Harald S. Müller
Michael Haist
Fernando Acosta
(Eds.)

Proceedings of
**The 9th *fib* International
PhD Symposium
in Civil Engineering**

Proceedings of

The 9th *fib* International PhD Symposium in Civil Engineering

Karlsruhe Institute of Technology (KIT)
22 – 25 July 2012, Karlsruhe, Germany

edited by

Harald S. Müller
Michael Haist
Fernando Acosta

Organising Institutions

Institute of Concrete Structures and Building Materials
Materials Testing and Research Institute MPA Karlsruhe
Karlsruhe Institute of Technology (KIT)

Impressum

Karlsruher Institut für Technologie (KIT)
KIT Scientific Publishing
Straße am Forum 2
D-76131 Karlsruhe
www.ksp.kit.edu

KIT – Universität des Landes Baden-Württemberg und
nationales Forschungszentrum in der Helmholtz-Gemeinschaft

KIT Scientific Publishing 2012
Print on Demand

ISBN 978-3-86644-858-2

Greetings

PhD studies mean the highest level of university education all over the world. Results of PhD studies are important not only for the PhD candidates but also for the universities themselves. Successful PhD researchers often become future professors at universities or recognized engineers by companies.

Every university has its own programme for PhD studies which can be considerably different at various universities. However, the purpose of them is always to support high level research.

The series of *fib* International PhD Symposia in Civil Engineering provides a special forum in addition to the PhD programmes to present the results of the ongoing research and to discuss them collecting advices on how to continue and finish the research.

In this forum only young colleagues (PhD candidates) are allowed to make presentations who have already started their research but did not submit their thesis yet (at the time of submitting the Abstract). Supervisors (anyhow professors and practicing engineers) are very welcome as Session Chair persons or contributors to the discussions.

Therefore, one of the unique features of these PhD Symposia is that a long obligatory discussion time (almost as long as the presentation) is included in the programme. In this way our young researchers are subjected to comments and questions of an international community.

In the *fib* International PhD Symposia many universities and institutes are represented from different parts of the world. It is always nice to realize how many researchers are active in various research fields and which are the most relevant research topics. Personal contacts of the participants taken during the PhD Symposia can help to develop future collaborations.

In this way we can fulfil the original objectives of the *fib* International PhD Symposia: (1) to provide a forum for PhD students in civil engineering to present the progress of their work; (2) to discuss the results of the ongoing PhD studies in order to support the future work; (3) to give the possibility for PhD students to establish contact for international communication and (4) to compare PhD studies in various countries.

I personally always enjoyed the unique atmosphere of previous PhD Symposia full with activities and enthusiasm of both the PhD candidates and the professors. Please enjoy it at KIT in Karlsruhe, Germany in 2012.

Finally, I take this opportunity to wish to our PhD candidates successful discussions in Karlsruhe and a successful PhD defense at your universities when you are ready with your research.

György L. Balázs
President of the International Federation for Structural Concrete (*fib*)

Preface

The *fib* International PhD Symposium in Civil Engineering is an established event in the academic calendar of doctoral students. It is held under the patronage of the International Federation for Structural Concrete (*fib*), one of the main international associations that disseminates knowledge about concrete and concrete structures. Previous venues of the *fib* International PhD Symposium in Civil Engineering were: Budapest (1996 and 1998), Vienna (2000), Munich (2002), Delft (2004), Zurich (2006), Stuttgart (2008) and Copenhagen (2010).

The Karlsruhe Institute of Technology (KIT) and its Institute of Concrete Structures and Building Materials are honoured to host the 9th Symposium from July 22 to 25, 2012 in Karlsruhe, Germany. The organizers are very happy to welcome 107 PhD-students from 21 countries presenting their work in the different fields of civil engineering both with a written paper contained in the proceedings at hand, as well as with an oral presentation during the conference. The topics of the conference – presented in different sessions – can be summarized as follows:

- load carrying and shear behaviour of concrete members
- load carrying behaviour and performance of masonry structures
- dynamics of structures and material behaviour under dynamic loadings
- innovative structures
- concrete technology and microstructure of concrete
- durability of concrete
- behaviour and performance of various materials and structures
- numerical simulation techniques
- fibre reinforced concrete
- monitoring and repair of building structures

The organizers of the symposium would like to thank the contributing PhD students for their efforts and their patience with us. We also thank the supervisors for their support for the symposium. It is an important part of academia to present research projects to an international audience and share information and experience with colleagues from the international research community. The organizers hope that you, the PhD students, have an enjoyable time in Karlsruhe and find the symposium a good chance to extend your professional network.

The review process in the *fib* PhD Symposia is very intense and requires much more time than for other conferences, owing to the educational nature of the symposium. This is why the organizers would like to thank the scientific committee, which consisted of 40 internationally recognized researchers from 19 different countries, for their tireless support. As for earlier years, the quality of the manuscripts ranged from good to excellent, illustrating the experience and training of the young researchers.

The organizers would also like to thank the PhD students and student assistants of the Institute of Concrete Structures and Building Materials of KIT, who supported the organising committee in the editing of the proceedings at hand and in the organisation.

Last but not least the organizers would like to express their gratitude to the sponsors of the symposium. Their significant support made the symposium possible.

We wish all participants a successful and worthwhile symposium and a rewarding career within the civil engineering community.

Harald S. Müller, Michael Haist and Fernando Acosta
Karlsruhe, June 2012

Scientific Committee

Andrzej B. Ajdukiewicz (Poland)

György L. Balázs (Hungary)

Josée Bastien (Canada)

Hans Beushausen (South Africa)

Wolfgang Brameshuber (Germany)

Rolf Breitenbücher (Germany)

Harald Budelmann (Germany)

John Cairns (United Kingdom)

Manfred Curbach (Germany)

Frank Dehn (Germany)

Marco di Prisco (Italy)

Michael Fardis (Greece)

Christoph Gehlen (Germany)

Mette Geiker (Denmark)

Carl-Alexander Graubner (Germany)

Petr Hajek (Czeck Republik)

Will Hansen (United States of America)

Josef Hegger (Germany)

Johann Kollegger (Austria)

Herbert Mang (Austria)

Peter Marti (Switzerland)

Peter Maydl (Austria)

Viktor Mechtcherine (Germany)

Aurelio Muttoni (Switzerland)

Franz Nestmann (Germany)

Hans-Wolf Reinhardt (Germany)

Koji Sakai (Japan)

Thomas Seelig (Germany)

Jürgen Schnell (Germany)

Karl Schweizerhof (Germany)

Surendra P. Shah (United States of America)

Lidia Shehata (Brasil)

Viktor Sigrist (Germany)

Johan Silfwerbrand (Sweden)

Lothar Stempniewski (Germany)

Luc Taerwe (Belgium)

Jean-Michel Torrenti (France)

Lucie Vandewalle (Belgium)

Jan L. Vitek (Czeck Republik)

Werner Wagner (Germany)

Joost Walraven (The Netherlands)

Organizing Committee

Harald S. Müller (Karlsruhe Institute of Technology, Germany)

Michael Haist (Karlsruhe Institute of Technology, Germany)

Fernando Acosta (Karlsruhe Institute of Technology, Germany)

Nico Herrmann (Karlsruhe Institute of Technology, Germany)

Participating Universities

Graz, University of Technology,
Austria

Technical University of Vienna,
Austria

University of Innsbruck,
Austria

Ghent University,
Belgium

KU Leuven,
Belgium

Laval University,
Canada

McGill University,
Canada

Université de Sherbrooke,
Canada

University of Zagreb Faculty of Civil
Engineering, Croatia

Brno University of Technology,
Czech Republic

Czech Technical University in Prague,
Czech Republic

VSB Technical University of Ostrava,
Czech Republic

University of Southern Denmark,
Denmark

Université Paris Est.,
France

Bauhaus-Universität Weimar,
Germany

Dresden University of Technology,
Germany

Federal Waterways Engineering and Research
Institute, Germany

Hamburg University of Technology,
Germany

Hochschule Ostwestfalen Lippe,
Germany

Karlsruhe Institute of Technology (KIT),
Germany

Leibniz Universität Hannover,
Germany

Leipzig University of Applied Sciences
(HTWK Leipzig), Germany

Ruhr University Bochum,
Germany

RWTH Aachen University,
Germany

Technical University Darmstadt,
Germany

Technical University of Braunschweig,
Germany

Technische Universität Kaiserslautern,
Germany

Technische Universität Kaiserslautern,
Germany

Technische Universität München (TUM),
Germany

Universität der Bundeswehr München,
Germany

University of Stuttgart,
Germany

University of Patras,
Greece

Budapest University of Technology and
Economics, Hungary

Indian Institute of Technology Madras,
India

Politecnico di Milano,
Italy

University of Brescia,
Italy

Hokkaido University,
Japan

Kagawa University,
Japan

Université du Luxembourg,
Luxembourg

Lodz University of Technology,
Poland

Silesian University of Technology,
Poland

University of Timisoara,
Romania

Technical University of Cluj-Napoca,
Romania

University of Stellenbosch,
South Africa

Technical University of Madrid,
Spain

Universitat Politècnica de València,
Spain

Ecole Polytechnique Fédérale de Lausanne,
Switzerland

ETH Zurich,
Switzerland

Delft University of Technology,
The Netherlands

Cardiff University,
United Kingdom

University of Bristol,
United Kingdom

University of Leeds,
United Kingdom

Contents

Dynamics of Structures and Material Behaviour under Dynamic Loadings (Session B-1)   **187**

Innovative Structures (Session B-2)   **255**

Concrete Technology and
Microstructure of Concrete (Session C-1)   385

Durability of Concrete (Session C-2) … 479

Load Carrying and Shear Behaviour of Concrete Members

Shear tests of reinforced concrete slabs and slab strips under concentrated loads

Eva O. L. Lantsoght

Department of Design & Construction – Concrete Structures
Delft University of Technology
Stevinweg 1, 2628 CN Delft, The Netherlands
Supervisors: Cor van der Veen and Joost C. Walraven

Abstract

In slabs subjected to concentrated loads, the shear strength checks are conducted for two limit states: 1) shear over an effective width, and 2) punching shear on a perimeter around the point load. In current practice, the shear strength at the supports is determined with models that do not consider the transverse redistribution of load that occurs in slabs, which results in underpredictions for the actual slab shear capacity. Currently, an experimental program is being conducted at Delft University of Technology to determine the shear capacity of slabs under point loads near to the support. This paper presents the results of the tests conducted in continuous slabs and slab strips. In addition to studying the influence of the slab width, the specimens are tested with two types of reinforcement (ribbed and plain bars). The results of the experiments are compared to strength predictions from current design models. Also, recommendations for the support effective width and an enhancement factor for considering the effect of transverse load redistribution are given.

1 Introduction

Shear in reinforced concrete one-way slabs loaded with a concentrated load near the support is typically checked in two ways: by calculating the beam shear capacity over a certain effective width and by checking the punching shear capacity on a perimeter around the load. The method of horizontal load spreading, resulting in the effective width b_{eff} of the support which carries the load, depends on local practice. In most cases (eg. Dutch practice) horizontal load spreading is assumed under a 45° angle from the centre of the load towards the support, Fig. 1 (left). The lower limit for the effective width is typically $2d$ for loads in the middle of the width and d for loads at the edge and corner of the slab. In French practice [1], load spreading is assumed under a 45° angle from the far corners of the loading plate towards the support, Fig. 1 (right). The punching shear (two-way shear) capacity in code formulas is developed for two-way slabs. Most empirical methods for punching shear have been derived from tests on slab areas around a column; a loading situation which is significantly different from a slab under a concentrated load close to the support.

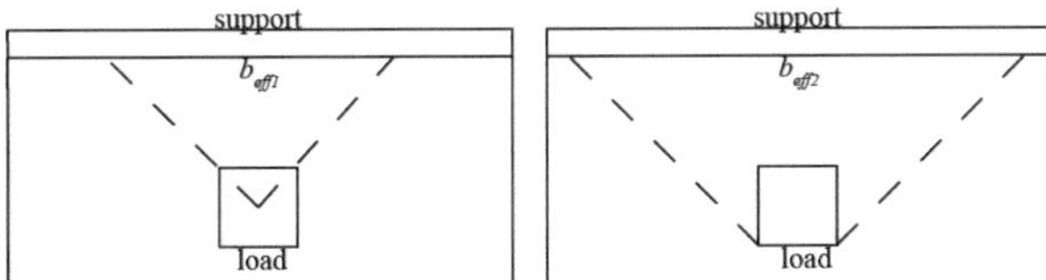

Fig. 1   Effective width (left) assuming 45° load spreading from the centre of the load: b_{eff1}; (right) assuming 45° load spreading from the far corners of the load: b_{eff2}; top view of slab.

2 Previous research

Recent research concerning shear in slabs has mainly focused on one-way slabs under line loads [2]. It was experimentally shown that one-way slabs under line loads behave like beams and that beam shear provisions lead to good estimates of their shear capacity. A database of 206 experiments on wide beams and slabs [3] shows that test data regarding the shear capacity of one-way slabs (b_{eff2} from Fig.1 (right) smaller than the total specimen width, b) under concentrated loads are scarce and only 22 experiments with $a/d < 2,5$ are available [4-7]. The majority of these experiments were car-

Proc. of the 9th *fib* International PhD Symposium in Civil Engineering, July 22 to 25, 2012,
Karlsruhe Institute of Technology (KIT), Germany, H. S. Müller, M. Haist, F. Acosta (Eds.),
KIT Scientific Publishing, Karlsruhe, Germany, ISBN 978-3-86644-858-2

ried out on small specimens ($d < 15$cm). A comparison between these results from the literature and EN 1992-1-1:2004 [8] "EC2" (combined with two different load spreading methods and $C_{Rd,c} = 0,15$ [9]) and the French National Annex [1]) and Regan's method [4] is shown in Table 1. The average (AVG), standard deviation (STD) and coefficient of variation (COV) of the results for the experimental value divided by the calculated value indicate that the traditional method of calculating the one-way shear strength (EC2 + b_{eff1}) underestimates the capacity. The French National Annex [1] and Regan's method [4] lead to the best results. The French National Annex allows shear stresses in slabs 2,27 times higher (for $k = 2$) than in beams as a result of transverse redistribution. Regan's method combines an enhancement factor from one-way shear with a punching perimeter from two-way shear. These results indicate that slabs can support higher concentrated loads as a result of their extra dimension. However, not enough experimental evidence is available to support this statement. Therefore, a series of experiments on slabs with $d = 265$mm is carried out.

Table 1   Comparison between test results from literature and design methods.

	EC2 [8] + b_{eff1}	EC2 [8] + b_{eff2}	EC2 + French NA [1]	Regan [4]
AVG	3,411	2,038	1,022	0,966
STD	1,005	0,501	0,294	0,204
COV	29,5%	24,6%	28,8%	21,1%

3   Experiments

3.1   Experimental setup

A top view of the test setup with a slab is presented in Fig. 2. The line supports (sup 1 and sup 2 in Fig. 2) are composed of a steel beam (HEM 300) of 300mm wide, a layer of plywood and a layer of felt [10] of 100mm wide. Experiments are carried out with a concentrated load close to the simple support (sup 1 in Fig. 2) and close to the continuous support (sup 2 in Fig. 2), where the rotation is partially restrained by vertical prestressing bars which are fixed to the strong floor of the laboratory. The prestressing force is applied before the start of every test, offsetting the self-weight of the slab. During the course of the experiment, some rotation could occur over support 2 due to the deformation of the felt and plywood and the elongation of the prestressing bars.

3.2   Specimens and Results

All slabs ("S") and slab strips ("B") have a thickness h of 300mm and an effective depth d of 265mm. The slabs are either loaded at the middle of the slab width (position M) at the simple and continuous support (two tests per slab), or consecutively at the east and west side (position S) at the simple and continuous support (four tests per slab).

Ribbed reinforcing bars with a diameter of 10mm (measured mean yield strength $f_{sy} = 537$MPa and measured mean ultimate strength $f_{su} = 628$MPa) and 20mm ($f_{sy} = 541$MPa and $f_{su} = 658$MPa) are used [10]. For S11 to S14, plain reinforcing bars with a diameter of 10mm ($f_{sy} = 635$MPa and $f_{su} = 700$MPa) and 20mm ($f_{sy} = 601$MPa and $f_{su} = 647$MPa) are used [10]. The flexural reinforcement is designed to resist a bending moment caused by a load of 2MN (maximum capacity of the jack) at position M (Fig. 2) along the width and 600mm along the span ($a/d = 2,26$). In practice, the amount of transverse flexural reinforcement for slabs is taken as 20% of the longitudinal flexural reinforcement. In the tested slabs, 13,3% of the longitudinal flexural reinforcement is used in S1 and S2; 25,9% in S3, S5 to S14 and 27,2% in the slab strips. In S4 the amount of transverse flexural reinforcement is only doubled as compared to S1 and S2 in the vicinity of the supports.

The properties and results of S1 to S9 and the slab strips can be found in [11]. The properties of S11 to S14 are given in Table 2, in which the following symbols are used:

f_c'	the cube compressive strength of the concrete at the age of testing the slab,
f_{ct}	the splitting tensile strength of the concrete at the age of testing the slab,
ρ_l, ρ_t	the longitudinal (ρ_l) and transverse (ρ_t) reinforcement ratios of the slab,
a	the centre-to-centre distance between the load and the support,
M/S	loading at the middle (M) or side (S) of the slab width, Fig. 2,
$b_{load} \times l_{load}$	the size of the loading plate.

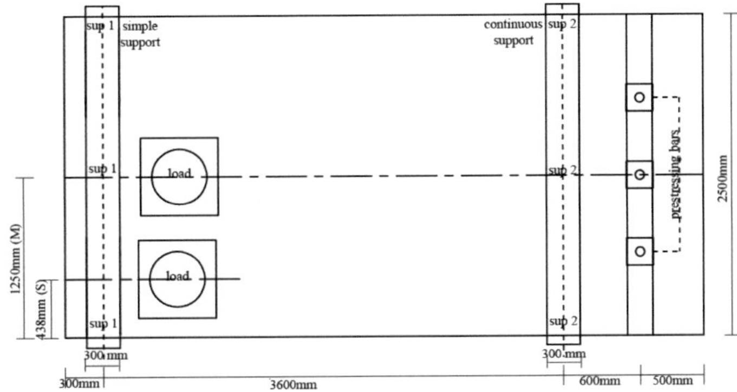

Fig. 2   Sketch of test setup, top view.

Table 2   Properties of slabs S11 to S14.

Slab nr.	b (m)	f_c' (MPa)	f_{ct} (MPa)	ρ_l (%)	ρ_t (%)	a/d	M/S	$b_{load} \times l_{load}$ (mm × mm)	test age (days)
S11	2,5	54,9	4,2	1,375	0,358	2,26	M	200 × 200	90
S12	2,5	54,8	4,2	1,375	0,358	2,26	S	200 × 200	97
S13	2,5	51,9	4,2	1,375	0,358	1,51	M	200 × 200	91
S14	2,5	51,3	4,2	1,375	0,358	1,51	S	200 × 200	110

The results of S11 to S14 are given in Table 3, in which the following symbols are used:

SS/CS   experiment at the simple (SS, sup 1) or continuous support (CS, sup 2), Fig. 2,
P_u the measured ultimate load,
WB/P/B   failure mode: wide beam shear (shear crack at the inside), punching shear or beam shear,
F_{pres} the force in the prestressing bars at failure,
V_{max} the resulting maximum shear force,
$V_{max,EC}$ the maximum shear force including reduction of the loads within $2d$ of the support [8].

Table 3   Results of slabs S11 to S14.

Test	SS/CS	P_u (kN)	Failure mode	F_{pres} (kN)	V_{max} (kN)	$V_{max,EC}$ (kN)
S11T1	SS	1194	WB + P	165	998	848
S11T4	CS	958	WB + P	307	886	766
S12T1	SS	931	WB + B + P	162	780	663
S12T2	SS	1004	P	173	839	712
S12T4	CS	773	WB + P + B	147	705	608
S12T5	CS	806	WB + B	158	735	633
S13T1	SS	1404	WB + P	157	1253	593
S13T4	CS	1501	WB + P	240	1411	706
S14T1	SS	1214	WB + P + B	133	1088	518
S14T2	SS	1093	WB + P + B	162	975	462
S14T4	CS	1282	WB + P + B	187	1207	605
S14T5	CS	1234	WB + P + B	142	1157	578

4 Results

4.1 Influence of the width

If the concept of an effective width can be applied to concrete slabs loaded in shear, then the shear capacity ceases to increase proportionally to the width after reaching a threshold value, the effective width. Increasing widths will lead to the same capacity, as only the effective width can carry the shear force from the load to the support [12]. For loads close to the support ($a/d < 2,5$) the results of S8 and S9 are compared to the results of the series of slab strips with different widths (BS1/0,5m – BX3/2m). The size of the loading plate, distance between the load and support and location of testing are variable. As shown in Fig. 3, the previously described threshold is achieved after an almost linear increase in capacity for an increase in width.

Table 4 gives the results for the effective width (b_{meas}) based on the experimental results, compared to the calculated widths based on the load spreading methods from Fig. 1 and from the ModelCode 2010 (b_{MC}) [13]. Remarkably, lower effective widths are found at the continuous support. The load spreading mechanism is thus influenced by the moment distribution in the shear span. The observed relation between the effective width and the size of the loading plate as well as the distance between the load and the support, are best reflected by b_{eff2}. The effective width from ModelCode 2010 gives too conservative results and does not correctly take the influence of the size of the loading plate into account.

Table 4   Effective width as calculated from the experimental results.

Test: $b_{load} \times l_{load}$, SS/CS, a/d	b_{meas} (m)	b_{eff1} (m)	b_{eff2} (m)	b_{MC} (m)
300mm × 300mm, SS, $a/d = 2,26$	2,12	1,10	1,70	0,99
300mm × 300mm, CS, $a/d = 2,26$	1,81	1,10	1,70	0,99
200mm × 200mm, SS, $a/d = 1,51$	1,25	0,70	1,10	0,63
200mm × 200mm, CS, $a/d = 1,51$	1,11	0,70	1,10	0,63
200mm × 200mm, SS, $a/d = 2,26$	1,63	1,10	1,50	0,98
200mm × 200mm, CS, $a/d = 2,26$	1,33	1,10	1,50	0,98

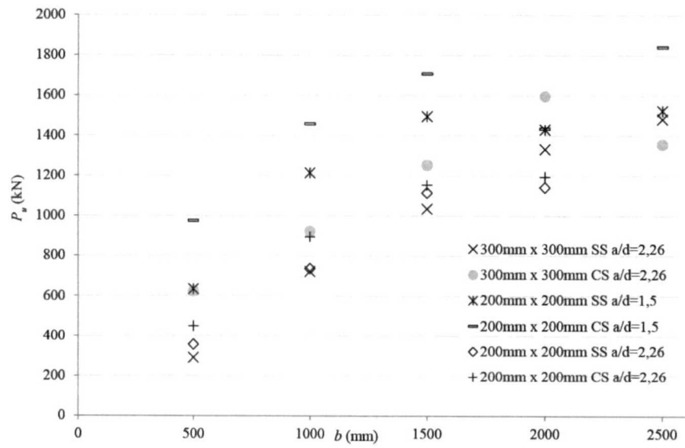

Fig. 3   Influence of overall width on shear capacity for the series discussed in Table 4.

4.2 Comparison to design models

All test results are compared to the following code methods: EN 1992-1-1:2004 [8] with $C_{Rd,c} = 0,15$ [9] with b_{eff1} and b_{eff2}, EN 1992-1-1:2004 [8] with the French National Annex for slabs [1] and Regan's formula [4]. The comparisons are based on the measured mean material properties. Safety and material factors equal 1. Punching shear was not the governing failure mode according to EN 1992-1-

1:2004 [8]. The critical perimeter as used in Regan's method is taken with 4 sides for loading in the middle of the width for the slabs, with 3 sides for loading near to the edge of the width for slabs and with 2 sides for the smallest slab strips.

Table 5 shows the comparison between the test results P_u or resulting shear forces $V_{max,EC}$ and the calculated values according to the considered methods. These results show that Regan's method and EN 1992-1-1:2004 with b_{eff2}, estimate best the shear capacity of slabs under concentrated loads close to the support. The French National Annex overestimates the shear capacity as 2,27 times higher stresses are allowed for slabs (for $k = 2$).

Table 5 Statistical properties from comparison between experimental data and calculated values.

Test data	$V_{max,EC}/V_{EC2,beff1}$			$V_{max,EC}/V_{EC2,beff2}$			$V_{max,EC}/V_{FR}$			P_u/P_{Regan}		
	AVG	STD	COV	AVG	STD	COV	AVG	STD	COV	AVG	STD	COV
All	2,71	0,58	21%	1,97	0,32	16%	0,85	0,15	18%	1,01	0,14	14%
Slabs	2,86	0,49	17%	1,98	0,25	12%	0,89	0,12	14%	1,03	0,15	14%
Ribbed	2,98	0,49	16%	2,01	0,27	13%	0,87	0,13	15%	1,08	0,13	12%
Plain	2,60	0,41	16%	1,91	0,18	9%	0,93	0,09	9%	0,92	0,14	15%
Slab strips	2,46	0,64	26%	1,95	0,42	21%	0,78	0,17	21%	0,97	0,13	13%

The influence of the considered effective width is reflected by the results of $V_{max,EC}/V_{EC2,beff1}$ and $V_{max,EC}/V_{EC2,beff2}$; using b_{eff2} agrees better with the experimental results. The average value (AVG) becomes smaller and more uniform: compare the range of 2,4 - 3,0 for b_{eff1} to the range of 1,9 - 2,0 for b_{eff2}. The standard deviation becomes smaller, as well as the coefficient of variation. These statistical parameters confirm that the French load spreading method (Fig. 1 (right)) is to be preferred for determining the effective width.

Comparing the row with the results of the slabs and the row with the results of the slab strips, shows a larger average capacity for slabs, which can be attributed to transverse load distribution. Therefore, in combination with [8] and b_{eff2}, an enhancement factor of at least 1,25 can be applied for slabs benefitting from transverse load redistribution and loaded close to the support. Also, the minimum effective width can be taken as $4d$ [14].

Kani [15] showed that plain bars result in higher shear capacities than deformed bars. As the concrete compressive strength and amount of transverse reinforcement of S1 and S11 were different, a direct comparison for the test results could not be made. The experiments mainly showed a difference in the cracking pattern, and a possibility for anchorage failure in the slabs with plain reinforcement. The influence of bond on the shear capacity of slabs under concentrated loads is thus studied based on the comparison to the code methods, Table 5. These results show that the average ratio of tested to predicted value is smaller for the plain bars as compared to the ribbed bars. Regan's method [4] slightly overestimates the capacity of slabs with plain bars. The higher calculated predictions according to [4], however, are the result of the increased amounts of transverse and longitudinal reinforcement.

Comparisons to non-linear finite element models [16, 17] show that predicting the experimental values strongly depends on the choice of the input parameters. A posteriori modelling leads to good results, but it is shown [16] that choosing a certain set of input parameters from modelling one experiment does not necessarily lead to an equally close modelling of another experiment. The recommended effective width b_{eff2} also most closely corresponds to the effective width based on the stresses at the support [16].

5   Conclusions and Recommendations

Transverse load redistribution leads to higher shear capacities in slabs as compared to beams. This conclusion is reflected in tests from the literature, as well as in the results from the discussed test series.

The French load spreading method is to be preferred. This conclusion is supported by data from the literature, the series of specimens with varying width, the comparison to code methods and results from non-linear finite element analysis.

The test data indicate that slabs reinforced with plain bars have a slightly smaller shear capacity. This conclusion does not correspond to the observations for beams with plain bars failing in shear.

It is recommended to assess the one-way shear capacity of reinforced concrete slabs by using EN 1992-1-1:2004 taking into account direct load transfer between the load and the support. This code method is to be combined with the effective width b_{eff2} (resulting from the French load spreading method) with a minimum effective width of $4d$, and an enhancement factor on the capacity of at least 1,25 for loads close to the support ($a/d < 2,5$) .

References

[1]   Chauvel, D., Thonier, H., Coin, A., Ile, N.: Shear resistance of slabs not provided with shear reinforcement. CEN/TC 250/SC 02 N 726, France (2007), 32 pp.

[2]   Sherwood, E. G., Lubell, A. S., Bentz, E. C., Collins, M.P.: One-way Shear Strength of Thick Slabs and Wide Beams. In: ACI Structural Journal 103 (2006) No. 6, pp.794-802

[3]   Lantsoght, E.O.L.: Shear in reinforced concrete slabs under concentrated loads close to the support – Literature review, Delft University of Technology (2011), 289 pp.

[4]   Regan, P. E.: Shear Resistance of Concrete Slabs at Concentrated Loads close to Supports. London, United Kingdom, Polytechnic of Central London. (1982), pp. 24.

[5]   Graf, O.: Versuche über die Widerstandsfähigkeit von Eisenbetonplatten unter konzentrierter Last nahe einem Auflager. In: Deutscher Ausschuss für Eisenbeton 73 (1933), pp. 10-16.

[6]   Richart, F. E., Kluge, R. W.: Tests of reinforced concrete slabs subjected to concentrated loads: a report of an investigation. In: University of Illinois Engineering Experiment Station Bulletin 36 (1939), No. 85, 86 pp.

[7]   Cullington, D. W., Daly, A.F., Hill, M.E.: Assessment of reinforced concrete bridges: Collapse tests on Thurloxton underpass. In: Bridge Management 3 (1996), pp. 667-674.

[8]   CEN: Eurocode 2 – Design of Concrete Structures: Part 1-1 General Rules and Rules for Buildings, EN 1992-1-1. Brussels, Belgium. (2004) pp. 229

[9]   König, G., Fischer, J.: Model Uncertainties concerning Design Equations for the Shear Capacity of Concrete Members without Shear Reinforcement. In: CEB Bulletin 224 (1995). pp 49-100.

[10]   Prochazkova, Z., Lantsoght, E. O. L.: Material properties – Felt and Reinforcement For Shear Test of Reinforced Concrete Slab, Stevin Report nr. 25.5-11-11, Delft University of Technology. (2011), pp. 28.

[11]   Lantsoght, E. O. L., van der Veen, C., Walraven, J.: Shear capacity of slabs and slab strips loaded close to the support. In: ACI SP – Recent Developments in Reinforced Concrete Slab Analysis, Design and Serviceability (2011), 17pp. (to appear)

[12]   Regan, P. E., Rezai-Jorabi, H.: Shear Resistance of One-Way Slabs under Concentrated Loads. In: ACI Structural Journal 85 (1988) No. 2, pp. 150-157.

[13]   SAG 5: Model Code 2010 – Final draft. (2011) 676 pp.

[14]   Lantsoght, E. O. L.: Voortgangsrapportage: Experimenten op platen in gewapend beton: Deel II: analyse van de resultaten, Delft University of Technology (2012), 288 pp.

[15]   Kani, G. N. J: The Riddle of Shear Failure and its Solution. In: ACI Journal Proceedings 61 (1964) No. 4, pp. 441-467.

[16]   Falbr, J. Shear redistribution in solid concrete slabs. MSc thesis. Faculty of Civil Engineering and Geosciences. Delft University of Technology. (2011), pp. 151

[17]   Voormeeren, L. O.: Extension and Verification of Sequentially Linear Analysis to Solid Elements. MSc thesis. Faculty of Civil Engineering and Geosciences. Delft University of Technology. (2011), pp. 95

Experimental symmetry bearings for punching shear tests on large reinforced concrete members

Karsten Winkler

Institute of Concrete Structures,
Ruhr-Universität Bochum,
Universitätsstraße 150, 44780 Bochum, Germany
Supervisor: Peter Mark

Abstract

Shear and punching shear tests on concrete members have reached their size-limitations since years, as financial and material amounts as well as requirements on testing infrastructure are rising with increasing member sizes. The existing lack of experimental results on large reinforced concrete members becomes more important, since there is a remarkable size effect concerning the effective depth's influence – especially for shear and punching shear failure.

Inspired by best practice for numerical simulations, an innovative test setup for shear and punching shear tests utilizing the conditions of symmetry is developed. Symmetric parts of test specimen are substituted by special modular bearing systems representing symmetric support. Fundamental investigations focus on engineering and realizing these bearing systems to ensure identical test performance of symmetrically reduced test specimen compared to the analogous full-size member. For this purpose, an almost frictionless sliding as well as a flexurally rigid performance at the symmetry bearing have to be realized.

1 Introduction and motivation

1.1. Experimental data available and size effects

Analysing currently available experimental data shows that experiments are designed and performed in considerably smaller scales than practically realized later on. In figure 1 (left) the experimental data of slender slabs without shear reinforcement failing in punching shear is plotted against effective depths of tested concrete members. The analysis demonstrates that 95 % of effective depths of test specimen are smaller than 20 cm. Simultaneously, there is a notable and continuing evolution of range of practical interest towards larger slabs. However, referring to experiments on slender slabs with large sizes, only one single slab with an effective depth of d = 45 cm tested by Guandalini [2] is available. This existing gap between effective depths of experimentally investigated and practically realized reinforced concrete members leads to one fundamental question:

How safe can load bearing capacities and mechanisms, derived from experiments on small concrete members, be extrapolated to real structures featuring considerably larger sizes?

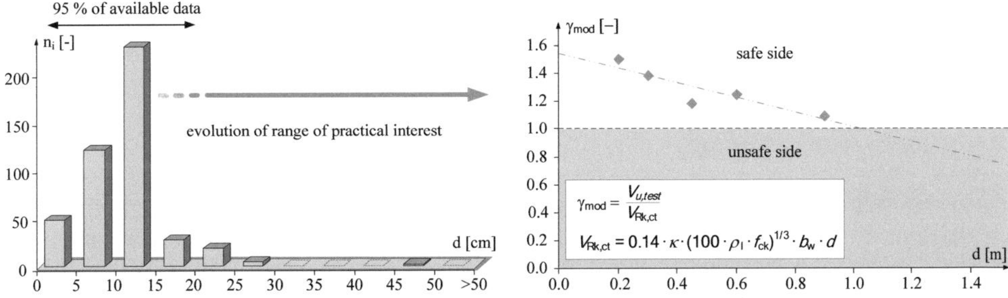

Fig. 1 Available experimental data of slender slabs ($\lambda \geq 3$) without shear reinforcement failing due to punching shear (left), model safety factors acc. to DIN 1045-1 [1] of own shear tests on reinforced concrete beams without shear reinforcement (right)

Proc. of the 9th *fib* International PhD Symposium in Civil Engineering, July 22 to 25, 2012, Karlsruhe Institute of Technology (KIT), Germany, H. S. Müller, M. Haist, F. Acosta (Eds.), KIT Scientific Publishing, Karlsruhe, Germany, ISBN 978-3-86644-858-2

The complexity of answering this question is based on the fact that existing size effects [3, 4] prohibit a direct extrapolation of the experimentally obtained bearing capacities. Especially for brittle failure modes that occur in shear and punching shear of concrete members without shear reinforcement, a significant non-linear influence of effective depth has to be considered. Current code provisions, such as DIN 1045-1 [1] and Eurocode 2 [5], use an approach based on fracture mechanics for approximating this size effect. This approach (eq.1) is uniformly used in design formulations for both shear and punching shear.

$$\kappa = 1 + \sqrt{\frac{200}{d[mm]}} \leq 2.0 \qquad (1)$$

To prove this approximation, own experiments on true-scaled concrete beams without shear reinforcement have been performed [6]. Therefore, effective depths of the test specimen varied from 20 cm to about 100 cm. Remaining parameters have been kept constant in order to fulfill the demanding requirements on true to scale experiments [3, 4]. In figure 1 (right) the experimental results are plotted in terms of model safety factors γ_{mod} [7, 8] against the effective depths of tested beams. The model safety factor γ_{mod} illustrates a safety margin depending on the 5 %-fractile of the total population of experiments on shear failure without shear reinforcement.

Although, the size effect approach acc. to Eurocode 2 and DIN 1045-1 is already included in model safety factors γ_{mod}, the analysis of the true-scaled series of experiments shows a trend to decreasing safety factors with increasing member sizes.

Besides other aspects [4], it should be considered carefully that current codes prescribe a minimum shear reinforcement for beam-like structures to ensure a ductile failure mode. By contrast, same code provisions tolerate a design of slabs subjected to shear or punching shear loads without shear or transverse reinforcement.

In summary, the combination of

- - range of practical interest, experimentally confirmed only in a small and restricted part,
- - continuous evolution towards increasing concrete member sizes and heights,
- - existing well-known but not really solved problem of size effects,
- - and pronounced brittle failure modes,

demonstrates the relevance of experimental research in the field of punching shear failure of large reinforced concrete slabs.

1.2 Experimental challenge: punching shear tests on thick reinforced concrete slabs

Using common test setups, the experimental challenge of testing slender slabs with large effective depths in punching shear becomes almost insuperable. Indeed, fundamental requirements on test setup and infrastructure for punching shear tests are significantly more challenging, than for comparatively simple tests on shear failure of beams. The realization of a true rotationally symmetric bearing of test specimen means to avoid a stiffness-dependent support of slab also for cracked states. Testing of slabs requires sufficient large slendernesses, which results in large specimen sizes and accordingly heavy weights of specimen.

At the same time, thick slabs and slabs with large effective depths lead to enormous testing loads that cannot be handled in international laboratories at this time. Further aspects of manufacturing, transportation, cuttings and removal have to be considered which cannot be (financially) legitimated in sum. Hence, one promising approach to obtain experimental results on thick concrete slabs failing in punching shear is developing a new testing method.

2 Innovative test setup

2.1 Symmetry conditions

At the institute of concrete structures at the Ruhr-Universität Bochum, an innovative concept for testing slabs loaded in punching shear is developed. The research program is financially supported by the German Research Foundation (DFG) and focuses on true to scale punching tests on slabs with large effective depths. The fundamental idea is inspired by best practice of numerical analyses [9].

To reduce time needs for numerical solutions, modeling techniques typically utilize symmetry where possible. In figure 2 (left), the finite element model of a reinforced concrete slab failing in punching shear is shown. Due to utilizing symmetrical conditions, only a quarter of the slab has been modelled.

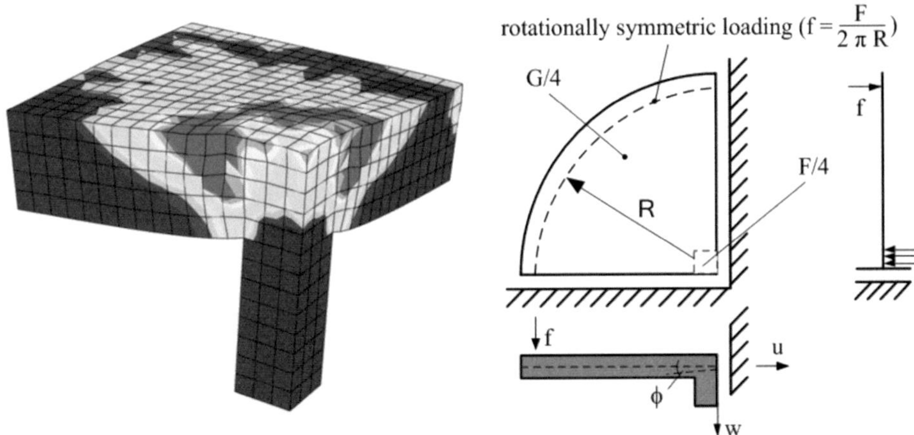

Fig. 2   Utilizing symmetry of system and load in numerical investigations [9] (left) and as basic principle for the innovative test setup (right)

Transferring this fundamental principle to experimental research and procedures means to substitute symmetric parts of test specimen by a symmetric support. In figure 2 (right), a schematic drawing of the principle operation mode of the test setup is given. Beneath realizing a flexurally rigid bearing combined with minimized deformations u, sliding with small friction of the specimen in vertical direction w has to be ensured. Accordingly, these characteristic features are equivalent to the main challenges discussed separately in section 3.

Specified for punching shear tests, utilizing the double symmetry leads to quartering of weight and test loads beneath a significant reduction of material and personal costs. Therefore, this concept allows testing considerably larger slabs than up to now. In figure 3, a visualization of the setup for a large concrete slab tested for punching shear is shown. The whole test setup consists of six bearing modules. Each one is composed of flexurally rigid reinforced concrete bearings and a slip-free embedded front end made of steel. The main advantage of the modular assembly is seen in flexible combination and setting up. This allows to obey individual requirements and sizes of test specimen. Furthermore, it benefits handling and storage of setup's components. In the remainder, the experimental realization is explained in detail.

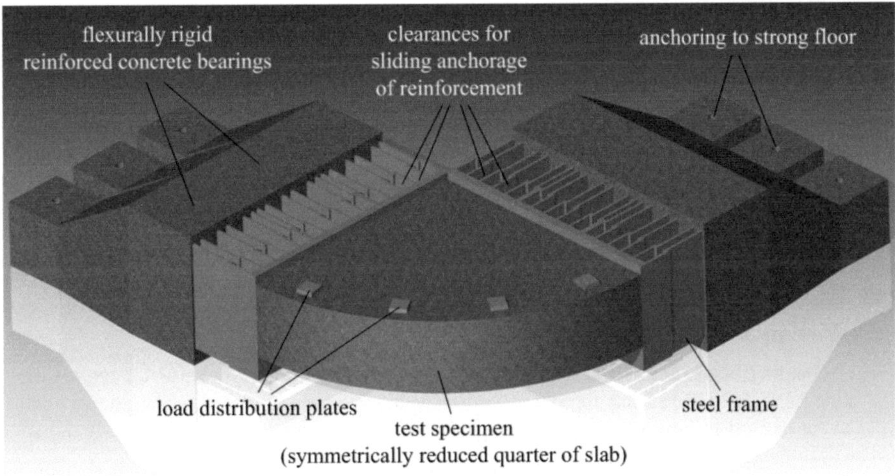

Fig. 3   Visualization of developed innovative test setup for punching shear tests on thick slabs

2.2   Experimental realization

The rotationally symmetric loading is applied through load distribution plates arranged axis-symmetrically along the outer edge of the specimen. On the inner edge, the testing slab is enclosed by a combination of steel segments divided by elastic interface layers also used as framework for placing of concrete. These layers prevent an unacceptable stiffening of the slab along the symmetrical support. A special sliding layer made of greased PTFE (polytetrafluoroethylene), also known as Teflon^{TM}, divides testing slab and front end of the bearing modules. Combined with the special mechanically finished and polished surface of the steel frame, an almost frictionless sliding is enabled.

The test specimen itself is connected and restrained by a unique socket construction (figure 4). The concrete-casted socket is connected to a threaded rod, which is anchored in between clearances of the steel frame. Anchor plates of the reinforcement as well as the reinforcement itself have to fulfill same displacements. Therefore, these anchor plates are supported on sliding layers made of greased PTFE, too. Additionally, this means that the spacings in the steel frame have to be sufficiently dimensioned for both vertical and horizontal amount of displacement of reinforcement's anchor plate.

In summary, the bearing modules carry the total reaction forces of the slab by direct contact on the compressed part and by restrained sliding anchors of reinforcement on the tension part. Finally, the reinforced concrete bearings are restrained and anchored in the prestressed strong floor of the testing laboratory.

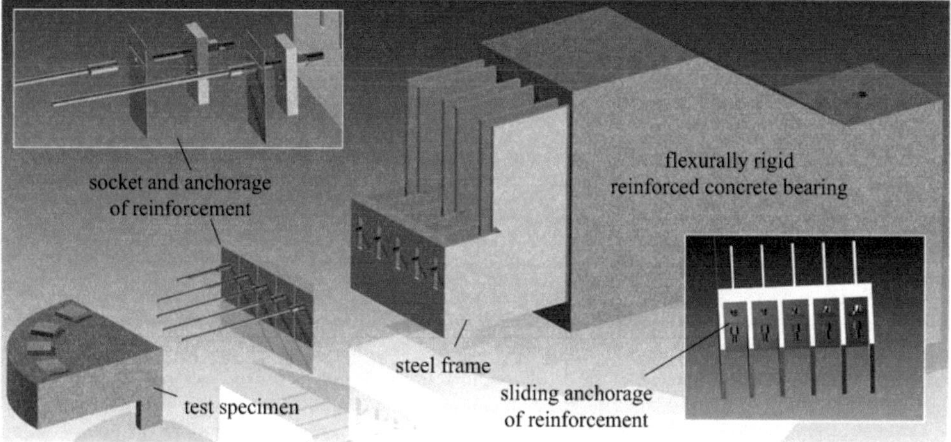

Fig. 4   Single bearing module and corresponding elements of test setup

3   First performance tests

3.1   Developing and optimizing component-by-component

First and foremost, the primary goal of the research program is developing and realizing symmetry bearings for experimental investigations on punching shear failure. Prior to first tests on large scale slabs, the development of the test setup itself is continuously supported by experiments on single major components. Therefore, smaller alternate samples of test specimen as well as symmetrically reduced concrete beams are taken to prove and verify functionality of elements and details. Especially the realization of almost frictionless sliding interface layers and the development of flexurally rigid support of the test members are in the focus of present work. Following, major results of performed sliding tests and first complete tests on reinforced concrete beams are presented exemplarily.

3.2   Friction and sliding tests

Simulating sliding of the actual test specimen, two anchorage plates are prestressed against the steel frame under defined prestressing force and vertically pushed then (figure 5). The anchor plates are similar to those used in future tests on reinforced concrete members to anchor the reinforcement bars between spacings of the steel frame.

Fig. 5   Sliding tests at steel frame with defined surface finish (left and middle), test specimen for sliding tests pasted with greased sheets of PTFE (right)

The test parameters include different thicknesses and sizes of PTFE-sheets, as well as varying positions of loading and different surface qualities and roughness of the sliding planes. Beneath a small sliding coefficient especially under high pressures, resulting deformations of the PTFE-sheets have to be minimized, to avoid unintentional rotations of the test specimen.

Considering the effective prestressing load as well as the dead load of the whole sliding system, the vertical load required to push the system is measured and results in a sliding coefficient of the system. Results show, that sliding combinations based on greased PTFE-sheets lead to sliding coefficients smaller than about 1 %. These excellent sliding coefficients can be reached even under high prestressing forces, corresponding to maximum reaction load of the reinforcement of future concrete tests. Additionally, the analysis of sliding tests provides prospects for optimization to even improve these sliding values in the future.

3.3   First experiments on symmetrically bisected reinforced concrete beams

The main goal of a first series of reinforced concrete tests on symmetrically reduced beams is testing the slip-free socket connection of the reinforcement bars in connection with sliding of test specimen and anchor plate at the symmetry bearing. Starting with three point bending tests, conventionally tested full-size members serve as reference for the performance of the identically reinforced, but symmetrically bisected beams at the innovative test setup (figures 6 and 7).

All test specimen are designed to fail in flexural tension, because yielding of tensile reinforcement marks an upper bound for future experiments on shear and punching shear. Therefore, a successful and consistent testing of this flexural limit state verifies also the functionality for shear tests.

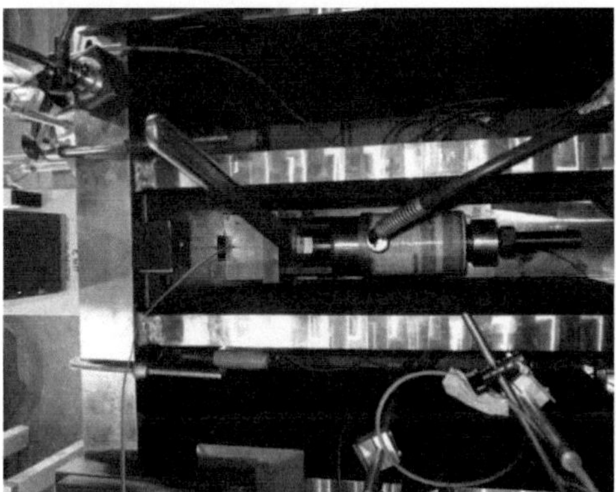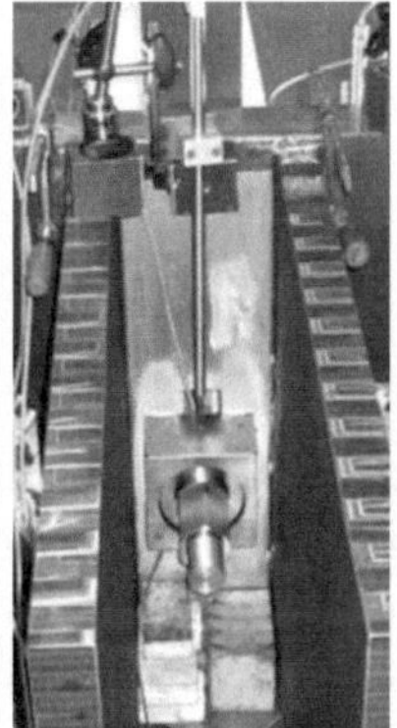

Fig. 6   Experiments on reinforced concrete beams at the new test setup: procedure of prestressing (left) and sliding anchorage of bending reinforcement (middle, right)

In figure 6, the experimental procedures for testing a symmetrically bisected concrete beam at the new test setup are shown. After adjusting of test specimen at the front end of the steel frame and placing sliding anchorages on the rear side, a threaded rod is screwed into the socket of the reinforcement bar. Following, the beam is restraint to the test setup by defined prestressing of the threaded rod (figure 6, left).

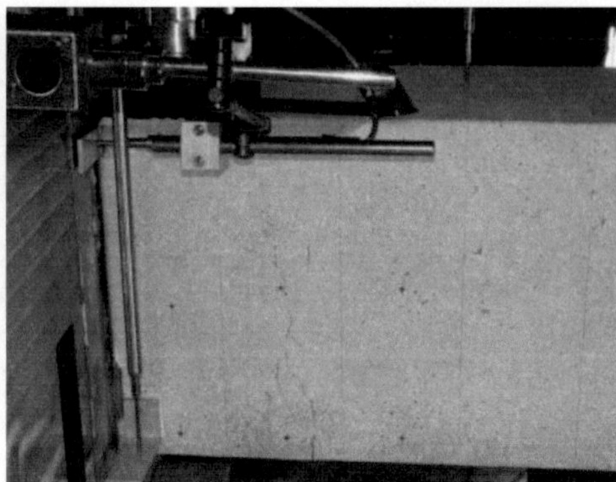

Fig. 7   Experiments on reinforced concrete beams at the new test setup: test setup (left) and sliding area of test specimen (right)

During the actual testing, displacements and deformations, especially at the interface to the bearing, are documented by conventional measuring devices like displacement transducers and strain gauges, as well as photogrammetric methods. Beneath the primary goal of testing socket connection and anchoring of the bending reinforcement, also the influencing effect of tensioning to the beam's deformation is analysed (figure 7).

In figure 8, a comparison of representative experimental load-displacement-curves of a symmetrically bisected beam (HBV-IIb-3) and corresponding full-size reference beams is given. Caused by extension of the restraining system, horizontal deformations and unintentional additional rotations of the test specimen at the sliding surface emerge. While the comparison of test loads at the beginning of yielding of reinforcement shows very good accordance between reference test and symmetrically reduced beam, corresponding vertical displacements of the latter are slightly larger, yet. The reason for this can be found in a too weak threaded rod as well as in a too low prestressing of specimen against the support.

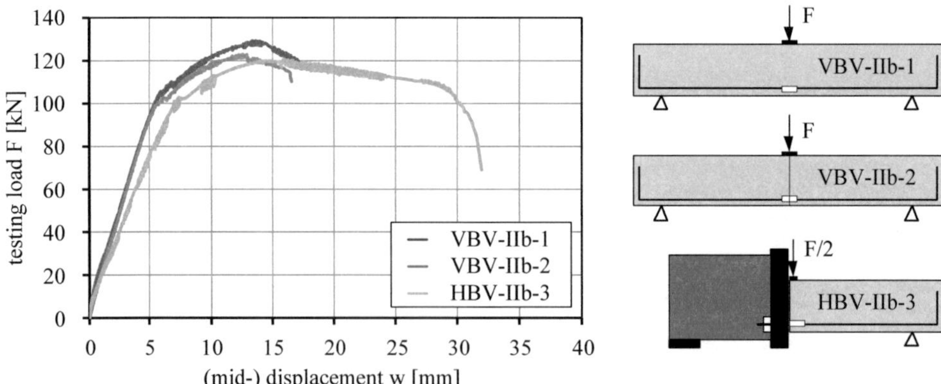

Fig. 8   Comparison of experimental load-displacement-curves of symmetrically bisected concrete beam (HBV-IIb-3) and corresponding full-size reference beams

4   Conclusions and outlook

An innovative test setup for reinforced concrete slabs failing in punching shear is presented. The article describes the basic concept of utilizing symmetry, development of experimental symmetry bearings as well as first results of performance tests. Beneath first satisfying results at the initial system, tests on symmetrically bisected concrete beams lead to valuable knowledge as well as optimization potential that will be integrated in further research and development work.

Upon finishing performance tests on the plane symmetrical bearing, two complete bearing modules will be assembled orthogonally, to start performance tests on symmetrically quartered slabs, until finally punching shear tests will be performed. The final verification will be performed by punching shear tests of two analogous designed and fabricated slabs: the full-size slab as reference tested in a common way, as well as the corresponding, symmetrically reduced quarter of slab tested at the new test setup.

The remarkable innovation potential of this innovative test setup is shown for axis-symmetrical punching shear tests in figure 9. Upon successful realization and verification, this innovative experimental approach will enable to expand the range of experimental investigations considerably.

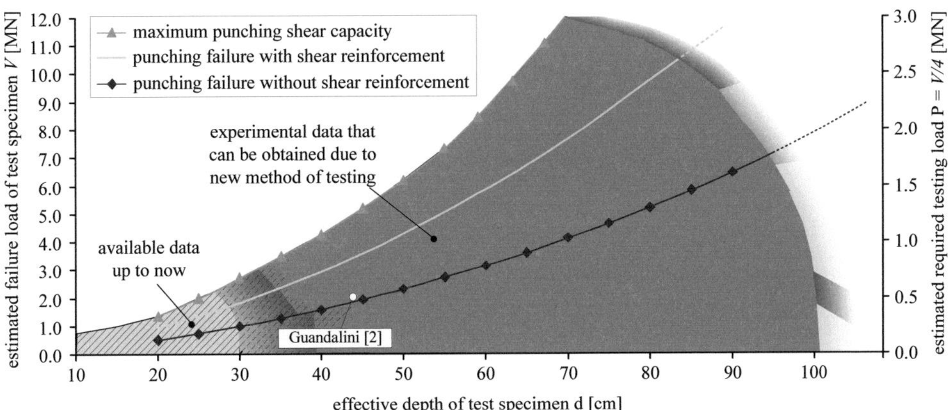

Fig. 9 Innovation potential of the innovative test setup: relationship between failure loads, required testing loads and effective depths of test specimen

Acknowledgement

The financial support of the research program by the German Research Foundation (DFG) is gratefully acknowledged. The author would like to express his gratitude to members of staff of the structural testing laboratory of the Ruhr-Universität Bochum, as well as student assistants at our institute, involved and participating in this project.

References

[1] DIN 1045-1: Tragwerke aus Beton, Stahlbeton und Spannbeton, Teil 1: Bemessung und Konstruktion. Beuth Verlag, Berlin, August 2008.

[2] Guandalini, S.: Poinçonnement Symétrique Des Dalles En Béton Armé, phd-thesis, EPFL Lausanne, 2005.

[3] Bazant, Z.P., Cao, Z.: Size Effect in Punching Shear Failure of Slabs. In: ACI Structural Journal 84 (1), pp. 44-53, 1987.

[4] Winkler, K.: Besonderheiten bei Bemessung und Konstruktion von Stahlbetonbauteilen mit großen Abmessungen. 51. Forschungskolloqium des DAfStb, pp. 125-136, 2011.

[5] DIN EN 1992-1-1: Eurocode 2: Bemessung und Konstruktion von Stahlbeton- und Spannbetontragwerken – Teil 1-1: Allgemeine Bemessungsregeln für den Hochbau, 2005.

[6] Winkler, K. und Mark, P.: Querkraftversuche an maßstäblich skalierten, schubunbewehrten Stahlbetonbalken, Berichte aus dem Lehrstuhl für Massivbau, Versuchsbericht 2011-1, Ruhr-Universität Bochum, Februar 2011.

[7] Reineck, K.-H., Kuchma, K. S. K., Marx, S.: Shear Database for Reinforced Concrete Members without Shear Reinforcement. In: ACI Structural Journal 100 (2), pp. 240–249, 2003.

[8]   Hegger, J., Görtz, S., Beutel, R., König, G., Schenck, G., Kliver, J., Dehn, F., Zilch, K., Staller, M. & Reineck, K.-H.: Überprüfung und Vereinheitlichung der Bemessungsansätze für querkraftbeanspruchte Stahlbeton- und Spannbetonbauteile aus normalfesten und hochfesten Beton nach DIN 1045-1, DIBt-Forschungsvorhaben IV 1-5-876/98, 1999.

[9]   Winkler, K., Stangenberg, F.: Numerical Analysis of Punching Shear Failure of Reinforced Concrete Slabs. 2008 Abaqus Users' Conference, May 19-22, 2008, Newport, RI, USA, 2008, pp. 244-257.

Shear fatigue behaviour of reinforced concrete elements without shear reinforcement

Juan Manuel Gallego

Department of Continuum Mechanics and Structures
Universidad Politécnica de Madrid
c/ Profesor Aranguren s/n, Madrid (28040), Spain
Supervisor: Luis Albajar and Carlos Zanuy

Abstract

Many concrete structures such as bridge approach slabs, wind towers, maritime structures or bridge deck slabs, are able to suffer fatigue because they are subject to a high number of repeated loads. In spite of this, structural concrete codes do not typically include a verification of shear fatigue safety or those which do still prefer empirical formulations instead of advanced mechanical models. Although empirical S-N curves provide a practical and simple way to estimate fatigue strength, a rational approach is more desirable in order to understand the shear fatigue process. For this reason, this paper proposes a rational S-N-based model to predict the number of cycles to diagonal cracking based on the stress state at the tip of flexural cracks.

1 Introduction

Existing experimental works [1-8] have shown that fatigue failure mode may differ significantly from static failure mode in concrete elements without stirrups. Elements designed to develop ductile flexural failure without transverse reinforcement are able to present brittle shear failure under repeated loading even before the fatigue fracture of longitudinal reinforcement occurs. Shear fatigue failure has been found in experiments where the maximum shear force under repeated load was significantly smaller than static shear strength. A deeper understanding of shear fatigue process is therefore necessary due to the brittle nature of this failure mode.

The motivation for the research presented in this paper was the study of the transverse shear fatigue behaviour of the lateral cantilever of reinforced concrete bridge deck slabs, which are normally built without shear reinforcement. The design of such elements is usually made on the basis of static strength and it is unusual that standards consider shear fatigue verification. It has to be noted that loads transferred to these elements by heavy vehicles have increased during last decades due to increasing transport demands. This fact should be considered in existing bridges that were designed for smaller loads and their safety margin against failure could be now lower than expected.

In order to investigate fatigue shear strength on one-way slabs without shear reinforcement, a mechanical model is proposed to predict the number of load cycles up to diagonal cracking. The diagonal cracking strength is formulated as a function of the stress state at the tip of existing flexural cracks. As a result of the mechanical model, a simplified but rationally based non-linear S-N curve is proposed.

2 Different modes of shear fatigue failure

Shear-related fatigue failures can be classified into two different groups. On the one hand, the so-called 'shear-compression fatigue' consists on the formation of a diagonal crack at the shear span and its progressive development towards the compression zone until failure takes place when the compression depth is too small to resist the applied force. This failure mode presents significant residual strength after diagonal cracking. On the other hand, the so-called 'diagonal cracking fatigue' is characterized by sudden failure as soon as a diagonal crack forms. Besides these two failure mechanisms, fatigue fracture of the reinforcement is as well possible depending on the stress oscillation.

The experimental evidence indicates that shear fatigue behaviour of elements without stirrups is a rather complex process. The first stage of the process is the formation of flexural cracks within the shear span. Such a crack pattern develops and stabilizes during the first cycles. The process continues until a diagonal crack forms from one of these flexural cracks. At that moment, some specimens

Proc. of the 9th *fib* International PhD Symposium in Civil Engineering, July 22 to 25, 2012,
Karlsruhe Institute of Technology (KIT), Germany, H. S. Müller, M. Haist, F. Acosta (Eds.),
KIT Scientific Publishing, Karlsruhe, Germany, ISBN 978-3-86644-858-2

collapse due to the full development of a diagonal crack across the beam's web (diagonal cracking fatigue) while other elements do not fail, presenting an important residual shear strength after the diagonal crack appears (shear-compression fatigue). When shear-compression fatigue takes place, the diagonal crack extends simultaneously into the compression zone and towards the support running horizontally at the level of the longitudinal reinforcement. Final fatigue failure is not due to insufficient capacity of shear resisting mechanisms (shear transferred by the compression zone, friction/aggregate interlock and dowel action), but rather to the progressive propagation of the diagonal crack into the compression zone until it becomes too small to resist the compression force acting on it. The crack pattern of a shear-compression fatigue failure is shown in Fig.1 (right), according to [8].

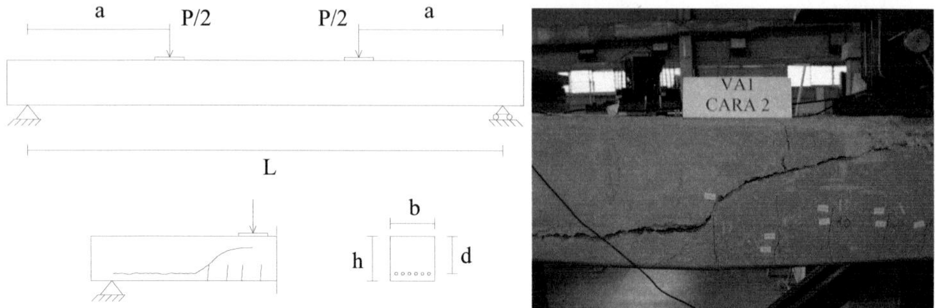

Fig. 1   (Left) Typical four-point bending configuration for shear fatigue tests; (Right) View of specimen VA1 [8] after shear-compression fatigue failure (a=1.35 m; b=0.30 m; L=4 m; d=0.26 m; a/d=5.4; P_{max}=120 kN; P_{min}=50 kN; ρ=2.51 %; f_c=25 MPa; Number of cycles up to diagonal cracking = 77000; Number of cycles up to final failure = 170718).

3   Evaluation of shear fatigue strength according to standards

Many codes of practice [9-11] still prefer empirical S-N curves to predict fatigue shear strength rather than advanced mechanical models. Eq. (1) is proposed by the Model Code draft [11] to estimate fatigue shear strength of concrete elements without stirrups. This linear expression allows to determine the amount of cycles that these elements can withstand up to failure (N), as a function of static shear strength (V_{ref}) and maximum applied shear force (V_{max}).

$$\log N = 10\left(1 - \frac{V_{max}}{V_{ref}}\right) \ (1) \qquad\qquad V_{ref} = k_v\sqrt{f_{ck}}zb \ (2)$$

According to Model Code draft [11], static shear strength can be determined by using Eq. (2), based on the modified compression field theory [12,13], where z is the inner lever arm, b is the web's width and f_{ck} is the concrete cylinder strength. Term k_v is a coefficient for concrete contribution that can be calculated through three different approximation levels depending on the required accuracy. Other codes of practice [9] establish different ways to evaluate fatigue shear strength, but all of them are based on empirical S-N curves. Code formulations present several drawbacks and even contradictions. For example, it does not seem logical that 1990 and 2010 Model Code versions [10,11] use the same S-N curve with different shear strength criteria. Moreover, significantly different results can be obtained if the three approximation levels are used in the last Model Code draft [11]. Besides, the use of such empirical S-N curves has two main weak points. On the one hand, due to the large inaccuracy that exists to calculate static shear strength that corresponds to the first point of the S-N curve where N is equal to zero. On the other hand, due to the S-N curve shape, especially when it comes to a high number of load cycles. Another great disadvantage is that most standards [9-11] do not provide criteria to distinguish between 'diagonal cracking fatigue' and 'shear-compression fatigue', so it is assumed that these codes only can assess the number of cycles up to complete failure.

4   Mechanical model to determine the number of cycles to diagonal cracking

A mechanical model is proposed to estimate the number of cycles to diagonal cracking from stress state at the tip of flexural cracks. A first attempt was proposed by the author and colleagues in [14]. The diagonal crack is considered to form from a flexural crack developed during the first cycles of the fatigue process. Such a crack will be referred to as critical flexural crack. According to the analysis of the experimental database [1-8], it will be assumed that, for four-point bending configuration tests, the position of the critical flexural crack is at a distance of d from the load point. The strain and stress distribution over the depth at the critical section before diagonal cracking is represented in Fig. 2, where point A lies on the neutral axis and point B lies on the tip of the flexural crack.

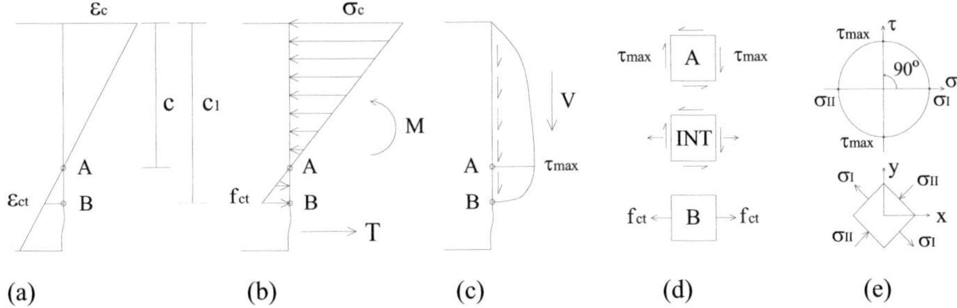

Fig. 2   Critical section analysis: (a) Plane strain distribution; (b) Normal stress distribution; (c) Shear stress distribution; (d) Stress state at three points of the critical section; (e) Mohr's circle and principal stresses orientation at the neutral axis (point A).

In order to simplify the mechanical model, it can be considered that the inclined crack starts at point A where the largest principal tensile stress is produced. By representing the stress state with the Mohr's circle, as shown in Fig. 2(e), it is possible to determine the principal tensile stress at the tip of the critical flexural crack (σ_I). The maximum shear stress at the critical section (τ_{max}) can be related with the principal tensile stress (σ_I) according to Eq. (3), obtained directly from Fig. 2(e). Assuming the second order parabolic distribution represented in Fig. 2(c), the maximum shear stress is related to shear force as indicated in Eq. (4). From Eqs. (3) and (4), Eq. (5) is obtained where the principal tensile stress (σ_I) depends on the applied shear force (V), the web's width (b) and the depth of the flexural crack tip (c_I).

$$\sigma_I = \tau_{max} \quad (3) \qquad V = \frac{2}{3}\tau_{max}bc_I \quad (4) \qquad \sigma_I = \frac{3}{2}\frac{V}{bc_I} \quad (5) \qquad c_I = c\left(1 + \frac{f_{ct}}{\sigma_c}\right) \quad (6)$$

The depth of the flexural crack tip (c_I) can be obtained from Eq. (6). This expression depends on the maximum normal stress (σ_c) at the critical section according to the distribution shown in Fig. 2(b), the tensile concrete strength ($f_{ct} = 0.3f_{ck}^{2/3}$ is assumed here) and the neutral axis depth (c). Even though a non-linear sectional analysis is required to obtain c_I and c, Eq. (7) can be used for practical purposes. From Eq. (7) it is possible to determine the neutral axis depth, where n is the modular ratio ($n = E_s/E_c$), d is the effective depth and ρ is the longitudinal reinforcement ratio.

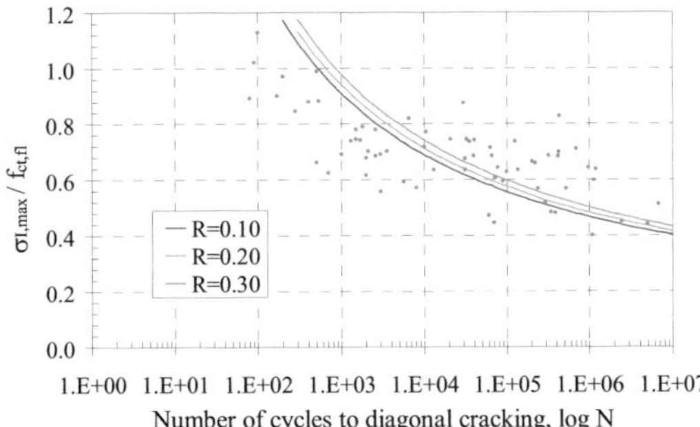

Fig. 3   Analysis of experimental tests (collected from the literature by the author) from the stress state at the tip of the critical flexural crack according to the mechanical model proposed.

In order to analyze the proposed model, the principal tensile stress for all the tests included in the database has been obtained according to Eq. (5) under maximum load. The results are plotted in Fig. 3 where the abscissa axis represents the amount of cycles up to diagonal cracking and the ordinate axis represents the principal tensile stress at the tip of the flexural crack. In order to include the size effect, this principal tensile stress has been normalized to the flexural tensile strength according to Eq. (8). According to Fig. 3, it is seen that the experimental test results would follow a non-linear S-N curve rather than the straight line proposed by other models.

$$c = n\rho d\left(-1 + \sqrt{1 + \frac{2}{n\rho}}\right) \quad (7) \qquad f_{ct,fl} = f_{ct}\left(1.6 - \frac{h[mm]}{1000}\right) \quad (8) \qquad \log N_{diag} = A(1-R)^B\left(\frac{\sigma_{I,max}}{f_{ct,fl}}\right)^C \quad (9)$$

From the comparison shown in Fig. 3 and using the least squares fitting method, the derivation of an exponential S-N curve has been investigated as indicated in Eq. (9). From the statistical analysis of the database, parameters A, B and C have been fitted to 2.631, -0.303 and -1.038, respectively. Eq. (9) can be used to calculate the number of load cycles that these elements can resist up to diagonal crack formation (N_{diag}). The minimum load level is included in the model ($R = \sigma_{I,min}/\sigma_{I,max}$).

5   Application to lateral cantilevers of highway box-girder bridges

The model proposed in the previous section is used in order to determine the number of cycles that a lateral cantilever of a highway box-girder bridge can withstand up to diagonal cracking. The element considered to simulate a lateral cantilever of a highway box-girder bridge is a cantilever beam with a 1.00 m width and a total length of 3.00 m (refer to Fig. 4). The longitudinal reinforcement at the top layer along the cantilever consists of 20 mm diameter bars with 0.20 m spacing. No shear reinforcement is provided. The element is subject to permanent loads owing to self weight and dead loads (pavement and security barrier) and also to a cyclic point load that may be applied at any section along the beam (except for the 1.00 m length close to the free edge) and simulates the cyclic action of a heavy vehicle wheel. The applied point load actually represents the fraction of the wheel load that is effectively carried by 1.00 m width.

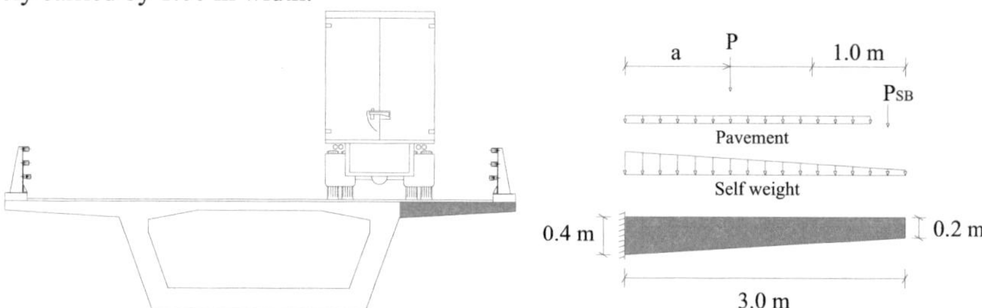

Fig. 4   (Left) Typical cross section of a highway box-girder bridge; (Right) Simplified model of lateral cantilever represented as a cantilever beam with a 1.00 m width.

Static shear failure is searched for an increasingly point load that can act at any section of the beam. Some standards [15,16] based on mechanical models [12,17] estimate static shear strength bearing in mind interaction between bending moment and shear force. Such mechanical models suggest smaller static shear strength in those areas of the cantilever with higher bending moment (at the clamped edge). In this case, the critical section (defined as the one with the least static shear strength) of the cantilever is always located at the clamped edge.

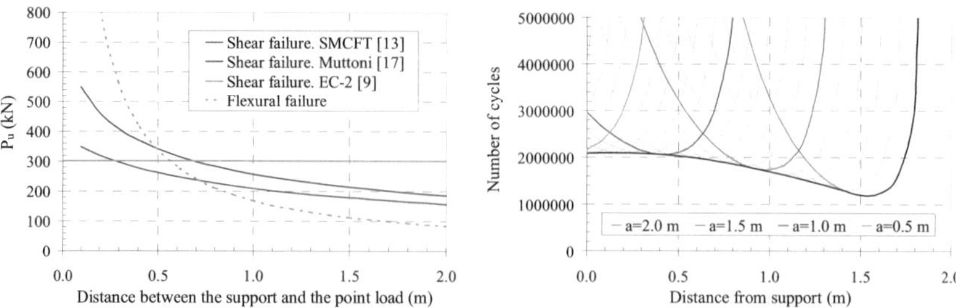

Fig. 5   (Left) Maximum admissible static load (P_u) for different load positions; (Right) Number of cycles to diagonal cracking at each section when the cantilever is subject to a cyclic load. The parameter 'a' is the distance between the cyclic load and the clamped edge.

According to these two models, the smallest value of static shear strength occurs when the load is as far as possible from support. For the static case, it is possible to plot the curves showed in Fig.5(left) that allow to determine the value of the point load that leads to failure of the cantilever as a function of the position occupied by such a load. In addition, the load value that produces flexural failure at the clamped edge is also plotted in Fig. 5(left). Static shear failure occurs when the distance between the point load and the support (abscissa axis) is less than a value that ranges between 0.45 and 0.75 m.

When the element studied above is subject to a cyclic action, there will be a certain value of the number of cycles that produces the diagonal crack for each value and position of the applied cyclic load. The critical section at which the diagonal crack forms is initially unknown and will be here investigated. To do it, it is necessary to determine, for each value and position of the applied point load, the number of cycles to diagonal cracking in each section of the cantilever. The critical section will be the one with the smallest value of number of cycles to diagonal cracking for a given position and magnitude of the load. Fig. 5(right) represents the number of cycles to diagonal cracking for a value of the point load of 120 kN as a function of the position of this load. The minimum value of each curve represents the numbers of cycles to diagonal cracking for each position of the point load and the corresponding position of the critical section where the diagonal cracks forms.

From the set of curves shown in Fig. 5(right), and for a given applied load, it is possible to establish an envelope from all the minimum values of the number of cycles to diagonal cracking at each section. The minimum number of cycles to diagonal cracking for the studied element occurs when the load is located at the furthest position from support. Contrary to static result where the critical section was always at the support, the position of the critical section varies along the element as a function of the magnitude and location of the cyclic point load. In this manner, for a given position of the cyclic load, each value of the load will be associated with a number of cycles to diagonal cracking. Therefore it is possible to plot a P-N curve of the cantilever that relates the number of cycles to diagonal cracking with the applied point load as Fig. 6 shows. A similar curve can be plotted considering the number of cycles that produce fatigue fracture of the longitudinal reinforcement. Hence, shear fatigue failure only occurs whenever the number of cycles that produce fatigue failure of longitudinal reinforcement exceeds the number of cycles to diagonal cracking. The fatigue strength of the reinforcement has been estimated by using the S-N curves for the steel provided by the Model Code [10].

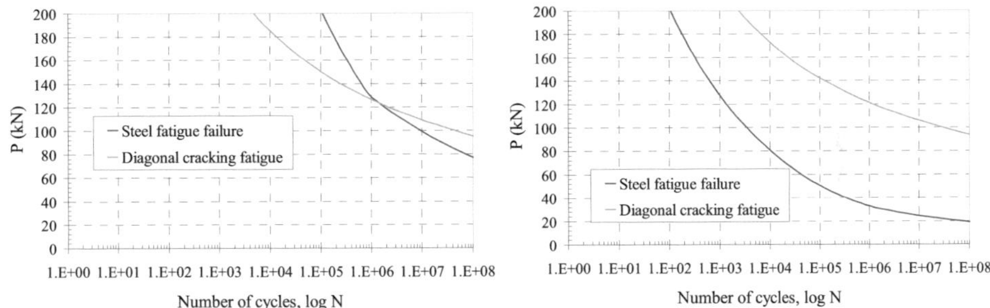

Fig. 6   P-N curve that relates the number of cycles up to fatigue failure to cyclic point load: (Left) Load at 0.5 m from support; (Right) Load at 2.0 m from support.

In the practical example developed, the critical section is located at support. Considering a typical real case of a highway box-girder bridge where the position of the most eccentric wheel is at 0.5 m from support, it is possible to determine (Fig. 5(left)) the load that produces static failure of the cantilever. In that case, static shear failure occurs before flexural failure when the point load is equal to 262 kN according to [13]. If applied cyclic point load is equal to 60 % of static shear strength (P=157 kN), Fig. 6(left) shows that the element can only withstand 60000 cycles up to diagonal cracking. In this case, the diagonal crack is formed before fatigue fracture of longitudinal reinforcement.

6   Conclusions

The study presented in this paper has dealt with the shear fatigue behaviour of reinforced concrete elements without shear reinforcement. Rather than mechanical models, most standards [9-11] currently propose empirical formulations to predict fatigue shear strength which has the problems mentioned in section 3. The study has focused on the estimation of the number of cycles to diagonal cracking. A predictive mechanically-based S-N equation has been obtained from the analytical study of the stress state at the tip of flexural cracks.

The proposed model has been used to study the transverse behaviour of lateral cantilevers of highway box-girder bridges subject to cyclic loads. The full development of this model aims to obtain a P-N curve of the lateral cantilever of bridge deck slabs whereby it is possible to predict the number of cycles to diagonal cracking. If this curve and the value of the traffic load are known, it is possible to determine the number of cycles to diagonal cracking. In addition, it has to be noted that diagonal cracking does not always imply the final collapse of such elements. Several experimental tests [1-8] have shown an important residual fatigue life after diagonal cracking. Accordingly, the research presented in this paper must be extended in order to study the reasons to explain why some elements have an additional strength after diagonal cracking.

References

[1]   Chang T.S., Kesler C.E.: Static and fatigue strength in shear of beams with tensile reinforcement. ACI Journal, 29 (12), 1958, pp. 1033-1057.

[2]   Chang T.S., Kesler C.E.: Fatigue behaviour of reinforced concrete beams. ACI Journal, 30 (2), 1958, pp. 245-254.

[3]   Stelson T.E., Cernica J.N.: Fatigue properties of concrete beams. ACI Special Publication, 5 (8), 1958, pp. 255-259.

[4]   Taylor R.: Discussion of a Paper by Chang T.S and Kesler C.E. Fatigue behaviour of reinforced concrete beams. ACI Journal, 55, 1959, pp. 1011-1015.

[5]   Higai T.: Fundamental study on shear failure of reinforced concrete beams. Proceedings of the Japan Society of Civil Engineers (JSCE), 279, 1978, pp. 113-126.

[6]   Ueda T.: Behaviour in shear of reinforced concrete beams under fatigue loading. Dissertation for the degree of Doctor of Engineering, University of Tokyo, 1982.

[7]   Johansson U.: Fatigue tests and analysis of reinforced concrete bridge deck models, Stockholm, Sweden, Royal Institute of Technology, 2004.

[8]   Zanuy C.: Análisis seccional de elementos de hormigón armado sometidos a fatiga, incluyendo secciones entre fisuras, Tesis Doctoral, Universidad Politécnica de Madrid, 2008.

[9]   CEN: Eurocode EC-2. prEN 1992-1-1:2004, 2004.

[10]   CEB-FIP.: CEB-FIP Model Code (1990). Lausanne, Switzerland, 1991.

[11]   FIB., Bulletin d'information No.56 and 57. Model code 2010. First complete draft, Lausanne, Switzerland, 2010.

[12]   Vecchio F.J., Collins M.P.: The modified compression field theory for reinforced concrete elements subjected to shear. ACI Journal, 83 (2), 1986, pp. 219-231.

[13]   Bentz E.C., Vecchio F.J., Collins M.P.: Simplified modified compression field theory for calculation shear strength of reinforced concrete elements. ACI Structural Journal, 103 (4), 2006, pp. 614-624.

[14]   Zanuy C., Albajar L., Gallego J.M.: Toward modelling the shear fatigue behaviour of reinforced concrete beams without shear reinforcement. 7th International Conference on Analytical Models and New Concepts in Concrete and Masonry Structures AMCM, 2011, Krakow, Poland, pp. 255-259.

[15]   Canadian Standards Association.: CSA A23.3 Design of concrete structures. Canadian standards association, Rexdale, ON, Canada, 2004.

[16]   SIA.: SIA 262. Code for concrete structures. Swiss society of engineers and architects, Zurich, Switzerland, 2003.

[17]   Muttoni A., Fernández Ruiz M.: Shear strength of members without transverse reinforcement as function of critical shear crack width. ACI Structural Journal, 105 (2), 2008, pp. 163-172.

Behaviour and strength of existing bridges with low amount of shear reinforcement

Michael Rupf

ENAC IIC IBETON,
Ecole Polytechnique Fédérale de Lausanne,
Bâtiment GC, Station 18, 1015 Lausanne, Switzerland
Supervisor: Aurelio Muttoni

Abstract

In the last decades, the assessment of the strength of existing structures has become a major issue in structural engineering. Prestressed concrete bridges are of particular relevance, due to the large number of these structures and to the significant changes occurred in the design approaches and traffic actions. In particular, a number of these structures built before the 1980's present insufficient amount of shear reinforcement or defective stirrup anchorage compared to current design standards. This, however, does not necessarily mean that these structures are actually unsafe and have to be retrofitted because code provisions are mostly oriented for design of new structures and their design provisions include a number of safe in-built hypotheses.

In this paper, the strength of prestressed girders with low amount of shear reinforcement or with defective anchorage is investigated by means of a test programme carried out at the Ecole Polytechnique Fédérale de Lausanne on 10 prestressed and 2 reinforced concrete girders (10 m long, 0.78 m high). The results show that if certain conditions are fulfilled, these structures can perform suitably and provide the expected strength according to plastic design approaches. For comparisons, the elastic-plastic stress fields method is used to predict the specimens' strength leading to excellent correlations between the measured-to-predicted behaviour and strength. Furthermore this approach allows a sound understanding of the various shear-carrying mechanisms developed in the girders and of the various failure modes observed.

Fig. 1   Tested beam SR23 after failure

1   Introduction

During the assessment of an existing structure, especially for prestressed concrete bridges, the requirements of the current design codes can often not be fulfilled. Many existing bridges show insufficient amount of shear reinforcement or defective stirrup anchorage compared to current codes. As a consequence, more accurate procedures like the elastic-plastic stress fields approach [1, 2] are required to ensure a safe design of the structure. Furthermore the improvement of current code provisions is desired to obtain a safe and economical design for the bridge girders.

Proc. of the 9th *fib* International PhD Symposium in Civil Engineering, July 22 to 25, 2012,
Karlsruhe Institute of Technology (KIT), Germany, H. S. Müller, M. Haist, F. Acosta (Eds.),
KIT Scientific Publishing, Karlsruhe, Germany, ISBN 978-3-86644-858-2

In this research project, the influence of low amount of shear reinforcement, the insufficient anchorage of the stirrups, and the presence of beam flanges on the behaviour of a structural element is analysed. The objectives are to show that an assessment for the mentioned bridges is possible and to give the guidelines for their verification in agreement with current code approaches. To gain a better understanding of the behaviour of such structures, a test series of ten prestressed concrete girders and two reinforced concrete girders has been performed at the Ecole Polytechnique Fédérale de Lausanne EPFL.

This paper presents an overview of the test series and shows the main results of the experiments. The experimentally obtained results have been compared to predictions using the elastic-plastic stress fields approach and using current codes of practice (Eurocode 2: 2004 [3], Model Code: 2010 [4]).

2 Test program

2.1 Test setup

The twelve tested single span beams with cantilever correspond to a multi span bridge with a span of about 40 m on a scale of 3/8. According to the test setup (figure 2), the maximal bending moment is acting together with the maximal shear force likewise the support regions of multi span bridges. In the tests, the applied force on the cantilever is chosen to be always half of the force in the span of the beam. Thus, the magnitude of the shear force is constant over the whole beam and corresponds to the applied force on the cantilever.

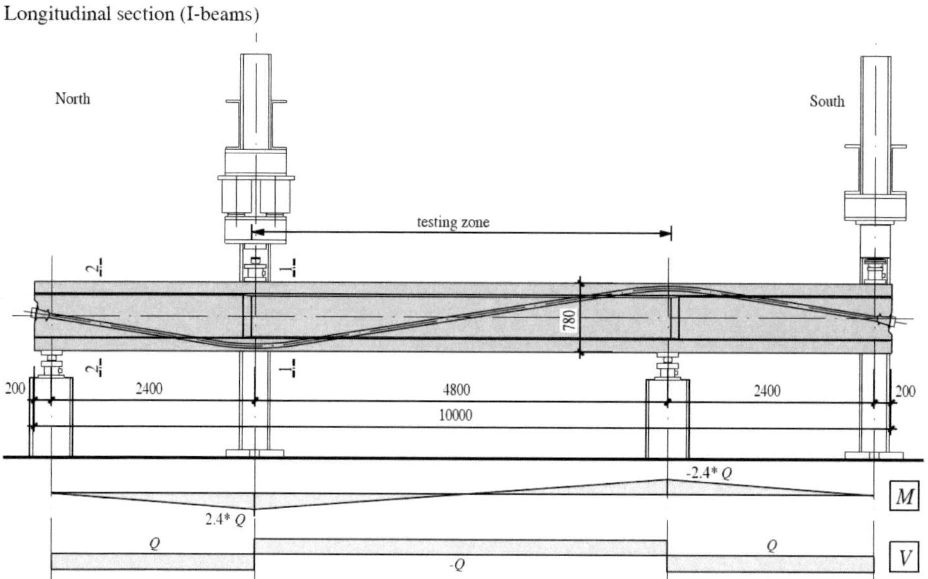

Fig. 2   Test setup: Longitudinal section with diagram of the bending moment and the shear force

Only the central part of the beam is used as test region (figure 2). The exterior parts of the beam had a larger width and amount of shear reinforcement and thus higher shear strength then the testing zone. With the conducted measurements the general behaviour of the beam is recorded on the whole length of the beam. Measurements of web deformations are limited on the testing zone.

2.2 Test specimen

2.2.1 Main parameters

The main parameters of the test series are the cross section, the amount of post-tensioning P/A, the shear reinforcement ratio ρ_w and the anchorage properties of the stirrups. The value P denotes the post-tensioning force and A the area of the cross section in the testing zone. In figure 3, the two types of cross sections are shown. The prestressing is introduced by one or two post-tensioning cables in the girder. Three beams contain open stirrups which would mean that the longitudinal reinforcement bars are not enclosed by the shear reinforcement (refer to table 1).

Table 1   Main parameters of the test series

Beam SR..	21	22	23	24	25	26	27	28	29	30	31	32
section	I	I	I	I	I	I	I	I	I	I	▯	▯
P/A [MPa]	2.5	2.5	2.5	2.5	5.0	5.0	5.0	-	2.5	2.5	3.0	-
ρ_w [%]	0.09	0.13	0.06	0.25	0.09	0.06	0.19	0.09	0.25	0.25	0.09	0.09
stirrup	▯	▯	▯	⊔	▯	▯	▯	▯	⊔	⊔	▯	▯

2.2.2   Geometry and material properties

The length, the height, and the web thickness in the central part of all tested beams are the same (figure 3). In the testing zone the shear reinforcement consists of stirrups or single bars with a diameter of 6 mm and a spacing between 150 mm and 300 mm. The post-tensioning consists of one or two cables VSL 6-4 in ribbed steel ducts and anchorage heads VSL-EC25. All the steel ducts are grouted with a high strength mortar after tensioning of the wires. The cable position follows the bending moment as indicated in figure 2.

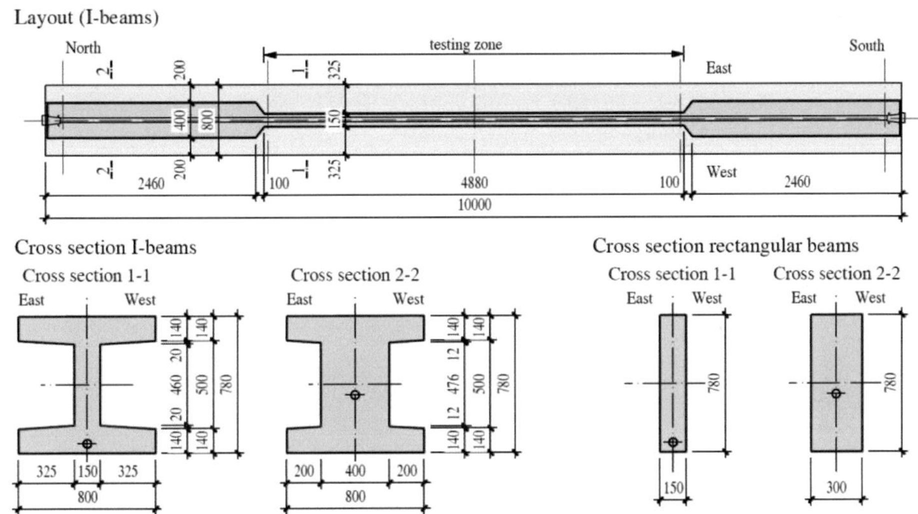

Fig. 3   Dimensions of the beams: Layout (top), cross section I-beams (bottom left) and cross section rectangular beams (bottom right)

Standard concrete without any additives and aggregates with a maximal diameter of 16 mm has been used. The concrete cylinder strength f_c at the testing day varies between 29.8 MPa and 37.8 MPa. For the beams with flanges the measured yielding strength of the shear reinforcement f_y is 580 MPa, the ultimate strength f_t is 630 MPa and the ultimate strain ε_u is 3.0 %. For the beams without flanges the yielding strength of the shear reinforcement f_y is 530 MPa, the ultimate strength f_t is 590 MPa and the ultimate strain ε_u is 5.5 %.

3   Test results

All tested beams failed in shear. Figure 4 shows the deflection under the loading point in the span versus the shear force. As expected, the larger the amount of shear reinforcement the larger the ultimate strength. The same applies to the increasing amount of post-tensioning force. The beams with flanges show a rather large deformation capacity, in spite of the lowest amount of shear reinforcement. The residual strength is between 60 and 70 % of the ultimate strength and the deformation could be increased without any big loss in this residual strength. In contrast to this observation, the beams without flanges behave more brittle and show a smaller loading capacity. After reaching the ultimate strength the beams fail suddenly with small strength afterwards.

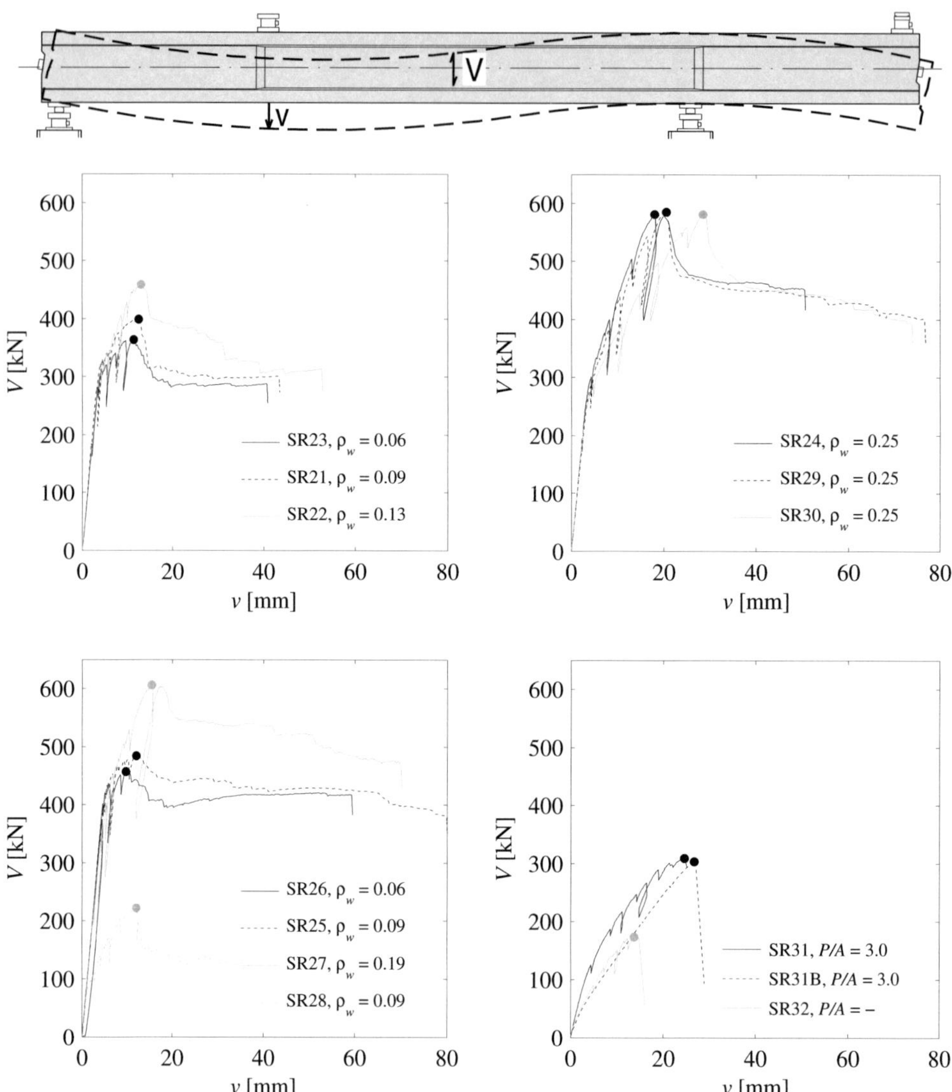

Fig. 4   Test results: scheme of the deformation (top), beams with flange and P/A = 2.50 MPa (centre left), beams with flange and open stirrups (centre right), beams with flange and P/A = 5.00 MPa or P/A = - (bottom left) and beams without flange (bottom right)

The numerical values of the ultimate strength $V_{R,test}$ are presented in table 2. It can be noted that the beam SR31 has been externally reinforced after failure and tested once again as SR31B.

4   Discussion of the test results

Before starting the test series, all beams were modelled to predict their strength and behaviour. This prediction has been done with a model using the elastic-plastic stress fields (EPSF) [2]. Table 2 and figure 5 (top left) present the resulting values. The comparison of the prediction with the test results over all the beams gives an average value $V_{R,test}/V_{R,pred}$ of 1.06 and a coefficient of variation of 0.05. The prediction of the EPSF method is thus in very good agreement with the test results.

Table 2   Resulting shear strength of the tests $V_{R,test}$ and predicted shear strength $V_{R,pred}$ using the elastic-plastic stress fields method

Beam SR..	21	22	23	24	25	26	27	28	29	30	31	31B	32
$V_{R,test}$ [kN]	399	459	364	579	484	457	606	222	585	581	309	303	173
$V_{R,pred}$ [kN]	370	430	355	560	470	445	580	220	560	540	265	265	175
$V_{R,test}/V_{R,pred}$	1.08	1.07	1.03	1.03	1.03	1.03	1.04	1.01	1.04	1.08	1.17	1.14	0.99

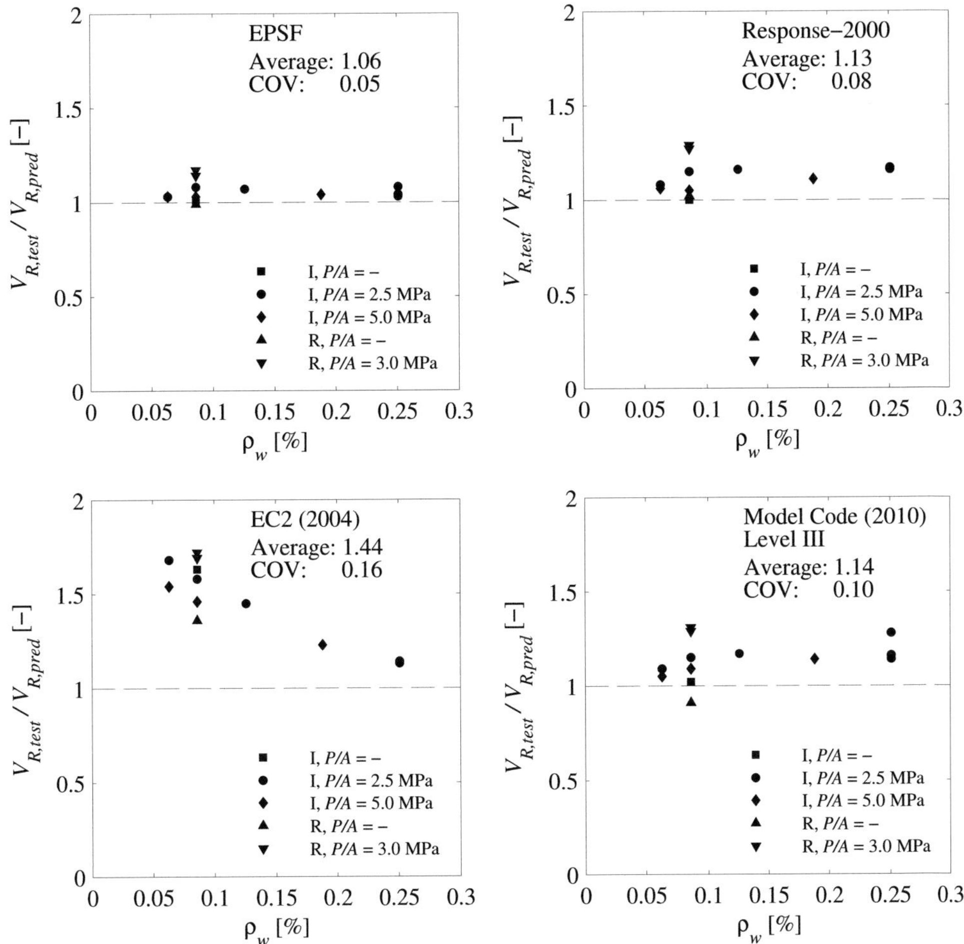

Fig. 5   Ratio of $V_{R,test}/V_{R,pred}$ for different models versus the shear reinforcement ratio: EPSF [2] (top left), Response-2000 [5] (top right), Eurocode 2 (2004) [3] (bottom left), and Model Code Level III (2010) [4] (bottom right), (I = beam with flanges, R = beam without flanges, COV = Coefficient of variation)

A comparison of the test results with the Eurocode 2 (2004) [3], the Model Code (2010) [4], and the program Response-2000 [5] is also given in figure 5. One can see that the Eurocode leads to conservative results for the tested beams and to a rather large coefficient of variation of 15 %. The Model Code (Level III) leads to less conservative results and to a coefficient of variation of 10 %. The program Response-2000 allows a sectional analysis based on the Modified Compression Field Theory [6]. With an overestimation of the ultimate shear force of 13 % the results are much better than the code predictions. The most accurate prediction is given by the EPSF method.

This comparison shows that design codes provide generally safe estimates as some shear-transfer actions are neglected (as the inclined component of the compression chord) and conservative values are given for some design parameters (angle of the compression struts, strength reduction factor of cracked concrete). A theoretical research on this topic is under work.

A more general view on the test results show that the presence of flanges is very beneficial and changes the behaviour of the whole structural element. For instance, they increased the ultimate strength of the reinforced beam without post-tensioning by 25 % (refer to specimens SR28 and SR32). Another beneficial aspect is the observed change in the deformation capacity. The beams with flanges showed large deformation capacity and residual strength whereas the beams without flanges showed a lower deformation capacity and less post-peak resistance.

5   Conclusion

This paper presents an investigation on the shear strength of prestressed reinforced concrete beams with low amount of shear reinforcement. The investigation is based on a test series of ten prestressed concrete girders and two reinforced concrete girders whose main results are presented in this paper. Its main conclusions are:

- The shear strength of the beams increase with larger amount of shear reinforcement and with increasing post-tensioning force.
- The shear strength of girders with flanges is significantly larger than the shear strength of beams without flanges, keeping the shear reinforcement ratio constant.
- The flanges provide rather large deformation capacity and residual resistance after peak load.
- Design codes generally provide safe estimates for the tested girders.
- The elastic-plastic stress fields method is applicable for structural elements like the tested beams and led to the best prediction.

References

[1]   Fernández Ruiz M., Muttoni A.: On Development of Suitable Stress Fields for Structural Concrete, ACI Structural Journal, Vol. 104 No. 4, pp. 495-502, Farmington Hills, USA, July 2007.

[2]   Fernández Ruiz M., Muttoni A., Burdet O.: Computer-aided development of stress fields for the analysis of structural concrete, fib Symposium Dubrovnik 2007, pp. 591-598, Dubrovnik, Croatia, May 2007.

[3]   European Committee for Standardization CEN: Eurocode 2 (2004), EN 1992-1-1, Design of concrete structures – General rules and rules for buildings, 225 pp., Brussels, Belgium, April 2004.

[4]   Fédération internationale du béton (fib): Model Code (2010), Final Draft, 653 pp., Lausanne, Switzerland, September 2011.

[5]   Bentz E.: Response-2000, Department of Civil Engineering, University of Toronto, Toronto, Canada, September 2001.

[6]   Vecchio F.J., Collins M.P.: The Modified Compression Field Theory for Reinforced Concrete Elements Subjected to Shear, ACI Journal, Proceedings Vol. 83 No. 2, pp. 219-231, Farmington Hills, USA, March 1986.

Shear capacity of concrete beams under sustained loading

Reza Sarkhosh

Faculty of Civil Engineering and Geosciences,
Delft University of Technology,
Stevinweg 1, 2628 CN Delft, the Netherlands
Supervisors: Joost Walraven, Joop den Uijl and René Braam

Abstract

The aim of this research is to predict the time-dependent mechanical behaviour of cracked concrete shear-critical beams without web reinforcement subjected to high levels of sustained loading. 42 concrete beams without shear reinforcement are tested under a sustained load close to their ultimate shear capacity, for periods ranging from 3 to 24 months. The beams are designed to fail in shear. The applied load is higher than 87% of the short term shear capacity. The crack opening displacement and crack length development are monitored in time. The experimental results show that there is no significant sustained loading effect on the shear capacity of concrete beams.

1 Introduction

Many attempts have been done to study the short-term shear capacity of concrete beams without shear reinforcement. Little attention has been paid to the behaviour of shear cracks under sustained load. The lack of experimental data has led to use a reduction coefficient in case of long-term loading in some codes. The Dutch Code for concrete structures (NEN 6720), uses a concrete strength in long-term loading that is 0.85 of the short-term strength. Since shear resistance is a function of the tensile strength of the concrete, also this aspect is then subjected to a sustained loading effect [1].

In this project the influence of long-term loading on the shear capacity of concrete beams without shear reinforcement is investigated. The goal is to quantify the possible shear capacity loss due to long-term loading. For that reason, several test series have been carried out on concrete beams subjected to high shear loads close to the short-term failure load, for periods ranging from 3 to 24 months, during which the deflection, crack development and crack widths are monitored.

When a concrete beam is subjected to a high sustained load, a flexural cracking pattern develops along the span. Various shear-carrying mechanisms may develop, e.g. aggregate-interlock and dowel action. These mechanisms induce tensile stresses in the concrete in front of the crack tip. Once the tensile strength of the concrete in these regions is reached, the existing flexural cracks progress in a diagonal direction or new ones are created. The development of a shear crack, however, does not necessarily imply the collapse of the member but in case of high sustained loading, the crack width and length will increase.

2 Background

Experimental investigations showed that the fracture process in concrete structures includes three different stages: crack initiation, stable crack propagation and unstable fracture [2]. In order to predict the crack propagation, several fracture models, like the fictitious crack model by Hillerborg et al. [3] and the effective crack model by Karihaloo and Nallathambi [4] have been proposed. In the models, different material parameters are introduced to describe the cracking properties of concrete materials. These fracture models that attempt to adapt LEFM to concrete structures can only be used to predict the unstable fracture (i.e. in front of crack tip) and cannot be used to predict crack initiation.

Zhou [5] proposed a time-dependent fracture model for concrete based on experimental tests on pre-notched concrete beams. The experimental procedure is based on two phases; in the first phase, the external load grows from zero to the nominal level (a fraction of the maximum load $P_{\max}$) under deflection control, while, during the second phase, the load is kept constant until creep rupture occurs (pre-peak sustained bending). Such tests are usually denoted as pre-peak sustained bending tests. Of course, in order to know the maximum load $P_{\max}$, a number of static tests have to be previously

Proc. of the 9th *fib* International PhD Symposium in Civil Engineering, July 22 to 25, 2012,
Karlsruhe Institute of Technology (KIT), Germany, H. S. Müller, M. Haist, F. Acosta (Eds.),
KIT Scientific Publishing, Karlsruhe, Germany, ISBN 978-3-86644-858-2

executed. To overcome this difficulty, various investigators prefer to use the so-called deformation-controlled post-peak tests where the creep phase starts beyond the peak load [6]. Furthermore, based on the inclusion of a standard rheological model for creep and relaxation into the fictitious crack model, the time dependency of crack opening is accommodated. Similar research is done by Barpi et. al. [7].

3 Experimental program

A total number of 42 concrete beams (divided in 7 groups, each group consisting of 6 beams) have been tested to investigate the behaviour of the beams under high sustained loads. The first group was only tested in short-term loading, to obtain the ultimate shear capacity, crack opening displacement (COD), type of failure, and to gain insight into the scatter of the results. In the other groups of beams 3 are tested in long-term and 3 in short-term loading, the latter as a reference of the ultimate shear capacity. The beams are reinforced longitudinally at the bottom to prevent bending failure, but no shear reinforcement is used. To provide sufficient anchorage capacity, steel plates are welded to both ends of the bars. The geometry of the beams and the loading scheme are illustrated in Fig. 1. The compressive strength, ultimate capacity P_{max} (short-term) and load ratio (ratio of the long-term load to the short-term ultimate capacity P_{max}) are shown in Table 1.

The measuring system installed on each beam consists of a load-cell, one LVDT at midspan to measure deflection and two diagonal LVDT's to measure the crack opening displacements (COD's). An additional measuring system with a manually operated LVDT is applied on beams in groups 6 and 7. The zero measurements are taken when the beams are only loaded by their self-weight. Thus, the influence of the self-weight is not incorporated in the measuring results of the LVDT's. After 28 days curing in a fog room at 99.99% RH and 20 °C, beams are stored and tested in a climate room at 50% RH and 20 °C.

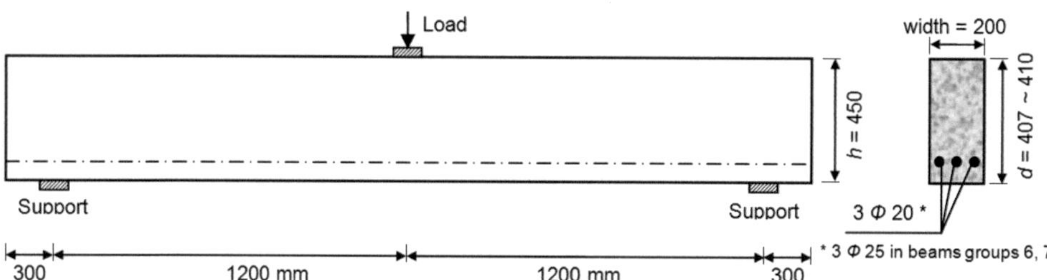

Fig. 1   Dimensions and cross section of the beams (all dimensions in mm).

Table 1   General properties of the tested beams (compressive strength is tested at 28 days).

Group	$f_{cm,cube}$ [MPa]	Type of test	P_{max} (mean) [kN]	Load ratio	Loading duration
1	30.3	6 short-term	184.7 (COV = 4.6%)	-	-
2	32.1	3 short-term	188.9 (COV = 3.2%)	-	-
		3 long-term	-	0.87	3 months
3	47.0	3 short-term	204.9 (COV = 1.4%)	-	-
		3 long-term	-	0.95	24 months
4	42.0	3 short-term	194.7 (COV = 4.9%)	-	-
		3 long-term	-	0.95	12 months
5	44.2	3 short-term	203.2 (COV = 3.1%)	-	-
		3 long-term	-	0.89	6 months
6	73.4	3 short-term	250.1 (COV = 2.7%)	-	-
		3 long-term	-	0.90	18 months
7	78.0	3 short-term	229.9 (COV = 6.7%)	-	-
		3 long-term	-	0.92	11 months

3.1 Short-term loading

In the short-term tests, P-COD curves are obtained for the reference beams. The COD's in Fig. 3 are related to the failure cracks and are measured along two diagonal LVDT's that are installed on the surface of the beam, see Fig. 2. These LVDT's are not measuring crack tip opening displacement or crack mouth opening displacement, which could be located in different areas of the beam in various tests, but they are monitoring the crack opening at the fixed location in every single test. That makes it easier to make a comparison with other beams.

Two types of failure are observed in the short-term tests; they are both flexural shear failure but the first type is failure due to a shear crack through (crossing) the potential compression strut and the second type is failure due to a crack not crossing the potential compression strut which is often associated with a relatively higher ultimate load, see Fig. 4. Specimens G1B4 (refers to Group 1-Beam 4), G2B2, G3B2, G4B1 and G4B2 have failed in Type II which is with a large midspan deflection and large crack opening displacement. The reason to have a larger deflection and a larger crack opening is that the crack tip ends under the loading plate and the stress concentration under the loading plate prevents the crack tip to open. So it is somehow protected under the loading plate, but more deflection leads to larger crack opening at the middle of the crack. The rest of the beams failed according to Type I which is with total crushing of the beam.

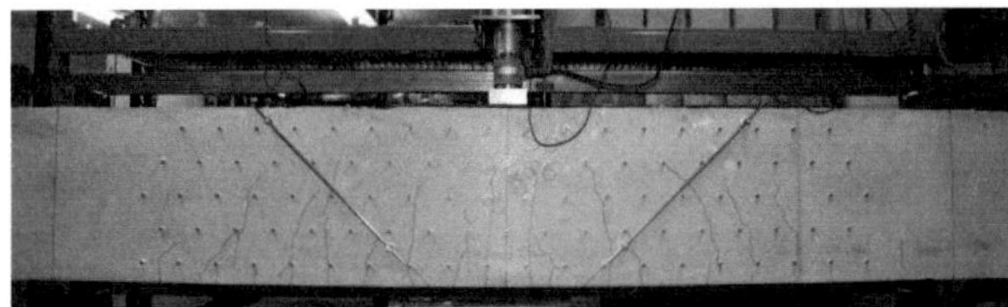

Fig. 2 Measuring devices installed on the surface of the beam to measure the strains and crack width during loading.

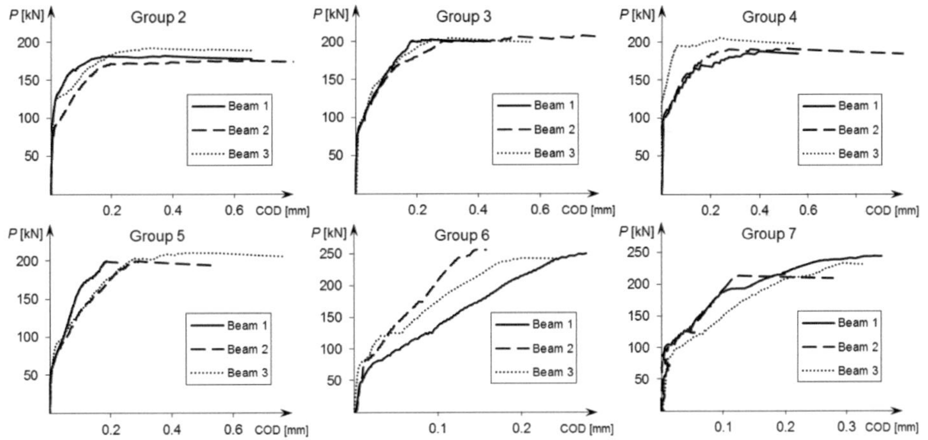

Fig. 3 P-COD curve of the reference tests in groups 2-7.

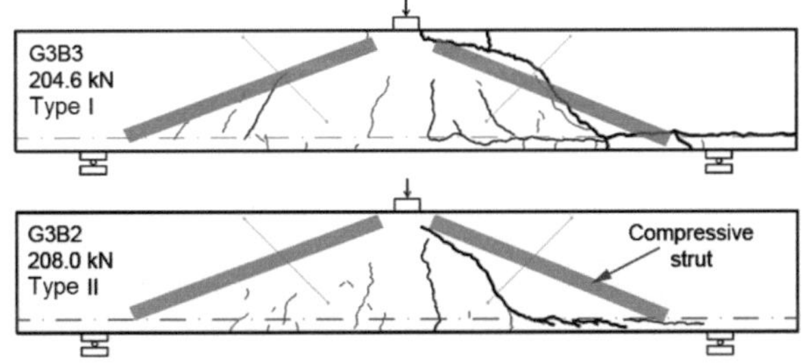

Fig. 4 Two types of failure; beam G3B3 failing in Type I and beam G3B2 failing in Type II.

3.2 Long-term loading

3.2.1 Test procedure and measurements

As presented in Table 1, three concrete beams in each group are subjected to long term loading (between 3 and 24 months). The main goal of the test is to study the behaviour of concrete beams under high sustained loads. Therefore, to apply a load close to the ultimate shear capacity, a load ratio of 0.87~0.95 of the mean value of P_{max} from short-term tests on reference beams in the same group is chosen. Of course, at a very high load ratio, the risk of failure during application of the load is increasing as it has happened in some cases. During the long-term loading tests, the crack opening development, the crack length development and the appearance of new cracks were measured.

3.2.2 Crack length development

The progress of every single crack on the surface of the beam is monitored in time. The surface cracks are categorized in two groups; major cracks and minor cracks. Minor cracks are very small cracks (less than 100 mm length) which can never lead to failure or be part of a failure crack, but are large enough (longer than 10 mm) to affect the stress redistribution in the beam. Since the concrete beams were older than 60 days at the time of loading, the shrinkage cracks had mostly appeared before loading, thus are not considered in these measurements.

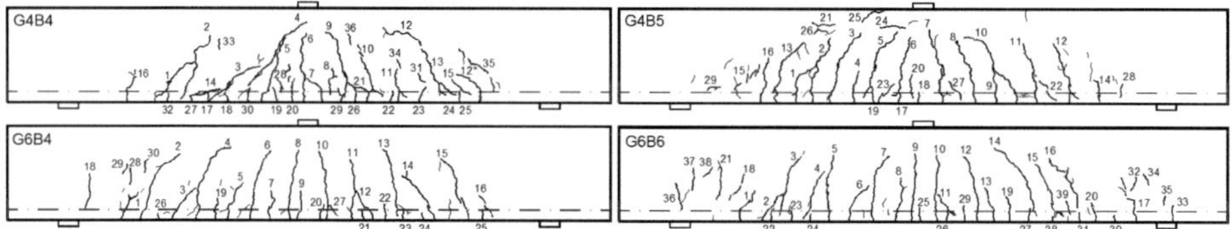

Fig. 5   Crack pattern in specimens S4B4, S4B5, S6B4 and S6B6

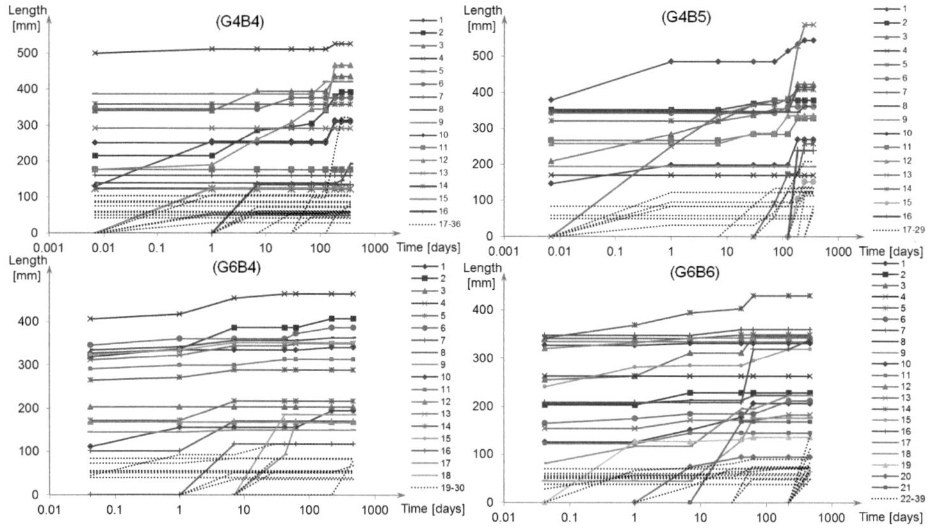

Fig. 6   Development of cracks length in time. Dot-lines represent the minor cracks.

Fig. 6 shows the development of the crack length in four beams. Clearly, some of the cracks are developing in time, while some have a constant length. The development of the cracks is not necessarily limited to the large cracks; sometimes small cracks show considerable progress, while there is no progress in the large cracks. To have a better overview of the crack length development, the total lengths of the cracks are shown in Fig. 7. It follows from Fig. 7 that the development of the cracks in beams group 4 is noticeably more than in group 6. As presented in Table 1, the mean compressive strength 42.0 MPa, while in group 6, the mean compressive strength is 73.4 MPa. As a result, cracks in beams with lower concrete strength show faster progress. Since this progress is in the first year, there is almost no appearance of new cracks in the second year of test.

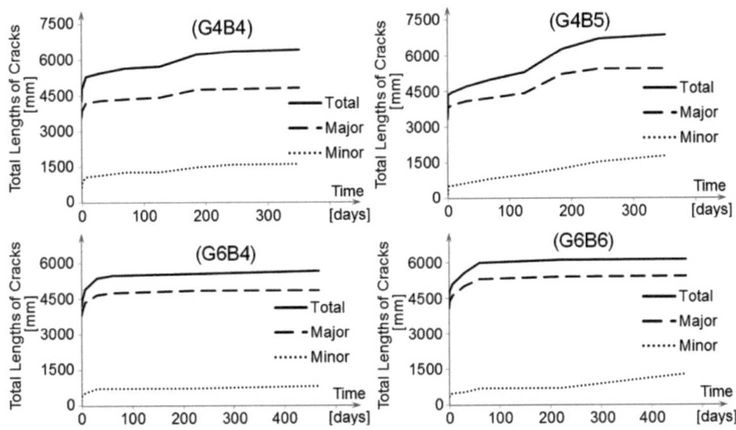

Fig. 7   Development of total crack lengths in time.

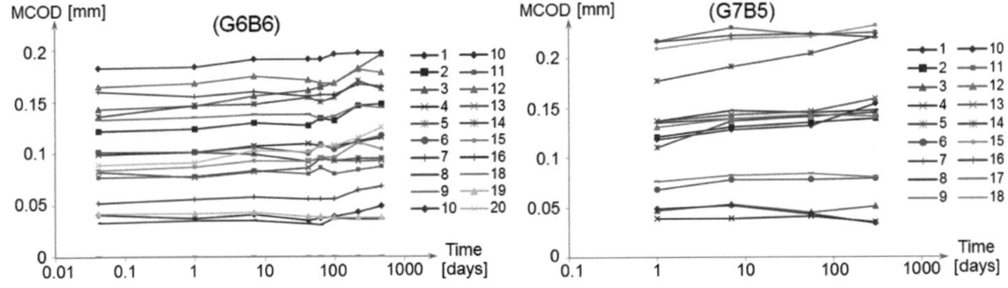

Fig. 8   Development of maximum opening displacement of major cracks in time.

3.2.3 Crack opening development

The width of major cracks are monitored in time. Measuring of the width of minor cracks was skipped since they were sometimes out of the measuring area, so it was impossible to measure the width. The opening of the cracks is measured perpendicular to the crack face in various positions, nevertheless the maximum opening of a crack is taken into account, which is in some cracks in the middle and in some others at the bottom of the crack. The maximum crack opening displacements (MCOD's) of two beams are shown in Fig. 8. Noticeably, in some cracks, MCOD increases in time while in some other cracks, remains constant. The cracks with a considerable increase in length have a significant opening in time (e.g. crack number 3 in specimen G6B6, see Fig. 6 and 8), however, there are always some cracks which propagate in time, but demonstrate a small change in crack width. The same happens when some cracks open in time, but there is no visible increase in length.

3.2.4 Appearance of new cracks

In the beginning of the loading, the crack pattern mostly appears in the middle of the beam and it does not reach the support area. As shown in Fig. 6, during sustained loading, new crack appear which are mostly near to the supports.

3.2.5 Ultimate shear capacity

In order to find out the effect of sustained loading (any reduction in strength), some of the beams were loaded to failure. Cube compressive strength tests were performed in time together with the beam tests to get insight into the concrete strength development. The ultimate capacity P_{max} of the beams was tested to investigate whether it is decreased due to long-term loading. Concrete beams at the end of the desired period of long-term loading, were overloaded to failure (without any load-reduction or unloading). In Fig. 9 (left), a comparison is made between the results of tests in groups 2 and 4 after long-term loading and the results of the reference tests together the estimated shear capacity based on the development of concrete strength. The ultimate bearing capacity of the beams after long-term loading corresponds to the estimated capacity (calculated numerically) and no effect of the previous sustained loading was observed.

Except in two beams, no failure was observed during long-term (sustained) loading. In two cases (G4B6 and G5B5), the beams failed shortly after initial loading, when they were still in the crack initiation stage. G4B6 failed in 3 hours and G5B5 failed in 48 hours. As shown in Fig. 9 (right), the crack opening displacement at both sides of the beam were stabilized after 1.5 hour, but suddenly when the load was constant, the shear crack on the right side opened and the beam failed.

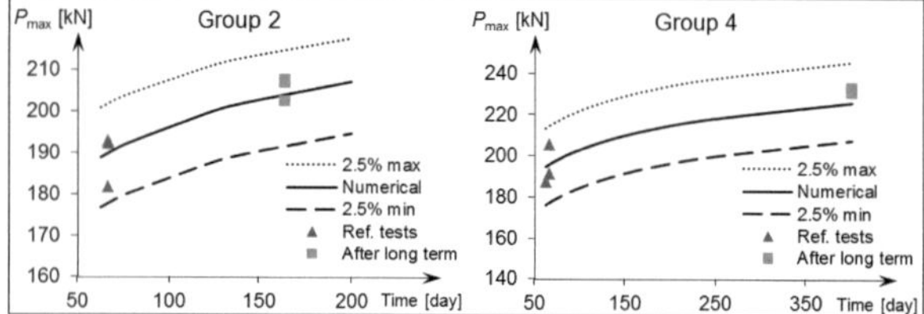
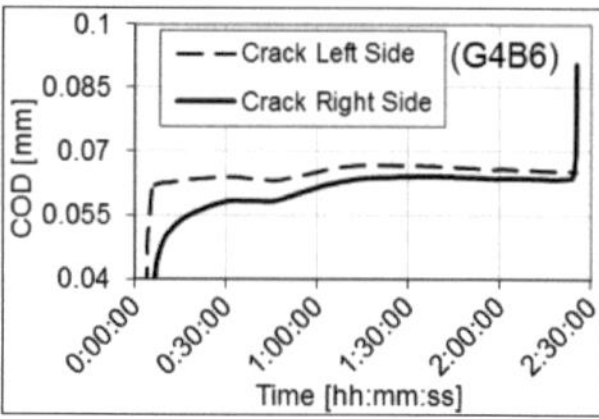

Fig. 9   Left: P_{max} of the beams at the end of long-term loading (squares) compared to P_{max} of reference tests in the same group (triangles) and estimated capacity based on the development of concrete strength. Right: Failure of specimen G4B6 shortly after loading.

4   Conclusions and outlook

The following conclusions can be drawn from this research:

- Shear tests have been carried out on concrete beams without shear reinforcement, subject to high levels of loading. The level of the sustained loading corresponded to values of 87 – 95 % of the average short-term shear capacity, which was determined in separate tests. The tests did not show a significant sustained loading effect.
- During sustained loading, slight propagation of shear cracks was noted. This crack propagation was more pronounced for lower concrete strength. However, it did not result in failure.
- In two cases shear failure occurred under sustained load: in one case 2.5 hours after loading and in the other case 48 hours after loading. In those cases, however, the sustained load was in the 95% confidence interval of the short-term shear capacity. For further extensions of the sustained loading results described in this work, numerical investigation is required.

Acknowledgment

The author would like to acknowledge his supervisors, Prof. Dr.ir. J.C. Walraven, Ir. J.A. den Uijl and Dr.ir. C.R. Braam, for their unparalleled support and guidance they gave throughout this research study and for providing a comfortable research group and well-equipped laboratory.

References

[1]   NEN. TGB 1990: Regulations for concrete - Structural requirements and calculation methods. Tech. Rep. NEN 6720, Nederlands Normalisatie-instituut, (1995)

[2]   Xu, S. and Reinhardt, H.W.: Determination of Double-K Criterion for Crack Propagation in Quasi-Brittle Materials, part I. Intl. J. of Fracture, 98(1999), Issue 2, pp. 111-149

[3]   Hillerborg, A., Modeer, M. and Petersson, P.E.: Analysis of crack formation and crack growth in concrete by means of fracture mechanics and FE. Cem. & Conc. Res. 6(1976), pp. 773–782

[4]   Karihaloo, B.L. and Nallathambi, P.: Notched beam test: Mode I fracture toughness. Fracture Mechanics Test Methods for Concrete, Report of RILEM Technical Committee 89-FMT (Edited by S.P. Shah and A. Carpinteri), Chapman & Hall, London, (1991), pp. 1–86

[5]   Zhou, F.P.: Time-dependent crack growth and fracture in concrete. PhD thesis, Report TVBM-1011, Lund University of Technology, Sweden, (1992)

[6]   Carpinteri, A., Valente, S., Zhou, F.P., Ferrara, G., Melchiorri, G.: Tensile and flexural creep rupture tests on partially-damaged concrete specimens. Mat. & Struc., 30(1997), pp. 269–276

[7]   Barpi, F., Chille, F., Imperato, L., Valente, S. Creep induced cohesive crack propagation in mixed mode. In: Durban, D., Pearson, J.R.A. (Eds.), Non-Linear Singularities in Deformation and Flow. Kluwer Academic Publishers, The Netherlands, (1999), pp. 155–168

Aspects regarding flexural behaviour of high strength reinforced concrete beams

Andreea Muntean

Faculty of Civil Engineering,
Technical University of Cluj-Napoca,
G. Baritiu-Str. 25, 400027 Cluj-Napoca, Romania
Supervisor: Cornelia Măgureanu

Abstract

This paper presents the flexural tests results of four high strength reinforced concrete beams. The research is a part of the experimental program of the author thesis regarding the long-term behaviour of high strength reinforced concrete beams subjected to flexure. In order to establish the sustained loading step for long term loading, similar beams were tested under short - term bending moment up to failure; this paper presents only aspects on behaviour of the short term loading beams from the author's thesis programme.

The short-term tested beams were realized with C70/85 high strength concrete class and Bst500S steel reinforcement. As a variable parameter, longitudinal reinforcement ratio was considered (2.01%, respectively 3.39%). Bending behaviour to the ultimate limit state (ULS) was recorded regarding experimental yielding of the longitudinal reinforcement and the beam; also failure values were measured (moment, deflection). A linear variation regarding the experimental deflection values can be noticed up to the moment that the beam reaches the yielding beam moment M_y, respectively for a value of the ratio M_y/M_u between 0.85 and 0.88. Also, the paper presents a study between experimental values and calculus values of deformations at yielding of the reinforcement and yielding of the beams. All design issues have been discussed and analysed in the perspective of EN 1992-1-1:2004 design code provisions. The increase of reinforcement ratio leads to an increase in design stiffness, therefore the difference between calculated values of deflection according to EN 1992-1-1 and realistic values of deflections is growing. The longitudinal reinforcement ratio influence is major in the design of deflection according to EN 1992-1-1, at a higher reinforcement ratio, higher element stiffness can be obtained; the experimental deflection values are set beyond 50% the calculated values.

1 Introduction

High strength concrete has, in recent years, been the focus of significant interest within the world's civil engineering community. Ever taller buildings and ever greater spans are sought in today's competitive market; such as an increase in the compressive strength of concrete is necessary. Of particular interest is the behaviour of high strength concrete elements subjected to flexure, because of the brittle failure that can occur, hence proper understanding of this behaviour is necessary in order to accurately predict bearing capacity and failure deflection. Ultimate limit states are often more critical for concrete structures. Consequently when design is undertaken, the ultimate limit state is designed for and then if necessary serviceability is checked for. However, element sizes ascertained in the pre-design stage usually ensure serviceability criteria are met. If actual deflections are required, then the structure must be analysed for the serviceability limit state, using design service loads. The deflections obtained will generally be short term values and will be multiplied by a suitable factor to allow for creep effects and hence give realistic long term values. The results of an investigation carried out on flexural behaviour of reinforced high strength concrete beams with different longitudinal reinforcement ration are presented in this paper. This paper also provides a study on the deformability of reinforced concrete elements subjected to bending moment action, in the stages of steel yielding and beam yielding.

Proc. of the 9th *fib* International PhD Symposium in Civil Engineering, July 22 to 25, 2012,
Karlsruhe Institute of Technology (KIT), Germany, H. S. Müller, M. Haist, F. Acosta (Eds.),
KIT Scientific Publishing, Karlsruhe, Germany, ISBN 978-3-86644-858-2

2 Experimental program

2.1 Concrete composition

The high strength concrete class studied was C70/85 for the reinforced elements subjected to short-term loadings. The concrete composition is presented in Table 1. The high strength concrete contains silica fume in 10% by weight of cement and water/binder ratio of 0.30. Physical-mechanical properties of the high strength concrete were determined at the age of 28 days on cubes of dimensions of 150mm×150mm×150mm and prismatic specimens of 100mm×100mm×300mm cast from the same batch as the reinforced elements.

Table 1   Concrete composition

Components	Quantity
Portland cement CEM I52.5R	480 kg/m^3
Silica fume	48 kg/m^3
Crushed gravel (4 - 16 mm)	1065 kg/m^3
Sand (0 – 4 mm)	710 kg/m^3
Water	129.6 l/m^3
Superplasticizer (Glenium ACE 30)	14.4 l/m^3
Water/cement ratio	0.27
$f_{cm,cube}^{28\ days}$ (MPa)	96.30
E_{cm} (MPa)	45217

2.2 Elements geometry and detailing

The experimental program was carried on a number of four simple reinforced concrete beams, tested at 4 point bending. The beams had a rectangular cross section of 125mm×250mm and a total length of L=3200mm. Different reinforcement ratios for longitudinal reinforcement was used as shown in Table 2 and Fig. 1, for each longitudinal percentage two beams were made. Transversal reinforcement consisted of Ø6/300 stirrups, with f_{yk}=500MPa, localized only on shear zone of the beam, in view of the main focus of the test was the behaviour under pure bending moment.

Table 2   Longitudinal reinforcement ratio for tested beams

Beams	Longitudinal reinforcement (A$_s$)	Reinforcement ratio for longitudinal reinforcement $\rho_l = \dfrac{A_s \cdot 100}{b \cdot d}$	Mechanical longitudinal reinforcement coefficient $\omega_s = \dfrac{A_s \cdot f_{yd}}{b \cdot d \cdot f_{cd}}$
AD 1-1, AD 3-1	3Ø14 Bst500S	2.01 %	0.164
AD 2-1, AD 4-1	3Ø18 Bst500S	3.39 %	0.276

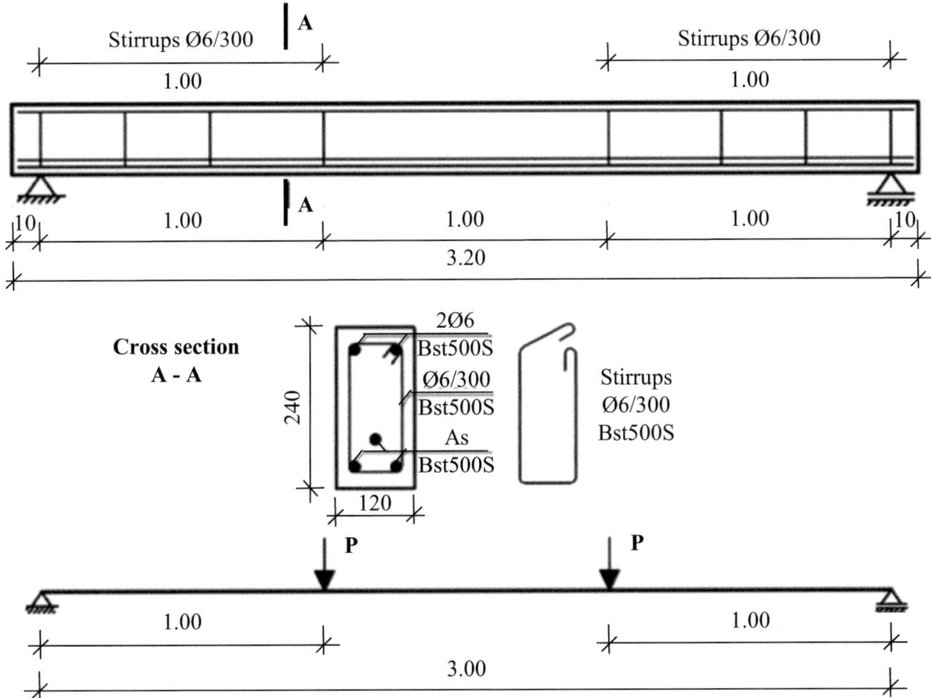

Fig. 1   Longitudinal reinforcement and cross section for short-term tested beams

2.3   Preparation and testing of elements

All elements were cast in metallic moulds. The beams and the control specimens were dismantled the next day, cured for 28 days in a moist environment and then air-dried in the laboratory prior to testing. Fig. 2 presents the beams short-term test set-up of four-point flexure loading over a clear span of 3000 mm. For each load increment, of 1/10 of bending design capacity, the concrete strains were measured with mechanical gauges (precision of 0.01mm) and digital micro comparator gauges (0.001mm precision), with 200 mm measurement base, applied on one side of the beam. The deflections along the reinforced concrete beam were measured not only with mechanical gauges (0.1mm precision), but also with LVDT displacements transducers (0.031mm precision). All strain and deformation readings were captured by a computer at pre-set load intervals until collapse. All beams were subjected to short-term loading to failure, in the ultimate limit state (ULS).

Fig. 2   Test set-up and failure mode

3　Test results and discussions

3.1　General behaviour of reinforced high strength concrete beams

Several distinctly different segments can be noticed to appear as a result of the events that took place during the loading history: first cracking (M_{cr}), yielding of tensile reinforcement ($M_{y0.2}$-$\Delta_{y0.2}$), and yielding of the beam (M_y-Δ_y), crushing of the concrete cover in the compression zone and at last the failure of compression zone (M_f). All these events can idealize a typical load - deflection curve as shown in Fig. 3. The experimental load - deflection curves, grouped according to the variable parameter considered, the longitudinal reinforcement ratio ρ_l, are presented in Fig. 4.

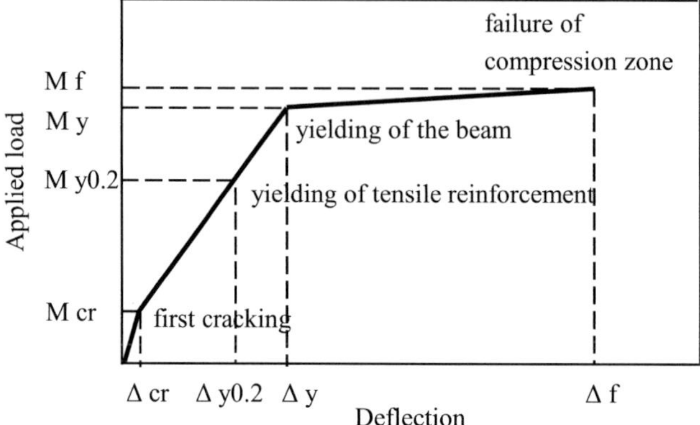

Fig. 3　Typical moment- deflection curve

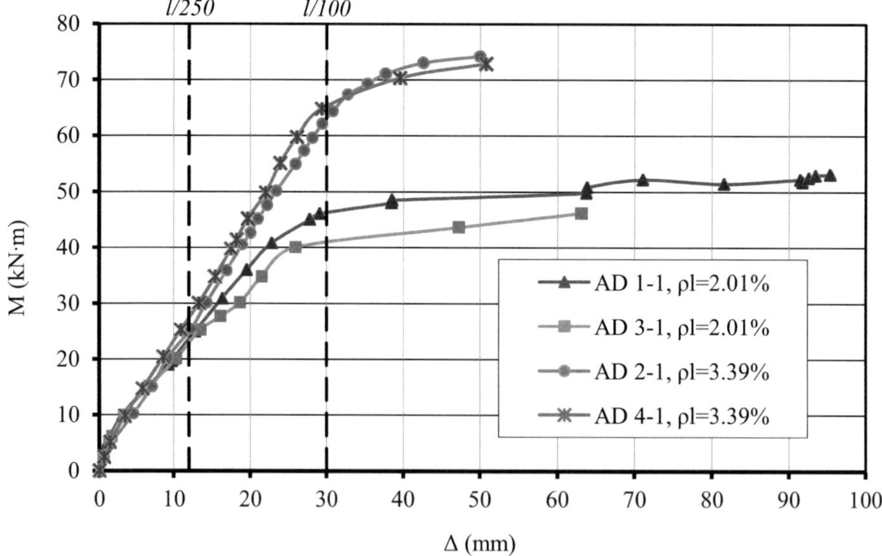

Fig. 4　Experimental curves moment - deflection

The variation law of the curves M - Δ can be considered approximately linear until the beams reaches the yielding moment M_y, respectively for a value of the ratio M_y/M_f between 0.85 and 0.88, regardless of the longitudinal reinforcement ratio used. After the value of 85%-88% of the ultimate bending moment, the loading moment stays constant, the beams presents good deformations up to failure, hence a good post-elastic behaviour.

The experimental test shows that beams with lower reinforcement ratio of 2.01% presented higher deflection ductility than the beam with 3.39% longitudinal reinforcement ratio and a decrease of the failure bending moment (M_f) with approximately 52%.

3.2   Deflection calculus according to SR EN 1992

When the maximum moment (M) in a beam does not exceed the cracking moment (M_{cr}), the beam is considered to be in the uncracked condition state. At the time that the bending moment (M) reaches M_{cr}, in the exterior layer of the tension zone vertical flexural cracks appear and propagate upward as the bending moment increases. Hence the flexural cracks reach the neutral axis and the section becomes fully-cracked. EN 1992-1-1, the European structural concrete codes, proposes for the deflection calculations a simplified method based on the determination of the curvatures and deflections of a concrete beam corresponding to its uncracked and fully-cracked conditions. Equation (1) was used to determine the design value of deflections of a reinforced concrete beam loaded at a level causing the beam to crack:

$$\Delta = \zeta \cdot \Delta_{II} + (1-\zeta) \cdot \Delta_{I} \tag{1}$$

$$\zeta = 1 - \beta \cdot \left(\frac{M_{cr}}{M} \right) \tag{2}$$

Where Δ_I and Δ_{II} are the deflection values corresponding to the fully-cracked and uncracked conditions of the beam, respectively, β is a coefficient accounting for the duration of loading or of repeated loading on the average strain. EN 1992-1-1 propose that β to be taken 1.0 for single short-term loadings and 0.5 for sustained loads or many cycles of repeated loading. Taking into consideration that long-term deflections of reinforced concrete beams are out of scope for the present study, the coefficient β is taken equal with the value of 1.0.

The design deflection can be obtained from the elements curvature calculated by summing the effects of uncrack section state and crack section state on curvature.

Fig. 5 and Fig. 6 presents the development of design deflections calculated according to EN 1992-1-1 and also the experimental values obtained for the beams AD 1-1 and AD 3-1 with 2.01% longitudinal reinforcement ratio, respectively for AD 2-1 and AD 4-1 beams having ρ_l = 3.39%. Calculated deflections were determined for the same loading steps as obtained from the experimental short-term tests.

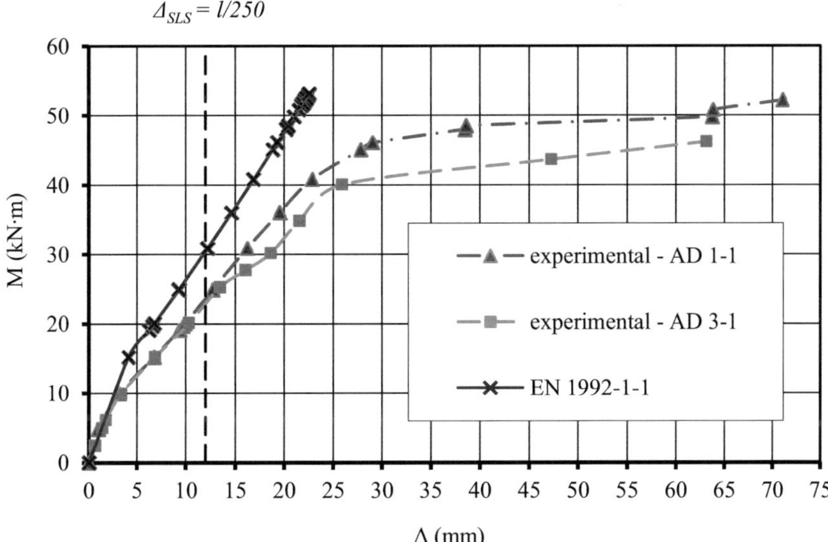

Fig. 5   Experimental and design deflection values for the beams with longitudinal reinforcement ratio of 2.01%.

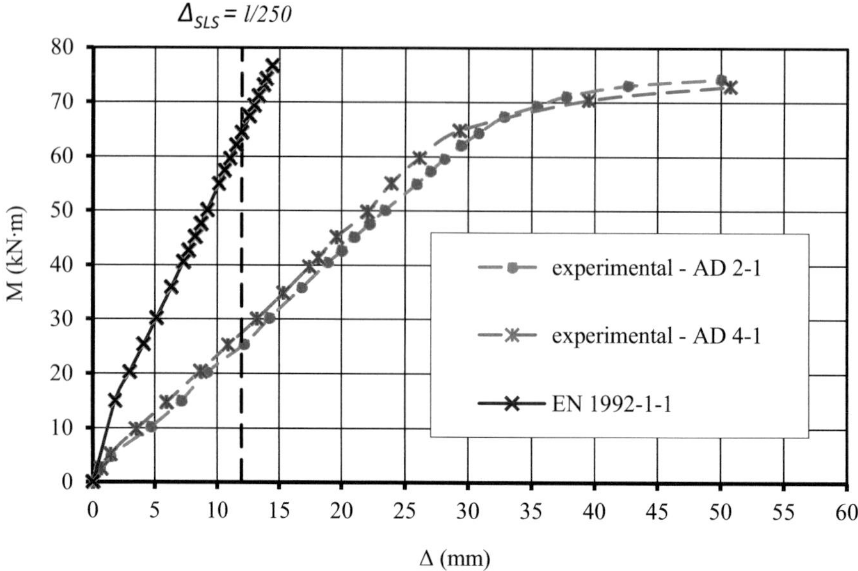

Fig. 6   Experimental and design deflection values for the beams with longitudinal reinforcement ratio of 3.39%.

Table 3   Experimental and calculated deflections for the same loading steps used in the tests

Results	Beam		AD 1-1	AD 3-1	AD 2-1	AD 4-1
	Longitudinal reinforcement ratio		ρ_l=2.01%		ρ_l=3.39%	
Experimental	Cracking moment	M_{cr} (kNm)	4.80	6.15	4.65	5.10
Experimental	At steel yielding	$M_{y0.2}$ (kNm)	40.80	27.68	42.60	39.75
		$\Delta_{y0.2}$ (mm)	22.80	16.01	20.00	17.40
		l/x	$l/132$	$l/187$	$l/150$	$l/172$
EN 1992-1-1		$\Delta_{y0.2}$ (mm)	16.86	10.637	7.671	7.094
		l/x	$l/178$	$l/282$	$l/391$	$l/423$
Comparison		$\Delta^{EN\,1992}/\Delta^{exp}$	0.74	0.66	0.38	0.41
Experimental	At beam yielding	M_y (kNm)	45.00	40.05	67.35	64.80
		Δ_y (mm)	27.76	25.90	32.80	29.30
		l/x	$l/108$	$l/108$	$l/91$	$l/102$
EN 1992-1-1		Δ_y (mm)	18.797	16.51	12.58	12.08
		l/x	$l/160$	$l/182$	$l/239$	$l/248$
Comparison		$\Delta^{EN\,1992-1-1}/\Delta^{exp}$	0.68	0.64	0.38	0.41
Experimental	Failure moment	M_f (kN)	53.10	46.20	76.65	74.40

Note: Effective span length l=3000mm, $l/x=\Delta$.

Fig. 5 presents the development of experimental and calculated values of deflections for the beams with longitudinal reinforcement ratio of 2.01%, with the assignation that experimental deflection reaches the serviceability limit deflection $\Delta_{SLS}= l/250$=12 mm when a load level of M/M_f=0.49-0.50 is obtained, whereas the calculated deflection reaches $\Delta= l/250$=12 mm at M/M_f=0.57-0.66. For the high strength reinforced concrete beams with 3.39% longitudinal reinforcement ratio (Fig. 6), design value of deflection in the serviceability limit state is obtained at a loading step of M/M_f=0.84-0.87 overestimating the experimental values, as the experimental deflection of 12mm is reached at a loading ratio M/M_f=0.35-0.41.

For beams with smaller longitudinal reinforcement ratio (ρ_l=2.01%), design deflections are close to the experimental values, the ratio $\Delta^{EN\ 1992-1-1}/\Delta^{exp}$ is kept approximately the same for the steel yielding as for beam yielding as shown in Table 3. Similarly the design values of deflection at steel and beam yielding were compared to the experimental values for the beams with a longitudinal reinforcement ratio of 3.39%, it was noticed that the calculated values were below 50% of the experimental values.

4   Conclusions

The following conclusions in relation to flexural behaviour of reinforced high strength concrete beams were drawn from this investigation. The steel reinforcement ratio ρ contribute to the load-deflection characteristics of the beams. Higher longitudinal reinforcement ratio ρ cause a decrease of the calculated deflections according to the equations given by EN 1992-1-1. Experimental deflections present a linear variation up to the yielding of the beam, at a value of the ratio M_y/M_u between 0.85 and 0.88. For a smaller longitudinal reinforcement ratio of 2.01%, the research shows that the experimental and design values of deflection are relatively close. Along with the increase of reinforcement ratio, it can be observed an increase in design stiffness, therefore the difference between calculated values of deflection and real values of deflections is increasing. Author remarks that the longitudinal reinforcement ratio influence is major in the design of deflection according to EN 1992-1-1, at a higher reinforcement ratio, higher element stiffness can be obtained, and also the experimental deflection values are more than 1.5 times the calculated values.

Acknowledgments

Paper prepared for the Project "Doctoral studies in engineering science in order to develop a society based on knowledge - SIDOC", Contract POSDRU/88/1.5/S/60078, with the aid of a type A research grant, code CNCSIS 1036 « High Strength/High Performance Concrete with Steel and Carbon Fibbers, and Rubber Powder. The Behaviour in seismic areas, aggressive environments, dynamic and wear actions. Environmental Ecology », director grants Cornelia Măgureanu, Professor and Ph.D.

References

[1]   EN 1992 -1-1 / 2004 Design of concrete structures - Part 1-1: General rules and rules for buildings, December, 2004.

[2]   L. Martin, J. Purkiss, Concrete design to EN 1992, Second edition, 2006.

[3]   Hegheş, B., Magureanu C., Rotation Capacity of Reinforced Concrete Elements, The 3rd Central European Congress on Concrete Engineering, Visegrad, Hungary, 17-18 September, 2007, pp. 301-308, ISBN 978-963-420-923-2.

[4]   Gilbert, R.I., Shrinkage, Cracking and Deflection – the Serviceability of Concrete Structures, Electronic Journal of Structural Engineering, 1 (2001), pp. 15 – 37.

[5]   Rashid, M.A., Mansur, M.A., Reinforced High-Strength Concrete Beams in Flexure, ACI Structural Journal/May-June 2005, pp. 462-471.

[6]   Hegheş, B. (2009), Ductility of high strength concrete, PhD. Thesis, Technical University of Cluj-Napoca, Romania, UT Press.

[7]   Doug Jenkins, Prediction of Cracking and Deflections, International Code Provisions and Recent Research, Concrete Solution 2009, Concrete Institute of Australia, 19 pp.

[8]   Tabassum, J., Analysis of current methods of flexural design for high strength concrete beams, master of engineering thesis, RMIT University, October 2007, 330 pp.

Shear capacity of reinforced concrete beams under complex loading conditions

Yuguang Yang

Faculty of Civil Engineering and Geosciences,
Delft University of Technology,
Stevinweg 1, 2628 CN Delft, the Netherlands
Supervisors: Joost Walraven and Joop den Uijl

Abstract

In this paper, the shear capacity of reinforced concrete specimens without shear reinforcement loaded under multiple point loads with both continuously and simply supported boundary conditions is investigated. Various experiments have been carried out in the laboratory. The test results are compared with those obtained on similar specimens loaded by a single point load in the same research project. Besides, the test results are compared with design formulas suggested in design codes such as Eurocode and *fib* Model Code 2010. Several recommendations are given with respect to the tested loading conditions.

1   Introduction

The shear capacity of reinforced concrete beams without transverse reinforcement is considered as a complex problem, being influenced by many factors. Among them, one pronounced phenomenon is that the loading condition has a significant effect on the shear capacity especially when the loading point is close to the support. This has been validated experimentally in the case of simply supported beams loaded by a single point load [1] and has been implemented in design codes, such as Eurocode [2]. However, when the structure is loaded under complex loading conditions, for example, in the case of statically indeterminate structures loaded by multiple point loads, the inner force distribution becomes more complicated. Several experiments reported in literature [3-6] have shown that uniformly distributed load and multiple point loads would clearly change the shear behaviour of a simply supported beam such as the ultimate shear capacity, the crack pattern etc. On the other hand, such kind of complex loading conditions is often found in many structures such as the multiple span concrete bridges.

Since reported experimental results with regard to complex loading conditions are quite limited in literature, additional experimental research is needed to get better insight into this problem. For that reason, a series of experiments with complex loading conditions have been designed and carried out in the Stevin Laboratory at Delft University of Technology upon request of the Dutch Ministry of Infrastructure and the Environment. As a part of the research program, the effect of single point load in case of continuously supported boundary conditions has been studied previously; see [7]. The research reported in this paper is a continuation of the research program.

2   Experiment Program

2.1   Test Setup

In total four tests have been carried out. The test setup is illustrated in Fig. 1. The loads are generated by two hydraulic actuators denoted P_1 and P_2 with constant ratio P_1/P_2. Load P_1 is split into two equal point loads by a steel beam. The distance between P_1 and P_2 is 2.4 m, and the centre to centre distance between the two loads split from P_1 is 1.2 m. The setup is designed in such a way that it is comparable with the other tests within the test program, in which the point load P_1 is applied directly onto the beam, see [8]. Since the forces of the two point loads within the main span are always half of P_1, P_1 is referred to when indicating their load levels. During the test, the forces P_1 and P_2 were monitored. Other measurements include the deflection of the specimen at several points,(see Fig. 1), the deformations of the side surfaces of the specimen with photogrammetry measurement and LVDT arrays.

Proc. of the 9th *fib* International PhD Symposium in Civil Engineering, July 22 to 25, 2012,
Karlsruhe Institute of Technology (KIT), Germany, H. S. Müller, M. Haist, F. Acosta (Eds.),
KIT Scientific Publishing, Karlsruhe, Germany, ISBN 978-3-86644-858-2

The specimens are 8 m long, 0.5 m high and 0.3 m wide. Thus the effective height of specimens d is about 450 mm. The mean compressive strength of the concrete measured at testing date is $f_{cm} = 68$ MPa. As indicated in Fig. 1, three Ø32 mm ribbed bars are used as top and bottom reinforcement in the specimens, which makes the reinforcement ratio 1.79%. Stirrups for positioning of the bars are only used outside the regions where shear failure may occur. The specimens are tested at both ends. The crack level of the untested end was carefully controlled, to make sure the shear behaviour of the untested end would not be influenced. If there is a shear crack in the longer span between the last point load and the support that might influence the second test, additional strengthening is applied with a steel frame. For more details of the test program the test report [8] is referred to.

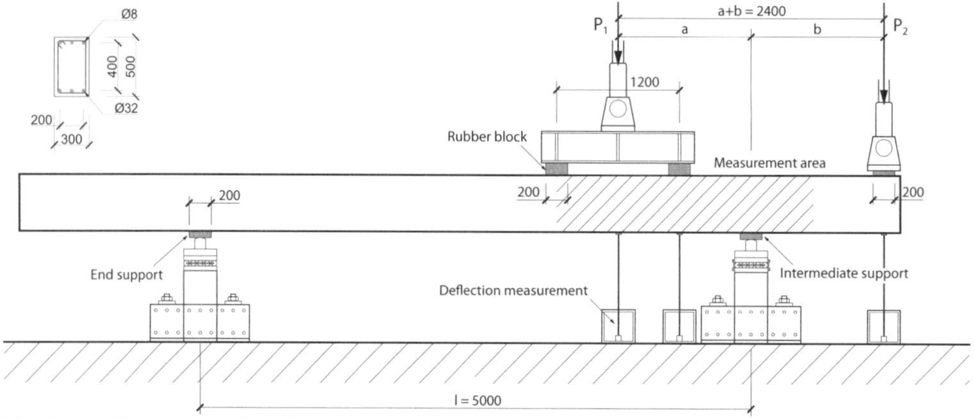

Fig. 1   Test setup and reinforcement configuration.

The configurations of the four tests are summarized in Table 1. The specimens are identified according to the rules explained in Fig. 3. Among the four tests listed, D17a151 and D18a121 are simply supported tests. Only P_1 was applied on the specimen. In order to properly indicate the variables and the locations across the specimens, they are subdivided into several spans by the loads and supports, the inner forces within the spans are numbered respectively. An example is given in Fig. 2 based on D17b154. The numbering system remains the same when P_2 is zero.

Table 1   Test configurations, definitions of the variables can be found in Fig. 2.

Test No.	a [m]	b [m]	a/d	M_1/M_0	P_1/P_2	V_1/P_1	V_2/P_1	V_3/P_1	V_4/P_1	M_2/M_0	M_3/M_0
D17a151	1.5	0.9	3.33	0	-	-	0.70	0.20	0.30	0.90	0.83
D17b154	1.5	0.9	3.33	1	1.46	0.68	0.82	0.32	0.18	0.20	0.83
D18a121	1.2	1.2	2.67	0	-	-	0.76	0.26	0.24	0.60	0.84
D18b152	1.5	0.9	3.33	1/3	3.17	0.32	0.76	0.26	0.24	0.70	0.83

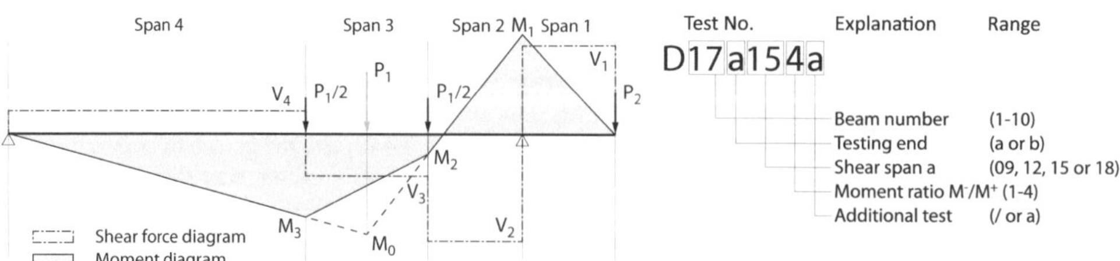

Fig. 2   Moment and shear force distribution and definition of variables; the diagram is based on test D17b154.

Fig. 3   Test numbering system

2.2 Experimental Results

The failure modes of all the four tests can be considered as diagonal tension failure, which is defined by the opening of one or more inclined cracks under the shear force in that span. As indicated in Table 1, the maximum shear force occurs within Span 2 in the tests, so that the cracking process can be monitored by the deformation measurement arranged there during the test. As expected, the first inclined cracks developed in the second span in all four tests. Unlike tests with single point load, in which the inclined crack normally stops at the edge of the loading point, the cracks extend into Span 3 along the top reinforcement in the specimens with multiple point loads. The crack patterns of the specimens after failure, illustrated in Fig. 4, show that the diagonal cracks stretched until the maximum moment, where the curvature of the specimen becomes zero. The shear force at which the inclined cracks developed is defined as inclined cracking force V_{cr}. If the beam has a diagonal tension failure, the value of V_{cr} normally equals to the ultimate shear force V_u. This was the case in test D17a151 and D18b152.

However, in tests D17b154 and D18a121, the opening of an inclined crack in Span 2 did not result in failure. Compressive struts developed between the first loading plate and the support in both cases. Because of that, the load applied on the specimen could be increased further, until diagonal tension failures occurred in a different span. The shear forces in the spans at P_{cr} and P_u are summarized in Table 2, the maximum M/Vd ratios in relevant spans are given as well. The spans in which the specific event occurred are shaded in the table.

Table 2   Shear forces over the spans at critical load levels. The shaded cells indicate the critical span; P_{code} is code prediction (upper one Eurocode, lower one Model Code).

Test	Variable	Span 1	Span 2	Span 3	Span 4	P_1	P_{code}
D17a151	M/Vd	-	2.00	9.76	6.46		
	V_{cr}	-	175.0	50.0	75.0	250.0	238.9 EC
	V_u	-	212.3	60.6	91.0	303.3	254.6 MC
D17b154	M/Vd	2.00	1.67	3.52	6.46		
	V_{cr}	156.1	188.3	73.5	41.3	229.6	234.5
	V_u	340.3	410.3	160.1	90.1	500.4	220.9
D18a121	M/Vd	-	1.33	6.56	6.46		
	V_{cr}	-	190.3	65.1	60.1	250.4	330.9
	V_u	-	465.0	159.1	146.9	611.9	234.5
D18b152	M/Vd	2.00	1.17	6.11	6.46		
	V_{cr}	104.5	248.3	85.0	78.4	326.7	376.2
	V_u	104.6	248.3	85.0	78.4	326.8	271.3

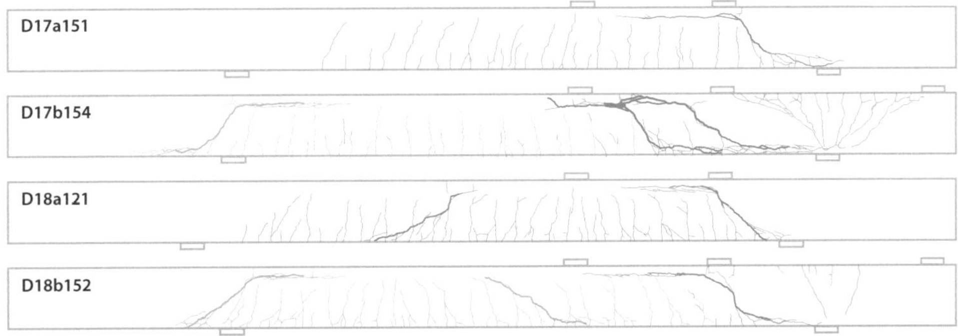

Fig. 4   Crack patterns after failure.

3 Evaluation of Design Codes

In many design methods, such as the Eurocode and the ACI Code, the formula calculating the shear capacity of reinforced concrete structures without transverse reinforcement is in principle based on the regression of tests on simply supported beams with single point load. Thus, they usually do not concern the influence of the moment distribution. The estimated shear capacities with these formulas could be conservative in certain cases. In Eurocode, a force reduction factor β is introduced, which is applied to the point loads that are within a distance of $a_v < 2d$ from the support. β equals to $a_v/2d$, so that the so-called shear slenderness effect found in simply supported beams [1] is taken into account. The formula works well in case of structures with only a sagging moment over the span. Once the moment distribution becomes more complex, for example in case of continuous beams, comparison between test results and code prediction showed that the Eurocode formula tends to underestimate the real capacity [7, 8]. Besides, the β factor is only allowed to apply to the forces in the vicinity of the supports. Once the maximum shear force is in the mid-span of the structure, it is not allowed to increase the shear capacity anymore. In the previous research with regard to continuous beams [8], Yang et al. suggest that the Eurocode formula shall relate to the inclined cracking load V_{cr} of the specimens without shear reinforcement. Besides, the value of a_v/d in the shear reduction factor $\beta = a_v/2d$ on the load side can be replaced with the maximum M/Vd. It can also be placed at the resistance side as an increase of the shear capacity under given load condition. In that way the value of V_{cr} in the span with the highest shear force (see Fig. 2) can be evaluated more accurately even in continuous structures. Formulas with better accuracy are quite welcome in the case of evaluating of the residual capacity of existing structures, where accurate evaluation of the loading capacity may save lots of structures from being demolished.

  The V_{cr} derived from the test program reported in this paper is plotted in Fig. 5 against the maximum M/Vd in the critical spans. The failure of tests D17b154 and D18a121, which occurred in Span 3 and Span 4, respectively, was also defined as diagonal tension failure. Therefore these test results are plotted in the graph as well. The single point load test results reported in [8] are plotted in the same figure as reference. All specimens presented in Fig. 5 have the same concrete mixture and the same reinforcement configuration. Besides, the Eurocode prediction of the mean shear capacity is plotted in the graph to be compared with the test results. The values of V_{EC} is calculated with the formula $V_{Rd,c} = C_{Rd,c}k(100\rho_l f_{ck})^{1/3}b_w d$. The value of f_{ck} is replaced by f_{cm}, and the value $C_{Rd,c} = 0.18/\gamma_c$ recommended in Eurocode is replaced by 0.15 according to the study of König [9], which was originally derived from a regression study on shear test results out of a selected test database covering large variety of test configurations. It is suggested to use this value in evaluating the average strength derived from experiments. As an additional check, a linear regression study has been carried out with the test data shown in Fig. 5. As a result, the evaluated $C_{Rd,c}$ based on the present test data is 0.144, which is very close to the recommended value.

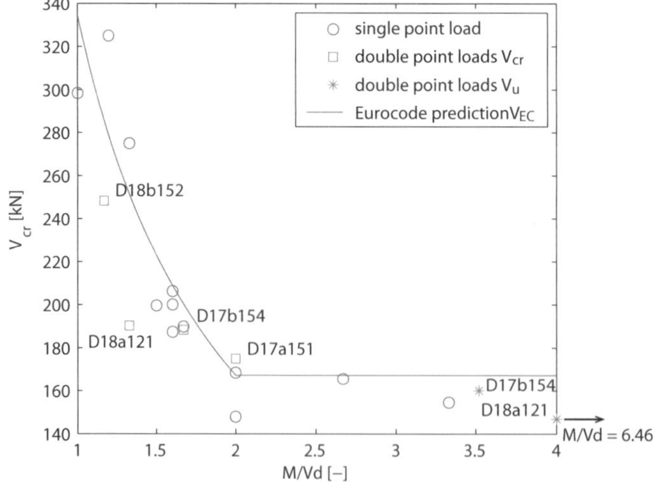

Fig. 5 Comparison of experimental results and prediction for mean shear capacity based on Eurocode, V_{cr} versus M/Vd.

In Fig. 5, it turns out that even under complex moment distributions, as the presented test program shows, the values of V_{cr} are still closely related to the maximum M/Vd in the concerned span. If the value of M/Vd is smaller than 2.0 the shear force required to open an unstable inclined crack can still be increased. On the other hand, once the flexural cracks are fully developed (maximum M/Vd is relatively large), the expected V_{cr} becomes less dependent to M/Vd. The test results derived from the present test series turns out to be quite consistent compared to the single load tests with same specimen configurations. Even in complex loading conditions, the Eurocode prediction reflects the test observations well, with the proposed adoption. The average difference between the individual results and the Eurocode prediction is about 11%.

On the other hand, the recently published Model Code 2010 [10] has introduced a quite different strategy with regard to the shear capacity, with which the shear capacity of a concrete structure can be calculated at three levels of approximation. A higher level formula asks for more complicated calculation and, as a reward, the prediction is supposed to be more close to the structures' real capacity. All three calculation levels recommend to check the critical cross section at $z = 0.9d$ from the support. With level 3 method, the calculation of the shear capacity of the structure is dependent on the average strain in the mid-height of the cross section ε_x, which is determined by the moment and shear force at the section. With this method, the influence of the moment distribution is automatically taken into account. It also means that for structures loaded under complex load conditions, the calculated shear capacity will depend on the location. Consequently, there might be more critical cross sections other than the one at $0.9d$ from the supports. However, the test results available with relatively complex loading condition are really limited to give solid conclusions on the choice of cross sections. With the presented test series it is still possible to evaluate the procedure specified in the new Model Code. The shear capacity of the specimens over the whole length is calculated with level 3 method. As an example, the results of tests D17a151 and D17b154 are plotted in Fig. 6. The shear forces along the specimen at P_{cr} are shown as a reference in the same graphs.

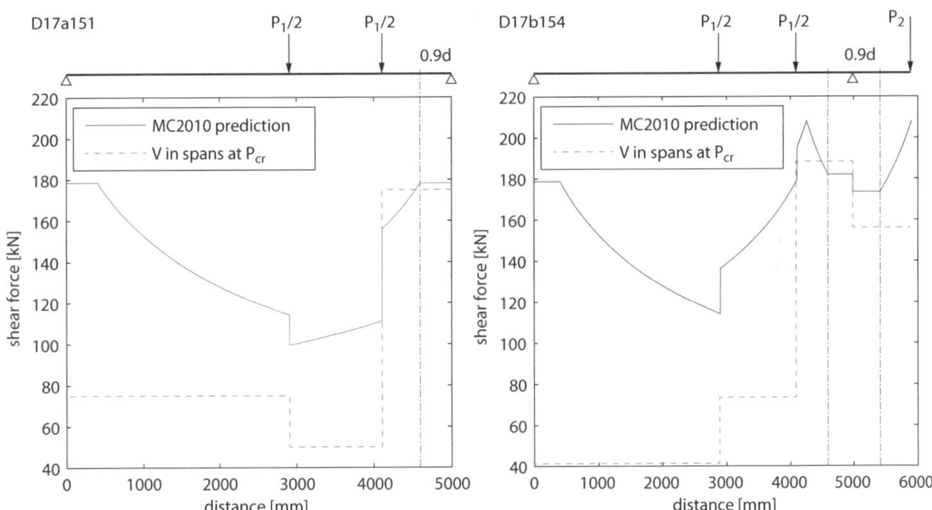

Fig. 6   Comparison of MC2010 prediction and tested results over specimen length

Since in both cases the spans with highest shear forces are still in the vicinity of the support, the rule with regard to the critical cross sections seems still valid. Similar as Eurocode, the predicted results are quite close to the measured inclined cracking load, see Table 2 as well. At that load level, the expected shear capacities of all the other spans are still higher than the shear forces. Thus, at least in the tested cases, only the spans with the highest shear force needs to be checked. On the other hand, Fig. 6 shows that the MC2010 predicted shear capacity increases with the decrease of the moment in the span; see Fig. 2 for the moment distribution. In the simply supported beam D17a151 the calculated shear force at the specified cross section is the highest one in the span, since according to the code the design shear force between the design cross section and the support shall be taken as the value calculated at that section. Other choices would yield more conservative predictions. In the continuously supported beam D17b154 the point of inflection is located in the critical span and the calculated shear capacity at the specified section becomes the lowest one in that span. Consequently, if the formula is able to give a close estimation at the design cross section, other sections in that span would show an overestimated prediction. In order to be consistent with the situation of simply supported beams, it would be more reasonable to set the control cross section of a continuous specimen at z from the point of inflection and assign the sections closer to the point of inflection with the same design shear force as the design cross section.

4   Conclusions

In this paper, the shear capacity of reinforced concrete specimens without shear reinforcement loaded under multiple point loads with both continuously and simply supported boundary conditions is investigated. Several experiments have been carried out in the laboratory. The test results are compared with those obtained in tests on similar specimens loaded by a single point load.

It turns out that the shear capacity of a structure is closely related to the moment distribution along its length. In order to get a more accurate estimation of the shear capacity, the moment distribution shall be taken into account.

In the case of Eurocode, it is suggested to replace the value of a_v/d in the shear reduction factor β with M/Vd.

With regard to Model Code 2010, the selection of the critical cross section shall be adjusted according to the point of inflection in continuous supported specimens.

References

[1]   Kani, G.N.J., The Riddle of Shear and Its Solution. ACI Journal, 1964. **61**(28): p. 441-467.

[2]   Eurocode 2, NEN-EN 1992-1-1 Design of Concrete Structures: General Rules and Rules for Buildings. 2004. p. 229.

[3]   van den Beukel, A. and T. Monnier, IBBC-TNO report B-84-522/62.5.0804. 1985, TNO-IBBC: Delft. p. 48.

[4]   Bryant, R.H., et al., Shear Strength of Two-Span Continuous Reinforced Concrete Beams with Multiple Point Loading. ACI Journal, 1962. **59**(44): p. 1143-1177.

[5]   Leonhardt, F. and R. Walther, Beiträge zur Behandlung de Schubprobleme im Stahlbetonbau. Deutscher Ausschuss für Stahlbeton, 1962. **151**: p. 83.

[6]   Rodrigues, R.V., A. Muttoni, and M.F. Ruiz, Influence of shear on rotation capacity of reinforced concrete members without shear reinforcement ACI Structural Journal, 2010. **107**(5): p. 516-525.

[7]   Yang, Y., J.A. den Uijl, and J. Walraven. *Shear Capacity of Continuous RC Beams without Shear Reinforcement*. in *fib Symposium* 2011. Prague.

[8]   Yang, Y. and J.A. den Uijl, Report of Experimental Research on Shear Capacity of Beams Close to Intermediate Supports. 2011, Delft University of Technology: Delft.

[9]   König, G. and J. Fischer, Model Uncertainties concerning Design Equations for the Shear Capacity of Concrete Members without Shear Reinforcement, in CEB Bulletin No. 224. 1995. p. 117.

[10]   FIB, CEB-FIP Model Code 2010 (Draft). 2010: Lausanne.

Shear behaviour of large and shallow fiber reinforced concrete beams

Antonio Conforti

DICATA – Department of Civil, Architectural, Environmental and Land Planning Engineering
University of Brescia
Via Branze 43, 25123 Brescia, Italy
Supervisor: Giovanni Plizzari

Abstract

Steel fibers are very important in beams under shear loading, as evidenced from several scientific papers reported into the last decade journals. This paper reports some recent results of an experimental campaign on FRC beams under shear loading tested at the University of Brescia, focusing on the size effect issue and the shear behavior of shallow beams. With the first regard, nine full scale beams, having a height varying from 500 to 1500 mm, were tested to analyze steel fibers influence on size effect. Concerning the shallow beams, eight beams (all having depth of 250 mm) with two different widths, fiber content were tested for evaluating the shear response of typical structural members utilized in Southern Europe in residential buildings. Results show that a relatively low volume fraction of fibers can significantly increase shear bearing capacity and ductility. The size effect issue is substantially limited and it is observed that, with a fairly tough FRC composite, it is possible to completely eliminate this detrimental effect. Shallow beams do not show the typical brittle failure also without any shear reinforcement and the effect of fibers is even more prominent than in deep beams.

1   Introduction

A significant change in the shear and punching shear provisions, included in the first draft of *fib* Model Code [1,2], referred to in the following as MC2010, which was officially presented during the fib Congress in Washington D.C., can be outlined. Four different approximation levels are proposed for designing shear members, incorporating statements and models well recognized such as Modified Compression Field Theory [3] and the variable truss angle theory. The equations have been defined with a basic structure that requires two significant parameters, namely the angle of inclination of the stress field, θ, and a coefficient for a concrete contribution k_v (more often referred as β). The MC2010 first draft provides three levels of approximation to calculate these terms with the first and third based on the MCFT and the second based on a strain-modified form of plasticity. As the level of approximation increases, the quality of the predictions improves, but with more complex calculations. Several reports published over the past 25 years [4,5,6] confirm the effectiveness of steel fibers as shear reinforcement. Fibers are used to enhance the shear capacity of concrete or to partially or totally replace stirrups in RC structural members. This relieves reinforcement congestion at critical sections such as beam-column junctions in seismic applications. Fiber reinforcement may also significantly reduce construction time and costs, especially in areas with high labour costs, and possibly even labour shortages, since stirrups involve relatively high labour input to bend and fix in place. Fiber concrete can also be easily deployed in thin or irregularly shaped sections, such as architectural panels, where it may be very difficult to place stirrups. This is of paramount significance for many secondary structural elements in which a minimum conventional reinforcement is not required for equilibrium. In this respect, nine experimental tests on full-scale SFRC beams (with a height up to 1.5 m) are firstly presented in this paper, which focus on the fiber's role in delaying shear crack localization, in mitigating the size effect and in allowing a stable crack development with associated load and ductility increases. Experimental results will be evaluated in terms of strength, ductility, shear cracking, collapse mechanism and effect of fibres. Secondly, further eight experiments on full-scale shallow beams will be briefly reported, aiming at investigating the shear behaviour of a very frequent structure in the residential buildings in Southern Europe, in which architectural needs require that beams have the same thickness of the floor (i.e. secondary floor beams and main beams should have the same depth).

Proc. of the 9th *fib* International PhD Symposium in Civil Engineering, July 22 to 25, 2012,
Karlsruhe Institute of Technology (KIT), Germany, H. S. Müller, M. Haist, F. Acosta (Eds.),
KIT Scientific Publishing, Karlsruhe, Germany, ISBN 978-3-86644-858-2

2   Experiments

All nine full-scale beams were tested under a three point loading system and a shear span-to-depth ratio a/d of 3. Beams were made with different amounts of steel fibers: 0, 50 and 75 kg/m^3 (corresponding to a volume fraction of 0, 0.64 and around 1%, respectively) and, for each fiber content, three beams with different depths were cast: 500 mm (beams H500), 1000 mm (beams H1000) and 1500 mm (beams H1500). All beams had the same width of 250 mm and gross cover (60 mm). Different effective depths were therefore obtained: 440, 940 and 1440 mm, respectively for specimens H500, H1000 and H1500.

In the case of shallow beams, all eight beams were tested under a four point loading system and a shear span-to-depth ratio a/d of 2.5. Beams were made with different amounts of steel fibers: 0, 25 (FRC25 samples) and 35 (FRC35 tests) kg/m^3 (corresponding to a volume fraction of 0, 0.32 and 0.45 respectively) and, for each fiber content, two beams with different widths were cast: 750 mm (series W750) and 1000 mm (beams W1000). Moreover, for each width, one beam with the minimum amount of transverse shear reinforcement as required by EC2 (2005) was also produced (4Ø6@150 mm stirrups for W750 MSR sample and 6Ø6@150 mm stirrups for W1000 MSR specimen). All beams had the same height of 250 mm and gross cover (40 mm), giving an effective depth of 210 mm.

Figure 1 illustrates the geometry of the specimens and the reinforcement details for all beams tested. In the case of deep beams, longitudinal reinforcement was positioned in two layers and the reinforcement ratio was approximately 1% for all test specimens. Eight longitudinal rebars having a diameter of 14, 20 and 24 mm were placed respectively in H500, H1000 and H1500 beams. For the wide shallow beams, the reinforcement ratio was again equal to 1%: 8Ø6 and 11Ø16, all disposed in one layer, were utilized respectively for series W750 and W1000.

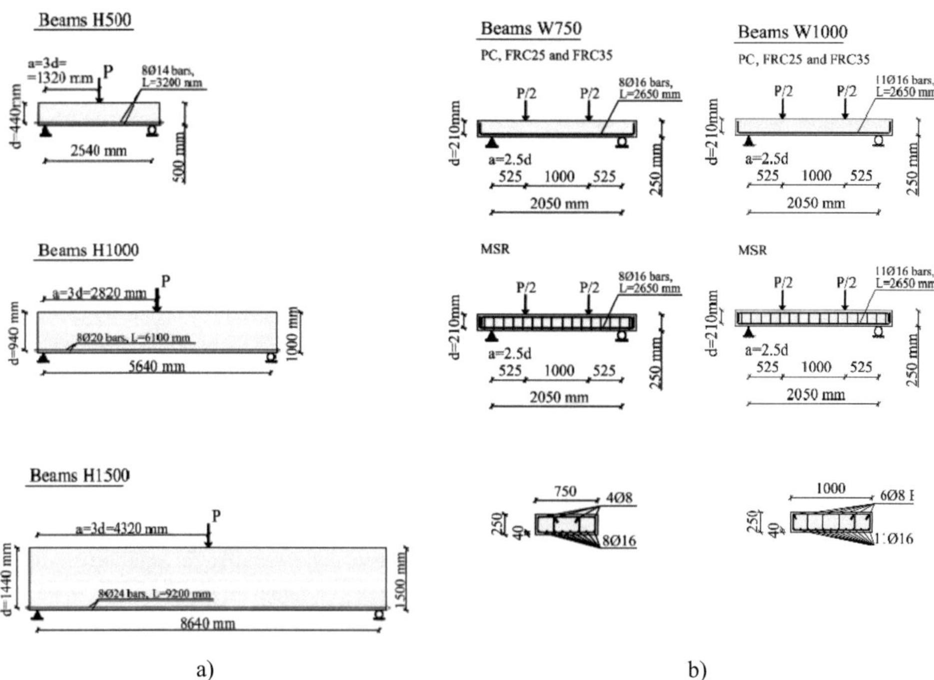

a)   b)

Fig. 1   Geometry and reinforcement details of deep (a) and shallow beams (b).

A normal strength concrete, provided by a concrete supplier, was utilized for all beams. In the case of deep beams, the mean value of the compressive strength was 38.7 MPa for PC series, 32.1 MPa for the FRC50 series and 33.1 MPa for the FRC75 set of tests. For the shallow beams, the mean value of the compressive strength was 40.5 MPa for PC and MSR series, 38.0 MPa for the FRC25 series and 36.9 MPa for the FRC35 set of tests.

Hooked end steel fibers, having a length of 50 mm, a diameter of 0.8 mm (aspect ratio L/Ø of 62.5) and a tensile strength of 1100 MPa were adopted for all FRC samples. The yielding and ultimate tensile strength of the longitudinal rebars were: 510 MPa and 588 MPa for Ø6 bars; 506 MPa and 599 MPa for Ø14 bars; 537 MPa and 630 MPa for Ø16 bars; 555 MPa and 651 MPa for Ø20 bars and 518 MPa and 612 MPa for Ø 24 bars, typical for S500 steel according to the current EC2 (2005).

Concerning the test setup, an electro-mechanical screw jack with a loading capacity of 1500 KN for all specimens was utilized. A displacement-controlled test was therefore guaranteed, allowing for a suitable test control during critical steps such as in the case of abrupt cracking phenomena or load drops.

With regard to the instrumentation, linear Variable Differential Transformers (LVDTs) were utilized for measuring deflections at midspan (front and back side) and support displacements. Also, potentiometric transducers were employed for measuring crack widths and strut shortening.

3 Experimental results on deep beams

The experimental results, reported in Figure 2 and Figure 3, represent the midspan-displacement and the shear-crack-width development as a function of the external load for H1000 and H1500 series. A shear failure was seen for all nine elements. However, in both the FRC shallowest beams (H=500 mm), the maximum flexure load was reached, with clear yielding of longitudinal rebar and a rather significant ductility for beam H500 FRC50.

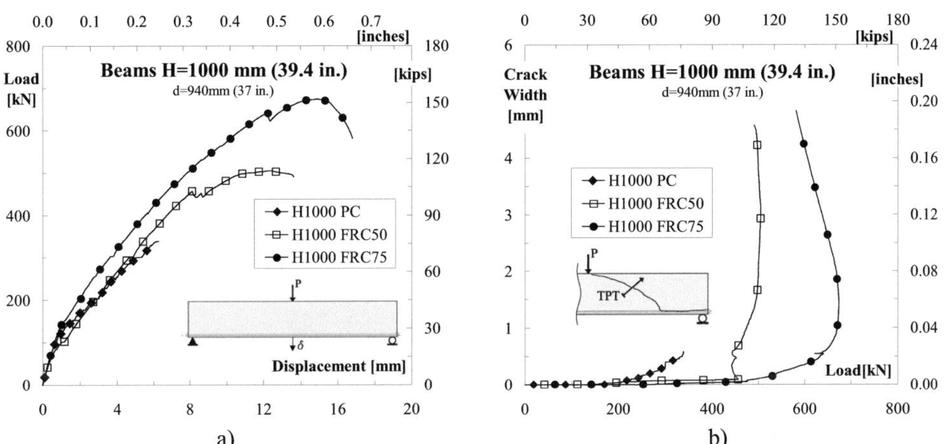

a)   b)

Fig. 2   Experimental curves load vs. deformation (a) and main shear-crack-width vs. load (b) of H1000 beams.

Especially from both H1000 and H1500 test series, a significant enhanced post-cracking stiffness is observed for FRC beams, due to stiffening effect, in tension, which is due to the bridging effect of fibers (residual tensile stress at a crack) and to the smaller crack spacing, both in flexure and in shear. With increasing depth, the difference between the two fiber contents tends to become higher, i.e. the lowest content of fibers has less positive impact in diminishing the scale effect.

The addition of fibers promoted a stable propagation and progressive development of several shear cracks, which led to a more ductile behaviour with vertical deflections 2-3 times greater than those recorded in the reference plain concrete beams, as clearly evidenced by the experimental plots. Concerning shear cracking, Figure 3 b) reports the crack development vs. the load, for the H1500 series: note that the crack width evolution is better controlled in FRC: in particular, evident shear cracking begins at 320 kN for the reference sample, whereas it occurs at 570 kN and 890 kN for the FRC50 and FRC75 beams, respectively. While the plain concrete member fails at the emergence of the first shear crack, with a maximum shear crack width of 0.2-0.25 mm, multi-cracking (in shear) was seen for the FRC samples, with single shear crack wider than 1-2 mm and, even more important, still steadily propagating. The same trend can be seen for sample H1000 (Figure 2 b).

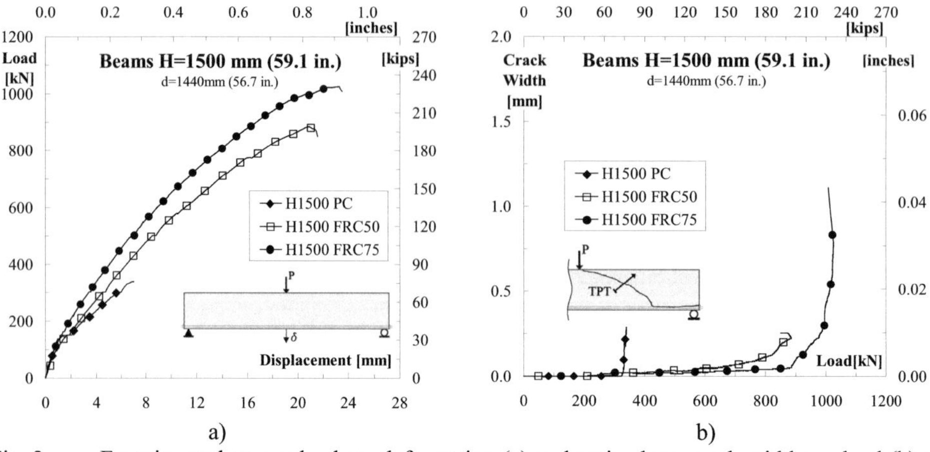

Fig. 3   Experimental curves load vs. deformation (a) and main shear-crack-width vs. load (b) of H1500 beams.

Figure 4 depicts the final crack patterns of specimens H1000 and H1500, with the indication of the progression (see different load levels in the pictures). Note, once again, a much more distributed crack pattern for FRC samples. For the same load level, both flexure and shear crack are fairly different in the three materials: for low loads, the crack pattern is much more developed in the reference elements and the cracking phenomenon tends then to quickly stabilize. On the contrary, in FRC samples, the cracking process, after a quite postponed appearance, becomes more dynamic so that it is quite difficult to properly define a stabilized crack stage: a more distributed crack patter forms, with narrower closely spaced cracks (this aspect is quite significant under a durability point of view). Moreover, the development of new cracks continues up to the failure.

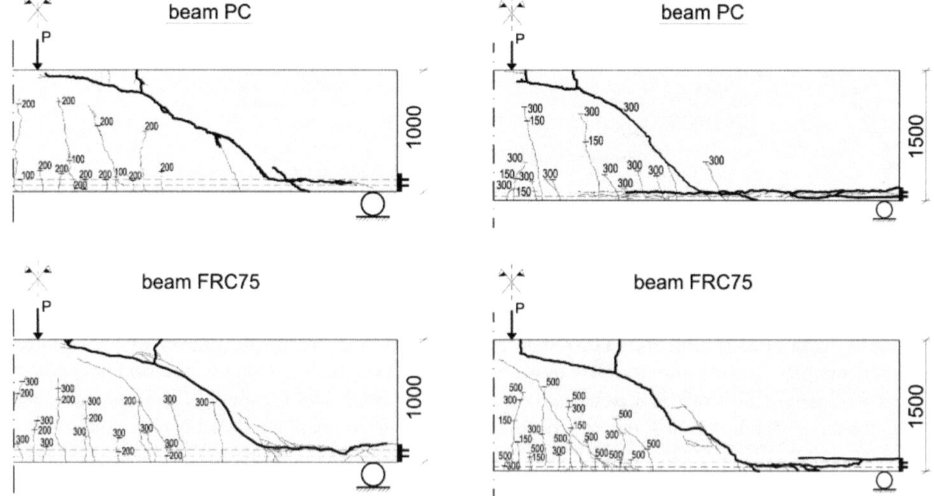

Fig. 4   Crack evolution for specimens H1000 and H1500 (loads in kN).

4   Experimental results on wide shallow beams

Figure 5 reports the load-displacement curves for all eight shallow beams: a shear failure was seen only for plain concrete members. All members containing either fibers or the minimum amount of transverse reinforcement reached the maximum flexure load, with clear yielding of longitudinal rebar and a significant ductility.

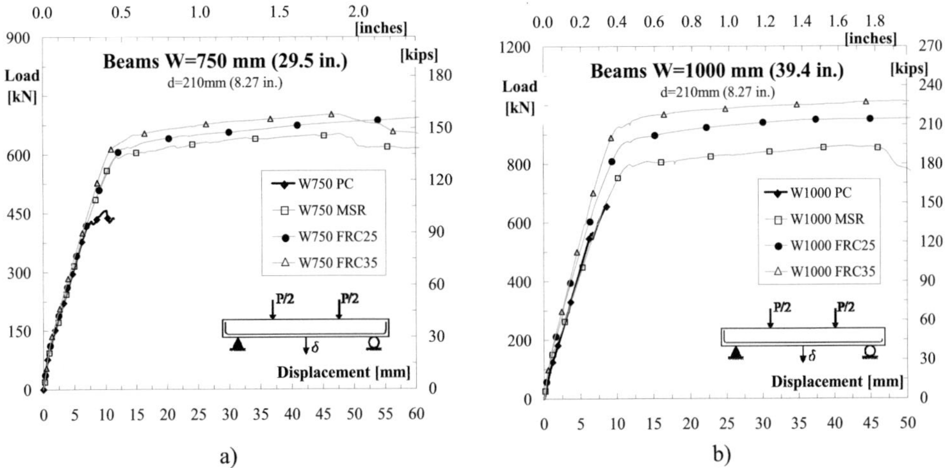

Fig. 5   Load-displacement curves for W750 (a) and W1000 specimens (b).

Table 1   Main experimental results about deep and shallow beams.

Specimen	Type of failure	$V_{u,exp}$ [kN]	v_u [MPa]	$v_u/(f_{cm})^{1/2}$ [-]	$M_{u,fl}$ [kNm]	$M_u/M_{u,fl}$ [-]	δ_u [mm]
H500-PC	Shear	116	1.05	0.17	254	0.60	3.70
H500-FRC50	Shear*	240	2.18	0.38	285	1.11	34.95
H500-FRC75	Shear*	235	2.13	0.37	293	1.06	9.14
H1000-PC	Shear	188	0.80	0.13	1210	0.44	6.26
H1000-FRC50	Shear	272	1.16	0.20	1325	0.58	13.61
H1000-FRC75	Shear	351	1.49	0.26	1356	0.73	16.78
H1500-PC	Shear	211	0.59	0.09	2511	0.36	7.03
H1500-FRC50	Shear	484	1.34	0.24	2791	0.75	21.58
H1500-FRC75	Shear	554	1.54	0.27	2864	0.84	23.46
W750 PC	Shear	237	1.50	0.24	169	0.74	11.37
W750 MSR	Flexure	335	2.12	0.33	169	1.04	61.28
W750 FRC25	Flexure	355	2.25	0.35	186	1.00	85.39
W750 FRC35	Flexure	360	2.28	0.36	189	1.00	97.03
W1000 PC	Shear	337	1.60	0.25	232	0.76	8.62
W1000 MSR	Flexure	441	2.10	0.33	232	1.00	51.64
W1000 FRC25	Flexure	488	2.32	0.37	262	1.00	80.93
W1000 FRC35	Flexure	517	2.46	0.39	266	0.97	89.20

* Shear failure mode took place, but maximum flexure load was reached.

Fibers were also able to increase the overall ductility under flexure, thanks to the positive effect in increasing the compression softening. The post-cracking stiffness was also higher. Moreover, 25 kg/m³ of fibers were able to completely substitute the minimum amount of transverse reinforcement in shallow beams. It results that either a minimum transverse reinforcement or a low content of fibers, in the case of shallow beams with a classical reinforcing ratio (around 1%), is able to bring the member up to a ductile flexure failure.

Looking at the cracking phenomenon, the collapse, as one would have expected, did not appear immediately after the first shear crack: two different shear cracks formed in the front and rear side and developed in a quite stable fashion, without connecting one to the other trough the width of the element. Clearly, the different behaviour at front and back side allowed for a more stable response of the member (the stress field, due to the significant width, is far to be simply 2D). Also, the height of these beams is very small, and in addition the central core of the beam results well confined by the surrounding concrete: this determines an extra resisting mechanism in shear. This is a peculiar behavior that deserves a number of thoughts for its comprehension and, eventually, for the definition of suitable design implications.

In conclusion, Table 1 reports, for all seventeen samples tested, the failure mode, the experimental shear capacity, the nominal shear stress and the ratio between the ultimate experimental moment, M_u, and the maximum flexural capacity of the member, $M_{u,fl}$ (the latter evaluated according to MC10 in the case of the FRC samples). One should note the strong size effect in plain concrete members, whereas a significant decay con be outlined in the case of FRC. Note, once again, the beneficial effects of fibers in moving the collapse mode from brittle to ductile in the case of shallow beams.

5   Concluding remark

In the present paper, the beneficial effects of providing steel fibers as spread shear reinforcement have been scrutinized. Fibers, even in relatively low amount, greatly influence the shear behaviour of beams, basically by delaying the occurrence of the shear failure mechanism and, eventually, by altering the collapse from shear to flexure, with enhanced bearing capacity and ductility. Fibers allow, if supplied in sufficient quantity, a well distributed crack pattern in the critical area under shear, delaying or even avoiding the formation of the single critical shear crack, which brings the member to a brittle failure. If this happens, it is associated with visible warning, cracking and deflections, unlike for plain concrete members. Steel fibers mitigate the size effect issue in shear, even if provided in relatively low amounts. Moreover, fibers might substitute the minimum shear reinforcement in shallow beams, typical structural typology, in Southern Europe, for residential buildings.

The Author would like to give his appreciation to the assistance of Prof. Giovanni Plizzari and Prof. Fausto Minelli. The help of PhD candidate Estefania Cuenca, Engineers Bianchi Francesco, Vezzoli Carlo, Bianchi Francesco and Bonomelli Mauro in performing the test program, in data processing and modelling is gratefully acknowledged.

References

[1]   *fib* Bulletin 55, Model Code 2010 – First complete draft, Volume 1 (chapters 1-6), 318 pages, 2010, ISBN 978-2-88394-95-6

[2]   *fib* Bulletin 56, Model Code 2010 – First complete draft, Volume 2 (chapters 7-10), 312 pages, 2010, ISBN 978-2-88394-096-3

[3]   Vecchio, F.J. and Collins, M.P. (1986) "The Modified Compression Field Theory for Reinforced Concrete Elements Subjected to Shear," ACI Journal, 83(2) pp. 219 231.

[4]   Choi, K-K., Park, H-G., and Wight, J. (2007) "Shear Strength of Steel Fiber-Reinforced Concrete Beams without Web Reinforcement", ACI Structural Journal, V. 104, No 1, pp. 12-22.

[5]   Dinh, H. H., Parra-Montesinos, G.J., and Wight, J. (2010) "Shear Behavior of Steel Fiber-Reinforced Concrete Beams without Stirrup Reinforcement", ACI Structural Journal, V. 107, No 5, pp. 597-606.

[6]   Minelli, F. (2005) "Plain and Fiber Reinforced Concrete Beams under Shear Loading: Structural Behavior and Design Aspects", Ph.D. Thesis, Department of Civil Engineering, University of Brescia, February 2005, pp. 422.

Shear capacity of concrete beams with FRP reinforcement

Martin Kurth

Institute of Structural Concrete,
RWTH Aachen University,
Mies-van-der-Rohe-Str. 1, 52074 Aachen, Germany
Supervisor: Josef Hegger

Abstract

Typical shear failure modes that can occur in steel-reinforced concrete members are diagonal tension failure, shear compression failure, and failure by crushing of the concrete struts. Further failure modes upon testing a concrete beam with fiber-reinforced polymers as internal reinforcement are possible. The FRP shear elements can fail due to either a dowel action rupture or a concentration of stress at a corner in the reinforcement.

This paper presents an experimental and theoretical study on shear performance of concrete I-beams with fiber-reinforced polymers (FRP) as internal reinforcement. A total of 24 beam tests were conducted, including tests without shear reinforcement and tests with glass fiber reinforced polymer (GFRP) stirrups. In all specimens GFRP bars were used as flexural reinforcement. The test variables were the ratio of shear reinforcement and the concrete strength. In the test without stirrups, diagonal tension failure occurred. Failure due to rupture of the GFRP stirrups was observed in the tests with lower shear reinforcement ratio of 0.75%. In the beam tests with higher ratios of 1.26% and 2.26% respectively, depending on the concrete strength, either GFRP stirrup rupture or web crushing failure occurred. Based on the results, a modified approach for calculating the shear capacity of concrete beams with FRP reinforcement was developed.

1 Introduction

Most research has been undertaken to investigate the behavior of FRP-reinforced members without transverse reinforcement. However, also studies on shear behavior of concrete members with FRP stirrups were conducted [1]. Due to the lower modulus of elasticity of the FRP-reinforcement compared to steel and the predominantly used rectangular cross-section of the test specimens, shear compression and diagonal tension failure occurred in the most cases. Failure by crushing of the concrete struts has only rarely been observed. This Paper evaluates the shear performance of large-scale FRP RC I-beams focusing on the maximum shear capacity for analyzing the failure modes and subsequently deriving shear capacity equations.

2 Experimental Study

2.1 Test specimens

The experimental program involved 24 shear tests on 12 concrete I-beams reinforced with FRP as flexural and shear reinforcement. The beam specimens were designed to fail in shear and were tested in two successive test phases shown in Figure 1.

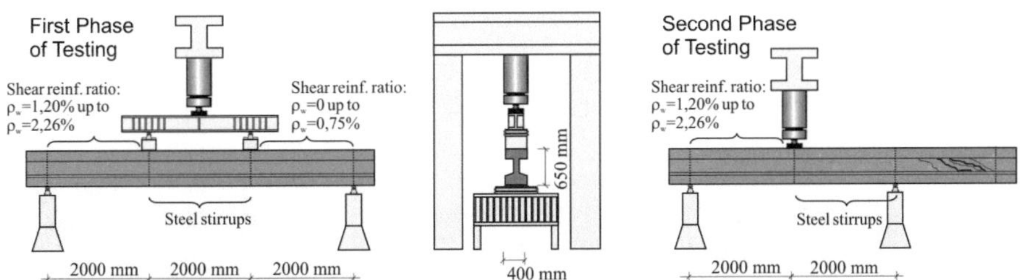

Fig. 1   Specimens and test set-up.

Proc. of the 9th *fib* International PhD Symposium in Civil Engineering, July 22 to 25, 2012,
Karlsruhe Institute of Technology (KIT), Germany, H. S. Müller, M. Haist, F. Acosta (Eds.),
KIT Scientific Publishing, Karlsruhe, Germany, ISBN 978-3-86644-858-2

In the first phase of testing (left diagram), two point loads were applied leading to rupture of the right-hand beam section with low shear reinforcement ratio or without shear reinforcement. In the second phase (right diagram), the left-hand section with a higher shear reinforcement ratio was tested in three-point bending. To prevent failure, the middle beam section was strongly reinforced with steel stirrups. Thus, a total of 5 tests without and 19 tests with shear reinforcement were conducted.

The beams had a total height of 650 mm and a web width of 100 mm. To ensure that a shear failure precedes a flexural failure, flexural reinforcement ratios ($\rho_l = A_{fl}/(b_w \cdot d)$) of 8.5 and 11.5 % respectively were required (six 32 mm resp. eight 32 mm GFRP bars) and steel reinforcement was added in the compression zone (Fig. 2). In the tests with shear reinforcement, transverse reinforcement consisting of either 12 mm GFRP stirrups or 16 mm GFRP headed bars was used.

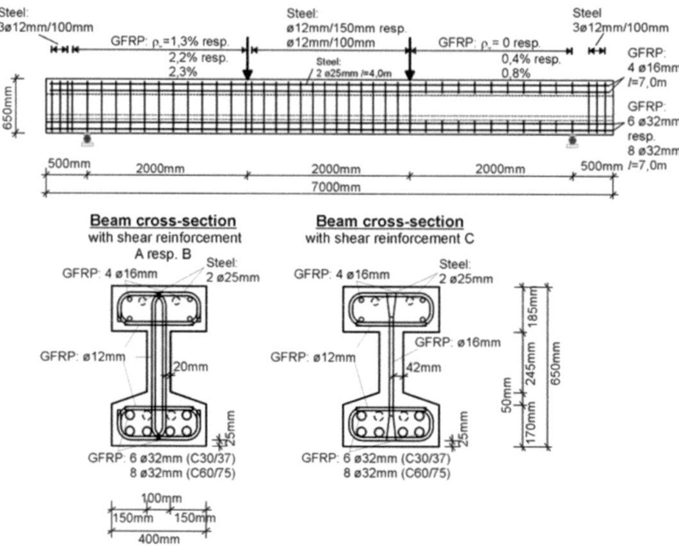

Fig. 2   Geometry and reinforcement details of test specimens.

2.2   Material properties

The stirrups and headed bars have a height of 600 mm. The mandrel diameter of the stirrups D is 84 mm (7ϕ where ϕ is the bar diameter). Figure 3 depicts the used shear reinforcement that is composed of three respectively four parts. The GFRP bars and stirrups exhibit surface deformations to enhance bond performance between reinforcement and surrounding concrete.

To manufacture the beam specimens, a ready mixed concrete with a maximum aggregate size of 8 mm was used. The properties of the test specimens summarized in Table 1 include mean values of compressive strength (f_{cm}), mean values of splitting tensile strength ($f_{ctm,sp}$) and moduli of elasticity of the concrete (E_{cm}), diameters and moduli of elasticity of the flexural and shear reinforcement ($\varnothing_f$, $\varnothing_s$, E_f, E_{fw}), spacings of shear reinforcement (s), effective depth of a cross-sections (d) and shear reinforcement ratios ($\rho_{fw} = A_{fw}/(b_w \cdot s)$).

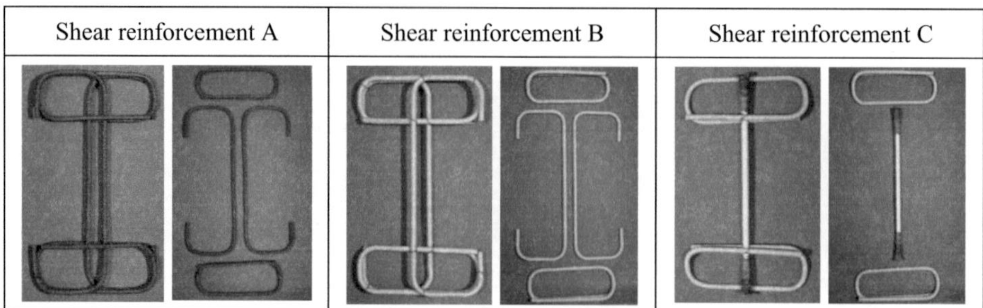

Shear reinforcement A	Shear reinforcement B	Shear reinforcement C

Fig. 3   Tested shear reinforcement.

Table 1   Concrete properties and reinforcement details of test specimens.

Test specimen	Concrete			Flexural reinforcement			Shear reinforcement			
	f_{cm} (Mpa)	$f_{ctm,sp}$ (Mpa)	E_{cm} (Mpa)	$\varnothing_f$ (mm)	d (mm)	E_f (Mpa)	$\varnothing_s$ (mm)	s (mm)	ρ_{fw} (%)	E_{fw} (Mpa)
S1AN-0-1	34.4	2.87	26200	6 x 32	566	59000	-	-	-	-
S1AN-1.3-2	38.4	2.87	27700	6 x 32	566	59000	12	180	1.20	56200
S2AN-0.8-3	34.1	3.00	25600	6 x 32	566	59000	12	300	0.75	56200
S2AN-2.3-4	33.5	2.84	24700	6 x 32	566	59000	12	100	2.26	56200
S3AH-0-5	80.4	4.40	35200	8 x 32	556	59000	-	-	-	-
S3AH-1.3-6	82.0	4.37	35200	8 x 32	556	59000	12	180	1.26	56200
S4AH-0.8-7	42.2	2.57	25500	8 x 32	556	59000	12	300	0.75	56200
S4AH-2.3-8	42.6	3.19	26000	8 x 32	556	59000	12	100	2.26	56200
S5BN-0-9	42.5	2.76	27300	6 x 32	572	62600	-	-	-	-
S5BN-1.3-10	42.8	3.20	27200	6 x 32	572	62600	12	180	1.26	57000
S6BN-0.8-11	30.4	2.58	24600	6 x 32	572	62600	12	300	0.75	57000
S6BN-2.3-12	30.7	2.65	25200	6 x 32	572	62600	12	100	2.26	57000
S7BH-0-13	74.9	4.16	33200	8 x 32	561	62600	-	-	-	-
S7BH-1.3-14	75.7	3.74	33300	8 x 32	561	62600	12	180	1.26	57000
S8BH-0.8-15	72.7	3.95	33400	8 x 32	561	62600	12	300	0.75	57000
S8BH-2.3-16	69.9	3.85	31200	8 x 32	561	62600	12	100	2.26	57000
S9CN-0-17	37.1	3.11	24900	6 x 32	572	62600	-	-	-	-
S9CN-1.3-18	37.7	3.03	25900	6 x 32	572	62600	16	160	1.26	63400
S10CN-0.7-19	32.0	2.44	23200	6 x 32	572	62600	16	270	0.75	63400
S10CN-2.2-20	33.6	2.44	25200	6 x 32	572	62600	16	90	2.22	63400
S11CH-0.4-21	67.6	3.20	33300	8 x 32	561	62600	16	450	0.45	63400
S11CH-1.3-22	71.0	3.08	32000	8 x 32	561	62600	16	160	1.26	63400
S12CH-0.7-23	73.1	3.71	31400	8 x 32	561	62600	16	270	0.75	63400
S12CH-2.2-24	73.9	3.62	30000	8 x 32	561	62600	16	90	2.22	63400

The mean values of the modulus of elasticity of the flexural and shear reinforcing elements provided in Table 1 were determined in the laboratory of the Institute of Structural Concrete (IMB) according to test method B.2 and B.5 respectively (see ACI440.3R 04 [2]).

2.3   Instrumentation, test set-up, and procedure

The strains in the flexural and shear reinforcement were measured using electrical resistance strain gauges. To measure the deflection of the beam, Linear Variable Differential Transducers (LVDTs) were used. The shear crack width was monitored during the test by a non-contact optical deformation measuring system (Aramis) and by using crack gauges. The results of the crack width and deformation measurements are not presented in this paper. In each test, the point loads were located at a distance of 2000 mm from the support, which corresponds to a shear-depth ratio of 3.50 up to 3.60. In the four-point bending tests, the load was applied through a spreader beam. The load was applied at a rate of 5.0 kN/min up to about 60% of the failure load. Then, displacement-controlled rates of 0.5 up to 1.0 mm/min were used to avoid a sudden shear failure. During the tests, all gauges and LVDT readings were recorded using data acquisition systems.

2.4   Particular experimental results

The results obtained for all of the beams tested are summarized in Table 2 and Table 3 which contain the shear force at the onset of diagonal cracking (V_{crack}), the maximum imposed shear force (V_{max}), the inclination angles of the diagonal shear cracks (β_r), average stirrup strains at failure ($\varepsilon_{fw,average}$), and mode of failure.

Table 2   Results of tests without shear reinforcement.

Test specimen	βr (°)	V_{crack} (kN)	V_{max} (kN)	Failure mode*
S1AN-0-1	24	91	126	DT
S3AH-0-5	19	122	184	DT
S5BN-0-9	23	85	136	DT
S7BH-0-13	42	118	205	DT
S9CN-0-17	40	86	122	DT

* DT is diagonal tension failure

Table 3   Results of tests with shear reinforcement.

Test specimen	βr (°)	$\varepsilon_{fw,average}$ (‰)	V_{max} (kN)	Failure mode*
S1AN-1.3-2	40	7.6	427	WC
S2AN-0.8-3	30	9.5	332	GR
S2AN-2.3-4	36	6.8	486	WC
S3AH-1.3-6	31	10.6	571	GR
S4AN-0.8-7	32	9.5	326	GR
S4AH-2.3-8	34	8.3	544	WC
S5BN-1.3-10	37	6.3	439	WC
S6BN-0.8-11	33	8.5	302	GR
S6BN-2.3-12	37	5.2	448	WC
S7BH-1.3-14	31	10.3	626	GR

Test specimen	βr (°)	$\varepsilon_{fw,average}$ (‰)	V_{max} (kN)	Failure mode*
S8BH-0.8-15	27	14.1	439	GR
S8BH-2.3-16	33	5.2	581	WC
S9CN-1.3-18	36	6.4	410	WC
S10CN-0.7-19	23	6.9	304	GR
S10CN-2.2-20	34	4.3	484	WC
S11CH-0.4-21	28	10.7	374	GR
S11CH-1.3-22	33	7.8	621	WC
S12CH-0.7-23	32	9.5	441	GR
S12CH-2.2-24	36	6.1	742	WC

* WC is web crushing failure, and GR is GFRP shear reinforcement rupture

2.4.1 Shear-carrying mechanism in the tests without shear reinforcement

At low load levels, only small vertical and slightly inclined flexural shear cracks appeared, so that shear transfer across cracks by mechanical interlock was ensured. As loading continued, large diagonal cracks developed and at the same time the deflections increased suddenly. The load was then carried by the concrete compression zone, and by the longitudinal bars in terms of dowel action. When the load was further increased, arch action occurred, and the horizontal component was ensured by the tensile flexural reinforcement. Finally, diagonal tension failure occurred, when the shear crack extended through the compressive zone towards the load points.

2.4.2 Shear-carrying mechanism in the tests with shear reinforcement

The crack pattern of the tested beams with FRP shear reinforcement corresponded to that of a steel-reinforced beam, so that the truss analogy can be assumed. All specimens with shear reinforcement ratios of $\rho_w = 0.45\%$ and 0.75% and those with $\rho_w = 1.26\%$ and concrete strength of $f_{cm} = 75.7$ and 82.0 N/mm² failed due to rupture of the GFRP shear reinforcement. Because the FRP stirrups do not yield, the shear reinforcement contributed to the shear resistance up to failure. In contrast to beams with rectangular cross-sections, the bend portions of the stirrups were sufficiently anchored in the flanges of the tested I-beams. Consequently, the stirrups did not fail due to a concentration of stress at a corner in the reinforcement. In fact, they failed at the straight portion. The specimens with higher shear reinforcement ratios $\rho_w = 2.22\%$ and 2.26%, and those with $\rho_w = 1.26\%$ and concrete strength up to $f_{cm} = 71.0$ /mm² failed suddenly due to crushing of the concrete struts within the web.

3   Shear design equations

The shear design equations were derived from the results of the experimental work and further tests results provided in literature. For FRP RC members without shear reinforcement, the design value of the shear resistance $V_{Rd,ct}$ (Eq. (1)) was empirically determined based on the Eurocode 2 [3].

$$V_{Rd,ct} = \beta \cdot \frac{1}{620} \cdot \kappa \cdot \left(100 \cdot \rho_l \cdot E_{fl} \cdot f_{ck}\right)^{1/3} \cdot b_w \cdot d \tag{1}$$

where $\beta = 3/(a/d)$ is the factor for increased shear resistance of concrete near the supports, $\kappa = 1 + \sqrt{200/d}$ is the size effect factor, ρ_l is the flexural reinforcement ratio, E_{fl} is modulus of elasticity of the flexural FRP-reinforcement, f_{ck} is the characteristic compressive cylinder strength of concrete, b_w is the width of web and d is the effective depth.

The design value of the total shear capacity of members with FRP shear reinforcement V_{Rd} was derived as the sum of the concrete contribution $V_{Rd,c}$ $(=V_{Rd,ct})$ and the contribution from shear reinforcement $V_{Rd,f}$ that was calculated using the truss analogy (Eq. (2)).

$$V_{Rd} = V_{Rd,c} + V_{Rd,f} \tag{2}$$

In the performed shear tests, a linear relation between the resultant of the axial stiffnesses $E\rho_{res}$ and the load bearing and deformability behavior of the specimens was observed. $E\rho_{res}$ (see Figure 4) includes the extensional stiffness of the flexural and shear reinforcement without the concrete contribution and can be expressed according to Equation (3).

$$E\rho_{res} = \sqrt{\left(\rho_l \cdot E_{fl}\right)^2 + \left(\rho_w \cdot E_{fw}\right)^2} \tag{3}$$

where ρ_l is the flexural reinforcement ratio, E_{fl} is modulus of elasticity of the flexural FRP-reinforcement, ρ_w is the shear reinforcement ratio and E_{fw} is modulus of elasticity of the FRP-shear reinforcement.

The components of the truss contribution truss angle $(\beta_r = \theta)$ and the limiting stirrup strain (ε_{fw}) depend on the resultant of the axial stiffnesses $E\rho_{res}$ as can be seen in Figure 5. The depicted diagrams include apart from the 19 performed shear tests further tests from literature.

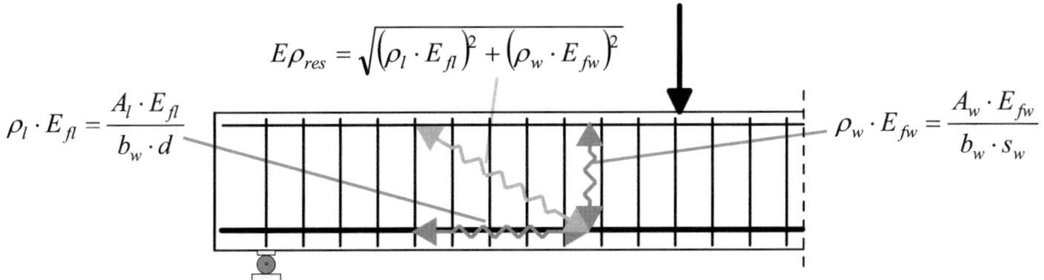

Fig. 4   Resultant of the axial stiffnesses $E\rho_{res}$.

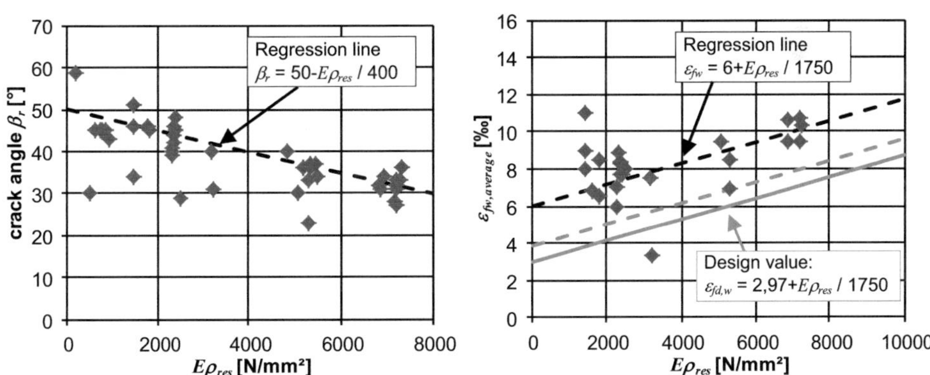

Fig. 5   Relationship between the resultant of the axial stiffnesses $E\rho_{res}$ and the crack angle (left) and the average stirrup strain at failure (right).

Based on this relationship, the design value of the contribution from shear reinforcement $V_{Rd,f}$ can be determined as follows:

$$V_{Rd,f} = a_{fw} \cdot f_{fd,w} \cdot z \cdot \cot(\theta) \leq \frac{b_w \cdot z \cdot \sqrt{f_{ck}} \cdot 1{,}5}{\cot\theta + \tan\theta} \tag{4}$$

where a_{fw} is the cross-sectional area of the shear reinforcement, $f_{fd,w}$ is the design tensile strength of the shear reinforcement ($f_{fd,w} \leq E_{fw} \cdot \varepsilon_{fd,w}$, with $\varepsilon_{fd,w}$ according to Figure 5, right, and $f_{fd,w} \leq f_{fd,w,exp}$, tensile strength derived from tests), z is lever arm of internal forces and θ is the truss angle ($=\beta_r$ according to Figure 5, left).

To verify these equations, 151 tests on FRP RC beams without shear reinforcement and 73 tests on FRP RC beams with shear reinforcement described above and cited in the literature have been taken into account. In Fig. 6, the ratios V_{test} / V_{calc} calculated by Eq. (1) as a function of the flexural reinforcement ratio and V_{test} / V_{calc} calculated by Eq. (2) as a function of the shear reinforcement rations are plotted. For all calculations mean values instead of design values were used. The derived shear equations show a good agreement with results from literature.

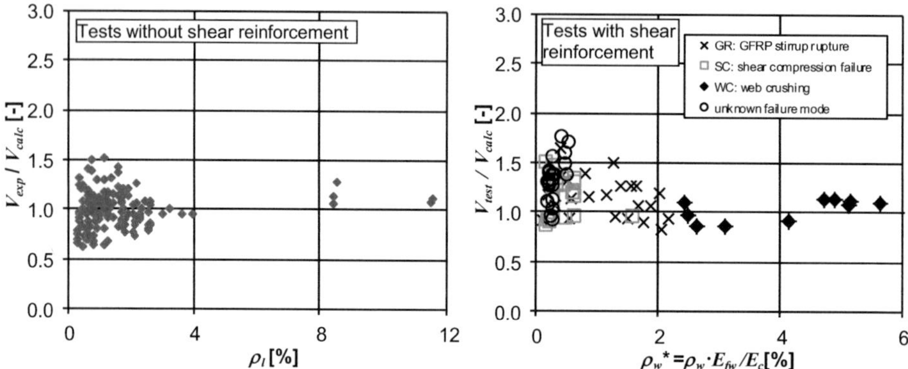

Fig. 6 V_{test} / V_{calc} as a function of the flexural (ρ_l) and the related shear reinforcement ratio (ρ_w^*), respectively.

4   Conclusions

A total of 24 beam tests were conducted, including tests without shear reinforcement and tests with glass fiber reinforced polymers (GFRP) stirrups. In all specimens, GFRP bars were used as flexural reinforcement. The test variables were the ratio of shear reinforcement and the concrete strength. In the tests without stirrups, diagonal tension failure occurred. Failure due to rupture of the GFRP stirrups and headed bars, respectively, was observed in the tests with lower shear reinforcement ratio of 0.75%. In the beam tests with higher ratios of 1.26% and 2.26%, respectively, depending on the concrete strength, either GFRP stirrup rupture or web crushing failure occurred. Based on the results, a modified approach for calculating the shear carrying capacity of concrete beams with FRP reinforcement was developed. The components of the truss contribution, truss angle ($\beta_r = \theta$) and the limiting stirrup strain (ε_{fw}) can thereby be appropriately determined by applying the resultant of the axial stiffnesses $E\rho_{res}$.

5   Acknowledgements

The author gratefully acknowledges the financial support of Deutscher Beton- und Bautechnik-Verein E.V. (DBV) and Arbeitsgemeinschaft industrieller Forschungsvereinigungen (AiF) and the provision of FRP reinforcement of FiReP Rebar Technology GmbH and Schöck Bauteile GmbH.

References

[1]   Hegger, J., Niewels, J., Kurth, M.: "Shear analysis of concrete members with fiber-reinforced polymers (FRP) as internal reinforcement", Proceedings of the 9th International Symposium on Fiber-Reinforced Polymer Reinforcement for Concrete Structures (FRPRCS-9) , Sydney, Australia, July, 13th-15th, 2009.

[2]   American Concrete Institute Committee 440, ACI 440.3R-04: "Guide Test Methods for Fiber-Reinforced Polymers (FRPs) for Reinforcing or Strengthening Concrete Structures", 2004

[3]   European Committee for Standardization, Eurocode 2: "Design of Concrete Structures - Part 1-1, General Rules and Rules for Buildings", EN 1992-1-1: 2004.

Experimental research on reinforced concrete dapped-end beams: shear strength and serviceability behaviour

Jaime Mata Falcón

Institute of Concrete Science and Technology (ICITECH),
Universitat Politècnica de València,
Camino de Vera s/n, 46022 Valencia, Spain
Supervisors: Luis Pallarés Rubio and Pedro F. Miguel Sosa

Abstract

Dapped-end beams are currently a common end in concrete structures, particularly in the field of precast concrete manufacturing. However, unacceptable diagonal cracks frequently occur at service loads in the vicinity of the re-entrant corner. There is a lack of experimental data and knowledge on the behaviour of dapped-end beams under service loads; the aim of this research is to analyze the influence of different variables on the crack width in dapped-end beams, such as the layout of steel bars, the amount of diagonal reinforcement and the geometric shape of the re-entrant corner.

An experimental program is carried out on 18 dapped-end beams with 3.00 m length and a cross-section of 25x60 cm, reduced to 25x30 cm at the ends.

Experimental results help to conclude that dapped-end beams are usually conditioned by cracking service limit state instead of ultimate state (load-bearing capacity). Smaller crack width is observed in specimens with diagonal reinforcement than those with only orthogonal reinforcement; increasing the amount of the diagonal reinforcement regarding to the orthogonal one, greater effect in reduction of crack width occurs. Change in geometric shape bevelling the re-entrant corner also leads to crack reduction.

1 Introduction

Dapped-end beams (DEB) are currently a common end in concrete structures, particularly in the field of precast concrete manufacturing. These elements appear in the construction of bridges and building structures. Dapped-end beams are a common and inexpensive way to carry out expansion joints.

From the structural point of view, the existence of geometric and static discontinuities, due to the change of depth and to the application of point load respectively, makes that concrete members with dapped-ends are classified as Discontinuity regions (D). A complex distribution of internal stresses appears in D regions, what disables Navier-Bernoulli's hypothesis. These regions are usually designed using the Strut-and-Tie Method (STM), proposed by Schlaich et al. [1].

STMs is used to design components under ultimate loads since is supported in the lower limit bound theorem, remaining untreated service requirements such as opening of cracks or deformations, as pointed out by Bulletín 45 of the fib [2]. This lack is especially important in the study of dapped-end beams, because at service conditions is generated a crack at the re-entrant corner under relatively low loads, being this fact particularly relevant in this case, since dapped-end beams are usually moisture concentration zones (Fig. 1), so the durability of the element may be compromised.

Among the first studies on dapped-end beams should be highlight the investigation of Reynolds [3] as the first major experimental contribution to the behaviour of these supports, which was concluded on the importance of the diagonal reinforcement. Mattock and Chan's investigations [4], lead to propose a formulation and design recommendations [5] that would be the most extended until the appearance of STM. Among the most recent researches on these elements should be highlighted the investigations of Clark and Thorogood [6] and Zhu et al. [7], since they are the only ones where it is discussed in detail the service crack width and it is proposed a methodology for its design.

Proc. of the 9th *fib* International PhD Symposium in Civil Engineering, July 22 to 25, 2012,
Karlsruhe Institute of Technology (KIT), Germany, H. S. Müller, M. Haist, F. Acosta (Eds.),
KIT Scientific Publishing, Karlsruhe, Germany, ISBN 978-3-86644-858-2

Fig. 1   Dapped-end beams in overpass of CV-500 road near Valencia, with accumulation of moisture

Given the problems indicated for dapped-end beams and due to the lack of detailed studies of cracking in service, is especially relevant the experimental campaign presented in this work, which focuses in both, the ultimate load behaviour and the serviceability behaviour.

2   Objective

The aim of this research is to study the influence on the cracking of the total amount of reinforcement, the existence of diagonal reinforcement in different amounts, the cover of the vertical main tie, the prestressed of the horizontal main tie and the bevelled of the re-entrant corner.

3   Methodology and parameters

3.1   Design of the testing method

Beam elements of 3.00 m length are tested, with 25x60 cm section, reduced to 25x30 cm at the supports. The width of the load bearing support is fixed at 15 cm. Anchor plates are disposed for the bottom reinforcement of the beam and the horizontal main tie, since the anchorage is not considered a variable of experimentation. The loading system and supports (Fig. 2) is such that it allows an independent testing of each of the two supports that consists each specimen.

The bearings are designed allowing the rotation and sliding, so can ensure that the load applied to the dapped-end beams has no horizontal component. This is achieved by a system of two plates of PTFE (Teflon) of 5 mm thickness, separated by a polished stainless steel plate of 5 mm thickness.

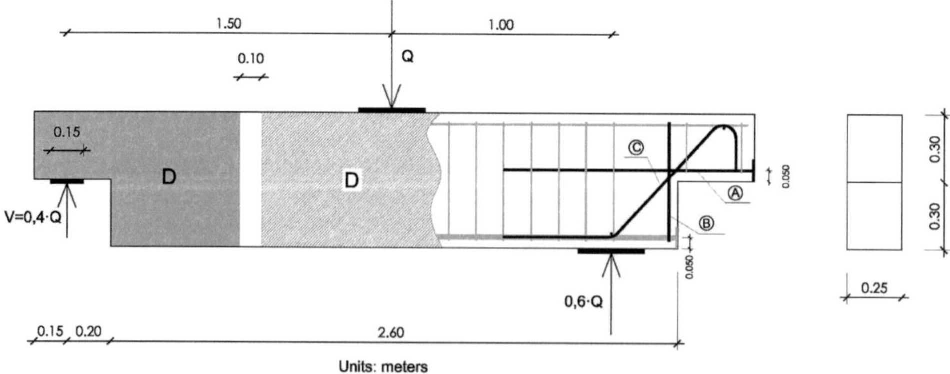

Fig. 2   Geometry of specimens, test setup and reinforcement codification

3.2 Instrumentation and equipments

The load is applied by a hydraulic actuator of 2500 kN with control of displacement speed constant, with a speed of 0.01 mm/sec. It is disposed a force transducer for controlling the applied load.

It is used a total of 21 LVDTs and strain gauges with 3 mm long glued to the steel bars (in a variable number depending on the type of reinforcement). In the concrete, near the re-entrant corner, are glued four strain gages with 50 mm long, which enable to accurately detect the cracking of the specimen.

To measure crack openings are used 2 Canon EOS 5D Mark II of 21.1 MPx resolution and 35x24 mm CMOS sensor with Canon EF 85 mm f/1.8 USM lens. It is also employed an optical microscope OPT-P2034-100M with 100 increases of Peak Optics for the redundant measuring of crack openings.

3.3 Parameters

The test results shown in this paper are part of a research project with a campaign of tests more extensive. In order to achieve the objectives pointed out above, 18 specimens have been carried out. The different parameters of the specimens are listed in Table 1, and described and discussed after it. The two dapped-ends of each beam have been built with the same characteristics in order to have a total of two tests for each specimen.

Table 1   Characteristics of the reinforcement of the different configurations tested

Specimen	Amount reinforc. -p-	Layout reinforcement	Reinforcement horizontal tie -A- (cm^2)	Reinforcement vertical tie -B- (cm^2)	N° layers vertical tie	Reinforcement diagonal tie -C- (cm^2)	Bevelled corner
DEB-1	49%	LR-O.1	3.93	2.58	1	0.00	No
DEB-2	49%	LR-O.2	2.36	2.58	1	0.00	No
DEB-3	49%	LR-O.3	3.93	1.01	1	0.00	No
DEB-4	49%	LR-O.1	3.93	2.58	3	0.00	No
DEB-5	49%	LR-O.2	2.36	2.58	3	0.00	No
DEB-6	49%	LR-D.1	2.36	1.51	1	1.57	No
DEB-7	100%	LR-O.1	8.04	5.40	3	0.00	No
DEB-8	100%	LR-O.2	4.52	5.40	3	0.00	No
DEB-9	100%	LR-D.1	4.52	3.14	1	3.05	No
DEB-10	71%	LR-O.1	5.65	3.71	3	0.00	No
DEB-11	71%	LR-O.2	3.39	3.71	3	0.00	No
DEB-12	71%	LR-D.1	3.39	2.26	1	2.26	No
DEB-13	100%	LR-D.2	3.14	2.26	1	4.27	No
DEB-14	100%	LR-D.2	3.14	2.26	1	4.27	Yes
DEB-15	100%	LR-D.3	1.57	1.01	1	5.15	No
DEB-16	100%	LR-D.3	1.57	1.01	1	5.15	Yes
DEB-17	100%	LR-D.4	8.04	1.01	1	5.15	No
DEB-18	100%	LR-D.4	8.04 - Prestessed	1.01	1	5.15	No

In order to analyse the ultimate load and the crack width in the re-entrant corner the next parameters are used in the study:

- The amount of reinforcement "p" related to the referenced specimen (taken as 100%). There have been studied three different amounts of reinforcements, the maximum (p=100%), an intermediate amount (p=71%) and the minimum amount of reinforcement (p=49%).

- The layout of reinforcement. A total of seven different layouts of reinforcement are tested, four of them with diagonal reinforcement. The first reinforcement (LR-O.1) corresponds to an orthogonal reinforcement designed to simultaneously reach the yielding of the vertical and the horizontal ties; the second one (LR-O.2) is a variation of LR-O.1 with 40% less in the horizontal tie; layout of reinforcement LR-O.3 is a variation of LR-O.1 reducing in this case the vertical tie in 60%. All the diagonal reinforcements mix the orthogonal ties with the diagonal tie; the first one (LR-D.1) is designed to carry about 50% of the load in the diagonal mechanism and 50% in the orthogonal mechanism; the second one (LR-D.2) carries 65% of the load in the diagonal and 35% in the orthogonal; the third one (LR-D.3) carries 80% of the load in the diagonal mechanism and 20% in the orthogonal; the last layout (LR-D.4) is a variation of LR-D.3 with a lot of more horizontal reinforcement, about five times more.
- Two different numbers of layers in the vertical tie are tested, one and three.
- In order to divide the crack generated in the re-entrant corner and consequently reduce its width, it is placed a bevel of 4x4 cm in two of the specimens.
- At last, in one specimen, the main horizontal tie is prestressed in order to generate a local compression concentration in the re-entrant corner that could lead to a delay of the cracking load.

4 Results and discussion

4.1 Results in ultimate state

In Table 2 are summarized the main results of the tests, with V_u representing the ultimate vertical load which supports the specimen and with w representing the crack width, shown for loads close to the service level, which is defined as the range of 40% of the breaking load. The results shown in Table 2 are derived from the average of the two tests done for each specimen.

Table 2   Summary of test results

Specimen	f_c (MPa)	f_{ct} (MPa)	V_u (kN)	w (mm) 35%·V_u	w (mm) 40%·V_u	w (mm) 45%·V_u
DEB-1	41.1	3.15	193.56	0.38	0.52	0.65
DEB-2	39.3	2.56	139,26	0.33	0.40	0.46
DEB-3	39.9	2.98	127.05	0.08	0.17	0.27
DEB-4	40.4	3.19	176,70	0.36	0.46	0.58
DEB-5	40.8	3.05	125.30	0.32	0.40	0.47
DEB-6	40.2	3.05	197.23	0.29	0.39	0.52
DEB-7	31.1	3.19	309.22	0.41	0.47	0.54
DEB-8	30.0	2.84	191.60	0.33	0.41	0.53
DEB-9	33.3	2.98	325.79	0.31	0.40	0.48
DEB-10	32.2	2.77	197.15	0.31	0.42	0.53
DEB-11	31.9	3.19	143.61	0.24	0.31	0.38
DEB-12	33.3	2.98	240.49	0.25	0.30	0.39
DEB-13	36.9	3.67	310.68	0.31	0.39	0.45
DEB-14	33.7	2.70	329.07	0.21	0.24	0.30
DEB-15	37.1	2.84	280.00	0.24	0.27	0.32
DEB-16	37.2	4.02	315.30	0.18	0.23	0.26
DEB-17	38.3	3.19	328.09	0.30	0.38	0.45
DEB-18	38.8	3.05	343.41	0.17	0.23	0.30

Analyzing the achieved ultimate loads (Table 2) it is shown that the specimens with maximum amount of reinforcement (p=100%) does not carry out double of ultimate load of those loads achieved by specimens with half amount of reinforcement (p=49%); only a load between 60% and 70% higher is reached. Also the ratio of ultimate loads is lower than the ratio of reinforcements in specimens with intermediate amount of reinforced (p=71%) regarding to those with minimum amount of reinforcement (p=49%). Between the maximum and the intermediate amount of reinforcement the ratio of ultimate loads is quite similar to the ratio of reinforcements.

The results of the ultimate loads pointed out are explained by a spalling of the upper concrete cover (Fig. 3 – left), that is generated for the maximum and the intermediate amounts of reinforcement. This spalling is generated in the nodal zone delimited by the diagonal strut, the vertical principal tie and the horizontal strut that connects with the continuity region of the beam. The loose of this concrete cover involves that a different Strut-and-Tie mechanism with a lower ultimate load is generated. This aspect has to be considered when proposing simplified Strut-and-Tie models for dapped-end beams with high amounts of reinforcements.

In most cases the ultimate load is reached with the plastification of all main ties (vertical, horizontal and diagonal).

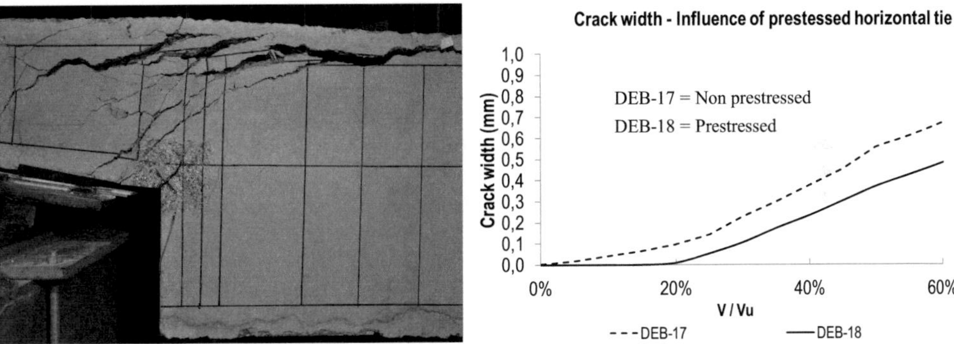

Fig. 3   (left) Detail of test DEB-11_2 at ultimate load – (right) Crack width. Influence of prestressed horizontal tie

4.2   Serviceability behaviour

The service load of the test is taken as around 40% of its ultimate load. At this load level is analyzed the influence of different variables in the crack width of the elements. For this loading level, crack widths are superior to 0.30 mm in most cases, indicating that the design of dapped-end beams is usually conditioned by cracking service limit state instead of ultimate state (load-bearing capacity).

Regarding to the amount of reinforcement there are not significant differences in cracking at service loads for the different specimens with different amounts analyzed.

In the case of prestressing the main horizontal tie (Fig. 3 - right) appears significant differences since the cracking load is delayed up around the 20% of the ultimate load. Therefore, for the service level, approximately taken as the 40% of the ultimate load, the crack width is considerably lower than the analogous specimen without prestressing.

Regarding to the presence of diagonal reinforcement, can be observed that its presence reduces cracking width at service loads (Fig. 4 - left). Increasing the amount of the diagonal reinforcement in reference to the orthogonal one, greater effect in reduction of crack width occurs.

The last parameter analyzed that influences serviceability behaviour is the bevelled of the re-entrant corner. By having a bevel of 4x4 cm on the corner are generated two cracks in the specimens, which lead to an appreciable reduction of crack width (Fig. 4 - right).

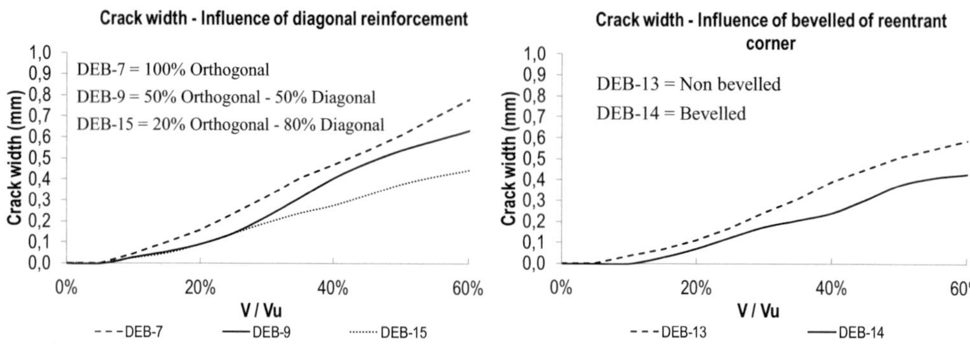

Fig. 4   (left) Crack width. Influence of reinforcement configuration – (right) Crack width. Influence of bevelled of re-entrant corner

5   Conclusions

An experimental program on 18 beams with dapped-ends is carried out to analyze the influence of different configurations and amounts of reinforcement. The following conclusions are given:

- It is generated for very low loads a crack at the re-entrant corner. Due to the crack widths obtained, the dimensioning of the reinforcement of dapped-end beams are usually conditioned by cracking service limit state instead of ultimate state (load-bearing capacity)
- Putting diagonal reinforcement it is considerably reduced the cracking in service. The cracking also decreases as increases the importance of diagonal reinforcement in reference to the orthogonal one.
- The bevelled of the re-entrant corner generates two cracks what reduces the crack width in service significantly.
- Prestressing the horizontal tie leads to a delay of the cracking load in dapped-end beams, reducing the crack width at service loads.

6   Acknowledgements

This research was funded through grants from the Ministerio de Ciencia e Innovación of Spain with the research project BIA2009-11369, part of the National Plan of I+D+i 2008-2011. The experimental program was developed in the Laboratory of Concrete of the Institute of Concrete Science and Technology (ICITECH) of the Universitat Politècnica de València (UPV).

References

[1]   J. Schlaich, K. Schäfer, y M. Jennewein, «Toward a consistent design of structural concrete», *PCI Journal*, vol. 32, n°. 3, pp. 74–150, jun. 1987.

[2]   fib bulletin No. 45, *Practitioners' guide to finite element modelling of reinforced concrete structures*. Lausanne, Suiza: , 2008.

[3]   G. C. Reynolds, *The strength of half-joints in reinforced concrete beams*. London, UK: Cement and Concrete Association, 1969.

[4]   T. C.-K. Chan, «A Study of the behaviour of reinforced concrete dapped end beams», Thesis, University of Washington, Seattle, Washington, USA, 1979.

[5]   A. H. Mattock y T. C. Chan, «Design and Behavior of Dapped-End Beams», *PCI Journal*, vol. 24, n°. 6, pp. 28–45, 1979.

[6]   L. A. Clark y P. Thorogood, «Serviceability behaviour of reinforced concrete half joints», *Structural Engineer*, vol. 66, n°. 18, pp. 295-302, sep. 1988.

[7]   R. R. H. Zhu, W. Wanichakorn, T. T. C. Hsu, y J. Vogel, «Crack width prediction using compatibility-aided strut-and-tie model», *ACI Structural Journal*, vol. 100, n°. 4, pp. 413–421, 2003.

A design method for beam members consisting of CFRP laminates

Hiroki Sakuraba

Graduate School of Engineering,
Hokkaido University,
Kita 13, Nishi 8, Kita-ku, Sapporo, Japan
Supervisor: Takashi Matsumoto

Abstract

The goal of this research is to develop a design method for beam members consisting of Carbon Fiber Reinforced Polymer (CFRP) Laminates. Specimens with four laminate structures which consist of orthogonal laminas are employed, and four point bending tests are carried out. Finite element analysis is conducted to examine a dominant stress to failure. Results of the bending tests are discussed with the ones calculated based on beam theory and on results of the finite element analysis. The calculated results show that the proportion of longitudinal laminas in laminate structure can design displacement by bending and that flexural strength can be determined when dominant stresses to failure and material strengths are known, which is affected by the effect of the interaction between longitudinal normal and in-plane shear stresses.

1 Introduction

1.1 The motivation of this research

Recently, Carbon Fiber Reinforced Polymer (CFRP) has been studied to develop durable beam members because of its superior properties such as high strength and stiffness, and non-corrosive nature. However, the flexural behavior of beam members consisting of CFRP has not been fully clarified because CFRP is an anisotropic and a brittle material.

Since CFRP is fabricated as a laminate composite, the way to stack and orient individual layers, called lamina, affects its material properties. The way to stack and orient laminas is called "laminate structure". Namely, different laminate structures govern the mechanical properties of CFRP. Therefore, it is important to understand the influence of different laminate structures to the flexural behavior of beam members consisting of CFRP.

Fig. 1 shows the outline of this research. Although beam members consisting of CFRP have been researched based on member-scale level [1], [2], the mechanical properties of CFRP are not fully utilized yet for a beam member. To achieve an efficient use of the mechanical properties of CFRP, a research based on laminate-scale level seems to be required.

1.2 The goal of this research

The goal of this research is to develop a design method for beam members consisting of CFRP laminates. Namely, this research presents the possibility to develop a design method based on laminate-scale level. In order to develop the design method, bending tests and structural analyses are needed with specimens which have different laminate structures.

1.3 Key steps

This research consists of three steps. First, specimens with different laminate structures are prepared to examine the flexural behavior of beam members consisting of CFRP, and four point bending tests are carried out. Second, structural analyses are conducted to understand the flexural behavior. Third, based on the two steps, design methods for displacement and flexural strength are developed.

Proc. of the 9th *fib* International PhD Symposium in Civil Engineering, July 22 to 25, 2012,
Karlsruhe Institute of Technology (KIT), Germany, H. S. Müller, M. Haist, F. Acosta (Eds.),
KIT Scientific Publishing, Karlsruhe, Germany, ISBN 978-3-86644-858-2

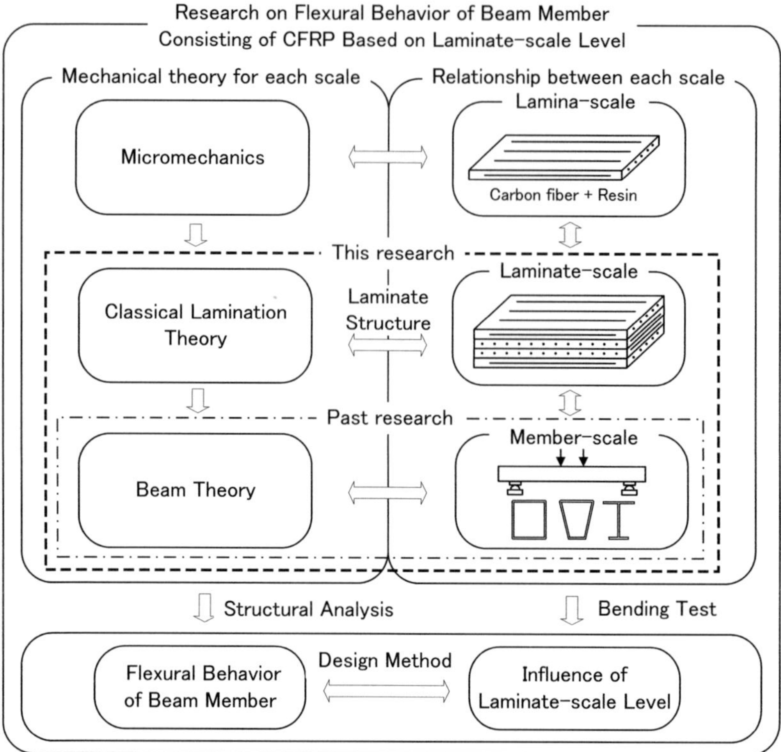

Fig. 1   Outline of this research.

Table 1   Laminate structures and material properties.

No.	Name	Laminate structure	E_1 (GPa)	G_{12} (GPa)	f_1^{T} (MPa)	f_1^{C} (MPa)	τ_{12}^{U} (MPa)
1	L9T1	$[0_2/90/0_9]/90/[0_9/90/0_2]$	98.1	3.50	2353	674	71.1
2	L1T1	$[0/90]_6/0/[90/0]_6$	81.0	3.56	1912	612	67.7
3	L1T2	$[90_3/0]_3/90/[0/90_3]_3$	39.0	3.53	715	247	67.7
4	L1T9	$[0/90_{11}]/0/[90_{11}/0]$	20.9	3.50	295	173	67.7

In this paper, specimens with four laminate structures which consist of orthogonal laminas are employed with the same loading condition and cross section, and four point bending tests are carried out. Also, finite element analysis is conducted to examine a dominant stress to failure. Finally, results of the bending tests are discussed with the ones calculated based on beam theory and on results of the finite element analysis.

2   Experimental Program and Finite Element Modeling

2.1   Test specimens and setup for bending test

Specimens with four laminate structures were employed to examine the effect of different laminate structures. One specimen was prepared for each laminate structure. The specimens consist of laminates with 25 laminas fabricated from carbon fiber and epoxy resin. The specimens have a square box cross section with 100mm height, 100mm width, and 5mm thickness, and a length of 1000mm.

Table 1 shows the four laminate structures and material properties in each specimen. The material properties were obtained from material tests in the past research [3]. Fig. 2 shows laminate coordinate

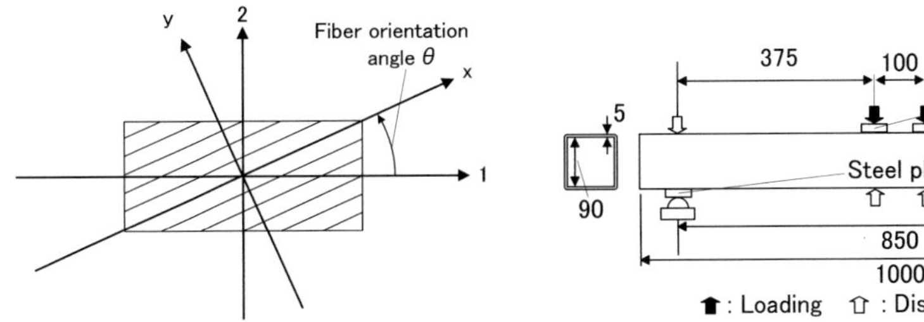

Fig. 2 Laminate coordinate system 1-2 and lamina coordinate system x-y. Fig. 3 Setup for bending test.

system 1-2 and lamina coordinate system x-y. The directions of 1 and 2 correspond to the longitudinal and transverse directions of the specimens, respectively. The rotation angle between direction 1 and x is defined as fiber orientation angle. In order to examine the influence of orthogonally stacked laminates to the flexural behavior of beam members consisting of CFRP, fiber orientation angles, 0° and 90°, were employed. For example, $[0/90]_6/0/[90/0]_6$, the first left side number means the fiber orientation angle of the first layer. The subscript six indicates that six [0/90] groups are continuously stacked. The four laminate structures are symmetric about the mid-plane. Specimen No.1, No.2, No.3, and No.4 are named L9T1, L1T1, L1T2, and L1T9, where L and T representing longitudinal and transverse, respectively. The numbers after L or T represent their proportions to the whole amount of laminas.

Fig. 3 shows a setup for bending test. The specimens were tested under four point bending and under load control. The span, shear span, and flexural span were 850mm, 375mm, and 100mm, respectively. Loading and support plates consisting of steel were placed to the loading and support points, respectively. The loading condition was determined so as to prevent shear failure at the web from arising. Also, in Fig. 3, white and black arrows show displacement gauges and loading points, respectively. The loading was applied until the specimens failed.

2.2 Material property and boundary condition for finite element analysis

L1T1 was analyzed with MSC.Marc to understand a dominant stress to failure. The analytical model utilized three dimensional solid elements. The loading and support plates were modeled as steel (E = 200 GPa; v = 0.3). L1T1 was modeled as orthotropic materials. The in-plane properties were obtained from material tests in the past research [3] (E_1 = 81.0 GPa; E_2 = 81.0 GPa; G_{12} = 3.50 GPa; v_{12} = 0.080), while the out-of-plane properties were assumed values (E_3 = 8.50 GPa; G_{23} = G_{31}= 3.20 GPa; v_{23} = v_{13} =0.465). E_3 was assumed from a transverse elastic modulus of a CFRP lamina [4], where subscript three representing the thickness direction. G_{23} and G_{31} were typical values of a lamina made of graphite-polymer composite [5]. v_{23} and v_{13} were estimated based on classical lamination theory [5]. L1T1 was simply supported, and a quarter analytical modeling was used because of a symmetrical condition of the bending test. Vertical loading was applied on top of the loading plate as pressure.

3 Results and Discussions

3.1 Displacement

The influence of the different laminate structures to displacement in the specimens is discussed. Fig. 4 shows total load-displacement relationships at the loading point in each specimen. Displacement at the loading point based on Timoshenko's beam theory [6] is calculated and is given by

$$\delta = \frac{P}{2E_1 I}\left\{\frac{a^3}{6}+\frac{a^2 b}{2}\right\}+\frac{Pa}{2G_{12}kA}\tag{1}$$

where P = total load; E_1 = longitudinal elastic modulus; I = second moment of area; a = shear span; b = flexural span; G_{12} = in-plane shear modulus; A = cross-section; k = shear correction factor (= A_w/A, A_w = cross section of the web). The first and second terms represent displacements by bending and by shear, respectively. Also, the first term corresponds to Bernoulli-Euler's beam theory. Longitudinal and in-plane shear moduli shown in Table 1 are used for the calculation. Since the shear moduli in each specimen show close values, displacements by shear among the specimens become almost equal at the same loading level.

As a result, the displacements based on Timoshenko's beam theory agree well with the experimental ones. On the other hand, the displacements based on Bernoulli-Euler's beam theory underestimate the experimental ones. At about 20kN level, L9T1 shows the smallest displacement, followed by L1T1, L1T2, and L1T9. The displacements in each specimen seem to depend on the proportion of the longitudinal laminas. Therefore, it is thought that the displacements by bending in the specimens can be designed based on the proportion of the longitudinal laminas. Moreover, in order to design displacement by shear, using laminate structure with diagonal laminas such as the one with fiber orientation angles of ±45° can be an important focus because laminate structure with diagonal laminas can govern in-plane shear modulus of a laminate.

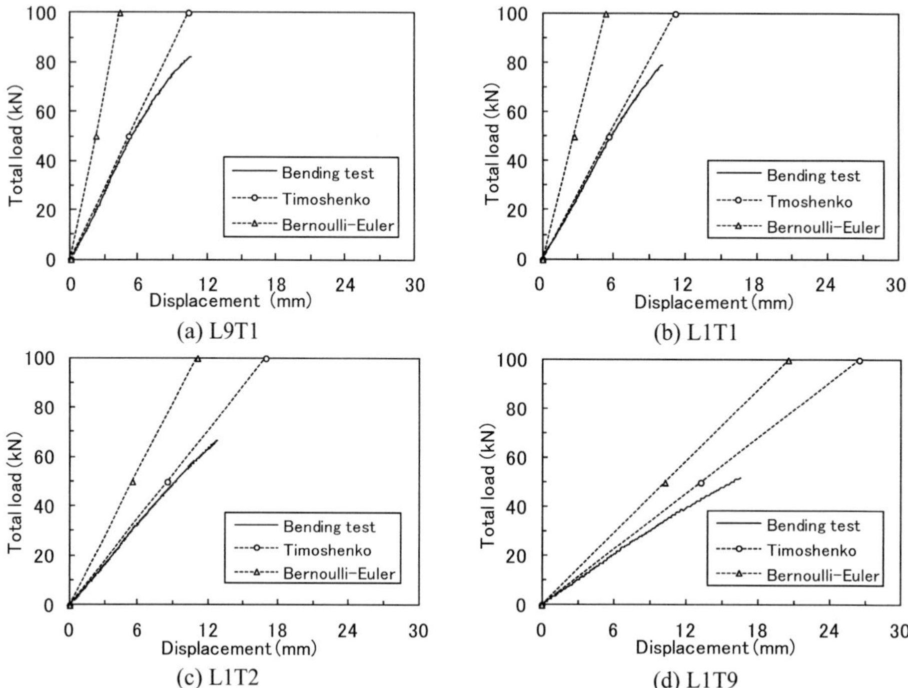

Fig. 4   Total load-displacement relationships at loading point in each specimen with calculated displacements based on Timoshenko's beam and Bernoulli-Euler's beam theories.

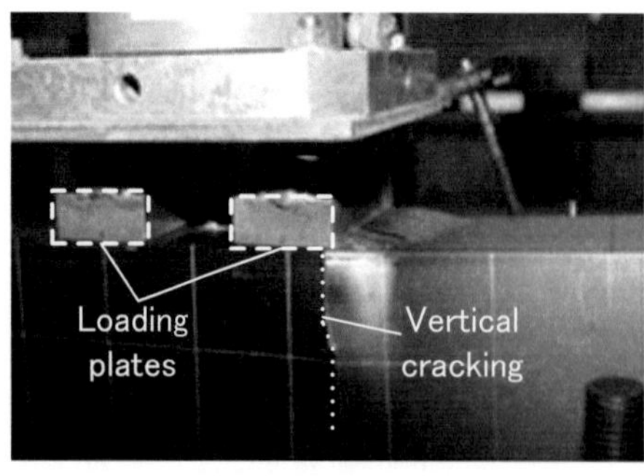

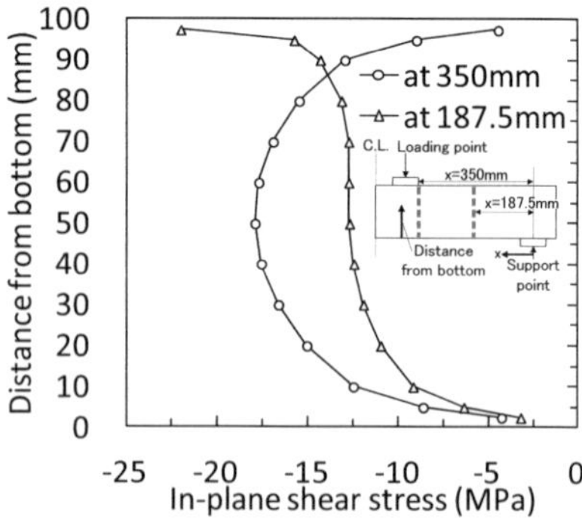

Fig. 5   Typical configuration of failure near loading plate.

Fig. 6   Distribution of in-plane shear stress in height direction of web in finite element analysis.

3.2   Dominant stress to failure

In order to develop an equation based on beam theory to calculate flexural strength, a dominant stress to failure in addition to longitudinal normal stress is examined by finite element analysis. In the bending tests, flexural strengths in L9T1, L1T1, L1T2, and L1T9 are 82.0 kN, 78.9 kN, 66.3 kN, and 51.5 kN, respectively. Upon reaching the flexural strengths, the specimens failed as a brittle behavior. Fig. 5 shows a typical configuration of failure in the specimens. The failure took place near the loading plate in the side of the shear span where vertical cracking arose as shown in Fig. 5.

In the analysis, a local shear deformation near the loading plate is observed. By resulting from the local shear deformation, a relatively high in-plane shear stress occurs. Fig. 6 shows distribution of in-plane shear stress in the analysis at 30kN in height direction of the web and at 350mm and 187.5mm in longitudinal direction which are shown in Fig. 6. The distribution at 187.5mm agrees well with that in beam theory. On the other hand, in the case of the distribution at 350mm, a relatively high shear stress arises near the top. This seems to be caused by the effect of the local shear deformation due to the loading plate, and the relatively high in-plane shear stress can be a dominant stress to failure. Thus, in addition to longitudinal normal stress, the relatively high shear stress seems to be needed for calculating flexural strength in the specimens.

3.3   Equation for flexural strength

Based on the results of the dominant stress, an equation to calculate flexural strength in the specimens is considered, and the calculated flexural strength is compared to the experimental one.

Longitudinal and in-plane shear stresses based on beam theory at the corner of the upper flange where the failure took place are given by

$$\sigma_1 = \frac{M}{I}\left(-\frac{h}{2}\right), \ \tau_{12}^0 = \frac{V}{I}\left(\frac{bh}{4}\right) \tag{2}$$

where σ_1 = longitudinal normal stress; M = bending moment; h = height of the specimen; τ^0_{12} = in-plane shear stress by beam theory; V = shear force; b = width of the flange. As the relatively high in-plane shear stress near the loading plate is observed in the analysis, the effect is considered to the equation for flexural strength. The in-plane shear stress is assumed by

$$\tau_{12} = \tau_{12}^0 + V/A \tag{3}$$

where τ_{12} = in-plane shear stress with additional in-plane shear stress due to the local shear deformation near the loading plate. The second term represents the additional in-plane shear stress.

In order to consider the effect of the interaction between the longitudinal and in-plane shear stresses, Tsai-Wu criterion [5] is employed. Tsai-Wu criterion consisting of the two stresses is given as

$$F_1\sigma_1 + F_{11}\sigma_1^2 + F_{66}\tau_{12}^2 = 1$$

$$F_1 = \frac{1}{f_1^T} - \frac{1}{f_1^C}, \ F_{11} = \frac{1}{f_1^T \cdot f_1^C}, \ F_{66} = \frac{1}{\left(\tau_{12}^U\right)^2} \tag{4}$$

where F_1 = Tsai-Wu's coefficient for longitudinal normal stress in first-order term; F_{11} = Tsai-Wu's coefficient for longitudinal normal stress in second-order term; F_{66} = Tsai-Wu's coefficient for in-plane shear stress in second-order term; f_1^T = longitudinal tensile strength; f_1^C = longitudinal compressive strength; τ_{12}^U = in-plane shear strength. The coefficients are calculated by using the material strengths shown in Table 1. When the left side equation becomes one, material failure takes place. It is assumed that the specimens fail when they satisfy Tsai-Wu criterion. By substituting equation (2) and (3) into equation (4), flexural strength in the specimens is determined. Also, as shown in Table 1, each specimen shows close in-plane shear strengths since they do not have diagonal laminas which contribute to in-plane shear strength of a laminate.

Table 2   Comparison of flexural strengths.

Specimen	L9T1	L1T1	L1T2	L1T9
Experiment (kN)	82.0	78.9	66.3	51.5
Calculation (kN)	85.1	80.6	59.4	49.2

Table 2 compares the experimental flexural strengths with the calculated ones. As a result, the calculated flexural strengths agree relatively with the experimental ones. In the calculated flexural strengths, L9T1 shows the highest fllexural strength, followed by L1T1, L1T2, and L1T9. Namely, the laminate structures with larger proportion of the longitudinal laminas exhibit higher flexural strengths. It is concluded that the specimens failed as a material failure with satisfying Tsai-Wu criterion and that the flexural strength can be determined when dominant stresses to failure and material strengths are known. Moreover, the influence of laminate structure with diagonal laminas to flexural strength can be a key issue when flexural strength is designed as laminate structure with diagonal laminas can govern in-plane shear strength of a laminate.

4   Conclusions

This paper presented the possibility to develop a design method for beam members consisting of CFRP, which is based on laminate-scale level. Specimens with four laminate structures which consist of orthogonal laminas were employed with the same loading condition and cross section, and four point bending tests were conducted. Results of the bending tests were discussed with the ones calculated based on beam theory and on results of finite element analysis. The conclusions are summarized as follows.

The displacements in the specimens agree well with the ones calculated based on Timishenko's beam theory, and the proportion of the longitudinal laminas in laminate structure can design displacement by bending. The calculated flexural strengths agree relatively with the experimental ones, which consider the effect of the interaction between longitudinal normal and in-plane shear stresses. The flexural strength can be determined when dominant stresses to failure and material strengths are known, and the proportion of the longitudinal laminas in laminate structure seems to be an important factor to design flexural strength.

For the next step, the author plans to investigate the influence of laminate structure with diagonal laminas to the flexural behavior of beam members consisting of CFRP under the same loading condition and cross section.

References

[1]   Mutsuyoshi, H., Aravinthan, T., Asamoto, S., and Suzukawa, K.: Development of New Hybrid Composite Girders Consisting of Carbon and Glass Fibers, COBRAE conference 2007 Benefits of composites in civil engineering, 2, university of Stuttgart, Germany, 2007.

[2]   Sugiura, K. and Kitane, Y.: Static Bending Test of Hybrid CFRP-Concrete Bridge Superstructure, *Proceedings of US-Japan Workshop on Life Cycle Assessment of Sustainable Infrastructure Materials*, Hokkaido University, Japan, 2009.

[3]   Sakuraba, H., Matsumoto, T., Hayashikawa, T., Inada, H, Yoshitake, K., Sugiyama, H., Goto, S., Ishizuka, Y., Suzukawa, K., and Matsui, T.: Study on the deformation and failure behaviour of concrete-filled CFRP box beams under bending and axial load, *Journal of structural engineering*, Vol.56A, pp.979-990, JSCE, 2010 (in Japanese).

[4]   Toray Industries: Fundamentals of carbon fiber technology and their application to Torayca products, Functional and Composite Properties, http://www.torayca.com/techref/index.html, 2011.

[5]   Hyer, M. W.: *Stress Analysis of Fiber-Reinforced Composite Materials*, pp.58-59, 387-419 502-508, WCB/McGraw-Hill, 1997.

[6]   Bank, L. C.: *Composites for construction*, pp.384-401, John Wiley & Sons, 2006.

Numerical simulation of the load-bearing behavior of thin-walled plate and shell structures made of TRC

Alexander Scholzen

Institute of Structural Concrete,
RWTH Aachen University,
Mies-van-der-Rohe-Straße 1, 52074 Aachen, Germany
Supervisors: Rostislav Chudoba, Josef Hegger

Abstract

Textile-reinforced concrete (TRC) is a cement-based composite material consisting of a fine-grained concrete matrix and non-corrosive high strength textile fabrics. Due to the orthogonal layout of the textile reinforcement TRC-composites exhibit a strong initial anisotropy. The cracking of the matrix leads to an additional damage induced anisotropy and to a pronounced strain-hardening response for tensile loading. In order to simulate the complex load-bearing behavior of TRC and yet allow for an efficient calculation of large structures the present work uses a smeared cracking approach with an implicit representation of the textile reinforcement. The implemented material model belongs to the class of microplane damage models. The microplane formulation gives the possibility to define directionally dependent damage functions reflecting the anisotropic characteristic of the material. The strain-hardening response under tensile loading can directly be captured by the damage functions. They can be calibrated using the results of TRC-tensile tests applying an automated calibration procedure. The validation of the model is performed based on TRC-slab tests supported at the corners by hinge supports exhibiting a fine biaxial crack pattern in the tensile zone.

1   Introduction

Textile reinforced concrete (TRC) is an innovative composite material that has been the topic of current research within the collaborate research centre "SFB 532" at RWTH Aachen University for the last years. A main advantage of TRC compared to conventional concrete is, that it allows for very thin-walled structures which only need a minimum amount of concrete cover due to its non-corrosive, high strength textile reinforcement. Furthermore, due to the flexibility of the reinforcement the realisation of double-curved shell geometries can be realized relatively easy. In order to demonstrate the potential of this new material an exhibition pavilion with a roof consisting of four double curved TRC-shells with a span of 7x7 m and a thickness of only 6 cm at its edges is currently realised at the campus of RWTH Aachen University.

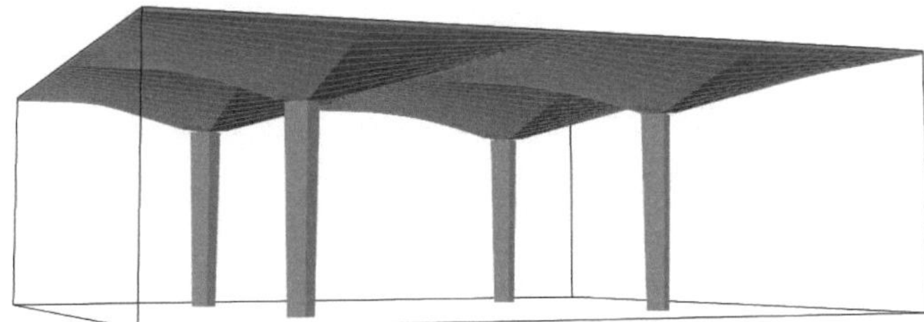

Fig. 1   Computer animation of a TRC roof structure currently realised within the collaborate research center "SFB 532" at RWTH Aachen University.

Proc. of the 9th *fib* International PhD Symposium in Civil Engineering, July 22 to 25, 2012,
Karlsruhe Institute of Technology (KIT), Germany, H. S. Müller, M. Haist, F. Acosta (Eds.),
KIT Scientific Publishing, Karlsruhe, Germany, ISBN 978-3-86644-858-2

In [1] the dimensioning of the TRC roof structure has been described in detail. It is based on a linear-elastic finite element analysis and an automatic dimensioning tool to evaluate the necessary number of reinforcement layers to resist the imposed stresses in the ultimate limit state.

The aim of the present work is to realistically simulate the load bearing capacity of thin-walled TRC-shell and plate structures. Therefore, a refined damage-based material model has been developed that is able to capture the anisotropic effects of TRC and yet allows for an efficient non-linear finite element analysis.

2 Anisotropic damage model for TRC

2.1 Basic principle of the applied microplane approach

The developed material model belongs to the class of microplane formulations [2, 3, 4] and is based on the microplane damage model introduced by Jirasek [5]. The basic concept of the microplane approach is to decompose the macroscopic strain ε (or stress σ) tensor into a set of microplane strain (or stress) vectors applying a geometrical projection into different spatial directions. Based on the value of the maximum normal strain e_{max} reached at each microplane the damage corresponding to this direction is evaluated. The damage state at each microplane is described by a scalar integrity variable denoted as ϕ_i. A value $\phi_i = 1$ reflects undamaged material and $\phi_i = 0$ a complete material degradation. Through the applied microplane approach a directional interpretation of the damage variables is explicitly given. Applying the principle of energy equivalence the damage contributions of all microplanes can be accumulated into an anisotropic fourth order damage tensor β placing the applied microplane damage model into the context of classical damage mechanics (Eq. 1; D_e denoting the elasticity tensor). For an undamaged material the damage tensor β evaluates to the unit tensor and the model reproduces the linear-elastic response for arbitrary values of the Poison's ratio [6].

$$\sigma = \beta : D_e : \beta^T : \varepsilon \tag{1}$$

For homogeneous materials an isotropic distribution of the damage function can be assumed. However, the specification of directionally dependent damage functions with specific characteristics in different directions makes it also possible to reflect an initially anisotropic material behavior. Elementary studies of the described model performed at the material point level for tensile loading and for varying orientation of the reinforcement with intuitively defined damage functions have been described in [7].

2.2 Automatic calibration of damage functions

In order to apply the material model in the investigation of real TRC structures the model needs to be calibrated for the present material configuration, i.e. concrete mixture and textile reinforcement. For this purpose, an incremental calibration algorithm has been implemented that determines the damage functions of the material model based on the experimentally obtained uniaxial tensile response of the composite material. The geometry of the tensile specimens is depicted in Fig. 2 (left).

In the calibration algorithm, the damage function of the material model is calibrated in such a way that the stress–strain curve returned by the numerical simulation of a material point subjected to pure tension reproduces exactly the measured response (Fig. 2 (right/top)). The resulting damage function is represented by a piecewise linear function with a resolution corresponding to the resolution of the time stepping procedure (Fig. 2 (right/bottom)). A detailed description of the calibration algorithm can be found in [8]. In the given example the stress-strain curve has been fitted assuming an isotropic distribution of the damage functions, which means that all microplane orientations have the identical form. In general, the textile reinforcement induces an anisotropic damage behavior along the yarn direction that needs to be taken into account. In the present case an alternating orientation of 10 layers of reinforcement has been used so that the anisotropy is not as pronounced.

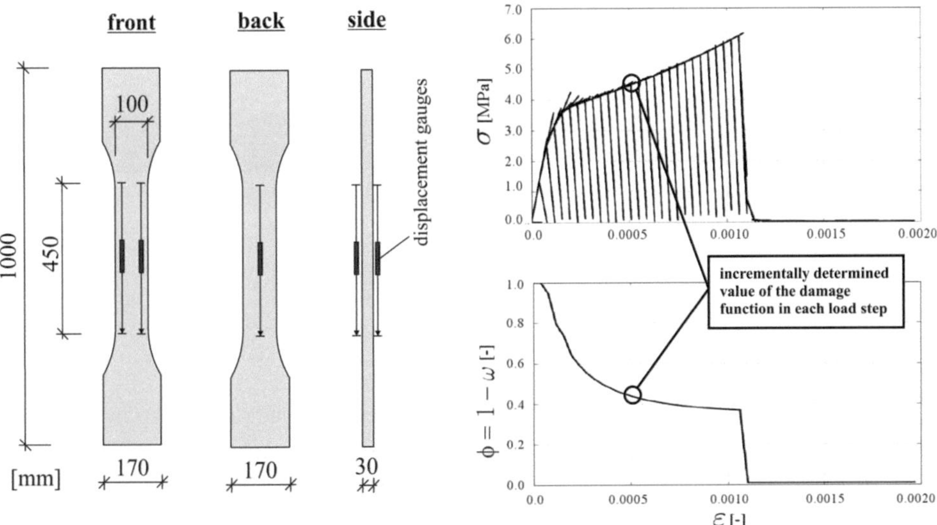

Fig. 2   Geometry of the tensile specimens (left) and measured stress-strain diagram with iteration steps (right/top) and calibrated damage function (right/bottom).

2.3   Implementation and application range

In most cases thin-walled TRC shell and plate structures exhibit a fine and regular crack pattern. For the validity of described material model this is a prerequisite for the use of a smeared cracking model.

Furthermore, the cracks in such structures develop mainly perpendicularly to the surface. Motivated by this fact, only a two-dimensional version of the described model has been implemented. In this form it has been embedded into volume elements that have been extended with the decomposition of the strain and stress tensors into "in-plane" and "out-of-plane" components referring to the shell geometry. The two-dimensional damage model is then applied for the evaluation of the "in-plane" damage, only. In the thickness direction, e.g. the "out-of-plane"-direction, linear elastic behavior has been assumed.

3   Validation of the model using TRC slab tests

3.1   Experimental results

In order to be able to validate the model with the damage function identified from the tensile tests, TRC slab tests have been conducted [9]. The setup for the slab tests is shown in Fig. 3. The thickness and the reinforcement ratio of the slabs was the same as for of the conducted tensile tests depicted in Fig. 2.

The edge length of the quadratic slabs was 1,25 m with a thickness of 3 cm. The loading was applied at the center of the slab. The hinge supports in the corners enabled free rotations. This test setup leads to a fine crack distribution at the bottom side of the slab as depicted for on quarter of the slap in Fig. 4 (left). While the center zone of the slab is cracked both in 0°- and 90°-direction, the pattern becomes coarser towards the edges. As stated above a fine crack pattern is a prerequisite for the application of the derived damage model.

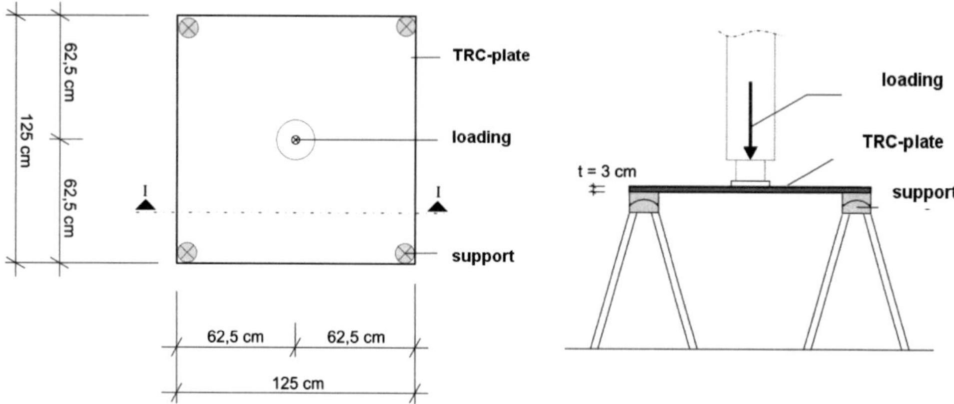

Fig. 3   Test setup of the conducted TRC slab tests: top view (left) and side view I-I (right)

3.2   Numerical results

The simulation has been performed for a quarter of the slab exploiting two planes of symmetry included in the setup. A discretization with 10 x 10 quadratic serendipity volume elements in the slab plane and two element layers in the thickness direction has been chosen. A 3x3x3 Gauss integration scheme has been applied yielding 9 integration points in the out-of-plane direction. In order to verify the finite element model, parametric studies have been conducted for varied element type, integration scheme and fineness of the discretization.

As stated before, the thickness and the reinforcementratio of the slab tests and the tensile tests are identical so that the calibrated damage function in Fig. 2 (right/bottom) could be directly used in the non-linear finite element analysis of the slab tests. This is not necessarily the case for varying reinforcement ratios

The damage distribution obtained from the simulation is shown in Fig. 4 (center) and Fig. 4 (right). The visualized field in Fig. 4 (right) shows the profile of the fracture energy, i.e. the amount of deformation energy dissipated per unit volume at the bottom side of the slab. The energy dissipation is a cumulative indicator for multiple cracking of the matrix oriented in all directions. Further details about the calculated damage pattern can be extracted from the principle direction of the damage tensor as depicted in Fig. 4 (center).

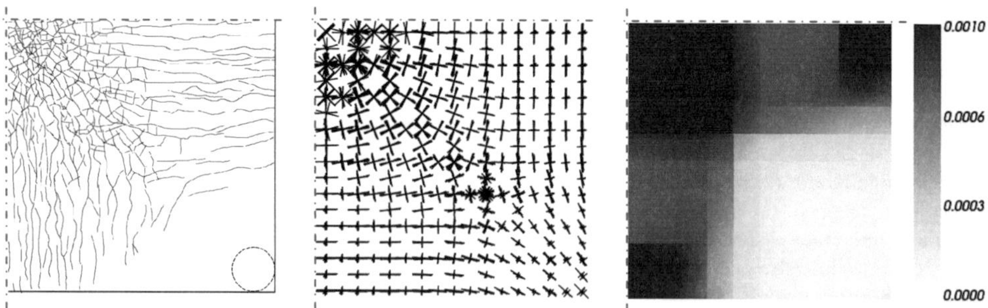

Fig. 4   Qualitative comparison of the final crack pattern (left) with the evolution of damage on the bottom side of the slab indicated by the principle damage direction (center) and the fracture energy [MN/m] (right).

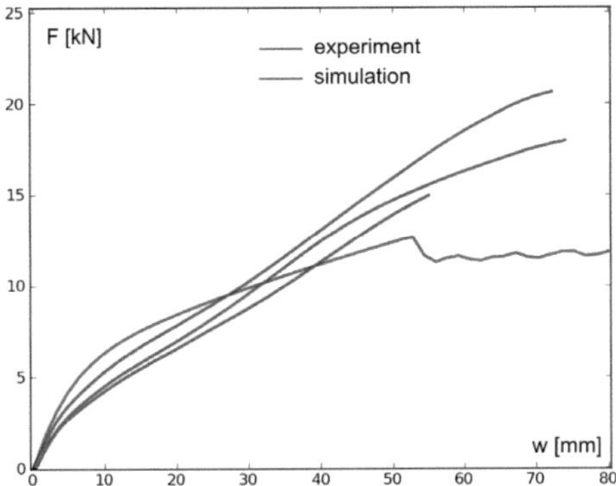

Fig. 5 Comparison between the load-deflection curves of the slab tests obtained in the experiments (gray curves) and in the simulation (red curve).

A comparison between the load-displacement curves as measured in the experiments with the non-linear finite element simulation is depicted in Fig. 5. It can be noted that the initial cracking strength of the slab is overestimated whereas the stiffness of the cracked slab and the ultimate load-bearing capacity is still underestimated. The reason for an overestimated level of crack initiation can be seen in the fact that the shown example has been calculated using a damage function that did not account for the initial anisotropy of the material. As a result the deflection profile obtained in the simulation was identical for the 0°- and 90°-direction.In contrast to this, the experimental response exhibited significant differences in the deflection along the 0°- and 90°-direction revealing a significant effect of the initial anisotropy of the composite due to the orientation of the textile fabrics.

The assumption of isotropy in the simulation has lead to a larger volume accumulating stresses reaching the damage threshold then in the experiment. This explains the overestimated level of loading at the initiation of the first macroscopic crack. These considerations motivate the next step in the elaboration of the subject, studying the effect of initially anisotropy captured by the directionally-dependent damage functions (cf. section 2.2) on the shape of the load-deflection curve of the slab.

Furthermore, due to the fact that TRC exhibits a very ductile behavior large deflections could be observed in the slab tests. Once a certain deflection is reached second order effects become more important and need to be included in the formulation. This could be a possible reason for the underestimation of the load bearing capacity of the slab in the numerical simulation.

4 Conclusions

Summarizing, the model yields plausible results for initially anisotropic cementitious composites. The model provides a valuable tool for the realistic simulation of the load-bearing behavior of TRC structures taking into account the evolution of damage.

The feasibility of the model has been demonstrated in the simulation of TRC-slab tests. Further adjustments of the introduced directional dependency of the damage specification for TRC including the coupling to the mesoscopic model are intended as the following steps [10, 11]. In particular, the kinematical behavior of the model during the localization process must be refined in order to reflect the meso-level damage mechanisms occurring in the tested material.

References

[1] Scholzen, A.; Chudoba, R.; Hegger, J.: "Modellierung und Bemessung von dünnwandigen Platten- und Schalentragwerken aus textilbewehrtem Beton" (in German), 6th Colloquium on Textile Reinforced Structures (CTRS6), Berlin, (2011), pp. 285-296.

[2] Carol, I.; Bažant, Z. P.: "Damage and plasticity in microplane theory". International Journal of Solids and Structures (1997), No. 34, pp. 3807–3835.

[3] Kuhl, E.; Steinmann, P.; Carol, I.: "A thermodynamically consistent approach to microplane theory. part ii. dissipation and inelastic constitutive modeling". Int. Journal of Solids and Structures (2001), No. 38, pp. 2933–2952.

[4]   Leukart, M.; Ramm, E., "A comparison of damage models formulated on different material scales". Computational Materials Science (2003), No. 28, pp. 749–762.

[5]   Jirasek, M.: "Comments on microplane theory", Mechanics of Quasibrittle Materials and Structures, Hermes Science Publications (1999), pp. 55–77.

[6]   Carol, I.; Jirásek, M.; Bažant, Z. P.: "A thermodynamically consistent approach to microplane theory. part ii. free energy and consistent microplane stresses". Int. Journal of Solids and Structures (2001), No. 38, pp. 2921–2931.

[7]   R. Chudoba, A. Scholzen, J. Hegger, "Modeling of reinforced cementitious composites using the microplane damage model in combination with the stochastic cracking theory", EURO-C, Schladming, Austria (2010), pp. 101-110.

[8]   Scholzen, A. ; Chudoba, R. ; Hegger, J.: "Damage based modeling of planar textile reinforced concrete structures", International Conference on Material Science and 64th RILEM Annual Week (MATSCI), Aachen, (2010), pp. 283-291.

[9]   Scholzen, A. ; Chudoba, R. ; Hegger, J.: "Calibration and Validation of a Microplane Damage Model for cement-based Composites applied to Textile Reinforced Concrete", International Conference on Recent Advances in Nonlinear Models – Structural Concrete Applications (CoRAN 2011), Coimbra (2011), pp. 417-428.

[10]   Konrad, M., Chudoba, R.; Kang, B., "Numerical and experimental evaluation of damage parameters for textile reinforced concrete under cyclic loading". In ECCM 2006 III European Conference on Computational Mechanics, (2006).

[11]   Konrad, M.; Jeřábek, J.; Vořechovský, M.; Chudoba, R., "Evaluation of mean performance of cracks bridged by multifilament yarns". In EURO-C 2006: Computational Modelling of Concrete Structures, Taylor and Francis Group, London (2006), pp. 873–880.

Analyses of the end-block of a high performance hollow core concrete slab with high initial prestressing stressed strands

Gábor Alajos Molnár

Faculty of Civil Engineering, Department of Structural Engineering,
Budapest University of Technology and Economics (BUTE),
Műegyetem rakpart 3-9. Kmf. 85., 1111 Budapest, Hungary
Supervisor: István Bódi and Gábor Kovács

Abstract

The aim of this article is to analyse the end-block of a high performance prestressed hollow core concrete slab. It especially focuses on the spalling cracks of the webs. The paper is trying to find the reasons of the spalling cracks in a design and also in a production point of view. From the side of the design, the results of the formulas calculating transmission length and spalling stress given by the Hungarian standards and FIB Bulletins, were compared with the results of the practical tests. From the side of the production the effect of the prestressing force to the angling of the end-cross-section of the slab was investigated. Existing additional spalling stresses causing by the tensed sawing blade was examined. Alternative practical solutions preventing the tensing of the sawing blade was researched.

1   Symbols

Latin upper case letters:

P_0	initial prestressing force just after release
M	prestressing moment
E_0	modulus of elasticity at time of release
I	second moment of the cross-section

Latin lower case letters:

l_{pt}	basic value of the transmission length
f_{bpt}	bond stress at time of release
$f_{ctd}(t)$	design tensile strength at time of release
b_w	thickness of an individual web
e_0	eccentricity of the prestressing steel
f_{ct}	tensile strength of the concrete at time of release
f	camber of the slab
l	length of the slab

Greek lower case letters:

α_1	coefficient depending on prestress dispersion
α_2	coefficient depending on the type of the reinforcement
$\varnothing$	nominal diameter of the strands
σ_{pm0}	strands stress just after release
$\eta_{p1}; \eta_1$	coefficients that take into account the type of strand and the bond situation at release
σ_{sp}	spalling stress
α_e	coefficient depending on the cross-section
φ_{creep}	creep factor of the concrete

Proc. of the 9th *fib* International PhD Symposium in Civil Engineering, July 22 to 25, 2012,
Karlsruhe Institute of Technology (KIT), Germany, H. S. Müller, M. Haist, F. Acosta (Eds.),
KIT Scientific Publishing, Karlsruhe, Germany, ISBN 978-3-86644-858-2

2 Description of the problem

In our work the horizontal cracks in the webs above the strands of the hollow core slabs are examined. These cracks can cause really serious problems, because they can reduce the shear capacity of the slab. Since the production technology of the slabs restricts the use of shear reinforcement, cracked webs will not resist the shear stress but the undamaged webs.

Fig. 1   Horizontal cracks in the webs above the strands of a HCS250 typed hollow core slab.

3 Description of the process of the examination

The goal of our paper is to find the reasons of these spalling cracks in a design and in a production point of view.

In a design point of view the transmission length and the spalling stress were calculated according to the Hungarian standards MSZ EN 1168; MSZ EN 1992-1-1 (Eurocode 2) and FIB Bulletin 6th. The calculations were done by using software Cslab made by Consolis Technology and by using our own-programmed application. The results of the two different applications were compared with each others. In the next steps the results of the calculations were compared with the behaviour of the 10 pieces test slabs. Using the experiences of the test slabs the own-programmed application was modified to be able to approach the real behaviour of the slabs.

In a production point of view the camber and the angling of the end-cross-section, which is the influence of the camber of the slab were considered. The camber and the end-cross-section angling were calculated also by using software Cslab and by using our own application. The calculated data were compared with the data measured on the test slabs. The own-programmed application was developed to be able to approach the real behaviour of the slabs. Knowing the actual value of the camber and the end-cross-section angling is important, because if this value exceeds a practical limit value, the concrete sawing blade will tense and will cause additional spalling stresses, which can be the production reason of these horizontal cracks. Finally, alternative practical solutions were researched for preventing the tensing of the sawing blade.

4 Description of the tested slabs

4.1 Geometry and Reinforcement

The type of the test slabs is HCS250, which means that it is a hollow core slab produced by slipformer technology width a depth of 250 mm.

The designed geometry of this slab can be seen on figure 2. The difference between the designed and the produced geometry is close to zero percent, so calculating with the designed geometry is more than sufficient.

The number of the strands in the examined cross-section is 10 from strand type of EN 10138-3-Y1860S7-12.5-I-F1-C1. The initial prestressing stress is 1200 N/mm^2.

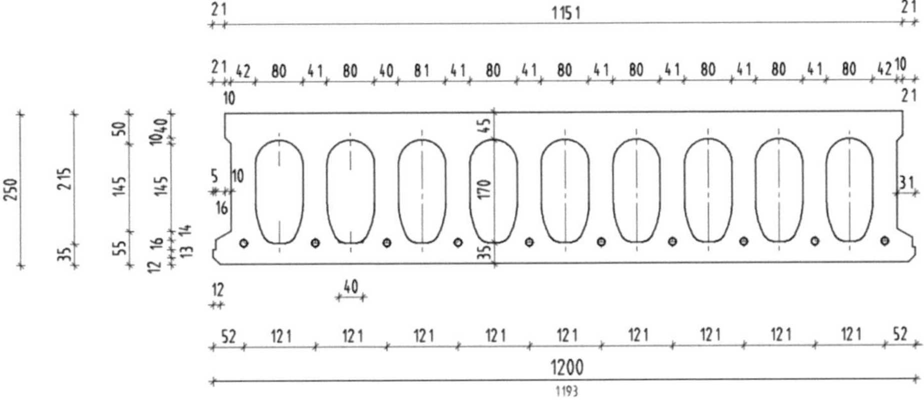

Fig. 2   Designed geometrical values of the cross-section of HCS250 product typed hollow core slab.

4.2   Used materials

The type of the used materials of the produced test slabs perfectly fits to the type of the materials using day by day in the precast industry:

- concrete: C40/50-XC2-12-F1-MSZ 4798-1:2004
- strand: EN 10138-3-Y1860S7-12.5-I-F1-C1

5   Design aspects of the spalling cracks

5.1   Calculation of the transmission length and the spalling stress

5.1.1   Theoretical basis

The transmission length was calculated according to the Hungarian standard MSZ EN 1992-1-1 from:

$$l_{pt} = \frac{\alpha_1 \times \alpha_2 \times \phi \times \sigma_{pm0}}{f_{bpt}} \tag{1}$$

During our calculation using eq. 1 sudden released 7-wire good bond conditioned strands were assumed. Considering the tendon stress the initial prestressing stress was decreased with the prestress losses existing before release. The bond stress can be calculated from the following expression, where the value of the tensile strength was considered at time of release:

$$f_{bpt} = \eta_{p1} \times \eta_1 \times f_{ctd}(t) \tag{2}$$

The spalling stress was calculated using the Hungarian standard MSZ EN 1168 and the FIB Bulletin 6th.

$$\sigma_{sp} = \frac{P_0}{b_w \times e_0} \times \frac{15 \times \alpha_e^{2,3} + 0,07}{1 + \left(\dfrac{0,8 \times l_{pt}}{e_0}\right)^{1,5} \times (1,3 \times \alpha_e + 0,1)} \tag{3}$$

According to the standards eq. 4 have to be verified in order not to have horizontal spalling cracks in the webs of the slab.

$$\sigma_{sp} < f_{ct} \tag{4}$$

As above already mentioned the calculations were done by using software Cslab made by Consolis Technology and by using our own-programmed application.

5.1.2 Numerical results

The final results of the calculations of the two different applications can be found in table 1.

According to the Hungarian standard MSZ EN 1168 the tensile strength of the concrete at the time of release must be considered, which 2.85 MPa is. According to FIB Bulletin 6th the 5% fractile characteristic value of tensile strength of the concrete at the time of release must be considered, which 2.0 MPa is.

Since both value of the tensile strength of the concrete is higher, than the calculated spalling stress the webs will not have to be cracked.

Table 1   Numerical results for the transmission length and the spalling stress of the two different applications.

	Transmission length	Spalling stress
Cslab	476 mm	1.96 MPa
Own application	488 mm	1.82 MPa

5.2 Comparison of the theoretical values with the real behaviour of the test slabs

10 pieces test slabs were produced by ASA Construction Ltd. (a company of CONSOLIS) for examining the spalling effect in a design and also in a production point of view. The measured values of the tensile strength of the test slabs at the time of release were at around 3.2 MPa. The summary of the behaviour of the test slabs can be seen in table 2.

Despite of having smaller calculated values for spalling stress than the tensile strength, 70% of the test slabs were cracked. In order to get reliable results from the theoretical calculations, the own-programmed application was calibrated using the test results. Mainly the values belonging to the bond conditions were modified considered the effect of the production technology, the concrete technology and the early development of the strength of the concrete, which is influenced by the curing process.

Table 2   Real behaviour of the test slabs considering the spalling cracks with the measured values for the camber and for the end-cross-section angling (l.a.c.: loaded against camber)

Number of test slab	Appearance of spalling cracks	Time of appearance	Camber	End-cross-section angling	Remark
TS_01	yes	just after lifting	16.4 mm	0.0044 rad	l.a.c.
TS_02	yes	just after lifting	17.0 mm	0.0046 rad	l.a.c.
TS_03	no	-	15.8 mm	0.0042 rad	l.a.c.
TS_04	yes	1 day later	16.2 mm	0.0044 rad	l.a.c.
TS_05	no	-	15.9 mm	0.0042 rad	l.a.c.
TS_06	yes	just after lifting	16.4 mm	0.0044 rad	l.a.c.
TS_07	yes	just after lifting	16.8 mm	0.0045 rad	l.a.c.
TS_08	yes	2 days later	16.7 mm	0.0045 rad	l.a.c.
TS_09	yes	just after lifting	15.9 mm	0.0042 rad	l.a.c.
TS_10	no	-	16.3 mm	0.0044 rad	l.a.c.

6   Production aspects of the spalling cracks

6.1   Calculation of the camber end the end-cross-section angling

6.1.1 Theoretical basis

For the theoretical calculation model of the camber the simplest analysis has been taken: full linear elastic/plastic behaviour. According to the practical experiences this assumption fits very well for hollow core slabs. The basic expression of this model is the following:

$$f = \frac{M \times \ell^2}{8 \times E_0 \times I} \times (1 + \varphi_{creep}) \qquad (5)$$

The end-cross-section angling was calculated using the value of the camber on the basis of a graphic model.

The calculations were done also by using software Cslab and by using our own application.

Finally, the horizontal displacement of the top point of the end-cross-section was calculated with a simple trigonometrically function using the value of the earlier calculated end-cross-section angling and knowing the depth of the slab.

6.1.2 Numerical results

The final results of the calculations of the two different applications can be found in table 3.

The practical limit value of the horizontal displacement of the top point is 1.0 mm. This value is a characteristic value of the concrete sawing blade using in the production. For the explanation of this effect please see figure 4 (left). Since the sawing blades can be different from plant to plant, this value must be revised in every new calculation belonging to a different precasting plant. Comparing the calculated values with the limit value, the sawing blade will not have to tense, so the webs will not have to crack.

Table 3   Numerical results for the camber, the end-cross-section angling and the horizontal displacement of the top point.

	Camber	End-cross-section angling	Horizontal displacement
Cslab	14.6 mm	0.0036 rad	0.9 mm
Own application	14.2 mm	0.0034 rad	0.85 mm

6.2   Comparison of the theoretical values with the real behaviour of the test slabs

On the one hand the camber and the end-cross-section angling values of the 10 pieces produced test slabs were measured at the time of release. The measured values were above the calculated values, as can be seen in table 2. In order to be able to approach the real behaviour of the slabs with the calculated values, our own-programmed application was calibrated using the measured data. Mainly the value of the modulus of elasticity and the creep factor was fitted to the local circumstances, which is first influenced by the production technology the concrete technology and the curing process.

On the other hand to examine the effect of the sawing blade 5 more pieces test slabs were sawed without loaded them against camber, because according to the 6.1.2 it is not necessary. The result was rankling. In two cases the sawing blade tensed so much so that it stopped, and damaged the slab. In three cases the sawing blade tensed, but it could saw the slab. After lifting the slab the horizontal cracks of the webs were perfectly visible. See figure 3. In spite of having smaller calculated values for the horizontal displacement than the limit value, the sawing blade tensed. The explanation is that the real values for the horizontal displacement of the 5 test slabs were above the calculated values and also above the limit values.

Fig. 3   Damaged slabs caused by the stopped tensed concrete sawing blade (left). Horizontal cracked webs of a slab caused by the tensed concrete sawing blade (right).

If the calculated value for the horizontal displacement is above the limit value two main alternative solutions are theoretically useable. Either of them, the sawing of the slabs must be delayed until a higher concrete strength is reached. The problem is that this solution decreases the productivity, so it is not really useable in the production. The other one is the loading of the slab during sawing against camber, which is a useable and productivity solution. See figure 4 (right).

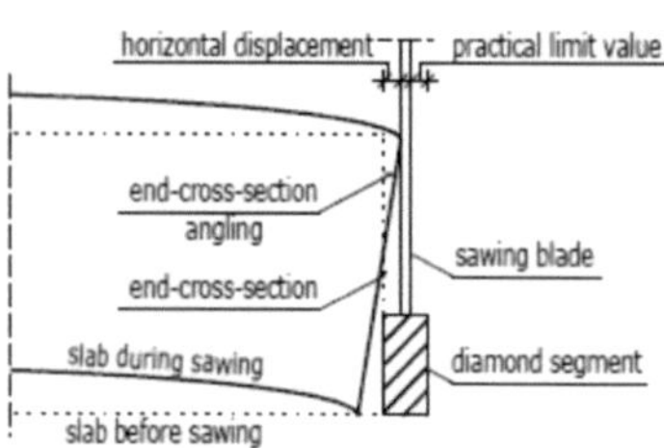

Fig. 4   Explanation of the effect of the concrete sawing blade (left). Loading of the slab during sawing against camber (right).

7   Conclusions

In our work the design and production reasons of the horizontal spalling cracks in the webs of the hollow core slabs, which can reduce the shear capacity of the slab, were examined. In a design point of view the transmission length and the spalling stress, in a production point of view the camber and the end-cross-section angling were calculated according to the Hungarian, the European standards and FIB Bulletins using software Cslab and using our own-programmed application.

10 pieces test slabs were produced for comparing the real behaviour of the slab, so the measured values with the theoretical calculated values.

In both cases despite of having fulfilled theoretical requirements, the webs of the slabs were spalling cracked. In order to get reliable results in the future approaching the real behaviour of the slabs from the theoretical calculations, the own-programmed application was calibrated using the measured values of the test slabs.

The conclusion is that the expressions given by the standards can't use perfectly without fitting them to the local circumstances. In our case first the values belonging to the bond conditions of the strands and the tensile strength of the concrete were modified considered the effect of the production technology, the concrete technology (e.g. the shape of the gravels, which influences the tensile strength of the concrete) and the early development of the strength of the concrete, which is influenced by the curing process.

To protect against the tensing of the sawing blade, which is the main production reason of the spalling cracks, the loading of the slab during sawing against camber was found as the only efficient solution.

References

[1]   Tassi, G.: Bond properties of prestressing strands. In: Proceedings of the FIP symposium, Jerusalem, Israel (1988) pp. 121-128.

[2]   Molnár, G., Bódi, I., Karkiss, B.: Analyses of the end-block of a high performance prestressed concrete beam with very thin web. In: Proceedings of the FIB symposium, Czech Republic, Prague (2011) Vol. 2, pp. 811-814.

[3]   FIB Guide to good practice: Special design considerations for precast prestressed hollow core floors. (1999)

[4]   Hungarian standard: MSZ EN 1992-1-1, Design of concrete structures (General rules and rules for buildings). (2005)

[5]   Hungarian standard: MSZ EN 1168, Precast concrete products (hollow core slabs) (2005)

Experimental and theoretical analyses of the load-bearing behaviour of slim biaxial hollow core slabs with flattened void formers

Christian Albrecht

Institute of Concrete Structures and Structural Engineering,
Technische Universität Kaiserslautern,
Paul-Ehrlich-Straße, D-67663 Kaiserslautern, Germany
Supervisor: Jürgen Schnell

Abstract

Hollow core slabs are used in reinforced concrete structures to save material and dead load. Furthermore, larger spans can be realised. For slab thicknesses from 20 to 35 cm a flattened type with rotationally symmetric flattened forms of hollow cores has been developed. For these slabs, the shear force resistance has not been analysed up to now. In addition, the local punching capacity of the upper floor slab section and other structural behaviours have not been tested for example.

Therefore the laboratory allocated to the institute did 16 tests for the shear force resistance and 80 tests for the local punching capacity up to today. There will be more tests for example about the special boundary conditions of the new slim hollow core slabs. After the experimental tests, the different load bearing behaviours will be analysed and design concepts will be developed.

1   Preamble

Hollow core slabs are used in reinforced concrete structures to save material and dead load. Furthermore, larger spans can be realised. Consequently, substantial effects on resource conservation and reduction of CO_2 emissions in the construction business can be achieved. Uniaxial hollow core slabs are state of the art.

During the last years biaxial hollow core slabs became more commonly used. Used types of hollow cores are for example "Airdeck", "Beeplate", "BubbleDeck", "Cobiax", "DONUT TYPE", "UBOOT" which are built in sphere, box or in rotationally symmetric formers. In Germany the only type with a general building approval [1] for slab thicknesses from 30 to 60 cm has a sphere formed hollow core. For slab thicknesses from 20 to 35 cm a new type with rotationally symmetric flattened forms of hollow cores has been developed.

Fig. 1   Biaxial hollow core slabs: hollow cores in a box form (left), hollow cores with a spherical form (middle) and new hollow cores with a rotationally symmetric flattened form (right).

Proc. of the 9th *fib* International PhD Symposium in Civil Engineering, July 22 to 25, 2012,
Karlsruhe Institute of Technology (KIT), Germany, H. S. Müller, M. Haist, F. Acosta (Eds.),
KIT Scientific Publishing, Karlsruhe, Germany, ISBN 978-3-86644-858-2

So far, there are no design rules for these types of hollow core slabs. The aim of the research project, supported by the Federal Ministry of Economics and Technology in Germany, is to develop a design concept for such hollow core slabs.

The following aspects of structural behaviour will be analysed:

- bending resistance
- shear force resistance
- local punching of the upper part of the floor slab section
- shear reaction at the interface between concrete cast at different times
- concreting

For all effects of action experimental tests were and will be done to develop a design concept. The article includes the analyses of the shear force resistance and the local punching capacity of the upper part of the floor slab section. The testing program, the testing results and the design concept for both aspects of the structural behaviour are shown.

2 Description of the slabs

The new hollow bodies consist of two shells being not been possible to re-open them. The bodies are fixed in holding cages of reinforcing steel. They will be produced in heights from 10 to 22 cm in steps of 2 cm for slabs from 20 to 35 cm.

Special features of the design are summarized in Table 1.

Fig. 2   Flattened Hollow body: Open (left), closed (middle) and in a cage (right).

Table 1   Special features of the design

Action	Special feature	Illustration
bending	reduced pressure zone	
shear force	reduced core area	
local punching	slim upper part of concrete floor slab section	
shear at pouring interface	reduced interface zone	
concreting	depending on the degree of flowability	

3 Shear force resistance

Caused by the embedded hollow bodies, the shear resistance is reduced. This is not a big problem for normal slabs. The question is what load bearing capacity is left in the weakened slab? The holding cages provide a positive influence on the shear resistance. Moreover, the spacing of the hollow bodies and the thickness of the upper part of the floor slab section are factors influencing the shear resistance.

3.1 Testing program

A testing program was established that covers the unfavourable situations. Four different situations were tested for three times and one reference test was done without hollow cores. We have examined the minimum and maximum slab thickness with two concrete strengths for each slab thickness (see Table 2). All tests were performed with minimum spacing of test bodies in both directions and the minimum upper part of the floor slab section minus an extra value for installation tolerances of 5 mm. In order to realize a practical longitudinal reinforcement ratio, we used a high-strength steel St 900/1100. At fig. 2 we see the testing specimen for the hollow bodies S-100. Two test runs were done with one testing body.

Table 2   Testing Program

Test No.	Slab thick-ness [cm]	Slab width [cm]	Hollow body height [cm]	Concrete strength class	Ratio for longi-tudinal rein-forcement	Number of tests with hollow cores	Reference test (no hol-low cores)
V-Q-10-20	20	103,5	10	C20/25	0,74%	3	1
V-Q-10-45	20	138	10	C45/55	0,95%	3	1
V-Q-22-20	35	103,5	22	C20/25	0,57%	3	1
V-Q-22-45	35	138	22	C45/55	0,74%	3	1

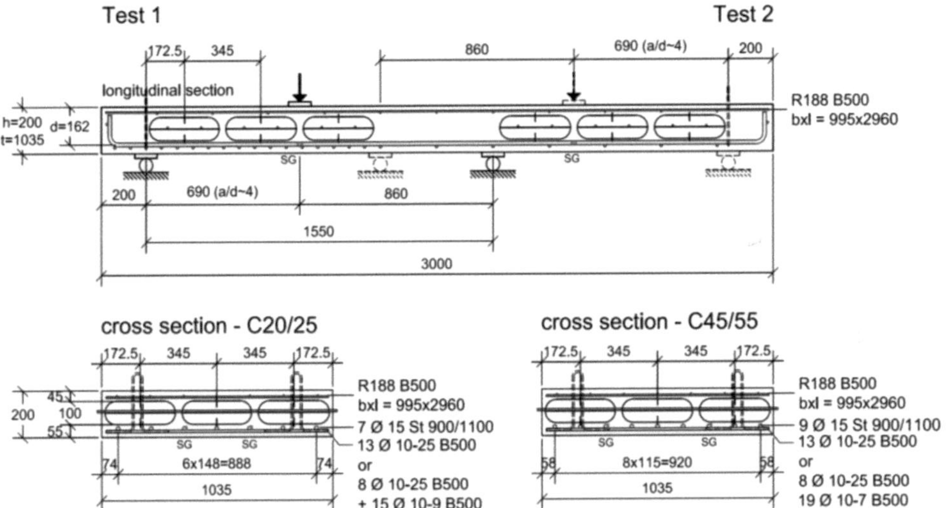

Fig. 3   testing specimen for the smallest slabs.

3.2 Testing results and design concept

The hollow core slabs showed no unusual structural behavior. The load carrying capacity was as expected lower than such of the regular slabs. As reference the formula 6.2a out of the Eurocode 2 [2] was used. The pre-factor $C_{Rk,c}$=0.15 represents the 5% quintile of the shear force capacity. In order to compare the testing results, the average of the capacity with $C_{Rm,c}$=0.2 is used. Longitudinal stresses were not taken into account. The ratio f shows the ratio of the experimental test load to the average shear force of regular slabs. At figure 3 you can see a chart with the ratio f to the concrete strength and a photo of the sheared off specimen with the shear crack running through and next to the hollow bodies.

$$f = \frac{V_u}{V_{Rm,c}} = \frac{V_u}{0.2 \cdot k \cdot \left(100 \cdot \rho_l \cdot f_{cm}\right)^{\frac{1}{3}} \cdot b_w \cdot d}$$

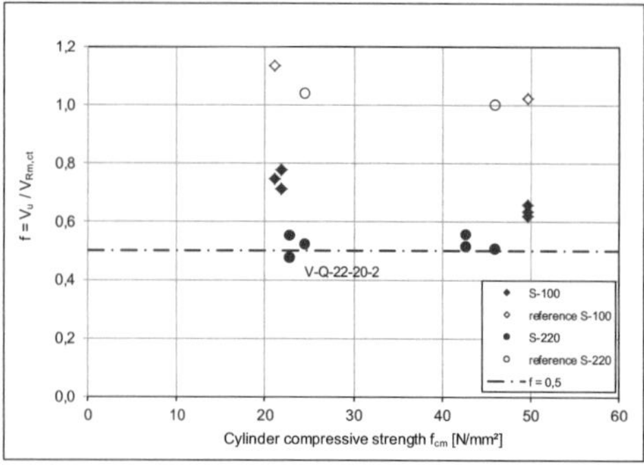
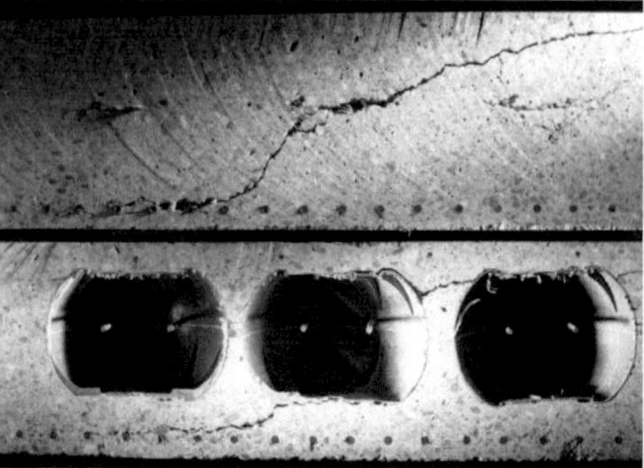

Fig. 4   Reducing factor versus concrete strength (left); sheared off specimens (right)

The aim of the investigations was to figure out a further pre-factor for the hollow core slabs produced with the new flattened type of hollow cores. This is to make the calculation easier for the structural engineers. Figure 3 shows that the smallest ratio to the average shear force capacity was come down to 0.5. It is recommended to design the hollow core slabs with the following formula. This means that the design load is carried out with a pre-factor $C_{Rk,c,Slim\ Line}$=0.075 instead of $C_{Rk,c}$=0.15. Following a statistical analysis similar to the evaluation of local punching, a factor of $C_{Rk,c,Slim\ Line}$ =0.09 would also be acceptable.

$$V_{Rd,c,\text{Slim Line}} = f \cdot C_{Rd,c} \cdot k \cdot \left(100 \cdot \rho_l \cdot f_{ck}\right)^{\frac{1}{3}} \cdot b_w \cdot d$$

$$with\ f = 0.5\ and\ C_{Rd,c} = \frac{0,15}{\gamma_c}$$

4   Local punching

Point loads with small contact areas occur both during construction by formwork support and during service lifetime. The upper part of the floor slab section above the hollow bodies shall be formed in such way that the local punching is excluded. Since there are no design rules for the local punching resistance for this geometry available, small-scale tests were carried out. The tests were anticipated to analyze the structural behavior and to establish a design concept. Local punching shear may occur in the compression and tension zone.

4.1   Testing program

There was carried out a series of experiments to the local punching in the pressure zone and in the tension zone. Being on the safe side, the experiments for the compression zone were performed without longitudinal normal stresses. In both series of experiments the following parameters were varied:

- - Concrete strength (C20/25 and C45/55)
- - height of the upper part of the floor slab section (ccb = 4.5 and 7.5 cm)
- - loading area (Aload = 5 x 5 cm and 10 x 10 cm)
- - hollow body height (h_{HK} = 100 mm and 180 mm)

The testing program for the local punching action without longitudinal normal stress includes a total of 72 experiments, of which each 18 tests were performed on one test body, wherein this has been tested from both sides (see Fig. 4).

The testing program for the local punching in the tension zone includes 16 tests. To produce a longitudinal tensile stress in the upper floor slab section, a bending moment was applied with a collet (see Fig. 4). The bending moment was chosen for a longitudinal tensile stress occurring in the reinforcement at 300 N/mm². For the test program all minimum and maximum parameters were combined.

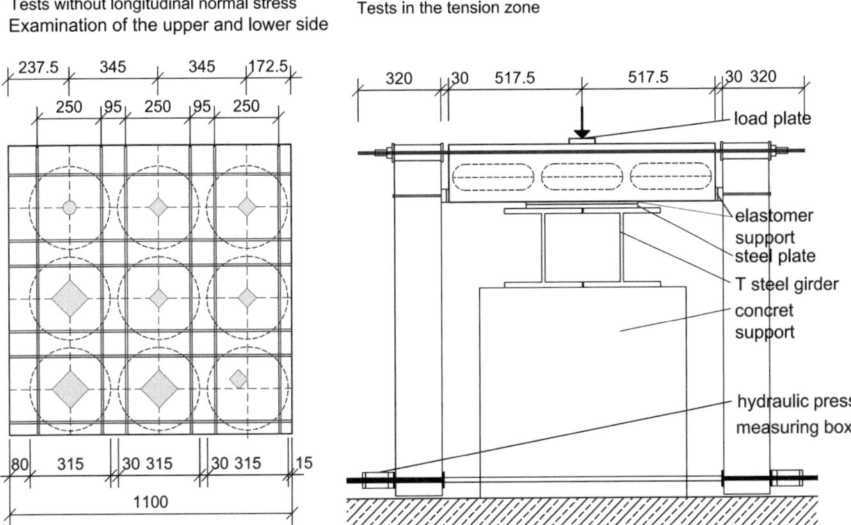

Fig. 5   Test setup: local punching without normal stress (left) and with tension (right)

4.2   Testing results and design concept

All punching tests resulted in a punching cone. The surface line of the cone was punching of the edges of the load footing plates starting tangential to hollow out (see Fig. 5).

We examined various approaches to measure the local punching shear capacity. The best correlation between calculated and measured fracture load could be achieved by a design model of the surface of the punching cone.

Fig. 6   Punching cones from above (left) and in section (right)

For comparison of the calculated failure loads with the measured failure loads, the quotient of the calculated ultimate load to failure load is calculated for all experiments. The design is performed with the 5%-quintile of the concrete strength, reduced by the material safety factor of 1.8, and a pre-factor determined.

$$f_{cal,M} = \frac{F_u}{F_{cal,Mantle(fcm)}}$$

$$F_{Rd,Mantle} = f_{5\%,M} \cdot F_{cal,Mantle(f_{ck},\gamma_m)}$$

with $f_{5\%,M}$ pre-factor from a statistical analysis of $f_{cal,M}$

The factor evaluating the quotient $f_{cal,\ M}$ is calculated as shown below. The factor $k_{s,\ 95\%}$ is used here to determine the lower limit of the one-sided limited statistical area proportion with unknown standard deviation of 5%-quintile at a confidence level of 75%.

$$f_{5\%,M} = m_x - k_{s;95\%} \cdot s_x$$

To determine the pre-factor mean value, standard deviation and 5%-quintile of the ratio of computational and experimental failure load for various groups in the test combinations are calculated and evaluated.

As a result of this analysis and other considerations such as limiting the maximum load and the consideration of site design equation, the following tolerances are recommended.

in the pressure zone without normal stresses:

$$F_{Rd,Mantle,pressure} = f_{5\%,M} \cdot F_{cal,Mantle(f_{cm},\gamma_m)} \leq 50kN$$

upper part of the floor slab section in the tension zone:

$$F_{Rd,Mantle,tension} = 0,7 \cdot F_{Rd,Mantle,pressure} \leq 35kN$$

5 Summary

Hollow core slabs with biaxial hollow bodies are used in reinforced concrete structures to save material and dead load. In addition, larger spans can be realized.

Affected by the hollow bodies a variety of structural actions have to be analysed. The testing program, the testing results and the design concept for the shear resistance and the local punching capacity are shown. Further structural reactions are still under investigation.

The studies indicate that the load-bearing capacity of slim biaxial hollow core slabs with flattened void formers are, as expected, lower than given by solid slabs. By considering the design concepts suggested the hollow core slabs can be designed without lack of safety.

References

[1]   Allgemeine bauaufsichtliche Zulassung Z-15.1-282: Hohlkörperdecke System "COBIAX", DIBt, Berlin, 5.2.2010.

[2]   DIN EN 1992-1-1:2011-01: Eurocode 2: Bemessung und Konstruktion von Stahlbeton- und Spannbetontragwerken – Teil 1-1: Allgemeine Bemessungsregeln und Regeln für den Hochbau; Deutsche Fassung EN 1992-1-1:2004 + AC:2010, Beuth Verlag GmbH, Berlin, 2011.

Compressive membrane action in concrete decks

Sana Amir

*Department of Design and Construction, Structural and Building Engineering,
Technical University Delft,
Stevinweg 1, 2628 CN Delft, the Netherlands
Supervisor: J. C. Walraven, C. van der Veen*

Abstract

One of the major challenges facing the designers today is to investigate if the old bridges are still safe for modern traffic. The current research deals with this question by taking into account compressive membrane action (CMA) in determining the capacity of reinforced and transversely prestressed concrete decks. CMA can significantly affect the flexural and punching shear strength of deck slabs but it is usually neglected for design and assessment purposes. Therefore, a flexural theory was used and a punching shear model was modified to fully utilize the effect of strength enhancement by CMA. Several experiments done by various researchers have been analysed by using these theories. It was concluded that considering CMA in the assessment shows that bridge decks can have a considerably larger shear capacity than assumed in the initial design. This is of high significance because in The Netherlands about 70 bridges have to be investigated, with very thin decks cast between the flanges of long prestressed beams. Using the actual design codes for the verification of the bearing capacity leads to values showing that the safety standards are not met. However, theoretical analyses show that nevertheless sufficient residual capacity might be available. In order to confirm the validity of the calculations large scale laboratory tests are carried out. Variables are the geometry of the deck, the confining effect on the punching shear capacity, and the role of transverse prestressing.

1   Introduction

In the Netherlands, there are a large number of bridges that were built around 50 years ago or even earlier using the design codes and construction methods of that time. Since then, not only the traffic loads have increased drastically but codes have also evolved with additional safety requirements being incorporated into them. It has been observed that bridge deck slabs in typical beam and slab type bridges have inherent strength due to presence of in-plane forces. This is typically defined as compressive membrane action (CMA) or arching action and it occurs in slabs with laterally restrained edges. This restraint induces compressive membrane forces in the plane of the slab enhancing the flexural and punching shear capacities. Therefore, it is possible that such bridges need not to be strengthened if CMA of concrete decks is taken into account while assessing their real capacities.

2   Past research

Traditionally concrete deck slabs have been designed for bending effects only, with the assumption that shear capacity is adequate. However, it has been generally observed that typical bridge deck slabs tend to fail in punching shear rather than flexure. A lot of research has been done in past on the flexural and punching strengths considering compressive membrane action focusing on reinforced concrete bridge deck slabs. CMA was first reported by Ockleston [1] during tests on a 3-storey building in South Africa. Subsequent research in the bending strength area was done by Wood [2] and Park [3]. Research conducted at Queen's University, Canada in the late 1960's has led to compressive membrane action been incorporated in the current Ontario Highway Bridge design Code (OHBDC) [4] and the New Zealand Code [5].

Another rational treatment of the compressive membrane action has been done in the UK Highway Agency standard BD81/02 [6] which resulted from the research done at Queen's University Belfast [7-10]. However the UK Highway method is valid more for rigidly restrained decks and therefore for deck slabs with low restraint, Taylor's approach [11] is used.

The most significant contribution in punching shear considering CMA was made by Hewitt and Batchelor (H&B model) [12] who modified the Kinnunen and Nylander (K&N) punching shear model [13] by including an empirical restraint factor to show the impact of boundary restraint. Some

Proc. of the 9th *fib* International PhD Symposium in Civil Engineering, July 22 to 25, 2012,
Karlsruhe Institute of Technology (KIT), Germany, H. S. Müller, M. Haist, F. Acosta (Eds.),
KIT Scientific Publishing, Karlsruhe, Germany, ISBN 978-3-86644-858-2

tests were done in Queen's University, Kingston, Canada [14] on transversely prestressed concrete decks with steel girders and the H&B model was used to evaluate the punching shear capacity. However, the tests were done on small scale models and till today there has been insufficient research done to include CMA in current codes for prestressed decks with precast concrete girders. Therefore, this research aims to investigate transversely prestressed concrete decks after studying reinforced concrete deck slab methods in detail.

3 Load capacity of reinforced concrete decks

3.1 Flexural capacity

The method used to calculate the flexural capacity considering CMA are Rankin and Long method. It is based on deformation theory and utilizes an elastic-plastic criterion for concrete. The loads carried by bending and arching are calculated separately and then added to give the ultimate load capacity [8].

3.2 Punching shear capacity

The methods used to calculate the punching shear capacity considering CMA are the UK Highway Agency standard BD81/02 [6], Taylor's approach [11] and the modified Hallgren model.

3.2.1 Modified Hallgren model

In 1996, Mikael Hallgren proposed a mechanical model of punching based on the model by Kinnunen and Nylander [13]. The ultimate tangential concrete strain was the failure criterion in the K&N model and was based on a set of semi-empirical expressions developed from the strains measured in punching shear tests and no size effect was considered. In the Hallgren model, the main modification was the ultimate tangential concrete strain derived from a simple fracture mechanics model reflecting both the size effect as well as the brittleness of the concrete [15]. The model did not take into consideration the lateral restraint, however, it was open for further development by introducing forces from the boundary restraint and prestressing.

Therefore, a modified form of the Hallgren model has been proposed in this paper and applied to relevant set of experimental data. In Fig. 1, boundary forces have been introduced into the Hallgren model of a slab with diameter or equivalent diameter, C and depth, h. An empirical restraint factor, η, proposed by Hewitt and Batchelor [12] is used in the Hallgren model to estimate the boundary forces.

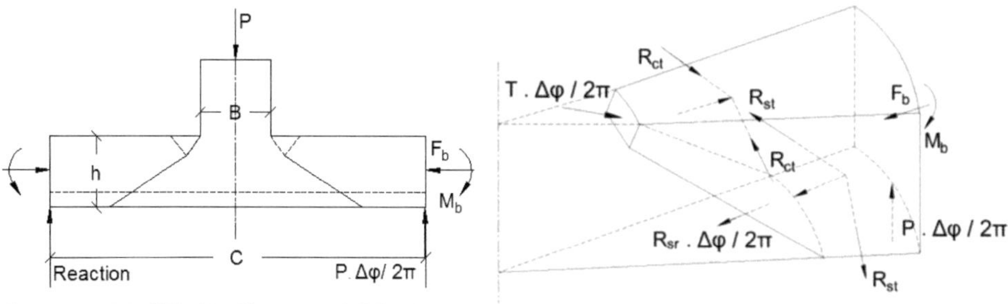

Fig. 1   Modified Hallgren Model for CMA

3.3 Application of CMA theories to experimental data

Several tests done by various researchers have been analysed by the CMA methods, both in flexure and shear but primarily focusing on punching shear as it is more commonly observed failure mode in rigidly restrained deck slabs.

Table 1 and Fig. 2 show comparison of test data (less than rigid restraint slabs) and the ultimate capacity calculated by various methods.

Table 1   Tests by S. E. Taylor et al [11]

Test Panel	P_t [kN]	P_{BS} [kN]	P_{BD81} [kN]	P_{taylor} [kN]	P_{mh} [kN]	P_t/P_{BD81}	P_t/P_{taylor}	P_t/P_{mh}
D1	185	49.4	341.1	219.8	219.8	0.54	0.96	0.84
D2	200	49.3	317.6	206.8	206.8	0.63	1.10	0.97
D5	150	38	268	164.5	164.5	0.56	0.99	0.91
D6	182	38.1	276	184.12	184.12	0.66	1.05	0.99
D7	135	38.1	280.9	155.29	155.29	0.48	1.46	0.87
D8	157	38.7	274	167.45	167.45	0.57	1.05	0.94
					Average	0.57	1.10	0.92
					St. deviation	0.06	0.18	0.057

Table 1 shows that Taylor's approach and the modified Hallgren model give good results as they incorporate deck slabs with less than rigid restraint. The UK Highway BD81/02 shows higher values confirming that it is more suitable for rigid boundaries. However, BS5400 gives very conservative estimates. Fig. 2 shows tests done on 1:3 scale model of an M-beam bridge deck. Both the modified Hallgren and BD81/02 show good estimation of failure loads with the latter being slightly conservative showing presence of adequate rigidity during tests. However, the flexural capacity method of Rankin and Long over estimates the failure load.

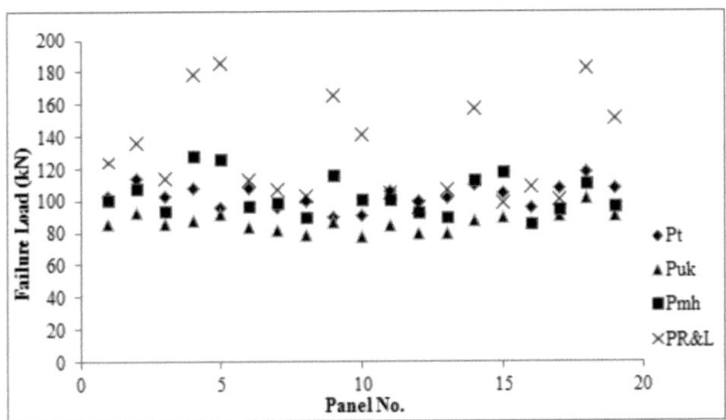

Fig. 2   Tests by Kirkpatrick et al [9]

Table 2 shows test results of experiments done on a real bridge. Each panel was intended to be loaded to three times the UK service wheel load (337.5kN), while not exceeding a midspan deflection of 2.5mm. The BS5400 code, BD81/02 and the modified Hallgren were used to predict the capacity. It is interesting to observe that while the code methods only gave the ultimate loads, the modified Hallgren method could be used to estimate the load corresponding to a certain deflection as well.

4   Punching shear capacity of transversely prestressed concrete decks

The methods used to calculate the punching shear capacity considering CMA are the New Zealand code [5] and the modified Hallgren model. An equivalent steel ratio was used in place of the pre-stressing steel ratio to find the capacity using the code. The modified Hallgren method was further adapted to include forces by prestressing.

Table 3 shows results by modified Hallgren model and the New Zealand code applied on tests done in Queen's University, Kingston, Canada. The method of superposition was used to calculate the punching load [14]. Both the methods show good results compared to the experimental loads and it can be observed that the punching load and hence the membrane action increases with the increasing transverse prestress level (TPL). It is interesting to note that the New Zealand code over estimates the capacity at lower transverse prestress levels.

Table 2   Tests by S. E. Taylor et al [16]

Test Panel	Deflection [mm]	P_t [kN]	P_{BD81} [kN]	P_{mh} [kN]	P_{BS} [kN]	P_{BD81}/P_{BS}	P_{mh}/P_{BS}	P_t/P_{mh}
A1	2.5	333	570.1	401	128.3	4.44	6.75	0.83
A2	1.5	428	600.8	426.4	178.3	3.37	5.70	1.00
B1	2.15	344	563.6	381	66.5	8.48	11.07	0.90
B2	1.15	428	610.4	445.2	92.3	6.61	9.60	0.96
C1	2.6	333	588	406	66.6	8.83	11.58	0.82
C2	1.2	428	588	427.5	92.2	6.38	9.24	1.00
D1	1.85	368	553.5	365	127.9	4.33	5.51	1.01
D2	1.75	428	568.3	412	177.3	3.21	5.35	1.04
E1	1.95	392	632.8	484	202.1	3.13	3.89	0.81
E2	1.6	428	648.7	484.7	280	2.32	3.36	0.88
F1	1.9	371	566.5	415	199.5	2.84	3.56	0.89
F2	0.75	428	601.2	464.2	275.2	2.18	3.19	0.92
						Average		0.92
						St. deviation		0.08

5   Future tests

In the Netherlands, about 70 bridges have to be investigated with very thin decks cast between the flanges of long pre-stressed beams. Using the actual design codes for the verification of the bearing capacity leads to values showing that the safety standards are not met. However, theoretical analyses show that nevertheless sufficient residual capacity might be available. In order to confirm the validity of the calculations large scale laboratory tests are carried out. Variables are the geometry of the deck, the effect of confinement or the restraint on the punching shear capacity, and the role of transverse prestressing.

Fig. 3 shows the test setup of the 1:2 scale model of the van brienenoord bridge near Rotterdam. However, the scale model is still in the design stage and many variables are yet to be determined. To ensure adequate confining effect and the failure within the slab portion, girders have been over designed and a suitable overhang is provided to the external girders for the development of compressive membrane forces. It is expected that a restraint factor, η, of atleast 0.5 will be observed during the tests.

6   Conclusions and recommendation

The UK Highway Agency BD81/02 gives good results for rigidly restraint deck slabs. However, when the restraint is low, the results are unsafe. Also, this method does not allow for the effect of varying reinforcement ratio. For such situations, Taylor's approach is a good tool especially because it incorporates both flexural punching and shear punching failures. The New Zealand code can be used for transversely prestressed decks. However, it gives better estimation when the TPL is high. Modified Hallgren model gives excellent results both for reinforced and transversely prestressed deck slabs, therefore it will be used for future tests as well.

It would be useful if the actual stiffness and restraint can be calculated as it could lead to better estimation of the compressive membrane action. Also, more tests on transversely prestressed decks could be done in future to verify results obtained from this research.

Table 3   Tests in Queen's University, Kingston, Canada [14]

Test Panel	A_p [mm²]	TPL [MPa]	P_t [kN]	P_{mh} [kN]	P_{NZ} [kN]	P_t/P_{mh}	P_t/P_{NZ}
SW-1A	0.0869	1.84	53.1	59.77	67.39	0.89	0.79
SE-1B	0.0869	1.84	53.04	59.77	67.39	0.89	0.79
CW-2B	0.105	2.15	54.82	64.16	70.45	0.85	0.78
CE-2B	0.105	2.15	57.26	64.16	70.45	0.89	0.81
NW-2A	0.1198	2.5	63.85	67.57	71.68	0.94	0.89
NW-2B	0.1198	2.5	48.7	67.57	71.68	0.72	0.68
CE-1B	0.14	2.91	74.43	72.08	74.74	1.03	1.00
CW-1A	0.14	2.91	65.82	72.08	74.74	0.91	0.88
SE-2B	0.1549	3.32	66.31	75.42	76.58	0.88	0.87
SW-2A	0.1549	3.32	72.97	75.42	76.58	0.97	0.95
NE-1B	0.176	3.88	80.54	80.15	79.65	1.00	1.01
NW-1A	0.176	3.88	77.52	80.15	79.65	0.97	0.97
CE-1A	0.19	4.37	94.12	83.42	80.87	1.13	1.16
NE-2A	0.19	4.37	92.28	83.42	80.87	1.11	1.14
NW-3B	0.19	4.37	80.11	83.42	80.87	0.96	0.99
CW-4B	0.19	4.37	82.66	83.42	80.87	0.99	1.02
SE-5B	0.19	4.37	87.3	83,42	80.87	1.05	1.08
SW-6A	0.19	4.37	92.23	83.42	80.87	1.11	1.14
					Average	0.96	0.94
					St. deviation	0.10	0.14

Notations

Φ	Angle of sector element of slab
B	Width of loaded area
F_b	Boundary restraining force
M_b	Boundary restraining moment
P, P_t	Failure load in tests or applied test load
P_{BD81}	Predicted ultimate capacity from BD81/02
P_{BS}	Predicted ultimate capacity from BS5400
P_{taylor}	Failure load according to Taylor's approach
P_{mh}	Predicted ultimate capacity from modified Hallgren model
P_{NZ}	Predicted ultimate capacity from the New Zealand code
$P_{R\&L}$	Predicted ultimate capacity from Rankin and Long method
R_{ct}	Horizontal force in concrete crossing the shear crack
R_{sr}	Horizontal force in reinforcement at right angles to the radial cracks
R_{st}	Horizontal force in reinforcement crossing the shear crack
TPL	Transverse Prestress Level

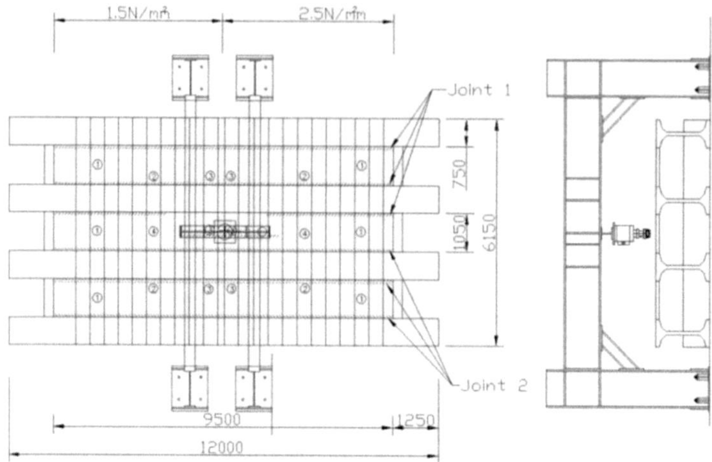

Fig. 3   The scale model test set-up (All linear dimensions are in mm)

References

[1]   Ockleston, A. J.: "Load Tests on a Three Storey Reinforced Concrete Building in Johannesburg," The Structural Engineer, V. 33, Oct. 1955, pp. 304-322

[2]   Wood, R. H.: 'Plastic and Elastic Design of Slabs and Plates," 1961, Ronald, New York

[3]   Park, R., and Gamble, P.: "Reinforced Concrete Slabs," 1980, John Wiley & Sons, UK

[4]   Ontario Ministry of Transport and Communications: Ontario Highway Bridge Design Code (OHBDC), Toronto, Ontario, 1979 (amended 1983 & 1992)

[5]   Transit New Zealand Ararau Aotearoa: New Zealand Bridge Manual, 2nd edition, 2003

[6]   UK Highway Agency, "BD 81/02: Use of Compressive Membrane Action in bridge decks," Design Manual for Roads and Bridges, V. 3, Section 4, part 20, Aug. 2002

[7]   Taylor, S. E., Rankin, G. I. B., and Cleland, D. J.: "Arching Action in High Strength Concrete Slabs," ICE Proceedings – Structures and Buildings, No. 146, Nov. 2001, pp. 353-362

[8]   Rankin, G. I. B., and Long, A. E.: "Arching Action Strength Enhancement in Laterally Restrained Slab Strips," ICE Proceedings – Struc. & Buildings, No. 122, Nov. 1997, pp. 461-467

[9]   Kirkpatrick, J., Rankin, G. I. B., and Long, A. E.: "Strength of Evaluation of M-Beam Bridge Deck Slabs," Structural Engineer, V. 62b, No. 3, Sept. 1984, pp. 60-68

[10]   Taylor, S. E., Rankin, G. I. B., and Cleland, D. J.: "Guide to Compressive Membrane Action in Bridge Deck Slabs," Technical Paper 3, UK Concrete Bridge Development Group/British Cement Association, June 2002, pp. 18-21

[11]   Taylor, S. E., Rankin, G. I. B., and Cleland, D. J.: "Real Strength of High-Performance Concrete Bridge Deck Slabs," ICE Proceedings–Bridge Engg., No. 156, June 2003, pp. 81-90

[12]   Hewitt, B. E., and Batchelor, B. deV,: "Punching Shear Strength of Restrained Slabs, ASCE J. of Structural Engineering, V. 101, ST9, 1975, pp. 1837-1853

[13]   Kinnunen, S., and Nylander, H.: Trans. Royal Inst. Technology, Stockholm, No. 158, 1960

[14]   Weishe, He.: "Punching Behaviour of Composite Bridge Decks with Transverse Prestressing," Ph.D. Thesis, Queen's University, Kingston, Canada, 1992

[15]   Hallgren, M.: "Punching Shear Capacity of Reinforced High Strength Concrete Slabs,"Ph.D Thesis, Royal Institute of Technology, S-11 44 Stockholm, Sweden

[16]   Taylor, S. E., Rankin, G. I. B., Cleland, D. J. and Kirkpatrick, J.: "Serviceability of Bridge Deck Slabs with Arching Action," ACI Structural Journal, V. 104, No. 1, Feb, 2007, pp. 39-4

Numerical and experimental analysis of a novel real-scale test set-up for the analysis of tensile membrane actions in concrete slabs

Dirk Gouverneur

Department of Structural Engineering,
Ghent University,
Magnel Laboratory for Concrete Research, Ghent, Belgium
Supervisors: Robby Caspeele and Luc Taerwe

Abstract

Research with respect to robustness of (concrete) structures has gained wide interest due to the partial collapse of some case examples, such as the famous collapse of the apartment building in Ronan Point (UK) in 1968. A very important property of concrete structures with respect to robustness is the rigid connectivity with neighbouring elements. When a continuous concrete slab is excessively loaded or when a certain support is lost due to an accidental event, membrane forces can be activated in order to establish a load transfer to the remaining supports, which can considerably enhance its load-carrying capacity compared to estimates obtained from small deformation theories. Thus, membrane actions can prevent a progressive collapse and increase the robustness of concrete structures.

Finite element methods allow to simulate the behaviour of concrete plates under these large deformations. Currently, however, different questions remain unsolved in order to properly assess the membrane actions under large deflections, i.e. the influence of the constitutive laws which are implemented, the modelling of connectivity, fracture mechanical aspects under tensile membrane actions, etc. As currently only limited research has been focussing on tensile membrane actions, a novel real-scale test set-up has been developed in order to assess these actions in real-scale concrete plates. The details of this test set-up will be explained and some experimental test results will be discussed. Finally, the results will be compared with numerical FEM analyses and the influence of different model assumptions will be evaluated.

1   Introduction

The interest in robustness of concrete structures emerged because of the complete or partial collapse of structures like the apartment building at Ronan Point (UK) in 1968 (Wei-Jian Yi et al. (2008)). Due to this accident, extensive research on progressive collapse has been established in order to develop related design guidelines. However, the explicit interest in the topic slightly decreased in the 1980s and the early 1990s. However, the bombing attacks on American buildings in several locations and the terrorist attacks on the World Trade Center in New York and the Pentagon in Washington in 2001, triggered a renewed interest in this research topic.

The design of reinforced concrete slabs is usually based on small deformation theories. Regardless of which design method is used, deflections and crack widths should remain within acceptable limits under service conditions. Despite the large number of slabs designed and built, the advantages of the plastic behaviour of slabs are not always recognized or sufficiently taken into account. The risk of a structural failure increases in accidental situations, typically combined with large deformations. In this case the structural interaction between the different elements becomes important in order to provide alternate load paths. According to Li et al. (2007) very large deflections may be induced in the floor slabs of a building without causing the collapse of the structure. It was found that this advantage is due to membrane action. Hence, such structural behaviour plays a considerable role in increasing the load-carrying capacity of floor slabs in accidental load situations compared to considering only flexural behaviour.

Previous experimental and theoretical research can be traced back to the early 1920s. Westergaard and Slater (1921) examined membrane behaviour in numerous full-scale tests on floor panels tested until structural collapse. However, an essential understanding of the membrane behaviour was provided in the 1960s when a significant amount of research was conducted following Ockleston's land-

Proc. of the 9th *fib* International PhD Symposium in Civil Engineering, July 22 to 25, 2012,
Karlsruhe Institute of Technology (KIT), Germany, H. S. Müller, M. Haist, F. Acosta (Eds.),
KIT Scientific Publishing, Karlsruhe, Germany, ISBN 978-3-86644-858-2

mark paper of 1955, on the observed membrane action in the tests of the Johannesburg Hospital (Bailey et al. (2008)). Consequently, Wood's (1961) research resulted in the first systematic analytical approach for membrane actions in slabs.

Fig. 1 illustrates a typical load – deflection curve taken from Park (1964).

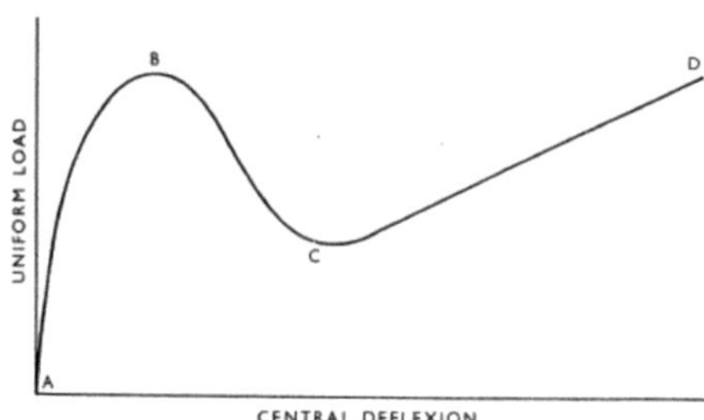

Fig. 1   Load-deflection curve of a slab with full edge restraint according to Park (1964).

In case of a rectangular slab panel with perfect edge restraints against lateral displacement due to the stiffness of surrounding panels, it was discovered that compressive membrane action is induced as a result of restraining the outward movement along the slab edges in the early part of the loading range. This may offer a much higher flexural load than the maximum load predicted by Johansen's (1962) Yield Line Theory. After the outermost flexural load has been reached in point B, the diagram shows a rapid decrease of the supported load with further deflection as a result of the reduction of compressive membrane forces. Moreover, near point C, membrane forces being in the centre bay of the slab reach the stage where they change from compression to tension. The slab boundary restraints start to resist inward movement of the edges. Accordingly, cracks extend over the full thickness of the slab and the yielding of steel reinforcement is established due to the elongation of the slab surface. It is observed that beyond point C the reinforcement acts as a tensile membrane that enables carrying the load under increasing deflections. The carried load increases once more until the reinforcement finally starts to fracture at point D. Already in 1964, Park stated that knowledge of the region CD in Fig. 1 is of importance since, a snap through will occur in point B unless the tensile membrane strength of the reinforcement is high enough to absorb the load.

However, Guice et al. (1989) state that the behaviour of reinforced concrete slabs is not entirely understood, in particular when large deflections occur. Researchers involved in case studies of failures in reinforced concrete slabs noted that an improved theoretical knowledge is necessary to characterise the behaviour of slabs under excessive loading.

2   Experimental program

2.1   Test set-up

In order to investigate the tensile membrane actions under accidental events in more detail, an experimental test set-up has been developed as illustrated in Fig. 2. The configuration basically consisted of a 160 mm thick and 1800 mm wide reinforced concrete slab strip. The total length of the specimen was 14.30 m, whereas the distance of the inner supports and the central support was 4 m changing to one span of 8 m at the removal of the latter.

2.1.1   Vertical support conditions

Heavily reinforced, transverse edge beams (650 mm x 500 mm x 3200 mm) at the lateral outer edges of the concrete slab realised the requested support conditions in vertical and horizontal direction. At these edge beams, the specimen was vertically supported on two concrete blocks at each side. Uplifting was prevented by providing a steel rod (diameter 14 mm) in the centre of the edge beam which ran through a vertical hole to a guide rail anchored in the laboratory floor (hence allowing horizontal displacements of the end beams).

Secondly, the slab was supported by a set of inner supports, consisting of concrete walls. The centre support, however, consisted of an I-beam placed on two concrete columns that in turn were positioned on an identical I-beam on the floor. In this way, the free space between the two I-beams could be used to locate two identical hydraulic jacks in order to simulate an accidental failure, namely the collapse of the central support. Steel rollers with diameter 50 mm were positioned between each support and the slab specimen.

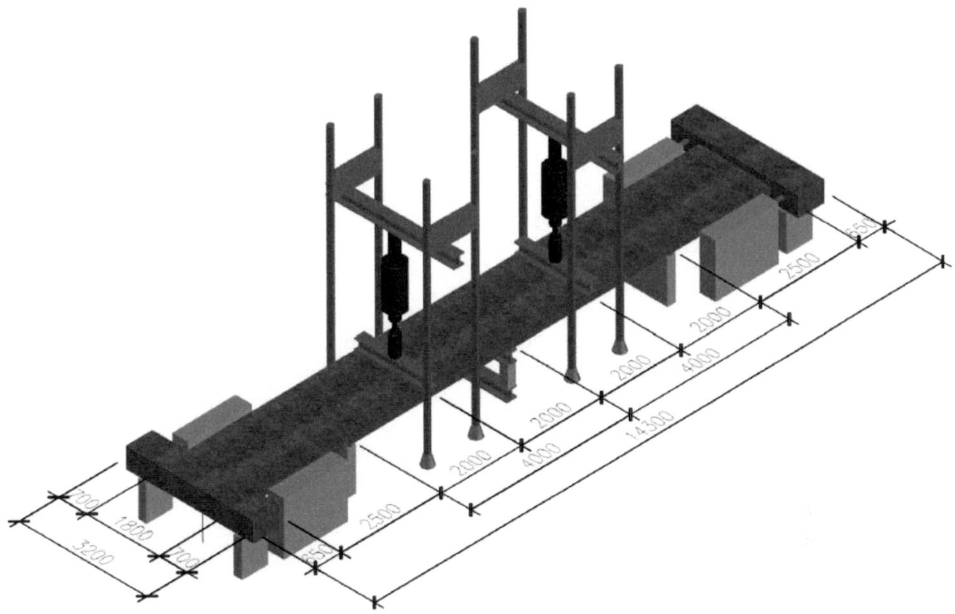

Fig. 2   Test set-up

2.1.2 Horizontal support conditions

In addition, four identical concrete anchor blocks (1300 mm x 400 mm x 1600 mm) established the horizontal reaction forces in order to generate tensile membrane actions. A gap in between the edge beams and the concrete blocks was used to fix 2 load cells on one side. On the other side, 2 aluminium cylinders with the same deformation behaviour as the load cells were placed in between the anchor blocks and the edge beam. The remaining small gap was filled up with thin steel plates at each connection point. Thus, the slab was allowed to elongate horizontally, yet was restricted to contract. Measurements of horizontal compression forces in the load-cells were recorded.

2.1.3 Reinforcement

The longitudinal reinforcement of the slab corresponded to a reinforcement ratio of $\rho = 0.50$ % (typical ratio for office buildings). The reinforcement bars were continuous over the entire length of the specimen and comprised of 16 bars diameter 10 mm for the bottom and top longitudinal reinforcement, respectively, with a concrete cover of 20 mm. Moreover, the top and bottom reinforcement bars were bent at each end in order to properly anchor this reinforcement in the edge beams. A picture of the reinforcement in the edge beam is given in Fig. 3 (a) and (b).

2.1.4 Loading and measurement equipment

The load was applied by using two hydraulic jacks acting on mortar embedded spreader I-profiles over the entire width of the slab. Fig. 2 shows the test set-up with the loading equipment. The loads were measured by pressure sensors. The vertical reaction force of the jacks was taken up by an adjustable steel frame that was anchored in the strong floor (see also Fig. 2). In the centre span of the specimen, another two jacks were positioned between the two I-beams in order to lift up the slab and remove the central support.

Various displacement transducers (LVDTs), strain gauges, potentiometers and dial gauges were located at various points in order to record vertical and horizontal displacements over the complete length of the specimen.

Fig. 3   Reinforcement edge beam   (a)   (b)

3   Numerical program

Considering accidental situations, the robustness related performance of reinforced concrete slabs is highly non-linear as a result of non-linear material behaviour and large deflections. The loading procedure of the test set-up as explained in the next section was numerically simulated by means of the FE code DIANA. The specimen had a rectangular cross section of 160 mm x 1800 mm and was symmetric to the central support regarding geometry and loading. Consequently, only one quarter of the specimen had to be modelled. The elements consist of isoparametric solid brick elements and a 3 x 3 x 3 integration scheme was used.

An interphase element enables to model the real boundary conditions between the load cell and the concrete specimen as elongation of the slab was not restricted. For the concrete, a fixed crack material model was used. For reinforcing steel, the actual stress-strain relationship based on laboratory tests was implemented. However, bond-slip behaviour between reinforcement and concrete was not yet taken into account. Fig. 4 shows the mesh of the finite element model.

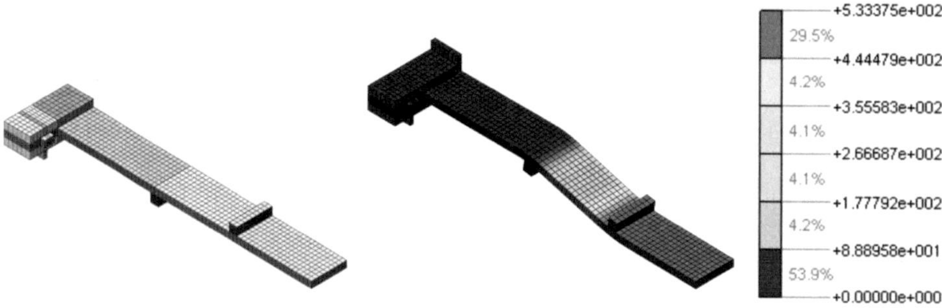

Fig. 4   Finite element model initial stage (left) and deformation right before failure [mm] (right)

4   Test procedure and observations

The loading procedure was split up into three different phases. During the first phase, the load was gradually increased up to a service load of 60 kN for the situation when the central support was present. Subsequently, the load was reduced to 0 kN again. Measurements have shown a gradual increase of deflections under the two line loads ending up with an approximate displacement of 2.50 mm at 60 kN. After unloading, deflections of 0.80 mm were measured.

In a second phase, the central support of the specimen was gradually removed in order to simulate a failure of the support and to obtain valuable data regarding the robustness of the specimen. In order to remove the support, the hydraulic jacks under the central support slightly lifted the slab up so that the concrete spacers could be removed and reaction forces were taken over by these jacks. Thereafter, the stroke of those jacks was reduced very slowly, ending in the disconnection of the central support and the specimen. Accordingly, the two inner spans of 4 m each, changed to one span of 8 m. A reaction force of 42 kN was measured under the central support.

The specimen was allowed to bend progressively in the middle span of the slab. Under this simulated accidental situation, stresses were redistributed and an alternate load path was established distributing the emerging forces to the remaining supports. In summary, the removal of the central support induced a deflection in the centre span of 14 mm.

Finally, in the third phase, the load was applied again by two line loads as specified in phase one. However, unlike in the first phase, load was applied displacement controlled on the specimen until failure. Fig. 5 illustrates the load-displacement diagram for the real-scale test as well as for the numerical simulation. The displacement measurements correspond to those under the line loads.

With regard to the laboratory test, the slab initially elongated under the increasing load. Measurements of horizontal forces could not be taken during this phase as the load cells between the concrete blocks and the specimen could only record compression forces. With increasing vertical deflections of the specimen, however, the slab started to contract again. At point 1 in Fig. 5 (left) the edge beam of the specimen and the load cells made contact again. From this point on, the slab was restrained against inward horizontal movements. Hence, horizontal forces could be measured in the load cells due to the so called *catenary action*. Considering the load-displacement curve obtained by the finite element calculation, this stage was reached slightly earlier compared to the real test situation.

As tensile membrane forces developed, the load-displacement curve was increasing again until a load of 160 kN per jack and a deflection of 330 mm in point 2 (Fig. 5, left). The top reinforcement bars over the left support started to fail consecutively until only the bottom reinforcement of the specimen was active over this support. Point 3 designates the rupture of the top reinforcement over the right support. From this point on, only the bottom reinforcement established the tensile membrane forces until the bottom reinforcement failed as well over the last mentioned support denoted with number 4 in Fig. 5 (left) at a load of 180 kN and a deflection under the load of 520 mm. At that point, the deflection in the central span of the slab panel amounted to 645 mm.

Thus, the ultimate load was found to be around three times the load that was calculated to be the service load for the two spans of 4 m. The finite element model predicted the load displacement curve until point 2 within an acceptable deviation. However, it failed to take the rupture of the top reinforcement bars over the support and hereby the development of very large cracks into account. The deflection when the failure of the slab occurred was predicted well.

Fig. 5 (right) gives an overview of the load-displacement curve whereas the deflection was measured under the load (left axis) in comparison with the development of horizontal membrane forces (right axis). As mentioned above, the finite element program predicted the activation of tensile membrane actions somewhat earlier. Nevertheless, the ultimate membrane force was predicted rather accurately with a value of 711 kN for the finite element model and 740 kN for the real test.

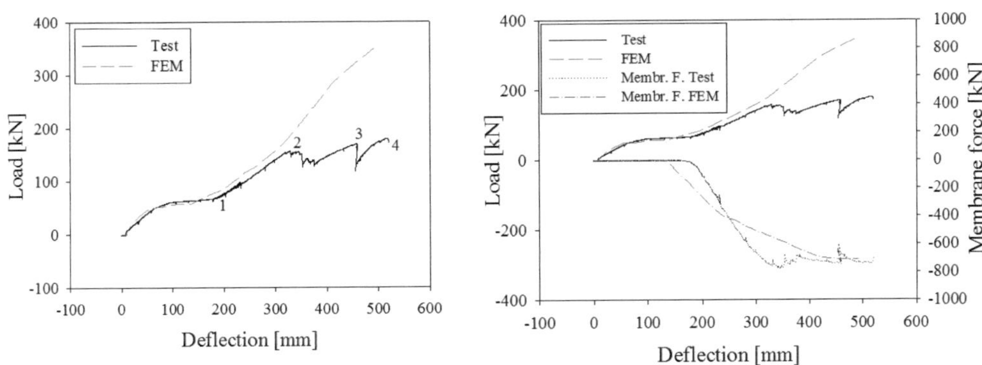

Fig. 5   Load-displacement curve (left) and load-displacement curve in comparison with development of membrane forces (right)

Fig. 6 displays the load displacement behaviour of the specimen calculated by the finite element program when the slab was restrained against inward horizontal displacements as well as for a horizontally unrestrained version of the specimen. According to the finite element model, the load-carrying capacity of the unrestrained version of the specimen was not enhanced significantly due to tension stiffening and, thus, the slab specimen deflected increasingly under approximately the same load. Thus, tensile membrane action can indeed favour the load carrying behaviour of reinforced concrete structures.

Finally, Fig. 6 (right) shows a picture of the laboratory test conducted within this study. It gives an impression of the slab strip's bending behaviour shortly before the collapse of the specimen.

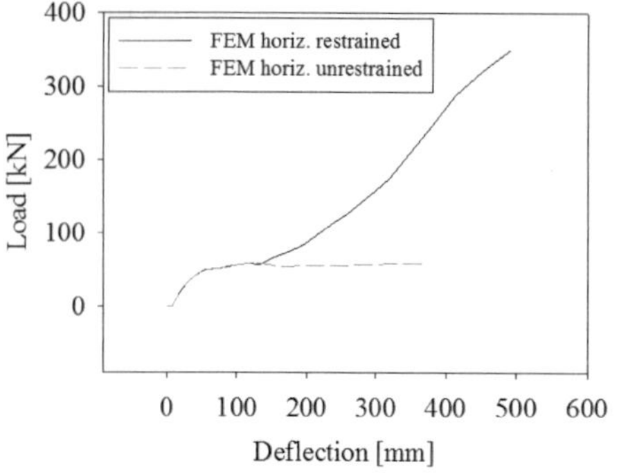

Fig. 6   Restrained slab model in comparison with unrestrained slab model (left) and picture of the laboratory test at a deflection of 635 mm (right)

5   Conclusions

The paper presents a novel real-scale testing procedure. The results as well as a numerical verification are presented. It has been demonstrated that the development of tensile membrane actions can indeed significantly favour the load-carrying capacity in accidental situations of a slab strip when very large displacements occur. The ultimate load was found to be around three times the load corresponding to the service load for the situation with central support. FE modelling is able to predict the increase in the load-carrying capacity due to tensile membrane forces. Although the numerical model is currently still being improved, the load-displacement curve until the rupture of the reinforcement bars as well as the ultimate membrane force was calculated within a reasonable tolerance.

References

[1]   Yi, W.-J., He, Q.-F., & Kunnath, S. K.: Experimental Study on Progressive Collapse-Resistant Behavior of Reinforced Concrete Frame Structures. In: ACI Structural Journal 105 (2008), pp. 433-439

[2]   Li, G.-Q., Guo, S.-X., & Zhou, H.-S.: Modeling of membrane action in floor slabs subjected to fire. In: Engineering Structures 29 (2007), pp. 880-887

[3]   Westergaard, H. M., & Slater, W. A.: Moments and Stresses in Slabs. In: American Concrete Institute Journal 17 (1921) No. 2, pp. 415-539

[4]   Ockleston, A. J.: Load Tests on a 3-Story Reinforced Concrete Building in Johannesburg. In: The Structural Engineer 33 (1955) No. 10, pp. 304-322

[5]   Bailey, C. G., Toh, W. S., & Chan, B. M.: Simplified and advanced Analysis of Membrane Action of Concrete Slabs. In: American Concrete Institute Journal 105 (2008) No. 1, pp. 30-40

[6]   Wood, R. H.: Plastic and elastic design of slabs and plates. In: Thames and Hudson (1961), London

[7]   Park, R.: Tensile membrane behaviour of uniformly loaded rectangular reinforced concrete slabs with fully restrained edges. In: Magazine of Concrete Research 16 (1964) No. 46, pp. 39-44

[8]   Johansen, K. W.: Yield-Line Theory. In: Cement and Concrete Association (1962), London

[9]   Guice, L. K., Slawson, T. R., & Rhomberg, E. J.: Membrane Analysis of Flat Plate Slabs. In: ACI Structural Journal 86 (1989) No. 1, pp. 83-92

Shear carrying capacity of old highway bridges according to the *fib* Model Code 2010

Patrick Huber

Institute for Structural Engineering,
Vienna University of Technology,
Karlsplatz 13/E212, 1040 Vienna, Austria
Supervisor: Johann Kollegger

Abstract

Austrian infrastructure management companies face the problem that their bridge stock contains numerous concrete bridges which were built between 1950 and 1995. If these old bridges are re-analysed according to current code provisions, in many cases the shear capacity is not sufficient. In the paper the results of comparative calculations on the shear capacity of old concrete bridges according to Eurocode 2 [2] and the new fib Model Code 2010 [5] will be presented. Moreover, recalculations of tests on fifty year old bridge girders were carried out. It can be shown that current standards lead to an underestimation of residual shear strength of prestressed concrete members. Therefore, a mechanically consistent shear model is needed, which is able to predict the shear strength of existing structures more accurately.

1   Introduction

The shear strength of concrete members has been researched for more than 100 years. A mechanically consistent model, which makes it possible to predict the shear capacity of existing bridges in a safe way, has still not be found. This fact must be considered as problematic because shear failures can occur without warning. The behaviour of concrete members in shear is a particularly complex phenomenon, which depends on many different physical parameters. The various bearing mechanisms in shear are known, but the mutual influence makes it quite difficult to formulate accurate equations.

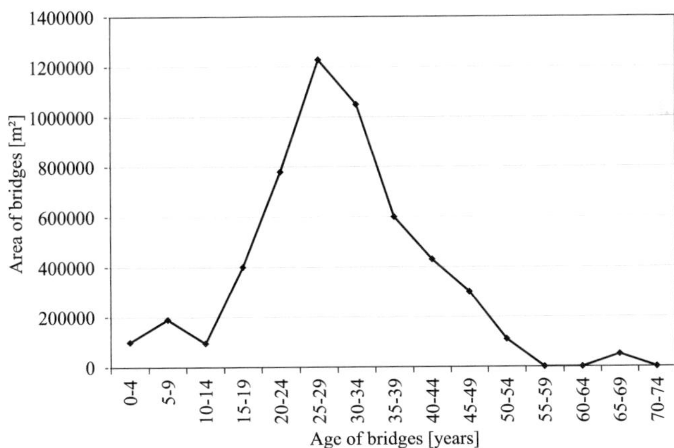

Fig. 1   Age structure of Austrian highway bridges in 2008 [11]

Austrian infrastructure management companies face the problem that the most part of their bridge stock contains bridges that are older than thirty years (see figure 1). They were designed according to ÖNORM B4200 [1], which required none or only little shear reinforcement. This code does not include any minimum shear reinforcement and the design shear resistance is based on principle stress calculations which are compared to allowable stresses. Today's Austrian standards are based on the design provisions of Eurocode 2. Moreover, traffic loads have been increased over the last 50 years.

Proc. of the 9th *fib* International PhD Symposium in Civil Engineering, July 22 to 25, 2012,
Karlsruhe Institute of Technology (KIT), Germany, H. S. Müller, M. Haist, F. Acosta (Eds.),
KIT Scientific Publishing, Karlsruhe, Germany, ISBN 978-3-86644-858-2

Therefore, assessment of existing bridges becomes more and more important for ensuring the bearing capacity and it will be one of the main tasks of structural engineers in the future.

2 Comparative Calculations of Existing Bridges Concerning Shear Strength

2.1 Considered Shear Design Approaches

Within the framework of a research order, comparative calculations were carried out to predict the shear capacity of existing bridge structures. The shear strength of these old bridges was recalculated according to ÖNORM EN 1992-1-1 (Eurocode 2) and the new Model Code 2010 design procedure.

All design approaches taken into consideration differ between members without or with shear reinforcement. The new Model Code 2010 represents various levels of approximation (LoA) for the design and assessment of concrete structures. The first LoA includes the simplest approach and can be used for preliminary calculations. In the higher LoA, physical parameters can be better estimated and thus the degree of accuracy increases. These equations can be used for designing new structures as well as for assessing existing structures such as old bridges.

2.1.1 Members without Shear Reinforcement

The shear resistance $V_{Rd,c}$ for members without shear reinforcement according to Eurocode 2 represents a semi empirical equation, which includes the parameters of compressive strength, geometric percentage of longitudinal reinforcement, width of a member, axial forces and the inner lever arm.

The first LoA of the Model Code 2010 is similar to the minimum shear resistance of Eurocode 2, since it only depends on compressive strength, width and inner lever arm. The shear resistance for members without shear reinforcement of the second LoA is influenced by the longitudinal strain ε_x at the mid-depth of the effective shear depth. These equations are based on the simplified modified compression field theory (SMCFT) [6].

2.1.2 Members with Shear Reinforcement

Eurocode 2, as well as the first LoA of Model Code 2010 design approach for members with shear reinforcement, are based on a variable angle truss model approach. The second LoA represents a generalized stress field approach, which also depends on the state of strain in a concrete member [7]. The third LoA is based on the already mentioned simplified modified compression field theory (SMCFT). The shear resistance V_{Rd} according to SMCFT is composed of a part provided by the shear reinforcement $V_{Rd,s}$ and a part attributed to the concrete $V_{Rd,c}$. Lower LoA do not include an implicit contribution of concrete.

2.2 Calculative Shear Strength of Existing Bridges

The comparative calculations aimed to predict the shear resistance of existing Austrian bridge structures. It shall be demonstrated that these old bridges, as far as shear strength is concerned, may be considered safe. Table 1 shows a summary of main parameters concerning these bridges. Seven reinforced and two prestressed concrete bridges were analysed. First, the calculation of the shear capacity was carried out separately for the main structure in longitudinal direction and subsequently for bridge deck slabs in transverse direction of the bridge.

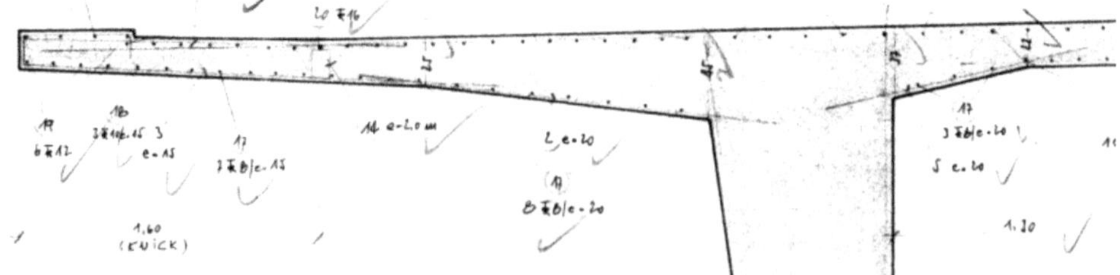

Fig. 2   Reinforcement drawing of a typical carriageway slab of a box-girder bridge

The assessment was carried out with safety values for load and materials according to today's standards. The extreme values for shear force were determined by dead load assumed by original static calculations and by traffic loads according to ÖNORM EN 1991-2 [4]. Dimensions of the structure, strength class of concrete and reinforcement etc. were taken from the original construction drawings, respectively static calculations. Due to the continuous hydration process, the compressive strength of concrete increases significantly. The effect of the subsequent hardening of concrete was also taken into account. In case of a detailed assessment, the compressive strength should be determined on drilled cylinders taken from the considered bridges. Additional to the shear strength, the detailing provisions of the shear reinforcement were also checked. The shear resistance was always calculated at control sections at a location d from the face of supports or at curtailments of reinforcement.

Table 1   Summary of parameters of selected existing bridges

Object	Construction	Height	Total Length	Bearing System	Spans	Cross Section	f_{ck}	f_{yk}
[-]	[-]	[m]	[m]	[-]	[-]	[-]	[Mpa]	[Mpa]
1	RC	0,8	12,3	Ss	1	Sl	18,3	400
2	RC	0,85	70	Cs	4	Sl	26,9	400
3	PC	2,5	39,6	Ss	1	Tb	40	500
4	RC	0,7	13,7	Ss	1	Sl	18,3	500
5	RC	0,75	29,2	Fr	3	Sl	26,9	500
6	RC	0,62	24,8	Ss	2	Sl	26,9	500
7	PC	2,2	100	Cs	3	Bg	40	500
8	RC	1,47	105,6	Cs	4	Bg	26,9	400
9	RC	2,4	33,5	Ss	1	Sl	26,9	400

Legend:

RC	Reinforced concrete	Ss	Single-span system	Sl	Slab
PC	Prestressed concrete	Cs	Continuous system	Tb	T-beam
		Fr	Frame	Bg	Box-girder

Table 2   Results of the shear capacity of bridge deck slabs in cross direction

Obj. No.	Pos.	EN 1991-2	EN 1992-1-1			MC 2010 Level I	MC 2010 Level II	
		d	V_{Ed}	$V_{Rd,c,min}$	$V_{Rd,c}$	η^*	$V_{Rd,c}$	$V_{Rd,c}$	η^{**}
[-]		[m]	[kN/m]	[kN/m]	[kN/m]	[-]	[kN/m]	[kN/m]	[-]
3	I	0,28	115	155	153	0,74	145	156	0,74
	I	0,17	119	106	106	1,12	97	138	0,86
7	C	0,37	95	154	165	0,57	146	119	0,65
	I	0,19	120	98	111	1,08	88	122	0,99
8	I	0,19	153	98	97	1,57	88	87	1,74
	I	0,30	152	127	116	1,19	119	96	1,27
	I	0,19	153	98	112	1,37	88	113	1,35
	I	0,30	152	127	169	0,90	119	190	0,80
9	I	0,17	132	87	103	1,29	80	123	1,07

Legend:

C   Cantilever slab   * $\eta = V_{Ed} / \max(V_{Rd,c,min}; V_{Rd,c})$

I   Inner slab   ** $\eta = V_{Ed} / \max(V_{Rd,c,(LoA\,I)}; V_{Rd,c\,(LoA\,II)})$

The calculated shear capacity, according to Eurocode 2, is considerably lower compared to old Austrian standards [1]. Deficiencies in the calculated shear strength occur especially at bridge deck slabs in transverse direction. This can be explained by the fact that carriageway slabs are thin and, in transervse direction, slightly reinforced concrete members (see figure 2). Table 2 shows the results of comparative calculations of the shear strength of carriageway slabs in transverse direction according to Eurocode 2 and Model Code 2010 provisions. Both design approaches contain two possibilities to reduce design shear force:

- reduction of point loads near supports
- favourable contribution of inclined tension and compression chords (V_{td} and V_{ccd}).

However, many of the calculated bridge deck slabs do not fulfil the requirements of the Eurocode 2 design rules, which raises the question whether today's design rules are too conservative. Tests also confirmed the discrepancy in shear capacity of bridge deck slabs [8,9]. The effect of compressive membrane action in slabs may lead to higher shear capacities than previously assumed in current design approaches. In most cases the new Model Code 2010 implies a higher calculated shear resistance than the Eurocode 2 design approach, although the ULS check for shear cannot always be fulfilled.

The recalculation of the slab bridges as well as webs of T-beams and box-girder bridges has shown that in most cases, due to the inclined bars which were common at the time, the calculated shear carrying capacity of bridges in the longitudinal direction presents no problem.

Another fact is that the detailing provisions in more or less all considered bridges do not fulfil the requirements. Non-compliance of the minimum shear ratio must be considered to be particularly problematic, since because of this the ductile behaviour of members cannot be regarded as certain.

3 Recalculation of Tests on Fifty Year Old Bridge Girders

Within the framework of the demolition of an underground railway station, the Institute for Structural Engineering at the Vienna University of Technology was given the opportunity to carry out large scale tests concerning the shear strength of three prestressed bridge girders. The test series were performed in 2008 [10]. The T-beams had the original length of 17.5 m, a slab width of 3.3 m and a variable height of 1.15 to 1.40 m. The web contained 14 parabolic tendons St 75/105 Rg Ø 26 mm. Close to the support the shear reinforcement consisted of Ø 12 mm with a distance of 250 mm.

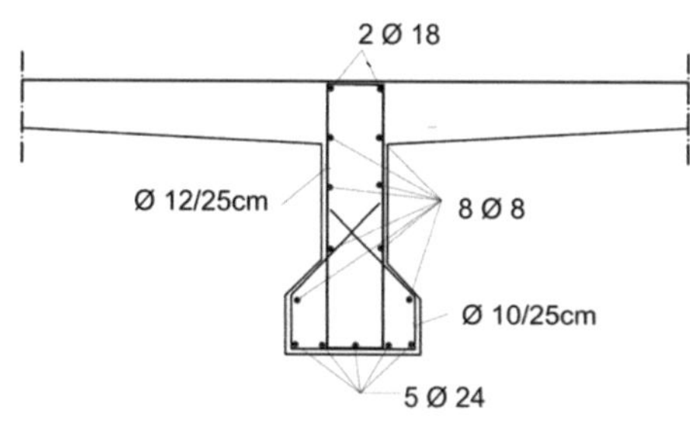

Fig. 3 Test Setup (left) and cross section and reinforcement of web close to the support (right)

Table 3 Mean values of material properties

	f_{cm} / f_{ym} [MPa]	E_{cm} [MPa]
Concrete	52,5	38000
Reinforcement steel	450	209500
Presstressing steel	780	210000

To obtain more information about the shear load carrying behaviour of the structure, the bridge girders were subjected to a four-point bending. The two loads were applied by hydraulic jacks in variable distances from 2.75 up to 3.5 m of the bearings.

In determining the theoretical shear carrying capacity safety factors were omitted and measured mean values of material properties were used (see table 3). The shear strength of the test beams was calculated at a location d from the face of supports, which is in dependence to current code provisions. The inclination of the tendons was on average 5 degrees at the critical section. Due to the fact that the bridge girders were more than fifty years old, the prestressing losses could not be ignored. They were determined according to Eurocode 2 with 30 per cent. The inner lever arm was taken to $z = 0.9 \cdot d$.

Table 4 outlines the recalculation results of test beams V2 und V3. Both tests displayed a typical flexure-shear failure. The theoretical shear carrying capacity in relation to the experimental shear failure load varies between 1.28 and 2.20. It is to be noted that at the load step $V_{Rm,test}$ the stirrups started to yield. This is in accordance to today's design approaches. It can be shown that the new Model Code 2010 design rules represent the best load prediction in the third Level of Approximation. When the stirrups started yielding, the maximum load had not yet been reached. At maximum load, the theoretical load carrying capacity in relation to the experimental load ranges between 2.6 and 3.1 according to the Eurocode 2 approach and in this case the Model Code 2010 predicts the shear resistance most accurately.

Table 4 Results of the recalculation of the test according to Eurocode 2 [2] and the new Model Code 2010 [5]

contribution of inclined tendons: $V_{pm} = 225$ kN	Test V2 $V_{Rm,test} = 1373$ kN* $a = 3{,}25$m		Test V3 $V_{Rm,test} = 1572$ kN* $a = 2{,}75$ m	
Code	$V_{Rm,calc}$	$V_{Rm,calc}/ V_{Rm,test}$	$V_{Rm,calc}$	$V_{Rm,calc}/ V_{Rm,test}$
	[kN]	[-]	[kN]	[-]
ÖNORM EN 1992-1-1	713,9	1,92	713,9	2,20
Model Code 2010: Level I	918,6	1,49	918,6	1,71
Model Code 2010: Level II	777,6	1,77	819,4	1,92
Model Code 2010: Level III	1069,7	1,28	1124,8	1,40

* $V_{Rm,test}$ is the maximum load during the test minus the favourable contribution of inclined tendons

4 Conclusions and Outlook

This paper has shown that the assessment of existing bridge structures according to current shear design rules can be regarded as particularly problematic. Especially bridge deck slabs often do not fulfil the ULS check for shear. Also the non-compliance of detailing provisions complicates the application of today's design approaches.

Recalculation of tests has shown that current standards lead to an underestimation of actual shear strength of prestressed concrete members. The predicted shear carrying capacity of current design approaches is too conservative. The favourable contribution of the inclined tendons seems to be much higher than previously assumed.

Therefore, a mechanically consistent shear model is needed, which is able to predict the shear strength of existing structures more accurately. Then it may be possible to avoid expensive strengthening measures or high costs for reconstruction of bridges, which do not show signs that confirmed the expectation of a serious insufficiency of shear bearing capacity. Hence, further shear tests will be carried out at the Institute for Structural Engineering at the Vienna University of Technology to obtain more information about the real structural behaviour of concrete structures according to shear.

5   References

[1]   ÖNORM B 4200: Stahlbetontragwerke: Berechnung und Ausführung, 4. Teil, Wien, 1957

[2]   ÖNORM EN 1992-1-1: Eurocode 2: Design of concrete structures - Part 1-1: General rules and rules for buildings, Vienna, 2011

[3]   ÖNORM EN 1992-2: Eurocode 2: Design of concrete structures - Part 2: Concrete bridges - Design and detailing rules, Vienna, 2007

[4]   ÖNORM EN 1991-2: Eurocode 1 - Eurocode 1: Actions on structures - Part 2: Traffic loads on bridges, Vienna, 2011

[5]   Model Code 2010, First complete draft, Volume 2. International Federation for Structural Concrete (fib), Lausanne, 2010

[6]   Bentz, E.; Vecchio, F. J.; Collins, M. P.: Simplified Modified Compression Field Theory for Calculating Shear Strength of Reinforced Concrete Elements, ACI Structural Journal, Vol. 103, No.4, July-August 2006, pp. 614- 624

[7]   Sigrist, V.: Generalized Stress Field Approach for Analysis of Beams in Shear, ACI Structural Journal, Vol. 108, No.4, July-August 2011, pp. 479- 487

[8]   Rombach, G., Latte, S.: Querkrafttragfähigkeit von Fahrbahnplatten ohne Querkraftbewehrung, Beton- und Stahlbetonbau, Vol. 104, Heft 10, 2009, S. 642 – 656

[9]   Tailor, S.E., Rankin, B., Cleland, D.J., Kirkpatrick, J.: Serviceability of Bridge Deck Slabs with Arching Action, ACI Structural Journal, Vol. 104, No.1, January-February 2007, pp. 39 - 48

[10]   Vill, M., Schweighofer, A., Kollegger, J.: Großversuche an Spannbetonbrückenträgern zur Beurteilung des Schubtragverhaltens, Beton- und Stahlbetonbau, Vol. 107, Heft 2, 2012

[11]   Brandtner, G.: alumniTalks011, Brücken verbinden, Vortragsunterlagen 06.06.2008: http://alumni.tugraz.at/tug2/alumnitalks/alumniTalks011_brandtner.pdf; Zugriff: 05.02.2012; 13:00 Uhr

Evaluation of shear tests using the generalised stress field approach

Britta Hackbarth

Institute of Concrete Structures,
Hamburg University of Technology,
Denickestraße 17, 21073 Hamburg, Germany
Supervisor: Viktor Sigrist

Abstract

The determination of the shear strength of structural concrete beams still is in discussion. Whereas for simple design cases code equations based on limit analysis may be used, the evaluation of existing structures demands more refined procedures. The Generalised Stress Field Approach (GSFA) enables a profound analysis and a general evaluation of a structure. It represents an equilibrium model with recognizable physical relevance of the used parameters but takes compatibility into account as well by adapting the parameters to the strain state.

To prove the suitability of the GSFA approach, in this contribution, results of 97 reinforced concrete beam tests are evaluated and compared to calculated data. The experimental findings are described and the influence of parameters such as e.g. the effective shear depth z is discussed. An iterative procedure finally yields the stress band inclination θ and the effective concrete compressive strength f_{ce}. The calculated shear resistances are compared to the test results and overall, a good correlation is obtained.

The inclination θ depends on the strain state and is usually limited to the range between a maximum value θ_{max} and a minimum value θ_{min}. As θ_{min} is associated with steel rupture, it might be dispensable when using very high ductile steel. In the case of normal and high ductility steel, a suitable strain-based definition of θ_{min} is found. Furthermore, it is investigated in which cases an additional shear resistance that is attributed to the concrete could be accounted for.

1 Introduction

Stress fields enable visualisation of the force flow and provide an established foundation for shear design and assessment. They form a widely used method for describing the stress state within a structure at ultimate. Within this framework, there are different approaches with varying degree of complexity and precision. Conservative results from calculations based on limit analysis are reasonable in the design process as the equilibrium solution is associated with low effort. The assessment of existing structures requires refined methods as the specific stress and strain states have to be considered for achieving a more precise result.

The Generalised Stress Field Approach is an efficient method for assessment cases which yields accurate results with manageable effort. It combines equilibrium with implicitly incorporated compatibility through strain dependence with reference to the Cracked Membrane Model [6]. The approach provides the background of the Level II and III calculations of the new Model Code [2]. In this contribution, the GSFA is evaluated with the help of 97 selected reinforced concrete beam tests.

2 Generalised Stress Field Approach

2.1 Theoretical foundations

The idea of the Generalised Stress Field Approach is to combine the plasticity based equilibrium models with compatibility considerations. Therefore, the generalisation comprises specific limits for the stress field inclination angle θ and the strain depending value of the effective concrete compressive strength f_{ce}. The approach is presented and discussed in several publications, e.g. in [5].

Proc. of the 9th *fib* International PhD Symposium in Civil Engineering, July 22 to 25, 2012,
Karlsruhe Institute of Technology (KIT), Germany, H. S. Müller, M. Haist, F. Acosta (Eds.),
KIT Scientific Publishing, Karlsruhe, Germany, ISBN 978-3-86644-858-2

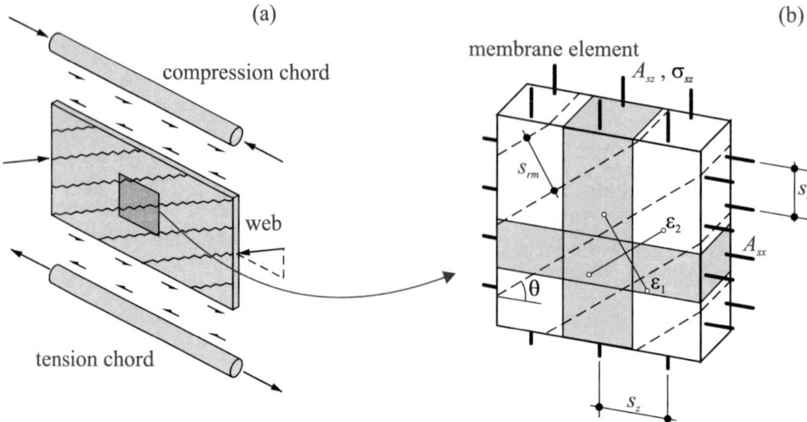

Fig. 1   (a) Splitting of a beam segment into compression chord, web and tension chord, (b) membrane element.

Figure 1 shows a concrete beam that is notionally splitted into tension and compression chord and the web. The latter consists of a conjunction of orthogonally reinforced membrane elements. From equilibrium and with the assumptions $\sigma_{sz} = f_{yz}$ and $-\sigma_2 = f_{ce}$, the well known design equations for the resistances attributed to the stirrup reinforcement V_{Rs} and the concrete struts V_{Rc} can be derived. With the help of $\tau_R = V_R / (b_w z)$ and $\rho_z = A_{sz} / (s_z b_w)$ the equations may be rearranged and solved for the shear strengths of the stirrups τ_{Rs} and the concrete τ_{Rc}:

$$\tau_{Rs} = \rho_z f_{yz} \cot\theta \tag{1}$$

$$\tau_{Rc} = f_{ce} \sin\theta\cos\theta \tag{2}$$

A_{sz} denotes the cross sectional area of the stirrups with a spacing s_z and yield limit f_{yz}. b_w represents the web width and z is the effective shear depth. The effective concrete compressive strength f_{ce} may be calculated from concrete cylinder strength f_c'

$$f_{ce} = \eta_c f_c' \left(\frac{30}{f_c'} \right)^{1/3} \tag{3}$$

where the factor $\left(30 / f_c' \right)^{1/3}$ diminishes the strength for $f_c' > 30$ MPa due to the more brittle behaviour of these concrete types in compression. The strength reduction factor η_c allows the strain state to be accounted for:

$$\eta_c = \frac{1}{1.2 + 55\varepsilon_1} \tag{4}$$

The principal strain ε_1 may be calculated as $\varepsilon_1 = \varepsilon_x + (\varepsilon_x - \varepsilon_2)\cot^2\theta$ from Mohr's circle of strains. Principal strain ε_2 is taken as the peak strain $-\varepsilon_{c0}$ at reaching f_{ce} where $\varepsilon_{c0} = 0.002$ may be used as a suitable approximation. The longitudinal strain ε_x can be calculated from the longitudinal strain in the concrete compression zone and the strain in the longitudinal reinforcement; the value at mid-depth of the web shall be decisive. This is an arbitrary definition but harmonises design equations and the evaluation of shear tests.

    Limits of the stress field inclination θ can be derived from Mohr's circle of strains where the relation

$$\tan^2\theta = \frac{\varepsilon_x - \varepsilon_2}{\varepsilon_z - \varepsilon_2} \tag{5}$$

may be further deduced. The lower limit θ_{min} is associated with stirrup rupture and therefore the maximum stirrup strain $\varepsilon_z = \varepsilon_{smu}$ applies. θ_{max} comprises web crushing at the onset of stirrup yielding, so $\varepsilon_z = \varepsilon_{smy}$ prevails. It takes the consideration of bond to calculate the parameters [5]. As a simplification and depending on the steel ductility, a linear approximation is recommended. For normal ductility steel, $\theta_{min} = 20° + 5000\varepsilon_x$ applies whereas $\theta_{min} = 16° + 4000\varepsilon_x$ is appropriate for high ductility steel. As the upper limit, $\theta_{max} = 35° + 5000\varepsilon_x$ may be used for both.

2.2 Comparison to experimental findings

2.2.1 Test beams

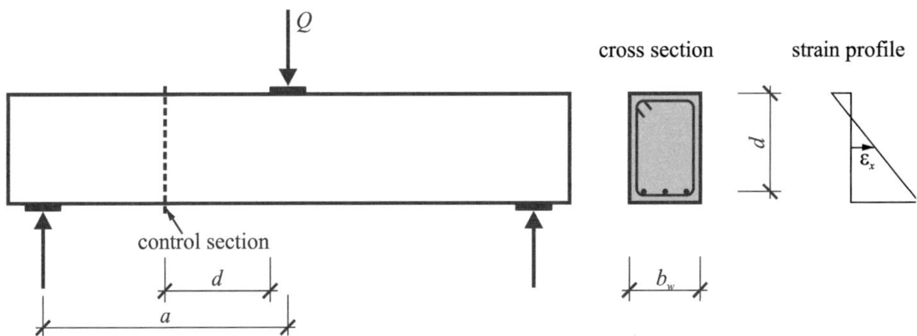

Fig. 2   Typical setup of beam tests.

In this contribution, 97 reinforced concrete beams of numerous experimental series [4] are analysed with help of the Generalised Stress Field Approach. T-beams as well as beams with a rectangular cross section are included. All beams are single span girders subjected to point loads Q at the distance a from the support axis. An example of a typical test setup is displayed in Figure 2. Shear strength is calculated for the control section at the distance d from the face of the loading plate of the point load.

In order to avoid disturbing influences, the considered shear tests were filtered. Hence, beams with flexural failure were excluded. This criterion comprises obvious flexural failure that is declared in the test report. In addition, the plastic bending resistance and the corresponding maximum point load Q_{plast} were computed for each of the beams; in the case of $Q_{test} / Q_{plast} \geq 0.95$, the beam was excluded. Moreover, a concrete compressive strength $f_c' \geq 15$ MPa was presupposed to represent a common range of concrete strengths. With the assumption of a minimum ratio of $a / d > 2.5$ short beams with possible strut or arch action were eliminated and via a ratio of stirrup spacing to effective depth $s_z / d \leq 0.65$ sound detailing rules were ensured. Finally, a minimum shear reinforcement ratio according to the new *fib* Model Code [2] was considered

$$\rho_{z,min} = 0.08 \frac{\sqrt{f_c'}}{f_{yz}} \tag{6}$$

2.2.2 Parameter variation and results

Within the GSFA, adoption of several parameters is possible to customise the approach. In order to evaluate the beam tests, the relevant parameters have to be determined.

For computing the effective shear depth z, different methods are possible. Calculation from the bending moment at the control section assuming a concrete stress block yields a realistic estimation with comparatively high effort. The approach $z = 0,9\ d$ is more simple and sufficient for a quick estimation on the safe side: z tends to be smaller and, as a consequence, θ as well as the ratio of observed and predicted shear strength increases. Another possibility for T-beams is $z = d - 0.5h_f$ where h_f denotes the depth of the compression chord. Depending on the geometry of the cross section, z is once again smaller than based on the two other methods. This is reasonable as the entire chord is assumed to account for the concrete compression zone this way. In summary, the difference in z is small, but noticeably affects the outcome. Expectedly, best correlation is obtained for the first method as it takes the actual configuration into account.

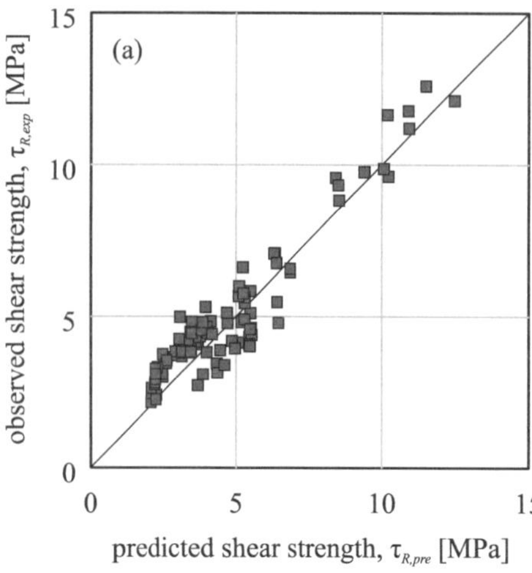
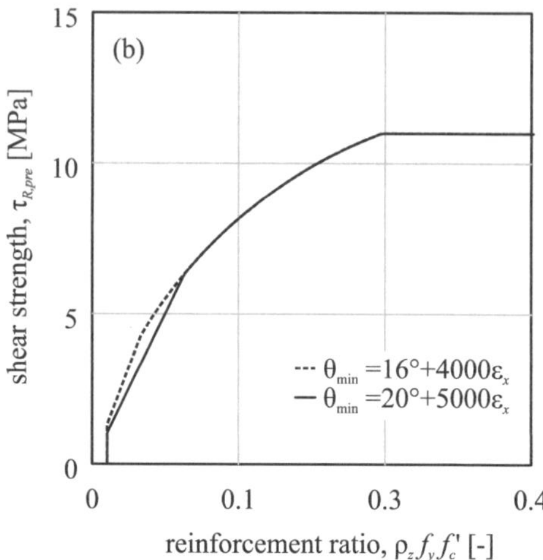

Fig. 3   (a) Comparison of predicted and observed shear strengths of 97 beam tests. The data refer to a specific selection of the database provided by K.-H. Reineck in [4]. A list of the 97 beam tests may be requested from the author. (b) Example curves for the shear strength depending on the mechanical shear reinforcement ratio and different approaches for θ_{min}.

An example for the change in shear resistance with increasing mechanical shear reinforcement ratio is depicted in Figure 3(b). Note that as the curve depends on material properties and the strain state there would be a specific one for each of the test beams. The three parts of the curve that are associated with specific failure modes can clearly be identified and in addition, the minimum stirrup reinforcement ratio is clearly apparent. The maximum shear strength is achieved with high stirrup reinforcement ratios and implies web crushing without yielding of the stirrups. This horizontal branch of the curve is associated with the upper limit θ_{max}. Instead of using the linear approximation mentioned above, the upper limit θ_{max} was calculated from equation (5) with help of $-\varepsilon_2 = \varepsilon_{c0} = 0.002$ and $\varepsilon_z = \varepsilon_{smy} = 0.8 f_{yz} / E_s$ [5].

For medium reinforcement ratios, the curved line implies concrete crushing with yielding of the stirrups. θ increases nonlinearly from θ_{min} to θ_{max} with increasing shear reinforcement ratio. By equating (1) and (2), failure stress of the curved line may be calculated from

$$\tau_R = \sqrt{\rho_z f_{yz} \left(f_{ce} - \rho_z f_{yz} \right)} \tag{7}$$

The lower limit θ_{min} represents stirrup rupture and therefore depends on the yield limit and the ductility characteristics of the reinforcing steel. The higher the ductility, the lower the minimum angle can be chosen. As a simplification, the lower θ-limit may be expressed as a linear approximation. The recommended linear approximation in the case of high ductility steel is

$$\theta_{min} = 16° + 4000\varepsilon_x \tag{8}$$

As displayed with a dashed curve in Figure 3(b), the inclination of this linear branch is increased compared to the outcome for normal ductility steel. Hence, the affected range of reinforcement ratios decreases and simultaneously, the shear resistance for a given shear reinforcement ratio increases.

The stirrup reinforcement ratio of many of the test beams is rather low and hence, θ tends to be rather small. Thus, the approach for θ_{min} importantly influences the outcome. In all these tests high ductility stirrup steel was used, therefore equation (8) may be applied for ribbed stirrups and leads to best correlation between observed and predicted shear strengths.

Whereas ribbed stirrups provide good bond conditions, in the case of plain stirrup reinforcement, concrete cracks and stirrup strains proceed (rather) independently due to the minor bond between steel and concrete. Plain stirrups were used in early test series as e.g. by Regan [3] or Bræstrup [1]. A large maximum steel strain follows from the used very high-ductile steel, thus, stirrup rupture was hardly ever controlling. Therefore, θ_{min} can be neglected. Note that all test beams nevertheless have calculated stress band inclinations greater than 14°. Another effect of the minor bond is a lower strain impact on the concrete. Consequently, f_{ce} and η_c, respectively should not be that much reduced as with equations (3) and (4). This imprecision is well balanced by neglecting θ_{min}.

Especially for small reinforcement ratios, an additional concrete contribution V_{Rc} [2] could be appropriate as with decreasing stirrup resistance concrete strength gains in importance. On a trial, this approach was used for those of the test beams with $\theta = \theta_{min}$ but the results did not improve due to the fact that θ_{min} hardly ever occurred because the GSFA allows for very small θ-values. In other words, a separate concrete contribution is not necessary for cases with $\rho_z > \rho_{z,min}$ as the effect is covered by the definition of θ_{min}. Test beams with $\rho_z < \rho_{z,min}$ were not considered.

Figure 3(a) compares the observed shear strengths $\tau_{R,exp}$ and the predicted shear strengths $\tau_{R,pre}$ calculated from the GSFA as described above. The observed shear strength refers to the experimental results including dead load. An average value $\tau_{R,exp} / \tau_{R,pre}$ of 1.11 and a coefficient of variation of 18% are evidence of a good correlation and prove the accuracy of the approach. The longitudinal strains ε_x are in the range between $0.19 \cdot 10^{-3}$ and $1.32 \cdot 10^{-3}$ and the inclination θ varies between $14.2°$ and $39.0°$. Most of the test beams cover the range of low and medium shear reinforcement ratios, the vast majority of the test beams refer to mechanical reinforcement ratios $\rho_z f_{yz} f_c' < 0.2$ which means that the upper limit of shear resistance is not equally approved.

3   Conclusions

Whereas classic shear design procedures neglect strain effects and yield conservative results, the Generalised Stress Field Approach implicitly incorporates compatibility. The approach forms a compromise between simple calculations based on limit analysis and a complex evaluation on the basis of the Cracked Membrane Model. As the results depend on the strain state, computation of the affected parameters is necessary and an iterative procedure is required. Compared to the results of 97 selected beam tests, the findings from the GSFA are remarkably consistent. For this reason, the method suits well for the assessment of structures. The influence of various assumptions for z and θ_{min} was elaborately discussed.

Available shear tests mostly cover low or medium stirrup reinforcement ratios. Evaluation of more test beams with high reinforcement ratios would be important for further validating the approach. Moreover, beams with $\rho_z < \rho_{z,min}$ might be evaluated with regard to the concrete contribution V_{Rc}. By means of prestressed test members, evaluation would be possible for a wider range of longitudinal strains. The influence of member size ought to be discussed because the findings from small test beams are not smoothly transferable to big bridge girders where the strain state in the webs is more membrane-like. Previous evaluations [6] have shown that the GSFA also covers the shear behaviour of panels quite well but however, also more panel tests should be analysed to further confirm this result.

References

[1]   Bræstrup, M.W.; Nielsen, M.P; Bach, F.: Shear tests on reinforced concrete T-beams: series V, U, X, B and S. Structural Research Laboratory, Technical University of Denmark, Lyngby, Denmark, 1980.

[2]   fib: Model Code for Concrete Structures. fib Bulletin, to be published in 2012.

[3]   Regan, P.: Shear in Reinforced Concrete: an experimental study. Department of Civil Engineering, Imperial College of Science and Technology, London, 1971

[4]   Reineck, K.-H.; Kuchma, D. A.; Fitik, B.: Extended Databases with Shear Tests on Structural Concrete Beams without and with Stirrups for the Assessment of Shear Design Procedures. Research Report. ILEK, University of Stuttgart and University of Illinois, Champaign, Urbana, July 2010

[5]   Sigrist, V.; Hackbarth, B.: A Structured Approach to the Design and Analysis of Beams in Shear. fib Workshop "Recent Developments on Shear and Punching Shear in RC and FRC Elements", Salò, Italy, October 2010, fib bulletin 57.

[6]   Sigrist, V.: Generalized Stress Field Approach for Analysis of Beams in Shear. ACI Structural Journal, Vol. 108, No. 4, July-Aug. 2011.

Post-installed shear reinforcement for concrete thick slabs

Mathieu Fiset

Centre de recherche sur les infrastructures en béton (CRIB),
Université Laval,
1065 Avenue de la médecine, Québec (Québec), G1V-0A6, Canada
Supervisors: Josée Bastien and Denis Mitchell

Abstract

Many concrete bridges may be associated to a simple thick slabs structural system. With the increase in traffic loads and material degradation, some of these structures need to be strengthened in shear. The goal of this research is to study the behaviour of slabs strengthened with post-installed shear reinforcement and to provide an analytical method of design. The studied reinforcement methods consist in installing rebars into pre-drilled holes in the slab with different anchor systems. The first experimental results showed that shear-strengthened slabs can have failure loads 46% higher than an unstrengthened slab but 29% lower than the Canadian code prediction for conventional stirrups. The VecTor2 finite element analysis tool will be used to study the parameters influencing the slabs behaviour. The first experimental results and associated numerical models outcomes will be presented.

1   Introduction

For simple structural systems such as thick slab bridges, it was commonly assumed that the concrete was able to resist shear stresses and that stirrups were therefore not required. However, due to the increase in traffic loads and material degradation, nowadays some of these thick slabs need to be strengthened in shear. In the past, shear strengthening methods have been examined and tested on beams. Among them, the addition of near surface mounted rods [1] and the addition of external carbon fiber reinforced polymer laminates [2,3] have been suggested. Although these methods can be effective on beams, the fact that the reinforcement is installed on either side of the concrete section raised the question of their effectiveness on the full width of large elements as slabs. More recently, the use of vertical rods anchored into thin slabs with epoxy adhesive to resist punching shear has proven to give good results [4]. However, very few studies were performed on thick slabs where the size effect can become important with regards to shear performance. Therefore, improved knowledge on shear strengthening methods for thick slabs through laboratory testing and numerical means has gained wide interest.

In the present paper, the strengthening methods under investigation consist in steel rebars introduced into pre-drilled holes with different anchor systems: epoxy adhesive, internal and external mechanical anchorage. Two series of tests performed on deep beams (slices of slabs) were conducted up to shear failure [5,6]. The first set of results showed that while shear-strengthened slabs can exhibit failure loads 46% higher compared with the ones of unstrengthened slabs, they showed failure loads 29% lower than the Canadian code prediction values [7] related to conventional shear reinforcement (stirrups). One of the main objectives of the underway research is to adequately predict the increase in shear strength of thick slabs strengthened by various methods and to provide basis for a normative strengthening design method in light of the experimental and numerical results. To achieve this, the proposed finite element models should be able to reproduce the behaviour of the tested beams. Once this is achieved, these models will help to perform parametric analysis to study the most important parameters influencing the slabs behaviour. The finite element analysis tool, VecTor2, will be used herein to study the influence of the relative rigidity of components, material properties and type of post-installed reinforcements including their anchorage and geometry. Some of the studied anchorage systems exhibited slippage during loading, a phenomena which should be taken into account by the numerical models.

Proc. of the 9th *fib* International PhD Symposium in Civil Engineering, July 22 to 25, 2012,
Karlsruhe Institute of Technology (KIT), Germany, H. S. Müller, M. Haist, F. Acosta (Eds.),
KIT Scientific Publishing, Karlsruhe, Germany, ISBN 978-3-86644-858-2

2   Review of experimental beams tests

Experimental tests were performed on two (2) series of beams (slices of slab), identified as the PP and BC series. The dimensional properties and strengthening rebar arrangements of the tested beams are summarized in Table 1 and Figure 1. For both series, each beam had a 4 meters free span, a 610mm width "b" and was designed to allow shear failure.

Table 1   Details of reinforced beams

Beam	Anchor	h [mm]	d [mm]	a/d	ρ [%]	s [mm]	s/d_v	A_v [mm²]	f_c [MPa]	E_c [MPa]
PP1	Epoxy	450	370	3.60	3.10	240	0.72	400	32.5	28 382
PP2	Epoxy	450	398	3.35	2.06	260	0.73	200	35.2	30 580
PP3	Epoxy	750	698	2.87	1.17	470	0.75	400	35.0	29 395
BC1	Stirrups	750	694	2.88	1.65	380	0.61	400	33.3	25 704
BC2	Epoxy	750	694	2.88	1.65	380	0.61	400	34.5	26 315
BC3	Epoxy	750	694	2.88	1.65	380	0.61	400	32.6	25 029
BC4	HSLG	750	694	2.88	1.65	380	0.61	292	31.5	24 144
BC5	Bolt	750	694	2.88	1.65	1000	1.60	1290	31.2	25 333

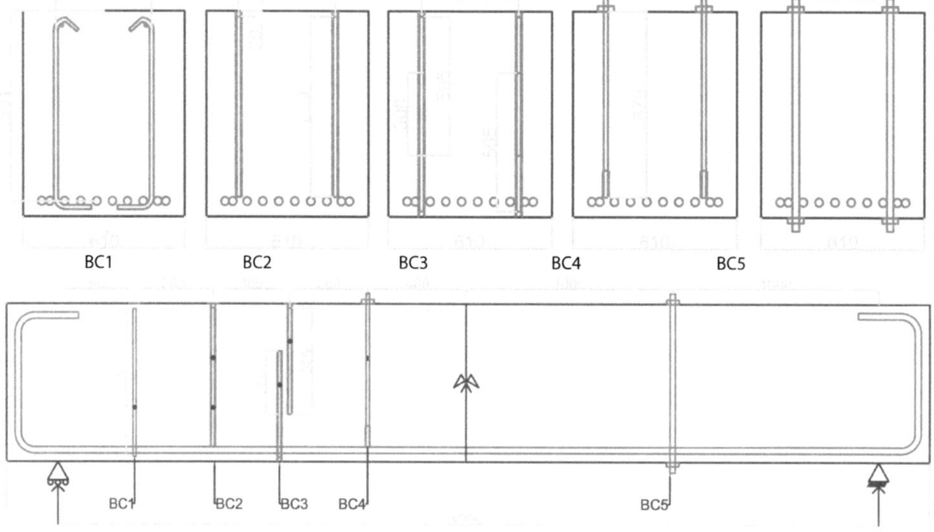

Fig.1   Specimens of BC series.

For all 3 PP beam categories (PP1, PP2 and PP3), 2 unstrengthened beams and 2 strengthened beams with the same overall dimensions were tested. In these PP series, the shear strengthening method consists in vertical post-installed rebars introduced into pre-drilled holes at specific locations along the beams and anchored by epoxy adhesive. This method is similar to that of the beam BC2 illustrated at Figure 1. The PP1 and PP3 specimens were strengthened with 2-15M rebars and PP2 specimens with 2-10M rebars respectively. The chosen spacing ratio of rebars, s/d_v, is close to the maximum value of 0.75 allowed by Canadian standards for conventional stirrups. While all specimen where simply supported, the PP1 and PP2 specimens were loaded at one-third of their span, whereas the PP3 specimens were loaded at mid-span.

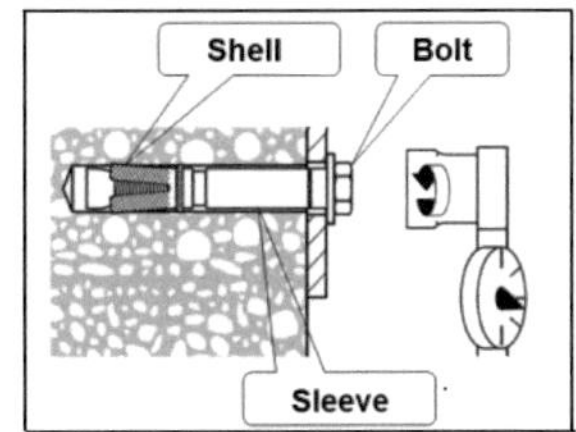

Fig.2   HSLG anchor used for BC4

The BC series differs from the PP series in terms of reinforcement ratios, transverse reinforcement spacing and strengthening methods. The BC1 specimen had stirrups as prescribed by Canadian standards and therefore was the only specimen of the BC series reinforced in shear before concrete casting. Similar to the PP3 specimens, the BC2 and BC3 specimens were strengthened with post-install 15M rebars anchored with epoxy adhesive into pre-drilled holes. The BC3 specimen has the particularity that its rebars overlap at mid height of the beam, over 300mm. The BC4 specimen was strengthened with vertical rebars inserted into pre-drilled holes from the top of the beam down to the location of the longitudinal rebars. The shear reinforcement is restrained with an anchor plate on the top face of the beam and with a mechanical anchorage at the bottom of the vertical rebars (Fig. 2). When opened, the shell of the mechanical anchorage exerts lateral pressure on the internal surfaces of the hole, which produces a frictional force and anchors the rebar. The BC5 specimen was strengthened with one pair of high strength bars inserted in pre-drilled holes and anchored on both top and bottom faces of the beam with an anchor plate. Each of the 5 BC beams were loaded at their mid span up to failure. Once a side of the beam reached its ultimate loading, this side of the specimen was strengthened in shear with external stirrups (Dywidag bars) and the beam was reloaded in order to reach the ultimate load of the specimen's other side.

2.1   Materials

The concrete mechanical properties presented in Table 1 were obtained according to standards ASTM-C39 and ASTM-C469. According to ASTM E08-04 and ASTM E111-04, the yield strength F_y, the ultimate strength F_u and the Young modulus E_s of the rebars were 472MPa, 660MPa and 178GPa respectively. The associate hardening strain was 23 mm/m and the ultimate strain was 114 mm/m. As specified by the manufacturer, the yield and the ultimate strength of the high strength steel bars of the beam BC4 were 642MPa and 800MPa, respectively. The maximum load for the expansive mechanical anchor was 84.5kN. The yield strength and the ultimate strength of the Dywidag bars used for the external shear strenthening of beam BC5 beam were 517MPa and 689MPa, respectively. A commercially available epoxy adhesive was used for all PP beam series as well as for beams BC1, BC2 and BC3. As specified by the manufacturer, the epoxy adhesive mechanical properties were: 12.4MPa bond strength (ASTM C882-91), 82.7MPa compressive strength (ASTM D-695-96), 1493MPa compressive modulus (ASTM D-695-96), 43.5MPa tensile strength and 2% elongation at failure (ASTM D-638-97).

3   Numerical model

The finite element numerical portion of the study is still underway. Up to now, the Vector2 numerical tool was used. This finite element software was developed at the University of Toronto for the analysis of two dimensional finite element models of concrete structures with rotating smeared crack. The analyses are based on the modified compression field theory (MCFT) [8] and disturbed stress field model (DSFM) [9]. With VecTor2, many options are available to model the material's behaviour. The basic options were initially selected. Therefore the steel behaviour is associated to a trilinear law as shown in Figure 3. For the tension behaviour of concrete, the σ-ε law is linear up to the tensile strength. Beyond this point, the tension softening effect is represented with a bilinear law. The tension stiffening effect is also included according to the model of Lee [10]. In compression, the cracked concrete behaviour includes compression softening effects and is modelled with equation (1).

As shown in Figure 4, the beam BC1 was modelled with 2D membrane elements. Because of the symmetry of the geometry and loading, half of the beam was modelled. Boundary conditions are imposed as follows, X displacements are blocked at mid-span and Y displacements are blocked at support. The supports and the loading plate surface were modelled to best represent the laboratory conditions. For the beam BC1, the longitudinal rebars and the stirrups were modelled with truss elements perfectly linked with the nodes of the finite element mesh. For the beams with epoxy adhesive, link elements (spring) may be used to model the potential slippage of the shear reinforcements.

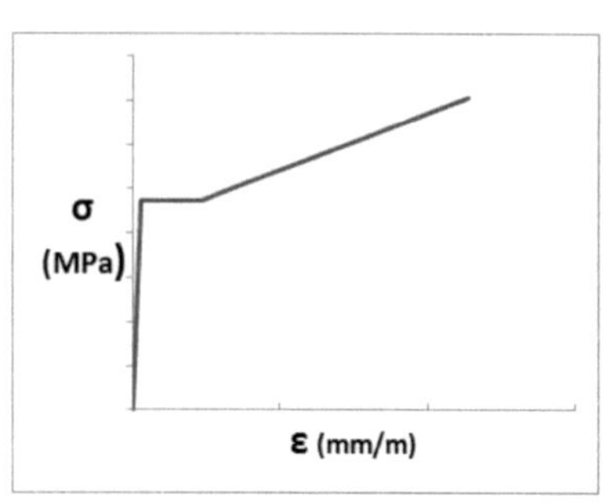

$$f_{c2} = -f_p \left[2\left(\frac{\varepsilon_{ci}}{\varepsilon_p}\right) - \left(\frac{\varepsilon_{ci}}{\varepsilon_p}\right)^2 \right] \tag{1}$$

Where:

$$f_p = \beta_d f_c' \tag{2}$$

$$\varepsilon_p = \beta_d \varepsilon_c' \tag{3}$$

$$\beta_d = \frac{1}{1 + 0.55 C_d} \tag{4}$$

$$C_d = \begin{cases} 0 & \text{if } r < 0.28 \\ 0.35(r - 0.28)^{0.8} & \text{if } r > 0.28 \end{cases} \tag{5}$$

$$r = \frac{-\varepsilon_{c1}}{\varepsilon_{c2}} \le 40 \tag{6}$$

Fig.3　　Behaviour of steel

4　Results and discussion

4.1　Experimental results

A summary of experimental results of the strengthened and unstrengthened beams is presented in Table 2 and Table 3 respectively. The designation of beam XXY-Z refers to the beam series XX, the beam specimens Y and the number of loaded beam (1 or 2 for PP series and Load or Reload for BC series). The predicted shear resistance V_{CSA} was calculated according to the Canadian standard [7]. This standard is based on the MCFT to define the concrete shear strength V_c and the shear resistance provided by shear reinforcements V_s. The shear strength attributed to concrete is the product between the tensile strength of concrete and a constant β, which is a function of the concrete strain and the crack spacing. The shear resistance attributed to the shear reinforcement is the load which leads to the yielding of the reinforcements that intercepts the main shear crack. For the beam BC3, the overlapping of the shear reinforcements is not considered for the calculation of the steel area A_v. δ_{ult} represents the ultimate deflection at mid-span.

Table 2　　Summary of results for the strengthened beams in shear

Beam	V_{exp} [kN]	V_{CSA} [kN]	V_{exp}/V_{CSA}	$V_{c\text{-}CSA}$ [kN]	$V_{s\text{-}CSA}$ [kN]	$V_{s\text{-}exp}$** [kN]	$V_{s\text{-}exp}/V_{s\text{-}CSA}$	δ_{ult} [mm]
PP1-1	476	603	0.79	214	389	262	0.67	13.6
PP2-1	293	420	0.70	243	177	50	0.28	6.7
PP2-2	321	420	0.76	243	177	78	0.44	9.7
PP3-1	504	705	0.71	355	350	149	0.43	12.2
PP3-2	519	705	0.74	355	350	164	0.47	11.4
BC1-L	740	779	0.95	352	427	388	0.91	10.6
BC1-R	801	779	1.02	352	427	448	1.05	39.5
BC2-L	756	783	0.97	357	426	399	0.94	11.9
BC2-R	783	783	1.00	357	426	426	1.00	22.1
BC3-L	956	776	1.23	349	427*	607	1.42	15.5
BC3-R	837	776	1.08	349	427*	488	1.14	37.4
BC4-L	593	796	0.74	362	434	231	0.53	11.5
BC4-R	604	796	0.76	362	434	242	0.56	16.7
BC5-L	731	903	0.81	334	569	397	0.70	12.0
BC5-R	983	903	1.09	334	569	649	1.14	24.3

* This value is calculated with A_v=400mm^2　　** $V_{s\text{-}exp}$=V_{exp}-$V_{c\text{-}CSA}$

Table 3   Summary of results for the unstrengthened beams

Beam	V_{exp} [kN]	V_{CSA} [kN]	V_{exp}/V_{CSA}	$V_{strengthened}/V_{unstrengthened}$	δ_{ult} [mm]
PP1-1	329	275	1.19	1.45	7.5
PP1-2	330	275	1.20	-	7.6
PP2-1	283	271	1.04	1.03	6.7
PP2-2	309	271	1.14	1.04	7.6
PP3-1	357	375	0.95	1.41	5.0
PP3-2	355	375	0.95	1.46	4.6

With the results shown in Table 2, it can be observed that the current Canadian standard, as expected, does not adequately predict the shear capacity of beams strengthened with post-installed shear reinforcement. However, Table 3 shows that, with an average V_{exp}/V_{CSA} of 1.08 for the unstrengthened beams, the Canadian standard adequately predict the shear strength of these beams. With a ratio V_{s-exp}/V_{s-CSA} between 0.28 and 0.67, the chosen rebar spacing and the anchorage system of strengthened beams PP1, PP2 and PP3 do not allow the development of the yield strength of rebars intercepting shear cracks. By comparing the BC series (BC1 through BC4), it can be seen that the selected strengthening method has a significant impact on the beam shear strength and the involved shear mechanisms. As an example, the expansive mechanical anchorage in beam BC4 has developed only 53% to 56% of the predicted steel strength. This situation is mainly due to the fact that the bars are linked to the concrete section at its top surface and at the expansion anchorage. The deformation of the bars is therefore redistributed across their full length rather than having a local deformation at crack locations which is the case with bonded rebars. The spacing ratio s/d_v has also a significant influence on the shear strength. This phenomenon can be examined with the help of PP3 and BC2 beams respectively. The spacing ratio of 0.75 for the beam PP3 has the effect of allowing main shear cracks to progress near the extremities of the shear reinforcements. The crack location enables the rebars to be properly anchored and to develop their full strength. The selection of a smaller spacing ratio of about 0.6, as for beam BC2, ensures that the rebars are anchored adequately as the main shear cracks intercept these rebars near their mid-length. This can also be observed by comparing the strengthened and unstrengthened beams of the PP series. With a little smaller spacing ratio, the ratio $V_{strengthened}/V_{unstrengthened}$ of the PP1 beams is greater than the PP2 beams. For the selected spacing of the beams PP2, the average of 1.04 means that the shear reinforcements have very small effect on shear strength. As anticipated, the code predictions are in good agreement with the experimental results involving bonded and adequately anchored rebars as it is the case for stirrups for which the code provisions are developed.

4.2   Numerical results and corroboration with experimental tests

For the loading stage of beam BC1, the finite element model and the experiments show a similar behaviour up to about 9.5mm of deflection. At the ultimate shear strength V_{exp} of 740 kN, the deflection δ_{ult} of the beam is 10.6mm, while the FE model predicts a deflection of 15.1mm for an ultimate shear strength of 830kN. For the reloading of the beam BC1, one can observe that the maximum experimental shear strength obtained is 801kN. According to the FE model, the ratios of shear strength V_{FEM}/V_{exp} are 1.12 and 1.04 for the loading and the reloading respectively. The Figure 4 compares the cracking pattern obtained experimentally with the one from the FE model. In a smeared crack model, each finite element (integration point) that reaches the tensile strength will exhibit a crack. However in Figure 4, the cracks with greater openings are illustrated with a bold line. As shown, the path of cracks is very well predicted by the FE model. The crack that leads to shear failure of the beam is the one that passes through the rows of stirrups S3 and S4. With the recorded strain of stirrups well above 23 mm/m, it has been experimentally observed that stirrups S3 and S4 were in the strain hardening behaviour state. The FE model supports this information: at ultimate, the maximum steel stress at the crack are about 493MPa and 481MPa for the stirrups S3 and S4 respectively, which is well above the yield strength of 472MPa.

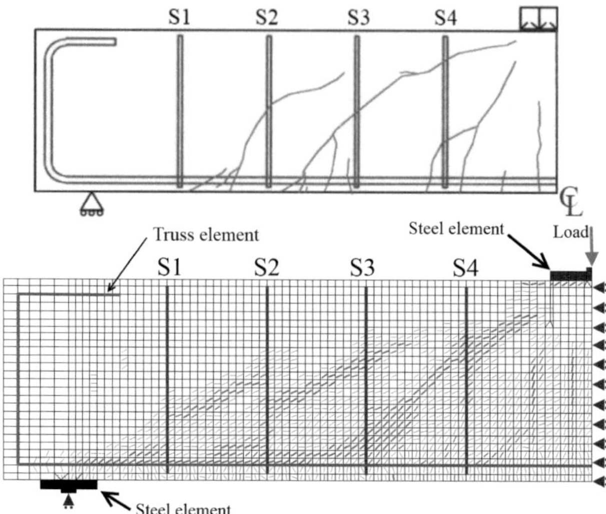

Fig.4   Cracking pattern of the beam BC1 (half beam). Experimental (top) and FE model (bottom)

5   Conclusions and future works

The main goal of this research is to adequately predict the increase of shear strength of thick slabs subjected to various methods of shear strengthening. For the matter, many slab specimens (beams) with different thickness, shear strengthening spacing ratios and strengthening methods were loaded up to failure in the laboratory. While the current Canadian standard is not suited for the shear strength prediction of post-installed shear reinforcements in general, it can nevertheless provide a first good estimate of the ultimate shear strength provided that the rebars are bonded to the section and are adequately anchored up to the beam shear failure which is not the case for all tested beams in Table 2. In this regard, finite element models can provide useful information. Using the VecTor2 software and taking advantage of the symmetry of loading and boundary conditions, half of the beam BC1 was modelled. This beam has been cast with standard stirrups with shear reinforcement spacing well under the maximum spacing ratio permitted by the Canadian code (0.61 versus 0.75).

While the numerical modelling part of the research is still underway, one of the first steps was to be able to replicate the BC1 beam behaviour. By comparing the experimental and numerical results associated to this beam, it appears that its behaviour and its ultimate strength are well evaluated by the model. Moreover, the experimental observations during the loading corroborate the crack pattern predicted by the model and its progression. As mentioned, the cracks' progression and pattern have a major impact on the shear strength of beams, especially those strengthened with rebars link to the concrete section with an epoxy adhesive. Therefore, the next step will be to model this type of shear strengthening and compare the numerical outcomes to the experimental results to perform, thereafter, an in-depth parametric analysis involving the parameters influencing the shear resistance. The testing program has demonstrated that many shear strengthening methods can be used. However, based on practical aspects and effectiveness, the use of additional rebars linked to the concrete section with an epoxy adhesive seems to be promising provided that geometric constrains are satisfied in order for the rebars to remain adequately anchored up to the shear failure.

References

[1]   De Lorenzis, L.; Nanni, A.: Shear strengthening of reinforced concrete beams with NSM fiber-reinforced polymer rods. In: ACI structural journal (2001). No. 1, pp. 60-68

[2]   Barros, J.A.O.; Dias, S.J.E.: Near surface mounted CFRP laminates for shear strengthening of concrete beams. In: Cement & concrete composites (2006) No. 28, pp. 276-292

[3]   Adhikary, B.B.; Mutsuyoshi, H.: Shear strengthening of reinforced concrete beams using various techniques. In: Construction and building materials (2006) No. 20, pp. 366-373

[4]   Fernández-Ruiz, M.; Muttoni, A.; Kunz, J.: Strengthening of flat slabs against punching shear using post-installed shear reinforcement. In: ACI structural journal (2010) No. 4, pp. 434-442

[5]   Provencher, P.: Renforcement des dalles épaisses en cisaillement (master thesis). In: Département de génie civil, Université Laval, Québec, Canada (2011)

[6]   Cusson, B.: Renforcement des dalles épaisses en cisaillement (master thesis). In: Département de génie civil, Université Laval, Québec, Canada (2012)

[7]   Canadian Standards Association: Canadian highway bridge design code (2006), 768 p.

[8]   Vecchio, F.J.; Collins, M.P.: The modified compression field theory for reinforced concrete elements subjected to shear. In: ACI Journal (1986) No.2, pp. 219-231

[9]   Vecchio, F.J.: Disturbed stress field model for reinforced concrete: Formulation. In: Journal of structural engineering (2000) No.9, pp.1070-1077

[10]   Lee, S.C., Cho, J.Y, Vecchio, F.J.: Model for post-yield tension stiffening and rebar rupture in concrete members. In: Engineering structures (2011) No.33, pp. 1723-1733

List of symbols

a	Distance between the load and one support
A_s	Area of all tension reinforcement
A_v	Area of all shear reinforcement within a distance s
d	Distance from extreme compression fibre to centroid of longitudinal tension reinforcement
d_v	Effective shear deep
f_{c2}	Compressive stress in concrete
h	Height of beam
s	Spacing of transverse reinforcement
V_{exp}	Total shear resistance calculated with experimental data
ε'_c	Concrete compressive strain corresponding to f'_c
$\varepsilon_{c1}, \varepsilon_{c2}$	Net concrete axial strain in the principal tensile (1) or compressive (2) direction
ρ	Ratio of tension reinforcement, equal to A_s/bd

Tensile strength of loop connections between precast bridge deck elements

Henrik. B. Joergensen

Institute of Technology and Innovation
University of Southern Denmark
Campusvej 55, 5230 Odense, Denmark
Supervisor: Linh C. Hoang

Abstract

Connections between precast concrete panels are often established by use of protruding U-bars, which are interleaved in an in-situ cast joint. Such connections are referred to as loop connections.

Loop connections can be used to create structural continuity in a bridge deck made of precast concrete panels. The capacity of a loop connection to transfer tension from panel to panel is governed either by yielding of the U-bars or by failure of the in-situ cast concrete. In present practice, calculation of tensile strength and prediction of failure mode are based on empirical formulas. Such formulas have the disadvantage that they can only be used when the design parameters are within the range covered by the empirical data.

The goal of this research work is to establish an analytical and design oriented method for strength calculation of loop connections. For this purpose, an upper bound plasticity model has been developed. From the model it is possible to derive analytical solutions which depend on a number of important design parameters. This includes U-bar spacing and diameter; loop diameter; strength of lacer bars and strength as well as cross sectional area of the in-situ concrete.

The model has been compared with experimental results. Satisfactory agreements have been found.

1   Introduction

This paper deals with a design-oriented model for calculation of the tensile strength of U-bar loop connections between precast concrete panels.

Loop connections are for instance used in steel-concrete composite bridges where they ensure structural integrity between the precast deck elements. Loop connections are established by use of protruding U-bars, which are interleaved in an in-situ cast joint. From this a flattened cylindrical concrete core confined by the interleaved U-bars is created.

In practice the engineers will always seek to design the connections in such a way that the yield strength of the U-bars is decisive for the tensile strength. However, yielding of U-bars is only possible if the connection is secured against premature failure in the concrete core. For the case of concrete failure only empirical formulas are available in the literature. Thus an application-oriented and analytical model is needed in this case. Such a model will make it possible to design the connection to avoid premature concrete failure.

The model presented in this paper is a plasticity upper bound model. This means that upper bound solutions for the ultimate load are developed on the basis of geometrical possible failure mechanisms. The model shows that the transverse reinforcement (also called lacer bars) has a great influence on the tensile strength of the connection. The model also shows that the tensile strength of the connection is greatly influenced by the distance between the U-bars and the cross-sectional area of the concrete core.

As usual when applying plastic theory to structural concrete, it is necessary to introduce the so-called effectiveness factors for reduction of the concrete compressive strength [1]. In this paper a modified version of the effectiveness factor for beam shear will be used as the failure in the concrete core of a loop connection is very similar to the shear failure in a continuous beam.

Proc. of the 9th *fib* International PhD Symposium in Civil Engineering, July 22 to 25, 2012,
Karlsruhe Institute of Technology (KIT), Germany, H. S. Müller, M. Haist, F. Acosta (Eds.),
KIT Scientific Publishing, Karlsruhe, Germany, ISBN 978-3-86644-858-2

The solutions derived are compared with tests found in the literature [2; 3; 4]. These tests confirm that the transverse reinforcement, the distance between the U-bars and the area of the concrete core influence the tensile strength of the connection in a manner similar to that predicted by the model.

2 Mechanism analysis and upper bound solution

2.1 Failure mechanisms with multiple yield lines

Consider a loop connection as shown in Fig. 1.

The U-bars protruding from each of the two precast elements are uniformly distributed at spacing $2a$. In the joint adjacent U-bars overlap by H and the distance between overlapping bars is a. A transverse lacer bar is placed in the center of the joint, which is filled with in-situ cast concrete. The joint is assumed to transfer a tensile force N from one precast element to the other.

In the case of concrete failure, the mechanism indicated in Fig. 1 is assumed to develop. The system of yield lines comprises of diagonal yield lines running between the ends of adjacent U-bars and splitting yield lines running from the end of each U-bar toward the nearest surface of a precast element. Finally, two yield lines are assumed to develop at the interfaces between the in-situ concrete and the precast elements. The system of yield lines divides the in-situ concrete into a number of segments that undergo rigid body movement.

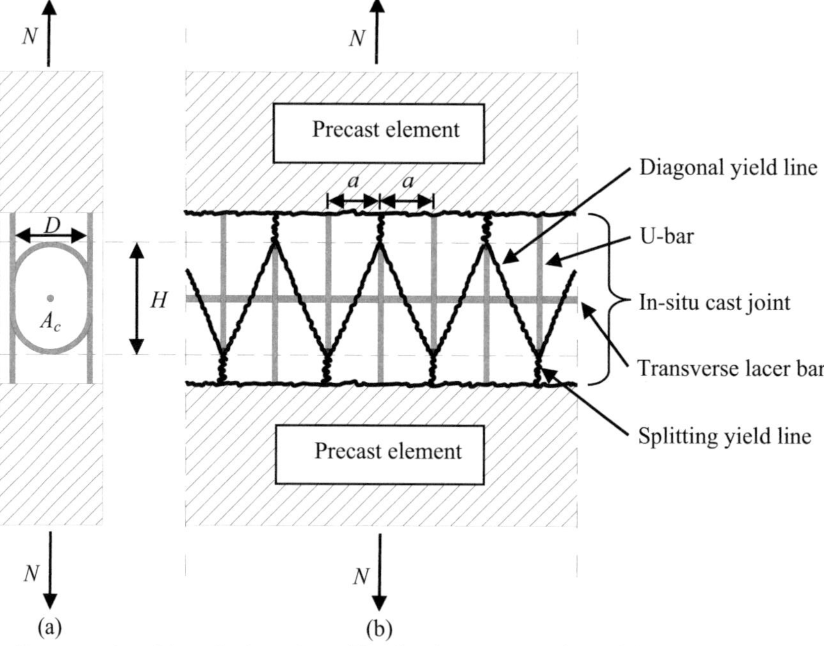

Fig. 1   Cross section (a) and plan view (b) of a loop connection with indication of failure mechanism

The plastic energy dissipated in the yield lines is calculated by considering concrete as a modified Coulomb material without tensile strength (i.e. $f_t = 0$).

The splitting yield lines undergo, due to symmetry, pure separation failures. Hence, no energy is dissipated in these yield lines as $f_t = 0$ is assumed. For the two yield lines formed in the interfaces between the precast elements and the in-situ concrete, it is assumed that the interfaces are smooth with a cohesion that is neglectable (at least when subjected to tension). So, in these two yield lines energy dissipation is also neglected.

It appears that only diagonal yield lines need to be considered. This means that it is justifiable to simplify the problem to that shown in Fig. 2, where the in-situ concrete outside the overlapping length H as well as the concrete cover is neglected. The diagonal yield lines then only cut through the concrete core with cross sectional area A_c. The relative displacement between the concrete segments shown in Fig. 2 may then be described as follows. In the direction of N, adjacent segments are displaced u_n relative to each other. This displacement is accompanied by a transverse displacement (in the t-direction) causing the segments to be pushed away from the centerline. As indicated in Fig. 2, adjacent segments are displaced u_t relative to each other in the transverse direction. This means that the absolute transverse displacement of the segments gets larger the further away a segment is from the center line. The transverse displacement will be counteracted by the transverse reinforcement, which is assumed to be rigid-plastic and thus in a state of tensile yielding.

Dowel action in the U-bars is neglected. The relative displacement vector $\boldsymbol{u} = (u_t , u_n)$ in the diagonal yield lines can now be determined by energy minimization.

It appears that it is only necessary to analyze yield lines emerging from one single U-bar. Then, the ultimate load of an entire connection can be determined simply by multiplying the results by the number of U-bars.

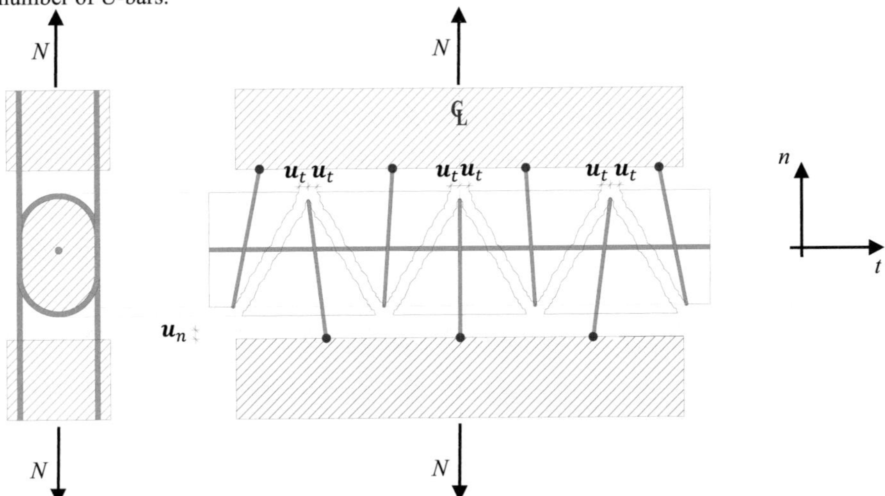

Fig. 2   Displacement of the concrete segments

2.2   Basic solution for one U-bar

Consider an internal U-bar as shown in Fig. 3. The tensile force, T_u, required to cause sliding failure in the two yield lines emerging from the end of the U-bar and resulting in the pull-out of one concrete segment will be determine in this section.

The displacement vector in the yield lines is denoted $\boldsymbol{u} = (u_t , u_n)$ and is inclined by an angle α to the yield line.

The internal work for both yield lines is calculated as follows:

$$W_I = 2\left(W_A A_{surface} + A_{sT} f_{yT} u_t\right) \tag{1}$$

The first term in this formula represents the dissipation in the yield lines and the second term is the reinforcement contribution, where A_{sT} and f_{yT} are the area and yield stress of the transverse reinforcement, respectively.

W_A is the dissipation per unit area, which for a modified Coulomb material appears as follows, [6];

$$W_A = \frac{1}{2} v f_c (1 - \sin\alpha)|\boldsymbol{u}| \tag{2}$$

Where v is the effectiveness factor (see below) and f_c is the uniaxial compressive strength. Finally $A_{surface}$ is the surface area of one diagonal yield line and may be calculated as follows:

$$A_{surface} = \frac{A_c}{\cos\beta} \tag{3}$$

Where the angle β is shown in Fig. 3 and A_c is the cross-sectional area of the concrete core and is calculated as:

$$A_c = \begin{cases} \frac{\pi}{4}\left(D + 2\phi_{loop}\right)^2 + \left(H - D - 2\phi_{loop}\right)\left(D + 2\phi_{loop}\right) & \text{for} \quad H \geq D + 2\phi_{loop} \\ \left(\frac{D}{2} + \phi_{loop}\right)^2 (\theta - \sin\theta) \quad ; \quad \theta = 2\,\text{Arccos}\left(1 - \frac{H}{D + 2\phi_{loop}}\right) & \text{for} \quad H < D + 2\phi_{loop} \end{cases} \tag{4}$$

Here, ϕ_{loop} is the diameter of the U-bars.

From geometrical considerations, see Fig. 3, the following relations apply:

$$|\boldsymbol{u}| = \frac{u_n}{\cos(\alpha - \beta)} \quad ; \quad u_t = u_n \tan(\alpha - \beta) \tag{5}$$

The external work is:

$$W_E = T_u u_n \tag{6}$$

An upper bound solution can be found by the work equation, $W_I = W_E$, where the relations in (5) replace $|\boldsymbol{u}|$ and u_t. As a result, an upper bound for T_u that depends on α is obtained. The solution is minimized with respect to α. The angle is found to be:

$$\alpha = \beta + \text{Arcsin}\left(\frac{1 - 2\Phi/\nu}{\sqrt{\left(\frac{a}{H}\right)^2 + 1}}\right) \tag{7}$$

Here Φ is the transverse reinforcement degree:

$$\Phi = \frac{A_{sT} f_{yT}}{A_c f_c} \tag{8}$$

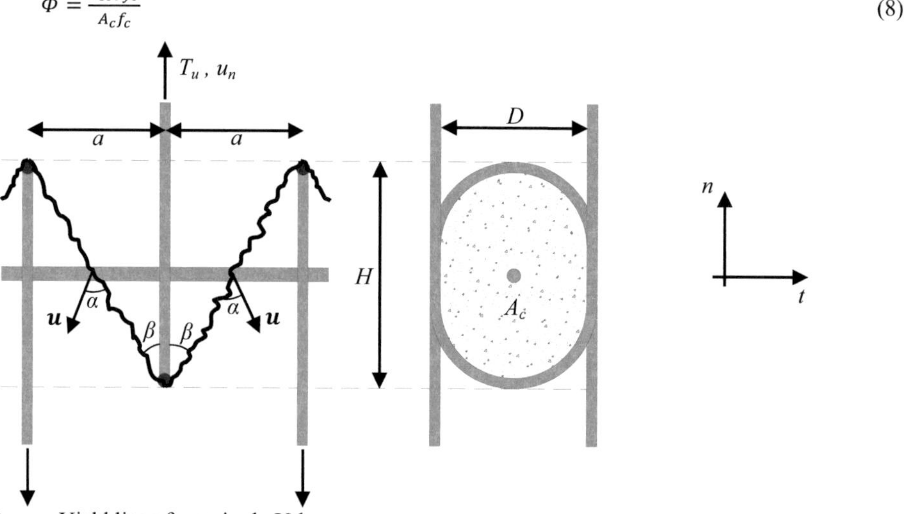

Fig. 3   Yield lines for a single U-bar

It is seen from Eq. (7) that $\alpha=\beta$ when $\Phi/\nu=\frac{1}{2}$ which means that the displacement vector is directed in the n-direction. This is the situation where there is no yielding of the transverse reinforcement, and the angle obviously cannot be smaller than this.

By inserting Eq. (7) into the upper bound solution, the optimal solution for a single U-bar is found to be:

$$T_u = \nu f_c A_c \left(\sqrt{4\frac{\Phi}{\nu}\left(1 - \frac{\Phi}{\nu}\right) + \left(\frac{a}{H}\right)^2} - \frac{a}{H}\right) \quad ; \quad \frac{\Phi}{\nu} \not> \frac{1}{2} \tag{9}$$

2.3   Tensile strength of symmetrical connections

The term symmetrical connection refers to the definitions of joints given in [3]. A symmetrical connection is a connection with the longitudinal reinforcement (the U-bars) placed symmetrically about the longitudinal centerline. This implies that the number of U-bars from one precast element is ξ while, from the other element the number of protruding U-bars is $\xi + 1$. Hence the tensile strength of an entire loop connection may be written as:

$$N_u = \min \begin{cases} \xi v f_c A_c \left(\sqrt{4\frac{\Phi}{v}\left(1 - \frac{\Phi}{v}\right) + \left(\frac{a}{H}\right)^2} - \frac{a}{H} \right) & ; \quad \frac{\Phi}{v} \not> \frac{1}{2} \\ \xi A_{se} f_{yl} \end{cases} \tag{10}$$

Here A_{se} is the cross sectional area of one U-bar and f_{yl} is the yield stress of the U-bars.

3   Comparison with test results

The theory has been compared to test results found in the literature. The tests with symmetrical connections include three series with 27 tests [2; 3; 4]. Details of the tests and calculations are summarized in [5]. In the published tests, no particular attention was paid to the anchorage of the lacer bars. Therefore when calculating the transverse reinforcement contribution, the yield stress f_{yT} is reduced by the factor l_i / l_b. Here, l_i is the distance from the end of the bar to its intercept with yield line number i and l_b is the anchorage length. Obviously f_{yT} is used when $l_i \geq l_a$. The anchorage length l_b is determined according to the Eurocode 2 [6]:

$$l_b = \frac{\phi_T}{4} \frac{f_{yT}}{2.25 f_{ct}} \quad ; \quad f_{ct} = 0.3(f_c - 8)^{2/3} \qquad \text{for} \quad \phi_T \leq 32mm \tag{11}$$

Eq. (11) is somewhat conservative because it does not take into account the confinement effects from the U-bars. The reduction factor for the strength of the transverse reinforcement is therefore expected to be in the interval between one and l_i / l_b. The calculations with and without the reduction factors l_i / l_b are compared with the tested strength of loop connections in Figure 4 (a) and (b), respectively.

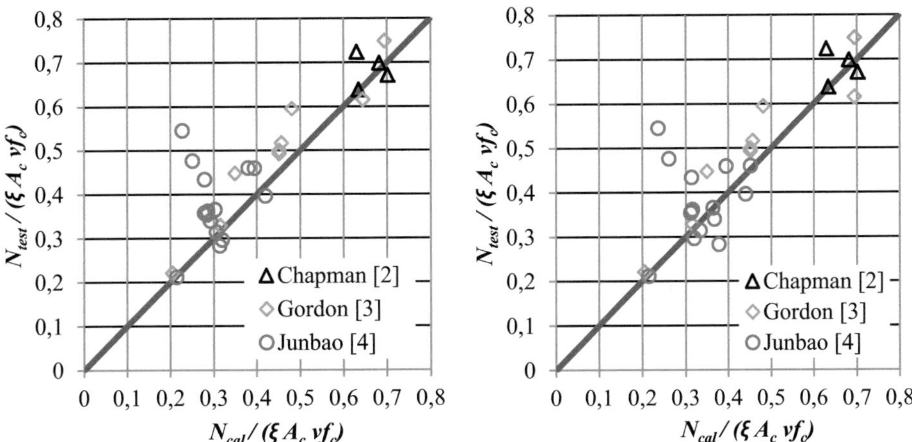

Fig. 4   Comparison of calculated and tested tensile strength of loop connections. **Left:** calculations with strength reduction from the anchorage length. **Right:** calculations without strength reduction of the transverse reinforcement.

As mentioned, an effectiveness factor v must be introduced when applying plastic theory to concrete structures. The considered failure mode consists of diagonal yield lines and therefore is similar to a beam shear mechanism. The formula for the effectiveness factor for beam shear, [1], is:

$$v_{beam} = \frac{0.88}{\sqrt{f_c}}\left(1 + \frac{1}{\sqrt{H}}\right)(1 + 26\,\rho_T) \not> 1 \tag{12}$$

Here f_c is in MPa, H in meters and $\rho_T = A_{sT} / A_c$.

It has been argued that the factor $(1+26\rho_T)$ accounts for dowel action in the critical cracks, see [7]. However, a contribution from dowel action is not possible when the tensile yielding capacity of the transverse reinforcement is fully utilized. For that reason the factor $(1+26\rho_T)$ is not included in the calculation of the effectiveness factor for a connection with U-bars. The effectiveness factor is then calculated by:

$$v = \frac{0.88}{\sqrt{f_c}}\left(1 + \frac{1}{\sqrt{H}}\right) \not> 1 \tag{13}$$

The test series from [4] consisted of 15 specimens having one U-bar overlapping two U-bars (i.e. $\xi =$ 1). The other two series consisted of multiple U-bar connections with three and two U-bars overlapping four and three U-bars, respectively (i.e. $\xi =$ 3 and 2).

The concrete compressive strength varied from 28.5 – 64.0 MPa. The aspect ratio a/H varied from 0.25–2.0. The inner bend diameter of the U-bars was $D = 47.7$mm in the test series from [2] and $D = 80$mm in the test series from [4]. In the test series from [3] the diameter D varied from 96-155mm.

A positive correlation is seen between the calculated and the tested ultimate tensile strength. However, when the overlapping length of the U-bars is very small ($H\!<\!<\!D$) the strength is underestimated by the theory (see the two tests from [5] having the largest deviation from the theory).

The mean value and the standard deviation for the ratio $N_{u,cal} / N_{u,test}$, when accounting for the anchorage length as described in Eurocode 2, are 0.88 and 0.17, respectively. However, if only the tests with $H - 2\phi_{loop} > D/2$ are taken into account (i.e. all specimens except the two tests with $H\!<\!<\!D$) the mean value and the standard deviation change to 0.92 and 0.13, respectively. If the strength is calculated using the yield strength without any reduction factor the mean value and standard deviation is calculated to 0.96 and 0.13, respectively (when not taking the two tests with $H\!<\!<\!D$ into account).

4   Conclusions

An upper bound solution for the tensile strength of loop connections between precast concrete panels has been presented. The solution covers the situation where concrete failure is governing. Positive correlation between the solution and test results found in the literature was seen. However, when the overlapping length of the U-bars is considerably smaller than the bend diameter of the U-bars, the tensile strength is underestimated by the model.

It was found from the upper bound solution that the transverse reinforcement has a significant influence on the tensile capacity. It was also found that the ratio a/H and the size of the concrete core confined by the overlapping U-bars are important for the tensile capacity.

5   Future work

Since only 27 tests could be found from the literature, and only 16 of these tests failed as concrete failure, there is a need for additional tests. The new test series should be made with varying transverse reinforcement degree (i.e. varying the area of the concrete core and the transverse lacer bars) to investigate if the effect of the transverse reinforcement is as high as the model predicts.

This paper deals with the situation of symmetric loop connections. Since asymmetric connections are also used in practice it could be interesting to find a solution for asymmetric connections as well.

Furthermore connections subjected to combined bending and tension forces should also be investigated since such combined loading is often seen in connections between bridge deck elements.

6   References

[1]   Nielsen, M.P., Hoang, L.C.: *Limit Analysis and Concrete Plasticity*, 3rd Edition, CRC Press LLC, 2011.

[2]   Chapman, C.E.: *Behavior of Precast Bridge Deck Joints with Small Bend Diameter U-bars*, Masters Thesis, University of Tennessee – Knoxville, 2010.

[3]   Gordon, S.R.: *Joints for Precast Decks in Steel Concrete Composite Bridges*, PhD Thesis, Heriot Watt University, 2006

[4]   Junbao, H.: *Structural Behaviour of Precast Component Joints with Loop Connection*, PhD Thesis, Department of Civil Engineering, National University of Singapore, 2004.

[5]   Joergensen, H.B., Hoang, L.C., Nielsen, L., Therkildsen, L.: *U-bar Loop Connections Subjected to Tension*, Internal Report, Institute of Technology and Innovation, University of Southern Denmark, 2011.

[6]   European Committee for Standardization. *EN 1992-1-1 Eurocode 2: Design of concrete structures – Part 1-1: General rules and rules for buildings.* Brussels: CEN, 2005.

[7]   Hoang, L.C., *Shear strength of lightly reinforced concrete beams*, Department of Structural Engineering and Materials, Technical University of Denmark, Report No. R-65, 2000.

Cracking risk in early-age RC walls

Agnieszka Knoppik-Wróbel

Faculty of Civil Engineering, Department of Structural Engineering,
Silesian University of Technology,
ul. Akademicka 5, 44-100 Gliwice, Poland
Supervisor: Barbara Klemczak

Abstract

Although early-age thermal–moisture effects are believed to be typical for massive concrete structures, thermal–shrinkage cracking may develop even in medium-thick structures if they are externally restrained. The goal of the research was to assess the cracking risk of such structures on the example of an RC wall cast against an old set foundation. A numerical model was prepared allowing of a multi-parameter investigation of thermal–moisture effects. The influence of concrete mix composition and curing conditions was analysed. The analysis confirmed that thermal–shrinkage cracking is a problem in early-age externally restrained concrete structures, yet it was proved that proper choice of mix composition and curing technology can reduce cracking risk to a great extent.

1 Introduction

1.1 Development of cracks in RC wall

Exothermic nature of cement hydration leads to development of significant temperatures in early-age concrete structures accompanied with moisture removal out of the structure. As a result, additional stresses occur in the element due to the consequent non-uniform volumetric changes. These stresses can reach significant values and may lead to early-age cracking of the structure, which has a significant influence on its durability.

The problem of thermal–shrinkage cracking of early-age concrete structures is well-known in massive concrete elements [1], however, cracks are also observed in newly constructed medium-thick concrete elements with limited possibility to deform [2] such as RC walls cast against an old set foundation (tanks walls [3], retaining walls, abutments [4]), walls being in contact with ground (tunnel walls) or walls cast in stages (massive containers walls [5]). Cracks, especially deep or through cracks, may adversely affect the serviceability, lifespan or even bearing capacity of a concrete structure. Cracking of liquid tanks walls endangers their tightness while in nuclear containers may promote leakage of radioactive elements into the environment during service life of the container. Early-age cracking also reduces the element's capacity and may lead to structural collapse as a result of unforeseen overpressure.

The thermal–shrinkage cracks in RC walls are non-structural cracks which occur because of a restraint of volume change. There are two main types of restraints: internal and external. The internal restraint is generated as a result of temperature and moisture gradients within the volume of the element and leads to formation of self-induced stresses. The external restraint is a limitation of deformation of the wall caused by more mature and stable concrete of previously cast foundation and as a result restraint stresses occur. Although considerable temperatures may develop in RC walls, because of relatively small thickness of the elements there are small temperature and moisture gradients in the elements sections. Nevertheless, walls (especially tanks walls) represent a high length-to-height ratio which causes relatively large restraints, thus developing stresses are of the restraint origin [6].

In newly constructed RC walls a uniform distribution of vertical cracks develops along the entire span of the structure. A typical pattern of cracking due to the edge restraint of a wall is shown in Fig. 1, assuming that the base is rigid. Restraint horizontal forces develop along the construction joint as a reaction to instantaneous deformation of the wall due to thermal–moisture effects. The observed cracks are vertical in the midspan of the wall and tend to splay towards free ends where a vertical force is required to balance the tendency of the horizontal force to warp the wall. A horizontal crack may occur at the construction joint at the ends of the wall due to this warping restraint. It is worth

Proc. of the 9th *fib* International PhD Symposium in Civil Engineering, July 22 to 25, 2012,
Karlsruhe Institute of Technology (KIT), Germany, H. S. Müller, M. Haist, F. Acosta (Eds.),
KIT Scientific Publishing, Karlsruhe, Germany, ISBN 978-3-86644-858-2

noticing that although cracks start to develop at the construction joint between the wall and the foundation, the maximum width of crack occurs at some height above the joint [4, 8].

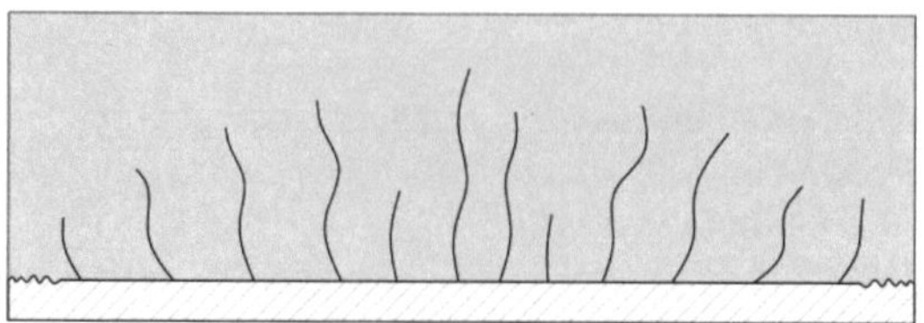

Fig. 1   Typical cracking pattern in an externally restrained RC wall.

1.2   Factors affecting the risk of early-age cracking and methods to reduce it

The goal of the research was to investigate the role of various factors in development of thermal–shrinkage stresses leading to crack formation of externally restrained medium-thick concrete structures on the example of an RC wall cast against an old set foundation.

The most important factor when analysing early-age thermal–shrinkage stresses in concrete elements is the temperature development in the element. The complex variables that affect the temperature increase rate and the maximum temperature as well as the temperature difference are [2]:

- thermal properties of concrete dependent on the amount and properties of concrete components, especially the amount and kind of cement (total amount of hydration heat, heat evolution rate, specific heat, thermal conductivity);
- conditions during casting and curing of concrete (initial temperature of concrete, type of formwork, the use of insulation or cooling);
- technology of concreting (e.g. segmental concreting);
- environmental conditions (ambient temperature, wind, humidity);
- dimensions and geometry of the concrete structure.

Potential solutions to minimise high temperatures and temperature differences in early-age concrete are referred to factors listed above. Currently used methods include optimal concrete mix design, concrete cooling before or after placement, the use of smaller placements as well as insulation.

Mechanical properties of maturing concrete such as strength development or elastic and viscous behaviour of concrete subjected to high levels of stresses and elevated temperatures are also of great importance in development of early-age stresses. Moreover, in case of restrained structures such as walls the degree of restraint also influences distribution of stresses [7, 8].

2   Research methodology

2.1   Strategy

The influence of various factors on development of thermal–shrinkage stresses and cracking risk in RC wall was investigated in a multi-parameter numerical analysis. A numerical model was developed at the Department of Structural Engineering of the Silesian University of Technology and implemented in a set of programs [9, 10]. The wall was modelled with use of the software and analysed considering different parameters, such as the ambient temperature, the difference between the ambient and the initial concrete mix temperature, the time of formwork removal and a fresh concrete mix composition – especially the type and amount of cement. The resultant values of stresses were compared.

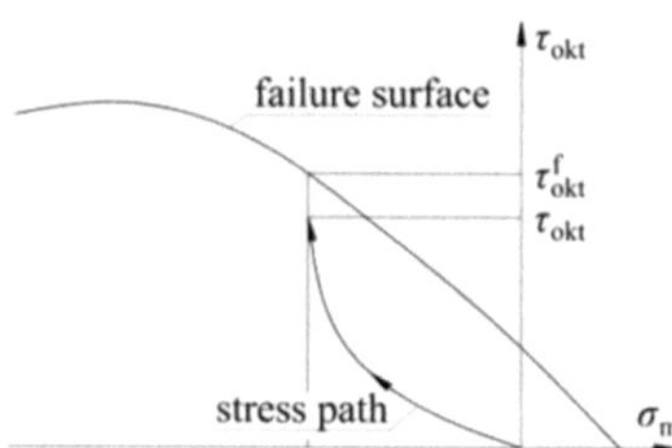

Fig. 2   Axiatoric section of failure surface.

The cracking risk was expressed by means of the damage intensity factor. The possibility of crack occurrence was defined on the basis of the location of the point representing the stress state with respect to the failure surface. This location was referred to as the damage intensity factor and described as $s_1 = \tau_{oct}/\tau^f_{oct}$ acc. to Fig. 2. The damage intensity factor equal to 1 is equivalent to the stress reaching the failure surface and signifies failure of the element. Character of this failure depends on the location where the failure surface is reached [9, 10]. In the further presented examples reaching the failure surface always occurred within the range of hydrostatic tensile stresses which was equivalent to crack formation in the plane perpendicular to the direction of the maximum principal stress.

2.2 Numerical model

The presented numerical model can be classified as a phenomenological model. The influence of mechanical fields on temperature and moisture fields was neglected, but thermal–moisture fields were modelled using coupled equations of thermodiffusion. The complex analysis of a structure consists of three steps. The first step is related to determination of temperature and moisture development, in the second step thermal–shrinkage strains are calculated and these are used as an input for computation of stress in the last step. For the purpose of determination of the stress state in early-age concrete structures a viscoelasto–viscoplastic model with a consistent conception was proposed. With respect to the engineering application of the theoretical models, the computer programs were developed: TEMWIL for determination of thermal–moisture fields, MAFEM_VEVP for determination of stress states and MAFEM3D user interface for data preparation and presentation of results [9, 10].

3 Analysis of the RC wall

3.1 Basic case

The analysed wall was assumed to have 20 m of length, 4 m of height and 70 cm of thickness, supported on a 4-m wide and 70-cm deep strip foundation of the same length. The wall and the foundation were assumed to be reinforced with a near-surface reinforcing nets of ⌀16 bars (RB400 steel). The wall was reinforced at both surfaces with horizontal spacing of 20 cm and vertical spacing of 15 cm. The foundation was reinforced with 20 cm × 20 cm spacing at the top and bottom surfaces. Due to double symmetry of the structure the model for finite element analysis was created for ¼ of the wall. Final geometry of the wall with a mesh of finite elements is presented in Fig. 3.

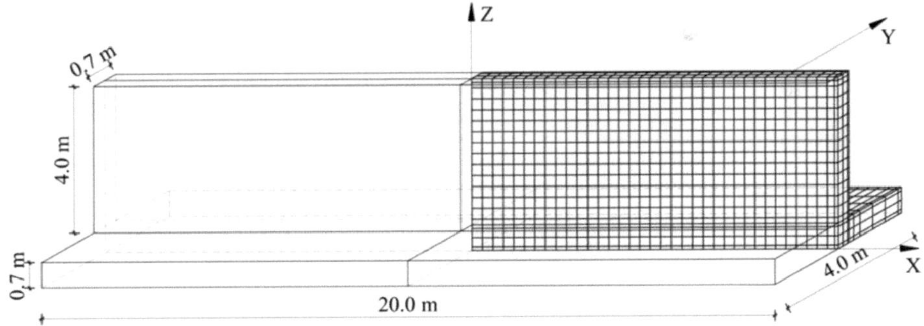

Fig. 3   Geometry of the wall with the finite element mesh.

The material properties of the foundation were assumed as for 28-day concrete, i.e. the compressive strength $f_{cm} = 35$ MPa, the tensile strength $f_{ctm} = 3$ MPa and the modulus of elasticity $E_{cm} = 32$ GPa. For the wall, development of material properties was assumed according to CEB-FIP MC90. Detailed material, environmental and technological parameters for the basic case were taken as:

- composition of fresh concrete mix: cement CEM I 42.5R – 375 kg/m^3, water – 170 kg/m^3, aggregate (granite) – 1868 kg/m^3;
- ambient temperature $T_z = 25°C$, initial concrete mix temperature $T_p = T_z = 25°C$;
- wooden formwork of 1.8 cm plywood at side surfaces, removed after 28 days, protection of top surface with foil.

The values of parameters taken in numerical analysis are presented in Table 1. The initial temperature of the foundation concrete was taken as equal to the ambient temperature T_z. The parameters connected with temperature and moisture migration in the wall after formwork removal were assumed as for a free, unprotected surface.

Table 1   Thermal–moisture parameters assumed in numerical analysis

thermal fields		moisture fields	
λ, [W/(m·K)]	2.52	q_v, [W/m^3]	acc. to equation:
c_b, [kJ/(kg·K)]	0.95		$Q(T,t) = Q_\infty e^{-at_e^{-0.5}}$ (1)
ρ, [kg/m^3]	2413	K, [m^3/J]	$0.3 \cdot 10^{-9}$
α_{TT}, [m^2/s]	$7.29 \cdot 10^{-7}$	α_{WW}, [m^2/s]	$0.6 \cdot 10^{-9}$
α_{TW}, [(m^2·K)/s]	$9.375 \cdot 10^{-5}$	α_{WT}, [m^2/(s·K)]	$2 \cdot 10^{-11}$
α_p, [W/(m^2·K)]	6.00 no covering[(2)] 5.80 foil 3.58 formwork	β_p, [m/s]	$2.78 \cdot 10^{-8}$ no covering[(2)] $0.10 \cdot 10^{-8}$ foil $0.18 \cdot 10^{-8}$ formwork
α_T, [1/°C]	0.00001	α_W	0.002

[(1)] the value of $Q_\infty = 508$ kJ/kg and coefficient a was described as $a = 513.62 t_e^{-0.17}$, which complies with test results of heat development for CEM I 42.5R

[(2)] without considering the influence of wind

Fig. 4 (left) shows stress development in time for the interior and the surface of the wall. A typical, two-phase (compressive–tensile) character was observed. It must be noted that greater stresses occurred in the interior of the wall because the wall was detained in formwork until it cooled down. The difference between the internal and surface stresses results from self-induced stress component caused by temperature and moisture differences in the section [6]. Fig. 4 (right) presents stress distribution in the internal mid-span cross-section of the wall after 19 days of curing. A characteristic stress distribution can be observed which results from the restraint exerted by the foundation [8].

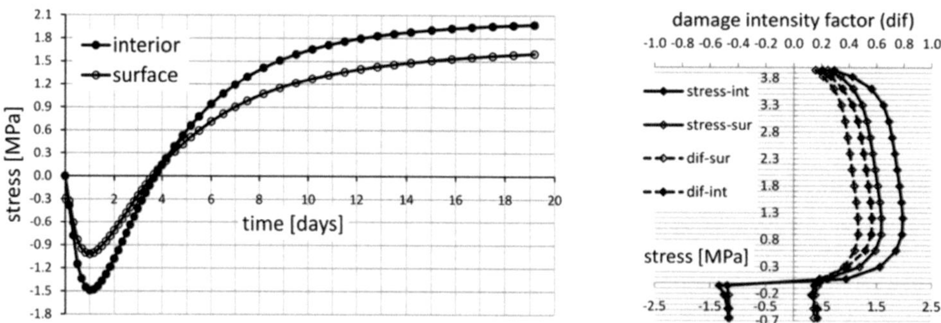

Fig. 4   Stresses in the RC wall: development in time (left) and distribution at the height of the wall, in its mid-span section after 19 days of concrete curing (right).

3.2   Parametric study

The impact of various parameters on cracking risk was investigated. As the measure of cracking risk the maximum value of damage intensity factor was taken. It should be noted that in each of the analysed cases the character of the observed stress development in time as well as the stress and damage intensity factor distribution at the height of the wall complied with the diagrams presented in Fig. 4. The maximum cracking risk was observed at the height of $0.9 \div 1.5$ m above the foundation; in more loaded walls it concentrated closer to the foundation joint.

Firstly, the influence of the ambient temperature and the temperature difference was checked. The following cases were taken into consideration: external temperature of 15, 20 and 25°C, without pre-cooling of concrete mix, and additionally in each case lowering of the mix initial temperature by 5 and 10°C. Diagrams in Fig. 5 (left) present juxtaposition of maximum damage intensity factor values in all the analysed cases. It is visible that lower ambient temperature posed smaller risk of cracking, although this difference was not significant. Pre-cooling of concrete mix was very beneficial for the structure. Lowering of the initial temperature by 5°C and 10°C reduced the cracking risk in the interior by as much as 17% and 30% while on the surface by 25% and 35%, respectively.

Then it was investigated how the time of formwork removal influences the cracking risk of the wall. Under the assumption of the external temperature being 15, 20 and 25°C and without initial cooling of concrete mix two cases were compared: the first in which the wall was detained in formwork for the whole analysed time and the second in which the formwork was removed after 3 days. Comparison of cracking risk is presented in Fig. 5 (right). Early formwork removal resulted in accelerated heat and moisture removal out of the structure and led to significant increase in stresses, especially in near-surface areas. That is why damage intensity factor increase was observed. Although not so considerable in the interior, the risk of cracking almost doubled on the surface of the wall.

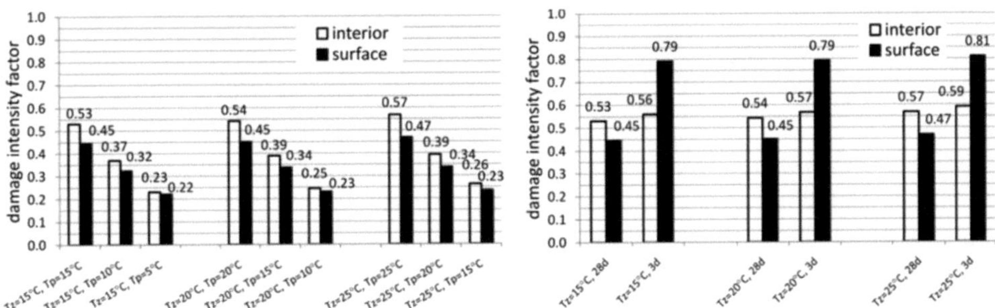

Fig. 5   Cracking risk: the influence of temperature (left) and time of formwork removal (right).

The influence of concrete mix composition was also analysed focusing on the type and amount of cement. In each case a mix was designed under the assumption that the final values of mechanical parameters of mature concrete were the same. Considered mixes are shown in Table 2.

Table 2   Composition and thermo-physical parameters of concrete mixes

	Mix 1	Mix 2	Mix 3	Mix 4	Mix 5
Cement	CEM II B-S 42.5R	CEM III/A 42.5N	CEM V/A 32.5R	CEM I 42.5R	
$[kg/m^3]$	375	375	375	325	425
a	$480.51t_e^{-0.115}$	$475.70t_e^{-0.093}$	$473.72t_e^{-0.091}$	$513.62t_e^{-0.17}$	
Q_∞ [kJ/kg]	466	469	396	508	
Water [kg/m^3]	170	170	151	147	193
Aggr. [kg/m^3]	1868	1868	1918	1973	1762
Adm.[kg/m^3]	-	-	8	8	7
λ [W/(mK)]	2.52	2.52	2.53	2.45	2.57
c_b [kJ/(kgK)]	0.95	0.95	0.92	0.99	0.92

Fig. 6 (left) shows the damage intensity factor development in time for concrete mixes with different types of cements while Fig.6 (right) presents the comparison of cracking risk depending on the type and amount of cement. For concrete mixes with the same type of cement there was a direct correlation: greater amounts of cement generated more heat and exerted greater stresses on the structure thus posing higher risk of cracking. When different types of cements were used, the relationship was not that straightforward because there are two factors of importance: total amount of heat together with its development rate and the rate of mechanical parameters development. Cements with lower hydration heat generate lower hardening temperatures, however, they usually have lower rate of mechanical parameters development. Therefore, until the concrete hardens, the value of the elasticity modulus of the element is lower comparing to the elements made of cements with fast strength development. It is a serious issue in restrained concrete structures where stresses arise mainly as a result of an external restraint such as a stiffer foundation (greater foundation-to-wall moduli ratio [7, 8]).

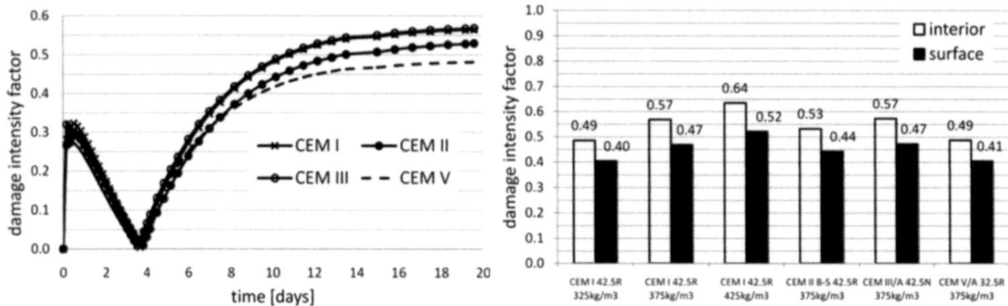

Fig. 6   Influence of mix composition on cracking risk in the wall: development in time for different cement types (left) and max. values for different cement types and amounts (right).

4   Conclusions

The phenomenon of crack development in young concrete structures is not only the problem of massive concrete structures but also externally restrained structures of smaller dimensions. The paper discusses the cracking risk of RC walls by means of a numerical analysis with the original software.

It was proved that adequate choice of technology and curing conditions can reduce the risk of cracking of RC walls. Application of lower hydration heat cements may also have a positive effect on reduction of heat generated in young concrete elements but special attention must be paid to the rate of strength development in the elements as there is a direct correlation between the two.

References

[1]   Bergstrom, S.G.; Byfors, J.: Properties of concrete at early ages. In: Materials and Structures Vol. 13 No. 75 (1980), pp. 265–274

[2]   Klemczak, B.; Knoppik-Wróbel, A.: Early-age thermal and shrinkage cracks in concrete structures – description of the problem. In: Architecture–Civil Engineering–Environment Vol. 4 No. 2 (2011), pp. 35–47

[3]   Zych, M.: Analiza pracy ścian zbiorników żelbetowych we wczesnym okresie dojrzewania betonu w aspekcie ich wodoszczelności (Analysis of work of reinforced concrete tanks walls in early phases of concrete curing in the view of their water tightness). PhD Thesis, Cracow Technical University (2011)

[4]   Flaga, K.; Furtak, K.: Problem of thermal and shrinkage cracking in tanks vertical walls and retaining walls near their contact with solid foundation slabs. In: Architecture–Civil Engineering–Environment Vol. 2 No. 2 (2009), pp. 23–30

[5]   Craeyea, B.; De Schutter, G.; Van Humbeeck, H.; Van Cotthem, A.: Early age behaviour of concrete supercontainers for radioactive waste disposal. In: Nuclear Engineering and Design Vol. 239 Issue 1 (2009), pp. 23–35

[6]   Knoppik-Wróbel, A.: The influence of self-induced and restraint stresses on crack development in reinforced concrete wall subjected to early-age thermal–shrinkage effects. In Proc.: 14th International Conference of PhD Students Juniorstav 2012, Brno, Czech Republic

[7]   ACI Committee 207: Report on Thermal and Volume Change Effects on Cracking of Mass Concrete (ACI 207.2R-07), ACI Manual of Concrete Practice, Part 1, American Concrete Institute, Detroit (2011)

[8]   Nilsson, M.: Restraint Factors and Partial Coefficients for Crack Risk Analyses of Early Age Concrete Structures. PhD Thesis, Luleå University of Technology (2003)

[9]   Klemczak, B.: Modelowanie efektów termiczno–wilgotnościowych i mechanicznych w betonowych konstrukcjach masywnych (Modelling of thermal–moisture effects in massive concrete structures). Wydawnictwo Politechniki Śląskiej: Monograph 183, Gliwice (2008)

[10]   Klemczak, B.: Prediction of Coupled Heat and Moisture Transfer in Early-Age Massive Concrete Structures. In: Numerical Heat Transfer. Part A: Applications Vol. 60 No. 3 (2011), pp. 212-233

Modeling coupling beams made of or retrofitted with HPFRCCs

Milot Muhaxheri

Department of Structural Engineering,
Politecnico di Milano,
Piazza Leonardo da Vinci 32, 20133 Milano, Italy
Supervisor: Liberato Ferrara

Abstract

The general aim of this work is to define the framework for predictive modeling of coupling beams made of or retrofitted with High Performance Fiber Reinforced Cementitious Composites HPFRCCs. This will be done by verifying the accuracy of different modeling techniques at different modeling scales. In details two approaches will discussed: 1) a "material level based approach and 2) a "structural element based" one. The first provides a detailed description at the material level and the latter a reasonable accuracy at the level of global response under complex loading paths and higher efficiency in terms of time running. Finally, the second model will be used to predict the performance of poorly designed coupling beams reinforced with HPFRCC layer.

1   Introduction

Coupling beams as interconnection of shear walls in earthquake-resistant structures have to be designed with the purpose of transmitting the forces resulting from earthquake excitation between the different shafts the wall consists of. In practice, achieving a realistic transfer of force requires careful arrangement of the reinforcement, which may result in details difficult to be executed. During the last decade several promising studies have been conducted (Canbolat et al. 2005; Lequesne 2011) aimed at improving the structural behavior of coupling beams using HPFRCC as an alternative replacement of the traditional concrete, allowing the reinforcement reduction and a simplified detailing. Such an alternative has also been shown to drastically improve energy dissipation capacity, besides highlighting a simpler constructability.

Progressive deterioration of existing r/c structures is highlighting the urgent need of repairing and strengthening structural elements in order to make the structure to perform as intended or anticipated or simply to make it complying with updated and, not seldom, more rigorous and stringent design provisions. Because of this, the use of High Performance Fiber Reinforced Cementitious Composites (HPFRCCs) has become an attractive solution within the research community, particularly for rehabilitation of slender beams and columns (Martinola 2010; Shin 2011). As a matter of fact reliable approaches to predict the behavior of HPFRCC coupling beams and transfer the garnered knowledge into reliable design prescriptions are so far lacking, also because of this still ongoing evolution of code prescriptions concerning structural applications of Fiber Reinforced Concrete.

This paper focuses on the predictive modeling of coupling beams made of or retrofitted with HPFRCCs. The problem has been first of all investigated at the "material level", employing the "Crush-Crack" damage model (di Prisco and Mazars, 1996; Ferrara and di Prisco, 2001) to model the behavior of HPFRCC, as experimentally identified by means of a dedicated experimental campaign (Ferrara et al. 2012). The tensile behavior of HPFRCC has been described by piecewise linear functions consistent with prescriptions contained in Model Code 2010. The same numerical approach has been then applied to model the behavior of HPFRCC coupling beams under monotonic loading, making reference to the experimental campaign performed by Canbolat et al. (2005) on 4 individual coupling beams, cast with either conventional reinforced concrete or HPFRCCs and different reinforcement arrangements. Continuum Damage Modelling allowed a deeper insight to be cast into local phenomena occurring in the tested specimens along the loading path.

Finally a fiber Timoshenko beam element, incorporating reliable description of unilateral effects of concrete cyclic behavior as well as suitable assumptions resulting from the previous modeling phase, has been adopted to predict the cyclic behavior of coupling beams made of or retrofitted with HPFRCCs. This work has to be intended as a preliminary step, though limited to a peculiar case,

Proc. of the 9th *fib* International PhD Symposium in Civil Engineering, July 22 to 25, 2012,
Karlsruhe Institute of Technology (KIT), Germany, H. S. Müller, M. Haist, F. Acosta (Eds.),
KIT Scientific Publishing, Karlsruhe, Germany, ISBN 978-3-86644-858-2

towards the incorporation of high performance fiber reinforced cementitious composites into design of new and retrofitting of either damaged or poorly designed earthquake resistant structures.

2   Material properties

The strengthening material herein investigated consists of a self leveling high strength mortar reinforced with 100 kg/m^3 steel fibers (l_f = 13 mm; d_f = 0.16 mm). Its tensile behavior has been characterized by using an innovative testing technique known as Double Edge Wedge Splitting (DEWS – Fig. 1 a; see di Prisco et al., 2010). To this purpose, 18 DEWS square tiles have been extracted from a slab, 1000 mm long 500 mm wide and 30 mm thick, cast with the material at issue, where the wedge-preordained fracture planes were either normal or parallel to the casting flow lines (Fig. 1b- Ferrara et al., 2012). The experimental results, shown in Fig. 1-c in terms of tensile stress vs. Crack Opening Displacement (COD), highlight that the same material can exhibit either hardening or softening post-cracking behavior in tension due to flow induced alignment of the fibres to the applied tensile stress. For numerical investigations the curves with highest and lowest tensile capacity have been chosen.

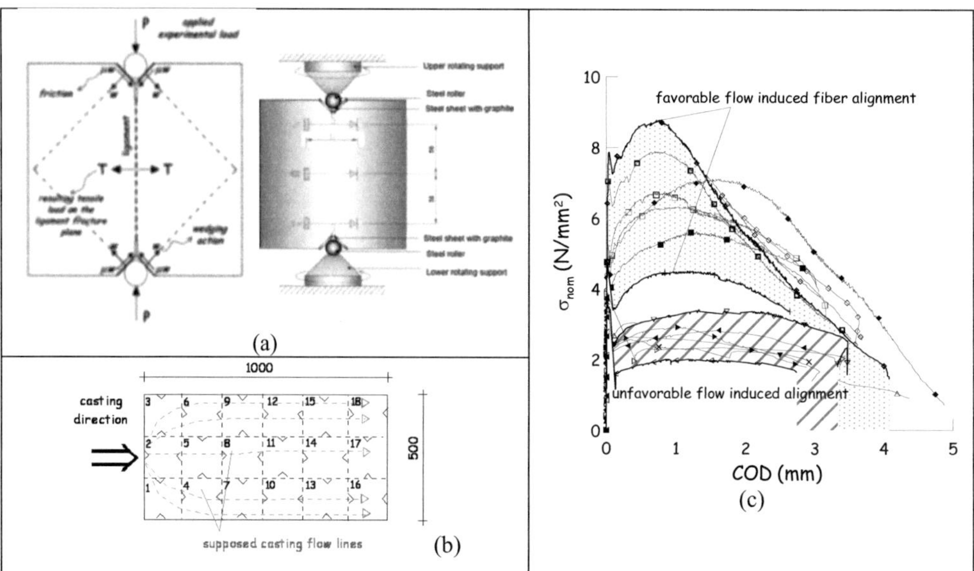

Fig. 1   DEWS test setup (a); cast slab and DEWS specimen cutting (b); stress-COD curves (c)

3   Constitutive laws for HPFRCC

Experimental results have shown that the orientation of fibers, triggered by the casting flow, may generate an orthotropic material behavior. Based on these considerations, a different procedure has been herein adopted to describe the mechanical response of cementitious composite in tension, whether strain softening (for unfavorable fiber orientation) (Fig. 2-a) or hardening (for favorable orientation) (Fig. 2-b). The constitutive curves are modeled by valuable reduced number of parameters assigning the points of peak (f_{ct}) and residual stresses ($f_{R,1}$ and $f_{R,3}$) at crack opening displacements equal to 0.25 mm and 1.25, such threshold being absolute or referred to the COD value at peak in case of a strain softening or hardening material respectively.

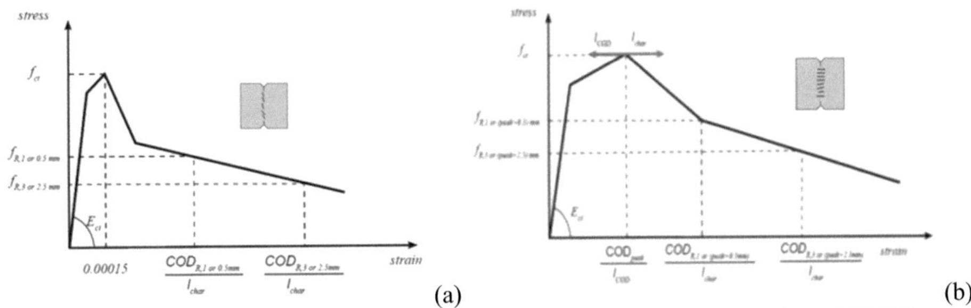

(a)   (b)

Fig. 2   Stress-strain relationship for: a) strain softening and b) strain hardening HPFRCCs

4   Numerical modeling approaches

4.1   HPFRCC behavior: continuum damage modeling

The behavior of the HPFRCC, experimentally identified as above, has been first of all back analyzed by means of the "Crush-Crack" damage model, implemented in finite element code Cast3M (di Prisco and Mazars, 1996; Ferrara and di Prisco, 2001). The model features the possibility of assigning any kind of constitutive law both in tension and compression, with no need to comply with pre-arranged built in options. The two options for tensile constitutive behavior of HPFRCC, as illustrated in (Fig. 2 a-b), have been employed to reproduce the experimental results obtained from DEWS tests, by employing the mesh shown in Fig. 3-a. Whereas in the pre-peak regime (i.e. before unstable crack localization), strains were calculated as the ratio between the measured crack opening and the gauge length, in the post-cracking regime the element size $h_{element}$ was used as a characteristic length parameter to convert crack opening into strain. This furthermore allowed mesh independent numerical solutions to be obtained (Figure 3b). The reliability of the identification procedure clearly appears from results shown in Fig. 3-c.

Interestingly, from damage patterns shown in Figures 4 a-b, it can be observed that whereas in the case of a strain softening material an immediate unstable localization occurs after the formation of the crack along the ligament, in the strain hardening case a significant spreading of the damage has been correctly predicted in the pre-peak regime, which corresponds to stable multiple cracking, consistently with experimental evidence. All along the softening stage, which is when a major crack unstably localizes and propagates along the ligament, the width of the damaged zone spreading no longer.

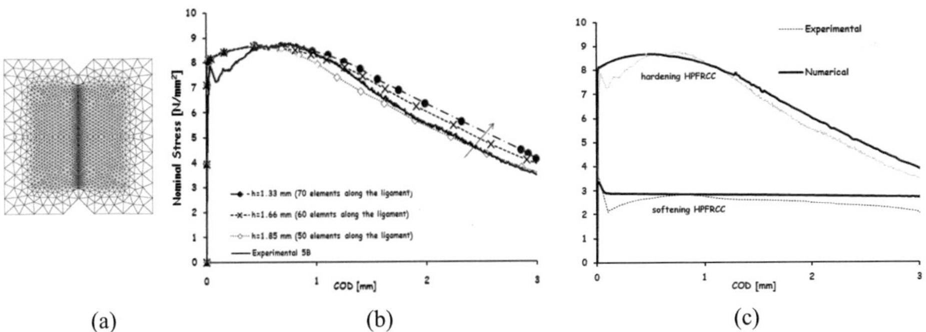

(a)   (b)   (c)

Fig. 3   Mesh employed for numerical analyses (a); mesh independence(b) and numerical vs. experimental comparison (c)

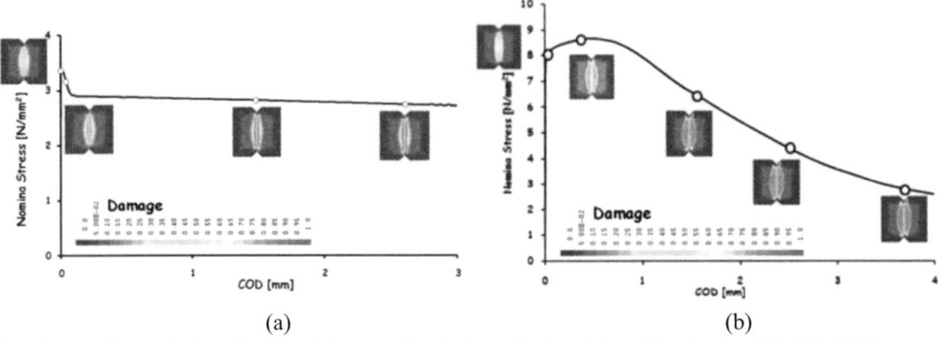

(a)   (b)

Fig. 4   Computed evolution of damage for strain softening (a) and hardening (b) HPFRCC

4.2   Coupling beams: damage modeling

The numerical model calibrated as above has been employed to reproduce the behavior of a HPFRCC coupling beam (Canbolat et al., 2005). The constitutive behavior of HPFRCC has been described as from experimental results provided by the authors, coherently with Fig. 2-b assumptions. A monotonic load path was considered, which matches reasonably well with the envelope of the experimental cyclic load-drift response (Fig. 5-a). The computed damage pattern at the maximum drift interestingly reproduces the experimental crack pattern (Fig. 5-b), furthermore highlighting the high level of damage which the material can tolerate before localization of a major unstable crack.

Numerical analyses further showed that, at 4% drift, which can be consistently assumed as the ULS deformation demand to such structures, the principal tensile strain nowhere exceeds a threshold corresponding to a crack opening equal to 0.5 mm beyond the peak strain. This supports the assumption of an elastic-perfectly plastic description of HPFRCC tensile behavior in further "structure level" modeling, with a "plateau" stress assumed equal to the residual stress corresponding to the aforementioned strain threshold, as hereafter detailed.

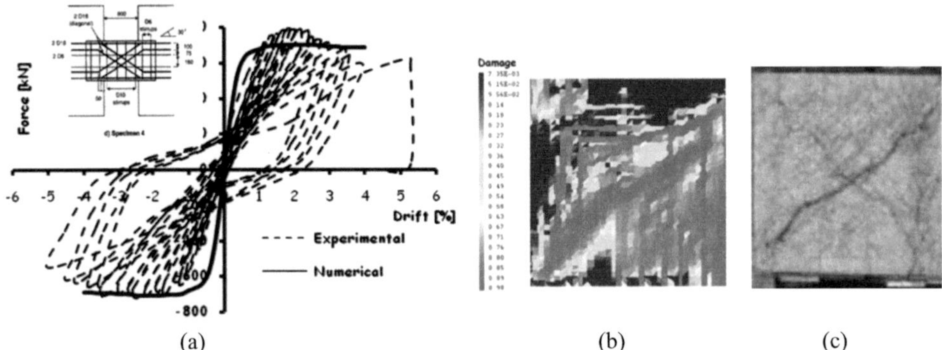

(a)   (b)   (c)

Fig. 5   HPFRCC coupling beam: a) load-drift experiments vs. numerical and b-c) damage pattern vs. experimental crack pattern experimental at 4% drift.

4.3   Coupling beams: Timoshenko-beam fiber modeling

A fiber beam model based on Timoshenko beam theory implemented in Cast3M (Guedes et al., 1994; Mazars et al., 2006) has been used for structure level modeling. Compressive behavior for both conventional concrete and HPFRCC has been described by means of a parabola, up to the peak stress, followed by a linear softening. The tensile behavior of conventional concrete has been described as linear elastic up to tensile strength, followed by a post-peak linear softening. Suitable description of the cyclic behavior is provided, in order to take into account the stiffness degradation and crack closure effects (Guedes et al., 1994). As for the behavior of HPFRCC in tension, a simplified bilinear elastic-perfect plastic stress-strain relationship has been adopted, according to aforementioned assumptions, with a "yield" strength equal to the residual strength corresponding to a crack opening value equal to 0.5 mm beyond the peak. Because of lack of data about the cyclic behavior of

HPFRCC, this has been assumed equal to that of a conventional concrete having the same compressive and tensile strength. Experimental evidence about the cyclic behavior of HPFRCC is very limited: anyway it has been shown that its cracking behavior under cyclic loadings is comparable to behavior under monotonic loading and there is no influence on the ultimate strain level (Mechtcherine, 2012). Steel behavior has been assumed to be elastic-hardening and perfectly bonded to concrete.

The numerical approach has been validated based once again on the experimental tests performed by Canbolat et al. (2005), with reference to coupling beams made of either conventional r/c or HPFRCC when subjected to reversal shear loading, under increasing amplitude cycles. The tested coupling beams has length and height equal to 600 mm, and width equal to 200 mm and 150 mm for r/c and HPFRCC respectively (they were designed at a ¾ scale to the full scale case study).

The compressive strength of regular concrete was assumed equal to 41 N/mm^2, as from experimental paper; other parameters needed for the analysis, such as tensile strength and fracture energy, were computed from MC2010 formulae. The yielding strength of the steel was equal to 450 N/mm^2, with a 726 N/mm^2 tensile rupture strength, once again accordingly to what reported by the authors. As for HPFRCC its compressive strength was assumed equal to 63 N/mm^2, and its post cracking tensile strength equal to 5.5 N/mm^2, both as reported. Perfect bond was assumed between steel and either concrete or HPFRCC. The influence of transverse reinforcement that provides confinement can be considered by assigning a non-zero residual strength in compression: actually 20 % of the compressive strength was chosen in this study. A relatively good agreement between numerical prediction and experiments is observed for the case of conventional concrete (Fig. 6-a), where also the influence of a residual compressive strength, in the aforementioned sense, can be appreciated.

With reference to HPFRCC coupling beams, the elastic-perfect plastic description of the material tensile behavior, has proven to be a reliable assumption to have a satisfactory agreement between experimental and numerical predictions (Fig. 6-b).

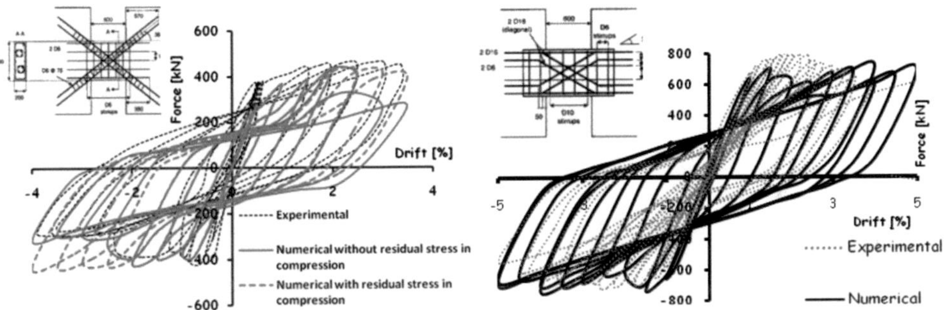

Fig. 6 Experiments vs. Numerical a) R/C coupling beam and b) HPFRCC coupling beam

5 Retrofitted coupling beams

With the same approach previously calibrated, numerical analyses have been performed in order to understand the effectiveness of using HPFRCC as a retrofitting solution for poorly designed coupling beams and design a dedicated experimental campaign. The beam dimensions were scaled by a 1:2 ratio to full scale size, to comply with the maximum capacity of available loading actuators to be employed in the experimental stage of this research: this resulted in a length equal to 450 mm and a rectangular cross section 300 mm high and 100 mm wide. Grade B500 longitudinal reinforcement was considered (3ϕ10 bars at the top and at the bottom). Class C20/25 concrete was chosen for the beam; the other parameters for the constitutive laws were obtained according to MC2010. The retrofitting HPFRCC material, consistently with tests described in § 2, has a compressive strength of 115 N/mm^2; the residual tensile strength at 0.5 mm crack opening is equal, as from experiments, to 7.11 N/mm^2 and 1.8 N/mm^2, respectively for favorable and unfavorable fiber orientation.

Perfect bond between the old concrete and the HPFRCC overlay has been assumed. This assumption, mainly due to the lack of suitable experimental reference data, is debatable, because of the influence that both the surface treatment of the existing structural element and its pre-existing damage conditions may have on the interface behavior

In Fig. 7 a-b the response of retrofitted beams is shown in terms of load-drift curves, in monotonic, for the different investigated retrofitting options. The jacketing has been modeled along two or three sides of the cross section, that might be the real situation during execution; two different repairing thickness have been evaluated, also considering the effects of the residual strength of HPFRCC, as it may be affected by the flow induced orientation of fibers. The effectiveness of the retrofitting is shown, also through the comparison with the performance of a r/c jacketing, which, assuming a ϕ6@200mm welded wire mesh skin reinforcement, should be at least 40 mm thick on each side.

6   Conclusions: addressing research needs

In this paper, a numerical framework to model the behavior under lateral loads of coupling beams retrofitted with HPFRCC jacketing, as instrumental to the design of a dedicated experimental campaign. The effectiveness of HPFRCC as a retrofitting material has been numerically assessed, predicting satisfactory performance which will be assessed through forthcoming experiments. A number of limitations still existing in the approach described above can be highlighted. The interface interaction between HPFRCC and existing concrete has not been considered, mainly because of the lacking test results to calibrate the interface model itself. A dedicated experimental campaign is ongoing to provide these data. Last but not the least, the orthotropy of HPFRCCs resulting from flow induced orientation of fibers and its outcomes on the cyclic behavior of the material, have to be both experimentally characterized and properly described by means of advanced constitutive models, in order to correctly predict the cyclic behavior of elements made of or strengthened with this kind of materials.

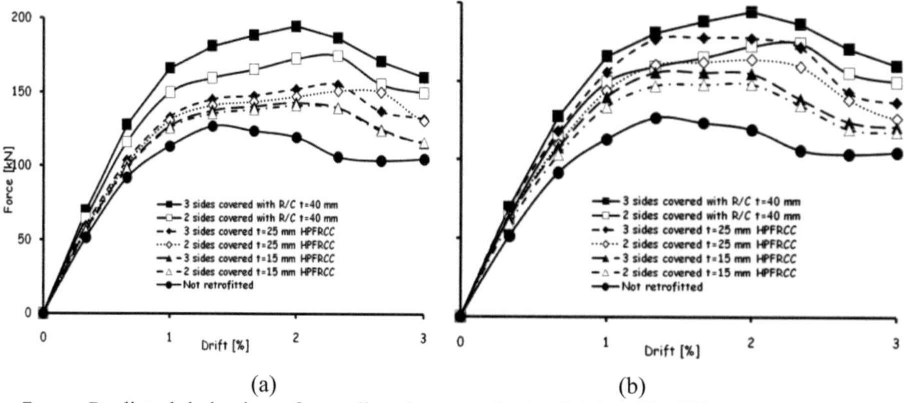

Fig. 7   Predicted behavior of coupling beams of retrofitted with different arrangements the HPFRCC layer: unfavorable (a) and favorable orientation (b).

Acknowledgements

The financial support is acknowledged of the Italian Department of Emergency Management through the ReLUIS project: Development of new materials for seismic retrofitting.

References

[1]   Canbolat, B.A., Parra-Montesinos, G.J., Wight, J.K.: Experimental study on the seismic behavior of high-performance fiber reinforced cement composite coupling beams. ACI Structural Journal 2005, 102(1): pp. 159-166.

[2]   di Prisco, M. and Mazars, J: Crush-crack: a non-local damage model for concrete. Journal of Mechanics of Cohesive and Frictional Materials, 1996, 1: pp. 321-347.

[3]   di Prisco, M., Lamperti, M.G.L., Lapolla, S. (2010) On Double Edge Wedge Splitting test: preliminary results. In: Oh B.H. et al. (eds.) Proceedings FraMCoS 7: pp. 1533-1540.

[4]   Ferrara, L., di Prisco, M.: Mode I Fracture behavior in concrete: Nonlocal damage modeling. ASCE Journal of Engineering Mechanics 200: pp. 678-692.

[5] Ferrara, L., Faifer, M.; Muhaxheri, M.; Toscani, S.: A magnetic method for non destructive monitoring of fiber dispersion and orientation in SFRCC: part 2: correlation to tensile fracture toughness. Materials and Structures, accepted 2011.

[6] Ferrara, L., Ozyurt, N., di Prisco, M.: High mechanical performance of fibre reinforced cementitious composites: the role of "casting-flow induced" fibre orientation. Materials and Structures 2011, 44: pp. 109–128.

[7] Guedes, J., Pegon, P., Pinto, A.V.: A Fibre/Timoshenko Beam Element in CASTEM 2000. Special Publication Nr. I.94.31, European Commission, JRC, Ispra (VA), Italy 1994.

[8] Lequesne, R.: Behavior and design of high-performance fiber-reinforced concrete coupling beams and coupled-wall systems. Ph.D. Thesis, University of Michigan, USA. 2011, pp. 298.

[9] Mechtcherine, V.: Towards a durability framework for structural elements and structures made of or strengthened with high-performance fibre-reinforced composites. Construction and Building Materials, 2012; 31: pp. 94-104.

[10] Martinola, G., Meda. A., Plizzari, G.A., Rinaldi, Z.: "Strengthening and repair of RC beams with fiber reinforced concrete", Cement & Concrete Composites 2010; 32: pp. 731-739.

[11] Mazars, J., Kotronis, P., Ragueneau, F., Casaux, G.: Using multifiber beams to account for shear and torsion - Application to concrete structural elements. Computer Methods in Applied Mechanics and Engineering. 2006; 195: pp. 7264-7281.

[12] Naaman A.E., Reinhardt H.W.: Proposed classification of HPFRC composites based on their tensile response. Materials and Structures 2006; 39: pp. 547-555.

[13] Shin, S.K., Kim, K., Lim, Y.M.: Strengthening effects of DFRCC layers applied to RC flexural members. Cement and Concrete Composites 2011; 33: pp. 328-333.

Reliability of highly stressed UHPC slender columns

Martin Heimann

Institut für Massivbau,
Technische Universität Darmstadt
Petersenstraße 12, 64287 Darmstadt, Germany
Supervisor: Carl-Alexander Graubner

Abstract

Ultrahigh-performance concrete (UHPC) has been used successfully since the early 1990s. Meanwhile concrete-technological developments have allowed production of UHPC with a compressive strength of more than 150 N/mm². Thus, the field of application of reinforced concrete constructions currently in use can be extended. In particular, the cross-sectional dimensions required for highly stressed compression members can be reduced. However, in many cases this positive development increased the trend to even more slender and therefore likelier to buckle structural systems. Hence, it is to be expected that under certain conditions the safety level of such construction elements decreases. Because of this, the reliability of slender UHPC structural members is being investigated in a research project, supported by the Deutsche Forschungsgemeinschaft (DFG), at the Technische Universität Darmstadt. The purpose of this project is the calibration of safety factors in order to achieve the target reliability for safe and efficient design. Realistic modelling of the load-bearing behaviour of slender UHPC columns requires material and geometric nonlinearities to be considered. Due to the load-dependent material behaviour, a sudden stability failure may occur long before the material's strength is exceeded. An adequate analysis of this phenomenon calls for realistic material laws. In this context a special FE-tool which takes into account this phenomenon was developed for stochastic simulations. As a first step, calculation results at cross-sectional level can be used to identify the sensitivity of the parameters. Therefore, specific moment-curvature-diagrams, which are based on stochastic simulation methods, are being developed to identify the influence of concrete compressive strength on the load-carrying behaviour of a structural member with rectangular cross-section. Moreover, the FE-Model is verified by experimental test results of slender columns. By applying the simulation method of adaptive importance sampling (AIS), preliminary results of the reliability analysis are presented in this paper. On the basis of the results of this research project, innovation potentials and limits of application of UHPC will be pointed out.

1 Modelling of slender UHPC columns

1.1 General

Realistic modelling of the load-bearing behaviour of slender reinforced concrete columns requires material and geometric nonlinearities to be considered. Material nonlinearity describes stiffness reduction with increasing load intensity while geometric nonlinearity results from second order effects. Due to the load-dependent material behaviour (moment-curvature-relationship) a sudden stability failure may occur long before the material's strength is exceeded. In this case, we talk about stability failure due to load-dependent stiffness reduction. Correct analysis of this phenomenon calls for realistic material laws both for the stress-strain-relationships of concrete and reinforcing steel as well as for the variation of the modulus of elasticity. The stochastic finite element program developed within this research project takes this phenomenon accurately into account and has been checked by comparison of numerical and experimental results.

1.2 Material

Concerning the probabilistic model of ultrahigh-performance concrete it has to be mentioned that in the past statistical information about the scatter of concrete compressive strength was rarely available. Differing from data in the assessment report UHPC [1] the results of investigations under the DFG priority programme SPP1182 point to a larger scatter with increasing compressive strength. Investigations of several experimental test series at the research laboratory for concrete structures at the Tech-

Proc. of the 9th *fib* International PhD Symposium in Civil Engineering, July 22 to 25, 2012,
Karlsruhe Institute of Technology (KIT), Germany, H. S. Müller, M. Haist, F. Acosta (Eds.),
KIT Scientific Publishing, Karlsruhe, Germany, ISBN 978-3-86644-858-2

nische Universität Darmstadt confirm these results. Based on the results of the experimental test series, idealized stress-strain-relationships were formulated for two different mixtures of UHPC (cp. Fig. 1). The scatter in the area of increasing stress-strain-relationship can be neglected, but in the area of decreasing stress-strain-relationship the scatter is distinctive. Therefore, mean stress-strain-relationships were formulated with limit points $\sigma_{cu}/\sigma_{c1}=0{,}125$ and $\varepsilon_{cu}/\varepsilon_{c1}=2{,}305$ for UHPC with fine grain aggregate and $\sigma_{cu}/\sigma_{c1}=0{,}183$ and $\varepsilon_{cu}/\varepsilon_{c1}=2{,}136$ for UHPC with basalt aggregate.

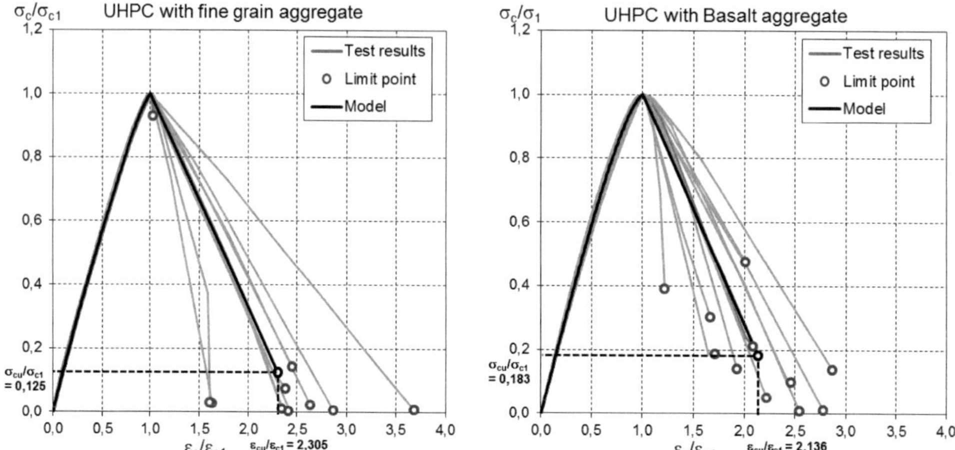

Fig. 1   Standardised stress-strain-relationship of UHPC with fine grain aggregate and with basalt aggregate

In addition to this assumption, the following relationship between the secant modulus of elasticity E_c at approximately $\sim 0.4{\cdot}f_c$ and compressive strength fc has been used.

$$E_c = \alpha_E \cdot f_c^{1/3} \tag{1}$$

The mean value of the parameter α_E was set to 9.350 [-] for both concrete mixtures. A lognormal distribution with a coefficient of variation of $v_{\alpha E} = 0.05$ [-] for UHPC with fine grain aggregate and $v_{\alpha E} = 0.045$ [-] for UHPC with basalt aggregate was used to fit the test results. The basic variable α_E thus represents the scatter of modulus of elasticity E_c independent of compressive strength f_c.

The yield strength of reinforcing steel f_y is described by a lognormal distribution with a coefficient of variation of $v_{fy} = 0.06$ [-]. The mean value for steel grade S500 amounts to $\mu_{fy} = 550$ [MPa]. The distribution type is normal according to the recommendation of the JCSS Probabilistic Model Code P.3 [2]. The nominal cross-sectional dimensions h_{nom}/b_{nom} correspond to their mean values $\mu_{h,b}$. The standard deviation depends on the size of the cross-sectional dimensions and is assumed to be $\sigma_{h,b} = 5$ [mm]. A normal distribution with a mean value equal to the nominal value and a standard deviation of 5 [mm] was used as stochastic model for the concrete cover.

1.3   Modelling of loads

Due to the strong scatter of live loads, their influence on the structural reliability is large compared to the influence of dead loads which can be easily predicted over the service life of the structure. Most standards and guidelines require that the characteristic value of the extreme value distribution of imposed loads does not exceed the 95%-ile in an observation period of 1 year. This means that the code value can statistically be exceeded only once in 50 years.

According to the Probabilistic Model Code P.2 [2] the weight density of a structural part is assumed to be Gaussian distributed and a coefficient of variation of $v_G = 0.1$ is chosen for the reliability analysis. In the case of imposed loads the distribution of the extreme values is represented by a Gumbel distribution (EV I) with a coefficient of variation of $v_Q = 0.4$.

1.4 Stochastic parameters

Table 1 shows the statistical parameters and adopted probabilistic density functions of the basic variables. UHPC-specific parameters were determined by experimental testing. All remaining parameters were modelled according to the Probabilistic Model Code [2]. Model uncertainty is one of the most important influence parameters. Initial results of the test evaluations suggest that due to the accuracy of the FE model developed, the coefficient of variation may be reduced to 0.07. According to the Probabilistic Model Code [2] a coefficient of variation of $v_m = 0.1$ was selected.

Table 1   Statistical parameters of the basic variables

class	Variable	Distribution	Mean value	Std.-deviation	Var.-coefficient
concrete	$f_c^{\,fine}$	LN	>150 N/mm²	-	0,06
	$f_c^{\,basalt}$	LN	>150 N/mm²	-	0,06
	$\varepsilon_{c1}^{\,fine}$	LN	-3,80 ‰	-	0,10
	$\varepsilon_{c1}^{\,basalt}$	LN	-3,92 ‰	-	0,075
	$\varepsilon_{c1u}^{\,fine}$	LN	-8,79 ‰	-	0,326
	$\varepsilon_{c1u}^{\,basalt}$	LN	-8,40 ‰	-	0,258
	$\alpha_{ct}^{\,UHPC}$	LN	0,45	-	0,15
	$\alpha_E^{\,fine}$	LN	9.350	-	0,049
	$\alpha_E^{\,basalt}$	LN	9.350	-	0,045
reinforcement	f_y	LN	550 N/mm²	-	0,06
	A_s	N	$A_{s,nom}$	-	0,02
geometry	h	N	h_{nom}	5 mm	-
	b	N	b_{nom}	5 mm	-
	c	N	a_{nom}	5 mm	-
load	G	N	G_m	-	0,1
	Q	Gumbel	$Q_k/1{,}746$	-	0,4
model uncertainty ξ		N	1,00	-	0,10

1.5 Finite element model

The calculation method acc. to Tran [5], developed for stochastic analysis, is based on a combination of finite element method, transfer matrices method, deformation-based method and an algorithm of cross-section calculation. The numerical method at cross-sectional level is based on an optimised Gaussian-Integral cross-section calculation which takes into account the UHPC-specific material law. The stress-strain-relation of concrete for tensile strength is implemented by the model of Quast [6].

The finite element method used at system level is adapted by the procedure of field transfer matrices which allows a significant reduction of computational time. The great benefit of this procedure lies in the flexibility of load arrangement and boundary conditions for slender columns. In the case of non-linear stiffness distribution in the column's longitudinal direction, a realistic determination of stiffness can be achieved by appropriately dividing the slender column lengthways. Division by iterative steps is done adaptively as a function of stiffness gradients. The system of equations is solved by means of the Newton-Raphson iteration method.

2 Verification of the finite element model

Accompanying the numerical calculations, experimental tests were carried out on slender columns. In order to test columns with a high ratio of slenderness ($\lambda \geq 80$) the cross-sectional dimensions have to be relatively small (h/b/l = 12cm/12cm/278cm) due to the capacity of the laboratory's mechanical equipment. The columns are loaded in a position-controlled component test by a hydraulic cylinder. During the test, deflections and compressions of reinforcement, concrete and structural system are measured with strain gauges and displacement sensors at more than 20 measuring points. In addition to these tests, investigations of compressive strength, flexural strength and modulus of elasticity of UHPC were carried out.

Based on recorded measurements of the experimental tests, the results of the FE-Model have been verified. Taking into account all measurable material and geometric parameters of the testing system, the recalculation shows an excellent approximation of the computational results to the experimental data (cp. Fig. 2).

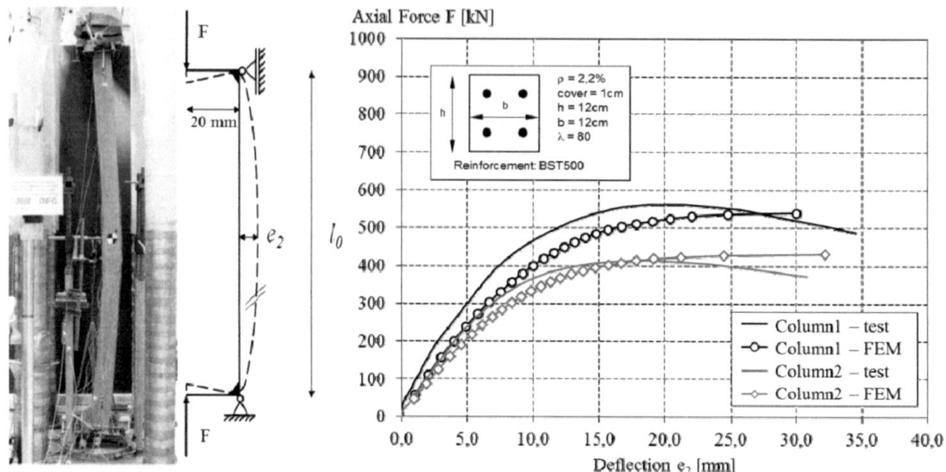

Fig. 2   Comparison of the computational results and the experimental data

3   Identification of important basic variables

3.1   General

In order to identify the sensitivity of important variables, a parametric study was carried out at cross-section level. The study was split into three parts. In part 1 specific moment-axial force diagrams were constructed to analyse the scatter of m-n-interaction. In part 2 moment-curvature-relationships were calculated to identify important variables for structural members made of UHPC. Finally the calculation results were compared with results based on material laws according to EC-2 Part 1-1 [3]. The study was carried out for structural members with rectangular cross-section and will be enhanced to members with circular and hollow cross-section. In the following section the results of part 3 will be presented in detail.

3.2   Parametric study at cross-section level

The analysis of part 3 is based on the concept of part 2 of our investigations. It must be mentioned here that the intention of part 2 was to identify important variables in the m- κ-behaviour of structural UHPC members. For this purpose, a calculation of scattering moment-curvature-diagrams was carried out with the help of stochastic finite element calculations (cp. Fig.3). The following variables have been modelled as random: concrete strength, yield and tensile strength of reinforcement, section height, cross sectional-width, concrete cover, diameter of reinforcing bar, maximum concrete strain and the modulus of elasticity of concrete. The diagram has been drawn for ν = -0.4. The solid line ellipses around the characteristic points (compressive yield point, concrete crack point, yield point, bearing capacity limit) define the 95%-fractiles. To start with, only one basic variable per sample was considered as random and different scattering diagrams were calculated for each variable.

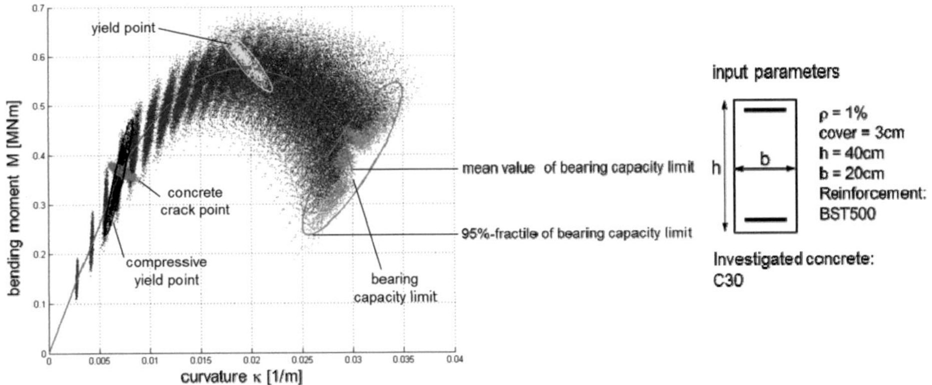

Fig. 3   Scattering moment-curvature-diagram ($\nu = -0.4$)

In order to compare these results and estimate the influencing variables, two parameters were calculated. The first parameter (V_{Rmax}) is the coefficient of variation of the bearing capacity limit. The second parameter (V_{Mmax}) is the coefficient of variation of the maximum moment.

Figure 4 illustrates exemplarily the coefficient of variation of the bearing capacity limit V_{Rmax} for NSC, HSC and UHPC. It can be seen, that at cross-sectional level the variations of concrete strength, concrete cover, cross-sectional width and section height are much more significant with UHPC compared to NSC and HPC. Considering all variables as random emphasizes the importance of concrete strength, because V_{Rmax} is approximately the same when all variables or concrete strength alone are taken as random.

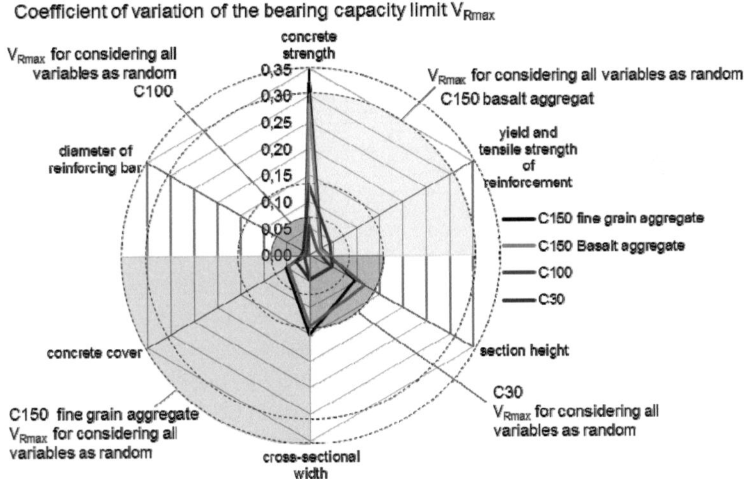

Fig. 4 V_{Rmax} for NSC, HSC and UHPC

4   Reliability analysis

4.1   General

As previously described, it has to be expected that the load-dependent material behaviour of UHPC may cause a sudden stability failure long before the material's strength is exceeded. Because of this phenomenon, a reduced reliability level is to be expected as already shown by Schmidt/Six [7] for HSC. This case of failure is not considered in the current design concept for concrete columns. In order to identify the reliability level, stochastic analyses at system level were carried out considering the characteristics of UHPC. The reliability analysis concentrates on short (slenderness ratio $\lambda = 0$)

and slender (slenderness ratio $\lambda = 100$) UHPC columns with a nominal compression strength of $f_{ck} = 150$ MPa.

4.2 Adaptive importance sampling method

A fundamental result of the reliability analysis is determination of the reliability index β. This requires calculation of the failure probability p_f . Due to the great number of basic variables as well as the non-linear limit-state function, failure probability p_f cannot be determined analytically. Therefore, with the help of the Monte Carlo simulation, an estimated value for p_f is determined, which tends to the exact answer given an appropriately large number of samples. Since in the case in question $p_f \approx 10^{-6}$ is to be expected, an extremely high number of samples are required. By contrast, the method of adaptively weighted simulations (adaptive importance sampling = AIS) allows an estimate of p_f with a markedly reduced sample size. The estimator for p_f can then be expressed as follows:

$$\hat{p}_f = \frac{1}{N} \sum_i^N I\left[g(x_i) < 0\right] \frac{f_x(x_i)}{h_x(x_i)} \tag{2}$$

where $I[g(x_i)]$ is the indicator function for failure related to the limit state function $g(x_i)$, $f_x(x_i)$ describes the original distribution and $h_x(x_i)$ the distribution of the importance sampling function (IS) of the basic variables x_i. By dividing the calculation of p_f into several steps, the estimator function can be adapted after every step. For this purpose equation 3 gives the adaptive expectation vector in the failure range.

$$\hat{m}_{x_j} = \frac{1}{\hat{p}_f} \frac{1}{N} \sum_i^N x_{j,i} \cdot I\left[g(x_i) < 0\right] \frac{f_x(x_i)}{h_x(x_i)} \tag{3}$$

Fig. 5 illustrates how the AIS method functions using an example with two normally distribution basic variables and X_{id} as design point.

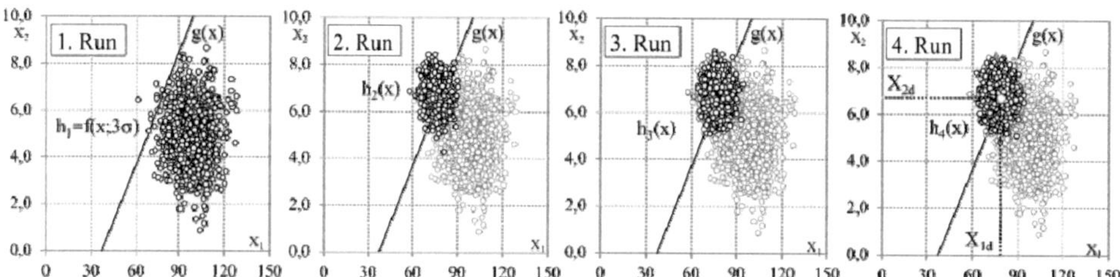

Fig. 5   Example of adaptive importance sampling

4.3 Preliminary results of the reliability analysis

Figure 6 shows preliminary results of the reliability analysis for a short (slenderness ratio $\lambda = 0$) and a slender (slenderness ratio $\lambda = 100$) column of NSC and UHPC. The geometric parameters of the column and the cross-section are given on the right-hand side. The results show precisely the expected effects of decreasing reliability with increasing slenderness. The minimum value of the reliability index can be determined, given a reference period of 1 year, as $\beta_{MIN,1a} = 4,1$. In the case of NSC the minimum value is adhered to both at cross-sectional and at system level. The UHPC slender column reaches lower reliability indices even at an eccentricity of $e/h = 0.7$, which fall short of the standard safety level required. This finding from initial results of the reliability analysis thus justifies further investigations of other systems with different column cross-sections.

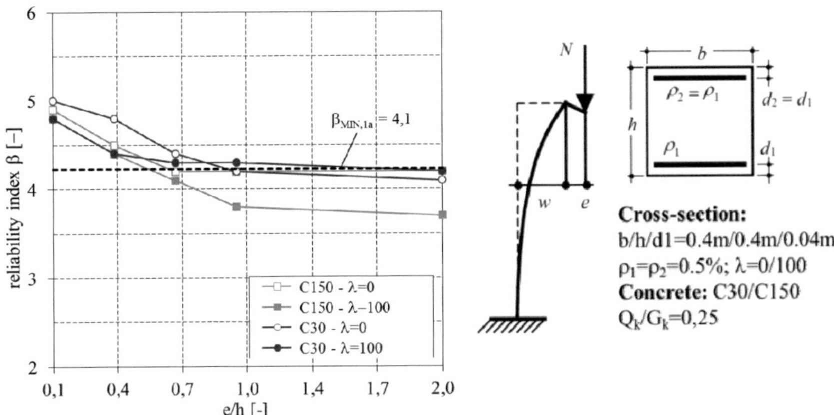

Fig. 6   Reliability index β depending on eccentricity e/h for an NSC and a UHPC column

5   Conclusions

This paper deals with probabilistic modelling and experimental testing of UHPC columns. The reliability analysis of slender UHPC columns considered new statistical parameters concerning UHPC compressive strength and concrete strain, as well as the modulus of elasticity. Stochastic simulations were carried out, taking into account the special characteristics of UHPC at cross-sectional level. As expected, the results of the parametric studies identified concrete strength as highly sensitive, while the variables of reinforcement are insignificant. After that, a finite element model for extending the results from cross-sectional to system level is presented. The finite element method used at system level based on the displacement method and is adapted by the procedure of field transfer matrices. The reliability analysis is realized by the Adaptive-Importance-Sampling-method (AIS) in combination with a stochastic finite element method. First results of a reliability analysis on slender columns show, that slender UHPC columns do not achieve the required safety level according to EN 1990 [4]. After completion and verification of the stochastic model at system level, the relevant influencing factors will be identified and a proposal for the design of slender UHPC columns will be presented.

References

[1]   DAfStb, Sachstandsbericht Ultrahochfester Beton (Heft 561). Berlin: Beuth Verlag, 2008

[2]   JCSS Probabilistic Model Code Part 3 Resistance Models, Joint Committee of Structural Safety, 2001

[3]   EN 1992-1-1 EUROCODE 2: Design of concrete structures - Part 1-1: General rules and rules for buildings, European Committee for Standardization, 2012.

[4]   EN 1990 EUROCODE 0: Basis of structural design, European Committee for Standardization, 2012.

[5]   Tran N.L.: Berechnungsmodell zur vereinfachten Abschätzung des Ermüdungsverhaltens von Federplatten bei Fertigträgerbrücken, Dissertation Heft 20, Technische Universität Darmstadt – Institut für Massivbau, Mai 2011

[6]   DAfStb, Programmgesteuerte Berechnung beliebiger Massivbauquerschnitte unter zweiachsiger Biegung mit Längskraft (Heft 415), Beuth-Verlag, Ed., 1990.

[7]   Six M., Schmidt H.: Probabilistic Modelling of HSC Slender Columns in High-Rise Buildings, IABSE Symposium 2008, Creating and Renewing Urban Structures, Chicago, USA, 2008

Verification of details of existing structures with the elastic-plastic stress field method

Galina Argirova

Structural Concrete Laboratory IBETON,
Ecole Polytechnique Fédérale de Lausanne (EPFL),
Station 18, CH-1015 Lausanne, Switzerland
Supervisor: Aurelio Muttoni

Abstract

The verification of critical details of existing structures is generally performed with strut-and-tie models. An alternative widely used in some countries, such as Switzerland or Denmark, is the rigid-plastic stress fields (RPSF) method (based on the lower bound theorem of plasticity) which allows for the consistent and systematic analysis of a variety of structural elements and details. It however requires a certain level of experience to be applied to a wider range of cases and in order to consider the various load-carrying actions. Therefore, the verification of existing structures requires a more systematic method. The elastic-plastic stress field method (EPSF) was developed at EPFL in order to overcome the limitations of the rigid-plastic stress field method. It takes into account the resistance of the concrete as a function of the deformations and has proven to give good estimates of the strength of reinforced concrete members.

In the framework of this project it is intended to study the applicability of the EPSF and the RPSF for the assessment of certain complicated details specific for concrete bridges. In order to do that, a large amount of performed tests available in the literature will be analysed with the two methods. Some of the details that are studied are zones of introduction of concentrated forces, cantilevers with dapped ends, diaphragms, and the extreme ends of beams with poor anchorage of the reinforcement which can simulate the behaviour of prefabricated beams. The results will then be used to develop guidelines for the choice of the coefficient accounting for the reduction of the compressive strength of the concrete due to transversal strains and the angle of the inclination of the compressive struts.

1   Introduction

Critical details in bridges can be regarded as the potentially weakest members of a structure, thus governing its strength. This is typically the case of geometrical discontinuity of the cross section or the reinforcement provided or the action of concentrated loads (fig 1). In order to better understand the behaviour of such regions a large number of tests have been performed over the decades and simple strut-and-tie models have been proposed by many researchers. More refined analysis of the discontinuity regions can be performed with RPSF accounting for the actual reinforcement layout. However, the concrete strength is still modelled based on rules. The EPSF allows for the detailed modelling of both the reinforcement and the concrete, and is a promising tool for such analysis.

Proc. of the 9th *fib* International PhD Symposium in Civil Engineering, July 22 to 25, 2012,
Karlsruhe Institute of Technology (KIT), Germany, H. S. Müller, M. Haist, F. Acosta (Eds.),
KIT Scientific Publishing, Karlsruhe, Germany, ISBN 978-3-86644-858-2

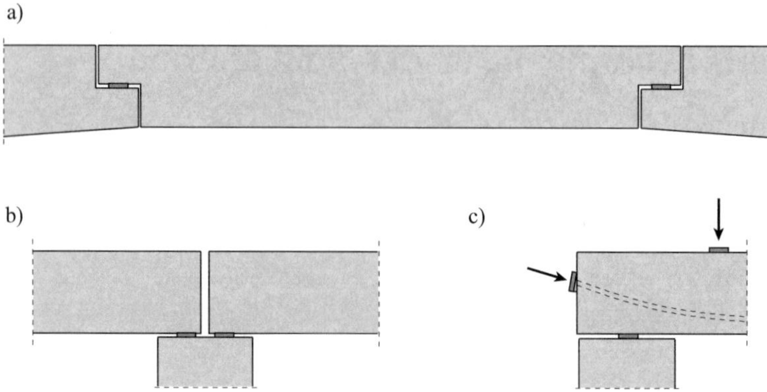

Fig. 1   Discontinuity regions typical for concrete bridges a) dapped end b) insufficient anchorage c) concentrated forces due to prestressing and traffic

It is the goal of this research, by using the EPSF method, to develop RPSF models to be solved by hand of common for concrete bridges details, to verify the applicability of the EPSF method for these details and to provide guidance for the development of more accurate strut-and-tie and RPSF models.

2   Numerical modelling with EPSF method

The Elastic Plastic Stress Field (EPSF) method was developed at EPFL, Switzerland and was implemented in a computer program - *jconc* [1], [2]. The access for students at EPFL and practicing engineers is free (http://i-concrete.epfl.ch) and the source code of the program can also be downloaded. This program provides an accurate and quick solution for 2D engineering static problems and for the analysis only two parameters are required - the strength and the modulus of elasticity of the materials.

The main assumptions of the elastic-plastic stress field method (EPSF)(fig. 2) are:

- Mohr-Coulomb yield surface for plane stress with tension cut-off
- elastic perfectly plastic behaviour of concrete in compression
- perfect bond between concrete and reinforcement

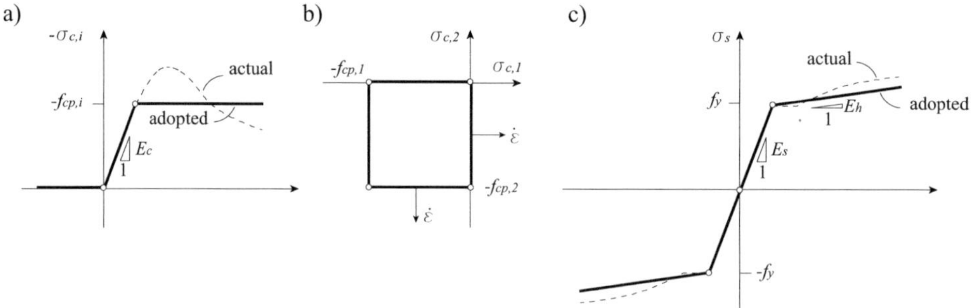

Fig. 2   a) constitutive model for the concrete b) Mohr-Coulomb yield surface c) constitutive model for the steel

To account for the brittleness of the concrete (fig 2a) the effective plastic concrete strength in compression is calculated as:

$$f_{cp} = f_c \cdot \eta_{fc} \tag{1}$$

with the coefficient η_{fc}, as introduced by Muttoni [3] :

$$\eta_{fc} = \left(\frac{30}{f_{ck}}\right)^{1/3} \leq 1 \tag{2}$$

The effect of the transversal strains ε_1 on the compressive struts in the concrete is accounted for with the coefficient k_c, as introduced by Vechio and Collins [4]:

$$k_c = \frac{1}{0.8 + 170 \cdot \varepsilon_1} \leq 1 \tag{3}$$

For the steel a bilinear constitutive law is introduced in the program with the possibility to account for strain-hardening.

3   Applicability of the EPSF method

In order to validate the applicability of the EPSF method a wide range of test specimens were selected. In the following paragraphs some of the selected test programs are discussed and the results obtained with the program *jconc*, based on the EPSF method are presented in Table 1.

Kuchma performed a total of 20 tests on high-strength prestressed bulb-tee girders [5]. The results from specimens G10E and G10W (which were strengthened with FRP sheets) and specimen 8E (in which an interface in the concrete was done by steel plates) are not discussed in the statistics because at present they are out of the scope of the program *jconc*.

Kaufmann and Ramirez performed tests on a total of 3 AASHTO Type II and 6 Type I prestressed girders [6]. The goal of the test program was to validate the current provisions for the use of high-strength concrete. The specimens failed due to insufficient anchorage, shear and bending.

Leonhardt and Walther tested a total of 15 T-beams with high amount of shear reinforcement and the anchorage length of the longitudinal reinforcement beyond the support was varied [7]. The results from specimens TA7 and TA8 are not included in the statistics because almost no longitudinal reinforcement was provided at the support.

Sagaseta and Vollum tested a total of 14 rectangular beams [8] (four of the beams were without shear reinforcement and they are not included in the final statistics). The main parameters varied in the test were the aggregate size and the loading arrangement. The goal of the experimental program was to investigate the effect of the compression zone and to compare the results with similar tests on T- and I-sections.

Table 1   Statistical analysis of the results from the analysis of specimens from literature with EPSF

Selection		
Bulb-Tee Girders	# specimens included	17
	Average	1.08
	COV, %	4.75
AASHATO Type I and II Girders	# specimens included	9
	Average	1.09
	COV, %	9.96
T-beams	# specimens included	13
	Average	1.10
	COV, %	6.58
Rectangular beams	# specimens included	10
	Average	1.04
	COV, %	7.97

For the assessment of the applicability of the EPSF method, the ultimate strength of the element was not the only parameter investigated. Additionally, a detailed analysis of the distribution of the angle of the compression struts θ and the coefficient k_c (Eq.3) is performed, as well as comparison with observed failure modes and measurements. The results of a test specimen from the test campaign [5] obtained with the program based on the EPSF method are presented in Figure 3. The mesh was refined so no change in the ultimate load was observed as the element size was reduced.

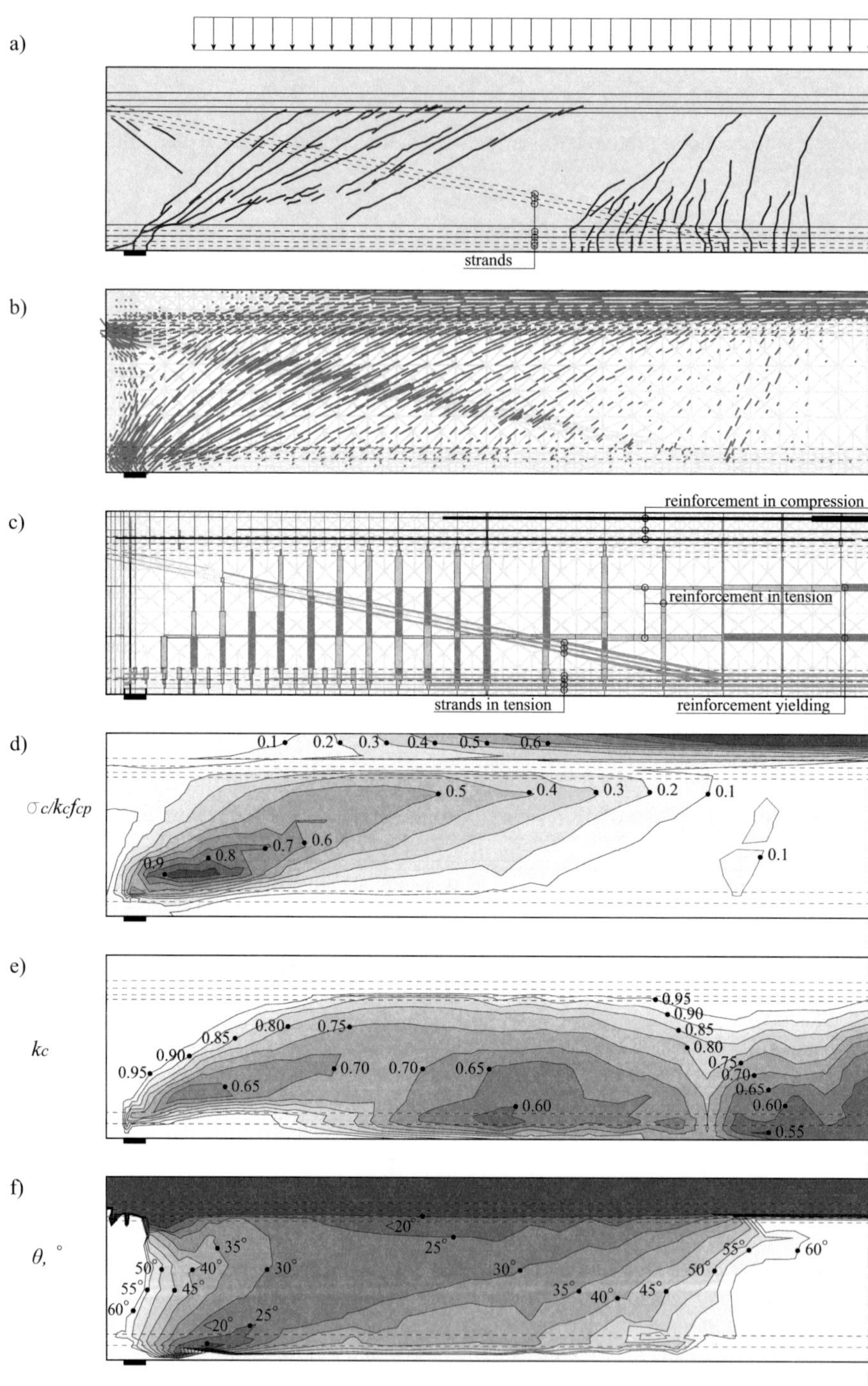

Fig. 3   Girder 1 East End, [3] a) cracking pattern and loads; b) direction of principal compressive stresses c) stresses in the reinforcement and the strands d) degree of use of concrete strength e) distribution of the coefficient k_c f) angle of principle compressive stresses in the concrete

4   Development of RPSF models

In order to present the development of the RPSF models an example of a specimen tested by Rainer and Elbadry [9] will be discussed. In this experimental program a total of seven dapped end specimens with inclined and straight shear reinforcement were tested. The ultimate load bearing capacity of all specimens was calculated with a strut-and-tie model and the average ultimate load predicted was approximately 70% of the actual one. The reinforcement layout and the strut-and-tie model used for the analysis of specimen A are presented in Figure 4.

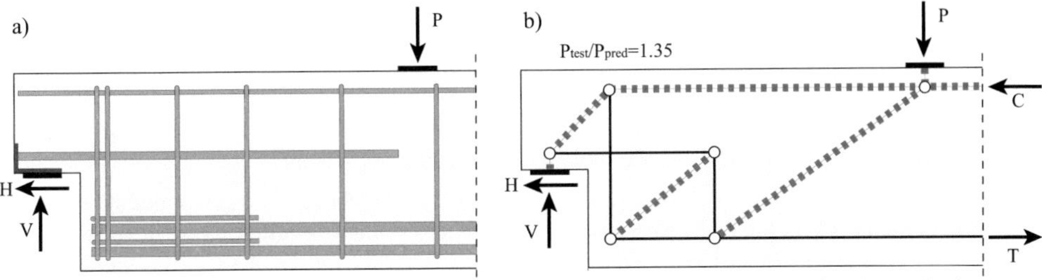

Fig. 4   Dapped End A a) reinforcement layout; b) strut-and-tie model used for the analysis

The specimen A was analysed with the program *jconc* and the results are presented in Figure 5a. It can be observed that the load transfer path obtained with the program differs from the one originally proposed in [9]. The additional concrete strut can be accounted for only if the detailed reinforcement layout is considered. The presence of this additional load-transfer action can be observed also in the crack pattern of the specimen.

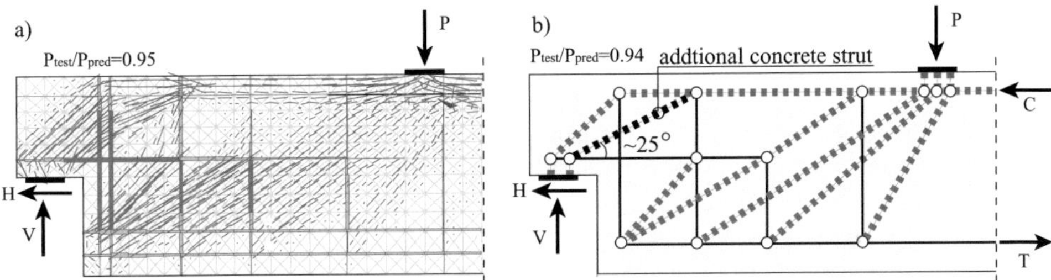

Fig. 5   Dapped End A a) model in *jconc*; b) refined strut-and tie model

Based on the refined strut-and-tie model (fig.5b) a RPSF model was developed (fig.7b), where a coefficient of k_c equal to 0.6 as usually adopted for such design was taken. The presence of wide crack openings in the region between the nib and the beam was observed during the test. From the analysis of the distribution of the coefficient k_c with *jconc* (Figure 6a), it can be concluded that the program can accurately predict the large transversal strains in this region which lead to wider crack openings (k_c=0.3).

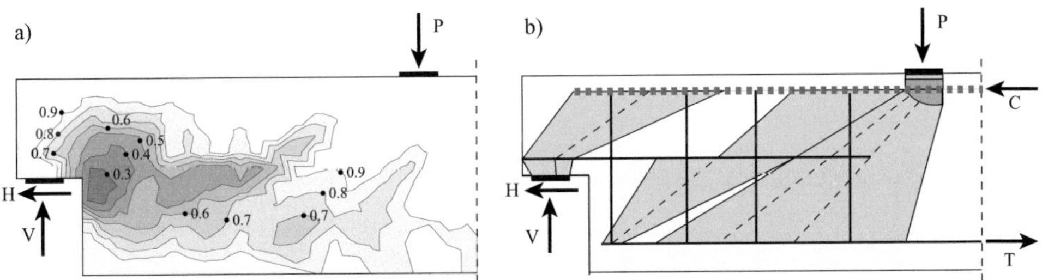

Fig. 6   Dapped End A a) distribution of the coefficient k_c; b) RPSF model

5 Conclusions

From the analysis of the results it was concluded that for certain critical details (such as insufficient anchorage of prestressed girders with sufficient shear reinforcement) the EPSF can accurately predict not only the load bearing capacity but also the mode of failure. Additionally, the computer program *jconc,* based on the EPSF method, is a valuable tool which allows for the development of rigid- plastic stress field models for common details.

Future work consists of analysis of a wider range of typical for reinforced concrete bridges details and investigation of different parameters such as concrete strength, reinforcement ratio and layout, type of loads, etc.

References

[1]   Fernandez Ruiz, M., Muttoni A., "On development of suitable Stress Fields for structural concrete", ACI Structural Journal, vol.104, no.4, 2007, pp. 495-502

[2]   Kostic, N., "Topologie des champs des contraintes pour le dimensionnement des structures en béton armé"(Typology of stress fields for the dimensioning of reinforced concrete structures) PhD Thesis, Lausanne, EPFL, 2009

[3]   Muttoni, A., "Die Anwendbarkeit der Plastizitätstheorie in der Bemessung von Stahlbeton", Institut für Baustatik und Konstruktion, ETH Zürich, IBK Bericht Nr. 176, Birkhäuser Verlag, Basel, 1990, 158p.

[4]   Vecchio, F.J., Collins, M.P., "The modified compression field theory for reinforced concrete elements subjected to shear", ACI Journal, Vol.83, No.2, 1986, pp.219-231

[5]   Kuchma, D., Kim, K.S., Nagle, Th.J., Sun Sh., Hawkins, N.M., "Shear tests on high-strength prestressed bulb-Tee girders: strength and key observations", ACI Structural Journal, vol. 105, 2008, pp. 358-367

[6]   Kaufman M.K., Ramirez J.A. "Structural behavior of high strength concrete prestressed I-beams", West Lafayette, Indiana: Purdue University, 1988.

[7]   Leonhardt F., Walther R., "Schubversuche an Plattenbalken mit unterschiedlicher Schubbewehrung", Deutscher Ausschuss Für Stahlbeton, Berlin, 1963

[8]   Sagaseta J., Vollum R.L. "Influence of beam cross-section, loading arrangement and aggregate type on shear strength", Magazine of Concrete Research, 2011

[9]   Rainer, H., Elbadry M.M., "Stud reinforcement in dapped ends of concrete bridge girders", The 2004 Concrete Bridge Conference, Charlotte NC, USA

Load Carrying Behaviour and Performance of Masonry Structures

Analytical model for cost optimization of climate-neutrally operated office and administration buildings

Achim Knauff

Institute for Concrete Structures, Building Materials and Masonry,
Technical University Darmstadt,
Petersenstraße 12, 64287 Darmstadt, Germany
Supervisor: Carl-Alexander Graubner

Abstract

This research study includes the development of an analytical model for the optimization of investment and operation costs of energy supply solutions for climate-neutrally operated buildings. Investigations presented in this study include photovoltaic, solar thermal and geothermal technologies. In addition, location conditions will be considered. The solution to the optimization problem will be developed in two steps. In a first step, the three technologies (photovoltaic, solar thermal and geothermal) are merged into a verified tool which enables the simulation of processes of renewable energy plants. The developed simulation tool will be used for modeling of realizable energy supply configurations for different locations. In a second step, dependencies between the costs for the technologies and their contribution to meet the total energy demand will be considered for each technology. These dependencies will be established with parameters which are as well independent as dependent from the local conditions. Based upon these considerations, this article will describe the implementation of an analytical model for the optimization of costs. All necessary intermediate steps will be explained using examples of geothermal plants.

1   Introduction

Motivation of this research study is the political climate protection targets of the European Union. By the year 2020 the "climate-neutral building standard" will be required for all new erected buildings. Already existing buildings have to be almost climate-neutral by the year 2050. The implementation of these demands means a more rational use of primary energy for buildings of tomorrow. Beside an energy-optimized building envelope and an intelligent building technology, a renewable energy supply and an interaction with the energy infrastructure at the site is particularly required. In addition, to ecological advantages of buildings which are operated in a climate-neutrally way, a holistic view of the target comprises economic aspects as well. Therefore, the goal of this research study is to find economic optimized energy supply solutions for operating office and administrative buildings in a climate-neutral way. Furthermore, related requirements of the building services engineering should be identified.

2   Methodology

This research study involves the development of an analytical model for the optimization of costs for climate-neutrally operated buildings in two steps. Step one includes the development and validation of a simulation tool for photovoltaic, solar thermal and geothermal plants. The development and validation required a diligent collection and evaluation of the basis of computation [1]. In order to model photovoltaic, solar thermal and geothermal plants by using a simulation tool, the plant-specific behavior of photovoltaic generators, inventers, solar collectors, stratified buffer storages, geothermal probes and heat pumps has to be described by algorithms. These algorithms can be used to determine the exact plant condition time stepwise [2]. The advantage of implementing all technologies in one simulation tool is that the basic conditions, such as climate data sets and the extent of used climate parameters, are identical. This is important for subsequent comparative examinations. The developed simulation tool can be then used to model renewable as well as realizable energy supply configura-

Proc. of the 9th *fib* International PhD Symposium in Civil Engineering, July 22 to 25, 2012,
Karlsruhe Institute of Technology (KIT), Germany, H. S. Müller, M. Haist, F. Acosta (Eds.),
KIT Scientific Publishing, Karlsruhe, Germany, ISBN 978-3-86644-858-2

tions for different locations. In addition, the simulation tool calculates and assesses the site-specific contribution of renewable energies which are available to cover the respective energy demand.

For the coverage of the total energy demand of a building it is necessary to find the optimal combination of the photovoltaic, solar thermal and geothermal technologies. Instead of performing many iterative calculations for the simulation, it would be an advantage to use parameterized equations. In order to describe a technology sufficiently by using a parameterized equation, all parameters have to be analyzed and all essential site-dependent and site-independent variables have to be determined. These parameter studies can be performed using the developed simulation tool. The impact of each identified variable has then to be evaluated. Results of these parameter studies, e.g. linear, potential or exponential correlation of the single variables and the respective impact for the energy demand, have then to be summarized within a non-linear equation.

In a further step, the correlation of the single variables and their impact on investment and operating expenses of the respective technology must be defined to be able to optimize all arising costs. The result is a relation between the percentage contribution of a technology to the coverage of the energy demand and the investment and operating expenses depending on the given parameters. The optimization of costs while using a combination of photovoltaic, solar thermal and geothermal technologies can be derived from considering the extreme values of the costs as a function of the proportion of the total energy demand delivered by each technology. By taking all these conditions into account for the implementation of an analytical model it would be possible to give site-specific recommendations for an economical optimized combination of renewable energies.

3   First results using the example of geothermal plants

The implementation of the analytical model requires the evaluation of the variables of each technology by performing parameter studies. Prior to these studies for photovoltaic, solar thermal and geothermal plants, boundary conditions have to be defined. The type of net energy demand, which differs for heating, ventilation, cooling and illuminating, as well as the total amount of the existing net energy demand have to be considered as boundary conditions. It also has to be taken into account that the net energy demand depends on the building size and the energetic standard. By using the developed simulation tool, the fraction of the net energy demand for heating can be calculated for solar thermal plants while for geothermal plants the fraction for heating and cooling can be calculated. Furthermore, the fraction of the net energy demand for illuminating as well for operating power for heating, ventilation and cooling can be calculated for photovoltaic plants. Prior to the evaluation, all boundary conditions must be specified to ensure well-structured parameter studies. For the simulation of a geothermal system an annual energy demand of a building (based on the net floor area) was considered as presented in Table 1. These values are comparable to the passive house standard.

Table 1   Annual energy demand for heating, ventilation, cooling and lighting; climatic region Germany as a function of the net floor area

	Dimension	Heating	Ventilation	Cooling	Lighting
Net energy demand	[kWh/(m²a)]	12,0	-	-	10
Generator net energy supply	[kWh/(m²a)]	13,9	-	-	10
Auxiliary energy	[kWh/(m²a)]	0,6	2,7	-	-

After the boundary conditions have been specified, the essential variables of each technology have to be identified. Additionally, parameter studies have to be performed to determine the impact of each variable. The essential variables for geothermal plants are location, number, length and order of the geothermal probes, as well as the performance number of the heat pump.

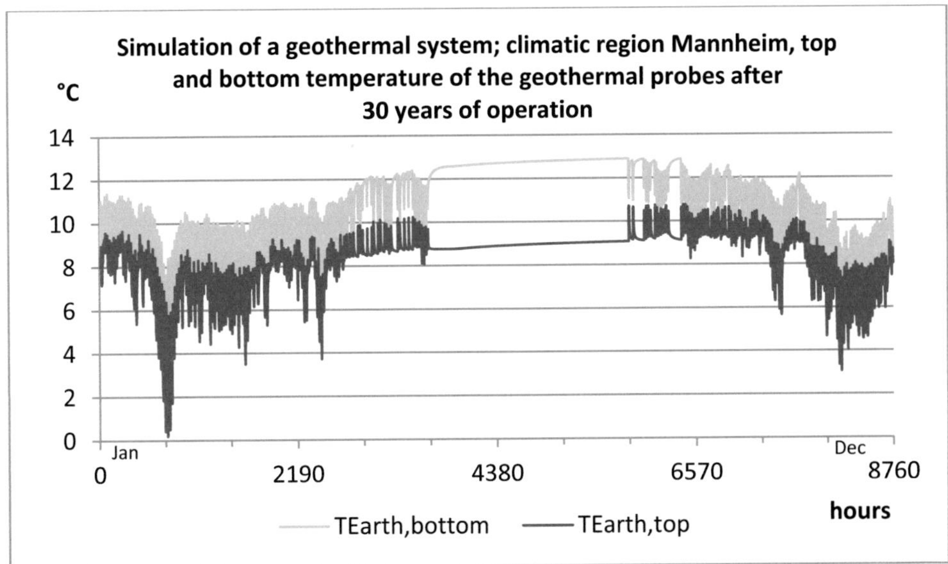

Fig. 1   Simulation of a geothermal system for the energy demand of a lowest-energy building with 5.000 m² net floor area located at the climatic region Mannheim; temperature at the top and bottom of the geothermal probes for 30 years of operation. The figure shows the modeled 30th year of operation.

The climate-dependent net energy demand, the temperature of the intact soil and the thermal soil properties belong to the site-specific parameters.The test reference year data set of the Geman Weather Service (Deutscher Wetterdienst) for the climate region Mannheim was used for the climate-specific parameters. Furthermore, the heat conductivity of $\lambda_E = 1,5$ W/(mK) and a combined value from heat capacity and density of $c_{p,E} \cdot \rho_E = 2,0$ MJ/(m³K) was applied for the soil properties. These values represent approximately the limiting case between a substratum existing of dry sedimentary rock and a substratum existing of normal bedrock or saturated sedimentary rock. Figure 1 shows the result of modeling for a building having a net floor area of 5.000 m² and a configuration of geothermal probes consisting of a probe field with 2 × 6 boreholes. The geothermal probes show a bore diameter of 125 mm and a length of H = 170 m each. The ratio of probes' distance (B) and length equals B/H = 0,1. The performance number of the heat pump amounts to $COP_{B0,W35} = 4,3$. The cold spell recorded in the used data set of the German Weather Service for the months January and February shows a significant impact and becomes to the design criterion as the operation of geothermal probes should be avoided within the freezing zone.

In a first parameter study, the useful energy demand, the respective fraction of the geothermal plant, as well as the number of probes and their placement are varied in order to analyze the single variables for geothermal plants. The required probe length, the actual performance number of the heat pump, the abstraction capacity of the geothermal probes and the necessary operating power can be derived as results from the simulation tool.

For the development of a parameterized equation an appropriate relation between the results and the parameter study has to be found. Especially, the placement of the geothermal probes within the probe field has to be considered: using the same number of boreholes the length of the probes varies with the order. A linear alignment of the probes results in less length than for a square or rectangular alignment. In order to compare results calculated for different borehole alignment a shape factor has to be implemented. It is proposed that this shape factor can be derived from by taking the actual heat transfer circumferences and a theoretical heat transfer circumferences and by building a geometrical ratio (cf. Fig. 2).

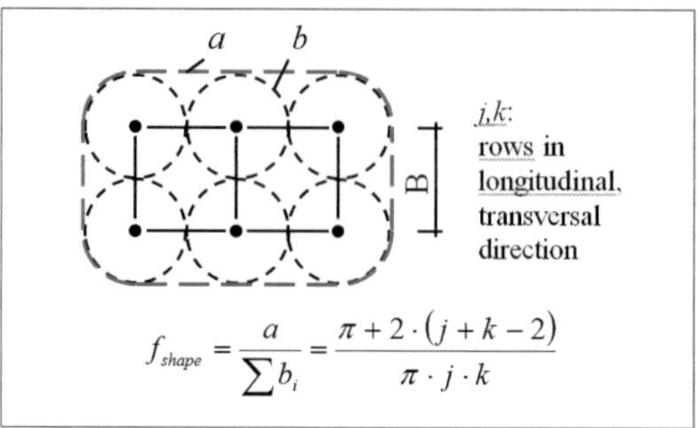

$$f_{shape} = \frac{a}{\sum b_i} = \frac{\pi + 2 \cdot (j + k - 2)}{\pi \cdot j \cdot k}$$

Fig. 2 Shape factor of borehole alignment. The shape factor was geometrically determined by comparing the heat transfer circumferences with a radius, or a distance respectively, from the probe field by the value of half the probe spacing.

By applying this implemented shape factor, the comparison of the results of the parameter study for different borehole placements is possible. Figure 3 shows the average annual abstraction capacities of probe fields as a function of the shape factors. Taking the shape factors into account, the highest determinateness was found by assuming a potential correlation between probe length-specific annual abstraction capacity and shape factor.

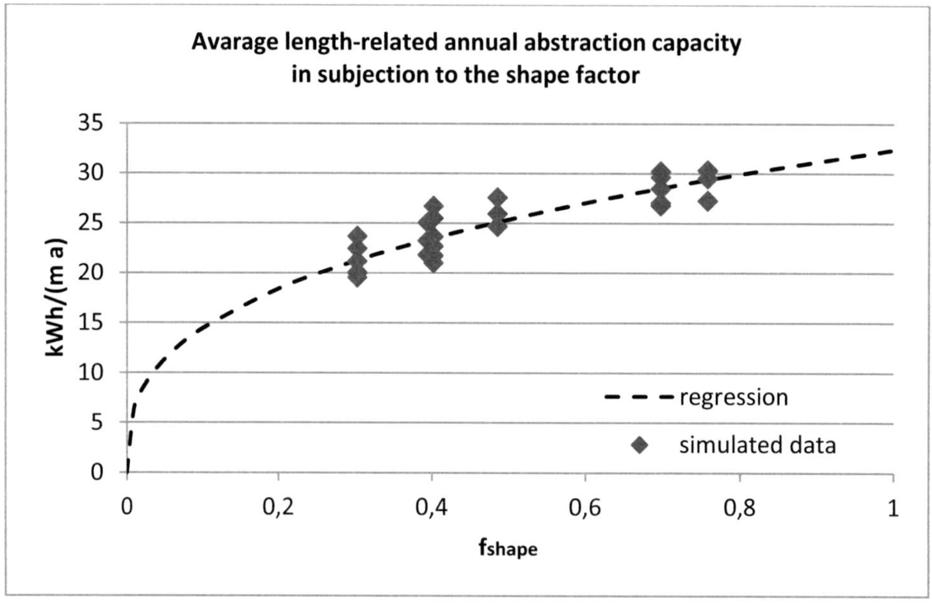

Fig. 3 Average length-related annual abstraction capacity against the shape factor; $\lambda_E = 1{,}5$ W/(mK), $c_{p,E} \cdot \rho_E = 2{,}0$ MJ/(m³K), B/H = 0,1, $COP_{B0,W35} = 4{,}3$

A further step will be to extend the parameter studies to the variation of site-specific parameters, the evaluation of their impact on the potential abstraction capacity and the determination of the respective course of function. The impact of the varied parameters on the course of function has to be analyzed and to be considered for the determination of the coefficient and exponent, in order to establish a generally valid function. The required geothermal probe length, dependent on the borehole alignment as well as on the needed annual abstraction capacity, can be then calculated. The needed annual abstraction capacity can be derived from the generator net energy supply and the performance number of the heat pump.

In the following, the correlation between the single variables and their impact on investment and operating expenses was defined in order to develop an analytical model for the optimization of costs, which can already be used for geothermal plants. Furthermore, the model will be extended to the photovoltaic and solar thermal technologies which will enable the user to determine the optimal combination of the fractions of photovoltaic, solar thermal and geothermal energy supply for covering the total energy demand.

4   Conclusions and Outlook

This research study is based upon the political climate protection targets of the European Union. A tool was developed in order to model the optimal energy supply for climate-neutrally operated buildings. The analytical model has the ability to give recommendation for the coverage of the total energy demand by using a combination of photovoltaic, solar thermal and geothermal plants. Beforehand, no elaborate simulations have to be carried out. Identical boundary conditions were applied for all technologies during the implementation of the model. This ensures that the weighting of technologies was accurately considered.

The impact of site-specific parameter was comprehensively analyzed and implemented so that by taking individual boundary conditions into account realistic result could be obtained. Besides the photovoltaic, solar thermal and geothermal technologies this model for renewable energy supply for buildings could be extended in future by additional plants, e.g. power-heat and power-heat-cooling coupling plants.

5   Acknowledgement

The author would like to thank Prof. Dr.-Ing. C.-A. Graubner for his support and for the fruitful discussions. The support of Bilfinger Berger AG is gratefully acknowledged.

References

[1]   Knauff, A.; Graubner, C.-A.: Increase of energy efficiency by the use of renwable energy. In: Darmstadt Concrete, Annual Journal on Concrete and Concrete Structures (24), Darmstadt 2009.

[2]   Knauff, A.; Graubner, C.-A.: Simulation of renewable energy. In: Darmstadt Concrete, Annual Journal on Concrete and Concrete Structures (25), Darmstadt 2010.

Diagnostics and stability analysis of stone masonry arch bridges

Gyula Bögöly

Budapest University of Technology and Economics
Department of Construction Materials and Engineering Geology
Műegyetem rkp. 3., Budapest, H-1111, Hungary
Supervisor: Péter Görög

Abstract

In Hungary there are no methods commonly used for controlling and investigating stone masonry arch bridges in bad and deteriorative condition. The old and simpler method does not give adequate results in many cases, and the new modern methods are usually very complicated or their practical adaptation is difficult.

The aim of my project is to find a way to assess simply and accurately enough the condition of such stone structures which is applicable in practice.

In order to see clearly the effects of each condition parameters on the stability preliminary investigations and modeling are necessary. My presentation demonstrates some case studies of Hungarian stone bridges through which the different stability analysis and the diagnostics will be compared. In conclusion there will be a few words about the future of this project, and the required roles further on.

1   Introduction

Stone masonry arches are one of the most ancient forms of the engineering structures. Their modeling and analysis bring up many questions even today. Although in these days new stone masonry arch bridges are not built, the maintenance and restoration of the old ones represent a special challenge at the present time as well. There are more than 1500 stone masonry arch bridges in Hungary [6]. These bridges were built mostly in the 18th and 19th century. Unfortunately these kinds of structures get low attention in Hungary and the condition of most of them increasingly deteriorates. Since the construction of these bridges the traffic and related load has increased significantly, thus these old bridges have to fulfill the new expectations (Table 1). Their stability control and the verification of their load-bearing capacity became necessary in many cases.

Table 1   Change of vehicular load [5]

150-200 years ago	Today	Ratio
six horse drawn artillery (~50 kN)	the heaviest possible load according to the Hungarian standard (800 kN)	16
passing farm-wagons (~2.53 m)	passing buses (~3.75 m)	1.5

2   Advancement of modelling

However the structural behaviour of masonry vaults seems to be simple at the first sight, in fact their mechanical behaviour and modelling is very complex. These inhomogeneous structures behave non-linearly under loading and their failure occurs in plastic state. These kinds of structures can be resisted only minimal tensions and this speciality influences especially the behaviour of the masonry structures. The stresses are usually low, thus the failure of the material is rare in masonry arches. Masonry vaults must satisfy three main structural criteria: strength, stiffness and stability. The structure must be strong enough to carry whatever loads are imposed, including its own weight. It must not deflect unduly, and it must not develop large unstable displacements, whether locally or overall [9].

In the history there were lot of different methods to control and design arches. In the antiquity and in the beginning of the Middle Ages thumbs of rules and simple geometrical rules were used. Afterwards methods advanced a lot. For instance Coulomb made a great progress in the practical use of

Proc. of the 9th *fib* International PhD Symposium in Civil Engineering, July 22 to 25, 2012,
Karlsruhe Institute of Technology (KIT), Germany, H. S. Müller, M. Haist, F. Acosta (Eds.),
KIT Scientific Publishing, Karlsruhe, Germany, ISBN 978-3-86644-858-2

"

statics with the development of his graphic method and Gaudi, the famous Catalan architect, elaborated an empiric method to design arches and domes [10]. Nowadays these methods are not sufficient because of the increased requirements, therefore more specific methods were developed such as thrust line analysis, rigid-block method, finite element and discrete element method. These methods are capable of taking into account many different influential effects, thus the failure load can be calculated with great accuracy in theory. On the other hand in practice these modern methods are not applicable so easily. In these models certain effects and attributes are taken into consideration by factual mechanical model parameters. These parameters often can be measured only with difficulty on existing structures, or it is difficult to characterize them with one factual value. The other primal trouble is the consideration of the failures. The aim of my project is to find a solution to these problems so that the condition of such stone structures can be assessed simply and accurately enough in practice. In order to see clearly the effects of each condition parameters on the stability previous investigations and modeling are necessary. In this article I will show some examples and give a conclusion from these preliminary analyses.

3 Analysed bridges

So far in the preliminary investigations I analysed 4 bridges with 3 different methods. Because of the limited size of this paper I cannot demonstrate detailed and complete analysis in all 4 cases, but briefly the descriptions of the bridges and the review of the analyses with conclusions will be presented.

3.1 Bridge over the Bükkös brook

The bridge is a 6.0 m span vehicular bridge and it was built in 1900 (Fig 1 left). It is located near Szentendre (North Hungary).The rise at mid-span of the segmental arch is 1.6 m (Fig 1 right). The bridge was made of Hárshegy sandstone. The sandstone is in a good condition despite of the fact that the blocks are constantly wet. The condition of mortar joints is disadvantageous, but there are no serious damages on the bridge.

Fig. 1   Bridge over the Bükkös brook (left), Basic dimensions to the MEXE method (right)

3.2 Bridge over the Derék brook

The bridge built in 1799 stands in the main road of Patak village (North Hungary). It is arching over a local brook with three 3.6 m long segmental barrels. The full length of the structure is 22 m. The bridge consists of different types of sandstones, thus the compressive strength of the blocks are quite different. The blocks and the mortar are also in a disadvantageous condition. There are serious mortal losses on the spandrel walls, and averagely 1 cm on the arches. There are no serious damages on the bridge. (Fig 2)

Fig. 2   Bridge over the Derék brook (left), Mortal loss on the bridge (right)

3.3   Bridge over the Lókos brook

It is a Baroque style three-span stone bridge in North Hungary which was built in 1790. For few decades the traffic uses a new bridge and the brook was regulated as well. Thus at the present time the bridge serves pedestrian traffic only. The full length of the structure is 17.8 m. The arches are 2.6, 2.8, 2.6 m long. The blocks were made of fine grained sandstone. The structure is in a good condition at all points. (Fig 3)

Fig. 3   Bridge over the Lókos brook (left), West side of the bridge (right)

3.4   Bridge over the Rédey-Nagy brook

The bridge stands in a North Hungarian village, Gyöngyöspata. It is a vehicular bridge, which provides passage possibility in the south-west border of the village. The bridge arches over the Rédey-Nagy brook with two 7.5 m long barrels (Fig 4 left). The bridge was built from different types of igneous lithologies, which are typical in this region. The main dimension stone of the bridge is andesite tuff with larger lapillis. Additionally there are pink vitreous rhyolite with flow structure, grey rhyolite with vuggy fabric, and black andesite. The blocks are carven, and their sizes are different. Due to the moisture there are several blocks which are weathered and their strength is decreased. These changes were measured in situ by using Schmidt hammer. The condition of mortar joints is disadvantageous. On several part of the bridge there is a mortar loss, with a depth of less than 5 cm. Beyond the mortar loss and the impact of the weathering some part of the structure is damaged. Several blocks of the barrel, of the spandrel wall and of the wing wall are missing partly or entirely. Part of the North East spandrel wall was damaged seriously, the stone blocks were disrupted and in the cracks the backfill can be seen [4] (Fig 4 right).

Fig. 4   Gyöngyöspata Bridge (left), Damaged, cracked spandrel wall (right)

4   Methods

The Hungarian standard of masonry arch bridges defines three different levels of the investigations. It begins with approximate calculations like MEXE method. On the 2nd level it suggests a simple 2D modelling (thrust line analysis, rigid block method), and the top level of structural analysis advises more difficult 2D and 3D modelling (rigid block method, FEM, DEM) [10]. The expected accuracy of the analysis determines which method is suggested to be used. During the preliminary investigations thrust line analysis were used with Archie-M (demo), and rigid block method with using the program developed by the University of Sheffield, named Ring. The traditional MEXE method is still widely used for assessing the carrying capacity of masonry arch bridges. For this reason approximate calculations were carried out by the MEXE method to compare the results.

The parameters, which were necessary for the models, were derived from site investigations and from laboratory tests. The site investigations included the inspections and recording of geometrical parameters, photo documentation. Lithotypes of the dimension stones were also determined and the differences between the strength of the blocks were measured in situ by using Schmidt hammer. Moisture content was also recorded on site. Some petrophysical properties such as compressive strengths, unit weights were identified under laboratory conditions and some model parameters were derived from suggestions of case studies. In the models of the bridges the above mentioned damages were taken into consideration.

4.1   MEXE method

The calculation is built on empiric coherences, and it takes in account the different effects by using modifying factors. It is a strongly approximate method, but it calculates fast and simply. It was developed in England during the World War II. The aim of this method was to calculate if a cruiser-tank can cross over a bridge or not. The provisional axle load (PAL) is obtained by using the values of the thickness of the arch barrel adjacent to the keystone, the span, the average depth of fill at the quarter points of the transverse road profile, between the road surface and the arch barrel at the crown, including road surfacing. The provisional axle load is obtained then modified by the modifying factors (span/rise factor, profile factor, material factor, joint factor) and the condition factor [10].

The MEXE method can not be used if the bridge is a multi-span bridge, the span is longer than 18 m, the skew of the bridge is bigger than 15 degree, the backfill above the extrados is bigger than 1 m, and the structure has remarkable damages. From the 4 analysed cases only 1 is acceptable at all points. The other 3 bridges are multi-span briges. Despite the calculations were accomplished in these cases as well because in theory if the piers of multi-span bridges are short (not slender), then the arches work separately. A pier is short if the heigth/width ratio is smaller then 2. This condition is true in these cases. The method has more deficiencies. It underestimates significantly the carrying capacity of long span bridges, and it overestimates the carrying capacity of short span briges.

4.2   Thrust line analysis

The basis of equilibrium analysis is set out by Heyman. It is based on the plastic theorems which were first developed in Hungary, but which were brought to Britain by Baker [14]. The thrust line analysis was carried out on the bridges by the Archie-M software. The program can take into account multi-span bridges, masonry strength, masonry unit weight, mortal loss, and the angle of friction of the fill unit. The program gives the line of the thrust according to a given load with a given position. Thus the load bearing capacity can be easily determined (Fig 5). The load carrying capacity of the bridge is adequate in case the line of the thrust does not leave the cross-section under the loads. With using this method the control can be carried out easily. In the case of the crossing over of the heaviest possible load according to the Hungarian standard (from now on load type "A") the load bearing capacity of the bridges was adequate (Fig 6). While a load is crossing over the bridge, there is a chance to see which part of the structure is under strain.

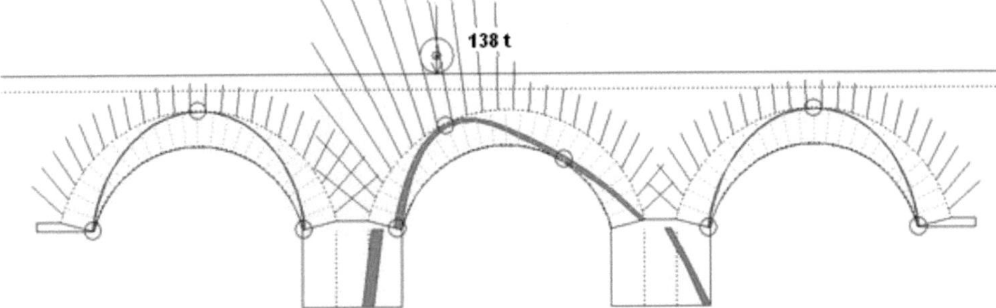

Fig. 5   Emerged thrust line if the failure load is at the worst position (Bridge over the Derék br.)

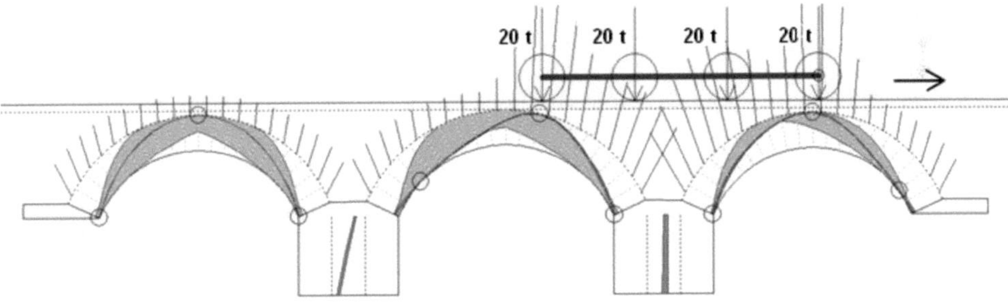

Fig. 6   Load type "A" is crossing over the Lókos Bridge

4.3   Rigid block method

The main principle of the method was set out by Heyman (1982) as well as Gilbert and Melbourne (1994). The method uses the upper-bound theory of plasticity in conjunction with geometrical compatibility criteria to obtain solutions to problems involving single- and multi-span arch [7].

Ring 3.0 can identify the critical failure load factor and associated failure mechanism and distribution of internal forces. This allows the safety of the bridge to be assessed. The software can take into account multi-span bridges, compressive strength of the blocks, masonry unit weight, mortal loss, the angle of friction and cohesion of the fill unit, the favourable effect of passive pressure, angle of dispersion and a few damages. The bond between the blocks is taken into account by the help of friction coefficients. The span/rise rate of masonry arch bridges influences significantly the emergent failure mechanism. The results of the models verified this as well. In case of lower rise arches (span/rise > 4.0) generally a 3 hinges mechanism emerges with sliding. In case of higher rise arches a 4 hinges mechanism is expected. And there is a third failure mechanism when there is a shear failure with sliding. On Figure 7, 8, 9 the analysis of the bridge over the Rédey-Nagy brook can be seen.

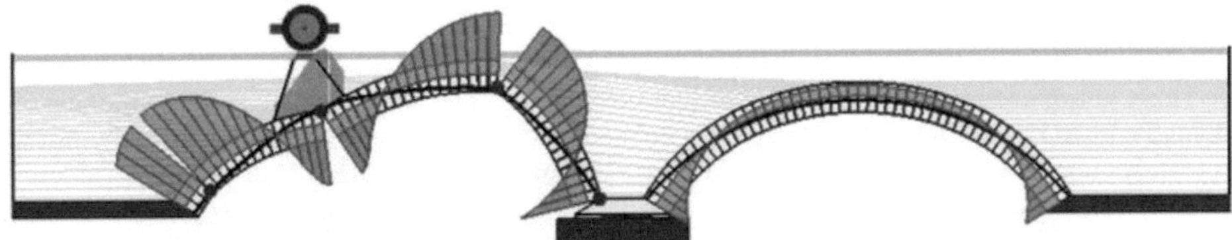

Fig. 7   Failure mechanism 2D and moment diagram

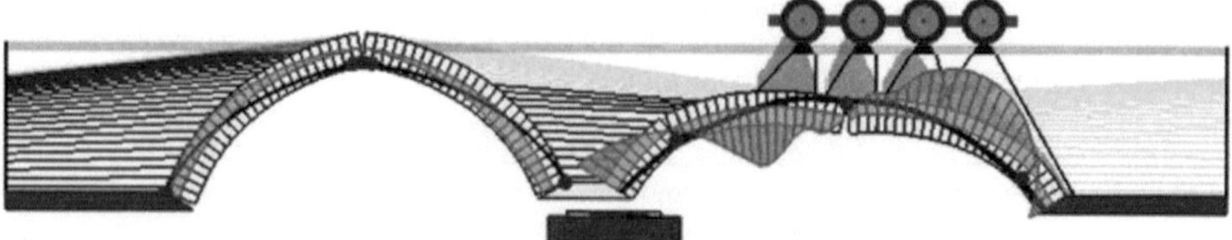

Fig. 8   Failure mechanism of repaired bridge by the load type "A" and shear force diagram

Fig. 9   Failure mechanism of repaired bridge 3D, and the moment diagram

5   Results and conclusions

To calculate the permissible axle load the value of the failure load has to be divided by a safety factor with a value between 2 and 3 according to the Hungarian standard [10]. The results of the different types of analyses were determined accordingly (Table 2).

Table 2   Results of the different analyses

Permissible axle load [t]	Bükkös	Derék	Lókos	Rédey-Nagy
MEXE method	52.4	37.1	35.6	23.7
Thrust line analysis	62.9	55.3	20.8	23.0
Rigid block method	109.6	70.5	26.6	23.2

In the table the results of the thrust line analysis and of the rigid block method are average values. These values due to the problematical attributes of the structures such as friction coefficients, dispersion of live load, passive pressures, inner components, damages, could change 5-20 % in either direction. If there is a new bridge and every parameter is known these models work perfectly. But in practise in case of an existing bridge it is almost impossible to measure some required parameters, for instance the friction coefficients between the blocks of the arch barrel. It is also difficult to characterize some properties such as the compressive strength or the masonry strength with one factual value, especially if the bridge consists of different types of stones with different level of weathering. Neither the loads nor the effects of the damages are simple. The diagnostic has difficulties also, because it is hard to see what is inside an old bridge. Although a backing, the surface fill depth, the properties of the backfill could change the outcome significantly.

It is noticeable that the MEXE method assesses the results quite well in case of the multi-span bridges with short pier as well. In fact it overestimates the carrying capacity of short span bridges. The difference between the levels of the investigations is conspicuous.

Although the masonry structures are ancient, we can see they bring up many questions at the present time as well. My purpose is to bring closer these methods to the practise and develop the diagnostics to get closer to an effective modelling.

These above used modern numerical design methods can be used primarily to determine the failure load with 2D modelling. With using these methods the stability control can be made quite fast and accurately enough to a simple control. To get more precise and detailed results or to get data about the stuctural behaviour under loading FEM and DEM methods can be used. In these methods almost every effect and circumstances can be taken into consideration but the applicability of these methods are quite difficult and the modelling takes a lot of time. Assessment of serviceability is becoming more and more important with increasing traffic volumes on masonry arch bridges. At present time there is, however, neither a suitable method for the serviceability assessment of masonry arches nor any criteria according to which such an assessment could be carried out [11]. These assessments of serviceability bring the more difficult 3D analyses into prominence. Therefore I plan to carry out 3D FEM and DEM modelling in the near future. At these preliminary analyses I wanted to compare the different methods to know what are they capable of and see clearly the differences between them. My intention is to compare the results of the numerical analyses with some factual behaviour of the structures. And I also plan to conduct experiments on a small scale laboratorial models which I hope eventually will help to reach my aim.

Acknowledgements

The work reported in the paper has been developed in the framework of the project "Talent care and cultivation in the scientific workshops of BME" project. This project is supported by the grant TÁMOP-4.2.2.B-10/1--2010-0009.

References

[1]   Andai, P.: A mérnöki alkotás története, (1959), pp 364.

[2]   Aydin, A., Basu, A.: Engineering Geology 81 (2005), pp 1-14.

[3]   Berkó, D.: Mélyépítés, 5 (2003) , pp 16-20.

[4]   Bögöly, Gy., Görög, P., Török, Á., Building Materials and Building Technology to Preserve the Built Heritage, WTA-publications (2011), pp 95-107

[5]   Gáll, I.: Régi magyar hidak, (1970), pp 3-46.

[6]   Gálos, M., Vásárhelyi, B.: Kő, VII. (2005), pp 21-25.

[7]   Gilbert, M., Melbourne, C.: The Structural Engineer, 72 (1994), pp 356-361.

[8]   Gubányi-Kléber, J., Vásárhelyi, B.: Mélyépítés, 1 (2004), pp 16-20

[9]   Heyman, J.: The stone skeleton, Cambridge University Press (1995), pp 3-15

[10]   Magyar Útügyi Társaság: Útügyi Műszaki Előírás 813/2005., (2006), pp 61-108

[11]   Orbán, Z., Gutermann, M.: Engineering Structures, 31 (2009), pp 2287-2298

[12]   Pegon, P., Pinto, V., Géradin, M.: Computers and Structures, 79 (2001), pp 2165-2181.

[13]   Santiago, H.: Nexus Network Journal, 8 (2006), pp 33-39

[14]   http://www.obvis.com/archie-theory/

Experimental testing of masonry structures

Marie Stará and Martina Janulikova

Department of Building Structures, Faculty of Civil Engineering,
VSB Technical University of Ostrava,
17. listopadu 15/2172, 708 33 Ostrava-Poruba, Czech Republic
Supervisors: Radim Cajka and Pavlina Mateckova

Abstract

Experimental testing of a masonry structure consists in building up the masonry corner, including the pre-stressing bars, into the prepared steel structure. The masonry structure of the brick corner is imposed on the vertical load and pre-stress is applied through the bars. The values of deformation from pre-stressing are sensed by potentiometer sensors. Mathematical modeling of the brick corner is based on the finite element method using appropriate software, and then the results are compared with those of experimental testing. The goal of testing is to contribute to improving the masonry structure numerical models while more closely approaching an accurate yet simple procedure for design of pre-stressed masonry.

1   Motivation

Due to underground mining as well as construction of tunnels and collectors in urban areas, terrain subsidence occurs with characteristic effects on building structures. In such an area it is often necessary to strengthen an affected structure with post-tensioning of walls and foundations, so that the structure is able to bear the load caused by horizontal terrain deformation and the effects of terrain curvature.

Post-tensioning of a damaged structure brings forth horizontal stiffening and increases stability,, preventing generation of new cracks and sometimes causing closure of existing cracks while generally extending structural service life.

The post-tensioning design procedure is elaborated for reinforced concrete structures, but for masonry structures is not given in detail [1]. Pre-stressing force design is especially problematic because it is necessary to appoint existing masonry material characteristics with regard to material strength and possible occurrence and extent of damage. The experiment in situ considers as a safe value 1/10 masonry compressive strength perpendicular to bed joints [2].

2   Experimental testing

Excessive damage of masonry is particularly expected in the corner of the object because of tri-axial stress and local deformation in the anchoring area, where there is extreme strain due to transfer of pre-stressing.

To understand better the behavior of masonry structures when post-tensioning occurs during reconstruction, experimental measurements have been performed on laboratory equipment. On such equipment it is possible to accomplish a number of tests for different input parameters, e.g., using various materials (brick, mortar), layout and number of pre-stressing cables, value of pre-stressing force, size and shape of anchor plate, value of vertical load, number of brick layers or the thickness of bed joint, brick bond and the support type of the masonry corner (simple, with slide joint and others).

2.1   Measurement procedure

Testing equipment, Fig. 1, has the plan dimensions 900×900 mm, wall thickness of 450 mm, height 870 mm. Masonry, pre-stressing bars and loads correspond to the real structure in a proportion of 1:1. During walling the specimen pre-stressing bars are placed into the bed joint at a height of 370 mm in one direction (direction A) and at a height of 520 mm in the second direction (direction B), Fig.1.

The masonry specimen is made of clay bricks CP $290 \times 140 \times 65$ and lime cement mortar M10. Extra specimens for laboratory strength testing were made both from bricks and from mortar.

Proc. of the 9th *fib* International PhD Symposium in Civil Engineering, July 22 to 25, 2012,
Karlsruhe Institute of Technology (KIT), Germany, H. S. Müller, M. Haist, F. Acosta (Eds.),
KIT Scientific Publishing, Karlsruhe, Germany, ISBN 978-3-86644-858-2

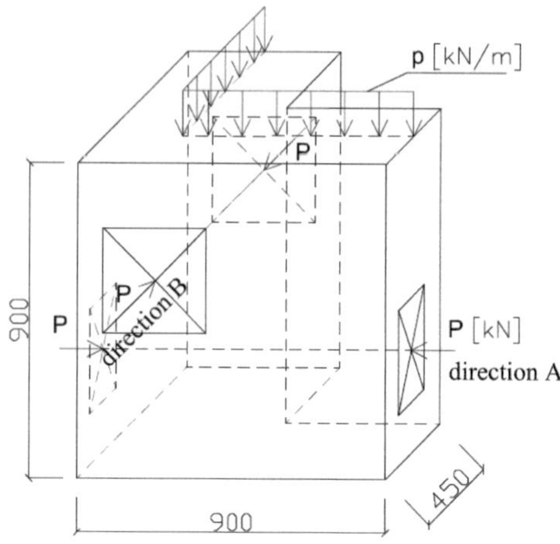

Fig. 1   Testing equipment

Strength of the mortar was determined to be f_m = 9.8 MPa according to CSN EN 1015-11 [3] and strength of bricks f_b = 20.88 MPa according to CSN EN 1052-1 [4]. Characteristic strength of masonry is then f_k = 7.6 MPa, [1].

Vertical load value was 0.12 MPa. Vertical load was applied using a hydraulic cylinder which was located between the I-profile and the steel plate with a thickness of 12 mm, strengthened with ribs, and placed in the mortar joint to spread the load as uniformly as possible. Pre-stressing was initiated with two pre-stressing steel bars and square anchoring plates with dimensions of 300 × 300, 200 × 200, and 150 × 150. Horizontal load was 50 kN and 100 kN. The specimen was loaded first in one direction and released, then in the second direction and released, and finally in both directions. Measurement of deformation was performed in the regular network of measuring points, with potentiometer sensors, identified from M20 to M25. Fig.2 shows the placement of measuring points in direction A; the location of measuring points was analogous in direction B.

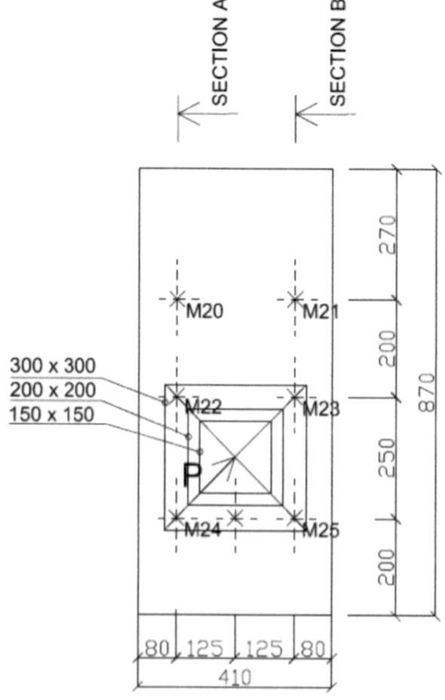

Fig. 2   Network of deformation measuring points in direction A

2.2 Measurement results

Fig.3 and Fig.4 show the graphs with measured deformations in direction A in section A; in section B, (according to Fig.2); the vertical load was 0.12 MPa and pre-stressing force was 100 kN. In both graphs the deformations are displayed for the arbitrary anchoring plates 150 × 150, 200 × 200, and 300 × 300. A horizontal line indicates the location of the pre-stressing force.

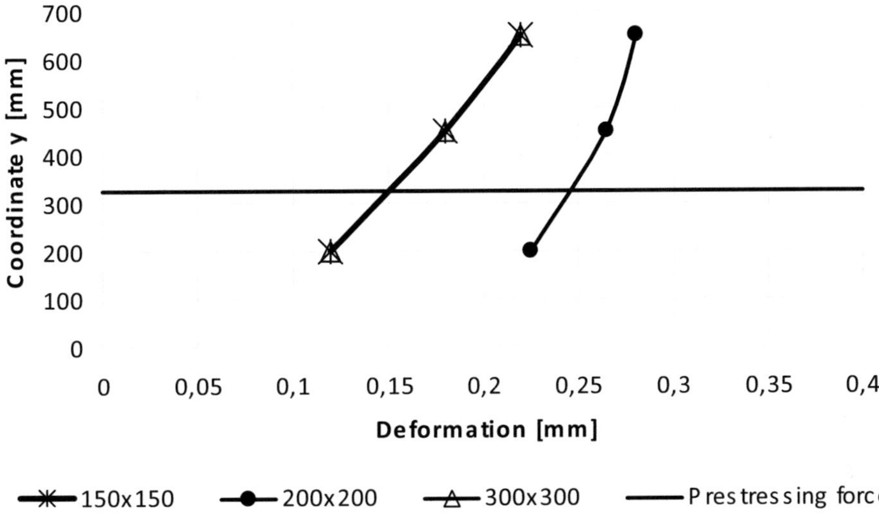

Fig. 3   Deformations for pre-stressing 100 kN in direction A, section A

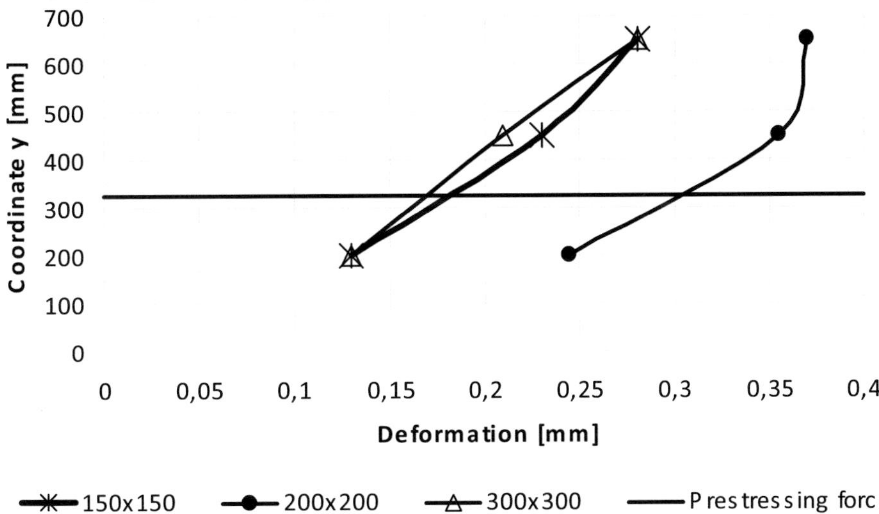

Fig. 4   Deformations for pre-stressing 100 kN in direction A, section B

In both graphs the deformation for the 300 × 300 anchoring plates is smaller than for the 200 × 200 plates, because greater stress occurs under the smaller anchoring plate. Deformations for the 150 × 150 plate are almost identical with the 300 × 300 plate. The same deformation could be caused mainly by the location of the potentiometer sensors, which are not in direct contact with the anchoring plate.

This could be due to the location of potentiometric sensors, which are not in direct contact with the anchoring plate 150 × 150 as well as the anchor 200 × 200 plate, see Fig. 2. Stress at a point with the potentiometric sensor is probably smaller than under the anchoring plate. The deformation could also be influenced with deformation of anchoring plates exposed to a concentrated load.

The results also show that the deformations are higher in the part above the anchoring plates. These waveforms of deformations could be caused by imperfect adherence of the anchoring plate to the masonry, although the masonry was underplayed with mortar.

In direction B there are similar resulting deformations.

3 Numerical model

Numerical models of a tested masonry corner were created in ANSYS software, which is used for a comparison with experimental measurements.

Masonry is a heterogeneous and anisotropic material which consists of bricks and mortar. Both of these components have different physical and material properties. Hence homogeneous properties in the entire masonry structure cannot be guaranteed. . In other words, creating a suitable model that expresses the real material and physical properties of masonry is difficult. The mathematical modeling process must take into account different material properties of basic components, geometrical arrangement of bricks, interaction between components, quality of manufacturing, environmental influence, etc.

There are two basic possibilities of modeling the masonry structure: a micro model, which portrays the actual arrangement of masonry units, including liaison and bed joints mortar, or a macro model without resolution of individual elements and masonry joints [5].

The tested masonry corner is performed in both ways, modeled by element SOLID45. Pre-stress is applied in the model using the 3D element LINK8, anchoring plates for insertion of tensile forces are modeled with the finite element SOLID45, Fig.5.

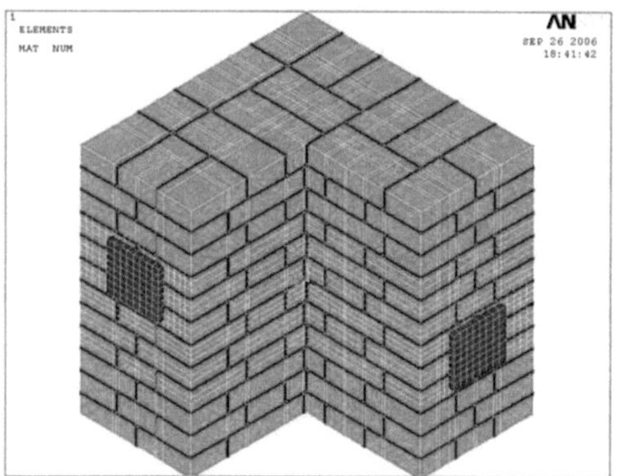

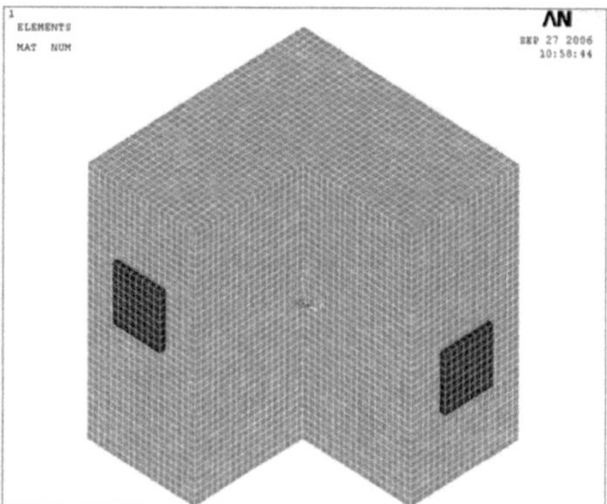

Fig. 5   Micro model (left) and macro model (right) of the post-tensioned masonry corner

The numerical model was created according to the experiment. We used CP 290 × 140 × 65 bricks and M10 mortar, input parameters of the micro model according to Tab.1 and parameters for the macro model according to Tab.2. Parameters for the macro model were derived from macro modeling of masonry, 3.1.

Section masonry, in Fig. 6, is located in place of pre-stressing bars, for better representation of the concentric preload and its effects on the resulting deformation of the wall. All input values were entered into the program as characteristic values (without the introduction of coefficients according to the applicable standards).

We found differences when comparing the two types of stress models, as was expected. The micro model usually demonstrated greater local maximum values than the macro model did in the areas of greatest stress. Deformation in these critical areas had the greatest influence on mortar and bricks that were in contact with the anchor plate. Only a small degree of deformation appeared in the other elements. This situation corresponds to practical experience where bricks and mortar in close proximity are crushed during the pre-stressing, while they are affected less by the preload at greater distance.

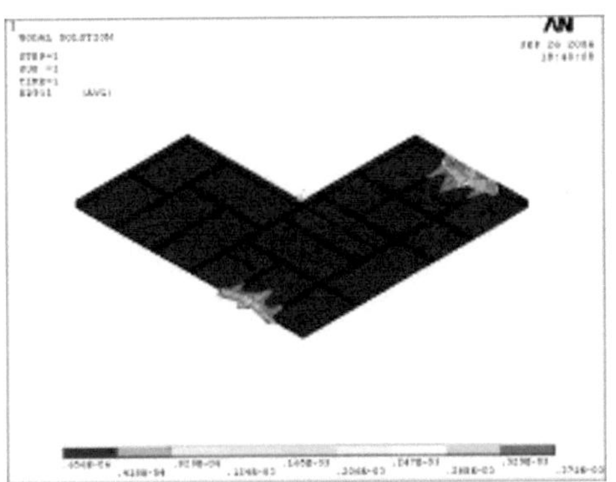 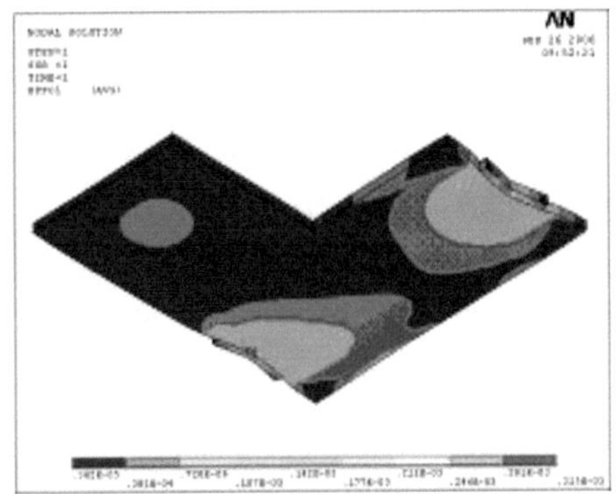

Fig. 6   Micro model (left) and macro model (right) – deformation in section in place of bottom pre-stressing bar

The homogenized model is, with its properties and resulting deformations, closer to the behavior of concrete. The deformations interfere with a larger area in the macro model, in micro model the deformations interfere only with local bricks. The resulting values, when compared with experimental measurements, are not the same but correspond in the decimal order,, which can be considered a very good result.

Table 1   Input parameters for illustrative example

Material	Modulus of Elasticity [GPa]	Shear modulus [GPa]	Poisson's ratio
Bricks 290/140/65	15		0.15
Mortar 10 MPa	10		0.2

Table 2   Output parameters for illustrative example

Material	Modulus of Elasticity [GPa]	Shear modulus [GPa]	Poisson's ratio
Masonry x direction	13.96	3.65	0.202
Masonry y direction	14.00	5.56	0.198
Masonry z direction	13.84	4.30	0.197

3.1   Homogenized properties for macromodel

Homogenization is performed by creating a fictitious material which replaces the actual material. The homogenized material is to express the orthotropic characteristics which correspond to behavior of masonry in two natural directions – parallel and perpendicular to bed joints, [6].

When creating homogenization of the material it is important to make a micro model first. Input values are the modulus of elasticity and the Poisson's ratio of bricks and mortar (Tab.1). Determination of the modulus of elasticity in three directions is done using a model of the wall, loaded on one side with constant deformations and supported on the other end. From the resulting response F, area A, length L and deformation u, the values of the modulus of elasticity for masonry E are determined in all three directions x, y and z, [6]. Shear modulus G is determined analogously to the modulus of elasticity.

$$E_x = \frac{F_x \cdot L_x}{u_x \cdot A_x} ; G_{xy} = \frac{F_{xy}}{\gamma \cdot A}$$

4　　Conclusions

In this paper the authors introduce the experimental testing of a masonry corner exposed to tri-axial load and partial measured deformations. Masonry as heterogeneous and anisotropic material requires a specific way of mathematical FEM modeling. The micro model and macro model of a tested masonry corner are introduced.

This experimental measurement simulates cases of strengthening and rehabilitation of masonry buildings. The reason the resulting deformation is so small is due to overly cautious selection of relatively small tensile force values, which were intentionally conservative for these early experiments.

The resulting deformation of the masonry can be influenced by imperfect adherence of the anchoring plate to the masonry or deformation of anchoring plates due to the concentrated load from pre-stressing.

In the upcoming year, the authors plan to build new masonry specimens with lower mortar and bricks strength, allowing the measured deformation to be larger and more distinctive.

We will develop further numerical models for the experimental measurements, which we expect to refine during the experimental testing so the new models will closely match the actual behavior of masonry. The result should be a simple (homogenized) model which allows us to avoid modeling the individual components of masonry. In other words, sufficient for obtaining accurate results without performing time-consuming experiments.

Acknowledgements

This outcome has been achieved with the financial support of SGS Grant, internal no. SP 2012/36.

References

[1]　　EN 1996-1-1 Eurocode 6 – Design of masonry structures – Part 1-1: General rules and rules for reinforced and unreinforced masonry structures. CEN Brussels 2005

[2]　　Bazant, Z.; Klusacek, L.: Statics at building reconstruction (in Czech). VUT Brno, 2004

[3]　　CSN EN 1015-11 Methods of test for mortar for masonry - Part 11: Determination of flexural and compressive strentgth of hardened mortar. CEN Brussels 1999

[4]　　CSN EN 1052-1 Methods of test for masonry - Part 1: Determination of compressive strength. CEN Brussels 1998

[5]　　Lourenco, P. B.: Computational strategies for masonry structures. Delft University Press, 1996

[6]　　Cajka, R.; Kalocova, L.: Modelling and Analysis of Post-Tensioned Masonry. In Proceedings of the Eleventh International Conference on Civil, Structural and Envirinmental Engineering Computing 2007, Malta 2007

[7]　　Bradac, J.: Effects of Undermining and Protection of Structures, vol. I and II. Dům techniky Ostrava. Ostrava, 1999. ISBN 80-02-01276-3

[8]　　Materna, A.; Brozovsky, J.: Constitutive model for two-dimensional modeling of masonry. In proceedings of the Eleventh International Conference on Civil, Structural and Envirinmental Engineering Computing 2007, Malta 2007.

[9]　　Mynarzova, L.: Static Analysis of Masonry Structures Dissertation thesis 2009. VŠB-TU of Ostrava 2009. ISBN 978-80-248-2064-4

[10]　　Cajka, R.: Strengthening of Historical Structures on Flooded and Undermined Territory. International Geotechnical Engineering,Saint Petersburg. 17-19. September 2003, Russian, ISBN 5-93093-204-2.

[11]　　Čajka, R.: Lifetime Enhancement of Historical Structures on Flooded and Undermined Territory. Integrated Lifetima Engineering of Buildings and Civil Infrastructures, and International Symposium ILCDES 2003. 1-3. December 2003, Kuopio, Finland, ISSN 0356-9403, ISBN 951-758-436-9.

Influence of biomineralization on mechanical and physical characteristics of Hungarian limestone

Péter Juhász

Department of Construction Materials and Engineering Geology,
Budapest University of Technology and Economics (BUTE)
Műegyetem rkp.3., H-1111 Budapest, Hungary
Supervisor: Katalin Kopecskó

Abstract

Bacteria induced calcium carbonate precipitation nowadays is a widely examined process being a possible alternative for traditional stone conservation methods. Thus application connected, in situ experiments are recommended by researchers engaged in bioconsolidation. In our experiment, method and materials developed by a French research group were applied on Hungarian porous limestone specimens in situ. Different from the original method samples were immersed into the liquid compound, instead of spraying the curing solution onto their surface. According to their apparent porosity characteristics and fabrics, samples were divided into four groups. Changes in mass-properties, compressive strength and Young's modulii of the specimens due to the biodeposition were measured, respectively. Cured specimens gave better results for all the examined properties, comparing to the non-cured ones. Due to previous recommendations and practical reasons, color changes induced by the biomineralization were also measured. Changes of the shade were evaluated by comparing images taken before, during and after the treatment. As result, some degree of darkening was observed.

1   Introduction

Biomineralization is based on the phenomena, that different bacteria strains are capable of producing calcium-carbonate crystals in adequate environment. Bacterially induced carbonate precipitation has been explored for the protection and consolidation of ornamental stone for more, than twenty years. Nowadays there are two main trends in biomineralization [1]. Biodeposition is an organic originated and highly compatible method for restoration and conservation of porous stone materials, especially porous limestone or lime-bound sandstone. It results in a deposition of a carbonate layer on the surface, and a few centimeters under the surface of building materials. Crystals produced during the precipitation integrate themselves into the matrix of the stone material in a high extent. This method was first used for conservation purposes by a French research group [2]. The other use of microbially induced carbonate precipitation (MICP) is biocementation, used for the generation of binder-based materials such as mortars and concrete.

In our experiment, method and materials developed by the French research group were applied on Hungarian porous limestone specimens in situ. Our aim was to determine the expectable changes in mass-properties, uniaxial compressive strength, Young's modulus and colour of the spe-cimens due to the biodeposition.

2   Materials and preparation of the specimens

2.1   Experimental circumstances

Upon previous recommendations of the researchers engaged in biomineralization, our experiment was carried out in situ [1]. With one exception we applied a one-week treatment on the limestone specimens following the instructions of the French Calcite Bioconcept company. Different from the original method, the samples were immersed into the liquid compound, instead of spraying the curing solution onto their surface. The treatment was carried out between the 23th and the 29th of August, 2011. Specimens were exposed to intensive sunlight during half of the day. Daily temperature fluctuated between 16 and 36 oC, while the temperature of the wet stone surfaces fluctuated between 25,0-42,5oC in the morning and 22,9 - 26,8°C at night.

Proc. of the 9th *fib* International PhD Symposium in Civil Engineering, July 22 to 25, 2012,
Karlsruhe Institute of Technology (KIT), Germany, H. S. Müller, M. Haist, F. Acosta (Eds.),
KIT Scientific Publishing, Karlsruhe, Germany, ISBN 978-3-86644-858-2

2.2 Bacterial isolates and culture media

In our experiment curing materials developed by the French company Calcite Bioconcept were applied on the porous limestone specimens. With the desired aim of obtaining homogenously, in-debt cured specimens, the samples were immersed into the liquid compound instead of spraying the curing solution onto their surface. The bacterium strain applied in the compound is Bacillus cereus, which is an aerobe, Gram-positive and non-pathogen soil-bacterium. Germination of this strain occurs between 15 and 50oC [3] , thus it is ideal for outdoor treatments in calmer weather. Bacillus cereus is a spore-forming bacterium, therefore it is capable for survival and revitalization depending on the accessibility to nutrition.

2.3 Stone material

The stone material used in our experiment is Hungarian porous limestone of two different types. The first one has ooidic texture with finer grains (a), the second one has an ooidic-bioclastic texture with coarse grains, bioclasts and larger pores [4]. Two blocks of the finer, as well as five blocks of the coarse lime-stone, altogether seven (marked with capital letters A-G) were drilled and cut in order to prepare the specimens.

2.4 Group and preparation of the samples

The porous limestone specimens were prepared according to the requirements of the MSZ EN 1926 standard (April, 2007). Thus diameters and heights of the specimens are 50±5 mm. The blocks were classified according to their textures (two types) and their mass density values (measured on the specimens). This classification resulted in the establishment of four groups: I. - blocks A and D; II.- blocks B, C and E; III. - block F; IV.- block G. Half of the specimens were cured, the other half was used as controls. In the uniaxial compressive strength test eventually only five blocks were involved.

3 Methods and experiments

3.1 Mass properties

During the measurement of mass properties we examined the changes of dry and water-saturated weight, apparent porosity and density. Measurements were carried out before and after the treatment, respectively. Subjecting the specimens to saturation with water and drying until constant weight could not affect the results. As result of the biomineralization we expected increase of dry mass and density, as well as decrease of the apparent porosity.

3.2 Uniaxial compressive strength and Young's modulus

Generally a stone material with high ultimate compressive strength is more durable, than other ones with lower values. Therefore in each group we measured the average uniaxial compressive strength of cured and non-cured specimens. Measurements and evaluation were done according to the regulations of the MSZ EN 1926 standard. Young's modulus (also known as modulus of elasticity) (E) was calculated according to the MSZ EN 14580:2005 standard. On each σ-ε diagram we marked the linear range, and Young's modulus was calculated from the values of this range. Idealized σ-ε diagrams of the specimens were drawn using the values of modulii of elasticity and of the ultimate strength. As result of the newly grown crystal-capsules due to the biomineralization, we expected strengthening of the stone matrix.

3.3 Colour changes

Appearance of the cured surface should not differ from the original, non-cured surface in a great extent. Using organic originated material, small degrees of discolouration or changes in shade are likely to occur. Nevertheless these changes should not result in complete discolouration, or darkening of the surface in an aesthetically unacceptable scale.

As the curing compound has a yellowish colour due to its urea-content, some degrees of darkening and colour-changes towards yellow were expected. Changes in colour and shade of the specimens were examined by images taken three times: before the treatment (Phase I.), after the treatment (Phase II.), and eventually after washing the cured specimens. Each image was taken in a dark room from the same distance, angle and lighting. Manual settings of the Kodak camera were: f/2.8, exposition time:

1/250 s, focus: 6mm, and without flash. In order to determine the changes of colours we analysed and compared the histograms of images in Photoshop CS3.

4 Results and discussion

4.1 Mass properties

During the consolidation process a few amount of new material (bacteria and nutrition) was added to the specimens. This resulted in the changes of mass properties, thus apparent porosity, density, and mass of the stones. Biomineralization induced the reduction of pore-diameters, thus increase in density and decrease in apparent porosity. The total amount of solid material used for the preparation of the curing liquid was 125 grams. This amount was distributed among 34 specimens and 3 smaller pieces of porous limestone. Consequently relatively low percentage of mass-increase was expected. In order to avoid miscalculations due to small grains remained unbound or unsolved in the porous matrix, specimens were washed in a vessel. Material losses during the measurement and transportation of the specimens are at a degree of tenthousandth, which is negligible. We measured increase in average masses. Relative differences were calculated by the following formula: $\text{Diff.}_{rel.} = (\mathbf{P}_{cured} - \mathbf{P}_{non\text{-}cur})/\mathbf{P}_{non\text{-}cur} *100$ [%], where $\mathbf{P}$ stands for the examined property (**Table 1**).

While dry mass of the specimens increased, the capability of water-uptake and the saturated mass decreased due to the biodeposition. An average amount of 0.459 grams of solid material in form of calcium-carbonate crystals was added to each specimens, which is equal to an average of 0.22 m/m % increase in mass. As result, mass-densities increased in a same extent. Change of the mass noticeably affected the apparent porosity, which decreased with 9.18 % (**Table 2**). Consequently the treatment is capable to provide higher resistance to the stone against water-related deterioration, such as freezing-thawing and salt crystallisation.

Table 1   Changes in dry and saturated masses between the non-cured, and the cured-and-washed phase.

Mark of block; Mark of group	Dry mass average [g]			Saturated mass average [g]		
	Non-cured	Cured and washed	Relative difference [%]	Non-cured	Cured and washed	Relative difference [%]
A	194,90	195,41	**+0,26**	230,53	228,11	**-1,05**
B	197,91	198,34	**+0,22**	229,09	227,73	**-0,59**
C	205,24	205,71	**+0,23**	233,67	231,29	**-1,02**
D	196,44	196,99	**+0,28**	231,84	229,15	**-1,16**
E	200,43	200,87	**+0,22**	229,38	227,79	**-0,69**
F	188,28	188,55	**+0,14**	221,43	221,36	**-0,03**
G	210,37	210,78	**+0,20**	237,23	235,13	**-0,89**
I. (A+D)	195,67	196,20	**+0,27**	231,19	228,63	**-1,11**
II. (C+D+E)	201,19	201,64	**+0,22**	230,71	228,94	**-0,77**
III. (F)	188,28	188,55	**+0,14**	221,43	221,36	**-0,03**
IV. (G)	210,37	210,78	**+0,20**	237,23	235,13	**-0,89**

Table 2   Changes in density and apparent porosity between the non-cured and the cured-and-washed phase.

Mark of block	Density average [g/cm^3]			Apparent porosity average [%]		
	Non-cured	Cured and washed	Relative difference [%]	Non-cured	Cured and washed	Relative difference [%]
A	1,638	1,643	**+0,31**	30,07	27,55	**-8,37**
B	1,647	1,651	**+0,21**	26,01	24,52	**-5,72**
C	1,695	1,699	**+0,23**	23,54	21,18	**-10,04**
D	1,628	1,633	**+0,29**	29,40	26,71	**-9,15**
E	1,664	1,668	**+0,22**	24,09	22,40	**-7,00**
F	1,525	1,535	**+0,67**	30,53	27,17	**-11,00**
G	1,739	1,742	**+0,20**	22,25	20,17	**-9,36**

4.2   Compressive strength

Uniaxial strengths of five blocks (A,C,D,E,G) were evaluated. For each block we measured higher average compressive strength for the cured specimens, than for the non-cured ones. Results are shown in **Table 3**, and plot in **Figure 1. B**. Connection between the type of texture and the ultimate strength of the limestone specimens is noticeable. Finer fabrics gave higher results, than more uneven ones with coarse structure. Standard deviations of the results are higher, than the differences of compressive strengths between the cured and non-cured specimens. Therefore experiments should be carried out on more homogenous porous materials, for example on mortar specimens.

Table 3   Differences between the average compressive strengths of the non-cured and the cured-and-washed blocks and groups of specimens

Mark of block or group, fabric	Compressive strenth [N/mm^2]		Relative difference [%]	Standard deviation [%]	
	Non-cured	Cured and washed		Non-cured	Cured and washed
A-fine	6,42	7,99	+24,59	7,44	30,75
C-coarse	3,84	4,54	+18,38	24,97	30,68
D-fine	6,35	6,53	+2,78	22,50	10,50
E-coarse	3,40	3,55	+4,40	14,34	18,85
G-coarse	4,21	4,54	+7,99	18,44	26,74
Average	**4,84**	**5,43**	**+11,63**	**17,54**	**23,51**
I. (A,D) - fine	6,39	7,26	+13,74	14,97	20,63
II.(C,E) - coarse	3,62	4,05	+11,81	19,66	24,77
IV.(G) - coarse	4,21	4,54	+7,99	18,44	26,74

4.3   Young's modulus

Young's modulii of the cured specimens are higher than non-cured ones. In four blocks of five the σ-ε diagrams shown higher modulus of elasticity. Exact values and differences compared to the values of the non-cured groups are shown in **Table 4**. Plot of the values for the block A is shown in **Figure 1. A**. An average relative difference of +24,196 %, thus increase was evaluated. Increase of the Young's modulus indicates that the stone material became more rigid, with the newly precipitated calcium-carbonate crystals. This also indicates that the capsules bound strongly to the original material, while they could hamper deformation inside the stone. Thus they decrease the specimen's overall rate of deformation during the loading. While unbound grains were washed out of the specimens, it can be also declared, that the solid material left inside the stone mostly transformed into rigid crystals around the bacteria.

Table 4   Changes of Young's modulus between non-cured and cured-and-washed specimens

Block	Young's modulus [N/mm²]		Relative difference
	Non-cured	Cured and washed	[%]
A	1524,05	2101,18	+37,87
C	623,77	1001,60	+60,57
D	1826,51	1986,17	+8,74
E	844,23	743,76	-11,90
G	628,12	783,26	+24,70

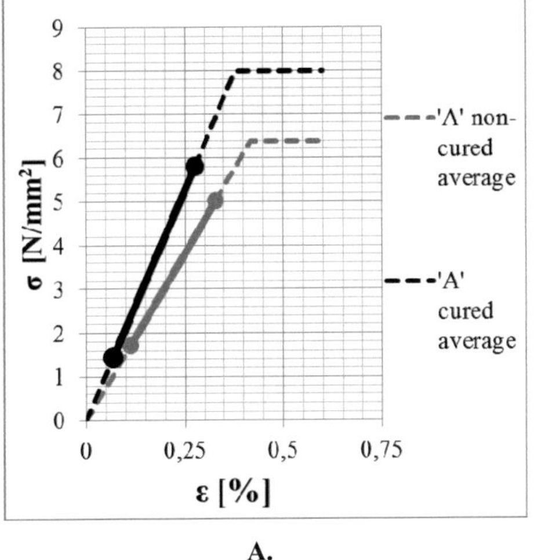

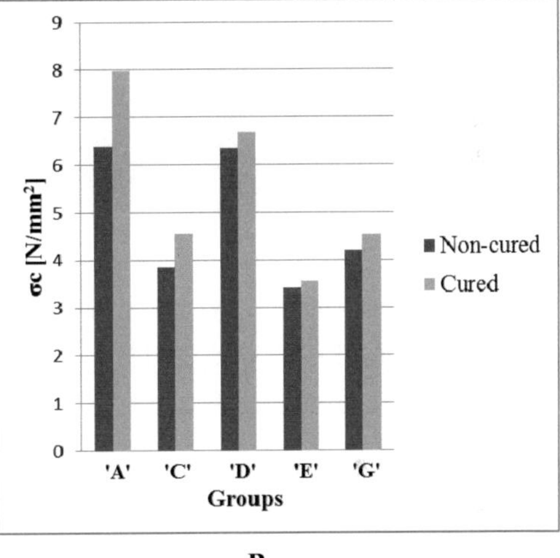

Fig. 1   **A.** Idealized σ-ε diagrams of the cured and non-cured specimens of block A. **B.** Changes of compressive strength in each group.

4.4   Colour changes

During and after the treatment colour changes of the specimens were noticeable. Original colour of our limestone specimens is light beige – yellowish, approximately RGB (153;124;94), with a mean luma component of 129. Changes of colour and their values are presented in **Table 5**. Original and altered colours of the specimens can be seen in the images of the second column. Mean values of the image histograms are also presented in grayscale in the third column. According to these columns the darkening and brightening of the specimen is noticeable. Histograms of all colour channels show the shift towards the darker shades (left side) comparing cured and non-cured specimens. After washing the specimens some degrees of widening of the histogram can be noticed with an almost similar mean value. This refers to almost no change in shade, but discoloration due to changes of pixel values, thus the shift of the RGB channels' histograms. Some degrees of darkening of the yellow colour can be seen in the last column. Yellow colours' histograms are plot where histograms of red and green channel overlap, without disturbance of the blue channel's histogram. While histograms are the density functions of shade of pixel values between 0 and 255 (0 for black, 255 for white), reading the last column darkening of the yellow component is noticeable. For more exact determination of the colour-changes, mathematical counting of the pixel values is necessary.

In practice color-changes between the non-cured and the cured-and-washed (final) phase is important. Considering faşcades exposed to rainwater, washing of the stone surfaces is possible. According to this measurement no change in shade, but higher significance of the yellow color should be expected comparing the non-cured, and the cured and washed phase.

Table 5   Colour of specimen, equivalent of mean value in grayscale, histograms of all the colours and change of the colour yellow.

State	Colour of specimen (~1cm*1cm)	Shade of mean value in gray-scale	Histogram of all color chan-nels	Histogram of luminosity	Changes of the yellow's shade
Non-cured			mean value:120,20	mean value:125,89	
Cured			mean value: 87,73	mean value:93,37	
Cured and washed			mean value: 118,14	mean value:125,39	

5   Conclusion

In our experiment, method and materials developed by a French research group were applied on Hungarian porous limestone specimens in situ. Our aim was to determine the expectable changes in mass-properties, uniaxial compressive strength, and colour of the specimens due to the biodeposition.

We measured an average of 0.22 m/m % increase of masse, which noticeably affected the apparent porosity, which relatively decreased with an average of 9.18 %. Relative increases of the compressive strengths (11,63%) and the Young's-modulii (24,20%) were evaluated. As result of the treatment some degree of darkening and changes of the yellow's shade were observed.

The treatment turned out to be capable to provide higher resistance to the stone. Further experiments will be carried out in order to determine the effects of biodeposition on the compressive strength and the durability of porous materials.

6   Acknowledgement

This work is connected to the scientific program of the "Development of quality-oriented and harmonized R+D+I strategy and functional model at BME" project. This project is supported by the New Hungary Development Plan (Project ID: TÁMOP-4.2.1/B-09/1/KMR-2010-0002).

7   References

[1]   De Muynck, W., De Belie, N., Verstraeteb, W., 2010. Microbial carbonate precipitation in construction materials: A review. *Ecological Engineering* **Vol.** (36) 118–136.

[2]   Adolphe, J.P., Loubičre, J.F., Paradas, J., Soleilhavoup, F., 1990. Procédé de traitement biologique d'une surface artificielle. European patent 90400G97.0. (after French patent 8903517, 1989).

[3]   Johnson, K. M., Nelson, C. L., Busta, F. F. 1983. Influence of temperature on germination and growth of emetic and diarrheal strains of Bacillus cereus in a broth medium and in rice. Journal of Food Science **Vol.** (48), Issue (1): 286–287.

[4]   Török Á. 2007. *Geológia mérnököknek*. Műegyetemi Kiadó, Budapest. 72.

Dynamics of Structures and Material Behaviour under Dynamic Loadings

Experimental results on fibre reinforced masonry structures subjected to seismic loads

Moritz Urban

Department of Reinforced Concrete and Building Materials,
Karlsruhe Institute of Technology (KIT),
Gotthard Franz Straße 3, 76131 Karlsruhe, Germany
Supervisor: Lothar Stempniewski

Abstract

Every year several big earthquakes like in Italy (L'Aquila 2009), Chile (2010) and Japan (Fukushima 2011) are causing high damages and collapses especially on masonry buildings. The 10 biggest earthquakes led to a total damage of 351 billion USD [4]. But also human losses like in L'Aquila where an earthquake with a comparable low 6,3 magnitude caused 260 losses, 1000 injured and 28.000 homeless are the main motivation to retrofit old masonry buildings.

The work is about different strengthening techniques with new materials like ductile polyurethane glues, cement based mortars and textiles with different fibre orientations. Two special developed systems for masonry retrofitting are tested on different kinds of laboratory specimen like 6-stone-masonry, 1.25 m x 1.25 m walls, 2,4 m x 2,6 m walls and a full 5,8 m high masonry building.

Results on a masonry building with the same material and the archetype like buildings in the region from L'Aquila showed in a shaking table test an increase in the maximum load capacity of more than 50% (peak ground acceleration) after strengthening.

1 Introduction

Fibre materials reinforcement are one out of several strategies to retrofit or strengthen existing masonry buildings. The benefit is the additional high tensile strength for masonry, while the weight compared to other materials is very low. This is the most important point for earthquake loads which are related to the mass of a building. The effects to masonry structures are trough the failure mechanisms very different and dependent on the orientation. The load bearing of walls can happen in the out of plane and in plane direction. Therefore all researchers divided the both loading scenarios in their experimental campaigns in the two different cases and conducted tests on out of plane specimens and in plane specimens.

In the first mentioned direction conducted Velazquez and Ehsani [11] wall tests with b/h ratio of 14 and 28 and reinforced with glass fibre stripes and epoxy glue. The maximum possible increase in the load bearing capacity was 650% compared to the unreinforced walls. Tumialan [10] conducted out of plane tests with hollow bricks and concrete block masonry which were reinforced with glass and aramid fibre reinforcement. He additionally determined the bonding with special putty. Further work was done by Hamilton and Dolan [7], Reinhorn and Madan [8], and Hamoush et al [6]. All of them used a static loading and stiff glues like epoxy resign.

Also for in plane loading different authors made several tests on walls. Schwegler [3] one of the first researchers in this field tested 1994 masonry walls with different fibre materials and different reinforcement strategies. Wallner [2] considered the existence of openings in walls with windows in his work. Other fibre reinforced materials were conducted by Hohlberg and Hamilton [9] and Münich [5] in static cyclic tests. Only El Gawady [12] tested his walls under dynamic conditions on a shaking table. He used hollow bricks, carbon stripes, glass and aramid fibre textiles. The possible increase in the load bearing capacity was between 30% and 150%.

Studies in the past showed the good possibilities of the fibre materials in the field of application masonry. But there is still missing the knowledge about the real behaviour under dynamic and full building conditions. This work shows two new reinforcement techniques which were tested under static, dynamic and full building conditions.

Proc. of the 9th *fib* International PhD Symposium in Civil Engineering, July 22 to 25, 2012,
Karlsruhe Institute of Technology (KIT), Germany, H. S. Müller, M. Haist, F. Acosta (Eds.),
KIT Scientific Publishing, Karlsruhe, Germany, ISBN 978-3-86644-858-2

2 Two new developed reinforcement systems

Two different fibre reinforcement systems have been developed at KIT. The first consists on a high ductile and breathable polyurethane glue on water base and a biaxial, alkali resistant glass fibre fabric. The adhesive is applied directly on the existing plaster surface on a wall in which the fabric is fixed. This has the advantage that the plaster must not be removed from the wall and no cleaning of the building after the work is needed. On the other side before the application the sufficient bonding capacity from the plaster to stone has to be found out. This system was developed for indoor use and the out of plane safety on non-structural walls. The system is called "eq-top" (see figure 1).

Fig. 1   Biaxial glass fiber textile – eq-top (left),   Cracking fibers (right)

The second fibre reinforcement system is applied directly to the stone surface in a special mortar textile sandwich configuration. The mortar as matrix for the bonding to the stone consists on three components: A = cement powder, B = epoxy resign, C = hardener. As fibre material a 425 g / m² heavy glass fibre/ polypropylene fabric was developed that takes into account all possible failure causes of masonry. It is woven in four fibre directions in angels of 0°, 90° and +/- 30° with the two different fibre materials. In a sandwich structure consisting of a layer of high-strength epoxy mortar, a layer fibre fabric and a covering layer of the same mortar, the whole system is applied flat to the outside surface of a building. The system is called "eq-grid" (see figure 2).

 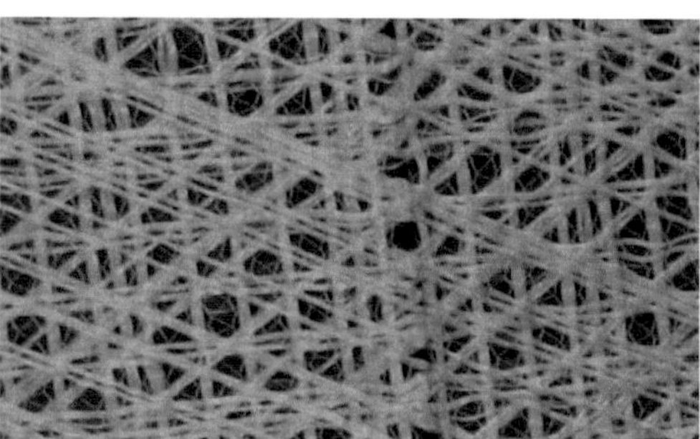

Fig. 2   Fabric in the first mortar layer (left),   Eq-grid fabric with four fibre directions (right)

3 Experimental campaign – Out of plane

The two most important parameters for the out of plane bending behaviour of masonry walls is the compression strength of the stones and the mortar, because unstrengthend masonry carries the load only over compression in the vertical and horizontal bending direction. In existing buildings it is not possible to change these parameters. Therefore the only solution is to increase the tensile strength with additional external fibre reinforcement systems. Like reinforced concrete the fibre reinforcement carries tensile forces and the stones together with the mortar the compression forces on the other side.

3.1 Static bending tests

While in the literature no normal force or only self-weight has been used, here was developed a test assembly with force controlled hydraulic pistons. The following figure shows the test configuration for small 6-stone-masonry specimens. The vertical piston is applied displacement controlled and running 0,05 mm/sec in the MTS 1000 machine. The force and the displacement are recorded and displayed in the next diagrams. The two horizontal pistons hold constant the given stress of 0.2, 0.4 or 0.8 MPa.

Fig. 3   Bending test in with bricks (left),   Sand lime brick specimen tested (right)

The results show a different behaviour dependent on the used stones. At Brick stones were observed an increase in the maximum force over 64%. The sudden collapse of the unreinforced masonry (URM) equal with or without plaster is for earthquake loads very unfavourable. Therefore much more important was the after cracking behaviour trough the reinforcement which the maximum displacement has nearly quadrupled. The failure occurred by stone cracking, while no damage of the textile was visible.

Sand lime bricks with their higher compression strength led not to a stone failure. Instead the textile was utilized till the fibre broke. The increase in the maximum force was four times the URM and the displacement by maximum force three times higher.

This pilot study on reinforced masonry (RM) showed clearly the benefits for the out of plane bending behaviour. Further dynamic studies are presented in the next chapter.

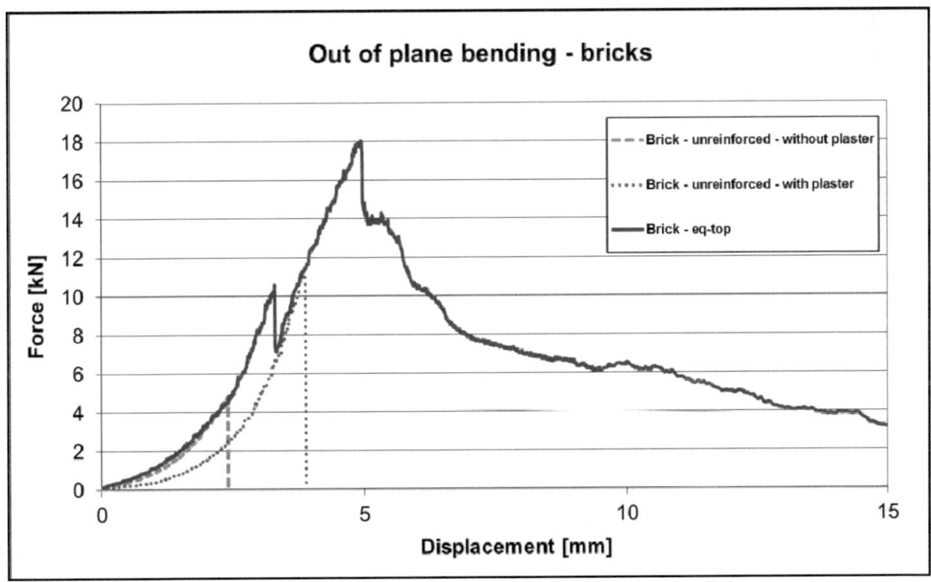

Fig. 4   Bending test results for brick stones. Unreinforced vs. eq-top reinforcement

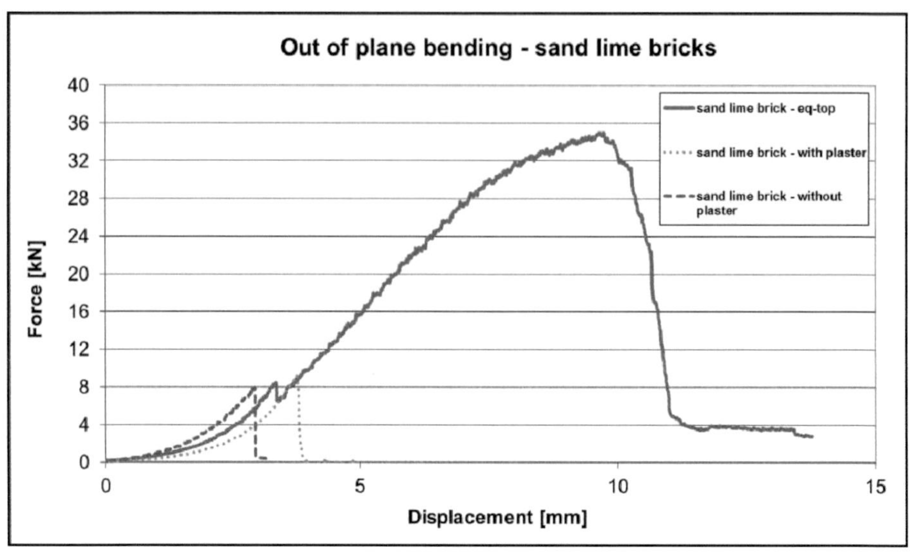

Fig. 5   Bending test results for sand lime bricks. Unreinforced vs. eq-top reinforcement

3.2   Dynamic bending tests

The dynamic testing was conducted on a uniaxial shaking table with the size 3 m x 3 m in a special constructed steel frame. Two tests were conducted in each case one 2,4 m x 2,6 m wall with only plaster and a second wall with plaster and the eq-top reinforcement. The difference in the second test was the inserted door opening and the two diagonal steel crosses for frame stiffening.

The stones were hollow bricks with tube orientation in the horizontal direction. To the stone surface was applied a typical in Istanbul used cement based plaster typically in Istanbul used. All walls were hold at three sides, while the top side could move freely.

In the first test cracks occurred early at the URM wall and led to a collapse of a part (see figure 9). The total collapse happened at 3.8 g acceleration in the wall. At the synchronous tested RM wall no crack after the maximum capacity of the shaking table was visible.

The door opening led to first cracks at the door corners and resulted in a collapse of the top part from the URM wall. While collapsing of the unstrengthend wall first cracks in the textile on the other wall were visible at the door corner location. The maximum accelerations at the URM wall was 2.4 g before collapse and 4.5 g at the RM wall.

Fig. 6   URM wall without opening is collapsing   URM wall with opening is collapsing (right)

4 Experimental campaign – In plane

4.1 Static cyclic wall tests

Scaled wall tests were conducted with different test specimens to select the best materials for the matrix-fibre-system. The shear loading in-plane was in the strong inertia force direction with a vertical stress of 0.2 MPa. For the horizontal cyclic displaced head beam the displacement and the horizontal force were measured.

Representative results of unreinforced masonry (URM) and reinforced masonry (RM) tests (1.25 m x 1.25 m) are shown below in figure 7. Both walls were built with sand lime bricks in 4 DF formate. The maximum resistance force of the URM wall was 98 kN and the maximum load of the RM structure was 232 kN. This is an increase of 136%. But the more important effect is the increase of ductility of more than 200%. Further dynamic studies are presented in the next chapter.

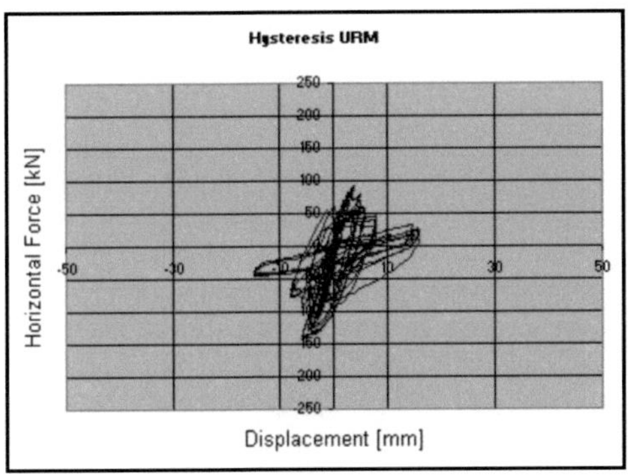

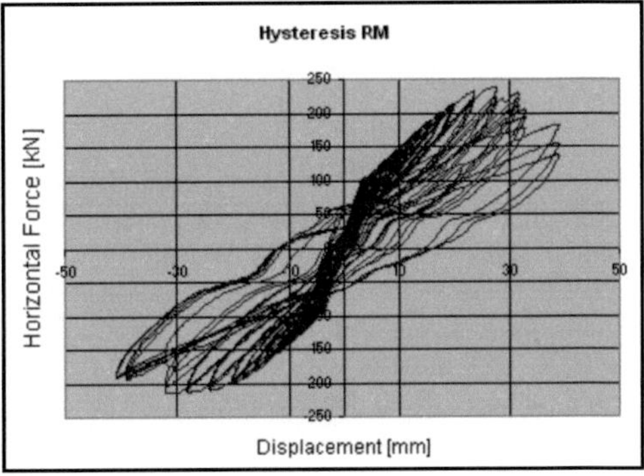

Fig. 7   Hysteresis for the URM wall (left),   Hysteresis for the RM wall (right)

4.2 Dynamic tests on a full building

Motivated by buildings damaged in the L'Aquila earthquake (2009) a building with the typical archetype with natural stones and a size from 5.8 m high, 5.8 m long and 4.4 m width was built on a shaking table in Pavia (Italy) at the EUCENTER. A uniaxial shaking table simulated an earthquake and in five increasing amplification steps from 0.05 g peak ground acceleration (PGA) to 0.4 g. After the last earthquake loading the building was nearby destroyed and had several big cracks (see figure 8).

The test of the URM structure exhibited common failure modes. The very soft wooden slab at the ground level led to out of plane bending failures in the front side (failure no. 1 and no. 2 in figure 8). Diagonal bending/tension cracks trough the mortar joints (no. 5) over the window locations were the most important failures, due to the fact that the front corner in figure 8 was shortly before collapse. Only the wooden beam held this part together with friction and the roof load. A joint sliding (no. 3), for the "in-plane" walls occurred in the cross between windows and doors and on the bottom between the doors (no. 6). The shear cracks at the "beam"-part (no. 5) resulted in the most moving point at the front and was the reason for the high deformation in this corner area. An existing eccentricity though the asymmetrical arrangement of the wall stiffness led to an additional torsion moment which was increased after cracks in the front side and shear point shift toward the walls without openings. This led to very high accelerations in the point A of about 0.9 g. High local deformations in this area caused different orientated cracks in this corner region (no. 4)

After repairing the displaced roof and the wooden slab as retrofitting tool the eq-grid system was used to repair all the cracks in the masonry which occured in the tests before. The tests of the repaired building showed no cracks after 0.4 g PGA, the maximum level of the URM building. Further tests at 0.5 g led to small micro cracks at the window locations. The maximum possible PGA for table was reached after 0.6 g. At this stage cracks in the glass fibres occured local under both windows at the front side which is seen in figure 8. Also delamination between the first layer mortar and textile was observed, but the total building was in good shape. Further information have been written in [1].

Fig 8   Cracked URM building (left),   Reinforced building in the laboratory (right)

5   Conclusions

Both presented fibre systems provide the engineer with a new tools for the seismic retrofit of unreinforced masonry structures. In this study the test results for out of plane, in plane, static, dynamic and on full masonry structures are presented, but much more knowledge was collected with additional small test samples and much more parameters. Together all results show the significant increase in strength and ductility which is the key for many old existing buildings in earthquake risk zones. The combination of both systems which are designed for indoor and outdoor application could be a new technology for the future in earthquake engineering.

References

[1]   Urban M., Stempniewski L.: Reinforcement and measurement method for earthquake damaged masonry buildings tested on a shaking table, 3rd ECCOMAS conference on computational methods on structural dynamics and earthquake engineering, Karlsruhe 2011

[2]   Wallner C.: Erdbebengerechtes Verstärken von Mauerwerk durch Faserverbundwerkstoffe, Dissertation, Karlsruhe, 2008

[3]   Schwegler G.: Verstärken von Mauerwerk mit Faserverbundwerkstoffen in seismisch gefährdeten Zonen, Dübendorf, 1994

[4]   Münchner Rückversicherung: Bedeutende Naturkatastrophen von 1980 – Juni 2010 – Die 10 teuersten Erdbeben für die Volkswirtschaft, 2010

[5]   Münich J. C.: Hybride Multidirektionaltextilien zur Erdbebenverstärkung von Mauerwerk – Experimente und numerische Untersuchungen mittels eines erweiterten Makromodells. Dissertation, Karlsruhe, 2010

[6]   Hamoush S.: Out-of-plane behavior of surface-reinforced masonry walls, 2002

[7]   Hamilton H. R., Dolan C. W.: Flexural capacity of glass FRP strengthened concrete masonry walls, 2001

[8]   Reinhorn A. M., Madan A.: Evaluation of tyfo-s fiber wrap system for out of plane strengthening of masonry, preliminary test report, 1995

[9]   Holberg A. M., Hamilton H. R.: Strengthening URM with GFRP composites and ductile connections, 2002

[10]   Tumialan J. G. et al: Flexural strengthening of URM walls with FRP laminates

[11]   Velazquez-Dimas J. I., Ehsani M. R.: Modeling out-of plane behaviour of URM walls retrofitted with fiber composites, 2000

[12]   El Gawady M.: Seismic on-plane behaviour of URM wall upgraded with composites, 2004

Seismic performance evaluation of buckling restrained braces and frame structures

Ádám Zsarnóczay

Department of Structural Engineering,
Budapest University of Technology and Economics (BME),
Műegyetem rkp. 3-9. Kmf. 85., Budapest H-1111, Hungary
Supervisor: László Gergely Vigh

Abstract

The objective of my research is the development of standardized specifications for design and verification of frame structures employing diagonal Buckling Restrained Braces (BRB). BRBs are displacement dependent anti-seismic devices capable of significant energy dissipation when subjected to extensive cyclic loading. A series of cyclic loading tests were performed at BME and an element level BRB model has been developed and verified with experimental data. It is being used to create numerical models of several braced frame archetypes. These frames are subjected to a set of ground motions using nonlinear incremental dynamic analysis. As a result, the probability of failure for each structural archetype is evaluated and eventually an appropriate design procedure is developed.

1   Introduction

Buckling Restrained Braces (BRB) are displacement dependent anti-seismic devices (Fig. 1). [1] BRBs are used as diagonal members of frames designed for lateral load resistance. The advantageous dissipative properties of the steel material are taken with these devices to the structural level, resulting in structures capable of significant energy dissipation. [2] Under seismic excitation the axial load on the diagonal BRB members of Buckling Restrained Braced Frames (BRBF) is resisted by the slender steel core of the braces. The cross-section of the steel core is reduced in its middle so-called yielding zone in order to ensure early yielding and thus efficient ductile behaviour. In order to prevent global buckling of such a slender beam under compressive loading, the steel core of the braces is surrounded by a concrete filled steel hollow section providing continuous lateral support. The steel core and the supporting concrete surface is decoupled by a layer of air, therefore the concrete section cannot participate in the axial load resistance.

The concept of BRBs had been developed in Japan at the end of the 1980s. It appeared in the United States at the end of the last century and by now its design is regulated in engineering standards as a displacement dependent dissipative lateral load resisting solution [3]. Eurocode 8 (EC8) [4] – the European Standard for Seismic Design of Structures – does not include specifications regarding the design of such devices. Instead they are controlled by a separate standard for anti-seismic devices, the EN 15129 [5]. This document focuses only on experimental analysis and quality control, leaving practicing European earthquake engineers without a simplified – behaviour-factor based – standardized design procedure for structures with BRBs. Thus, currently these dissipative structural systems practically have to be verified with nonlinear analysis as per EC8. This requires more computational resources and engineering expertise, which limits the applicability of BRBs and other innovative anti-seismic devices.

The objective of the presented research is to propose a design procedure of BRB-based structural solutions. The procedure shall be compatible with the existing elastic modal response spectrum analysis and capacity design principles in EC8. Therefore an appropriate behaviour factor (represents the energy dissipation capability) and a set of capacity design rules regarding the design of elements and connections in Buckling Restrained Braced Frames (BRBF) is to be developed. Evaluation of the design procedure is based on a numerical BRB model validated by laboratory experiments. Capacity design rules and the corresponding behaviour factor are determined by performing global nonlinear static and dynamic analyses and calculating the probability of failure for typical building configurations (so-called archetype structures).

Proc. of the 9th *fib* International PhD Symposium in Civil Engineering, July 22 to 25, 2012,
Karlsruhe Institute of Technology (KIT), Germany, H. S. Müller, M. Haist, F. Acosta (Eds.),
KIT Scientific Publishing, Karlsruhe, Germany, ISBN 978-3-86644-858-2

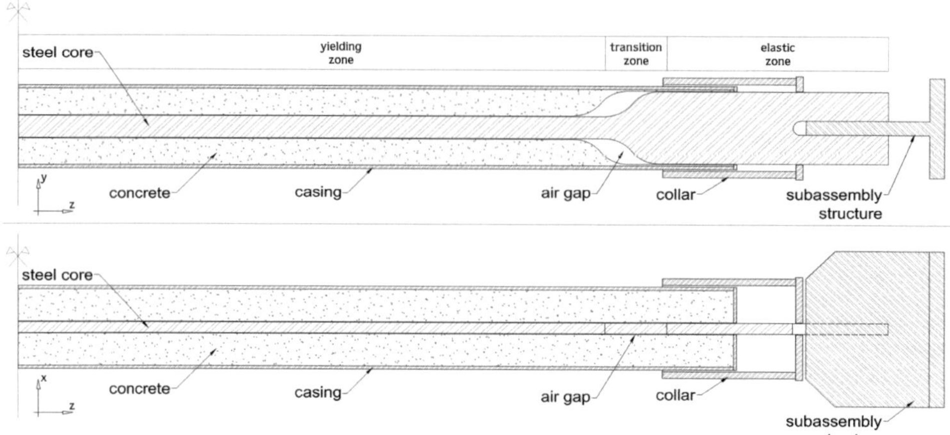

Fig. 1 Main components of buckling restrained braces are illustrated on longitudinal sections

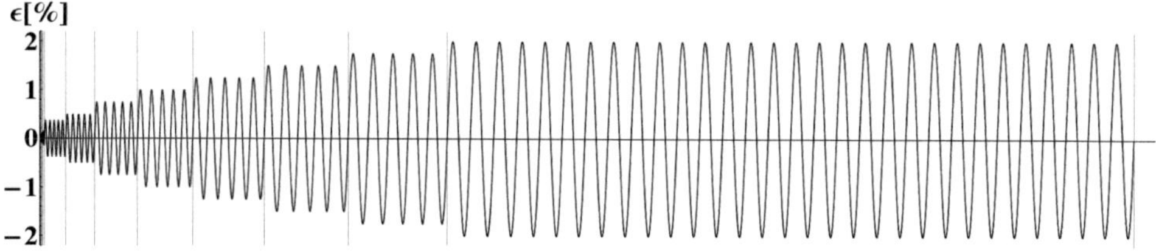

Fig. 2 Displacement dependent load protocol used for testing the EWC 800 specimens

2 Element level numerical model

2.1 Laboratory tests

A total of six laboratory tests have been performed on BRB specimens at the Budapest University of Technology and Economics (BME). Table 1 summarizes the main characteristics of the specimens. The quasi-uniaxial cyclic load tests were designed in accordance with EN 15129 specifications. Recommendations of ECCS [6], AISC [3] and US professionals familiar with the BRB system [7] were also taken into account. Braces were loaded in a custom loading frame by hydraulic jacks using a displacement controlled loading protocol (Fig. 2). Note that the top of the specimen is moved 50 mm in the horizontal direction, therefore the brace is not perfectly vertical (Fig. 3 (left)). This measure takes the second-order moments into account, which act on BRBs due to the impeded rotation of their connections. The loading protocol significantly surpasses the requirements of EN 15129 to enhance the evaluation by subjecting BRBs to a more diverse set of load cycles, thus simulating actual seismic excitation in a more realistic manner. Fig. 3 (right) shows experimental force-displacement curves for two tested specimens. Results confirm the stable, plastic hardening behaviour with significant ductility and energy dissipation capabilities.

Table 1 Main characteristics of tested BRB specimens

specimen(s)		EWC 500 A EWC 500 B	EWC 800 A EWC 800 B	600 BCE	825 BCE
total brace length [mm]		2960	2960	2760	3120
yielding zone	length [mm]	2000	2000	1802	2198
	cross-section [mm]	20 x 25	20 x 40	15 x 40	15 x 55
core material		S235	S235	S235	S235
actual load bearing capacity tension/compression [kN]		199.5/232.1	323.7/442.0	230.9/270.9	317.3/392.3

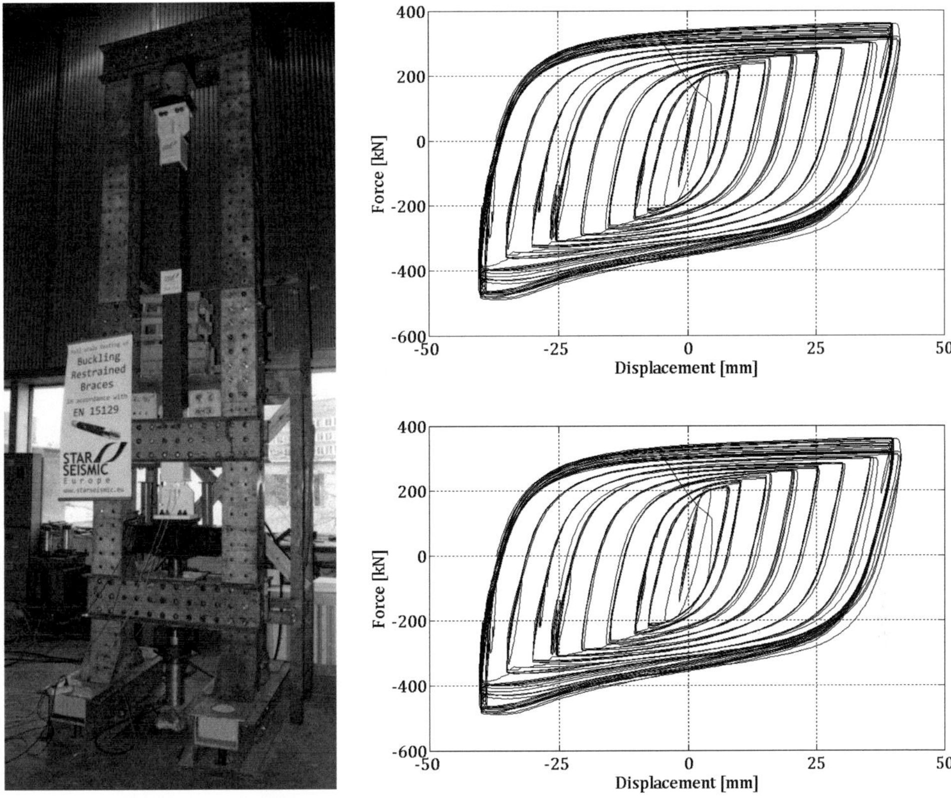

Fig. 3   Photo of the loading frame (left) and two typical experimental force-displacement relationships (hysteresis loops) (right)

2.2   Numerical BRB model

The objective at this phase was to develop a numerical model in a Finite Element Modelling (FEM) environment that can represent BRB behaviour accurately while sparing computational resources. A complex BRB model made of solid elements has already been developed for detailed analysis, but it was too expensive computationally to be applicable in the global analysis of frame structures. Therefore a simple beam model was made, where the BRB characteristics are simulated with a complex nonlinear material applied at simple beam elements (Fig. 4). Due to its simplicity, this latter approach can be used for the analysis of large frame structures with reasonable computational capacity, while accurately capturing the following important behavioural characteristics:

- Bauschinger effect
- plastic strain hardening under tension and compression
- cyclic hardening for a large number of load cycles at different displacement levels
- asymmetric hardening under tension and compression
- monotonic loading curve

The numerical model was developed in the OpenSEES FEM environment [8]. A complex material is created by connecting three separate materials in parallel. The parallel placement means that strain in the finite element equals strain in each material component, while stress is the sum of the stresses in each material component. The components are described in Table 2.

The model was calibrated with data from one experiment and validated with the other experimental results. The experimental and numerical hysteresis curves on Fig. 5 illustrate how accurately the developed model is able to follow real behaviour.

Table 2   Individual material components of the complex artificial BRB material

material type	characteristics
Steel 02	- strain hardening models behaviour under monotonic loading - isotropic hardening models asymmetric cyclic behaviour - configurable smooth transition from elastic to plastic behaviour
Hardening	- symmetric combined kinematic and isotropic hardening - isotropic hardening is used to model the symmetric part of cyclic hardening - kinematic hardening is not used
Pinching4	- designed to represent degradation under cyclic loading - used to model the pinching (increasing stiffness) effect under large strains in compression - practically negligible tensile resistance

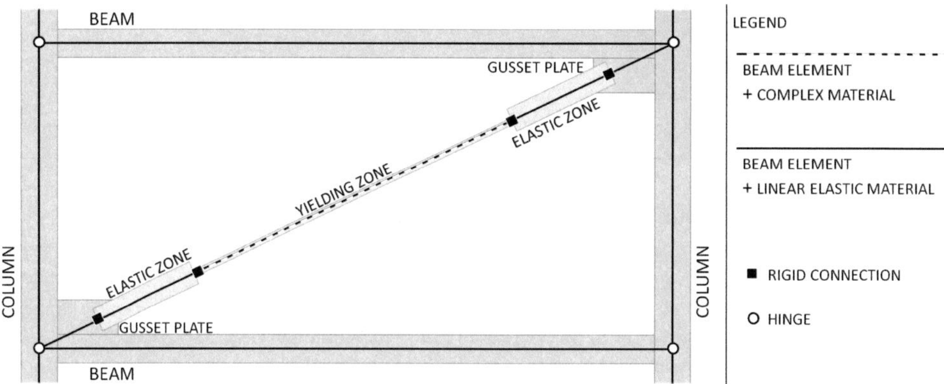

Fig. 4   Numerical model of BRB with simple beam elements

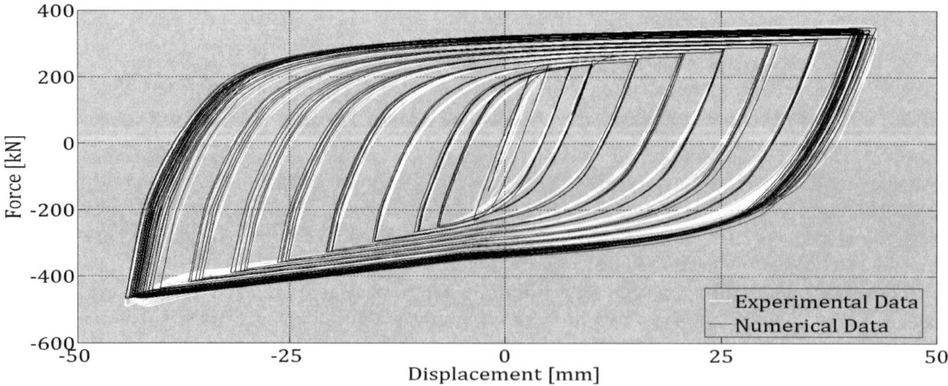

Fig. 5   Comparison of numerical and experimental hysteresis curves to validate the BRB model

3 Design procedure development

The developed element level numerical model is used to analyse the global behaviour of frame structures with diagonal BRB elements. The methodology for this analysis is based on the recommendations of FEMA 965 [9]. The objective is to develop a framework that can evaluate different design procedures. The key steps of this procedure are explained in the following.

3.1 Assembly of structural archetypes

First, the typical structural configurations and their main parameters need to be assembled. Accurate description of a frame structure for global analysis requires 10-15 attributes like number of stories, bay size, storey height, steel material type, brace topology etc. The numerous parameters define a large problem space where a discrete number of specific archetypes that represent the typical applications of the system shall be identified. The archetypes are analysed in detail in the following steps. The number of selected archetypes is limited by time and computational constraints and is usually not more than 30.

3.2 Assembly of ground motion sets

The next step is the assembly of a statistically independent set of ground motion records that is representative of the seismic characteristics of the region under examination. Currently a set of 22 records included in FEMA 965 is used for the analysis while the framework is still under development. This set will be revised by the author with the help of seismologists to better describe European ground motion features.

3.3 Seismic design procedure

A design procedure is defined and each structural archetype shall be designed with it. The procedure is based on elastic modal response spectrum analysis included in EC8. Dissipative structures are designed with the help of system-specific capacity design specifications. The applicable seismic load reduction due to energy dissipation (the behaviour factor in practice) depends on how strictly these specifications are followed. Currently there are no specific capacity design rules for BRBFs in EC8, thus the initial design procedure is based on design on conventional braced frames, but modified to take the special characteristics of BRBs into account.

3.4 Incremental Dynamic Analysis (IDA)

During the IDA each designed archetype is subjected to each ground motion record. IDA is a nonlinear time-history analysis, therefore unlike modal response spectrum analysis it provides accurate internal load levels, stresses and strains at each time step and it can determine the plastic deformations of the structural members. IDA is incremental, because the intensity of each ground motion is increased incrementally from a sufficiently low level (where all of the structural members remain elastic during the entire earthquake event) to a maximum that the structure can resist without exceeding force or deformation limits of the applicable standard.

Phenomena leading to structural damage and eventually collapse shall either be included in the numerical analysis (e.g.: P-Δ effect), or evaluated after each analysis separately (e.g. BRB failure due to extensive plastic deformation). Fig. 6 shows an example of IDA results. The spectral acceleration levels (acceleration experienced by a single degree of freedom system with the same natural period as the BRBF under consideration) are plotted versus the maximum interstorey drift values from each analysis on Fig. 6 (left). Each curve on the figure represents the structural response to different intensities of one ground motion record. Note how increasing ground motion intensity introduces inelastic behaviour (increasing drift increments for the same acceleration increment). Also note how the same acceleration level leads to very different responses and thus different interstorey drift levels. This highlights the inherent uncertainty when acceleration alone is used to define an earthquake event.

This uncertainty is expressed by calculating the fragility curve of the structure that describes the probability of failure for the given archetype depending on spectral acceleration. (dashed curve on Fig. 6 (right)) The fragility curve is a cumulative distribution function of a lognormal distribution with the same median value (highlighted on Fig. 6) as the analysis results. Variance of the fragility curve due to the uncertainty in earthquake events is currently taken as 0.4 according to the recommendation of FEMA 965 [9].

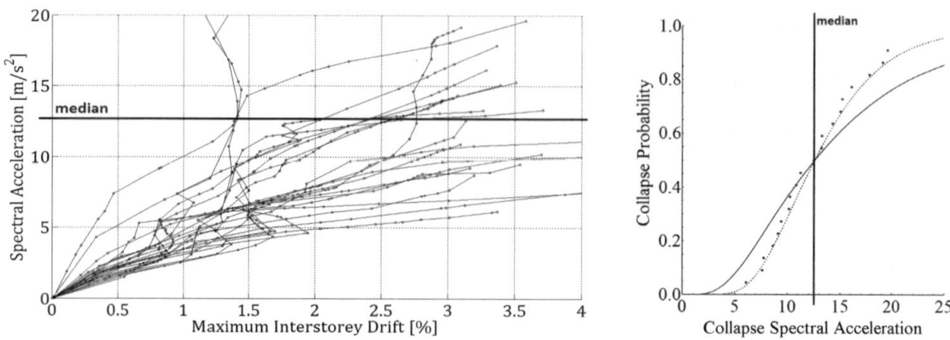

Fig. 6   Acceleration-interstorey drift relationship of one archetype for different ground motion records (left) and the calculated fragility curve (right)

Other sources of uncertainty in the methodology (numerical model, experimental data, design procedure etc.) are taken into account by increasing the variance of the lognormal distribution. The solid line on Fig. 6 (right) shows how an increased variance affects the results. Note the significant increase in the probability of failure at low acceleration levels.

Fragility curves are used for design procedure evaluation. An appropriate design procedure shall ensure that the probability of failure for the designed structure at the design spectral acceleration level is below a certain threshold that is in agreement with acting European Standards. The design procedure shall be improved in an iterative process until it reliably fulfils this requirement.

4   Concluding remarks

The presented IDA methodology for global BRBF analysis is being implemented into an application based on the OpenSEES environment. This application is going to be used to analyse the seismic performance of several BRBF configurations. The expected result is a proposal for a standardized BRBF design procedure and the applicable seismic design parameters (e.g.: behaviour factor).

5   Acknowledgements

The results discussed above are supported by the grant TÁMOP-4.2.2.B-10/1—2010-0009. The authors would like to express their gratitude to Star Seismic Europe Ltd. for the specimens and the BRB design details.

References

[1]   Watanabe A., Hitomi Y., Saeki E., Wada A., Fujimoto M.: Properties of Brace Encased in Buckling-Restraining Concrete and Steel Tube. In: Proceedings of Ninth World Conference on Earthquake Engineering (1988) IV. pp. 719-724

[2]   López, W.A., Sabelli, R.: Seismic Design of Buckling Restrained Braced Frames. In: Steel Tips (2004)

[3]   AISC: ANSI/AISC 341-06 Seismic Provisions for Structural Steel Buildings (2005)

[4]   CEN: EN 1998-1:2008 Eurocode 8: Design of structures for earthquake resistance – Part 1: General rules, seismic actions and rules for buildings, (2008)

[5]   CEN: EN 15129:2010 Anti-seismic devices (2010)

[6]   ECCS: Recommended Testing Procedure for Assessing the Behaviour of Structural Steel Elements under Cyclic Loads (1986)

[7]   Sabelli, R.: Recommended Provisions for Buckling-Restrained Braced Frames. In: Engineering Journal (2005) Q4, pp. 155-175

[8]   UC Regents: The Open System for Earthquake Engineering Simulation (OpenSEES)

[9]   ATC: FEMA P965 Quantification of Building Seismic Performance Factors (2009)

Seismic performance evaluation of RC bridge piers subject to combined earthquake loading and material deterioration in aggressive environment

Mohammad M. Kashani

Earthquake Engineering Research Centre (EERC), Department of Civil Engineering
University of Bristol,
Queens Building, University Walk, Bristol, BS8 1TR, United Kingdom
Supervisors: Adam J. Crewe and Nicholas A. Alexander

Abstract

The aim of this research is to investigate the effect of different degrees of corrosion on the ultimate strength, ductility and failure mode of Reinforced Concrete (RC) bridge piers subject to earthquake loading using experimental and analytical studies on medium scale bridge piers at the Earthquake Engineering Research Centre (EERC) of the University of Bristol.

The experimental studies in this project have three stages: a) testing of corroded reinforcement to investigate the effect of corrosion on the nonlinear stress-strain behaviour of reinforcing bars under monotonic and cyclic loading, b) monotonic compression tests on corrosion damaged well-confined RC short columns to investigate the effect of corrosion on confined concrete behaviour and the overall stability of corroded bars under high compression loads in the column and c) reaction wall tests on scaled bridge piers to investigate the response of corrosion damaged bridge piers subject to cyclic loading. Based on the experimental tests at stages (a) and (b) new constitutive material models are being developed to take into account the effect of long-term material deterioration. Finally these material models will be incorporated in the numerical simulation of experimental tests at stage (c) by developing a multi-mechanical nonlinear finite element code using fibre-section discretization technique.

A selection of the experimental results of the monotonic tests on the corroded reinforcing bars in tension and compression (including buckling) and its effect on inelastic section response (moment-curvature) of corrosion damaged RC bridge piers are reported in this paper.

1 Introduction

Over the past couple of decades researchers have made significant efforts to improve the general knowledge of seismic design and analysis of structures in earthquake prone regions. As a result of these efforts the Performance-Based Earthquake Engineering (PBEE) framework has been developed which is currently very well stabilised among researchers and practising engineers around the world. Based on the observed response of existing structures in recent large earthquakes, it is clear that most structures designed and constructed after 1990 have performed well. However there are a significant number of older major infrastructure artefacts that are located in an aggressive environment and are suffering from the material aging and deterioration. In order to provide a durable and reliable solution, the functionality, safety and the service life of various infrastructure artefacts must be estimated based on scientific and engineering knowledge. Therefore for existing deteriorated infrastructure a rational maintenance plan should be employed in accordance with their current condition. Climate change, sustainability issues and limited funding in the recent years means there is a need for a new bridge management system that can capture all this complexity into a single comprehensive framework. This has led researchers around the world to consider extending the existing performance based earthquake engineering framework to include the long term material performance.

Among the different deterioration mechanisms, corrosion of reinforcing steel is the most common reason for the premature deterioration of RC structures in a chloride laden environment. This leads to loss of the steel cross section and weakening of the bond and anchorage between concrete and reinforcement, which directly affects the serviceability and ultimate strength of concrete elements within a structure. This will increase the probability of failure of deteriorated structures in big earthquakes. It can also result in significant permanent damage in smaller earthquakes which will increase the whole

Proc. of the 9th *fib* International PhD Symposium in Civil Engineering, July 22 to 25, 2012,
Karlsruhe Institute of Technology (KIT), Germany, H. S. Müller, M. Haist, F. Acosta (Eds.),
KIT Scientific Publishing, Karlsruhe, Germany, ISBN 978-3-86644-858-2

life cycle cost (WLCC) of the structure. In the following sections a solution procedure to this complex problem is discussed.

2 Effect of Corrosion on Nonlinear Stress-Strain Behaviour of Reinforcing Steel

2.1 Effect of Corrosion of Tensile Properties of Corroded Bars

In order to simulate the corrosion of embedded steel reinforcement in concrete a total of 5 reinforced concrete specimens were cast. Two specimens dimensioned 200x150x500, designed for tension tests, incorporated 8mm and 12mm diameter reinforcing bars and three specimens dimensioned 250x250x500, designed for buckling tests, incorporated 12mm diameter reinforcing bars. An accelerated corrosion process was employed for corrosion simulation. This comprised an electrochemical circuit using an external power supply. The reinforcing bars act as an anode in the cell and an external material acts as the cathode. A schematic arrangement of test setup is shown in Fig. 1 (left) and reinforcing bars after corrosion are shown in Fig 1 (right). The percentage of corrosion is calculated using averaged measured mass loss along the reinforcement in relation to the uncorroded bars and the stress is calculated based on the mean reduced diameter relative to average mass loss (Mean Stress) [1]. The properties of reinforcement used in the experiment are E_s = 210GPa, f_y = 524MPa, ε_u = 0.0466 for 8mm diameter bars and 0.06 for 12mm diameter bars and the average f_u/f_y = 1.2.

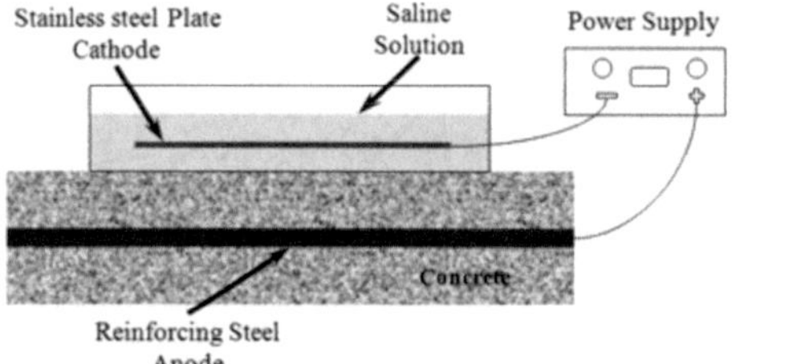

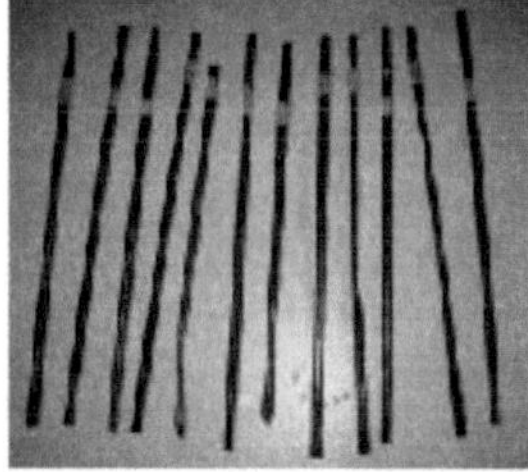

Fig. 1   Accelerated corrosion setup in the laboratory (left) and reinforcing bars after corrosion (right)

The results of tension tests show that corrosion levels up to about 15% don't have a significant effect on stress-strain curves. However, once the corrosion level is greater than 15% a significant drop occurs in plastic deformation capacity and the residual capacity of the corroded bars. This is similar to the results from previous studies which used British manufactured reinforcement [1]. Fig. 2 shows the observed stress-strain curves of tension tests based on average reduced areas and the relevant code requirements for ductile design (BS 4449-2009). It should be noted that a 100mm gauge was used to measure the bar extension in these tests. In many cases the rebar fracture occurred within this gauged section but because the point of fracture depends on the pitting corrosion the fracture sometimes occurred outside the gauge, as indicated in the Fig. 2, which affected the strain recorded at failure.

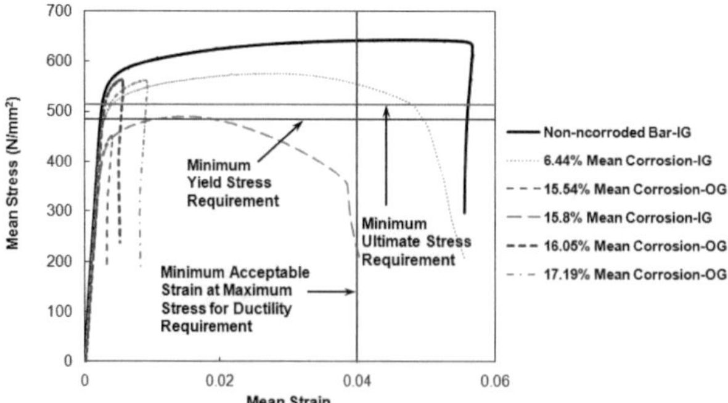

Fig. 2   Mean Stress-Strain Curves of Tension Tests for the 8mm Diameter Bars (IG = Failure Inside the Gauge, OG = Failure Outside the Gauge)

2.2   Effect of Corrosion on Inelastic Buckling of Corroded Bars

A total of 57 compression tests were carried out on corroded bars with different effective lengths. The buckling lengths of bars (slenderness ratios) considered in the experiment were chosen based on the ratio of spacing of horizontal ties in common construction of RC columns (L) to bar diameter (D) known as L/D ratio. The L/D ratios tested in this experiment are 5, 6.5, 8, 9.5, 12, and 13.5.

As expected reinforcement of the group of L/D=5 and 6 showed more stable behaviour compared to those with greater L/D ratios. Almost all the rebars were quite stable up to the yield stress and instability started after yielding which was then followed by buckling and softening behaviour. Unsymmetrical corrosion resulted an uneven reduction of cross section along the length of the corroded bar. This produced an imperfection which caused an additional stress in the bar due to load eccentricity. Therefore a corroded reinforcement yields earlier and its buckling load is reduced compared to uncorroded reinforcement. In addition, the cross sections of corroded bars are not circular any more. This unsymmetrical shape of cross section creates strong and weak axes in the reinforcement and changes the radius of gyration of corroded bars in different axes. This directly affects the theoretical buckling load of bars. As a result, different buckling collapse mechanisms were observed in each test relative to the distribution of pitting corrosion along the bar. Fig. 3 shows an example of the response of 12mm diameter corroded bars in compression with slenderness ratio of L/D=12.

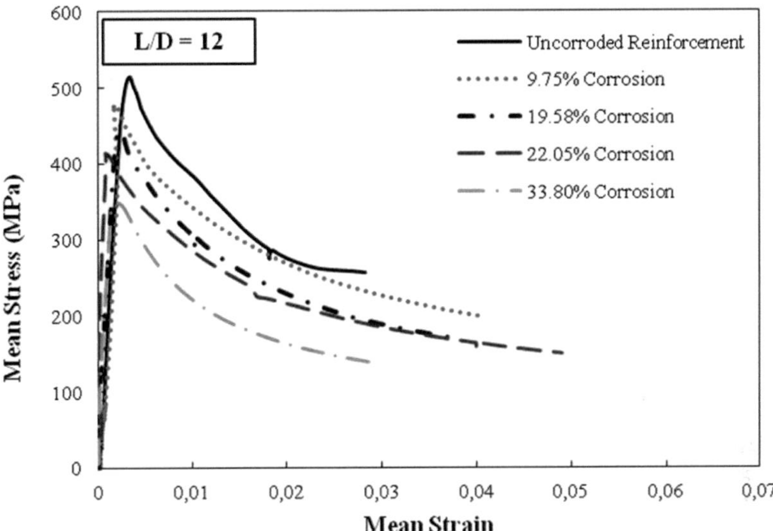

Fig. 3   Mean Stress-Strain Curves of Corroded Bars in Compression

3   Effect of Corrosion on Confined Concrete and Stability of Reinforcement

Reinforced concrete bridge piers exhibit inelastic response when they are subjected to large lateral forces during major earthquakes. Therefore a very important design/assessment consideration is the ductility of structures when subjected to seismic-type loading. This is because the seismic design philosophy relies on energy absorption and dissipation by post-elastic deformation for survival in major earthquakes. It is well known that confinement associated with hoop reinforcement will increase the ductility and energy absorption capacity of RC bridge piers. In addition, as discussed in the previous section, buckling of vertical bars is a very important damage parameter in seismic design and it can be avoided by providing adequate hoop reinforcement.

However the corrosion of vertical bars and hoop reinforcement can change the behaviour of confined concrete and result in instability in the vertical bars under high compression loads. Fig.4 shows buckling of vertical bars during an experiment on scaled bridge piers in the US [2]. The effect of corrosion damage on confined concrete and stability of vertical bars has not been explicitly studied to date. Accordingly in this research some experimental tests on short confined columns are introduced. The accelerated corrosion of columns is now completed and the columns will be tested in the near future.

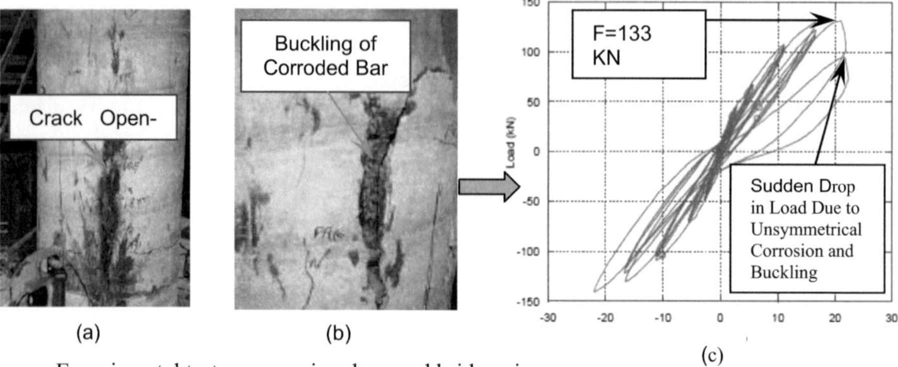

Fig. 4   Experimental test on corrosion damaged bridge pier

The interaction between core and cover concrete, vertical bars and hoop reinforcement is shown in the predicted model of the test specimens in Fig. 5. The results of the experiments will be used to modify the existing analytical confined concrete models to include the corrosion damage. This will then be incorporated in the analytical model that is explained in the following sections of this paper.

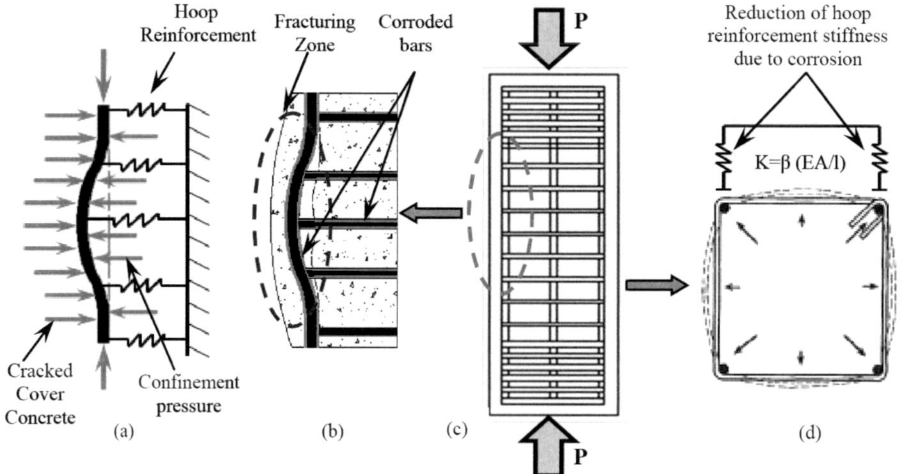

Fig. 5   Predicted Behaviour of corrosion damaged well confined RC column: a) Buckling Model b) Damage Model c) Test Specimen and d) Corrosion induced Volumetric Loss of Confining Steel

4   Experimental Modelling of Corrosion Damaged RC Bridge Piers

Columns are first subjected to an accelerated corrosion process by applying anodic current of specified intensity and time as previously shown in Fig. 1. Only the part of column which is immediately above the base (800mm above base level) is immersed in 5% NaCl solution in a tank. A data acquisition system is set up to monitor the current and voltage applied to the column. After completion of accelerated corrosion the columns are cast into a reaction block. The columns are 250mm by 250mm in section and 2500mm high and are designed to EC2 criteria. One column is a control specimen and the next three are being corroded at different levels. The lateral load will be applied by a 50KN actuator as shown in Fig. 6. Lateral deflection and at the top of the column, rotation at the base and strains in the concrete and steel will be measured through internal and external stain gauges and external displacement transducers.

5 Analytical and Numerical Modelling

In recent years, for nonlinear analysis of RC structures subject to seismic loading, a lot of attention has been given to the development of the fibre element technique. In this method the member cross section is discretized into a number of steel and concrete fibres at selected integration points. The material nonlinearity is then considered through a uniaxial constitutive material model of steel (tension and compression) and concrete (confined core concrete and unconfined cover concrete). More recently, other researchers have attempted to investigate the effect of reinforcement corrosion on the behaviour and response of RC bridges subject to seismic loading through nonlinear finite element analysis using fibre-based section discretization technique [3]. However, due to lack of available data and constitutive nonlinear material models of corroded reinforcing bars their analytical model couldn't represent the real behaviour of corroded structure. This is more critical in compression zones where buckling of bars and confined concrete behaviour make an important contribution to the response of the structure.

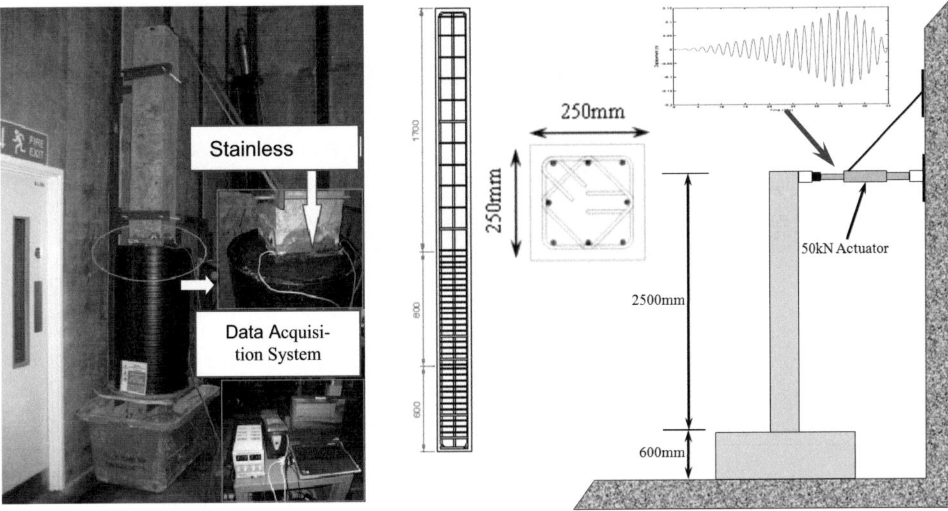

Fig. 6   Experimental Test Procedure and Setup

In the fibre-based section discretization technique the element response is greatly influenced by the moment-curvature response of cross section. Therefore, a moment-curvature analysis has been done for the proposed experimental specimens for both uncorroded and corroded conditions. The moment-curvature analysis is developed based on the fibre-base section discretization technique using MATLAB computer code. An unconfined concrete model is used for cover concrete and a confined concrete model is used for the core concrete [4]. The mechanical properties and cross sectional area of longitudinal reinforcement in tension have been modified using existing mathematical models available in the literature and experimental data [2]. The Dhakal-Makeawa buckling model has been modified based on the experimental tests on corroded reinforcement in compression and is included in the code to model the effect of reinforcement buckling [5] as shown in Fig. 7. The effect of corrosion induced cracking of cover concrete in the compression zone is also considered in the analysis [6]. This analysis is the basis of a multi-mechanical nonlinear finite element code for push-over analysis of corroded bridge piers using a fibre element technique which can account for corrosion damage.

6 Discussion and Conclusion

The effect of corrosion on the stress-strain behaviour of corroded reinforcement has been investigated experimentally. The experimental results show that corrosion has a significant effect on the ductility and residual strength of corroded bars in tension. Corrosion also changes the buckling collapse mechanism of compression reinforcement. Furthermore, unsymmetrical corrosion will cause yielding of the weakest cross section along the length of the bar followed by premature buckling. This is a very complex behaviour which also has interaction with cover concrete and horizontal ties (which may also be corroded). Further experimental and analytical work is needed to study this interaction.

The effect of corrosion on the inelastic response of a corroded RC section has been studied through moment-curvature analysis of the section. The analysis shows that corrosion has a significant effect on the ductility of the section under large strain demands as shown in Fig. 8. This is an important finding and further study is needed in order to provide guidelines for practicing engineers. The effect of corrosion should also be included in seismic assessment codes.

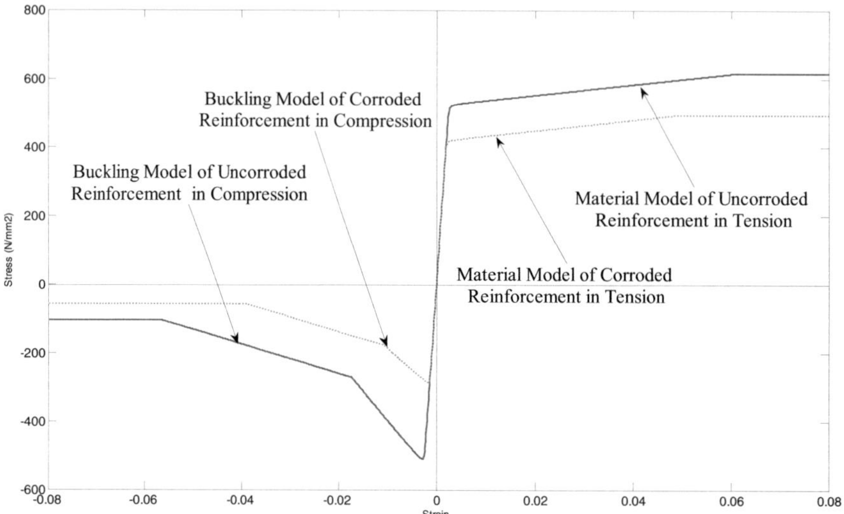

Fig. 7   Non-linear uniaxial material model of corroded reinforcing steel

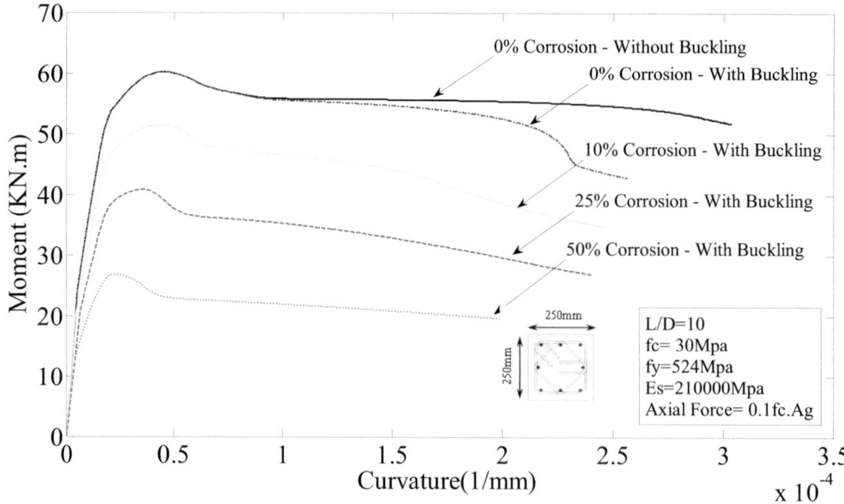

Fig. 8   Moment-Curvature plot of corrosion damaged RC section

References

[1]   Du, Y. G., Clark, L. A. and Chan, A. H. C. "Residual capacity of corroded reinforcing bars." Mag. of Con. Res., (2005) 57(3), 135–147.

[2]   Aquino W., "Long-term Performance of Seismically Rehabilitated Corrosion-Damaged Columns" PhD Thesis, University of Illinois at Urbana-Champaign (2002).

[3]   Ghosh J., and Padgett J. E. "Aging considerations in the development of time- dependent seismic fragility curves." Journal of Structural Engineering (2010) 136 (12), 1497–1511.

[4]   Park, R., Priestley, N. and Gill, W. "Ductility of Square-Confined Concrete Columns" Journal of Structural Division, (1982), 929-951.

[5]   Dhakal, R., and Maekawa, K. "Modeling for postyield buckling of reinforcement." Journal of Structural Engineering (2002) 128(9), 1139–1147.

[6]   Coronelli D. and Gambarova P. "Structural Assessment of Corroded Reinforced Concrete Beams: Modelling Guidelines." Journal of Structural Engineering (2004) 130(8), 1214-1224.

Inelastic seismic analysis of as-built and retrofitted RC frame structures considering joint distortion

Akanshu Sharma

Institute for Construction Materials,
University of Stuttgart,
Pfaffenwaldring 4, 70569 Stuttgart, Germany
Supervisor: Rolf Eligehausen

Abstract

In this work, a new model for predicting the inelastic shear behaviour of joints is suggested that can be used with existing commercial programs and can reasonably accurately capture the shear behaviour of joints. An extension to pivot hysteretic model parameters is developed that can, with only two parameters, capture the hysteretic behaviour of RC frame structures including joint distortion. Both the models have been validated against experimental results at joint and structural level. The model is further extended for analysing the joints retrofitted with a practical retrofitting scheme using fully fastened haunch elements. All the models have been validated against experimental results.

1   Introduction

Non-conforming reinforced concrete (RC) frame structures under the action of seismic forces often display undesirable failure modes such as shear failure of beam-column joints combined with bond failure of reinforcing bars of beams framing into the joints. The deficiencies of joints are mainly caused due to (i) inadequate design to resist shear forces due to inadequate transverse and vertical shear reinforcement and (ii) due to insufficient anchorage capacity in the joint. Under the action of seismic forces, beam-column joints are subjected to large shear stresses in the core. The axial and joint shear stresses result in principal tension and compression that leads to diagonal cracking and/or crushing of concrete in the joint core. These stresses in the joint core are resisted by the so-called strut and tie mechanism [1]. To prevent the shear failure of the joint core by diagonal tension, joint shear reinforcement is needed, which is therefore prescribed by the newer design codes [2, 3, 4]. Moreover, these codes prescribe a large anchorage length of the bars terminating in case of exterior joints, so that a bond failure may be avoided. However, a vast majority of the structures worldwide consist of structures designed prior to the advent of seismic design codes. For the correct assessment of such structures against seismic forces, modelling the inelastic behaviour of joints is essential.

Various models for the joints are proposed by several researchers but the models are generally too complex to be implemented at the structural level, and therefore joints are mostly modelled as rigid. Similarly, associating a hysteretic rule that is not too complex but is still realistic enough is very important. In this work, practically usable and accurate models are reported to realistically model the inelastic response of the structures considering joint distortion. A model to consider the inelastic behaviour of the joints of poorly detailed structures is developed and presented. A practical hysteretic rule based on the extension of "Pivot hysteretic model" is developed for members and beam-column joints and the same is also reported. The analytical models are validated against the experimental results and the efficiency of the models in predicting the seismic response of structures considering joint distortions is shown.

Fully fastened haunch retrofit solution (HRS) has been recently investigated as a viable alternative for retrofitting of beam-column joints of reinforced concrete (RC) frame structures. Certain guidelines for design of haunch elements are available in literature [5]; however, it is of utmost importance to have numerical models to predict the behaviour of the retrofit considering various failure modes possible. In case of fully fastened HRS where the haunch elements are connected to the beams and columns using post-installed anchors, the numerical modelling of joint core as well as of anchors is highly important. This paper presents a method of modelling the beam-column joints of RC frame structures retrofitted with fully fastened HRS. The model has been explained and the results of numerical analysis have been validated with experimental ones.

Proc. of the 9th *fib* International PhD Symposium in Civil Engineering, July 22 to 25, 2012,
Karlsruhe Institute of Technology (KIT), Germany, H. S. Müller, M. Haist, F. Acosta (Eds.),
KIT Scientific Publishing, Karlsruhe, Germany, ISBN 978-3-86644-858-2

The paper focuses on the numerical modelling procedure and hence detailed description of the experiments is out of scope of the paper. However, cross-references are mentioned for each experiment and the detailed information can be obtained from the same.

2 Joint Model

The model proposed in this work uses limiting principal tensile stress in the joint as the failure criteria so that due consideration is given to the axial load on the column. The spring characteristics are based on the actual deformations taking place in the sub-assemblage due to joint shear distortion. Due to the joint shear deformation, γ_j, the column experiences a relative shear displacement of $\gamma_j h_b$, where h_b is the total depth of the beam. This deformation can be divided into two as $\Delta_c = \gamma_j h_b/2$ for the column half above the beam centre line and $\Delta_c = \gamma_j h_b/2$ for the column half below the beam centre line. Also, the beam experiences a rotation of γ_j due to which the beam tip displacement is equal to $\Delta_b = \gamma_j L_b$, where L_b is the length of the loading point from the face of the column. The contribution of joint shear deformation to overall storey drift is considered as shown in Fig. 1 where shear springs in the column portion and a rotational spring in the beam region are assigned [6]. Thus, in a frame analysis by software packages, in addition to hinges assigned at the ends of the members (beams and columns), a joint core may be modelled and hinges are provided in the core region to consider the shear deformations of the joint. The failure criteria for a typical joint is shown in Fig. 2.

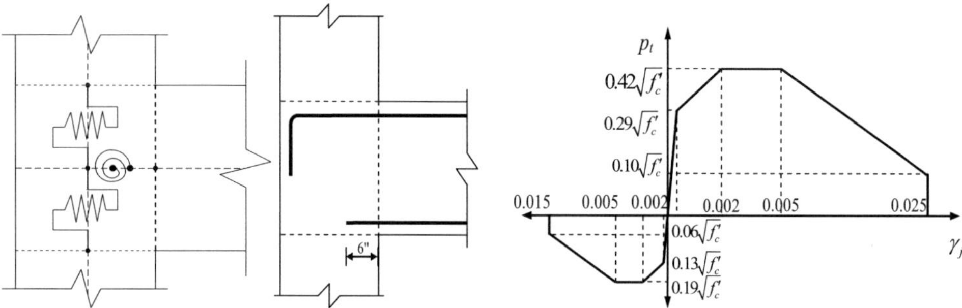

Fig. 1   Joint Springs   Fig. 2   Principal tensile stress-shear deformation relation

The proposed model was validated with experiments. Nonlinear static pushover analysis was performed to get the load-displacement behaviour of the joints that was compared with the envelope of the hysteretic loops obtained from the experiments. Fig 3 shows the results of a full-scale beam-column joint tested with beam bars on top and bottom face bent-in [7], while Fig. 4 shows the same for another joint [7] with top beam bar bent in and bottom beam bar embedded to 150mm inside the joint. The beam section was 300mm X 400mm and the column section was 350mm X 300mm. The total height of joint sub-assembly was 3.0m and the beam length was 1825mm. The complete details of the experiment can be obtained from the reference [7]. Two types of analysis were performed, one considering the joint springs as per the joint model and another assuming joint as rigid.

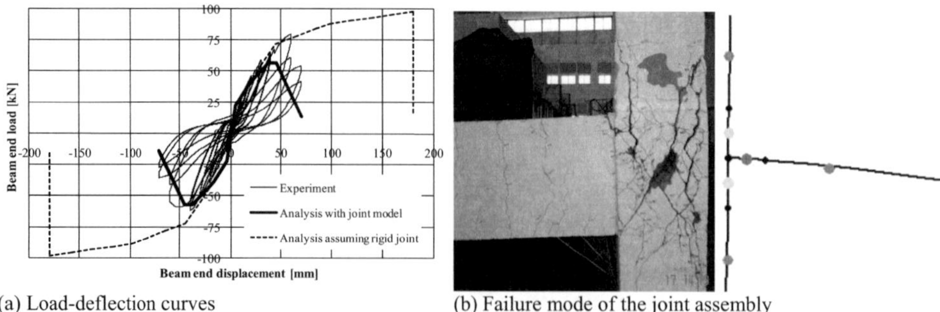

(a) Load-deflection curves   (b) Failure mode of the joint assembly

Fig. 3   Results of joint with beam bars bent in

In Fig. 3(b) and 4(b), the experimental and numerical failure modes are given. The red coloured circles show the hinge in collapse region, while the green one shows the hinge to be in near collapse region. The pink coloured circles show the hinge state close to yield. Fig. 3 and 4 clearly show that analysis without considering joint nonlinearity may lead to quite un-conservative results whereas the results obtained for "with joint model" case are very close to experimental results.

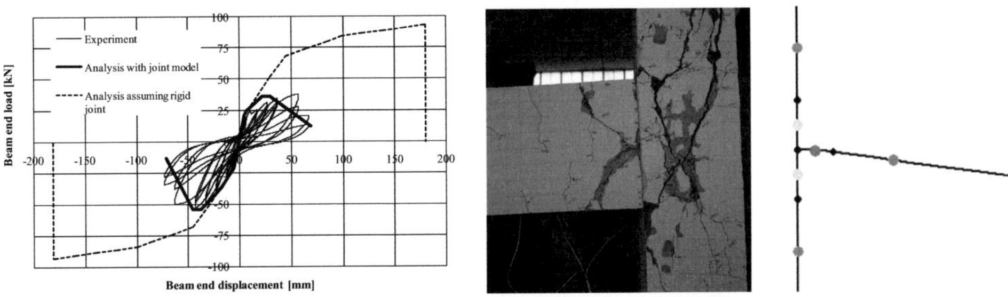

(a) Load-deflection curves　　　　　　　　(b) Failure mode of the joint assembly

Fig. 4　　　Results of joint with top beam bars bent in and bottom bars straight

After validation of the model at joint level, the same model was applied at the structural level. A full-scale experiment was performed by author [8] on RC frame structure under monotonically increasing lateral pushover loads. Fig. 5 shows the general geometric arrangement of the structure. Typical beam sizes were 230mm X 1000mm, while the column sizes varied from 300mm X 700mm to 400mm X 900mm. The slab thickness was 130mm. Inverted triangular lateral load pattern was applied on the structure and the test was performed under load control till failure. The major failure modes observed included flexure and flexure-shear failure of beams and columns, torsional failure of transverse beams, joint shear and bond failures. The complete details of the experiment can be obtained from the reference [8].

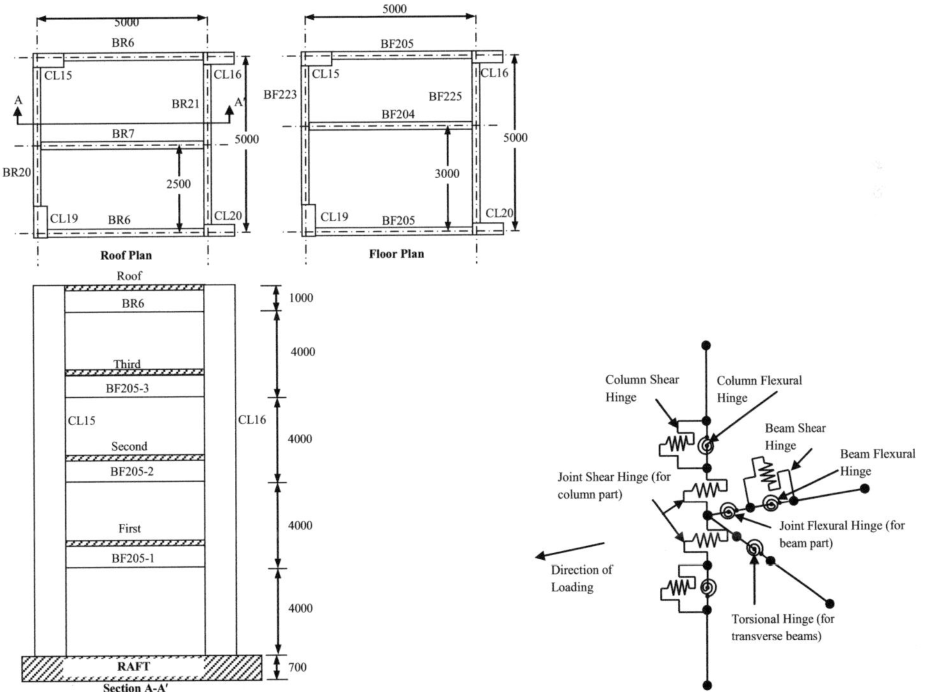

Fig. 5　　　Geometry of structure tested　　　　　Fig. 6　　　Numerical modelling of joint of structure

In the numerical model, flexural hinges, shear hinges, torsion hinges and joint hinges following the joint model were modelled. The typical joint of a structure with springs is shown in Fig 6. Three different analyses were performed for the model. In the first case, only the flexural and shear hinges were considered for the members. In the second case, additionally torsion hinges were considered and in the third case, joint hinges in addition to flexural, shear and torsional hinges were considered. Fig. 7 shows the comparison of experimental and numerical results for the structure. It can be observed that the model with the joint hinges modelled along with other hinges (flexural, shear and torsion) yields the best results not modelling the joint hinges can yield to significantly un-conservative results.

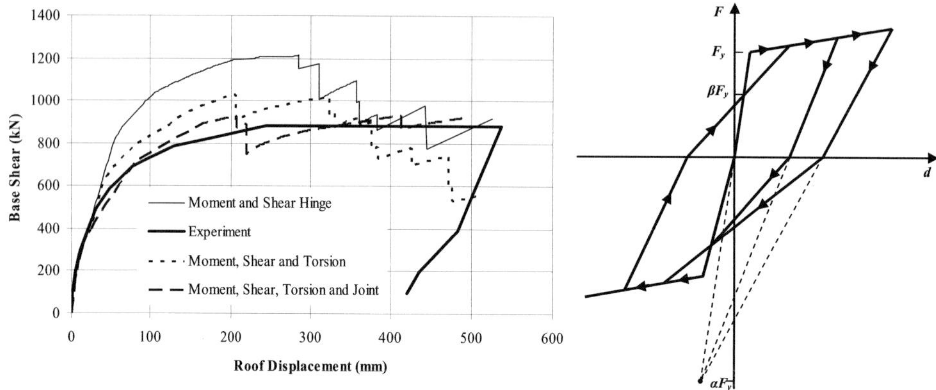

Fig. 7	Validation of model at structural level	Fig. 8	Pivot hysteresis law

3   Pivot hysteresis law

In order to capture complete inelastic behaviour of structures subjected to seismic loads, modelling of hysteretic behaviour of members as well joints is essential. The pivot hysteresis law that was originally developed for circular bridge columns [9] was extended in this work for application on rectangular members and beam-column joints [10]. The experimental observations show that the unloading, back to zero force from any displacement level is generally guided towards a single point in the force-displacement plane, on the idealized stiffness line. Also, all force-displacement paths tend to cross the elastic loading line at approximately the same point. The first point was named as "primary pivot point" and the second point was named as "pinching pivot point". Thus, the model basically needs to define only two parameters, α and β (Fig. 8).

Based on curve fitting of the test data of large number of experiments, using regression analysis and the method of minimum sum of the square of residues, the equations defining 'α' and 'β' for rectangular sections were obtained as [10]

$$\alpha = 0.170 \, k_\alpha + 0.415 \tag{1}$$
$$\beta = 0.485 \, k_\beta + 0.115 \tag{2}$$
$$k_\alpha = p_t / ALR \tag{3}$$
$$k_\beta = (ALR)^{0.25} \times (p_{sh})^{0.2} \tag{4}$$

p_t = percentage longitudinal reinforcement,
p_{sh} = percentage volumetric shear reinforcement
ALR = Axial load ratio

Similarly, the relations for 'α' and 'β' parameter for poorly detailed joints with beam bars bent in were obtained as

$$\alpha = 4.4 - 25.38 * (ALR); \text{ for } 0 \leq ALR \leq 0.13, \text{ and} \tag{5}$$
$$\alpha = 1.0 \text{ for } ALR > 0.13 \tag{6}$$
$$\beta = 0.125 + 0.44 * (ALR) \tag{7}$$

To demonstrate the efficacy of the model in predicting the cyclic hysteretic behaviour of non-seismically detailed RC structures, the results on a frame structure as tested by Calvi et al [11] were used. Fig 9 shows (a) experimental hysteretic loops as obtained for the structure and (b) shows the analytical hysteretic loops. This shows that the analytical predictions for the hysteretic behaviour are very close to the experimentally observed ones. Thus, the hysteresis law with suggested parameters is able to capture the hysteretic behaviour of members, joints as well as of frame structures quite well.

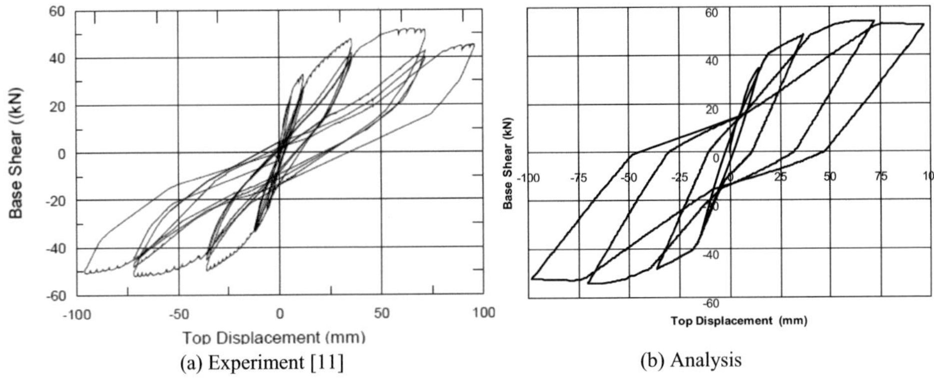

(a) Experiment [11]　　　　　　　　(b) Analysis

Fig. 9　　Validation of hysteresis law at structural level

4　　Modelling of haunch retrofit solution

Fully fastened haunch retrofit solution (HRS) was tested by author to verify its efficacy in preventing the brittle joint shear failure [12]. Fig. 10 shows (a) experimental setup, and (b) modelling of joint retrofitted with HRS as tested by the author [12]. The geometry of the joint was same as that shown in Fig. 3 and the HRS was connected to the joint members using post-installed bonded anchors. In numerical model, in addition to the springs for joint panel and member hinges, axial springs were used to model the anchor group behaviour, with the typical characteristics as shown in Fig. 11.

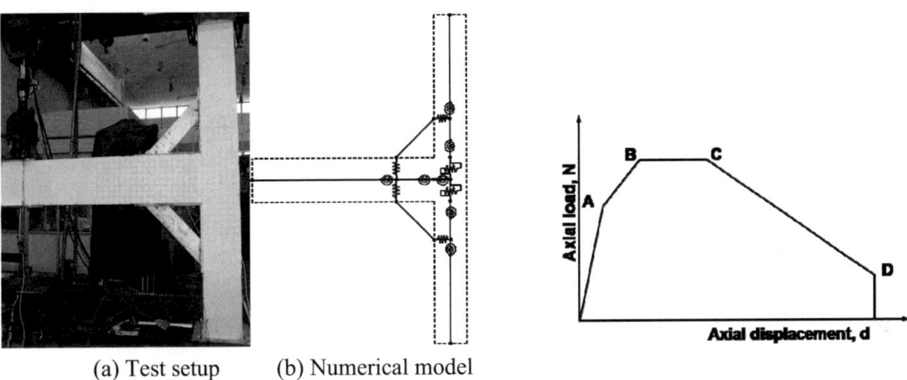

(a) Test setup　　　(b) Numerical model

Fig. 10　　Joint with fully fastened HRS　　　　　Fig. 11　　Modelling characteristics of anchor group behaviour

To develop the characteristics for anchor springs, the following assumptions are made:

- Only tension loads are critical for the anchorage system.
- Anchor system spring loaded in compression can be considered as stiff and linear elastic
- The ultimate load carrying capacity and failure mode can be calculated using CC Method [13]
- Unless special provisions are made, cracked concrete shall be assumed
- The model is applicable only for anchors suitable for cracked concrete

The value of 'N' corresponding to point B and C is equal to load carrying capacity (N_u) evaluated by the CC method [13], while that for point A is considered as 0.8 N_u and for point D as 0.2 N_u. The values of 'd' corresponding to these points depend on the type of failure predicted by the CC method. To consider the behaviour of anchorage system in cracked concrete, the peak load carrying capacity of the anchorage system in cracked concrete can be considered as 70% of the peak load carrying capacity of the anchorage system in uncracked concrete, i.e., $N_{u,cracked} = 0.7*N_{u,uncracked}$ [13]. Fig. 12 shows the comparison of experimental and numerically obtained results of the joint retrofitted with fully fastened HRS [12] using bonded anchors. The comparison shows that the model is able to nicely predict the behavior of joints retrofitted with fully fastened HRS.

5   Conclusions

In this work, models are presented to (i) predict inelastic shear behaviour of joints, (ii) capture the hysteretic behaviour of RC frame structures including joint distortion, and (iii) analyse joints retrofitted with a practical retrofitting scheme using fully fastened haunch elements. All the models are validated against experimental results and the comparison shows their efficacy in realistically predicting the seismic behaviour of as-built and retrofitted structures considering joint distortion.

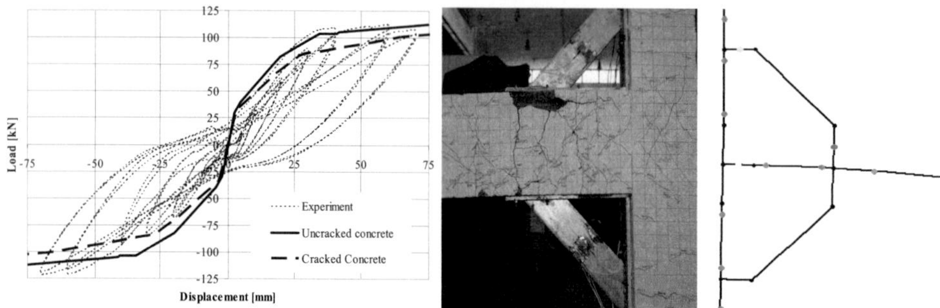

(a) Load-deflection curves   (b) Failure mode of the joint assembly

Fig. 12   Results of joint retrofitted with fully fastened HRS tested by Genesio and Sharma [12]

References

[1]   Paulay T, Priestley MJN: Seismic design of reinforced concrete and masonry buildings. John Wiley publications New York, 1992.

[2]   ACI 318M-11: Building Code Requirements for Reinforced Concrete", ACI Detroit (2008)

[3]   NZS 3101: Design of Concrete Structures. Standards Association of New Zealand (1995)

[4]   IS 13920: Indian Standard ductile detailing of reinforced concrete structures subjected to seismic forces — Code of Practice. Bureau of Indian Standards, New Delhi, 2002

[5]   Pampanin S, Christopoulos C, Chen TH. Development and validation of a metallic haunch seismic retrofit system for existing under-designed RC frame buildings. Earthquake Engineering and Structural Dynamics (2006) 35, pp.1739-1766

[6]   Sharma A, Eligehausen R, Reddy GR: A new model to simulate joint shear behavior of poorly detailed beam–column connections in RC structures under seismic loads, part I: exterior joints. Engineering Structures (2011) 33(3), pp. 1034-1051.

[7]   Genesio G, Sharma A: Seismic retrofit solution for reinforced concrete exterior beam-column joints using a fully fastened haunch - Part 2-1: As-built joints (2010) Test report, No. WS 221/07 - 10/01. University of Stuttgart, Germany.

[8]   Sharma A, Reddy GR, Babu RR, Revanna D, Vaze KK: Experimental Seismic Assessment of Full Scale Non-seismically Detailed RC Structure by Pushover Method. Transactions, Structural Mechanics in Reactor Technology SMiRT 21, 2011, New Delhi: Paper ID#316

[9]   Dowell RK, Seible F, Wilson EL: Pivot Hysteresis Model for Reinforced Concrete Members. ACI Structural Journal (1998) 95(5), pp. 607-617

[10]   Sharma A, Eligehausen R, Reddy GR: Pivot hysteresis model parameters for RC columns, joints & structures. Accepted for publication in ACI Structural Journal

[11]   Calvi GM, Magenes G, Pampanin S: Experimental Test on a Three Storey R.C. Frame Designed for Gravity Only," 12th European Conf on Earthquake Engg, 727, 2002

[12]   Genesio G, Sharma A: Seismic retrofit solution for reinforced concrete exterior beam-column joints using a fully fastened haunch - Part 2-2: Retrofitted joints (2010) Test report, No. WS 221/08 - 10/02, University of Stuttgart, Germany.

[13]   Eligehausen R, Mallée R, Silva J. (2006) "Anchorage in Concrete Construction", Wilhelm Ernst & Sohn, Berlin.

Residual seismic displacements of RC oscillators

Liosatou Eftychia

*Department of Civil Engineering,
University of Patras,
Rio, Patras, 26504, Greece
Supervisor: Michael N. Fardis*

Abstract

Residual seismic deformations are computed for cyclic force-deformation relations typical of RC structures. 729 historic seismic motion records are used as input and lognormal probability distributions are fitted to the residual deformation expressed as ratio to the peak inelastic or the 5%-damped elastic deformation of the response. The force-deformation relation and the reduction of the lateral force resistance have a strong influence on the probability distributions. The ratio of residual deformation to the peak 5%-damped elastic deformation is affected by the fundamental period of the system in the short period range. Velocity pulses in the ground motion also influence the residual deformation ratio in certain of the cases studied. Theoretical results were compared to residual displacements in several hundred cyclic or seismic tests on concrete members or systems that entailed abrupt changes in the response due to local or global failure.

1   Introduction

With the emergence of performance-based seismic design, assessment or retrofitting as a focal point of research, codification and practice, the interest of the structural earthquake engineering community in residual deformations has increased. Collapse prevention and life safety do not monopolize our attention anymore; reparability and usability of our structures are also important. Residual deformations are therefore important, because it is hard to reverse them if they are large.

In this paper the question of the magnitude and prediction of residual deformations of SDoF systems is investigated with the emphasis on the effect of hysteresis rules that represent the cyclic degradation of stiffness, strength and energy dissipation typical of reinforced concrete structures, and on the effect of the presence of velocity pulses in near field records.

2   Calculation of residual deformations of SDoF systems with hysteresis typical of RC structures

The strong dependence of residual deformations on the hysteresis model (notably on the rules governing the unloading/reloading stiffness and the post-yield degradation of strength) has emerged as the most notable outcome of past studies (Riddell and Newmark (1979), Mahin and Bertero (1981), Pampanin et al. (2002)). The most comprehensive of them have dealt with the standard case of a bilinear oscillator (Garcia and Miranda (2005, 2006), which suits well steel structures but is far from representative of RC ones. The present study addresses exclusively concrete systems and aims to capture a range of cyclic behaviors typical of them. To this end, it has adopted the versatile modelling approach of Erberik and Sucuoglu (2004) and Erberik (2008, 2011), which has been tuned to different categories of RC structures, ranging from new earthquake resistant ones to existing substandard construction, with intermediate situations. Erberik has categorized RC buildings as follows:

    - Group A comprises those which exhibit very little cyclic strength degradation, without pinching and are representative of recent good quality RC construction conforming to modern codes.

    - Group B stands for RC structures with moderate cyclic strength degradation and limited pinching. It represents RC construction of fair quality with certain engineered earthquake resistance, but without full conformity to modern-day seismic codes.

    - Group C includes RC systems with dramatic cyclic strength degradation, clearly pinched loops and therefore limited energy dissipation capacity. They are typical of RC structures without engineered earthquake resistance and/or serious deficiencies.

Proc. of the 9th *fib* International PhD Symposium in Civil Engineering, July 22 to 25, 2012,
Karlsruhe Institute of Technology (KIT), Germany, H. S. Müller, M. Haist, F. Acosta (Eds.),
KIT Scientific Publishing, Karlsruhe, Germany, ISBN 978-3-86644-858-2

The afore-mentioned features of the corresponding modelling are adapted to fit within an extension of the Bouc-Wen model by Sivaselvan and Reinhorn (2000). The model is modified for the present purposes to suit the modeling of strength and stiffness degradation and of pinching according to Erberik (2011). The model with degradation of strength and stiffness but without pinching consists of two springs in parallel: an elastic spring and a hysteretic one with: (a) smooth transition from elastic loading to a horizontal monotonic post-yield branch, and (b) straight unloading at a stiffness.

Strength degradation is modelled via a reduction of the force at the initially horizontal monotonic post-yield branch of the hysteretic spring. Pinching is modelled by adding a three-parameter "slip-lock spring" in series to the hysteretic spring. Its tangent flexibility is proportional to the range of inelastic deformations experienced times a Gaussian function of the spring's force.

Implicit in the use of the above hysteresis model to the end of the seismic response is the assumption that there is no abrupt failure that may cause the system to deviate from the gradual loss of stiffness or strength in the Erberik model.

Non-linear time-history analyses of SDoF oscillators with the above hysteresis model are carried out under 729 historic records. According to Mavroeides and Papageorgiou (2003) and Mavroeides et al (2004), 36 of these records exhibit a velocity pulse. The oscillators have periods, T, from 0.1 to 2.5s (at intervals of 0.1s) and a yield force equal to the peak elastic force that they would develop for 5% damping divided by a force reduction factor, R, with values from 2 to 8 at intervals of 1. For consistency with the elastic response and its peak force and displacement, viscous damping equal to 5% of critical is used in addition to the intrinsic hysteretic one of the model. One set of runs is carried out for each one of Erberik's groups.

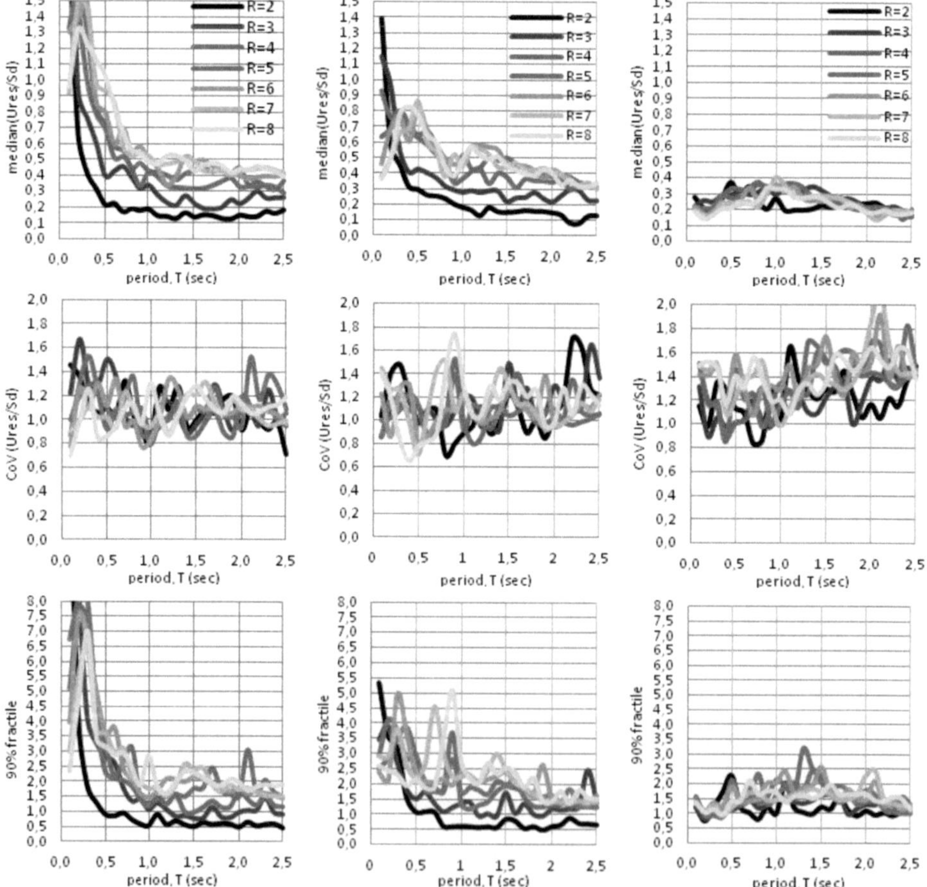

Fig. 1   Residual-to-peak-elastic displacement ratio for 36 records with velocity pulse. Median (*top*); coefficient of variation (*middle*); and 90%-fractile of lognormal distribution (*bottom*) - (*Left column*) Group A; (Middle column) Group B; (*Right column*) Group C.

3 Probability distribution of ratio of residual to peak inelastic or elastic deformations

As computed residual deformations have much larger record-to-record scatter than peak inelastic values, for each set of parameters (i.e., the oscillator's T and R, its Group per Erberik and the presence or not of a pulse), a lognormal distribution is fitted to the time-history results for the suite of records.

Figures 1 and 2 depict the median (top row), the coefficient of variation (middle row) and the 90%-fractile (bottom row) of the ratio of residual deformation, u_{res}, to the peak elastic displacement of the oscillator, $S_d(T)$, (i.e., the 5%-damped elastic spectral displacement). Each column (left, middle, right) refers to one of Erberik's Groups A, B and C. Figure 1 refers to the 36 records with a velocity pulse per Mavroeides and Papageorgiou (2003) and Mavroeides et al (2004); Figure 2 is for the 693 records without a pulse. Figures 3 and 4 repeat the exercise for median and coefficient of variation of the ratio of residual deformation, ures, to the peak inelastic displacement of the response, u_{max}.

To see whether the difference between Figures 1 and 2 and Figures 3 to 4 is not due to the velocity pulse per se but to the distance from the source, analysis results have been also grouped according to the closest distance from the fault. No systematic difference can be identified between the results for distances of less than 5 km from the source, or from 5 to 20 km, or beyond 20 km, which complies with Garcia and Miranda (2005, 2006) and Kawashima et al (1998). Therefore, the pulse and not the distance seems to be the reason for the difference.

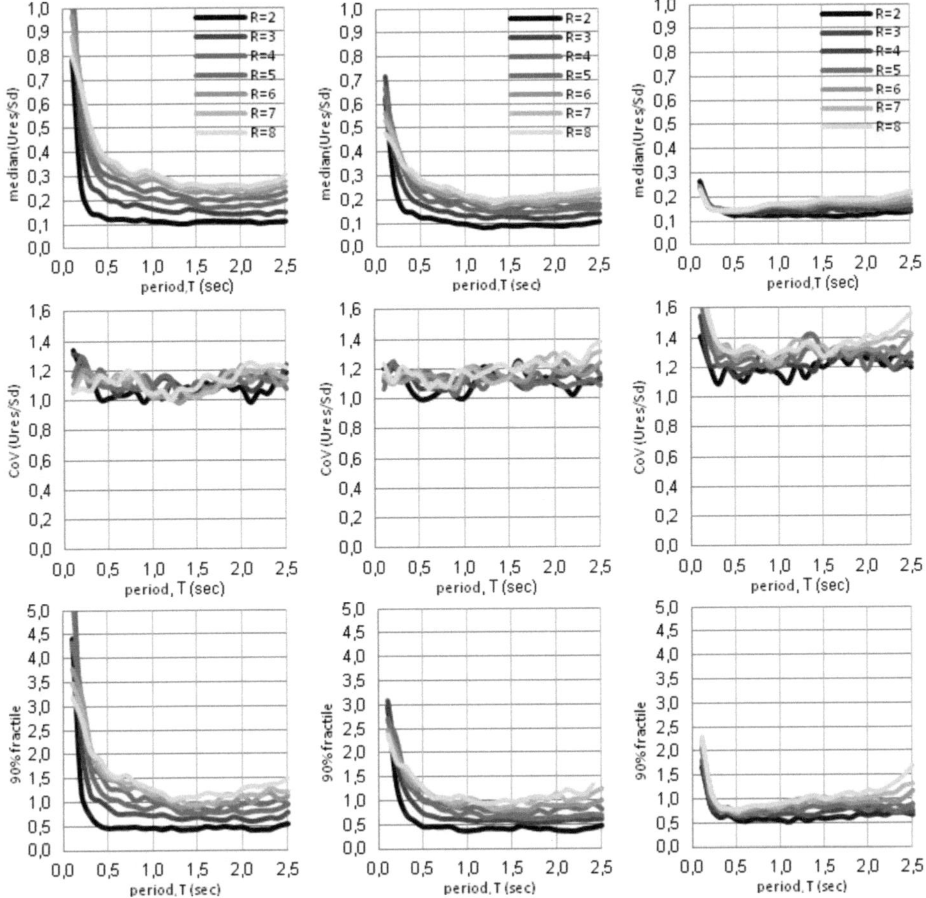

Fig. 2 Residual-to-peak-elastic displacement ratio for 693 records without velocity pulse Median (*top*); coefficient of variation (*middle*); and 90%-fractile of lognormal distribution (*bottom*) - (*Left column*) Group A; (Middle column) Group B; (*Right column*) Group C.

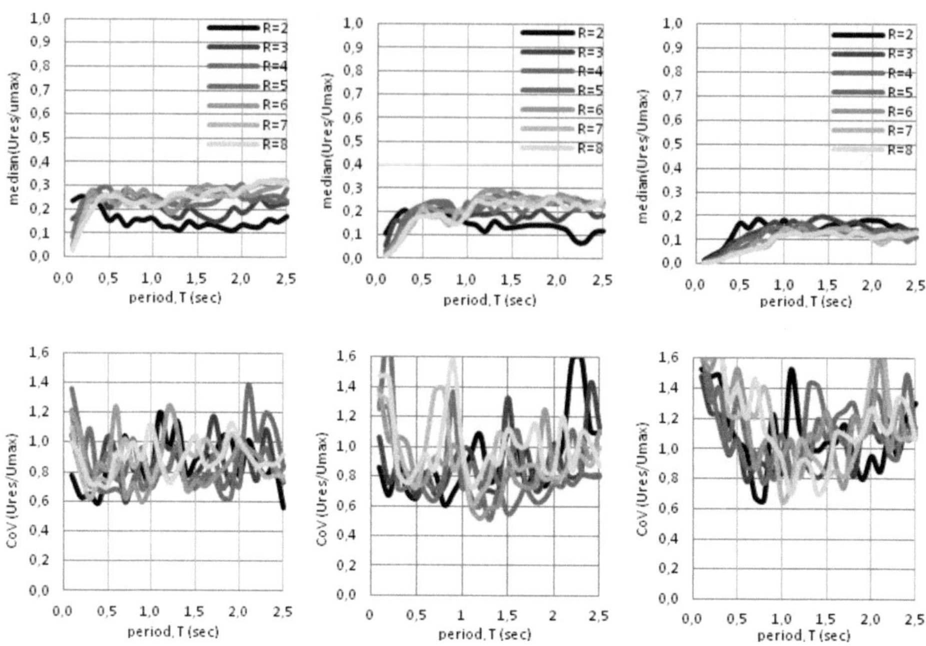

Fig. 3   Residual-to-peak-inelastic displacement ratio for 36 records with velocity pulse Median (*top*); coefficient of variation (*middle*); and 90%-fractile of lognormal distribution (*bottom*); - (*Left column*) Group A; (Middle column) Group B; (*Right column*) Group C.

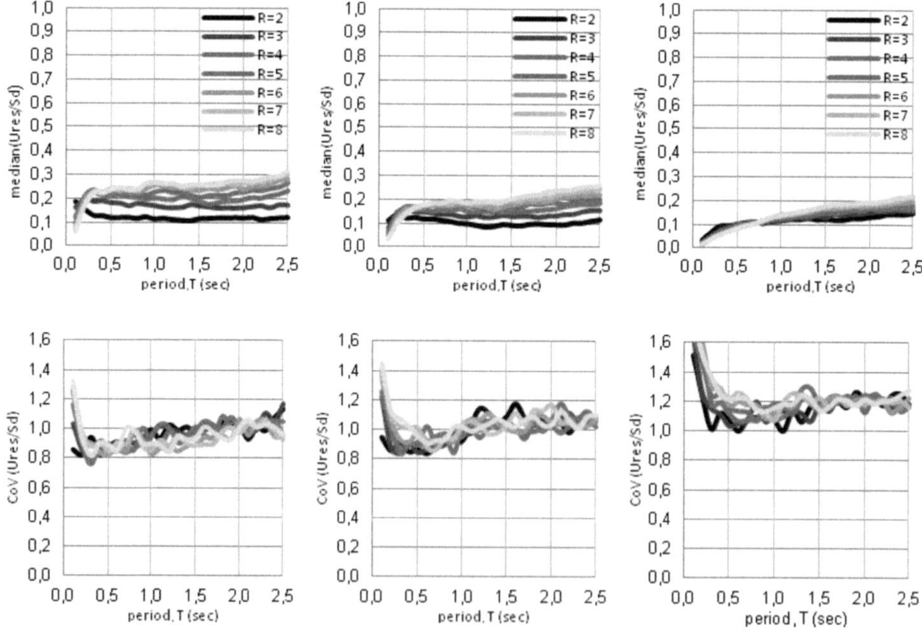

Fig.4   Residual-to-peak-inelastic displacement ratio for 693 records without velocity pulse Median (*top*); coefficient of variation (*middle*); and 90%-fractile of lognormal distribution (*bottom*); (*Left column*) Group A; (Middle column) Group B; (*Right column*) Group C.

4   Residual displacement in cyclic tests of RC members leading to abrupt failure

It has been assumed in the present analyses of the seismic response that the Erberik model, with its gradual loss of stiffness or strength with cycling, holds until the end of the response. This precludes an abrupt failure that may cause the system to deviate from its hitherto pattern of behavior. Indeed, RC members designed and detailed for earthquake resistance normally do not fail abruptly in cyclic loading. Their failure is normally in flexure and is gradual, governed by the progressive deterioration of the compression zone. Such behavior is consistent with the present modelling. However, sometimes at a certain point during a cyclic test the macroscopic pattern of hysteretic behavior changes abruptly. Normally, changes are due to fracture (of a significant part) of the tension or shear reinforcement, or to almost full disintegration of the compression zone in flexure or of the web in diagonal compression, etc., which are typical of members not well designed and detailed for earthquake resistance. The modelling adopted in Erberik (2011) and herein for Group C does not capture such abrupt failures. It is more in tune with a gradual strength decay that may or may not exceed the conventional ultimate deformation, normally identified with a 20% drop in peak resistance.

An abrupt change in the macroscopic pattern of behavior is often associated with a nearly complete loss of lateral force resistance. Hence, it may lead to a residual displacement that well exceeds the predictions of a hysteretic model that provides only for a gradual decay of stiffness, strength and energy dissipation capacity. To check whether this may be the case, a total of about 270 cyclic tests were identified in the databases of flexural or shear failures in (Biskinis and Fardis 2010) or (Biskinis et al 2004), respectively, as having an abrupt change in the hitherto pattern of behavior, leading to an almost full loss of lateral force resistance. The ratio of residual deflection to the maximum one experienced before this type of failure has a mean and a median of about 0.64 and a coefficient of variation (CoV) of 34%. The ratio of such an experimental post-failure residual deformation to the predicted ultimate one according to Biskinis and Fardis (2010) and Part 3 of Eurocode 8 has mean and median of about 0.68 and CoV of 68%. (The larger mean is because the conventional ultimate deformation at a 20% drop in peak resistance is normally lower than the peak deformation just before abrupt failure; the larger CoV is due to the scatter of the prediction). However, these experimental post-failure residual deformations are on the high side, because a cyclic test is typically stopped soon after such a failure. In the cases it continues, the amplitude of imposed deflections does not diminish, as in a real seismic response. Indeed, residual-to-maximum displacement ratios in shaking table or pseudo-dynamic tests simulating such a response are in the order of the present analytical results, notwithstanding occurrence of events normally associated with abrupt member failure. Further study of this aspect is due.

5   Conclusions

Results allow the following conclusions for the ratio of residual displacement to the 5%-damped spectral displacement:

The median and 90%-fractile of the ratio are almost independent of the period, T for Group C while for Groups A or B their dependence on T resembles the pattern of the ratio of inelastic-to-elastic displacements (with lower values, of course).

There is a marked reduction of residual displacements from Erberik's Group A to B to C. Therefore, the better stability of the Group A loops and their higher energy dissipation is of no benefit for residual displacements. By contrast, systems belonging to Group C seem to profit from the self- centering effect of the pinching. The CoV of residual displacement is almost the same across the three groups, with very high values.

On average, residual displacements are a small fraction of the peak elastic ones. However, owing to quite high values of the coefficient of variation (CoV) at a 90%- fractile level they may well exceed the peak elastic ones.

The presence of a velocity pulse increases residual displacements essentially uniformly across Erberik's groups and periods T. For Groups A and B the increase is more marked for R- values of 3 or more. The velocity pulse increases also slightly the CoV, maybe owing to the smaller sample, which increases also the CoV's volatility across periods.

Increasing the relative lateral strength, R, i.e., the level of inelasticity, from 1 to 2 or 3 increases the residual displacement ratio; a further increase in R has a limited effect on the residual deformation of systems belonging to Groups A or B but almost no further effect on those of Group C. The sensitivity to the R- value is larger for records with a velocity pulse.

The same conclusions apply qualitatively for the ratio of residual to peak inelastic displacement, but with every effect and influence quantitatively reduced:

The median ratio is almost independent of the period, of the exact R-value, for R greater than 2 for Groups B and C or above 3 for Group A and of the presence of a pulse.

Owing to the larger CoV-value if there is a pulse, the 90%-fractile of the residual to peak inelastic displacement ratio increases due to the pulse.

The Group per Erberik has a marked effect yet much less than on the absolute value of residual displacement, as peak inelastic displacements decrease also from Group A, to B, to C.

The analysis discounts an abrupt failure that may cause the system to deviate from the gradual loss of stiffness or strength in the Erberik model and lead to nearly full loss of lateral force resistance (as happens in members not well designed and detailed for earthquake resistance, owing to fracture of a good part of the tension or shear reinforcement, or to almost full disintegration of the compression zone in flexure or of the web in diagonal compression, etc.). Cyclic tests with such failures lead to residual displacements which, on average, are 4- to 5-times higher than the present analysis results. However, this large magnitude is not confirmed by shake table or pseudodynamic tests under realistic ground motions and may be due to the imposed cyclic displacement history.

References

[1] Biskinis, D.E., Fardis, M.N.: Flexure-controlled ultimate deformations of members with continuous or lap-spliced bars. In: Structural Concrete (2010) No. 11(2), pp. 93-108

[2] Biskinis, D.E., Roupakias, G.K., Fardis, M.N.: Degradation of shear strength of RC members with inelastic cyclic displacements. In: ACI Structural Journal (2004) No. 101(6), pp. 773-783

[3] Erberik, M.A.: Fragility- based assessment of typical mid- rise and low- rise PC buildings in Turkey. In: Engineering Structures (2008) No. 30, pp. 1360- 1374

[4] Erberik, M.A.: Importance of degrading behavior for seismic performance evaluation of simple structural systems. In: Journal of Earthquake Engineering (2011) No. 15, pp. 32-49

[5] Erberik, M.A., Sucuoğlu, H.: Seismic energy dissipation in deteriorating systems through low-cycle fatigue. In: Earthquake Engineering and Structural Dynamics (2004) No. 33, pp. 49–67

[6] Mahin, S.A., Bertero, V.V.: An evaluation of inelastic seismic response spectra. In: ASCE Journal of Structural Division (1981) No.107, pp. 1777-1795

[7] Kawashima, K., MacRae, G.A., Hoshikuma, J., Kazuhiro, N.: Residual displacement response spectrum. In: ASCE Journal of Structural Engineering (1998) No. 124(5), pp. 523-530

[8] Mavroeidis, G.P., Papageorgiou, A.S.: A mathematical representation of near-fault ground motions. In: Bulletin of the Seismological Society of America (2003) No. 93, pp.1099-1131

[9] Mavroeidis, G.P., Dong, G., Papageorgiou, A.S.: Near-fault ground motions, and the response of elastic and inelastic single-degree-of-freedom (SDOF) systems. In: Earthquake Engineering and Structural Dynamics (2004) No. 33, pp. 1023-1049

[10] Pampanin, S., Christopoulos, C., Priestley, M.J.N.: Residual deformations in the performance-seismic assessment of frame structures. In: Research report No. Rose-2002 /02, (2002) Rose School, Pavia, Italy

[11] Riddell, R. and Newmark, N.M.: Statistical analysis of the response of nonlinear systems subjected to earthquakes. In: Civil Engineering Studies, Structural Research Series (1979) Research Report No. 468, University of Illinois, Urbana-Champaign, Ill.

[12] Ruiz- Garcia, J., Miranda, E.: Residual displacement ratios for assessment of existing structures. In: Earthquake Engineering and Structural Dynamics (2006) No. 35, pp. 315-336.

[13] Ruiz- Garcia, J., Miranda, E.: Performance-based assessment of existing structures accounting for residual displacement. In: The J.A. Blume Earthquake Engineering Center, Stanford University, Stanford (2005), CA Report No. 153

[14] Sivaselvan, M., Reinhorn, A.: Hysteretic models for deteriorating inelastic structures. In: ASCE Journal of Engineering Mechanics (2000) No. 126 (6), pp. 633-640

Dynamic behaviour of rockfall protection galleries

Christina Röthlin

Institute of Structural Engineering,
ETH Zurich,
Wolfgang-Pauli-Strasse 15, 8093 Zurich, Switzerland
Supervisor: Thomas Vogel

Abstract

Rockfall protection galleries usually consist of a reinforced or prestressed concrete slab covered by a cushion material. The advantage of the cushion layer is the ability to dissipate the impact energy.

The goals of this research are preparing principles of examination and design of remedial interventions as well as structural design and dimensioning of rockfall protection galleries.

This paper describes the planned research, which focuses on analytical, numerical and experimental investigations. With the help of the Discrete-Element-Method (DEM), the behaviour of the cushion material due to rockfall impact will be numerically simulated in order to analyse the loading characteristics. A full-scale test is proposed to be carried out on a replicated rockshed. Findings of analytical investigations of the dynamic response of structures by an equivalent single-degree-of-freedom (SDOF) analysis are given. The presented approximate methods are important, because rockfall impacts involve significant uncertainties, particularly with regard to loading characteristics, where rigorous solutions obtained by complex methods are often not justified.

1 Motivation and relevance

In mountainous regions, rockfall protection galleries are significant elements of the infrastructure and assure efficient functionality of the traffic system and the safety of its participants.

The existing Swiss Guideline for the design of rockfall protection galleries [1] provides empirical formulations for calculating the maximum penetration depth and the maximum design forces acting on the structure. Significant improvements have been made with modeling techniques for the structural design of rock sheds in the last decades, resulting in advanced numerical and analytical models for evaluating and predicting the structural behaviour. At the Institute of Structural Engineering, ETH Zurich, research in the field of rockfall protection galleries started in 2006. Schellenberg developed a mass-spring-damper-model [12], which is based on a two-degrees-of-freedom-system developed by Eibl et al. [6] for analysing aircraft impacts on reinforced concrete structures. This model, which takes into account global bending failure and local punching failure, was adopted and enhanced for rockfall protection galleries by adding a further degree of freedom for the behaviour of the cushion material. Ghadimi Khasraghy examined the dynamic load-bearing behaviour of rockfall protection galleries by means of finite element analysis and proposed an analytical model with an uncoupled evaluation of the dynamic behaviour of the cushion layer from the structure [7].

The above-mentioned projects, [7] and [12], indicate that there are significant modelling uncertainties mainly related to the dynamic behaviour of the cushion layer. The possibility of uncoupling of the dynamic behaviour of the cushion layer and the structure has to be verified for specific design situations.

The following research questions have been arised:

- Are the developed models applicable for rockfall protection galleries with arbitrary boundary and loading conditions?
- Can uncertainties related to the dynamic behaviour of the cushion layer due to impact be reduced through systematic analysis of the most significant parameters?
- Can the dynamic behaviour of the cushion layer be evaluated independently from the structural behaviour?

An optimal cost-efficiency-analysis is necessary in order to predict the actual structural safety of rockfall protection galleries. For this purpose, it is necessary to:

Proc. of the 9th *fib* International PhD Symposium in Civil Engineering, July 22 to 25, 2012,
Karlsruhe Institute of Technology (KIT), Germany, H. S. Müller, M. Haist, F. Acosta (Eds.),
KIT Scientific Publishing, Karlsruhe, Germany, ISBN 978-3-86644-858-2

- comprehend the influence of the characteristics of the cushion layer (material, thickness, layer structure) on the load-time-function, dependent on the loading position, -area and impact velocity,
- validate the existing numerical models with full-scale tests, and
- prepare principles of examination and design for remedial interventions as well as for dimensioning and structural design of rockfall protection galleries.

2   Research program

The approach and methodology of this research was defined with regard to optimal achievement of the goals. The main focus of this research addresses numerical, experimental and analytical investigations.

2.1   Discrete Element Method

The models for rockfall protection galleries developed in recent studies at ETH Zurich, show large uncertainties regarding the modelling of the cushion material [7], [12]. In order to better understand the behaviour of the cushion material due to rockfall impacts, numerical simulations of boulder impacts on sand cushion shall be performed by means of a 3D Distinct Element Code (PFC[3D], Itasca Consulting Group, Inc, 2003) [8], which is based on the Discrete Element Method (DEM) [5]. The mechanical behaviour of the granular material used in the modelling approach is characterised by its porosity, a simplified grain-size curve, the assumption of round particles (no rotation) and an appropriate choice of the mechanical micro-macro relations [4].

First, field tests [10] are reproduced; then, the numerical model is used to study different characteristics of the cushion layer such as porosity (ratio of the volume of voids to the total volume of the soil), water content, thickness, layer structure, etc. on the time history of the impact force, penetration depth and velocity of the boulder, and the vertical stresses acting on the concrete plate as a function of the distance from the impacting location for different falling heights and masses of the boulder.

The aim of the numerical analysis by DEM is to establish practice oriented correlations for the impact force, penetration depth and the stresses acting on the structure depending on the cushion characteristics and different falling heights and masses of the boulder.

2.1.1   Blind prediction of field tests

A blind prediction of field tests on the absorbing performance of sand cushion layers conducted in Japan [10] by means of the 3D Distinct Element Code (PFC[3D]) is planned to be carried out. Relationships between impact force and time, transmitted force and time, and penetration depth and time were examined in the field tests. The modelling approach [4] adopted for the simulation uses spherical elements with simple elasto-plastic contact constitutive laws, characterised by the interparticle friction angle (ϕ_μ) and normal and tangential contact stiffnesses (k_n and k_s) as shown in Fig. 1, and relies on the proper reproduction of the mechanical material behaviour characterised by the porosity and the grain-size curve. The DEM model is intended to be as simple as possible in order to minimise the micromechanical parameters needed in the calibration process rather than attempting to reproduce the real grain-to-grain interaction.

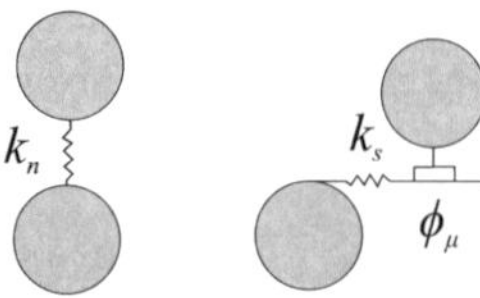

Fig. 1   Contact model in PFC[3D] (k_n: normal contact stiffness, k_s: tangential contact stiffness, ϕ_μ: interparticle friction angle).

The field tests were carried out on a rigid concrete plate (5 m x 5 m x 0.5 m) covered by a cushion layer of sand. The plate was supported by a steel structure, and its deformability can be disregarded. A top view of the experimental device and the shape of the boulder are displayed in Fig. 2. The mass of the falling weight was 5 t; Impact tests have been performed from 1.0 m up to 15.0 m. Hence, the energy domain ranges between 49 kJ and 736 kJ. Three cushion thicknesses, 0.3 m, 0.5 m and 0.7 m, were used. The measuring devices used for recording the data were accelerometers mounted on the falling block, with 29 load cells placed on the concrete plate according to the arrangement shown in Fig. 2.

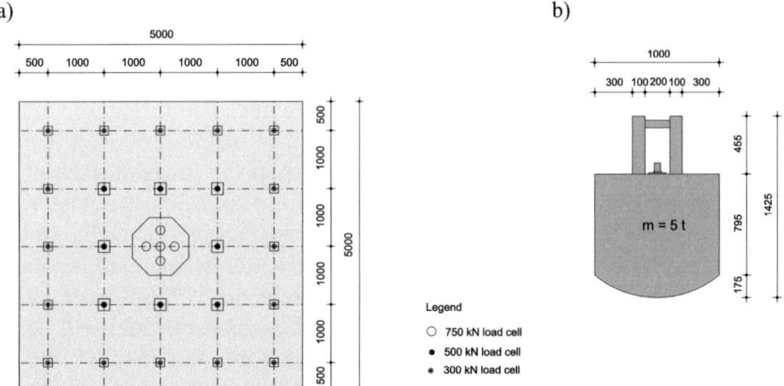

Fig. 2   Experimental setup: a) top view and b) dimensions of the falling weight. Dimensions in mm.

2.1.2 Parameter study and prediction for full-scale test

Once the test results are reproduced by the numerical model, parameters such as porosity, thickness and interparticle stiffness and friction angle of the cushion material, and falling height, mass and impact angle of the boulder will be adjusted and their influence on relations, such as impact force vs. time, impact force vs. displacement, transmitted force vs. time, loading area vs. time as well as contact stresses acting on the gallery will be evaluated.

Additionally, a prediction of the planned full-scale test (see section 2.2) with respect to the dynamic loading characteristics will be conducted prior to the test.

2.2 Full-scale test

A full-scale test is planned to be carried out on a replicated rockfall protection gallery. The concrete slab will be supported by a wall at the back side and by three columns at the front side as shown in Fig. 3. The span of the gallery is 8.7 m and the thickness of the slab is 0.7 m. Three different cushion systems, sand, gravel and a three-layered system (TLAS) [9] consisting of a polystyrene layer, a reinforced concrete slab and a sand-layer, will be tested.

The slab will be designed for an impact mass of approximately 2 tons, falling from a height of about 5 m. The thickness of the sand cushion is 0.9 m. The dimensioning of the gallery will be obtained by a 3D elastic structural analysis.

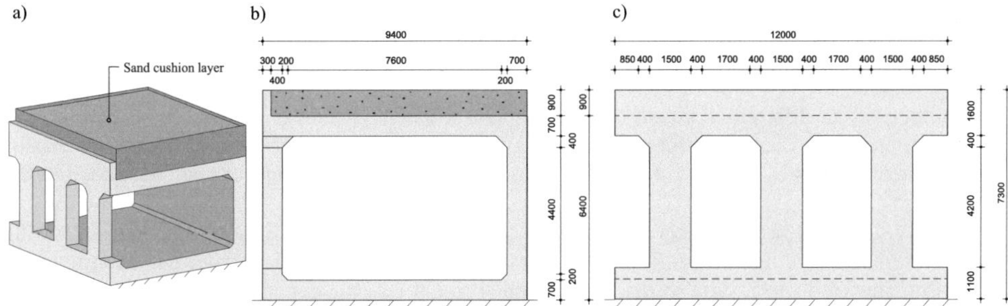

Fig. 3   a) General view of the gallery, b) elevation and c) sectional view. Dimensions in mm.

Emphasis will be placed on crack propagation, the behaviour of the reinforcing bars (strain), the stress distribution on the concrete plate, and the deceleration of the impacting block during penetration into the cushion material.

The following measurement devices are going to be used. Accelerometers will be mounted on the impacting boulder to record the accelerations along three orthogonal directions. The accelerations will be used to compute the forces acting on the boulder and may be integrated to obtain the velocity and penetration depth of the boulder. The forces obtained by the accelerations (force = mass x acceleration) correspond to the impact forces exerted by the boulder acting on the surface of the cushion layer. Accelerations are also measured in the reinforced concrete slab (top and bottom slab). Gages will be used to measure the strains at the upper surface of the reinforced concrete slab and within the bending reinforcement. It is not appropriate to equip the lower surface of the reinforced concrete plate because cracking would arbitrarily influence the strain measurements. Loading cells will be placed on the concrete slab in order to evaluate the stress distribution on the slab.

The testing program will be conducted as follows: Firstly, the gallery will be loaded within the elastic, cracked range of the structure for different loading schemes with all three cushion materials tested one after another. The thickness of the single-layered cushion material will be varied. As the experimental results correlate with the prognoses, the expected ultimate load will be applied on the gallery covered by sand. If the prognoses do not comply with the measured data, the prognoses will be revised. The structural and the deformation behaviour will be investigated up to failure.

The evaluation of the structural behaviour due to impact and the detailed test procedure are now in progress.

3 Analytical investigations

3.1 Elasto-plastic single-degree-of-freedom-system

It was shown in [3] and [11], that the ratio of the load duration to the natural period of the structure (t_F/T) has a significant influence on the structure's response behaviour. Depending on this ratio, one can evaluate if the response at the margin is similar to a very short loading pulse or, at the other extreme, similar to that in static domain behaviour. For a ratio t_F/T between these two limits, a dynamic analysis is required [11]. In the following, these theoretical solutions will be examined for a SDOF system with an elasto-plastic resistance function (Fig. 4) for two different load-time-functions with comparable pulse shapes for rockfall impacts.

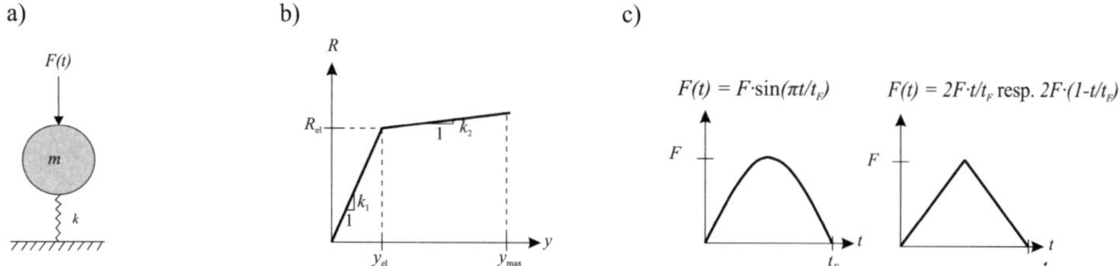

Fig. 4   Scheme a) Mass-spring-model (damping neglected), b) Bilinear resistance function with hardening and c) Load-time functions (sinusoidal load and triangular load).

The solution of the differential equation of motion for a SDOF system is obtained by the numerical integration procedure 'central difference method'. The derived response spectra for undamped elasto-plastic systems with a bilinear resistance function, shown in Fig. 5, account for responses beyond the linear range.

Usually, the structural designer is interested only in the maximum displacement and not in the complete response as a function of time. From these maximum response charts, one can determine the ratio of maximum displacement to the elastic displacement y_{max}/y_{el}, dependent on the ratio t_F/T and the ratio of yield resistance to peak load R_{el}/F. Each curve represents a different yield resistance to peak load ratio R_{el}/F. The quotient R_{el}/F has the character of an elastic/plastic coefficient of impact. For very short durations of impact (t_F very small compared to T), the structural responses (y_{max}/y_{el}) for a sinusoidal load and a triangular load are similar. However, for ratios t_F/T approaching 1.0 the influence of the pulse shape becomes more important and hardening (e.g. $k_2 = 0.3k_1$) has an influence on the maximum response of the structure (y_{max}/y_{el}).

Approximate design methods were analysed for structures loaded with rectangular and asymmetrical triangular short-duration loading [2], [3]. It was shown that, by means of energy consideration, the complicated response behaviour of reinforced concrete structures due to very short loading pulses can be simplified. This limiting case is valid if the deflection of the structure is small during the loading duration. In this case, the inner resistance of the structure R can be neglected, since it increases only with increasing deflection.

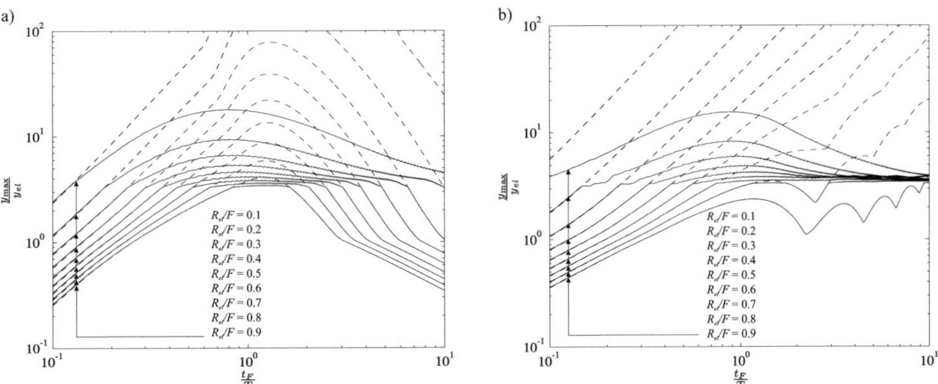

Fig. 5 Maximum response of elasto-plastic SDOF system (undamped) with (dashed line, assumption $k_2 = 0.3k_1$) and without hardening (solid line, $k_2 = 0$) a) sinusoidal load and b) triangular load. Modified after [3].

Consequently, the differential equation can be simplified; the inertia forces are in equilibrium with the short-duration load. Integration of the differential equation leads to the impact velocity, which is a function of the impulse I (area under the load-time function). The required resistance is obtained by equalizing the kinetic energy with the total energy absorbed by the system (deformation energy equal to the area under the resistance curve).

$$R_{el} = \frac{I\omega}{\sqrt{2\mu - 1}} \text{ (elastic resistance)} \tag{1}$$

Whereas μ is the ductility, defined as the ratio of the yield deflection to the maximum deflection and ω is the eigenfrequency. By rearranging the equation above for a half-sinusoidal load and an equal-sided triangular load respectively, the ratio F/R_{el} can be obtained (impulse loading):

$$\frac{F}{R_{el}} = \frac{T}{4t_F}\sqrt{2\mu - 1} \text{ (sinusoidal load)}, \quad \frac{F}{R_{el}} = \frac{T}{\pi t_F}\sqrt{2\mu - 1} \text{ (triangular load)}.$$

The other limiting case of an inelastic response is the maximum deflection of a system subjected to a long-duration loading relative to the natural period (quasi-static loading):

$$\frac{F}{R_{el}} = (1 - \frac{1}{2}\frac{y_{el}}{y_{max}})$$

The approximate solutions (impulse loading, quasi-static loading) for half-sinusoidal loading and equal-sided triangular loading are given in Fig. 6. The evaluation of the percentage difference between the exact and approximative solution for very short loading shows that for $t_F/T < 0.35$ the accuracy for an equal-sided triangular loading lies in the range of 10%. The accuracy for a half-sinusoidal loading lies in the range of about 20% for $t_F/T < 0.25$. Consequently, in these mentioned ranges, approximative solutions for very short loading pulses are appropriate, since the deformation of the structure during the loading duration is small.

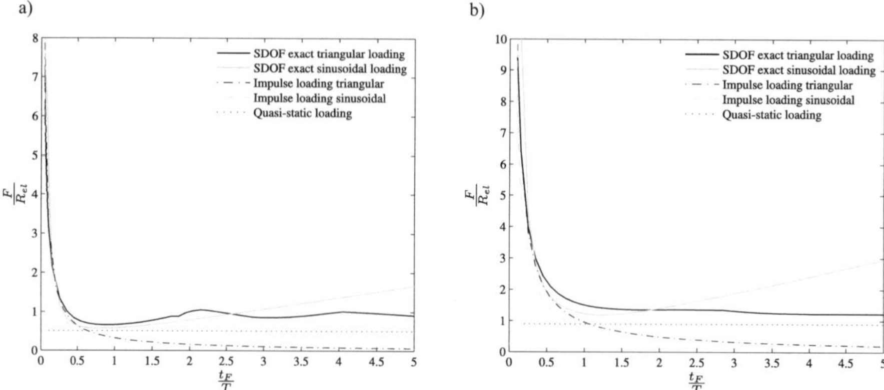

Fig. 6   Comparison of approximate (impulse loading, quasi-static loading) and exact elasto-plastic SDOF system solution for a half-sinusoidal loading and equal-sided triangular loading a) Ductility $\mu = 1$ and b) Ductility $\mu = 5$. Modified after [2].

4   Conclusions

The presented approximate solutions obtained by analysis of an SDOF system are important, because rockfall impacts involve significant uncertainties, particularly with regard to the dynamic behaviour of the cushion material. Therefore, rigorous solutions obtained by complex methods are often not justified. The possibility of uncoupling the dynamic analysis of the cushion layer and the structural behaviour due to impact was shown with the help of approximate methods. Uncoupling is justified for ranges of t_F/T where the approximate impulse loading solution is equal to the exact solution obtained by the SDOF system (small deformations during loading time). Considering rockfall protection galleries, however, characteristics of the boulder (mass, shape, velocity and angle of incidence) as well as of the cushion material (thickness, layer, stiffness, etc.) have an influence on the duration of the impact and the shape of the load-time-curve. The structural response behaviour, depending on the location of impact, geometry and support conditions, influences the natural period of the structure. Hence, all these parameters have an influence on the ratio t_F/T and need to be analysed in more detail.

References

[1]   ASTRA, SBB, (2008): Einwirkungen infolge Steinschlags auf Schutzgalerien, Richtlinie Ausgabe 2008 V2.03, BBL, Verkauf Bundespublikationen, Bern.

[2]   Basler, E. (1966): Der Druckstoss und seine Auswirkungen auf Bauwerke: Vortrag. Schweizerische Bauzeitung, Vol. 84. Zürich.

[3]   Biggs, J. M. (1964): Introduction to structural dynamics. New York: McGraw-Hill.

[4]   Calvetti, F. (2008): Discrete modelling of granular materials and geotechnical problems. European Journal of Env. and Civil Eng., 12; pp. 951-965.

[5]   Cundall, P. A., Strack, O. D. L. (1979): A discrete numerical model for granular assemblies. Géotechnique, 29, pp. 47-65.

[6]   Eibl, J., Keintzel, E., Charlier, H. (1988): Dynamische Probleme im Stahlbetonbau, Teil II, Stahlbetonteile und -bauwerke unter dynamischer Beanspruchung, Heft 392, Berlin, Deutscher Ausschuss für Stahlbetonbau, 1988.

[7]   Ghadimi Khasraghy, S. (2011): Numerical Simulation of Rockfall Protection Galleries. Doctoral thesis, No. 19931, ETH Zurich.

[8]   Itasca Consulting Group, Inc (2003). PFC3D (Particle Flow Code in 3 Dimensions), Version 3.0. 656 pp.

[9]   Kishi, N., (1999). Absorbing performance of sand cushion and three-layered absorbing system. 3rd Asia-Pacific Conference on Shock and Impact loads on Structures.

[10]   Kishi N. (2012). Field test on absorbing performance of sand cushion layer. Outline. Structural Research Team, CERI for Cold Region, Japan. Unpublished.

[11]   Petersen, Ch. (1996): Dynamik der Baukonstruktionen. Friedr. Vieweg & Sohn Verlagsgesellschaft mbH, Braunschweig/ Wiesbaden.

[12]   Schellenberg, K. (2008): On the design of rockfall protection galleries. Doctoral Thesis, No. 17924, ETH Zurich.

Vortex-induced vertical vibrations on bridges

Abraham Sánchez Corriols

School of Civil Engineering,
Technical University of Madrid,
Prof. Aranguren s/n, 28040 Madrid, Spain
Supervisor: Miguel A. Astiz

Abstract

The present research work focuses on the study of vortex-induced vertical vibrations of bridge sections through some basic configurations: rectangular single section, twin boxes in tandem arrangement, and H-shaped section. The main purpose is to identify their susceptibility to this type of excitation as well as to propose an analytical expression that provides a good estimation of the maximum oscillation amplitudes. The excitation mechanisms are described and some basic recommendations to prevent or minimize such effects are given. Numerical simulations of existing bridges, which have suffered from such excitation phenomena, are carried out using a CFD code based on the Vortex Particle Method. Results are compared to those obtained in wind tunnel tests on sectional and full-bridge models.

Such methods are gaining popularity in the field of civil engineering because of their growing reliability provided by the implementation of new computational methods and the availability of more powerful computational resources.

1 Introduction

In January 2006 a vortex-induced vertical vibration episode was observed on the Alconétar Bridge (220 m of main span, Cáceres, Spain) after the completion of one arch. Two rectangular variable depth boxes in tandem arrangement braced by X-trusses form each of the two arches. The wind began to blow at an average horizontal speed not exceeding 30km/h and continued for more than two hours with almost no turbulent component. The maximum amplitude reached up to 80cm in the vertical axis and seriously threatened the structural stability of the arch, which could resist the wind action through the braces. Aerodynamic studies were conducted – sectional models on wind tunnel tests - and a number of deflectors were added along the whole arch to avoid such problem. Vibration episodes do not have occurred again since the bridge was completed.

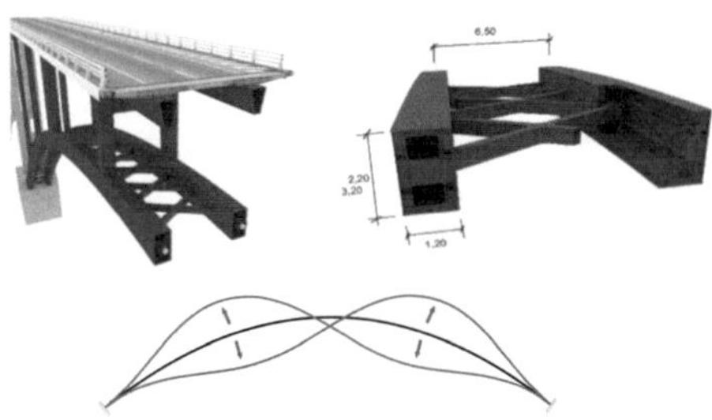

Fig. 1 The Alconétar Bridge: overview during erection (left), diagram of the whole unit formed by arch, metal piers and deck (top, right) and antimetrical vibration mode (bottom, right) [1]

Proc. of the 9th *fib* International PhD Symposium in Civil Engineering, July 22 to 25, 2012,
Karlsruhe Institute of Technology (KIT), Germany, H. S. Müller, M. Haist, F. Acosta (Eds.),
KIT Scientific Publishing, Karlsruhe, Germany, ISBN 978-3-86644-858-2

2   Vortex shedding excitation

When the airflow impinges on a bluff body with a dimension D normal to the velocity vector **U**, two separated shear layers are generated giving rise to wake instability along the body. This results in the generation of vortical structures on both sides of the body, which are shed with a frequency f_v and transported downstream. To balance the pressure difference caused by the alteration of the flow in the wake, the body begins to oscillate in transversal direction according to its natural vibration frequency f_n which depends on its mass and stiffness. If the vortex shedding frequency, which depends on the geometry of the body and the wind speed, is within a range of values close to the natural vibration frequency, coupling occurs and their formation and release is "adapted" to the body transversal oscillation according to its natural vibration frequency. This resonance phenomenon is known as "lock-in".

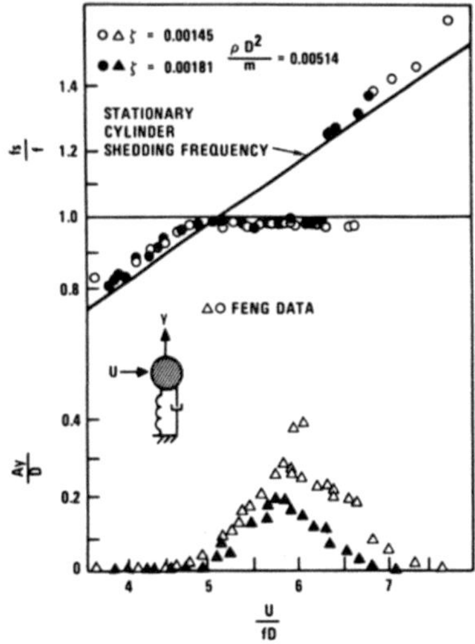

Fig. 2   "Lock-in" characterization of VIV ([2], p. 60)

$$St = \frac{f_v D}{U}$$

St is the Strouhal number, which relates vortex shedding frequency f_v with horizontal component of wind speed U, and transversal dimension of the body, D. For coupling to occur, the average wind speed should be maintained for a certain time. Also the absence of a turbulent wind flow favours the process. If oscillations reach high amplitudes – close to D –, vortex shedding behind the body becomes unstable and resonant effect disappears. Vortex shedding can be considered as self-limiting phenomenon.

2.1   Rectangular section

For small aspect ratios - B/D -, the boundary layer is detached from both corners of the body and the Karman vortex street pattern arises immediately behind the body. This type is the so-called leading edge vortex shedding (LEVS). For side ratios greater than 4, streamlines cannot go over the section width and are reattached to its surface, giving rise to smaller vortex formations that configure a distinct shedding wake pattern known as impinging leading edge vortex (ILEV). Finally, for side ratios above 10, vortex shedding phenomenon is governed by the trailing edge of the body, resulting in smaller and weaker vortex structures that are shed downstream.

For side ratios between 2 and 4, there is a transition zone where streamlines reattach intermittently and more than one vortex shedding pattern can be observed. The vortex shedding frequency does not have a well-defined value in this range and shows a clear jump in the Strouhal number as depicted in Fig. 3, given by Deniz & Staubli [3]. Different cases of vortex-induced excitation associated with three basic geometries are compared attending to the Strouhal number. Numerical simulations of twin boxes in tandem arrangement with an aspect ratio B/D=½ are carried out using a CFD code based on the Vortex Particle Method developed by Morgenthal [4]. Reynolds numbers are close to $2 \cdot 10^6$. Also the Strouhal numbers of sectional wind tunnel tests of H-shaped section at low Reynolds numbers (between 60 and 1000) are shown herein.

2.2   Rectangular section in tandem arrangement

Fig. 4 shows the CFD simulation of vortex shedding process in a rectangular prism with an aspect ratio B/D=½, and its interaction with another twin section situated downstream. The vortex pattern for both sections corresponds to LEVS. For a specific spacing ratio, vortices shed from upstream prism coalesce with those being formed on the downstream prism leading to a resonance phenomenon. Conducting some wind tunnel tests and using numerical simulation techniques two different phases have been distinguished:

1. Vortex formation and shedding (Fig. 4, 1 to 5). This phase lasts half of the vortex shedding period:

$$t_1 = \frac{T_v}{2} = \frac{1}{2f_v}$$

where T_v=vortex shedding period; f_v=vortex shedding frequency.

Vortex size is similar to the prism from which it has been shed. Its centre is situated at a distance of s_1=B=D/2 from the trailing edge.

2. Vortex convection in the flow wake. (Fig. 4, 5 to 9) The horizontal component of vortex convection speed is defined by the following expression:

$$v_c = 4StU = 4\left(\frac{f_v D}{U}\right)U = 4f_v D$$

where St=Strouhal number; U=horizontal wind speed component.

Convection flow speed depends on the wind speed and the Strouhal number, being the latter dependent of the aspect ratio B/D.

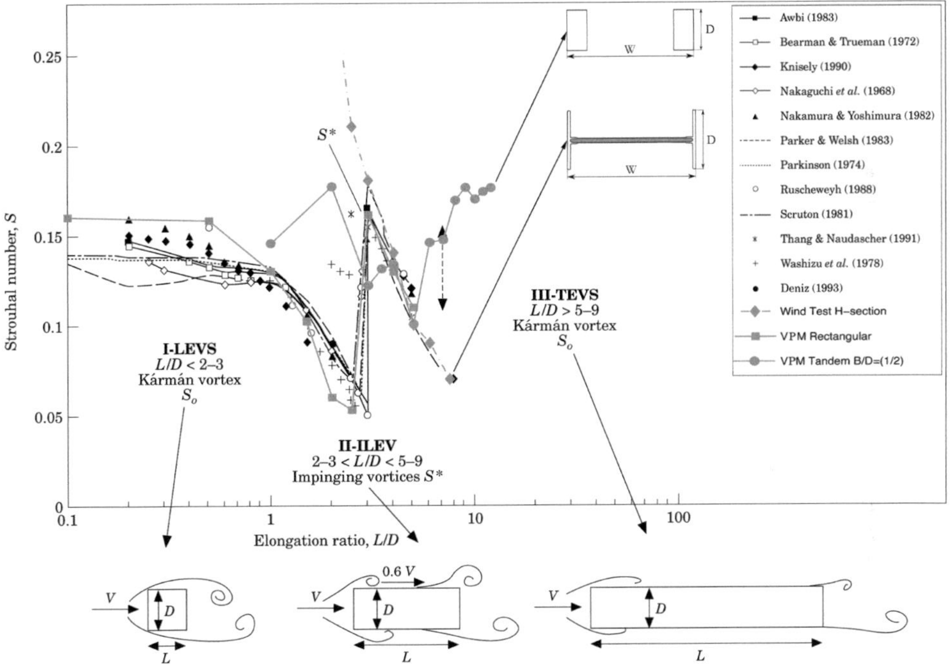

Fig. 3   Effect of aspect ratio from various section geometries on the Strouhal number, St, after [3] with present results added.

In order for the vortex shed from the upstream prism to engage in phase, i.e. synchronously, with the one being formed in the downstream prism, the overall time must be equal to the vortex shedding period, T_v. If the first phase takes t_1=T_v/2, the second phase should last t_2=T_v/2. The distance covered by the vortex in the horizontal axis is:

$$s_2 = v_c t_2 = 4f_v D\left(\frac{1}{2f_v}\right) = 2D$$

Then, the total between the trailing edge of the upstream prism and the leading edge of the downstream prism to make the coupling effective should be:

$$s_T = s_1 + s_2 = \frac{D}{2} + 2D = 2.5D \qquad (W - 2B) = 2.5D$$

There is a range between 2D and 3D where it is possible to have a coupling effect, being the vortex shedding frequency close to the natural frequency of vibration. This fact must be taken into account in the design of rectangular sections in tandem arrangement. If vortex coalescence occurs on every double vibration period, $2T_v$, instead of one, T_v, the spacing is:

$$t_2 = 2T_v - t_1 = 2T_v - \frac{T_v}{2} = \frac{3T_v}{2} \quad s_2 = v_c t_2 = 4f_v D\left(\frac{3}{2f_v}\right) = 6D \quad s_T = s_1 + s_2 = \frac{D}{2} + 6 = 6.5D$$

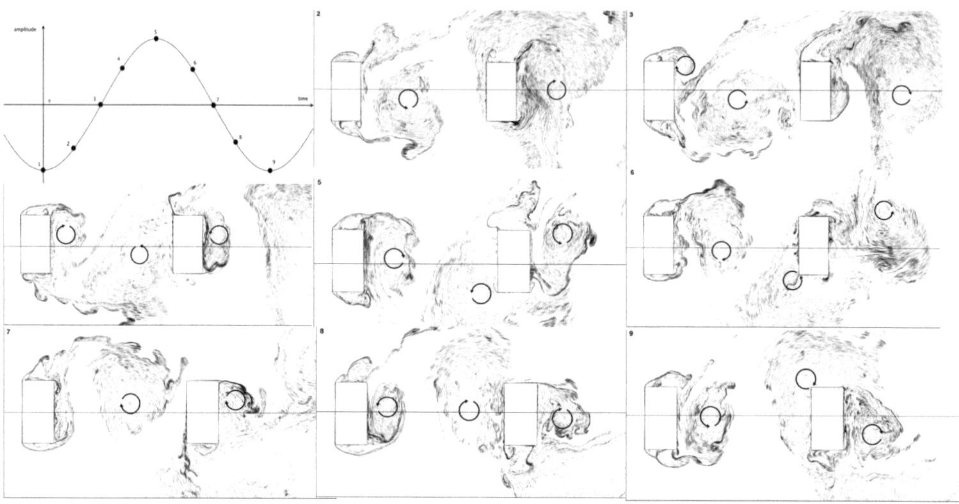

Fig. 4   VPM flow visualization of two rectangular prisms in tandem

However, the coalescence between the two prisms is no longer produced as effectively since they are very separated and the tandem effect disappears as distance increases. The upstream prism dominates the vortex shedding pattern as isolated and the downstream prism acts as an obstacle against the vortex street in the near-wake, thus being the coupling process more difficult to occur. On the other hand, if both sections are located in close proximity – less than 3D – it behaves like a single prism with a side ratio B/D = W/D and the vortex flow pattern is of ILEV type.

2.3   H section

This represents a transition between rectangular section and prisms in tandem arrangement. In this case, the vortex shedding pattern presents a peculiar behaviour as described below and depicted in Fig. 5. For small side ratios the central spacing between section borders do not have any influence and H-shaped section behaves as an equivalent rectangular section. For side ratios greater than 2, vortex shedding mechanism responds to the ILEV type. That is due to the appearance of convection cells in the upper and lower gap, which maintains the detached boundary

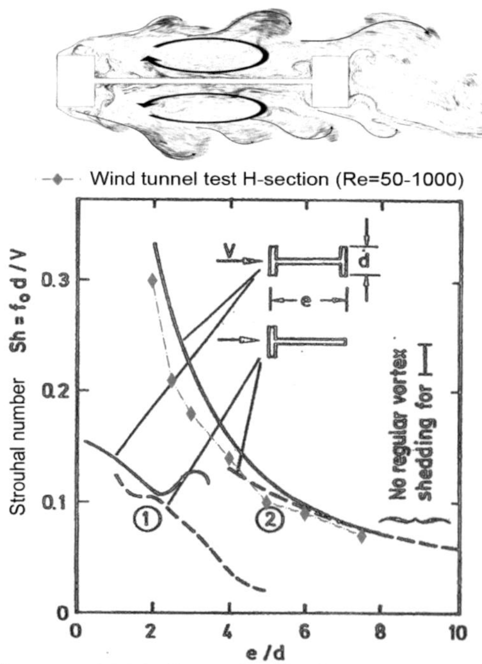

Fig. 5   VPM flow visualization and Strouhal number for different H-section side ratios (After [5], p. 129), with present results added

layer and favours the formation of impinging leading edge vortices along the surface (ILEVcc). This effect is less important as aspect ratio increases. Furthermore, according to Fig. 5, with side ratios between 2 and 4, a transition zone between normal ILEV and ILEVcc pattern gives rise to a dual vortex shedding frequencies. Thus, resonant effects may occur for two different horizontal wind speeds.

3 Case study. The Alconétar Bridge

Previous considerations are especially important and help to explain vortex-induced vertical vibrations on Alconétar Bridge. The box depth corresponding to the maximum amplitude value is D=2.41m, the box width is B=1.20m and the total width W=7.74m, thus giving a side ratio of W/D=3.21. The first vertical vibration frequency is f_n=0.70Hz. According to Fig. 4, this configuration belongs to the transition zone, thus having two different Strouhal numbers: St_u = 0.205 corresponding to the upper curve and St_l = 0.125 to the lower curve. Critical resonant wind speeds associated are:

$$St_u = \frac{f_v D}{U} \quad \Rightarrow \quad U_{cr,u} = \frac{f_v D}{St_u} = \frac{0.70 \cdot 2.411}{0.205} = 8.23\,\text{m/s} \approx 30\,\text{km/h}$$

$$St_l = \frac{f_v D}{U} \quad \Rightarrow \quad U_{cr,u} = \frac{f_v D}{St_l} = \frac{0.70 \cdot 2.411}{0.125} = 13.50\,\text{m/s} \approx 48.6\,\text{km/h}$$

Measured wind speeds during the Alconétar Bridge vibration episode did not exceed from 30km/h which corresponds to a Strouhal number close to 0.205. If wind speeds around 50 km/h had been achieved, another resonant situation would have happened, the latter according to the vortex shedding pattern associated to a Strouhal number of 0.125.

The Alconétar Bridge does not have a conventional H-shaped section because X-braces connecting the two boxes do not occupy the whole space in between. It can be thought that vortex shedding mechanism typical from tandem configuration, having a spacing ratio of (W-2B)/D=2.21. The Strouhal number is 0.14, corresponding to a side ratio B/D=½, and hence the critical wind speed is:

$$U_{cr} = \frac{f_v D}{St_u} = \frac{0.70 * 2.411}{0.14} = 12.1 \text{ m/s} \approx 43.4 \text{ km/h}$$

Furthermore, a perfect coalescence effect between the vortex detached from the box upstream and the one being formed on the box downstream could exist, since the gap between the twin boxes is very close to 2.5D. However, some arguments suggest that the vortex shedding pattern is closer to H-shaped section rather than tandem configuration:

- Braces partially occupy the gap between the two boxes, interceding in the vortex convection process and thus hindering the coalescence phenomenon. In addition to that, the varying depth of the boxes results in different shedding frequencies, not excited by the same wind speed.
- Critical wind speed below 30 km/h, as measured during the excitation episode, does not correspond to the critical wind speed estimated for tandem configuration.
- Aerodynamic sectional tests were carried out resulting in a measured Strouhal number of 0.15, corresponding to a tandem configuration flow pattern. However, it is necessary to point out that three-dimensional effects must be considered: the fact that the wind speed is not homogeneous along its axis, the varying depth (D=3.20m in abutments and D=2.20m in the central point) and the bracing configuration.

4 An analytical model for VIV

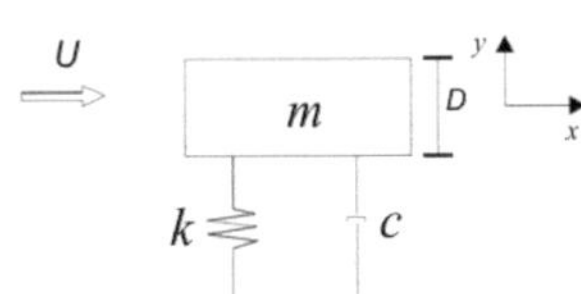

Let us consider a one-degree-of-freedom simple oscillator. The equation that describes the behaviour of the physical system in terms of its vertical movement as a function of time is:

$$m(\ddot{y} + 2\zeta_s \omega_0 \dot{y} + \omega_0^2 y) = F_y(t) = \frac{1}{2}\rho U^2 D C_Y(t)$$

where m=effective mass; ζ_s=structural damping ratio; ω_0= natural angular frequency; ρ=air density; U=wind speed;

D=dimension normal to wind flow; and $C_Y(t)$=force lift coefficient. The main purpose is not only to find a solution to the previous equation, but also to provide a good expression that defines $C_Y(t)$. There are a number of analytical models regarding experimental results from vortex-induced vertical vibrations to estimate the maximum amplitude thresholds [6]. The model proposed by Vickery & Basu [6,7] and the other by Ruscheweyh [7] are two of the most well known. The model used herein by the author is the one proposed by A. Barrero [6]. It has been found to give quite good amplitude estimations. This is the so-called nonlinear model based on amplitude-modulated delay-time quasi-static hypothesis [6].

$$r \approx \frac{U_{r0}^2}{4Sc}\left(\frac{|a_1|}{U_{r0}} + C_{YT}\right)\cos r$$

where r=y/D=non dimensional amplitude, y=maximum oscillation amplitude, U_{r0}^2 =1/(2πSt)=reduced wind speed, Sc =mζ/ρD^2=Scruton number, a1 is a factor depending on the drag coefficient and C_{YT} is a term depending on lift coefficient. This simple two-dimensional model uses aerodynamics parameters obtained through static wind tests and considers the nonlinear evidence of such phenomenon.

5   Conclusions

Vortex-induced vertical vibration mechanisms have been identified in some bridge sections. Alconétar Bridge, composed by two steel boxes, is studied herein. The more homogeneous the bridge is, the more representative the sectional wind tunnel tests are. Specifications previously mentioned cause the vortex shedding mechanism to have a significant three-dimensional component which must be taken into account. As a consequence, a full-bridge wind tunnel test should be conducted to consider all of these properties and experimentally corroborate the above results arisen.

It is also important to study the influence of the aspect ratio of the horizontal strip that braces the two sections in tandem arrangement, since this parameter determines the force coefficient and therefore influences greatly in the oscillation amplitude.

The behaviour of bluff bridge sections against vortex-induced vibrations could be simplified attending to the basic mechanisms presented herein, thus predicting their effects and avoiding them with the corresponding countermeasures. The effect of barriers on a bridge deck could be compared to the H-section configuration with an impinging vortex pattern developing on the upper surface.

Presently some research works are being carried out in order to propose a newer analytical model to account three-dimensional effects such as the correlation length parameter used by Ruscheweyh and others [6,7]. Simulation techniques based in the Vortex Particle Method are also a helpful tool to continue the present work.

References

[1] Astiz, M.: Wind-Induced Vibrations of the Alconétar Bridge, Spain. Structural Engineering International (2010) No. 2, pp. 195-199

[2] Blevins, R.D.: Flow-Induced Vibrations, 2nd ed., Krieger Publishing Co., Malabar, FL, USA (2001), Chap. 3, pp. 43-103

[3] Deniz, S., Staubli, Th.: Oscillating Rectangular and Octagonal Profiles: Interaction of Leading- and Trailing-Edge Vortex Formation. Journal of Fluids and Structures (1997) No. 11, pp. 3-31

[4] Morgenthal, G., Walther, J.H., An immersed interface method for the Vortex-In-Cell algorithm. Journal of Computers and Structures, No 85, 2007, pp. 712-726, Elsevier Science Ltd, The Netherlands.

[5] Naudascher, E., Rockwell, D.: Flow-Induced Vibrations. An Engineering Guide 1st ed. (reprint), Dover Publications, Inc., Mineola, NY, USA (2005), Chaps. 3, 4, 6 and 9.

[6] Barrero, A.: Dinámica de Osciladores Aeroelásticos, PhD Thesis, Technical University of Madrid (2008)

[7] Simiu, E., Scanlan, R.: Wind Effects on Structures. John Wiley & Sons, Inc., New York, USA (1996), Chaps. 6 and 14.

Influence of loading frequency on the fatigue behaviour of high-strength concrete

Nadja Oneschkow

Institute of Building Materials Science,
Leibniz Universität Hannover,
Appelstraße 9a, 30167 Hannover, Germany
Supervisor: Ludger Lohaus

Abstract

Load cycles which occur throughout the service life of building structures are characterised by different loading frequencies and magnitudes of loads. The influence of loading frequency on the fatigue behaviour of high-strength concrete has not yet been sufficiently investigated. In addition, most research was concerned with the effect on the numbers of cycles to failure without including detailed analyses of strain development. The fatigue behaviour of a high-strength concrete under compressive loading is investigated with respect to loading frequency within the scope of a research project. The test results are analysed considering the numbers of cycles to failure and particularly with regard to the strain development of concrete and thus damage development. In this paper, selected test results and the corresponding analyses are presented.

1 Introduction

New developments of lightweight structures and recent developments in the wind energy sector have led to increasing magnitudes of fatigue relevant loads, making fatigue design more important. Fatigue loading, especially with high maximum stress levels $S_{c,max}$, or rather high amplitude levels S_a, causes considerable damage. For this range of fatigue loading, also called low-cycle fatigue, different researchers have found that the loading frequency or rather the rate of loading has a significant influence on the numbers of cycles to failure for normal strength concretes [1], [2]. This increases the susceptibility of concretes in the low-cycle fatigue range when exposed to lower loading frequencies. Only limited data exists for high-strength concrete concerning the influence of frequency. Furthermore, the topics covered in most literatures concentrate mainly on the influence of frequency on the numbers of cycles to failure. Systematic analysis of strain development under fatigue loading has hardly been conducted. Even today, the root causes and mechanisms of failure leading to lower numbers of cycles to failure with decreasing loading frequency are not well-established. Thus, the design concepts of CEB-FIP Model Code 90 [3] and of fib-Model Code 2010 [4] do not cover frequency effects sufficiently.

Within the scope of a research project [5], the static, cyclic, uniaxial, and multi-axial material behaviour of a high-strength concrete was investigated using small-scale test specimens [6]. One of the main focus areas of these investigations was the fatigue behaviour under uniaxial compressive loading. Furthermore, the influence of the loading frequency on the fatigue behaviour was of special interest. As a first step towards a more mechanism-orientated analysis of fatigue behaviour, investigations are being carried out to determine the effects of loading frequency on the numbers of cycles to failure and on the strain development of concrete. In this paper, selected test results and analyses concerning the influence of loading frequency on the fatigue behaviour of high-strength concrete are presented.

2 Experimental program

2.1 Concrete and specimens

The experimental investigations on fatigue behaviour under uniaxial compressive loading were conducted on a high-strength concrete with an average 28-day compressive strength of $f_{ck,cube} = 115$ MPa after storage in water. The maximum grain size was 8 mm. Cylindrical specimens with the dimensions $d/h = 60$ mm/180 mm were used for the tests. The specimens were cast to a height of h ≈ 250 mm and later cut at both ends to the required height of 180 mm to achieve a more uniform concrete quality along the height of the specimens. The PVC-formworks were removed after 48 hours and the cylinders were then stored in standardized conditions (20°C/65% R.H.) until testing. It is absolutely crucial, especially in fatigue testing, to ensure a centric loading in order to attain a uniform stress distribution. Very slight deviations can affect the test results to a high extent. In order to achieve a more uniform stress distribution, the loading surfaces of the cylinders were first plane-parallel ground and later polished.

2.2 Experimental setup

A servo-hydraulic universal testing machine with a 1 MN actuator was used for the experimental investigations presented in this paper. The test frequencies applied were f_P = 0.1, 1.0 and 10.0 Hz. The fatigue loading was applied using a force-controlled system in such a way that, at first, the axial force was increased constantly up to the mean stress level, and afterwards, the sinusoidal fatigue loading was applied. The full amplitude was applied in the first load cycle. The minimum stress level was kept constant with $S_{c,min}$ = 0.05 for all test series, while the maximum stress level was varied between $S_{c,max}$ = 0.95, 0.90 and 0.80. During the tests, the axial deformations were measured continuously using three laser distance sensors which were positioned at 120° from one another. In addition, the temperature of the specimen's surface, the axial force and the axial stroke of the actuator were measured. The sampling rate was chosen in the range of 60 to 600 Hz, depending on the test frequency. All experimental tests were conducted using specimens with a minimum concrete age of 28 days. The static compressive strength of the high-strength concrete was tested just before conducting the fatigue tests, using three specimens from the same batch having the same geometry as the specimens used in the fatigue tests. The reference compressive strength which is required to determine the axial forces was calculated as the mean value of those static compressive strengths. Since the maximum test duration for each set of test series was 10 days, the strength increment due to subsequent hardening was negligible and thus it was not deemed necessary to determine the compressive strength at the end of each test series. Table 1 presents an overview of the experimental tests conducted.

Table 1   Overview of the experimental investigations conducted

Batch	Frequency f_P [Hz]	Number of tests for maximum stress levels $S_{c,max}$		
		0.80	0.90	0.95
B12	0.1	7	6	6
	1.0	6	-	-
B14	1.0	7	6	6
B13	1.0	6	-	-
	10.0	6	-	-

Each test frequency was investigated using the specimens from one batch in order to examine the fatigue behaviour for different maximum stress levels $S_{c,max}$, thereby eliminating undesirable batch-related influences. In addition, the maximum stress level $S_{c,max}$ = 0.80 with a frequency of f_P = 1.0 was applied for all three batches, B12, B13 and B14, in order to make comparisons of test results between these batches.

3   Results and analyses

3.1   General remarks

Most of the research projects dealing with the fatigue behaviour of concretes focus mainly on the investigation of the numbers of cycles to failure as one criterion for analysing the effects of frequency. The developments of strains under fatigue loading have often been measured with varying degrees of accuracy. However, research work with regard to systematic analyses of the developments of strains with respect to the influence of loading frequency can hardly be found in literature. Assuming that the development of strains under fatigue loading is connected to the induced damage, the analysis of these fatigue strains with respect to the influence of loading frequency might be the first step in the direction of a mechanism-oriented investigation of fatigue behaviour and fatigue damage. On the basis of this principle, the experimental investigations have been planned, conducted and analysed taking the numbers of cycles to failure and the development of strains into consideration.

3.2 Numbers of cycles to failure

Here, in contrast to other usual test procedures, the total amplitude was applied with the first load cycle. Therefore, the numbers of cycles to failure were counted, starting with the first load cycle. In Figure 1, the correlation between the maximum compressive stress level $S_{c,max}$ and the numbers of cycles to failure N_f is shown. For simplicity, the numbers of cycles to failure are plotted on the axis of abscissa on a logarithmic scale. The test results are presented as single values as well as mean values. The minimum stress level in all tests was $S_{c,min} = 0.05$. Statistical analyses of the results obtained for $S_{c,max} = 0.80$ show that the differences between the three batches are not significant. Therefore, the mean value at this stress level was calculated considering the test results of all three batches.

The Woehler curves from fib-Model Code 2010 [4] and CEB-FIP Model Code 90 [3] are also included as reference. The background information regarding the Woehler curves in [4] is described in [7] and [8]. This material model was developed based on experimental investigations conducted on an ultra-high-strength concrete with a test frequency of $f_P = 10$ Hz. The Woehler curves found in [3] are based on research work from [9]. Those experimental investigations were conducted on concretes with compressive strengths of $f_c < 100$ N/mm² and a test frequency of $f_P = 1.0$ Hz. As specified in both codes, the Woehler curves are applicable for maximum stress levels of $S_{c,max} < 0.9$ and frequencies of $f_P > 0.1$. Besides, [3] is only valid for concrete grades of up to C80, while [4] is valid for concrete grades of up to C120. Despite these requirements, both curves are used for comparison with the experimental test results obtained for the high-strength concrete.

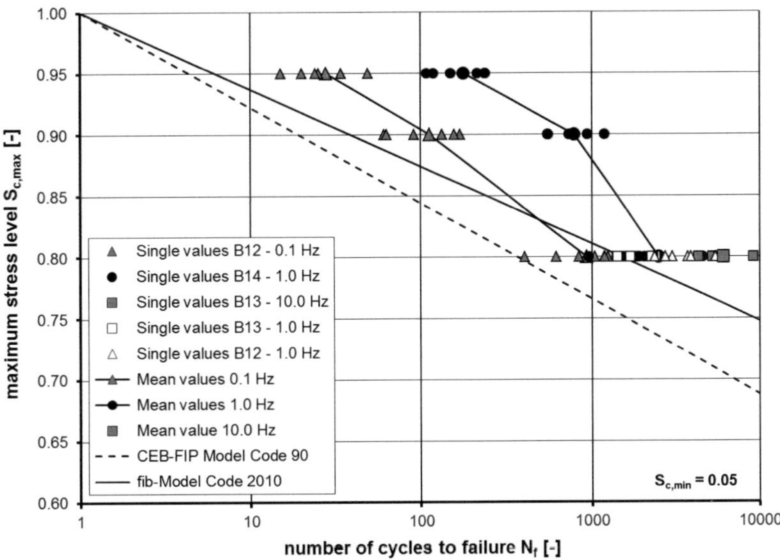

Fig. 1   Numbers of cycles to failure for different loading frequencies and maximum stress levels

Fig. 1 clearly demonstrates the influence of frequency. It is obvious that the number of cycles to failure increases with decreasing maximum stress level $S_{c,max}$ and with increasing loading frequency. In addition, the influence of frequency seems to be smaller for $S_{c,max} = 0.80$ than for the two higher maximum stress levels. Insofar, the results obtained for the high-strength concrete correspond to those for normal strength concretes presented in [1] and [2].

Comparison of the test results with the Woehler curve in [3] shows, that all numbers of cycles to failure – single values and mean values, even those for $f_p = 0.1$ Hz – are higher than the curve specified. The Woehler curves in [3] are, therefore, deemed to be quite conservative. The gradient of the curve in [4] is smaller than in [3]. Hence, the number of cycles to failure is higher for the same maximum stress level. The test results lie to the left side of the curve in fib-Model Code 2010 for a maximum stress level of $S_{c,max} = 0.80$ and a test frequency of $f_p = 0.1$ Hz. All mean values and most of the single test results for $f_p = 1.0$ Hz and $f_p = 10.0$ Hz are higher than the values given in [4]. The test results for $S_{c,max} = 0.90$ and $S_{c,max} = 0.95$ are underestimated by both approaches. This inaccuracy is typically related to the common approach in approximating the logarithmic numbers of cycles to failure by a straight line intersecting at $S_{c,max} = 1.0$ [5]. Considering the limitations of applicability, the Woehler curve of [4] approximates the test results very well.

3.3   Development of strains

It is assumed that the development of strain under compressive fatigue loading can give reference to the development of damage. Therefore, a precise analysis can be helpful in acquiring new knowledge about the development of damage. Here, the main objective is to investigate the influence of loading frequency and maximum stress levels on the fatigue behaviour, respectively damage development under compressive fatigue loading. Precise measurements of deformations with high sampling rates are thus prerequisites for this analysis. The accuracy with regard to strain measurement has increased considerably in the last years. Moreover, nowadays, higher computing capabilities enable the gathering and handling of large measurement data.

Prior to the analyses, the measurement data were processed in such a way that the axial displacements recorded at the peak points of maximum stress ($\sigma_{c,max}$) and minimum stress ($\sigma_{c,min}$) were gathered in the first step. Afterwards, the corresponding strain developments at these maximum and minimum stress levels were generated and the data were further evaluated. In Fig. 2, the strain developments with respect to normalized numbers of cycles to failure N/N_f are presented exemplarily for $S_{c,max} = 0.80$, $S_{c,min} = 0.05$ and $f_p = 1.0$ Hz. In general and similar to normal strength and ultra-high-strength concretes, the strain development can be divided into three phases. The first and third phases with the rapid strain increase each amount to approximately 5 % of the service life and are thus shorter than those observed for normal strength concretes [5], [6].

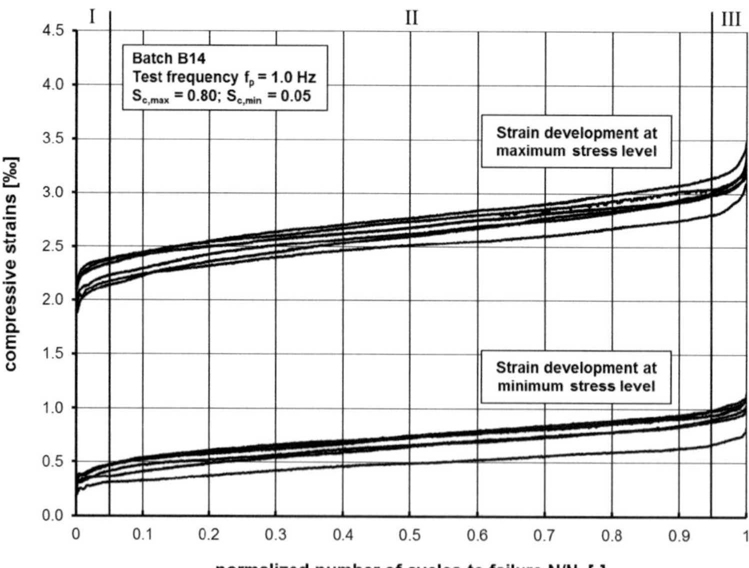

Fig. 2   Development of strains ($S_{c,max} = 0.80$; $S_{c,min} = 0.05$; $f_p = 1.0$)

The relevant parameters that describe the characteristics of strain developments have to be defined prior to any detailed analyses and to comparison for different loading conditions. Different parameters are currently under investigation. In this paper, the results of the analyses of the strain gradient in the second phase of the strain development (from now on referred to as "secondary strain gradient") are presented in relation to the loading frequencies investigated. Following [7], the secondary strain gradients at maximum stress levels are evaluated for the range $N/N_f = 0.20$ to $N/N_f = 0.80$ (cf. Eq. 1).

$$\varepsilon_{II} = \left(\frac{\varepsilon_{0.80 \cdot N_f} - \varepsilon_{0.20 \cdot N_f}}{N_{0.80 \cdot N_f} - N_{0.20 \cdot N_f}} \right) \qquad \text{Eq. 1}$$

The logarithmic secondary strain gradients are calculated as follows:

$$\log \varepsilon_{II} = \log \left(\frac{\varepsilon_{0.80 \cdot N_f} - \varepsilon_{0.20 \cdot N_f}}{N_{0.80 \cdot N_f} - N_{0.20 \cdot N_f}} \right) \qquad \text{Eq. 2}$$

In Fig. 3, the logarithmic secondary strain gradients calculated at different maximum stress levels with their numbers of cycles to failure are presented with respect to the test frequencies investigated in double-logarithmic scale. The different maximum stress levels are distinguished by means of the data point colour. A strong linear correlation between secondary strain gradient and number of cycles to failure was found for the test frequencies $f_P = 0.1$ Hz and 1.0 Hz. The two almost parallel regression lines are also shown in Fig 3. Furthermore, with increasing maximum stress levels $S_{c,max}$, the logarithmic secondary strain gradient decreases and thus, the secondary strain gradient increases. Similar strain gradients can be identified for $S_{c,max} = 0.90$, $f_P = 0.1$ and $S_{c,max} = 0.95$, $f_P = 1.0$, or rather for $S_{c,max} = 0.80$, $f_P = 0.1$ and $S_{c,max} = 0.90$, $f_P = 1.0$. Thus, the same strain gradient might be obtained for decreasing frequencies by increasing the maximum stress level. The experimental tests using specimens of batch B12 and B13 for $S_{c,max} = 0.80$ and $f_P = 1.0$ confirm the fact that the results obtained are not only due to batch dependency.

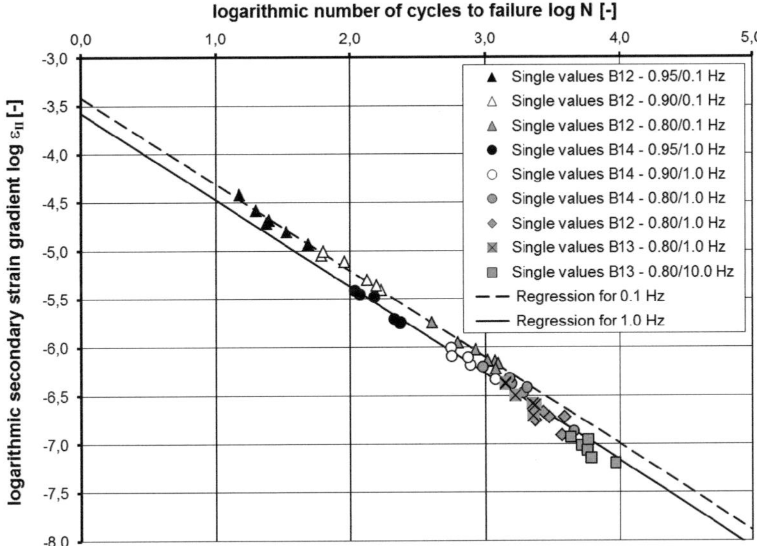

Fig. 3   Logarithmic secondary strain gradient at different loading frequencies and maximum stress levels

Sparks and Menzies [2] report on static and fatigue tests on three normal strength concretes with different coarse aggregates. The fatigue tests were conducted applying a triangular loading at different rates of 0.5 N/(mm²·s) and 50 N/(mm²·s). The loading frequency was adjusted depending on the maximum stress level in such a way that the loading rate remained the same for all tests. The maximum stress level ranged between $S_{c,max} = 0.90$ and $S_{c,max} = 0.70$, based on static tests at a loading rate of 5 N/(mm²·s). The minimum stress level was maintained at $S_{c,min} = 0.33$ based on static tests at a loading rate of 0.25 N/(mm²·s). Similar to the results presented here, a frequency dependency of numbers of cycles to failure, as well as a linear correlation between the secondary strain gradient and the numbers of cycles to failure, is presented in [2]. However, contrary to the results presented here, it is stated in [2] that the loading rate has no influence on the secondary strain gradient. The reason for this might be the triangular cyclic loading. Unfortunately, the results shown in [2] cannot be assigned to the different loading rates and thus a verification of these test results is not possible.

4 Conclusions and outlook

Experimental results concerning the influence of loading frequency and maximum stress level on the fatigue behaviour of high strength concrete are presented in this paper. Within the scope of this research, their effects on the numbers of cycles to failure and especially on the development of strains are under investigation. Only selected test results and analyses are included in this paper. It is shown that the number of cycles to failure increases with a decreasing maximum stress level and increasing loading frequency. This influence seems to decrease for a maximum stress level of $S_{c,max} = 0.80$. Therefore, these results are compatible to results documented in literature for normal strength concretes. Furthermore, the strain development is generally analysed and with respect to the secondary strain gradient as one reference parameter. A strong linear correlation exists between the logarithmic secondary strain gradient and the logarithmic number of cycles to failure. In addition, the secondary strain gradient depends on the loading frequency and the maximum stress level. A higher frequency leads to smaller secondary strain gradients. A similar effect can be obtained by decreasing the maximum stress level.

The test results are currently under further examination. In addition to the first test results presented here, further analyses including additional parameters are underway. Additional tests will be conducted with respect to the influence of rate of loading. Based on these investigations, new knowledge will be obtained concerning the influence of frequency or rather rate of loading on the fatigue behaviour of concrete. Hereby, the basis shall be established for including the influence of frequency into the fatigue standards.

References

[1]   Reinhardt, W., Stroeven, P., den Uijl, J., Vrencken, J.: Einfluss von Schwingbreite, Belastungshöhe und Frequenz auf die Schwingfestigkeit von Beton bei niedrigen Bruchlastwechselzahlen. Betonwerk + Fertigteil-Technik, Heft 9, 1978, pp. 498 – 503

[2]   Sparks, P. R., Menzies, J. B.: The effect of rate of loading upon the static and fatigue strength of plain concrete in compression. Magazine of Concrete Research, Vol. 25, No. 83, 1973, pp. 73 – 80

[3]   CEB – Comité Euro-international du Béton: "CEB-FIP Model Code 90". Bulletin d'Information, No. 213/214, Thomas Telford Ltd., London, 1993

[4]   fib – International Federation for Structural Concrete: Model Code 2010, Final draft, September 2011

[5]   Grünberg, J., Oneschkow, N.: Gründung von Offshore-Windenergieanlagen aus filigranen Betonkonstruktionen unter besonderer Beachtung des Ermüdungsverhaltens von hochfestem Beton. Abschlussbericht zum BMU-Verbundforschungsprojekt, Leibniz Universität Hannover, 2011

[6]   Oneschkow, N.: Hochfester Beton unter ein- und mehraxialer dynamischer Beanspruchung. Proceedings of the 51. Forschungskolloquium des DAfStb, Kaiserslautern, 2010

[7]   Wefer, M.: Materialverhalten und Bemessungswerte von ultrahochfestem Beton unter einaxialer Ermüdungsbeanspruchung. Dissertation, Leibniz Universität Hannover, Institut für Baustoffe, 2010

[8]   Lohaus, L., Wefer, M., Oneschkow, N.: Ermüdungsbemessungsmodell für normal-, hoch- und ultra-hochfeste Betone. Beton- und Stahlbetonbau, Vol. 106, No. 12, Ernst & Sohn, 2011, pp. 836 – 846

[9]   Petković, G., Stemland, H., Rosseland, S.: High Strength Concrete SP 3 - Fatigue, Report 3.2 Fatigue of High Strength Concrete. SINTEF Structures and Concrete, Trondheim, August 1992

Analysing the concrete behaviour under projectile impact by means of fracture mechanics and surface analytic methods

Steve Werner

Institut für Werkstoffe des Bauwesens,
Universität der Bundeswehr München,
Werner-Heisenberg-Weg 39, 85579 Neubiberg, Germany
Supervisor: Karl-Christian Thienel

Abstract

The behaviour of normal strength concrete under projectile impact has not been investigated in detail until now. A generally accepted criterion has not yet been established to describe this local damage of concrete. Thus, the main goal of this research project is to describe the aforementioned behaviour for normal strength concrete. Therefore, different concrete mixtures were used with varying w/c-ratios and maximum aggregate diameters. This investigation is based on an energy balance, which considered different energy forms such as kinetic energy and fracture energy. The kinetic energy of the projectile before and after perforating a concrete specimen was determined in projectile impact tests. Furthermore, the surface area of the resulting craters and fragments was analysed utilizing a combination of a 3-dimensional laser scanner and a photo-optical particle analyser. In order to obtain the fracture energy E_V the specific fracture energy of the concrete mixtures G_F was determined using a standard fracture test. Additionally, the fracture areas of the broken specimen were analysed. This surface analysis was used to describe the fracture behaviour. One finding of this study is that a design code has to consider the concrete mix design for a reasonable description of the impact behaviour of normal concrete specimens under impact. Especially the maximum aggregate diameter is decisive for the impact resistance.

1   Introduction

Concrete behaviour under projectile impact has rarely been determined until now. Often, the compressive strength is the only concrete property considered in a design process of protective barriers [1]. In recent years some studies have shown that other parameters such as the concrete composition need to be considered in a design code. These studies [e.g., 1-4] were mainly focused on high strength concrete mixtures. A basic study of the behaviour of normal strength concrete is still missing. Therefore, five different concrete mixtures were developed to determine the effects of w/c-ratio and maximum aggregate diameter with respect to the resistance of concrete against projectile impact.

Many empirical formulas were generated to describe the impact behaviour. An overview can be found in Li et al. [5]. Usually these formulas use one (or more) of the following four measurements to quantify the impact behaviour: penetration depth, scabbing limit, perforation limit and / or ballistic limit. Another arising effect is not considered in these models: the fragmentation of the specimen within the range of the scabbing craters. Fragments occur as a result of the projectile impact. These fragments are ejected of the specimen and form fragment clouds. Dinovitzer et al. [6] developed a model to predict the number of fragments caused by impact. They used an energy balance which considered different energy forms such as kinetic energy of the projectile and fragments, deformation energy and heating energy. In our study an energy balance was utilised as well. The change of energy forms is described as a result of different concrete mixtures. Thereby, the focus is on the fracture energy E_V, which creates fracture surface areas. By means of standard tests and newly developed test procedures the fracture energy and the fracture surface was determined.

Proc. of the 9th *fib* International PhD Symposium in Civil Engineering, July 22 to 25, 2012,
Karlsruhe Institute of Technology (KIT), Germany, H. S. Müller, M. Haist, F. Acosta (Eds.),
KIT Scientific Publishing, Karlsruhe, Germany, ISBN 978-3-86644-858-2

2 Experimental Investigation

2.1 Concrete Mixture and Properties

The basic concrete was a normal strength concrete with a w/c-ratio of 0.60 and a maximum aggregate diameter of 16 mm. The binder was German CEM I Portland cement with a minimum compressive strength of 42.5 N/mm^2 at an age of 28 days. The aggregate was limestone gravel from a local quarry, which followed the grading curve A/B 16 according to DIN 1045 [7]. Further concrete mixtures were developed by changing only one parameter of the basic concrete mixture at a time: the w/c ratios 0.45 and 0.35 were chosen and the other maximum aggregate diameters were 8 mm and 4 mm, respectively. Details of the concrete mix designs are given in Table 1.

Table 1   Mix design

Test series	W/C [/]	Water [kg/m^3]	Cement [kg/m^3]	Sand 0/4 [kg/m^3]	Gravel 4/8 [kg/m^3]	Gravel 8/16 [kg/m^3]
BC	0.60	185	310	847	364	680
C045	0.45	166	370	847	364	680
C035	0.35	148	425	847	364	680
C8	0.60	185	310	1268	619	0
C4	0.60	185	310	1879	0	0

Ten specimens were cast for each test series. The specimens were stripped after one day and stored subsequently in lime-saturated water for 27 days until testing.

The mechanical properties were measured at an age of 28 days. Compressive strength and Young's modulus were determined in each case on three cylindrical specimens (d/h = 150/300 mm) according to DIN EN 12390 [8]. The results of these tests are presented in Table 2.

Table 2   Mechanical properties

Test series	Density ρ [kg/dm^3]	Compressive strength f_c [N/mm^2]	Young's modulus E [GPa]
BC	2.489	46.7	32.6
C045	2.485	57.8	36.7
C035	2.523	75.6	39.8
C8	2.404	48.8	29.6
C4	2.277	49.4	27.2

2.2 Fracture Mechanics Investigation

The fracture mechanics investigation aimed at determining the specific fracture energy G_F [e.g., 9, 10]. Therefore, the fracture behaviour of notched beams was studied in three point bending tests according to a method described by RILEM [11]. The specific fracture energy is defined as the amount of energy necessary to create one unit area of a crack. According to RILEM [11], G_F can be calculated as the quotient of the energy W_0 represented by the area under the load-deformation curve and the projection of the fracture zone on a plane perpendicular to the beam axis.

Additionally, the fracture area was measured with a three-dimensional laser scanner. On the one hand the fracture area measured was used to determine more accurately the specific fracture energy. This specific fracture energy $G_{F,3D}$ is calculated as the quotient of the energy W_0 and the fracture area measured A_{3D}. On the other hand the fracture area was used to describe in a more precise way the fracture behaviour of concrete.

2.3 Impact Investigation

The impact investigation was conducted in the ballistic laboratory at the Universität der Bundeswehr München. A detailed description of the test set-up used is given in Werner et al. [12].

The concrete specimens were slabs with a quadratically loaded surface area of 30 x 30 cm^2. This size was necessary to minimize edge effects. The slab thickness of 5 cm was only small in order to ensure a perforation.

The parameters measured to describe the impact load are the projectile velocities before and after perforating the specimens. They are used to calculate the kinetic energy of the projectile at these two points in time (E_{in} and E_{out}). The velocity of the projectile before impacting was approx. 870 m/s for all tests. It was measured with a photoelectric barrier. After perforating the specimen the velocity of the projectile was determined with a double exposed picture of a digital camera. The velocity could be calculated from the time difference of two flashes (here: 0.1 ms) and the distance covered by the projectile within this time difference (Fig. 1).

The fracture energy E_V of the specimen is described as the product of specific fracture energy $G_{F,3D}$ and fracture surface area A_{3D}. The fracture area consists of different parts: the crater surfaces on the front and rear side of the specimen and the fragment's surface.

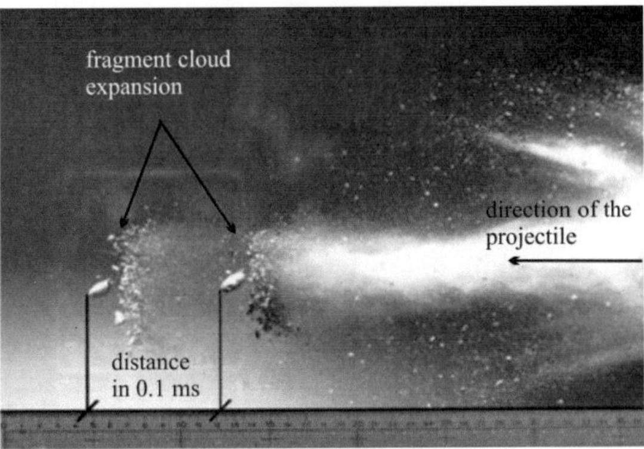

Fig. 1   Double exposed picture of a projectile after perforation

2.4 Surface Determination

Three different methods were used to determine the fracture areas which consist of different parts. The applied methods were a laser scanning system, a particle analysing system and a photographic technique. A detailed description of the different methods is given in Werner et al. [12].

The laser scanning system was applied to investigate the fracture areas that result from the fracture mechanics tests. Additionally, the crater areas on the front and rear side of the specimens, which were caused by projectile impact, were scanned as well. This system uses laser triangulation (see, Winkelbach et al. [13]).

Millions of fragments occur as a result of the projectile impact. The surfaces of the fragments provide a main part of the fracture area. An automatic computer particle analyzer (CPA) was used to determine the number, the size and the shape of the fragments. The CPA is based on digital processing of images taken by a high resolution digital line scan camera. Additionally, a mathematical model was used to idealise the surface area of the fragments. It is predicated on an elliptic elementary form.

Since the original surface in the range of the craters before impacting was not caused by fracture, it was subtracted from the total fragments surface. These surface areas were detected using the photographic technique.

3 Experimental Results

3.1 Energy of the projectile

The kinetic energy of the projectile before perforating the specimen E_{in} is nearly the same for all test series. It ranges between 3579 J and 3643 J.

The energy after perforating the specimen E_{out} is one of the decisive parameters to evaluate the resistance of the material against projectile impact. Figure 2 shows the mean values of E_{out} and their normal distribution depending on the w/c-ratio (left) and the aggregate size (right). It can be seen that the energy increases with an increasing w/c-ratio and decreases with an increasing maximum aggregate diameter. A linear regression between w/c-ratio respectively maximum aggregate diameter and E_{out} led to linear coefficients of determination of the mean values: $R^2_{wc} = 0.902$ and $R^2_{as} = 0.995$.

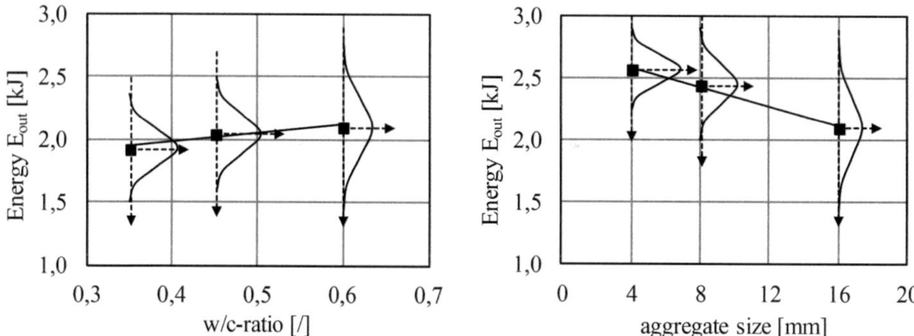

Fig. 2   Energy of the projectile after perforating the specimen E_{out} depending on the w/c-ratio (left) and maximum aggregate size (right).

3.2 Fracture area and fracture energy

The mean values of the fracture area of the perforated specimens are presented in Figure 3, which shows the normal distributions as well. A correlation between w/c-ratio and fracture area is not definite, but a small increase of the fracture area can be seen with an increasing w/c-ratio. The increase goes along with a higher number of fragments while the crater areas decrease for higher w/c-ratios.

The fracture area increases significantly with an increasing maximum aggregate diameter. This is a result of a higher number of fragments and a larger crater area.

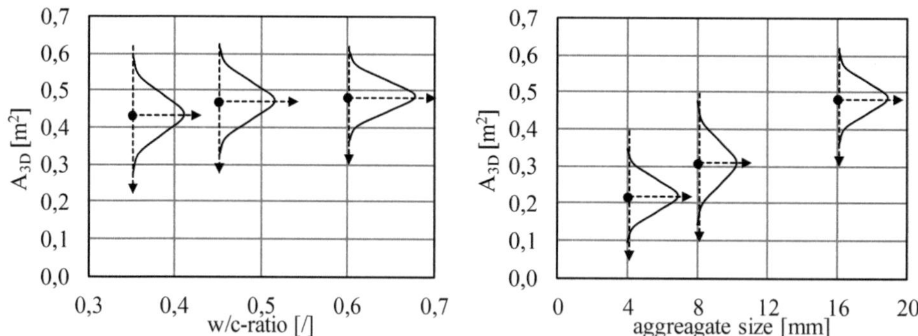

Fig. 3   Fracture area A_{3D} of the specimen depending on the w/c-ratio (left) and maximum aggregate size (right).

The specific fracture energy measured in the three point bending tests increased with decreasing w/c-ratio and with increasing aggregate size, respectively. These correlations can be seen in the calculation of G_F according to RILEM as well as in the calculation of $G_{F,3D}$ with respect to the fracture area A_{3D}. Figure 4 shows the mean values of G_F and $G_{F,3D}$ and their normal distributions.

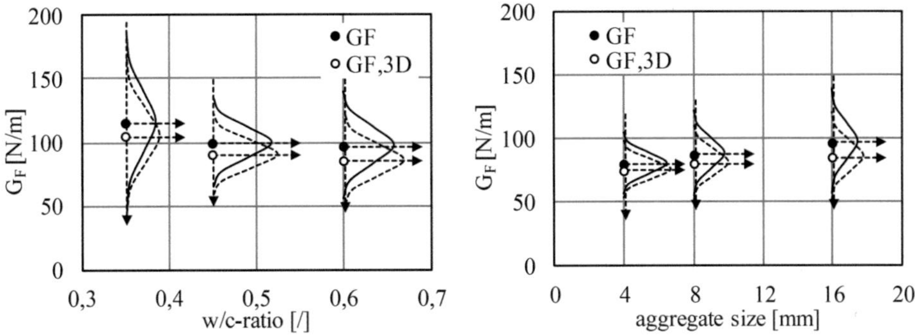

Fig. 4 Specific fracture energy G_F and $G_{F,3D}$ depending on the w/c-ratio (left) and maximum aggregate size (right).

The specific fracture energy was measured under static test conditions but is known to be dependent on the load velocity, especially high dynamic load velocities (see e.g., [14-16]). As a consequence, a correction value must be considered for the implementation of this parameter into the evaluation of projectile impact tests: $k_{dyn} = G_{Fdyn}/G_{Fstat}$. Its dimension is not yet known for a velocity of 870 m/s. Thus, the absolute value of the fracture energy E_V can only be shown depending on k_{dyn}. Figure 5 shows the quotient of E_V and k_{dyn} depending on the w/c-ratio and the maximum aggregate size. The normal distributions of the test series are also presented. A clear correlation between w/c-ratio and fracture energy cannot be seen. The fracture energy increases with an increasing maximum aggregate diameter. The correlation is linear ($R^2 = 0.893$ for the individual values, $R^2 = 1$ for the mean values).

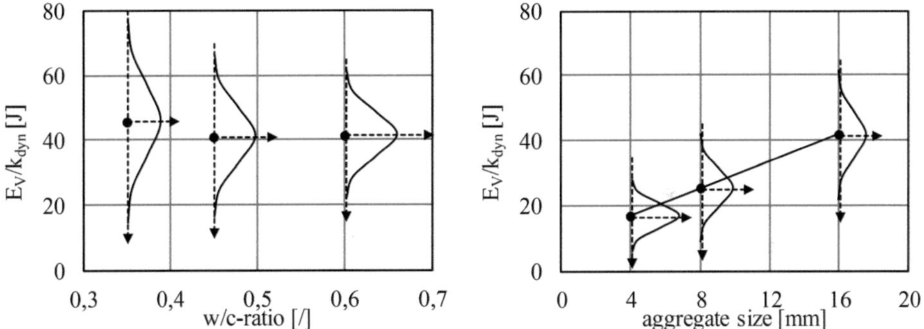

Fig. 5 Quotient of fracture energy and dynamic correction value E_V/k_{dyn} depending on the w/c-ratio (left) and maximum aggregate size (right).

4 Discussion

The correlation of the energy after perforating the specimen and the w/c-ratio is not definite. However, it is well known that the resistance of concrete against impact increases with an increasing compressive strength of concrete. The w/c-ratio is an important parameter to change the compressive strength. Therefore, the results shown in our investigation could be expected. One reason for the increase can be seen in a more compact cement matrix and less pores. More energy is necessary to fracture this matrix.

The correlation between E_{out} and the maximum aggregate diameter is definite: E_{out} decreases for increasing diameters. Concrete mixtures with a larger aggregate diameter have a more distinctive crack pattern as a result of the load. More branched cracks dissipate more energy of the projectile. A similar increase of resistance against impact was shown by Kustermann et al. [17] for high strength concrete using the crater volume as a resistance parameter.

The impact on specimens with a w/c-ratio of 0.35 caused less fragments compared to specimens with a higher w/c-ratio. These fragments exhibit a coarser distribution in addition. As a result, the fracture area is smaller than the fracture area of concrete mixtures with a higher w/c-ratio. The coarser distribution could be a result of the higher specific fracture energy G_F. A higher specific fracture energy means that more energy is necessary to create a fracture area. The influence of the correction of G_F to $G_{F,3D}$ is low, but the variation of the mean value decreases. The fracture energy E_V of the impact tests does not correlate with the w/c-ratio.

The fracture area of the impact tests correlates well with the maximum aggregate diameter. The fragment size distributions are equal. Thus, the higher fracture area of concrete mixtures with a higher aggregate diameter is a consequence of more fragments. In this study the concrete mixtures with smaller aggregate diameters have a higher porosity. The pores can operate as expansion spaces during the penetration of the projectile. As a result, the craters of the specimens are smaller and fewer fragments occur. The specific fracture energy G_F and $G_{F,3D}$, respectively declines with an decreasing aggregate diameter as a result of the dissipated energy (see, Mechtcherine [18]). The fracture energy E_V increases as a result of the increasing fracture area and the increasing specific fracture energy.

The correlation found between the aggregate diameter and fracture energy is valid for the experimental set-up used: $E_V/k_{dyn} = 2.0 * D_{Max} + 8.5$, where D_{Max} is the maximum aggregate diameter measured in mm.

5   Conclusion and outlook

This study shows that a design code has to consider the concrete composition in order to describe impact behaviour of projectiles in specimen of normal strength concrete. Especially the maximum aggregate diameter is decisive for the impact resistance.

Additionally, the use of fracture mechanics basics and an analysis of the fracture area led to a closer insight into impact tests. The following conclusions were drawn:

- Crack branches consumed energy which led to smaller kinetic energy of the projectile after perforation. More distinctive crack branches were found in concrete mixtures with larger aggregate diameters.
- A smaller w/c-ratio has a higher specific fracture energy which provides a smaller number of fragments in impact tests. Therefore, the whole fracture area decreases although the crater areas increase.
- A larger aggregate diameter led to higher fracture energy.

Future work will focus on further concrete components such as different grading curves, age of concrete, storage of the specimen, type of cement etc.

References

[1]   Dancygier, A.N., Yankelevsky D.Z. and Jaegermann Ch.: Response of high performance concrete plates to impact of non-deforming projectiles. In: Int. J. Imp. Eng. 34 (2007), pp. 1768-1779

[2]   Zhang, M., Shim, V., Lu, G. and Chew, C.: Resistance of high-strength concrete to projectile impact. In: Int. J. Imp. Eng. 31 (2005), pp. 825-841

[3]   Børvik, T., Gjørv, O.E. and Langseth, M.: Ballistic perforation resistance of high-strength concrete slabs. In: Con. Int. 29 (2007), pp. 45-50

[4]   Frew, D., Hanchak, S., Green, M. and Forrestal, M.: Penetration of concrete targets with ogive-nose steel rods. Int. J. Imp. Eng. 21(1998), pp. 489-497

[5]   Li, Q., Reid, S., Wen, H. and Telford, A.: Local impact effects of hard missiles on concrete targets. In: Int. J. Imp. Eng. 32 (2005), pp. 224-284

[6]   Dinovitzer, A., Szymczak, M. and Erickson, D.: Fragmentation of targets during ballistic penetration events. In: Int. J. Imp. Eng. 21(1998), pp. 237-244

[7]   Deutsches Institut für Normung: Tragwerke aus Beton und Stahlbeton. 2008. Beuth

[8]   Deutsches Institut für Normung: Prüfung von Festbeton. 2009. Beuth

[9]   Bažant, Z. P. and Planas, J.: Fracture and size effect in concrete and quasibrittle materials. (1998), CRC Press LLC, ISBN: 084938284X.

[10]   Wittmann, F. H. (Ed.): Fracture mechanics of concrete. (1983), Elsevier, ISBN: 0444421998.

[11]   RILEM: Determination of the fracture energy of mortar and concrete by means of three-point bend tests on notched beams. Mat. Str. 18 (1985), pp. 287-290

[12]   Werner, S. and Thienel, K.-C.: Influence of impact velocity on the fragment formation of concrete specimens. In: Proceedings of the II International Conference on Particle-Based Methods, Barcelona, Spain, October 26 - 28, 2011. Onate, E. and Owen, D.R. J., Ed.

[13]   Winkelbach, S., Molkenstruck, S. and Wahl, F.: Low-Cost Laser Range Scanner and Fast Surface Registration Approach. In: Pattern recognition. 28th DAGM Symposium, Berlin, Germany, September 12 - 14, 2006

[14]   Vegt, I., Weerheijm, J. and van Breugel, K.: The rate dependency of concrete under tensile impact loading. Fracture energy and fracture characteristics. In: Proceedings of the 13th Int. Symposium on the Interaction of the effects of munitions with structures. May 11 - 15, 2009

[15]   Zhang, X., Ruiz, G., Yu, R. and Tarifa, M.: Fracture behaviour of high-strength concrete at a wide range of loading rates. In: Int. J. Imp. Eng. 36 (2009), pp. 1204–1209

[16]   Brara, A. and Klepaczko, J.: Fracture energy of concrete at high loading rates in tension. In: Int. J. Imp. Eng. 34 (2007), pp. 424–435.

[17]   Kustermann, A., Zimbelmann, R.K., Keuser, M. and Grimm, R.: Hochfeste Bindemittel und Zuschlagstoffe für hochfeste Betone unterschiedlicher Güte für Schutzanlagen der militärischen Sonderinfrastruktur. (2005). Neubiberg

[18]   Mechtcherine, V.: Bruchmechanische und fraktologische Untersuchungen zur Rissausbreitung in Beton. Dissertation. (2000). Karlsruhe

Numerical simulation of the impact of fluid-filled projectiles using realistic material models

Ferdinand Borschnek

*Materials Testing and Research Institute (MPA Karlsruhe),
Karlsruhe Institute of Technology (KIT),
Gotthard-Franz-Str. 3, 76131 Karlsruhe, Germany
Supervisor: Harald S. Müller*

Abstract

A number of investigations were performed to find out the damage potential of large passenger aircraft impacts. Due to technical difficulties and high costs associated with experimental tests, these events are studied with the help of computational approaches. In the present work small-scale impact tests are modeled using the explicit finite element code LS-DYNA. For the aluminium alloy the material and failure model by Johnson and Cook (1985) is used. The modeling of the tank filling is carried out by the methods Smooth Particles Hydrodynamics (SPH) and Arbitrary Lagrangian Eulerian (ALE). Thus, a numerical model validated on the experiments is generated in order to provide an enhanced confidence level for future calculations of real impact scenarios. The complex numerical model and first results of the ongoing studies are presented.

1 Introduction

The threat scenario of safety related structures has changed with the events of September 11, 2001. A check-up of current security measures for nuclear power plants was therefore necessary. Until today, according to the German building codes for (nuclear) power plants only the load-time diagram of a relatively small fast-flying military aircraft has to be considered [1, 2]. So the increased damage potential of large masses, as for example the impact of a large airplane, requires detailed investigation. Based on this background small-scale tests with aluminium projectiles were performed at the Materials Testing and Research Institute (MPA Karlsruhe) of the Karlsruhe Institute of Technology (KIT). The geometry of the projectiles and their stiffness and mass distribution along the longitudinal axis were adapted to the real conditions of a commercial aircraft (Airbus A 340-600 [3]). The focus of the investigations was placed on the influence of speed and impact angle of the aircraft on the force-time function. Also different levels of the tank filling were examined. A detailed overview about the investigated parameters is given in [3-6].

The aim of the investigations is the development of a realistic numerical description of the complex impact behaviour of fluid-filled projectiles. Based on the simulation of small-scale experimental impact tests, the behaviour of structural models under impact loading is investigated using arbitrary projectile geometries.

2 Modeling

2.1 Fundamentals

For a realistic modeling of an impact scenario some simplifications and assumptions are necessary. Figure 1 shows the changeover from reality of the passenger plane to the experimental projectiles according to Kreuser et al. [3] and Ruch et al. [4]. The relevant experimental boundary conditions are also mentioned in the figure. First, the shown wide-body passenger airplane can be divided into four different sections with varying mass and stiffness distribution: the tip of the aircraft (section 1), subsequently a long cylindrical part (section 2), the stiff middle area where the wings are attached (section 3) and the tail of the aircraft (section 4). In section 3 of the projectile a cylindrical container simulating the additional mass of the tank-filling of the plane can be added. According to [3] section 4 of the large capacity airplane has only a minor influence on the impact load-time function. Therefore, a separate modeling of the aircrafts tail in the experimental and numerical investigations was desisted from.

Proc. of the 9th *fib* International PhD Symposium in Civil Engineering, July 22 to 25, 2012,
Karlsruhe Institute of Technology (KIT), Germany, H. S. Müller, M. Haist, F. Acosta (Eds.),
KIT Scientific Publishing, Karlsruhe, Germany, ISBN 978-3-86644-858-2

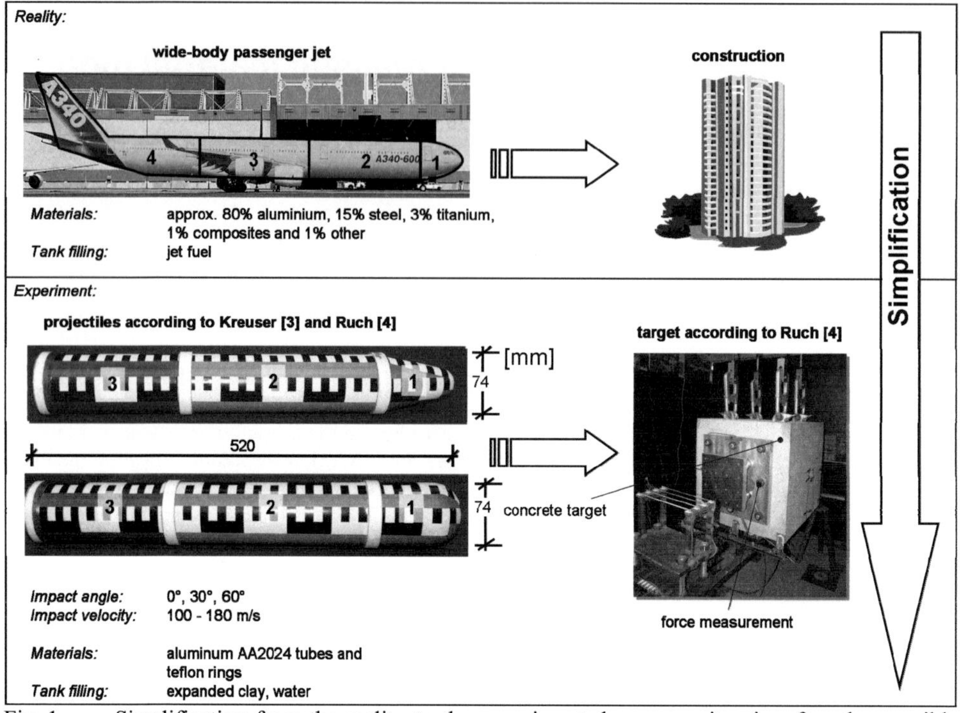

Fig. 1   Simplification from the reality to the experiment: large-capacity aircraft and a possible target building (top), test projectiles and concrete target (bottom)

2.2   Numerical model

For the numerical modeling the explicit finite element program LS-DYNA is used [7]. The calculations were performed on the distributed memory parallel computer HP XC3000 (hc3) with 356 eight-way compute nodes [8] of the Steinbuch Centre for Computing (SCC) of the Karlsruhe Institute of Technology (KIT). Each node has two Quad-Core Intel Xeon E5540 processors with 2.53 GHz frequency and 24, 48 or 144 GB main memory. The contact-impact problem was modeled on a common PC and solved on the HP XC3000 using the LS-DYNA version R4.2.1 ls971d respectively R5.1.1 ls971d with double precision. The calculations are performed on the HP XC3000 as Shared Memory Parallel (SMP) system and the run time is about 4 days.

The aluminium projectiles were modeled and the finite element mesh was generated with the program Hypermesh. Screenshots of the generated geometries are shown in Figure 2. All components of the projectiles are modeled; these are the sections 1-3, a pressed in stiffener in the transition section, additional a closing cover plate and three teflon rings. These serve as guidance in the acceleration phase of the experiments. The components are realized in a geometric full-model, which is able to reproduce the observed asymmetrical folds of the experiments. A water-filled tank is placed inside section 3.

The FE-model consists of four-node fully integrated shell elements with five integration points over the thickness. To model the material properties and failure an appropriate deformation and failure model for the aluminium material is necessary. The *MAT_JOHNSON_COOK, which is based on the deformation and failure model by Johnson and Cook [9] was selected. Therefore the input material parameters and the deformation and failure parameters according to [10] are being used. For the material properties of the teflon rings the *MAT_PLASTIC_KINEMATIC model with a failure criterion is used. In the fluid-like materials such as the tank-filling great deformations and fluid-structure-coupling-problems can occur. These problems can lead to numerical instabilities in the mesh and therefore special approaches in order to deal with these problems are necessary.

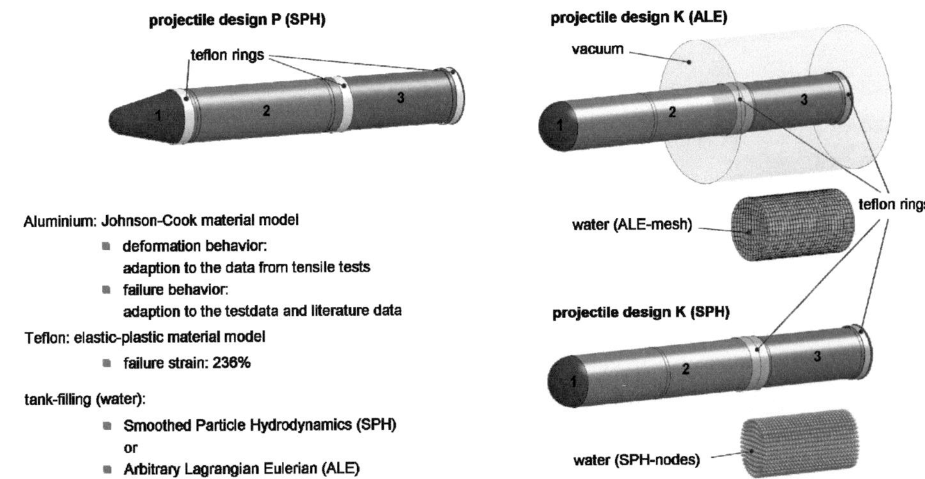

Fig. 2   Numerical model: projectile design P according to Kreuser [3] (left); projectile design K according to Ruch [4] (right); in addition the two different methods for the tank-filling SPH and ALE are shown.

Given that the Lagrangian formulation which is used for the modeling of the structural elements of the projectile is not suitable to model fluids, two fundamentally different methods were chosen to simulate the tank-filling. The first method is a coupling of a Lagrangian with an Eulerian formulation the so-called Arbitrary Lagrangian-Eulerian method (ALE). In this method the advantages of both approaches (Lagrangian and Eulerian) are combined. Therefore the mesh is discretised in Lagrangian form and is considered as fixed in space. If now in some areas of the mesh an unfavorable geometry appears, the nodes of the mesh are moved to avoid a numerically unstable mesh.

The second method is the so-called Smoothed Particle Hydrodynamics (SPH) method, a mesh-free method. Herefore the simulated fluid is separated in a multitude of discrete particles, which have no geometric connection to each other. Each of the particles is associated with physical parameters as mass, density, and pressure. The field functions such as mechanical deformation or pressure at a particle is approximated as the summation over nearest neighbouring particles through a smoothing function a so-called kernel function. This function establishes a relation for the discrete particles which allows the individual particles to act as a continous material (fluid). For the material properties of the tank-filling i. e. the water, values from literature are used. *MAT_NULL is selected in LS-DYNA to specify an ideal fluid so that no deviatoric stresses are computed. Additionally an equation of state (EOS) is required to simulate the water. An EOS determines a relation between the pressure, the density and the energy. In this case *EOS_LINEAR_POLYNOMIAL was selected. The target is modeled as a volumetric cuboid with rigid steel material properties and eight-node solid (brick) elements with one integration point being used for the mesh. The acceleration phase of the projectile is not shown in the numerical simulations. A constant velocity is assigned to all projectile nodes.

In the simulation a contact/impact interaction is necessary. For contact between the projectile and the target *CONTACT_AUTOMATIC_SINGLE_SURFACE with Option SOFT=1 was chosen. Also self contact is considered between the individual parts by this contact formulation. Contact between the SPH part, the projectile parts and the target, respectively, is adjusted by *CONTACT_NODES_TO_SURFACE.

3   Results

In Figure 3 (left) an example of the experimentally determined impact load-time functions is shown. It can be observed that the peak load of the experiment is smaller than the numerically calculated force.

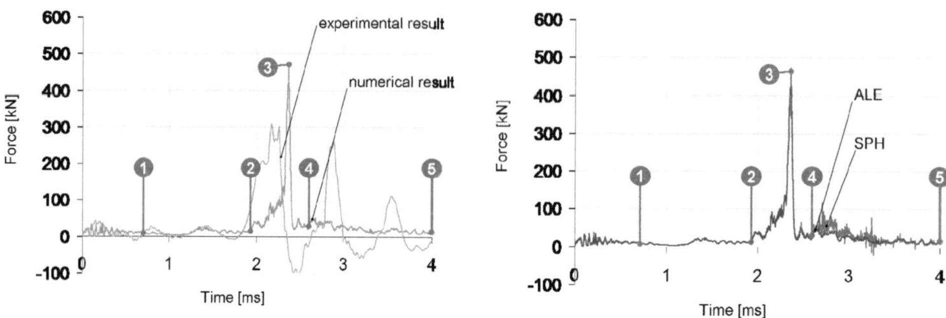

Fig. 3   left: comparison of experimental and numerical results (projectile design K); right: comparison of the two approaches ALE and SPH (projectile design K)

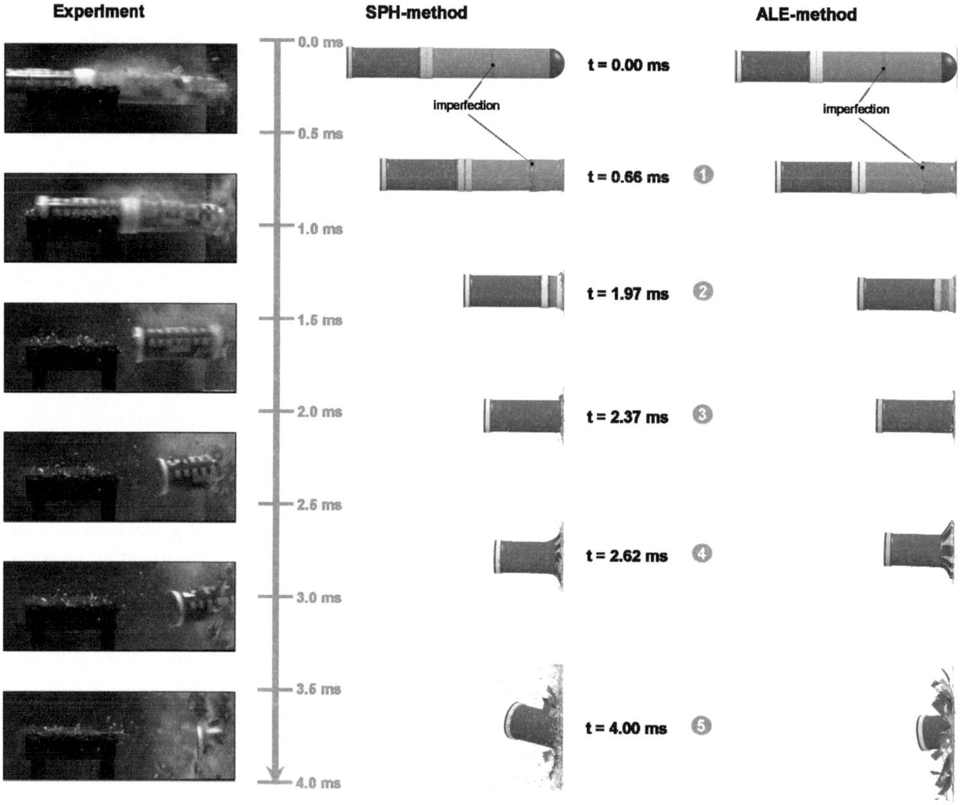

Fig. 4   Comparison between experiment and numerical simulation; left: experiment; middle: SPH-method; right: ALE-method according to [10]

This discrepancy results only from the fact, that the experimental values were cutoff at approximately 250 kN because of the limit of the transmitter. The data, which is published in [10], show a characteristic curve progression.

The first contact of the projectile with the measuring platform (t = 0 ms) leads to inversion of the hemisphere (section 1) and continuous bursting of the cylinder (section 2). Then a load drop at about t = 0.66 ms is registered which is explained by the failure of a production-related imperfection. Furthermore a continuous bursting of the projectile is regarded until the stiff section 3 with the tank-filling reaches the target (about the instant of time t = 1.7 ms). This phase of the projectile impact delivers also the maximal load-entry to the target.

In addition to the results the numerically determined load-time-functions of the impact of water-filled projectiles can be seen in Figure 3. It is shown for the ALE-model as well as for the SPH-model. Figure 4 displays the short-takes of the relevant impact occurrences corresponding to the load-time function. The principal characteristic of the curves that start similarly to the experiments with a continuous load input is visible. This load input is due to the impact of the projectile tip and folding of the cylinder. At t = 0.66 ms the aforementioned imperfection fails and a visible load drop is noticed in the diagram. At t = 1.97 ms the impact of the first teflon ring can be seen, which has shifted during the acceleration phase and is accompanied by a slight increase in the load curve. The maximum load is again recorded during the impact of the stiff and bulky section 3 at t = 2.37 ms. There are no significant differences in the two modeling methods for the tank-filling with respect to the load-input into the target. Only during the bursting of the tank (section 3) there is a slight difference between the corresponding curves. Differences in the previously performed calculations result in the burst behavior of section 3 (see figure 5).

Both methods show the characteristic fracture appearance of the third part, the so-called "peeling off". But the fracture appearance of the SPH-method is more asymmetric. This method allows in contrast to the ALE method for a droplet or mist formation of the fluid.

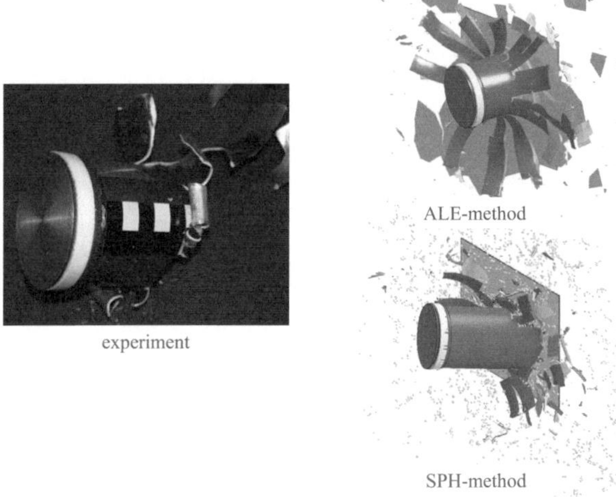

Fig. 5   Fracture performance of section 3; left: experiment; right: numerical results

In addition to the modeling of the tank-filling the variation of the impact angle of 30°, 60° and 80° and the varying impact velocity from 100 m/s to 180 m/s are of particular importance. The key aspect of current experiments is to investigate the influence of these impact angles on the load-time-function.

4   Summary and Conclusion

In this work numerical models to analyse the impact of fluid-filled projectiles onto a target have been developed by using LS-DYNA. The key aspects of the work are the detailed investigation of the numerical model with different velocities and/or angles of impact and the simulation of the tank-filling with different feasible numerical methods. Within this context, projectile impacts (with the available methods (ALE and SPH)) were simulated. Some work steps are not yet completed and will be supplemented further with the influence of the effects of the target structure. This is necessary to obtain an effective numerical model able to describe the different influences on the load-time-function of a wide-body aircraft impact.

So far, reasonable quantitative correlation between numerical predictions and experimental results were obtained. The performance of such a complex computational model depends on the knowledge of the properties and failure phenomena of the individual materials. But also the correct treatment of contact has an influence on the quality of the numerical results. Despite these complexities the so far obtained results are quite promising.

The presented research project is conducted on behalf of the Gesellschaft für Anlagen- und Reaktorsicherheit (GRS) mbH and is financially supported by the Bundesministerium für Wirtschaft und Technologie (BMWi). (Reactor safety research project number: 1501374)

Supported by: Project executing organization:

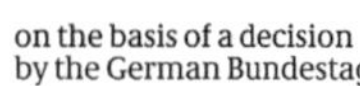

References

[1]   Kerntechnischer Ausschuss (KTA), KTA 2202; Schutz von Kernkraftwerken gegen Flug-zeugabsturz und äußere Explosionen, Grundsätze und Lastannahmen, Regelentwurfsvorla-gen 8/80

[2]   Kerntechnischer Ausschuss (KTA), KTA 2203; Schutz von Kernkraftwerken gegen Flug-zeugabsturz; Auslegung der baulichen Anlagen (bei vorgegebenen Lastannahmen), Regel-entwurfsvorschlag Dezember 1983

[3]   Herrmann, N.; Kreuser, K.; Stempniewski, L.: An experimental Approach to determine Load-Functions for the Impact of Fluid-Filled Projectiles; 76th Shock and Vibration Sym-posium, October 30 – November 3, Destin, FL, USA, 2005

[4]   Ruch, D.; Herrmann, N.; Müller, H. S.: Abschlussbericht zum Vorhaben 150 13 10 Simula-tion des mechanischen Verhaltens von steifen Strukturen beim Anprall flüssigkeitsgefüllter Stoßkörper, Karlsruher Institut für Technologie (KIT), Materialprüfungs- und Forschungs-anstalt, MPA Karlsruhe, Karlsruhe, Germany, 2010

[5]   Hyvärinen, J.; Hakola, I.; Saarenheimo, A.; Ruch, D.; Herrmann, N.; Schimpfke, T.; Siev-ers, J.: *Structure Mechanics Simulation of Phenomena during High Energetic Impact;* EU-ROSAFE 2007 – Towards Convergence of Technical Nuclear Safety Practices in Europe, Berlin, Germany, 2007

[6]   Ruch, D.; Herrmann, N.; Müller, H. S.: *Evaluation of Load-Time Functions due to Impact of soft Missiles;* Proceedings of the 17th International Conference on Nuclear Engineering; ICONE17, July 12-16, Brussels, Belgium, 2009

[7]   LS-DYNA, Keyword User's Manual, Volume I, June 2009, Version 971 / release 4 Beta, Livermore Software Technology Corporation (LSTC)

[8]   http://www.scc.kit.edu/scc/docs/HP-XC/ug/erstinfhc.pdf

[9]   Johnson, G.R.; Cook, W. H.: A Constitutive Model and Data for Metals Subjected to Large Strains, High Strain Rates and High Temperatures. In: Proceedings of the 7th International Symposium on Ballistics, The Hague, Netherlands, page 541-547, 1983

[10]   Ruch, D.: Bestimmung der Last-Zeit-Funktion beim Aufprall flüssigkeitsgefüllter Stoßkör-per, PHD thesis, Karlsruhe Institute of Technology (KIT), KIT Scientific Publishing, Karls-ruhe, Germany, 2010

Innovative Structures

Sandwich panels with thin folded and curved concrete layers

Alexander Stark

Institute of Concrete Structures,
RWTH Aachen University,
Mies-van-der-Rohe Str. 1, 52074 Aachen, Germany
Supervisor: Josef Hegger

Abstract

The construction and design of multiple folded and curved sandwich panels prestressed with non-metallic tendons to gain long-span roof or façade structures with a high corrosion resistance are to be investigated. The experimental test program is presented within this paper. The bond behaviour between pre-tensioned CFRP (carbon fibre reinforced polymers) tendons and concrete (textile reinforced concrete: TRC, ultra-high performance concrete: UHPC) will be analysed. Pull-Out tests and experiments on the transfer length will be carried out. Special attention has to be paid to the anchorage system due to sensitivity to lateral pressure of the CFRP tendon. The technology to fabricate curved and folded slender sandwich shells will be developed with regard to concrete, core foam and connecting devices. Tension and shear tests of sandwich panels with and without connecting devices will be conducted. Finally bending and shear tests of prestressed folded and curved sandwich panels will be carried out.

1   Introduction

In a priority programme [1] founded by the German Research Foundation (DFG SPP 1542: Concrete light. Future concrete structures using bionic, mathematical and engineering form-finding principles) prestressed sandwich panels with thin folded and curved concrete layers are investigated. The concrete layers are either made of UHPC (ultra-high performance concrete) or TRC (textile reinforced concrete) and prestressed with CFRP (carbon fibre reinforced polymers) tendons. To achieve this goal the following questions have to be clarified:

- Which parameters describe the bond behaviour of pre-tensioned CFRP tendons in high strength and ultra-high performance concrete? Is the Hoyer-effect applicable to CFRP tendons which are sensitive to lateral pressure? Which thickness of the concrete layers and what spacing of the tendons is required?
- Which technology is necessary for the fabrication of curved and folded sandwich structures in terms of concreting, core foam and connection devices?
- How can a shear resistant connection of the concrete layers be set up which causes only small restraint forces?
- How does the bending and shear carrying capacities change due to the profiled cross-sections under short and long time loading? Which shape of cross-section leads to minimised time dependant deformations and to a high sandwich action at the same time?

2   Materials

For construction of the test specimens materials are chosen which mechanical properties are known from previous research projects which were partly conducted at the Institute of Concrete Structures of RWTH Aachen University [2][3]. Two different types of concrete mixtures are used. On the one hand an ultra-high performance concrete (UHPC) with an uniaxial compression strength of about 180 N/mm² and on the other hand a textile reinforced concrete (TRC) with a fine grained matrix (maximum grain size smaller than 5 mm) exhibiting an uniaxial compression strength of about 80 N/mm². Straight micro fibres of high strength will be added to either mixtures to enhance the brittle behaviour.

Proc. of the 9th *fib* International PhD Symposium in Civil Engineering, July 22 to 25, 2012,
Karlsruhe Institute of Technology (KIT), Germany, H. S. Müller, M. Haist, F. Acosta (Eds.),
KIT Scientific Publishing, Karlsruhe, Germany, ISBN 978-3-86644-858-2

For the core foam prefabricated PUR (polyurethane) and XPS (extruded polysterene) plastic slabs are used. For connection of the concrete layers the application of connecting devices in terms of lettuce beams, made of GRP (glas fibre reinforced polymers), CFRP or stainless steel, are planned, which are rigid to tension, compression and shear.

For TRC a fabric is used, similar to a reinforcement grid in common concrete structures. Coated textiles normally exhibit a better bond behaviour since the single rovings are bonded to each other where no bond to the concrete exists [4]. Either AR-glass or carbon textiles will be used.

CFRP tendons are generally applicable because of a high tensile strength of about 2000 N/mm² and corrosion resistance [5] which is necessary due to the lack of concrete covering in slender structures. One indented strut and two different diameters of a seven-wire tendon are used for pre-tensioning and are provided by Tokyo Rope Mfg., Co., Ltd.. The trade name is CFCC™ (carbon fibre composite cable). The material proporties given by the manufacturer are listed in table 1.

Table 1   Material properties of CFCC

Designation [-]		Diameter [mm]	Effective cross-sectional area [mm²]	Guaranteed capacity* [kN]
●	U 5.0∅	5	15.2	38
✿	1x7 7.5∅	7.5	31.1	76
	1x7 10.5∅	10.5	57.8	141

* the guaranteed capacity applies to strands using an anchor system provided by Tokyo Rope

Since the anchor system developed by Tokyo Rope is not used, tests with different shapes of anchors are carried out (chapter 3.2).

3   Experimental tests on the bond behaviour

3.1   Anchorage behaviour of strands

The load carrying behaviour of concrete structures prestressed with steel strands is mainly governed by the prestress load and the bond between concrete and tendon. The resulting bond stress can be divided into three parts:

- A constant part resulting by basic friction,
- a stress dependent part (Hoyer-effect) which increases with the transfer of pre-tensioning
- and a slip dependent part which is independent of the prestressing ("lack of fit" which results from the non-uniform geometry of the tendons).

The bond stress is not constant along the transfer length l_{bq} (fig. 1). Releasing the pre-tensioning causes a difference between steel and concrete strain, hence the slip and the lateral stresses rise. At the concrete edge nearly the full pre-tensioning has to be transferred which results in a high lateral stress between concrete and steel. Thus all three bond parts are activated. Since the pre-stressing of the concrete increases along the transfer length, the bond stress decreases. At the end of the transfer length the bond is mainly governed by the base value, since the most stresses have been transferred from steel to concrete.

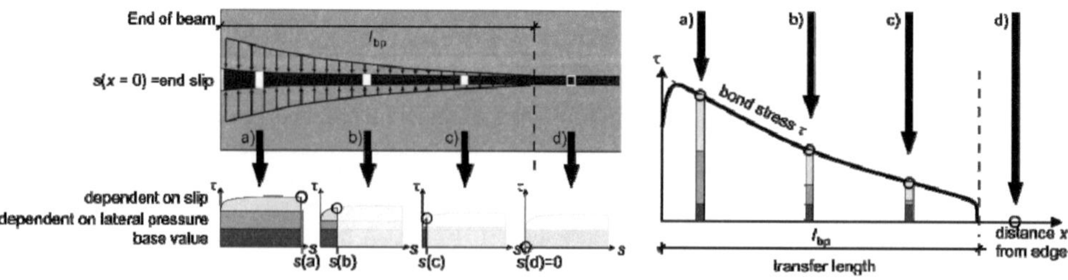

Fig. 1   Schematical stress distribution along the transfer length of a tendon [6]

Concrete structures prestressed with pre-tensioned steel tendons have been widely studied in NSC (normal strength concrete) [7], HSC (high strength concrete) [8] and in UHPC [6].

For CFRP tendons investigations on the bond behaviour have been performed by many researchers, e.g. [9][10]. But there are no experiments in literature known to the author with very small concrete covering and use of UHPC or HSC (besides [5]) with a fine grained matrix. Hence Pull-out tests and experiments on the transfer lengths are carried out.

3.2 Anchorage system

The high tensile strength of CFRP tendons ($f_{CFRP} \geq 2000$ N/mm²) and corrosion resistance makes an application highly attractive in slender concrete structures. Since CFRP is very sensitive to lateral pressure special attention has to be paid to the anchorage device. Standard devices for steel tendons do not seem to be applicable since concentrated forces occur. Nevertheless these ones are analysed as well.

To overcome this issue, different types of anchorage systems are analysed. Three examples are shown in fig. 2.

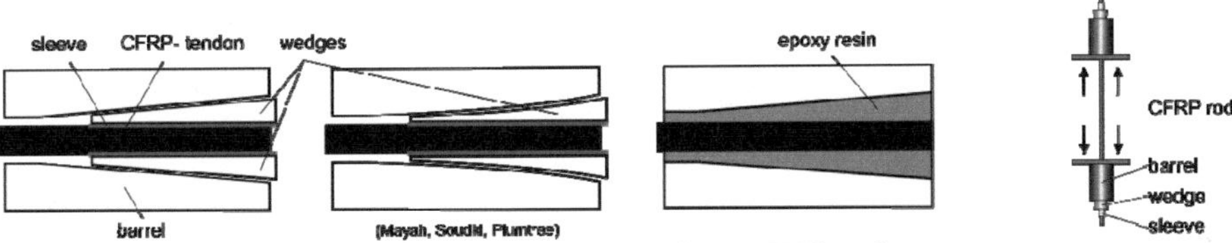

Fig. 2   Examples of anchorage systems (central figure: [10]) and test setup

Tension tests with 1 m CFRP rods and anchorage on either sides of the tendon are carried out (fig. 2). Different diameters (5 mm, 7.5 mm and 10.5 mm) and shapes (seven wire strand and single indented rod) of CFRP tendons are investigated (tab.1). Load and slip are measured. The anchorage type which leads to sufficient results is used for subsequent investigations.

3.3 Pull-Out tests

Pull-Out tests (fig. 3) are conducted according to CEB/FIB/RILEM [11]. The position of the tendon in the concrete (TRC and UHPC) cube as well as the prestressing force are varied to determine the influence of concrete covering and the Hoyer-effect to the bond-slip relationship and the bond strength.

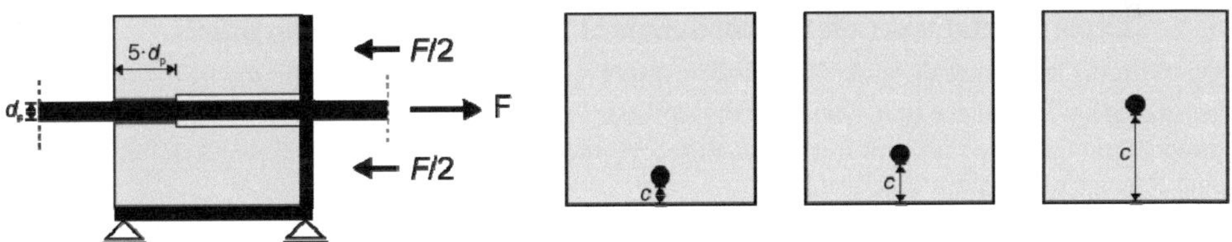

Fig. 3   Pull-Out test [8]

Table 2 shows an overview of the varied parameters of the Pull-Out tests. The main parameters to be investigated are the type of concrete, the concrete covering, the concrete strength (age) and the type of pre-tensioned member. Each test will be conducted three times with the same parameters and three strain stages (0 %, 50 %, 100 %), i.e. nine tests are carried out for each parameter variation.

Table 2   Parameters of Pull-Out tests

Concrete	Pre-tensioning member	Concrete covering	Concrete age	Strain stages
[-]	[mm]	c / d_p [-]	[d]	[%]
UHPC/TRC	U 5∅ 1x7 7.5∅ / 10.5∅	1 / 2 / 3	1 / 3	0 / 50 / 100

The strut and tendons are pre-tensioned inside a testing rig. Afterwards the test cubes are concreted. After one and three days respectively the first three tests are carried out. After decreasing the prestress force of about 50 % the next three tests are conducted. At the end the last three tests are performed with full release (100 %), i.e. full lateral strain of the strut and tendons respectively.

3.4 Tests on the transfer length

Experiments on the transfer length (fig. 4) are carried out to gain the minimum value of concrete covering as well as distance between CFRP tendons without any cracks in the transfer zone.

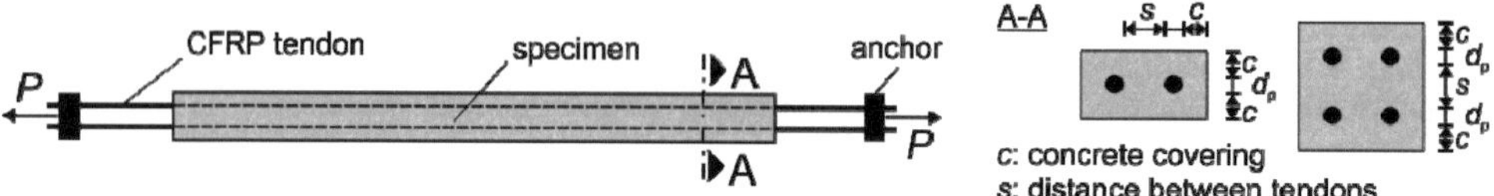

Fig. 4   Test setup to determine the transfer length (small beam tests)

The varied parameters of the small beam tests are listed in table 3. Besides the type of concrete, the concrete covering, the concrete strength (age) and the type of pre-tensioned member the spacing s of the struts and tendons is varied. To determine the minimum concrete covering c two strands are used, while four strands are required to determine the minimum spacing s. The concrete age is always three days.

Table 3   Parameters of small beam tests

Concrete	Pre-tensioning member	Concrete covering	Spacing	Number of strands
[-]	[mm]	$c\,/\,d_\mathrm{p}$ [-]	$s\,/\,d_\mathrm{p}$ [-]	[-]
UHPC/TRC	U 5∅ 1x7 7.5∅ / 10.5∅	1 / 2 / 3	2 / 3 / 4	2 / 4

The specimens are fabricated in a rig similar to the Pull-Out test specimens. After three days the prestress force is released in steps of 20 % as in [6]. Measuring the concrete strains along the longitudinal axias of the small beams at each load step one can obtain the transfer length from the strain differences. Additionally the slip at the end of the specimens is measured continuously with displacement transducers.

4   Experimental tests on the behaviour of prestressed concrete shells

Shells and folded cross-sections gain their stiffness and load bearing capacity due to the spatial load bearing behaviour of the thin-walled cross-sections. For large spans the filigree cross-sections do not provide enough space for reinforcement, thus they have to be prestressed to prevent deformations in consequence of cracking and shrinkage.

The load carrying behaviour of prestressed concrete shells is to be investigated in bending tests as depicted in figure 5. First of all fabrication tests of simply folded or curved cross-sections are conducted. A hinged or moulded formwork is used. These tests will give information about the influence of the type of cross-section, the prestress level and the slenderness ratios $L\,/\,h$ and $h\,/\,d$.

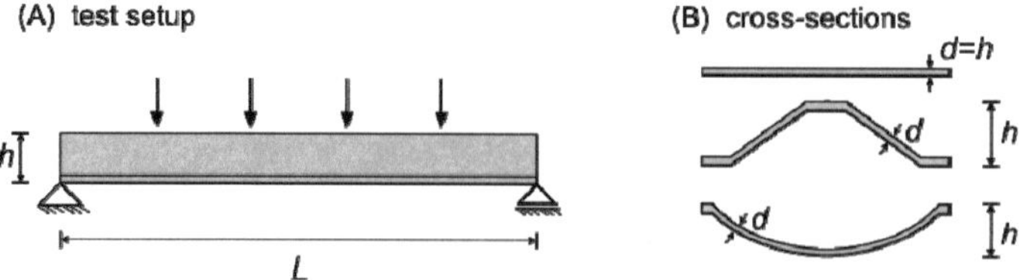

Fig. 5   Bending tests of prestressed concrete shells

Additionally the results of the bending tests will give information about the difference in load carrying capacity compared to sandwich panels which are assembled with these concrete shells.

Table 4   Parameters of bending tests of prestressed concrete shells

Prestress-level [%]	Beam slenderness L / h [-]	Cross-section slenderness h / d [-]
30-70	10-30	1-25

The parameters shown in table 4 are finally chosen after the push-out and small beam tests have been carried out (chapter 3).

5   Experimental tests on the behaviour of prestressed sandwich layers

The tension and shear carrying capacity of sandwich panels is mainly governed by the bond strength between core foam and concrete as well as the stiffness of the core foam and the connecting devices. As a result of profiled cross-sections the relationship of shear deformation g, shear modulus G and shear stress t for planar stress states is not valid, since the shear stress distribution is variable over the cross-sections width. The resulting shear stiffness of applied core foams has to be derived in terms of shear flow. The test setups are depicted in figure 6.

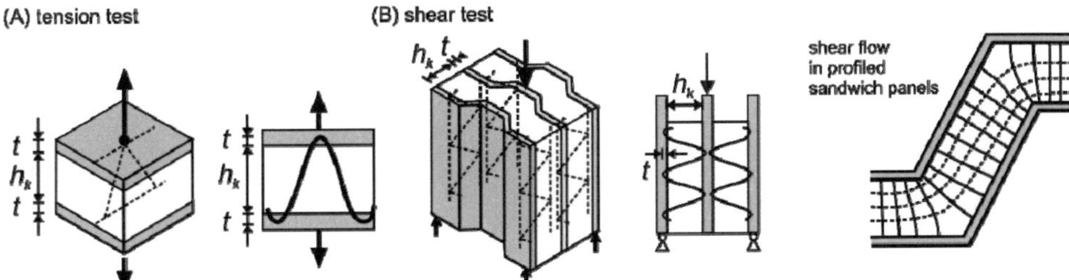

Fig. 6   Tension and shear test of sandwich panels as well as shear flow of profiled cross-sections

For the tests the cross-sections geometry, the type and height h_k of the core foam and the number and type of connecting devices are varied. With the results of these tests the bending tests of prestressed sandwich panels are set up. Table 5 shows the variation of parameters.

Table 5   Parameters of tension, shear and bending tests on sandwich panels

Test [-]	Concrete [-]	Core foam [-]	Height h_k [mm]	Connecting device [-]	Depth of anchoring t [mm]
tension shear bending	UHPC/TRC	without PUR XPS	100 200	without ⋀⋀⋀ ⋀⋀⋀	Variation after defining the layer thickness

The anchoring depth t is varied after defining the thickness of the concrete layers from the bending tests of concrete shells (chapter 4).

The load bearing capacity of prestressed sandwich panels is investigated in bending tests (fig. 7). The degree of sandwich action in comparison to prestressed concrete shells (chapter 4) is analysed as well as the shear carrying capacity of the core foam and connecting devices is compared to the results of the shear tests.

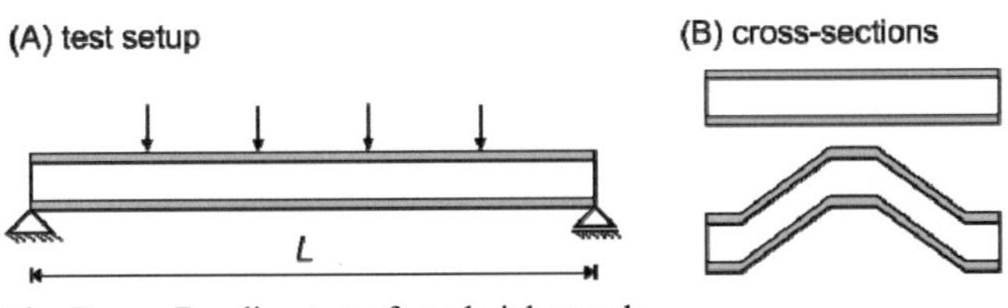

Fig. 7   Bending test of sandwich panels

The prestress level is altered on basis of the tests described before. Experiments without any connecting devices will give information about the shear carrying capacity of the core foam under the effect of different profiled cross-sections and core foams heights h_k.

6 Summary and Outlook

In this paper the main steps to reach the desired goal of a folded or curved sandwich panel are drawn. The investigations will give information about the bond behaviour of pre-tensioned CFRP tendons in TRC and UHPC, the fabrication techniques as well as the load carrying capacity of prestressed simply folded or curved sandwich panels. Additionally to the experimental investigations computational simulations by means of finite elements will be conducted and compared to the test results.

After completion of this goal multiple folded sandwich panels will be investigated.

Acknowledgement

This research project is part of the SPP 1542 founded by the German Research Foundation (DFG). The author acknowledges the support.

References

[1]   German Research Foundation, priority program (DFG SPP 1542), Leicht Bauen mit Beton (Concrete light. Future concrete structures using bionic, mathematical and engineering form-finding principles)

[2]   German Research Foundation, priority program (DFG SPP 1182), Nachhaltig Bauen mit UHPC (Sustainable Construction with UHPC)

[3]   German Research Foundation, collaborative research centre SFB 532, Textilbewehrter Beton - Grundlagen für die Entwicklung einer neuartigen Technologie (Textile Reinforced Concrete - Development of a new technology)

[4]   Schätzke, C.; Schneider, H.N.; Joachim, T.; Feldmann, M.; Pak, D.; Geßler, A.; J. Hegger, A. Scholzen: Doppelt gekrümmte Schalen und Gitterschalen aus Textilbeton", 315-328, 6th Colloqium on Textile Reinforced Structures (CTRS6), Berlin, 285-296 (2011).

[5]   Katz, A.; Aronoff, J.; Frostig, Y.: Bond Failure Between CFRP Rebars and High Strength FRC. The 8th International Symposium on Fiber Reinforced Polymer Reinforcement for Concrete Structures (Patras 2007), pp. 1-8

[6]   Bertram, G.; Hegger, J.: Anchorage behaviour of strands in ultra-high performance concrete. 8th International Symposion on Utilization of High-Strength and High-Performance Concrete (Amsterdam 2008), pp. 921-926

[7]   Russel, B; Burns, N.: Measurement of Transfer Length on Pretensioned Concrete Elements. In Journal of Structural Engineering (1997) , pp. 541-549

[8]   Nitsch, A.: Spannbetonfertigteile mit teilweiser Vorspannung aus hochfestem Beton. In: Schriftenreihe des Lehrstuhls und Instituts für Massivbau der RWTH Aachen (2001)

[9]   Mayah, A.; Soudki, K.; Plumtree, A.: Experimental and Analytical Investigation of a Stainless Steel Anchorage for CFRP Prestressing Tendons. In: PCI Journal (March-April 2001), pp. 88-100

[10]   Mayah, A.; Soudki, K.; Plumtree, A.: Simplified Wedge Anchor System for FRP Rods. The 8th International Symposium on Fiber Reinforced Polymer Reinforcement for Concrete Structures (Patras 2007), pp. 1-8

[11]   RILEM/CEB/FIP: Bond Test Reinforcing Steel, 2. Pull-Out Test 1987

Finding new forms for bearing structures by use of origamics

Jan Dirk van der Woerd

Institute of Structural Concrete,
RWTH Aachen University
Mies-van-der-Rohe-Straße 1,52074 Aachen, Germany
Supervisors: Rostislav Chudoba and Josef Hegger

Abstract

Today in building industry different, partly contrary interests are competing within the design of the load bearing structure of a new building. These usually contradicting interests and requirements are for example the function, economic issues, an attractive design or the maximum exploitation of construction materials. One possible strategy how to achieve harmony between the contradicting requirements might be the application of the folding technique inspired by the Japanese paper folding art origami. The term for technical application of origami is origamics. The idea is to take a thin-walled flat plate made out of concrete and fold it along a pre-defined pattern of creases. With this technique it is possible to create complex spatial structures made out of cement-based materials without complex formwork. The aim of the research is to develop this technique for light-weight concrete structures and its improvement onto a cost-effective level.

The elaborated work has been split into three parts. The first part is the identification and verification of suitable foldable crease patterns by use of a parametric mathematical model, which has been developed for this purpose. The second part is focussing on production issues and design of details. Within the third part structural analysis is performed for evaluation of the load bearing capacity.

In this paper the idea of creating new load bearing structures by folding is presented. The results of a feasibility study are shown. The parametric mathematical model and the work on identifying adequate material combinations are presented. An outlook on the structural analysis is given.

1   Introduction

In the design of the load bearing structure of a building often several different interests with partly contrary requirements are competing with each other. The main interests are:

(1)   High utilization of the construction volume for the desired function;

(2)   Financial constraints: Economic production and maintenance;

(3)   Attractive design;

(4)   High exploitation of the construction material;

For accomplishing these requirements different concepts and methods are available for architects and structural designers. These conceptual instruments can be distinguished within global and local design principles. Forming the global shape according to the principle "form follows force", it is expected to achieve a high utilization ratio of the used materials (Requirement 4). This design principle is the basis of the design of shell structures. Due to their light appearance, shell structures are mostly regarded as elegant and architecturally attractive (Requirement 3).

The focus of designing at the local level lies on the form and layout of efficient cross-sections, which can be produced in a cost-effective way (Requirement 2) and lead to space saving load bearing structures (Requirement 1).

The two first mentioned requirements can be found in the almost exclusive rectangular forms of the realized buildings in our days. The reason for the overweight of the local design principles can be found in the strong emphasis on an economic fabrication (Requirement 2) and a maximum utilization of the construction volume (Requirement 1).

Proc. of the 9th *fib* International PhD Symposium in Civil Engineering, July 22 to 25, 2012,
Karlsruhe Institute of Technology (KIT), Germany, H. S. Müller, M. Haist, F. Acosta (Eds.),
KIT Scientific Publishing, Karlsruhe, Germany, ISBN 978-3-86644-858-2

2   Classical design approaches and new design and production technique

In the field of concrete structures two design approaches can be found. The first approach is shell structures and the second one is folded plates. The global shape optimization has been used for the design of light-weight shell structures [1]. The local design mainly focuses on the form and layout of cross-sections as with folded plates.

The two approaches will be shortly discussed concerning their load bearing mechanism and their fabrication. Out of these two approaches a new approach is going to be derived by the use of the technique of folding. This new approach also requires the development of new production methods.

2.1   Shell structures

The geometry of shell structures is chosen in a way that the dominant constant loads (dead load) are carried by membrane forces [2], [3]. However their load bearing capacity for variable loads is poor, since they differ from the form determining constant loads. A critical aspect is the expensive production of the formwork. Mostly the formwork is unique and so it can only be used one time.

2.2   Folded Plates

The design of folded plates is focusing on the local design of the cross-sections. By placing plates in saw tooth shape the statically effective depth is enlarged.

Folded plate structures can be produced in two ways. The first way is to use prefabricated elements, which are connected on site to the final structure. By the use of prefabricated elements the variety of different global shapes is limited, since for production reasons the segment shape is limited to simple, mainly rectangular forms. The size of the elements is also limited due to transportation reasons. The second way is to produce the elements on-site. Simple regular structures produced with in-situ concrete are cheap. But for complex folded plate structures the effort for producing the framework is rising very fast. Here the same problem arises concerning high cost of the formwork like for the shell structures. Only regular and simple structures can be produced with little costs.

2.3   New design approach by use of origamics

Neither of the classic design approaches, shells and folded plates, fulfils all requirements which have been mentioned in the introduction. For a better fulfilment of the requirements local and global design principles should be connected within a new design approach. Variable loads are the main problem for shells if they are large compared to the constant loads. Variable loads can lead to bending moments within the shell. The load bearing behaviour of shells concerning bending moments is poor. A possible strategy to engage this problem is to increase the shell thickness, but this is conflicting with an economic material usage. By folding it is possible to increase the statically effective depth with a lower amount of material increase than with increasing the shell thickness. The technique of folding is quite common in nature, e.g. leaves of palms or seashells [4]. Steel profiles are also stiffened by folding like for example sheet piles [5].

For realization of the generally folded concrete structures a specialized production technique has been devised to introduce crease patterns in thin-walled concrete plates. The folding technique is demonstrated in Figure 1. Spacers are used to introduce crease lines with the required rotation capacity. The spacers avoid concrete getting at these places. The reinforcement is passing through. After folding the creases are fixed with grout.

Using this technique, generally curved shapes can be constructed similar to the way how a piece of paper is folded in origami. The technique of origami serves as a source for identifying suitable crease patterns for load bearing structures. The flexibility of the technique can be used to design shapes that combine the global and local design principles for statically loaded structures leading to high exploitation of the material. In the next section the production of a prototype is described.

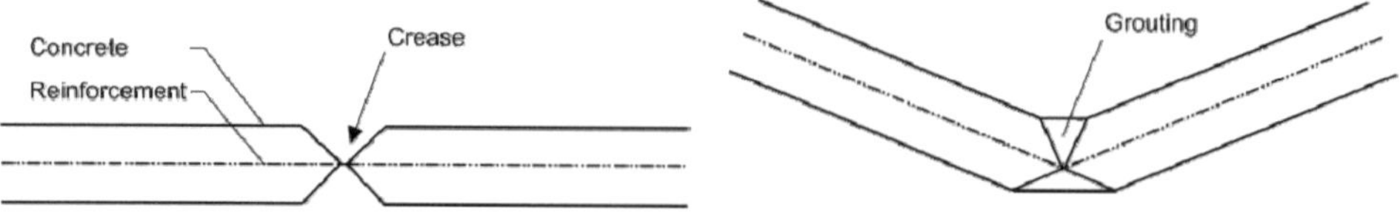

Fig.1   Sketch of the principle of folding: (left) before folding, (right) after folding and fixing

3   Feasability study

For demonstrating the advantages and the basic principles of the new production method a feasibility study has been performed. As a prototype a thin-walled arc was chosen, which was going to be folded out of a thin concrete plate. The basis crease-pattern is a regular rhombus crease-pattern [6]. This crease pattern is well-known and has already often been used in architecture and it is rigid foldable.

The size of the concrete plate was 1,40 m x 0,80 m (length x weight) with a thickness of 6 mm. As reinforcement steel wire with a diameter of 0,6 mm was placed in a rectangular mesh. As concrete UHPC was used. The workflow and the individual production steps can be seen in Figures 2 and 3.

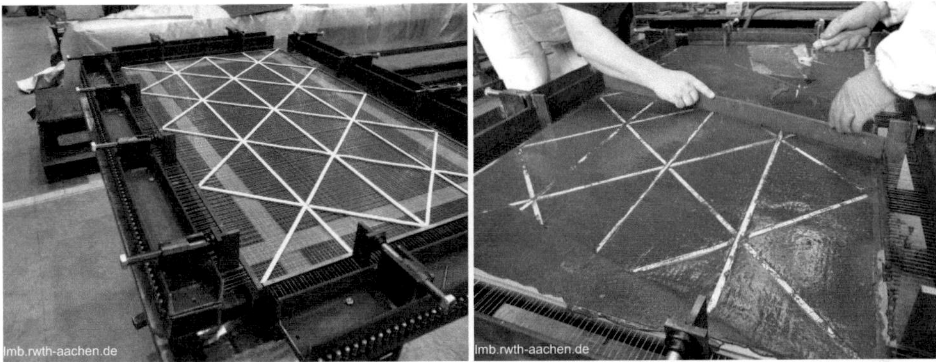

Fig. 2   (left) Reinforcement with spacers; (right) Placing of concrete

Fig. 3   (left) Concrete after hardening and removal of the spacers; (right) Completed structure after folding and filling the creases

In the first step the reinforcement is placed into the formwork and the wires are tensioned (Fig. 2 left). Along the future creases gaskets have been placed on the reinforcement to create spaces within the concrete during concreting (Fig. 2 right). After hardening of the concrete the gaskets have been removed (Fig. 3 left) and the crease pattern can be clearly seen. For getting the plate out of 2-D into 3-D shape the plate has been lifted at five intersections of the crease lines and the remaining creases have been folded by hand. After reaching the final shape, the creases have been grouted with concrete, to achieve a stable connection between the segments. However the focus within the study lied on production issues the thin-walled arc has a high stiffness in vertical and horizontal direction. With this feasibility study it was shown, that there is a high potential with the proposed design and production method. A new application area for innovative cement-bounded composite materials like textile reinforced concrete or concrete with carbon reinforcement is unclosed.

4   Road map of the work for the new design method

For development of the new design method several problems have to be solved. It seems convenient to split the process into several parts. One can distinguish between three major parts.

- Finding suitable crease patterns
- Production and material issues
- Structural analysis and assessment of the load bearing capacity

Each of the parts influences the others and it is necessary to redo them several times. The problem of finding the optimal solution for a load bearing structure can be approximated iteratively by going through the parts in a circular way.

4.1   Form finding by use of a mathematical model

A major task is the finding of rigid foldable crease patterns. A crease pattern can be regarded like a map for the folding process, wherein a 2-D structure is turned into 3-D. Depending on which side the angle formed by a fold and its attached facets opens the resulting folds are distinguished in mountain or valley folds. Several different rigid foldable crease patterns and their resulting folding patterns have already been identified and are known in literature [6]. Two basic folds are the accordion and the reverse fold. A lot of the known folding patterns can be constructed by repeated use of the reverse fold.

A common approach to find new patterns and forms is to try-out by hand [6],[7]. However this approach is very time-consuming and it cannot be excluded, that the founded forms are strongly influenced by the preferences of the researcher. For more complex forms and geometries it is nearly impossible to find a crease pattern just by trying. There are too many relations and constraints. The use of a mathematical model greatly simplifies the identification of suitable crease patterns.

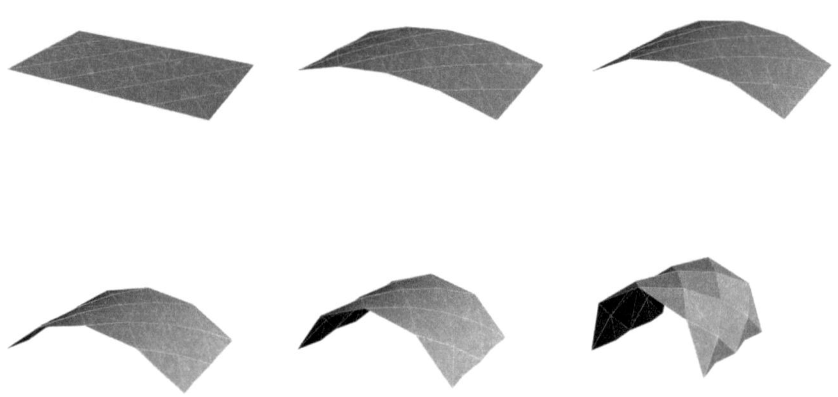

Fig. 4   Mathematical simulation of the folding process of the structure of the feasibility study

Several constraints have to be fulfilled by a mathematical folding model [8]:

- Neither the creases nor the segments stretch
- The segments stay flat and do not crinkle
- The segments and the crease lines do not rip or have holes
- The segments are not allowed to penetrate each other

With the open source programming language Python a parametric mathematical model has been implemented for simulating and visualisation of the folding process. The model consists of several modules with different tasks. There are modules for creating the crease pattern, for simulating the folding process or for visualisation. The three first mentioned constraints have been implemented within the mathematical formulation. The last constraint is ensured by visual control.

The simulation of any arbitrary rigid foldable crease pattern is possible with this model. However the crease pattern and the boundary conditions, e.g. fixed nodes for support have to be chosen that way, that the structure is rigid foldable. Rigid foldable means, that if instead of the creases hinges can be placed and the attached segments are rigid the structure can be folded [9]. The conditions for global rigid foldability are yet not completely known. A solution to this problem is to use small crease pattern elements that are rigid foldable and construct large crease patterns by parallel or symmetric repetition. The simulation of the structure of the feasibility study can be seen in Figure 4.

4.2 Production and material issues

For limiting the weight it is desired to make the concrete plate thin. For durability reasons steel reinforcement needs a certain concrete coverage which leads to relatively thick diameters. Within the feasibility study steel wire was used as reinforcement with a diameter of only 0,6 mm, but the folding was hard. Textile reinforcement cannot corrode and needs no concrete coverage due to durability reasons. It was decided to skip metallic reinforcements and to focus on textile fabrics. With the use of textile reinforcement a suitable matrix had to be found.

Several small prototypes have been produced for identifying adequate material combinations. Instead of UHPC normal concrete with a small grain size has been tried, since its plasticity is better. For the reinforcement different glass and carbon textile fabrics have been used. As adequate combination textile reinforced concrete has been identified.

Besides the variation of material combinations, attention has been spent in reduction of production steps and in detailed design of the creases. One of the latest prototypes can be seen in Figure 5.

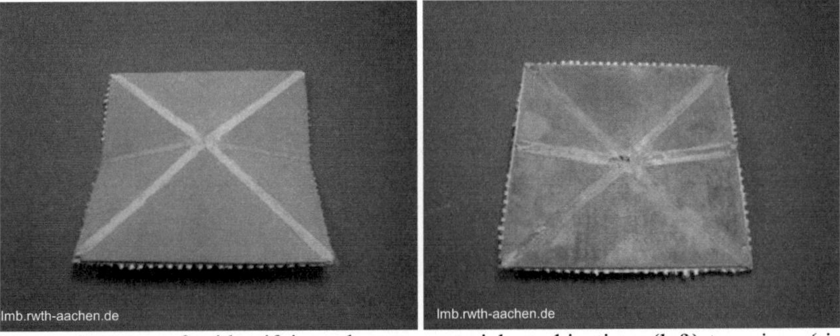

Fig.5   Prototype for identifying adequate material combinations (left) top view, (right) bottom view

It is planned to perform strength tests for getting information about the stiffness and bending capacity of the creases. These data will be used for detailed structural models simulating the load bearing behaviour.

4.3 Structural analysis and assessment of the load bearing

In the third step structural analysis is going to be performed for assessing the load bearing capacity of larger structures. With the evaluation it is intended to identify parameters for designing the crease patterns for a good load bearing behaviour. Some of such parameters are mentioned by Trautz[5], like the height of folds and their frequency. The height of a fold for example is directly connected to its statically effective depth.

The findings of this assessment for the static parameters will be included in the detail design of the crease patterns. The major task is to combine the static parameters with parameters which determine the final global shape of the structure. The static parameters are mainly influencing the local design of the final structure and the geometric parameters influence the global design.

It is intended to connect the mathematical simulation of the folding process with a structural analysis to accelerate the assessment and the identification of good parameters for the crease patterns.

5 Summary

A design method considering global and local design principles by use of folding and inspired by origamics has been presented. For realizing this method a new production method has been developed. Thin-walled concrete plates are equipped with a crease pattern and are folded to complex spatial structures. Within a feasibility study the new production method was tested. For the simulation of the folding process a parametric mathematical model has been developed. With small prototypes the production method has been improved and as adequate material textile reinforced concrete has been identified. An outlook has been given on the procedure of the structural assessment of the load bearing capacity.

With the technique of folding it is possible to build powerful bearing structures out of cement bounded composite materials like textile reinforced concrete. The characteristic of these structures are little material consumption, an attractive design and an economic production without the use of complicated formwork. It is expected that this technique will expand the range of application of cement bonded composite materials. For example the application of folded concrete structures within façade systems, which are dominated by metal and plastic based construction materials. Another possible application is the use as "lost formwork" for complex geometries.

References

[1]   Ramm, E., Mehlhorn, G.: On shape finding methods and ultimate load analyses of reinforced-concrete shells, In: Engineering Structures, vol. 13 (1991), no. 2, pp. 178-198

[2]   Schmidt, H.: Von der Steinkuppel zur Zeiss-Dywidag-Schalenbauweise. In: Beton- und Stahlbetonbau 100 (2004), Heft 1, pp. 79-92

[3]   Koch, K.: Bauen mit Membranen: der innovative Werkstoff in der Architektur. München; Berlin; London; New York: Prestel (2004)

[4]   Trautz, M., Ayoubi, M.: Das Prinzip des Faltens in Architektur und Ingenieurbau. In: Bautechnik 88 (2011), Heft 2, pp. 76-79

[5]   Werner, H.: Falten im Leichtbau. In: Archplus (2011), Heft 2, pp. 82-89

[6]   Buri, H.: Die Technik des Origami im Holzbau – Faltwerke aus BSP-Elemente. In: Tagungsband 5.Grazer Holzbautagung (2006), pp. N-1 – N-13

[7]   Buri, H.: Origami-Faltkunst für Tragwerke. In Detail (2010), Heft 10, pp.1066-1069

[8]   Hull, T.: Origami3: Third International Meeting of Origami Science, Mathematics, and Education. A K Peters, Ltd., 2002.+

[9]   Tachi, T.: Geometric Considerations for the Design of Rigid Origami Structures. In: IASS Symposium Shanghai China (2010), pp. 771-782

Discretisation of light-weight concrete elements using a line-geometric model

Markus Hagemann

Daniel Klawitter

Institute of Geometry,
Dresden University of Technology,
Zellescher Weg 12-14, 01069 Dresden, Germany
Supervisor: Daniel Lordick

Abstract

The requirements for concrete construction are found mainly on the capacity and its weight. These issues can be already considered in the design process. Therefore integration to CAD systems is necessary. Since we restrict ourselves to ruled surfaces, which are particularly interesting in terms of manufacturing, the use of a line-geometric point model is justified. We can adapt well-known subdivision schemes that allow a high degree of design flexibility. This makes it possible to interpolate a set of straight lines by ruled surfaces. Finally, parameters are presented that allow an intuitive modification or which can be optimized with respect to static issues.

1   Introduction

The realisation of concrete elements primarily depends on static and mould. The aim is to map both factors of influence to a more complex mathematical model, where a consistent variation of these elements is possible. Under the aspect of mould and static properties the class of ruled surfaces is distinguished. This includes the amount of surfaces that can be generated by the motion of a line. In civil engineering this class of surfaces has not yet gained the attention it deserves. The production is not as complicated as it is for free form surfaces, for example, ruled surfaces can be manufactured by heat wire cutting.

The well-known line-geometric models provide an elegant mathematical method for the description of ruled surfaces, especially for the modification of these surfaces. We want to use the model of the dual unit sphere, because helicoids are of particular importance. Besides that, helicoids are minimal-surfaces, *i.e.*, they minimize the surface area and hence the weight of (concrete) solids. This is the reason why they attract our special attention. The aim of this contribution is the development of algorithms, which are easy to handle. We develop a ruled surface interpolation algorithm for lines, where the design parameters are evident. The algorithm was implemented, using the grasshopper plugin for Rhino3D.

The paper is organized as follows: In section 2 we give a brief introduction into the mathematical background. Therefore, dual numbers are presented and special line coordinates are introduced. Afterwards, we show how displacements can be treated. Furthermore, subdivision algorithms for points in the plane and especially on the sphere are presented. In section 3 we transfer these methods to the dual unit sphere. Thereby, we obtain an interpolation method for lines with interesting properties. Last but not least, we discuss the occurring design parameters.

2   Mathematical background

2.1   Dual numbers and spear coordinates

The ring of dual numbers is an extension of the real numbers. Every dual number has the form

$$z = a + \epsilon b,$$

where $a, b \in \mathbb{R}$ and ϵ is the dual unit with $\epsilon^2 = 0$. Addition and multiplication are defined with respect to the dual unit:

$$z_1 + z_2 := (a_1 + a_2) + \epsilon(b_1 + b_2), \qquad z_1 \cdot z_2 := (a_1 \cdot a_2) + \epsilon(a_1 b_2 + a_2 b_1).$$

Proc. of the 9th *fib* International PhD Symposium in Civil Engineering, July 22 to 25, 2012,
Karlsruhe Institute of Technology (KIT), Germany, H. S. Müller, M. Haist, F. Acosta (Eds.),
KIT Scientific Publishing, Karlsruhe, Germany, ISBN 978-3-86644-858-2

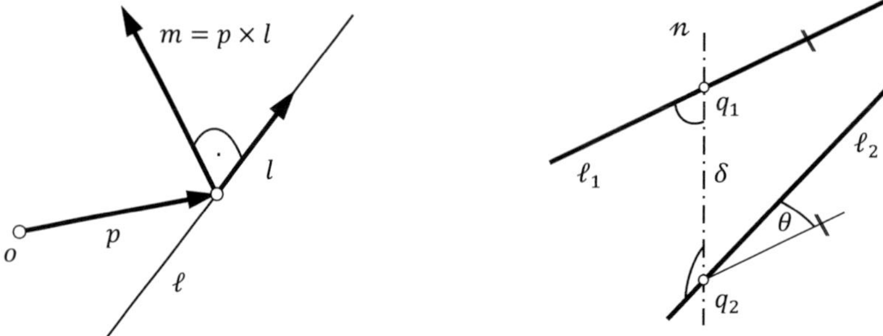

Fig. 1   Interpretation of the line coordinates (left). Composition of the dual angle (right).

Note that a pure dual number, *i.e.*, a dual number with vanishing real part, has no inverse. Known functions can be extended to the dual number calculus by Taylor expansion, cf. [7]. Therefore we have

$$\cos(a + \epsilon b) = \cos(a) - \epsilon b \sin(a), \qquad \sqrt{a + \epsilon b} = \sqrt{a} + \epsilon b \frac{1}{2\sqrt{a}}.$$

An arbitrary line in the three dimensional space can be described by a point p and a directed vector l. Since this representation is not unique for a line, some restrictions are necessary. This includes the normalisation of l and a special choice of p on the line so that the vectors p and l are perpendicular. Using l in combination with the momentum $m := p \times l$ instead of p to describe a line leads to the well-known Plücker coordinates (Fig.1). Note that l and m are perpendicular, too.

Let l and m be the real and the dual part of a dual vector ℓ, then the norm of ℓ is defined by

$$\|\ell\| = \sqrt{\|l\|^2 + 2\epsilon\langle l, m\rangle} = \|l\| = 1.$$

Therefore we can suppose that all lines are elements of a sphere in a three dimensional vector space over the dual numbers with radius 1. Due to the fact that the orientation of l is negligible for describing lines, it is obvious that two points on the dual sphere correspond to the same line. To avoid this, it is expedient to establish oriented lines, called spears. Thus, there exists a one-to-one correspondence between oriented lines of $\mathbb{R}^3$ and points on the dual unit sphere $S_\mathbb{D}^2$, which can be used to map ruled surfaces to curves on $S_\mathbb{D}^2$, and vice versa, see [4]. The presented description of spears offers a geometric interpretation of the inner and outer product; cf. [1]. Both products are defined as the extension of the usual ones to dual vector calculus.

$$\langle \ell_1, \ell_2 \rangle = \langle l_1, l_2 \rangle + \epsilon(\langle l_1, m_2 \rangle + \langle l_2, m_1 \rangle)$$

$$\ell_1 \times \ell_2 = l_1 \times l_2 + \epsilon(l_1 \times m_2 + m_1 \times l_2)$$

Let q_1 and q_2 be the intersection points of two spears ℓ_1 and ℓ_2 with their common normal n. As mentioned above, the momentum vector of a line can be obtained as the cross-product of a point on the line and its direction vector. Hence we have

$$\langle \ell_1, \ell_2 \rangle = \cos(\theta) - \epsilon \sin(\theta)\, \delta = \cos(\theta + \epsilon\delta) = \cos(\phi).$$

Therefore, a dual angle ϕ is defined, which consists of the angle θ between the two spears and their shortest distance δ. (Fig.1) Thus the cross product of two dual unit vectors equals the common normal of the two spears ℓ_1 and ℓ_2 multiplied by the sine of their dual angle:

$$\ell_1 \times \ell_2 = \sin(\theta + \epsilon\delta)\,(l_n + \epsilon m_n) = \sin(\phi)\, n. \tag{1}$$

The orientation of n depends on the order of the factors.

2.2   Displacement of lines

Now, we describe transformations, especially displacements of spears. At first we consider the Rodrigues rotation formula. The formula can be involved by a vector of $\mathbb{R}^3$ that corresponds to the axis and its norm to the angle of rotation. Accordingly, the rotation matrix can be determined by using the exponential map for skew-symmetric matrices. It can be shown that this formula is also advantageous to describe displacements of lines, see [6].

Therefore the formula needs to be extended to dual vectors. We apply the Rodrigues rotation formula to two lines ℓ_1, ℓ_2 and their common normal n. This results in a dual matrix

$$R(t) = I + \sin(\phi t)\,\mathcal{N} + (1 - \cos(\phi t))\mathcal{N}^2,$$

where $\mathcal{N}$ is a skew-symmetric dual matrix representing the cross product of n. Applying this to the initial line ℓ_1 yields

$$\ell(t) = R(t) \cdot \ell_1 = \ell_1 + \sin(\phi t)\,(n \times \ell_1) + (1 - \cos(\phi t))\big(n \times (n \times \ell_1)\big)$$
$$= \sin(\phi t)\,(n \times \ell_1) + \cos(\phi t)\,\ell_1.$$

Considering (1), the equation can be transformed to

$$\ell(t) = \frac{\sin((1 - t)\phi)}{\sin(\phi)}\,\ell_1 + \frac{\sin(\phi t)}{\sin(\phi)}\,\ell_2. \tag{2}$$

It should be pointed out, that this is only valid for regular dual angles, *i.e.*, angles between the lines ℓ_1, ℓ_2 which are not purely dual. In the case of parallel lines an arbitrary normal of both lines can be chosen.

In conclusion, these equations give the opportunity to determine all lines of a helicoid including ℓ_1 and ℓ_2, whose screw parameter is minimal. Note that the definition of the pitch, *i.e.*, the ratio of translation and rotation, is depending on the orientation of the spears.

2.3 Subdivision algorithms for curves

In computer graphics the use of subdivision algorithms for designing free-form curves or surfaces has established itself. These algorithms base on the linear interpolation of two points p_0, p_1 in $\mathbb{R}^3$.

$$p(t) = (1 - t) \cdot p_0 + t \cdot p_1.$$

The easiest and most popular algorithm is named after its developer de Casteljau and allows the evaluation of Bézier curves (Fig. 2) by specifying an ordered set of control points p_i. In contrast to this a cubic Hermite spline is determined by defining two control points p_0, p_1 and two tangent vectors t_0, t_1 at these points, respectively. Due to the degree of this curve four control points $\tilde{p}_i$ can be extracted from p_i and t_i, whose Bézier curve is equivalent to the Hermite spline. By this construction a piecewise G^1-interpolation is possible.

For our purpose, we want to study Kochanek-Bartels splines (also called TCB-splines), a special form of cubic Hermite splines. The advantage of these splines is their flexibility and their very intuitive design parameters. Moreover, we can trace their generation back to the evaluation of Bézier curves using the de Casteljau algorithm. Everything that needs to be done is to determine the points of the control polygon.

The concept of subdividing a line segment can be easily transferred to the sphere S^2. Instead of lines, geodesics have to be used, which are great circles on the sphere. Hence, the adaption of linear algorithms is not a difficulty and is already known as "spherical linear interpolation" (short: slerp), see [5]. Let p_0 and p_1 be points on the sphere S^2 and θ their including angle, then the formula for all points on the great circle, that contains p_0 and p_1, is given by

$$p(t) = \frac{\sin((1 - t)\theta)}{\sin(\theta)} \cdot p_0 + \frac{\sin(t\theta)}{\sin(\theta)} \cdot p_1. \tag{3}$$

In the following, this equation enables us to transfer the concept of TCB-splines to the sphere.

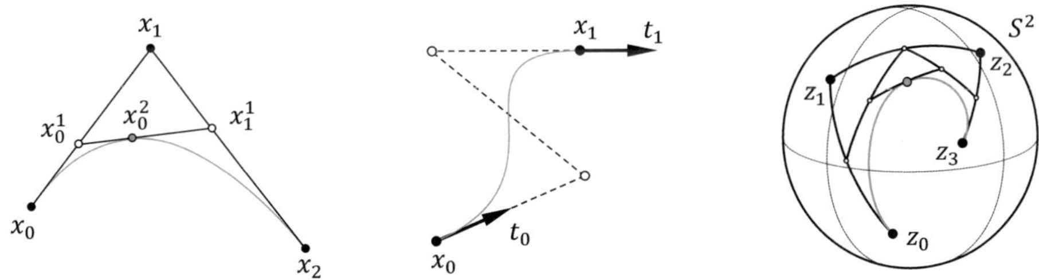

Fig. 2   The de Casteljau algorithm (left); cubic Hermite Interpolation (middle); spherical linear interpolation of four points using the de Casteljau algorithm (right).

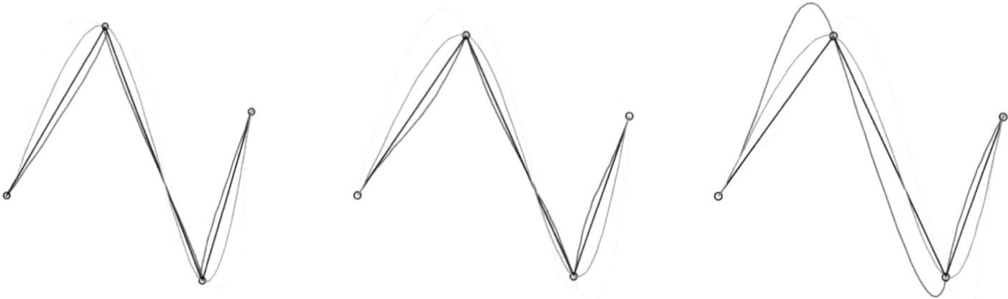

Fig. 3 TCB-splines for a given polygon (black) with a variation of tension, continuity and bias (f.l.t.r.) to the values -0.8 (green), 0 (red), 0.8 (blue).

3 Designing curves on the dual unit sphere

The similar structure of equation (2) and (3) is an important aspect. That means more specifically that we can transfer spherical linear interpolation to the dual unit sphere for generating ruled surfaces. Furthermore, the de Casteljau algorithm allows an adaption of cubic Hermite splines. Note that tangential elements on $S_{\mathbb{D}}^2$ have no precise meaning. The ultimate goal is the application of Kochanek-Bartels splines which offer a high degree of flexibility. The difficulty is now to find an analogy to finite differences on the dual unit sphere and to declare functional parameters for modifying the resulting ruled surface.

The question of derivatives can be answered mathematical correct by using the theory of Lie groups, cf. [3]. However, the algebraic result is intuitive and corresponds to the osculating great circle of the curve. This, in turn, is related to a helicoid having contact of first order and thus the design parameters affect essentially this helicoid. Let $\ell_0, \ell_1, \ell_2 \in S_{\mathbb{D}}^2$, then the axis a_1 and the pitch σ_1 of the finite difference surface between ℓ_0 and ℓ_2 at ℓ_1 are given by

$$l_n^* + \epsilon m_n^* := n_{01}\phi_{01} + n_{12}\phi_{12}, \qquad \sigma_1 = \frac{\langle l_n^*, m_n^* \rangle}{\langle l_n^*, l_n^* \rangle}, \qquad a_1 = m_n^* - \sigma_1 l_n^*.$$

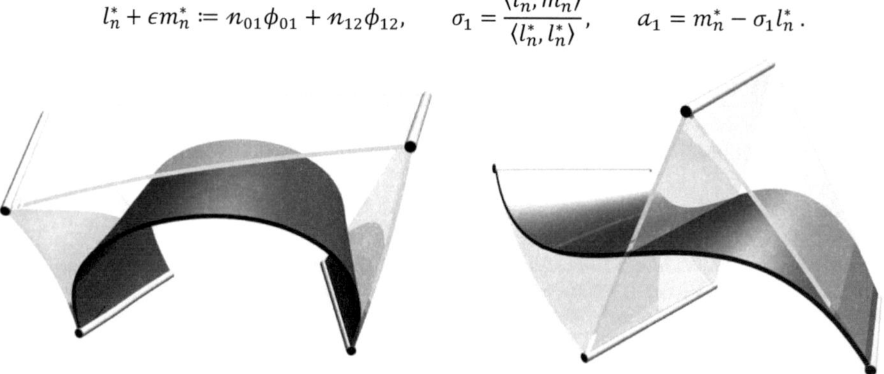

Fig. 4 Subdivision scheme for ruled surfaces (grey) with helicoids (transparent), control lines (orange) of third degree.

3.1 Design parameters and their influence on interpolants

In analogy to the modification of splines by changing control points or tangents, we want to establish appropriate parameters to control the design of the ruled surface patches. This has the advantage of being very user-friendly and intuitive. It also offers the opportunity to define a ruled surface analytically. A fundamental question, which has to be answered before starting the design process, is that of the continuity of the ruled surface patches at their boundary control lines. It is obvious, a higher degree causes dependencies of design parameters. Since cubic splines are used, the ruled surface patches are defined by 4 lines. Two lines of them are interpolated and therefore the degree of freedom for connecting two surface patches depends on the remaining lines.

This degree is equal to 8 for each surface. In the following we shall mainly restrict ourselves to the case of G^1-continuity which means an identical set of tangent planes along the two boundary lines of adjacent ruled surfaces.

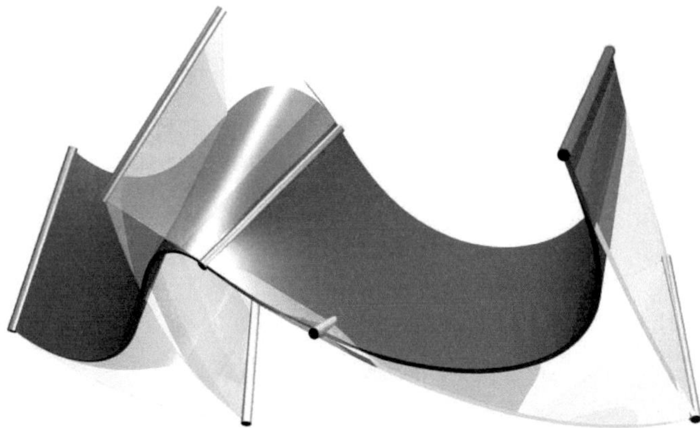

Fig. 5   Subdivision scheme for ruled surfaces (grey) with helicoids (transparent), control lines (orange) of third degree.

3.1.1 Continuity

Due to the fact that we use helicoids the type of contact of adjacent ruled surfaces is already known. They are determined by the interpolated ruling itself and the first adjacent control line of the ruled surface patch. In other words, G^1-continuity is guaranteed by choosing three control rulings on the same arbitrary helicoid.

3.1.2 Tension

A convenient pendant to the tension of splines can be realised by the choice of arbitrary lines on the helicoid. This naturally ensures the continuity and allows the modification of the tension of the surface.

3.1.3 Bias

In addition to the tension and the continuity, we can define an equivalent to the bias of curves. Considering ruled surfaces, the modification of this parameter corresponds to a rotation of the helicoid around the interpolated line. Moreover, the evaluation of the axis also results in a displacement along the ruling. Generally, the set of axes of the helical surfaces forms a Plücker conoid whose axis is the interpolated line.

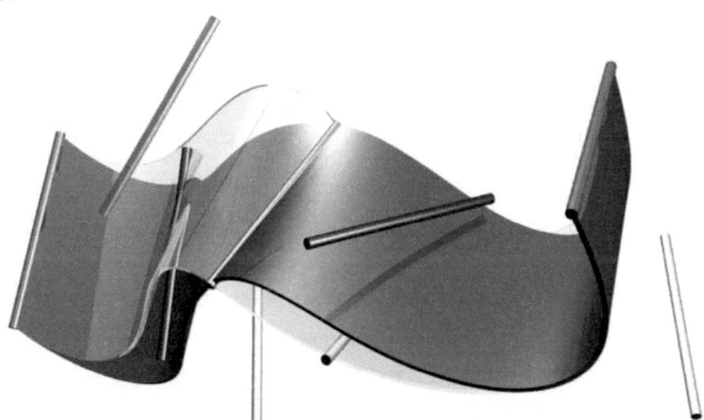

Fig. 6   Modification of the bias yields new control lines (blue); resulting ruled surface (grey); initial ruled surface (transparent), cf. fig. 5 (grey).

3.1.4 Shift

Due to the fact that the resulting surface needs to be continuous, the explanation of each modification can be reduced to the interpretation of the behaviour of the helicoid. The degree of freedom for all involved helical surfaces is equal to 3. Therefore we search an axis and a screw parameter for the helicoid with first-order contact. In contrast to the bias parameter, which allows a rotation of the axis, the shift parameter causes a translation of the axis along the interpolated boundary line.

3.1.5 Screw parameter

Finally, we can introduce a variation of the screw parameter, which should be greater than zero because the inversion of the orientation can already be realised by the bias. The parameter affects primarily the ratio of rotation and translation along the axis of the helicoid.

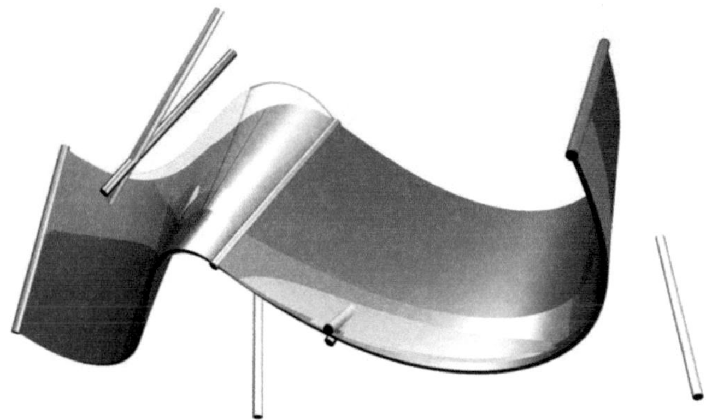

Fig. 7   Modification of the screw parameter results in new control lines (blue); ruled surface (grey); initial ruled surface (transparent), cf. fig. 5 (grey).

4   Conclusion and further research

The advantage of ruled surfaces for concrete structures and their reinforcement is that straight lines are contained in the surface. Moreover, this property is necessary for prefabrication using heat-wire cutting. The presented design method offers intuitive parameters for modification. This degree of freedom can further be used for best-fit approximations or static optimisation, following the idea of "form follows force".

Obviously, continuity of second order involves more control lines of the ruled surface patches. Because of the de Casteljau algorithm we can expect dependencies from each patch which imply important constraints to the flexibility of the resulting surface.

Acknowledgement

The results of this paper are part of the research project "Line Geometry for Lightweight Structures", funded by the DFG (German Research Foundation) as part of the SPP 1542.

References

[1]   Bottema, O.; Roth, B.: Theoretical Kinematics. Dover Publications, Inc. (1990)

[2]   Hoschek, J.; Schwanecke, U.: Interpolation and approximation with ruled surfaces. In: The Mathematics of Surfaces 8 (1998) 213-231

[3]   Karger, A., Novák, J.: Space kinematics and Lie groups. Gordon & Breach Science (1985)

[4]   Pottmann, H.; Wallner, J.: Computational Line Geometry. Springer (2001)

[5]   Shoemake, K.: Animating rotations with quaternion curves. In: Computer Graphics 19 (1985)

[6]   Sprott, K.; Ravani, B.: Kinematic generation of ruled surfaces. In: Advances in Computational Mathematics 17 (2002) 115-133

[7]   Veldkamp, G.: On the use of dual numbers, vectors and matrices in instantaneous, spatial kinematics. In: Mechanism and Machine Theory 11 (1976) 141-156

Methods for transforming flat concrete plates into double curved shell structures

Benjamin Kromoser

Institute for Structural Engineering,
Technical University of Vienna
Karlsplatz 13, A-1040 Vienna, Austria
Supervisor: Johann Kollegger

Abstract

Over the last few years, different methods for building shells by using a combination of pneumatic formwork and post-tensioning technologies have been investigated at Vienna University of Technology.

The idea behind all these construction methods is to build ice and concrete shells with double curvature originating from an initially plane plate.

The various construction methods will be presented and the experiments on spherical shells with base diameters of approximately 10m will be described. The advantages and disadvantages of individual methods will be explained and a proposal for a new construction method will be presented.

A patent application for the new construction method has been submitted by Vienna University of Technology in November 2011. A description of the method and a feasibility study will be presented in this paper.

1   Motivation

Domes are esthetic, efficient, impressive and also fascinating because of their shape, thinness and structure. A particular feature of double curved shells is their effective load transfer. Loading is mostly carried by normal forces in the middle surface of the shell which causes a very uniform and efficient distribution of the stresses in the entire cross section. Thus, large spans can be achieved while the stresses in the structure are kept relatively small [1].

2   Aim of the research work

The transformation of a flat plate into a double curved shell can be accomplished if an orthotropic construction material were available, which would permit large strains at low stresses in one direction, and which would have much better strength properties in the perpendicular direction. Searching for such a material has been the concern of many scientists in the past.

Extensive research work has already been carried out in regard to this topic. But still there is no method for building shell structures which could compete with current building methods. The Institute of Structural Engineering at Vienna University of Technology aims at inventing a new construction method for building concrete and ice domes in a quick and easy way.

3   Methods already tested

Different methods for building shells originating from a flat starting plate have been investigated at Vienna University of Technology. In the experiments, circular flat plates with a diameter of 13m were transformed into dome structures. The sections 3.1, 3.2, 3.3 give an overview of the different building methods and the results obtained.

3.1   Polystyrene wedge method [3,4]

The new construction method will be explained based on the example of a reinforced concrete shell built in Vienna in July 2005.

The simple shape of a dome with a span of 12m and a height of 2.16m was chosen for the construction of the shell. A flat reinforced concrete plate with 32 polystyrene segments was produced

Proc. of the 9th *fib* International PhD Symposium in Civil Engineering, July 22 to 25, 2012,
Karlsruhe Institute of Technology (KIT), Germany, H. S. Müller, M. Haist, F. Acosta (Eds.),
KIT Scientific Publishing, Karlsruhe, Germany, ISBN 978-3-86644-858-2

(see Fig. 1a). The circle in the plan was approximated by a polygonal line with 30 straight parts and two concrete anchor blocks for the post-tensioning tendons. Along the circumference a formwork had to be provided in order to form the boundary of the reinforced concrete shell with a thickness of 50mm. Two unbonded tendons were placed along the edge of the plate and two stressing anchorages were installed at each of the two anchor blocks. Fig. 1c shows a section through the RC shell which was obtained after the stressing of the tendons. Fig. 1d shows a section of the circumference of the flat plate. The two unbonded strand tendons were also used to tighten the two layers of foil at the circumference of the plate.

An inexpensive product, which is used in agriculture, was used as foil. The thickness of the poly-ethylene foil was equal to 0,12mm and the tensile strength was equal to 3.3N/mm².

During the transformation of the flat plate into the shell the diameter was reduced from 13m to 12m. As indicated above, this construction method is only applicable if large strains in the ring direction occur during the forming process. The shell consisted of reinforced concrete in the meridian direction, whereas in the ring direction, the stiffness properties were governed by the elastic-plastic behaviour of the polystyrene segments. During the shaping process the polystyrene segments were compressed from 144mm to 38mm at the circumference. The shape of the polystyrene segments was determined by hand calculations in such a way that during the forming process a reduction of the segment width to approximately 25% of the original width would produce the intended shape of a dome with a 2.16m elevation.

During the transformation of the plate into the shell, the radius of curvature was decreasing from an infinite radius for the plate to a radius of 9.365m for the final shape. This radius of curvature had to be equivalent to the thickness of the shell divided by the sum of the absolute values of the strains at the top and bottom surface. For a given thickness of 50mm and a radius of 9.365m the sum of these strains had to be around 0.5%. Since the tensile strength of concrete is smaller than the compressive strength, cracking occurred during the shaping process. Reinforcement bars with 5 mm diameter at 70mm distance were used as reinforcement in order to obtain many distributed cracks with small crack widths.

The photographs in Fig. 2 demonstrate the shaping process of the flat plate into the shell with double curvature. In the experiment, the plate was lifted additionally to the tendons with the aid of a pneumatic formwork to compensate the dead weight of the shell. A maximum pressure of 6 mbar was measured. In the experiment the centre of the plate was lifted by about 20mm with the aid of the pneumatic formwork. Then the tendons were stressed simultaneously at both anchor blocks up to forces of 40kN. Further stressing of the tendons resulted in a reduction of the circumference and a rise in the centre of the shell. The tendon force remained approximately constant during the experiment. When the final elevation was achieved the tendons were locked at the anchorages.

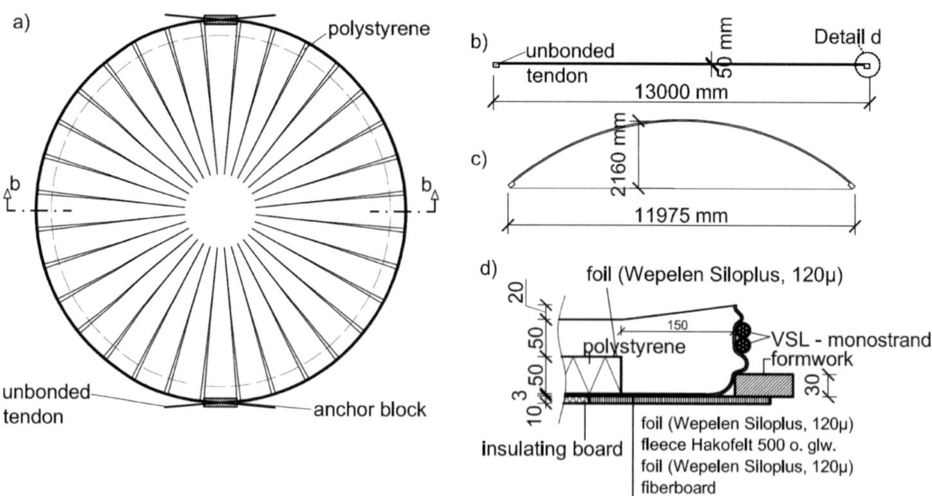

Fig. 1:   a)Plan view of concrete plate, b)section b-b of concrete plate, c) section of concrete shell, d) detail d

Fig. 2:   Pictures of the shaping process

The results in in Fig. 3 show that the measured shape of the RC-shell after locking the tendons is approximately the same as the theoretical shape of the circular dome. The difference between the measured elevation of 2.12m and the theoretical elevation of the spherical dome is only 40mm.

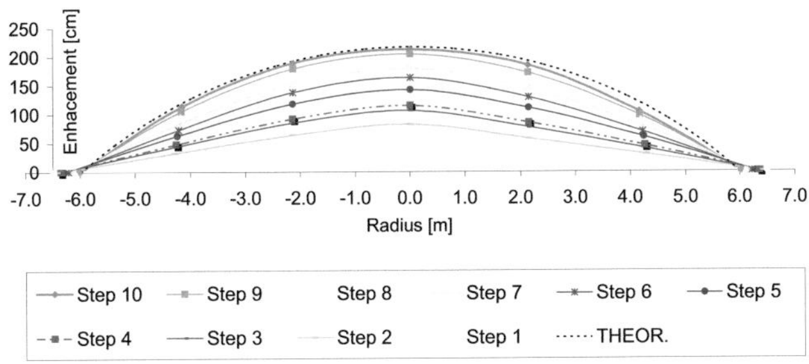

Fig. 3:   Enhancement during the shaping process

In December 2005 an ice shell was built using the construction method described above. An ice plate with a diameter of 13m and thickness of 200mm was produced and was then reinforced with glass fibre fabric and transformed into a dome with a height of 1,8m (Fig. 4).

Fig. 4:   Ice dome in Obergurgl, Austria a) completed structure, b) view from the inside (Photographs: Günter R.Wett)

3.2   Pneumatic formwork method [2]

The basic idea of the different shell building methods is more or less the same. A strong material is to be transformed into a dome structure compensating the shortening in ring direction with a "soft material". Instead of using polystyrene wedges, the pneumatic formwork method uses air as shortening material. First the elements are placed on a plane working surface. In order to transform the plate into a shell, the pneumatic formwork is placed under the precast elements. While air is inflating the pneumatic formwork the elements lift and the plane plate is transformed into a shell.

3.3   Segment lift method [2]

Even though the elements used are still of the same type, the lifting process is completely different. The production of the plane plate, subdivided in 16 elements, is the same as depicted in 3.1. Before the elements are lifted, they have to be formed until the necessary curvature is accomplished. So far two different methods to "bend" the elements are proposed. The elements can either be lifted by means of a pneumatic formwork placed underneath the ice elements before producing or they can be lifted with the aid of a lifting device and supported by wooden planks. By maintaining this condition, creep deformations are added to this instant deformation caused by dead load. Subsequently, the shaped segments are equipped with temporary tension chains and then lifted and assembled in order to form a shell. The tensions chains carry the tension forces while the segments are lifted to a dome. Fig.5 shows the different steps for building a double curved shell according the segment lift method.

Fig.5:   Different production steps of the segment lift method [2]

4   Presentation of a new shell construction method

The research on the different methods mentioned above revealed valuable results but still those methods are not able to rival conventional construction methods. Furthermore, there are some details which have to be improved regarding the design of the new structure. The experiments proved that the segment lift method is efficient but has one disadvantage: too many construction steps. No doubt an improvement compared to conventional formwork methods but the method is still too complicated and expensive for being used by the building industry. Based on the previous research, a new construction method will be investigated at Vienna University of Technology. The following section describes the characteristics of the new construction method.

4.1   Analysis of the already tested methods

The polystyrene wedge method, explained in section 3.1, can only be used for shells with a small curvature. If this construction method were to be used for building a dome structure or an ice shell it would be necessary to achieve a higher curvature. When analyzing the first already tested method we found out that the polystyrene wedges have to be replaced by another material. This material should have the following properties:

- The circumference of the structure is reduced up to 30%, therefore the wedges should "disappear" while the structure is being built.
- The inner surface of the ice shell built in 2005, shown in Figure 4, contained residues of the polystyrene wedges and of the tape used for fixing the wedges to the pneumatic formwork. On the one hand, it is difficult to clean the construction site after the ice has melted and, on the other hand, the optical impression of the inner surface is affected. It should be possible to remove all the materials used as formwork after finishing the building.
- A large structure with a big curvature requires wedges with a big width. The usage of the polypropylene wedges may lead to an increasing stability problem. If the width of the wedges is too large the wedges might pop out due to a buckling failure.

In summary, a compressible, removable stable material is needed to build the wedges between the different elements of the shell.

4.2   The pneumatic wedge method [5]

This method is based on the idea of using pneumatic wedges to fill the gaps between the elements. The pneumatic formwork is easy to compress and can be removed without leaving any residuals. Inflatable structures always cause tension in the membrane and thus it is impossible for the pneumatic wedges to lose their stability during the shaping process of the shell.

5   Outlook

5.1   Research work on the new construction method

The Institute for Structural Engineering at Vienna University of Technology plans to research this new method. In course of the research work two concrete shells will be built in Amstetten, Austria. Prior to the building of a big shell with a diameter of 16m, a smaller one with a diameter of 8m and a thickness of 50mm will be erected.

The following description is referring to the smaller shell with 8m diameter. It is expected that a height of 2,2m can be reached in the middle of the shell. The cast in situ reinforced concrete plate is subdivided in 16 parts. Two unbonded tendons will be used to tighten the two layers of the foil at the circumference of the plate. Steel ropes are to be chosen for the reinforcement. The low modulus of elasticity (60.000N/mm²) of the ropes at the first load will allow the transformation of the plate into a dome with a small radius of curvature. The position, and the very high yield strain of the reinforcement, is essential for attaining the high curvature. The effective depth amounts to 36mm. Figure 6a shows the plane concrete plate prior to the shaping process. Figure 6b shows a section of the plane plate and Figure 6c shows a section of the fully shaped structure. The pneumatic formwork wedges will be compressed from 350mm to 20mm. During this process the air pressure in the pneumatic formwork as well as in the individual wedges and the post-tensioning force in the tendons can be controlled. In the course of the research project it will be investigated which levels of air pressure and post tensioning prove to be ideal for the shaping process. After finishing the shaping process the pneumatic formwork for lifting the segments and the pneumatic wedges will be removed. For concluding the shaping process, the gaps between the segments will be filled with grout.

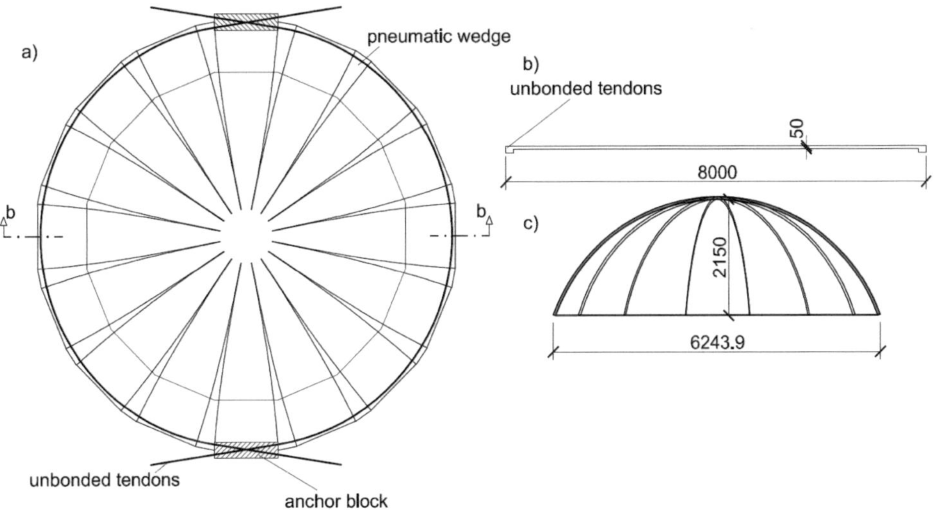

Fig. 6:   a) Plan view of concrete plate, b) section b-b of concrete plate, c) section of concrete shell

5.2 Research work on the different production materials

As already mentioned, shells will be built of concrete and ice. Especially the choice of ice as construction material is a complex matter.

In the course of the preparation for the new experiments the properties of ice will be investigated. Test series should help to understand how the material works and to find parameters to limit the range of the compressive strength in combination with pre-cast ice. The properties of ice will be investigated by pressure tests with different additives. Additional creep experiments will be carried out at different temperature levels.

6 Conclusion

The motivation to research this topic has already been explained in section 1. The building of shell structures is only of marginal importance to the building industry. Nevertheless, it is a very interesting field of technical engineering. At a time when nature is the template for architectural ideas and when resources are scarce, we are looking forward to build thin shells of concrete and ice using the materials in their best operating range.

References

[1]   Billington, D.P., Thin shell concrete structures, McGraw-Hill Book Company, 1965

[2]   Dallinger, S., Ice Domes – Development of construction methods. Phd – thesis, Vienna University of Technology, 2011

[3]   Kollegger, J., Preisinger, C., Verfahren zur Herstellung von zweifach gekrümmten Schalen, Austrian Patent Application, 2004

[4]   Kollegger, J., Preisinger, C. Method for the production of double –bent shells, European Patent EP 1706553, 2008

[5]   Kollegger, J., Kromoser, B., Verfahren zur Herstellung von zweifach räumlich gekrümmten Schalen, Austrian Patent Application, 2011

Electronic controlled adaptive formwork for freeform concrete walls and shells

Matthias Michel

Hochschule Ostwestfalen Lippe
Detmolder Schule für Architektur und Innenarchitektur
Emilienstraße 45, 32756 Detmold
Supervisor: Ulrich Knaack

Abstract

Concrete walls in general require two form work sides that are exposed to considerable rheological concrete pressure during casting process and the initial curing phases. For realization of complex geometry i.e. double curvature, additional technological complexity is the consequence. Free form wall formwork is nowadays dominated by the use of rigid milled foam blocks or router cut wooden rib and bulkhead style systems, resulting in a considerable amount of technical and financial effort in realization.

A reusable formwork with the ability to change its topology multiple times according to a given geometry has the potential to reduce the economic and ecologic impact of free formed concrete walls thus allowing an extended formal freedom in architectural design.

While adjustable form work tables for horizontally cast shells are presently researched on be at various places [1,2,4], the researcher's objective is a reusable two surface system that engages electronically controlled hydraulic or mechanic actuators to change the topology of its two upright formwork surfaces in a defined curvature range.

1 Introduction

The Idea of making use of an adjustable mould origins back to the 1960s when we find illustrations by the architect Renzo Piano, showing a pin cushion style mould table featuring an array of individually driven actuators. The actuators were covered with an elastic material to interpolate between the actuator tips. With the upcoming of digitalization concepts, patents were focusing on Computer controlled actuators systems that are elastically deforming a flexible formwork surface, among those of Spybroek and Kosche [3,5].

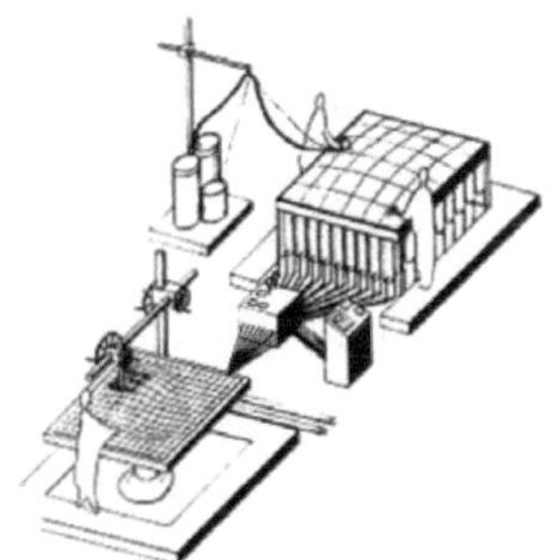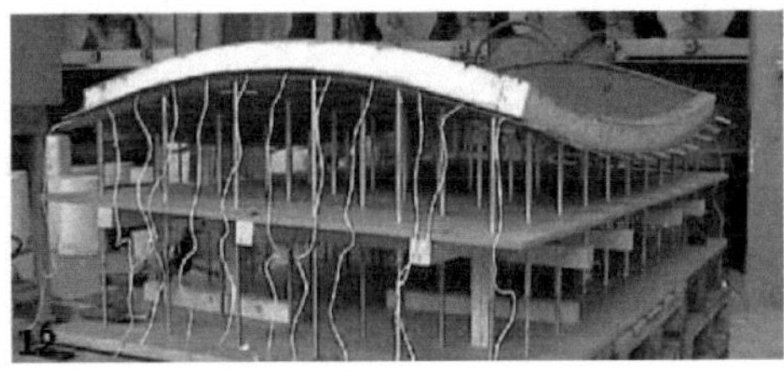

Fig. 1 Left: Sketch of Renzo Piano, 1966 – Right: static mould mockup of Shippers

Known working systems are installed at North Sail Inc. as a mould for double curved mylar-carbon based sails, the company Adapa Aps, Aalborg, is using an actuator driven mould table to form thin concrete panels [4]. A manual adjusted mould table for concrete casting is researched on by Shippers, TU Delft, resulting in publications recently [2]. The research of Riedbergen and Vollers, TU Delft, lead to a patent for an flexible passive mould for deforming glass panels [1]. Riedbergen developed a working mockup of an actuator driven active mould and undertook initial evaluations [7].

Proc. of the 9th *fib* International PhD Symposium in Civil Engineering, July 22 to 25, 2012,
Karlsruhe Institute of Technology (KIT), Germany, H. S. Müller, M. Haist, F. Acosta (Eds.),
KIT Scientific Publishing, Karlsruhe, Germany, ISBN 978-3-86644-858-2

According to the authors overhead research project, the present research focuses on a systems for walls, making use of two mould surfaces that are withstanding high rheologic pressure. However basic aspects are applicable to both kinds of systems, vertical and horizontal.

2   Concept Sketch

The adjustable formwork for walls will have at least on set of hydraulic or mechanical driven actuators to elastically imprint a specified topology into a formwork surface. This process is electronically preprocessed and controlled. As walls require an enclosed shuttering, two layers of formwork surfaces are being deformed, using either a set of coupling anchors to slave one surface to the other or two independent surfaces with actuators on either side.

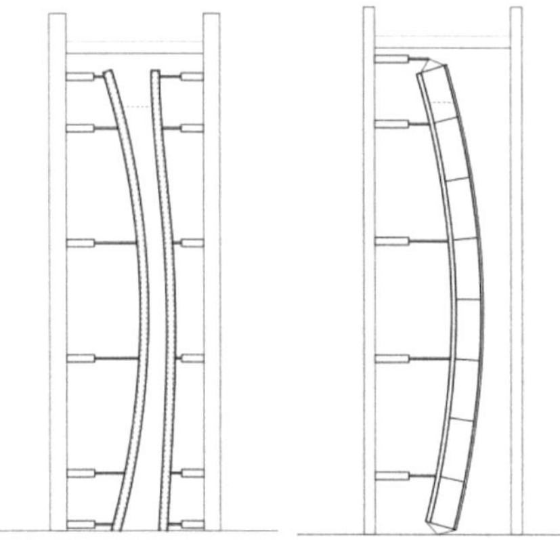

Fig. 2   Schematic sketch of different versions of flexible wall shutting systems - left: single actuator set with anchored outer mould surface - right: two individual systems of varying wall strength

The actuators will require to be mounted on a rigid structure giving the required stiffness and serving as the mould's backbone. The actuator may be arranged as a regular grid or as irregular pattern where an actuator's position is individually determined by the intended topology. The shuttering's surfaces will be elastically deformed but needs to withstand the rheologic pressure of the fresh concrete as well. To comply with these contradicting requirements of global elastic deformability and local stiffness, for at least some kinds of configurations a dense anchoring of the two surfaces may be provided, so the concrete pressure is short cut.

The mould shall be capable to integrate a certain degree of reinforcement, as free formed curved walls are unlikely to be free of bending action. The concept aims to deform also the reinforcement bars during actuator movement, but for the research the reinforcement diameter will be limited to dimensions that will not become the dominating factor for configuration of the formwork.

3   Research goals

The research identifies functionally different working principles with their individual geometric and mechanical constraints. It evaluates potentially suitable formwork surface materials and their deformation strategies in respect of elastic behavior versus rheological interaction. For the design of the active components the corresponding logic and mechanical system requirements are identified. A method of system configuration based on essential parameters of influence shall predict a formwork configuration and characteristic results will be tested using a working scale mockup. A vital part of the above program is a simulation- and control environment by which various configurations are simulated for evaluation of geometry, structure and logic.

The major objectives of the research in detail

- Geometric principles
  - Surface representation and approximation
  - Evaluation of operating principles
- Operational aspects
  - Methods of mould control and software preprocessing
  - Actuator systems and it's interfacing and control

- Material behavior and evaluation
  - o Principles of elastic deformation and methods of surface approximation
  - o Deformable Materials in conjunction with rheologic interaction
- Implementation and testing
  - o Implementation aspects like edge design, reinforcement, structural aspects, modularity
  - o Testing and mockups

During the sequential research work, the geometric analysis as well as interface and simulation has been accomplished first while the material research is pending at the time if writing. The following texts reports about the present work on geometry and the simulation environment for an adjustable wall form work.

Fig. 3   Spline ruler with spline weighs

4   Basic Aspects

Free formed surfaces can be understood as a dense sequence of spline curves, representing the surface. The spline is nowadays firmly associated with the CAD technology but it originates from ship building culture when the word described a thin and flexible ruler, which was elastically deformed to interpolate between defined points resulting in a smooth curvature. The spline was held in place by spline weights, acting as control points. The ruler's path can by described by a spline interpolation of with minimal bending energy and amount of curvature. The areas with maximum bending are located at the control points while the sections where curvature changes and bending is neutral will be found between the control points.

In relation to our mould a spline ruler can be understood as a fiber of the elastic mould surface material and the physical spline weights resemble the control points of the spline, represented by the actuator tips imposing deformation to the mould.

While splines in computer programs my encounter an arbitrary sequence of curvatures, the materialized spline is limited by its mechanical properties. In additions to geometric limits, for approximation of a digitally created curve or non-planar surface by bending using physical components the following parameters are of influence:

- Control point spacing
- Bending and shear stiffness of the mould surface's material
- Physical limits of curvature
- Exposure to load by the work piece or the liquid filling

The above points influence the configuration of a mould system and its ability to approximate a given topology. During the yet to be started process of material evaluation a correlation between the parameters shall be identified to allow individual configuration, and will be verified by simulation and testing.

5   Principles of Operation

Deformation of members can be analytically determined based on the differential equations of Euler/Bernoulli as long as the supports (or control points) and the member (in our case the mould surface) are not influenced by effects of II. order theory. However, large deformations are wanted and either affect the position of supports or - if they are immovable – result in tensile forces in the mould surface material under actuator movement.

According to Shippers' approach, the mould surface is not connected to the actuator tips at all, but held in place by gravity and thus is able to slide freely. For an upright wall formwork the mould surface has to be firmly attached to the actuators and it is obvious that the necessary lateral movement can be accomplished by letting the actuators tilt.

The above effect in combination with the complex geometric relationships and resulting misalignments that come into play with assignment of physical extensions and material width make an analytical approach difficult and limiting. Thus a numerical approach is regarded as the appropriate method to analyze and predict the mould's behavior.

To identify a method of controlling the actuators minimizing tolerances in the physical mould, linear as well as nonlinear effects have to be dealt with, which will result from geometric dependencies, effects of large deformations as mentioned and imposed loads of the fresh concrete. To handle the above, a parameterized geometrical model is engaged to cover the geometric part. This model can be freely configured by adjusting predefined parameters and feedback is given in real time. By an interface to a nonlinear numeric solver, effects of large deformation are iteratively input and the model is being updated after each loop until a convergence is reached.

The coupling of interactive parameterization and numeric solution makes it feasible to simulate the mould's behavior covering a broad range of its complexity. The environment allows to simulate the effect of varying configuration parameters and finally to control a mockup directly by sending the calculated actuator strokes to the hardware. So the parametric-numerical model qualifies as a basic research instrument.

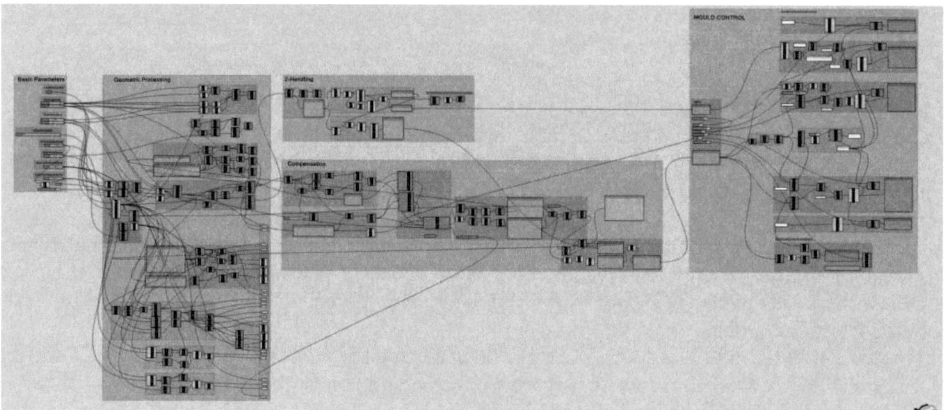

Fig. 4 Wiring diagram of the parametric model, clustered in functional units

The following process was created by using the Grasshopper environment for Rhino. At the time of writing, the interface software processes the data in the following sequence:

- Rasterization of the given NURBS geometry
- Calculation of the initial Stroke of the actuators under strictly vertical movement
- Nonlinear numerical processing of a framework for tilting actuators
- Compensation of the stroke length for inclined actuators
- Generation of actuator commands to send to the data bus for directly driving the mould

6 Geometric Processing

The geometric dependencies are first illustrated for a horizontal mould and drawn with horizontal x-axes. Further steps will be illustrated for upright mould surfaces.

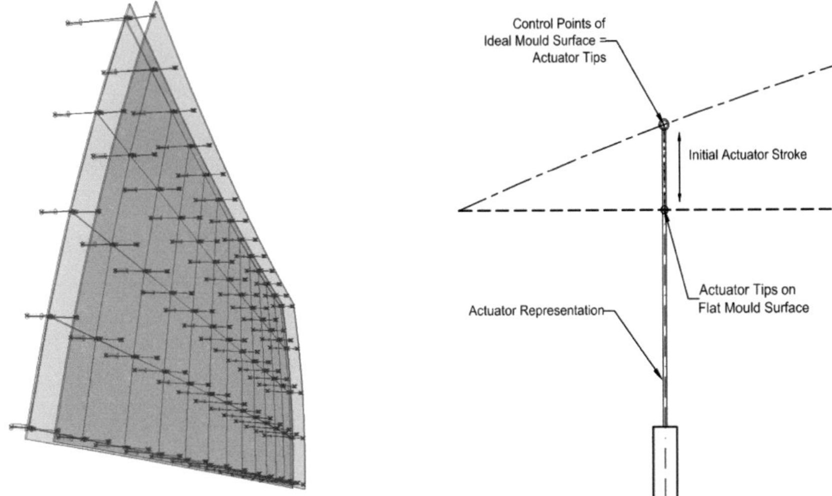

Fig. 5   Left: Parametric model in rendered in sketch mode - Right: Idealized functional scheme

Initially the cross section curve representing the mould surface is allowed to perform according to spline interpolation: Control points are moved parallel to the y-Axis, and are part of the fiber, that may encounter elongation relative to the cord. The x-Axis components of control point distances remain constant. This situation may be represented by differential equations.

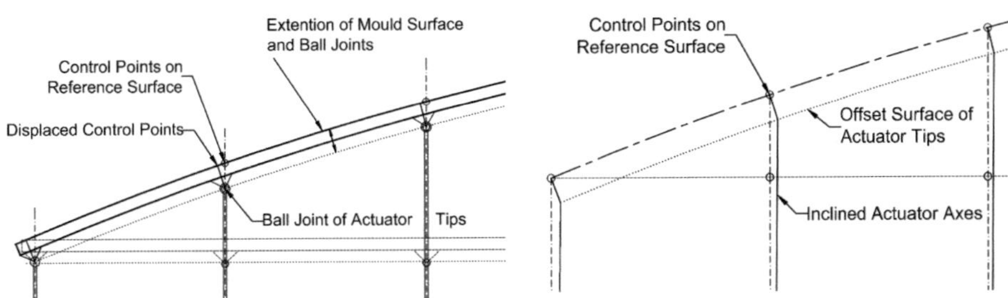

Fig. 6   Functional scheme with material width and offsets

For the physical implementation, material with and offset of the ball joints of the actuator tips is added to the system. The ball joints lie on a surface that is an offset from the reference surface by the amount of material width and ball joint height. The necessity to offset ball joints may not be inevitable but is included in the considerations as it later may render necessary for subsequent material options.

By the offset of the actuator tips, the control points on the reference surface either will become displaced leaving the actuator axis strictly parallel to the y- Axis or on the other hand the actuators have to be allowed to incline to keep the control point unchanged. For the following considerations the latter will be assumed.

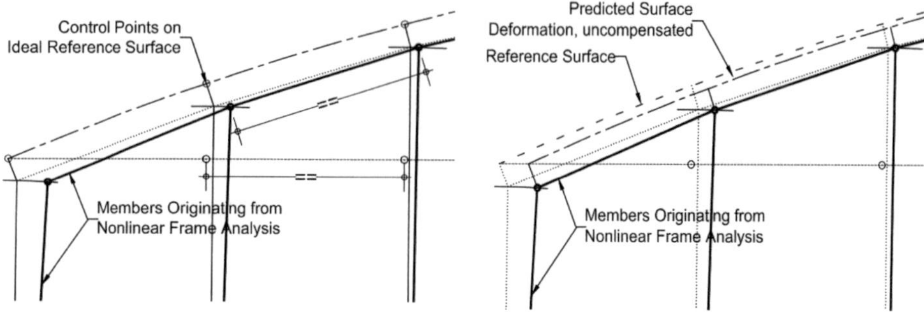

Fig. 7   Functional scheme with beam representations of numerical model

As mentioned, it is expected that in the physical implementation the actuators tips are rigidly attached to the mould surface and not allowed to slide along the surface. A further premise is that the physical mould surface has adequate elastic bending capabilities but is inelastic in membrane direction.

Under those conditions the resulting curvature of the mould surface will be incorrect for initially calculated actuator strokes as the chord length reduces. The actuators that are assumed to be mounted hinged at their base will follow the surface by tilting. To anticipate this behavior, a numerical space frame model is coupled to the geometry. This model consists of the actuator axes and an interconnecting framework of the actuator tips. It is processed iteratively in a numerical solver and results lead to an updated geometry in the parametric model.

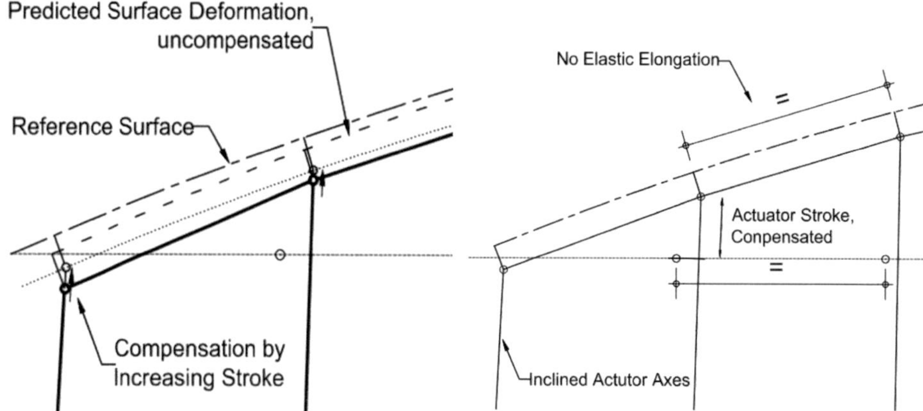

Fig. 8   Left: Principle of compensation – Right: Resulting system

To regain the correct curvature of the mould surface the actuators will have to extend or retract until their tips will be repositioned on the originating surface. This compensation is processed iteratively because by the adjustment of stroke a lateral movement is implied by the interconnecting members, requiring recalculation and adaption of the geometry. After a state of visually observed convergence, the geometry resulting has constant distance of the reference points and features inclined actuators. The amount of additional stroke for compensation is finally calculated and can be sent to the hardware.

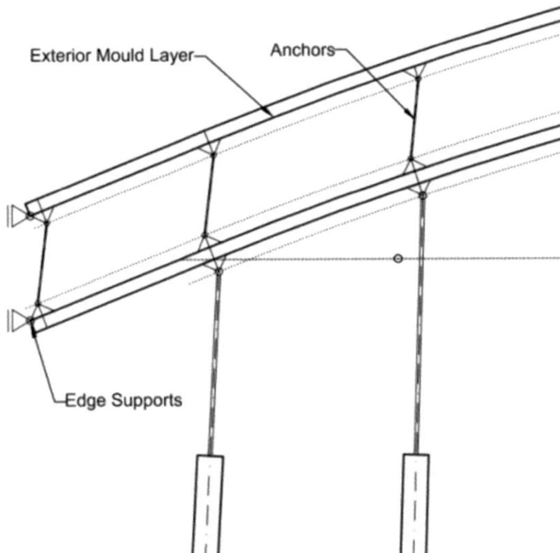

Fig. 9   Scheme of wall formwork with two surfaces and anchoring

For a corresponding wall formwork different layouts is can be imagined. To cope with the high concrete pressure and still preserve the ability to deform of the shuttering surfaces elastically, a system with two anchored surfaces is most suitable to solve the problem as the anchors shortcut concrete pressure and slave the exterior mould surface to the inside that is driven by actuators. It is not likely that spacing of actuators and anchors relates one to one, though for the illustrations in this report it does for simplicity reason.

The dependencies of the additional layer can be solved geometrically. However it is obvious that the presence of material width and offsets bring in bending action for the mould surface, resulting from actuator forces that are imposed for deforming the mould surfaces and its reinforcement. Also by the curvature of the mould surfaces the anchors become inclined. The anchor force resulting from concrete pressure causes opposite tangential forces in the mould surfaces. The mould layers may be subject lateral movement and have to be supported on the edges.

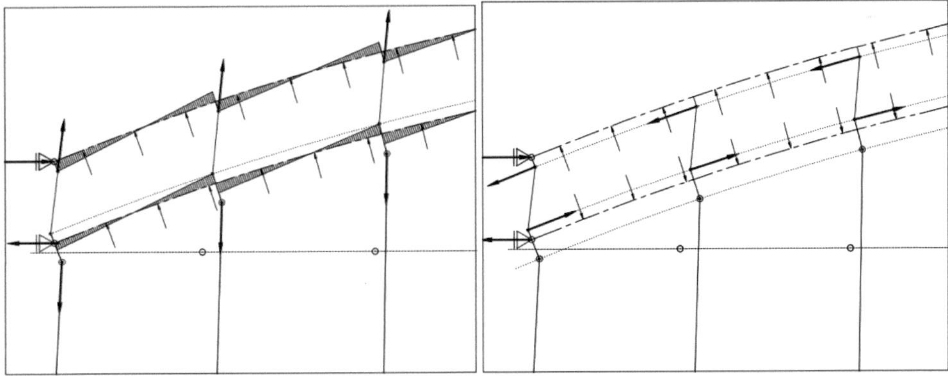

Fig. 10   Effects resulting from material offsets under concrete pressure and actuator force

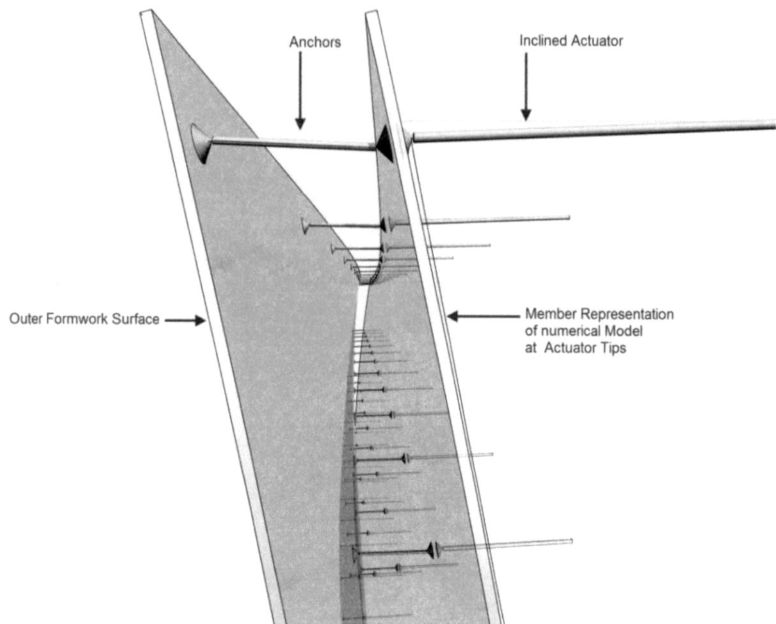

Fig. 11   Deformed   parametric   model   with   abstract   component   representations (dimensions not to scale)

7   Conclusions

By coupling parametric geometry processing and nonlinear numeric analysis, the geometric dependencies of an actuator driven mould system were successfully simulated. The technique is suitable to implement more complex solvers for mechanical analysis and to cover effects of structural deformation and to give information about the structural limits of a specific mould surface material. The system qualifies itself as a platform and tool for further research, but also serves as an interface to physical actuators.

For further research the evaluation of mould materials will be in special focus along with identification of a method of configuration of a specific mould system. The architecture of the active components will have to be outlined. Their requirements will be integrated into the parameter research as well as implementation aspects like edge design, reinforcement placing and deformation that will round up the research that was started in September 2011 and completes in 2014.

Acknowledgment

The Research is conducted by Prof. Dr.-Ing. Alexander Graubner, Chair of Concrete and Masonry Structures, TU Darmstadt and Prof. Dr.-Ing. Ulrich Knaack, Chair for Design and Construction, Hochschule Ostwestfalen Lippe, and is promoted by DFG in connection with DFG Special Research Project 1542 „Leicht bauen mit Beton"

Research title of the common overhead project: "Fundamentals of Development for Adaptive Formwork for Free Formed Concrete Members"

References

[1]   K.J. Vollers and D. Rietbergen. Patent number 2000699, Werk-wijze en inrichting voor het vormen van een dubbelgekromd paneel uit een vlak paneel. NL Octrooicentrum, 2008.

[2]   R.Shippers, B. Janssen, Manufacturing Double Curved Elements in Precast Concrete using Felxible Mould in: Proceedings fib Symposium Prague, 2011.

[3]   Kosche, F., "Schaltisch und Verfahren zur Herstellung von doppelt gekrümmten Bauteilen", German Patent DE19823610B4, 2005

[4]   Ch. Raun, M. Kristensen, P.H. Kirkegaard, "Dynamic Double Curved Mould System" in: Gengnagel et al. Computational Design Modelling, Proceedings Design Modelling Symposium, Berlin 2011

[5]   L. Spuybroek, "NOX: machining architecture", Thames and Hudson, London, 2004

[7]   K. Huyge and A. Schoofs, Precast Double Curved Concrete Panels, Delft University of Technology, 2009

Ultra-lightweight concrete members inspired by bamboo

Patricia Sawicki

Institut für Baustoffe, Massivbau und Brandschutz, Fachgebiet Massivbau
Technische Universität Braunschweig
Beethovenstr. 52, 38106 Braunschweig, Germany
Supervisor: Martin Empelmann

Abstract

Inspired by the optimised structure of bamboo culms, ultra-lightweight concrete members with small wall thicknesses are developed in this research project. On the one hand bamboo is very lightweight and elastic due to cavities; on the other hand it possesses a high stability as a result of diaphragms. Within the approach of light concrete constructions, which is the background of the research program SPP 1542 of the German Research Foundation (DFG), bamboo provides an excellent archetype for ultralightweight concrete members. In accordance with the principle "Form follows force", the material is placed in the optimal position for load transmission and stability. The resulting thin walled concrete elements have to be investigated in regard to their structural behaviour and load-bearing capacity.

1 Bamboo as an archetype for slender structural elements

In the interdisciplinary field of biomimetics innovative strategies for technical problems are developed from comparable analogies of the nature. With regard to sustainable and ecological constructions, the decrease of dead load and an optimised material usage have to be aspired. In this case, especially for slender compression members, the load-bearing capacity and the stability of the structure become the decisive design factors.

An impressive archetype for very slender, hollow columns can be found in nature: Guadua angustifolia. This is a bamboo which is preferred for construction and used for amazing buildings like the bamboo pavilion (architect: Simon Velez, Columbia) at EXPO 2000 (fig. 1).

Fig. 1 Bamboo pavilion at EXPO 2000

Proc. of the 9th *fib* International PhD Symposium in Civil Engineering, July 22 to 25, 2012,
Karlsruhe Institute of Technology (KIT), Germany, H. S. Müller, M. Haist, F. Acosta (Eds.),
KIT Scientific Publishing, Karlsruhe, Germany, ISBN 978-3-86644-858-2

Within the approach of "Concrete light" (SPP 1542), the Fachgebiet Massivbau of the Institut für Baustoffe, Massivbau und Brandschutz (iBMB), TU Braunschweig, works on the project "Ultra-lightweight concrete members", in which the load bearing and fracture behaviour of bamboo-like concrete columns is reviewed and optimised.

2   Reflection on the structure and compression failure of bamboo

Bamboo is an exceptional product of evolution: Its culms are able to reach a height of more than 30 m only having diameters of 12 cm. This remarkable slenderness results from the efficient anatomical structure of bamboo.

Every stem is divided into single segments (internodes), which are connected together at plain nodal points (fig. 2, left). The wall of the internodes features a thickness of only 1 to 2 cm and is longitudinally pervaded by vascular bundles which cause the typical hardness as well as the distinctive elasticity of the culms (fig. 2, right). Besides, the thin diaphragms of the nodes entail an additional stiffening of the bamboo plant (fig. 3, left).

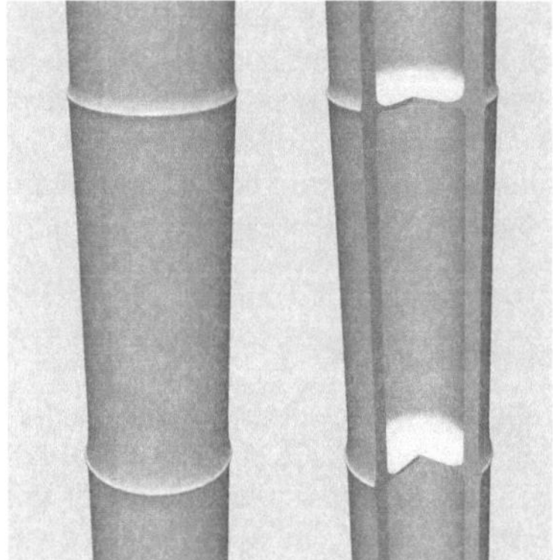
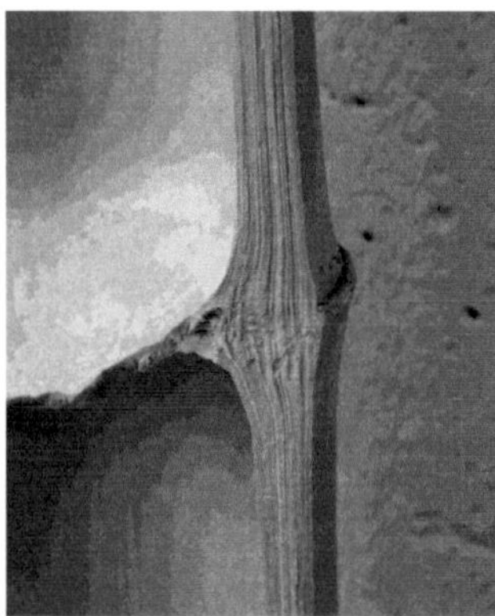

Fig. 2   Cross section of a bamboo stem and detail view of the orientation of fibres in the wall

A common, quantitative relationship between diameter, wall thickness and spacing of the nodes does not exist as the period of growth is mainly influenced by the environment. Though, the archetype of the nature demonstrates clearly that the nodes are spaced according to the principle, "Form follows force": The intervals are closer at the base of the restrained stem than at the upper culm (fig. 3, right).

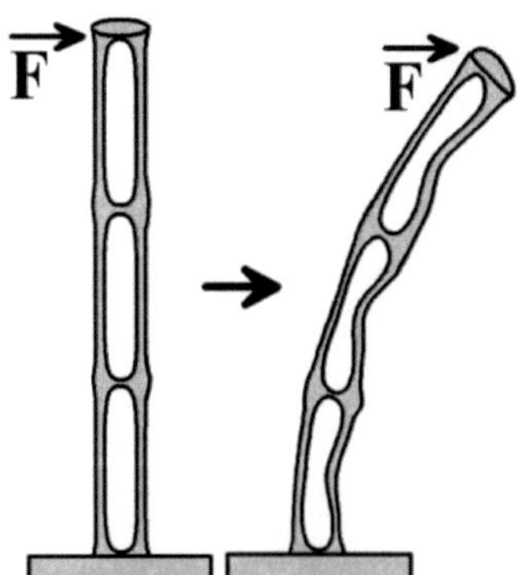

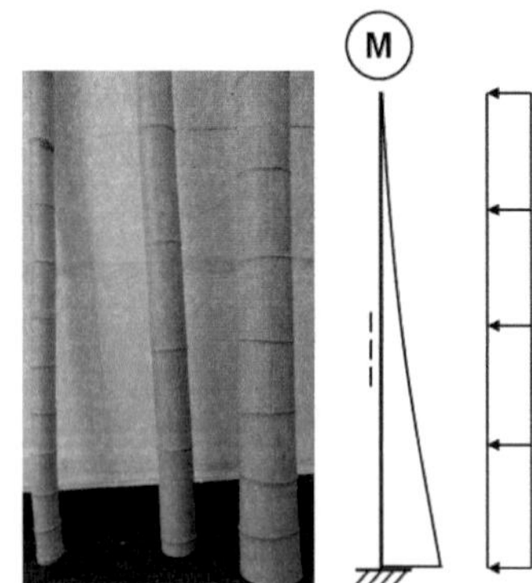

Fig. 3   Response of a bamboo stem to a horizontal force (modelled after [1]) and spacing of the nodes according to the principle "Form follows force"

In experimental series at the iBMB, specimens of bamboo stems with diameters of 8 to 14 cm were loaded with a centrical normal force until collapse to investigate general mechanisms of failure.

Thereby, the position of the nodes was varied so that segments without nodes resp. with only exterior or internal nodes were inspected (fig. 4, left).

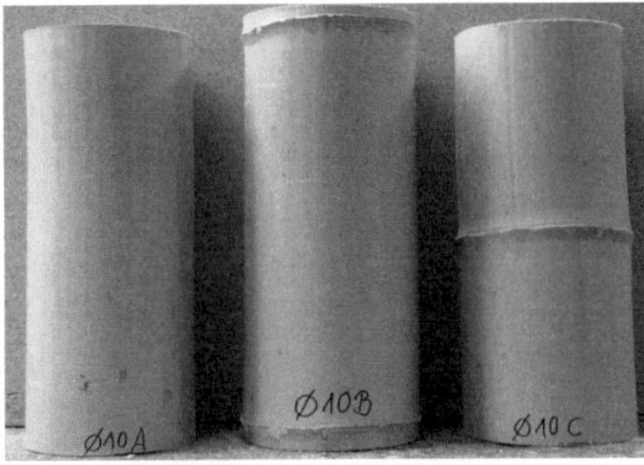

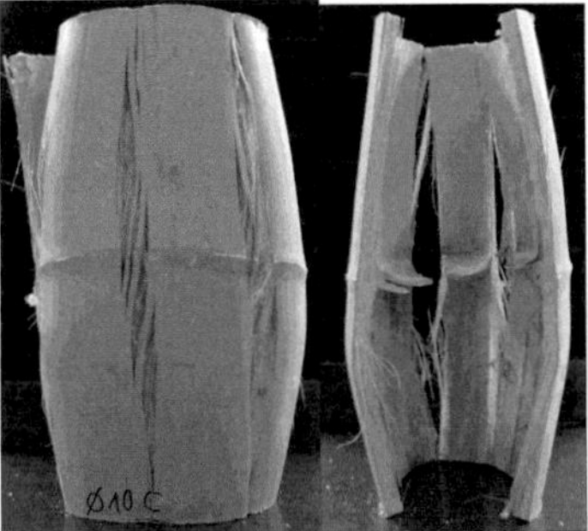

Fig. 4   Bamboo specimens before and after testing

Every specimen showed lamellar fracture behaviour due to pure cross tension (fig. 4, right). Before the critical strength was reached, a ductile behaviour could be observed, which was identified by a considerable plateau in the load-deformation curve (fig. 5). However, the number and position of the nodes does not decisively affect the amplitude of the bearing load. The reason for this is that under centrical compression the cross-section resistance is the leading parameter. The influence of stiffening becomes more important when influences of 2nd order theory or eccentrical forces have to be considered.

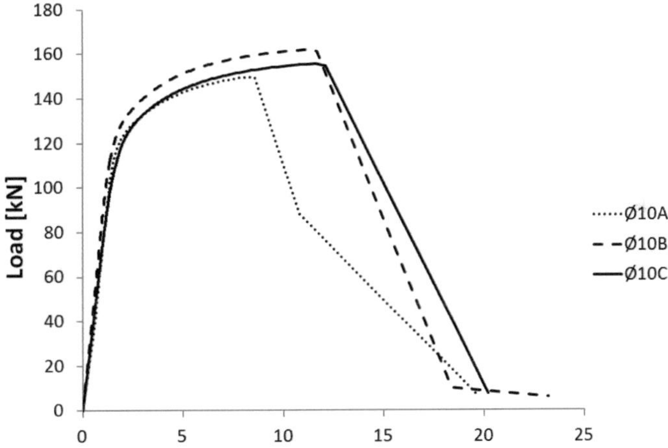

Fig. 5   Load-deformation-curve of the test series shown in fig. 4

3   Hollow, thin-walled concrete members made of UHPC

Basically, the cross section of a cylindrical ring is well suitable for material reduction in structures: On the one hand self-weight is minimised in contrast to solid profiles and on the other hand flexural strength is relatively little decreased due to the second moment of inertia. Additionally, the arrangement of nodes raises the buckling resistance of the slender structures.

3.1   Formwork concept

For fabrication of the thin-walled components the biomimetical principle is adopted. This means that profiling and segmentation are used for structural design to avoid instability.

In a first step, additionally to the conventional external formwork, a core of very light-weight PU-foam is employed to create the block-outs of the concrete tubes (fig. 6). It remains within the element as permanent formwork. The profiling of diaphragms – in analogy to the bamboo nodes – is achieved by accumulation of fitted foam sheets, which are threaded on a rod and prestressed with nuts to secure the position of the panels during concreting.

Currently, further investigations are undertaken in regard to the shaping of the inner mould.

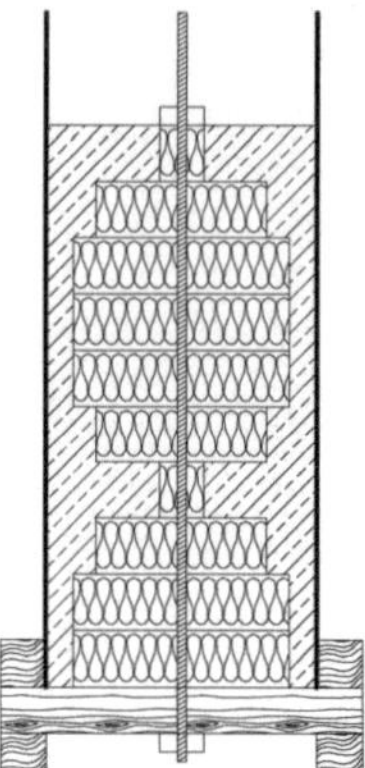

Fig. 6   Formwork arrangement for the concrete columns and construction of the PU-core as interior formwork

3.2   Reinforcement concept

Another focus of this research project is the development of an adequate concept of reinforcement for thin-walled structures that generates an approximately similar load bearing capacity under tensile and compressive forces. To establish the desired small wall thickness of a few centimetres, 3 different innovative alternatives of reinforcement will be examined.

With type "A" the use of micro-reinforcement is tested. It is placed plane within the walls (fig. 7, left). The single layers are laid directly one upon the other.

Type "B" amplifies the experience with conventional reinforcement. In this case, high-strength reinforcement bars, e.g. S670/800, shall be used. To ensure a sufficient concrete cover, e.g. for bonding, the bars are embedded in ribs (Fig. 7, middle). Therefore, in the interior PU-core rills are reamed in the concerning locations. These longitudinal ribs feature an additional positive effect on the stability of the thin shell.

In Type "C" (fig. 7, right) a perforated metal sheet is integrated, which is fixed in the centre of the tube wall. By the potential use of steel fibres a strong interlock between concrete and the sheet can be aspired, if necessary.

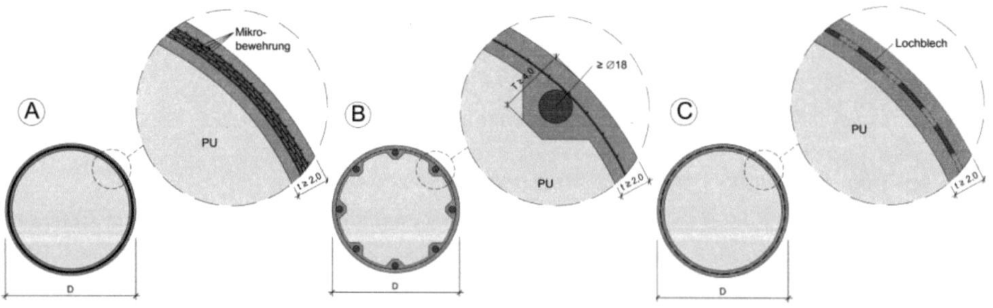

Fig. 7   Reinforcements concepts

3.3   Concrete composites

Hollow concrete columns have to feature a higher strength than solid concrete members to obtain a similar load bearing capacity for centrical compression. In competition with common columns a compressive strength of at least $f_{ck} \geq 60$ N/mm² is aspired for the bamboo-like structures.

To realise the delicate profile without air voids, the plasticity and the texture of the concrete play a decisive role. A concrete or mortar with fine-grained microstructure is required which exhibits self-compacting and self-venting properties and is not prone to separation.

Hence, various compounds were examined which require only the addition of aggregate, water and admixtures and guarantee a constant concrete quality. Generally, all tested concrete mixtures met the requirements and the properties of the hardened concrete were suitable (table 1).

Table 1   Selected testing results of the concrete specimens

Concrete	Dyckerhoff NANODUR[®]	Dyckerhoff FLOWSTONE[®]	PAGEL[®]-Verguss V1[®]/10
Age	33 d	33 d	28 d
Cylinder strength (water immersion)	116.5 N/mm²	75.0 N/mm²	84.4 N/mm²
Cylinder strength (air immersion)	121.6 N/mm²	77.5 N/mm²	78.9 N/mm²
Cylinder strength (position-controlled)	115.4 N/mm²	76.6 N/mm²	75.6 N/mm²
Strain at failure (position-controlled)	4.36 ‰	2.67 ‰	5.29 ‰
Splitting tensile strength (air imm.)	-	3.5 N/mm²	2.4 N/mm²
Density (air immersion)	2.33 kg/dm³	2.35 kg/dm³	2.22 kg/dm³

4   Structural behaviour of thin-walled concrete columns

The iBMB has got experience with the load-deformation behaviour of thin-walled tubes made of ultra high strength concrete. For example, within a research project – funded by the German Research Foundation in the research program SPP 1182 – design rules for concrete tubes with a wall-thickness of 5 cm were established to assure a robust load bearing capacity (fig. 8).

Fig. 8   Specimens of the sub-project TE 587/2-1 (SPP 1182, [2])

In the research project at hand the wall thickness shall be reduced to a minimum of around 2 cm. Thereby, the wall thickness is smaller than today's common concrete covering. The bond of the reinforcement has to be explored as well as the influence of profiling, which is in use for steel constructions but not for concrete structures due to the lack of experience.

The missing knowledge impedes the usage of ultralightweight hollow concrete members and until now advantages, like the maximal portable element size or the assembly effort in precast constructions, cannot really be used.

5   First tests on centrical compression elements

To investigate the bamboo-like structures, first, concrete columns without reinforcement were tested. The design of the PU-cores resembled fig. 6. The diameter of all specimens remained constant at 30 cm, the wall thickness of the internodal sections ranged from 2.4 to 3.0 cm. The results of the compression tests can be seen in table 2.

Table 2   Experimental results for centrical compression

Specimen	S 1.1	S 1.2	S 1.3
Concrete	NANODUR®	FLOWSTONE®	PAGEL V1®/10
Mean wall thickness	2.9 cm	2.4 cm	3.0 cm
Fracture load	1603.5 kN	1028.9 kN	1377.3 kN
Compressive strength	64.9 N/mm²	49.4 N/mm²	54.1 N/mm²

The corresponding compressive strength of the columns amounts only to approx. 60% of the tested cylindrical strength (cf. table 1). This decrease is most probably caused by the different cross section and the small element thicknesses. In order to quantify such scale effects, further tests have to be performed.

The crack patterns of the concrete specimen resemble the formation of cracks in the tests with the bamboo culms. The cracks disperse circumferential in uniform distances (fig. 9, left). Thus, the structure of bamboo was successfully adapted (fig. 9, right).

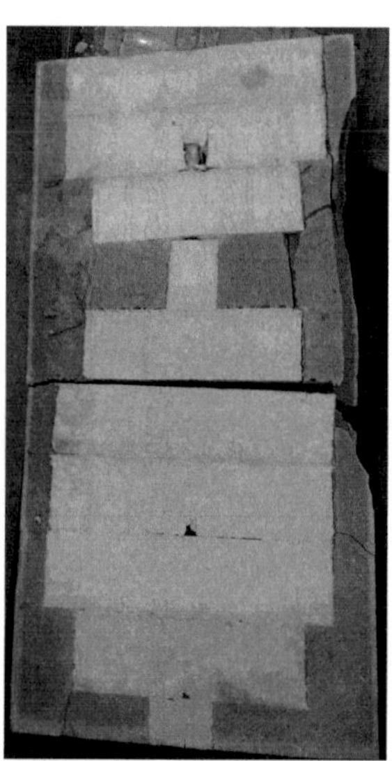

Fig. 9   Tested concrete column S 1.1 in the test bench and as longitudinal section

6   Future activities

On the basis of the aforementioned investigations on the shape and the scaffolding as well as the concrete mixture, further tests are planned in order to study the influences of the wall thickness, the reinforcement concept and the applied load (centrical and eccentrical forces). Besides, to align the geometry to the flow of forces and to quantify the influence of thin walls, numerical computations will be executed by nonlinear finite element analyses. In this way, efficient and optimised structural components will be developed.

References

[1]   Klein, B.: Leichtbau-Konstruktionen – Berechnungsgrundlagen und Gestaltung, 2009

[2]   Empelmann, M.; Müller, C.: DFG-Ergebnisbericht. Dünnwandige Rohrprofile aus UHPC, Forschungsvorhaben innerhalb des SPP 1182, 2012

Shape-optimised Parabolic Trough Collectors made of micro reinforced Ultra High Performance Concrete

Patrick Forman

Institute of Concrete Structures,
Ruhr-University Bochum,
Universitätsstraße 150, 44780 Bochum, Germany
Supervisor: Peter Mark

Sören Müller (Co-Author)

Institute of Concrete Structures and Structural Design,
Technical University of Kaiserslautern,
Paul-Ehrlich-Straße, Gebäude 14, 67663 Kaiserslautern, Germany
Supervisor: Jürgen Schnell

Abstract

Parabolic trough collectors are a special kind of Concentrated Solar Power (CSP) Systems. Solar irradiation is focused on line-like absorber tubes to heat up synthetic oil to be converted in subsequent conventional steam cycles to electricity. Existing collector structures usually consist of steel frameworks point-wise supporting curved mirror-elements. In addition to the self-weight of the trough the main load contribution for the design arises from orientation-dependent wind actions on the structure. A maximum power generation is achieved, if the geometric shape of the trough is ideal. So, unintended initial deformations or deformations induced by load, temperature constraints, shrinkage or creep have to be minimised. Hence, a shape-optimised shell structure made of Ultra High Performance Concrete (UHPC) and micro reinforcement merging both tasks – support and mirror surface – is assumed to be most effective.

At the Ruhr-University Bochum multiple stress analyses using continuum-based finite shell elements are performed in order to iteratively find an optimal shape taking into account constraints from admissible deformations, geometry and materials.

At the Technical University of Kaiserslautern research concerning the shrinkage of UHPC is done as unintended deformations result from this effect. One aim of the project is to fabricate the troughs within the formwork as short as possible. For that purpose the suitability of a heat treatment of UHPC is examined and its interaction to shrinkage.

In the contribution first results of the collaborative optimisation process of shapes and materials are presented.

1 Introduction

1.1 Concentrated Solar Power plants

In Concentrated Solar Power (CSP) plants (Fig. 1), whose number and importance will gain enormously in the next years, thousands of parabolic troughs are needed.

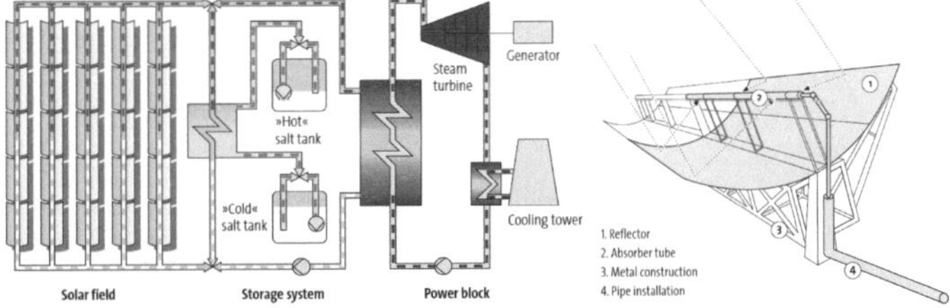

Fig. 1   Basic prinicple of CSP systems (left) and a parabolic trough to concentrate solar radiation (right) [9]

Proc. of the 9th *fib* International PhD Symposium in Civil Engineering, July 22 to 25, 2012,
Karlsruhe Institute of Technology (KIT), Germany, H. S. Müller, M. Haist, F. Acosta (Eds.),
KIT Scientific Publishing, Karlsruhe, Germany, ISBN 978-3-86644-858-2

Up to now parabolic troughs, which are built up of a mirror and a substructure, are being carried out as steel structures, which brings a number of disadvantages about:

- The manufacturing costs are high.
- The assembly time is high.
- The mirrors have a low stiffness, which leads to deformations due to wind and dead load. Torsion also is of importance because the troughs are rotated along their longitudinal axis to follow the sun and to be cleaned. Every minimal distortion of the reflector causes the deflection of sun rays of the absorbing pipe and by this reduces the efficiency of the system.
- The power plants are often located in areas with corrosive environmental conditions, e. g. deserts with soils with high salinity levels. This entails high maintenance efforts and connected costs.

These disadvantages can be avoided, if one succeeds in replacing the reflectors and the steel framework by thin concrete shells.

1.2   UHPC shells

Shells made of UHPC with micro reinforcement can be carried out extremely slender and elegant. They need only a few centimeters of thickness and therefore have a minimal material consumption while achieving a high manufacturing efficiency and durability. The extraordinary slenderness entails tight restrictions concerning the fabrication and geometries of the shells, which elementary can only be achieved through interaction of numerical calculations and experiments. Furthermore all levels of consideration from material to system (multilevel) have to be taken into account. A particularly applicable example are the shells of parabolic trough collectors of Concentrated Solar Power Systems, as they have to meet high, interactive demands regarding low deformation and precision under exposed environmental conditions. Because of their exceptional high repetition as a precise precast segment, these trough collectors are suited for detailed baseline investigations of free formings.

The aim of this project is the consistent development of structure optimized UHPC-shells using a multilevel approach, strictly following the principle "form follows force", thus being led only by the flow of forces and deformation restrictions. To realize this, optimizations of the concept, the shell design with bracings and details, oscillation analyses resulting from wind and complex nonlinear instability calculations are designated. Minimized fabrication tolerances in high precision formworks, shrinkage reduced concrete technology and novel manufacturing techniques are to be focused. Realistic uncertainties in geometry, material and impacts are identified and comprised, just like degradation ramifications during the life cycle.

1.3   Creep and Shrinkage

When designing constructional elements made of UHPC autogenous shrinkage as well as shrinkage due to self-desiccation is to be considered. The autogenous shrinkage is of bigger influence and is very dependent on the cement being used and the existing amount of cement paste. The research presented in [10] shows a smaller degree of shrinkage, if HS cement and an increased contingent of coarse aggregate are used. It is possible to reduce the autogenous shrinkage by the addition of shrinkage absorbers [3], whereas the addition of micro steel fibres does not reduce the shrinkage significantly [1]. Considerable internal stresses can also occur in structural elements without external restrictions of deformation, because of the restrictions of deformation resulting from reinforcement. Through utilisation of heat treatment – in particular at the span 70 - 90 °C – the shrinkage due to self-desiccation can be mostly anticipated [2, 6].

Another effect to be considered is the creepage of concrete. Creep tests with unreinforced cylinders d/h = 100/300 mm are presented in [11]. Presently systematic investigations are carried out by Müller [5].

1.4   Heat treatment

At using heat treatment a slow heating up and cooling down of the concrete are very important, so that the improvement of the microstructure is not compensated by thermally caused microcracks. According to [2] the heat treatment accelerates the strength development so much that with a heat treatment at 90 °C for 48 hours a higher strength than after 28 days immersion in water at 20 °C is reached.

1.5 Autoclaving

Autoclaving could be an option for aftertreatment. Following the principle of the pressure cooker a high water vapour saturation pressure is built up under high pressures. By autoclaving the hydration of the binders is improved substantially, which manifests itself in a low content of unhydrated cement clinker [4]. High compressive and tensile strengths occur. At present sufficient practical experience in aftertreatment with temperatures > 90 °C does not exist yet but shall be gained within this project.

2 Accuracy of the collector's surface

The performance of parabolic trough collectors depends on the efficiency of the mirror, the absorber tube and the mirror surface being true to form. While the efficiency of the mirror and the absorber tube is already fixed due to manufacturing, high demands on the accuracy of the collector's surface have to be set. The primary aim of a parabolic trough collector is to concentrate the solar irradiation – perpendicular to the collector's aperture – along the focal point F, where line-like absorber tubes are arranged.

2.1 Geometric properties of the parabola

The shape of a parabola (see Fig. 4)

$$z = -ay^2 + z_0 \tag{1}$$

is defined by the coefficient a (a < 1 – shrunk; a > 1 – stretched). Furthermore, this coefficient can be described as a function of the focal length.

$$a = 1/(4f) \tag{2}$$

The focal length corresponds to the focus F, where every light ray parallel to the axis of symmetry is concentrated. This holds true for each point on a parabola recognising that incident and reflection angles are equal. To retrieve a general geometrical representation the focal length f can also be related to the aperture width w and the rim angle φ.

$$f(w, \varphi) = \frac{w}{8 \tan \varphi} \left(2 \pm 2\sqrt{1 + \tan^2 \varphi} \right) \tag{3}$$

Results of first shape analyses (Fig. 2) illustrate an increasing parabola height for wide rim angles, whereas the focal length decreases. Furthermore, an effective parabolic shape should always have a balanced ratio between height and focal length to limit the torsional loading through self-weight and to avoid an excessive high focal point. Already realised typical parabolic trough collectors posses rim angles of 80° [10].

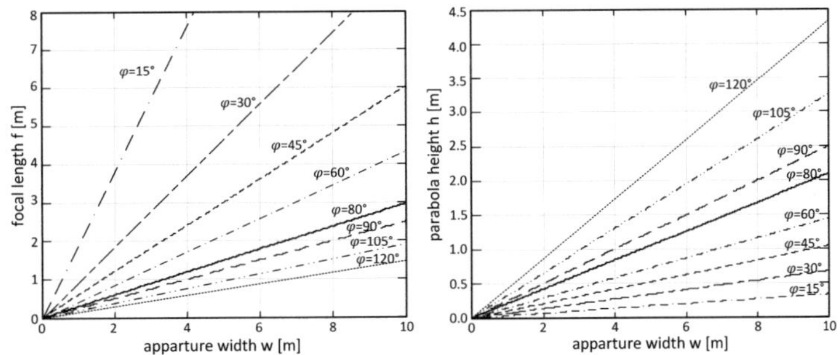

Fig. 2   Focal length f (left) and parabola height h (right) for varying aperture width w and rim angle φ

2.2 Performance losses caused by collector's deformation

In order to ensure a persistent and optimal power generation, losses – caused by unintended initial as well as additional deformation induced by other loads – have to be minimised. Therefore, a criterion to express the deviation of the deformed collector's surface from an ideal parabolic shape has to be derived, which enables to evaluate the performance. In order to fulfill the predefined aim to concentrate the solar irradiation in the focal point of an ideal parabola, the collector's accuracy can be described by a function of the distance with respect to the deflected, distorted solar irradiation form of the absorber tube.

Fig. 3 illustrates a derived model to estimate the performance of a deformed collector's surface. Thus, the surface – disturbed by deformations – is compared to an ideal parabola. Due to deformation each reflected light ray has a certain deviation to the ideal focal point. This leads to a course of distance along the aperture width. The distances are determined and weighted with respect to the inner and outer radius of the absorber tube. Related to a typical absorber tube consisting of an inner tube, in which synthetic oil is heaten up and an outer tube containing a vacuum to reduce the losses by heat transfer, the distribution of the efficiency factor is estimated in a simplified way to

$$\eta(x) = \begin{cases} 1, & if\ dist(x) \leq r \\ 0, & if\ dist(x) > r \end{cases}. \tag{4}$$

Generally, other distributions of effectivity might be assumed on the realistic technical performance of the absorber tube. Only parts of the collector's surface whose distance of the reflected rays to the focal point is inclosed by the radius of the inner tube r contribute to the local efficiency.

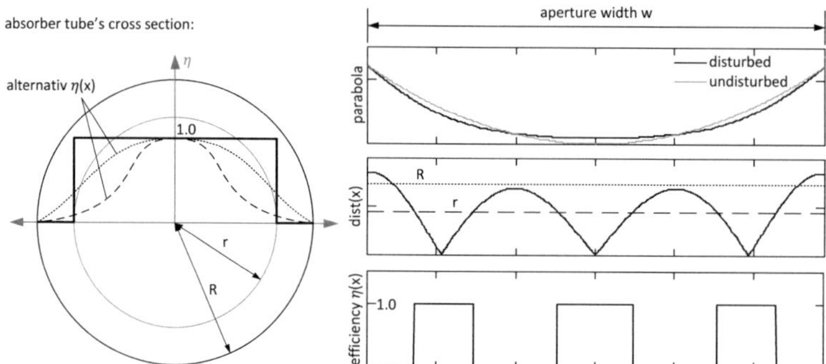

Fig. 3   Estimated efficiency of the inner and outer radius of an absorber tube (left), disturbed versus undisturbed shape of the parabola (top), distance of the deflected light rays form the focal point (middle) and effectiveness of the deformed parabola (bottom)

The performance of the entire deformed collector's surface is obtained by integrating the course of efficiency $\eta(x)$ over the aperture width w.

$$\eta = \int_w \eta(x)dx \tag{5}$$

3   Finite Element model

To achieve a slender, low weight shell structure made of micro reinforced UHPC and a parabolic surface mostly free from deformation, a shape optimisation is applied. It iteratively uses FE supported structural analyses and interactive optimisation cycles. The Finite Element model incorporates continuum-based shell elements. Due to the limited deformations of the parabolic trough surface, a linear-elastic analysis – with stress restrictions through material constrains – assures a sufficient accuracy. Fig. 4 shows the initial cross section and a Finite Element model of the trough.

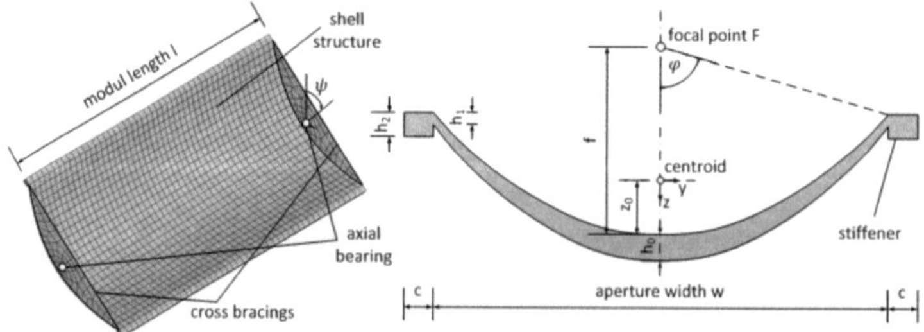

Fig. 4   FE-Model with continuum-based shell elements (left) and model's cross section (right)

In a first simplified starting modus the shell thickness decreases from the lowest point of the parabola to the open borders. Deviated stiffeners insure a sufficient stability there. However, the non-uniform thickness distribution has to be tested for their properties concerning heat treatment and time-variant deformations of creeping and shrinkage. The lateral supports as well as the centre of rotation coincide with the centre of the cross section. Additional cross bracings are modelled to gain more stiffness and to transfer the loads from the shell to the boundaries. The main geometric parameters have been chosen on the basis of the existent steel framework collector EuroTrough [8] having an aperture width $w = 5.77$ m, a modul length $l = 12.00$ m and a focal length of $f = 1.72$ m.

The optimisation process has to cover a superposition of several loading conditions. They occur in the daily sun tracking with different loading conditions. Deflection-dependent stress-distributions from self-weight (Fig. 5) as well as wind actions on the structure have to be taken into account for the design. Additionally, stresses induced by temperature constraints, shrinkage and creep are considered.

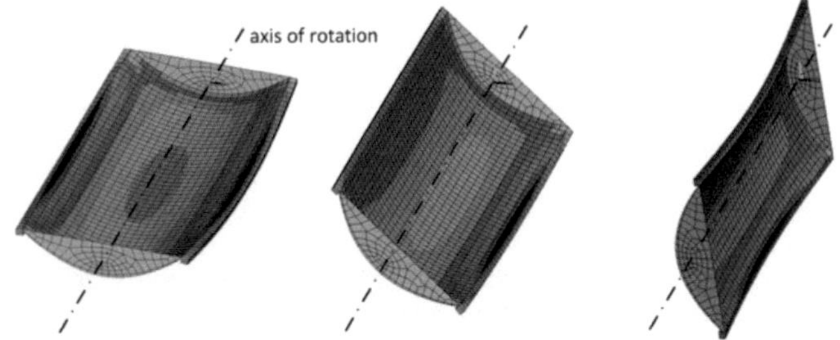

Fig. 5 Stress plot under self-weight for a deflection angle ψ=0° (left), ψ=40° (middle) and ψ=80° (right)

4 Experimental tests

At present experiments to investigate the general deformation behaviour of shell structures made of UHPC are prepared at the Technical University of Kaiserslautern. It is planned to analyse the influence of several parameters such as shell thickness, shell symmetry, aftertreatment, concrete mix, reinforcement and stripping time.

4.1 Specimens and concrete mix

Currently test specimens are fabricated at the Technical University of Kaiserslautern. Their geometry is shown in Fig. 6.

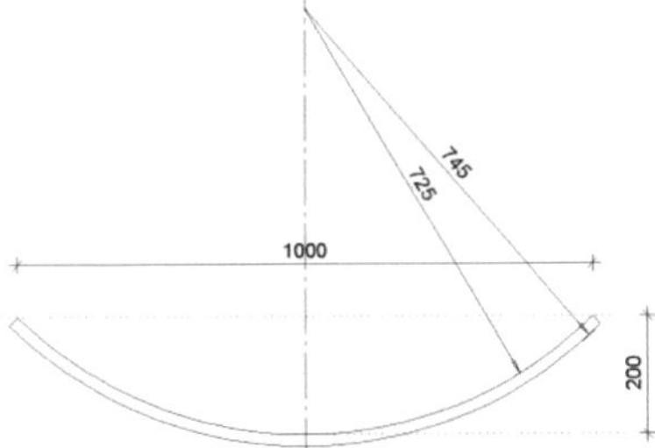

Fig. 6 Geometry of test specimens with different shell thicknesses (10, 20, 30, 50 and 100 mm) and a shell width of 80 mm will be tested

The following concrete mix was developed at the Technical University of Kaiserslautern and will be used to start off with:

- Cement CEM I 42.5 HS	831.57 kg/m³	- Silica dust QM 1600	205.35 kg/m³
- Water	73.80 kg/m³	- Silicoll EMSAC	249.48 kg/m³
- Sand 0/2	693.99 kg/m³	- Superplasticizer ACE 30	35.76 kg/m³
- Sand 0.125/0.5	223.87 kg/m³		

4.3 Measurement technology

One aim of the here presented experiments is to quantify the deformations which emerge in the concrete while hydrating. This is planned to be realized by juxtaposing the shell geometry a few hours after casting and the geometry at certain ages. At present different measuring methods (e.g. photogrammetry, see Fig. 7) are being compared to detect the most suitable one for this task.

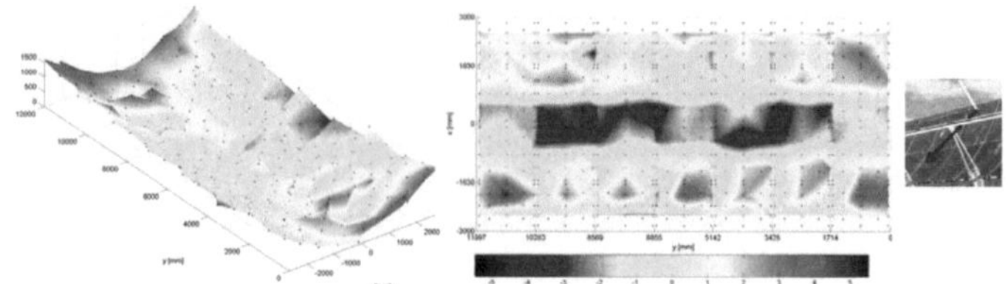

Fig. 7   Unintentional deformations of the reflectors, measured with photogrammetry: vertical deviation (left) and inclinations in home position (right) [7]

5   Conclusion

Thin concrete shells made of micro reinforced UHPC replacing the steel framework and the reflectors of already existing parabolic troughs seem to be most effective. To determine the efficiency of the collector a performance model for the disturbed surface is introduced, which weights the reflected light rays based on their distance to the focal point. Hence, the performance of the collector depends on the accuracy of the surface. Deformations through creep and shrinkage additional to self-weight and wind actions as well as temperature constraints have to be considered. Experimental tests at TU Kaiserslautern will include heat treatment and autoclaving to minimise initial deformations and enable a short-time fabrication. At Ruhr-University Bochum a first simplified Finite Element model employing continuum-based shell elements is used to predict the deformations and therefore the accuracy of the collector's surface. Optimisation cycles using superposition of several loading conditions through sun tracking will lead to an optimal shape.

References

[1]   Association Francais de Genie Civil (AFGC): Documents scientifiques et techniques: Betons fibres a ultra-hauts performances, Janvier 2002.

[2]   Deutscher Ausschuß für Stahlbeton (DAfStb): Sachstandsbericht Ultrahochfester Beton, Heft 561, Beuth, Berlin, 2008.

[3]   Dudziak, L., Mechtcherine ,V.: Reducing the cracking potential of Ultra-High Performance Concrete by using Super Absorbent Polymers (SAP), Institute of Construction Materials, TU Dresden, Germany, In: Advances in Cement-based Materials, G. van Zijl and W.P. Boshoff (eds.), Taylor & Francis Group, 2010.

[4]   Fontana, P.: Mikrostruktur und mechanische Eigenschaften von wärmebehandeltem und autoklaviertem ultrahochfestem Beton. In: Fachtagung „Brücken aus ultrahochfestem Beton", Berlin, 9. Juli 2009.

[5]   Müller, H., Burkhart, I.: Materialgesetze für das Spannungs-Dehnungs-Zeitverhalten von ultra-hochfestem Beton, Forschungsvorhaben Universität Karlsruhe KIT (Mu 1368/9) im Rahmen des DFG-Schwerpunktprogrammes 1182 Nachhaltiges Bauen mit ultra-hochfestem Beton.

[6]   Philip, U., Dehn, F., Schreiter, P.: Temperatureinfluss auf die Phasen- und Gefügebildung von UHPC, in: Ultrahochfester Beton, Bauwerk-Verlag, Berlin, 2003, S. 79-88.

[7]   Pottler, K.; Lüpfert, E.; Johnston, G. H. G.; Shortis, M. R.: Photogrammetry: A powerful tool for geometric analysis of solar concentrators and their components, Proceedings of ISEC, 2004, ASME, Portland, Oregon.

[8]   Schiel, W.: World's biggest solar power plant under construction in Spain, Schlaich Bergermann & Partner, 2007.

[9]   Solar Millennium: The parabolic through power plants Andasol 1 to 3, self-published Solar Millennium, 2008.

[10]   Tue et. al.: Das Verbundrohr als Innovationsmotor für hybrides Bauen, Schlussbericht an Bilfinger Berger AG, Universität Leipzig, 2004.

[11]   Tue, N. V., Ma, J., Orgass, M.: Kriechen von Ultrahochfestem Beton (UHFB), in: Bautechnik, Band 83, Heft 2, 2006, S. 119–124.

Concrete columns formed by nature

Katrin Schwiteilo

Institute of Concrete Structures, Faculty of Civil Engineering,
Technische Universität Dresden,
01062 Dresden, Germany
Supervisor: Manfred Curbach

Abstract

This project is part of the Priority Program 1542 "Concrete light. Future concrete structures using bionic, mathematical and engineering formfinding principles", which is funded by the Deutsche Forschungsgemeinschaft (DFG). The aim of the project described in this paper is the development of reinforced concrete columns whose shape is determined by the force flow to carry loads more efficiently. The results might be material savings, less power consumption in the production process and hopefully more aesthetic structures in the built environment of human beings. In this paper the planned methods and the first steps have been presented. The structural behaviour of columns and the modes of failure were analysed. As a first result, shapes of haunches were designed and evaluated regarding a decrease in failures due to load application and perhaps slab punching. Finally, some further steps were presented.

1   Introduction

Concrete is flowable during building time and, consequently, it is possible to create nearly every form (Fig. 1). Nonetheless, concrete structures are normally designed with flat and parallel surfaces like plates, beams and other cubic elements. The reason for this is the simple and low cost fabrication and that there is no code available yet for the structural design of concrete columns with a variable cross section in vertical direction. But how can typical concrete elements be optimised? Columns with a constant rectangular cross section have areas which do not make full use of the material's load-bearing capacity. It is the aim of this project to design more slender, economic and aesthetic columns by making them more efficient. This means that an attempt is made to construct a column whose form is determined by the main forces.

Fig. 1   Flowable concrete before hardening (left) and sculptures in concrete by V. Mixsa (right)

Nature shows how this could work. Slender columns could be developed by analysing natural compression structures with the help of computational models and experimental research. The planned key steps are

- Searching for an inspiration from compression structures in nature,
- Analysing the main load situations and the reasons for failure of typical reinforced concrete columns,

Proc. of the 9th *fib* International PhD Symposium in Civil Engineering, July 22 to 25, 2012,
Karlsruhe Institute of Technology (KIT), Germany, H. S. Müller, M. Haist, F. Acosta (Eds.),
KIT Scientific Publishing, Karlsruhe, Germany, ISBN 978-3-86644-858-2

- Developing more efficient column shapes for each failure mode and, if necessary, experimental testing,
- Combining these shapes to a few column forms,
- Choosing of different modern concrete types, for example reinforced HPC with or without textiles or UHPC with fibres,
- Dimensioning of the chosen column shapes with the selected material combination,
- Planning of the test set-up and performing experimental tests,
- Evaluating and recalculating the experiments and
- Developing a draft for a dimensioning code for columns with a variable cross section in vertical direction.

The first step is to find a form of columns with a variable cross section that can carry more loads than a column with a constant cross section and to optimise the load flow.

2 Possibilities

Software to optimise the shape of structure component, such as the Computer Aided Optimization method [1], is available. However, this FEM software can only optimise the structure for one load situation. Apart from this, it needs a lot of time and processing power.

To make columns more efficient it is necessary to know how they carry loads and why typical columns fail. In addition to this, the choice of the material can have an influence on the structural behaviour. Accordingly, it is for example relevant whether reinforcement is used or not and at which load level plastic deformation starts.

In this project cast-in-place concrete columns with fixed supports at head and bottom shall be investigated (Fig. 2a). Columns are part of the load-carrying structures that are basically loaded by axial forces from dead and life loads of the system. Theoretically, these idealised loads would generate constant compression in the homogenous material of concrete independent of the form chosen for the cross section. Therefore, there is no point in shaping the columns in vertical direction because the material capacity is constant, except for the connecting point of column and slab.

However, normally there is no perfectly centric load situation in a structural system (Fig. 2b). The contiguous slabs, for example, do not necessarily have to be subjected to equal loads and thus would not deform in the same way. This generates changeable bending moments in the centre column. A similar load situation can be observed in exterior and corner columns. The difference between the centre column and the others is the ratio of the maximum bending moments to the normal forces which is smaller in the centre column.

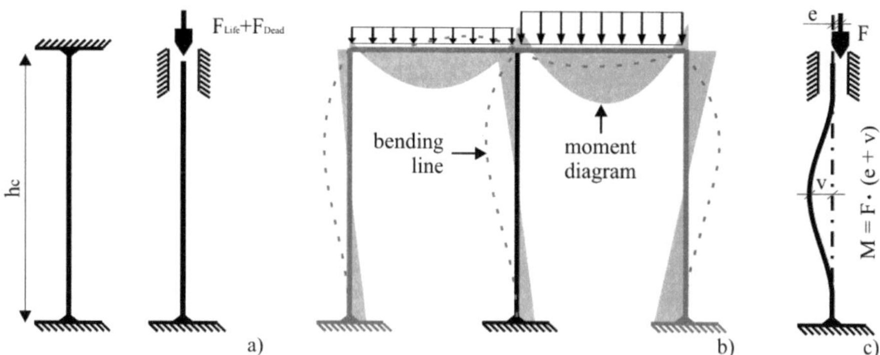

Fig. 2 Static system of the considered column with idealised fixed support (a), example for an asymmetric load situation (b) and principle of the increase of the internal forces due to deformation under axial loads (c)

Another bending moment could be introduced in a column under axial loads by the stability problem. Typical concrete columns are subjected to a number of inhomogeneities in geometry, material and structure which might, for example, occur in the support or the shape of the column. Therefore spontaneous deformation due to a stability problem could not be a problem. The inhomogeneities lead to deformations before spontaneous failure occurs. However, these deformations can increase the internal forces of the column and, consequently, lead to large deformations and earlier failure (Fig. 2c).

On the basis of the described load situations and the support conditions, the inner structural behaviour can be described. In typical columns under compression, the stress is uniformly distributed over the cross section. Furthermore, usually there should not be any tension in the longitudinal direction of a column.

One mode of failure is material failure which is caused by reaching the ultimate compressive stress in one part of the column. This can be calculated very well for the non-linear material of concrete and reinforced concrete.

However, in the connection point between column and slab – fixed or not – the stress can be enlarged (Fig. 3). This is a well-known problem in load application. Therefore, more transverse reinforcement is usually placed in these areas than in the rest of the columns. Furthermore, the problem of slab punching is probably related to this concentrated stress at the point of connection.

So what are the possibilities to strengthen columns by forming in vertical direction? A look at the column's inner stress behaviour can give an answer to this question. Toward this aim, an unreinforced concrete column fixed at both endings has been modelled with a 2D-FEM software. The value of the applied load is not important for this investigation. The application and the direction of the forces are relevant to show how the stress in a column is allocated. The aim of this investigation would be a completely harmonised stress allocation at the surface of the column. As this is not realistic because of the bending moment, it is desirable to at least normalise the stress at the surface as much as possible.

Under axial load the stress concentrates in the connecting point of column and slab (Fig. 3a). Possibly the maximum stress acts like a predetermined breaking point and, thus, can be the reason for the problems of load application and slab punching. It is obvious that the maximum stress of a column under axial and bending load is also located in this connecting point because it is also the place of the largest bending moment (Fig. 3b). The horizontally displaced support at the column head generates the bending moment in the column. Both pictures show the stress in the upper part of the column in direction of the x-axis and in the lower part in direction of the y-axis.

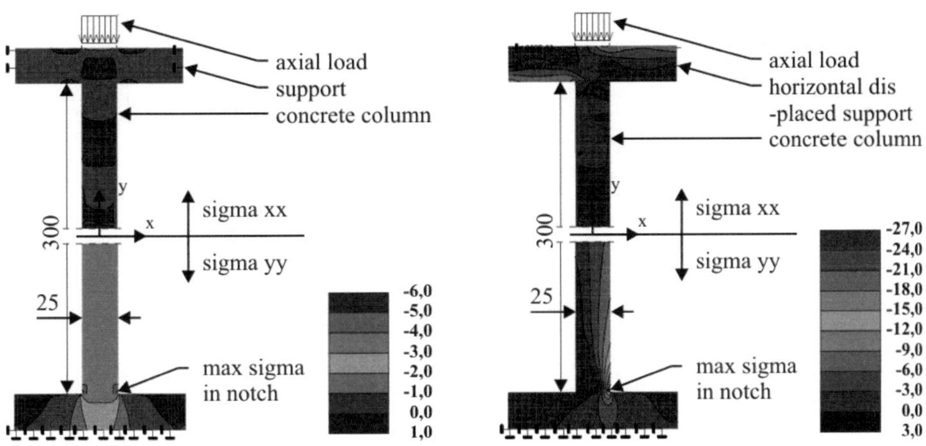

Fig. 3a   Inner stresses in a column under axial load

Fig. 3b   Inner stresses in a column under axial and bending load

To eliminate the concentrated stress at the fixed ends of the column, they can be enlarged. The optimum would be a constant stress level in vertical direction along the surface. The result would be the disappearance of any predetermined breaking point. Apart from this, the extension of the column ends should also have a positive influence on the stability problem.

For one load situation it is possible to find an optimum shape to meet this axiom of constant stress [1], by considering the point of reaching the material strength and the subsequent stress rearrangement. But is it even possible to develop the optimal shape if there are variable load situations, such as different directions or conditions?

3 Forming to reduce the maximum stress

This point will be examined first because it is a problem which is relevant for all loads – be they axial or not. The maximum stress occurs in the notch where slab and column are connected. Accordingly, it is not only the load that creates inner stress, but also the form how elements are connected which could make a difference. Otherwise the stress at this point would be significantly smaller. The idea to form the head of columns to create an improved load flow from the slab into the column is not new [2]. Robert Maillart, for example, used the column head enlargement for round or hexagonal columns in some of his projects (Fig. 4). He then enlarged the head of the columns linearly or rounded it out to transfer the loads of the slab into the column. What is the effect of such a haunch and which is the best form to harmonise the stress in the column?

Fig. 4   The Zurich warehouse and the filter building at Rorschach [3]

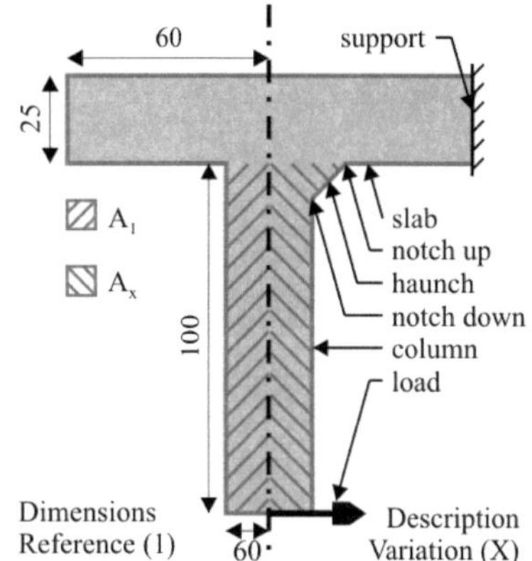

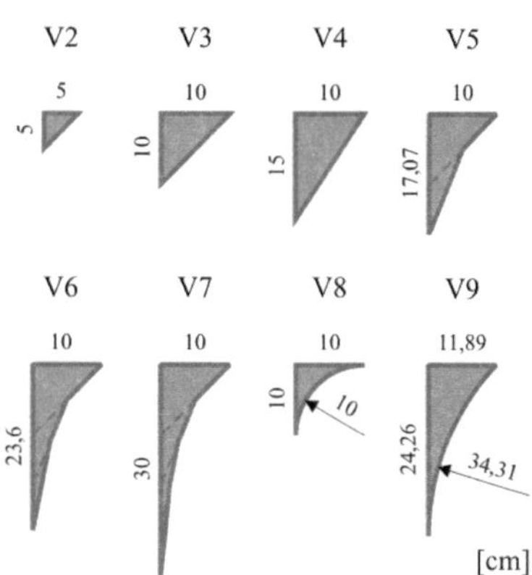

Fig. 5   Dimensions, loading and description   Fig. 6   Numeration and dimensions of the
         of the half simulated steel column   investigated haunches

The investigation starts with a very simple linear haunch which is also often used by Maillart [2]. Its basic idea is similar to the principle of trees or bones which put material in the place of the maximum stress [1]. One simple way to do this is a small triangle related to the direction of shear. A 2D-FEM model of a half column has been created to compare the stress allocation qualitatively. The model which is based on a homogeneous and isotropic material (in this case steel) has the typical dimensions of a reinforced concrete column (Fig. 5). Among different simple haunches (V2 - V7) – adding a triangle in the notch with the maximum stress – a typical quadrant (V8) and a simple modification (V9), which are often used in the construction with steel, were tested (Fig. 6).

To evaluate the results, the maximum stress reached in the column (Max) was compared. But in the stress plots it became evident, that the maximum stress is not always in a notch. As a result, the stress in the notch which connects slab and column (notch up) and in the notch where the haunch begins (notch down) were also compared (Fig. 5). The notch down was always the notch subject to the larger stress. Furthermore the area increase of the columns ($\Delta A = A_x / A_1 - 1$) has been compared in relation to the decrease of the stress. The results are shown in Fig. 7.

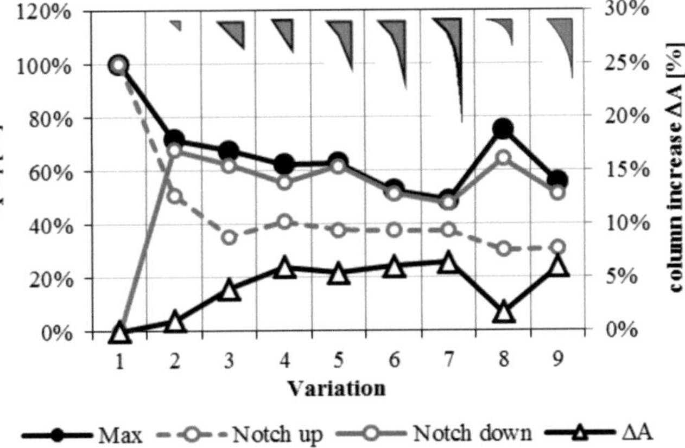

Fig. 7   The left diagram shows the maximum stress in the column (Max), the stress in the upper notch (Notch up) and in the lower notch (Notch down) of each haunch (2 - 9) in relation to each of these stresses in the reference (1) without a haunch marked by circles ordered to the left axis. The line with the triangles gives the increase of the column area through the haunches ordered to the right axis.

All of the chosen haunches reduce the maximum stresses in the column while they never exceed a half column area increase of 7 %. The maximum stress seems to be a function of the height of the haunch. This is logical because the bending moment at the starting point of the haunch is smaller when it is closer to the middle of the column. The function is approximately linear. However, in several haunches, such as haunch V4, the reduction of stress was more significant in the notch down. This might be due to the fact that the maximum stress is not always located directly in the notch down but sometimes also above or under it.

The maximum stress and the stress in the notch down can already be reduced by about 30 % through haunch V2 which is an isosceles triangle with a 5 cm side length. The doubling of this triangle, which is haunch V3, reduces the stresses only by 4 % more. Consequently, the bigger size cannot exclusively result from the smaller bending moment, as this would lead to a linear increase of the stress. The haunch V8 in the shape of a circle with a radius of 10 cm – the notches being in the same point like in V3 – reduces the maximum stress only by 25 %.

4   Conclusions

In summary, the implementation of a haunch reduces the maximum stress in every case. In general, this was to be expected because in the case of notch up the cross section increases and in case of the notch down the bending moment decreases. Both of these lead to lower stress in each notch. However, if the notches of differently formed haunches are at the same point, the stresses in the notches should be the same. As can be seen in the comparison of V3 and V8, this is not the case. Consequently, it can be concluded that the shape of the haunch has to have an influence too.

The area increase of the column was always smaller than 7 % in the 2D view. Just a bit more material reduces the maximum bending stress in columns by 30 % to 50 % and harmonises the stress allocation. The question whether this could for example reduce the necessary stirrups in the column head or the problem of the slab punching should be investigated by 3D simulation and experiments. Nonetheless, the assumption seems reasonable.

A statement about the best haunch shape seems to be difficult because of the different haunch sizes, additional areas and resulting differing bending moments at the notches. Therefore the factor C was created to compare the different situations, formula (1).

$$C = \frac{\sigma_x}{\sigma_1} \cdot \frac{A_x}{A_1} \qquad (1)$$

σ_x ... maximum stress in the haunches V2 to V9

σ_1 ... maximum stress in the reference V1

A_x ... area of the column with the haunches V2 to V9

A_1 ... area of the reference V1

The reduction of the stress, the increase of the area and the coefficient C for the columns with different haunches are shown in Table 1. The largest reduction of stress and the smallest C occurred in V7, which is also the shape with the most harmonised stress allocation. But also the very small haunch of V2 and V3, which could be built very easily, reduces the stress by nearly 30 %. Here the maximum stress is located in the column close to the notch down. In the circle section of V9 the maximum stress is located in the haunch close to the notch down. The reduction in V9 amounts to more than 55 % for the stress and almost 60 % for the coefficient. Consequently, the haunches of V2 or V3, V6 and V9 will be further analysed in this project.

Table 1   Reduction of the stress, the increase of the area and the coefficient C for the columns with different haunches

	V1	V2	V3	V4	V5	V6	V7	V8	V9
σ_x / σ_1 [%]	100,0	71,4	67,4	62,3	62,8	52,7	48,9	75,2	55,6
A_x / A_1 [%]	100,0	101,0	104,0	106,0	105,4	106,1	106,4	101,7	106,0
C [%]	100,0	72,2	70,1	66,1	66,2	55,9	52,0	76,5	58,9

5   Outlook

The presented influence of the shape of the column's head can offer a solution to the problems of load application and slab punching because the load application from the slab into the column is smoother. Consequently, if the high stress is the reason for more transversal reinforcement in the column head, these stirrups can perhaps be set at a larger distance such as in the middle. Both of this should be investigated through experiments. Further questions have to be investigated, for example how the longitudinal reinforcement is to be continued in the chosen haunches or how these haunches can be built.

Furthermore, the increased deformation and the resulting stress in the middle of the column because of theory 2nd order are the next part of the investigation. It is likely that the cross section in the middle of the column will get larger. But the enlarging of the fixed ends also has a stabilising influence on the stiffness of the column. Additionally, the investigation now has to be upgraded for an in-homogeneous and an-isotropic material like concrete and to the interaction with the reinforcement in a 3D model.

References

[1]   Mattheck, C.: Design in Nature. Springer Verlag, 1998, Berlin

[2]   Gesellschaft für Ingenieurbaukunst: Robert Maillart, Beton – Virtuose. Katalog zur Ausstellung des Instituts für Baustatik und Konstruktion der ETH Zürich. Band 1. Vdf Hochschluverlag AG, 1998, ETH Zürich

[3]   Billington, C. P.; Robert Maillart and the art of reinforced concrete. Verlag für Architektur Artemis, 1990, München

Numerical simulation of single-span lightweight concrete sandwich slabs

Michael Frenzel

Institute of Concrete Structures,
Technische Universität Dresden,
George-Bähr-Str. 1, 01069 Dresden, Germany
Supervisor: Manfred Curbach

Abstract

It is the aim to develop light and efficient structures by dividing the force flow into compression and tension and assign these to the appropriate materials – pressure to concrete and tension to reinforcement. A first step to reach this goal is to numerically and experimentally investigate the bending load bearing capacity of sandwich elements made of modern high performance concretes (compressive zone), a lightweight core layer and steel or textile reinforcements (tensile zone). This paper presents a numerical, nonlinear analysis of a single-span sandwich element and indicates the potential of weight reduction based on the effective application of different concretes.

1   Introduction

Ceiling and floor slabs of concrete building constructions are frequently erected with site concrete or as a combination of pre-cast panels with poured in-situ topping. Solid or T-beam cross sections are often executed because they can be fabricated fast and economically. However, from the structural point of view, these slab designs are quite inefficient because the ultimate bearing capacity of the reinforced concrete is reached in only a few sections of the mainly flexurally loaded elements. The optimisation potential of uniaxial spanned ceiling structures with regard to saving material and weight is investigated within the Priority Program 1542 "Concrete light. Future concrete structures using bionic, mathematical and engineering form finding principles" [1], which started in July 2011 and is funded by the German Research Foundation (DFG).

Efficient concrete structures may be realised if the different acting compressive and tensile stresses can be assigned to appropriate carrying materials. During the last years, new concretes and reinforcements such as (ultra) high performance concretes ((U)HPC), infra- and high performance lightweight concretes (ILC, HPLC) [2], glass and carbon fibre bars, sheets and textiles were developed and investigated. As a result, a wide range of high performance materials is available for the development of light slab structures. There are two approaches traced:

- cross-sectional material separation: a thin layer of a concrete with higher compressive resistance is applied to areas of intensive compressive stresses resp. an appropriate reinforcement serves to absorb tensile forces. Less loaded areas are filled with a lightweight material of low bearing capacity. Nevertheless, it needs to be high enough to ensure enough resistance against shear forces, especially because shear reinforcement is avoided at slabs due to its extensive application.
- form finding for the main load-bearing direction: Due to functional reasons the top surface of the slab needs to be plane. Accordingly, only the bottom can be adjusted to the force flow.

In this paper, the optimisation potential of a common single-span slab with regard to its height and self-weight through a cross-sectional material separation (sandwich structure) is presented.

2   Initial situation

The considered one way steel reinforced concrete slab has a span of 5000 mm (Fig. 1) and a width of 1000 mm. It is part of a meeting room. The aim is to determine the slab's required thickness with the objective of designing the structure as light as possible while complying with the most decisive requirements from EURO CODE 2 [3]. Three different variants are considered and compared:

Proc. of the 9th *fib* International PhD Symposium in Civil Engineering, July 22 to 25, 2012,
Karlsruhe Institute of Technology (KIT), Germany, H. S. Müller, M. Haist, F. Acosta (Eds.),
KIT Scientific Publishing, Karlsruhe, Germany, ISBN 978-3-86644-858-2

- solid cross section with the commonly used concrete C 25/30 (reference slab, No. 1)
- solid cross section with the lightweight concrete LC 25/28 (No. 2)
- three-layered slab made of infra- and lightweight concrete, ILC and LC 25/28 (No. 3 & 4)

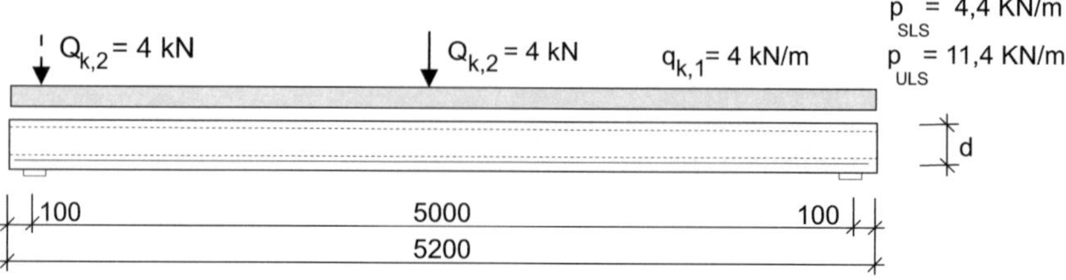

Fig. 1   Structural system of analysed floor slab

The infralightweight concrete (ILC) was developed by the Chair of Conceptual and Structural Design of the University of Technology Berlin [4], [5] during the last years and is currently one of the lightest structural lightweight concretes. It has a dry density of 760 kg/m³ and a mean cube compressive strength of 7.4 N/mm² [6]. At present time, it can only be used as core layer material and not in the tensile zone because the bond behaviour between steel and concrete and the ability to protect steel from corrosion still needs to be substantially investigated.

The steel reinforcement is covered by a concrete layer of 20 mm, by a lightweight concrete layer of 25 mm (exposition class XC 1) [3]. Considering a reinforcement diameter of 10 mm and an additional minimal concrete cover of 10 mm for bond, a total bottom layer thickness of 45 mm is required for the sandwich element. The minimal thickness of the top layer depends on the maximum aggregate size and is part of the calculation results. Only the longitudinal, structurally required reinforcement is provided. Shear reinforcement is not applied and the normatively prescribed minimum reinforcement areas are neglected.

3   Loads and requirements

The slabs are loaded by their self-weight, an assumed additional dead and two imposed loads. Table 1 shows their values, safety and combination factors according to EURO CODE 1 & 2 [3], [7], which need to be applied for calculations of the serviceability and the ultimate limit state (SLS resp. ULS). The considered, normatively required imposed load $Q_{k,2}$ (single load) is replaced by a line load ($q_{k,2}$) to reduce the number of different load positions. The value for the line load of 1.6 kN/m is chosen in such a way that the resulting bending moment at middle span and the shear force at the support are identical to the values caused by the single load placed at middle span or over the support.

The basis for the calculations is the quasi-permanent load combination at the SLS and the persistent and transient design situation at the ULS [3]. The following five requirements are defined as decisive: SLS- limitation of deformations ($f \leq 1 / 250 = 20$ mm), ULS – Bending in compression and tension and shear. Furthermore, steel reinforcement and concrete should be utilised economically and efficiently. This can be achieved if the steel is plastified at ULS while high concrete stresses occur. Besides, the plastification of steel fulfills the requirements of a ductile failure mode.

To be able to comply these requirements with a numerical, non-linear analysis, a global material safety factor of 1.27 is applied for concrete and steel [8], [9] instead of the commonly used partial material factors. Both the load and material safeties are added to the characteristic loads. When applying the load combinations, design loads of p_{SLS} = 4.4 kN/m at SLS and p_{ULS} = 11.5 kN/m at ULS are considered additionally to the slab's self-weight (see Fig. 1).

Table 1   Characteristic loads, safety factors and combination factors

	Load		Dim.	Value	SLS $\gamma_{load} / \gamma_{glob}$	SLS Ψ_2	ULS $\gamma_{load} / \gamma_{glob}$	ULS Ψ_0
1	Self-weight slab	$g_{k,S}$	kN/m	variable	1.0/1.0	1.0	1.35/1.27	1.0
2	Add. dead load	$g_{k,A}$	kN/m	1.0	1.0/1.0	1.0	1.35/1.27	1.0
3	Imposed load 1	$q_{k,1}$	kN/m	4.0	1.0/1.0	0.6	1.50/1.27	1.0
4	Imposed load 2	$Q_{k,2}, q_{k,2}$	kN, kN/m	4.0, 1.6	1.0/1.0	0.6	1.50/1.27	0.7

4   Material properties

4.1   Uniaxial Stress-strain relations

Mean material stress-strain relations can be used for the determination of deflections at SLS to get quite realistic simulation results [3]. Consequently, mean compressive strengths were taken into account for the concrete material law. The tensile resistance is based on characteristic values (5 % - fractile values) because the ultimate tensile stress of concrete varies comparatively much. For the same reason characteristic yield and ultimate tensile stresses are used for the description of the material law of the reinforcing steel. This is a conservative assumption. No long-time dependent deflections caused by e.g. creeping have been taken into account for the following structural analysis.

Compared to SLS, modified material laws need to be applied for the determination of the load bearing capacity at ULS due to the demanded safety factors. The used global material safety concept is utilised both for concrete and steel under compression and tension. The calculative maximum material resistances, which are the basis for the material laws at ULS, are determined with equation (1).

$$f_{calc} = f_d \cdot \gamma_{glo} = \frac{\alpha \cdot f_k}{\gamma_{mat}} \cdot \gamma_{glo} \tag{1}$$

f_{calc}...calculative stress, f_d...design stress, f_k...characteristic stress, γ_{glo}...global safety factor (1.27), γ_{mat}...material factor (concrete 1.5, steel 1.15), α...reduction factor (appr. for all concretes 0.85)

Fig. 2 shows the applied stress-strain relations for the concretes C 25/30, LC 25/28, the infralightweight concrete (ILC) and the steel reinforcement BSt 500. The ascending non-linear gradient of concrete compressive stresses is described by equation (2) [3]. The decisive values are summarized in table 2. The material parameters of the ILC were taken from experiments executed by EL ZAREEF [6]. After reaching the compressive stress peak, a linear descending softening law is considered based on compressive displacements and energy dissipation. They are localized in a plane normal to the direction of compressive principle stress and are independent of the size structure [10]. An appropriate und commonly used value for the maximum plastic displacement $w_{pl,max}$ of normal concrete is 0.5 mm [11]. A value of 0.2 mm is assumed for both lightweight concretes. According to MARKESET [12] the plastic displacement of lightweight concrete should be less than 0.3 mm. The linear descending softening laws are shown in figure 2 for the ultimate limit state stress-strain relations.

$$\sigma_c = -f_{cm} \cdot \frac{\dfrac{1.05 \cdot E_{cm}}{f_{cm}} \cdot \varepsilon_c - \left(\dfrac{\varepsilon_c}{\varepsilon_{c1}}\right)^2}{1 + \left(\dfrac{1.05 \cdot E_{cm} \cdot \varepsilon_{c1}}{f_{cm}} - 2\right) \cdot \dfrac{\varepsilon_c}{\varepsilon_{c1}}} \tag{2}$$

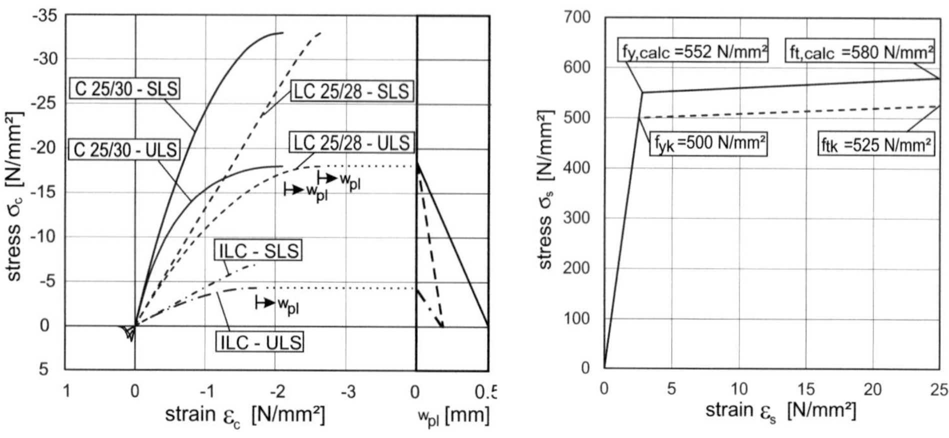

Fig. 2   Stress-strain relations of the used concretes (left) and the reinforcing steel (right)

Table 2   Material properties of concrete

Material	$f_{cm} / f_{ck} / f_{ccalc} / f_{ctk,0,05} / f_{ctcalc} / E_{cm}$ [1]	Dim.	ε_{c1} [1]	Dim.	ρ_{sup} / ρ_k	Dim.
C 25/30	-33.0 / -25.0 / -18.0 / 1.8 / 1.3 / 31000	N/mm²	-2.1	‰	- /2.4	t/m³
LC 25/28	-33.0 /-25.0 / -18.0 / 1.4 / 1.0 / 12550	N/mm²	-2.6	‰	1.4/1.45	t/m³
ILC	*-6.9 / -6.0 / -4.3 / 0.55/ 0.4/ 4000*	N/mm²	-1.7	‰	*0.9/0.95*	t/m³

[1] ...definition of symbols see [3], ρ_{sup}...superior calculative weight, ρ_k...characteristic weight, *cursive values are assumed*

4.2   Bond-slip relations

The bond between concrete and reinforcement ensures the functional capability of reinforced concrete. A certain bond stress is necessary to guarantee a sufficiently small crack width at SLS and the required anchorage of the reinforcement at ULS. The joint between the reinforcement bar and the concrete carries bond stresses in dependence of the relative displacement (slip) between both materials. Due to the small concrete cover, failure by splitting of the unconfined concrete is assumed with good bond conditions. The ascending part of the bond-slip relation may be described for normal concrete with equation (3) [13], and for structural lightweight concrete with equation (4) [14].

The bond stresses may be determined through the cylindrical compressive strength of concrete. The considered bond-slip relations are based on the characteristic strength values. Figure 3 shows the curves for the C 25/30 and LC 25/28. As noted in [13], the maximum bond stress of unconfined concrete is reached with a slip of 0.6 mm. This value is assumed for both concretes.

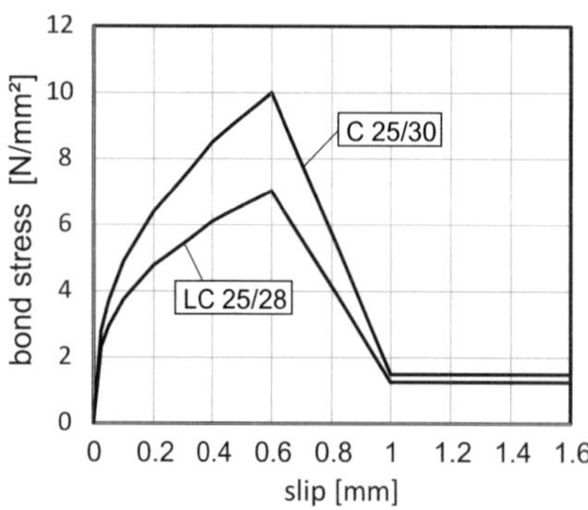

$$\tau(s) = 2 \cdot \sqrt{f_{ck}} \cdot \left(\frac{s}{0.6} \right)^{0,4} \quad \text{for C 25/30} \tag{3}$$

$$\tau(s) = 0.6 \cdot f_{lck}^{0.82} \cdot \left(\frac{s}{1.0} \right)^{0,4} \quad \text{for LC 25/28} \tag{4}$$

Fig. 3   Bond-slip relations

5   Finite-Elemente Model

The following analysis was carried out with the Atena 2D[1] program. The slabs were modeled in a two-dimensional concept room. Due to the symmetrical support and loading, only one-half of the slab was analysed. The results are transferable to the entire element. Figure 4 shows the cracked FE-model of a three-layer sandwich element with the reinforcing steel at ultimate limit state. The different concrete layers are rigidly bonded. It is assumed that concrete is a homogeneous material.

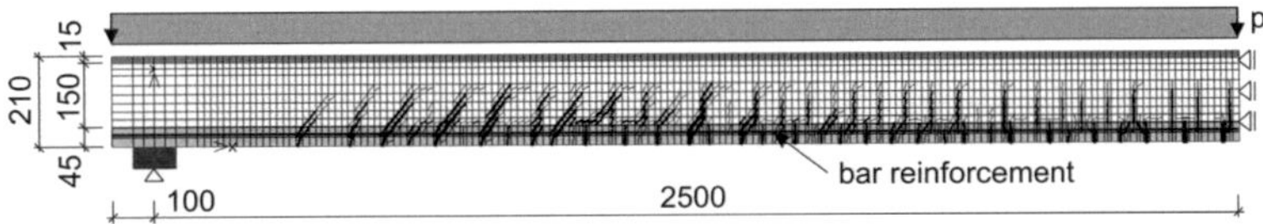

Fig. 4   FE-model for concrete sandwich elements

The implemented material model includes the following effects of concrete behaviour [10]:

- non-linear behaviour in compression, including hardening and softening,

- linear reduction of compressive strength after cracking,

- linear stress-strain relation in tension,

- fracture of concrete in tension based on nonlinear fracture mechanics,

- biaxial strength failure criterion,

- tension stiffening effect.

A finite concrete element fails if a combination of principal stresses fulfills the biaxial strength failure criterion. The implemented biaxial failure function for concrete, based on experiments by KUPFER [15] is schematically displayed in Fig. 5 [10]. The most important input material parameters are given in table 2.

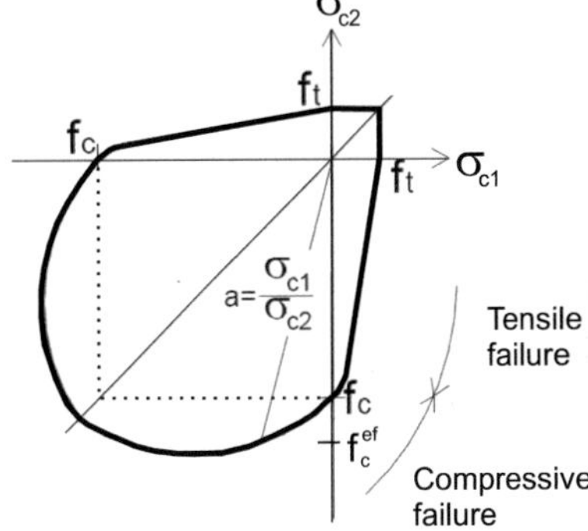

Fig. 5 Biaxial failure function of concrete [10]

f_c …uniaxial compressive failure stress
f_c^{ef}…biaxial compressive failure stress
f_t …tensile failure stress
σ_c …concrete stress

6   Results and Conclusions

The slab No. 1 (C 25/30) meets the requirements of the SLS and ULS with a height of 180 mm, the slab No. 2 (LC 25/28) and the sandwich element No. 3 of 210 mm. The sandwich element consists of a 15 mm top layer, a 45 mm bottom layer made of LC 25/28 and a 150 mm thick ILC-core. Table 3 furthermore shows the required area of reinforcement and the slab's characteristic self-weight. Slab No. 3 fulfills all requirements with only 53 % of slab No. 1's self-weight. The reduction of self-weight would be even bigger if it was designed of the same height as slabs No. 2 & 3 (h = 210 mm).

Table 3   Results

No.	h/d [mm]	a_s [cm²/m]	g_{self} [kN/m]	Red. [%]
1	180/155	7.14	4.3	100
2	210/180	5.24	3.0	70
3	210/180	5.24	2.3	53
4	180/150	10.28	2.0	47

h…slab height, d…effective depth, a_s… area of reinforcement, g_{self}…slab's self-weight

---

[1] Atena 2D, Version 4.2.2.0, Červenka Consulting Ltd., Prague, Czech Republic

Figure 6 shows the middle deflections which were determined in dependence of the bending moment. Furthermore, the required bearing moments at SLS and ULS are highlighted. The deflection-moment course may be described by three parts: ① shows a large curve increase until the concrete cracks. The resulting loss of stiffness leads to a lower curve increase in part ②. When the steel starts to plastify, the slab can hardly take up any additional loads ③.

A further weight reduction can be achieved with a sandwich element height of 180 mm (No. 4). However, a comparatively high (uneconomic) percentage of reinforcement has to be applied to meet the requirements of the SLS and ULS. While the reinforcement of the slabs No. 1 to 3 plastifies just before or when reaching the ultimate limit load, which guarantees a ductile failure mode, slab No. 4 fails brittle due to concrete

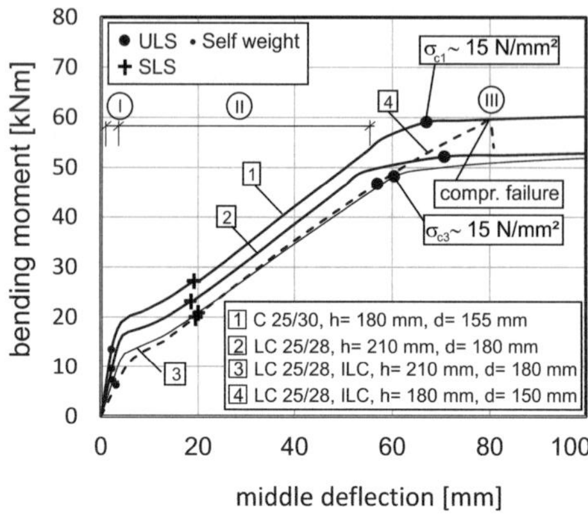

Fig. 6   Deflection-bending moment curves

failure in the slab's top compression zone. This type of failure is undesired. The top edge compressive stresses of 15 N/mm² of the slabs No. 1-3 indicate a well utilized concrete because the maximal calculative stresses are 18 N/mm² (see table 2).

The theoretical investigations have shown exemplarily that an effective and lightweight structure may be developed through the combination of different lightweight concretes. However, the structure may not be executed more slender than that of commonly designed ones because the deflections of the structure when cracked at SLS are mainly limited by the effective depth. Figure 6 shows that the slabs remain uncracked under self-weight. The application of lightweight sandwich elements is the more effective, the bigger the part of the self-weight is compared to the entire structure loading. Furthermore, lightweight concrete is preferable to normal concretes when due to the loss of weight the cracked cross section can be converted to an uncracked one at the serviceability limit state.

7   Outlook

The cross-sectional material separation is assigned to multi-spanned slabs. Particularly, the force-flow orientated shape finding along the longitudinal axis (Fig. 7) is focused. Besides, the bond behaviour between different concrete layers is analysed. After developing the theoretical background, experiments are executed to confirm the effective load bearing behaviour of sandwich structures.

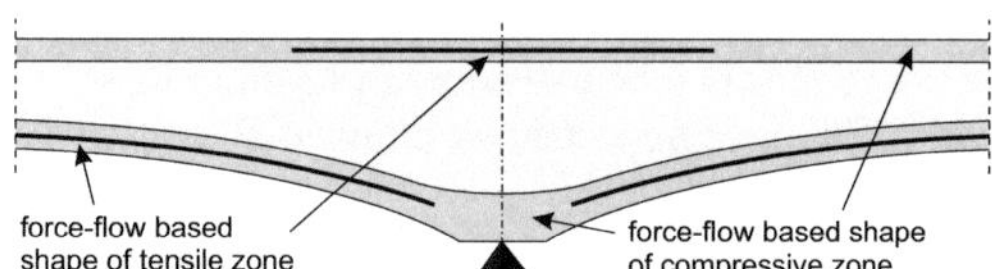

Fig. 7   Force-flow shaped sandwich slab

References

[1]   SPP 1542. http://spp1542.tu-dresden.de/ (15.02.2012)

[2]   Müller, H. S.; Reinhardt, H.-W.: Beton. In: Bergmeister, K.; Fingerloos, F.; Wörner, J.-D. (eds.): BetonKalender 2009. 1. ed. Berlin : Ernst & Sohn, 2009. – ISBN 978-3-433-01854-5, pp. 1–150

[3]   EN 1992-1-1 (EC2): Eurocode 2: Design of concrete structures – Part 1-1: General rules and rules for buildings; German version EN 1992-1-1:2004 + AC:2010, 2011

[4]   Schlaich, M.; El Zareef, M. E.: Infraleichtbeton. In: Beton- und Stahlbetonbau vol. 103 (2008), Nr. 3, pp. 175–182

[5]   Schlaich, M.; El Zareef, M.: Infra-lightweight Concrete. Proceeding of FIB International Conference, Tailor Made Concrete Structures - New Solutions for our Society, 2008, Amsterdam, Netherlands, P. 159.

[6]   El Zareef, M. E.: Conceptual and structural design of buildings made of lightweight and infra-lightweight concrete. Berlin, TU Berlin, Dissertation, 2010

[7]   EN 1991-1-1 (EC1): Eurocode 1: Actions on structures – Part 1-1: General actions – Densities, self-weight, imposed loads for buildings; German version EN 1991-1-1: 2002, 2010

[8]   CEB-FIP Comitè Euro-International du Bèton (ed.): CEB-FIP Model Code 2010 - First complete draft, CEB Bulletin No. 55. 1. ed. Lausanne : International Federation for Structural Concrete (fib), 2010. – ISBN 978-2-88394-095-6

[9]   Zilch, K.; Zehetmaier, G.-M.: Bemessung im konstruktiven Betonbau. 1. Auflage. ed. Berlin : Springer-Verlag Berlin Heidelberg, 2006. – ISBN 3-540-20650-7

[10]   Červenka, V.; Jendele, L.; Červenka, J.: Atena Program Documentation Part 1-Theory, 2009.

[11]   Mier, J. G. M. van: Multiaxial Strain-softening of Concrete, Part I: fracture. In: Materials and Structures vol. 19 (1986), Nr. 111

[12]   Markeset, G.; Hillerborg, A.: Softenning of concrete in compression localization and size effects. In: Cement and Concrete Research vol. 25 (1995), Nr. 4, pp. 702 – 708

[13]   CEB-FIP Comitè Euro-International du Bèton (ed.): CEB-FIP Model Code 1990, CEB Bulletin No. 213/214. 1. ed. London : Thomas Telford Services Ltd, 1993.

[14]   CEB-FIP Comitè Euro-International du Bèton (ed.): FIB Lightweight Aggregate Concrete, Recommended Extension to Model Code 90, CEB Bulletin No. 8. 1. ed. Lausanne : International Federation for Structural Concrete (fib), 2000. – ISBN 2-88394-048-7

[15]   Kupfer, H.; Hilsdorf, H. K.; Rush, H.: Behavior of concrete under biaxial stress. In: ACI Journal vol. 58 (1969), pp. 656–666

Shape optimized, filigree rods out of UHPC and non-corrosive CFK-reinforcement for variable spatial frameworks

Michael Henke

Department of Concrete Structures,
Technische Universität München (TUM),
Theresienstraße 90, 80333 München, Germany
Supervisor: Oliver Fischer

Abstract

Except of columns, concrete load-bearing systems are up to now usually characterized by transferring loads as surface elements or massive beams with comparatively unbalanced material utilization. Besides weight reduction and resource efficiency, the transparency can also be increased significantly by resolving these systems and designing filigree rod-like structures that are following the flux of forces. By using innovative materials (e.g. UHPC) and manufacturing processes combined with skillful designing – on system level (arrangement of nodes and rods) and component level (shape optimized rods and nodes in concern of material consumption) – it is possible for concrete structures to advance into fields of application that have so far almost exclusively been reserved for steel constructions. In addition, the use of reinforcement elements made of fiber-reinforced polymers and high-performance concrete involving a high density results in almost non-corrosive structural elements linked with low maintenance effort and a high durability.

The main objective of this research project is to develop general engineering models and design principles based on theoretical and experimental investigations for future applications of filigree variable spatial frameworks out of concrete. For that purpose first of all a modular concept is studied theoretically, with laboratory tests problems that are not yet sufficiently solved, will be ensured scientifically and finally the mechanical models will be validated through tests of the structural components.

1 Introduction

So far the term "concrete construction" is generally associated with solid and heavy constructions. In the majority of cases the minimization of material consumption is considered to be of subordinate importance. The initiators of the DFG (Deutsche Forschungsgemeinschaft) priority programme "Concrete light. Future concrete structures using bionic, mathematical and engineering formfinding principles" share the opinion that by further developments and researches of the material concrete, significantly lighter and more elegant concrete structures can be built in the future. Therefore numerous research projects are financed by the DFG within a total period of six years subdivided into two funding periods lasting three years each.

The research project presented in this paper focuses on filigree rod-like structures built up of prefabricated, shape optimized components (strut, tie and node) that are joined together modularly. With regard to weight reduction, load-bearing capacity and durability, the rods and nodes are made out of ultra-high performance concrete (UHPC) and are reinforced or tensioned exclusively with non-metallic elements. The force-fitted connection of the members is ensured inside of a node, whose geometry is fitted to the local load paths. For an improved introduction and transfer of the tensile and compression forces of the rods, a multiaxial state of stress should be obtained by using an expansive high-performance mortar combined with a spatial, stiff confinement. Whereas the use of UHPC with its high compressive strength is obvious for the highly stressed struts, it is not apparent for the ties at the first glance. However by using UHPC a higher extensional stiffness of the uncracked tie can be obtained, which is with regard to serviceability of high relevance. In order to ensure this stiffness permanently, the ties are tensioned.

Proc. of the 9th *fib* International PhD Symposium in Civil Engineering, July 22 to 25, 2012,
Karlsruhe Institute of Technology (KIT), Germany, H. S. Müller, M. Haist, F. Acosta (Eds.),
KIT Scientific Publishing, Karlsruhe, Germany, ISBN 978-3-86644-858-2

The priority programme started in summer 2011 and hence only a few results are already available. Because of that in the subsequent sections mainly the planned programme of work within the first funding period is presented.

2   Work programme

The focus of research within the first period is on the shape, the structural behaviour and on special details of the components strut and tie. Besides, connection possibilities of the rods are investigated theoretically and experimentally, fundamental mechanical considerations concerning shape and structural behaviour of the confined nodes are made and adequate preliminary tests on this topic are carried out.

2.1   Strut

2.1.1   Theoretical investigations

The main focus of the theoretical investigations is on the shape optimization of the strut. Slender confined solid cross-section at the ends of the strut and a middle section fitted to avoid a failure due to loss of stability (if applicable realized as hollow cross-section) are the basis for the optimization process, figure 1. Based on that, extensive numerical analyses are carried out with the main objective of approaching the two predominant failure criteria stability and bearing capacity. In this process, effects like time dependent material behaviour, multiaxial states of stress or changes of cross-section are taken into consideration. In order to be able to describe these effects properly, existing results of research projects (e.g. [1]) are used and effects that have not yet been sufficiently described are investigated by the experimental tests presented in the subsequent section.

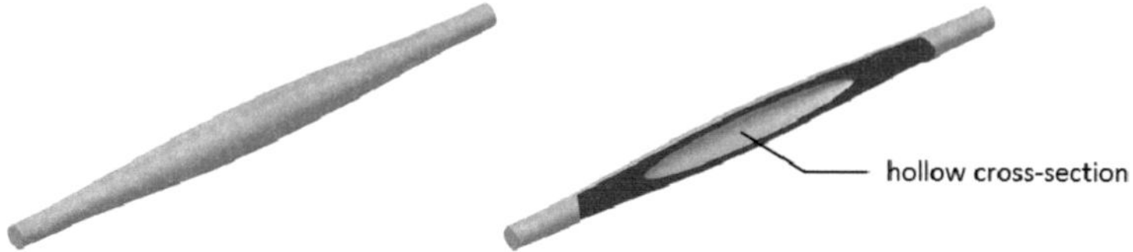

Fig. 1   Schematic sketch of a shape optimized strut

2.1.2   Experimental programme

In order to be able to describe the time dependent nonlinear behaviour of the highly stressed confined ends of the strut, creep test are carried out, figure 2. Due to the fact that the creep deformations of concretes are influenced by various parameters, such as the age of loading, the specimen size and the stress level, the relevant parameters with regard to this research project are varied in the experimental programme, table 1. In order to investigate the influence of longitudinal reinforcement and confinement on the creep deformation, some tests with confinement and/or reinforcement are carried out. In addition to that, shrinkage tests with the same specimen diameters are performed in order to determine the creep deformations out of the total deformations.

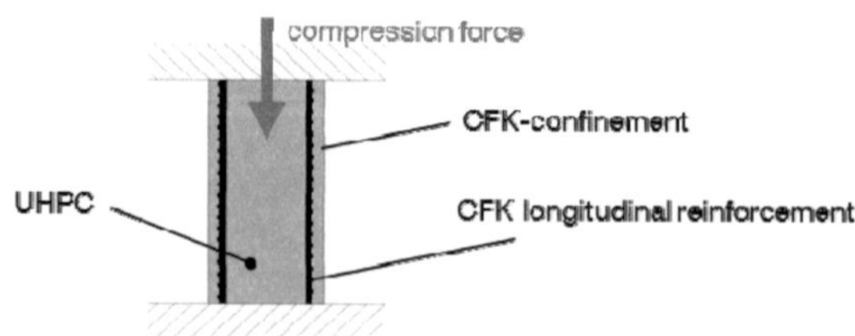

Fig. 2   Sketch of a creep test

Table 1   Creep experiments

Diameter [mm]	Length [mm]	Stress level	Longitudinal reinforcement	Confinement
75	225	0,30 / 0,60 / 0,70	yes / no	yes / no
100	300	0,30 / 0,60 / 0,70	yes / no	yes / no

Apart from these tests of particular details, component tests on struts are carried out to determine all influences of the manufacturing process and their effects on the structural behaviour. So far the scheme of the experimental programme includes the variation of the geometry of the specimens, of the ratio of moment to normal force and of the reinforcement arrangement, figure 3 and table 2.

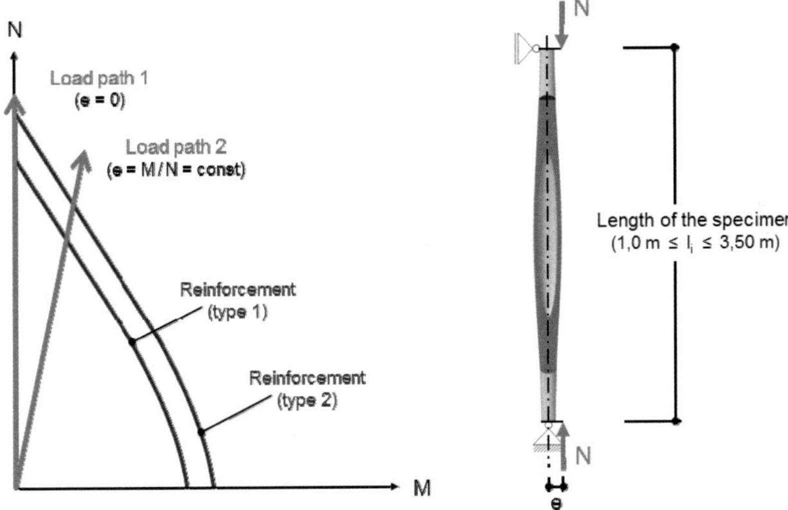

Fig. 3   Scheme of an interaction diagram and sketch of the component tests

Table 2   Overview of the planned component test on struts

Planned eccentricity (e = M/N)	Type of strut	Length of the strut	Reinforcement
e = 0	S / H	l_1 / l_2 / l_3	type 1 / type 2
e > 0	S / H	l_1 / l_2 / l_3	type 1 / type 2

It is planned to test three different strut lengths between 1,0 m and 3,5 m. The exact lengths and the exact cross section geometries that are finally tested are determined on the basis of the shape optimizing analyses mentioned in 2.1.1 and according to boundary conditions given due to manufacturing or testing. Comparatively short struts are presumably optimized and tested as rods with a solid cross-section (S), whereas for longer struts a middle part with a hollow cross-section (H) is probably considered. In order to verify the influence of the reinforcement on the loading capacity, the component tests are carried out with two different reinforcement concepts (type 1 and type 2).

2.2 Tie

The main objective concerning the tie is to develop a truss member, which remains uncracked in the serviceability limit state in order to prevent high deformations of the total structure due to a lack of stiffness of the components. Therefore the tie is pretensioned by single CFK-rods. The necessary amount of tendons and the dimensions of the tie are determined by theoretical investigations. In order to minimize the dimensions of the node, the dispersion lengths of the CFK-tendons should be as short as possible. The bond behaviour of these tendons mainly depends on the diameter and the surface conditions of the tendon. After a preselection of suitable CFK-tendons, pull-out tests according to Janovic [2] are carried out that, in contrast to usual pull-out tests, allow an investigation of the demolition effect of the reinforcement. With regard to different thermal expansion coefficients of the concrete and the tendons, some of the pull-out tests are performed with an additional temperature stress, table 3.

Table 3   Overview of the planned pull-out tests

Type of the CFK-tendon	Concrete cover	Temperature
A1 / A2 / A3	c_1	T_0
	c_2	$T_0 / T_1 / T_2$
	c_3	T_0

Based on the results of the pull-out tests a suitable type of CFK-tendon and a suitable concrete cover is chosen. Afterwards additional bond tests according to the DIBT-guideline [3] are carried out in order to determine the transmission length, table 4.

Table 4   Overview of the planned tests for the determination of the transmission length

Type of the CFK-tendon	Concrete cover	Temperature
B	c	$T_0 / T_1 / T_2$

Besides, fundamental investigations on the anchorage possibilities of the ties are performed. Figure 4 shows a scheme of the planned anchorage tests. Apart from the already mentioned multiaxial state of stress inside of the node, which is ensured by an expansive high-performance mortar combined with a spatial, stiff confinement, the surface of the tie is profiled for an improved transfer of the tensile force.

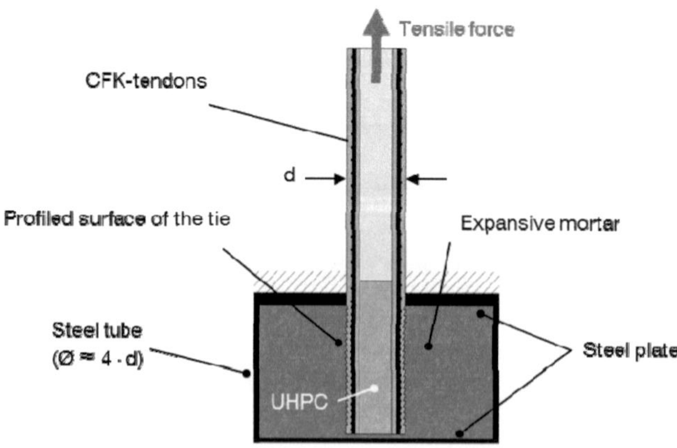

Fig. 4   Scheme of the anchorage tests

Initially the anchorage tests will be carried out with two different surface profiles (C1 and C2), which are determined by theoretical considerations. Based on the results from these test additional test are performed using the more efficient surface profile (C), table 5. By variation of the temperature, the level of lateral compression and the specimen diameter, the influence of these parameters on the anchorage behaviour is investigated.

Table 5   Overview of the planned anchorage tests

Surface profile	Specimen diameter	Temperature	Lateral compression
C1	d_1	T_0	p_1
C2	d_1	T_0	p_1
C	d_1 / d_2	$T_0 / T_1 / T_2$	p_1 / p_2

In the ultimate limit state the tie will have cracks transverse to the rod axis and therefore in certain cross-sections the complete tensile force has to be absorbed by the tendons. In order to investigate the nonlinear behaviour of the cracked ties, component tests are carried out and the stress-strain-diagram is determined. It is planned to vary the arrangement of the tendons, the diameter of the specimens and the temperature, table 6.

Table 6   Component tests on ties

Tendon arrangement	Diameter of the tie	Temperature
D_1	d_1 / d_2	$T_0 / T_1 / T_2$
D_2	d_1 / d_2	$T_0 / T_1 / T_2$

3   Acknowledgement

The author gratefully acknowledges the financial support provided by the Deutsche Forschungsgemeinschaft (DFG) within the priority programme "Concrete light. Future concrete structures using bionic, mathematical and engineering formfinding principles".

References

[1]   Speck, K.: Beton unter mehraxialer Beanspruchung – Ein Materialgesetz für Hochleistungsbetone unter Kurzzeitbelastung. Dissertation, Technische Universität Dresden, 2008

[2]   Janovic, K.: Bericht über den neuen konsolenförmigen Ausziehkörper als Vorschlag für ein allgemeingültiges Verbundprüfverfahren. Bericht Nr. 1349 vom 20.11.1979 des Lehrstuhls für Massivbau, Technische Universität München, 1979

[3]   Deutsches Institut für Bautechnik: Richtlinien für die Prüfung von Spannstählen auf ihre Eignung zur Verankerung durch sofortigen Verbund – Stand Juni 1980. Mitteilungen des Instituts für Bautechnik 6/1980

Pilot tests with lightweight woodchip concrete in composite slab constructions

Andreas May

Faculté des Sciences, de la Technologie et de la Communication
Université du Luxembourg
6, rue Richard Coudenhove-Kalergi, L-1359 Luxembourg, Luxembourg
Supervisor: Daniéle Waldmann

Abstract

Lightweight concrete based on woodchips is a material with several characteristics which can be used on one site for load carrying structures when dense lightweight concrete matrixes are required and on the other side porous lightweight concrete matrixes can be realized for mechanical requests. Lightweight woodchip concrete for bearing constructions can make a contribution to reduce the construction's weight. Sufficient compression strength but a reduced elasticity modulus of this concrete requires an application in composite constructions. Therefore an interesting aspect could be an application in composite slabs. In the context of a research project at the University of Luxembourg composite slabs with lightweight woodchip concrete are investigated. The results of the first static tests on small elements with various element heights and sheet thicknesses are presented.

1 Introduction

Lightweight concrete based on renewable products in constructive elements is able to reduce the self-weight of the construction. Although adequate compression strength of the woodchip concrete can be realized, the elastic modulus is, compared to other lightweight concrete with similar density class, lower, due to the renewable aggregates themselves. According to this, the renewable aggregates must be considered as imperfections in the concrete matrix. Thus, for designing constructive elements a composite partner becomes necessary. A relative young researcher's field in the last years is the use of lightweight concrete in composite slabs [1]. New in this context is the application of a lightweight woodchip concrete. For this purpose, small composite slabs are made by using composite sheets with a re-entrant profile and additional embossments to increase the composite actions. These small elements are tested to obtain first information about the composite behaviour, the composite actions, the slippage, the deflections and the load bearing behaviour. These first results are used for the planning of representative composite slabs related to EC4 [2].

2 Present research project

A given lightweight woodchip concrete should be implemented in constructive elements, and finally a parameter study will be applied via the finite element method. A number of pilot tests with different mixtures are realized to determine an adequate slab system. The present used composite sheet is a re-entrant profile with additional embossments. In this paper an interpretation of the results of small pilot tests will be presented.

2.1 Dimensions of the elements and varied parameters

The chosen composite sheet is kept constant for all elements while the sheet thickness and the height of the slabs are varied. The width and the lengths of the elements are constant as well as the span, see hereunto table 1. Additionally a required reinforcement related to EC 4 and EC 2 [3] is embedded. The variations of the parameters can be identified in the elements name. For example the slab name "4-120-1.25" describes: "4" the slab's number, "120" the height in mm and "1.25" the sheet thickness in mm. In the following sections the presented names are used and in the context of a better identification in the following figures, the different sheet thicknesses are coloured in black (1.25 mm) and grey (1.0 mm).

Proc. of the 9th *fib* International PhD Symposium in Civil Engineering, July 22 to 25, 2012,
Karlsruhe Institute of Technology (KIT), Germany, H. S. Müller, M. Haist, F. Acosta (Eds.),
KIT Scientific Publishing, Karlsruhe, Germany, ISBN 978-3-86644-858-2

Table 1   Dimensions of the small slabs and sheet thickness

height [mm]	length [mm]	width [mm]	span [mm]	composite sheet SHR 51/150/	sheet thickness [mm]
120	1100	450	900		1.0/ 1.25
160	1100	450	900	© Montana AG	1.0/ 1.25
200	1100	450	900	2011	1.0/ 1.25

2.2   Experimental set-up and measuring equipment

The small composite elements are tested by a three point bending test via displacement control. In figure 1 the equipment is shown in a lateral view (left) and a top view (right). The deflections are measured via displacement transducers at three points. At the third points of the span, onto the concrete side, two measuring points are installed, and finally at the steel sheet side in the middle point of length. The strains are measured by strain gauges on the concrete and sheet surfaces. The relative displacements between the concrete and the sheet are measured above the sheet profile at the lateral

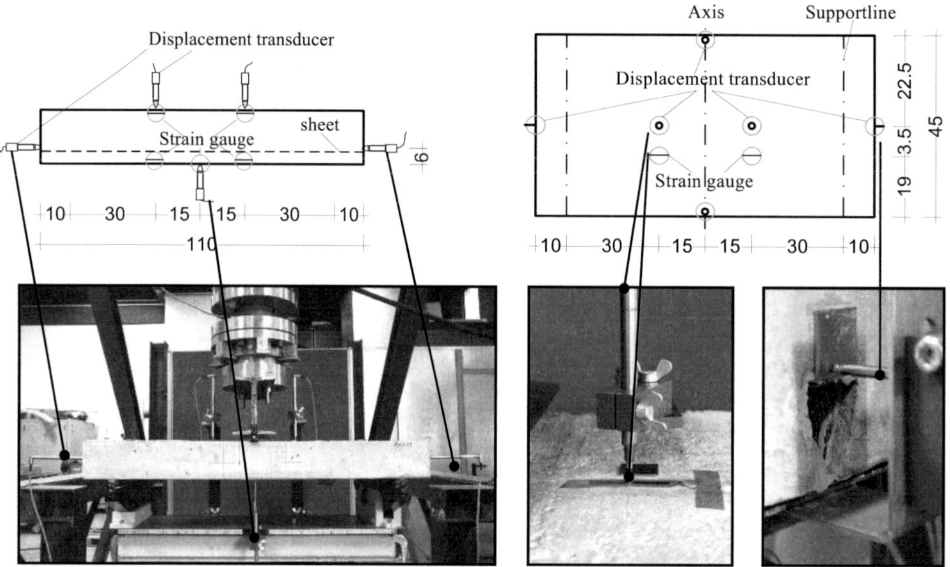

Fig. 1   (left) Lateral view, (right) Top view of the measuring equipment

2.3   Material characteristics

The material characteristics are defined according to DIN 12390 [4] and DIN 1048 [5], thereby the mixture should be conform to the requirements related to EC 4. In table 2 the material parameters of the here used concrete are listed, therein the minimum compression strength is not achieved. EC 4 as well as the accreditation requires minimal compression strength of LC20/22. The compacting of a lightweight concrete must be done in a careful manner [6]. At the one hand the freshly mixed concrete needs due to a low density more compacting energy or compacting time, but on the other hand local demixing occurs faster due to more compacting energy respectively compacting time. The compacting of the elements and the specimen in the concrete forms are different. Thus, to compare the compression strength of the default samples with the present strength of the elements, some core samples according to DIN 12504 [7] are tested.

Table 2   Material characteristics

Compression strength, cylinder, 28d	13.1 [N/mm²]	Compression strength, core sample, d> 90d	13.3[N/mm²]
Splitting tensile strength, 28d	1.55 [N/mm²]	Bending tensile strength, 28d	2.8 [N/mm²]
Elastic Modulus, 28d	5200 [N/mm²]	Density class, oven-dry	1,2

3   Present test results

A representative load-displacement diagram and the corresponding load-slippage diagram of two different tests are shown in figure 2. The elements were realized with a height of 200 mm. The black line represents a sheet thickness of 1.25 mm, while the grey line shows the behaviour for a sheet thickness of 1.0 mm. In both diagrams the y-axis describes the forces and the x-axis the deflection in the middle of the slab for the load displacement diagram as well as the relative displacement between the concrete and the steel sheet. For the bond behaviour diagram the influence of the height of the slab as well as the influence of the sheet thickness can be discussed. The ultimate load is higher when using a sheet with a thickness of 1.25 mm, but the beginning of slippage occurs earlier than for a sheet thickness of 1.0 mm, and this is representative for all conducted tests. All in all a ductile material behaviour can be observed by using a composite sheet with a re-entrant profile and additional embossments [8], see also section 2.2. By regarding the bearing load and the slippage load, the bending behaviour is nearly the same for these presented slabs. Furthermore, in figure 2 (left) the influence of the neoprene support can be detected. Therein, the origin of the disturbance of the linear curvature (dashed line) can be explained with the application of an elastic neoprene stripe.

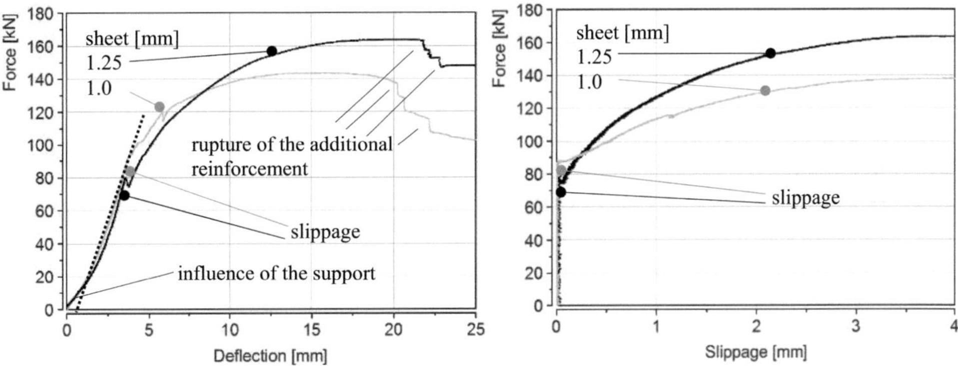

Fig. 2   (left) Load displacement-diagram, (right) Bond behaviour

3.1   Analysis of the results

The results of the load bearing behaviour of all conducted tests are resumed in figure 3 (right), and therein the y-axis of the bar diagram represents the loading and the x-axis represents the different heights of the slabs. Furthermore, the different coloured bars represent the results for different sheet thicknesses. In this diagram it can be observed, that independent of the heights of the slabs, the bearing load is increasing with the sheet thickness. In figure 3 (right), the y-axis represents the load at the beginning of slippage while the x-axis also describes the height of the slabs. In this illustration it can be seen that the slippage forces, for all varied element heights, are higher by using the thinner steel sheet of 1.0 mm.

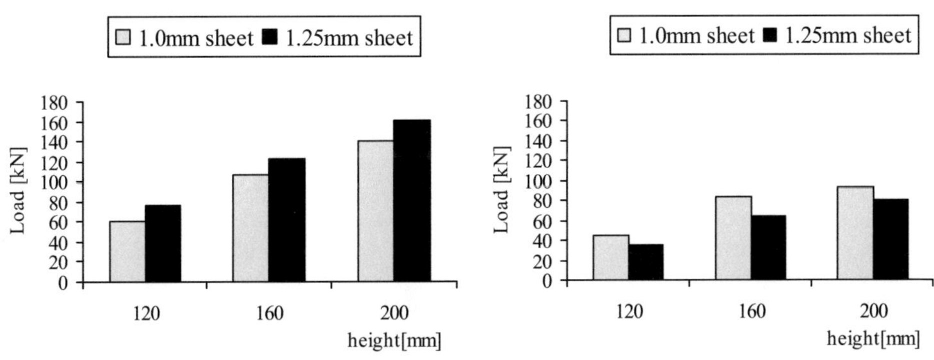

Fig. 3   (left) Load-bearing behaviour, (right) Load by beginning of slippage

3.2 Ductility criterion

The ductility criterion related to EC 4 is fulfilled, when the loading after slippage, until the limit load $F_{,max}$, can be increased about 10 percent: $F_{max} \geq 1.1*F_s$. Thereby the slippage is defined with a relative displacement about 0.1mm. Although the dimensions of the elements are not comparable with slabs and the weak material characteristics are not conform to the standards, the ductility criterion is fulfilled. In figure 4 the loading increment, from start of slippage till ultimate load, is shown. The ductility criterion is fulfilled with an adequate reserve. The increase of the ultimate load due to an increase of the steel thickness as well as the decrease of the slippage load due to an increased sheet thickness can be observed in figure 3.

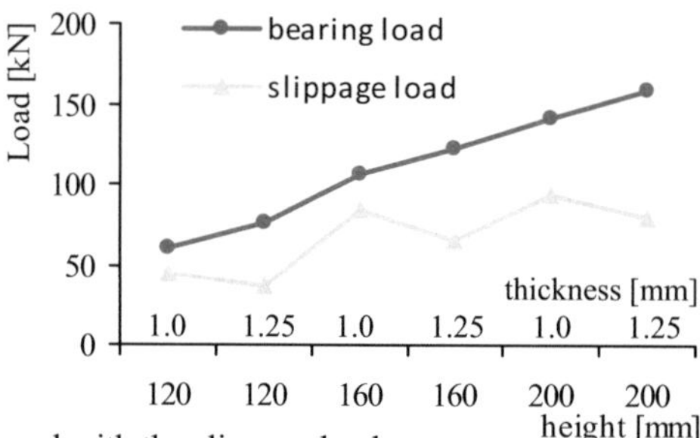

Fig. 4   Limit load compared with the slippage load

3.3 Interpretation of the results

To interpret the results, the composite action will be split into the mechanical and the frictional bond. First only the mechanical bond will be discussed. By reaching the slippage force, relative displacements between the concrete and the sheet occur. The embossments can be regarded as barrier for the concrete which must be first overcome. In the contact zone of the embossment and the concrete, additional forces are developed, see figure 6 (left). Relative displacements leads to two possible reactions: The concrete resists and the embossments get deformed, or the concrete get deformed while the embossments remain intact. The last case can occur when the compression strength is lower than a concrete C12/15 [9]. Although a LC12/13 is present, in these tests the concrete get deformed what can be explained by the low elasticity module. In figure 5 the relative displacement between the concrete and the steel sheet is shown. The left figure shows the intact embossments while the right figure pictured the deformed concrete surface at the contact zone.

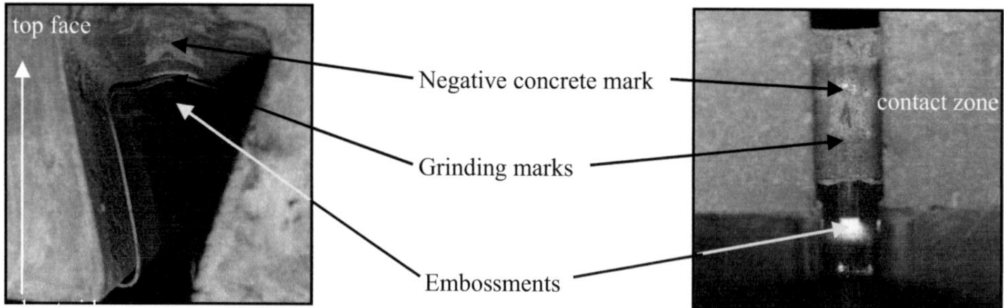

Fig. 5   (left) Intact sheet embossments due to reduced elasticity module of the lightweight concrete formulation , (right) Grinding marks at the contact zone on the concrete site.

The results in section 2.1 show that the forces, when slippage is activated, are higher for a sheet thickness of 1.0 mm as for a sheet thickness of 1.25 mm, see figure 3 (right). Therefore, with the present results it can be interpreted, that an increase of the sheet thickness does not lead necessarily to an increase of the mechanical bond due to the embossments. Consequently, the origin lies in the frictional bond. The frictional bond occurs when the transverse strain of the steel sheet is activated during the loading. Then, the re-entrant form get deformed and the concrete between the steel is clamped, see hereunto figure 6 (left). An adequate frictional bond due to frictional forces can only be obtained, when the sheet stiffness is elastic enough and when the stiffness of the concrete is high enough [10]. The elastic module of the applied concrete is very low and in combination with a sheet thickness of 1.25 mm, the stiffness of the steel sheet will be increased. Consequential, adequate frictional forces due to transverse strains between the sheet walls are not able to be developed in an adequate manner. Therewith the frictional bond became the crucial point to increase the slippage load.

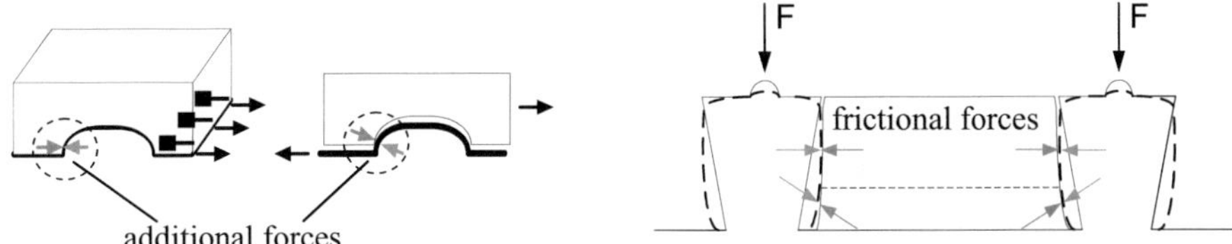

Fig. 6   (left) Additional forces in the contact zone, (right) Frictional forces due to transverse strains resulting of the deformation of the sheet

4   Crack pattern due to shrinking effects and loading

Before the elements are tested crack pattern due to shrinking could be observed at nearly all elements. The corresponding height of these cracks had a maximal length of two third of the element height. Only these cracks opened while loading, hence the shrinking cracks have determined the crack pattern in this pilot tests. In figure 7 a representative load displacement diagram with the corresponding crack pattern is shown. While the cracks width increased the crack length also increased. In accordance with [1] only one crack pattern occurred after reaching the ultimate load and it is related to the longitudinal cracks at the elements with a height of 120 mm. These cracks also started at the top of the re-entrant steel sheet and continued up to the top of the element, see figure 8. No longitudinal cracks are developed at the elements with a height of 160 mm or 200 mm.

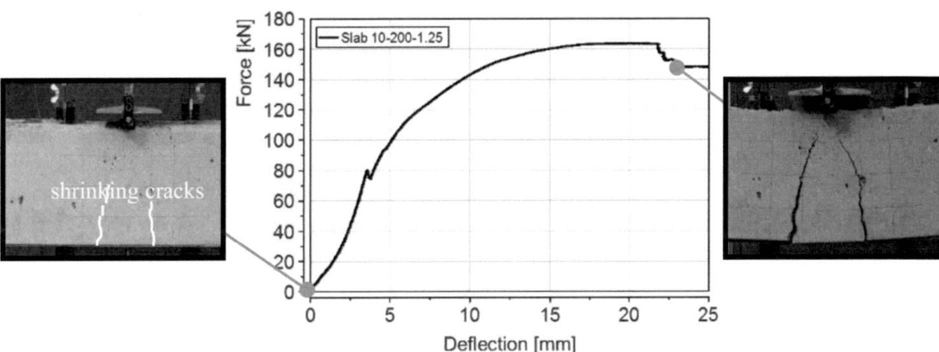

Fig. 7   Crack pattern due to shrinking and crack evaluation due to loading

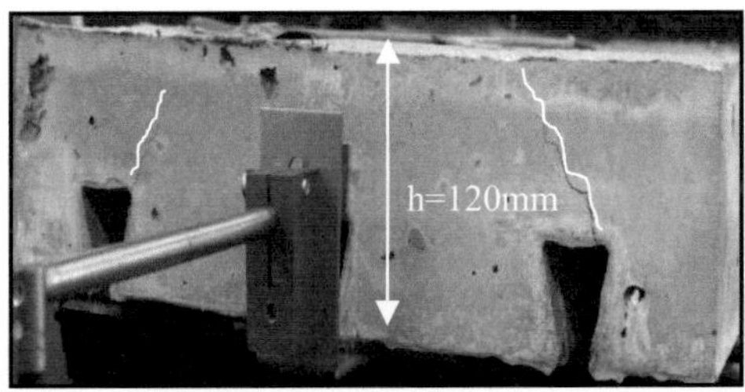

Fig. 8   (left) Beginning of the longitudinal cracks at the top of the steel sheet, (right) To the top continued longitudinal cracks

The pilot tests are done via displacement control and after reaching the tensile strength of the concrete, the steel sheet is directly activated. However, no influence of the concrete's crack forces can be obtained in the curvature of the load-time diagram. In figure 9 the corresponding load-time diagrams of the slabs with a height of 120, 160 and 200 mm are shown and a continuous curvature until the beginning of slippage can be seen.

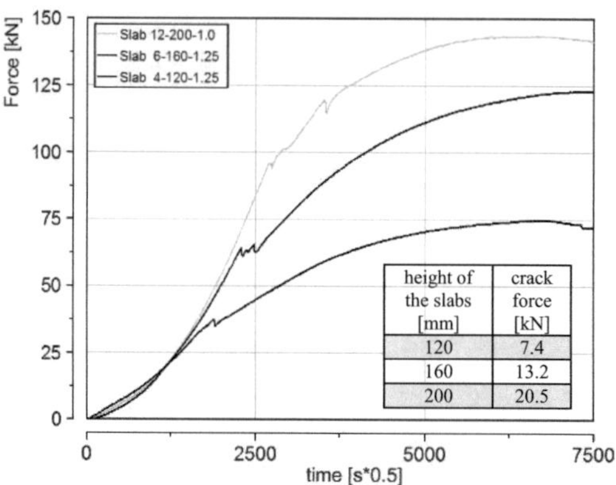

height of the slabs [mm]	crack force [kN]
120	7.4
160	13.2
200	20.5

Fig. 9   Load-time diagram of the slabs with all present heights and the corresponding crack forces

5   Conclusions

By regarding the load bearing behavior and the composite action in these pilot tests, an application of lightweight woodchip concrete in composite slabs seems to be possible; however an improvement of the mixture to obtain more useful material characteristics becomes necessary. The influence of the height of the elements as well as the influence of the sheet thickness could be demonstrated in these test series. Recapitulatory, an increase of the ultimate load is caused with an increase of the sheet thickness as well as the height of the elements. Furthermore, due to an increase of the elements height, the beginning of slippage can be delayed. Though, the beginning of slippage has occurred earlier by using a thicker steel sheet due to the low elastic module of the lightweight woodchip concrete. This research results should be taken into account when designing a woodchip lightweight concrete in composite slabs, so the requirements related to the serviceability state due to slippage and deflection, can be a crucial point. A test series related to EN 1994-1, Annex B will start at the University of Luxembourg soon. With the standardised test set-up the design value of the longitudinal shear resistance will be determined [11].

References

[1]   Kurz, W., Mechterine V., Brüdern, A., Hartmeyer, S., Kessler, C.: Leicht Bauen mit Verbunddecken im Wohnungs- und Gewerbebau. In: Fraunhofer IRB Verlag (2009)

[2]   DIN EN 1994-1, Eurocode 4: Bemessung und Konstruktion von Verbundtragwerken aus Stahl und Beton - Teil 1-1: Allgemeine Bemessungsregeln und Anwendungsregeln für den Hochbau. In: Beuth-Verlag (2004)

[3]   DIN EN 1992-1, Eurocode 2: Bemessung und Konstruktion von Stahlbeton- und Spannbetontragwerken - Teil 1-1: Allgemeine Bemessungsregeln und Regeln für den Hochbau. In: Beuth-Verlag (2011)

[4]   DIN EN 12390-1-7, Prüfung von Festbeton. In: Beuth-Verlag (2001)

[5]   DIN 1048-5, Prüfverfahren für Beton: Festbeton, gesondert hergestellte Probekörper. In: Beuth-Verlag (1991)

[6]   Faust, T.: Leichtbeton im Konstruktiven Ingenieurbau. In: Ernst & Sohn (2002)

[7]   DIN EN 12504-1, Prüfung von Beton in Bauwerken, Teil 1: Bohrkernproben- Herstellung, Untersuchung und Prüfung unter Druck. In: Beuth-Verlag (2000)

[8]   Bode H. ,Sauerborn, I.: Zur Bemessung von Verbunddecken nach der Teilverbundtheorie. In: Stahlbau 61, Heft 8, pp 241-250 (1992)

[9]   Kuhlmann, U.: Stahlbau Kalender- Verbundbau. In: Ernst & Sohn (2010)

[10]   König, G. ,Faust, T.: Konstruktiver Leichtbeton im Verbundbau. In: Stahlbau 69, Heft 7, pp 528-533 (2000)

[11]   Minnert ,J.,Wagenknecht, G.:Verbundbau-Praxis, Berechnung und Konstruktion", In: Bauwerk (2008)

Fresh concrete pressure on inclined or curved formwork

Björn Freund

Institut für Massivbau,
Technische Universität Darmstadt,
Petersenstraße 12, 64287 Darmstadt, Germany
Supervisor: Carl-Alexander Graubner

Abstract

Architects increasingly use free formed concrete units as design elements. Lightweight shell structures, for instance Florante Submarino in Valencia, are built up of a thin concrete membrane with variable inclinations. The pressure of fresh concrete is the most important load for the design of the formwork and has important effects on the construction costs. In Germany the calculation of the pressure on formwork has to be determined according to DIN 18218 [1]. The standard indicates rules for vertical forms and formwork with a maximum deviation of 5 degree to the vertical axis. Inclined formworks are analyzed and designed according to simplified models, which normally lead to an oversize of the formwork construction.

This paper deals with the pressure of fresh concrete on inclined and curved formworks. The related research project "Entwicklung der Grundlagen zur Entwicklung adaptiver Schalungssysteme für frei geformte Betonbauteile" (Basics for the development of adaptive formwork systems for free-formed concrete structures) belongs to the DFG Schwerpunktprogramm (SPP) 1542 (Priority Programme 1542 of the German Research Foundation): „Leicht bauen mit Beton" (Light concrete structures). Extensive numeric investigations, based on a FE-model, are conducted. Further on the experimental setup for large-scale tests are shown, as well as the test program.

1   Introduction

Several engineering structures such as water-tanks as well as buildings with special architecture consist of inclined concrete units with variable inclinations. For such structures the pressure on formwork is often determined according to conservative methods, such as fluid pressure. Standards like German DIN 18218 [1] or the French CIB-CEB-FIB Bulletin d'Information N° 115 [2] do not give explicit information for the determination of pressure of fresh concrete for inclined concrete members. Ast and Fröhlich [3] propose a simplified model for the fresh-concrete-pressure on inclined formwork based on the design load of DIN 18218. In the American ACI 347-04 [4] and the British CIRIA Report 108 [5] special rules for the calculation of the uplift pressure on the inclined upper formwork are given. These regulations are based on the fluid pressure.

The present load-models for the determination of the design-pressure of fresh concrete by inclined or curved formwork are relatively conservative. Hence, the formwork systems are not economically designed or concreting is unnecessarily slow. Both reasons lead to relatively high construction costs of such structures. One main target of the DFG SPP 1542 is to develop construction methods that allow lightweight concrete structures to be built easier and more often. In the project the concrete pressure on inclined and curved formwork will be analyzed in details. The obtained results can be used to develop new systems for such shaped concrete members. Therefore, the results of this project will support a simplified construction process of lightweight and aesthetically shaped concrete structures.

In order to analyze fresh concrete pressure on formworks, currently a new numerical model is developed, based on the experiences of numerical modeling at the TU Darmstadt [6]. The already existing model is described in Chapter 2. The new model should consider the friction between the fresh concrete and the formwork by inclined and curved shapes. Another part of the research project includes full-scale tests on concrete walls with different inclinations. The pressures of the fresh concrete are measured with pressure sensors. The test procedure is described in Chapter 3. The results of the full-scale tests will be used to verify the numerical model. Further, extensive parametric studies are

Proc. of the 9th *fib* International PhD Symposium in Civil Engineering, July 22 to 25, 2012,
Karlsruhe Institute of Technology (KIT), Germany, H. S. Müller, M. Haist, F. Acosta (Eds.),
KIT Scientific Publishing, Karlsruhe, Germany, ISBN 978-3-86644-858-2

planned. Finally, a simplified but realistic design model to calculate the pressure of fresh concrete on inclined and curved formwork will be developed.

2   Numerical modeling and preliminary results

At the TU Darmstadt a two-dimensional numerical model was already developed based on the finite element method. In this model the fresh concrete is modeled with plane elements. The entire fresh concrete consists of 16 layers with 16 plane elements. In the simulated installation of a new layer element the material parameters of the older layer elements are adjusted according to the time respective casting rate. The material behavior of the fresh concrete is regarded according to Mohr-Coulomb theory. This corresponds to a soil mechanics model and considers the internal friction angle φ.

The different internal friction angles were calculated based on the ratio between the horizontal and vertical pressure measured in small-scale material tests and the approach of the active pressure in soil mechanics. The stiffness of the formwork is modeled with springs (k = 20,000 kN/m per m² of formwork).

Connected to the springs, a continuous beam is modeled which has the stiffness of a 3 cm thick wooden board. The contact between fresh concrete and formwork is modeled with nonlinear contact springs. The contact springs can transmit normal and shear stresses but no tension. Figure 1 schematically shows the numerical model.

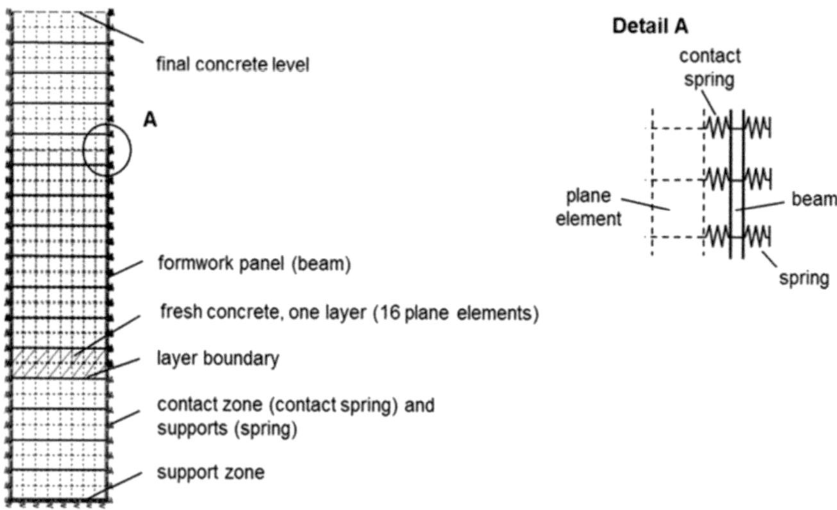

Fig. 1   Numerical model

The placing of the fresh concrete can be simulated in two different ways. The first possibility is the stepwise installation of all 16 plate layers. The material parameters will be adjusted according to the age of the single plane elements. The stress states from the previous calculation steps will be considered in the following step. The second way is the installation of all 16 layers at the same time. This case can be considered as a one-time step. Therefore, a stress state of a previous time step cannot be considered. The material parameters of the single plane elements are the same like the stepwise placing in the final state. The numerical model can calculate the pressure of fresh concrete on vertical formwork with and without friction between fresh concrete and formwork. Furthermore, the numerical model can calculate the pressure of fresh concrete on formworks which are inclined on one or both sides. In this case the simulation of friction between the fresh concrete and the formwork is not possible up to now.

The updated model, which is currently being developed should be able to calculate inclined and curved formwork and should also consider the friction between the fresh concrete and the formwork. Beyond, the model should be able to simulate dynamic impact on the material properties. Furthermore the influence of the casting from below on the formwork pressure will be analyzed including the influence of the reinforcement.

3   Large scale tests

For the large-scale tests two formwork systems will be built. Both formworks consist of two opposite formwork panels and have a width of 1.25 m. The thickness of the wall can be adjusted from 0.10 m to 0.50 m. One vertical formwork has a height of 3.50 m and the other formwork can be adjusted to different degrees of inclination. If the inclination angle α is 0°, the vertical height of the formwork is 5.00 m. If the inclination α is 45°, the vertical height of the formwork is approx. 3.50 m. Figure 2 shows the test formwork which can be used for different inclinations.

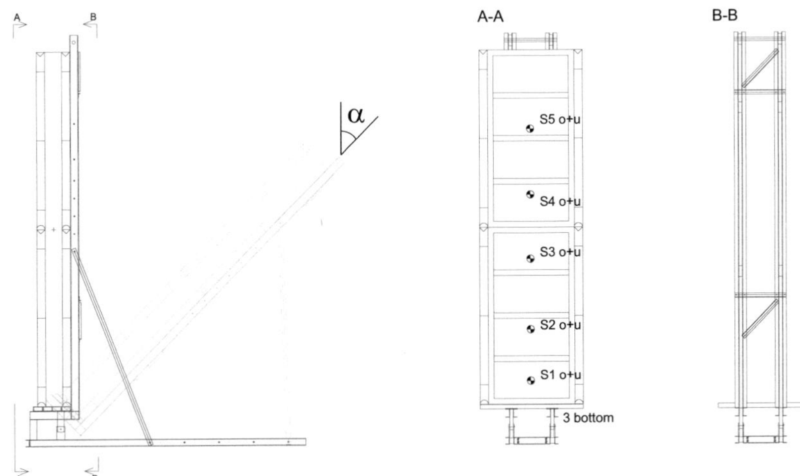

Fig. 2   Test stand of the adjustable formwork for different inclinations. From left side: (1) Side view, (2) Section A-A: Front view with formwork; (3) Section B-B: Back view without formwork

In order to measure the pressure distribution along the top and bottom panels of the formwork, 5 pressure sensors are placed on each side. The reducing effect of friction between concrete and formwork or reinforcement should be particularly investigated. Therefore, at the bottom of the formwork a pressure sensor will be placed. By subtracting the measured vertical load from the weight of the concrete, the friction between concrete and formwork panel or reinforcement can be determined. In all tests, the pressure is measured until the final setting time of the concrete occurs. The vibration of the fresh concrete changes the material properties significantly and can lead to an increase of the fresh concrete's pressure. This influence will be also investigated.

Table 1 shows the planned test series. Totally 10 test pairs are provided. The vertical formwork and the second inclined formwork are filled with concrete at the same time and with the same casting rate. The standard concrete has a consistency class F5 (flow-table-test) according to DIN EN 206-1. Investigated parameters are the following:

- Inclination angle
- Casting rate
- Thickness
- Final setting time
- Consistency class
- Reinforcement

The tests will be conducted in the second and third quarter of 2012. In addition to the large-scale tests, the following properties of fresh concrete will be determined:

- Spread in flow-table test,
- Final setting time according to Vicat, setting-bag test and with an ultrasonic-meter
- Relative yield stress and viscosity with rheometer BT2

Table 1   Overview of large-scale tests

Test No.	Consis-tency class	Casting rate v	Inclination Wall 1 / 2	Reinforce-ment	Thickness d	Final set-ting time t_E
Dim.	[-]	[m/h]	[°]	[-]	[cm]	[h]
1	F5	2	0/45	Yes	20	7
2	F5	2	0/22,5	Yes	20	7
3	F5	4	0/22,5	Yes	20	7
4	F5	4	0/45	Yes	20	7
5	SCC	2	0/45	Yes	20	10
6	F3	2	0/45	Yes	20	7
7	F5	2	0/45	No	20	7
8	F5	2	0/55	Yes	20	7
9	F5	2	0/45	Yes	10	7
10	F5	2	0/45	Yes	20	10

4   Conclusions and Outlook

Within the research project a realistic numeric model for the calculation of the fresh concrete's pressure on formwork will be developed. Parallel to the numerical analyses, the test setup and the program for the full-scale test is defined. The results of the full-scale tests will be used to verify the numerical model. Based on intensive parametric studies, a realistic analytical model for the calculation of the pressure of fresh concrete on inclined and curved formworks will be elaborated and a proposal for a simplified design model proposed.

References

[1]   DIN 18218:2010-01: Frischbetondruck auf lotrechte Schalungen. Beuth Verlag, Berlin 2010.

[2]   CIB-CEB-FIP Bulletin d'Information No 115: Manuel de technologie "coffrage". Paris 1977.

[3]   Ast, G.; Fröhlich, K.-C.: Seitenwanddruck bei geneigten Schalungen. Holzbau-Taschenbuch, Band 1, 9. Auflage. Ernst&Sohn Verlag, Berlin 1996, Seiten: 875 – 921

[4]   ACI 347-04: Guide to Formwork for Concrete. ACI – American Concrete Institute, Farmington Hills 2004.

[5]   CIRIA Report 108: Concrete pressure on formwork. CIRIA – Construction Industry Research and Information Association, London 1985.

[6]   Proske, T.: Frischbetondruck bei Verwendung von Selbstverdichtendem Beton. Dissertation, Institut für Massivbau der TU Darmstadt, Fachgebiet Massivbau, Prof. Dr.-Ing. C.-A. Graubner, Darmstadt 2007.

Adhesive joints for structural elements of high performance concrete (HPC)

Sebastian Oster[1], Inga Shklyar[2], Heiko Andrä[2]

Institute of Concrete Structures and Structural Engineering,
Technische Universität Kaiserslautern (1),
Paul-Ehrlich-Straße, Building 14, 67663 Kaiserslautern, Germany
Supervisor: Christian Kohlmeyer

Fraunhofer ITWM (2),
Fraunhofer-Platz 1, 67663 Kaiserslautern, Germany

Abstract

As a first step of the research project „Design and Optimization of Adhesive Joints for Plate and Sheet Structures of High Performance Concrete", tension tests on micro reinforced specimens of HPC with different joint geometries were carried out. In addition reference tests with specimens without a joint were tested. As one possible manufacturing technique to realize the joint geometry, the water jet cutting technique was used. A reactive powder concrete (RPC) adhesive was used to glue the different parts together. In all tests the load bearing capacity is only limited by the tensile strength and the cross section of the reinforcement. The joints affect the load bearing capacity only in the way that the cross section of the reinforcement is reduced but not because of the low bond tensile strength of the adhesive. Furthermore numerical calculations were carried out to investigate the stress distribution.

1   Introduction

„Form follows Force" is the motto of the „Schwerpunktprogramm 1542" called „Concrete Light. Future Concrete Structures Using Bionic, Mathematical and Engineering Formfinding Principles". One aim, among other things, is the finding of constructions which are as filigree and material-saving as possible. To achieve this aim, shell structures are particularly well suited. Because of the reason that these structures can not be cast in one construction step, inevitably the question arises: how can single members be joined to an entire structure. As a possible solution of this problem, the use of adhesive joints as a method of a continuous connection can be applied. The drawback of adhesive joints of structural concrete elements is the relatively low adhesive strength of the near-surface concrete layers, assuming that the strength of the adhesive is sufficient. Because of this reason, the overall aim of this research project „Design and Optimization of Adhesive Joints for Plate and Sheet Structures of High Performance Concrete" is the optimization of the adhesive joint geometry, in a way that the tensile stresses caused by the applied loads are reduced to a minimum. This optimization will be carried out in co-operation with the Fraunhofer ITWM (Institute for Industrial Mathematics). Depending on the different load cases, the joint geometry will be optimized by using a shape and topology optimization method with regard to the resulting stresses.

As first experimental investigations, tension tests on specimens with three different joint geometries (see Fig. 1) were carried out. The specimens were reinforced with micro steel meshes [1] and the joint geometry was cut out with the water jet cutting technique (see Fig. 2). First numerical calculations were carried out by the Fraunhofer ITWM to investigate the stress and strain distribution of these initial experiments.

2   Specimens and test set-up

All specimens for the tension tests were cast in one piece with a thickness of 30 mm, a width of 144 mm and a length of 600 mm. All specimens were reinforced with ten layers of micro steel meshes (see Fig. 3) with a wire diameter of 1.0 mm and a grid spacing of 12.6 mm.

Proc. of the 9th *fib* International PhD Symposium in Civil Engineering, July 22 to 25, 2012,
Karlsruhe Institute of Technology (KIT), Germany, H. S. Müller, M. Haist, F. Acosta (Eds.),
KIT Scientific Publishing, Karlsruhe, Germany, ISBN 978-3-86644-858-2

Fig. 1   The three different joint geometries, a) geometry A: rectangular teeth, b) geometry B: triangular teeth, c) geometry C: undercut geometry with pre-wetted surface

For the manufacturing of the specimens, a fine-grained, self-compacting high performance concrete (HPC) was used (see Table 1). The mechanic properties of the concrete at 28 days were as follows: mean compressive strength $f_{cm,cube} = 140$ N/mm², mean elastic modulus $E_{cm} = 39,500$ N/mm² and mean flexural tensile strength $f_{ctm,fl} = 12.5$ N/mm² (equivalent bending tensile strength $f_{ctm} = 6.4$ N/mm² [4]).

Fig. 2   Detail of the surface of the teeth cut with the water jet technique

Fig. 3   Formwork with micro steel mesh reinforcement

Five days after concreting , the joint geometry was cut into the specimens by ultra-high pressure water jet technology. In this way, two parts which fitted well together were obtained. The concrete and the steel wires were both cut by water jetting. However the water jet seemed to be deflected by the wires. This can be noticed by the slightly widened cut at the bottom side (see Fig 2). For the dimensions of the joint geometry see Fig. 5. In order to compare three reference specimens were cast. The reference specimens were not cut into two pieces.

The two parts of the specimens with joint were glued together again with a reactive powder concrete (RPC) adhesive. The composition of the used cement based adhesive is similar to the composition developed by Zilch [2], [3] (see Table 2). Before applying the adhesive on the joint the concrete surface was pre-wetted. At all specimens, the thickness of the adhesive layer was 3 mm. The mechanic properties of the adhesive at 28 days were as follows: mean compressive strength $f_{cm,cube} = 120$ N/mm², and mean flexural tensile strength $f_{ctm,fl} = 10.0$ N/mm² (equivalent bending tensile strength $f_{ctm} = 5.2$ N/mm² [4]).

Table 1   Composition of HPC

Raw material	[kg/m³]
cement CEM I 42,5 R-SR	831,6
water	73,8
coarse sand ($d_{max} < 2$ mm)	694,0
fine sand ($d_{max} < 0.5$ mm)	223,9
quarz powder	205,4
silica suspension (incl. 50% water)	249,5
superplasticizer (PCE)	35,8

Table 2   Composition of RPC adhesive

Raw material	[kg/m³]
cement CEM I 42,5 R-SR	730,0
water	203,3
fine sand (0.06/0.2 mm)	995,8
quarz powder	262,8
silica dust	189,8
superplasticizer (PCE)	48,3

For the tension tests a test set-up was developed (see Fig 4). At all tests the displacement was measured with three or four displacement transducers (see Fig. 4 and 5). The measuring length of the displacement transducers for the calculation of the strains can be found in Fig. 5. In some of the tests, strain gauges (gauge length 10 mm) were placed on the teeth. All tests were carried out displacement controlled. The displacement rate up to the yield point was 0.1 mm/min. Afterwards it was 1.0 mm/min.

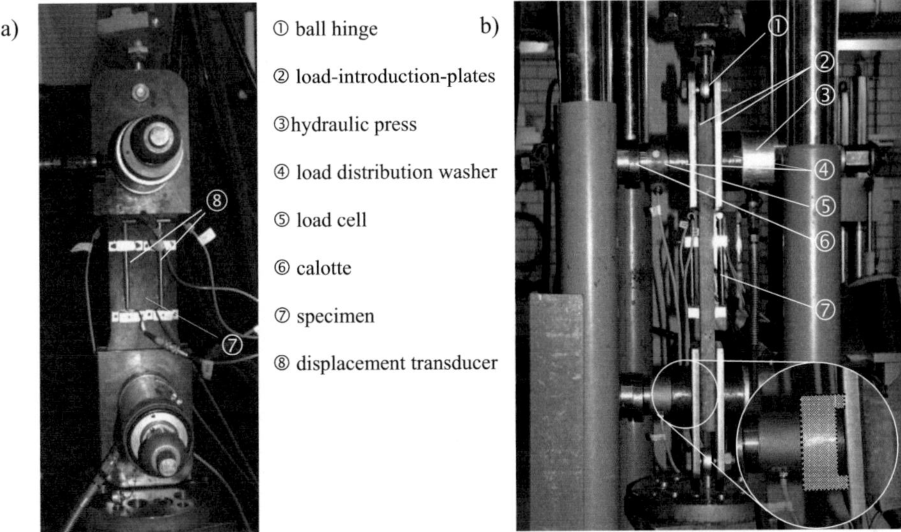

Fig. 4   a) Front view of the test set-up, b) Side view

3   Test programme

In this first experimental series tension tests on micro reinforced specimens with steel meshes were carried out. This is only one kind of reinforcement which is supported to investigate in this project. An alternative possibility is the use of thin steel bars. A well suited manufacturing technique was the water jet cutting method. Additionally, specimens with a cast joint geometry were produced. However, this method was more time-consuming. These specimens have not been tested so far. Three basic types of joint geometries (see Fig. 5) were tested: a rectangular geometry (A), a triangular geometry (B) and an undercut geometry (C). As reference tests, specimens without joints were tested.

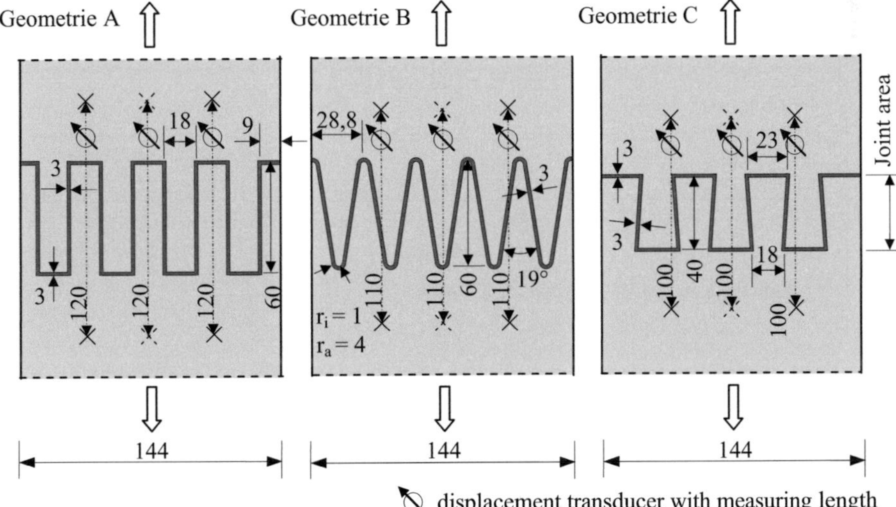

Fig. 5   Dimensions of the examined joint geometries in [mm] and location of the displacement transducers

4　　Test results

In Fig. 6 the load-strain curves of the tests are shown. The mean values of the displacements of the three or four displacement transducers were converted into strains.

The increase of load until the first crack was linear, for all tests. After the first crack, the stiffness decreases. In the reference tests the cracks developed along the whole length of the specimen. The crack spacing was nearly equal to the mesh spacing. At all specimens with joints, the first crack appeared perpendicular to the load at one end of the joint area (see Fig. 7; ①). The cracks run through the face surface of the teeth and the reinforced concrete cross section of the teeth. With increasing load, further cracks develop in the undisturbed area of the specimens. Within the joint area, no further cracks develop until the maximum load is reached. The maximum load in the reference tests was about 44 kN (see Table 3). The load is limited by the strength and cross section of the reinforcement. The load bearing capacity of all specimens with joints is limited by the reinforcement cross section at the ends of the joint areas, where the first cracks appear.

For example, the weakened cross section of the undercut geometry is about 25 % compared to the undisturbed area. Therefore, the load bearing capacity is also nearly 25 % of the reference tests. In the tests with the triangular and rectangular geometry, the load bearing capacity of the remaining reinforced cross section is about 55 % and 30 %, respectively.

Concluding the joint affects the load bearing capacity only in the way that the reinforced cross section is reduced. It is not reduced due to the low bond tensile strength of the adhesive.

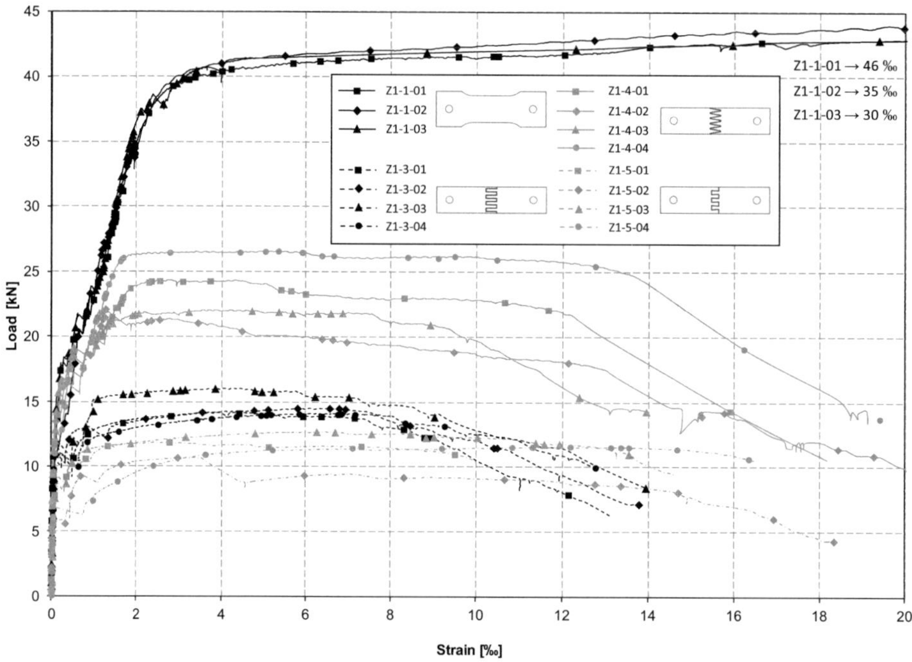

Fig. 6　　Load-strain curves of the tested specimens

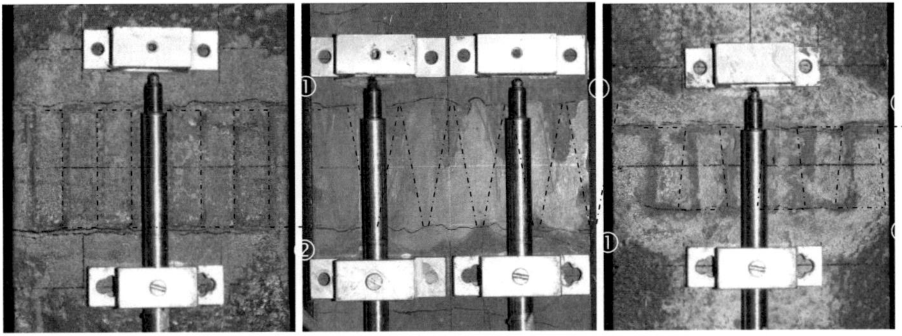

Fig. 7   Typical crack pattern at the maximum load of the different joint geometries (A-C)

Table 3   Maximum loads and tensile stresses at the first crack[2]

Specimen	Joint geometry [1]	Maximum load	Maximum load (mean value)	Tensile stress at first crack [2]	Tensile stress at first crack [2] (mean value)
		F_u [kN]	$F_{u,m}$ [kN]	$\sigma_{c,cr}$ [N/mm²]	$\sigma_{c,cr,m}$ [N/mm²]
Z1-1-01		44,98		4,17	
Z1-1-02	no joint	45,20	44,62	2,45	3,56
Z1-1-03		43,68		4,05	
Z1-3-01		14,11		2,72	
Z1-3-02	A	14,52	14,65	2,53	2,54
Z1-3-03		15,91		2,57	
Z1-3-04		14,06		2,35	
Z1-4-01		24,31		2,94	
Z1-4-02	B	21,31	23,54	2,91	2,84
Z1-4-03		22,01		2,74	
Z1-4-04		26,54		2,76	
Z1-5-01		11,71		1,95	
Z1-5-02	C	10,70	11,63	1,29	1,73
Z1-5-03		12,53		2,10	
Z1-5-04		11,57		1,58	

[1] see Fig. 5

[2] stresses calculated with real concrete cross section
series Z1-2 not yet tested

5   Numerical Investigations

Additionally to the tests, numerical calculations with the Finite-Element Software ABAQUS were undertaken by the Fraunhofer ITWM.

For the simulation a linear material model was used for the adhesive layer as well as for the HPC with micro steel meshes. As elastic modulus 34,000 N/mm² was used for the RPC adhesive. For the composite material consisting of micro steel meshes and concrete, effective orthotropic elastic moduli were calculated by an numerical homogenization method (FFT-Solver for the Lippmann-Schwinger-Equation). HPC and the reinforcement with the percentage of reinforcement according to the tests and an elastic modulus of 210,000 N/mm² were smeared. According to the tests an elastic modulus of 36,000 N/mm² for the concrete was used. The calculated effective isotropic elastic modulus is 39,000 N/mm².

The joint geometries with the dimensions according to the tests, were modelled by using symmetry conditions. The model consists of three-dimensional hexahedron finite elements with an edge length of 1 mm. In preliminary examinations also tetrahedron elements were used. The mesh size was varied from 1 mm to 3 mm and the results were compared.

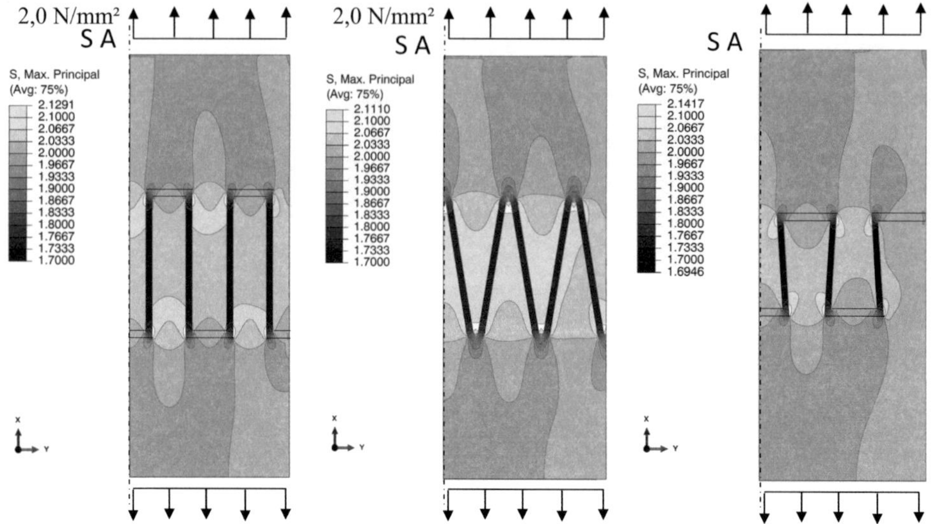

Fig. 8   Maximum principal stress distribution for the different joint geometries (A-C) at applied stress of 2 N/mm²

Fig. 8 shows the distribution of the maximum principal stress at an applied stress of 2 N/mm² for the different joint geometries (A-C).

Because of the linear material model and the small difference of the elastic moduli of the two materials, reinforced concrete and adhesive, there is only a small difference between highest and lowest stress value. The calculation shows that the stresses in the parts of the adhesive layer, which are parallel or nearly parallel to the load direction, are lower than in the rest of the specimen.

6   Conclusion and Outlook

The undertaken tension tests were the first step of the research project. The load bearing capacity of the tests with the joints is lower than in the tests without joints. It is only limited by the cross section of the reinforcement.

Due to this reason the reinforcement has to be improved in order to achieve a failure of the adhesive. One possibility is to replace the micro steel meshes with steel bars. In the research project the influence of the surface preparation of the joints is aimed to be investigated. Furthermore the manufacturing technique should be varied as well as other load cases (bending, shear force) should be observed. Additionally to the RPC adhesive, an epoxy resin adhesive should be investigated

The overall aim is the optimization of the adhesive joint geometry, in a way that the tensile stresses caused by the applied loads are reduced to a minimum. Finally design recommendations should be set up, based on the experimental and numerical investigations.

References

[1]   Hauser, S.; Wörner, J. D.: DUCON, ein innovativer Hochleistungsbeton. In: Beton- und Stahlbetonbau 94 (2), pp. 66-75, (3), pp. 141-145, 1999.

[2]   Mühlbauer C.; Zilch K.: Fügen von Bauteilen aus UHPC. In: Tagungsband zum 50. Forschungskoloooquium des Deutschen Ausschuss für Stahlbeton, 8. -9. Oktober, 2019

[3]   Mühlbauer C.; Zilch K.: Glued Joints of Ultra High Performance Concrete Structures. In: Proceedings of the 3rd fib International Congress 2010, Washington D.C., May 29 – June 2, 2010

[4]   Deutscher Ausschuss für Stahlbeton e.V. (editor): Sachstandsbericht Ultrahochfester Beton. In: Heft 561 des Deutschen Ausschuss für Stahlbeton, Beuth Verlag, Berlin 2008

Development of new jointing systems for lightweight UHPC structures

Jeldrik Mainka[1], Stefan Neudecker[1], Christoph Müller[1], Lukas Ledderose[1], Wibke Hermerschmidt[2]

[1] *Institute for Structural Design (ITE) /* [2] *Institute for Building Materials (iBMB)*
Department Architecture and Civil Engineering
Technical University of Braunschweig,
Pockelsstraße 4, 38106 Braunschweig, Germany
Supervisors: Harald Budelmann and Harald Kloft

Abstract

This DFG SPP 1542 project aims to combine new developments in the advanced material technology of Ultra-High-Performance-Concrete (UHPC) with the innovations in digital design and fabrication as CNC controlled mould production. The main goal is to develop innovative jointing systems regarding the specific properties of this new material. The research of the project is structured into two main mutually supplementing areas:

The first phase and subject of this paper is the fundamental research and development of nodes for prefabricated spatial tubular frameworks; in particular the jointing system between the node and the tubular elements. The second phase will be the research on connecting principles of two-dimensional bended elements for shell structures.

First steps taken for the fundamental research are:

- Digital form finding process with Rhino 3D and Grashopper
- Draft of a geometrical reference model of a prototype node of a spatial tubular framework
- Coherent studies for suitable concrete and reinforcement strategies and mould technology
- Study of stress distribution in the node with Ansys FEM calculations in association with the digital form finding process. The calculation is part of the development of a transferable digital process chain for force flow optimization testing and optimizing the CNC - controlled mould production
- Construction of a 1:1 prototype node

1   Introduction

Architects and engineers of the modern like Luigi Nervi, Antonio Gaudi and Heinz Isler, created freeform architecture based on the self-adaptive material behaviour. In the further development of free-form architecture, purely computer-generated forms without regard to materiality where transferred into reality. This approach resulted in ambitious architectures but with very inefficient material use, high dead loads and therefore with no regard to any sustainability. Nowadays the need for sustainable construction and the will to continue to create free-form systems requires a new appropriate approached regarding efficient material use. One possible approach is to generate a structural system based on a geometry, which uses the entire material capabilities and reduces dead loads to save the available resources. The combination of high performance materials such as UHPC with force optimized structural systems is of a crucial importance for further resource-efficient designs. Based on optimized prototypes this research aims to create fundamental parameters for a new innovative planning and fabrication process based on UHPC. One of the overall goals is the development of applications for customized material effective freeform load-bearing structures made out of UHPC. While the production of UHPC members for these structures based on quality reasons required has to be prefabrication in factories, a modular system of members is required. The jointing of these prefabricated members has to be inherent in this system and is therefore at the focus of this research project.

Proc. of the 9th *fib* International PhD Symposium in Civil Engineering, July 22 to 25, 2012,
Karlsruhe Institute of Technology (KIT), Germany, H. S. Müller, M. Haist, F. Acosta (Eds.),
KIT Scientific Publishing, Karlsruhe, Germany, ISBN 978-3-86644-858-2

2   Digital Formfinding

The architectural constructions built in the 1950's by Pier Luigi Nervi can be seen as an early approach of freeform design based on the principle of "form follows force". One of the outstanding projects was the „Palazzetto dello Sport"in Rome, built in the years 1956/1957. (Fig.2.1) Nervi transformed the highly efficient and solid shell structure of the cupola to a network of Y- beams at the edge (Fig.2.2). Using standard concrete as the main building materials at that time provokes a question. How could it be built more efficient in our times with UHPC and modern fabrication methods? Inspired by this geometry we designed the first reference nodes based on formal principles of the Y-beams used by Nervi at the „small sport palace".

Figs. 2.1 and 2.2   Pier Luigi Nervi - small sport palace in Rome [1]

In a first step the node was described by a sophisticated, geometrical model which was transferred in a parametrical script in Rhinoceros 3D (using Grasshopper). By adaption of directional vectors and variable section geometries, single nodes or arrays of different nodes (Fig.2.3) can be automatically generated. Jointing elements were variable generated by the chosen way of construction. To be easily adapted for the specific needs of the research process the script assembly was laid universal and highly modular. Components for the variable generation of surfaces, alternative jointing principles, more precise geometrical description, integrated FEM software, integrated evolutionary optimization processes and integrated fabrication controlling where already implemented in the script.

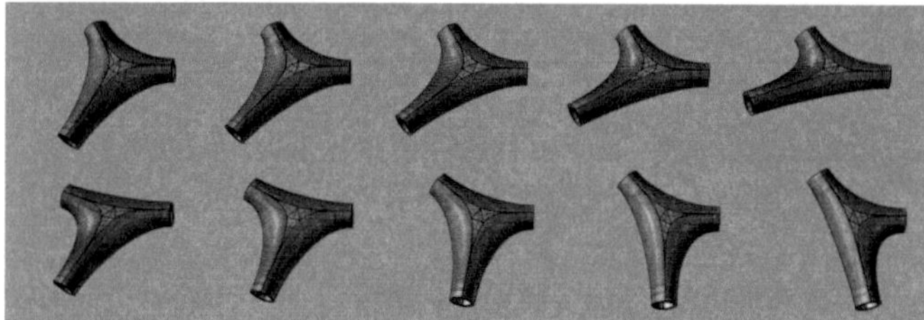

Fig. 2.3   At the ITE designed parametric arrays of nodes

The resulting node components can also be applied to complex networks, described by directional vectors. In this way not only single elements but also complex spatial structures can be generated automatically. Figs. 2.4 and 2.5 show examples of these complex spatial structures, as hexagonal systems with different Y- node systems.

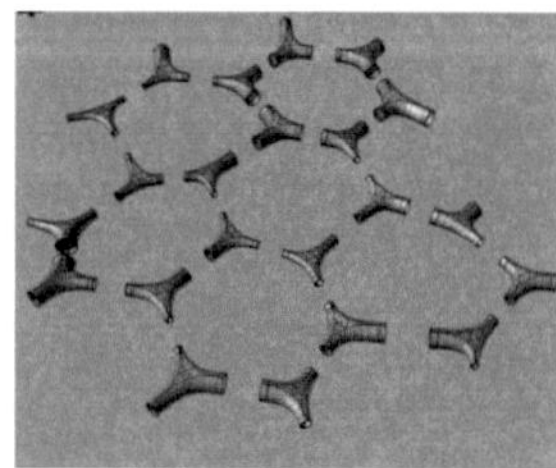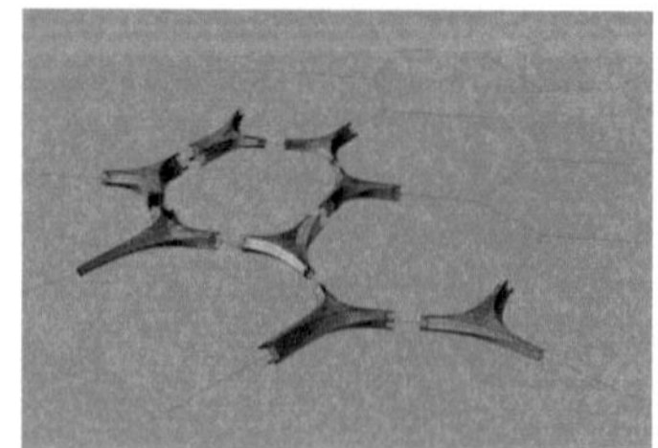

Fig. 2.4   Spatial structure with tubular nodes   Fig. 2.5   Spatial structure with ruled surface nodes

At this point of the project, the consideration of precise force flow has not yet been part of the geometrical optimization. At the current stage, the capabilities of the parametrical generation in the structural design are therefore not fully used. To get an efficient load bearing structure with optimized node geometries, the force effects have to be parametrically implemented in the geometrical design. An integrated fabrication process will then generate members as „serial unicums ". In this case, not the fabrication of the spatial form is repeated, it is the chain of processes itself. This vision can only be realized systematically and so the on-going research is focused on the production of a first single element where the fabrication processes are proven.

3   Structural Detailing

3.1   Geometrical Conditions

A parametric model of the node was generated with Rhinoceros 3D that allowed the free setting of angle and diameter, giving the possibility to create node arrays as shown in Fig 3.1. This model was used to generate the geometric and structural design of a first prototype node, using simple setting of three times 120 ° - angles as shown in Fig. 3.2. The chosen settings were seen as favourable for the first FEM calculations and for the CNC controlled mould production, due to the resulting multiple symmetries. As defined in the pre-dimensioning requirements, the tubular members of the framework were designed out of UHPC pipes with a diameter of 100mm and a wall thickness of 15mm. Upon this and the requirement to be bending-stiff, the connections on the node sides where designed as an external sleeve with a wall thickness of 30mm and an insertion depth of 100mm.

These requirements are close to the research of the interdepartmental working group the iBMB (also TU Braunschweig) "Ultra-lightweight concrete members inspired by bamboo" within the framework of the DFG SPP 1542

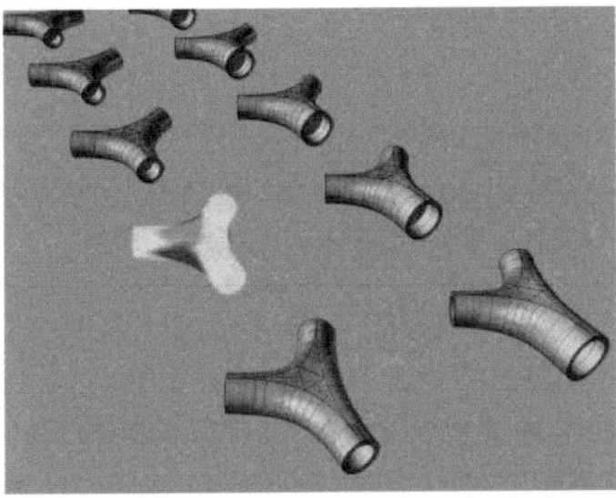

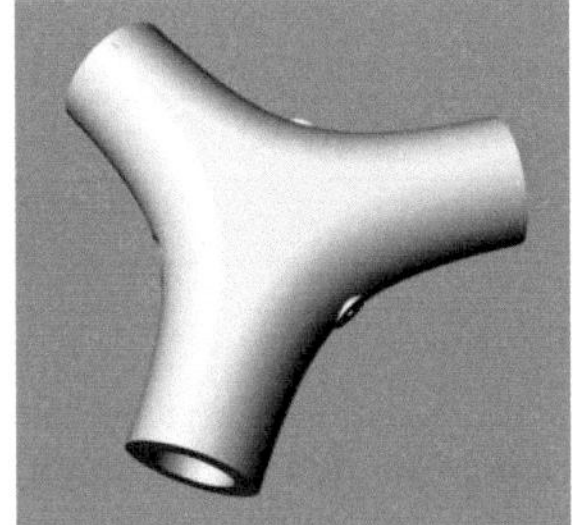

Fig. 3.1   Parametric node arrays with different angles and diameters

Fig. 3.2   Prototype node with three times 120°angles

3.2   Material Technology of UHPC

To reduce the cross section and the weight of the construction elements, an ultra high performance fiber reinforced concrete (UHPFRC) with a compressive strength of 150 MPa is used. The high compressive strength is reached by using high cement content, a low water/cement ratio and an optimized packing density of the aggregates. As the concrete is designed for thin elements, a maximum grain size of 2 mm is used. The ductility and the tensile strength are improved by an amount of 2.5 Vol-% steel fibers with a length of 9 mm and a diameter of 0.15 mm.

Table 1   Mixture for 1 m³ concrete

UHPFRC type FK1-2.5		
cement CEM I 52,5 R	kg/m³	595
quartz sand	kg/m³	1029
quartz flour 1	kg/m³	314
quartz flour 2	kg/m³	119
silica fume	kg/m³	69
steel fibers	kg/m³	192
water	kg/m³	156
super plasticizer	kg/m³	40
w/c-ratio	-	0.3
slump flow	cm	80

A good workability of the fresh concrete is of crucial importance for the fabrication of thin elements. By using a moderate amount of silica fume and adding a super plasticizer, based upon polycarboxi-latether, a nearly self-compacting consistence and a slump flow of 80 cm can be achieved. Table 1 shows the used mixture for 1 m³ concrete, the mechanical properties of the hardened concrete after 28 days are given in tab.2.

Table 2   Mechanical properties of the hardened concrete after 28 days

compressive strength	fcc,28	MPa	150.7
tensile strength	fct,28	MPa	9.93
static Young's modulus	Ecc,28	MPa	46,700

3.3   Pre-stressing System

Although the performance of UHPC approaches steel under compression, the tensile strength of the used UHPC material is only about 6% of its compressive strength even under addition of steel fibers. Therefore, the allocation of reinforcement is essential to handle the tensile forces in order to achieve an optimized performance for lightweight components. For the reinforcement layout of nodes and tubular components, two different approaches are considered:

Principle A is an internally enclosed reinforcement concept for every single node and tubular member, considering all load cases. Therefore, every element would require a separate reinforcement respectively an internal pre-stressing system in order to allow a force-fitted connection of the elements.

Principle B is a force-fitted connection of nodes and tubular members realized by externally applied tension on the internally passing tendons.

Beside the combination of UHPC with steel fibers, this externally applied tension has the advantage that the UHPC in the node and particularly in the connection area is compressed. This allows taking advantage of the benefits of UHPC material under compression in an optimal way.

Due to the advantages of the externally applied tension, with respect to the connection area / joining points and the possibility to suppress tensile stresses resulting from moment loads, the design possibilities of Principle B were pursued in the further study.

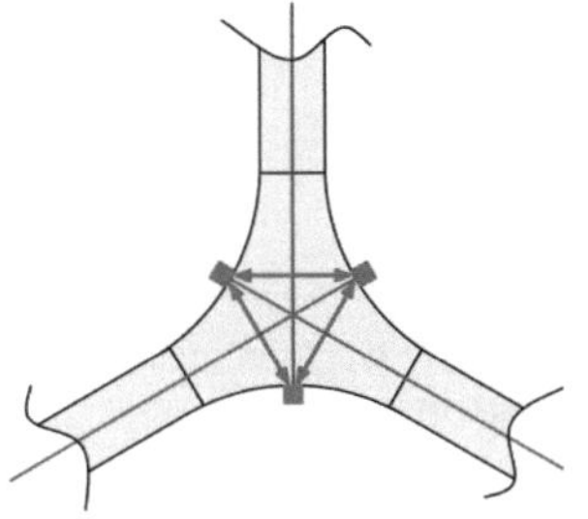

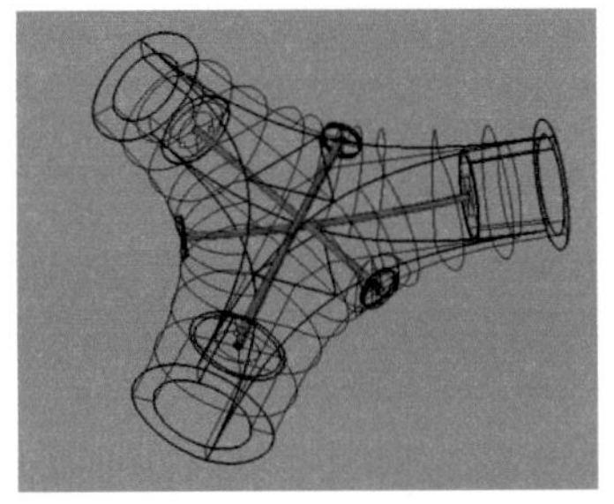

Fig.3.3   Pre-stressing concept of the created node

Fig.3.4   Node concept transferred into a 3D Rhinoceros model

Figure 3.3 shows schematically the pre-stressing concept of the created triple tension node, which has been transferred into a Rhinoceros 3D model as shown in figure 3.4. On closer examination, the tendon profiles (red line) in the 3D node model show a geometric problem. If the tendons on the inside of the node are laid out in a rectilinear way, (as in the concept) the tendons intersect in the centre of the node. To solve this geometric problem, two of the tendons are running curved around the third rectilinear running tendon.

Although deviation forces occur inside the node due to the tension in the curved tendons, this approach is considered appropriate for the time being. The influence of the occurring deviation forces will be examined in more detail later. An alternative approach is the division of the tendons into pairs of rectilinear running tendons. This would require to be distributed to rectilinear tendon pairs over at least five layers in the cross section of the node resulting in technical problems for applying the tension.

4   Numeric Analysis with FEM-Method

The first evaluation of the stress distribution in the prototype nodes was performed with the software, Ansys' using the finite element method FEM. To facilitate the calculations analysis in the first instance, the following simplifications where assumed. The entire node was assumed to be a solid body without voids and the nonlinear deformation behaviour regarding the occurrence of cracks was neglected. Of primary interest was the distribution and visualization of stress in the selected node geometry.

4.1   Input Parameter

As described in the initial considerations, it is assumed that the node is bending resistant. The expected load were normal force, moment and a combination of normal force and moment as shown in Fig.4.1.

The magnitude of the load was derived from the load bearing capacity of the connecting UHPC tubular members, but was increased to gain a sharp visual distinction of the significant stress gradients

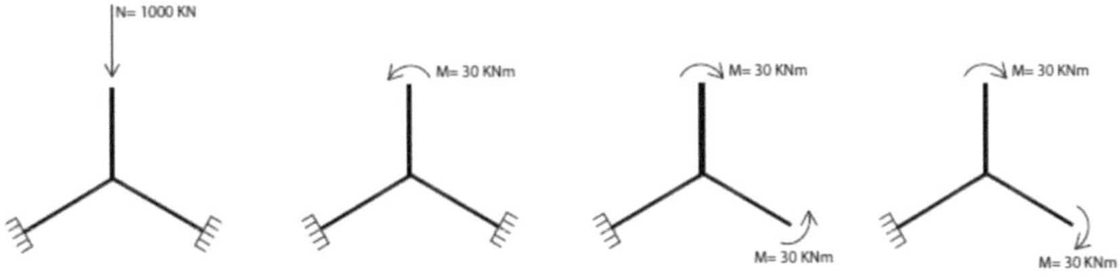

Fig.4.1   On prototype, node considered load cases

The material parameters where provided by the iBMB institute of building materials of the TU Braunschweig which is the partner of the ITE in this research project. See also paragraph 3.2. The static Young's modulus was set to 46.7 [GPa]. The Poisson's ratio was assumed to be 0.18. Based on the assumption that UHPC has a linear elastic behaviour until the failure of the material, Ansys calculated the modulus of shear to be 19.92 [GPa] and the bulk modulus to be 2.45 [GPa]

As described earlier the node was created in 3D with Rhinoceros 3D, imported to Ansys and meshed to approx. 50,000 tetrahedrons.

4.2   Interpretation of results

Based on the described approach a purely qualitative statement of the structural behaviour is possible. Fig 4.2 shows graphically the results for the main principal stresses in the node.

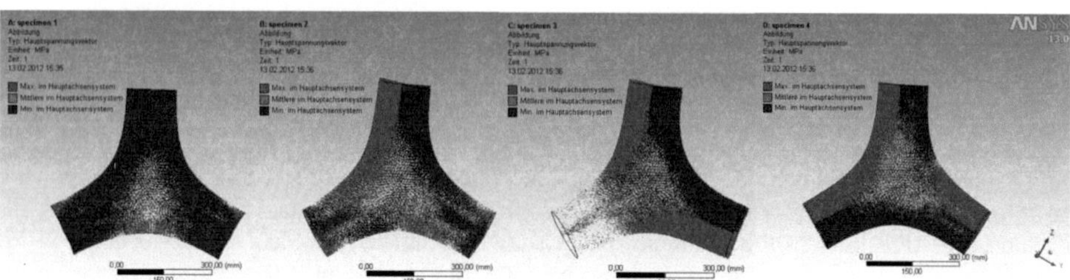

Fig.4.2   Main principal stresses in the node based on considered load cases (red tension, blue compression)

As expected, it can be seen that the stress due to bending moments was evidently higher than due to normal force. According to reinforced concrete design rules, it is necessary to have sufficient reinforcement in the location of the tensile force. To avoid this, the alternative is to use a pre-stressing system to suppress potential cracks in the tension area as already described. This would use the advantage of UHPC with its availability to suffer high compression force in a better way.

Furthermore, Fig. 4.2 shows that under pure compressive axial forces a tension zone arises in the rounding of the node. This area is therefore only partly suitable for anchoring of a tendon profile. As an alternative, the tendon could run in a curve to the longitudinal axes to suppress this round area. A curved tendon profile, however, results in a deviation force, which would adversely affect the existing nodes in design. This shows that the presented node with its geometrical dimensions has a high optimization potential especially in the tension zones. In the course of further research, a more accurate material model for the calculations with ANSYS will be elaborated and verified. Especially the behaviour of steel fibres such as the determination of cracks and notches will be in the focus of the research.

5   Fabrication of a 1:1 prototyp joint

A first prototype was developed in order to gather basic information about the digital production process as well as the general workflow of the fabrication of UHPC-elements. For this purpose, the digital model of the node (Fig 5.1) was first split and then transformed into two convex half shells that were embedded in a solid block (Fig 5.2).

For the development of the final geometrical design of the mould, it is necessary to consider the accurate process of assembling the half shells as well as the following concreting of the actual node. Therefore, it is suggested to design the segments of the mould in the digital process in way that the real segment can be assembled accordingly to a puzzle.

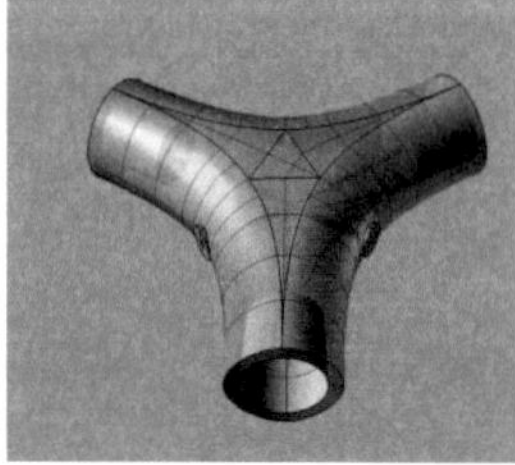

Fig. 5.1   Digital 3D-model of
Prototype-node

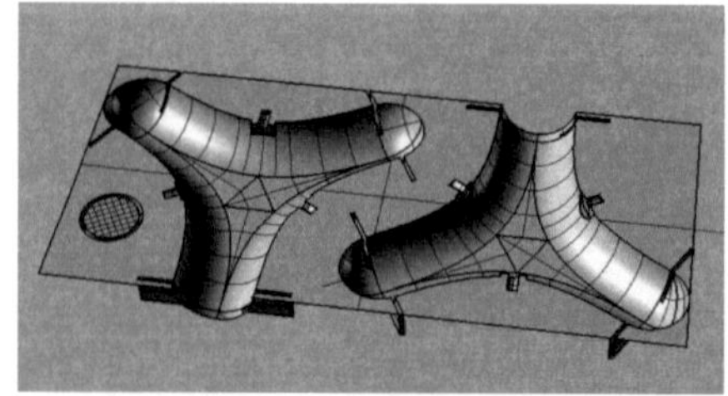

Fig. 5.2   Digital model of the mould consisting
of two convex half shell

Fig. 5.3 shows the retracted node and the CNC-milled mould segments, that consists of expanded polystyrene XPS 300-SF (material: 300kPa = 30 to/m2 compression strength at 10% compression strain and a coefficient of elasticity of 12 N/mm2 according to EN 826.) The chosen material can be CNC-milled in a precise way.

Using a portal milling machine and an 8mm milling head, the cutting time of the first two milling sequences was about 6 hours. However, the operational hours can be reduced by choosing a bigger milling head and optimizing the integrated milling path. The surface was finally grinded manually and sealed with silicone gel. By this approach the surface is not glossy but absolutely watertight in order to prevent bleeding of the concrete.

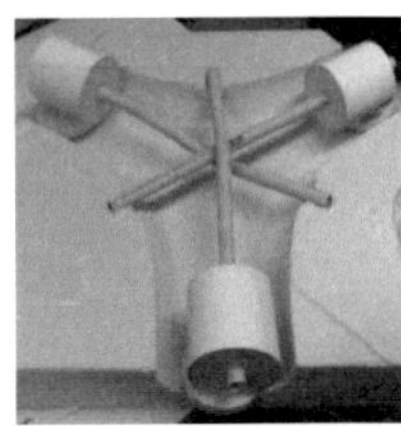

Fig. 5.3   Retracted prototype-node and elements of the mould made from expanded polystyrene

Fig. 5.4   built-in parts for the connecting sleeves

Afterwards the built-in parts for the connecting sleeves and tendons were adjusted to the mould, as shown in figure 5.4. The geometry of the tendons was realized by thermoforming the PVC ductwork (cf. 3.3). As shown in Fig. 6.1, the two shells of the mould were glued together using the silicone gel and then interlocked by all-thread rods. These were inserted through control holes that were congruently milled in the two shells. The CNC-milled and in PVC pipes embedded placeholders of the connecting sleeves were screwed on closing plates made from acrylic glass. These packages were inserted in the shells and bolted together.

Due to the results of the pre-tests executed by the iBMB that investigated the air removal of the selected UHPC type FK1-2.5, it was decided to cast the node in an upright position. Compared to the lying position, the horizontal areas are less, which allowed a better de-ventilation of the concrete.

To concrete the node, the Concrete Laboratory of the iBMB prepared the UHPC mix, as described in Paragraph 3.2, and was poured into the mould as documented in fig 6.2.

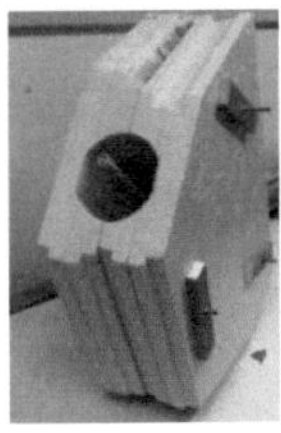 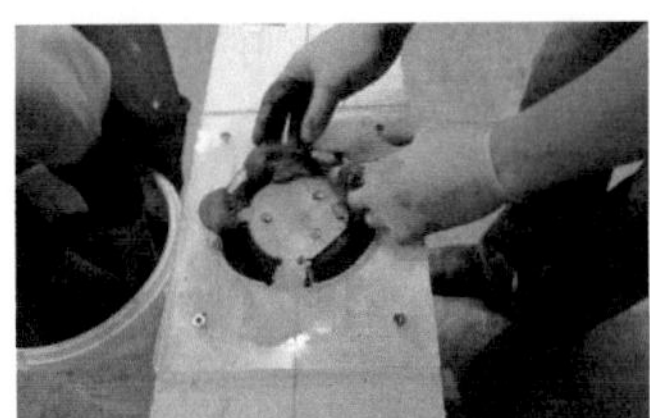 

Fig. 6.1   Prepared mould   Fig. 6.2   Filling the UHPC   Fig. 6.3   Rising air bubbles

In order to support the de-ventilation (Fig.6.3) of the concrete, the entire mould was shaked slightly for about 15 minutes. In contrast to the high frequency shaking used by standard concrete, a slower oscillation frequency of 2 Hz, as recommended in [2], was applied. Besides the accumulation of bubbles on the free surface, this treatment supported a proper de-ventilation of the UHPC and resulted in an almost pore-free surface on the entire node.

6   Conclusions

The fabrication of the first UHPC node as shown in Fig. 6.4 and Fig 6.5 makes no claim to deliver comprehensive research results, however it was possible to gain fundamental conclusions of the digital workflow and the general producibility.

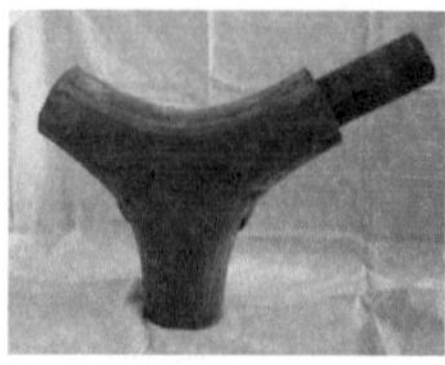
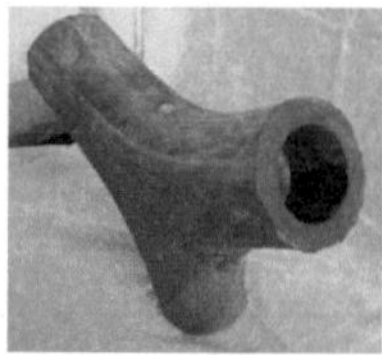

Fig. 6.4   Complete UHPC   Fig. 6.5   Connecting sleeves with joined
prototype node   bearer

By passing through the entire production chain, including the digital design of the node, CNC-milling of the mould and the concreting of the node, the upcoming problems showed the limits of the producibility. Using a three-axes-CNC-mill, the team gained experience about concrete mould, in particular about the realization of complex undercuts and exact edges. These problems need to be investigated in more detail. The production in most instances depends on the radius of the milling head, the geometrical location of the shells and therefore on the segmentation of the individual elements of the mould. Based on these experiences we intend to rather use a 5-axes-CNC-mill for more complicated moulds.

Based on the investigations described above, further research about the load bearing behaviour of the node elements are intended in order to optimize the form of the node according to the flow of forces. Besides the field of research concerning the production of the moulds and predictability of the node, the further investigations will focus on the interaction between the flow of forces and the inherent geometry. Therefore, the digital model of the material UHPC will be refined and verified by tested samples. Additionally, the influence of the radius of chamfers in thin-walled beams made from UHPC will be tested with special interest in the bending tension strength. The results of this research will be used for finding the optimized form of the node and its connection members.

Besides of the research on UHPC nodes itself, a further focus will be the investigation and basic evaluation of effective connections between UHPC nodes and the tubular members. Different principles of connecting, such as gluing and bracing or even the use of form-optimized built-in parts made of metal will be assessed.

References

[1]   P. Desideri, Pier Luigi Nervi, Zürich : Verlag für Architektur Artemis, Zürich , 1882.

[2]   B. Schmidt, Sachstandsbericht Ultrahochfester Beton, Beuth, Berlin: 2008.

[3]   Kloft/Budelmann, DFG Forschungsantrag - Entwicklungen neuartiger Verbindungen für geometrisch komplexe Flächen und Stabtragwerke, Braunschweig, 2011

Implants for Highly Efficient Load Introduction

Martin Kobler

*Institute for Lightweight Structures and Conceptual Design (ILEK),
University of Stuttgart, Pfaffenwaldring 7 & 14, 70569 Stuttgart, Germany
Supervisor: Werner Sobek*

Abstract

The development of ultra-high performance concrete offers a variety of new possibilities. Provided with steel fibres, regular reinforcement is no longer necessary; this allows very thin elements with thicknesses of only 20 mm. Due to complex mixing, concreting and sealing procedures, prefabrication is necessary for constant high performance. These precast elements need to be assembled to structures on-site. Point connections are easy to handle and ensure a constant high capacity of the load transmission area, but lead to a concentration of stresses in the connection area. An integral element (called "implant") for the introduction of high local loads was developed. This implant guarantees peak-free, homogeneous stress-fields and allows maximum utilization of the whole structural UHPC-member. Developed by analytical and numerical investigations, the load bearing behaviour could be finally verified by a series of experimental tests, which proved the high performance and efficiency of the load introduction.

1 Introduction

1.1 General

Ultra-high performance concrete (UHPC) has the compressive strength of regular steel but only 1/3 of its specific weight; therefore it can be considered as a very efficient lightweight material. In addition to its low weight combined with the high compressive strength, UHPC has another key advantage: just as regular concrete it can be used to build structures in almost any shape and. Compared to regular concrete, the highly increased performance significantly extends the range of possible forms and constructions – such as thin-walled shells and spacial structures. As achieving a thickness of only 20 mm is possible ([1]-[3]), UHPC is predestined as material for thin compression members.

Specific requirements for the production of UHPC elements (ingredients, mixing and sealing process [5], heightened attention during concreting) make prefabrication highly advisable. Only stationary production can make UHPC elements economic and therefore competitive.

1.2 Connecting UHPC elements to structures

Only individual components of a structure can be prefabricated, the structure itself still has to be assembled on-site. This leads necessarily to element connections that require special attention. For a discussion of possibilities to connect precast UHPC elements see [8], [9], [12]. There are two general ways of connecting precast elements that differ substantially from each other: continuous connections and point connections. In the context of the load path, continuous joints are optimal: stresses are transferred directly from one structural element to the other ([6], [7]). But, unfortunately, it is very difficult to produce continuous joints for UHPC elements in the required homogeneous quality.

Only point connections meet the requirements of economic production, easy handling on-site and the future necessity of deconstruction – but they logically cause a concentration of stresses in the connection area which will govern the design of the UHPC element. Avoiding stress concentrations and completely eliminating the resulting stress peaks by a specially designed implant will prepare the ground for highly efficient and optimized UHPC structures.

2 Implants for the introduction of high concentrated forces

2.1 Basic principles for efficient load introduction

At a certain distance from the point of load application, the compressive stresses reach an equal distribution [11]. The application of local forces thus creates singularities; these singularities cause

Proc. of the 9th *fib* International PhD Symposium in Civil Engineering, July 22 to 25, 2012,
Karlsruhe Institute of Technology (KIT), Germany, H. S. Müller, M. Haist, F. Acosta (Eds.),
KIT Scientific Publishing, Karlsruhe, Germany, ISBN 978-3-86644-858-2

stresses in the area of load application that are much higher than in areas of homogeneous utilization. For an efficient structure, a homogeneous distribution of stresses is necessary in order to avoid stress peaks that may cause early failure [10]. Only such a homogeneous distribution of stresses allows an optimal utilization of the structural components: failure stress is reached at any point at the same time.

2.2 Problem and solution statement

Trying to reach maximum utilization in an unaffected area of the element leads inevitably to problems in the singularity area as there is no high margin for increased stresses. The compressive stresses at a minimum distance to the point of load application have to be almost constant. A thin UHPC-plate with a thickness of 20 mm, a width of the implant of 250 mm, a compressive strength of 180 N/mm² for the UHPC and a high utilization of more than 80 % of this strength are the boundary conditions for the implant development (see [12] for detailed information).

The objective of connecting thin UHPC elements by implants demands a significant reduction of the cross-section in the connection point. The high compressive strength of UHPC together with the designated utmost utilization requires implant materials with a much greater strength than regular steel, i.e. high strength steel with a yield stress of almost 900 N/mm².

The extreme utilization of the precast element generates tensile stresses orthogonal to the compressive trajectories which already exceed the tensile strength of UHPC at only 14 % of the required load. The implant has to carry not only these tensile stresses by suitable tension elements adhering to the directions of the principle tensile stresses, but also the compressive stresses. The best solution for the latter is an adaptation of the saw-tooth-connections developed for normal strength concrete [11] to the needs of UHPC and the special requirements of the implant. The interlocking component (fig. 1) is of major influence for the homogenization of stresses ([10], [12]). Therefore, the stiffness of the interlocking component was investigated in detail [12]; these investigations identified a linear decreasing stiffness and a reduced modulus of elasticity by the use of titanium alloys as the optimum with regard to the homogenization of stresses. Fig. 1 shows and names the implant components schematically.

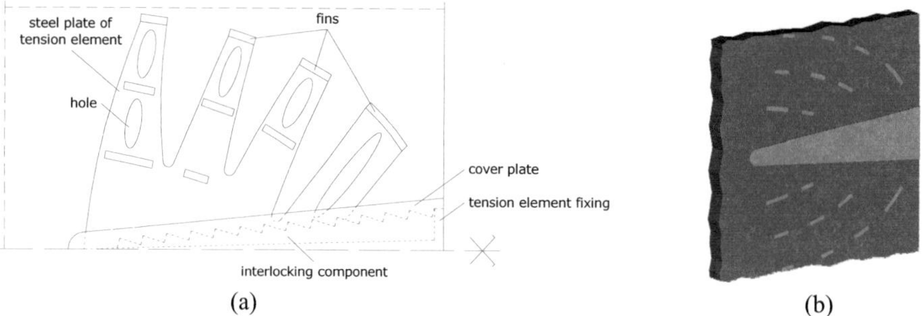

(a)   (b)

Fig. 1:   Implant components and integration into UHPC plate
           a) schematic illustration and names; b) visualisation of implant in UHPC plate

2.3 Load bearing behaviour of implants

The implant developed by numerical, analytical and conceptual investigations [12] bases on the following material properties:
- implant: high strength steel, strength 890 MPa, Youngs modulus 210 GPa
- interlocking component: titanium alloy, strength 1.000 MPa, Youngs modulus 110 GPa
- UHPC-plate: strength 180 MPa (compression), 10,6 MPa (tension), Youngs modulus 55 GPa

Fig. 2 opposes the normalised stress distributions for one symmetric half of the UHPC plate with different materials for the interlocking component, as identified by numerical investigations. Obviously, the discrepancy to the ideal homogeneous stress distribution is only half as big for interlocking components made of titanium than for those made of steel. As failure always starts in the maximal stressed area, the use of interlocking titanium components will finally allow higher utilization of the whole structural UHPC member.

Based on finite element analysis the maximum load for a steel implant with interlocking titanium components is 749 kN, compared to only 603 kN when using steel also for the interlocking components.

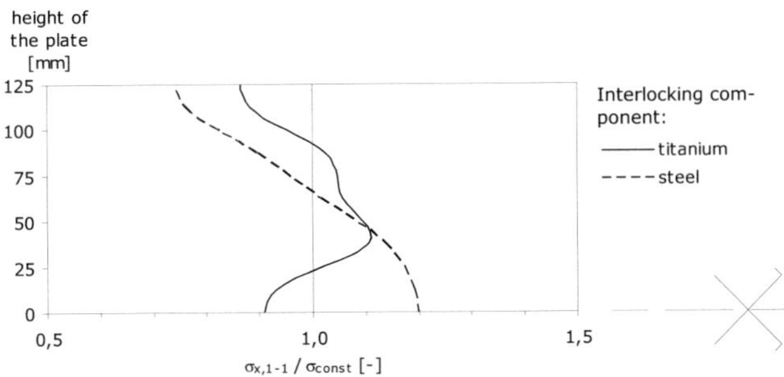

Fig. 2:   Normalised stress distribution at 220 mm from point of load application

3   Test series for verification

3.1   Basics and objective

In order to verify the analytical and numerical investigations a first test series with 6 specimens was carried out. The test specimens were thin UHPC plates; for the first test series 3 specimens were provided with implants, 3 were produced without, in order to obtain reference values for the efficiency of the load introduction with implants. The most important objective of the first series was to validate the manufacturing processes in general (implant and thin UHPC plates) and to confirm the consistency of the theoretical investigations with the test results.

A further test series with more comprehensive measuring instrumentation is currently being carried out in order to validate also the distribution and flow of forces influenced by the implant.

3.2   Assembly of the implants

The individual parts of the implant were cut out of steel plates by high pressure water jet. The fins (fig. 3b) are placed in the appropriate cut-outs in the tension element (fig. 3a) and welded. Afterwards, the tension elements provided with the fins are welded to the tension element fixing (fig. 3c); during this process, welding distortion has to be minimized by an appropriate welding sequence and careful handling. The interlocking components are positioned and fixed during concreting by instant adhesives (fig. 3c+d). Eventually (set in UHPC and subjected to compressive forces) they are kept in place by the compressive forces. The materials of the individual parts are shown in table 1.

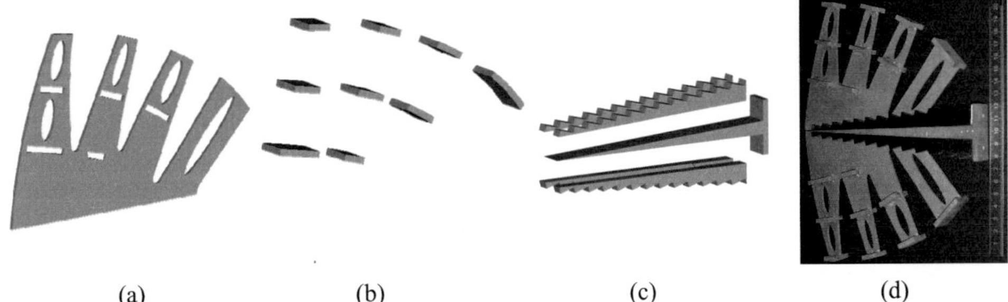

(a)   (b)   (c)   (d)

Fig. 3:   Single components of an implant (schematically): a) tension element; b) fins; c) tension element fixing & interlocking components; d) implant ready for concreting

Table 1: Materials & dimensions of the implant components

implant component	material	strength	thickness
tension element	high strength steel "Optim 900 QC"	900 MPa	2,5 mm
fins	high strength steel "Optim 900 QC"	960 MPa	4 mm
tension element fixing	high strength steel "Dillimax 890 T"	890 MPa	20 mm
interlocking component	titanium AL6V4	900 MPa	8 mm

3.3 Manufacturing of the test specimens

The UHPC used for the test specimens was Ductal® FM with a steel fibre content of 2 Vol.-%. The mechanical properties for the material of this series of specimens were identified on reference blocks and resulted in the following average values:
- Compressive strength: 206,3 N/mm²
- Modulus of elasticity: 59.771 N/mm²

The specimens with the dimensions of 600 mm x 420 mm x 20 mm were concreted in a formwork for 6 plates (fig. 4a). Approximately 20 hours after concreting the specimens were struck and put into water quench for 48 hours at 94°C. Finally the UHPC plates were cut orthogonally and abraded at the edge opposite to the implant. This edge defines the contact area of the UHPC plate with the test stand in the following compressive test.

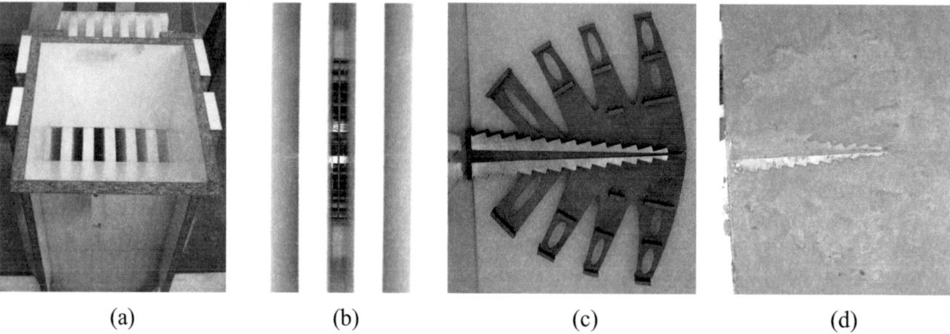

(a)   (b)   (c)   (d)

Fig. 4: Framework, concreting & striking
a) concreting rig; b) & c) implant in framework; d) implant in UHPC plate after striking

3.4 Compressive tests

The UHPC plates were prepared as described and afterwards loaded until failure. To prevent early collapse due to buckling, the thin plates were put into a steel test frame, where screws at certain points avoid out-of-plane deflections (see fig. 5a). Friction effects at those points were prevented by the use of PTFE pads fixed on the screws and plates.

The compressive tests were carried out on 6 test specimens (3 with implants, 3 without). The press was controlled by displacement at a speed of 0,05 mm per minute. All specimens experienced the initial crack at a press load of 150 to 250 kN in longitudinal direction in the middle of the UHPC plate. In the further test procedure, the specimens provided with implants showed additional analogue cracks in front of the area influenced by the implant. These cracks had no influence on the load-bearing behaviour; the cracks were hardly visible to the naked eye, so the crack-width did not exceed 0,1 mm.

Table 2: Ultimate load of specimens

specimen	type	failure load	stress[*]
C1-1	without implant	301 kN	334 N/mm²
C1-2	with implant	664 kN	738 N/mm²
C1-3	without implant	266 kN	295 N/mm²
C1-4	with implant	622 kN	691 N/mm²
C1-5	without implant	197 kN	219 N/mm²
C1-6	with implant	629 kN	699 N/mm²

*[)] compressive stress at the point of load introduction

Specimen C1-6 showed exactly the load bearing behaviour observed in the numerical investigations: cracks occur in the direction of the compressive trajectories along the interlocking component, followed by concrete spalling until the failure load is reached (fig. 5b). The test specimens C1-4 and C1-

2 showed an initial crack starting at the tip of the implant with a marginal crack width widening in further test procedure (fig. 5c), probably due to tolerances in implant production. These tolerances may have caused a load concentration on the tip of the implant resulting in orthogonal tensile stresses exceeding the tensile strength of UHPC. For the next series of specimens these tolerances are reduced significantly.

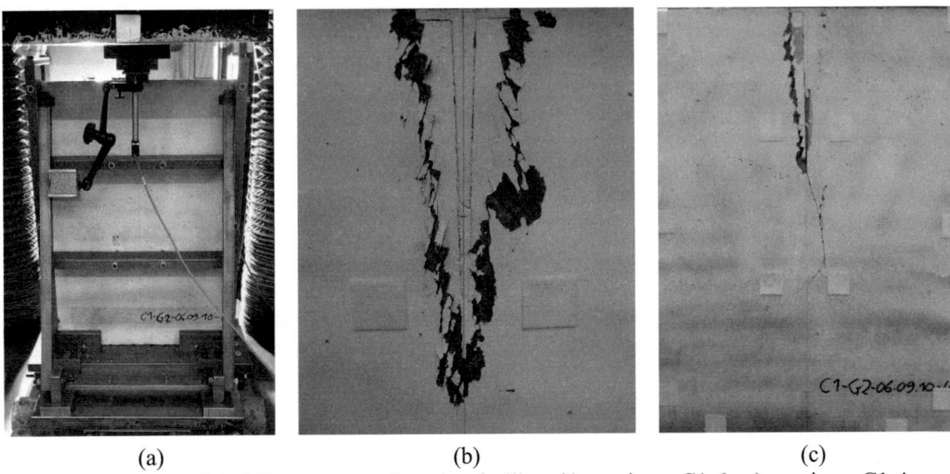

(a)	(b)	(c)

Fig. 5:   Test stand & failure pattern: a) testing facility; b) specimen C1-6; c) specimen C1-4

The load displacement diagram in fig. 6 demonstrates impressively the effectiveness of the implant: with the implant the maximum load is 3 times higher than without the implant (table 2) and the ductility has also greatly increased as shown clearly by the pronounced progression of the curve after failure. Since UHPC is a brittle material, the increased ductility is of major importance because it may offer the possibility to reduce the steel fibre content and thus improve the total energy balance of UHPC. Another important advantage of the implant shows the comparison of the maximum loads in table 2: the deviation of the maximum loads reached for the specimens with implants is only 4 % from the average value, whereas it is almost 23 % for the specimens without implants.

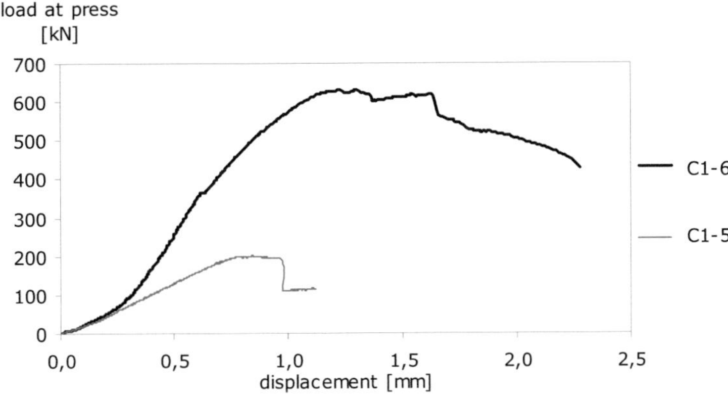

Fig. 6:   Load-displacement-diagram: C1-6 with implant; C1-5 without implant

By looking at specimen C1-2 it is obvious that the stresses transferred in the point of load introduction are 3,6 times greater than the already high compressive strength of UHPC (table 2). Compared to the numerical investigations, the maximum loads reached are 11 – 17 % lower. The most probable reason for this effect is some small production tolerances; for the next series of specimens these tolerances are already reduced as much as possible.

4   Conclusion & acknowledgement

The implant developed – based on conceptual, numerical and analytical investigations – reaches a homogenization of stresses in the area of load introduction. This homogenization increases the load bearing capacity of the load introduction area provided with implants significantly and allows to reach very high ultimate loads while at the same time reducing variation.

The test series verifies the applicability of the implant and the possible high ultimate loads. The homogenization of the load transfer by the adjusted stiffness of the load-inducing components is verified indirectly by the estimated effects. Another additional series of tests using numerous strain gauges is being carried out at the moment and will further analyse this effect.

The high concentration of stresses in the connection points provided with implants allow efficient structures made of thin UHPC members. In addition to the economic and competitive application of UHPC implants will allow to use the enormous potential of the high performance material.

The authors would like to thank the Lafarge Company, Paris (France), for the extraordinary and great support of the research in the field of UHPC at ILEK.

References

[1]   Fehling, E.; Bunje, K.; Schmidt, M.; Schreiber, W.: Ultra High Performance Composite Bridge across the River Fulda in Kassel. p. 69-75 in: Ultra High Performance Concrete. Proceedings of the International Symposium on Ultra High Performance Concrete. Schmidt, M.; Fehling, E.; Geisenhanslücke, C. (Hrsg.). Universität Kassel, Schriftenreihe Baustoffe und Massivbau, Heft 3, 2004.

[2]   Schmidt, M.; Bunje, K.; Fehling, E.; Teichmann, T.: Brückenfamilie aus Ultra-Hochfestem Beton in Niestetal und Kassel. Beton- und Stahlbetonbau 101 (2006), Heft 3, p. 198-204.

[3]   Vicenzino, E.; Culham, G.; Perry, V.H.; Zakariasen, D.; Chow, T.S.: First Use of UHPFRC in Thin Precast Concrete Roof Shell for Canadian LRT Station. PCI Journal, Vol. 50 (2005), No. 5, p. 50-67.

[4]   Holschemacher, K.; Dehn, F.: Ultrahochfester Beton – Stand der Technik und Entwicklungsmöglichkeiten. p. 1-12 in: Ultrahochfester Beton. Innovationen im Bauwesen. Beiträge aus Praxis und Wissenschaft. König, G.; Holschemacher, K.; Dehn, F. (Hrsg.). Berlin, Bauwerk Verlag, 2003.

[5]   Dehn, F.: Qualitätsgerechte Herstellung, Verarbeitung und Nachbehandlung von Ultra-Hochleistungsbeton. Betonwerk + Fertigteil-Technik 71 (2005), Heft 2 (Kongressunterlagen 49. Ulmer Beton- und Fertigteil-Tage), p. 22-23.

[6]   Sobek, W.; Schäfer, S.: An der Nahtstelle – Fügen von Bauteilen aus unterschiedlichen Werkstoffen. deutsche bauzeitung 130 (1996), Heft 1, p. 106-114.

[7]   Witta, E.: Kraftschlüssige Verbindungen und ihre Rückwirkungen auf das Bauwerk. Betonwerk + Fertigteil-Technik 41 (1975), Heft 11, p. 527-530.

[8]   Fehling, E.: Konstruieren von Tragwerken aus Ultra-Hochleistungsbeton – eine Herausforderung. Betonwerk + Fertigteil-Technik 71 (2005), Heft 2 (Kongressunterlagen 49. Ulmer Beton- und Fertigteil-Tage), p. 26-27.

[9]   Behloul, M.: Les micro-bétons renforcés de fibres. De l'éprouvette aux structures. XIVième Journées de l'AUGC, Clermont-Ferrand. Prix Jeunes Chercheurs „René Houpert", 1996.

[10]   Wiedemann, J.: Leichtbau – Band 2: Konstruktionen. Berlin, Springer Verlag, 1989.

[11]   Schlaich, J.; Schäfer, K.: Konstruieren im Stahlbetonbau. p. 721-895 in: Beton-Kalender 1998, Teil II. Berlin, Ernst & Sohn, 1998.

[12]   Kobler, M.; Sobek, W.: The Introduction of High Forces into Thin-Walled UHPC Elements by the Use of Implants. p. 683-690 in: Ultra High Performance Concrete. Proceedings of the Second International Symposium on Ultra High Performance Concrete. Schmidt, M.; Fehling, E.; Stürwald, S. (Hrsg.). Universität Kassel, Schriftenreihe Baustoffe und Massivbau, Heft 10, 2008.

Composite structures made of ultra-high performance concrete and fiber-reinforced polymers

Franz Xaver Forstlechner

Institute for Structural Design
Graz, University of Technology
Technikerstraße 4/4, 8010 Graz, Austria
Supervisor: Stefan Peters

Abstract

The ambition of this study is the development of innovative composite structures made of steel-fiber reinforced Ultra-High Performance Concrete (UHPC) and Fiber-Reinforced Polymers (FRP). The new composite construction method is characterized by low self-weight, high durability and simplicity, and could represent a robust alternative to prefabricated steel, timber or normal concrete elements. Within the study, two different types of composite structures are investigated: composite structures consisting of UHPC and structural profiles made of Glassfiber-Reinforced Polymers (GFRP), as well as UHPC structures with reinforcement made of Carbonfiber-Reinforced Polymer (CFRP) lamellae. For evaluation of feasibility, a comprehensive theoretical and experimental research into the bond and bending behavior of UHPC-FRP composites is carried out and several structural building applications like beams, shells and façade elements are investigated.

1 Introduction of Materials

The use of UHPC in structural design is a relatively new trend and started at the end of the 1980th in Canada and France. Its structural behavior is currently being scientifically tested all over the world in combination with different kinds of reinforcement. The enormous compressive strength and the excellent bond properties allow the design of extremely slim and filigree structures comparable with steel constructions, as well as the use of high strength reinforcement. However, the combination of UHPC and FRP, which are characterized by outstanding tensile strengths, has not intensively been investigated in science up to now.

FRP are also rather new materials in building industry and mainly used for tension cables, post strengthening of constructions and reinforcement in aggressive environments. Structural FRP products are manufactured in a pultrusion process, which ensures high fiber content and constant quality. They represent a corrosion resistant and light alternative to steel components and the standard product range covers all kinds of profiles like angels, tubes and double-t beams.

1.1 Ultra-High Performance Concrete

UHPC is characterized by compressive strengths up to over 200 N/mm², high resistance against environmental influences and great freedom in geometric shape due to its good flow and self-compacting properties. The superior performance is realized by low water-cement ratio, optimization of grain mixture and heat treatment after solidification. Unfortunately, elastic modulus and tensile strength do not increase at the same rate as compressive strength and the material becomes increasingly brittle. Furthermore, the material shows growing linear stress-strain behavior until material failure and a steeply falling working line after concrete cracking (see Table 1) [1].

Proc. of the 9th *fib* International PhD Symposium in Civil Engineering, July 22 to 25, 2012,
Karlsruhe Institute of Technology (KIT), Germany, H. S. Müller, M. Haist, F. Acosta (Eds.),
KIT Scientific Publishing, Karlsruhe, Germany, ISBN 978-3-86644-858-2

Table 1   Mechanical properties of normal concrete C 25/30 [2] and UHPC with 2.5 vol.-% steel fibers [3]

Characteristic values	C 25 /30	UHPC
Specific weight [kN/m³]	25	25
Elastic modulus [N/mm²]	31,000	46,000
Compressive strength [N/mm²]	25	200
Breaking strain [‰]	3.5	4.8
Tensile strength [N/mm²]	2.6	10
Coefficient of thermal expansion [10^{-6}/K]	10	11

For improvement of mechanical properties, UHPC is usually reinforced with short steel fibers. They do not have a significant influence on mechanical properties of non-cracked concrete, but with increasing crack formation, the structural behavior is positively influenced: they improve the flexural strength by transferring tensile forces over the crack and prevent the explosive failure mode under compression, which is typical for high strength concrete without fibers [1].

1.2   Fiber-Reinforced Polymers

FRP are composite materials that consist of fibers and polymer matrix. They have a distinctively orthotropic behavior, in which the fibers control the mechanical properties in fiber direction and the matrix normal to. The matrix fixes the fibers in space and introduces loads, the fibers have the function to transfer loads in span direction.

Glass- and carbon-fibers are most widespread among all FRP. They have high tension and compressive strength and high breaking elongation. Glass-fibers are furthermore incombustible and have good resistance to chemical and biological aggressions. Their weakness is the low elastic modulus. Carbon-fibers have even more outstanding mechanical properties, but are enormously expensive as well (see Table 2). The matrix is commonly made of unsaturated polyester, vinyl ester or epoxy resins. Polyester and vinyl ester are cheaper than epoxy, however, they do have lower strength properties [4].

Table 2   Mechanical properties of different materials [5], [6], [7]

Characteristic values	Steel S 235 JR[5]	GFRP pultruded[6]	CFRP rebar[7]
Specific weight [kN/m³]	78.5	2.1	1.6
Elastic modulus [N/mm²]	210,000	41,000	165,000
Tension strength [N/mm²]	360	620	3.100
Breaking strain [%]	26	1.6	1.7
Coefficient of thermal expansion [10^{-6}/K]	11.7	6.7	0.7
Thermal conductivity [W/mK]	50	0,25	-

2   Problem Definition & Aim of the Study

The fundamental problem with using UHPC as a structural member is the handling of its low tensile strength and the utilization of its enormous compressive strength. In accordance with normal concrete, there are actually three ways to deal with tensile stresses: reinforcing the tensile stressed areas, pre-stressing the construction to suppress tensile stresses or choosing a construction with low bending stresses and/or improving concrete's tensile strength by using special types of cement.

In this study, the first approach is followed, by reinforcing the UHPC with FRP. The reason behind the decision is that this method is technically simple and does not have a strong influence on the structure's form. The aim of this study is the development of efficient UHPC-FRP composite structures with a balanced ratio between compressive and tensile strength, which are able to utilize the high resistance of both materials, as shown in Figure 1.

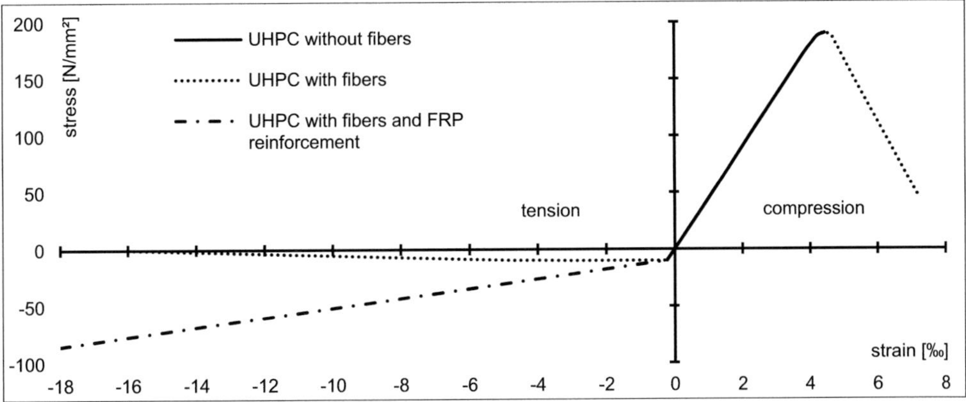

Fig. 1   Qualitative stress-strain relationship of UHPC with different kinds of reinforcement under centric tensile and compressive load [3]

3   Why UHPC and FRP

Within this study, two different types of composite structures are investigated: *composite structures made of UHPC and GFRP structural profiles*, and *UHPC structures with reinforcement made of CFRP lamellae*. Since GFRP are rather cost efficient, they can be applied to composite structures intensively in form of structural profiles. CFRP, however, are more cost intensive and therefore must be used very economically in form of thin lamellae. In both cases, the material combination appears useful for several reasons:

- Due to the good bond properties of UHPC, shear force transmission between UHPC and FRP is solely done by friction and adhesion, which is improved by roughening the FRP-surface. No additional mechanical joining means are required.
- The compact UHPC protects the FRP components against harmful environmental influences and improves the fire resistance. FRP has the advantage of being corrosion resistant compared to steel.
- By using closed FRP profiles, hollow parts for weight reduction can be realized easily. Light constructions with big static height provide benefits for wide spanned constructions, since UHPC's elastic modulus is rather low compared to its compression strength.
- The flexible fabrication of varying FRP cross sections and the free formability of UHPC allow the realization of customized and statically optimized structures.
- The joining of FRP components is rather problematic because of its low strength normal to fiber direction. By the combining FRP with UHPC, new alternatives for the introduction of single loads become possible.

4   Building Applications

In the following chapter, different building applications are investigated, which seem to be particularly appropriate for UHPC-FRP composite structures. Up to now, UHPC applications in architecture are rather rare and limited to columns, pre-stressed structures and non-load bearing façade applications without considerable tensile stresses. By the approach of reinforcing UHPC with FRP, new opportunities in design are provided, and the high performance of both materials is better utilized.

4.1   Free Formed Shell Constructions

Free formed shell structures appear frequently in contemporary architecture and are usually realized with steel grids. Prefabricated thin UHPC elements could represent an interesting alternative, since the material can be poured into almost any form without energy intensive processes (see Figure 2). Centric reinforcement made of CFRP-lamellae increases the load capacity without enlarging the material thickness and can be bent into curved formwork easily due to its low bending stiffness. Furthermore, centric CFRP reinforcement might be able to reduce or completely substitute the UHPC's steel fibers, whose alignment is difficult to control.

Fig. 2   Prefabricated UHPC learning pavilion at the campus of the TU Graz – student work by Stephanie Jordan, Nikolaus Pfusterschmied and Felix Zmölnig

4.2   Slim/Light Prefabricated Concrete Elements

Uniaxial spanned elements made of UHPC and GFRP-profiles are another promising application (see Figure 3). The GFRP functions as reinforcement and lost formwork at the same time and enables the production of efficient and light structural elements. For the manufacturing of non-standardized GFRP cross sections, available profiles can be joined together with glue. The conglutination is relatively simple, since the adhesive gap runs parallel to fiber direction and solely transfers shear forces.

The developed composite elements can be used for beam structures and columns as well. Compressive forces are absorbed by the UHPC cover, tension forces as a result of bending loads and buckling problems by the GFRP.

Fig. 3   Left: Quadratic hollow profile made of UHPC and GFRP tube elements – student work by Philipp Kramer and Mathias Schmid; Right: Light floor element made of UHPC and GFRP tube elements – student work by Magdalena Lang and Romana Streitwieser

4.3   Façade Application

Another promising application of UHPC-FRP composites are light façade elements, which consist of thin UHPC plates and a thermally insulated core. By frictional connection of plates and insulation, sandwich elements with high degree of stiffness and stability can be realized. High quality concrete plates enable architecturally demanding surface design and coloration, the good thermal insulting properties of GFRP can be utilized to avoid heat bridges.

5   Feasibility Studies

For evaluation of the intended applications´ feasibility, a comprehensive research into the bond and bending behavior of UHPC-FRP composites was carried out and is still running. In the following, the results of a pull-out test series with UHPC and CFRP-lamellae, and the results of a bending test series using thin UHPC-plates with centric CFRP-lamellae are presented.

5.1   Bond behavior

The described pull-out tests were carried out in October 2011 at the Laboratory for Structural Engineering at the University of Technology Graz (see Figure 4). The research target was to explore the bond behavior of UHPC and FRP in principle, and to find out the influences of surface roughening. The tests were carried in accordance to RILEM Recommendations [8].

The concrete cubes had dimensions of 20x20x20 cm and were manufactured with Ceracem, a UHPC premix developed by the companies Sika and Eiffage, which has maximum aggregate grain size of 7.0 mm and a steel fiber content of 2.0 vol.-%. [9]. The test objects were stripped after 24 hours and stored in the laboratory for at least 28 days without water storage or thermal after treatment. The CFRP-lamellae Carbodur S1014 (see Table 2) were used as reinforcement elements and were cut to cross sections of 15 mm x 1.4 mm.

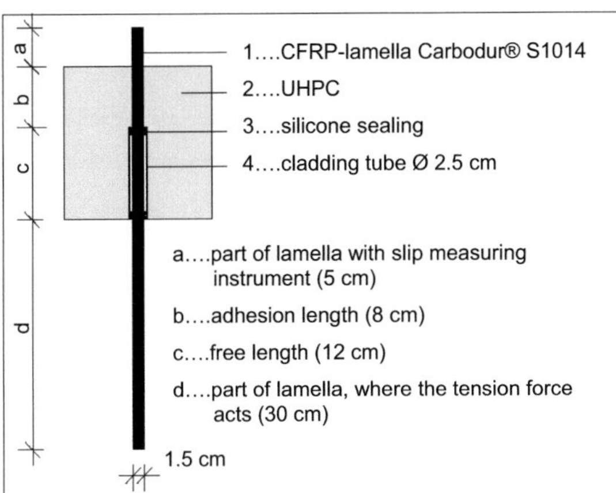

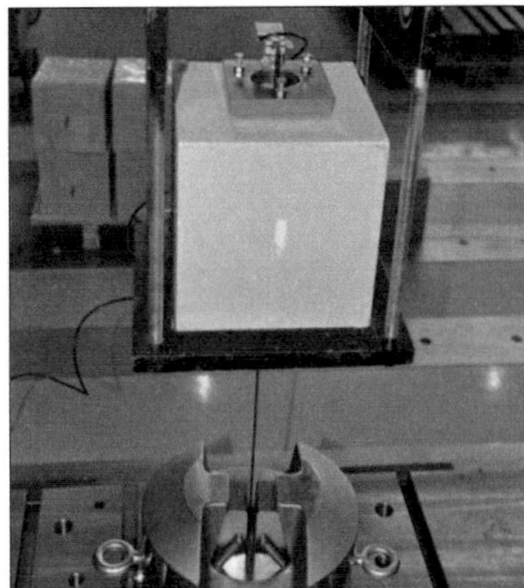

Fig. 4   Experiment setup of the performed pull-out tests

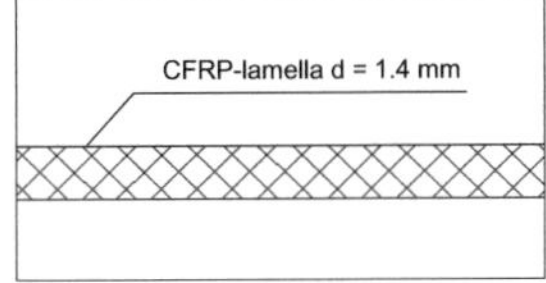

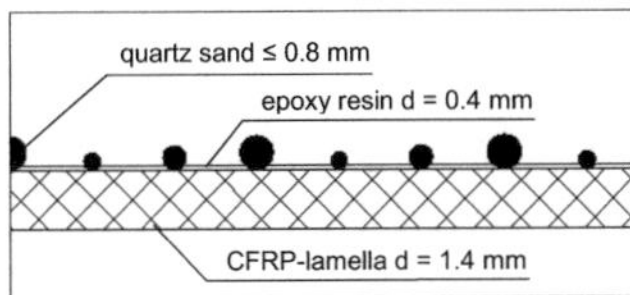

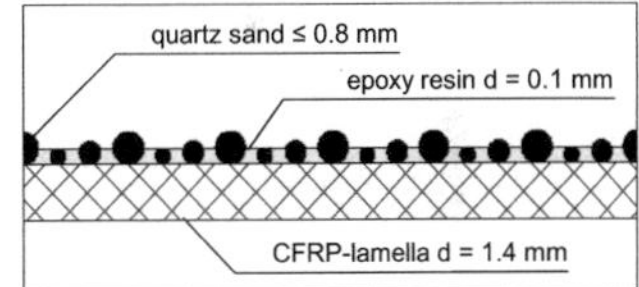

Fig. 5   Investigated surface characteristics: smooth, fine sanded and coarsely sanded lamellae

The tests were carried out with three varyingly rough lamellae surfaces: smooth (test series 1), fine sanded (test series 2) and coarsely sanded (test series 3), as shown in Figure 5. The smooth lamellae did not have any after-treatment, whereas the surface of the fine and coarsely sanded lamellae was manually coated with Sikafloor-156, a 2-component epoxy resin, and covered with quartz sand, which had maximal grain size of 0.8 mm. The fine sanded variant had an approximate resin thickness of 0.1 mm and the surface was only partially covered with sand grains. The coarsely sanded lamellae had an approximate resin thickness of 0.4 mm and the surface was fully covered with sand grains.

Table 3 and Figure 6 show that fine sanded surface (test series 1) achieved the best bond strength results by far with an average maximum bond stress of 8.30 N/mm². The smooth and coarsely sanded surface achieved an average of 2.51 N/mm² (test series 1) and 1.99 N/mm² (test series 3), which demonstrates that the surface roughening does not necessarily lead to bond strength improvement. The coefficient of variation, a measure of results´ dispersion, was 0.32 for smooth, 0.08 for fine sanded and 0.27 for coarsely sanded surface. Consequently, the fine sanded lamellae also achieved the lowest dispersion.

The test series 1 with fine sanded lamellae reached a maximum pull-out force of approximately 20 kN with a bond length of 8 cm. On the assumption, that the maximum chargeable bond stress remains constant when the bond length is enlarged, the projected anchorage length of the investigated CFRP lamellae is 25 cm.

Table 3   Mean values and coefficients of variation (COV) of maximum bond stress τ_{max} and corresponding slip s_{max}

Type of outer surface	Number of tests	τ_{max} (MPa)	COV	s_{max} (mm)	COV
Test series 1	3	2.51	0.32	0.08	1.35
Test series 2	3	8.30	0.08	0.04	0.12
Test series 3	3	1.99	0.27	0.27	0.35

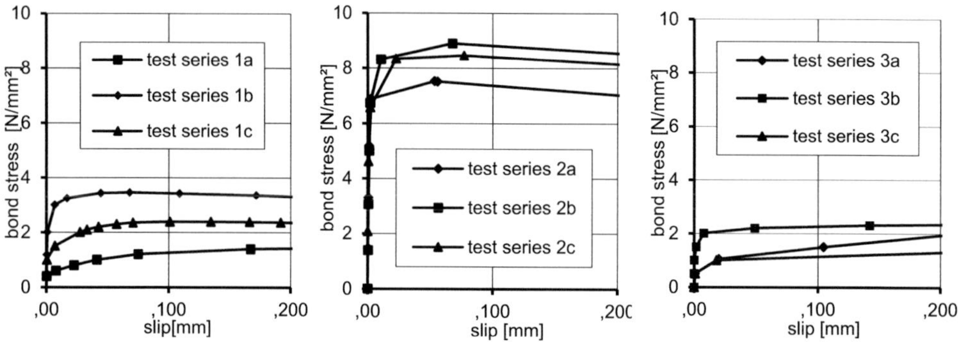

Fig. 6   Bond stress-slip relationship of test series 1-3

5.2   Bending Behavior

For the investigation of principle bending behavior of thin walled UHPC plates with steel-fibers and centric CFRP lamellae, 4-point bending tests were carried out in February 2012 at the Laboratory for Structural Engineering. The plate elements had a thickness of 2.5 cm and spanned a length of 60 cm (see Figure 7). The CFRP surface was roughened with fine sand, as described chapter 5.1.

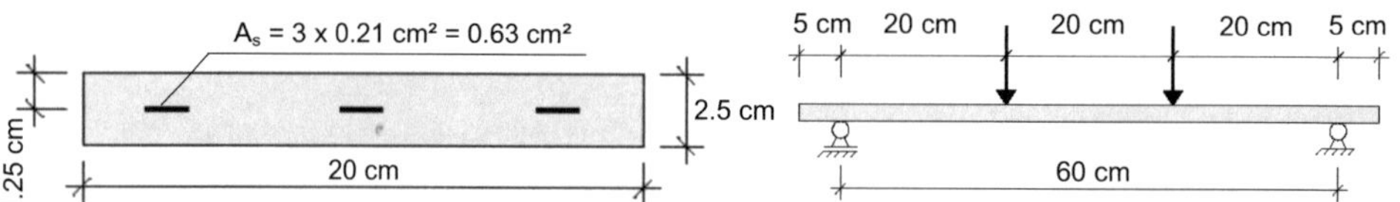

Fig. 7   Left: Cross section of the UHPC plate used for test series 1; Right: Experiment setup for the 4-point bending test

In order to get comparison values, the tests were carried out with two different types of reinforcement: UHPC plates with steel fibers and 3 centric CFRP-lamellae (test series 1), and UHPC plates with steel fiber reinforcement only (test series 2). The test's intention was to investigate the influence of centric CFRP-reinforcement on stiffness, crack formation and maximum load capacity of thin UHPC plates.

The used UHPC had a maximum aggregate grain size of 5.0 mm, 1.0 vol.-% steel fiber content and a water/cement-ratio of 0.285. The target compression strength was 160 N/mm². The plates were stripped after 24 hours and stored in the laboratory for at least 28 days without water storage or thermal after treatment.

Figure 8 and Table 4 show that UHPC plates with centric CFRP reinforcement (test series 1) had an almost 8 times higher breaking load than those without (test series 2). Consequently, the reinforcement causes a significant increase of load bearing capacity. The bending stiffness EJ^I in the non-cracked condition was 150,000 (test series 1) and 120,000 kNcm² (test series 2), the concrete cracked at a moment of 35 kNcm (test series 1) and 31 kNcm (test series 2). Thus, the centric CFRP reinforcement did not in-crease bending stiffness in non-cracked condition and crack moment substantially.

The bending stiffness in the cracked conditions EJ^{II} was 41,000 kNcm², which is approximately 27% of the bending stiffness in the non-cracked condition. The theoretically determined bending stiffness in the cracked condition without taking account of tension stiffening effect and fiber reinforcement is 12,000 kNcm², which is 3.4 times lower than the experimentally determined value. Hence, the good bond properties between CFRP and UHPC and the steel fiber reinforcement do have a remarkable influence on bending stiffness of the investigated plate.

In test series 1, the average crack distance on the underside of the plate was approximately 3.0 cm. This value also confirms the good bond properties, which were described in chapter 5.1. In test series 2, only one single crack appeared after exceeding the concrete's tensile strength, which kept on growing with increasing load.

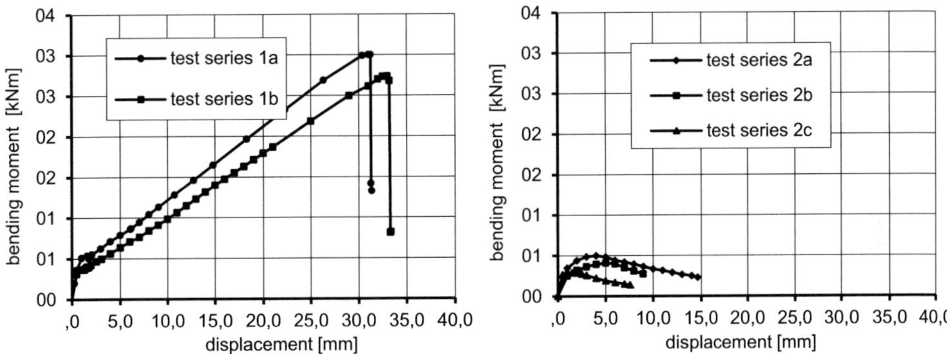

Fig. 8   Bending moment-displacement relationship of test series 1 and 2

Table 4   Mechanical properties of investigated plate

Characteristic values	Test Series 1	Test series 2
Number of test	2	3
Max. bending moment	287 kNcm	40 kNcm
Crack moment	35 kNcm	31 kNcm
Bending stiffness EJ^{I}	150,000 kNcm²	120,000 kNcm²
Bending stiffness EJ^{II}	41,000 kNcm²	-
Mean crack spacing	3.0 cm	-

6   Conclusions

Structures made of UHPC and FRP are a new kind of composites, which have only rarely been investigated up to now. Both materials are characterized by high mechanical and environmental resistance, and the good adhesion properties of UHPC make the combination with FRP reasonable.

The investigated construction method could be applied successfully to thin walled UHPC structures with CFRP lamellae for reinforcement. Due to the easy formability of UHPC, the realization of geometrically complex structures with prefabricated elements should be thought about. Furthermore, the combination of UHPC with structural GFRP profiles could generate light and robust alternatives to steel and timber constructions. Another interesting aspect is the utilization of GFRP's low thermal conductivity and the creation of thermally insulated elements for façade applications.

For evaluation of feasibility, pull-out tests with CFRP lamellae and UHPC were carried out, which show similar results as with normal concrete and circular FRP rods described in literature [11].

Proper sanding of the outer surface of FRP causes a remarkable increase of bond strength due to better chemical adhesion and friction coefficient. The achieved results are partially even larger than those of steel reinforcement and normal concrete. However, the increases in bond strength are related to an increase of brittleness.

For investigation of flexural behavior, 4-point bending tests were carried out. The results show an enormous increase of bending load capacity due to centric CFRP reinforcement and a linear elastic behavior until material failure. The results correspond with the behavior of over-reinforced cross sections in accordance to ACI 440.1R-06 [12], where the material failure is characterized by concrete crushing. Both performed studies suggest that a technical feasibility is given, although further extensive experimental and analytical investigations are required.

7   Acknowledgement

Our special thanks go to the Laboratory for Structural Engineering at the TU Graz and their head Dr. Bernhard Freytag, who enabled us to perform the experiments. Our thanks also go to the company Sika and Dr. Guenther Grass for the provision of materials. Furthermore, we would like to thank the company Exel Composites for supporting our research work and seminars with architecture students.

References

[1]   König, G.: Hochleistungsbeton (2001), Ernst & Sohn Verlag, Berlin

[2]   ÖNORM EN 1992-1-1: Eurocode 2: Bemessung und Konstruktion von Stahlbeton- und Spannbetonbauten (2009), Österreichisches Normungsinstitut

[3]   Deutscher Ausschuss für Stahlbeton: Sachstandsbericht UPHC (2008), Beuth Verlag, Berlin

[4]   Schürmann, H.: Konstruieren mit Faser-Kunststoff-Verbunden (2007), Springer Verlag, Berlin Heidelberg

[5]   Peters, S.: Glasfaserverstärkte Kunststoffe für den Fenster- und Fassadenbau (5/2009), In: *Innovative Fassadentechnik*, Ernst & Sohn Verlag, pp. 70-75

[6]   Bank, L.: Composites in construction (2005), John Wiley & Sons, New Jersey

[7]   Sika® CarboDur® Lamelle – Pultrudierte CFK-Lamellen für die Bauteilverstärkung (2009), Product Data Sheet, Company Sika®, Versionsnummer 2

[8]   RILEM Recommendation RC 6: Bond Test Reinforcing Steel. 2. Pull-Out Test (1982).

[9]   Maeder, U.: CERACEM a new high performance concrete: characterization and applications (2004), in M. Schmidt (ed): *International Symposium on Ultra High Performance Concrete, Kassel, 13-15 September 2004*, University Press, Kassel

[10]   Cosenza, E. et al.: Behaviour and Modelling of Bond of FRP Rebars to Concrete (1997), in Journal of Composites for Construction, pp. 40-51

[11]   ACI 440.1 R-06.2006: Guide for the Design and Construction of Structural Concrete Reinforced with FRP Bars (2006), American Concrete Institute, Farmington Hills, Michigan, USA

Model of cable supported bridges

Jan Koláček

Institute of Concrete and Masonry Structures,
Brno University of Technology, Faculty of civil Engineering
Veveří 331/95, 602 00 Brno, Czech Republic
Supervisor: Jiří Stráský

Abstract

This work deals with curved cable stayed and suspension pedestrian bridges.

In the first part, the equilibrium on a single-sided suspended section is studied. Distinguished are the sections with and without prestressing. The section with prestressing is analysed in detail on a cable stayed structure whose deck is in a form of semicircle with a radius of 30.00 m and on a suspension structure as an internal arrangement – both on the same plan. The notice is given to the verification of the equilibrium in the cross section, to finding of an initial state and to static and dynamic analyse.

The second part of the work deals with the verification of all findings. A physical model was built up in a scale 1:10 that included the both variants. Performed static and dynamic loading test verified the theoretical results.

1   Equilibrium in transversal direction with single-sided suspension

If a deck of asymmetrical section is suspended single-sided only, then its section is stressed by a torsional moment. That is why a section and its arrangement should be designed so that the torsion caused by the deck self-weigh and dead load would be minimal. The torsion of the deck will probably happen when the deck is stressed by a variable load (i.e. by pedestrians), but this stress should not influence the structure significantly.

1.1   Sections without prestressing

For the section without prestressing (Fig. 1 – left) holds that the position of suspension point (SP beginning point of a suspender) does not depend on a force magnitude but only on the section geometry (characteristics), specifically on the horizontal distance between the centre of shear and gravity axis and on the magnitude of angle α of the suspenders. At the changing angle α the point of suspension will change, too. Fig. 1 – right shows that the suspenders at any angle direct always into one point. This point is called a centre of suspenders and is labeled as C_z. Fig. 1 – (right) does not exhibit the whole cross section, but only its important points. In a simplified way it can be stated that the centre of suspenders will originate as an intersection of the vertical plane leading through the centre of gravity with the horizontal plane leading through the centre of shear.

This theory was verified on a simple study of single-sided suspended section.

1.2   Equilibrium of section with prestressing

Application of prestressing has several advantages. In the case of a deck formed either as a concrete slab or a concrete box it is possible to create a pressure reserve and thus minimise the flexural stress of extreme fibres of the deck. Another advantage of prestressing is utilization of the radial effects of prestressing cables for the regulation of equilibrium in the cross section in the case of a curved deck.

With the help of the prestressing moment it is possible to locate the suspension point almost everywhere. On the contrary of the section without prestressing, it does not depend on the mutual position of the gravity axis and centre of shear. Nevertheless, it is important so that the suspender's plan and radial resultant of the cables will be perpendicular to a curve of the deck. It will happen when the deck curvature is created by a circle.

Proc. of the 9th *fib* International PhD Symposium in Civil Engineering, July 22 to 25, 2012,
Karlsruhe Institute of Technology (KIT), Germany, H. S. Müller, M. Haist, F. Acosta (Eds.),
KIT Scientific Publishing, Karlsruhe, Germany, ISBN 978-3-86644-858-2

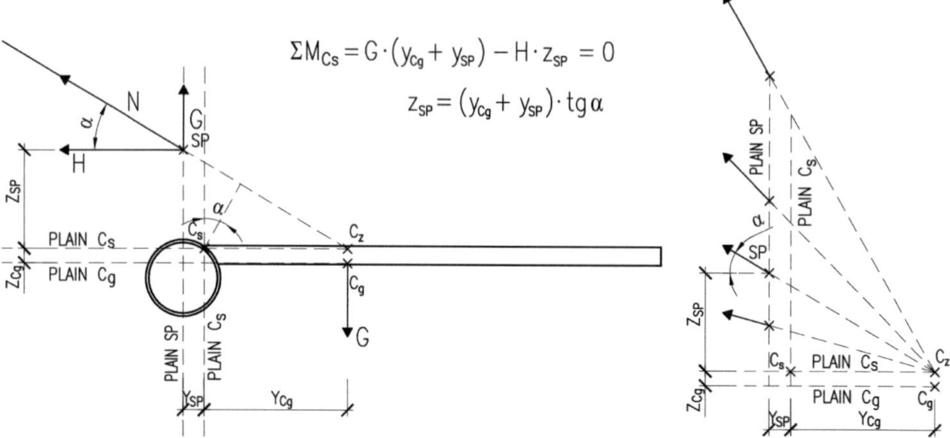

$$\Sigma M_{Cs} = G \cdot \left(y_{Cg} + y_{SP}\right) - H \cdot z_{SP} = 0$$

$$z_{SP} = \left(y_{Cg} + y_{SP}\right) \cdot \mathrm{tg}\,\alpha$$

Fig. 1   Equilibrium of the forces in a section for angle α (left) and for any angle α (right).

2   Study of curved cable stayed structure

Objective of the study was to design a curved cable stayed structure with the application of all the above mentioned piece of knowledge and methods. The study consisted of a design of a bridge deck with prestressing cables, finding of an initial state and static and dynamic analyse.

2.1   Description of the structure

The deck was formed by a curved concrete stress ribbon stiffened with a steel pipe embedded at the inner edge. The pedestrian bridge of a span 60 m was in a form of a semicircle of a radius 32 m in the deck axis. It completed exactly an angle 180°. Both the deck and steel pipe were fixed into the anchor blocks. The steel pipe was supplemented with L-shaped members that formed railing posts. Prestressing cables were situated in the steel pipes led between the railing posts and in the floor beams situated under the deck. The structure was suspended with 14 stay cables on a pylon situated in the middle of the circle.

The cross section was designed in combination of the steel pipe and concrete slab. The steel pipe transfers the longitudinal torsion and bending, slender concrete slab resists to the stress in the transverse direction. The solution considered the slab to be composite with the steel pipe. The railing was welded to the pipe, together with the T-shaped steel floor beam. For leading of the prestressing cables, the holes were left both in the railing posts and floor beams. The position of prestressing cable in the railing was common with the position of suspension point SP.

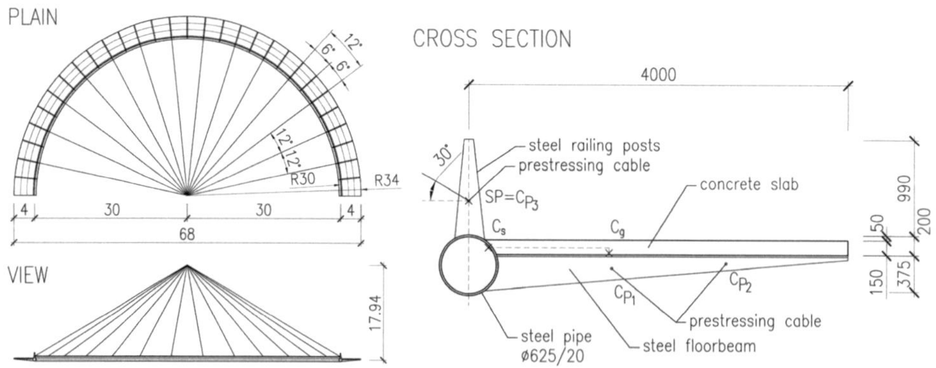

Fig. 2   Plain and view (left) and cross section (right) of cable-supported footbridge.

2.2   Verification of the equilibrium

The verification of the section was carried out in two phases. The position and diameter of prestressing cables were designed in the first phase. In the second phase was designed the position of the suspending point according to the conditions of equilibrium. For future designs it will be enough to determine the balance in a cross-section only in the suspending points about one half of a span with eventual small correction. The suspending points for others suspenders will be designed on the same level then. This structure is not sensitive on small deviations (up to 10 mm) of the suspending points and so an averaging of the suspending point position of all suspenders (i.e. the same level of the suspending point) has an importance in the simplification of design and construction.

2.3   Initial state of the structure

An aim of finding an initial state of the cable-stayed structure was to find out such the forces and strains so the vertical deformation and rotation at the suspending points were zero. For methods for finding such a shape were tested. The best results were reached by a method of vertical deflections equilibrium. This iteration method needs two random steps at the start. In the third step the strain in each suspender was adjusted so the linear extrapolation or interpolation of the vertical deformation of the upper surface of the pipe was zero at the suspending point SP.

2.4   Results of static and dynamic analyse

The 3D calculation model using software ANSYS was set up for the static and dynamic analyse. Both the response of the stress induced by pedestrians in various positions and temperature were modelled only, since these variable loads are the most frequent. Wind load wasn't considered. The dynamic analyse included the calculation of a shape of the structure (Table 1) and due to very low first bending frequencies also the response of the structure to the harmonic excitation.

Table 1   Comparing of cable stayed and suspension structure with regard to natural frequencies.

Study		Cable stayed structure			Suspension structure		
Mode		Frequencies [Hz]	Mode Shape		Frequencies [Hz]	Mode Shape	
vertical	1st	1.05	B	$f_{(1)}$	0.97	A	$f_{(1)}$
	2nd	1.09	A	$f_{(2)}$	1.11	B	$f_{(2)}$
swing	1st	3.40	C	$f_{(7)}$	3.86	C	$f_{(10)}$
torsion	1st	4.84	D	$f_{(10)}$	5.10	D	$f_{(12)}$
MODE A		MODE B		MODE C		MODE D	

3   Study of a plan curved suspension structure

A scope of this study was processing of the finding of an initial state of the suspension structure similar to the study of the cable stayed footbridge. The process was verified on a timber model in a scale 1:100 with the static and dynamic analyse followed.

3.1   Description of the structure

The footbridge structure is formed by curved concrete stress ribbon - the same as the cable stayed structure. The structure is of a span 60 m, in the plan of a semicircle with a radius of 32 m in the deck axis. It completes exactly an angle 180°. Both the deck and steel pipe are fixed into the anchor blocks.

Prestressing cables are situated in the steel pipes led between the railing posts and in the floor beams situated under the deck. The suspension structure supported by 12 suspenders attached to the suspension cable anchored in a pylon that is situated in the middle of the circle.

A height of the pylon was derived from the cable-stayed variant, i.e. the vertical distance from the suspending point to the pylon's top was the same. In the transversal direction was designed a combination of the steel pipe with concrete slab. The concrete slab was considered to be composite with the steel pipe. All the members of the structure were of the same dimensions and materials, only the suspending point was moved to the steel pipe top (see Fig. 3 – right). The prestressing cable in the railing was in such a height so the balance in the section was fulfilled.

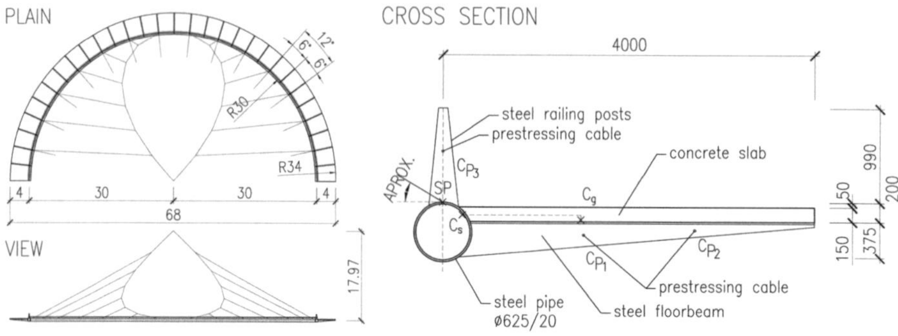

Fig. 3   Plain and view (left) and cross section (right) of suspension footbridge.

In opposite to the cable stayed structure, both the position of suspending point and prestressing force are designed and the position of prestressing cable in the railing is searched for.

3.2   Finding of an initial state

The initial state of the suspension structure is the same - vertical deformations and rotations of the structure at suspending points have to be zero. The suspension cables is assembled of 12-0.6" strands. There are two possibilities of arrangement of suspension cable that we can call an internal (see Fig. 3 – left) and external arrangement (see Fig. 4 – right).

Since the structure with internal arrangement has not been built so far, it was decided to study this structural arrangement. The initial state was determined in two steps, vertical state and horizontal one. The vertical state was determined in the way that similar to the way used in straight suspension structures.

The horizontal state was determined by iteration. At first the curved deck was considered as infinitively stiff and vertical positions of the suspenders were taken from the first step. During the iteration the horizontal movement of the suspension cable and strain in the cable and suspenders were modified. The final state was the state in which horizontal was minimum. After that the deck received the actual stiffness and iteration was repeated.

Fig. 4   Internal (left) and external (right) arrangement - semicircle curve of the bridge deck.

Due to the fact that the suspenders are not perpendicular to the deck curvature, the suspenders loads the deck not only by radial forces, but also by tangent ones. However, since the deck's gravity central is close to the suspension, the corresponding stresses are not significant. The optimum plan curvature, in which the suspenders are always perpendicular to the structure, was not found. The curvatures of the shapes of second degree parabola and ellipse were also studied. The best arrangement gave the second degree parabola with a rise of 30 m gave the best results. However, since the differences between this curvature and a semicircle curve was not significant, the semicircle curve was used.

For the verification of the described analytical process simple timber models were prepared - see Fig. 4. The models proved that the described process is correct.

3.3 Results of static and dynamic analyse

The structure was checked for the same loads as the cable stayed structure. Though the suspension structure has lower bending stiffness, its correct function was proved by the dynamic analyse [1].

4 Model in the scale 1:10

The results from previous studies were verified on a model in a scale 1:10. Both structures were put together so the horizontal reactions from the pylon were eliminated. It also made possible to compare both the structures.

4.1 Description of the model

The static model combines both structures that are anchored at common anchor blocks. Its span is 6.0 m, the radius in deck axis is 3.2 m.

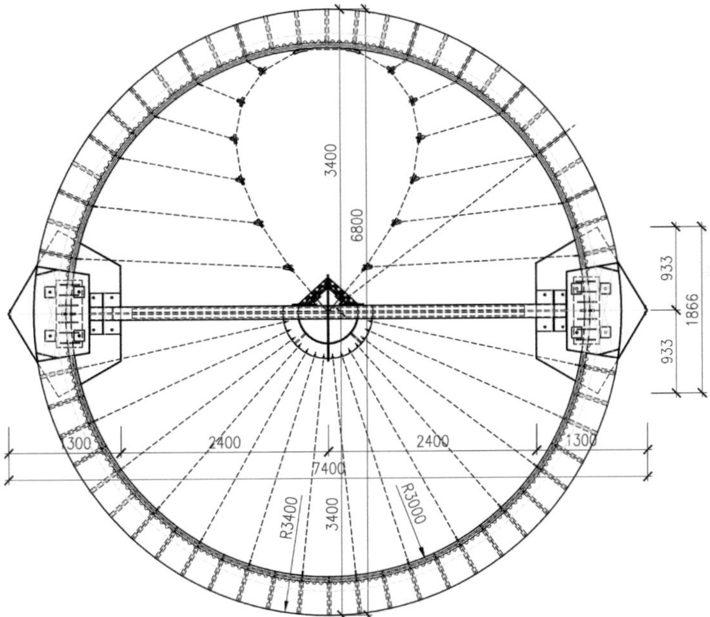

Fig. 5   Plain of the model (M1:10).

The anchor blocks are supported by concrete pedestal used for testing another structure. The pedestals are connected by steel girders and PT bars. The pedestals also support an A shape pylon that is formed by two legs mutually connected by a tension tie. The external cables overlapped at the anchor blocks.

The composite deck is assembled of a steel pipe of 63 mm diameter and the deck slab of thickness of 20 mm. The thickness of the pipe is 3.6 mm. The connection of the pipe with concrete slab is done by a steel wavy line welded to the pipe and cast in the slab and by a shear members placed on the floor beams. Since it is not possible to place the reinforcing bars into the 20 mm thick plate, the concrete is reinforced by fibres. To compare their function one half is from steel fibres, the second one from glass fibres. A design of both concrete mixes was influenced by the requirements on minimum concrete shrinkage.

To guarantee a model similarity, both structures have to loaded by additional load formed by steel bars and concrete blocks. The value of this load was determined in such a way that resultant stresses in the actual structures and in the model are same. Additional load is suspended on steel floor beams and steel pipe.

The model was analysed by a program system ANSYS with a similar modelling as the actual structures (see Fig. 6 – left). At first both structures were analysed individually, later the structures were connected together.

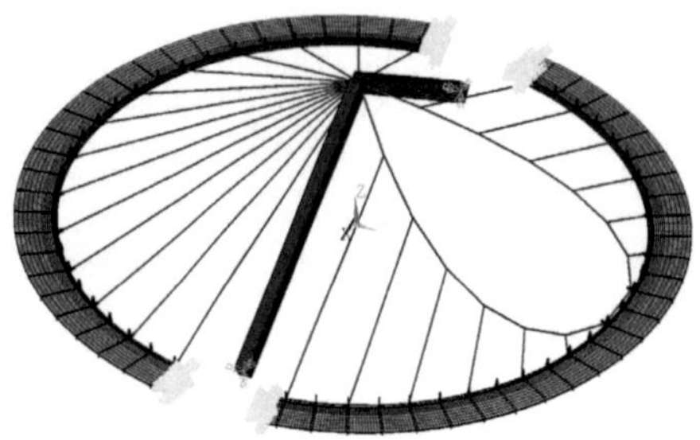

Fig. 6   The analysis model M1:10 (left) and completed structure (right).

4.2   Construction sequences

Since it is supposed that the actual structure will be erected on the temporary supports, the model was built similarly. The construction process was as follows:

- erection of the concrete pedestal,
- erection of the pylon and decks steel structure. Each erected structure is supported by two temporary supports and pedestals,
- erection of the stay cables, suspension cables and suspenders,
- placing of the formwork of the deck slab and anchor blocks,
- erection of the additional load that guarantee the model similarity,
- tensioning of the stay cables, suspension cables and partial tensioning of the radial cables, removing the temporary supports,
- reinforcing of the concrete blocks, casting the blocks and deck slab,
- when concrete reached 80% of the design strength the stay and suspension cables will be tension to their final values (Fig. 6 – right),
- loading the structure by a design and ultimate loads.

4.3   Loading tests and results

At first, basic loading tests verifying the function and correctness of calculation were carried out on the model. Then, both structures were individually tested on the ultimate limit strength.

The test results demonstrated very good correspondence with the calculation model.

5   Conclusions

From up to now results it can be stated that the semicircle curved deck suspended at internal edge only is functional for common loads acting on the footbridges, namely both for the simple cable stayed structure and for suspension structure in the internal arrangement.

The cable stayed structure has an advantage both of the easy geometry and calculation of the initial shape. In the deck none additional stresses originate since the suspender is always perpendicular to its tangent.

The suspension structure has more aesthetical appearance but the finding of the initial shape of cables and suspenders is more complicated task. Since the finding of the deck curve with the suspenders to be always perpendicular to it was not successful, the origin of additional tensile stress should be admitted. However these stresses will be in this case reduced by the longitudinal prestressing. These forces and the longitudinal moments originated from them are not so great due to short distance of the suspending point with the centre of gravity of the section.

This outcome has been achieved with the financial support of the Ministry of Education, Youth and Sports of the Czech Republic, project MSM No. 0021630519 „Progressive Reliable and Durable Structures" a Ministry of Industry and Trade of the Czech Republic, project FI-IM/185 „New, Cost-effective Structures from High-strength Concrete"

References

[1]   Strasky, J.: Stress ribbon and cable-supported pedestrian bridges. Thomas Telford Publishing (2005), ISBN 0 7277 3282 X, 229 pages.

Global model quality evaluation of coupled partial models for restraint effects in reinforced and prestressed concrete structures

Bastian Jung

Research Training Group "Model Quality" (Ger.: Graduiertenkolleg "Modellqualitäten")
Bauhaus-Universität Weimar,
Berkaer Straße 9, 99425 Weimar, Germany
Supervisor: Guido Morgenthal

Abstract

In the general design process of structures the engineer needs to decide which models are suitable for simulating realistically and efficiently the physical processes determining the structural behaviour. The theoretical knowledge, but also the experience from prior design processes will influence the model selection decision. It is thus often a qualitative choice of different models. The goal of this paper is the quantitative evaluation for coupled numerical partial models to assess the Global Model Quality. The evaluation is applied to the simulation of a semi-integral prestressed concrete girder bridge. The material behaviour, the creep and the shrinkage both for the superstructure and piers, furthermore geometrical nonlinearities and temperature distributions are considered as partial models. The results show that the Global Model Quality is strongly dependent on the sensitivity and quality of each model class with its corresponding partial models.

1 Evaluation method for Global Model Quality assessment

Global models (GM) for numerical simulation approaches utilise different model classes (M) with subordinate partial models (PM). Material descriptions, creep and/or shrinkage models are named here as possible M for concrete structures. Interactions and couplings of their PM are necessary for determining an appropriate structural behaviour. Therefore the following evaluation method enables to assess the Global Model Quality. For detailed information the author recommends Keitel et al. [1].

1.1 Sensitivity according a model class

The first step is to quantify whether the class has an influence on a certain target value. This is evaluated by using Sensitivity Analysis [2] which in general is the study of how the output of a model (Y) is related to the model input (X). By using discrete random variables for selecting the model class, the Sensitivity Study in this case is not an estimation of uncertainty, but a quantified value of the influence of the M class (X_i). The First Order Sensitivity Index is

$$S_i = \frac{V(E(Y|X_i))}{V(Y)}. \tag{1}$$

This index S_i illustrates the exclusive influence of model X_i. According to interactions in complex engineering problems higher order Sensitivity Indices are needed. The Total Effect Index is defined as

$$S_{Ti} = 1 - \frac{V(E(Y|X_{\sim i}))}{V(Y)}. \tag{2}$$

A finite number of possible model class combinations n_{comb} are necessary for the indices:

$$n_{comb} = 2^{n_M} \tag{3}$$

with n_M random variables (model classes). A measure of interaction between X_i and other model classes is the difference between S_i and S_{Ti}. High values of these Sensitivity Indices highlight a significant influence of this partial model class on the response of the global model. Models with values

Proc. of the 9th *fib* International PhD Symposium in Civil Engineering, July 22 to 25, 2012,
Karlsruhe Institute of Technology (KIT), Germany, H. S. Müller, M. Haist, F. Acosta (Eds.),
KIT Scientific Publishing, Karlsruhe, Germany, ISBN 978-3-86644-858-2

smaller than a given threshold (here: $S_{Ti} \leq 0.03$) shall be neglected for the next evaluation method step. No further investigations about their Partial Model Quality are performed.

1.2   Sensitivity according the choice of partial models

This second method step quantifies the importance of selecting a partial model from one model class. It is also based on Sensitivity Studies [1,2]. The choice of each PM within a model class is controlled by X_i. The Total Effect Sensitivity Index indicates how this choice leads to a variation of the global model response according to a certain output value. Low values show, that the different partial models in the same model class give a similar contribution to the structural response value and do not affect significantly these response value. These indices are used as weighting factors for the importance of the quality of a PM in a model class.

1.3   Quality of coupled partial models

The Global Model Quality of coupled partial models is quantified by a path through a graph (MQ_{GM}) with the vertex as the quality of the partial model MQ_{PM} and the edges as the coupling quantities. A number between 0 and 1 express this quality. 0 signifies a poor and 1 a high MQ_{PM}. These quantitative values come from the evaluation of the PM itself, using Uncertainty, Complexity or Robustness criteria's [3]. Assuming a perfect data coupling between each model classes the model quality of a global structural model is defined as [1]:

$$MQ_{GM} = \sum_{i=1}^{n_{M,red}} \frac{S_{Ti}^{MC} * MQ_{PM_j}}{\sum_{i=1}^{n_{M,red}} S_{Ti}^{MC}} . \tag{4}$$

PM_j is one partial model of class M_i. The number $n_{M,red}$ is the count of non-negligible partial model classes on the global response, determined by method step one. This Global Model Quality Evaluation method is applied to a reinforced and prestsressed concrete bridge below.

2   Application to semi-integral concrete bridge

2.1   Geometry, material properties and loading

The geometry of the longitudinal and vertical direction of the bridge and the prestressing steel is shown in Fig.1. The cross section of the superstructure and the piers is shown in Fig. 2 and the material properties are listed in Table 1.

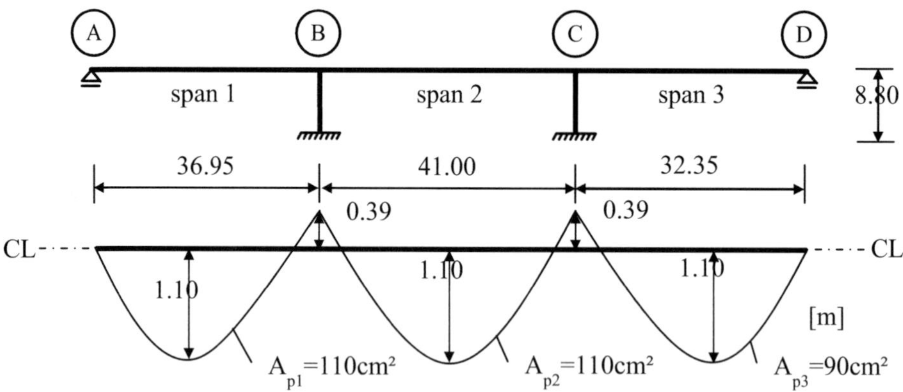

Fig. 1   Bridge and prestressing geometry

The classical decoupled connection between the piers and the superstructure is adjusted to a coupled semi-integral bridge. Therefore the overall structural load-deformation behaviour is affected by the interaction within the piers and superstructure, particularly in case of restraint effects. Hence, the interference between the partial models is investigated.

The structural behaviour is simulated under quasi-permanent loading [4] for 100 years of service life (design life):

- dead load (G):   superstructure 142 kN/m, pavement 24 kN/m, piers 6.75 kN/m
- prestressing (P) σ_p=1295 MN/m²
- imposed traffic (Q_{k1}):   UDL 46.375 kN/m, TL 400 kN (span 1)
       $\psi_{2,1}$=0.20
- temperature load (Q_{k2}): T_0=10°C, T_{min}=-24°C, $T_{e,min}$=-16°C, ΔT_N=-26 K, ΔT_M=-8.8K
       $\psi_{2,2}$=0.50

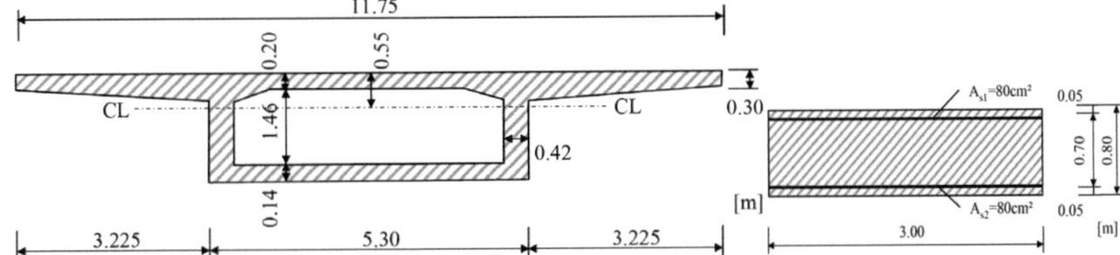

Fig. 2   Cross section superstructure and piers

Table 1   Material parameters

	Unit	Deck	Column
Concrete		C50/60	C35/45
CEM	-	II 52,5N	II 42,5N
E_{c0m}	[MN/m²]	38,500	33,300
E_{cm}	[MN/m²]	32,800	28,300
f_{cm}	[MN/m²]	58	43
f_{ctm}	[MN/m²]	4.1	3.2
Steel		Y1770	B500B
E_s	[MN/m²]	190,000	200,000
f_y	[MN/m²]	1,500	500

2.2   Considered partial models

The material description (Model Class A) for the concrete compression range is modelled with linear-elastic relation between strains and stresses. Because of the prestressing and the quasi-permanent loading the compression stresses are smaller than $0.40f_{cm}$. Therefore linear-elastic material behaviour can be assumed. In the range of tensile concrete parts the concrete can either sustain stresses until f_{ctm} (A-1) or cracking shall be considered with the application of a tension stiffening model as $\beta_{ct}*f_{ctm}$ until $\varepsilon_{ct} \leq \varepsilon_{ys}$ (A-2).

In order to describe the time-dependent increase of the creep compliance two creep models (Model Class B) are investigated. These are the model according Model Code 90-99 (B-1: MC 10) [5] and Gardner and Lockman (B-2: GL2000) [6].

The time-dependent shortening of concrete due to the hydration process and the drying of the cement is simulated by the shrinkage models (Model Class C). The chosen shrinkage models correspond to the creep models: Model Code 90 (new edition 2010, B-1: MC 10) [5] and Gardner and Lockman (B-2: GL2000) [6].

Geometrical nonlinearities (Model Class D) can affect displacement values and section forces. The nonlinear kinematic (D-1) and the p-Δ (D-2) approaches are considered in this model class.

Restraint effects in concrete structures may occur as a result of imposed deformations such as thermal actions (Model Class E). In the standard code EN 1991 [9] specific values are stated for temperature conditions and temperature distributions. One possibility to take thermal actions on bridges into account is the approach with constant temperature (ΔT_N) and linear shifting values over the cross section height (ΔT_M). Beyond, thermal actions can also be considered by the temperature (ΔT_N) and nonlinear varying values (ΔT) over the cross section height. Combination factors for the concurrent occurrence of both temperature parts are included to account for their coincident probability. Four temperature distributions are considered as partial models in the model class temperature:

- TEMP1   E-1:   constant with linear shifting $0.35*\Delta T_N + \Delta T_M$
- TEMP2   E-2:   constant with linear shifting $\Delta T_N + 0.75*\Delta T_M$
- TEMP3   E-3:   constant with nonlinear distribution $0.35*\Delta T_N + \Delta T$
- TEMP4   E-4:   constant with nonlinear distribution $\Delta T_N + 0.75*\Delta T$

The creep $\varepsilon_{c,cr}(t)$, shrinkage $\varepsilon_{c,sh}(t)$ and temperature $\varepsilon_{c,t}(t_0)$ strains are expressed by additional strain components of the concrete, which leads to the total strains of the concrete:

$$\varepsilon_{c,tot}(t) = \varepsilon_{c,el}(t) + \varepsilon_{c,pl}(t) + \varepsilon_{c,cr}(t) + \varepsilon_{c,sh}(t) + \varepsilon_{c,t}(t_0) \tag{5}$$

with $\varepsilon_{c,el}(t)$ and $\varepsilon_{c,pl}(t)$ as the time-dependent elastic and plastic strains.

2.3   Structural response values for the quantification

In case of the first evaluation method step, the Sensitivity is quantified for the vertical deformations in all spans, horizontal deformations at each bridge axis, concrete compression and prestressing steel tensile stress in the superstructure, concrete and reinforcement stresses in the piers and axial and bending moment section forces at different positions. The 8 model classes lead to 256 model combinations (2^8) independent of the target values for the structural behaviour.

2.4   Sensitivity according the model class

The discrete random variables control, whether the model class is activated or deactivated. In terms of the material behaviour either tension stiffening or purely linear-elastic material is modelled, in terms of creep or shrinkage either creep or shrinkage strains are computed or neglected, in terms of geometric nonlinearity either the second order or the first order kinematic is used and in terms of temperature either temperature strains occurring from constant and shifting parts are considered or zero. Table 2 shows the First Order and Total Effects Sensitivity Indices for a selection of target values.

It is obvious that the material description, the shrinkage of the piers and the kinematic have a negligible influence on the results. They are not studied further in the assessment of model quality.

Table 2   Sensitivity indices for the model classes according target values, first row for each target value: First Order Effect S_i^M, second row for each target value: Total Effects S_{Ti}^M

Model Class	$\sigma - \varepsilon$ super-struct. A	$\sigma - \varepsilon$ piers A	creep super-struct. B	creep piers B	shrink-age super. C	shrink-age piers C	geom. kine-matic D	tem-pera-ture E
Vertical dis.	0.000	0.000	**0.975**	0.001	0.000	0.012	0.000	0.000
span 1	0.000	0.000	**0.976**	0.012	0.001	0.012	0.000	0.001
Horizontal dis.	0.000	0.000	**0.076**	0.000	**0.850**	0.000	0.000	**0.074**
axis C	0.000	0.000	**0.076**	0.000	**0.850**	0.000	0.000	**0.074**
Concrete stress	0.000	0.004	**0.054**	**0.265**	0.002	0.022	0.000	**0.514**
superstr. span 2	0.000	0.023	**0.098**	**0.355**	**0.072**	0.025	0.000	**0.580**
Concrete stress	0.000	0.003	0.014	**0.296**	**0.548**	0.004	0.000	**0.045**
pier axis C	0.000	0.013	0.016	**0.382**	**0.620**	0.004	0.000	**0.059**
Bending moment	0.000	0.003	**0.132**	**0.170**	0.001	0.023	0.000	**0.424**
right axis B	0.000	0.016	**0.316**	**0.230**	**0.047**	0.025	0.000	**0.622**
Axial force super-	0.000	0.002	0.006	**0.164**	**0.732**	0.001	0.000	**0.047**
str. right axis B	0.000	0.010	0.008	**0.208**	**0.770**	0.001	0.000	**0.056**
Bending moment	0.000	0.003	0.018	**0.207**	**0.663**	0.000	0.000	**0.057**
pier bottom axis C	0.000	0.016	0.020	**0.252**	**0.700**	0.001	0.000	**0.070**
Axial force pier	0.000	0.001	0.000	**0.091**	**0.631**	0.016	0.000	**0.217**
bottom axis C	0.000	0.004	0.024	**0.110**	**0.647**	0.016	0.000	**0.245**

The creep phenomenon increases the strains for the quasi-permanent loading for 100 years design life. The vertical displacements in the superstructure are almost exclusively sensitive to this model class. Non activated creep modelling will reduce the predicted vertical displacement significantly. The creep phenomenon regarding the vertical displacements can thus not be neglected.

In case of horizontal displacements the major impact occurs from the shrinkage model class. Shrinkage strains must be included without any reduction factors. Temperature strains for the quasi-permanent loading are reduced by the combination factor $\psi_{2,2}=0.50$. This leads to:

- Shrinkage MC10: $\varepsilon_{c,sh}(36510d)$ $= -4.204E\text{-}4$
- $\Delta T_N=-26$ K: $\varepsilon_{c,t}(10d)= -26K*1.0E\text{-}5*0.5$ $= -1.300E\text{-}4$

and therefore to higher sensitivity of the shrinkage phenomenon according the horizontal displacement.

The difference between S_i and S_{Ti} (>0.05) such as the concrete stress in span 2 clarify a strong interaction between model classes. The deformation behaviour of the piers and superstructure affect each other and therefore the coupling of their model classes has a strong influence on the structural response. The influence of choice of different partial models in each model class is quantified for the grey marked rows of Table 2.

2.5 Sensitivity according the model choice

The analysis of the Total Effect Sensitivity Index enables the quantification of the model choice importance (comparable as weighting factors). For example, the prognosis of the models MC10 and GL2000 for creep and shrinkage are different and the influence of it can be computed by Sensitivity Analysis. Table 3 shows these weighting factors, which quantify the impact of model selection according to the chosen structural response values.

Table 3   Total Effect Sensitivity Indexes S_{Ti}^{MC} for the model choice according the important model classes for different target values, -... are unimportant model classes

Model Class	σ - ε super-struct. A	σ - ε piers A	creep super-struct. B	creep piers B	shrink-age super. C	shrink-age piers C	geom. kine-matic D	tem-pera-ture E
Horizontal dis. axis C	-	-	0.405	-	**0.496**	-	-	0.099
Concrete stress superstr. span 2	-	-	0.121	0.341	0.007	-	-	**0.622**
Bending moment right axis B	-	-	0.285	**0.490**	0.010	-	-	0.252

2.6 Global Model Quality

The Partial Model Quality for the creep models is analysed by uncertainty analysis including model- and parameter uncertainty and is stated in [1]. The Partial Model Quality of the shrinkage models is assessed on the variation of the error of the prediction. This uncertainty is $CV_{MC10}=0.481$ and $CV_{GL2000}=0.433$ [8]. In relation to the lowest model uncertainty of $CV_{B3}=0.374$ the Partial Model Quality is defined as:

- MC10: $MQ_{PM,MC10}=$ $0.374/0.481$ $= 0.78$
- GL2000: $MQ_{PM,GL2000}=$ $0.374/0.433$ $= 0.86$

Linear and nonlinear temperature models [7] are quantified by their prognosis of the induced strains. The complexity of the nonlinear temperature distributions is higher in comparison to the linear approaches. It can be assumed, that their Partial Model Quality is highest (of the considered) and the linear distributions are quantified relatively by the model outputs for the concrete stress in span 2, which is selected of the Global Model Quality Evaluation.

The important model classes with their respective partial models are shown in Fig. 3. The unimportant model classes are excluded for the Global Model Quality Evaluation. The influence of the model selection in every model class is expressed by the Total Effect Sensitivity Index (bottom of Fig. 3). The mentioned Partial Model Qualities are expressed in the vertices. The coupling (edges) is without any loss of data information.

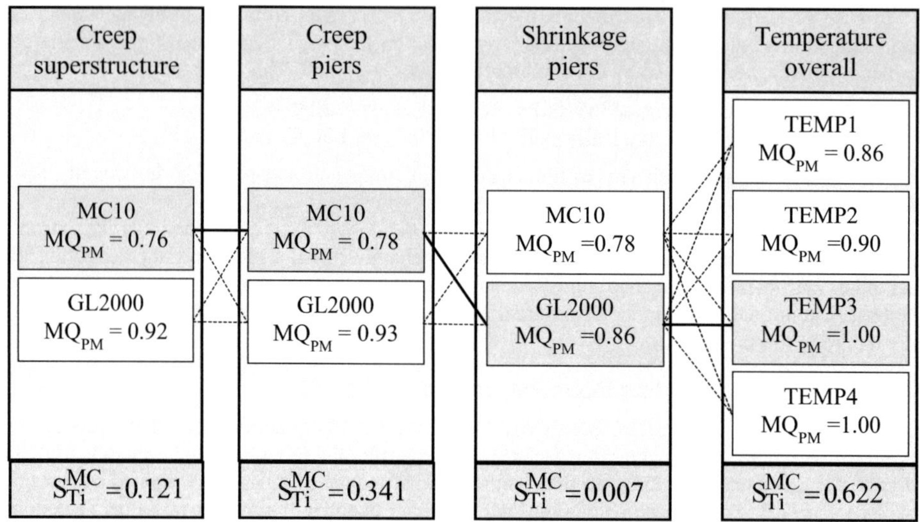

Fig. 3   Global Model Quality Evaluation according the concrete stress in span 2 for the application of a semi-integral bridge

The grey highlighted partial models express one admissible path throw the graph (32 possible combinations). This combination of partial models will lead to the following Global Model Quality:

$$MQ_{GM} = \frac{0.121*0.76 + 0.341*0.78 + 0.007*0.86 + 0.622*1.00}{0.121 + 0.341 + 0.007 + 0.622} = 0.90 \qquad (5)$$

Conclusions

The evaluation method for accessing the Global Model Quality for coupled partial models [1] is applied on a semi-integral bridge. Sensitivity Analyses quantify in a first step the influence of the phenomenona (model classes) like creep, shrinkage, material description, geometrical nonlinearities and temperature distributions. They depend on the structural output value (displacements, stresses, section forces). In a second method step the impacts of the model choice of a partial model in the same model class are analysed. Global Model Quality is evaluated by a path through the graph of partial models whereby each possible combination of the models will lead to a changed Global Model Quality. The structural application of a semi-integral bridge shows the applicability of the evaluation method and quantifies the important model classes and the model selection process. The Global Model Qualities are useful to compare different simulations in a quantitative manner.

References

[1]   Keitel, H. et al.: Evaluation of coupled partial models in structural engineering using graph theory and sensitivity analysis. In: Engineering Structures (2011) No. 33, pp. 3726-3736

[2]   Saltelli, A. et al.: Global sensitivity analysis. In: The Primer. John Wiley and Sons (2008)

[3]   Lahmer, T. et al.: Bewertungsmethoden für Modelle des konstruktiven Ingenieurbaus. In: Bautechnik Sonderheft (2011) No.6

[4]   DIN-Fachbericht 101: Actions on bridges. Beuth (2009)

[5] CEB-Comité Euro-International du Beton: Structural concrete: textbook on behaviour, design and performance – updated knowledge of the CEB/FIB Model Code 90. tech. rep. (1999) and CEP-FIP Model Code 2010 – First complete draft (2010)

[6] Gardner, N.; Lockman, M.: Design provisions for drying shrinkage and creep pf normal-strength concrete. ACI Mater (2001) No. 98, pp. 159-165

[7] EN 1991-1-5: Actions on structures – General actions – Thermal actions. Beuth (2010)

[8] Bazánt, Z. P. et al.: Unbiased statistical comparison of creep and shrinkage prediction models. In: ACI Material Journal (2008) No. 6, pp. 610-621

Structural challenges of textile reinforced FRC panels

Thomas Henriksen

Architectural Engineering & Technology,
TUDelft, Julianalaan 134, 2628 CR Delft, The Nederlands
Supervisor: Ulrich Knaack

Abstract

The aim of the research is to investigate the current developments in the use of Fibre Reinforced Concrete (FRC) as a cladding material for high-end architectural applications, and to look into widening its structural applications. In close collaboration with at least one leading FRC manufacturer, it is proposed to explore the possibilities of new applications and also evaluate the current state of the art in production capabilities.

As a cladding material FRC has not been exploited for new applications in the same way as structural glass. In particular, curved, and especially double-curved and free-form, FRC is significantly cheaper to produce than similar forms in glass, and it is possible to create a similar architectural language with both materials.

The goal of this research is to develop a systematic approach for FRC and to develop this to allow designers to understand and make better use of FRC in buildings, both as cladding and for more structural applications. The methods used to achieve this goal will include analysing the results of tests conducted on new FRC applications.

In particular, this paper explores the current application of textile reinforced FRC panels and the structural properties of textile reinforced FRC panels are investigated. The ability to fix FRC panels to a sub-structure is also investigated and the advantages between a mechanical and a bonded solution are discussed.

1   FRC as cladding material

FRC has been developed over the last hundred years into the material it is today. The idea of flexible elements mixed into mortar or similar brittle materials is not new but has been used for centuries. African clay huts are reinforced by using straw in the mortar, with old examples use masonry with animal hairs [1]. The first modern example of FRC include asbestos with further development of FRC made in the 1960s where steel and glass fibres were mixed into the concrete [2][3]. The most commonly used FRC today includes a mix of glass fibres, which have been coated to make them resistant to the alkali process in the concrete, and polypropylene fibres. A comprehensive description of today's FRC and understanding of the material can be found in [4].

FRC as a cladding material and the issues related to using thin fibre reinforced panels as cladding is the main topic of this paper. Currently, FRC is mostly used as a cladding material; it is also used to a lesser extent for structural elements. FRC as a cladding material is widely used in USA. The main method of production is using the spraying process, which allows for the simple production of decorative elements with only minimal flaws in the surface.

Recently FRC panels have been used in high-end architecture. Two of the best known projects are the Expo 2006 Bridge in Zaragoza, and the Soccer City Stadium in South Africa, as shown in Figure 1 and Figure 2.

Proc. of the 9th *fib* International PhD Symposium in Civil Engineering, July 22 to 25, 2012,
Karlsruhe Institute of Technology (KIT), Germany, H. S. Müller, M. Haist, F. Acosta (Eds.),
KIT Scientific Publishing, Karlsruhe, Germany, ISBN 978-3-86644-858-2

Fig. 1　Shows the Soccer City stadium in South Africa for the 2010 football World Cup. Picture © Rieder Smart Elements.

Fig. 2　Shows the Zaragoza bridge, designed by ZHA for the 2006 Picture © Rieder Smart Elements.

In collaboration with Rieder FibreC, new applications of FRC are being discussed. Rieder FibreC has developed an extrusion process for FRC panels. The panels have two layers of embedded textile reinforcement, which increases the tensile capacity of the panels and allows for thinner applications. The use of textile reinforcement in FRC is relative new. The application of textile reinforced concrete (TRC) has been described by [5]. The question that arises is; can FRC be developed into structural elements, without having to cast the concrete in situ? With the use of TRC, this might be possible.

2　Applications of FRC

Concrete as a material has fascinated people ever since ancient times. The most famous example of historic concrete is the Pantheon in Rome [6]. The Pantheon is a concrete shell structure without any reinforcement that fully utilises the compression strength of the concrete. However, most modern big dome structures have been achieved by in situ casting. Examples are the Torrojas dome in Spain [7] and, lately, the work of Heinz Isler [8]. Heinz Isler perfected the fabrication of thin shell domes, forming the shapes as pure compressions shells using simple techniques and a limited amount of reinforcement. These concrete shell structures are rarely built today because of the excessive cost of the formwork.

FRC is currently very popular in high-end architecture. As mentioned above, FRC has traditionally been used for flat cladding panels, however, buildings are now being built with cladding formed from double-curved geometry FRC. The ability to shape the concrete panels in a simple manner offers a wide range of applications. Spraying is one current way to form FRC panels in a complex geometry. If the sprayed method is applied correctly, the quality of the panels can be controlled and a high surface quality achieved. This method has been partly applied on the Heydar Aliev Cultural Center in Azerbaijan [9]; but due to the cost, the complex, doubly-curved wall panels were produced with glass fibre reinforced plastic (GFRP). This is an indication that the technology for producing cost-efficient, double-curved FRC panels is not in place. However, it is evident that the current architectural trend is to use free-form FRC to achieve the architectural intent.

Structurally, FRC panels have been exploited very little. Over the past decade, the structural use of glass has developed dramatically compared to the structural use of FRC. A similar development of FRC is now on-going, the principal challenge being to improve the tensile capacity of FRC. If the current structural use of FRC is assessed, it can be seen that that a combination of concrete and woven fibres gives the additional strength that makes FRC attractive for use as a structural element. The way forward to achieve additional strength is therefore to add textile reinforcement into the concrete. TRC is being researched at the moment by leading universities with the aim of improving its strength [5][10][11][12].

3　Structural challenges of FRC panels

FRC panels are normally relatively thin to allow for handling on site. The thin panels, usually 13mm – 20mm, have little room for reinforcement. The panels therefore have textile reinforcement. The reinforcement is placed on both sides of the plates. The textile reinforcement cast into the panel is shown in Figure 3.

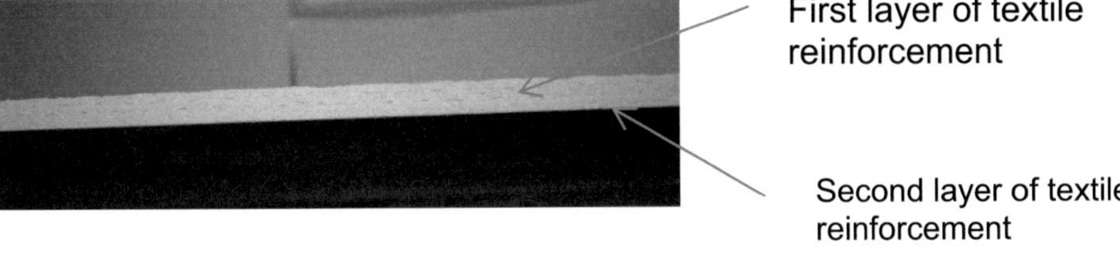

Fig. 3   FRC panel with textile reinforcement

The textile reinforcement ensures an increase in the bending strength of the panels. The panels have been tested for the maximum bending stress using a 3-point bending test [13]. The test result of the bending tests is shown in Figure 4 and Figure 5.

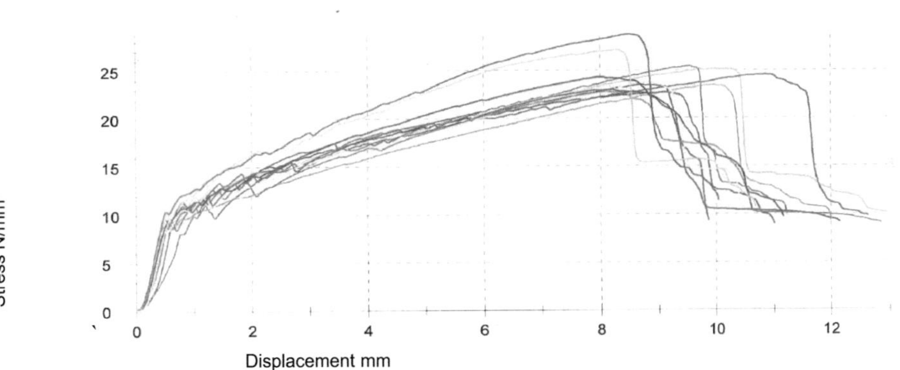

Fig. 4   Stress/Displacement graph for 3 point bending test of TRC panels

Legende	Nr	Prüfseite/-richtung	E_{mod} N/mm²	F_{max} N	$ß_{max}$ N/mm²	dL bei F_{max} mm	Betonversagen N/mm²	a_0 mm	b_0 mm
	max				---				
	min				22,50				
	1.1	Rückseite ZZ / Längsfaser	11600	3609	23,41	8,7	8,90	13,6	250
	1.2		11000	3540	23,31	10,0	8,84	13,5	250
	1.3		11800	3425	22,55	9,1	9,50	13,5	250
	2.1	Rückseite ZZ / Querfaser	12100	3921	25,07	10,0	10,37	13,7	250
	2.2		10200	3767	24,44	10,9	9,19	13,6	250
	2.3		9030	3950	25,25	9,5	9,37	13,7	250
	3.1	Sichtseite ZZ / Längsfaser	11300	3698	23,64	8,9	8,94	13,7	250
	3.2		12400	3477	22,89	8,1	8,77	13,5	250
	3.3		5810	3495	22,67	8,1	9,88	13,6	250
	4.1	Sichtseite ZZ / Querfaser	12000	3781	24,18	8,0	9,61	13,7	250
	4.2		11900	4293	27,05	8,3	9,81	13,8	250
	4.3		12200	4486	28,68	8,5	10,37	13,7	250

Fig. 5   Test results from the 3 point bending test.

From the test it can be seen that the panels perform in a linear manner until approximately 9 MPa. At 9 MPa the first cracks start to propagate from the surface and the first fibres and the textile reinforcement is activated. The panels then exhibit post-cracked behaviour until failure of the panel at approximately 23 MPa. From this it can be concluded that the thin TRC has an initial characteristic bending strength of 9 MPa, and an ultimate breaking strength of 23MPa. The thin plates are tested 28 days after production. The failed test specimen is shown in .

Fig. 6   Test specimen after breakage.

In the final application the design is limited to the initial bending strength. This is mainly due to the architectural requirements for the panels to remain un-cracked. Given the small thickness and an initial breaking strength of 9MPa the panels have a limited span dependent on its application. In this respect it is necessary to research ways of increasing the initial bending strength (without cracks forming in the surface) without changing the surface texture of the panel. Further research will look into ways of increasing the initial bending stress. An option which is being considered is to pre-stress the matrix similar to pre-tensioned concrete.

Because of the relatively low bending strength of the FRC panels it is necessary to use a substructure if the FRC panels is to have a certain size. Given the limited thickness of the panels due to handling of the panels a fixing connection to a sub structure is very important, This makes it possible to limit the deflection of the panels. Mechanical fixings of FRC panels have been used, since the limited thickness of the panels does not allow for many fixing types. Because of the thin panels there is no space to embed reinforcement into the panel. Initially the panels were fixed to the sub structure with screws penetrating the panel, this was done on the Zaragoza bridge. But from architectural point of view visible fixings is not accepted. Therefore a fixing method using Anchors were used. However, the Anchors have very little capacity, and can only transfer 1.2 kN in tension and 3.2 kN in shear per fixing [14]. For a panel with the dimensions; 2.1m x 1.3 m, the number of Anchors were analysed to accommodate wind suction on the panel equivalent to 5.8 kN/m^2. The calculation also takes into account the dead load of the panel. The support configuration for a panel is shown in figure 6. In figure 7 the maximum stress in the panel is shown.

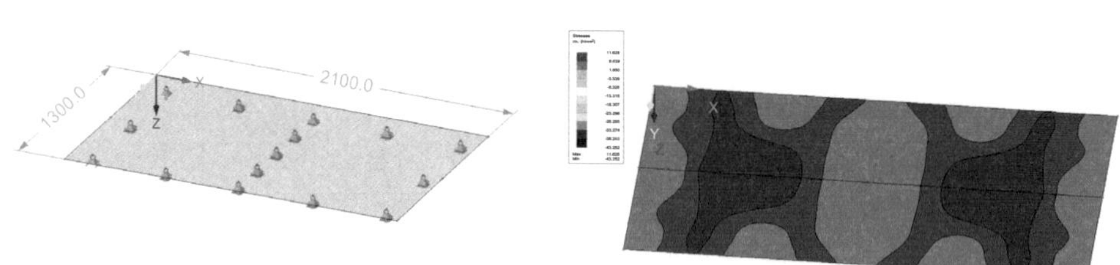

Fig. 7   Support condition for a panel, where each support point picture a mechanical anchor. Maximum tension force is 1.6kN

Fig. 8   Maximum principle stress 11.6 MPa, in the plate with the support condition shown in figure 6.

The maximum tension force from the analysis is 1.61 kN, which is above the maximum pull out capacity. In Figure 9 and Figure 10 are shown the anchor which is used in the FRC panels to fix it to the substructure.

Fig. 9   Picture showing the KIEL anchor.

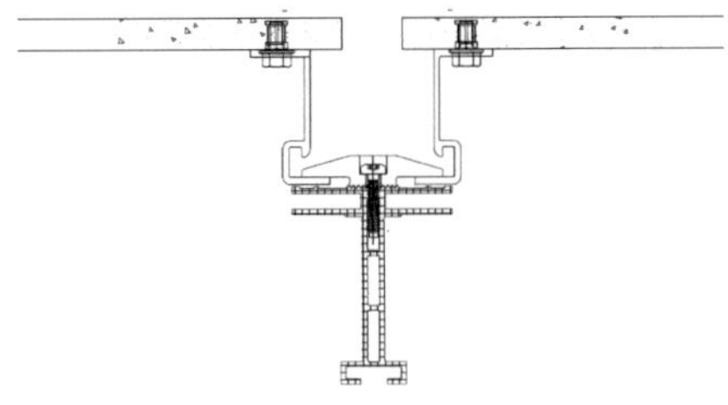

Fig. 10   Shows the connection between the FRC panel and the sub-structure with mechanical anchors.

From the calculation is it seen that the distance between the anchors is very small and still the required tension capacity is not met. Therefore an alternative method of fixing the FRC panels is to bond a profile to the FRC panel using an adhesive. The bonding would be able to replace the mechanical fixing in terms of live loads on the panel. The bonded solution met all the requirements in terms of structural loads. The bonded fixing was tested under various conditions, however, one problem occurred when testing the panels in extreme weather conditions. In terms of the temperature elongation of the sub-structure the FRC panel started to crack as shown in figure 10.

Radial cracks propagating from the edge of the panel due to the temperature elongation of the support structure.

Fig. 11   FRC panel subjected to temperature variations

When using thin FRC panels a sub-structure is necessary for panels of a certain size. If the live-loads on the panels is significant then a mechanical anchor will in most case not be sufficient. The alternative method of bonding the panels to the sub-structure leads to problems with the difference in temperature elongations between the FRC panel and the sub-structure. Further developments are necessary to find an improved method of fixing the panel to the sub-structure for large size panels greater than 1.2m x 1.2m. Alternatively as mentioned above, the surface strength of the FRC panels can also be improved.

4   Production of FRC

The production of FRC is equally an important topic to discuss, since the production method of the FRC limits how the FRC can be utilised. The current FRC production is aimed at façade panels and therefore mainly focuses on the surface quality. As previously mentioned, the predominant production method is a spraying process, for large quantities of which it is, however, more difficult to control the quality, in terms of colour, strength and surface quality.

An alternative method of production is premixed FRC, with which the quality of the concrete can be controlled more easily. However, the tensile capacity of the panels is very limited. This can be increased if ultra-high performance concrete is used, and, as mentioned above, textile reinforcement can be applied to enhance further the tensile capacity. In the FibreC product, the tensile reinforcement is part of the FRC panels. The disadvantage of premixed concrete is the control of the surface quality especially if the panels are cast as normal precast concrete elements.

In both production methods FRC panels face similar challenges in relation to high-end architectural applications. Currently some of the limitations in production of the concrete panels are the following:

a. Producing the moulds for the concrete, if all panels are different.
b. Flat panels with an edge return if all panels are different
c. Single curved panels
d. Doubly curved panels
e. Free form panels

The difficulty of the production and the cost increases from a to e. If the panels need an edge return, which usually is required architecturally, it is usually necessary at present to produce a thick panel. These can be either cast from a premix or sprayed. In both cases this increases their weight, which also influences the substructure. Today it is possible to produce complex geometry panels with ruled surfaces. The moulds for the panels can be produced with a wire cutter. For the Vienna design week an Art installation was produced with FRC elements which had a double curvature. The Art of work is shown in Figure 12 and Figure 13..

Fig. 12   shows an Art of Work for the Vienna designed by Reed Kram and Clemens Weisshaar.

Fig. 13   shows the panels with different shapes

The current state of the art development in regards to production methods is computer controlled moulds. Adapa; which is a spin-off company from Aalborg University have developed a flexible mould which can deal with free form shapes. The technology was presented in Berlin in 2011 [14]. At TUDelft a similar concept is being exploited [15]. The flexible moulds will in the future enable more complex forms to be generated. However the current challenge for FRC in relation to the flexible moulds is the necessary curing time. This at the moment is an obstacle in terms of how many panels can be produced using the flexible moulds. A machine which can generate the flexible surface is shown in Figure 14. In Figure 15 is shown an FRC panels placed on the flexible mould 2 hous after it has been cast.

Fig. 14   shows a machine which can produce flexible surface

Fig. 15   shows a FRC panel on the flexible mould.

In the long term, it should be possible to print the FRC directly with the correct build-up of textile reinforcement. Similar technology is being utilized in the production of sails for high-end sailboats. It is the intention of the research to get a better understanding of structural application of FRC, similar to initial research has been made by [5]. The research will look at utilising the FRC panels as self-supporting shell elements.

5 Conclusion

FRC as material has a long history; however its application in high end architecture is relatively new. When working with FRC, relatively limited information is available in terms of design guidelines for the material. Several companies have advanced the material but in different directions. All the different developments have the ultra high performance concrete in common. In this paper the Structural challenges of textile reinforced FRC panels has been investigated in terms of maximum strength of the panels and possibilities to strengthen the panels with secondary support structure. The limitations of the mechanical anchors have been discussed and an alternative bonded solution has been investigated. In both cases there a significant limitations which means that none of the solutions would work in all circumstances. A solution is to improve the surface strength of the FRC panels. This topic will be investigated further in this research.

Additional research will also be undertaken into the development of FRC as a cost-effective, free-form material, specifically looking at how to progress the use of free form moulding.

6 Acknowledgement

I would like to thank my professors for the support in my research, additionally I would like to thank © Rieder Smart Elements, and especially Wolfgang Rieder.

References

[1] C. D. Johnston: "Fiber-Reinforced Cements and Concretes (Advances in Concrete Technology)" Taylor & Francis, 1 edition, 2000

[2] A. Naaman: "Development and Evolution of Tensile Strain-Hardening FRC Composites" 7th RILEM Symposium on FRC- BeFIB 2008, India

[3] S. Shar el Al: "Fiber-Reinforced Cement-Based Composites: A Forty Year Odyssey" 6th RILEM Symposium on FRC- BeFIB 2004, Varenna, Italy

[4] A. Bentur and S. Mindes: "Fibre Reinforced Cementitious Composites", Second Edition Tayler & Francis, 2007

[5] J. Hegger, S Voss: "Application and Dimensioning of Textile Reinforced Concrete", FRPRCS-8, University of Patras, Patras Greece, July 16-18, 2007

[6] R. Mark, P. Hutchinson: "On the Structure of the Roman Pantheon", Art Bulletin, Vol. 68, No. 1 (1986), p.24

[7] E Torroja "The Structures of Eduardo Torroja: An Autobiography of an Engineering Accomplishment", Literary Licensing, LLC (Sep 2011)

[8] J. Chilton: "Heinz Isler (Engineer's Contribution to Architecture)", Thomas Telford Ltd; First Edition edition (13 Nov 2000)

[9] S. K. Bekiroglu, "Assembling Freeform Buildings in Precast Concrete", Symposium Precast 2010, TUDelft, Holland 2010

[10] S.Dallinger, J Kollegger, "Thin post-tensioned concrete shell structures", Tailor made concrete structures – Walraven & Stoelhost, London, 2008

[11] H. N. Schneider et al, Textile reinforced concrete – application and prototypes, ICTRC, Aachen, Germany 2006

[12] T Tysmans et al. "Shell Elements of Architectural Concrete using Fabric Formwork – Part B: Case Study", FRPRCS-9 Sydney, Australia 2009.

[13] Test protocol, FibreC, Rieder Boton werk, March 2012.

[14]   ETA 06/0220, KEIL undercut anchor KH for glass fibre concrete skin panels "FibreC",

[15]   C. Raun, "Dynamic Double-Curvature Mould System", Design Modelling Symposium Berlin, 2011

[16]   R. Schipper, "A flexible mould for double curved pre-cast concrete elements", Symposium Precast 2010, TUDelft, Holland 2010

Concrete Technology and Microstructure of Concrete

Effect of internal vibration and concrete rheology on properties of reinforced concrete

Guillaume Grampeix

IFSTTAR
Université Paris Est.,
58 Boulevard Lefebvre, Paris 75732 Paris Cedex 15, France
Supervisors: Nicolas Roussel and Jerome Dupoirier, Eiffage

Abstract

At building sites, internal vibrating pokers are traditionally used to compact fresh concrete and enable the steel bars/concrete composite to reach its maximal efficiency. Effects of vibration on properties of reinforced concrete have been extensively studied in the fifties.

In order to identify what can be called an adequate vibration, this study focuses on the effect of vibrating pokers of standard amplitude and frequency on the properties of reinforced concrete of various consistencies (from low slump concretes to almost self compacting).

Our experimental results suggest that vibration and consistency have a very strong influence on bond strength and surface quality.

1   Introduction

Internal vibrating pokers are traditionally used to compact fresh concrete and enable the steel bars/concrete composite to reach its maximal efficiency. Effects of vibration on properties of reinforced concrete have been extensively studied in the fifties. It can be reminded here that, at the beginning of the 20th century, concrete was deposited in shallow lifts and rammed into place by heavy tampers. Around 1930, along with the development of reinforced concrete, vibration was introduced in order to improve both the filling of the forms and the bond between concrete and steel bars. The history of this technology can be found in [1]. However, in the last two decades, modern concrete mix design has gone through major changes with the industrial production of *e.g.* high strength or high fluidity concretes. In parallel, the number of components entering mix design of concrete has increased along with the use of organic admixtures. Many cases were reported recently where the applied vibration characteristics (frequency and amplitude) did not seem to be compatible with castings of these modern concretes. Nowadays national or international technical recommendations, [2] – [4], mostly concern vibration duration as a function of concrete workability. These recommendations are based on empirical and experimental approaches from the first half of the 20th century, in which concrete was considered as a material dominated by direct frictional contacts between the coarsest aggregates [1] and [5]. For these materials, vibration is said to be able to break the contact network between these coarse grains. However, in the case of modern fluid concretes containing higher amounts of fine particles, it was shown recently that consistency is strongly affected by the rheology of the constitutive cement paste [6].

In order to identify what can be called an adequate vibration for this type of modern concrete, this study focuses on the effect of vibrating pokers of standard amplitude and frequency on the properties of reinforced concrete of various consistencies (from low slump concretes to self compacting concrete (SCC)). In a first part, we investigate the effect of vibration on compressive and tensile strength measured on standard cylindrical samples. Then, we measure the effect of vibration on the distribution of air bubbles located either at the surface or in the bulk of the sample. We finally study the effect of vibration on the bond strength between reinforcing steel bars and concrete.

Proc. of the 9th *fib* International PhD Symposium in Civil Engineering, July 22 to 25, 2012,
Karlsruhe Institute of Technology (KIT), Germany, H. S. Müller, M. Haist, F. Acosta (Eds.),
KIT Scientific Publishing, Karlsruhe, Germany, ISBN 978-3-86644-858-2

2   Materials and mix design

All concretes were designed from a Calcia cement CEM III/A 52.5 CE PM-ES-CP1 NF (Blaine specific surface of 3890 cm^2/g [7]) and plasticizer Plastiment HP from Sika. The water/cement ratio was kept constant and equal to 0.60. Variations in the volume of constitutive cement paste (or total volume of aggregates) allowed for a variation in consistency. The mix proportions are gathered in Table 1.

Table 1   Concrete mix design

Consistency	S1	S3	S4	S5	SCC
Amount of cement (kg.m^{-3})	280	300	320	340	380
W/C (-)	0,60	0,60	0,60	0,60	0,60
% (in cement weight) Plastiment HP (Sika)	0,41	0,41	0,41	0,41	0,85
Slump (S)/Slump Flow Test (SF) (mm)	40 (S)	130 (S)	180 (S)	220 (S)	650 (SF)
Semi crushed sand (Seine) 0/4mm (kg.m^{-3})	796	775	755	737	695
Semi crushed gravel (Seine) 8/20mm (kg.m^{-3})	973	948	923	901	849

3   Vibration effect on standard mechanical strength

Cylinder samples were cast following NF EN 12390-2 [8]. Then, vibration durations were chosen according to NF P 18-422 [9] for a 25 mm diameter poker. It can be noted that, although SCC shall not be vibrated, we did not spot any segregation in the case of the SCC tested here for a 25s vibration. For each tested concrete mix design, samples were cast with or without internal vibration. After 28 days, mechanical compressive and tensile strength were measured according to NF EN 12390-3 [10]. The values in Fig. 1 are the average of three samples. It can be concluded that vibration does not affect mechanical strengths as measured following NF EN 12390-3 except for consistency class S1.

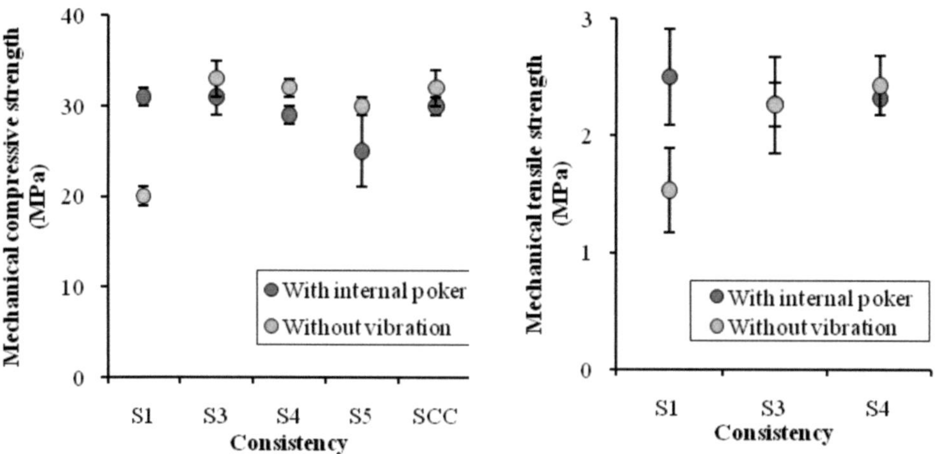

Fig. 1   Mechanical compressive strength on 16/32 samples at 28 days (left). Mechanical tensile strength on 16/32 samples at 28 days (right).

These results suggest that measurement of mechanical strength does not seem to be the appropriate way to appreciate the efficiency of the vibration (Cf. below).

4   Density

We also measured the apparent density for each specimen in the fresh state and compared it with the theoretical value computed from mix design assuming 2% of entrained air, see Fig. 2. Moreover, at 28 days, samples were drilled and we measured the density of the internal extracted bulk (100mm diameter) and of the external ring, see Fig. 3.

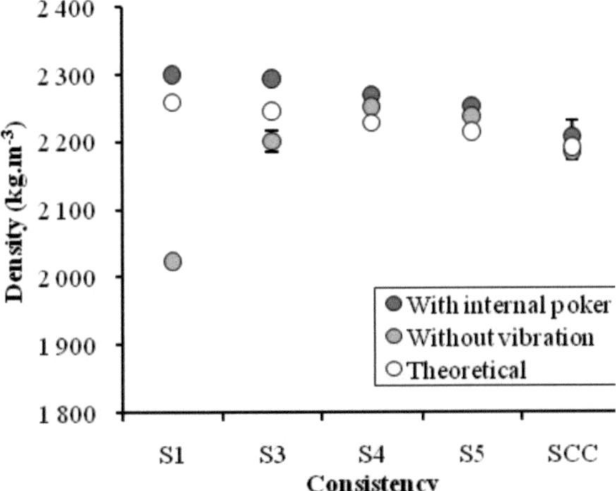

Fig. 2   Fresh concrete density as a function of consistency.

It can be concluded from Fig. 2 that vibration has almost no effect on the density in the fresh state except for concretes belonging to class S1 and S3. From Fig. 3, we conclude that vibration does not seem to strongly modify bulk density for the concretes tested here whereas, close to the interface, in the external ring, a strong drop in density is measured. Close to this interface, gravity alone is not able to fill correctly the mould whereas in the bulk (*i.e.* far from the interface) the concrete own weight is sufficient to create an apparently homogeneous material. We suggest that this peculiar effect is at the origin of the inability of mechanical tests to detect the influence of vibration on the samples as these standard tests mainly test the bulk of the samples where the stresses generated by the mechanical load are the highest.

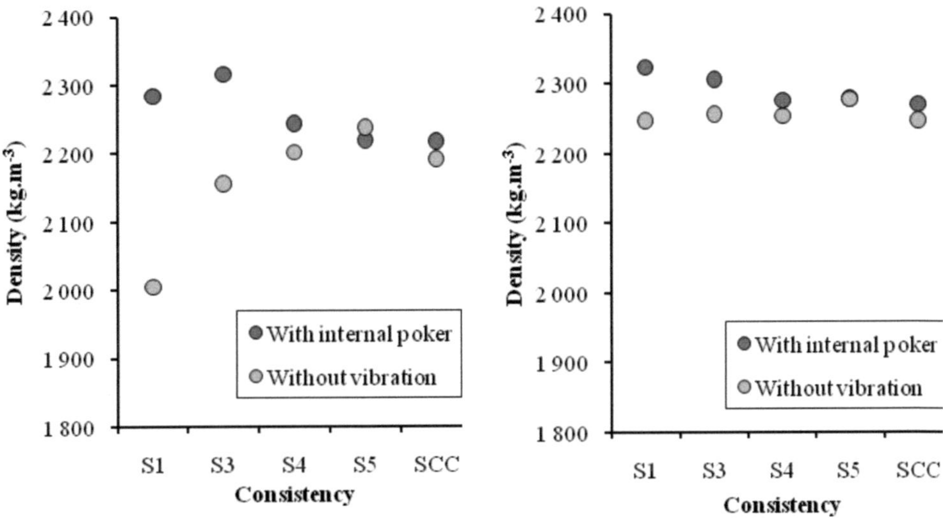

Fig. 3   External ring concrete density at 28 days (left). Internal extracted bulk concrete density at 28 days (right).

5   Effect of vibration on surface porosity

In order to characterize surface quality, we developed an image analysis technique. We painted the concrete samples surfaces, see Fig 4. The paint projection was inclined in order to prevent the paint from entering the bubbles and cavities. The enhanced contrast generated by this pre-treatment gives access to surface porosity and average diameter of surface defaults. We plot these results in Fig. 5. It can be concluded that vibration has a strong effect on surface porosity for all concretes except the two most fluid ones (S5 and SCC). Simultaneously, vibration tends to decrease the average size of the surface defaults. This positive influence decreases when the fluidity of the concrete increases.

Fig. 4   Examples of surface quality before painting treatment. Concretes samples casted without internal vibration

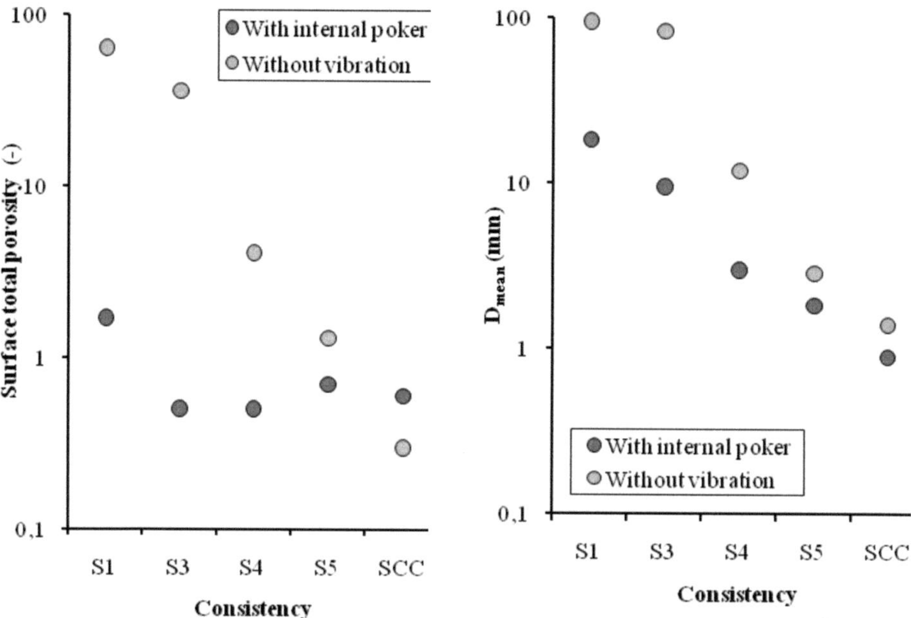

Fig. 5   Surface total porosity measured by image analysis on concrete samples at 28 days (left). Mean diameter of surface defects (right).

6   Effect of vibration on bond strength

Two types of samples were cast. The first one was typical of a beam or slab casting with a dominant influence of vertical bottom form surface on the flow of the material whereas the second one was typical of a wall casting with a dominant influence of an horizontal form surface. In both cases, concrete was poured from the top, see Fig. 6. The diameter of the steel bar was 14 mm. At 28 days, we tested bond strength for each concrete mix design. A jack pulled out the steel bar while the concrete sample was fixed. A strength sensor allowed the measurement of bond force peak. We compare, in

Fig. 7, the dimensionless ratio between peak bond strength between steel bar and concrete with vibration and without vibration as a function of concrete consistency.

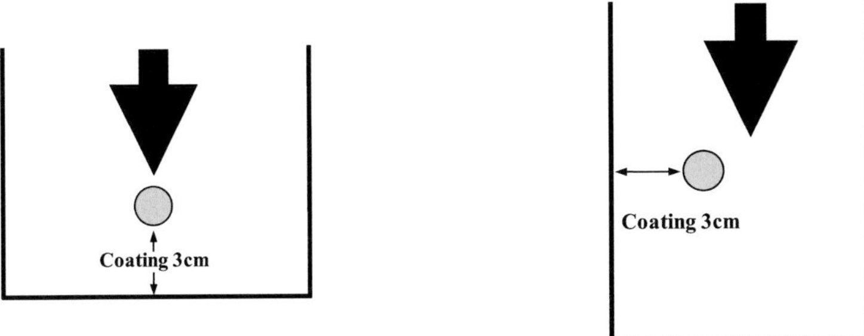

Fig. 6   Typical beam casting (left). Typical wall casting (right). Arrows indicate the casting direction.

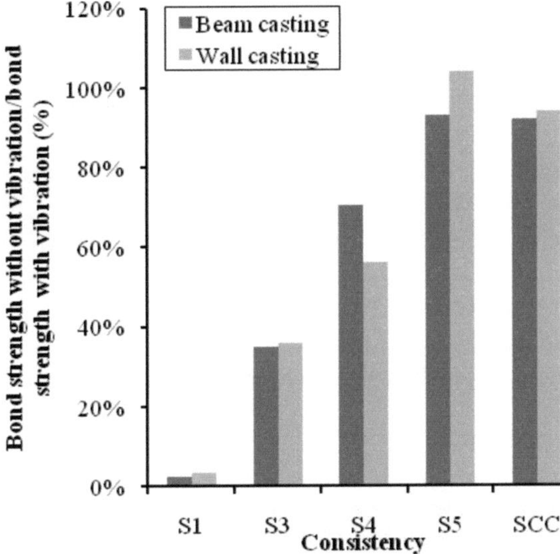

Fig. 7   Dimensionless ratio between bond strength between steel bar and concrete without and with vibration as a function of concrete consistency. The two types of casting are shown.

We conclude from Fig. 7 that vibration is necessary to reach proper bond strength for all concretes except the two most fluid ones (S5 and SCC). It can be noted that effect of vibration for this property is extreme as, for consistency S1 and wall casting, some reinforcing steel bars fell off during the removal of samples from their moulds.

6   Conclusions

In order to identify what can be called an adequate vibration, this study focused on the effect of vibrating pokers of standard amplitude and frequency on the properties of reinforced concrete of various consistencies (from low slump concretes to almost self compacting).

Our experimental results suggest that traditional pokers do not have significant influence on mechanical strength and average densities of concretes no matter their consistencies. They however show that vibration and consistency strongly influence bond strength and surface quality.

From a practical point of view, the following conclusions can be drawn:

- Concrete of class S1 and S3 have to be vibrated as all properties strongly decrease when no vibration is applied. The bond strength between steel bars and concrete almost drops to zero in the case of S1 concretes.

- If non visible and non-exposed to chemical aggressive environment, it could be tempting to apply no vibration to concretes of class S4. However, a decrease in bond strength of 50% should then be taken into account in the structural design. The surface quality of the obtained concrete would moreover be poor and it can be expected that the access of aggressive chemicals to the steel bars would be facilitated. It seems therefore safer to recommend a correct vibration of these concretes.

- Concretes of class S5 do not seem to need a vibration and, as expected, vibration does not improve the properties of SCC. It could moreover be expected to be detrimental to vibrate SCC as some segregation could be induced. It was however not the case for the SCC studied here.

7   Acknowledgment

The authors wish to thank the Federation National des Travaux Publics (FNTP) and Federation Francaise du Batiment (FFB) for their financial help. The authors acknowledge FNTP and FFB Committees reviewers for their constructive comments to the text.

References

[1]   L'Hermite, R. : Idées actuelles sur la technologie du béton. In Institut technique du bâtiment et des travaux publics. (1955)

[2]   Geoffray, J-M. : Béton Hydraulique - Mise en Oeuvre. In Les Techniques de l'Ingénieur. (2008). C 2229.

[3]   Ecole   Française   du   béton. :   La   vibration   EFB.   In. http://www.efbeton.com/IMG/ppt/LA_VIBRATION_EFB_V12_10_07.ppt. (2007)

[4]   CimBéton: La vibration des bétons. In Collection Technique CimBéton (1998)

[5]   Forssblad, L : Investigation of Internal Vibration of Concrete. In Civil Engineering and Building Construction Series Acta Polytechnica Scandinavica 29. (1965), p. 32

[6]   Yammine, J., et al: From ordinary rheology concrete to self compacting concrete: A transition between frictional and hydrodynamic interactions. In Cement and Concrete Research 38. pp. 890-896. (2008)

[7]   Calcia cement CEM III/A 52.5 CE PM-ES-CP1 NF product profile from Calcia in http://www.ciments-calcia.fr/NR/rdonlyres/577276F7-AEB9-4E95-AE53-61CD6D1CA382/0/CEMIIIA525Lou525LLHCEPMESCP1NF.pdf

[8]   AFNOR: Nome Européenne. NF EN 12390-2. Essai pour béton durci. Partie 2: Confection et conservation des éprouvettes pour essais de résistance. (2001)

[9]   AFNOR: Nome Française: NF P 18-422 Béton Mise en place par aiguille vibrante. (1981)

[10]   AFNOR Norme Européenne: Norme. NF EN 12390-3 Essai pour béton durci Parie 3 : Résistance à la compression des éprouvettes. (2003)

Rheological behaviour of cement based grouting materials

Raphael Breiner

Institute of Concrete Structures and Building Materials (IMB),
Karlsruhe Institute of Technology (KIT)
Gotthard-Franz-Str. 3, 76131 Karlsruhe, Germany
Supervisor: Harald S. Müller

Abstract

The reliable and durable sealing of hydraulic engineering structures in karst areas has an outstanding significance for their serviceability and structural safety. As it usually takes extensive efforts to prevent or minimise seepage with grout curtains, the controlling of the rheological behaviour of grouting materials is the key factor to achieve the best possible tightening of porous karst rock. In this paper comprehensive rheological investigations are presented which provide the database for the development of a reliable and physically consistent prediction model. It quantifies the influence of additives as well as of admixtures on the rheological properties of cement suspensions. It is crucial not only to describe the relative effect of the additives on the rheological suspension behaviour in an empirical model but to clarify the interactions between the numerous source materials used in geotechnical engineering. Consequently the practical model verification in a pilot scale will provide a comprehensible connection between geology and rheology.

1 Introduction

Within the scope of a German-Indonesian joint project, funded by the German Federal Ministry of Education and Research (BMBF), a hydropower plant with an underground concrete barrage (see Figure 1) was initialised, planned and built during the years of 2002 to 2006. During the dry season, it provides an urgently required water supply for the karst region Gunung Sewu in central Java, Indonesia [Müller et al. 2008]. Within an on-going German-Indonesian follow-up project funded by the BMBF, the hydro power plant is currently embedded into the frame of an "Integrated Water Resources Management" (IWRM), which couples all aspects of water supply, distribution, usage and treatment in an overall concept [Oberle et al. 2005].

The Institute of Concrete Structures and Building Materials (IMB, Subproject 5) of the Karlsruhe Institute of Technology (KIT) focuses in particular on the warranty of durable watertight and functional hydraulic constructions. This includes the construction of a grouting curtain in pervasively karstified limestone under difficult climatical, spatial and geological boundary conditions [Breiner et al. 2011]. The objective of the works is the effective and economical sealing of the karstic rock surrounding the concrete barrage to minimise the amount of incidental seepage water and to guarantee the secure and durable long time operation of the plant.

The reliable and durable sealing of hydraulic engineering structures in karst areas has an outstanding significance for their serviceability and structural safety. However, it usually takes extensive efforts to prevent or minimise seepage due to the inherent structure of karst rock which includes a heterogeneous pore size distribution and a non-uniform connectivity. Numerous case studies show the difficulties of karst rock injections which can be dealt with in different ways [Kreutzer 1997; Linortner 2009]. The key to a successful sealing of such structures is a detailed adaption of the sealing material – and here especially its rheological behaviour at the fresh state – to the geological boundary conditions of the karst rock. Mostly, the optimisation of the grouting materials is done based on empirical approaches, e.g. the study of relevant literature, the experience of the executing engineer and own, often extensive preliminary explorations and tests. Systematic scientific investigations which link the karst geology with the rheological behaviour of grouting suspensions are missing. Further, the raw adjustment of the viscosity of the injection material is usually carried out by an adjustment of the water/cement-ratio while often neglecting the possibilities of modern concrete technology.

Proc. of the 9th *fib* International PhD Symposium in Civil Engineering, July 22 to 25, 2012,
Karlsruhe Institute of Technology (KIT), Germany, H. S. Müller, M. Haist, F. Acosta (Eds.),
KIT Scientific Publishing, Karlsruhe, Germany, ISBN 978-3-86644-858-2

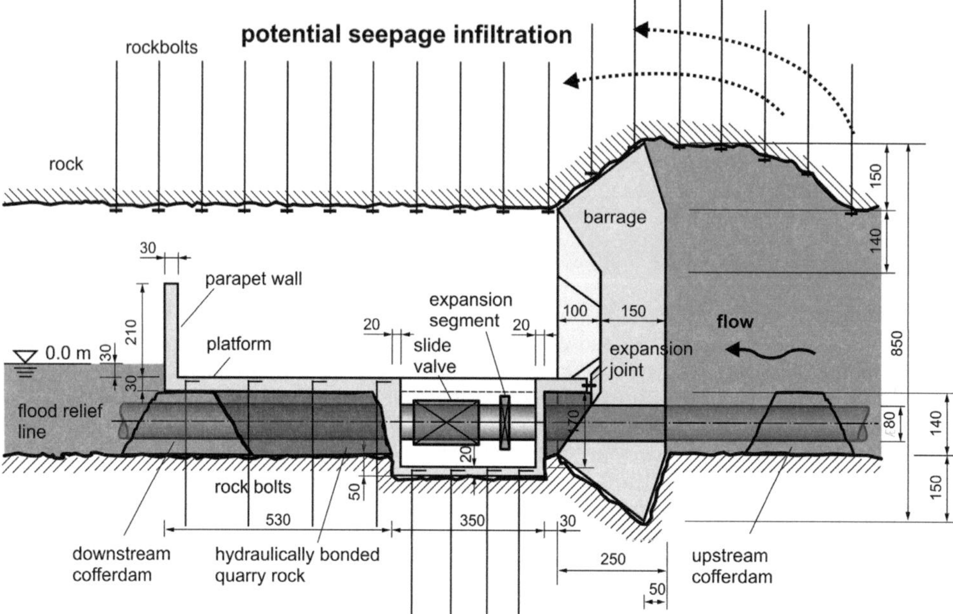

Fig. 1   Drawing of the underground concrete barrage (longitudinal section, without components of the hydraulic power plant) [Bohner et al. 2009]

Before this background the presented work focusses on the development of cementitious suspensions with optimized workability as well as applicability which can be used by means of a cement injection for the durable tightening of porous karst rock. The key factor to succeed both from a technical and economical point of view is the sound knowledge of the materials rheological properties which may differ in a wide range depending on the physical properties of the selected source materials.

The flow characteristics of pure cement suspensions are already well understood and can be predicted correspondingly by means of a physico-rheological model [Haist 2009]. However in modern concrete technology different additives and admixtures are normally added to the cement suspension to optimize its behaviour both at the fresh and the hardened state. The effects of these additions on the rheological properties of the cement pastes are qualitatively known without a doubt, e. g. summarized in [Ramachandran 2001]. However, there is at present no prediction model available which can be used to characterize the dependencies also in a quantitative way to enable in the case of need a fast, simple and reliable material development considering only the physical properties of the source materials and the boundary conditions of the karst rock to be filled.

2   Scientific approach

The main objective of this research project is the development of a reliable and physically consistent prediction model. It is intended to quantify the influence of additives as well as admixtures on the rheological properties of cement suspensions with focus on their physical and chemical interactions.

Based on a literature survey as well as various experiences gained during a recent field study [Breiner et al. 2011], source materials will be selected and carefully characterised considering their usage as grouting materials. Preliminary tests to develop and optimise the rheological measurement techniques are followed by a sensivity analysis to identify the main acting parameters. Systematic rheological parameter studies shall reveal especially the quantitative dependencies regarding the additives interactions in the rheological behaviour of cement suspensions. These will be combined in a physico-rheological model to predict the suspensions' behaviour considering only the physical properties of the raw materials used. Finally the practical model verification in a pilot plant scale will prove the applicability.

3   Materials and methods

Preliminary tests were conducted on cement suspensions prepared with a well cement class G according to DIN EN ISO 10426-1. It has a density of 3.2 kg/dm³ and a Blaine-value of 3700 cm²/g. As the usage of liquefying additives is a question of central importance for the flow behaviour of cement suspensions, five superplasticizers with different mechanisms of action were tested covering the whole range of commercially available products (see table 1).

Table 1   Physical properties of the tested superplasticizers

Notation	Mechanism of action	Density [kg/dm³]	Solids content [%]	Dosage [% by mass of c]
FM 1	Naphtaline sulfonate, sterical	1.20	39	0 - 0.3
FM 2	Polycarboxylatether, electrostatic	1.07	31	0 - 0.3
FM 3	Acetonformaldehydsulfit	1.18	30	0 - 0.3
FM 4	Polycarboxylatether, modified	1.10	31	0 - 0.1
FM 5	Polycarboxylatether, modified	1.06	30	0 - 0.1

With regard to the control of the sedimentation behaviour of grouts the clay mineral bentonite is used to stabilize cement suspensions by increasing the suspensions thixotropy. Further, bentonite binds free water due to its stratification structure and thus prevents a bleeding of the suspension. Consequently two bentonite minerals were examined: An activated Ca-bentonite with a density of 2.70 kg/dm³ and a Na-bentonite with a density of 2.71 kg/dm³.

To determine the physical-chemical interactions of different cement suspensions a measuring system consisting of a high-end rheometer Haake MARS combined with a measuring cell specifically developed for cement suspensions and mortars is used (see Figure 3). The measuring cell consists of a cylindrical vessel with an adjustable wall serration to account for different grain size diameters and to prevail a sliding of the cement suspension in the contact face to the wall. The cell filled with cement suspension is installed into the rheometer and defined shear stresses are consequently applied with a paddle-shaped rotor to determine the rheological properties. In addition the base of the measuring cell is equipped with an electroacoustic measuring device, which enables a real-time detection of the electrostatic charges of the suspension particles (Zeta-potential), the size of the forming agglomerations as well as of temperature and conductivity of the suspensions [Haist 2009].

To detect the rheological and electroacoustic properties of the cement suspensions even under high temperature and high pressure conditions which may occur e.g. during karst grouting or well cementing processes, a novel measuring unit was developed (see Figure 2). It enables rheological measurements and consequently a material optimisation up to 150 °C and 350 bar while maintaining a very good reproducibility and it can be integrated into the present experimental setup without major efforts.

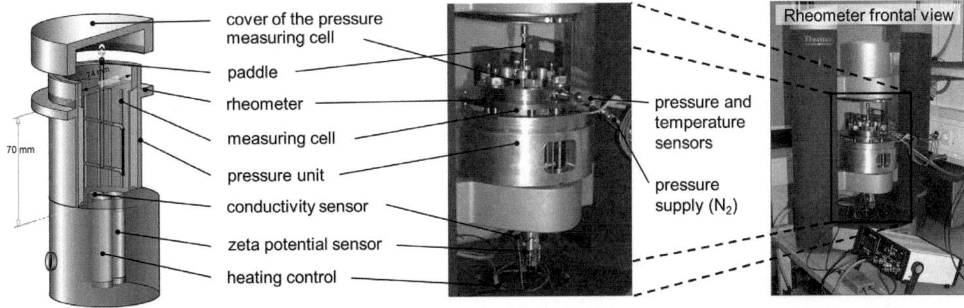

Fig. 2   Rheometer including pressure measuring cell with electroacoustic measuring device: schematical (left), laboratory in detail (middle) and laboratory overview (right)

The pressure cell can be heated up with an electrical heating or alternatively cooled down with a liquid temperature control unit. The pressurisation is realised with nitrogen (N_2) which can be exactly regulated over a pressure reducing valve to counteract for pressure increases e.g. in the case of sudden temperature variations or chemical reactions. Both the pressure and the temperature are registered automatically with an electronical measuring setup.

4   Results

To judge the effectivity of the tested superplasticizers in connection with the present cement, rheological measurements with graded dosages of the respective superplasticizer were carried out while maintaining a temperature of 35 °C. The water/cement-ratio of the cement suspension amounted in each case to 0.3. The influence of the type and amount of superplasticizer (left) and the addition of bentonite (right), respectively, on the yield stress of the cement suspension is illustrated in Figure 3. The shown yield stress was determined by regression from the experimental data with decreasing shear stress according to the Herschel-Bulkley model.

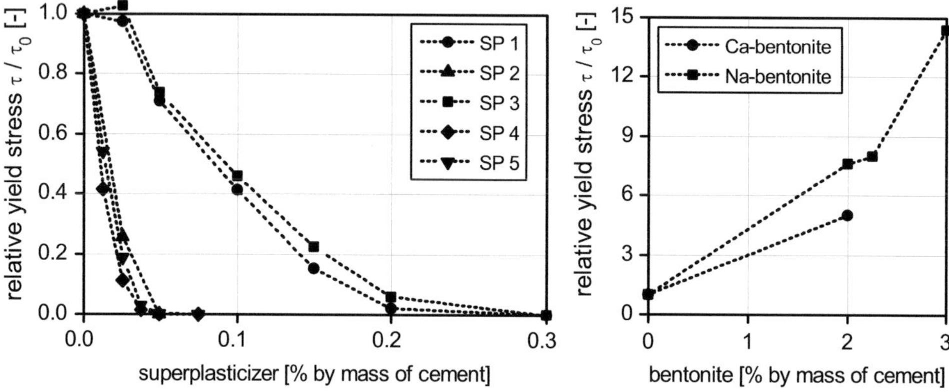

Fig. 3   Influence of type and amount of superplasticizer on the yield stress (left) and influence of the bentonite content on the yield stress (right)

Further the influence of the suspension age and of higher temperatures on the superplasticizer efficiency was studied in long-time measurements with selected dosages to exclude an unexpected reduction of the liquefying effect and to compare the behaviour of the selected additives. For this purpose the cement suspensions were mixed with the previously determined amount of superplasticizer to reach the saturation point (see Figure 4) and consequently rheologically characterized at various points of time.

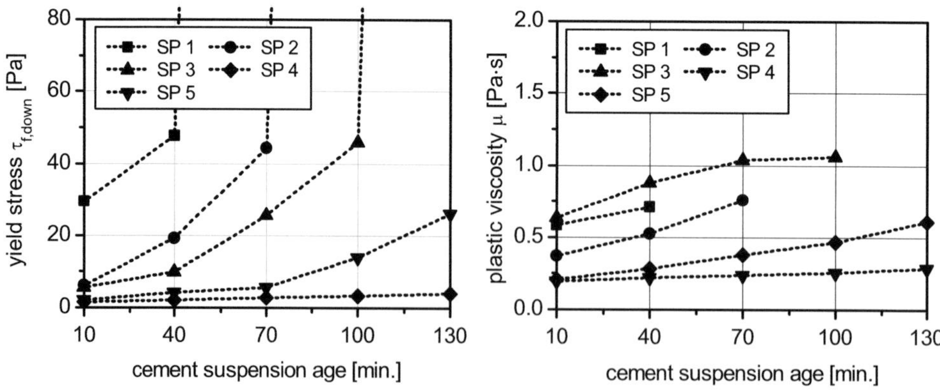

Fig. 4   Development of the yield stress (left) and the plastic viscosity (right) over time depending on the superplasticizer type at a constant temperature of 70 °C

Measurements were performed 10, 40, 70, 100 and 130 minutes after the water was added to the dry cement at a constant temperature of 70 °C. Figure 5 shows the change of the yield stress (left) and the plastic viscosity (right) over the whole measurement time for the cement suspensions depending on the superplasticizer type.

Though superplasticizers (SP) based on Polycarboxylatether (PCE) are believed to possess only a limited thermal stability, SP 4 and 5 show a good liquefying effect even at 70 °C. Actually SP 4 exhibits a stable behaviour over the whole measurement time, the yield stress and the plastic viscosity remains almost constant. However the behaviour of SP 3, which was classified as temperature-stable by its producer, could not be proven in the measurements. SP 3 shows an increase of the yield stress which advances even more with time. After 130 minutes the yield stress exceeded the stress limit of the instrument and was therefore classified as inappropriate for the desired purpose.

5 Discussion

The different behaviour of the cement suspensions due to the varying mechanism of action of the tested superplasticizers could be clearly observed in the experimental data (see figures 4 and 5). However, besides the well-known rheological properties yield stress and plastic viscosity, it is important to include also other characteristic values such as the shear modulus $|G^*|$ or the creep viscosity η_{creep} to account for the elastic properties of the suspensions and viscous sliding processes respectively. This helps on the one hand to simulate the deformation behaviour when only small shear rates are prevailing and on the other hand to characterise the suspensions' sedimentation behaviour.

The rheological properties already can be calculated for pure cement suspensions depending on the physical properties of the cement and the mix design [Haist 2009]. By a comparison of the separate suspensions with or without superplasticizers, the influence of the additives on the characteristic rheological values can be determined in a quantitative way, too. This could be probably further simplified by a classification (see Figure 3, left) which will be implemented in a physically sound model.

The model is based on a combination of the rheological elements spring, damper and Saint-Venant friction element, whose properties in turn are to be related to the material composition again. It will be possible in full knowledge of the underlying constitutive material laws to predict the rheological behaviour of a liquefied cement suspension in the whole shear rate range based only on the physical and chemical properties of the additives such as e. g. the length or the density of the PCE's side chains.

The rheological model described above is planned to be incorporated into a systematic approach to simplify the planning of the grouting process in general and the karst injection process in particular. Based on commonly used laboratory experiments to determine the karst porosity and in field Lugeon-tests (LU-test) to estimate the rock permeability, it is possible to derive an equivalent pore size diameter [Smith et al. 1976]. If in addition the desired penetration depth of the respective injection step and the maximum injection pressure are known from geotechnical investigations or are defined by the contractor, the equivalent pore size diameter can be used to calculate the required rheological properties of a cement suspension based on simplified equilibrium considerations [Wittke 1968, Cambefort 1969]. The combination of these results with the rheological model to be developed in this thesis enables the definition of a pore size range which can be reached, i.e. filled with cement suspension, while using the available raw materials.

As karst rock occurs all over the world, the respective physical properties of local cements and additives will be an important input parameter for the model to enable a fast, simple and reliable material development resulting in a practicable mixture composition.

6 Summary and conclusion

From a technical and economical point of view, the controlling of the rheological behaviour of grouting materials seems to be the key factor to achieve the best possible tightening of porous karst rock.

It is crucial not only to describe the relative effect of the additives on the rheological suspension behaviour in an empirical model but to clarify the interactions between the numerous source materials used in geotechnical engineering. Based on this a physically sound model will be derived which is modular constructed and reveals an enormous optimisation potential in the development of novel, cement based grouting suspensions.

Up to now there are only a few experiences regarding the rheological behaviour of cement suspension under high temperatures and nearly no experiences in respect of high pressures known in literature. In particular concerning the deformation behaviour when small shear-loadings are prevailing (e.g. start of pumping or sedimentation processes) no defined results are available. The recent research will close this gap and thus account for the controlling of grouting systems.

As approximately 20 % of the worlds' population live on carbonate rock, the question of grouting in karst is not only a problem of academic relevance. Consequently the practical model verification in a pilot plant scale enables a reliable application of cementitious grouting suspensions for the injection of karst rock which will be demonstrated during a scheduled underground injection campaign in Indonesia in spring 2013. Thus the concept offers a comprehensible connection between geology and rheology. Moreover its modular format combined with the physical sound knowledge of the underlying dependencies offers a wide range of further application possibilities e.g. in the petroleum and deep drilling industry.

7 Acknowledgements

The financial support by the German Federal Ministry of Education and Research (BMBF) under the grant number 02WM0881 is gratefully acknowledged by the author.

8 References

[1]   Bohner E., Fenchel M. and Müller H. S.: Konzeption und Herstellung eines unterirdischen Betonsperrwerks zur Trinkwassergewinnung auf Java. In: WasserWirtschaft 99 (2009) No. 7-8, pp. 47-52

[2]   Breiner, R., Bohner, E., Fenchel, M., Müller, H. S., Mutschler, T., Triantafyllidis, T.: Grouting of an underground concrete barrage in karst limestone. In: Asian Trans-Disciplinary Karst Conference (2011), Haryono, E. et al. (Hrsg.), Faculty of Geography, Gadjah Mada University, Yogyakarta, Indonesia, pp. 181-190

[3]   Cambefort, H.: Bodeninjektionstechnik (1969), deutsche Bearbeitung: Klaus Back, Bauverlag GmbH, Wiesbaden und Berlin, Germany

[4]   Ford, D., Williams, P.: Karst Hydrogeology and Geomorphology (2007), Wiley, England

[5]   Haist, M.: Zur Rheologie und den physikalischen Wechselwirkungen bei Zementsuspensionen (2009), Dissertation, Institute of Concrete Structures and Building Materials, Karlsruhe Institute of Technology, Karlsruhe, Germany

[6]   Kreuzer H.: Bogenmauer Francisco Morazán in Honduras – Erweiterung des Dichtungsschleiers im Karstgestein. In: Felsbau 15 (1997) No. 1, pp. 38-44

[7]   Linortner J.: Design and construction of the grout curtain for the Ermenek hydropower plant. In: Geomechanics and Tunnelling 2 (2009) No. 5, pp. 421-429

[8]   Müller H. S., Fenchel M., Bohner E. and Mutschler T.: Bau eines Höhlenkraftwerkes zur Trinkwassergewinnung auf Java, Teil 2: Konzeption und Realisierung des Sperr-werkes unter Berücksichtigung örtlich verfügbarer Baustoffe und Technologien. In: Symposium Baustoffe und Bauwerkserhaltung, Betonbauwerke im Untergrund – Infrastruktur für die Zukunft (2008), Müller H. S., Nolting U. and Haist M. (eds.): University of Karlsruhe Scientific Publishing, Karlsruhe, Germany, pp. 121-137

[9]   Oberle P., Kappler J. and Unger B.: Integriertes Wasserressourcen-Management (IWRM) in Gunung Kidul, Java, Indonesien. Schlussbericht zur Machbarkeitsuntersuchung im Auftrag des BMBF (2005), Institut für Wasser und Gewässerentwicklung, Bereich Wasserwirtschaft und Kulturtechnik, University of Karlsruhe, Karlsruhe, Germany

[10]   Ramachandran, V. S.: Concrete admixtures handbook – properties, science and technology (2001), Norwich, New York, USA

[11]   Smith, D. I., Atkinson, T. C. and Drew, D. P.: The hydrology of limestone terrains. In: The Science of Speology (1976), Ford, T. C. and Cullingford, C. H. D. (eds.), Academic Press, London, England, pp. 179-212

[12]   Wittke, W.: Zur Reichweite von Injektionen in klüftigem Fels. In: Felsmechanik und Ingenieurgeologie, Supplement IV (1968) pp. 79-89

Influences of the grout composition on the dewatering behaviour of annular gap grouts

Bou-Young Youn

Chair of Building Materials,
Ruhr-University Bochum,
Universitätsstr. 150, 44801 Bochum, Germany
Supervisor: Rolf Breitenbücher

Abstract

The main objective of this research study was to investigate the correlations between the individual components of different grout compositions and their dewatering behaviour under defined pressure conditions and filter mesh as well as the development of shear strength fundamentally. Therefore, tests on different grout compositions with variations of parameters, inter alia cement:fly ash-ratio, were performed. For this purpose, a filter press was developed to determine the amount of filtrate water, including temporal changes. From the results, influences of the varied parameters on the workability and on the development of required shear strength, achieved by the dewatering behaviour, were analyzed and discussed.

1   Introduction

The hollow space between tunnel lining and bedrock, caused during tunnel driving with segment lining, must be filled with a so-called annular gap grout, to embed the tunnel lining in an adequate position and minimize the settlements of the ground surface. The decisive requirements on such annular gap grouts are on the one hand optimal flow properties lasting for several hours, on the other hand a rapid development of shear strength immediately after grouting, corresponding to that of the surrounding soil. This can be achieved by pressing out the excess water of the grout into the soil. Thus, two contradictory requirements are demanded on annular gap grouts. Hitherto, grouts have been exclusively defined on empirical basis. Studies about specific investigations on grout compositions with regard to dewatering and infiltration behaviour into the surrounding soil are missing so far.

Therefore, it is necessary to investigate the correlations between the individual components of different grout compositions and their dewatering behaviour as well as development of shear strength under defined boundary conditions, e.g. pressurization, permeability of filter mesh. Based on this experimental description of dewatering and also infiltration behaviour of grouts, it should be possible, to perform this material in accordance with a real design concept and ensure a diagnosable and unerring annular gap grouting in praxis.

2   General requirements on annular gap grouts

During tunnel driving the annular gap remains between tunnel lining and bedrock, with a width of 13 until 18 cm [1]. This gap must be backfilled with a suitable grout composition immediately after installation of the segments, to minimize settlements of the ground surface and to ensure a durable stabilization of the tunnel lining [2, 3, 4]. In consideration of the geological and hydro-geological conditions, following general requirements on annular gap grouts can be compiled:

- Workability lasting for several hours, meaning high flowability and sedimentation stability at the same time [1]
- Optimal pumpability between 12 h and 24 h [1, 5]
- Very fast development of shear strength and stiffness modulus, corresponding to those of the surrounding soil [3, 6]
- Adequate and durable embedment of the tunnel lining and
- Minimization of settlements of the ground surface [2, 3, 4]
- Protection of the segmental lining against concrete-aggressive agents in water and soil [7, 8]
- High erosion stability in case of high flow velocities of water and
- High filter stability in case of very permeable soil [6, 9]

Proc. of the 9th *fib* International PhD Symposium in Civil Engineering, July 22 to 25, 2012,
Karlsruhe Institute of Technology (KIT), Germany, H. S. Müller, M. Haist, F. Acosta (Eds.),
KIT Scientific Publishing, Karlsruhe, Germany, ISBN 978-3-86644-858-2

Further requirements regarding compressive strength are not demanded, so that cementitious constituents are not necessary for the mix design or just in small amounts [6]. In this context, annular gap grouts can be classified in three different types [2]:

- Active grout: complete hydration of the binder, generally based on Portland cement with an amount > 200 kg/m³
- Semi-inert or reduced active grout: small amount of cement between 50 kg/m³ and 200 kg/m³; rapid stiffness, followed by a very slow strength development
- Inert grout: no cement, at most 50 kg/m³; cement is usually replaced by e.g. hydraulic lime or fly ash

Typical parameters concerning the requirements of grouts are summarized in Table 1.

Table 1   General requirements on annular gap grouts [1, 5, 7]

Requirements	Parameter		
Workability/ consistency	Spread flow diameter Flow diameter a Flow diameter a	(t = 0h) (t = 0h) (t = 8h)	15 ± 5 cm 20 ± 5 cm 15 ± 5 cm
Strength development	Shear strength τ_y		≥ 2.0 MPa (un-drained grout) ≥ 32.0 MPa (drained grout)
	Stiffness modulus E_s		1,000 MPa (corresponding to that of a common soil in the scope of tunnelling with segments)

3   Experiments

Within the scope of this research work, the workability and dewatering behaviour of different grout compositions for annular gap grouting were observed under defined pressure conditions and filter mesh. In addition, the development of shear strength was examined at dewatered specimens.

The investigations were carried out on grout compositions, which are commonly practiced for a major traffic tunnel, as shown in Table 2.

Table 2   Grout compositions (active, semi-inert, inert) for a major traffic tunnel [1]

Components	Dimensions	Active	Semi-inert	Semi-inert	Inert
Cement CEM I 42,5 R	[kg/m³]	194	120	60	0
Sand 0-1 mm	[kg/m³]	169	169	169	169
Sand 0-2 mm	[kg/m³]	674	674	674	674
Gravel 2-8 mm	[kg/m³]	454	454	454	454
Grain size distribution	[-]	B8	B8	B8	B8
Bentonite slurry (Concentration 6%)	[kg/m³]	153	183	166	183
Fly ash	[kg/m³]	194	268	328	420
Water	[kg/m³]	207	177	164	135

4   Test setup and test procedure

The investigations of the dewatering behaviour of grout were initially carried out by using a standardized test procedure, called "filter press test", according to DIN 4126, as shown in Fig. 1 (left). Thereby, the amount of filtrate water is determined under defined pressure conditions, filter permeability and test duration.

In a further step, a test setup was developed to simulate the annular gap with its width up to 18 cm. By way of derogation from DIN 4126, an entirely perforated base plate of steel was assembled, to ensure a uniform dewatering over the whole bottom of the grout material (Fig. 1,middle).

Fig. 1   Filter press test setup according to DIN 4126 (left) and a modified filter press setup (middle, right)

A filter paper with a defined permeability (quantitatively very fast, pore size 20-25 µm) was applied onto the perforated base plate. The grout material with a very fluid consistency was poured into the pot with a defined volume. Afterwards, a Perspex plate was put onto the grout surface to achieve an even pressurization. The steel pot was hermetically sealed and a defined pressure of 3.5 ± 0.25 bar was applied for about 30 min, corresponding to the typical parameters of tunnel driving with a segment length of 2.00 m [10]. By means of a plastic funnel, assembled at the bottom of the steel pot, the filtrate water was collected in a measuring glass (Fig. 1, right). The amount of water was weighed by a precision scale and documented per minute, to comprise the temporal changes.

5   Test results

5.1   Workability

The workability, represented by the development of consistency or flow diameter at different points in time after grout mixing, is shown in Fig. 2. As expected, the flow diameter decreased considerably in the course of time, the higher the cement content of the grout composition was. Consequently, the active grout composition with the highest cement content of 194 kg/m³ exhibited the smallest flow diameters, in comparison to the semi-inert and inert grout compositions. The stiffening consistency of the active grout material was attributed to the hydration progress of cement.

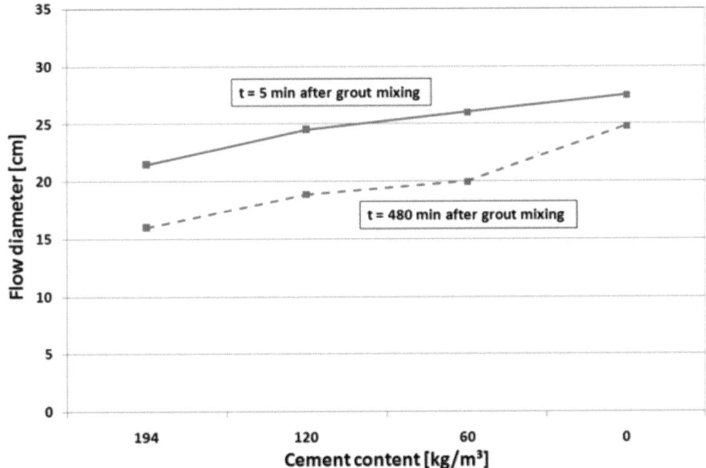

Fig. 2   Development of consistency of different grout compositions (active, semi-inert, inert)

5.2 Dewatering behaviour according to DIN 4126

Grout materials for annular gap grouting are distinguished by their very high water content to achieve the very fluid consistency for the required workability lasting for several hours, pumpability and complete backfilling of the annular gap. Concurrently with the grouting, the very fluid material must harden or consolidate by dewatering in terms of pressing out the excess water into the surrounding soil under defined pressure conditions. Fig.3 shows the total amount of filtrate water under pressurization of 7.0 ± 0.35 bar after test duration of 7.5 min, according to DIN 4126.

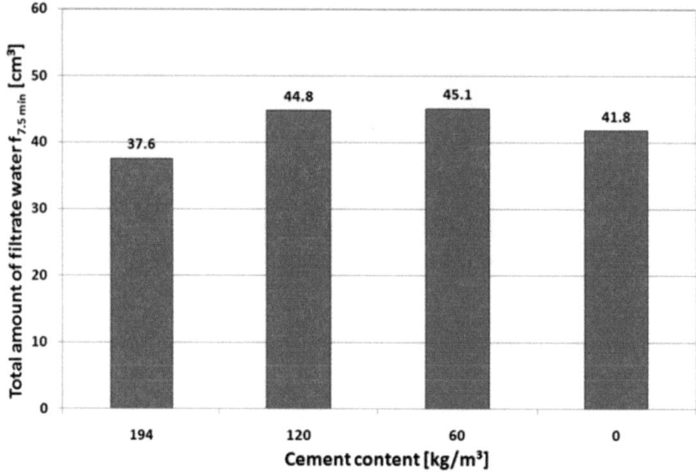

Fig. 3 Total amount of filtrate water, depending on cement content, according to DIN 4126

It can be clearly seen, that the active grout composition (194 kg/m³) displayed the smallest amount of filtrate water at the end of the filter press test, while the semi-inert grout compositions (60 kg/m³, 120 kg/m³) attained around 20 % higher amount of filtrate water. The dewatering behaviour is regulated by the water retention of the amount of ultra fines (0/0.125 mm) and fines (0/0.25 mm) as well as the initial water of the grout composition. With a higher amount of ultra fines, inter alia cement of the active grout composition, the specific surface increased and thus the physical bonding as well as the chemical reaction of cement with part of the mixing water.The relatively smaller amount of filtrate water of the cement-free grout composition, in comparison to the semi-inert grout compositions, was predominantly attributed to the smaller amount of initial water.

5.3 Dewatering behaviour by means of a modified filter press test

Further investigations on the dewatering behaviour of semi-inert grout materials (60 kg/m³) were carried out by means of a modified filter press test, to simulate the practical conditions of shield tunnelling. These filter press tests were conducted with a pressurization of 3.5 ± 0.25 bar and duration of 30 min, and, by comparison, also on the basis of DIN 4126. The curve shapes of both test series are almost identical, but the results reveal significant differences in the amount of filtrate water, as seen in Fig. 4. The amount of filtrate water under lower pressurization (3.5 ± 0.25 bar) was around 50 % lesser than that under higher pressurization (7.0 ± 0.25 bar). Throughout the entire filter press test with a pressurization of 3.5 ± 0.25 bar, a continuous rise of the amount of filtrate water could be observed. In contrast, a higher pressure (7.0 ± 0.35 bar) led into a significant increase in the first two minutes. In the further course of pressurization, the amount of filtrate water increased steadily, but with reduced intensity, which can be explained by the consolidation progress of the material, associated with the disconnection of the fluid transportation. Hence, a higher pressurization would have been needed for further dewatering.

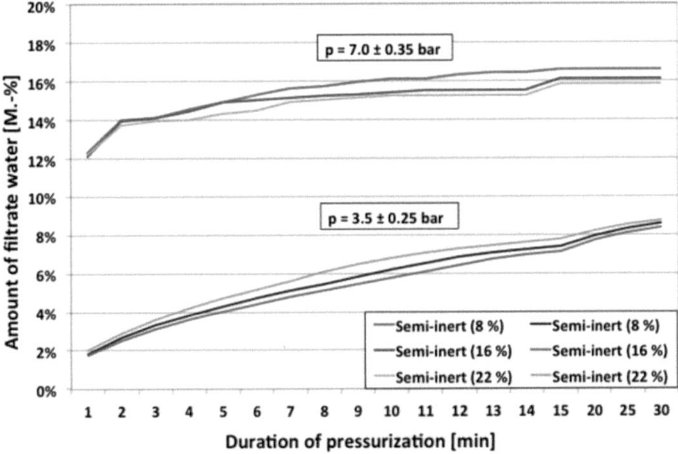

Fig. 4   Temporal changes of filtrate water under different pressurization

5.4   Development of shear strength at dewatered specimens

The dewatering behaviour of a grout material influences its development of shear strength. The shear strength tests were carried out by means of a shear vane, which was installed with a defined depth in the grout specimen before starting the filter press tests. The shear strengths were determined in different depths of the dewatered specimens, to draw possible conclusions from the values on the dewatering behaviour of the grout material. The results of the shear strength tests in different depths of dewatered grout specimens with a cement content of 60 kg/m³ are shown in Fig. 5.

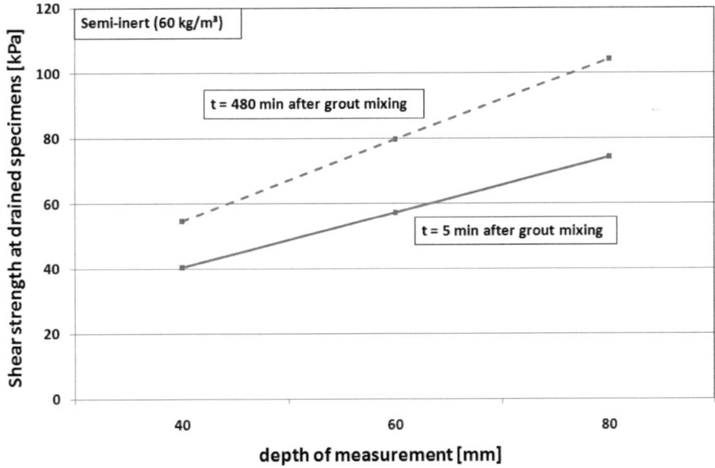

Fig. 5   Development of shear strength in different depths of dewatered grout specimens

As expected, the highest shear strengths were determined in greater depths of dewatered grout specimens, which is explained by the formation of different dewatered layers. At the end of the filter press tests, deeper layers were almost completely dewatered, whereas the upper part of the specimen was still watery, as shown in Fig. 6.

Fig. 6   Formation of different dewatered layers during the filter press test

6   Conclusions

The results of this research work demonstrated, that grout compositions with higher cement contents led into a significant decrease of consistency and smaller amounts of filtrate water and thus, higher strength values, attributed to the hydration progress of cement. The results of the investigations of shear strength in different depths revealed higher values in greater depths due to the consolidation progress of the grout material and the disconnection of the fluid transportation in the grout structure. During the filter press test under pressurization deeper layers dewatered more significantly than the upper ones.

7   Acknowledgement

Financial support was provided by the German Science Foundation (DFG) in the framework of project B3 of the Collaborative Research Center SFB 837. This support is gratefully acknowledged.

References

[1]   Thewes, M., Budach, C.: Grouting of the annular gap in shield tunnelling – an important factor for minimization of settlements and production performance. In: Proceedings of the 35th ITA World Tunnel Congress (2009)

[2]   EFNARC: Specification and Guidelines for the use of specialist products for Mechanised Tunnelling (TBM) in Soft Ground and Hard Rock (2005), pp. 17-25

[3]   Babendererde, L., Holzhäuser, J., Babendererde, S.: Verpressen der Schildschwanzfuge hinter einer Tunnelvortriebsmaschine mit Tübbingausbau. In: Tunnelbau-Taschenbuch (2002), pp. 228-254

[4]   Girmscheid, G.: Baubetrieb und Bauverfahren im Tunnelbau. Ernst & Sohn (2001), p. 458

[5]   Linger, L., Cayrol, M., Boutillon, L.: TBM's backfill mortars – Overview – Introduction to Rheological Index. In: Taylor Made Concrete Structures (2008), pp. 271-276

[6]   Bäppler, K.: Development of a two-component dynamic grouting system for tailskin injected backfilling of the annular gap for segmental concrete lining in shielded tunnel boring machines. Dissertation at the Colorado School of Mines, Mining and Earth System Engineering (2006)

[7]   Astner, M.: Spezialbaustoffe im Bauvorhaben "Hofoldinger Stollen". In: Zement + Beton 1 (2005) No. 4, pp. 12-15

[8]   Maidl, B., Herrenknecht, M., Anheuser, L.: Maschineller Tunnelbau im Schildvortrieb (2001)

[9]   Wittke, W., Druffel, R., Erichsen, C., Gattermann, J., Kiehl, B.: Statik und Konstruktion maschineller Tunnelvortriebe (2006)

[10]   Herrenknecht, M., Thewes, M., Budach, C.: The development of earth pressure shields: from the beginning to the present. In: Geomechanics and Tunnelling 4 (2011) No. 1, pp. 11-35

Subsoil strengthening by using jet-grouting technology

Anna Juzwa

Faculty of Civil Engineering, Department of Geotechnics
Silesian University of Technology
Akademicka 5, 44-100 Gliwice, Poland
Supervisor: Joanna Bzówka

Abstract

The weak subsoil conditions like an anthropogenic, organic and cohesive soil, contribute to the application of subsoil strengthening.

One of the most popular technology of subsoil strengthening is jet-grouting. Jet-grouting process consists in high-pressure injection of an inject (the cement grout usually) stream to the soil massif. The stream cuts and crushes the soil structure. Performance is started by an insert of drilling rods with high-pressure nozzle at the end, in a subsoil at expected depths. During pulling the rod, the inject is flowing from the nozzle under a high pressure. Soil and cement particles fill space in the range of erosion stream, creating the binding rigid block, called *soilcrete*. The blocks may be shaped into eg. columns.

The main motivation of my research project is preparing the explanation of interaction between jet-grouting columns and subsoil. It can be useful for designers to optimizing the procedure of dimensioning columns. The key goal is preparing the numerical model, which can be useful to illustrate the real conditions of subsoil strengthening by using jet-grouting columns.

The computational model describes the interaction between group of jet-grouting columns and soil. Presented model will reflect the essence of jet-grouting technology, and after checking the results with *in situ* tests it will be also used to verify engineering methods of dimensioning columns. The main element of this analysis is selection and calibration of computational model of the "group of jet-grouting columns - subsoil" interaction. The starting point for creating a model is the concept of dividing the model into three zones: jet-grouting columns, subsoil and the contact layer, formed between the columns and the soil massif.

1 Description of jet-grouting technology

The necessity of foundation of buildings structures on weak subsoil contributes to the application of diverse geotechnical methods for subsoil strengthening. Depending on subsoil types, their parameters and stratification, the groundwater level and the load value, there are some techniques used. One of most popular is jet-grouting method. The method was invented in the 1970s in Japan. Since this time it has been developed and used all around the world. The method brought new sense into traditional injection geoengineering methods and became the greatest invention for subsoil strengthening.

Performance of jet-grouting elements consists in high-pressure injection to the subsoil of an inject stream, which cuts and disintegrates the ground mass, forming after cement binding so-called cement-soil solid [2]. The element performance starts from drilling to the planned depth a borehole, 100 to 180 mm in diameter. Once the borehole is drilled, the drilling rod is lifted by a rotating – sliding motion. At the same time the soil around the drilling rod is cut by a very strong stream of water ejecting from nozzles. In the case of drilling rod with three nozzles this effect is increased by an envelope of compressed air pressed from the second nozzle. The cutting of soil by a water stream destroys its natural structure and this way increases many times its, very low in natural state, water permeability. The stream of cement grout ejecting from the nozzle at very high pressure reaching 80 MPa penetrates the formed soil pores without difficulty, pushing outside part of previously injected water and mixing with the soil.

Jet-grouting method is recommended for subsoil strengthening, foundation under structures which could be settled too much, as vertical and horizontal water-proof cut-off walls. It is used to form

Proc. of the 9th *fib* International PhD Symposium in Civil Engineering, July 22 to 25, 2012,
Karlsruhe Institute of Technology (KIT), Germany, H. S. Müller, M. Haist, F. Acosta (Eds.),
KIT Scientific Publishing, Karlsruhe, Germany, ISBN 978-3-86644-858-2

engineering structures as follows: columns, sealing screens, lamel walls, cell walls, tight palisades etc. [2, 3]. The main advantage of jet-grouting structures is high bearing capacity. Structures may reach substantial depths. The performance is fast and does not cause vibrations in the surroundings. Works can be provided very close to existing structures. The equipment has small dimensions and weight.

For subsoil strengthening are mainly used columns (fig. 1). A cement–soil column is formed, 0.6 to 1.5 m in diameter, depending on the soil type and inject pressure. Before cement grout binding, it is possible to insert reinforcement to the column.

Fig. 1   The pit of a jet-grouting column (J. Bzówka [3])

2   Dimensioning of jet-grouting columns

2.1   Classical method

Many scientist recognize foundation columns similar to concrete piles and the method of pile load capacity is using in engineering calculation. Polish authors calculate columns in accordance with formulas of Polish standard [4]. For every column:

$$Q_r \leq m \cdot N$$

(1)

where:

Q_r – computational column load – compressing or uplifting,

N – computational load capacities of the compressing or uplifting columns,

m – correcting coefficient.

The computational load capacities N depends on load resistance of soil under the base, frictional resistance along the shaft and dimensions (surface area of base and shaft) of columns.

This engineering method of dimensioning of jet-grouting columns has empirical origin and remains away from soil mechanics and present opportunities to create numerical models.

2.2   Numerical model of jet-grouting columns cooperated with subsoil

The problem of interaction between single jet-grouting column and subsoil was thoroughly analysed by Bzówka [2]. The author prepared some numerical models and calibrated them so as to obtain a precisely presentation of the system settlement and of the effort of its material within a wide range of load. A single column represents an idealized form, not occurring in practice. The analyses presented in this paper provide a basis for further simulations allowing for the most realistic reflection of the calculation model of interaction between a group of jet-grouting columns and the subsoil. The final model should be reflected the system settlement and the effort of its material under changing value of load [1].

2.2.1 Creation of adequate model

The computational model should properly describe mechanical parameters of materials, like different stiffness of soil layers and soilcrete, plastic character of deformation under load action and especially non-linearity of contact zone. According to Bzówka [1] computational model should:

- Applying heterogeneous elements of model (jet-grouting columns, subsoil and the contact layer, formed between the columns and the soil massif) to particular zones.
- Describing adequate constitutive models expressing elastic-plastic behaviour of the materials to particular zones.
- Using computational method for numerical solving boundary problems of mechanics of continuum, eg. method of finite elements.

Two homogeneous materials, reflecting column and soil surroundings, are considered for initial model. These zones are divided by contact layer. An elastic-ideally plastic material model with boundary condition by Coulomb-Mohr and not associated flow law is attributed for each zone.

2.2.2 Calibration of materials parameters

For every soil layer following parameters should be known: modulus of elasticity E, Poisson's ratio v, internal friction angle Φ and cohesion c. The material of columns is heterogeneous. Parameters of columns depend on surrounding soil, cement grout, performance techniques (like pressure, velocity of injection). To estimate the column parameters, tests of triaxial compression on the samples were carried out. Figure 2 shows the pit of the jet-grouting columns prepared to taking samples. The tests [1] were performed on 20 cylindrical samples, taken from columns sections with two different soil component: medium sand and silty clay.

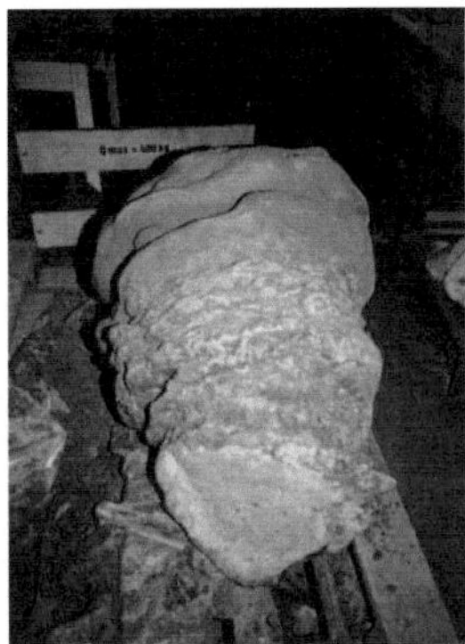

Fig. 2   The jet-grouting column excavated in the building site (left) and in the laboratory after cleaning (right) (J. Bzówka [1])

The maximum compressive stress obtained in tests by Bzówka [1, 3] equaled 28 to 34 MPa. The values of modulus of elasticity E and Poisson's ratio v were estimated on the basis of « stress-strain » characteristic. As input for computational program the following values of parameters were taken for *soilcrete*: $E = 9888$ MPa, $v = 0{,}186$, $\Phi = 59{,}3°$ and $c = 1772$ kPa, for both soil component.

More difficult problem is evaluation of contact zone parameters. For calculation was taken the solution proposed by Bzówka [1]: 33% reduction of estimated parameters for soil massif (E, v, Φ, c).

2.2.3 Creation of numerical model

The finite element method (FEM) is used to perform numerical analysis. The model is calculated in Z_Soil computational program, which uses the most robust numerical algorithms to simulate an elastic-ideally plastic material model with boundary condition by Coulomb-Mohr and not associated flow law.

For creation of numerical model there was adopted hypothesis which assumes similarity between an interaction between groups of jet grouting columns with the subsoil and a single jet grouting column with the subsoil. It is proposed to transfer a single jet grouting column model to a group of columns. The jet-grouting column is modelling by cylindrical surface with diameter d and length l. The surrounding subsoil is modelling by 3D space. The appropriate dimensions of modelled soil massif should equal: *11d* horizontally and *(l+5d)* vertically. The area of defined materials is divided into eight-noded, quadrilateral, isoparametric elements. The boundary conditions were assumed as typical: sliding support along the vertical side-edge and node fixed support for the bottom edge of the model.

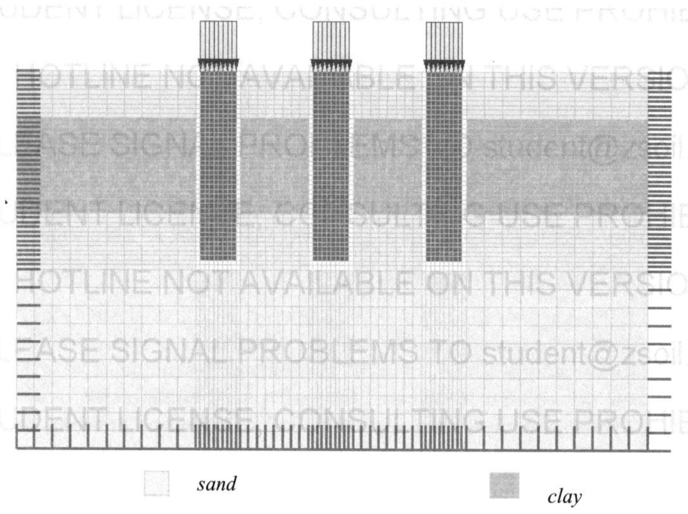

Fig. 3   The cross-section of modelling structure (Z_Soil)

Figure 3 shows a cross-section of the computational model: the jet-grouting column with diameter 0,8 m and lengtht 4,0 m, located in spacing equals 2,5 m are founded in stratified subsoil. For subsoil the following values of parameters were taken: sand E = 55,5 MPa, v = 0,3, Φ = 31,8° and c = 1 kPa and for clay E = 33,8 MPa, v = 0,3, Φ = 18,0° and c = 30 kPa. The columns are loaded by 3m height embankment. The scheme of deformation, direction and values of displacement are shown in Figures 4 to 7.

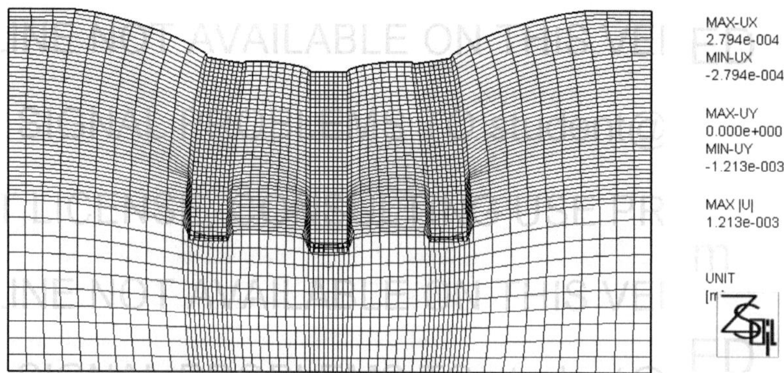

Fig. 4   The deformation of model (Z_Soil)

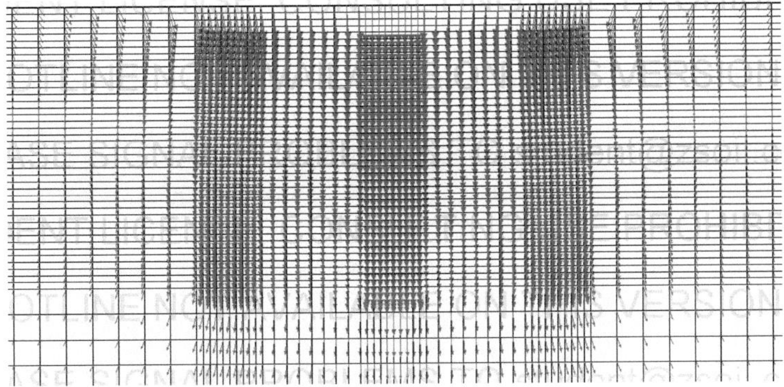

Fig. 5   The plan of displacement vectors (Z_Soil)

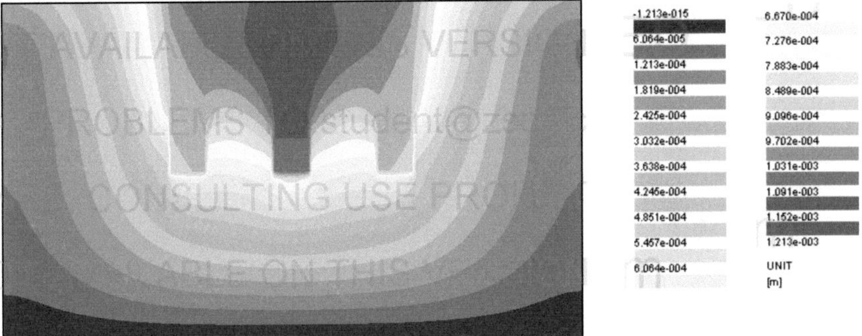

Fig. 6   The map of total displacement of model (Z_Soil)

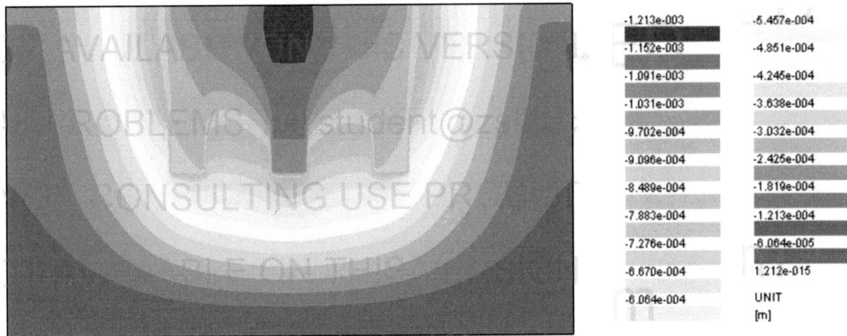

Fig. 7   The map of the main displacement along vertical Y axis (Z_Soil)

3   Conclusions

For the best calibration of numerical model it is planned to conduction of *in situ* research, like: wide range of the subsoil tests, *soilcrete* samples tests (grout material tests, triaxial compression test, etc.), trial load tests of groups of jet-grouting columns. Based on these results, comparative analyses of the theoretical and experimental curves "load-displacement" will be carried out.

Results of *in situ* tests will validate the numerical method of dimensioning jet-grouting columns. The research will form the basis for the development of a knowledge base concerning the essence of interaction between group of jet grouting column and the weak subsoil and the emerging importance of the contact layer between the columns and massive soil. It will also allow to introduce numerical techniques into dimensioning of jet grouting columns, which will be important not only for scientist but also for engineering applications.

References

[1]   Bzówka J.: Calibration of three-zone model of jet-grouting pile-soil system. In: Studia Geotechnica et Mechanica (2003) Vol. XXV, No. 3-4, pp. 3-51

[2]   Bzówka J.: Computational model for jet-grouting pile-soil interaction. In: Studia Geotechnica et Mechanica (2004) Vol. XXVI, No. 3-4, pp. 47-89

[3]   Bzówka J.: Interaction of jet grouting columns with subsoil. Monograph. (2009) Gliwice (in Polish)

[4]   PN-83/B-02482 Polish Standard: „Building Foundations. Load capacity of piles and pile foundations"

Complete Recycling of Demolished Concrete with Low Environmental Loads

Wakiro Nishigori and Koji Sakai

Faculty of Engineering,
Kagawa University,
2217-20 Hayashicho, Takamatsu, Kagawa, 761-0396, Japan
Supervisor: Koji Sakai

Abstract

Amounting to above 30 million tons every year, demolished concrete lumps are demanded to be recycled into concrete as aggregate. Studies on recycled aggregate so far have focused on the quality enhancement of recycled coarse aggregate, and thus apparatuses that impose heavy environmental loads have been used in many cases. Also, recycled fine aggregate and recycled powder are yet to be put to practical use. With this as a background, a rubbing recycling machine with low environmental loads was used in this study. The authors also devised a technique involving aggregate modification to investigate major technical problems related to recycled aggregate concrete, such as strength development and drying shrinkage, with the goal of completely recycling demolished concrete lumps, including recycled fine aggregate and recycled powder.

1 Introduction

While the emissions of construction by-products have tended to decrease in recent years, demolished concrete lumps have continued to exceed 30 million tons annually, with their proportion to total construction by-products tending to increase in Japan. Though the recycling ratio of concrete lumps is as high as nearly 100%, it depends highly on the demand from road construction including base course material. This demand is beginning to decrease recently due to reduced road construction, urging recycling of concrete into concrete. The enormous amount of debris generated after the Great East Japan Earthquake that struck in March 2011 also demands prompt disposal. With the pressing need to effectively utilize these materials as well, it is imperative to accelerate the development of technology for producing and using recycled aggregate concrete.

Meanwhile, Japan set a medium-term goal of a 25% reduction of greenhouse gas emissions by 2020 with respect to the level of 1990 as a measure to cope with global warming[1]. Control of CO_2 emissions is expected to be more important for the construction sector from now, particularly regarding concrete, one of major materials for construction. Therefore, it is vital, when utilizing demolished concrete for concrete structures, to adopt processes imposing low environmental loads.

Nevertheless, recycled aggregate concrete tends to show lower strength and greater drying shrinkage than those of normal concrete. In an attempt to solve these technical problems, studies related to recycled aggregate for concrete so far have focused on enhancing the quality of recycled coarse aggregate, resulting in technologies imposing heavy loads on the environment. Also, recycled fine aggregate and powder generated during production have not yet been put to practical use. Recycling of concrete lumps will hardly be achieved unless such fines are effectively used.

With this as a background, a rubbing recycling machine was used in this study to produce recycled aggregate with low environmental loads. Also, the authors devised a technique involving not only quality enhancement during production but also post-production modification of recycled aggregate. Based on this technique, major technical subjects related to recycled aggregate concrete, such as strength development and drying shrinkage, were investigated with the goal of achieving complete recycling of concrete including recycled fine aggregate and recycled powder.

Proc. of the 9th *fib* International PhD Symposium in Civil Engineering, July 22 to 25, 2012,
Karlsruhe Institute of Technology (KIT), Germany, H. S. Müller, M. Haist, F. Acosta (Eds.),
KIT Scientific Publishing, Karlsruhe, Germany, ISBN 978-3-86644-858-2

2 Production of recycled aggregate

2.1 Original concrete

Tables 1 and 2 give the materials and mix proportions of original concrete used for producing recycled aggregate, respectively.

Table 1   Materials of original concrete

materials	density(g/cm^3)*	water absorption(%)	fineness modulus
Fine aggregate (S)	2.56	2.2	2.75
Coarse aggregate (G)	2.60	2.1	6.70

*density in saturated surface-dry condition

Table 2   Mix proportions and fresh properties of original concrete

W/C (%)	S/a (%)	Mix proportions (kg/m^3)				super-plasticizer	air entraining agent	Slump (cm)	Air content (%)
		W	C	S	G				
65	49	185	285	855	900	CX1.35%	CX0.2%	18.5	4.9

2.2 Outline of the rubbing recycling machine

Photo 1 (left) shows the appearance of the rubbing recycling machine used in this study. Photo 1 (right) shows the inside features of the machine. The inside of the machine is separated into two lanes by a steel cylinder. Threaded steel rods 24 mm in diameter are vertically arranged in each lane. When the rods rotate, particles charged into the machine roll down through clearances between the rods, collectively flowing along the lanes. During this process, mortar adhering to aggregate surfaces is removed by abrasion with the rods and between aggregate particles.

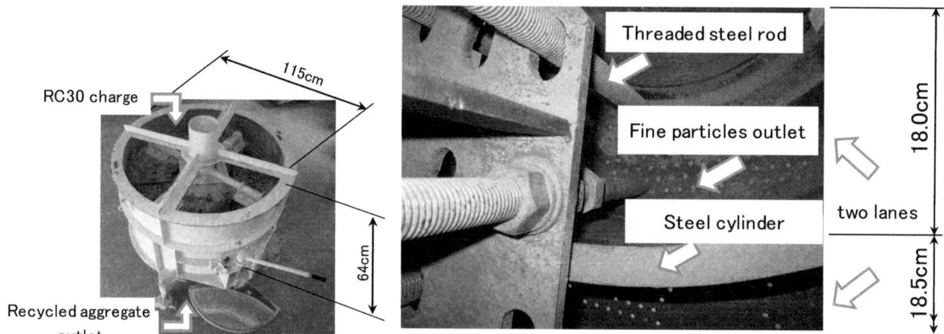

Photo 1   The rubbing recycling machine

2.3 Recycled aggregate production flow

Figure 1 shows the flow of recycled aggregate production. Original concrete blocks were coarsely crushed at an age of 56 days using a hydraulic breaker and crusher and then crushed to a medium size with an impeller breaker to produce recycled crusher-run RC30.

RC30 was then subjected to abrasion in a rubbing recycling machine, and particles greater than 5 mm were classified into recycled coarse aggregate. Figure 2 (left) shows the grading of the resulting recycled coarse aggregate, which is slightly finer than crushed stone 2005 (Japanese Industrial Standard (JIS A 5005)), being in the range of the grading of crushed stone classified as crushed stone 1505 (JIS A 5005). As for particles finer than 5 mm, those finer than 0.15 mm were further classified into recycled powder. Particles from 0.15 to 5 mm were treated with a ball mill, and those 0.15 mm or greater and finer than 0.15 mm were classified as recycled fine aggregate and recycled powder (blended with formerly classified recycled powder) , respectively.

Table 3 gives the treatment conditions by the rubbing recycling machine in this study. These were determined based on test results by Ogita et al.[2] to produce aggregate of maximum quality while imposing minimum environmental loads.

2.4　Improvement of recycled fine aggregate grading

The grading of 0.15 to 5 mm particles obtained by rubbing is as shown in Fig. 2 (right) before treatment, being slightly coarser than the standard grading of fine aggregate. Treatment was therefore carried out using a ball mill with the conditions given in Table 3.　Namely, not only a rubbing recycling machine but also a ball mill was used to adjust the grading of recycled fine aggregate. Figure 3 also shows the grading after treatment, which is proven to be in the standard grading for fine aggregate.

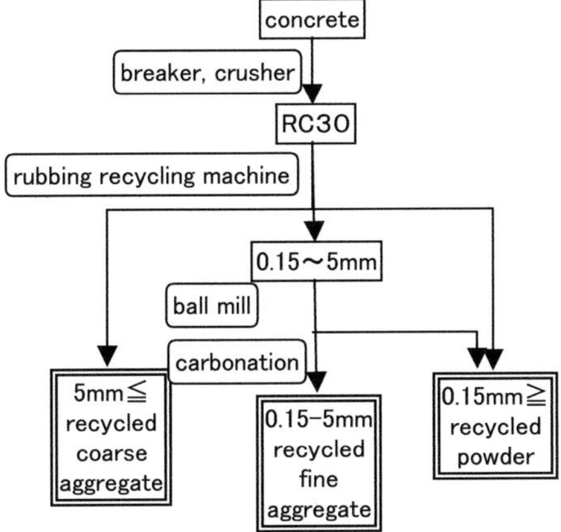

Fig. 1　　Flow of recycled aggregate production

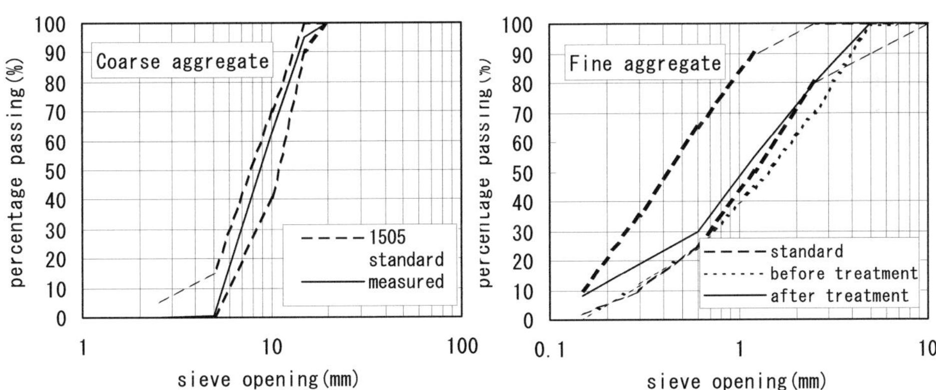

Fig. 2　　Grading of the resulting recycled aggregate

Table 3　　Treatment conditions

	conditions	
rubbing recycling machine	distance between each rod	26.5mm
	time for treatment	20min
	mass of each charge	50kg
ball mill*	time for treatment	15min
	speed of rotation	40rpm
	mass of each charge	10kg

* mass of material : mass of ball =1:1

2.5 Improvement of the density and water absorption of recycled fine aggregate

Production and use of recycled fine aggregate have not been practiced on a commercial basis. This is because the small particle size makes it difficult to physically improve recycled aggregate. That is, it hampers removal of cement paste from the surfaces of fine aggregate. The water absorption of recycled fine aggregate is generally high, presumably due to the effects of the quality and microcracks of cement paste adhering to its surfaces. Treatment using a ball mill conducted in this study led to no appreciable improvement in the density and water absorption of recycled fine aggregate.

For this reason, it was decided to attempt improvement in the density and water absorption by carbonation in addition to physical abrasion. This is to fix CO_2 into adherent cement paste by carbonation to fill microcracks, thereby modifying the surface properties.

Carbonation of recycled fine aggregate was carried out using a carbonation-accelerating chamber with a temperature, relative humidity, and CO_2 concentration of $20°C$, 60%, and 5%, respectively. Carbonation time was determined by testing. Each sample of recycled fine aggregate was spread as thin as possible in a container and stirred once every day to make the sample uniform. Figures 3 show the changes in a rate of mass increase, oven-dry density, and water absorption, respectively, over time. The rate of mass increase due to carbonation subsided by the end of the fourth day and leveled off thereafter. The oven-dry density and water absorption rapidly increased and decreased, respectively, for the first two days, but the changes grew slower thereafter.

Table 4 gives the JIS-requirements for the oven-dry density and water absorption of recycled fine aggregate. Figures 3 (center and right) demonstrate that the carbonation process improved the oven-dry density and water absorption to the JIS-required ranges for recycled fine aggregate classified as M (JIS A 5022), while the ball mill treatment alone led to a quality level equivalent to recycled fine aggregate L (JIS A 5023). It can therefore be said that the carbonation process in this study is effective in modifying recycled fine aggregate.

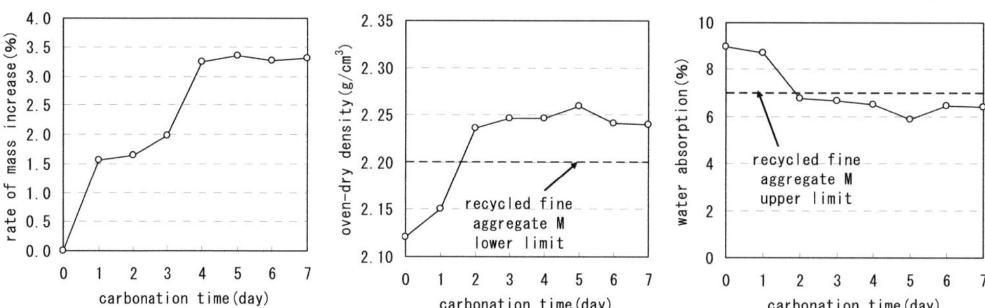

Fig. 3   Changes in a rate of mass increase, oven-dry density, and water absorption

Table 4   JIS-requirements for the oven-dry density and water absorption of recycled fine aggregate

	L	M	H
oven-dry density(g/cm³)	—	2.2 ≦	2.5 ≦
water absorption(%)	13.0 ≧	7.0 ≧	3.5 ≧

3   Tests on recycled aggregate concrete

3.1   Materials

Table 5 gives the materials used for concrete testing and their qualities. The recycled coarse aggregate meets the requirements for recycled coarse aggregate M (JIS A 5022). On the other hand, the recycled fine aggregate did not achieve an improvement comparable to Fig. 3, slightly deviating from the standard for recycled fine aggregate M (JIS A 5022), though it was subjected to carbonation for 3 days by being spread to a depth of 3 cm in a container and homogenized by stirring 3 times every day.

The recycled powder were also subjected to carbonation for 1 hour, but showed little change in the mass.

Note that the paste content by mass of the recycled fine aggregate and recycled powder was measured by dissolving paste in hydrochloric acid, with the results being approximately 22% and 48%, respectively.

3.2 Trial mix and mixing of concrete

Table 6 gives the mix proportions of trial mixtures. The target slump and air content were 8 ± 1.5 cm and $5 \pm 1\%$, respectively. The water-cement ratio (55%) and unit water content were fixed, while the type and unit weight of admixtures were changed to adjust the slump and air content.

Crushed andesite sand (NS) and recycled fine aggregate (RS) were used as fine aggregate, and crushed andesite stone (NG) and recycled coarse aggregate (RG) were used as coarse aggregate. Recycled powder (RP) were included in place of part of fine aggregate. Note that a maximum recycled powder content of 60 kg/m^3 was adopted, because 60 kg of recycled powder was generated from 1 m^3 of original concrete in the production of recycled aggregate. Tests were also conducted using a half of that proportion, 30 kg/m^3, of recycled powder.

Concrete was produced using a forced type twin-shaft mixer with a capacity of 50 liters in a laboratory at $20°$C and 60 % R.H. The batch size was 30 liters. The mixing procedure was as follows: Dry-mix cement, recycled powder, fine aggregate and coarse aggregate for 15 sec, add water and chemical admixtures, and mix for 105 sec.

Table 5   Materials used for concrete testing

materials	density(g/cm^3)	water absorption(%)	solid content(%)	fineness modulus
Normal fine aggregate (NS)	2.63*	1.84	62.5	2.85
Recycled fine aggregate (RS)	2.36*	7.74	67.2	3.45
Recycled powder (RP)	2.40	5.30	-	-
Normal coarse aggregate (NG1)	2.62*	1.89	57.4	6.54
Normal coarse aggregate (NG2)	2.62*	1.60	57.1	7.00
Recycled coarse aggregate (RG)	2.57*	3.40	59.9	6.37

*density in saturated surface-dry condition

Table 6   Mix proportions of recycled aggregate concrete

W/C (%)	S/a (%)	Mix proportions (kg/m^3)								water-reducing agent	super-plasticizer	air entraining agent	Slump (cm)	Air content (%)
		W	C	NS	RS	RP	NG1	NG2	RG					
55	46	171	311	823	-	-	481	481	-	CX1.1%	-	CX0.05%	8.5	5.7
55	46	171	311	823	-	-	-	-	944	CX1.0%	-	CX0.05%	9.5	5.8
55	46	171	311	-	739	-	-	-	944	CX1.0%	-	CX0.05%	7.9	5.9
55	46	171	311	-	709	30	-	-	944	-	CX0.3%	CX0.05%	7.5	5.8
55	46	171	311	-	680	60	-	-	944	-	CX0.6%	CX0.05%	8.2	5.5

3.3 Test items and methods

The following concrete tests were conducted in this study:

3.3.1 Slump and air content

Tests for slump and air content were conducted in accordance with JIS A 1101 and JIS A 1128, respectively.

3.3.2 Compressive strength and elastic modulus

Tests for compressive strength and elastic modulus were conducted on specimens fabricated by demolding one day after placing and water-curing at $20°$C until testing in accordance with JIS A 1108 and JIS A 1149.

3.3.3 Length change

Length change tests were conducted in accordance with JIS A 1129-3. Specimens were demolded one day after placing, water-cured at $20°$C until an age of 7 days at which the datum length was measured.

Specimens were then air-cured in a room with constant temperature and humidity at 20°C and 60% R.H. for a specified period. Note that a test age in these tests represents the number of days from the time of datum length measurement.

4 Test results

4.1 Unit weight of chemical admixture

As given in Table 6, unit weight of the chemical admixture necessary for attaining the target slump and air content of recycled aggregate concrete, with or without recycled fine aggregate, were equivalent to that of normal concrete made using normal coarse and fine aggregates. This is presumably because the particle shape of both recycled fine and coarse aggregates, of which solid content were higher than those of normal aggregate (see Table 5), were better.

Meanwhile, the target slump was not achieved by an air-entraining and water-reducing agent, when recycled powder were used, requiring a superplasticizer. A higher recycled powder content required a higher dosage of a superplasticizer.

4.2 Compressive strength

4.2.1 Test results

Figure 4 (left) shows the results of compressive strength tests at 7 and 28 days.

4.2.2 Effect of the use of recycled aggregate

In regard to the effect of recycled aggregate on the compressive strength shown in Fig. 4 (left), the compressive strength at 7 days is slightly lower than normal concrete when recycled coarse aggregate is used (NS-RG) and even lower when both recycled coarse and fine aggregates were used (RS-RG), but the difference from normal concrete is only several percent. On the other hand, significant differences are recognized at 28 days. The compressive strength with NS-RG is 7% lower than that of normal concrete, while the strength loss is as large as 20% with RS-RG.

The strength loss with NS-RG may be attributed to the effect of paste adhering to the surfaces of recycled aggregate. As to the case of RS-RG, the strength loss can be attributed to the effect of above-mentioned microcracks in addition to the greater amount of paste adhering to recycled fine aggregate than to recycled coarse aggregate. In other words, the carbonation process was able to fill microcracks, improving the density and water absorption, but did not lead to modification in terms of strength.

Also, such strength losses due to the effect of aggregate appear to be more significant under high strength conditions.

4.2.3 Effect of recycled powder

Figure 4 (left) reveals the compressive strength gains of specimens containing recycled powder (RP). Thus the compressive strength of specimens with RS-RG-RP is related to the recycled powder content in Fig. 4 (right). This figure demonstrates that the compressive strength increases as the recycled powder content increases, with the strength with 60 kg/m^3 powder (RP 60) being comparable to that of normal concrete (dashed lines in the figure). This is presumably due to the microfiller effect of recycled powder, suggesting a possibility that the effective use of recycled powder may compensate for strength losses, which are inevitably associated with the use of recycled aggregate.

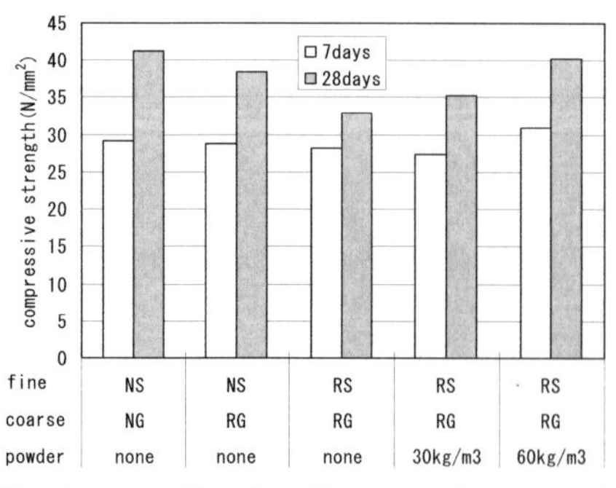

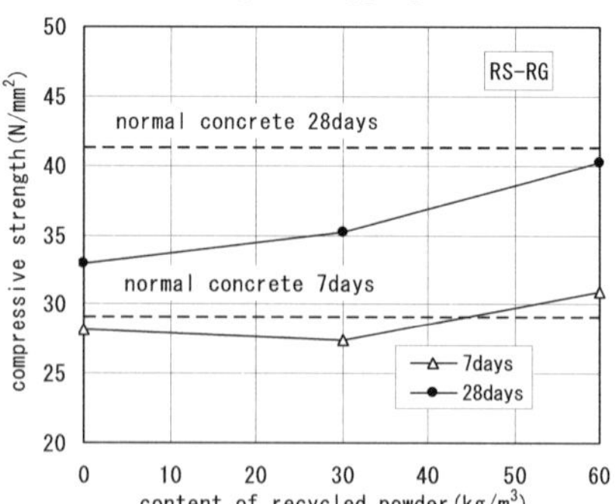

Fig. 4   Results of compressive strength tests

4.3   Elastic modulus

Figure 5 (left) shows the measured static modulus of elasticity at 7 and 28 days. The relationship between the compressive strength and elastic modulus is shown in Fig. 5 (right) along with this relationship appearing in the Standard Specifications for Concrete Structures (Design)[3] (hereafter referred to as Standard Specifications).

As seen from Fig. 5 (left), a tendency similar to compressive strength is observed for elastic modulus. The elastic modulus of specimens with NS-RG is lower than that of normal concrete, and the modulus is even lower with RS-RG. When recycled powder are contained, the ratio of increase in the elastic modulus is not so significant as the case of compressive strength.

When focusing on the relationship between the compressive strength and elastic modulus shown in Fig. 5 (right), the relationship for normal concrete and concrete with NS-RG nearly coincides with the relationship shown in the Standard Specifications. On the other hand, the elastic modulus of concrete with RS-RG tends to be low with respect to the compressive strength. This is presumably due to the brittleness of recycled fine aggregate containing cement paste particles.

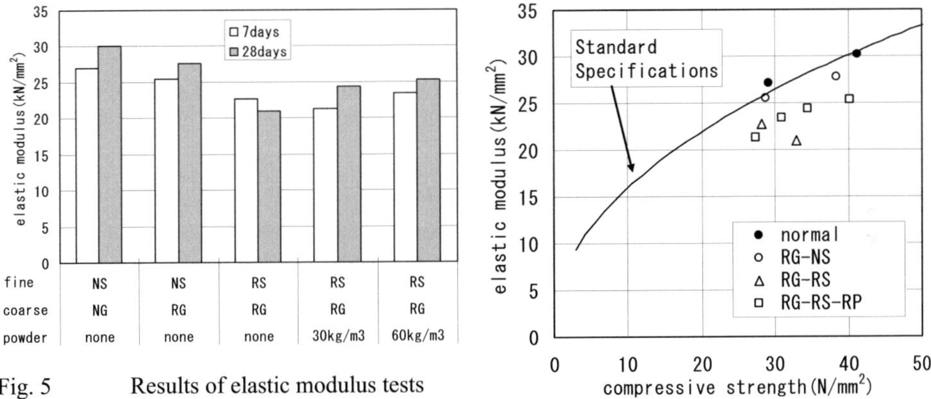

Fig. 5   Results of elastic modulus tests

4.4   Length change due to drying shrinkage

Figure 6 shows the results of length change tests.

The length change of recycled concrete due to drying shrinkage is generally larger than that of normal concrete. When comparing concretes containing recycled aggregate, specimens with NS-RG, RS-RG, and those containing 30 kg/m³ recycled powder (RP 30) show similar length changes. That of RP 60 is smallest among recycled aggregate concretes. In other words, the length change is increased by the use of recycled aggregate but can be curbed by using recycled powder. It is inferred that the use of recycled powder has a microfiller effect, but detailed investigation is necessary for elucidating the mechanism.

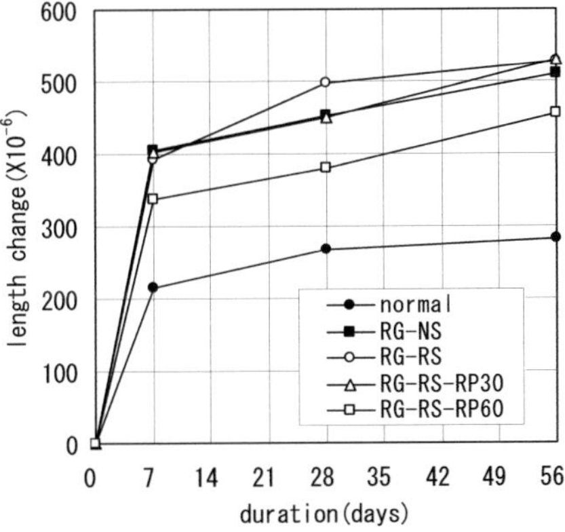

Fig. 6   Results of length change due to drying shrinkage

5   Conclusions

A rubbing recycling machine imposing low environmental loads was used in this study to produce recycled aggregate, which was then subjected to quality improvement including treatment using a ball mill to adjust its grading and accelerated carbonation to improve its density and water absorption. Concrete tests were then conducted using the resulting recycled aggregate with the aim of completely recycling concrete on the premise of using the whole amount of not only recycled fine and coarse aggregates but also recycled powder. The results are summarized as follows:

5.1   Production of recycled aggregate

The production of recycled coarse aggregate that meets the requirements for recycled coarse aggregate M was possible by the use of a rubbing recycling machine of a low environmental load type.

It was difficult to produce recycled fine aggregate that meets the requirements for recycled fine aggregate M by the use of the rubbing recycling machine alone.

5.2   Modification of recycled fine aggregate

The 0.15 to 5 mm particles of recycled fine aggregate produced using the rubbing recycling machine were treated in a ball mill to adjust the grading. The standard grading for recycled fine aggregate was achieved by this process. Recycled fine aggregate was also subjected to modification by carbonation. The requirements for recycled fine aggregate M were thus satisfied in terms of density and water absorption.

5.3   Proportioning of recycled aggregate concrete

The chemical admixture dosage necessary for achieving the target slump and air content of recycled aggregate concrete was nearly the same as that for normal concrete even when the whole amount of recycled coarse and fine aggregates was used.

When recycled powder were used, a superplasticizer was necessary, with its demand increasing as the powder content increased.

5.4   Quality of recycled aggregate concrete

The compressive strength of recycled aggregate concrete was lower than normal concrete when recycled coarse aggregate and normal fine aggregate were used, and even lower when both recycled coarse and fine aggregates were used. With the inclusion of recycled powder, however, the compressive strength increased as the recycled powder content increased. The compressive strength with 60 kg/m^3 recycled powder was comparable to that of normal concrete.

The elastic modulus of recycled aggregate concrete showed tendencies similar to compressive strength, being lower than normal concrete when recycled coarse aggregate and normal fine aggregate were used and even lower when both recycled coarse and fine aggregates were used. In contrast to compressive strength, the elastic modulus gains due to the inclusion of recycled powder were not as large as the case of compressive strength.

The length change of recycled aggregate concrete due to drying shrinkage was greater than that of normal concrete but was curbed by the inclusion of recycled powder, with the difference narrowing as the powder content increased.

This study appears to mark a major step forward for achieving the target of complete recycling of concrete from demolished concrete lumps into concrete.

References

[1]   Ministry   of   the   Environment   (Japan):   Challenge   25, http://www.challenge25.go.jp/about/index.html

[2]   Ogita, I. et al.: A Fundamental study on the efficient production of recycled coarse aggregate from demolished concrete. 38th Forum for Crushed Stone (2011)

[3]   Japan Society of Civil Engineers: Standard Specifications for Concrete Structures (Design) (2008), p. 44

Influence of the microstructure of low density mineralized foams on their thermal conductivity

Albrecht Gilka-Bötzow

Institute of Concrete and Masonry Structures and Building Materials,
Technische Universität Darmstadt,
Petersenstr. 12, 64287 Darmstadt, Germany
Supervisor: Harald Garrecht

Abstract

Our research wants to make utilizable mineralized foams with densities lower than $250\,kg/m^3$ for thermal insulating. It should be suitable on the one hand for applications prefabricated panels and on the other hand as a part of complex wall systems. Based on empiric results, this paper describes in particular structural-physical characteristics of mineralized foams with lower densities. The Dependency between the thermal conductivity and the density of mineralized foam is characterized in detail. In the light of the illustrated results there are some considerations about the influencing factors that cause the microstructure of mineralized foams which has effect on their thermal conductivity.

1   Introduction

So called foamed concrete undoubtedly is a material with attractive characteristics for certain purposes. Research has been carried out on this material for decades [1]. Most of the analyzed mixes reach densities about $1200\,kg/m^3$ [2] and are nowadays used mainly outside Europe for the manufacture of structural components in monolithic constructions. Recent research conducted in Germany was about mineralized foam with densities between 300 and $1200\,kg/m^3$ [3, 4]. With low density foams for example it is possible to realize thermal insulation layers as a part of an exterior insulation and finish systems (EIFS). Finding out the lowest possible density of mineralized foams that still secures determined and reproducible characteristics remains a challenge. In a two year industrial research program the main characteristics as well as a robust production form of mineralized foam with densities under $250\,kg/dm^3$ were identified [5].

2   Material and methods

Low density mineralized foam consists of lime and more than 80 % pores. The lime was reformulated of cements (CEM I, II and III; standard strength 42.5 and 52.5), pozzolanic and non reactive additives and water. The amount of water was 40 % in reference to the whole quantity of fines. A great variety of mineralized foam mixtures were analyzed amongst others in terms of thermal conductivity, density, compressive strength and shrinking.

A two step production process was used in the studies. Firstly the mineral lime components are mixed by a special slurry mixer in order to reach a viscous flowing consistence without using superplasticizers that might interact with the foaming agents. Afterwards, prefabricated foam based on protein foaming agents is mixed under the fluid phase. Starting with the density of approximately $400\,kg/m^3$ there is only binder and no fine aggregates has to be added [6]. The fresh mortar shows an good workability. The strength development occurs, in contrast to aerated concrete, by the hydration of the cement under normal storage conditions. Once hardened the material is still processable which offers advantages in a possible prefabrication production process.

The focus of our past work was on optimizing mineralized foam regarding low thermal conductivity by finding new mixtures with lower densities. These were supposed to be able to minimize the density and simultaneously considering the necessity of a robust production method. Conductivity was measured with the guarded hot plate method for every mixture on three temperature levels (10, 25 and 40 °C). Subsequent the thermal conductivity of the dry material at 10 °C was extrapolated. It is possible to reach thermal conductivities under $0.05\,W/(m \cdot K)$.

The minimum value for plaster systems of $0.4\,N/mm^2$ for 28 day compressive strengths is obtainable by mineralized foams with minimum densities of approximately $150\,kg/m^3$. Materials with lower

Proc. of the 9th *fib* International PhD Symposium in Civil Engineering, July 22 to 25, 2012,
Karlsruhe Institute of Technology (KIT), Germany, H. S. Müller, M. Haist, F. Acosta (Eds.),
KIT Scientific Publishing, Karlsruhe, Germany, ISBN 978-3-86644-858-2

compressive strengths are not approvable as thermal insulating layers. The shrinkage value is about 0.2 % after one year. The explicit comportment of mineralized foam under fire load was not investigated. But it is generally known that the material is incombustible. Thermal conductivity also depends on the temperature. Nevertheless, under higher temperatures mineralized foam should have a nearly unvaried conductivity because of its high porosity and its low conductivity under normal temperatures [7, 8]. If there is no development of expanded fissuring or if the material is protected by another airtight layer, mineralized foam could assume insulating functionalities at higher temperatures.

3 Data base

The investigations showed that the dependency on the density of thermal conductivity is not linear. In fact its gradient corresponds approximately to the known function of thermal conductivity for densities from 400 kg/m³ and higher of non fines lightweight concrete (Fig. 1) [9, 10]. Therefore it does not make much sense to attempt lower thermal conductivities by just minimizing the density of the material. This approach would lead to low compressive strengths.

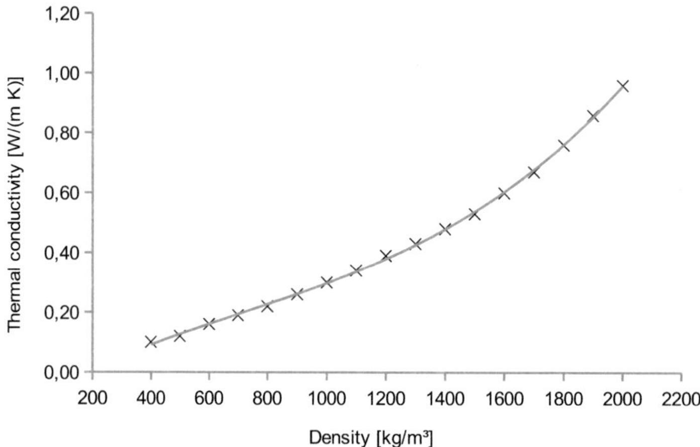

Fig. 1   Gradient of thermal conductivity in relation to density for non fines lightweight concrete [9]

Based on our great number of test results we derived a polynomial correlation between thermal conductivity and the density of mineralized foam for limited range of densities from 125 to 250 kg/m³. With this function $\lambda(\rho)$ it is possible to assign a variation v from the measured thermal conductivity λ_m to the calculated one on the base of the measured density ρ_m. Figure 2 shows the variation expressed as a percentage of measured density related to the calculated one for a number of mineralized foam mixtures. The variation shows the possible impact of influences beyond the density on thermal conductivity of mineralized foam up to 20 %. It is noticeable that this obtains for every here reconsidered density level. The probable maximum error of measurement for a hot plate apparatus is 3,07 % [11].

$$v = \frac{\lambda(\rho_m)}{\lambda_m}$$

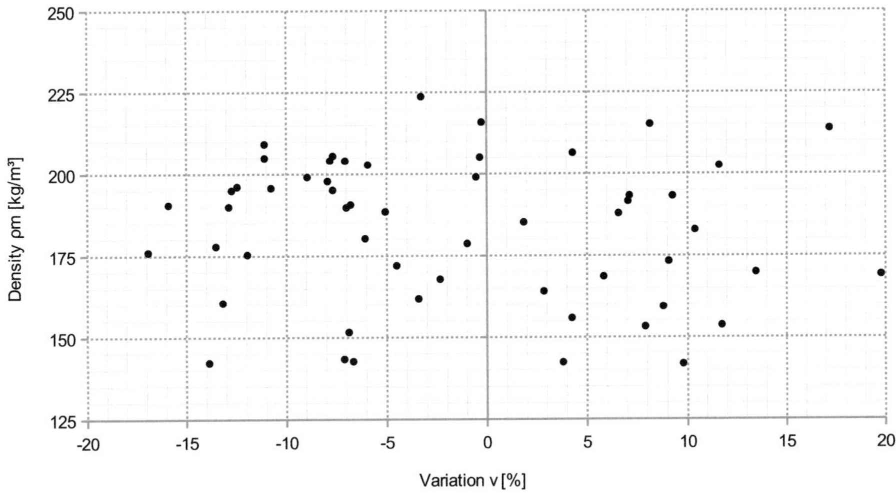

Fig. 2 Relationship for different mineralized foam mixtures between the variations from measured thermal conductivity to the calculated one and the measured densities

4 New challenges emerging from the findings

The empirical studies showed that there are three factors influencing the microstructural development of mineralized foam. The rheology of the lime is a determining factor for the stability of the fresh foam and the pore structure of the hardened material. Normally in an emulsion the gravitational force respectively the buoyancy effects a phase separation of fluid and gaseous phase. If the fluid phase has a certain viscosity with a pronounced flow limit the buoyancy has to be higher than the flow limit to make ascent the gaseous phase (Fig. 3, left) [12]. Foams additionally have addition sucking forces in the lamella which act in direction of the so called plateau until the bubble is destroyed. Viscous fluids in the lamella contrast this effect (Fig. 3, right) [13]. A fresh foam mortar remains more stable if the viscosity of the lime is able to resist the gravitational force and the sucking forces in the lamellas. Very liquid slurry is not able to preserve the intermixed gas pores due to its low viscosity. The air ascends in the suspension and at the same time flows downwards the pore membranes. A stiff or plastic lime destructs the foam due to mechanical stress and a not saturated water demand of the fines. If the ideal consistency of mineralized foam is adjusted, more flowable slurries lead to bigger pores and thinner membranes (Fig. 4), because bubbles joining when they get destroyed. We are still searching for a suitable method to carve out the differences of rheology between the various usable slurries.

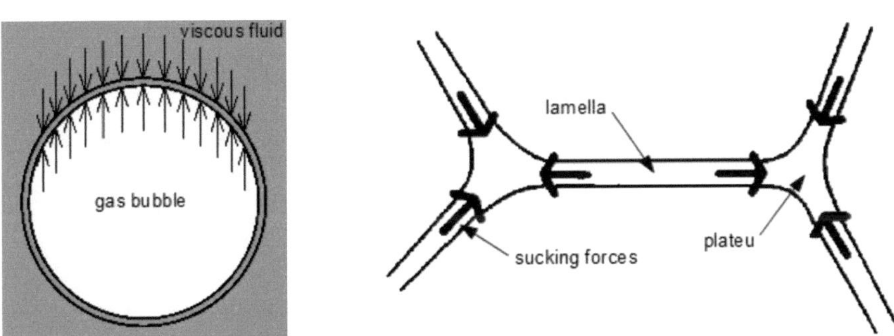

Fig. 3 Model of an emulsion with equilibrated buoyancy and forces caused by the viscosity (left), model of sucking forces acting in foam lamella [13] (right)

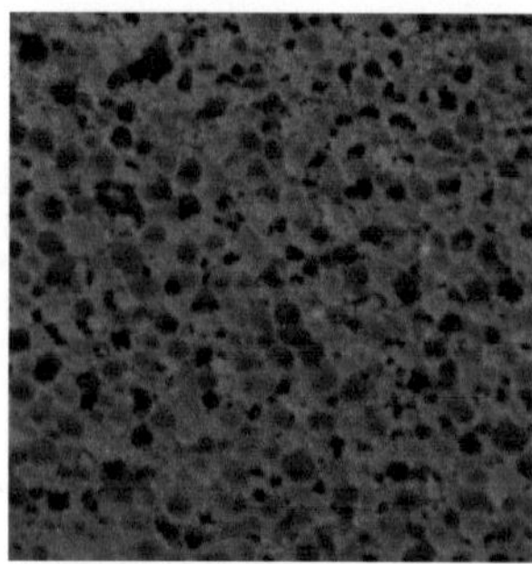 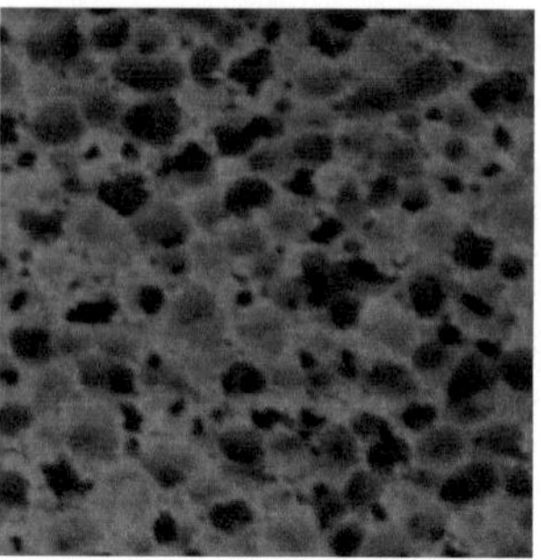

Fig. 4   Details (30 x 30 mm) from mineralized foam made with slurries with higher (left) and lower (right) viscosity.

The microstructure of the mineralized foam is also defined by the composition of the lime. The binding and non binding fines form a more or less compact wall structure around the pores depending on their granulometry and their chemical activity. This circumstance has an influence on the thermal conductivity of the hardened material.

The third important factor for the microstructural development of mineralized foam is the interaction with materials in contact while the hardening of mineralized foam. As part of a composite material especially the stability of the fresh foam mortar is influenced by the surficial area and the absorbency of the contact material. These factors determine the demand of water of the composite partner. If the demand has to be covered with water from the foam there is a high risk of destroying the mortar. Accordingly for every material combination has to be found the right mixture for the foam mortar or rather an appropriate conditioning of the contact surface.

5   Conclusions

The previous considerations and the conducted empirical studies showed that there is a complex correlation between thermal conductivity and density of mineralized foam with lower densities. Starting from microscopic examinations we will try to appreciate the mechanism of the treated material. In this context, it seems promising to study especially the influences of the lime characteristics on the microstructure of mineralized foam.

References

[1]   Ramamurthy, K., Kunhanandan Nambiar, E.K., Indu Siva Ranjani, G.: A classification of studies on properties of foam concrete. In: Cement and Concrete Composites 31 (2009) pp. 388–396

[2]   Karl, S.: Leichtzuschlag-Schaumbeton als Konstruktionsleichtbeton mit abgeminderter Rohdichte. TU Darmstadt (1979)

[3]   Pott, J.U.: Entwicklungsstrategien für zementgebundene Schäume. Universität Hannover (2006)

[4]   Just, A.: Untersuchungen zur Weiterentwicklung von chemisch aufgetriebenen, lufthärtenden, mineralisch gebundenen Schäumen. Shaker, Aachen (2008)

[5]   Harald Garrecht, Albrecht Gilka-Bötzow: Wandfertigteile aus Leichtbeton - Ist die massive Bauweise mit der EnEV 2009 noch möglich? In: BFT International (2010) pp. 112–115

[6]   Reul, H.: Handbuch Bauchemie. Verl. für Chem. Industrie Ziolkowsky, Augsburg (1991)

[7]   Bergmeister, K., Fingerloos, F., Wörner, J.-D.: Beton-Kalender 2011. Schwerpunkte: Kraftwerke, Faserbeton (A1, Gb). Ernst & Sohn (2010)

[8]   Brandverhalten von Porenbetonbauteilen. Bundesverband Porenbetonindustrie e.V., Hannover (2008)

[9]   DIN EN 1520:2007-10, Vorgefertigte bewehrte Bauteile aus haufwerksporigem Leichtbeton, Beuth Verlag

[10]   Gilka-Bötzow, A.: Mineralisierter Schaum. In: Beiträge zum 51. DAfStb Forschungskolloquium 2 (2010) pp. 709–720

[11]   Deutsches Instiut für Normung: DIN EN 1946-2:1999-04, Technische Kriterien zur Begutachtung von Laboratorien bei der der Durchfühung der Messungen von Wärmeübertragungseigenschaften - Messungen nach Verfahren mit dem Plattengerät, Beuth Verlag

[12]   Brauer, von H.: Grundlagen der Einphasen- und Mehrphasenströmungen. Sauerländer, Aarau [u.a.] (1971)

[13]   Mollet: Formulierungstechnik - Emulsionen, Suspensionen, Feste Formen. Wiley-VCH (1999)

Exploitation of TiO$_2$ nanoparticles in concrete industry

Petr Bílý

Department of Concrete and Masonry Structures,
Faculty of Civil Engineering, Czech Technical University in Prague,
Thákurova 7, 16629 Prague 6, Czech Republic
Supervisor: Alena Kohoutková

Abstract

This paper reviews state of the art in the field of titanium dioxide nanoparticles (n-TiO$_2$) utilization in concrete industry. By using n-TiO$_2$-additised cement, we can obtain structures with photocatalytic, self-cleaning and biocide surfaces. Presence of n-TiO$_2$ also influences cement hydration process. Principles of these phenomena are described, several examples of first applications in real construction are presented. Plans for further own research are briefly outlined.

1   Introduction

Nanotechnologies belong to the most prospective areas of contemporary science. In accordance with this idea, the author of this paper decided to focus on possible utilization of various nanomaterials in concrete industry in his doctoral study. This paper deals specifically with titanium dioxide nanoparticles effect on concrete properties.

Generally speaking, titanium dioxide (TiO$_2$) has a wide range of applications in paints, sunscreens, glass making, food-colouring or jewellery. Photocatalytic properties of TiO$_2$ were discovered in 1972 by Fujishima and Honda [1]. They exposed TiO$_2$ electrode in an aqueous solution to strong light, which lead to decomposition of water to hydrogen and oxygen. Initially they wanted to utilize this phenomena to extract hydrogen, a clean energy source, from water using sunlight, but experiments showed that the efficiency of the method was too low for commercial use.

In 1989, Fujishima, Hashimoto and Watanabe realized that photocatalysts could be used another way – to decompose trouble-making materials. After they covered the walls and floor of a hospital operating room with tiles coated with TiO$_2$ paint, concentration of bacteria and pollutants in the room fell sharply. Finally, in 1995, scientists from Toto's Research Institute employed photocatalysis to create superhydrophillic glass surfaces with a self-cleaning function [1].

As concrete is a material whose properties are often being modified by admixtures, it is no surprise that these discoveries quickly attracted concrete researchers and since the end of 1980s they experimented with TiO$_2$-enriched cement matrices. In these days, the focus is on titanium dioxide nanoparticles (n-TiO$_2$), because they exhibit increased reactivity compared to ordinary TiO$_2$ thanks to their high specific surface area.

2   Principles of photocatalysis and self-cleaning effect

The process of decomposing air pollutants by UV-radiation from sunlight is a natural phenomenon called photolysis. It helps us to get rid of smog in our cities. This reaction proceeds very slowly, but we can accelerate it by a catalyst. Then we call it photocatalysis.

The catalyst is usually a semiconductor – in our case n-TiO$_2$. Being activated by UV-radiation, an electron is transported from valence-band to semiconductor-band of an atom. When oxygen (O$_2$), gets in contact with such an electron, a radical with high oxidation ability called active oxygen or superoxidion (O$_2{}^-$) is created. In case of water (H$_2$O), hydroxyl radicals (HO$^\bullet$) are generated. Both are highly reactive agents which are able to oxidize most organic compounds and also pollutants such as nitrogen oxides (NO$_x$) [2], [3]. The process of NO$_x$ oxidation on n-TiO$_2$ additised surface is graphically illustrated by fig. 1.

Proc. of the 9th *fib* International PhD Symposium in Civil Engineering, July 22 to 25, 2012,
Karlsruhe Institute of Technology (KIT), Germany, H. S. Müller, M. Haist, F. Acosta (Eds.),
KIT Scientific Publishing, Karlsruhe, Germany, ISBN 978-3-86644-858-2

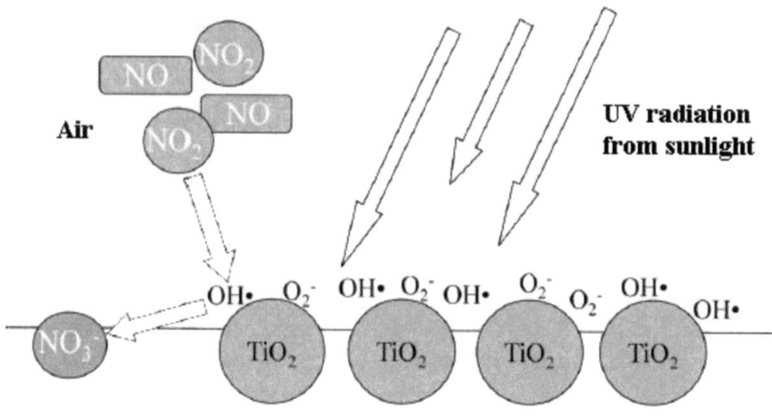

Fig. 1   NO$_x$ removal process scheme, adapted from [4].

Denominating electron as e^-, electron hole with positive charge as h^+ and energy of a photon as hv, we can describe the reactions exactly by following equations [4]:

Photocatalysis:

$$TiO_2 + hv \longrightarrow TiO_2 * (e^- + h^+)$$

$$O_2 + e^- \longrightarrow O_2^-$$

$$2H_2O + 2h^+ \longrightarrow 2OH\bullet + H_2$$

Oxidation:

$$NO + 2OH\bullet \longrightarrow NO_2 + H_2O$$

$$NO_2 + OH\bullet \longrightarrow NO_3^- + H^+$$

$$NO_x \xrightarrow{O_2^-} NO_3^-$$

At this point, it is apposite to stress that the UV-radiation is necessary for the reaction to take place and that n-TiO$_2$ serves only as a catalyst and therefore is not consumed during the reaction. To maintain long-term efficiency, the reaction products (nitrates, NO$_3^-$) should be from time to time washed off the surface, for example, by rain.

Self-cleaning effect is another benefit we can get from photocatalysis. UV-light partially removes oxygen atoms from the surface of the n-TiO$_2$. The areas where oxygen atoms are missing are hydrophilic, while the areas with no oxygen atoms taken away are hydrophobic. As both types of areas exist side by side on the surface (their size is several hundreds or thousands nm^2), water droplets do not remain spherical but became flat forming a uniform film as water spreads through the hydrophilic areas. If a dirt is already present on the surface, the water penetrates under the dirt and removes it [1].

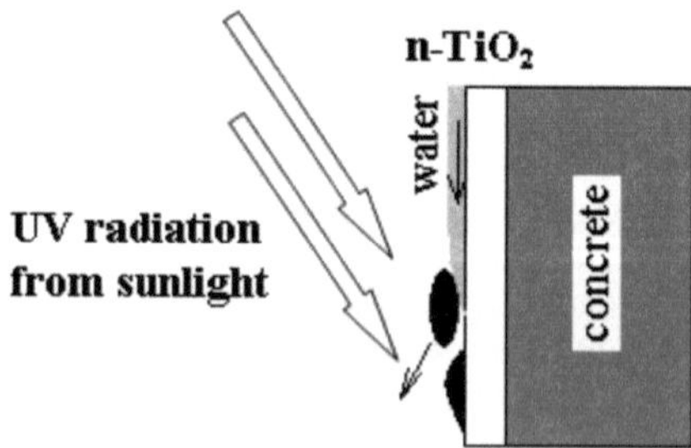

Fig. 2   Self-cleaning effect of n-TiO$_2$ additised concrete surface.

3 Photocatalytic concrete in practice

Photocatalytic, self-cleaning and biocide n-TiO$_2$-additised concrete surfaces were already applied in real structures, which makes them pioneering material of nanotechnology in concrete. So far, there are two companies producing commercially available cement with n-TiO$_2$ admixture: Mitsubishi Corp. with its NOxer and Italcementi SA, manufacturer of TioCem.

The price of these cements is significantly higher than the price of ordinary portland cement, 5 to 10 times depending on supplier and exact type of cement used. However, the impact on the total price of the structure is much milder, because in most cases special cement is not used for structural purposes, i.e. not the whole volume of the element contains n-TiO$_2$-additised cement. Usually, this type of cement is used in precast elements like acoustic barriers, cladding panels, paving blocks etc., where it is possible to make only surface layer of several milimeters or centimeters from photocatalytic material. As a result, primary costs of the elements are increased by some 10 to 20 %, which can be justified by positive impact on the environment, architectural needs and reduction of long-term maintenance costs [5], [6]. Following examples should serve as a support for this statement.

Probably the most well-known application of n-TiO$_2$-additised cement is Dives in Misericordia church in Rome built in 2003. American architect Richard Meier, the author of this structure, wanted to secure long-time whiteness of the church as a symbol of purity and perfection – very challenging demand in the conditions of three-million agglomeration with heavily polluted air. Periodical measurements of coloring of selected panels have shown that the white color is very stable , the only problem was found on windward side of the building, where the panels became slightly yellow due to the effect of fine Saharan sand that is often carried by wind from North Africa to Italy [7].

Fig. 3   Dives in Misericordia church (left) and Umberto I. road tunnel (right), reprinted from [7].

Also in Rome, the reconstruction of road tunnel Umberto I. was performed in 2007. The tunnel reveal was treated with cement painting containing n-TiO$_2$, the lights with high ratio of UV radiation were installed. The aim was to reduce the emissions of NO$_x$. Subsequent study showed that the emissions were reduced by almost 50 % [7].

Fig. 4   Construction of a road with photocatalytic surface in St. Louis (left). The lower layer made of standard application concrete was experimentally covered by 5 cm of photocatalytic concrete (right). Reprinted from [9].

Very promising is application of photoactive cement in acoustic barriers along the roads and interlock pavement blocks. Trial "photocatalytic pavements" were constructed in Paris, Bergamo and Malmö in previous years. Results of laboratory experiments and in-situ measurements summarized in [8] present reduction of NO_x concentration between 25 and 80 % in the surroundings of the pavement or acoustic wall, depending on UV-light intensity, air circulation rate and concentration of the pollutants.

In St. Louis, Missouri, USA, 500 meters of a road were paved by 5 cm thick layer of photocatalytic concrete in 2011 (see fig. 4). The influence on air quality will be monitored for one year by experts from Iowa State University and the University of Missouri at Kansas City. If successful, this in-situ experiment could lead to massive application of photocatalytic surfaces in transportation structures.

4   Influence of n-TiO$_2$ on cement hydration

Titanium dioxide nanoparticles are generally considered to be inert additives to cement. However, recent studies have shown their effect on the rate of hydration and dimensional stability of concrete.

Early age hydration rate is important because of potential influence on setting time, dimensional stability and strength development. It is well-known that fine fillers accelerate cement hydration when added to cement up to 15 % cement replacement levels [11]. Jayapalan et al. investigated influence of increasing dosage of n-TiO$_2$ on cement hydration rate [10], [11] and chemical shrinkage [10]. They used two n-TiO$_2$ powders of different particle sizes and surface areas (see table 1) to produce cement pastes with varied n-TiO$_2$ content. To study hydration rate, the pastes were poured into calorimetric capsules and subjected to isothermal calorimetry. The rate of hydration was measured every 60 seconds as power (mW) and normalized per gram of cement. Chemical shrinkage tests were carried out according to American standard ASTM C 1608-07.

Table 1   Characteristics of n-TiO$_2$ powders used by Jayapalan et al., adapted from [11].

Type of powder	Crystal size [nm]	Agglomerate size [μm]	Surface area [m^2/kg]	pH	Purity [%]
T1	20 – 30	1.5	45 – 55	3.5 – 5.5	>97
T2	15 – 25	1.2	75 – 95	3.5 – 5.5	>95

Results have shown that the rate of hydration was proportional to the dosage of n-TiO$_2$, smaller particles were found to accelerate the reaction more than larger ones (see fig. 5). Chemical shrinkage tests performed on the paste containing T2 powder also proved direct dependence on the dosage of the additive (fig. 6).

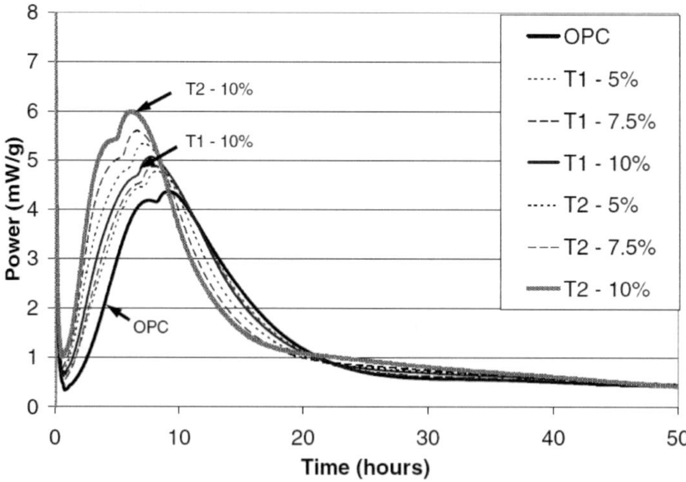

Fig. 5   Hydration rate time development measured by Jayapalan et al., reprinted from [11]. OPC = ordinary portland cement, percentages express the cement replacement level by n-TiO$_2$.

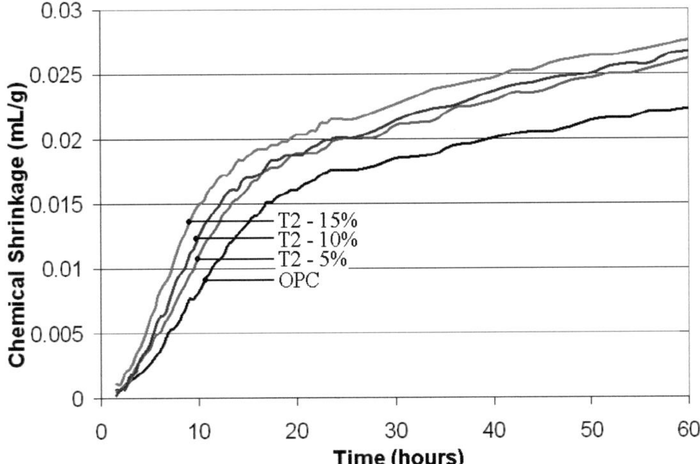

Fig. 6   Chemical shrinkage time development measured by Jayapalan et al., adapted from [10]. OPC = ordinary portland cement, percentages express the cement replacement level by n-TiO$_2$.

The conclusions are notable. They imply that apart from giving concrete surface photocatalytic and self-cleaning properties, addition of n-TiO$_2$ could also affect mechanical properties and chemical durability of concrete, both in favourable and adverse ways.

5   Plans for further research

Previous chapter showed that the research cannot be focused only on the main reasons for which an additive is added to concrete – such an attitude could lead to neglect of important facts. Even if the filler is non-reactive, which is the case of n-TiO$_2$, it affects hydration process for three reasons: modification of particle size distribution, heterogeneous nucleation and cement dilution. Fine particles are able to fill the voids between cement grains and calcium-silicate-hydrate (CSH) crystals, modify porous system and thus, for example, reduce segregation and bleeding in fresh concrete. They serve as nucleation sites promoting cement hydration. On the other hand, as they are used as a replacement of cement, they decrease the amount of hydratable binder (dilution effect) with possible negative impact on mechanical properties.

The role of n-TiO$_2$ in cement hydration with respect to the three previously mentioned phenomena does not seem to be well documented in literature, although we can assume that they are quite important for the process. Hydration rate acceleration and resulting increase in chemical shrinkage was studied in general in [10] and [11]. With respect to the fact that n-TiO$_2$-additised concrete is usually applied only in the surface areas of elements, the author believes that the investigation of differential shrinkage of concrete element's constituent layers is one of the topics that should be studied more in detail. In case when shrinkage in surface area is significantly higher than in the rest of an element, formation of cracks could take place. This could further lead either to separation of layers and devaluation of an element or to easier penetration of chemical agents into concrete. Such effects would be extremely unwelcome, because n-TiO$_2$-additised surfaces are usually applied in places where they are subjected to chemically aggressive environment (pavements, roads, facades etc.). For the same reason, effect of various dosages of n-TiO$_2$ on the ability of concrete to endure actions of chlorides, nitric acid, sulphuric acid, temperature gradients and other factors that affect structures in everyday life should be theoretically and experimentally examined. Without thorough research in this area and with growing number of application of n-TiO$_2$, one day we could face unpleasant problems with maintenance of degraded photocatalytic surfaces.

At the moment, the author is in literature survey stage of his doctoral thesis, evaluating all possible applications of nanotechnologies in concrete industry. Abovementioned ideas on research opportunities in the area of n-TiO$_2$ are one of many others that will be considered before final decision on narrowing the aim of the thesis will be taken.

6 Conclusions

The reasons for writing this paper were to summarize basic information on titanium dioxide nanoparticles influence on concrete and to outline opportunities for author's own research. The author believes that the first purpose was met by explaining principles of photocatalytic and self-cleaning effect, pointing out the question of n-TiO_2 impact on cement hydration and noticing several examples of practical applications. He also hopes that people interested in nanotechnologies in concrete, but without deeper knowledge of the area, might find the paper interesting and helpful for their familiarising with this issue. The second purpose was fulfilled by working out this essay and should serve as a basis for further independent research.

7 Acknowledgements

The author is indebted to the support of Studentská grantová soutěž ČVUT 2011 (SGS 2011) under grant SGS11/106/OHK1/2T/11 „Engineered Concretes and Cement Composites".

References

[1]   Fujishima, A.: Discovery and applications of photocatalysis – Creating a comfortable future by making use of light energy. In: Japan Nanonet Bulletin, 44th Issue (2005)

[2]   Bolte, G.: Innovative Building Material – Reduction of Air Pollution through TioCem®. In: Nanotechnology in Construction 3 – Proceedings of the NICOM 3 (2009), pp. 55 – 61

[3]   Folli, A.: Inovativní fotokatalytický cement obsahující nanočástice TiO_2. In: Beton TKS 6/2011, pp. 28 – 32

[4]   Allen, G.C. et al.: Photocatalytic oxidation of NO_x gases using TiO_2: a surface spectroscopic approach. In: Environmental Pollution 120 (2002), pp. 415 – 422

[5]   Questions and answers on photocatalytic products, Italcementi Group (2006). Available online at http://blog.antaeus.com/downloads/ta_qna.pdf

[6]   Interview with Ing. Ondřej Hranička, managing director of Liadur s.r.o. (2011). Available online at http://www.bezsmogu.cz/clanky/rozhovory/rozhovor-hranicka/

[7]   http://www.bezsmogu.cz

[8]   Hubertová, M., Matějka, O.: Protihlukové stěny Liadur s technologií TX Active. Časopis Stavebnictví 09/2009, pp. 20 – 23

[9]   Highway Research Project Paving Way to Cleaner Environment. Available online at http://www.txactive.us/pdf/7654-MoDOT_Profile.pdf

[10]   Jayapalan, A.R. et al.: Influence of Nano-Anatase Titanium Dioxide on Cement Hydration: Experiments and Modelling. Available online at http://blogs.cae.tntech.edu/hydration-kinetics/files/2009/07/jayapalan_poster.pdf

[11]   Jayapalan, A.R. et al.: Effect of Nano-sized Titanium Dioxide on Early Age Hydration of Portland Cement. In: Nanotechnology in Construction 3 – Proceedings of the NICOM 3 (2009), pp. 267 – 273

Conceptual modelling of C-S-H formation and the occurrence of Hadley grains in hardened cement paste

Richard J. Aquino

Faculty of Civil Engineering and Geosciences
Delft University of Technology,
Stevinweg 1, Delft 2628 CN, The Netherlands
Supervisors: Klaas van Breugel, Eduard A.B. Koenders

Abstract

The current state of knowledge of C-S-H, the main binding phase of Portland cement-based system, is limited despite decades of investigations. The C-S-H is nanogranular, nearly amorphous and fractal structure in nature. C-S-H is still complex to understand and more research is needed in both experimental and computational aspects. The mechanism of C-S-H formation and densification at length scale of 100nm to 1µm remain fundamentally unknown. Hence, the main goal of this research work is to contribute to the investigation of the main mechanism behind the C-S-H formation and the causes of possible densification.

A new conceptual model using colloidal approach is being developed in order to clearly define the growth mechanism of C-S-H, eventually, be able to describe and explain the densification process. The proposed growth and densification mechanism are as follows: the basic building block (BBB) of C-S-H is assumed spherical in shape and nanometer in size known as C-S-H particles. The C-S-H particles flocculate to form into larger units called clusters or globules [2]. At initial reaction, the C-S-H particles precipitate and gradually form into clusters outside the original boundary of the cement particles. At early-age, Hadley grains [5] will form and eventually be filled-up, first, by C-S-H particles then as clusters or globules. The C-S-H clusters will continue to attract each other to form the following C-S-H morphologies, i.e. LP, LD, HD and UHD. These four C-S-H morphologies differ in porosity or packing density. The proposed mechanism is expected to explain the existence of LP that appears in the early period of hydration, LD, HD and UHD that appear in the later period. Furthermore, this conceptual model is envisaged to improve the existing hydration models to bridge the gap between micro- and nano- level.

1   Introduction

Portland cement, the main ingredient of concrete, is the most common and ubiquitous material used in civil engineering works. It works like magic when combined with water. After a few hours, the liquid solution transforms into solid with strength that continues to increase in time. The calcium silicate hydrate (C-S-H) is the main hydration product and binder phase. Taplin [1] classified C-S-H as inner product (Ip) formed inside the original boundary of anhydrous cement and outer product (Op) formed in space originally occupied by water. Jennings [2] used the terms high density (HD) C-S-H and low density (LD) C-S-H that correspond to Ip and Op, respectively. In addition, loose-packed density (LP) C-S-H [3] and ultra-high density (UHD) C-S-H [4] are also used.

The resulting stone-like product is a porous material that makes it susceptible to many microscopic problems such as calcium leaching, transport of deleterious materials, and corrosions. Most of these problems occur via the intrinsic pore system that influences many engineering properties. The intrinsic pores are comprised of capillary and gel pores that depend on the initial water/cement (w/c) ratio of the mixture. As a consequence of increasing the w/c ratio, the porosity increases while the material's strength decreases and more ingress of deleterious materials are expected.

The capillary pores are defined as the space originally occupied by water in a mixture of cement and water. The hydration products will occupy the capillary pores hence they will decrease during hydration process. In addition, an often neglected type of pore and most unrecognized feature of hydrated cement paste microstructure is made available during the course of hydration process that forms in space occupied originally by cement. These pores are known as hollow shell pores or

Proc. of the 9th *fib* International PhD Symposium in Civil Engineering, July 22 to 25, 2012,
Karlsruhe Institute of Technology (KIT), Germany, H. S. Müller, M. Haist, F. Acosta (Eds.),
KIT Scientific Publishing, Karlsruhe, Germany, ISBN 978-3-86644-858-2

'Hadley grains' [5]. Many Hadley grains are completely hollow while others contain remnant anhydrous cement particles [6]. Further, some hollow shells formed as 'Williamson grain' [7] are later infilled by inner products that grow towards the middle of anhydrous cement until they are completely filled. The hollow shells are not artifacts caused by drying or particle pull-outs hence they are considered a significant feature of the microstructure [6] that accounts to 10% or less of the total porosity in cement paste and even increases up to 20% with the presence of silica fume [8]. In addition, Diamond [9] reported that a significant portion of larger pores that are found in concrete was derived from Hadley grains. Furthermore, the properties and performance of cement and concrete depend significantly on the system's porosity.

Nowadays, computer modelling and simulation of cement hydration are progressively advancing. These models include CEMHYD3D [10], DuCOM [11], HYMOSTRUC-3D [12] and μic [13]. Each model has advantages and disadvantages but all does not include a feature of the hollow shell pores. Hence, this study aims to develop a conceptual model of hydrating cement paste microstructure including hydrates formation and morphology. The model incorporates the often neglected but important mode of hydration called Hadley grain. The model will be implemented in HYMOSTRUC-3D to further improve the microstructural model to give realistic results in terms of engineering properties such as the strength. Moreover, the model including C-S-H formation and morphology will be studied further in the sub-micro level using colloidal approach. In the end, the HYMOSTRUC-3D model is envisaged to be a useful tool to study and develop new cementitious materials by incorporating the important mechanism in cement hydration.

The technique employed in this study is to use experimental evidence with the use of calorimeter and microscopy. From the results, conceptual models and later, mathematical models will be developed.

2 Experimental

2.1 Materials

Portland cement CEM I 42.5N was used in this study. All the pastes were mixed at w/c ratio of 0.50 and hydrated at 8, 10, 12 and 24 h, and at 7 and 28 days. The pastes were mixed using electric mixer for 5 mins. After mixing, the pastes were cast into a plastic cylindrical container and sealed. Mixing and storage took place at room temperature. The specimens were freeze-dried using liquid nitrogen and the frozen water was then removed through sublimation. The specimens were subjected to vacuum-impregnation with epoxy resin and oven-dried for 24 h to harden the epoxy resin. After hardening, a series of silicone carbide paper of increasing (fineness) mesh sizes 500, 800, 1200 and 4000 were used to grind the specimens followed by polishing using diamond particles of sizes 6, 3, 1 and 0.25 μm. Ethanol was used during grinding and polishing of the specimens to prevent further hydration of the paste. Finally, the polished specimens were cleaned through ultrasonic-bath for 5 mins. to remove debris possibly embedded in pores during the process of grinding and polishing.

2.2 Isothermal Calorimetry

The heat development at 20°C was measured using TAM Air isothermal calorimeter. Cement paste of 10 g was put into the ampoules and immediately placed in the calorimeter. The heat evolution was measured for 25 h.

2.3 ESEM

All the specimens were ground and polished flat for microscopy. The examination was carried out using Philips XL30 ESEM with an accelerating voltage kept at 20 kV throughout. The backscattered electrons (BSE) imaging detector of the ESEM were used. The BSE image provides valuable information in terms of average atomic number by setting the image contrast and brightness properly. The same image is used to quantify the microstructural characteristics of hardened cement paste. The grey-scale image gives a high atomic number for bright areas to lower average atomic values in darker areas. Anhydrous cement appears white, CH brighter grey, C-S-H grey and pores/epoxy as black.

3 Conceptual Model

A conceptual model of hydrating cement particle incorporating Hadley grain formation is being developed as shown in Fig. 1. The proposed growth and densification mechanism are as follows: The basic building block (BBB) of C-S-H is assumed spherical in shape and nanometer in size known as C-S-H particles. The C-S-H particles flocculate to form into larger units called clusters or globules [2]. At initial reaction, the C-S-H particles precipitate and gradually transform into clusters outside the original boundary of cement particles. At early-age, Hadley grains [5] will form and eventually be filled-up, first, by C-S-H particles then act as clusters or globules. The C-S-H clusters will continue to attract each other to form the following C-S-H, i.e. LP, LD, HD and UHD. These four C-S-H morphologies differ in porosity or packing density. The proposed mechanism is expected to explain the existence of C-S-H morphologies, i.e. LP that appears in an early period of hydration, LD, HD and UHD that appear in a later period.

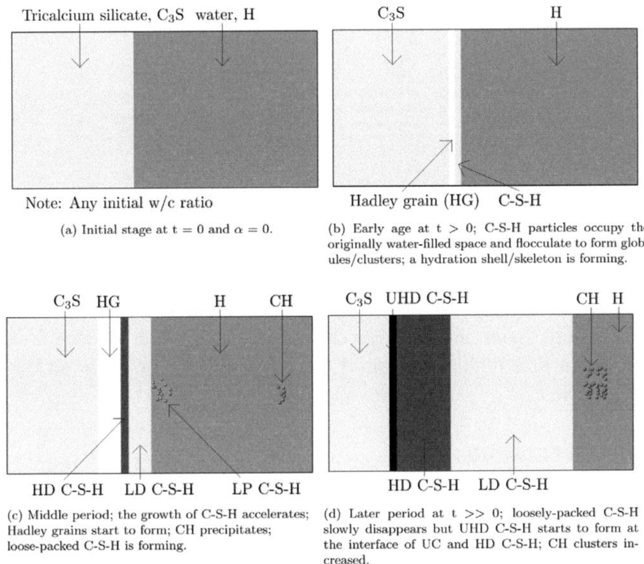

(a) Initial stage at t = 0 and α = 0.

(b) Early age at t > 0; C-S-H particles occupy the originally water-filled space and flocculate to form globules/clusters; a hydration shell/skeleton is forming.

(c) Middle period; the growth of C-S-H accelerates; Hadley grains start to form; CH precipitates; loose-packed C-S-H is forming.

(d) Later period at t >> 0; loosely-packed C-S-H slowly disappears but UHD C-S-H starts to form at the interface of UC and HD C-S-H; CH clusters increased.

Fig. 1   A conceptual model of hydrating cement particle incorporating Hadley grain formation.

The HYMOSTRUC-3D generates the virtual microstructure of hydrated cement paste at microscale level assuming cement particles as spherical in shape. The program inputs are particle size distribution (PSD), chemical composition, w/c, and hydration temperature. The corresponding outputs are virtual microstructure (Ip, Op, CH, pores, and clinker phases), degree of hydration, RH, etc. During hydration, mass balance will be satisfied at all times. This implies that C-S-H formed in the capillary pores instead of forming as inner product will be denser when Hadley grains occur. The conceptual model will be implemented in HYMOSTRUC-3D. In order to be compatible, the flowchart in Fig. 2 will be used. A thin middle hydration shell is introduced for Hadley grain and Williamson grain to form.

The thin middle hydration shell is assumed to occur during the acceleration period and becomes stable at the end of final set to act as skeleton for the Hadley grain to form. The Hadley grains are observed to form only on the surface of calcium silicates. The hydration mechanism can be divided into two methods, i.e. without (conventional) and with (extended) thin middle hydration shell. HYMOSTRUC-3D uses two layers of hydration shells that represent the inner and outer products. The conceptual model will extend HYMOSTRUC-3D by introducing a third layer at the middle called the thin middle hydration shell. The hydration will start at time, t = 0, immediately after mixing C$_3$S and water. Depending on a certain criteria to be defined during the course of the main author's research, cement particles will hydrate without and with thin middle hydration shell. The conventional method follows the classical cement hydration of HYMOSTRUC-3D. The extended method introduces a thin middle hydration shell that will act as skeleton for the Hadley grain to occur. In the formation of Hadley grain, the hydration products will form in the capillary pores instead at space originally occupied by cement particles. Here the cement particles will be partially and completely hollow shells during hydration process. After the Hadley grain formation commences, C-

S-H will continue to precipitate in the capillary pores and some will precipitate inside the thin middle hydration shell known as 'Williamson grain'.

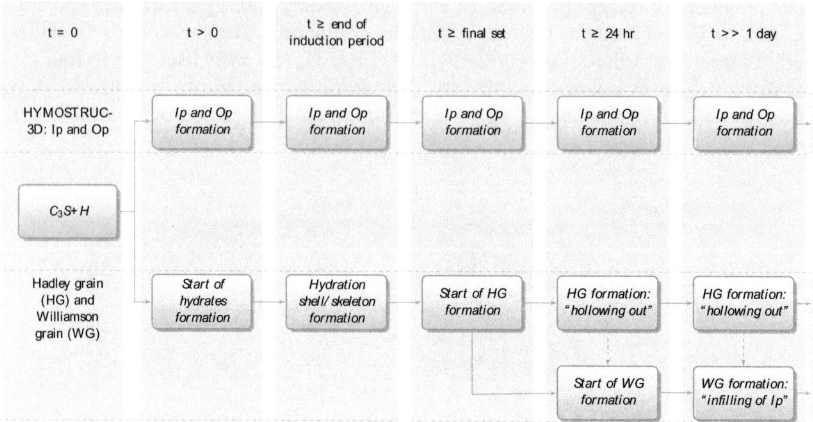

Fig. 2   The new HYMOSTRUC-3D process with and without thin middle hydration shell incorporating the 'hollowing out' and 'infilling of Ip' mode of hydration. The conventional hydration contains inner and outer products formation while the novel/extended hydration incorporates a thin middle hydration shell that acts as the skeleton for the formation of Hadley grains.

4   Results and Discussion

4.1   Calorimeter curve

The result of calorimetry is presented as the rate of heat evolved per gram of solid material in Fig. 3. The calorimeter curve is usually divided into 5 stages, i.e. pre-induction, induction, acceleration, deceleration and slow periods. An initial burst of heat occurred during the pre-induction period followed by an almost constant release of heat where an initial C-S-H forms. C-S-H and CH precipitate rapidly during the acceleration period where the thin middle hydration shell is assumed to form initially. Before the peak or after the final set occurs, it is assumed that the thin middle hydration shell becomes stable to act as skeleton for the Hadley grain to commence formation as typified by physical separation of the phase product and anhydrous cement. Hadley grain formation continues rapidly during the deceleration period and slows down.

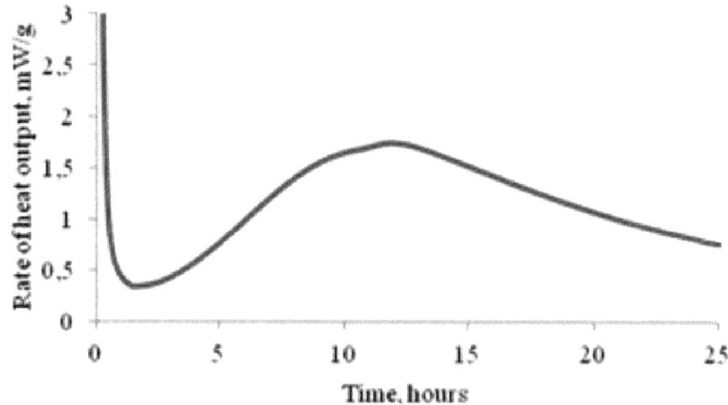

Fig. 3   The rate of heat evolution of the Portland cement (CEM I 42.5N) paste of w/c=0.5.

4.2   Cement microstructure

4.2.1   At 8 h

A backscattered electron image of the 8-hour old Portland cement paste is shown in Fig. 4a. The resolution is not satisfactory good to expose any fibrous C-S-H that grows on the surface of the calcium silicates as what is usually reported. The fibrous structure will appear in later periods.

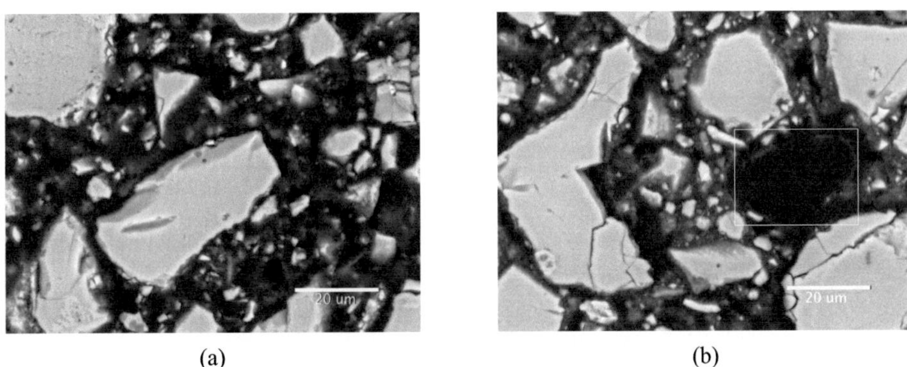

(a)   (b)

Fig. 4   BSE micrographs of the Portland cement hydrated for (a) 8 h and (b) 10 h.

4.2.2   At 10 h

Ten hours is about the peak of the calorimeter curve as shown in Fig. 3. Again, the resolution is not satisfactory but there is a clear contrast between the anhydrous cement and hydration products as revealed in Fig. 4b. Further, a completely hollow cement particle is apparent as shown in the framed area. Hence, Hadley grain formation might have started already between 8 and 10 h of hydration. This period represents approximately the final setting of cement using Gillmore method ($\leq$ 10 h) and Vicat method ($\leq$ 6.25 h) [14]. It can be assumed that a stable thin middle hydration shell is needed before the hollowing out process will commence that may occur after or during the final set.

4.2.3   At 12 h

The formation of C-S-H on the surface of Alite and CH precipitation are obviously revealed in Fig. 5. Close examination of the micrographs visibly shows a thin physical separation at the interface between the fibrous C-S-H and anhydrous cement. This separation very likely represents an early stage of Hadley grain formation.

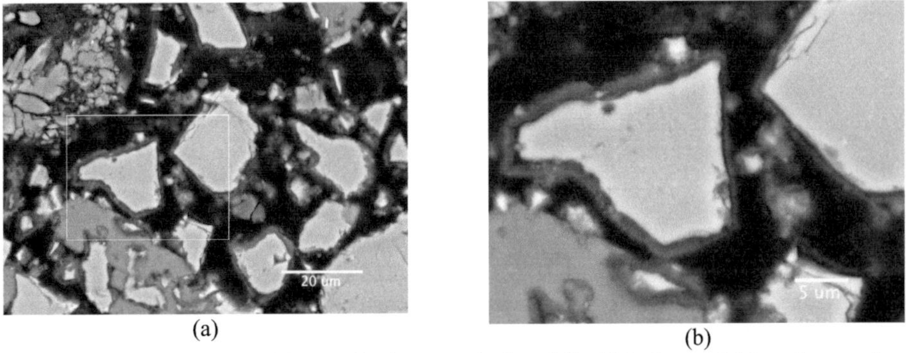

(a)   (b)

Fig. 5   BSE micrographs of the Portland cement hydrated for 12 h. Image (b) is a close-up image of the framed area on (a).

4.2.4 At 24 h

After 24 hours of hydration, fully developed Hadley grains are revealed in Fig. 6. Complete and partial hollowing out of the cement particles are clearly observed with almost constant thickness of thin middle hydration shell about < 1 μm. The remnant anhydrous cement core is shown in Fig. 6b.

4.2.5 At 7 days

A partially hydrated large cement particle and a remnant anhydrous cement core are revealed in Fig. 7. The open space between the thin middle hydration shell and remnant anhydrous cement is around 2 μm and hydration shell is less than 1 μm with fibrous morphology outside. Further, a remnant of thin middle hydration shell is visible on the large hydrating cement particle next to dense inner product. Smaller size cement particles that are completely hollowed are also observed.

(a)

(b)

Fig. 6   BSE micrographs of the Portland cement hydrated for 24 h. Image (b) is a close-up image of the framed area on (a).

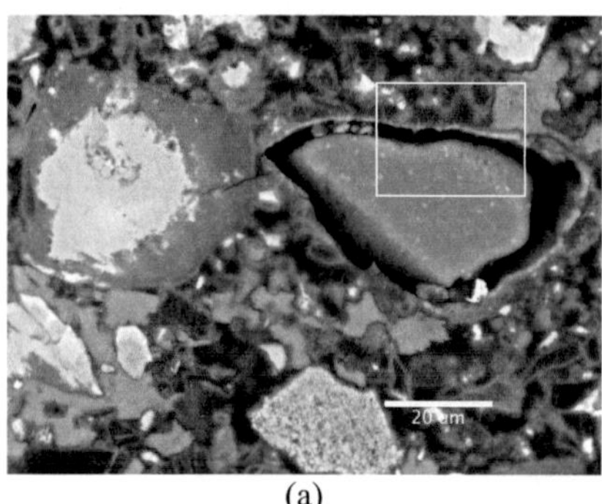
(a)

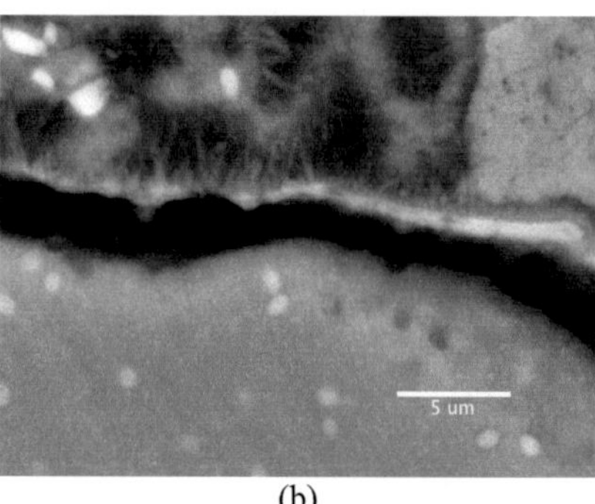
(b)

Fig. 7   BSE micrographs of the Portland cement hydrated for 7 days. Image (b) is a close-up image of the Hadley grain (framed area) in (a).

4.2.6 At 28 days

Fig. 8 shows a backscattered electron image of the 28 days Portland cement paste. Different phases can be recognized, i.e. anhydrous cement, CH, C-S-H, capillary pores and Hadley grain pores. It is clearly revealed that Hadley grain formation is favoured in porous region than the dense region of the microstructure. Even at this period, completely hollowed cement particles are apparent. This implies that porosity of the microstructure as Hadley grain type of pores increases. Furthermore, the C-S-H that precipitates in the capillary space instead of forming as inner product may densify if mass balance is observed during the hydration process.

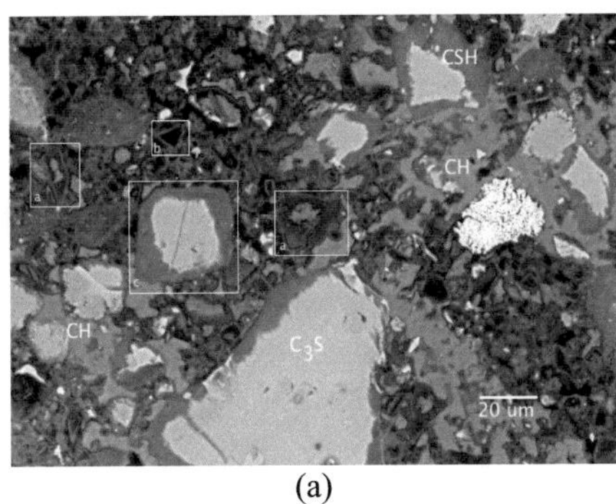
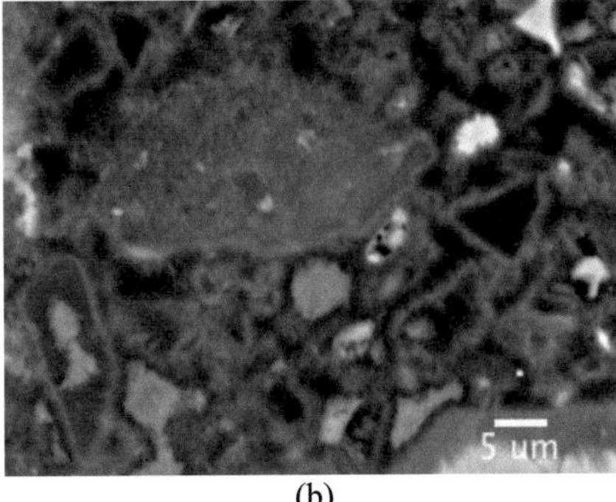

(a)   (b)

Fig. 8   BSE micrographs of the Portland cement hydrated for 28 days. Left image is composed of partially *a* and completely *b* hollowed hydrating cement particles. Image (b) is an enlarge area of particles with partial and complete Hadley grains.

5   Conclusion

The hydrated Portland cement paste microstructure has been shown to have plenty of hollow shells or Hadley grains. The occurrence of Hadley grains is abundant during the middle period of hydration and will stay as hollow or be filled by the hydrates at the later time. From these experimental evidences, the Hadley grain pores is considered an important mode of hydration [8] and prominent feature of the microstructure. The current cement microstructural models do not include the Hadley grain phenomenon. It is the aim of this study to include this feature in the HYMOSTRUC-3D model to give more realistic results using the conceptual model developed. In order to grasp the effect of these pores, it is apparent that the nature and formation of Hadley grain must be fully understood.

Acknowledgements

This work was supported by the Seventh Framework Program theme 4 NMP of the European Commission under the <u>CO</u>mputationally <u>D</u>riven design of <u>I</u>nnovative <u>CE</u>ment-Based materials (CODICE) project (www.codice-project.eu) under grant agreement no. 214030. Also, the main author would like to thank Arjan Thijssen for the assistance using the ESEM.

References

[1]   Taplin, J. (1959). "A method for following the hydration reaction in Portland cement paste." <u>Australian J Appl Sci</u> 10: 329-345.

[2]   Jennings, H. (2000). "A model for the microstructure of calcium silicate hydrate in cement paste." <u>Cement and Concrete Research</u> **30** (1): 101-116.

[3]   Howind, T. (2011). Mechanical properties mapping of cement paste microstructures. 3rd CODICE workshop on Multiscale Modelling and Experiments for Cementitious Materials. Istituto degli Innocenti Piazza SS. Annunziata 12 – Florence, Italy.

[4]   Ulm, F. J. and M. Vandamme (2009). "Probing Nano-structure of C-S-H by Micro- mechanics Based Indentation Techniques." Nanotechnology in Constr 3, Proceedings: 43-53.

[5]   Hadley, D. W. (1972). The Nature of the Paste-Aggregate Interface. <u>School of Civil Engineering</u>, Purdue University. **PhD**.

[6]   Kjellsen, K. O., H. M. Jennings, et al. (1996). "Evidence of hollow shells in the microstructure of cement paste." <u>Cement and Concrete Research</u> **26**(4): 593-599.

[7]   Williamson, R.B. (1972). "Solidification of Portland Cement." <u>Progress in Materials Science</u> 15(3): 189-286.

[8]   Kjellsen, K. O., B. Lagerblad, et al. (1997). "Hollow-shell formation - An important mode in the hydration of Portland cement." <u>Journal of Materials Science</u> **32**(11): 2921-2927.

[9]   Diamond, S. (1999). "Aspects of concrete porosity revisited." Cement and Concrete Research 29(8): 1181-1188.

[10]   Bentz, D. (1997). "Three-dimensional computer simulation of portland cement hydration and microstructure development." Journal of the American Ceramic Society 80(1): 3-21.

[11]   Maekawa, K. et al (2003). "Multi-scale Modeling of Concrete Performance - Integrated Material and Structural Mechanics. " Journal of Adv Concrete Technology 1(2): 91-126.

[12]   van Breugel, K. (1997). Simulation of hydration and formation of structure in hardening cement-based materials. Delft University of Technology, The Netherlands. **PhD**.

[13]   Bishnoi, S. (2008). Vector Modelling of Hydrating Cement Microstructure and Kinetics. Ecole Polytechnique Federale de Lausanne, Switzerland. **PhD**.

[14]   http://pavementinteractive.org/index.php?title=Portland_Cement_Setting_Time

Comparative study on autogenous phenomena of Portland cement paste with and without internal curing

Alexander Assmann

Department of Construction Materials,
University of Stuttgart,
Pfaffenwaldring 4, 70569 Stuttgart, Germany
Supervisor: H.-W. Reinhardt

Abstract

Autogenous phenomena were studied on two plain Portland cement pastes with w/c ratios of 0.30 and 0.36 and one water-entrained cement paste with a basic w/c ratio of 0.30. Water-entrainment was provided by the addition of superabsorbent polymers (SAP) and extra water. The amount of entrained water corresponds to w/c = 0.06. The cement pastes were investigated with regard to autogenous deformation, internal relative humidity, early-age elastic properties and degree of hydration. The investigation shows that water-entrainment due to SAP provides effective internal curing. As a result, self-desiccation shrinkage is prevented and the degree of hydration is increased.

1   Introduction

Concrete with low w/c ratio, i.e. high performance concrete with w/c < 0.30, is subject to a significant drop in the internal relative humidity (RH) during sealed hydration. It is caused by a shortage of free water since the low amount of water is insufficient to achieve complete hydration of the cement. A phenomenon closely related to the change in RH is autogenous shrinkage. It is defined as the bulk deformation of a sealed cementitious system, devoid of external forces. Before setting, autogenous shrinkage can be regarded as equal to chemical shrinkage, which is a consequence of hydration. After setting, when the fluid cement paste begins to form a solid skeleton, autogenous shrinkage is mainly driven by self-desiccation. Restraint autogenous deformations result in the development of tensile stresses. Consequently concrete with low w/c ratio is more sensitive to early-age cracking than traditional concrete. Due to the dense microstructure of the matrix, conventional curing procedures are not able to counteract autogenous phenomena effectively.

Superabsorbent polymer (SAP) particles can be used as admixture for water-entrainment in cement-based materials in order to prevent self-desiccation. Within the first minutes of mixing, they absorb and store a certain quantity of the mixing water, forming water-filled macropore inclusions. During hydration, the water is released to the surrounding cement. As a result, air filled macropores remain. Detailed information on the application of SAP in concrete construction can be found in [1].

This paper deals with the investigation of autogenous phenomena in Portland cement pastes with and without internal curing due to SAP. Measurements were performed on autogenous deformation, internal relative humidity, elastic material properties and degree of hydration.

2   Objectives

The investigations presented in this paper attempt to attain two goals:

- The demonstration of the potential of SAP as admixture for internal curing.
- The provision of data for modelling of the self-desiccation shrinkage of Portland cement paste as a function of the internal relative humidity according to an approach presented by Lura [2].

3   Materials and methods

3.1   Used materials

One cement paste with water-entrainment due to addition of 0.25 % SAP by mass of cement and two plain cement pastes with w/c ratios of 0.30 and 0.36 were investigated. The basic w/c ratio of the SAP-modified cement paste was 0.30. The amount of entrained water corresponds to w/c = 0.06. It is

Proc. of the 9th *fib* International PhD Symposium in Civil Engineering, July 22 to 25, 2012,
Karlsruhe Institute of Technology (KIT), Germany, H. S. Müller, M. Haist, F. Acosta (Eds.),
KIT Scientific Publishing, Karlsruhe, Germany, ISBN 978-3-86644-858-2

in compliance with the amount of entrained water theoretically needed to attain optimized internal curing of a sealed system, as proposed by [3]. In this case the relative volume of initially entrained water is equal to the expected volume of chemical shrinkage at the maximum degree of hydration. The SAP used was provided by BASF. Its water absorption capacity in the cementitious environment was found to yield 24 g of water per 1 g of dry polymer. The cement used was of type CEM I 42.5 R. Mixing was done by a pan-type mixer. The vessel had a volume of 1.5 litres.

3.2   Measurement of autogenous deformation

Free autogenous deformation of cement paste was measured according to [4] using corrugated tubes made of polyethylene as a mould. The watertight moulds were filled with cement paste on a vibrating table. Afterwards they were placed horizontally in a dilatometer. The length of the sample was approximately 380 mm and the diameter 30 mm. Two samples of each cement paste were tested simultaneously for 28 days. The measurements every 10 minutes were started half an hour after mixing. The linear strains were measured by means of linear variable differential transformer with an accuracy of ±5 µstrain. The dilatometers were placed in a thermostatically controlled room at 20 ±0.2 °C.

3.3   Measurement of internal relative humidity

About 12 g of fresh cement paste were filled into a plastic tray placed in the measuring cell of a Rotronic HygroLab3 station. The station is equipped with an AW-DIO sensor for RH and temperature measurement. It was located in a thermostatically controlled room at 20 ±0.2 °C.

Assuming thermal equilibrium between sensor and sample, the measured relative humidity is regarded as internal relative humidity. In the first hours after mixing, the RH measurements are affected by the development of heat of hydration in the paste. Condensation may occur if the sample has a higher temperature than the sensor itself. On this account, the RH measurements started 24 to 36 hours after mixing. The sensor was calibrated and, if necessary, adjusted before each measurement. Three saturated salt solutions with known constant RH in the range 80 - 95 % RH were used.

3.4   Determination of elastic material properties

The elastic material parameters were determined by using the testing device FreshCon. FreshCon was developed for the continuous monitoring of the hardening of cementitious materials in the 1990s at the University of Stuttgart. The method behind FreshCon is ultrasound (US) transmission testing. The present version of the device allows combined measurements with pressure (P) and shear (S) waves [5]. Figure 1 shows a picture of the entire measurement device.

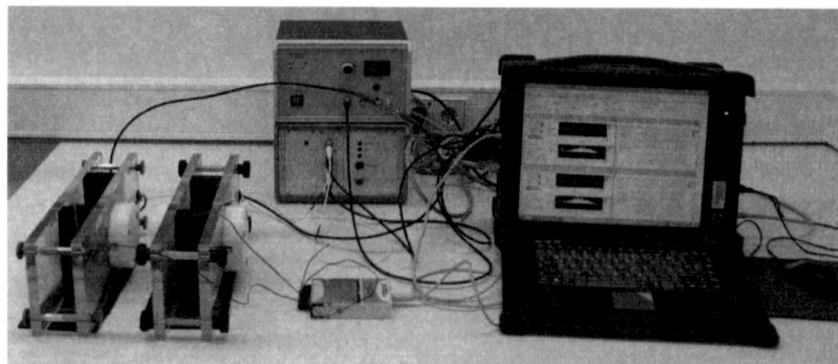

Fig. 1   Picture of the FreshCon system.

Two containers made of perspex, one for measuring pressure waves and one for measuring shear waves, were filled with about 500 cm³ of fresh cement paste. Each container is equipped with two sensors: one US-transmitter connected to the amplifier generating the signals and one receiver picking the onsets of the signals. Incident and transmitted pulses are analogous signals, digitalised by means of an A/D converter. The system is triggered automatically with a recording interval of 10 min. The signals were evaluated by software SMARTPICK [6] in order to determine the speed of the primary wave transmitted through the material. For determination of the S-wave onset time, the time signal is transformed into a time-frequency domain by means of a continuous wavelet transformation. This is necessary since the shear wave transducer also induces a pressure wave into the material. The picker for the determination of the P-wave onset bases on the Akaike Information Criterion (AIC) adapted for ultrasonic signals [7]. The continuous monitoring was aimed at the first 7 days of hardening. The dynamic modulus of elasticity E_{dyn} can be calculated on the basis of the P-wave velocity v_P and the S-wave velocity v_S according to equation 1.

$$E_{dyn} = (2 + 2\sigma_{dyn})v_S^2 \rho_C \text{ and } \sigma_{dyn} = \frac{0.5 \cdot v_P^2 - v_S^2}{v_P^2 - v_S^2} \tag{1}$$

With ρ_C = density of the fresh cement paste, and σ_{dyn} = Poisson's ratio.

3.5 Determination of the degree of hydration

The device Netzsch STA 409 PC was used in order to determine precisely the amount of chemically bound water in hardened cement paste samples by means of controlled heating. It allows combined measurements of two thermo analytical methods: differential scanning calorimetry (DSC) and thermo gravimetric analysis (TGA). Detailed information on the testing procedure and the evaluation methods can be found in [8]. In simplified terms, the chemically bound water is assumed to be driven off at temperatures between 105 and 1050 °C. The weight loss of the sample Δm_S within this temperature range leads to the degree of hydration α by the relation given in equation 2.

$$\alpha = \left[(\Delta m_S - \Delta m_C \cdot c) / 0.23 \right] / c \tag{2}$$

With c = mass fraction of cement in the sample, and Δm_C = loss on ignition of the plain cement.

It was estimated that one g of full hydrated Portland cement binds approximately 0.23 g of water chemically. At the time of testing, the cement paste samples were 6 hours to 28 days old. The cement pastes had been filled in plugged test glasses directly after mixing to maintain sealed hydration.

4 Results and discussion

The measured free autogenous deformations for all three cement pastes are provided by figure 2 (left). The curves show the average of two tested samples. The deformations were zeroed at the time of maximum deformation rate as shown exemplarily by figure 2 (right). This time marks the transition period in which the plastic cement paste starts to form a solid skeleton [9]. The skeleton gives stiffness to the paste leading to higher deformation resistance and a consequent decrease of the deformation rate. From that moment, restraint autogenous deformation would induce tensile stresses in the system. The initial deformations of the pastes, which occur before time zero, are mainly due to chemical shrinkage. Own measurements showed good agreement between the time of maximum deformation rate and the time of final setting determined by the Vicat needle test.

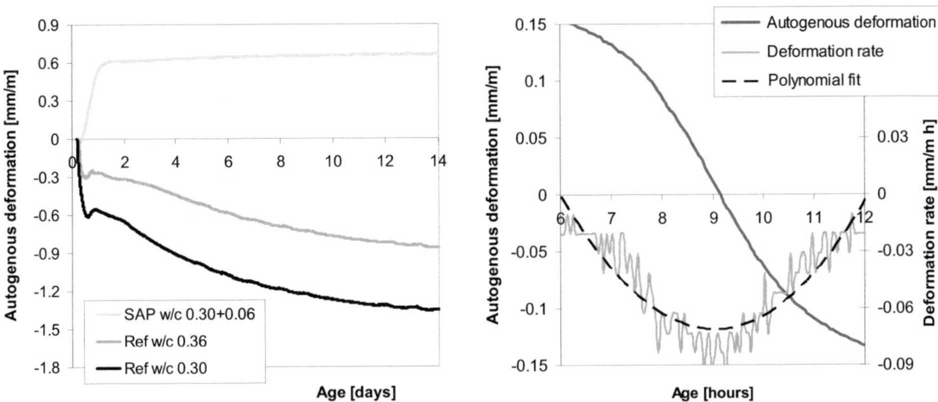

Fig. 2   Free autogenous deformation curves (left); Example for determination of time zero (right)

Both reference pastes show high autogenous shrinkage. With decreasing the w/c ratio, the shrinkage is increasing. After two weeks the pastes show total shrinkage of 850 and 1350 µstrain, respectively. However, approximately 15 hours after mixing, the shrinkage rates are temporarily decreased. For a short period of 6 hours, the deformation even turns into expansion. This observation is associated with an increase in the heat evolution due to hydration.

In contrast to the shrinkage of the reference pastes, the water-entrained cement paste even shows expansion right after setting. This is seen as a result of the internal curing water from SAP. It is well known that a continuous water supply to the hydrating cement leads to expansion as a consequence of absorption of water from the cement gel. Even for plain cement pastes exposed to water, this expansion may amount to 1000 - 2000 µstrain [10]. The observed expansion of the water-entrained cement paste shown in figure 2 is about 600 µstrain in the first day of sealed hydration. The expansion increased by less than 70 µstrain in the following two weeks.

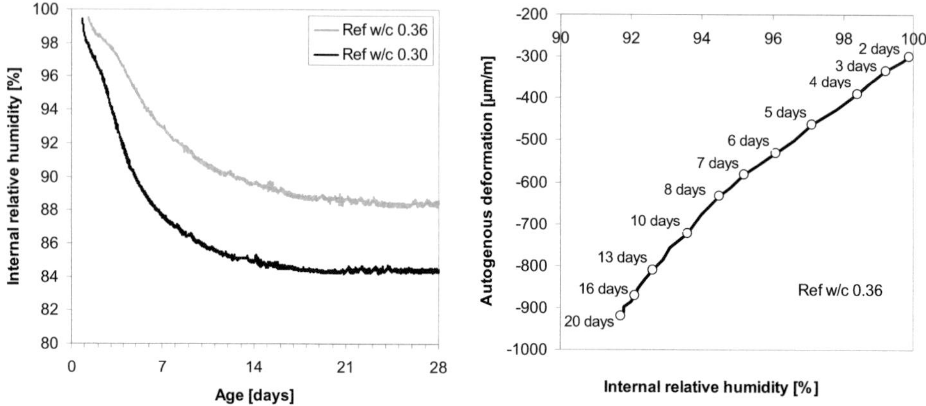

Fig. 3   Change in relative humidity of reference pastes up to 28 days (left); Free autogenous deformation of paste Ref w/c 0.36 in dependence on internal relative humidity (right).

Figure 3 (left) provides the development of internal RH with hydration time for both reference pastes up to an age of 28 days. The RH decreased directly after the measurements were started and equilibrium between the sensor and the paste was reached. The significant lowering of the autogenous RH is due to hydration reactions and is referred to as self-desiccation. With decreasing the w/c ratio the RH change is faster and steeper. After three weeks the internal RH of the reference pastes seem to approach limit values of 88 and 84 % RH, respectively. The result of the RH measurement of the water-entrained cement paste is not shown in the diagram. The values for internal RH ranged permanently near saturation, i.e. > 98 % RH. However, the most of the time the sensor measured 100 % RH. The relative humidity near saturation is strongly influenced by changes in temperature. Obviously the sensor was affected by condensation due to insufficient thermostatic control. Nevertheless, the measured values show that internal curing due to supply of water by SAP prevents self-desiccation at least for 28 days. If the amount of initially entrained water was decreased, self-desiccation and consequently self-desiccation shrinkage would be expected, even though delayed.

The relation between measured internal relative humidity and autogenous shrinkage of cement paste Ref w/c 0.36 is presented in figure 3 (right). According to the diagram autogenous shrinkage is directly linked to self-desiccation. Within the first two days of sealed hydration, autogenous shrinkage occurs although the internal relative humidity has decreased only slightly. This strong deformation at the beginning is mainly due to the low modulus of elasticity of the cement paste at early ages.

The dynamic modulus of elasticity determined by ultrasonic transmission testing is given in Figure 4 (left). In the time period when the cement paste begins to forms a solid skeleton, the dynamic modulus of elasticity is rapidly increasing. When decreasing the w/c ratio, the values become higher. As a consequence of water absorption by SAP, the water-entrained cement paste shows earlier stiffening. But already one day after mixing, the elastic material properties are very similar to the one of the reference paste Ref w/c 0.36 with same total amount of water. At this time, the modulus of elasticity has reached at least about 75 % of the value found at the end of the measurement.

If restraint hardening of the reference pastes was considered, tensile stresses would be induced in the system. Figure 4 (right) shows the results of a theoretical approach where the stress development was calculated by using Hooke's law according to equation 3:

$$\sigma = E_{dyn} \cdot \varepsilon \tag{3}$$

With σ = theoretical tensile stress due to complete restraint, E_{dyn} = measured dynamic modulus of elasticity, ε = measured free autogenous deformation.

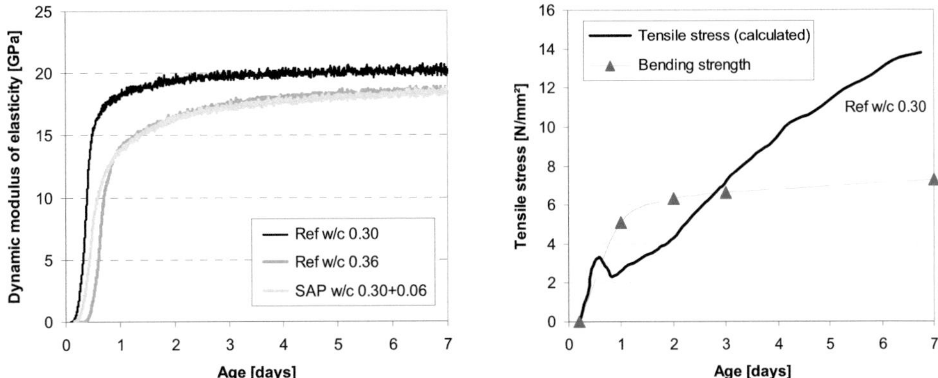

Fig. 4   Dynamic modulus of elasticity curves (left); Calculated stress development due to theoretical restraint of paste Ref w/c 0.30 (right).

Indeed, the calculated stress development is based on a strong simplification, e.g. creep is not taken into account and is larger than the static modulus of elasticity. Reasonable values for the tensile stress are expected to be lower. But, nevertheless, it is likely that the tensile stresses induced could reach the tensile strength of the material even at very early age. As a consequence, the reference paste would be subjected to cracking. In contrast to this finding, expansion of the water-entrained cement paste would completely prevent the development of the tensile stresses when restraint.

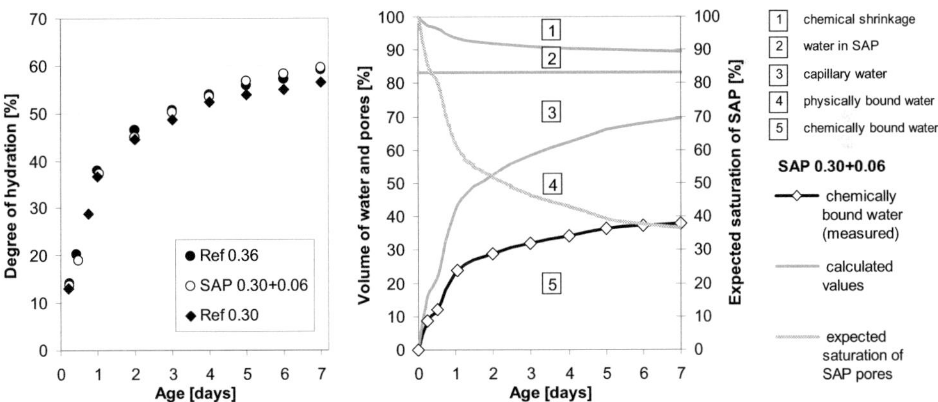

Fig. 5   Degree of hydration of cement pastes resulting from DSC-TG measurements (left). Phase distribution of water-entrained cement paste and expected saturation of SAP pores (right).

The degree of hydration up to an age of 7 days is given by figure 5 (left). Within the first two days, hydration is almost the same for all three cement pastes. This is in good agreement with the findings of the RH measurements. The first significant change in RH of the reference pastes occurs after two days. Later on, the degree of hydration is increasing with increasing w/c ratio. The water-entrained cement paste and the plain cement paste with same total amount of water show the same hydration at least for 7 days. That means, concerning hydration, it is not relevant whether the water is provided by SAP or is ordinarily added.

Figure 5 (right) shows the phase distribution of water and pores in the cement paste with internal curing by SAP. The volume of chemically bound water was determined by DSC-TG measurements. On the basis of Powers theory, the volumes of physically bound water, capillary water and chemical shrinkage were calculated. As a first estimation, it is assumed that the initially water-filled macropore inclusions are emptied due to chemical shrinkage. The expected saturation of SAP pores is also presented in figure 5 (right).

5 Conclusions

In this paper autogenous deformation, internal relative humidity, elastic material properties and degree of hydration were examined for two plain Portland cement pastes and one paste with internal curing due to addition of dry SAP particles. The following conclusions could be drawn:

- The main part of the autogenous deformations of the plain cement paste occurred around setting and was related to chemical shrinkage. Due to gain in stiffness by the formation of a solid skeleton the autogenous shrinkage rate is reduced. After two days of sealed hydration, autogenous deformations were accompanied by a drop of the internal relative humidity.
- The water-entrained cement paste showed expansion right after setting. This expansion was due to internal curing water provided by SAP. As a result, autogenous shrinkage and self-desiccation were completely prevented for at least 28 days.
- The degree of hydration for the plain cement paste and the water-entrained cement paste was the same, provided that the total amount of water is the same. That means it is not relevant whether the water is supplied by SAP or is added ordinarily to the mix.

6 Outlook

For modelling self-desiccation shrinkage of Portland cement paste as a function of internal relative humidity, further information is required. Besides internal relative humidity and elastic material properties, information on the degree of saturation of the pores and the surface tension of the pore water is needed. More research is under way in order to provide these data.

7 Acknowledgement

The German Research Foundation (DFG) is gratefully acknowledged for the financial support on this research. The Materials Testing Institute (MPA) Stuttgart is thanked for the assistance in the Fresh-Con tests. The company BASF Construction Chemicals GmbH is thanked for the provision of SAP.

References

[1] RILEM, State-of-the-Art Report of the RILEM Technical Committee 225-SAP: Application of superabsorbent polymers (SAP) in concrete construction. Mechtcherine, V. & Reinhardt, H.-W. (Eds.), Springer, Dordrecht, 2012.

[2] Lura, P.: Autogenous deformation and internal curing of concrete, Ph.D. thesis, Technical University Delft, Netherlands, 2003.

[3] Jensen, O.M. & Hansen, P.F.: Water-entrained cement-based materials I. Principles and theoretical background. In: Cement and Concrete Research 32 (2001), pp. 647-654.

[4] ASTM International, Standard ASTM C 1698-09. In: Annual Book of ASTM Standard, C-09 Vol. 04.02 - Concrete and Aggregates, Section 04 – Construction. West Conshohocken, USA

[5] Krüger, M.; Große, C.U.; Lehmann, F. & Reinhardt, H.-W.: Zuverlässige Qualitätssicherung von Frischbeton mit Ultraschall - das FreshCon-System. In: Messtechnik im Bauwesen 2011 (2011), pp. 88-92.

[6] Krüger, M. & Lehmann, F.: SmartPick - User Manual, Rev. 1.50, Smartmote, 2010.

[7] Kurz, J.H.; Große, C.U.; Reinhardt, H.-W.: Strategies for reliable automatic onset time picking of acoustic emissions and of ultrasound signals in concrete. In: Ultrasonics, Vol. 43 (2005), pp. 538-546.

[8]   Assmann, A. & Reinhardt, H.-W.: Effect of water-filled macropore inclusions on the hydration of Portland cement detected by thermo gravimetric analysis. In: Leung, C.; Wan, K.T. (Eds.): Advances in Construction Materials through Science and Engineering, RILEM Publications Pro 79 (2010), p. 60 (abstract book), full text on CD-ROM.

[9]   Fontana, P.: Einfluss der Mischungszusammensetzung auf die frühen autogenen Verformungen der Bindemittelmatrix von Hochleistungsbetonen, Schriftenreihe des DAfStb, Heft 570, Beuth Verlag, Berlin, 2007.

[10]   Neville, A.M.: Properties of concrete, 4th edition, Longman, Harlow, 1995.

Theory for the early age plastic cracking behaviour of concrete

Riaan Combrinck

Civil Engineering Department,
University of Stellenbosch,
Banhoekweg, Stellenbosch (7600),South Africa
Supervisor: William P. Boshoff

Abstract

Plastic settlement cracking (PSeC) and plastic shrinkage cracking (PShC) are the most prominent forms of cracking in concrete at early ages before the concrete has obtained significant strength. PSeC is caused by the differential vertical settlement of a concrete mix, whereas PShC is caused by the horizontal contraction of the concrete mix due to evaporation. Although the mechanisms causing PSeC and PShC are known, the underlying concepts remain incompletely understood. This is believed to be due to the tendency of most researches to investigate PShC and PSeC individually, which often leads to unexpected results due to the interaction between PShC and PSeC. This study investigated the phenomena of both PShC and PSeC as well as the combined effect of PShC and PSeC through several tests, each with different combinations of PSeC and PShC severity. The test results are used to create a theory that can explain the behaviour of early age plastic cracking for any condition. The theory distinguishes between the following six different cracking behaviours: *no cracking, delayed cracking, classical PSeC, classical PShC, settlement induced PShC* and *crack jumping.*

1   Introduction

The two earliest forms of cracking in concrete is plastic settlement cracking (PSeC) and plastic shrinkage cracking (PShC). Both these cracking types occur while the concrete is still in the plastic phase. PSeC is caused by the differential settlement of the concrete i.e. the concrete settles at different amounts in different parts of the concrete body. This causes a tensile stress to develop directly above the boundary of the two different settlement zones which ultimately leads to cracking [1]. PShC is caused by the loss of water from the concrete surface through evaporation which causes a negative capillary pressure build up in the concrete paste that ultimately leads to cracking [2]. PShC occurs once the concrete surface has dried out and ends once the concrete has gained enough strength to resist the capillary pressure build up. Concrete elements with a large exposed surface, for example highway pavements and floor slabs are especially vulnerable to evaporation and therefore PShC.

A study in Korea found that 40 % of cracks in concrete structures involve PShC and about 40 % of these cracks are associated with PSeC [1]. This indicates a clear connection between PSeC and PShC which can be explained by considering some of the influencing factors that PSeC and PShC have in common. Firstly, the settlement responsible for PSeC also results in bleeding water at the surface of the concrete which delays the capillary pressure build up and therefore PShC. Bleeding is the thin film of water on the concrete surface due to the settlement of the solid particles in the concrete that displaced the water to the surface [3]. Secondly, the vertical restraint required for the differential settlement and therefore PSeC is also required for the horizontal restraint needed for PShC. Without the horizontal restraint the concrete body would only undergo a uniform shrinkage without any cracks. Finally, both types of cracks normally occur at the same position. This is because PSeC generally occurs first and creates a weak spot in the concrete which is naturally exploited as the position of further crack growth once plastic shrinkage starts.

Although the mechanisms causing PSeC and PShC are known, the underlying concepts remain incompletely understood [4]. The author believes this is due to the tendency of most researches to investigate PShC and PSeC individually which often leads to unexpected results due to the interaction between PShC and PSeC. This paper describes the phenomena of both PShC and PSeC as well as the combined effect of PShC and PSeC based on observations made from test results. Finally, a theory is proposed that can explain the behaviour of early age plastic cracking for any condition.

Proc. of the 9th *fib* International PhD Symposium in Civil Engineering, July 22 to 25, 2012,
Karlsruhe Institute of Technology (KIT), Germany, H. S. Müller, M. Haist, F. Acosta (Eds.),
KIT Scientific Publishing, Karlsruhe, Germany, ISBN 978-3-86644-858-2

2 Test framework

Two different mould types were used for the tests. The first mould was newly developed to give significant PSeC and is called the Settlement Mould as shown in Fig. 1 (left). The mould is made of PVC and Perspex panels. The side panels are transparent to allow observation of settlement cracks. The crack is forced by differential settlement at the boundary between the deep and swallow sections of the mould. The mould also contains two triangular inserts. The bigger triangle is situated at the variation in depth and assists in creating two clearly defined settlement zones. The smaller triangle is placed near the end of the shallow part of the mould and provides the horizontal restraint needed for PShC. The second mould is based on moulds used for fresh concrete testing as proposed by ASTM C 1579 [5] and is called the Shrinkage Mould as shown in Fig. 1 (right). The mould is rectangular and contains three triangular inserts. In addition to this normal mould, several tests were also carried out with the same mould, but with a 20 mm steel rod as centre restraint. The concrete cover to the centre triangle of the normal Shrinkage Mould was 40 mm while the concrete cover for the rod Shrinkage Mould was 20 mm. For both Shrinkage Moulds a single crack is induced by the centre restraint while the two small triangles at the sides of the mould act as the horizontal restraint required for the PShC.

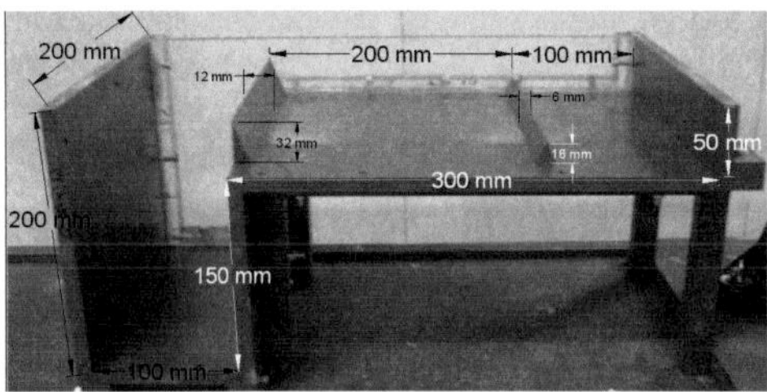

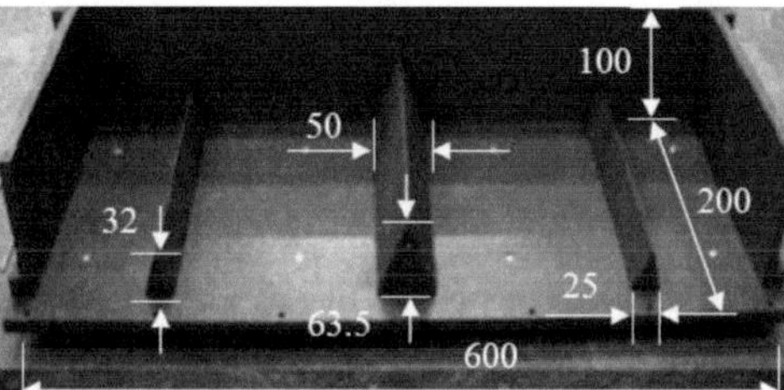

Fig. 1   Settlement Mould (left) and Shrinkage Mould (right).

The tests were conducted in an environmental chamber designed to create stable hot, dry and windy conditions. The chamber was used to create an extreme and a normal environmental condition called Climate E and Climate N respectively. Climate E had an air temperature of 40 °C, relative humidity of 20 % and wind speed of 33 km/h whereas Climate N had an air temperature of 23 °C, relative humidity of 55 % and wind speed of 2.2 km/h. This resulted in evaporation rates of 1 kg/m²/h and 0.2 kg/m²/h for Climate E and Climate N respectively. Two conventional concrete mixes were used for the tests, one with high and the other with low bleeding or settlement characteristics. The high bleeding mix (HB-Mix) and low bleeding mix (LB-Mix) achieved a bleeding rate of 0.82 kg/m²/h and 0.53 kg/m²/h respectively during the first hour. The HB-Mix contained a coarse natural sand and low cement content, while the LB-Mix contained a crushed Greywacke dust sieved through a 2 mm sieve and a high cement content. This resulted in the HB-Mix containing significantly less fines than the LB-Mix, which ultimately was responsible in giving the HB-Mix a higher bleeding. During the tests the following measurements were also taken: The capillary pressure was measured with electronic pressure sensors [2]. The initial and final setting times of a sieved concrete paste were measured with a Vicat needle apparatus [6]. The crack area was calculated using high resolution photos scaled with CAD software. The bleeding was measured by weighing extracted bleeding water every 20 minutes [3]. The evaporation was measured by continuously weighing a square mould filled with concrete. More details on the environmental chamber and measurements methods can be found in [7].

3 Classical PSeC and PShC

PSeC is caused by the differential settlement of the concrete within a concrete body. PSeC stops once settlement ends which corresponds closely to the initial setting time of concrete [7]. Furthermore, the crack position is normally somewhere above rigid inclusions such as reinforcing steel. In practice these cracks are only visible from the surface and might include crack growth due to PShC. This leads to the question whether PSeC originates from within or at the surface of the concrete. To answer this question a simple test was conducted with the LB-Mix in the Settlement Mould at Climate N except with no wind. In addition, the concrete surface was continuously kept wet to eliminate any capillary pressure build up and therefore preventing PShC. The test revealed a small crack that originated at the tip of the triangular insert at the boundary between the two settlement zones. The crack did not reach the surface and was only visible through the Perspex side panel as shown in Fig. 2 (left). Fig. 2 (right) shows the crack of the same mix at Climate E where evaporation was allowed. This crack was clearly visible at the concrete surface and notability followed the same path as the crack shown in Fig. 2 (left). This verifies that PSeC creates weak points were PShC will form. Due to the poor quality of the photos in Fig. 2 the outline of both the triangular restraint and crack pattern is drawn in by hand. From the results it is evident that PSeC originates from the bottom of the concrete at the point that differentiates between different settlement zones. The origin of PSeC is from the bottom upwards while the origin of PShC is from the surface downwards. This means that PSeC may be present in a concrete element even if it is not visible from the surface, especially in conditions with low evaporation rates and therefore none or only minor surface cracks due to PShC. The hidden PSeC are weak spots in the concrete element which are soon exposed once drying or thermal shrinkage commences.

PShC is caused by the build up of capillary pressure due to the evaporation of surface bleeding water. The higher the evaporation rate the higher the risk for PShC. Fig. 3 show the results conducted with the normal Shrinkage Mould for the HB-Mix at Climate E. The figure illustrates classical PShC behaviour with capillary pressure build up once the surface bleeding water has evaporated. As evaporation continues capillary pressure increases until air entry pressure is reached which can be identified by the significant drop in the capillary pressure. Air entry signifies the start of any possible cracks and corresponds closely to the initial setting time of concrete. The first cracks appear close to the initial setting time and is called crack onset. The majority of the cracking occurs and stabilises before the final setting time of the concrete. This describes classical PShC behaviour [7].

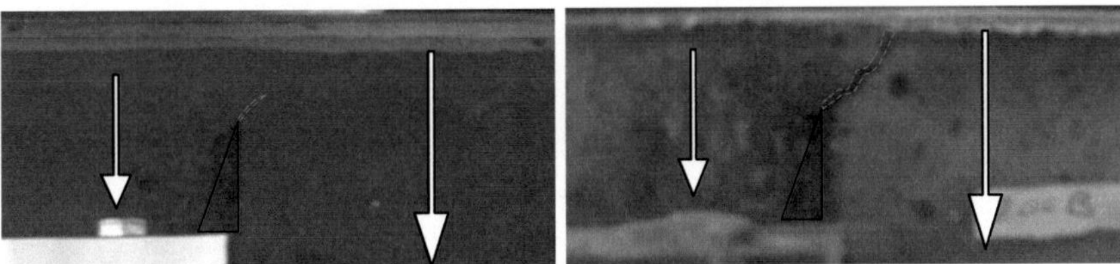

Fig. 2   Side view of a crack (indicated as dotted red line) in the Settlement Mould for LB-Mix at Climate N with no capillary pressure due to evaporation by keeping the surface continuously wet (left) and of the same mix at Climate E were evaporation was allowed (right).

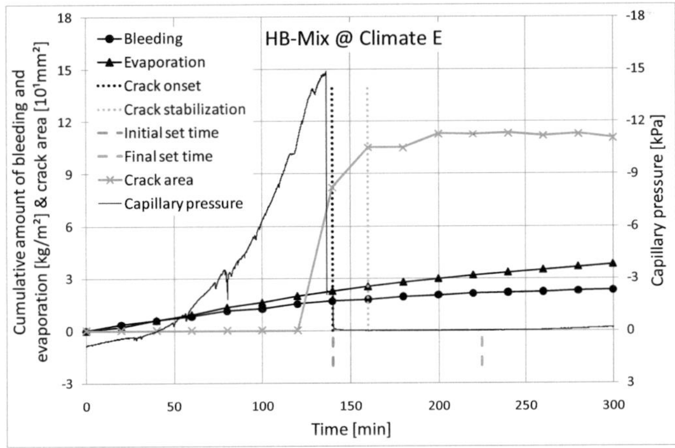

Fig. 3   Classical PShC behaviour for the HB-Mix in the normal Shrinkage Mould at Climate E.

4   Theory of early age plastic cracking behaviour

The previous section described the so-called classical PSeC and PShC behaviour. With this in mind a theory was developed that can explain early age plastic cracking behaviour due to PShC and PSeC respectively and more importantly the combined effect of PSeC and PShC. In this paper the term early age plastic cracking refers to any cracking due to PSeC and/or PShC while the concrete is still in the plastic phase and excludes cracking due to drying or thermal shrinkage. A graph of the proposed theory for early age plastic cracking behaviour is shown in Fig. 4. The figure shows the PSeC on the x-axis and the PShC on the y-axis, both scaled according to the severity of the cracking from zero to one. With zero meaning no chance and one meaning a severe chance for cracking. The assignment of crack severity values to a specific mix and condition will not form part of this paper.

According to the theory, cracks can be classical PSeC or PShC as well as a combination of the two cracking types. In addition, when PSeC and PShC behaviour is combined at different intensities several types of cracking behaviour may occur. These cracking behaviour accounts for the interaction between PSeC and PShC and is based on the cracking behaviour observed during the tests. The following six cracking behaviours were observed: *no cracking, delayed cracking, classical PSeC, classical PShC, settlement induced PShC* and *crack jumping. Classical PSeC* and *PShC* has already been discussed, while a discussion of the other four cracking types follows in the next sections.

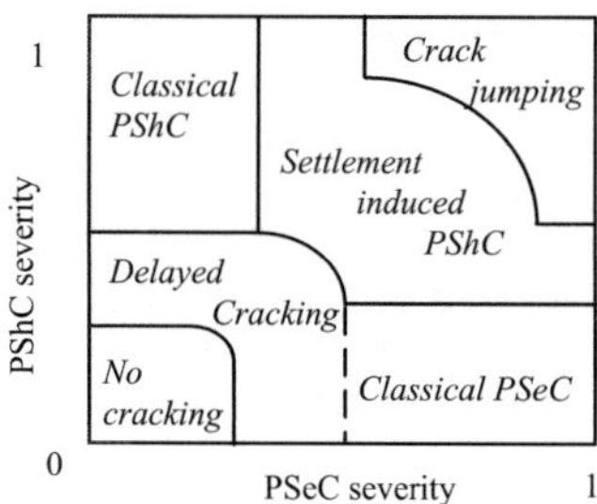

Fig. 4   Theory of early age plastic cracking behaviour as a graph.

4.1   No cracking and delayed cracking

In certain cases the severity or potential for both PSeC and PShC is insignificant and *no cracking* is expected. The HB-Mix in the normal Shrinkage Mould at Climate N showed no cracking due to the low evaporation rate and high bleeding rate, which resulted in no capillary pressure build up and therefore no PShC. Furthermore, the normal Shrinkage Mould does not cause significant differential settlement and therefore insignificant PSeC.

Delayed cracking does not fall within the normal PShC behaviour and only becomes visible once all early age plastic cracking has thought to have ended. Although these cracks are formed while the concrete is still in the plastic phase they are not visible and can therefore be described as delayed or hidden. These cracks or weak spots will be exposed as soon as drying or thermal shrinkage commences. This may included all classical PSeC which starts from below the concrete surface or any PShC that have been closed by finishing operations or is still filled with bleeding water. Fig. 5 shows a delayed crack which was observed for the LB-Mix in the normal Shrinkage Mould at Climate N.

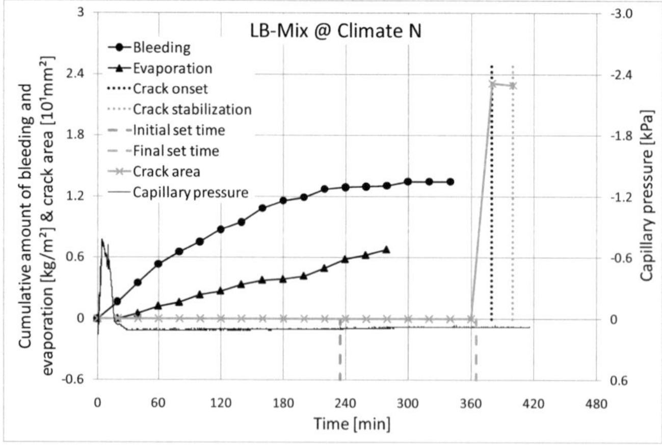

Fig. 5   Delayed cracking behaviour for the LB-Mix in the normal Shrinkage Mould at Climate N.

4.3 Settlement induced PShC

Settlement induced PShC requires both severe PSeC conditions in terms of significant differential settlement as well as conditions with enough evaporation to cause PShC. The rod Shrinkage Mould with 20 mm concrete cover at Climate E resulted in the ideal conditions to cause settlement induced PShC. Fig. 6 show the results of the HB-Mix in the 20 mm concrete cover rod Shrinkage Mould at Climate E. The first crack was observed long before the initial setting time of concrete as well as before any significant capillary pressure build up and therefore also air entry. This indicates that the crack is indeed a settlement crack caused by the severe differential settlement induced by the near the surface location of the centre rod. The figure also shows that the crack continues to grow once the capillary pressure starts to build up, which indicates that the crack is widening due to PShC. The settlement crack therefore created a weak spot which allowed PShC to commence much earlier than for normal PShC behaviour. It is also evident from the figure that for settlement induced PShC the time of air entry no longer signifies the start of cracking as for normal PShC. This suggests that PSeC is not a function of capillary pressure build up and air entry due the evaporation of surface bleed water. Finally, the majority of the crack growth ceased long before the final setting time.

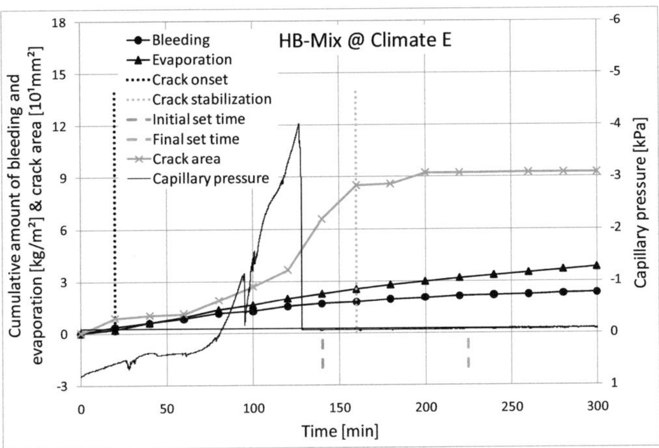

Fig. 6   Settlement induced PShC for the HB-Mix in the rod Shrinkage Mould at Climate E.

4.4 Crack jumping

The final and most severe cracking is called *crack jumping* and is due to the interaction between PSeC and PShC. Crack jumping requires the utmost severe conditions for both PSeC and PShC and was observed for the LB-Mix in the Settlement Mould at Climate E as shown in Fig. 7. The figure shows an extremely rapid crack growth just after the first crack was observed long before the initial setting time. This is called crack jumping and represents more than half of the total crack growth as measured two hours after the final setting time. As for settlement induced PShC, the first crack occurred long before the initial setting time and is therefore PSeC. The main difference between settlement induced PShC and crack jumping is the amount of capillary pressure that has built up at crack onset. For settlement induced PShC no significant capillary pressure is present at crack onset which causes the crack to grow gradually as the pressure builds up. In comparison, a significant amount of capillary pressure due to evaporation is already present at crack onset for crack jump behaviour. This means that the entire concrete body is under severe stress just before crack onset. Once the crack or weak spot is formed due to differential settlement, there occurs a global stress relieve over the entire concrete body which localises all the deformation at the crack position.

In addition to the global stress relieve, the magnitude and the rate of the crack growth can also be explained by the plastic nature of the concrete paste at the time of crack onset which allows crack growth in both the horizontal and vertical direction. This means that the crack growth can be due to both the vertical PSeC and horizontal PShC opening the crack at the same time and is called bi-directional cracking. Bi-directional cracking can only continue if the settlement of the concrete body is present and therefore ceases close to the initial setting time of concrete. Bi-directional cracking is only present at settlement induced PShC and crack jump behaviour.

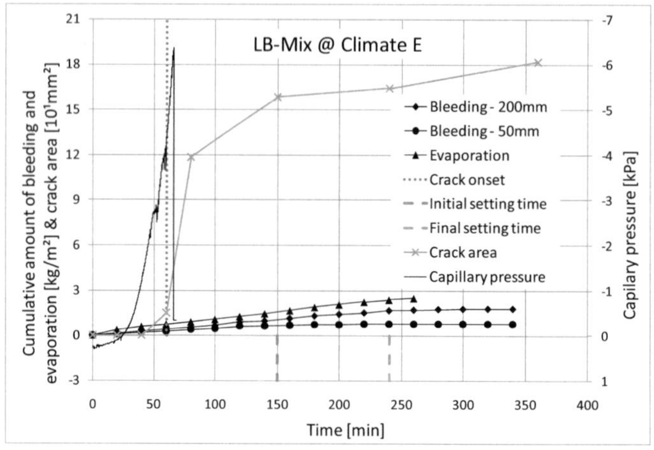

Fig. 7   Crack jump behaviour for the LB-Mix in the Settlement Mould at Climate E.

5   Conclusions

In this paper a theory was developed for the early age plastic cracking behaviour of concrete. The theory distinguishes between the following six types of cracking behaviour:

- *No cracking*, for conditions with low PSeC and PShC severity.
- *Delayed cracking*, these are minor cracks that only become visible after the early age cracking has thought to have ended and occurs for conditions with a low PSeC and/or PShC severity.
- *Classical PSeC*, these cracks are formed at the bottom of the concrete at the point of differential settlement and may or may not be visible at the concrete surface during the plastic stage, but will become visible as soon as drying and thermal shrinkage commences.
- *Classical PShC*, these cracks occur during the initial and final setting times under conditions with sufficient evaporation but negligible differential settlement.
- *Settlement induced PShC*, these are premature PShC induced by PSeC for conditions where capillary pressure only develops after crack onset and may involve bi-directional cracking.
- *Crack jumping*, these are rapid forming and stabilising bi-directional cracks which are induced by severe PSeC for conditions where significant capillary pressure has already developed at the time of crack onset.

References

[1]   Kwak, H.-G. and Weiss, W.J.: Experimental and Numerical Quantification of Plastic Settlement in Fresh Cementitious Systems. In: Journal of Materials in Civil Engineering 22 (2010) No. 10, pp. 951-966

[2]   Slowik, V., Schmidt, M. and Fritzsch, R.: Capillary pressure in fresh cement-based materials and identification of the air entry value. In: Cement & Concrete Composites 30 (2008), pp. 557-565

[3]   Josserand, L. and De Larrard, F.: A method for concrete bleeding measurement. In: Materials and Structures 37 (2004), pp. 666-670

[4]   Dao, V.T.N., Dux, P.F., Morris, P.H. and O'Moore, L.: Plastic shrinkage cracking in concrete. In: Australian Journal of Structural Engineering 10 (2010) No. 3, pp. 207-214

[5]   ASTM C 1579.: Standard Test Method for Evaluating Plastic Shrinkage Cracking of Restrained Fiber Reinforced Concrete. ASTM International, West Conshohocken, (2006)

[6]   SANS 50196-3.: Methods for testing cement Part 3: Determination of setting times and soundness 2nd Edn. Standards South Africa, Pretoria, (2006)

[7]   Combrinck, R.: Plastic shrinkage cracking in conventional and low volume fibre reinforced concrete. Stellenbosch, The University of Stellenbosch, MSc-thesis, (2011)

Modeling of concrete creep based on microprestress-solidification theory

Petr Havlásek

Department of Mechanics,
Faculty of Civil Engineering, Czech Technical University in Prague,
Thákurova 7, 16629 Prague 6 - Dejvice, Czech Republic
Supervisor: Milan Jirásek

Abstract

Creep of concrete is strongly affected by the evolution of pore humidity and temperature, which in turn depend on the environmental conditions and on the size and shape of the concrete member. Current codes of practice take that into account only approximately, in a very simplified way. More realistic description can be achieved by advanced models, such as model B3 and its improved version that uses the concept of microprestress. The value of microprestress is influenced by the evolution of pore humidity and temperature. In this paper, values of parameters used by the microprestress-solidification theory (MPS) are recommended and their influence on the creep compliance function is evaluated and checked against experimental data from the literature. Certain deficiencies of MPS are pointed out and its modified version is proposed and validated.

1   Introduction

In contrast to metals, concrete exhibits creep already at room temperature. This phenomenon results into a gradual but considerable increase of deformation at sustained loads and needs to be taken into account in design and analysis of concrete structures. The present paper examines an advanced concrete creep model, which extends the original B3 model [1] and uses the concepts of solidification [5, 6] and microprestress [2, 3, 4]. The main objective of the paper is to clarify the role of non-traditional model parameters and provide hints on their identification. The creep tests performed by Fahmi, Polivka and Bresler [7], covering creep of both sealed and drying specimens under elevated and variable temperatures, are used as a source of experimental data and are compared with the results of numerical simulations.

All numerical computations have been performed using the finite element package OOFEM [8, 9, 10] developed mainly at the CTU in Prague by Bořek Patzák.

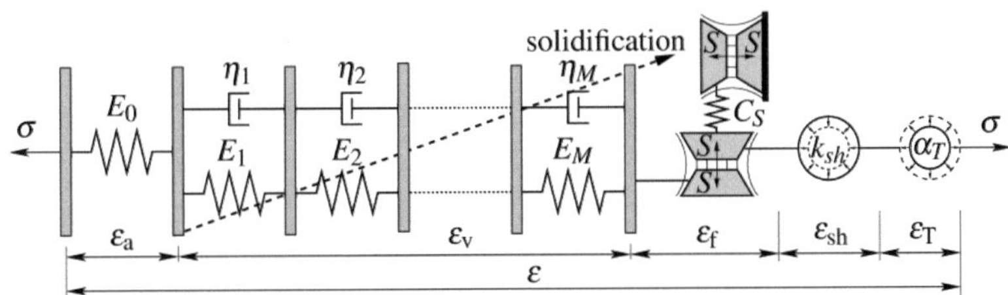

Fig. 1   Rheological scheme of the complete hygro-thermo-mechanical model

2   Description of the material model

The complete constitutive model for creep and shrinkage of concrete can be represented by the rheological scheme shown in Fig. 1. It consists of (i) a non-aging elastic spring, representing instantaneous elastic deformation, (ii) a solidifying Kelvin chain, representing short-term creep, (iii) an aging dashpot with viscosity dependent on the microprestress, S, representing long-term creep, (iv) a shrinkage unit, representing volume changes due to drying, and (v) a unit representing thermal expansion. In the experiments, shrinkage and thermal strains were measured separately on load-free specimens and subtracted from the strain of the loaded specimen under the same environmental conditions. It should be noted that even after subtraction of shrinkage and thermal strain, the evolution of mechanical strain

Proc. of the 9th *fib* International PhD Symposium in Civil Engineering, July 22 to 25, 2012,
Karlsruhe Institute of Technology (KIT), Germany, H. S. Müller, M. Haist, F. Acosta (Eds.),
KIT Scientific Publishing, Karlsruhe, Germany, ISBN 978-3-86644-858-2

is affected by humidity and temperature. Dry concrete creeps less than wet one, but the process of drying accelerates creep. Elevated temperature leads to faster cement hydration and thus to faster reduction of compliance due to aging, but it also accelerates the viscous processes that are at the origin of creep and the process of microprestress relaxation.

The microprestress is understood as the stress in the microstructure generated due to large localized volume changes during the hydration process. It builds up at very early stages of microstructure formation and then is gradually reduced by relaxation processes. Additional microprestress is generated by changes of internal relative humidity and temperature. This is described by the non-linear differential equation [2].

$$\frac{dS}{dt} + \psi_s(T,h)c_0 S^2 = k_1 \left| \frac{d(T\ln h)}{dt} \right| \tag{1}$$

in which T denotes the absolute temperature, h is the relative pore humidity (partial pressure of water vapor divided by the saturation pressure), c_0 and k_1 are constant parameters, and ψ_s is a variable factor that reflects the acceleration of microprestress relaxation at higher temperature and its deceleration at lower humidity (compared to the standard conditions). Owing to the presence of the absolute value operator on the right-hand side of (1), additional microprestress is generated by both drying and wetting, and by both heating and cooling, as suggested in [2].

3   Numerical simulations

In this section, experimental data of Fahmi, Polivka and Bresler are compared to results obtained with the MPS theory, which reduces to the standard B3 model in the special case of basic creep. All examples concerning drying and thermally induced creep have been run as a staggered problem, with the heat and moisture transport analyses preceding the mechanical one. The available experimental data contained the mechanical strains (due to elasticity and creep), with the thermal and shrinkage strains subtracted.

The four parameters of the B3 model describing the basic creep, q_1, q_2, q_3 and q_4, were determined from the composition of the concrete mixture and from the compressive strength using empirical formulae according to [1]. The result of this prediction exceeded the expectations; only minor adjustments were necessary to get the optimal fit (see the first part of the strain evolution in Fig. 2 (left). The following values were used: $q_1 = 19.5$, $q_2 = 160$, $q_3 = 5.25$ and $q_4 = 12.5$ (all in 10^{-6}/MPa). The MPS theory uses three additional parameters, c_0, k_1 and c, but parameter c should be equal to $c_0 q_4$. It has been found that the remaining parameters c_0 and k_1 are not independent. What matters for creep is only their product. In practical computations, k_1 can be set to a fixed value (e.g. 1 MPa/K), and c_0 can be varied until the best fit with experimental data is obtained; in all the following figures c_0 is specified in MPa^{-1}day^{-1}. All other parameters were used according to standard recommendations.

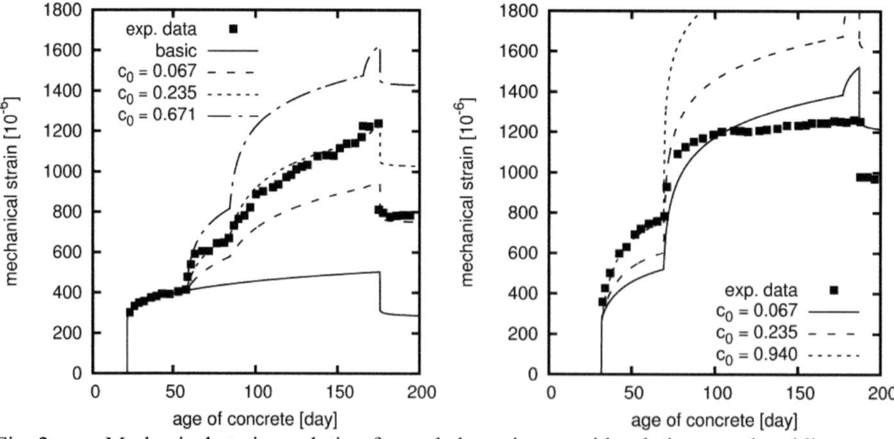

Fig. 2   Mechanical strain evolution for sealed specimens with relative pore humidity assumed to be 98%, loaded at time t' = 21 days (left), and for drying specimens at 50% relative environmental humidity, loaded at time t' = 32 days (right); both specimens are loaded by compressive stress 6.27 MPa

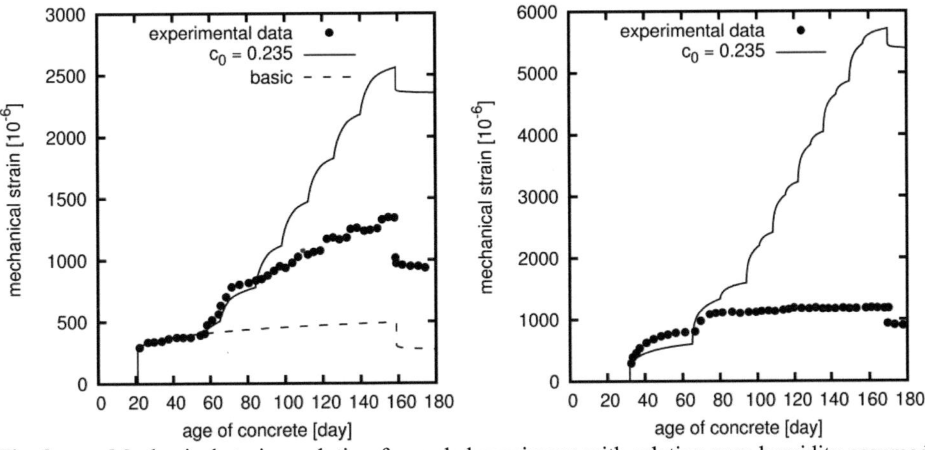

Fig. 3   Mechanical strain evolution for sealed specimens with relative pore humidity assumed to be 98%, loaded at time t' = 21 days (left), and for drying specimens at 50% relative environmental humidity, loaded at time t' = 32 days (right); both specimens are loaded by compressive stress 6.27 MPa and subjected to cyclic variations of temperature

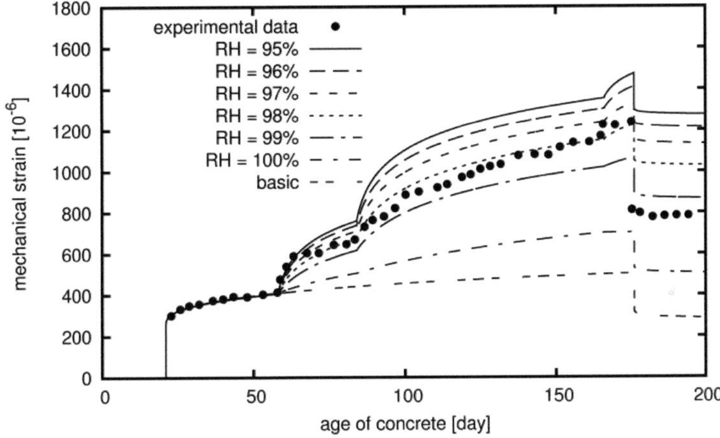

Fig. 4   Mechanical strain evolution for sealed specimens, loaded by compressive stress 6.27 MPa from age 21 days, with the assumed relative humidity of pores varied from 95% to 100%; parameters of MPS theory: $k_1 = 1$ MPa/K, $c_0 = 0.235$ MPa^{-1}day^{-1}

A really good fit of the first experimental data set (98% RH, i.e., h = 0.98) was obtained for $c_0 = 0.235$ MPa^{-1}day^{-1}; see Fig. 2 (left). The agreement is satisfactory except for the last interval, which corresponds to unloading. It is worth noting that the thermally induced part of creep accounts for more than a half of the total creep (compare experimental data with solid curve labeled *basic* in Fig. 2 (left). Unfortunately, with default values of the other parameters, the same value of c_0 could not be used to fit the second experimental data set, because it would have led to overestimation of creep; see the dashed curve in Fig. 2 (right). In the first loading interval of 37 days, creep takes place at room temperature and the best agreement would be obtained with parameter c_0 set to 0.940 MPa^{-1}day^{-1}; see the dash-dotted curve in Fig. 2 (right). However, at the later stage when the temperature rises to 60°C, the creep would be grossly overestimated. A reasonable agreement during this stage of loading is obtained with c_0 reduced to 0.067 MPa^{-1}day^{-1} (solid curve in Fig. 2 (right)), but then the creep is underestimated in the first interval in Fig. 2 (left).

For the last two testing programs with cyclic variations of temperature, the agreement between experimental and computed data is reasonable only until the end of the second heating cycle (solid curves in Fig. 3). For the sealed specimen, the final predicted compliance exceeds the measured value almost twice, for the drying specimen almost five times. In order to obtain a better agreement, parameter c_0 would have to be reduced, but this would result into an underestimation of creep in the first two testing programs.

Another deficiency of the model is illustrated by the graphs in Fig. 4. They refer to the first set of experiments. As documented by the solid curve in Fig. 2 (left), a good fit was obtained by setting parameter $c_0 = 0.235$ MPa^{-1}day^{-1}, assuming that the relative pore humidity is 98%. The pores are initially completely filled with water; however, even if the specimen is perfectly sealed, the relative humidity slightly decreases due to the water deficiency caused by the hydration reaction. The problem is that the exact value of pore humidity in a sealed specimen and its evolution in time are difficult to determine. In simple engineering calculations, a constant value of 98% is often used; unfortunately, the model response is quite sensitive to this choice (see Fig. 4). The source of such a strong sensitivity is in the assumption that the instantaneously generated microprestress is proportional to the absolute value of the change of $T \ln(h)$; see the right-hand side of (1). Rewriting (1) as

$$\frac{dS}{dt} + \psi_S(T,h)c_0 S^2 = k_1 \left| \ln h \frac{dT}{dt} + \frac{T}{h}\frac{dh}{dt} \right| \tag{2}$$

we can see that at (almost) constant humidity close to 100%, the right-hand side is proportional to the magnitude of temperature rate, with proportionality factor $k_1|\ln(h)| \approx k_1(1-h)$.

4 Improved material model and its validation

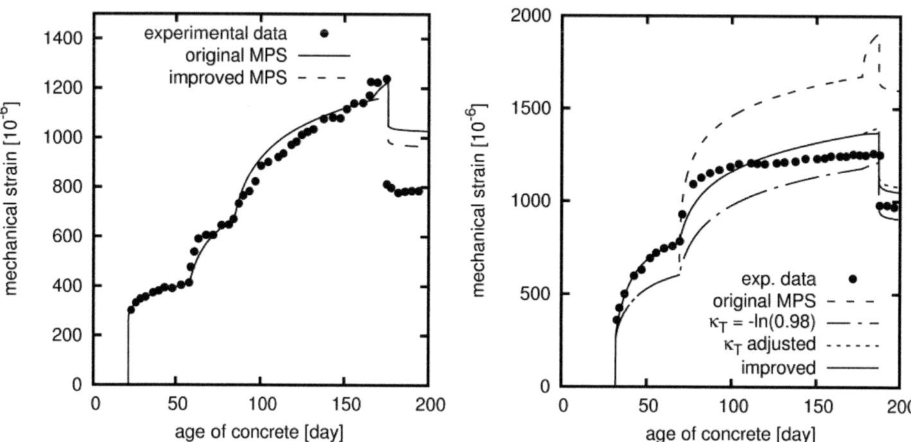

Fig. 5   Mechanical strain evolution for sealed specimens loaded at time t' = 21 days (left), and for drying specimens loaded at time t' = 32 days (right); both specimens are loaded by compressive stress 6.27 MPa and subjected to cyclic variations of temperature

As a simple remedy to overcome these problems, the microprestress relaxation equation (1) is replaced by

$$\frac{dS}{dt} + \psi_S(T,h)c_0 S^2 = k_1 \left| \frac{T}{h}\frac{dh}{dt} - \kappa_T k_T(T) \frac{dT}{dt} \right| \tag{3}$$

$$\text{with} \quad k_T(T) = e^{-c_T(T_{max}-T)} \tag{4}$$

in which κ_T and c_T are new parameters and T_{max} is the maximum previously reached temperature. With $\kappa_T = 0.02$, the creep curves in Fig. 4 plotted for different assumed pore humidities would be almost identical with the solid curve that nicely fits experimental results. Introduction of a new parameter provides more flexibility, which is needed to improve the fit of the second testing program in Fig. 2 (right), with combined effects of drying and temperature variation. For sealed specimens and monotonous thermal loading, only the product $c_0 k_1 \kappa_T$ matters, and so the good fit in Fig. 2 (right) could be obtained with different combinations of κ_T and c_0.

The results are shown in Fig. 5 for sustained thermal loading and in Fig. 6 for cyclic thermal loading. In these plots, the curves labeled *original MPS* show results obtained with standard MPS. Data series

$\kappa_T = -\ln(0.98)$ were obtained with $c_0 = 0.235$ MPa^{-1}day^{-1}, $k_I = 1$ MPa/K, $\kappa_T = 0.020203$ and $c_T = 0$. Data series κ_T *adjusted* correspond to parameters $c_0 = 0.235$ MPa^{-1}day^{-1}, $k_I = 4$ MPa/K, $\kappa_T = 0.005051$ and $c_T = 0$. Note that in the case of constant relative humidity (Fig. 5 (left) and Fig. 6 (left)) these series coincide with data series *original MPS*. The best agreement with experimental data is obtained with $c_0 = 0.235$ MPa^{-1}day^{-1}, $k_I = 4$ MPa/K, $\kappa_T = 0.005051$ and $c_T = 0.3$ K^{-1}; these series are labeled *improved*. In Fig. 5 (left), only a small change can be observed compared to data series *original MPS*; these differences arise when the temperature ceases to be monotonous. For the sealed specimen Fig. 5 (left), this change is detrimental, but the deterioration is negligible compared to a substantial improvement in the case of cyclic thermal loading (Fig. 6).

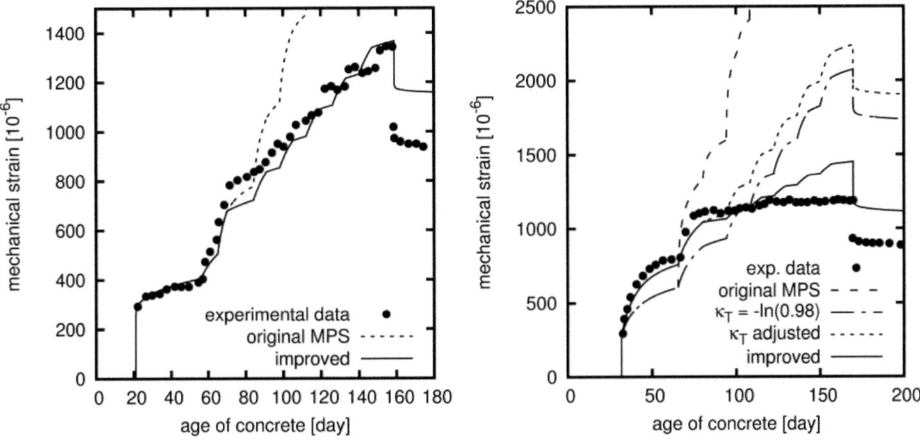

Fig. 6   Mechanical strain evolution for sealed specimens loaded at time t' = 21 days (left), and for drying specimens loaded at time t' = 32 days (right); both specimens are loaded by compressive stress 6.27 MPa and subjected to cyclic variations of temperature

5   Conclusions and further work

The material model based on the MPS theory has been successfully implemented into the finite element package OOFEM and has been used in simulations of concrete creep at variable temperature and humidity.

For sealed specimens subjected to variable temperature, the results predicted by the MPS theory are very sensitive to the assumed value of relative pore humidity (which is slightly below 100% due to self-desiccation). In order to overcome this deficiency, a modified version of the model has been proposed and successfully validated. Excessive sensitivity to the specific choice of relative humidity has been eliminated. Also, it has become easier to calibrate the model because thermal and moisture effects on creep are partially separated.

The original MPS theory grossly overestimates creep when the specimen is subjected to cyclic temperature. A new variable k_T has been introduced in order to reduce the influence of subsequent thermal cycles on creep. This modification does not affect creep tests in which the evolution of temperature is monotonous.

Even though the existing material model has been improved, it is not yet applicable to a general experimental setup. Comparing the results of experimental measurements of Pickett [11] and the data obtained from FE simulations (see Fig. 7), it is apparent that the experiment is well fitted in the case of basic creep (first part of series B) and monotonous drying (series C). Even the effect of curing before loading and drying is captured satisfactorily (series E and F). On the other hand the influence of drying and wetting cycles on creep is not described correctly (series D), e.g. at the end of the experiment (t = 140 days) the computed total deflection exceeds measured data more than two times. Further adjustments will need to be addressed in future work.

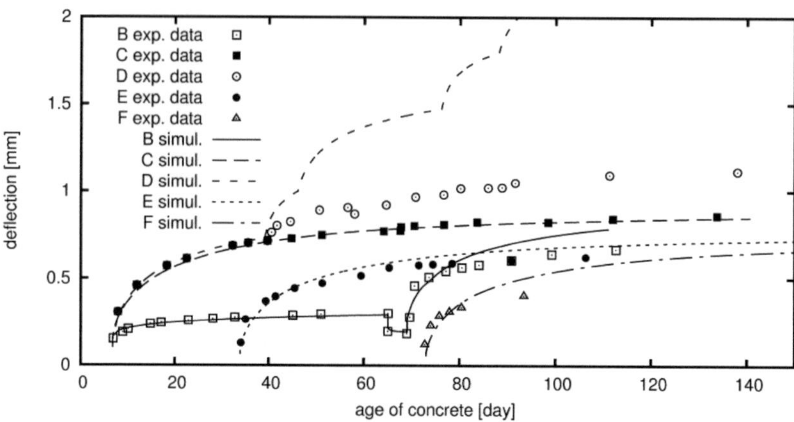

Fig. 7 Time evolution of deflection of prismatic beams subjected to three and four point bending (points refer experimental data, lines to data obtained from FE simulations; B = sealed conditions until reloading, C, E, F = cured until loading, then drying, D = drying and wetting cycles)

Acknowledgements

Financial support for this work was provided by projects SGS "Numerické modelování v mechanice konstrukcí a materiálů", reg. number OHK1-009/12 and 103/09/H078. The financial support is gratefully acknowledged.

References

[1]   Bažant, Z.P. and Baweja, S.: Creep and shrinkage prediction model for analysis and design of concrete structures: Model B3. In: Adam Neville Symposium: Creep and Shrinkage – Structural Design Effects (2000)

[2]   Bažant, Z.P., Cedolin, L. and Cusatis, G.: Temperature effect on concrete creep modeled by microprestress-solidification theory. In: Journal of Engineering Mechanics, 130 (2004) No. 6, pp. 691-699

[3]   Bažant, Z.P., Hauggaard, A.B. and Ulm, F.: Microprestress-solidification theory for concrete creep I: Aging and drying effects. In: Journal of Engineering Mechanics, 123 (1997) No. 11, pp. 1188-1194

[4]   Bažant, Z.P., Hauggaard, A.B. and Ulm, F.: Microprestress-solidification theory for concrete creep II: Algorithm and verification. In: Journal of Engineering Mechanics, 123 (1997) No. 11, pp. 1195-1201

[5]   Bažant, Z.P., and Prasannan, S.: Solidification theory for concrete creep. I: Formulation. In: Journal of Engineering Mechanics, 115 (1989) No. 8, pp. 1691-1703

[6]   Bažant, Z.P., and Prasannan, S.: Solidification theory for concrete creep. II: Verification and application. In: Journal of Engineering Mechanics, 115 (1989) No. 8, pp. 1704-1725

[7]   Fahmi, H.M., Polivka, M. and Bresler, B.: Effects of sustained and cyclic elevated temperature on creep of concrete. In: Cement and Concrete Research, (1972) No. 2, pp. 591-606

[8]   Patzák, B.: OOFEM home page. http://www.oofem.org, (2000)

[9]   Patzák, B., and Bittnar, Z.: Design of object oriented finite element code. In: Advances in Engineering Software, 32 (2001) No. 10-11, pp. 759-767

[10]   Patzák, B., Rypl, D. and Bittnar, Z.: Parallel explicit finite element dynamics with nonlocal constitutive models. In: Computers Structures, 79 (2001) No. 26-28, pp. 2287-2297

[11]   Pickett, G.: The effect of change in moisture-content on the creep of concrete under a sustained load. In: Journal of the ACI, 13 (1942) No. 4, pp. 333-355

Time dependent behaviour of concrete in tension

Nguyen Ngoc Toan

Faculty of Civil Engineering & Geosciences
Delft University of Technology
Stevinweg 1,PO-box 5048,2628 CN Delft ,The Netherlands
Supervisor: J.C. Walraven and C.R. Braam

Abstract

Direct tensile test were conducted under various loading rates in order to study the time dependence of tensile strength of concrete. The loading rates applied vary to such an extent that failure of the concrete specimens in tension occurs after time periods ranging from 3-5 seconds to 3-4 days. Concrete in strength classes C28/35 and C53/65 which hardened under various curing conditions were tested. The influence of the loading rate on the tensile strength and the appropriateness of a sustained loading factor are discussed. In this paper, test results for the strength class C28/35 are introduced.

1   Introduction

The bearing capacity of old concrete bridges is now a subject of discussion in the Netherlands. To resist heavier and more intensive traffic loading, strengthening might become requisite for those structures. However, theoretical service life might be as well be prolonged by using more advanced calculation techniques that use new insights into material and structural behaviour. The tensile strength of concrete is one of the parameters that will be studied in more detail.

In calculating the bearing capacity of old concrete structures, effect of time should be taken in to account. It is known that when subjected to long term loading tensile strength of concrete is reduced. Fig 1 (left) shows the principle to determine longterm tensile strength of concrete of Al-Kubaisy and Young [1]. The specimens were subjected to constant load level ranging from about 60%-95% of short term tensile strength of concrete. This study showed that sustained loading reduces tensile strength of concrete with about 30%. In VBC (Dutch Design Code for Concrete structures) therefore a factor of 0.7 is introduced.

T.Shkoukani [2] suggested that to determine the long-term direct tensile strength of concrete, it is necessary to have an appropriate definition of short-term tensile strength. In his study, sustained tensile load reduces tensile strength by about 25% if the short-time strength is determined with a failure time of 100 seconds. This value is 30% in the case short-time tensile strength is determined with a much higher failure time of 10 seconds.

Under sustained tensile load, tensile stress causes formation and extension of micro-cracks which lead finally to unstable crack propagation. However, strength of concrete continues to increase due to hydration of concrete if water is available. Which mechanism is prevailing of those two opposite effects is still a question.

In the Euro code for concrete structures the sustained loading factor is subject of national choice. The advisory value is $\alpha ct = 1.0$. The reason for this choice is the consideration that the strength values for concrete are determine after 28 days, and that at the time that substantial loading of the structure may be expected the ongoing hydration after 28 days has compensated a loss of concrete strength due to sustained loading. It should be noted, however, that this consideration only applies to the design of new structures. For the assessment of the bearing capacity of old structures, the actual concrete strength is mostly determined on drilled cylinders. Now, however a compensating effect of hydration may not be expected and the effect of sustained loading should be seriously considered.

In order to increase the insight in to the effect of sustained loading in tension, direct tensile tests were carried out under various loading rates in order to study time dependent behaviour of concrete in tension and more comprehensively understand the fracture mechanism of concrete.

Proc. of the 9th *fib* International PhD Symposium in Civil Engineering, July 22 to 25, 2012,
Karlsruhe Institute of Technology (KIT), Germany, H. S. Müller, M. Haist, F. Acosta (Eds.),
KIT Scientific Publishing, Karlsruhe, Germany, ISBN 978-3-86644-858-2

2　Test program

2.1　Test set up

The tensile testing machine was developed and adapted over the years at Stevin lab. The load was generated from a 100 kN actuator. A load cell was used to measure the load transmitted to the specimen. The specimen was connected to the testing rig through two steel plates. The whole system was connected to supports by a hinge joint. To prevent the rotation of the loading platen a guiding system was added to the testing rig. Although ball bushings were used, a small friction was still exited in the system. It led to some error in calculating the test results. To reduce this disadvantage, four load cells were added to the system. A guiding system was developed to ensure that the specimen is located at the centre of the loading platen. Fig 1 (right) shows the test set up for the direct tensile test.

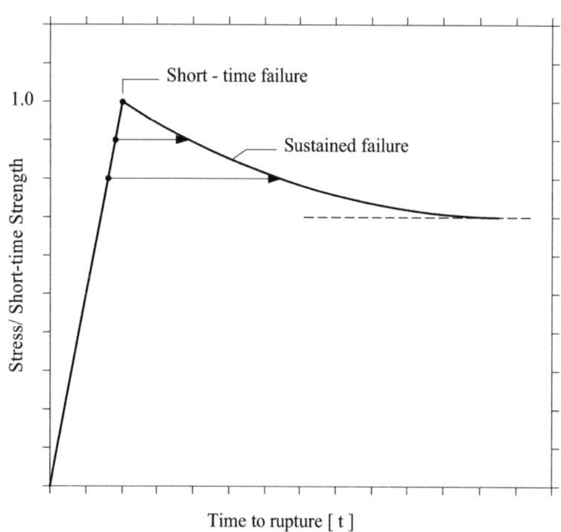

Fig. 1　　Al-Kubaisy and Young sustained loading test principle and setup of axial tensile test

2.2　Material

Concrete of strength class C28/35 was used. Aggregate size ranged from 4-16mm. Cement CEM III/B 42.5 N was used for the concrete strength class. Details about concrete components and the properties of concrete and the dry density after 28 days are given table 1.

Table 1　　Properties of concrete

Strength class	Cons. class	Aggregates (mm)	Cement	28 days compression strength (MPa)	28 day splitting tensile strength
C28/35	S3	4-16	CEM III/B 42.50N	38.17	3.53

2.3　Specimen's preparation

To produce specimens, concrete blocks were cast. They were cast in vertical direction. The block were kept in the mould after casting, covered with a plastic sheet and stored in the laboratory for 6 days. They were then demoulded and moved to the fog room at 100% RH.

About 20 days after casting, cylindrical specimens 100mm in diameter and 200mm in length were drilled out. The cores were stored in a climate-conditioned room (at 20° C, 60% RH) until the date of testing.

To investigate the effect of humidity content on the tensile strength of concrete a series of specimens in C28/35 strength class were oven-dried until totally dry. It takes about 38 days to complete this process. Those specimens were left to cool down in the climate room for several hours.

Subsequently that they were sealed. At first a layer of plastic was applied, follow by a layer of aluminium foil. Finally the specimens were covered by another plastic layer. Any change in the weight of the sealed specimens was documented. The results showed that the sealing technique was successfully applied.

2.4   Test conditions

Table 2 shows the testing condition for the direct tensile test. Six different rates of loading were applied in order to investigate the effect of loading rate to the direct tensile strength of concrete (LR1-LR6). In test series NR, the concrete strength of class C28/35 was cured under 20°C and 60% RH. The specimens of series DR were carried out on oven dried specimens. They are sealed during testing process.

Table.2   Conditions of direct tensile test

Specimen	Loading rate (kN/s)	Testing environment conditions	Time until tensile rupture	Maximum stress reached (MPa)
LR1.NR	6	Lab environment	3.5 sec	2.75
LR2.NR	1	-	22 sec	2.79
LR3.NR	0.1	-	3.37 min	2.54
LR4.NR	0.015	-	22.2 min	2.56
LR5.NR	0.001		4.8 hour	2.2
LR6.NR	0.000065		3.2 day	2.32
LR1.DR	6	Sealed	2.4sec	1.75
LR2.DR	1	-	16.8 sec	2.02
LR3.DR	0.1	-	2.6 min	1.85
LR4.DR	0.015	-	19.3 min	1.69
LR5.DR	0.001	-	3.8hour	1.67
LR6.DR	0.000065	-	2.83 day	1.64

3   Results and discussion

Figure 2 shows the results of the NR test series. Six specimens were tested per loading rate. The axial tensile strength of concrete tends to reduce with decreasing loading rate.

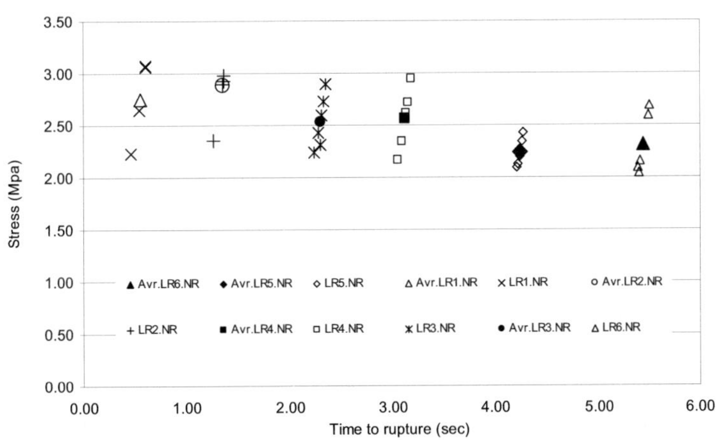

Fig. 2   Direct tensile strength as a function of time to rupture in log scale - DR test series

The test results of the DR series are shown in figure 3. Axial tensile strength tends to decrease as well with decreasing loading rate. Comparing the mean values of the tensile strength of the NR and the DR test series, it seems that the dried specimens are slightly less sensitive to the loading rate than the normally cured specimens. This difference in loading rate sensitivity can be explained on the basic of Stefan Effect due to the free water in the specimens of the NR series test. This effect is described as follows: when a thin film of a viscous liquid is trapped between two perfectly plane and parallel plates that are separated apart at a certain velocity of displacement, an opposing force are created. This force is proporational to the velocity of separation. In hydrated concrete, the wall of the pores can be considered as the plates. The free water in the pores works as the viscous films. From this perspective, the Steffan Effect is depended on the volume of the free water in concrete. This explains why the loading rate effects are larger in the NR test series.

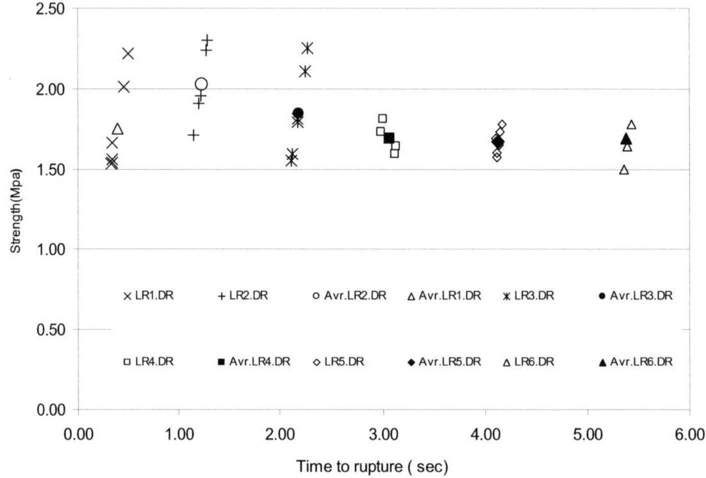

Fig. 3   Direct tensile strength as a function of time to rupture in log scale - DR test series

Table 3   The critical strain of the NR and DR test series.

Loading rate (kN/s)	6	1	0.1	0.015	0.001	0.000065
Test Series						
NR	0.000208	0.000213	0.000197	0.000209	0.000223	0.000223
DR	0.00024	0.000284	0.00025	0.000268	0.000248	0.000237

Figure 4.1 and 4.2 show the stress-strain relations of the NR test series. Figure 5.1 and 5.2 show the stress-strain relation of the DR test series. The critical strain in both test series is summarized in table 3. In this test series, the specimens were ruptured at almost the same critical strain.

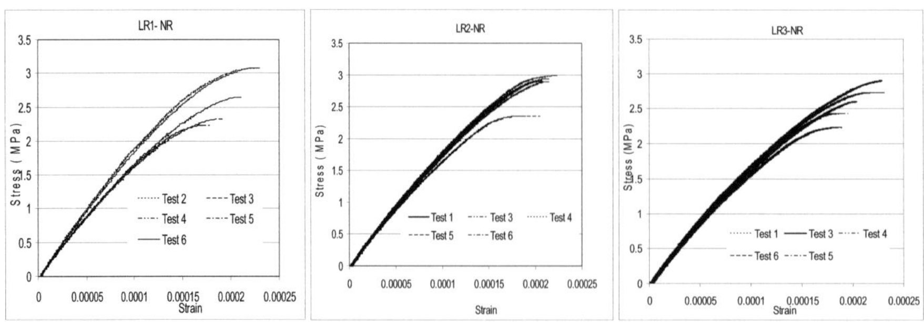

Fig. 4.1   Stress-strain relation of test series NR at loading rate LR1, LR2 and LR3

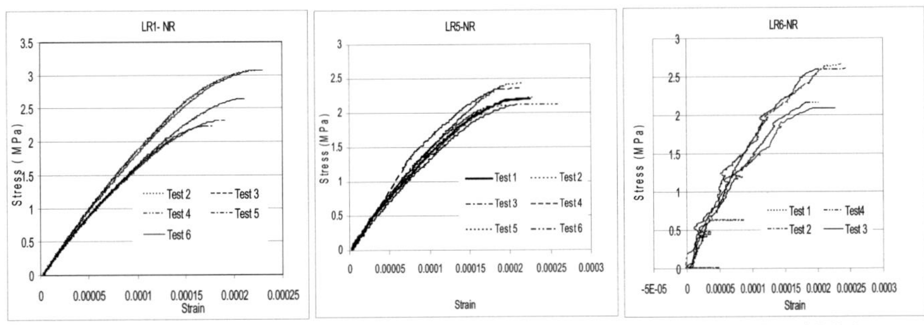

Fig. 4.2   Stress Stress-strain relation of test series NR at loading rate LR4, LR5 and LR6

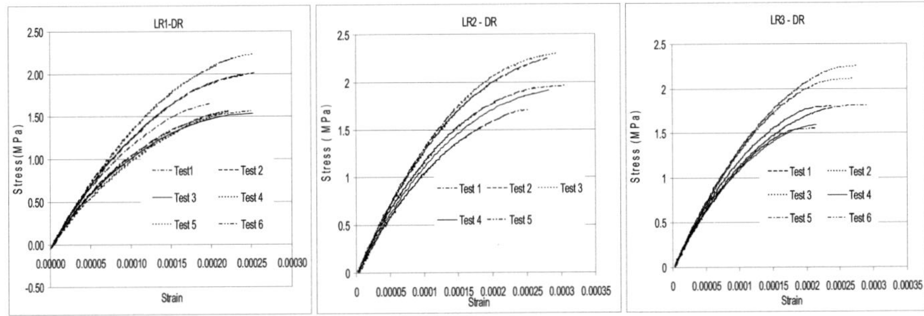

Fig. 5.1   Stress-strain relation of test series DR at loading rate LR1, LR2 and LR3

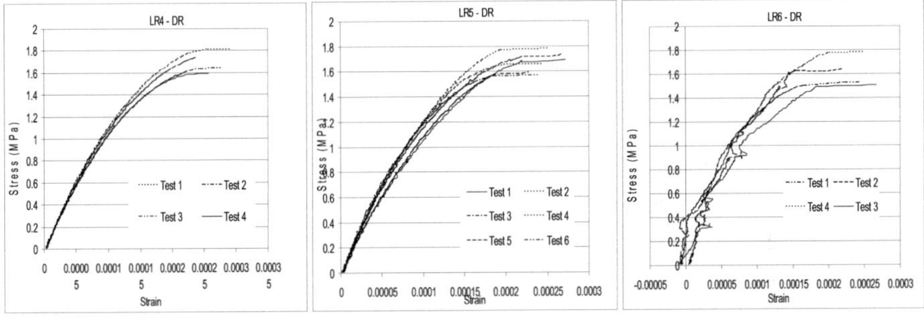

Fig. 5.2   Stress Stress-strain relation of test series DR at loading rate LR4, LR5 and LR6

4   Conclusions

To study the time dependent behaviour of concrete under axial tensile load, axial tensile tests under various loading rate were carried out. On the basis of the test results, the following conclusions can be drawn:

- The direct tensile strength of concrete decreases with decreasing on loading rate.
- Dried specimens are slightly less rate sensitive than specimens cured under normal condition.
- In determining long-term tensile strength of concrete, this behaviour should be taken into account to produce a more appropriate sustained loading factor
- To define a sustained loading factor agreement should be reached about the loading rate of the "short-time' reference test.
- The critical strain can be used as a fracture criterion.

References

[1]   M.A. Al-Kubaisy and A.G.Young: Failure of concrete under sustained tension - Magazine of Concrete Research 27(1975), pp.171-178

[2]   Shoukani: Behaviour of Concrete under concentric and eccentric sustained tensile loading - Damstardt Concrete, 4 (1989), pp. 223–224

[3]   Hans-Wolf Reinhardt, Tassilo Rinder: High strength concrete under sustained tensile loading - Otto-Graf-Journal Vol. 9(1998), pp 123-134

[4]   Y.Aoki and T.Shimomura: Effects of loading rate and drying condition on uniaxial tensile stress-strain relationship and cracking strength of concrete - Creep, Shrinkage and Durability Mechanics of Concrete and Concrete Structures-Tanabe et al @2009 Taylor & Francis group, London, ISBN 978-0-415-48508-1, pp. 207-214

Modelling explosive spalling of concrete under fire

Josipa Bošnjak

Institute of Construction Materials,
University of Stuttgart,
Pfaffenwaldring 4, 70569 Stuttgart, Germany
Supervisor: Joško Ožbolt

Abstract

The paper summarizes the results of the research project on explosive spalling of concrete. A fully coupled 3D thermo-hygro-mechanical model for concrete was employed in numerical simulations of explosive spalling. Because of the local character of this failure mode, the influence of local material and geometrical parameters was studied. Concrete was modelled at macro and meso scale and results were compared with the experiments. It was demonstrated that the meso scale model gives more realistic results than the macroscopic model. Experimental part of the program consisted of fire and permeability tests, both aimed at comparing the performance of high strength concrete with and without addition of polypropylene fibres. Fire tests were performed on relatively small slabs under one-sided fire. It was shown that only concrete without fibres exhibited explosive spalling. Permeability tests at high temperatures (on hot specimens) demonstrated a constant increase of permeability with rising temperature for concrete without fibres, whereas concrete containing fibres exhibited a sudden jump in permeability around 150°C. This confirmed that permeability is a major parameter that controls explosive spalling and validated the numerical model.

1   Introduction

One of the largest concerns for concrete under fire load is the occurrence of explosive spalling. This is especially pronounced in case of high performance concretes (HPC) with relatively dense matrix. Explosive spalling occurs at temperatures around 200-250°C, thus at an early stage of fire. The implications for concrete structures are: (i) spalling of concrete cover exerts the reinforcement to direct fire and (ii) concrete cross section is reduced. Explosive spalling has been widely investigated in the past decades [1,2], but the mechanism of spalling is still not fully explained. Addition of polypropylene fibers is a recognized method for mitigating explosive spalling. Fibers melt at around 160°C leaving free path for the water vapor to escape, i.e. increasing the concrete permeability. The majority of experimental measurements of permeability were in the past performed after cooling down the specimen to room temperature [3].

Numerical modelling of explosive spalling was in the past carried out by several research groups. The use of thermo-hygro-mechanical models [4, 5] made it possible to investigate the transport processes in concrete under fire conditions and their interaction (coupling) with mechanical properties of concrete. One such model was developed at the Institute of Construction Materials, University of Stuttgart [6]. The model was used in the numerical part of the project and it will be briefly discussed in the following chapters. The objective of the numerical study was to validate the existing model through comparison of numerical and experimental results, and to investigate the influence of different local inhomogeneities on explosive spalling of high strength concrete.

In the framework of the above mentioned project, fire tests were performed to compare the performance of concrete with and without polypropylene fibres under two different heating rates (ISO 834 and ZTV-ING fire). The permeability tests were performed at temperatures ranging from 20°C to 300°C. The measurements were performed on heated specimens, i.e. in hot state.

2   Numerical modelling

The thermo-hygro-mechanical model used in this study is relatively simple single phase phenomenological model for concrete. It is formulated in the framework of continuum mechanics under the assumption of validity of irreversible thermodynamics [6]. The response of the model is controlled by the following state variables: temperature, pore pressure (moisture), stresses and strains. In the numer-

Proc. of the 9th *fib* International PhD Symposium in Civil Engineering, July 22 to 25, 2012,
Karlsruhe Institute of Technology (KIT), Germany, H. S. Müller, M. Haist, F. Acosta (Eds.),
KIT Scientific Publishing, Karlsruhe, Germany, ISBN 978-3-86644-858-2

ical model temperature, moisture and pore pressure are coupled with stresses and strains, i.e. thermo-hygral part of the model depends on damage of concrete. Moreover, the relevant macroscopic mechanical properties of concrete (Young's modulus, tensile strength, compressive strength and fracture energy) are temperature dependent.

2.1 Coupled heat and moisture in concrete

The general approach for the solution of the problem of coupled heat and mass transfer in a porous solid, such as concrete, is well known within the framework of irreversible thermodynamics. However, there are a number of complex details, therefore, for the practical applications the model must be simplified. After introducing simplifications and assuming for a moment that the moisture flux ($\mathbf{J}$) and heat flux ($\mathbf{q}$) in concrete are independent of the stress and strain, the following is valid [7]:

$$\mathbf{J} = -\frac{a_p}{g}\operatorname{grad} p \tag{1}$$

$$\mathbf{q} = -b\operatorname{grad} T \tag{2}$$

with p = pore pressure, T = temperature, b = heat conductivity, $a = a_p/g$ = permeability, which is in the present model taken as a function of temperature according to the proposal of Bažant and Thonguthai [7], with g = gravity constant.

The governing equation for mass conservation is written in Eq. (3), where w = water content, t = time and wd = total mass of water released into the pore by dehydration. In the present model dehydration is not accounted for. The balance of heat is given by Eq. (4), where = mass density and C = isobaric heat capacity of concrete, Ca = heat sorption of free water and Cw = heat capacity of water, which is in the present model neglected.

$$\frac{\partial w}{\partial t} = -\operatorname{div}\mathbf{J} + \frac{\partial w_d}{\partial t} \tag{3}$$

$$C\rho\frac{\partial T}{\partial t} - C_a\frac{\partial w}{\partial t} - C_w\frac{\partial T}{\partial t}\mathbf{J}\operatorname{grad} T = -\operatorname{div}\mathbf{q} \tag{4}$$

Boundary conditions at concrete surface can be defined as:

$$n \cdot \mathbf{J} = \alpha_W\left(p_0 - p_E\right) \tag{5}$$

$$n \cdot \mathbf{q} = \alpha_G\left(T_0 - T_E\right) \tag{6}$$

where α_w = surface emissivity of water, α_G = surface emissivity of heat, T_0 and p_0 are temperature and pore pressure at concrete surface and T_E and p_E are temperature and pore pressure of environment.

The constitutive laws for p, w and T follow simplified suggestions proposed by Bažant and Thonguthai [7]. To describe the state of pore water in concrete, one has to distinguish between three different states: (i) non-saturated concrete, (ii) saturated concrete and (iii) transition from non-saturated to saturated concrete. For more detail see also [6].

The model is implemented into a 3D FE code using direct integration scheme. To assure stability of the time integration, a backward difference method is used. Since the controlling parameters are coupled, the linear differential equations (3) and (4) have to be solved iteratively.

2.2 Thermo-hygro-mechanical coupling

To account for the influence of temperature on the strain development in concrete, the total strain tensor for stressed concrete exposed to high temperature is decomposed as [6]:

$$\varepsilon = \varepsilon^m(T,\sigma) + \varepsilon^{ft}(T) + \varepsilon^{tm}(T,\sigma) + \varepsilon^c(T,\sigma) \tag{7}$$

where ε^m = mechanical strain tensor, ε^{ft} = free thermal strain tensor, ε^{tm} = thermo-mechanical strain tensor and ε^c are strains due to the temperature dependent creep of concrete. For more detail see [6]. The mechanical strain tensor ε^m, that comes into the 3D constitutive law for concrete (microplane model), is calculated as $\varepsilon^m = \varepsilon - (\varepsilon^{ft} + \varepsilon^{fm} + \varepsilon^c)$. The mechanical strains are then used to calculate the effective stresses increments $\overset{\cdot}{\sigma}$ (stress in solid phase of concrete matrix) and macroscopic stresses increments $\dot{\sigma}$ from the microplane constitutive law [6]:

$$\dot{\sigma} = \mathbf{D} : \dot{\varepsilon}^m + \dot{\sigma}^p \tag{8}$$

in which $\mathbf{D}$ = tangent material stiffness tensor obtained from the microplane model, $\dot{\varepsilon}^m$ = increment of the mechanical strain tensor and $\dot{\sigma}^p$ increment of pore pressure, which is calculated from the increment of volumetric pore pressure $\dot{\sigma}^v = n\dot{p}$, with $\dot{p}$ = increment of pore pressure. Note that according to definition pore pressure p is negative. The internal parameters of the microplane model are modified such that the macroscopic response of the model fits temperature dependent mechanical properties of concrete [6].

It is known that permeability and porosity of concrete are relevant parameters that control transport processes in concrete. On the other hand, both porosity and permeability are strongly influenced by damage, i.e. for higher level of damage, porosity and permeability increase. To account for this, permeability and porosity of concrete are assumed to be strain dependent [6].

To account for finite strains the co-rotational stress tensor together with Green-Lagrange finite strain tensor are used in the formulation of microplane model. The finite strain formulation is needed in order to investigate the influence of the geometrical instabilities (buckling) of a concrete layer on the explosive type of spalling of concrete cover. The objectivity of the mechanical part of the analysis with respect to size of the finite element is assured by the crack band method.

In the finite element analysis the mechanical and non-mechanical parts of the model are treated separately, however, in every time (load) step the relevant state variables that control model response are continuously updated. In this way the interaction between both parts of the model is implicitly accounted for.

3 Numerical modelling of explosive spalling

Numerical investigation was performed on relatively small concrete specimens under assumption of plain strain conditions, see Fig. 1 (right) and Fig. 3 (left). Concrete, mortar matrix, aggregates and interface between aggregate and matrix were all discretized using hexahedral three-dimensional elements assuming linear strain field. Material properties, fire loading and boundary conditions were same as in the experiments (see Section 4). Considering the symmetry of the test setup only half of the specimen was modelled. The analysis was performed for two different fire curves: ISO 834 and ZTV-ING fire (see Fig. 1).

3.1 Macro scale modelling

In the first part of the study concrete was modelled as a macroscopic material with following properties: compressive strength f_c = 88 MPa, tensile strength f_t = 4.8 MPa, Young's modulus E = 42.3 GPa and fracture energy G_F = 0.09 N/mm. These properties correspond to the concrete used in experiments. Free thermal strains and load induced thermal strains were accounted for.

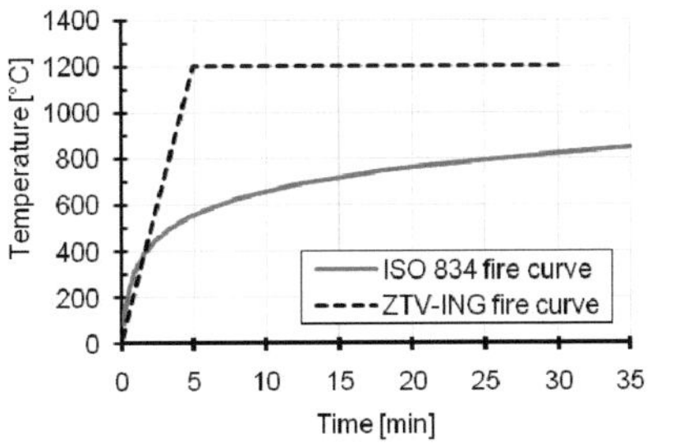

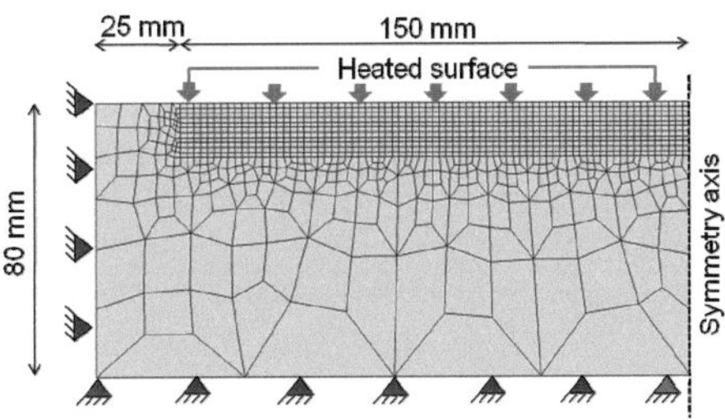

Fig. 1   Fire curves (left); Model geometry - macro scale (right)

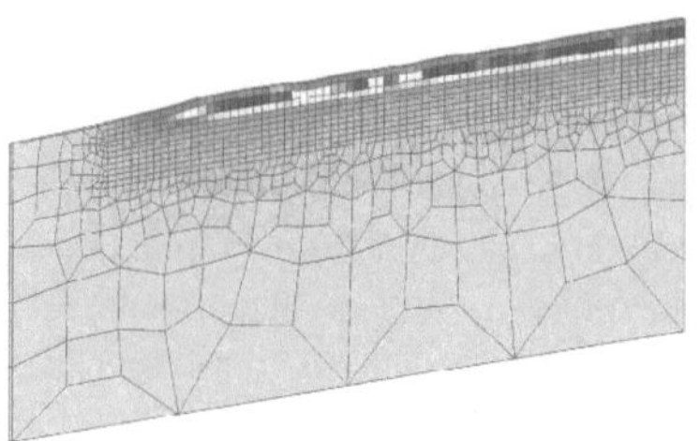 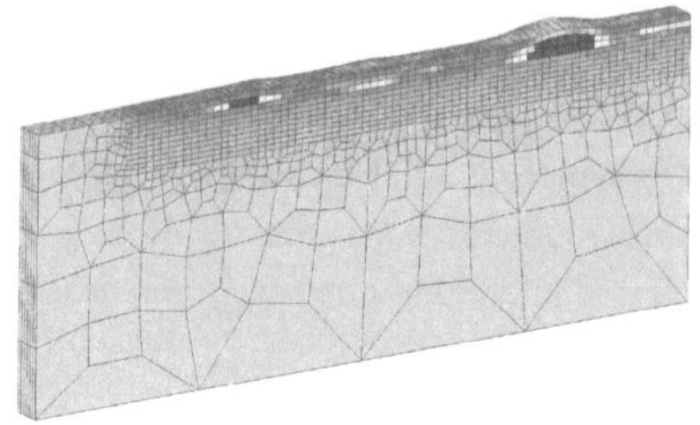

Fig. 2   Crack pattern (max. principal strains) in case of uniform heating (left) and non uniform heating (right)

Parametric studies were performed under variation of concrete permeability, fire load and external compressive load. The occurrence and time of spalling were mainly influenced by the concrete permeability. Spalling occurred within a certain range of permeability. After increasing permeability by two orders of magnitude no spalling was observed, thus confirming the experimental results.

However, the failure mode was not realistic, since the whole heated surface exploded instantaneously, as shown in Fig. 2 (left). Explosive spalling occurred after 11 min of exposure to the ISO 834 fire, whereas experimentally tested specimens spalled after 8 minutes of fire.

The effect of external compressive load is presented in Fig.3 (right). The results indicate that compressive load delays explosive spalling, whereas experiments [8] showed that compressive loads enhance spalling. These results confirmed that macroscopic consideration cannot fully capture this failure mode.

Explosive spalling is a local phenomenon, and it is therefore required to consider any local variations in terms of material properties, geometry or load. Concrete is a heterogeneous materials and its local properties can vary from the average properties. Furthermore, concrete is a composite of aggregates and cement matrix and at elevated temperatures these two constituents exhibit different behaviour. All these influences should be accounted for.

To confirm the effect of inhomogeneities, a macro scale model with 4 rows of elements was exposed to non uniform heating. Fire load on concrete surface was taken as variable across the specimen thickness. This variation resulted in localization of damage and spalling across the specimen length, as shown in Fig. 2 (right). This result illustrated the effect of the local variations on explosive spalling. Furthermore, failure occurred after 8 minutes of fire exposure, as it did in the experiments.

3.2   Meso scale modelling

In the second part of the numerical study concrete was discretized as three-phase material consisting of mortar matrix, coarse aggregates and interfacial transition zone. Only the concrete close to the free surface was modelled as three-phase material. The properties of the three phases were: mortar matrix $f_c = 79$ MPa, $E = 43.2$ GPa, $f_t = 4$ MPa, $G_F = 0.075$ N/mm; aggregates (basalt) $f_c = 200$ MPa, $E = 60$ GPa, $f_t = 18$ MPa, $G_F = 0.13$ N/mm; interfacial transition zone: $f_c = 79$ MPa, $E = 21.6$ GPa, $f_t = 0.5$ MPa, $G_F = 0.04$ N/mm. These properties correspond to the properties of experimentally tested concrete. Free thermal strains were considered for aggregates and mortar separately. Load induced thermal strains of mortar were not accounted for.

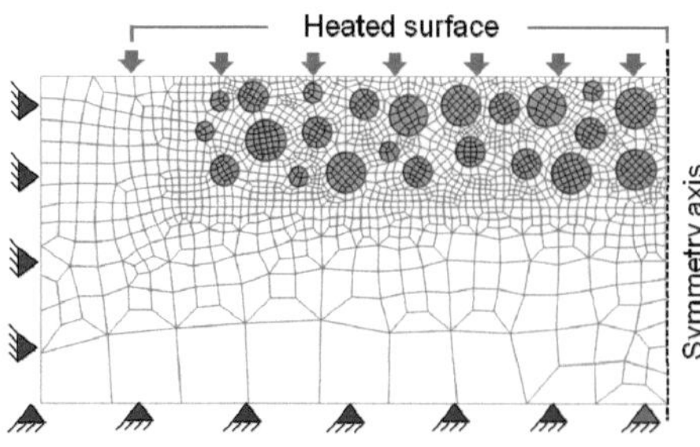

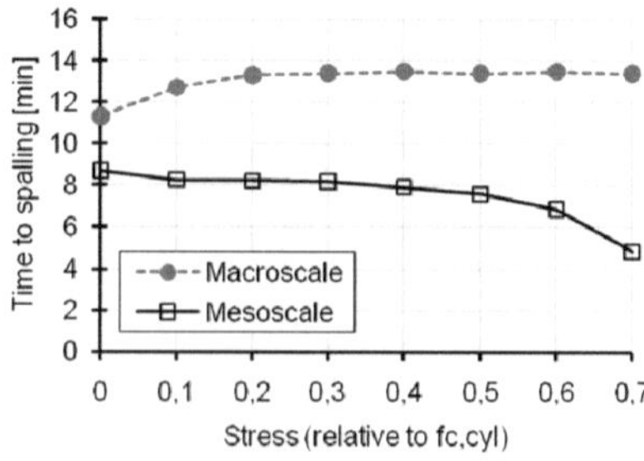

Fig. 3   Model geometry- meso scale (left); Influence of external load on spalling (right)

Typical failure mode for a not loaded specimen is shown in Fig. 4. Two different aggregate configurations were considered. The failure is localized as it was in the experiments. Cracking initiates in the interface zone between mortar and aggregates. Due to the relatively high tensile strength of aggregates, the cracks progressed only through cement mortar. It can be observed that the configuration of aggregates also influences the failure mode.

The effect of external load is shown in Fig 3 (right). Unlike in the case of macroscopic model, meso scale model indicates that compressive load enhances explosive spalling. With increase in the degree of external load spalling occurs earlier and also in thicker layers. These results are consistent with the available experimental data [8].

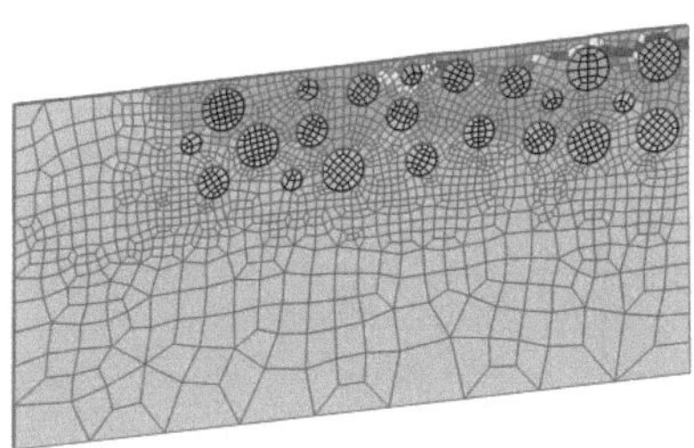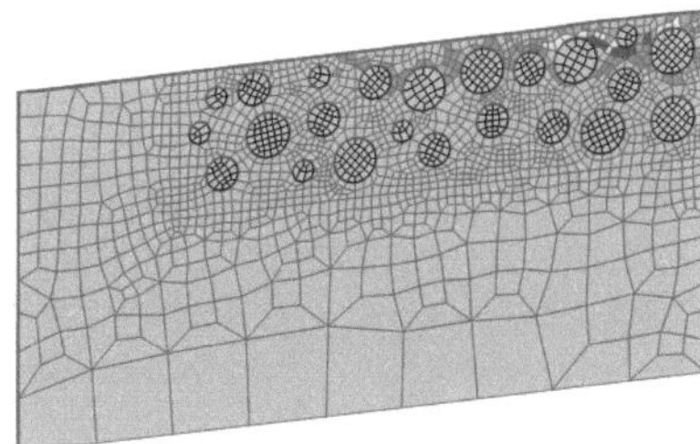

Fig. 4   Crack pattern for two different aggregate distributions

All parametric studies were performed on 2D models. However, a 3D FE analysis on macro and meso level was performed as well. Dimensions of the model were 50 x 50 x 50 mm. Failure modes obtained from the macro and meso scale analyses are shown in Fig. 5. Material properties and fire load were same as in the 2D analysis. As can be seen, macroscopic model yields an unrealistic failure of the entire heated surface, whereas 3D meso scale model results in a localized failure. The time of spalling for both models showed similar values as for 2D analysis, thus showing that the results of 2D analysis are realistic and can be used to investigate explosive spalling.

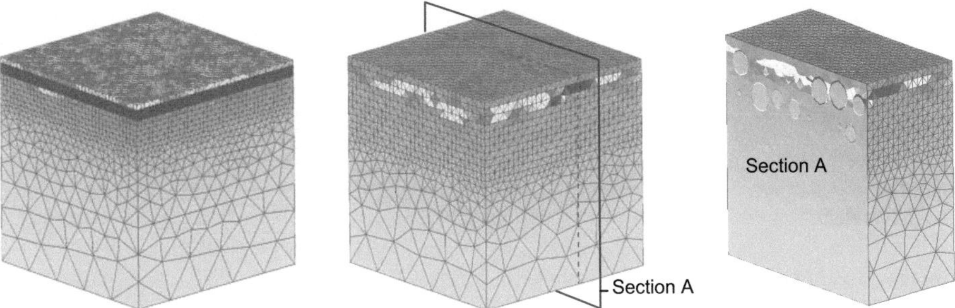

Fig. 5   Crack pattern: 3D macro scale model (left) and 3D meso scale model (mid and right)

4   Experimental investigations

4.1   Fire tests

Relatively small concrete slabs 700x700x300mm with and without addition of polypropylene fibres (content = 1kg/m^3) were exposed to one-sided fire. Concrete quality corresponded to the concrete class C80/88. The fire curves used were: ISO 834 fire curve and ZTV-ING tunnel curve (Fig.1 left). Concrete without fibres exhibited severe spalling in both loading cases, whereas concrete with polypropylene fibres exhibited almost no spalling. In the case of ISO 834 fire curve specimens spalled after around 8 minutes, whereas for the ZTV-ING curve the spalling began already after 2 minutes of fire exposure. In both loading cases explosive spalling occurred successively in thin layers. The experiments clearly showed that addition of polypropylene prevents explosive spalling.

4.2 Permeability tests

In the framework of the project a new simple test setup for permeability measurement at high temperatures was developed and validated against widely used RILEM method. Concrete permeability was investigated for temperatures between 20°C and 300°C. Test setup is shown in Fig. 6 (left). The gas used in the experiments was nitrogen and method of measurement was the pressure decay method. The test results are plotted in Fig. 6 (right). Concrete without addition of polypropylene showed a constant increase in permeability with rising temperature, whereas concrete with polypropylene fibres exhibited a sudden jump in permeability at around 150°C. This temperature corresponds to the melting point of the fibres. These results confirm the governing influence of permeability on explosive spalling and the validity of our numerical model.

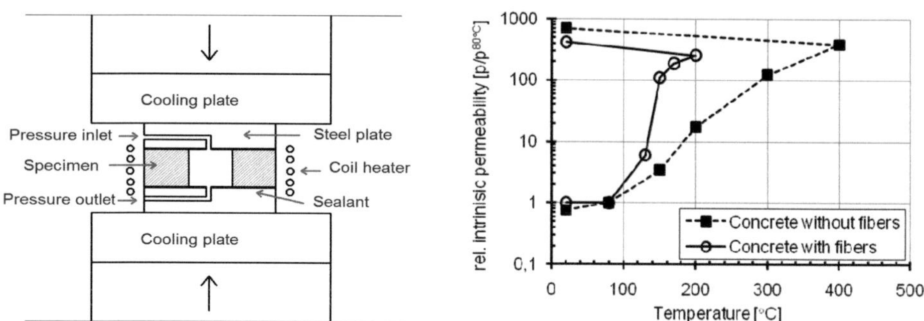

Fig. 6   Test setup (left); Relative intrinsic permeability as a function of temperature (right)

5 Conclusions

Because of the complex nature of the experiments on concrete at high temperatures, only limited data could be obtained experimentally. Numerical analysis, on the other hand, allowed for a detailed investigation of the influence of single parameters. Numerical results showed that the permeability is the major material property that governs spalling. Explosive spalling is a local phenomenon and all possible local variations in loading, geometry and material properties should be taken into consideration. Analyses at macro and meso scale were performed and it was shown that only meso scale model can realistically predict explosive spalling. Fire experiments have demonstrated that explosive spalling can be effectively reduced by adding polypropylene fibres to concrete. Permeability tests showed that at approximately 150°C there is a progressive increase of permeability with increase of temperature. This confirmed that permeability is major parameter that controls explosive spalling and validated the numerical model.

Acknowledgments

The author greatly acknowledge financial support of Deutsche Forschungsgemeinschaft (DFG Project AOBJ: 575356).

6 References

[1]   Phan, L.T.: Pore pressure and explosive spalling in concrete. In: Materials and Structures 41 (2008), pp 1623-1632

[2]   Sullivan, PJE.: Deterioration and explosive spalling of high strength concrete at elevated temperature. In: RILEM International Workshop Aging of Concrete Structures. Cannes, France (2000), pp 1–12

[3]   Kalifa, P., Chene, G. and Galle, C.: High-temperature behaviour of HPC with polypropylene fibres: From spalling to microstructure. In: Cement and Concrete Research 31 (2001), pp 1487-1499

[4]   Bažant, Z.P. and Kaplan, M.F.: Concrete at High Temperatures: Material Properties and Mathematical Models, Harlow, Longman (1996)

[5]   Gawin, D, Pesavento, F. and Schrefler, B.A.: Towards prediction of the thermal spalling risk through a multi phase porous media model of concrete. In: Computational Methods in Applied Mechanics and Engineering 195 (2006), pp 5707-5729

[6]   Ožbolt, J., Periškić, G., Reinhardt, H.-W. and Eligehausen, R.: Numerical analysis of spalling of concrete cover at high temperature. In: Computers and Concrete 5(4) (2008), pp 279-293

[7]   Bažant, Z.P. and Thonguthai, W. : Pore pressure and drying concrete at high temperature. In: ASCE Journal of Engineering Mechanics 104 (1978), pp 1059-1079

[8]   Boström, L., Wickström, U., Adl-Zarrabi, B.: Effect of specimen size and loading conditions on spalling of concrete. In: Fire and Materials Fire 31 (2007), pp 173–186

Effects of thermal load to the mechanical properties of polymer concrete

Orsolya Ilona Németh

Department of Structural Engineering,
Budapest University of Technology and Economics,
Műegyetem rkp. 3-9., Budapest 1111, Hungary
Supervisors: György Farkas and Éva Lublóy

Abstract

Polymer concrete is defined as a type of concrete in which the binding material is some kind of polymer. Consequently, its mechanical properties and possible applications are essentially influenced by the change of temperature and the load caused by fire.

Our study introduces the changes caused by thermal load in the mechanical and chemical properties of a polymer concrete of given composition.

During the tests, the polymer concrete specimen of given composition was subjected to load in four stages (20°C, 100°C, 200°C, 300°C), then the residual flexural strength, the residual compression strength and the modulus of elasticity was determined.

The changes occurring as a result of thermal load were compared to the measured strength values of cement concrete. Up to 300°C the measure of decrease of the compression strength of polymer concrete did not pass the same measure of the analogous cement concrete.

1   Fire resistance of polymer concrete

The binders of polymer concrete are organic materials, whose heat resistance is much lower than that of inorganic materials (e.g. stone, cement and metal). High temperature results in the degradation and, finally, loss of strength of the resin [2].

The physical features of organic polymers show a sudden change already at a slight temperature increase. The temperature at which a polymer transfers from its rigid, glassy condition into a much more flexible and plastic condition, is called glassing temperature (T_g). Glassing temperature may vary widely and depends on the molecular structure of the final polymer. The glassing temperature of polymers usually used in polymer concretes ranges from 10 to 200°C. Resistance to high operating temperatures is the task of the polymer used. Only very special PCs can resist temperatures over 150°C and such materials are usually very expensive or unavailable [1].

The features of some resins change drastically if temperature reaches or approaches or exceeds the softening temperature of resin (HDT – heat distortion temperature). At this temperature the resin starts to soften, deform or possibly flow as a result of the load. During the production of polymer concrete strength values must also be given for lower and higher temperatures than the expected operating temperature. In the case of structural applications, the highest expectable ambient temperature should be below the softening temperature. Where performance is expected even at increased temperatures as well, a detailed test in a temperature range including the expectable exposure temperature is recommended [1].

Polymer concrete mixes do not tolerate the impact fire if their resin content is 10% or more. Most polymer concretes with a resin content higher than 10% require a fire-retarding additive if the prevention of inflammability is an expectation. A sudden change in the mechanical features of polymers occurs when the temperature reaches the softening temperature of the resin used [1].

2   Material composition

The following tests were carried out to determine the properties for the present recipe (Tab. 1). The construction material made by the recipe will here be referred to as "UP polymer concrete", meaning polymer concrete whose binder is unsaturated polyester. Tab. 1 also includes the components of comparative (reference) cement concretes.

Proc. of the 9th *fib* International PhD Symposium in Civil Engineering, July 22 to 25, 2012,
Karlsruhe Institute of Technology (KIT), Germany, H. S. Müller, M. Haist, F. Acosta (Eds.),
KIT Scientific Publishing, Karlsruhe, Germany, ISBN 978-3-86644-858-2

Table 1   Components of polymer concrete and of comparative cement concretes

			Cement concrete	
	"UP" Polymer concrete		**Mix 1 (kg/m³)**	**Mix 2 (kg/m³)**
Binder	**16w%**	**POLIMAL 144-01 unsaturated polyester**	350 Portland cement	445 Portland cement
Water		0	151	144
Aggregate 0-2 mm	**38w%**	**particle-size dried bulk graded quartz gravel**	912	818
Aggregate 2-4 mm	**38w%**	**particle-size quartz sand**		
Aggregate 4-8 mm		0	485	363
Aggregate 8-16 mm		0	544	636
Other components	**3w%**	**Trigonox 44 B catalyst**	1.4 plasticiser	8.9 plasticiser
		CO-1 Cobalt initiator		
	5w%	**Calcium-Carbonate**		

3   Experimental methods and results

3.1   Derivatographic measurement

Derivatographic measurement is a simultaneous thermo-analytical method which simultaneously produces TG (thermo-gravimetric), DTA (differential thermo-analysis) and DTG (derivative thermo-gravimetric) signs. A small amount of the sample was powdered, put in a skillet of an inert material (corundum or platinum), and annealed in a furnace chamber at an even heat-up speed (in the so-called dynamic mode). Meanwhile the changes in the mass of the sample (TG curve) were measured by an analytical balance and the changes of enthalpy taking place in the sample compared to the temperature of an inert material in the furnace chamber (DTA curve) were measured by thermo-elements. The appliance produces the first derivative of the TG curve, i.e. the DTG curve, in an analogous manner, which determines the place and extent of the processes accompanying the change of mass on the temperature scale. The above three curves and the test result also containing the temperature (T °C) sign and taken in function of the measurement time (t min) are called the derivatogram. The derivatogram can be shown in the function of temperature (T °C) as well. A derivatograph Q-1500 D appliance was used for the measurements (Fig. 1 (left)). The parameters of the derivatographic measurement were as follows:

- reference material: aluminium oxide,
- heat-up speed: 10°C/minute,
- temperature range: 20-1000°C,
- measured mass of sample: 200 mg,
- TG sensitivity: 50 mg,
- corundum skillet.

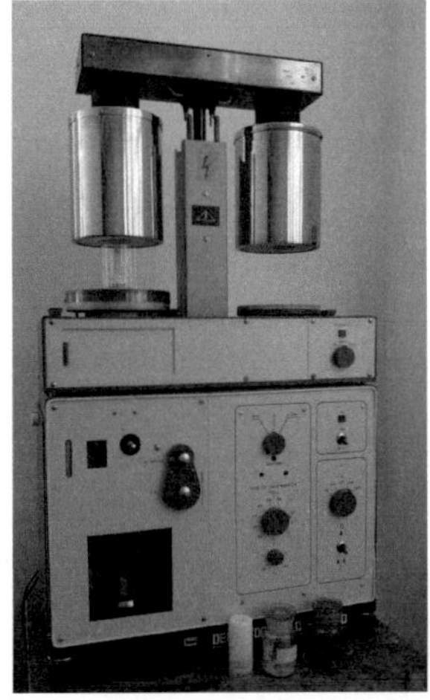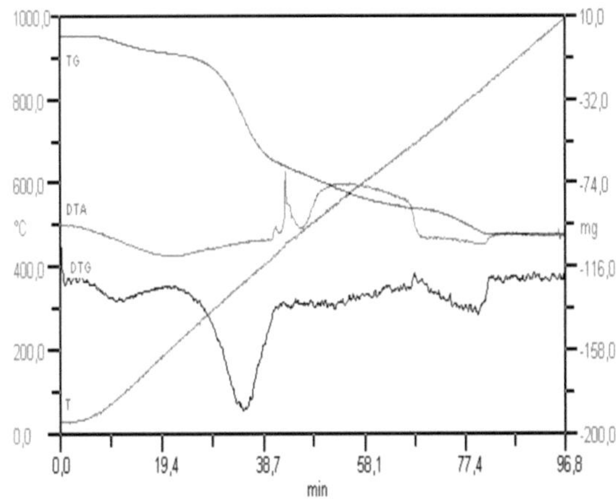

Fig. 1   Derivatograph Q-1500 D (left) and the derivatogram about UP polymer concrete (right)

Results of the derivatographic test are summarized in Fig 1(right). It can be read from the DTA and DTG curves that the binder of the polymer concrete undergoes a softening process between 300°C and 400°C and the phenomenon is accompanied by a change of mass.

3.1 Fire resistance test

In the case of tests at high temperatures or tests following heat effects the speed and manner of heat load are essential issues. Based on data described in literature and in the standards, several kinds of fire curves are used for the experiments. In this case a heating up curve approximating the normative fire curve was used, i.e. the one applicable to buildings and halls of architectural engineering. Tests were carried out without applying any direct flame effect. The strength of the specimens was examined after cooling down, at room temperature.

3.1.1 Ocular inspection

The specimens were inspected after the heat load.

It can be clearly observed that as a result of a heat load of 200°C and 300°C, respectively, the specimens got discoloured (Fig. 2 (left)). Extent of discolouration: as a result of the temperature increase, the specimens became darker. The inner layer was not discoloured during the two-hours heat load (Fig. 2 (right)).

Fig. 2   Discoloured specimens after heat load (left) and the discoloration of the outer layer (right)

During the 400°C heat load the specimens caught fire without any lighting effect and continued to burn freely until they got carbonized (Fig. 3 (left and right)). This behaviour is similar to that of wood, however, when wood was taken out of the furnace, after a time it went out by itself in the open air, which behaviour is due to the carbonized layer developed. In the case of concrete with plastic binder, the burning of the plastic provides enough energy for the continuation of burning.

Fig. 3   Burning of UP polymer concrete specimens (left and right)

3.1.2 Compressive strength tests

The edge length of the specimens – taking into account the maximum grain size of the additive and the prescriptions of the standard – was 150 mm. After heat load and cooling down the specimens were broken in an ALPHA 3-3000S type breaker. The thermal steps used were 20°C, 100°C, 200°C, and 300°C. The loading speed of the crusher was 11.4 kN/s. At 20°C the following compressive strengths were measured: Mix1: 64 N/mm^2, Mix2: 89 N/mm^2, UP Polymer concrete: 98.7 N/mm^2.

The relative residual values of compressive strength are provided in Figure 4. Based on Figure 4 it can be stated that:

- the extent of the strength reduction of polymer concrete does not exceed, up to a heat load of 300°C, the strength reduction of cement concrete of the same strength;
- in the case of concrete of a lower strength (Mix1), the extent of strength reduction was higher than in the case of polymer concrete;
- over 300°C the UP polymer concrete started to burn and continued to burn freely until the specimens got carbonized, practically until their strength was considered to be 0 N/mm^2.

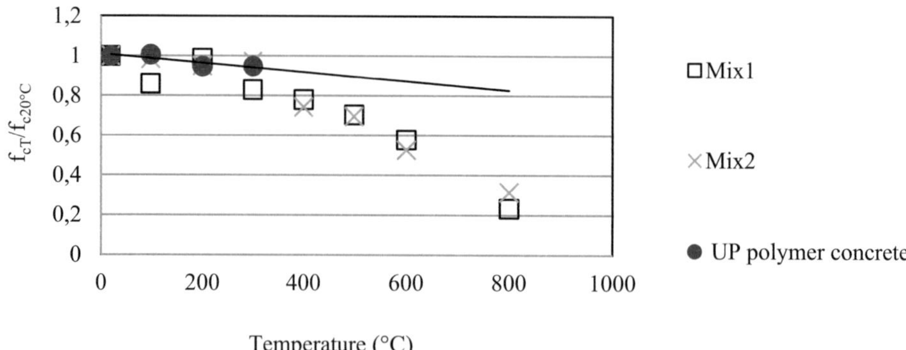

Fig. 4   Relative residual values of compressive strength dependent temperature [3] (all points represent an average of three measurement results)

3.1.3 Flexural strength tests

Flexural strength was measured on 70*70*250 mm prisms. The specimens were broken after heat load and cooling down. A three-point flexural strength test was carried out by the standard VPM machine, loaded at a speed of 11.4 kN/s. The distance of the supports was 200 mm. At 20°C the compressive strength of UP polymer concrete was 28.5 N/mm^2.

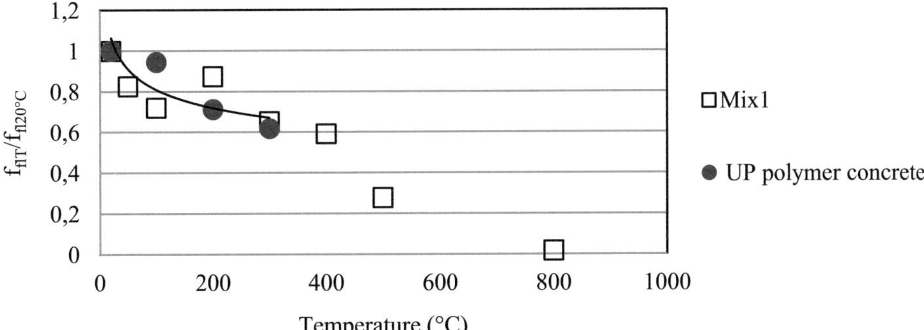

Fig. 5 Relative residual values of flexural strength dependent temperature (all points represent an average of three measurement results)

The relative residual values of flexural strength are provided in Figure 5. Based on Figure 5 it can be stated that:

- the extent of the flexural strength reduction of polymer concrete is higher than the extent of compressive strength reduction;
- the extent of the flexural strength reduction of polymer concrete exceeds, up to a head load of 300°C, the strength reduction of concrete.

3.1.4 Measurement of the modulus of elasticity

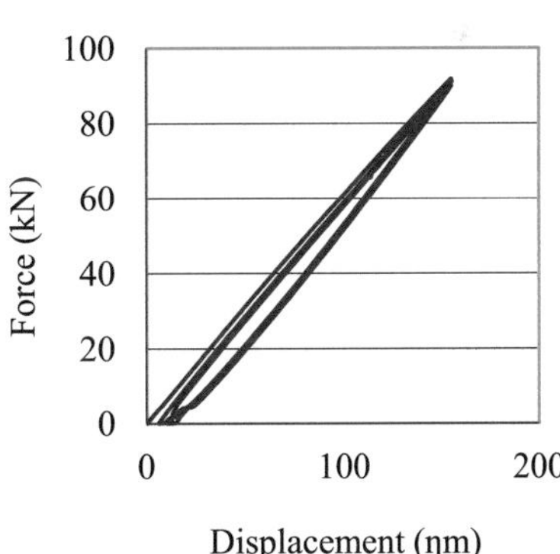

Fig. 6 Specimen with the measuring instrument in the breaker (left) and typical force-displacement curve (right)

Flexural strength was measured on 70*70*250 mm specimens. After the formerly discriminated heat load and cooling down, the specimens were loaded with the tierce of the fracture force and the load-displacement results were set down (Fig. 6 (left and right)). The relative residual values of the modulus of elasticity were counted from these diagrams as shown in Figure 7. At 20°C the modulus of elasticity of UP polymer concrete was 19400 N/mm^2.

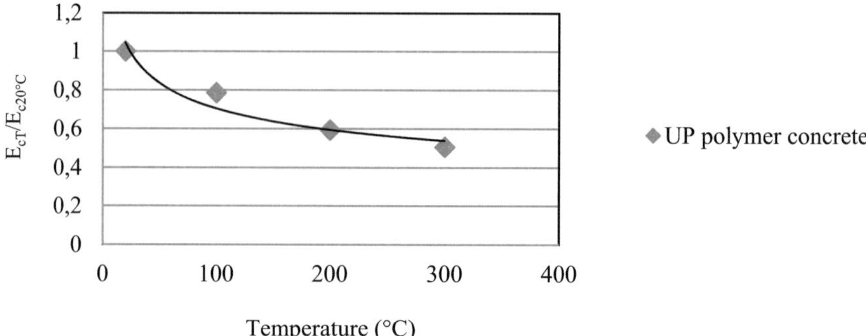

Temperature (°C)

Fig. 7   Relative residual values of the modulus of elasticity dependent temperature (all point represent an average of three measurement results)

Based on Figure 7 it can be stated that:

- the extent of the modulus of elasticity reduction of polymer concrete is higher than the extent of compressive strength reduction; analogous with cement concrete;
- the tendency of the extent of the flexural strength reduction of polymer concrete and the extent of the modulus of elasticity reduction of polymer concrete are similar.

4   Summary

This study examined the effect of high temperatures on UP polymer concrete. Experiments were conducted to detect the behaviour of UP polymer concrete at these temperatures. The observations and measurements made during the experiments yield the following conclusions:

- the extent of the strength reduction of polymer concrete does not exceed, up to a heat load of 300°C, the strength reduction of concrete of the same strength;
- the tendency of the extent of the flexural strength reduction of polymer concrete and the extent of the modulus of elasticity reduction of polymer concrete are similar, but they show more significant decrease up to a head load of 300°C.

The UP polymer concrete had no fire resistance. Our future research will be extended in this direction.

Acknowledgements

The work reported in the paper has been developed in the framework of the project „Talent care and cultivation in the scientific workshops of BME" project. This project is supported by the grant TÁMOP-4.2.2.B-10/1--2010-0009.

References

[1]   ACI Committe 548.: Polymer Concrete – Structural Applications State-of-the Art Report, (1996), pp. 8-9

[2]   Balaga, A., Beaudoin, J.J.: CBD-242. Polymer Concrete. In: Division of Building Research, National Research Council Canada, Canadian Building Digest, (1985)

[3]   Balázs, Gy. L.- Lublóy, É.: Concrete at high temperature, The Third International fib Congress incorporating the PCI Annual Convention and Bridge Conference, Washington D.C., United States, (2010)

Durability of Concrete

A three-phase model for concrete

Matthias Aschaber

Institute for Basic Sciences in Civil Engineering,
University of Innsbruck,
Technikerstraße 13, A-6020 Innsbruck, Austria
Supervisor: Günter Hofstetter

Abstract

Drying shrinkage of a concrete overlay influences the structural behaviour of a strengthened concrete structure. Hence, a three-phase model for concrete, based on the theory of porous media, is developed to simulate the change of moisture distribution and the resulting drying shrinkage in a realistic way. Because shrinkage strains start to develop after concrete setting, the hydration reaction, modelled by an evolution equation for the degree of hydration considering the influence of temperature and relative humidity must be taken into account. Comparisons of computed results from the numerical simulation of a brick-shaped concrete specimen strengthened by a concrete overlay with a thickness of 9 cm with test data, demonstrate the capabilities of the numerical model.

1   Introduction

Drying shrinkage of concrete is characterized by the time-dependent volume decrease due to moisture migration and moisture transfer to the environment caused by a change in relative humidity [11]. Due to the variable moisture distribution in a structure, shrinkage of the dryer near-surface regions will be restrained by the moist inner regions. Additional restraint effects are encountered, when a concrete structure is strengthened by adding a concrete overlay, since for the latter shrinkage is restrained at the interface with the original concrete [4]. For commonly encountered values of relative humidity the physical origin of drying shrinkage is related to the increase of the capillary pressure in the porous concrete during the drying process [1]. Hence, a physically based model of drying shrinkage relies on a multi-phase formulation, in which concrete is considered as a porous material, consisting of a solid skeleton and voids, filled by liquid water and gas [5, 8].

2   Multi-phase concrete model

Concrete is modelled as a porous material with the voids filled by liquid water and/or gas. Hence, a multi-phase concrete model consists of the solid phase (s), the water phase (w) and the gas phase (w). The latter phase is assumed to be a mixture of two ideal gases, dry air (ga) and water vapour (gw). By means of the hybrid mixture theory [6] the macroscopic balance laws are derived from the microscopic balance laws averaged over a representative elementary volume, whereas the constitutive models are introduced directly at the macroscopic level.

The set of governing equations consists of (i) the rate form of the linear momentum equation, (ii) the mass balance equations of the individual phases, (iii) the enthalpy equation, (iv) the linearized kinematic relations and (v) constitutive relations for the mechanical, thermal, hygric and chemical behaviour of the individual phases. They are formulated in terms of state variables, consisting of the displacement vector $\mathbf{u}$ of the solid phase, the temperature T , the gas pressure p^g and the capillary pressure

$$p^c = p^g - p^w \tag{1}$$

representing the difference of the gas pressure p^g and the water pressure p^w [5].

From the large number of constitutive relations required for the multi-phase concrete model only those related to drying of the overlay will be reviewed briefly, following [5]. Further constitutive equations for the multi-phase concrete model are described, e.g., in [3], [5], [8].

Proc. of the 9th *fib* International PhD Symposium in Civil Engineering, July 22 to 25, 2012,
Karlsruhe Institute of Technology (KIT), Germany, H. S. Müller, M. Haist, F. Acosta (Eds.),
KIT Scientific Publishing, Karlsruhe, Germany, ISBN 978-3-86644-858-2

Within the considered range of relative humidity between 100 % and 50 % the water in the pores is present as capillary water. The capillary pressure p^c is determined from the pore relative humidity φ by the Kelvin-Laplace law

$$p^c = -\rho^w \frac{RT}{M_w} \ln \varphi \, ,\tag{2}$$

with M_w, R and ρ^w denoting the molar mass of water, the universal gas constant and the intrinsic density of the water phase, respectively. The sorption/desorption isotherm provides the relationship between relative humidity φ and mass water content w at sorption equilibrium. The latter relationship can be converted to a relationship between the degree of water saturation S_w and capillary pressure p^c, which can be approximated by the van Genuchten equation [10] as

$$S_w = \left[1 + \left(\frac{p^c}{a} \right)^{b/(b-1)} \right]^{-1/b} \, ,\tag{3}$$

with a and b denoting material parameters for fitting measurement data.

Flow of the fluid phase π, $\pi = w, g$, relative to the solid phase s is described by Darcy's law

$$nS_\pi \mathbf{v}^{\pi s} = \frac{k^{r\pi} \mathbf{K}_\pi}{\mu^\pi} \left(-\nabla p^\pi + \rho^\pi \mathbf{g} \right) \, ,\tag{4}$$

with $nS_\pi \mathbf{v}^{\pi s}$ denoting the artificial relative velocity, and n, $k^{r\pi}$, $\mathbf{K}_\pi$ and μ^π representing the porosity, the relative permeability, depending on the degree of saturation of the respective fluid phase, the intrinsic permeability tensor and the dynamic viscosity of the fluid, respectively. Assuming concrete as an isotropic material yields $\mathbf{K}_\pi = K_\pi \mathbf{I}$ with K_π as the intrinsic permeability and $\mathbf{I}$ as the second order unit tensor. The relative water permeability k^{rw} can be approximated by the equation proposed by van Genuchten [10]

$$k^{rw}(S_w) = \sqrt{S_w} \left(1 - (1 - S_w^b)^{1/b} \right)^2 \, ,\tag{5}$$

where the parameter b is the same as in equation (3).

The diffusion inside the gas phase is described by Fick's law

$$nS_g \rho^\pi \mathbf{v}_{diff}^\pi = -\rho^g \mathbf{D}_g^\pi \nabla \left(\frac{\rho^\pi}{\rho^g} \right) \, ,\tag{6}$$

with $\pi = gw$ or $\pi = ga$, and

$$\mathbf{D}_g^{gw} = D_0 f_s \mathbf{I} \, ,\tag{7}$$

is the diffusivity tensor with $D_0 = D_0(p^g, T)$ as the free water vapour diffusion coefficient of the air and $f_s(S_w, n)$ as the resistance factor, depending on the porosity and the degree of water saturation. Commonly, the resistance factor is determined by the analytical formula proposed by Millington [7]

$$f_s(S_w, n) = n^{4/3} (1 - S_w)^{10/3} \, ,\tag{8}$$

However, Baroghel-Bouny [2] showed that the exponents in (8) are not appropriate for low values of water saturation and, hence, they should be determined by fitting test data.

Within the framework of the multi-phase concrete model the effective stress concept is applied, splitting the total stress $\boldsymbol{\sigma}$ into the effective stress $\boldsymbol{\sigma}'$ and the solid pressure p^s, i.e.,

$$\boldsymbol{\sigma} = \boldsymbol{\sigma}' - \alpha p^s \mathbf{I} \, .\tag{9}$$

The effective stress is related to the deformation of the solid matrix by a constitutive law, e.g. the generalized Hooke's law $\boldsymbol{\sigma}' = \mathbf{C} : \boldsymbol{\varepsilon}^{tot}$, and α is the Biot coefficient, which accounts for different bulk moduli of the solid phase and the solid skeleton. The solid pressure, exerted by the pore fluids on the solid phase, is defined as

$$p^s = p^g - p^{atm} - \chi(S_w)p^c \; , \tag{10}$$

where p^{atm} is the atmospheric air pressure and $\chi(S_w)$ represents the Bishop parameter. In (9) and (10) pressures are defined as positive quantities, whereas stresses are positive in tension.

Inserting (10) into (9) assuming the pore gas pressure at atmospheric conditions, i.e. $p^g = p^{atm}$, yields

$$\boldsymbol{\sigma}' = \boldsymbol{\sigma} - \alpha\chi(S_w)p^c \, \mathbf{I} \; . \tag{11}$$

Since drying causes an increase of the capillary pressure, from (11) together with the constitutive relations a volumetric compaction of the concrete is obtained. Hence, departing from a fully saturated state, characterized by $p^c = 0$, the drying shrinkage strain at a partially saturated state with $p^c \neq 0$, is determined by means of the bulk modulus K of the solid skeleton as

$$\boldsymbol{\varepsilon}^{sh} = -\frac{\alpha\chi(S_w)}{3K}p^c \mathbf{I} \; . \tag{12}$$

3 Numerical study

3.1 Determination of material parameters

For determining material parameters and for validating the multi-phase concrete model test data are indispensable. An extensive experimental study has been carried out in [9]. This testing program contains drying shrinkage tests on concrete slices, concrete prisms and on larger brick-shaped specimens with added overlays. The water desorption isotherm was determined by drying concrete slices with dimensions of 110 x 110 x 6 mm at constant temperature and at constant relative humidity until the sorption equilibrium was achieved. The latter was determined at relative humidities of 100%, 90%, 80%, 65% and 50% respectively. Additionally, the drying shrinkage strains of the thin concrete slices were measured using a high-precision strain-measuring extensometer. With the desorption isotherm the parameters a and b of the van Genuchten equation (3) and, with the additional information regarding the time to achieve the sorption equilibrium, the transport properties, i.e. the intrinsic water permeability K_w in (4) and the exponents of the resistance factor in (8), were determined. Since in (12) both α and $\chi(S_w)$ are unknown, the term $\alpha\chi(S_w)$ in (12) was determined from the measured relationship between the drying shrinkage strain and relative humidity [5].

3.2 Validation

The tests on prismatic specimens with dimensions of 100 x 100 x 56 mm served for determining the mass water content evolution and distributions in the samples along the length of a Multi-Ring-Sensor (MRS) during one-dimensional drying at a relative humidity of 65%. They were determined by a calibrating curve for the respective concrete, relating electrolytic resistances, measured by Multi-Ring-Sensors, to the mass water content. In Figure 1 the mass water content distribution along a single MRS is plotted for selected time instants during moist curing and drying. The initial constant mass water content distribution at the end of the moist curing gradually changes with progressing drying time to a distribution with decreasing values towards the surfaces due to moisture migration and moisture transfer to the environment.

The computed distributions of the mass water content are shown by the dashed lines in Figure 1. Comparison of the measured and computed mass water content distributions shows satisfactory agreement. Hence the determined water desorption isotherm, the intrinsic water permeability tensor in (4) and the diffusivity tensor in (6) are sufficiently accurate.

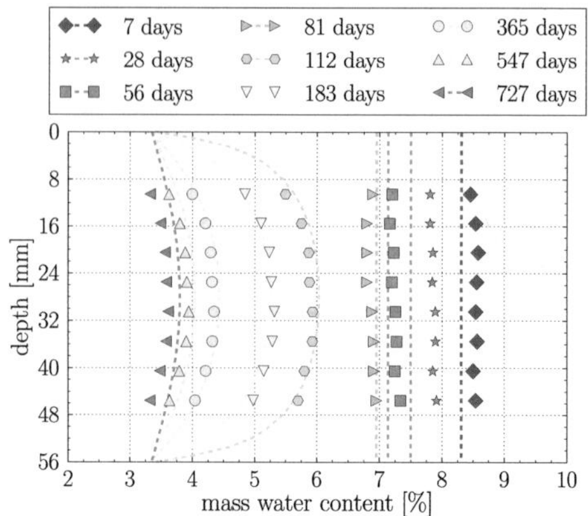

Fig. 1   Measured and calculated mass water content plotted along the specimen for selected time instants during moist curing and drying.

The strengthening of existing concrete structures by concrete overlays was investigated experimentally by adding a concrete overlay with a thickness of 90 mm to a brick-shaped concrete specimen with dimensions of 800 x 300 x 300 mm [9]. The latter was moist cured for 81 days and, subsequently, the lateral surfaces were sealed and the top and bottom surface with dimensions of 800 x 300 mm each were exposed to drying at 65% relative humidity. After more than two years of drying the top surface was roughened and wetted by a high-pressure water blasting unit and, subsequently, a concrete overlay was placed at the top surface of the brick-shaped specimen. For both, the brick-shaped specimen and the overlay a normal strength concrete was employed.

The mass water content distributions in the original brick-shaped specimen was measured for depths ranging from 10 mm to 87 mm below the top surface for several time instants during moist curing and subsequent drying. Measurements of the mass water content were continued during roughening and wetting of the top surface of the brick-shaped specimens and after placement of the overlay. The measured distributions of the mass water content in the brick-shaped specimen immediately before roughening and wetting (corresponding to the specimen age of 823 days) and afterwards are displayed in Figure 2 (a). As can be seen in Figure 2 (a) before wetting the mass water content in the near-surface region is considerably lower than in the interior region, which is the consequence of having exposed the top-surface of the brick-shaped specimen to drying for more than two years. The wetting produces a sharp increase of the mass water content in the uppermost 40 mm of the specimen. Subsequently, the mass water content in the near-surface region is decreasing slowly with progressing time. In addition, the mass water content distributions were measured in the overlay at several time instants during hardening and drying of the overlay, the latter due to exposing its top surface to a relative humidity of 65%. The evolution of the measured longitudinal strain at the top surface of the overlay is displayed in Figure 2 (b) by the continuous line.

The numerical model of the brick-shaped specimen and the overlay consists of 2856 3D finite elements with quadratic interpolation of the displacements and linear interpolation of the temperature and the capillary pressure. The gas pressure is assumed to be equal to the atmospheric air pressure.

In the first step of the analysis drying of the brick-shaped specimen is simulated, followed by the numerical simulation of wetting of the top surface of the brick-shaped specimen by applying a high water pressure to its top surface. The computed distributions of the mass water content before and after wetting are shown by the dashed lines in Fig 2 (a). Compared to the measured values, the computed mass water content is somewhat higher, however, with a similar distribution. It is emphasized that it is difficult to numerically simulate the high-pressure water blasting.

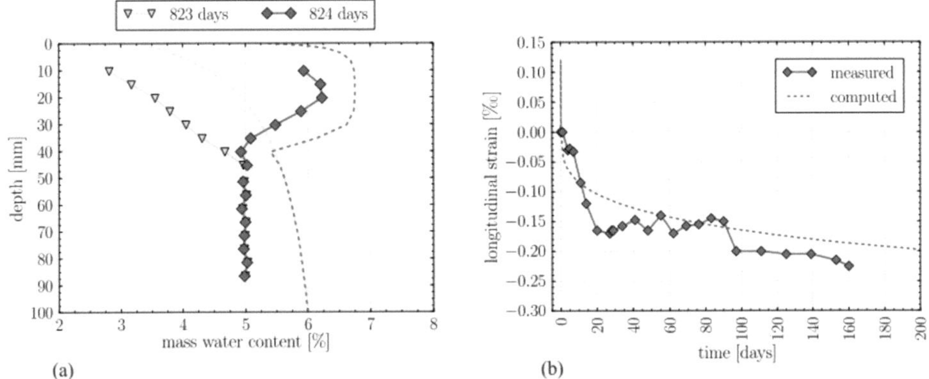

Fig. 2   (a) Measured mass water content (continuous lines) and computed mass water content (dashed lines) in the brick-shaped specimen before (823 days) and after (824 days) roughening and wetting of the interface. (b) Measured evolution (continuous line) and computed evolution (dashed line) of the longitudinal strain at the top surface of the concrete overlay.

In the second step of the analysis the concrete overlay is applied and hardening of the latter is simulated. After 12 hours of moist curing, its top surface is exposed to drying at a relative humidity of 65%. The computed evolution of the longitudinal strain at the top surface of the overlay is depicted by the dashed line in Figure 2 (b). It follows from (12) that in the employed multi-phase concrete model drying shrinkage results from the capillary pressure, which is generated by a decrease of the degree of water saturation during drying. Since the capillary pressure exerts a hydrostatic pressure on the solid skeleton, it also produces creep strains. However, the latter are not taken into account in the present investigation.

Finally, the shrinkage strains, computed according to (12), and the computed total as well as effective vertical stress are plotted for the overlay after 14 days of hardening and drying in Figure 3. It also shows the deformations of the overlay (magnified by a factor of 1000). As can be seen in Figure 3 (c), effective vertical tensile stresses are predicted at the interface between old and new concrete along the outer boundary of the specimen, in particular, at the outer edge. If the effective tensile stresses exceed the tensile strength, debonding of the overlay from the old concrete will be initiated.

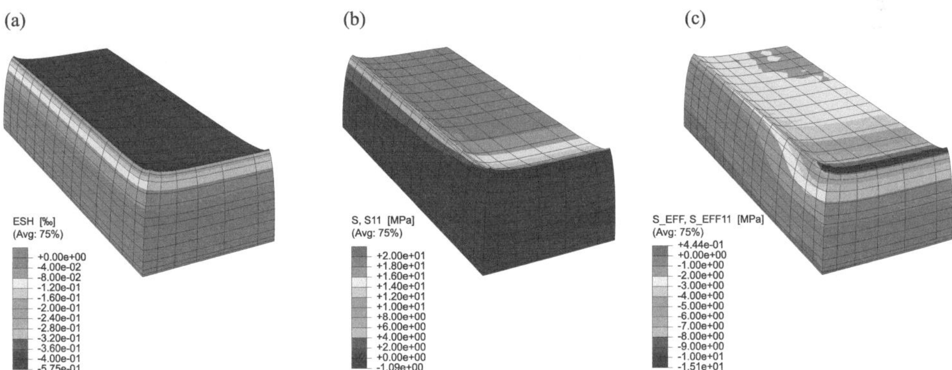

Fig. 3   Computed response of the overlay after 14 days of hardening and drying: (a) drying shrinkage strains [‰], (b) total and (c) effective stress [MPa] in longitudinal direction

4   Summary and conclusions

A multi-phase concrete model, in which drying shrinkage is predicted from the capillary pressure, generated by the decreasing degree of water saturation during drying, was applied to the numerical simulation of laboratory tests, aiming at the investigation of the effects of drying shrinkage of concrete overlays. Since in the employed multi-phase concrete model drying shrinkage results from the capillary pressure, the latter also produces creep strains [14]. Hence, in the next step the impact of creep on the overlay behaviour will be investigated.

References

[1]   Baroghel-Bouny, V. & Mainguy, M. & Lassabatere, T. & Coussy, O. 1999. Characterization and identification of equilibrium and transfer moisture properties for ordinary and high-performance cementitious materials, Cement and Concrete Research, 29: 1255-1238.

[2]   Baroghel-Bouny, V. 2007. Water vapour sorption experiments on hardened cementitiuos materials Part II: Essential tool for assessment of transport properties and for durability predictions, Cement and Concrete Research, 37: 438-454.

[3]   Cervera, M. & Oliver, J. 1999. Thermo-Chemo-Mechanical Model for Concrete. I: Hydration and Aging, Journal of Engineering Mechanics, 125: 1018-1027.

[4]   Feix, J. & Andreatta, A. & Niederegger, C. & Fritsche, G. &. Hofstetter, G. & Niederwanger, G. & Theiner, Y. & Cordes, T. 2010. Composite constructions used for load-bearing structures and roadways of bridges, research report, published in Scientific Series of Research (Straßenforschungsheft Nr. 589), Federal Ministry of Traffic, Innovation and Technology, Vienna, Austria, in German.

[5]   Gawin, D. & Pesavento, F. & Schrefler, B. A. 2006. Hygro-thermo-chemo-mechanical modeling of concrete at early ages and beyond; part I: hydration and hygro-thermal phenomena, part II: shrinkage and creep of concrete, International Journal for Numerical Methods in Engineering, 67: 299-363.

[6]   Hassanizadeh, M. & Gray, WG. 1979. General conservation equations for multi-phase systems: 1. Averaging procedure, Advances in Water Resources, 2: 131-144.

[7]   Millington, R.J. 1959. Gas Diffusion in Porous Media, Science, 130: 100-102.

[8]   Pesavento. F. & Gawin, D. & Schrefler, B. A. 2008. Modeling cementitious materials as multiphase porous media: theoretical framework and applications, Acta Mechanica, 201 :313-339.

[9]   Theiner, Y.: Evaluation of the effects of drying shrinkage on the behaviour of concrete structures strengthened by overlays, Tiroler Wissenschaftsfonds (TWF), 2011, 1-120.

[10]   Van Genuchten M.T.1980. A closed-form equation for predicting the hydraulic conductivity of unsaturated soils, Soil Science Society of America, 44: 892-898.

[11]   Wittmann, FH. 1985. Deformation of Concrete at Variable Moisture Content, in Mechanics of Geomaterials, Ed.: Z. Bazant, John Wiley & Sons, 425-459.

Relationship between early stage microstructure and long term durability of slag-containing cements

Mark Whittaker

School of Civil Engineering, Institute of Resilient Infrastructure
University of Leeds
Woodhouse Lane, LS2 9JT
Supervisor: Leon Black

Abstract

This study examines the relationship between slag reactivity, paste microstructure and long term durability; particularly with regards to sulphate attack. This is achieved by combining standard engineering tests on slag cement blends with comprehensive characterisation. The engineering property measured to-date is the compressive strength. Characterisation has been achieved by simultaneous thermal analysis (STA) and scanning electron microscopy (SEM). The degree of hydration of both the slag and cement components has also been followed by chemical shrinkage and calorimetry, plus image analysis of SEM micrographs.

1   Introduction

Whilst concrete may be considered a durable material, it can be prone to damage by external sulphate attack. External sources of sulphate, e.g. in groundwater where sulphate concentrations may vary from $0.02\text{-}3\text{g.L}^{-1}$, may react with aluminate phases within the hydrated cement paste to form ettringite and gypsum, whilst the associated cations may be a source of further damage. Sources of aluminium typically include unreacted C_3A, monosulphates and other AFm phases (hydroxyl and carbonate) [1]. These reactions are expansive and the re-formed ettringite may not be accommodated in the hardened cement paste, resulting in tensile forces. This, in turn, may lead to expansion then ultimately cracking and spalling of the cement. One potential solution is blending cement with ground granulated blast furnace slag (ggbs).

Being a latent hydraulic material slag can react with water in an alkaline environment such as cement, potentially improving performance whilst reducing the environmental burden. Slags typically replace up to 40% clinker, but specialist applications may require up to 90% replacement. The substitution of cement with slag may lead to a denser matrix opposing ingress of the deleterious species and thus increasing resistance as well as reducing the availability of calcium ions needed for ettringite and gypsum formation. Slag reactivity is affected by its particle size, glass content, composition and the type of activator used [2, 3]. Furthermore, slags are typically richer in aluminium than cement clinker; and since it is the aluminate content in the clinker which is involved in sulphate attack, then an increase in total aluminium content within a blend could potentially increase the risk of sulphate attack. Sulphate resistance shows no real improvement at low levels of replacement, sometimes even having an adverse effect, but high levels of replacement almost always lead to improvement [4,5].

This study probes the relationship between slag reactivity, paste microstructure and long term durability; particularly with regards to sulphate attack. This is achieved by combining standard engineering tests on slag cement blends with comprehensive characterisation. Compressive strength data will be complemented by linear expansion due to sulphate attack and air permeability. Characterisation by simultaneous thermal analysis (STA) and scanning electron microscopy (SEM) has been supported by X-ray diffraction (XRD). The degree of hydration of the slag and cement components has been followed by chemical shrinkage, isothermal calorimetry and image analysis of SEM micrographs.

2   Experimental Method

This project uses combinations of 2 cements, CEM I 42.5R and CEM I 52.5 N referred to as C1 and C2 respectively with three slags (A, B and C) of varying compositions. Slag replacement of cement is 40% or 70%, with the effect of additional sulphate also studied by replacing 3 wt% of the binder in the form of anhydrite. These latter blends are denoted with a $. This results in 28 different mixes as

Proc. of the 9th *fib* International PhD Symposium in Civil Engineering, July 22 to 25, 2012,
Karlsruhe Institute of Technology (KIT), Germany, H. S. Müller, M. Haist, F. Acosta (Eds.),
KIT Scientific Publishing, Karlsruhe, Germany, ISBN 978-3-86644-858-2

shown in Table 1. Full characterisation thus far has been limited to 5 mixes; C1, C140Sb, C140Sc, C170Sc and C140Sc$, with selected characterisation (by chemical shrinkage and isothermal calorimetry) performed on the complete matrix. Slags B and C were chosen for complete analysis based on their aluminium contents, having the lowest and highest aluminium contents respectively.

Characterisation has been performed on pastes with w/b=0.5. Engineering properties have and will be measured on mortars using a mix design ratio of 1:3:0.5 of binder:sand:water.

Table 1 Matrix Design, with samples chosen for characterisation marked in bold

	S_a		S_b		S_c	
	40%	**70%**	**40%**	**70%**	**40%**	**70%**
C_1	C_140S_a	C_170S_a	$\mathbf{C_140S_b}$	C_170S_b	$\mathbf{C_140S_c}$	$\mathbf{C_170S_c}$
C_2	C_240S_a	C_270S_a	C_240S_b	C_270S_b	C_240S_c	C_270S_c
$C_1\$$	$C_140S_a\$$	$C_170S_a\$$	$C_140S_b\$$	$C_170S_b\$$	$\mathbf{C_140S_c\$}$	$C_170S_c\$$
$C_2\$$	$C_240S_a\$$	$C_2\text{-}70S_a\$$	$C_240S_b\$$	$C_270S_b\$$	$C_240S_c\$$	$C_270S_c\$$

2.1 Engineering Properties

Expansion due to external sulphate attack has been measured according to ASTM C452-06 [6] and C1012/1012M-09 [7] with some deviation. The sample size was reduced to 200mm in length, with no changes in mix design as suggested by the standards to satisfy workability conditions when working with blended cements. All samples were exposed to sulphate solutions 14 days of curing, regardless of their strengths. Furthermore the solution concentration was reduced to $3g.L^{-1}$ to better reflect field conditions, and renewed every fortnight with a risk of calcium leaching. The net expansion from sulphate attack will be determined by curing parallel samples in a saturated CH solution. Samples will be polished prior immersion to remove any surface carbonation occurring during curing or storage.

Strength was assessed on mortars as described in BS EN 196-1:2005 [9], with samples cured in a water bath until testing.

2.2 Characterisation

Paste samples 20mm in diameter and 50mm long were prepared for characterisation. These were placed in a sulphate solution ($3g.L^{-1}$) after 14 days of hydration, with reference samples placed in a saturated CH solution. Care was taken to analyse the outer layer ($<500\mu m$) of the samples placed in a sulphate solution. Samples prepared for TG and SEM were freeze dried prior analysis.

The CH content of the blends was measured using a STA 780 Series simultaneous thermal analyser. 15-18mg of sample was placed in a crucible and heated to 1000°C at a rate of 20°C/min under a nitrogen atmosphere

Chemical shrinkage was measured by dilatometry [10, 11] following initial work by Geiker [12]. 50g of paste was hand-mixed for 2 minutes, with 15g taken and placed in a plastic beaker. A thin layer of water (5g) was carefully poured onto the paste, taking care to minimise surface disturbance. The beaker was then filled to the rim with paraffin oil and sealed with a bung through which a 1mL pipette was attached. Chemical shrinkage was measured by following the drop in height of the oil. Images were captured with a 10MP camera every 5 minutes for a period of 28 days.

Isothermal calorimetry was conducted using a TAMair twin channel calorimeter. Samples were prepared by mixing 50g of paste (w/b=0.5) for 2 minutes of which 6g were placed in a plastic ampoule. Reference samples prepared with quartz in place of slag were prepared to account for particle effects. Measurements were taken every 5 minutes for 28 days.

Image analysis allowed quantitative determination of the degree of hydration of cement and slag, using grey-level histograms to identify unreacted material, hydration products and pores [13]. Backscattered electron mages were captured using a JEOL JSM-5800LV at 800x magnification with a 15KeV accelerating voltage. 50 images were taken for analysis with chemical mapping in respect to magnesium performed to isolate the slag particles.

3 Results and Discussion

Figure 1 shows the heat evolution rate (left) and total heat evolved (right) of the 5 mixes mentioned in §2. The addition of slag to cement affects cement hydration, with both quantity and type of slag having an effect. Slag addition leads to the appearance of a new exotherm located just after the main alite hydration peak, with the location dependent on the slag composition. The addition of sulphate, a known slag activator, delays the appearance of the new peak, whilst increasing the slag content accelerates its appearance, such that it almost coincides with the alite hydration peak [14].

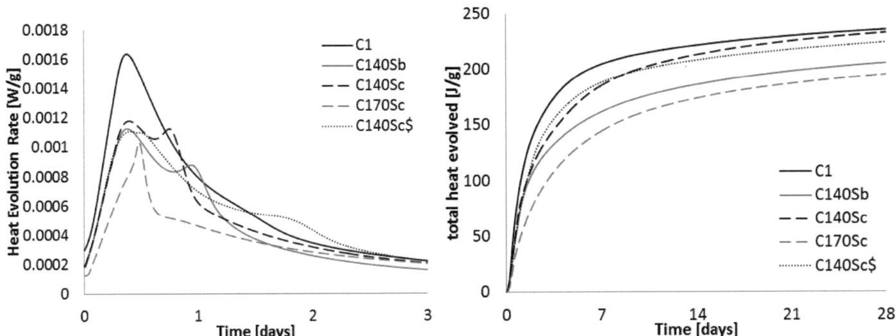

Fig. 1   Heat Evolution Rate [left] And Total Heat Evolved [right]

Blended mixes evolved less heat than pure cement pastes, as shown in Figure 1 on the right. After 28 days, slag C had evolved more heat than slag B, and higher slag replacement levels reduced the heat further. The addition of sulphate (C140Sc$) increased the total heat evolved during the early stages of hydration but not at later ages. The contribution of the slag towards heat evolution could be determined by comparisons with the quartz blend; the quartz being non-reactive yet providing nucleation sites and acting as a diluent. This contribution is determined by first normalising the heat evolution to their respective cement contents. The difference between the quartz and slag blends is due to slag hydration, which can then be normalised to the slag content (Fig 2, left). The slag reactivity is demonstrated in Fig 2 (right) showing a higher reactivity of Slag C over slag B at 40% replacement. The apparent increase in slag reactivity with increased slag content could be explained by a greater availability of water for slag hydration. Early stage hydration appears to be endothermic, however, perhaps due to precipitation of the slag.

As with isothermal calorimetry, chemical shrinkage allows measurement of the degree of hydration of cements and blends. Chemical shrinkage occurs because water occupies a lesser volume when bound to a solid than in a free state. As hydration continues, more water is bound, resulting in paste shrinkage. Slag addition reduces shrinkage, with the same trends seen as when measuring heat evolution during hydration. Mix C140Sc however exhibited an unusually high shrinkage. Plotting the total heat evolved against the total shrinkage (Fig. 3, right) gave a linear relationship, with $R^2>96\%$ for all mixes except C140Sc, suggesting error when collecting data. Future work will aim to determine the shrinkage of the slag component alone.

These two methods allow quantitative measurement of the total degree of hydration. The results are complemented by SEM backscattered electron image analysis and XRD with Rietveld refinement (not shown), to enable quantification of the degree of hydration of the cement and slag alike. To date results are limited, but preliminary results have enabled the degree of hydration of pure cement pastes to be determined and correlated with equivalent calculations by isothermal calorimetry and dilatometry.

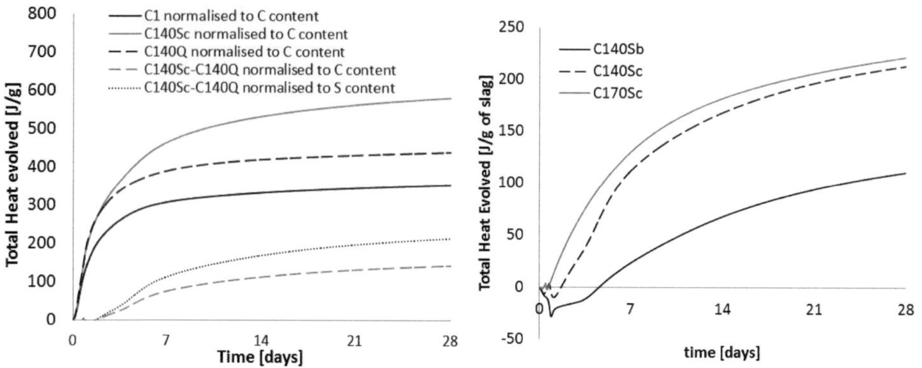

Fig. 2 Heat Evolution Plots Used To Determine The Heat Evolved By The Slag Component [left] and Total Heat Evolved Due to Slag Hydration Normalised To The Slag Content

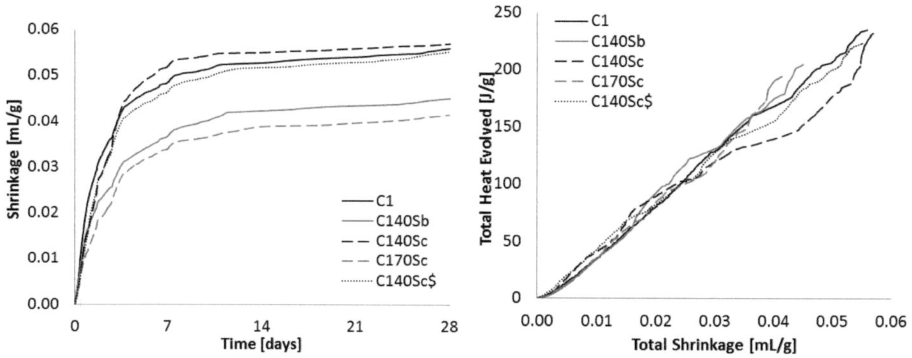

Fig. 3 Total Chemical Shrinkage [left] And Its Linear Relationship with the Total Heat Evolved During Hydration

Figure 4 shows a SEM-BSE image of a pure cement paste hydrated for 14 days with its associated grey level histogram. The histogram can be split into 4 different sections including pores, C-S-H and other hydrates, CH and anhydrous material. It is then possible to calculate the degree of hydration by measuring the change in area of the anhydrous fraction over time. The degree of hydration is shown in Figure 5.

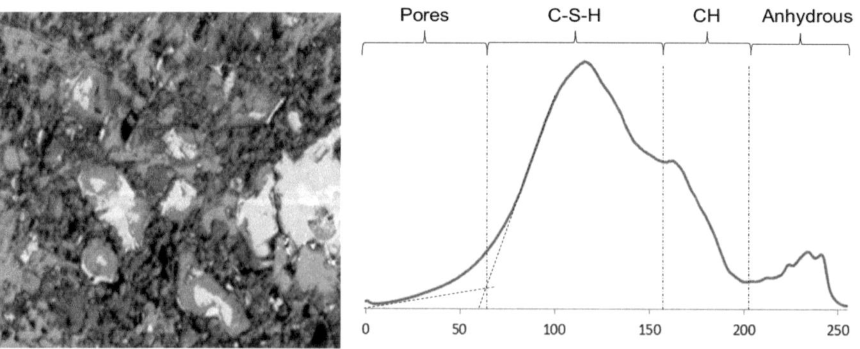

Fig. 4 SEM-BSE of a Pure Cement Paste Hydrated For 14 days (left) And Associated Greylevel Histogram (right)

The area fraction of the anhydrous material, V_f, before hydration can be easy calculated and has been estimated at 37.56% for the pure cement paste. It is also possible to follow the formation of the other phases as shown in Figure 5 (right) where a reduction in both the anhydrous and pore contents is observed, with a continual increase in CH, C-S-H and various hydrates. Although not shown here, the degree of hydration of the slag can also be estimated by first isolating the slag particles by chemical mapping and again by greyscale segregation; with slag hydration revealing a characteristic darkened rim upon hydration.

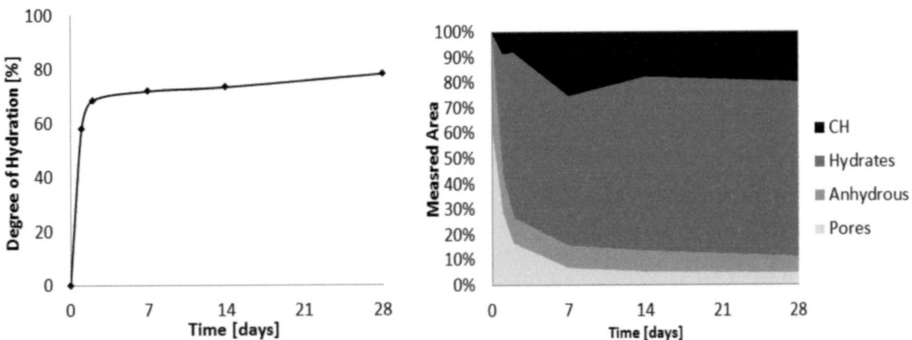

Fig. 5   Measured Degree of Hydration Of A Pure Paste (left) And The Total Area Of the Other Hydrates (right)

Figure 6 (left) shows the compressive strength evolution over time. Replacement of cement with slag led to a reduction in strength up to 28 days, with the effect being greater at early ages, particularly at high replacement levels, e.g. 80% reduction in strength from C1 to C170Sc. After 28 days of hydration the strengths were comparable for all the blends containing 40% slag, all marginally lower than the pure cement system; with now only 18% difference between C1 and C170Sc. The addition of sulphate also appears to aid early stage hydration of the blended system C140Sc$.

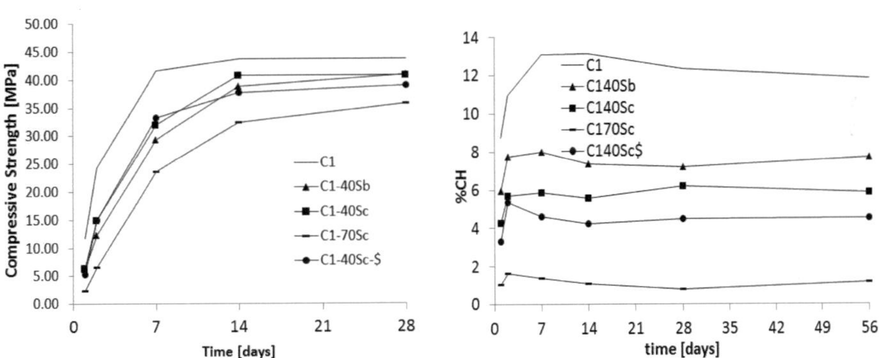

Fig. 6   Strength [Left] and CH [right] Evolution During Hydration

Figure 6 (right) shows the CH evolution during hydration in a CH solution. The pure cement paste contained the most portlandite. Increased replacement of cement with slag reduced the CH content. Slag hydration consumes portlandite [15] and it has been suggested that there is a maxima in the CH content during the early stages of hydration. No such trends have been found here, agreeing with previous observations by Luke and Glasser [16]. For the same level of replacement however, the blend using Slag C consumed more CH than slag B. This may be attributed to its higher reactivity. The addition of sulphate either further increased slag reactivity or reacted with CH to form gypsum marked with an increase.

4 Conclusions

This paper has primarily focused on qualitative determination of the degree of hydration as shown by calorimetric and shrinkage data, together with SEM-BSE image analysis. Slag addition greatly affects the normal hydration of cement. Firstly, by reducing the overall heat evolution and chemical shrinkage, and secondly resulting in a second prominent exotherm during isothermal calorimetry, as seen in figure 1 (left).

Further work will include permeability and expansion data due to sulphate attack as well as XRD with Rietveld refinement combined with SEM-IA to complement calorimetric and shrinkage data.

References

[1] Irassar, E.F., Sulfate Attack on Cementitious Materials Containing Limestone Filler – a Review, Cement and Concrete Research, 39, pp 241-254, 2009

[2] Esacalante-Garcia, J.I., Gomez, L.Y., Johal, K.K., Mendoza, G., Mancha, H., Mendez, J., Reactivity of Blast-Furnace Slag in Portland Cement Blends Hydrated Under Different Conditions, Cement and Concrete Research, 31, pp. 1403-1409, 2001

[3] Escalante-Garcia, J.I., Fuentes, A.F., Gorokhovsky, A., Fraire-Luna, P.E., Mendoza-Suarez, G., Hydration Products and Reactivity of Blast-Furnace Slag Activated by Various Alkalis, Journal of the American Ceramic Society, 86, pp. 2148-2153, 2003

[4] Fernandez-Altable, V., Leinekugel-Le-Cocq-Errien, A.Y., Le Saout, G., Scrivener, K., Effects on Slag on Aluminate Distribution and Sulphate Resistance, Proceedings of the 17th International Conference on Building Materials, ed. J. Stark, vol. 1, pp 691-696

[5] Gollop, R.S., Taylor, H.F.W., Microstructural and Microanalytical Studies of Sulfate Attack, V, Comparison of Different Slag Blends, Cement and Concrete Research, 26(7), pp. 1029-1044, 1996.

[6] ASTM C452-06, Standard Test Method For Potential Expansion of Portland-Cement Mortars Exposed to Sulfate, 2006

[7] ASTM C1012/C1012M-09, Standard Method for Length Change of Hydraulic-Cement Mortars Exposed to a Sulphate Solution, 2009

[8] Cabrera, J.G., Lynsdale, C.J., A New Gas Permeameter for Measuring the Permeability of Mortar and Concrete, Magazine of Concrete Research, 40, pp. 177-182. 1988

[9] British Standard Institution, BS EN 196-1:2005 Methods of Testing Cement – Strength Test: BSI 2005

[10] Costoya Fernandez, M.M., Effect of Particle Size on the Hydration Kinetics and Microstructural Development of Tricalcium Silicate, 2008, PhD Thesis

[11] Kocaba, V., Development and Evaluation of Methods to Follow Microstructural Development of Cementitious Systems Including Slags, PhD Thesis, 2009

[12] Geiker, M., Studies of Portland Cement Hydration By Measurement of Chemical Shrinkage, PhD Thesis, 1983

[13] Scrivener, K.L., Backscattered Electron Imaging of Cementitious Microstructures: Understanding and Quantification, Cement and Concrete Composites, 26, pp 935-945, 2004

[14] Escalante-Garcia, J.I., Sharp, J.H., The Effects of Temperature on the Early Hydration of Portland Cement and Blended Cements, Advances in Cement Research, 12, pp 121-130, 2000

[15] Richardson, I.G., Groves, G.W., Microstructure and Microanalysis of Hardened Cement Pastes Involving Gound Granulated Blast-Furnace Slag, Journal of Material Science, 27, pp. 6204-6212, 1992

[16] Luke, K., Glasser, F.P., Internal Chemical Evolution of the Constitution of Blended Cements, Cement and Concrete Research, 18, pp. 495-502, 1988

Development and Testing of CO_2-resistant Well Cements

Astrid Hirsch

Institute of Concrete Structures and Building Materials,
Karlsruhe Institute of Technology (KIT),
Gotthard-Franz-Str. 3, 76131 Karlsruhe, Germany
Supervisor: Harald S. Müller

Abstract

The following paper gives a brief outline on current research activities regarding the testing and development of CO_2-resistant well cements. Within the presented work, several approaches were studied in order to improve the durability of well cements exposed to supercritical CO_2 and to carbonic acid formed from CO_2 and brines from formation waters, respectively. Among these was the addition of puzzolanic and inert fillers to the cement aiming for reduced cement permeability. Another approach consisted in adding salt to the mixing water and thus reducing the solubility of CO_2 in the pore fluid. In the third approach barium silicate cements were developed and tested for a possibly increased corrosion resistance. All cements were optimized for their rheological performance and hardened cement samples were examined thoroughly. To evaluate their corrosion resistance, samples were exposed to carbonic acid in autoclave experiments. Long-term autoclave tests are still ongoing.

1 Introduction

To protect global climate, the carbon dioxide emission from power plants is to be reduced worldwide in the future. To justify a further use of fossil fuels in power plants, the Carbon Capture and Storage (CCS) technology is considered as an alternative. This means that CO_2 has to be separated from flue gas and subsequently be stored in deep geological reservoirs. One key requirement for storing CO_2 securely underground is to guarantee long term tightness of the boreholes in the reservoir.

In the oil and gas industry well cements have been employed as borehole seals for decades. However, it has to be expected that conventional well cements are not chemically stable when in contact with CO_2 and carbonic acid, respectively, which may form from formation waters and CO_2. Carbonic acid and hardened cement react with each other dissolving portlandite initially and forming calciumcarbonate ($CaCO_3$) subsequently. A continuous carbonic acid attack leads to anew dissolution of the freshly formed calciumcarbonate, possibly causing a complete deterioration of the hardened cement and thereby of the borehole seal. If this is the case long-term reservoir integrity is not guaranteed.

Based thereon, rheologically optimised CO_2-resistant cement types should be developed to enable a durable sealing of the CO_2-reservoirs. The following approaches were looked into:

- **reference well cement**

- **approach 1: salt cements**

  reduced solubility of CO_2 in pore fluid of hardened cement due to salt addition

- **approach 2: modified well cements**

  reduced permeability of hardened cement due to addition of puzzolanic and inert fillers, respectively

- **approach 3: barium cements**

  substitution of Barium for Calcium in cement clinker in anticipation of sparingly soluble phases in hardened cement and therefore slowing down of corrosion

The main subjective of the presented research activities was to investigate these newly developed cements thoroughly and to compare them to conventional well cements. Thereby, autoclave experiments were of special importance, since they could simulate the cement corrosion under temperature and CO_2-pressure conditions, as may be found in geological CO_2-reservoirs.

Proc. of the 9th *fib* International PhD Symposium in Civil Engineering, July 22 to 25, 2012,
Karlsruhe Institute of Technology (KIT), Germany, H. S. Müller, M. Haist, F. Acosta (Eds.),
KIT Scientific Publishing, Karlsruhe, Germany, ISBN 978-3-86644-858-2

2 Cement development

2.1 Salt cements and modified well cements

A large number of experiments was carried out to choose the most suitable raw materials for the cement development. Miscellaneous cements, quartz powders, fly ashes, salts and additives were tested.

Table 1   Composition and porosity of the developed cement types

cement type	composition[*]	$w/c_{equ.}$ [-]	porosity[***] [vol.-%]
Class G	Class G well cement	0.40	30 (16)
SaltCem-24	Class G well cement, NaCl (24 % by weight of water), $CaCl_2$, superplasticiser, latex suspension	0.40	30 (22)
SaltCem-36	Class G well cement, NaCl (36 % by weight of water), $CaCl_2$, superplasticiser, latex suspension	0.40	34 (28)
MWC-FA	Class G well cement, 3 types of quartz powder, fly ash, superplasticiser, latex suspension	0.44	18 (12)
MWC-1	Class G well cement, 3 types of quartz powder, superplasticiser, latex suspension	0.46	19 (10)
MWC-2	Class G well cement, 3 types of quartz powder, superplasticiser, latex suspension	0.46	14 (10)
BaCem 1 and 2	barium silicate clinkers	0.40 and 0.50	not yet tested

[*] water contained in all cement types

[**] dertermined 28 days after production

[***] first value for non-corroded cement (second value for cement carbonised for 4 weeks at 40 °C and 90 bar)

In Table 1 all developed cement types are listed. For the purpos of comparison a reference well cement only consisting of conventional well cement (i.e. Portland cement with defined composition) and water was produced (Class G).

Two salt cements (see Table 1, SaltCem-24, SaltCem-36) were developed. For their production NaCl brine was used instead of water. The solubility of CO_2 in brine is lower than in water, therefore a slowing down of the carbonation of the hardened cement is expected.

Three modified well cements were developed within the project (see Table 1, MWC-FA, MWC-1, MWC-2). These cement types are characterised by a substitution of cement by quartz powder and fly ash, respectively. The addition of quartz powders and fly ash is intended to lead to an optimized grain-size distribution of the modified well cements and thus increase durability of the hardened cements in raising turtuosity and reducing porosity and permeability. In addition the added quartz powders are inert and therefore sparingly soluble. An increased durability of the modified well cements is expected.

During the development of all cement types listed above, the main objective was to design materials as durable and fluid-tight as possible. Fluid-tightness of a borehole seal though requires an optimal placing of the cement slurry in the borehole with a minimum of flaws. In order to achieve this, the rheological properties of the developed cements were optimised in numerous tests. In comparison with the reference well cement, all salt cements and modified well cements show strongly improved workability. This part of the research however is not dealt with in detail in this paper.

Following the rheological optimisation, hardened cement samples were produced of all the cement types. These samples were used in corrosion experiments and for further mechanical, physical and chemical analysis (see chapter 3).

2.2 Barium cements

As special barium silicate based cements are not available worldwide, barium silicate clinkers were synthesised and subsequently investigated. It is known from literature [1] that hydraulic barium silicates exist, corresponding to the hydraulic phases $2CaO \cdot SiO_2$ and $3CaO \cdot SiO_2$ (C_2S and C_3S) in portland cement. Thus, the synthesization of the pure phases $2BaO \cdot SiO_2$ and $3BaO \cdot SiO_2$ (B_2S and B_3S) was investigated.

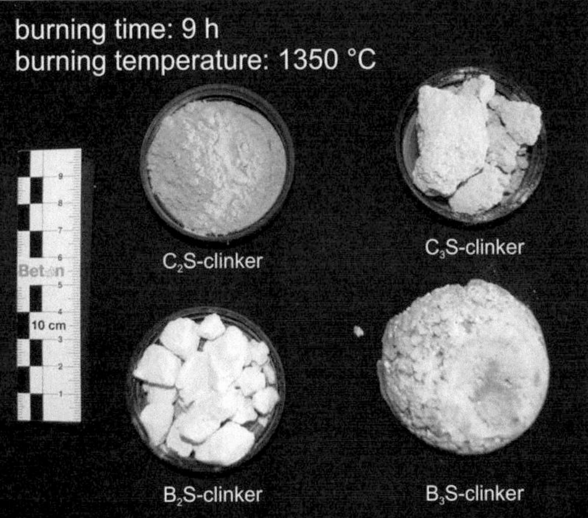

Fig. 1   Kiln and crucible fort the burning of cement clinker (left); several burnt barium silicate clinkers and calcium silicate clinkers (right)

Although not pure B_2S and B_3S, several barium silicate clinkers could be produced (BaCem in Table 1, below referred to as barium cements) from barium carbonate and amorphous silica. The raw materials were burnt at 1350 °C for 9 hours and subsequently ground. The so generated powders showed hydraulic properties and reacted with water to form white solids with defined strength. Characteristic for the barium cements was an extremely rapid setting, which caused very poor workability and compactability of the material. Several samples of hardened barium cement were examined in corrosion experiments.

3   Corrosion experiments

For the realistic simulation of the CO_2-attack in boreholes, a high-pressure-high-temperature autoclave system was constructed and built. To determine the time-dependent cement corrosion, samples of all developed cement types were exposed to carbonic acid under high pressures and temperatures in the autoclaves. Following the corrosion experiments, carbonated cement samples and non-corroded control samples were thoroughly examined and compared with each other (see chapter 3.2 and 3.3).

3.1   Autoclave System and experimental procedure

Figure 2 shows the autoclave system. It consists of a pressure generator with a control panel (left) and of a framework with two autoclaves (right). Two identical autoclaves, each 0.99 dm³ of volume, are operated independently. They can be heated up to 150 °C. In the start of each experiment hardened cement samples and brine or water are filled into the autoclaves before they are closed. The target pressure, up to 300 bar, is subsequently built up, using fluid CO_2. During experiments continuous temperature and pressure control as well as CO_2 supply are guaranteed. Furthermore temperature, pressure and CO_2 consumption are documented digitally. The key advantage of the autoclave system lies in the possibility to investigate the chemical processes in the autoclave during experiments by fluid sample extraction (see Chapter 3.2) and replacement of attack medium during tests at nearly constant pressures and temperatures.

Numerous experiments were carried out. In most, cement samples were covered with water or brine to guarantee carbonic acid attack throughout the experiment. At specified intervals, a defined volume of less than 2-3 cm³ of attack medium were extracted for further analysis. For test durations exceeding seven days, half of the attack medium was replaced by new water or brine. This was done to prevent slowing down of the corrosion process due to saturation of the attack medium with dissolved solids.

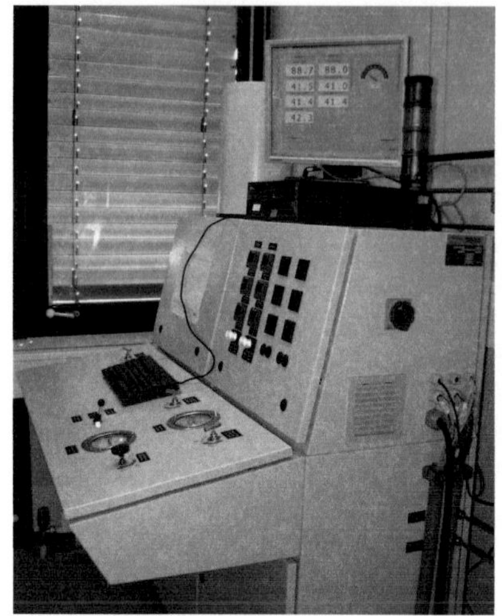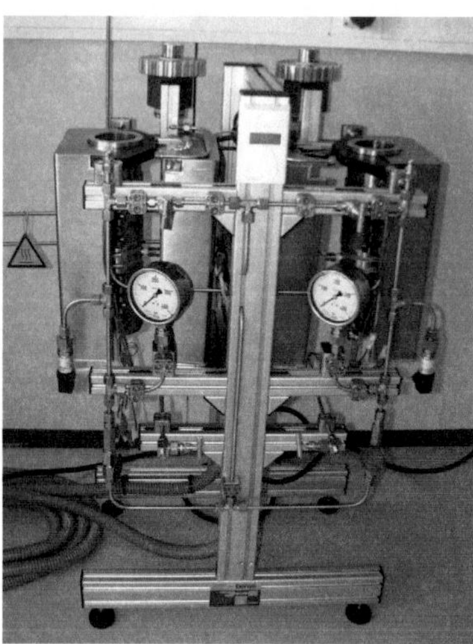

Fig. 2   Autoclave system; pressure generator with a control panel (left); framework with two autoclaves (right)

3.2   Analytical methods

The identification of the minerals found in non-corroded control samples and in carbonated samples was performed by means of X-ray powder diffraction.

Changes of pore space distribution due to carbonation are of vital importance for the durability development of the hardened cements. An exact description of the pore space was gained by the use of mercury porosimetry. Again non-corroded control samples and carbonated samples were examined.

To measure the carbonation depth, samples were split lengthways after the corrosion experiments. The carbonation front was always clearly visible, even more so when the sample surface was wetted with phenolphthalein. So the carbonation depth could easily be measured on high resolution digital scans.

The interaction between the attack medium and hardened cement samples was evaluated by the examination of fluid samples extracted during corrosion experiments. This is of importance because the leaching out of material indirectly indicates the durability of the cement. Fluid samples were analysed primarily via atomic absorption spectrometry. The contents of calcium, silicon, sodium, magnesium, potassium, chloride and sulphate were determined. For each date, the complete quantity of an element (e.g. calcium) leached out until this point was calculated.

3.3   Results of corrosion experiments

So far only experiments with a duration of up to 28 days have been carried out. Long term tests are still ongoing. For most of the experiments the test conditions were set to 40 °C and 90 bar, guaranteeing supercritical conditions. These pressure and temperature conditions were chosen to resemble the ones found in real CO_2-Reservoirs (e.g. Ketzin, Germany).

3.3.1   Mineralogy

X-ray powder diffraction showed considerable amounts of portlandite in non-corroded samples of the reference well cement and the salt cements, whereas all tree modified well cements only contained little portlandite. This can be explained by the fact that in these cements a part of the originally formed portlandite is consumed by the puzzolanic reaction with fly ash and SiO_2 from fine quartz powders.

All non-corroded cement samples contained remains of non-hydrated cement. Modified well cements showed very high SiO_2-peaks, whereas small amounts of the chloride containing Friedel's salt were identified in salt cements.

In all cement samples that were carbonated for 28 days, considerable amounts of calcium carbonates were detected. Furthermore, the formation of calcium carbonate was always coupled with a decrease of the portlandite content. This confirms the assumption, that most of the calcium in the newly precipitated calcium carbonates is derived from the dissolution of portlandite. Thus, in the originally portlandite rich cement types (reference well cement and the salt cements) much more calcium carbonate was detected than in the modified well cements, formerly poor in portlandite.

3.3.2 Porosity

In table 1 the porosities of non-corroded cement samples and of samples carbonated for 28 days are given. Porosities were measured using mercury porosimetry.

Porosities of non-corroded samples of reference well cement and salt cements were comparatively high (30 to 35 vol.-%). Due to their optimized grain-size distribution non-corroded modified well cements have reduced porosities of only about 10 to 15 vol.-%.

28 days long carbonation experiments lead to a porosity reduction of about 5 to 10 vol.-% for all examined cements. The reason therefore is primarily the precipitation of calcium carbonate in the pore space of the hardened cement. As discussed in chapter 3.3.1 this effect is particularly pronounced for the reference well cement and the salt cements.

3.3.2 Carbonation depth

Figure 3 shows two carbonated samples of hardened cement. The carbonation front is clearly visible and marked by an arrow. Figure 3 (left) shows a barium cement sample that was carbonated for 7 days at 40 °C and 90 bar. Caused by the wetting of the sample with phenolphthalein, the non-corroded inner part of the sample is displayed darker than the carbonated outer part. With a carbonation depth of 4 mm, the barium cements corrodes much faster than all other cement types, for which the maximum carbonation depth under the same conditions is 1.8 mm (see figure 4, left). The reason therefore is seen in the extremely poor workability and compactability of the barium cements (see chapter 2.2), which obviously leads to a highly permeable pore system.

Fig. 3

BaCem, w/c = 0,4, carbonised for 7days at 40°C and 90 bar (left); MWC-FA, w/c = 0,44, carbonised for 7days at 40°C and 90 bar (right)

In contrast, figure 3 (right) shows the much smaller carbonation depth of only 2.5 mm after 28 days at 40 °C and 90 bar for the modified well cement MWC-FA.

Figure 4 (left) shows the development of the carbonation depth fort he first 28 days at 40 °C and 90 bar for all developed cement types with the exception of the barium cements. Already after 28 days, it can be seen that all curves level off. For comparison a $\sqrt{t}$–curve is added to figure 4 (left), as carbonation of concrete is often described by $\sqrt{t}$–curves in literature [2]. The curves for the reference well cement and the salt cements are more or less parallel, whereas modified well cements exhibit slightly higher carbonation depths. Again, the reason is probably the lower portlandite contents of modified well cements and therefore less formation of in the pore system. In comparison, for the reference well cement and the salt cements the precipitation of calcium carbonates seems to cause clogging of pores and thereby slowing down the carbonation.

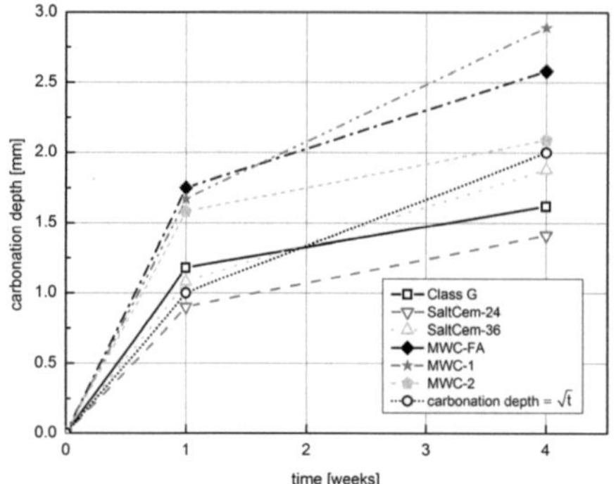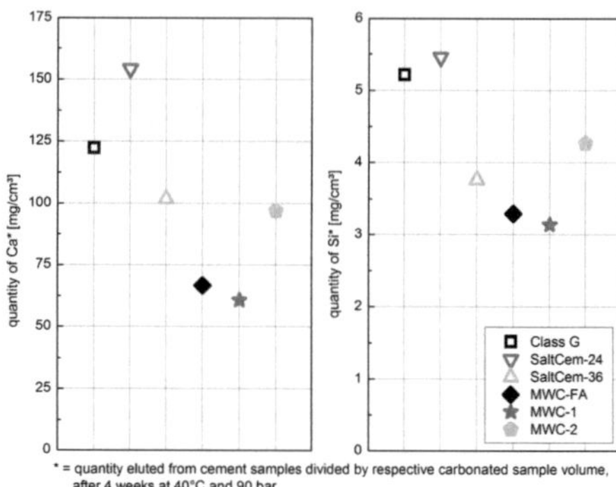

Fig. 4 development of carbonation depth for the first 28 days at 40 °C and 90 bar (left); dissolved amount of calcium and silica in relation to carbonised sample volume, after 28 days at 40 °C and 90 bar (right)

3.3.4 Leaching tests

The low durability of the barium cements is confirmed in the leaching tests. After 7 days up to three times as much silica is dissolved from barium cement as from all other cement types.

Figure 4 (right) shows the amount of dissolved calcium and silica (after 28 days at 40 °C and 90 bar) in relation to carbonised sample volume, that was calculated from the measured carbonation depths. It can be seen, that modified well cements are leached less intensive than the reference well cement and the salt cement SaltCem-24, even if the former exhibit higher carbonation depths. This effect seems to be a result of the smaller porosity and permeability of the modified well cements.

3 Conclusions and summary

In the presented project seven new cements of three types have been developed: salt cements, modified well cements and barium silicate cements. All cements were thoroughly investigated as slurries and in the hardened state. For the simulation of the CO_2-attack in a borehole, a high-pressure-high-temperature autoclave system was constructed, so that cements samples could be tested under realistic conditions. To the knowledge of the authors such equipment has not been available in market so far.

In comparison with the reference well cement all developed cements, with the exception of the barium silicate cements, show strongly improved workability and same or better durability against borehole conditions.

For the carbonic acid attack on reference well cement and salt cements, a fast clogging of pores by calcium carbonate was observed that led to a strong reduction of porosity and slowing down of the carbonation process. However, it has not yet been determined, whether long term carbonic acid attack may lead to dissolution of the newly formed calcium carbonate.

Due to their reduced portlandite contents modified well cements show less clogging of pores and deeper carbonation than the reference well cement and the salt cements. However, as a consequence of their very low porosity and permeability as well as their low leaching tendency, it is expected that under long term carbonic acid attack, modified well cements will be at an advantage.

At present, long-time tests with durations longer than one year are being carried out aiming at a definite estimation of the developed cements' long-term durability.

In addition, important basic information on the production of barium silicate cements and their physico-chemical properties was gained within the project. Further experiments with barium silicate cements are planned.

References

[1] Braniski, A.: Die Eigenschaften der silicatischen Bariumzemente in Abhängigkeit vom Aufbau ihrer Klinkerminerale. In: Zement-Kalk-Gips 21 (1968), N0. 2, pp. 91-98

[2] Stark, J., Wicht, B.: Dauerhaftigkeit von Beton – Der Baustoff als Werkstoff. Birkhäuser, Basel, 2001

Prediction model for the carbonation of cement reduced green concretes

Stefan Hainer

Institut für Massivbau,
Technische Universität Darmstadt (TUD),
Petersenstraße 12, 64287 Darmstadt, Germany
Supervisor: Carl-Alexander Graubner

Abstract

Green concretes with a low amount of Portland cement clinker were developed to reduce the global warming potential of concrete structures. The relevant criterion which often prevents the application of the cement reduced concretes for elements in exterior environmental climate conditions is normally the adequate carbonation resistance of the concrete. A new prediction model for the carbonation resistance based on the raw materials should support the further development and application of cement reduced concretes.

1   Introduction

The major environmental impact of concrete is caused by the CO_2-emission during the cement production. Recent studies revealed that the use of superplasticizer, high reactive cements and the reduction of the water volume enables a significant reduction of the Portland cement clinker in the mixture. Furthermore a possibility is the substitution of the cement by secondary raw materials like fly ash, furnace slag or limestone powder. Unfortunately this option is limited by the availability of these resources. For the practical application of green concretes with very low cement content, questions arise regarding the durability including the carbonation progress of the concrete to avoid corrosion of the steel reinforcement. Using the environmental performance evaluation the ecological advantages were identified.

During the hydration of the cement phases, solid calcium hydroxide $Ca(OH)_2$ is formed. Additionally a highly alkaline solution (KOH/NaOH) is enriched in the pore water [1]. The alkalinity ensures the passivation of the steel reinforcement to avoid corrosion. In the course of time, the calcium hydroxide $Ca(OH)_2$ and eventually the alkali hydroxides KOH/NaOH are attacked by the carbon dioxide from the air. The reaction of $Ca(OH)_2$ and CO_2 to calcium carbonate $CaCO_3$ and water (the carbonation process) is connected with a reduction in the pH of the pore water solution. The pH in the pore solution should not decrease below approximately 9-10 [1]. Otherwise the passivation layer of the steel reinforcement will diminish and the risk of corrosion in wet conditions and the spalling of the concrete will increase.

In cement reduced concretes the potential amount of $Ca(OH)_2$ decreases in proportion to the cement clinker reduction. In addition, the use of pozzolanic reacting additives reduces the calcium hydroxide accumulation [2]. This raises questions about whether cement reduced concretes provide sufficient resistance against carbonation induced corrosion.

2   Experimental program

At the Technische Universität Darmstadt, intensive studies on the topic of "Green Concrete" have been carried out as part of research projects in the field of Sustainable Concrete Structures. Different kinds of concrete with reduced cement content were developed for the ready-mix and precast industry especially for building structures.

Cement reduced concretes are intended to be used for interior structures (exposure class XC1) as well as for exterior structures (exposure classes XC4, XF1 and XA1). The German standard DIN 1045-2:2008-08 defines the requirements for the concrete mix design depending on the exposure class. For the application in exterior structures the minimum cement content is 270 kg/m³, whereas the minimum strength class is C25/30 and the maximal water-cement ratio is 0.60. For interior structures the minimum cement content is 240 kg/m³, the strength class C16/20 and the maximal water-

Proc. of the 9th *fib* International PhD Symposium in Civil Engineering, July 22 to 25, 2012,
Karlsruhe Institute of Technology (KIT), Germany, H. S. Müller, M. Haist, F. Acosta (Eds.),
KIT Scientific Publishing, Karlsruhe, Germany, ISBN 978-3-86644-858-2

cement ratio 0.75. For the evaluation of the concrete performance, reference concretes based on the concrete mix design according to DIN 1045-2:2008-08 were included in the test program. Referring to this Table 1 illustrates the mix design of reference concretes with a cement content of 240 kg/m³ (exposure class XC1) and 270 kg/m³ (exposure class XC4, XF1 and XA1).

Starting with the reference concrete, the cement content was reduced from 270 kg/m³ to 100 kg/m³ (see Table 2). The cement was substituted incrementally by additives. At the same time the water volume was reduced. The lowest value was 125 kg/m³. To maintain a sufficient workability, the paste volume was fixed at a constant value of 300 l/m³ by the addition of fly ash und limestone powder. The concrete consistency was adjusted by changing the dosage of superplasticizer to reach a table flow of about a = 550 mm. A cement CEM I 52.5 R was generally used. Additionally, a cement CEM I 32.5 R was included. Subsequently, the influence of a high performance CEM III/A 52.5 R with reduced clinker content and a high amount of blast furnace slag was analysed. The limestone powder and fly ash additives were used in the ratios 0, 50, and 100% by volume.

Table 1   Mix design of the reference concretes

Mix design Mass per m³ concrete		B270 CEM I 52.5 R SFA10 w165 DIN	B270 CEM I 42.5 R SFA10 w165 DIN	B270 CEM I 32.5 R SFA10 w165 DIN	B240 CEM I 52.5 R w180 DIN	B240 CEM I 42.5 R w180 DIN	B240 CEM I 52.5 R SFA160 w180	B240 CEM I 52.5 R SFA160 w145
Cement	kg	270	270	270	240	240	240	240
Fly ash	kg	10	10	10			160	160
Limestone	kg							
Water	kg	162	162	162	180	180	179	142
SP	kg	2.8	1.9	3.0		1.3	1.7	4.0
Aggregates	mm	0-16	0-16	0-16	0-16	0-16	0-16	0-16
w/c	[-]	0.61	0.61	0.61	0.75	0.76	0.75	0.60
w/c$_{eq}$	[-]	0.60	0.60	0.60	0.75	0.76	0.66	0.53

Table 2   Mix design of the cement reduced concretes (selected mixtures)

Mix design Mass per m³ concrete		B175- CEM I 52.5 R- SFA160- w145	B150 CEM I 52.5 R SFA250 w145	B150 CEM I 52.5 R SFA125 KSM145 w145	B150 CEM I 52.5 R KSM289 w145	B150 CEM III/A 52.5 R SFA250 w145	B150 CEM III/A 52.5 R SFA125 KSM145 w145	B150 CEM III/A 52.5 R KSM289 w145
Cement	kg	175	150	150	150	150	150	150
Fly ash	kg	225	250	125		250	125	
Limestone	kg			145	289		145	289
Water	kg	142	142	142	142	143	142	142
SP	kg	3.9	3.1	4.0	5.1	2.4	3.3	4.5
Aggregates	mm	0-16	0-16	0-16	0-16	0-16	0-16	0-16
w/c	[-]	0.83	0.96	0.97	0.97	0.96	0.97	0.97
w/c$_{eq}$	[-]	0.73	0.85	0.85	0.97	0.85	0.85	0.97

3   Experimental results

It was shown that even concretes with cement content lower than 125 kg/m³ were able to meet the usually required strength for the strength class C20/25 and a sufficient workability. Even concretes with cement content lower than 150 kg/m³ were able to meet the defined strength requirements for the strength class C30/37 (Figure 1). However, a significant reduction in the water content as well as the addition of reactive powders like fly ash or a higher strength cement class was necessary to achieve this strength. Secondary, a higher demand of superplasticizer was required.

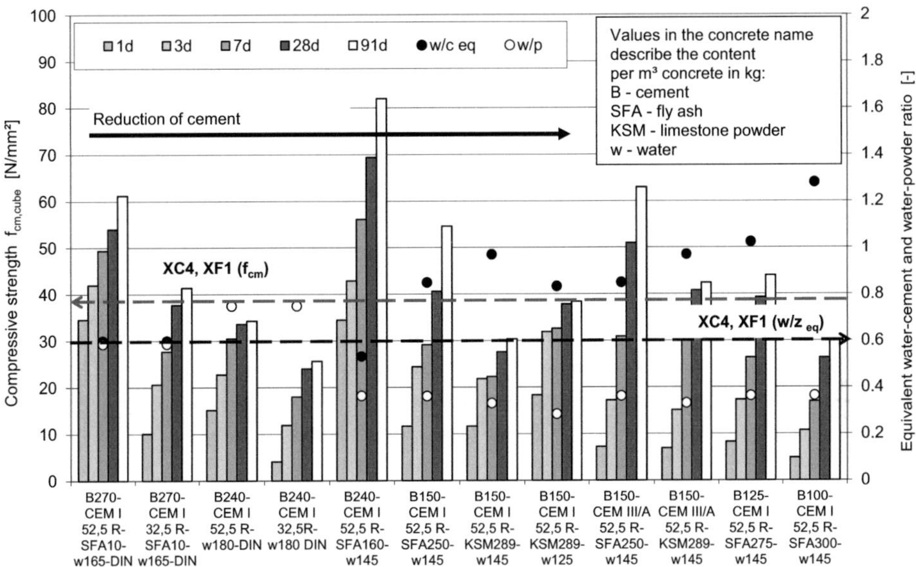

Fig. 1   Strength development of the concretes.

Figure 2 shows the measured carbonation depth as a result of the ACC-test. The concrete mix B270-CEM I 32.5 R-w165-DIN according to the German standard has a relatively high carbonation depth of approximately 6 mm. The concrete with CEM I 52.5 R has a significantly lower carbonation depth.

Compared to the carbonation depth of the concrete with CEM I 32.5 R, approximately the same value was measured for the concretes with only 150 kg/m³ CEM III/A 52.5 R. In contrast, the concrete mixes with 150 kg/m³ CEM I 52.5 R had a considerably higher carbonation depth than the reference concrete. However, requirements for exterior structures can be fulfilled by the reduction of water or a slight increase in cement. The influence of the concrete additives on the carbonation was remarkable. Notwithstanding the consumption of calcium hydroxide fly ash reduced as expected, the carbonation even more than limestone powder. A reduction of the cement content to 125 kg/m³ and 100 kg/m³ tends to result in values which are significantly higher than the carbonation depths of the reference concrete. However, concretes may be applied in interior structures.

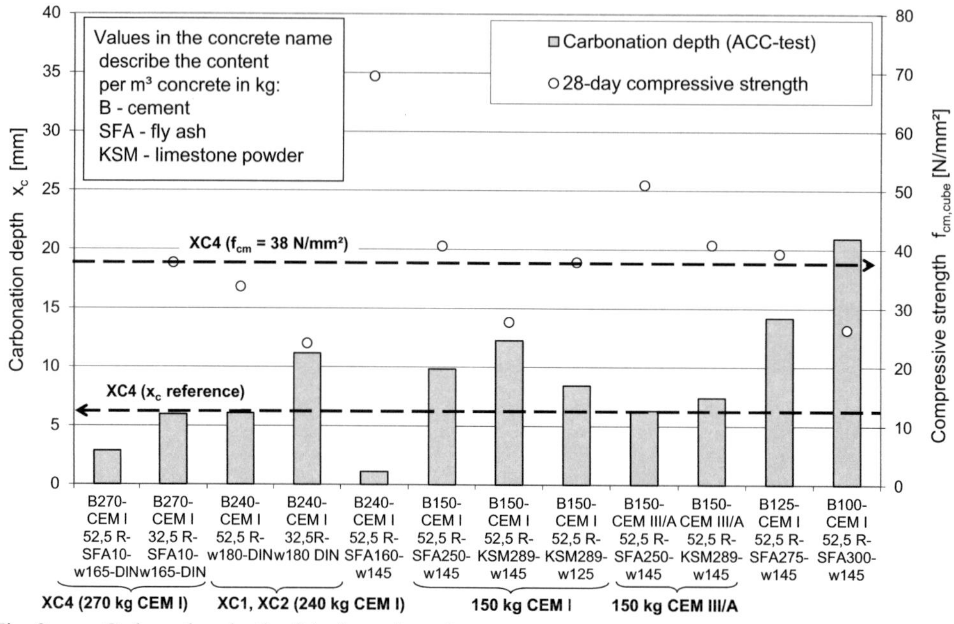

Fig. 2   Carbonation depth of the investigated concretes.

4   Prediction model

In order to ensure the durability of concrete structures, the national standard DIN 1045-2 in accordance with EN 206-1 provides limiting values for the minimum cement content, the compressive strength and the water-cement ratio. Reactive additives such as fly ash can be taken to some extent into account for the calculation of the normative cement content and the water-cement ratio, using an efficiency factor (k-value). Based on the k-value concept, concretes with the same equivalent water-cement ratio show an equivalent performance, even if the mixture composition varies significantly. If the efficiency factor is k = 1, the additive will contribute to the durability in the same way as cement. The maximum amount of concrete additives that can be taken into account is limited according to EN 206-1.

As part of the investigations performed at the TU Darmstadt, it was determined whether this approach is also applicable for concretes with highly reduced cement content or if modifications are necessary. Assuming that concrete additives contribute differently to the concrete, compressive strength and carbonation resistance, the k-values as well as the concept were modified. In contrast to the existing concepts, this approach takes the influence of the cement strength class and the "inert" additives (< 0.25 mm) into account directly (equation 1). For the calibration of the k-values and constants, the previously presented test results were used. The k-values are given in Table 3. Based on the equivalent water-cement ratio for the carbonation, the carbonation depth for cement reduced concrete with a high amount of additives can be predicted (equation 2).

The w/c-ratio directly correlates to the carbonation depth in the accelerated test. The carbonation depth of the accelerated carbonation test [3] resembles the carbonation in a conditioned environment (65% r. h., 20°C) for a time period of approximated 2.7 years. In addition, an adequate carbonation resistance can be proved by the equivalent w/c-ratio.

$$\left(\frac{W}{C}\right)_{carbo} = \frac{W}{C \cdot k_C + FA \cdot k_{FA} + P \cdot k_P} \tag{1}$$

$$x_c = \left(\frac{W}{C}\right)_{carbo} \cdot 35.8 - 15.9 \tag{2}$$

Table 3   K-values for approach 2

Component	Symbol	Quantity	k-value	Symbol k-value
CEM I 32.5			1.00	
CEM I 42.5	C	100%	1.10	k_C
CEM I 52.5			1.20	
Fly Ash	FA	max. 33% from C	0.40	k_{FA}
Powder	P	100%	0.03	k_P

Figure 3 highlights the measured carbonation depths of the accelerated test compared to the calculated equivalent w/c-values based on the described approach. The correlation of the two variables can be described linearly. With the regression lines (Figure 3) the carbonation depths of the accelerated test can be calculated based on the equivalent w/c-ratio of the mix design. This means that only the material weights of the individual concrete additives are the input parameters. In particular, a simplified prediction of the resistance against carbonation-induced corrosion of reinforcement is given by the equivalent w/c-ratio. For instance, the equivalent w/c-ratio must not exceed 0.75 for exposure class XC4 and 0.60 for XC1. To guarantee the durability, the limiting values with respect to the cement content and strength class are no longer required.

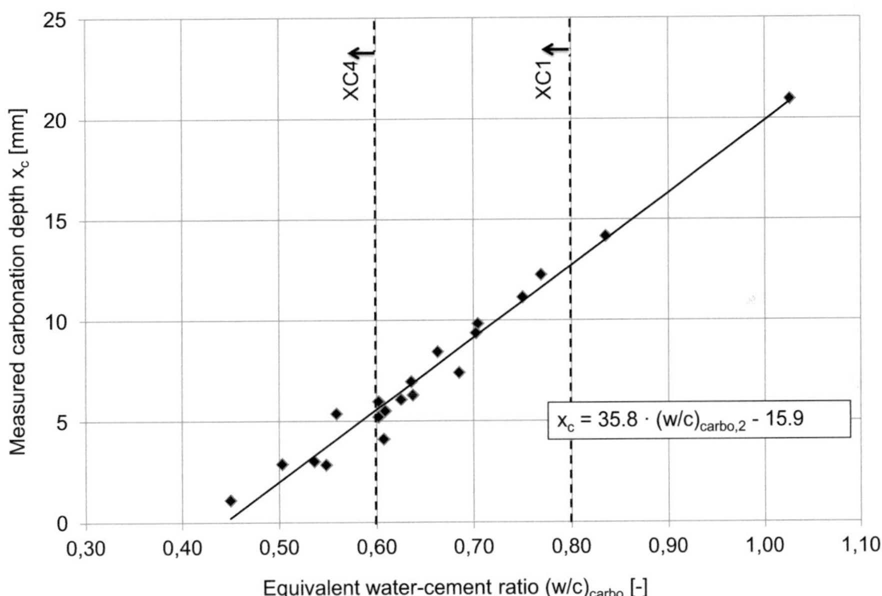

Fig. 3   Correlation between the calculated equivalent w/c-ratio of the carbonation and measured carbonation depth of the accelerated test.

5   Comparison and verification

Figure 4 presents the measured and calculated carbonation depths. Using both equations (1) and (2) it is obviously possible to predict the carbonation depths. The chosen k-values are calibrated for the materials used in the investigations at TU Darmstadt and are not necessarily applicable for all such materials.

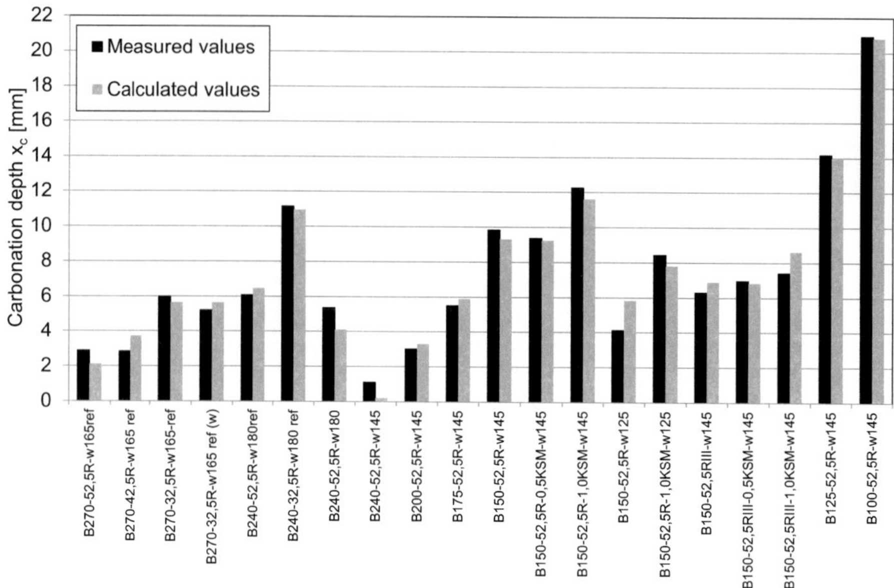

Fig. 4   Comparison of the calculated and measured carbonation depth

6   Conclusions

Based on the experimental studies presented here, the following conclusions can be drawn: Applying the equivalent performance concept of concrete demonstrates that highly reduced cement concretes are able to fulfil the requirements for the carbonation resistance according to EN 206-1 and DIN 1045-2:2008-08 for exposure classes XC1 to XC4. For the application of cement reduced concretes in Germany, a building authority approval for the production of reinforced concrete structures is necessary if the mixture design does not meet the national standards, especially the minimum cement content of 240 kg/m³ (for dry and wet conditions) and 270 kg/m³ (for moderate exposed concrete members). The existing k-value concept was improved, focused on the compressive strength and the carbonation depth. For these concrete properties, a prediction model based on the experimental tests was developed. It considers the cement strength class, the reactive and inert concrete additives and the absolute cement and water content. Additional approaches are publicized in [4]. According to the existing design procedure, the equivalent w/c-ratio can be used for the simplified verification of the resistance against the corrosion induced by carbonation. However, the durability against other exposure (e.g. sulphate attack, freeze-thaw) should be verified. The results should also be proved based on analytical methods which reveal the fundamental chemical reactions.

References

[1]   Stark, J.; Wicht, W.: Dauerhaftigkeit von Beton - Der Baustoff als Werkstoff. Bau Praxis. 2001, Birkhäuser Verlag, ISBN 3-7643-6344-4.

[2]   Meng, B.; Wiens, U. Wirkung von Puzzolanen bei extrem hoher Dosierung – Grenzen der Anwendbarkeit. Tagungsband der 13. Internationalen Baustofftagung. 1997, Weimar, 1-0175-1-0186.

[3]   fib bulletin 34: Model Code for Service Life Design. February, 2006: 110 S., fib Task Group 5.6, International Federation for Structural Concrete (fib).

[4]   Proske, T.; Hainer, S.; Graubner, C.-A.: Carbonation of cement reduced Green Concrete, International Congress on Durability of Concrete, 2012.

Influence of compaction on chloride ingress

Jure Zlopasa

Microlab, Faculty of Civil Engineering and Geosciences,
Technical University of Delft,
Stevinweg 1, 2628 CN Delft, the Netherlands
Supervisors: Eduard Koenders and Aad van der Horst

Abstract

Experiences from practice show the need for more of an understanding and optimization of the compaction process in order to design a more durable concrete structure. Local variations in compaction are very often the reason for initiation of local damage and initiation of chloride induced corrosion. Poor compaction is often manifested locally in concrete structures in the form of large voids, profuse honeycombing and heterogeneity. Controlling compaction in practice today depends largely on the operator and his experience. The reliability of such a practice goes with many uncertainties. Depending on the exposure conditions, cement based materials maybe attacked by aggressive substances which can influence the performance of a concrete structure. Therefore the change in chloride ingress was investigated as a function of compaction time and sample depth. In this paper the Rapid Chloride Migration method, (RCM), was used and test results show a change of the diffusion coefficient with time of compaction. Along with the RCM results, the compressive strength was measured and the results are presented as well.

1   Introduction

Durability of concrete depends on the long-term quality of a microstructure, which depends, on its turn, on the used materials, quality of construction, quality of design, the exposure conditions, and the on-site execution. When placed in a form, concretes may contain from 5 up to 20 % of entrapped air by volume. Presence of air voids in hardened concrete turned out to have a significant effect on mechanical properties and can be controlled by compaction. Compaction has a very good correlation with several important properties such as strength, permeability (and hence durability) and shrinkage [1]. It has been found that for every 1 % of air voids there is a 5 % of decrease in compressive strength [2]. Air voids are expelled by compaction. Compaction can be achieved by vibration, centrifugation, rodding, tamping, or combination of these actions [3].The compaction consists of two processes; in the first stage the particles are set in motion and concrete is "liquefied" and slumps to fill the forms, while during the second stage air bubbles escape from the concrete element. There is a distinction between the two stages because they don't start and finish at the same time. The consequence of inadequate compaction is appearance of sand streaks, honeycomb and as mentioned excessive entrapped air voids. If the workability of the concrete is constant, the skill and consistency of workers in operating the vibrators (i.e., compaction effort) will significantly influence the degree of compaction attained and with this the long-term durability [4]. Because concrete contains 65-75 %vol. of aggregate and all of the air voids reside in the cement paste, a small amount of air entrainment causes a significant change of the microstructure of the paste, and to its pore structure in particular[5]. The focus of this research was on the relationship between time of compaction, transport and mechanical properties of concrete a specimen. Transport properties are measured by the Rapid Chloride Migration, RCM, test and the mechanical properties are presented in terms of compressive strength after 7 and 28 days. Because the chloride transport is of great significance, but in real life this process lasts for many years, the RCM test is used which has shown that it can adequately represent chloride transport in cement-based materials [6].

2   Materials and experimental procedure

Concrete samples with the same mix design, table 1, were casted. In order to view the influence of compaction of the properties of concrete mix design, mixing, vibration frequency (65 Hz), vibration amplitude (0,087 mm), curing conditions and temperature were kept constant. Samples were cast in cube moulds 150mm×150mm×150mm. The compaction was done on a vibration table. After the

Proc. of the 9th *fib* International PhD Symposium in Civil Engineering, July 22 to 25, 2012,
Karlsruhe Institute of Technology (KIT), Germany, H. S. Müller, M. Haist, F. Acosta (Eds.),
KIT Scientific Publishing, Karlsruhe, Germany, ISBN 978-3-86644-858-2

compaction was done the samples were stored in the fog room at 20 °C. The specimens were tested for compressive strength after 7 and 28 days and chloride migration coefficient, which was performed by the RCM test. The whole experimental procedure is outlined in figure 1.

Table 1   Mix proportions

CEM III/B 42.5N [Kgm^{-3}]	370
Water [Kgm^{-3}]	185
Aggregate [Kgm^{-3}]	1794

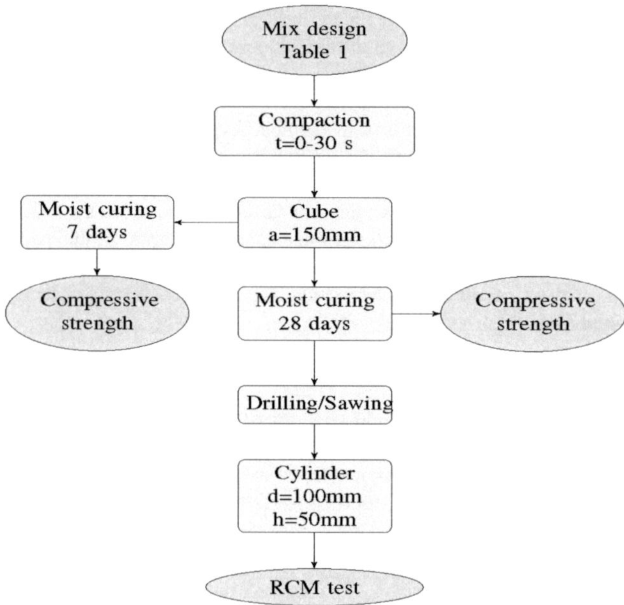

Fig. 1   Experimental scheme

In order to evaluate the difference in transport properties of the samples determination of the chloride coefficient in concrete using the RCM is used. The method is prescribed in NT Build 492 [7]. The principle behind this method is the application of external electrical potential axially across the concrete specimen which forces the chloride ions to migrate into the specimen. A sketch of the RCM test is shown in figure 2. Sampling was done by drilling the cubes and slicing the cylinders in order to obtain 100 mm diameter and 50 mm height sample. The samples were then preconditioned by putting them in vacuum for 3 hours and then, with the vacuum pump still running, the samples were covered with saturated Ca(OH)$_2$ solution. After 1 hour the vacuum pump was turned off and air was allowed to enter the container and the specimens were kept in the Ca(OH)$_2$ solution for 18 ± 2 hours. After the preconditioning the samples were mounted in the RCM setup which consists of a rubber sleeve in which the sample is placed and anolyte solution (0.3 M NaOH) was poured in the sleeve. The specimen was then put into a catholyte solution (2 M NaOH). The electrodes were immersed in anolyte and catholyte solutions and connected with the power supply.

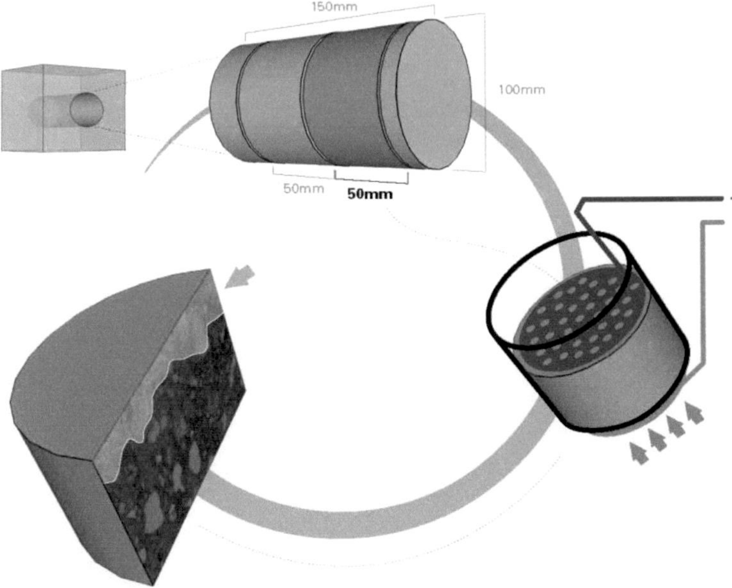

Fig. 2   Schematic representation of the RCM test

The initial current through the specimen at 30 V is recorded and the voltage was adjusted according to the standard which then states the duration of the test as well. Initial and final temperatures were recorded. After the stated duration of the test the specimens were removed from the RCM setup and sliced axially into two pieces. On the freshly split surface 0.1 M $AgNO_3$ was sprayed and after a few minutes white silver chloride started to precipitate. The white precipitated silver chloride represents the penetration depth of the chlorides which was measured by a calliper. For each measurement three samples were made.

3   Results and discussion

In figure 3 the change in compressive strength after 7 and 28 days is shown. There is a difference in compressive strength with time of compaction. This is due to the presence of the air voids that are entrapped in the specimen. Air voids are reduced by the time of compaction which is shown as an increase of the compressive strength. There is a big rise in the compressive strength from no compaction to 10 seconds of compaction, after that increase in compressive strength is shown. So it could be said that most of the air inside of the concrete has escaped.

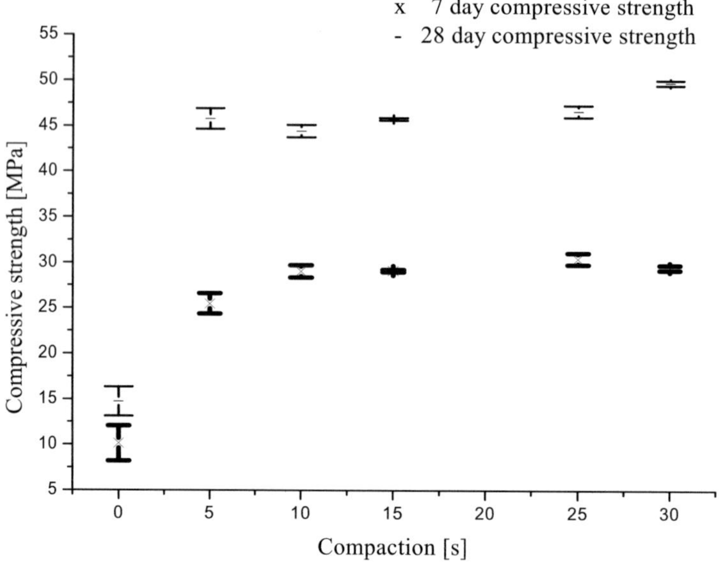

Fig. 3   7 and 28 day compressive strength as a function of compaction duration.

Also shown in the figure 3 is less scatter in the attained compressive strength results for compaction from10 to 30 seconds. That shows a more homogenous and consistent microstructure and results.

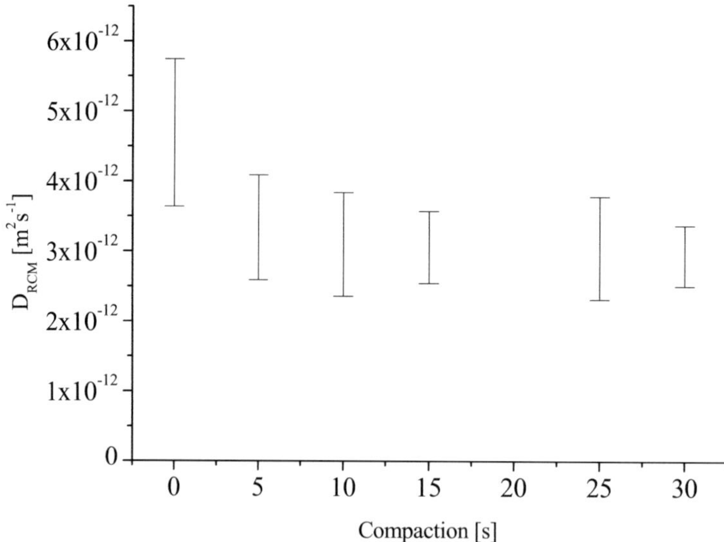

Fig. 4   Chloride migration coefficient measured from the chloride penetration depth after RCM test as a function of compaction time

From the penetration depths, measured after silver chloride, AgCl, precipitation, chloride non-steady state migration coefficient was calculated. It can be seen in figure 4 and table 2 that the migration coefficient decreases with time of compaction. The reason for the lowering in the migration coefficient can be found in change in aggregate to cement ratio which changes during compaction, and decrease of volume fraction of air voids. Wong et al. [5] found that the microstructure of the air void-paste interface is similar to that of the aggregate-paste interfacial transition zone. Kreijger [8] showed that the "skin" of a concrete consists of three layers, and during compaction the biggest change in composition, aggregate to cement ratio, is within few cm. This skin concrete layer that is rich in cement paste has different transport properties than the bulk concrete. That difference could also be enhanced due to chloride binding by cement hydration products. That is one of the downfalls of RCM method in which up to 2 cm of "concrete skin" is cut off.

Table 2   Summary of the results measured for compressive strength after 7 and 28 days, and RCM test after 28 days.

Sample	Compressive strength [MPa]		D_{RCM} [m^{2}s^{-1}]
	7 day	28 day	
CC0	10.13 ± 1.92	14.76 ± 1.60	$4.69 \times 10^{-12} \pm 1.05 \times 10^{-12}$
CC5	25.46 ± 3.28	45.79 ± 1.12	$3.34 \times 10^{-12} \pm 0.75 \times 10^{-12}$
CC10	29.05 ± 1.94	44.44 ± 0.68	$3.10 \times 10^{-12} \pm 0.74 \times 10^{-12}$
CC15	29.12 ± 0.87	45.76 ± 0.13	$3.06 \times 10^{-12} \pm 0.51 \times 10^{-12}$
CC25	30.45 ± 1.82	46.62 ± 0.65	$3.05 \times 10^{-12} \pm 0.73 \times 10^{-12}$
CC30	29.50 ± 0.71	49.74 ± 0.28	$2.94 \times 10^{-12} \pm 0.43 \times 10^{-12}$

4   Conclusions

This study was undertaken to investigate the importance and the influence of compaction on the chloride migration coefficient and compressive strength. Specimens with constant mix design were characterized and the findings are outlined below.

RCM test shows sensitivity towards time of compaction, this sensitivity is greater while there is a higher volume fraction of air inside the concrete specimens. For compaction longer than 10 seconds the part that changes most is the outermost part where the segregation happens with the change in aggregate to cement ratio which in the standard RCM test is cut off.

Compressive strength showed an increase with time of compaction which is a consequence of en-trapped air leaving the concrete specimen due to vibration. The biggest difference was until 10 seconds of vibration, later little difference was observed. The difference was more pronounced on the standard deviation rather than the compressive strength.

Up to thirty seconds of compaction leads to a more uniform and homogenous concrete element, in which the properties differentiate less than compared with no compaction.

5   Acknowledgements

The authors like to acknowledge the Dutch National Science foundation STW. The research conducted within this project is financed by STW as part of the IS2C program (www.is2c.nl), number 10962.

References

[1]   Ayton, G., A Recipe for Compaction of Concrete, In: 7th International Conference on Concrete Pavements (2001)

[2]   Heaton, B. S., Relationship in Concrete between Strength, Compaction and Slump, In: Constructional Review 39 (1966) No.2, pp.16-22

[3]   ACI 309R-96: Guide for Consolidation of Concrete. In: ACI Manual of Concrete Practice (2005)

[4]   Stewart, M.G.: Concreting Workmanship and Its Influence on Serviceability Reliability. In: ACI Materials Journal 94 (1997) No.6, pp.501-509

[5]   Wong, H.S., Pappas A.M., Zimmerman, R.W., Buenfeld, N.R.: Effect of entrained air voids on the microstructure and mass transport properties of concrete. In: Cement and Concrete Research 41 (2011), pp. 1067-1077

[6]   Bouwmeester - van den Bos, W.J., Schlangen, E.: Influence of curing on the pore structure of concrete, In: Tailor Made Concrete Structures (2008), pp.65-70

[7]   NT Build 492, 1999, Concrete, Mortar and Cement-based repair materials: Chloride Migration Coefficient from non-steady-state migration experiments. Nordtest Finland.

[8]   Kreijger, P.C.: The skin of concrete: Composition and properties. In: Materials and Strucutres 17 (1984) No. 4, pp.275-283

Determining sulphate movement in concrete exposed to evaporative transport

Julie Ann Hartell

Department of Civil Engineering and Applied Sciences,
McGill University
Montreal, Canada
Supervisor: Andrew J. Boyd

Abstract

In order to properly identify ongoing sulphate attack on concrete and assess the levels of damage a structure has sustained, it is important to understand how the attack progresses relative to its environment: sulphate exposure parameters and ambient environmental exposure conditions. Currently, much debate revolves around the path that sulphates follow in partially exposed concrete sitting in sulphate laden soils. Simulating a realistic attack mechanism and reproducing field conditions in the laboratory are added challenges in sulphate attack research. This research utilized large scale concrete blocks similar in dimension to residential foundation elements, cast from a 0.45 water to cement ratio concrete mixture. The specimen was partially exposed (i.e. partly submerged) to a 5% sodium sulphate solution for a period of two years. During the exposure period, dual degradation mechanisms co-existed, both chemical and physical, as seen in ongoing attacks in the field. Chemical analysis was used to create a two-dimensional sulphate profile within the blocks to determine the effect of an evaporative surface above the immersion line on accelerating movement of sulphate into and through the concrete. This process may lead to paste softening and expansive cracking along the transport path and surface scaling due to salt crystallization at the evaporation surface. To establish this profile, a method of sampling the deteriorated concrete block without changing the chemical composition of the cementitious matrix was devised. Samples were taken at regular intervals throughout the cross-section of the block and prepared for analysis of sulphate content using inductively coupled plasma optical emission spectrometry (ICP-OES).

1   Introduction

The culprits involved in external sulphate attack (ESA) and their related degradation mechanisms have been extensively reviewed in literature; Mehta 1992, Skalny et al. 2002 and Stark 2002 to name a few. In the field, sulphate related damage to concrete begins at any surface exposed to a sulphate-rich solution, which may come from an underground, marine or industrial environment; and from sulphate containing salts that form on the concrete surface (i.e. salt weathering). Arguably, there are many ways of reproducing the effects of ESA seen in the field under laboratory conditions; each case leading to different results due to a wide spectrum of deterioration processes dependant on various exposure conditions. (Santhanam et al 2001)

It is often argued that specimen size and shape affect the outcome of durability tests. An increase in surface reaction (i.e. a small specimen with a large surface area such as a cube) is one way to accelerate the attack's development. Doing so, however, may change the course of the reaction kinetics. (Cohen and Mather 1991) Traditional small scale simulations can rapidly provide information on specific durability issues associated with sulphate attack but they do not provide an overall realistic degradation pattern since the entire specimen is subjected to the aggressive agent. In reality, the core of a larger scale concrete element may not be affected at all, thus changing the mechanical behaviour of the concrete. This study avoids the problems with small scale samples by utilizing large blocks similar in dimension to actual structures in the field.

Since virtually all structures subjected to ESA are partially buried or submerged, such as house foundations, simulating a partial exposure in the lab instead of complete submersion was found to be more representative of actual structural or surface damage observed in the field by producing both chemical and physical attack modes. It is believed that the change in moisture gradient at the immersion line creates a wicking transport mechanism that draws the soluble sulphates through the porous concrete, which then recrystallize near the evaporative surface. (Boyd and Mindess 2004) On the

Proc. of the 9th *fib* International PhD Symposium in Civil Engineering, July 22 to 25, 2012,
Karlsruhe Institute of Technology (KIT), Germany, H. S. Müller, M. Haist, F. Acosta (Eds.),
KIT Scientific Publishing, Karlsruhe, Germany, ISBN 978-3-86644-858-2

other hand, claims have also been made that the sulphates only travel up, or near, the outer surface via capillary suction and that the surface damage above the immersion line is essentially superficial. (Hime 2005) Reported field case studies, both in Southern California, demonstrate opposing conclusions on sulphate transport mechanism and their related chemical interactions with concrete.

Brown and Doerr's (2000) study on cores taken from housing foundations which had not been in direct contact with the soil revealed that multiple phases had formed in the concrete as a result of the transport of sulphate, and other typical compounds found in aggressive soils, to significant distances throughout the cementitious matrix. However, a Novak and Colville (1989) study analyzed efflorescent salts filling cracked foundation slabs and the concrete material adjacent to these cracks. X-ray diffraction results indicate that no ESA by-products where found in the concrete material with the exception of trace amounts of gypsum.

This research attempted to reproduce the exposure conditions seen in the field in order to demonstrate the true path of the sulphates as they move upward toward the evaporation plane. Prior work has indicated that based solely upon visual inspection, the chemical changes causing paste softening and expansive cracking in the bulk cementitious matrix can be overshadowed by superficial physical damage at the evaporation plane, but mechanical testing distinctly illustrates the more significant damage hidden beneath this scaling. (Hartell et al. 2011) Confirming the path that sulphates take in larger scale specimens is thus critical toward understanding the degradation of concrete subjected to ESA in the field. Inductively coupled plasma optical emission spectrometry (ICP-OES) was the chosen analytical technique to meet this objective.

2 Experimental Program

2.1 Sample Preparation

In order to recreate a realistic field exposure mimicking that of foundation structures, concrete blocks measuring 900 mm long x 240 mm thick x 485mm high were cast and immersed to a depth of 150mm in a 5% sodium sulphate solution. Only one mixture, 0.45 w/c, was analyzed in this study; representing the prescribed value in the Canadian standard to resist a severe exposure type. Table 1 provides further details on the mixture design.

Table 1   Concrete Mixture Design

Mixture		0.45 w/c
Cement – Type I	[kg/m^3]	507.7
Water	[kg/m^3]	228.6
Fine Aggregate	[kg/m^3]	855.1
Coarse Aggregate	[kg/m^3]	733.4
Admixture - Superplasticizer	[L/m^3]	0.8

After two years of exposure, the block was removed from the solution and sealed with paraffin wax in order to minimize further interactions with the ambient surroundings. A section was then removed for chemical analysis. This section, measuring 100 mm in thickness, was cut from the mid section of the block, using a concrete chain saw. Then, a concrete bench saw was used to accurately cut the section in half along its vertical axis to produce two 100 mm long x 120 mm thick x 485 mm high samples, each containing one exposure face and one inner-core face. All cuts were performed dry to prevent soluble sulphate compounds from washing out of the concrete samples. Even though dry-cutting can induce friction heat and change the chemical structure of the sulphate phases present, this was not considered to be an issue since only the total amount in sulphur was desired. Whether it was present in the form of ettringite, gypsum or even thaumasite was of no interest for the purpose of this study, only the distribution and quantities of sulphur were evaluated.

The two dimensional sulphate profile was created by accurately milling the surface of the concrete prism layer by layer (Fig. 1). The powdered concrete was collected using a high pressure vacuum system. Each sample represented a portion of concrete measuring 100 mm long x 10 mm thick x 20 mm high.

Fig. 1   Sectional milling of concrete prism.

Because of the large scale of the specimen, the entirety of the prism was not processed. Powdered samples were prepared for the entire submerged portion and the region immediately above the immersion line containing the evaporative plane. This included the physically damaged zone (extending approximately 60 mm above the immersion line) and a visually intact zone (extending 50 mm above that). Reference samples were also taken from the top end of the concrete block to determine the initial 'background' sulphur content of the concrete due to the presence of sulphates in the cement (Fig. 2). It was assumed that the upper portion of the concrete block had not been affected by ESA since it was situated more than 250 mm above the evaporative plane.

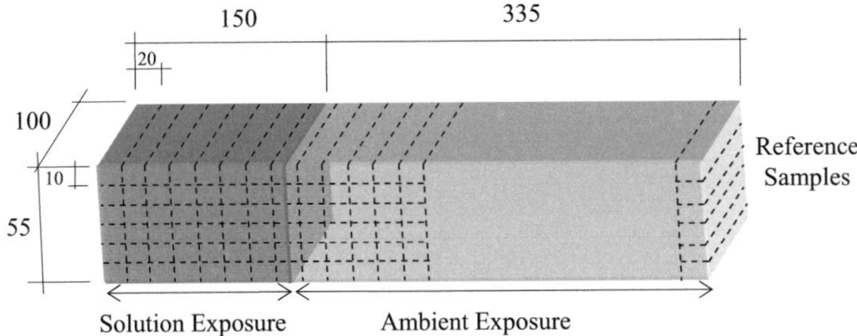

Fig. 2   Details of powdered concrete sample location and profiling of concrete prism

The outer layer of concrete (measuring 5 mm in depth) was not used in this study because of surface geometry inconsistencies (i.e. loss of mass due to salt crystallization).

2.2   Analytical Procedure

The amount of sulphur in each concrete powder sample was determined using an analytical technique called inductively coupled plasma optical emission spectrometry (ICP-OES). This type of emission spectroscopy uses inductively coupled plasma to generate excited atoms which produce electromagnetic radiation at wavelengths characteristic to a particular element and then, in this work, examining the intensity of the emission produced by the sulphur element in solution. Herein, the 182.040 emission line was selected to conduct the quantitative analysis on only the sulphur element in order to minimize spectral interference from other elements which might be present in the sample, such as silicon.

Individual sample solutions were prepared according to the following procedure. First, 0.5 g (± 0.001 g) of a concrete powder sample was selected and diluted into 25 ml of deionised water. Slowly, 5 ml of hydrochloric acid was added to the solution, heated just below its boiling point and allowed to digest for a few minutes. Then, the acid solution was further diluted to 50 ml with hot deionised water to obtain an approximate 10% v/v HCl solution. The latter was further digested using a sonicator bath and then filtrated to minimize the presence of solid particles in the solution. Finally, approximately 10 ml of filtrate was collected for ICP-OES analysis.

From the known quantity of dissolved concrete powder in the solution (10 g/l ± 0.05 g) and the obtained concentration of sulphur (ppm) determined by ICP-OES, the mass of sulphur (kg) over the mass of concrete (kg) ratio was determined, S/C. These results were then used to create a quantitative map of the sulphur concentration throughout the concrete matrix, which should be instructive as to sulphate movement mechanisms within the concrete element.

3. Results and Discussion

The graphical results displayed in Figure 3 represent a two-dimensional sulphur concentration profile within the analyzed section of the 0.45 w/c concrete block. The values are expressed as a ratio between the sulphur quantity in kilograms per kilogram of concrete. Each region of varying colour corresponds to a range of sulphur concentration. Both axes represent the concrete prism's longitudinal (i.e. height) and transversal (i.e. depth of penetration) dimensions in millimeters.

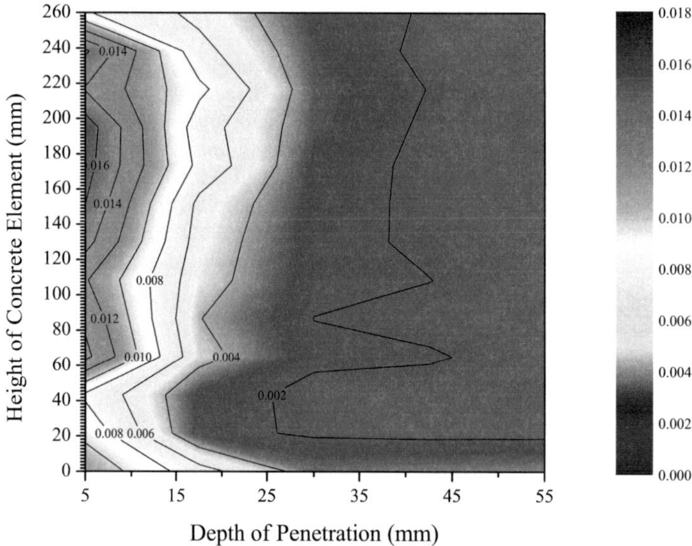

Fig.3   Sulphur concentration profile of 0.45 w/c concrete prism, S/C.

From the ICP-OES analysis on the reference samples situated at the upper end of the block exposed to ambient conditions (465 to 485mm), it was determined that the average background sulphur concentration in the concrete was below 0.002 kilograms per kilogram of concrete. Therefore, the initial range shown in Figure 3 (0.000 to 0.002) represents the area of the cementitious matrix that was not affected by ESA.

Starting with the solution exposed half, the depth of penetration was greater for the bottom portion of the block (0 – 20 mm) because of the two-dimensional ingress front. The bottom face of the block was not sealed to prevent solution penetration. However, between 20 and 60 mm from the bottom there is a drop in sulphate concentration from the surface to the core; this being the lowest recorded penetration depth. The values presented herein assume that the concrete was uniform in composition throughout the block. On the other hand, this decrease may be a consequence of aggregate segregation, thus changing the concrete's density in the lower section of the block. However, the continuing trend as it approaches the immersion line might suggest otherwise.

The sulphur concentration and the depth of penetration seem to increase as the immersion line (150 mm) is approached. It is thought that the presence of an evaporation plane draws greater amounts of solution further within the porous concrete to compensate for the continuous change in moisture gradient above the immersion line. From 60 to 140 mm, the penetration depth reaches its greatest, approximately 45 mm into the bulk. It should be noted that this exceeds the depth of the average covercrete where structural rebar is present in residential foundation structures. Chemical interactions between the sulphates and the cementitious matrix may lead to diminished mechanical properties of the concrete, which may become critical at these depths as it will affect the bond between the concrete and the reinforcing steel. Looking at the results above the immersion line, the concentration values continue to increase within the evaporation plane but, the overall depth of penetration decreases.

It seems that the combined action of capillary suction and wicking, due to the presence of continuous evaporation, creates a transport mechanism where the pore solution is drawn upwards through the concrete matrix, but not just along the outside surface of the block. As the solution travels, more and more sulphates are available to react with the matrix where various compounds may form. This corroborates with Brown and Doerr's (2000) findings. Also, as the soluble sulphates migrate toward the surface of the concrete, the solution evaporates leaving the sulphates behind to crystallize. This phenomena causes crystallization pressures which lead to surface disintegration. The highest concentration values recorded were situated within 5 mm of the block's surface and between 160 and 220 mm up from its bottom. This portion of the prism corresponds to the area damaged by salt crystallization. The amount of sulphur peaks around the 220 mm mark, which is directly above the upper edge of crystal formation due to efflorescence. Then, the concentration values decrease for the upper remainder of the block (220 to 260 mm).

It was noticed that as the attack progressed in time the area affected by salt crystallization spread upwards from the immersion line. One hypothesis is that a barrier was formed at the surface of the concrete element by the crystal growth. Therefore, the solution was wicked higher to evaporate above this pore filled concrete. Thus, with time the physically damaged area grew and the actual evaporation plane continuously moved upward.

As the concrete degrades physically and chemically the pore structure is altered further, accentuating the transport mechanisms described above. The rate at which the sulphates penetrate and circulate is unknown from this investigation but is currently being analyzed for various exposure types in separate research. Further analytical and mechanical testing techniques are also being devised for determining the type of damage being induced along their path of travel.

4 Conclusions

The primary outcome from this research is a better understanding of how sulphates move through a concrete element in the field. The results of ICP-OES testing corroborate past research investigations where soluble sulphates migrated through the bulk of the porous concrete to then crystallize once the solution evaporates at the surface, leaving higher concentrations of sulphate salts behind in the pores. After an exposure period of two years the solution does not appear to travel along the outside surface through capillary suction alone, but penetrates a significant distance into the concrete. Moreover, the presence of salt crystal growth at the surface seems to form an barrier to evaporative transport, thus drawing more sulphates from other regions of the concrete by continually displacing the evaporative front.

The use of ICP-OES was found to be an accurate and viable method for evaluating sulphur concentration and distribution throughout the matrix. Knowing the sulphate transport mechanisms, this research offers insight into how and where to sample a deteriorated structure, with respect to its exposure conditions, in order to appropriately identify the extent of damage it has sustained.

References

[1]   Mehta, P.K.: Sulfate attack on concrete – A critical review. In: Materials Science of Concrete III, J. Skalny ed., American Ceramic Society (1992), pp. 105-130

[2]   Skalny, J., Marchand, J. and Odler, I.: Sulfate Attack on Concrete. In: Modern Concrete Technology Series, Spon Press, London (2002)

[3]   Stark, D.C.: Performance of Concrete in Sulfate Environments. In: RD129, Portland Cement Association, Skokie, Illinois (2002)

[4]   Santhanam, M., Cohen, M. D. and Olek, J.: Sulfate Attack Research – Whither Now?. In: Cement and Concrete Research 31(2001), pp. 845-851

[5]   Cohen, D.M. and Mather, B.: Sulfate Attack on Concrete – Research Needs. In: ACI Materials Journal, (1991) January-February, pp. 62-69

[6]   Boyd, A.J. and Mindess, S.: The Use of Tension Testing to Investigate the Effect of W/C Ratio and Cement Type on the Resistance of Concrete to Sulfate Attack. In: Cement and Concrete Research 34 (2004), pp. 373-377

[7]   Hime, W.G.: A Discussion of the Paper: "The Use of Tension Testing to Investigate the Effect of W/C Ratio and Cement Type on the Resistance of Concrete to Sulfate Attack by Andrew J. Boyd, Sidney Mindess. In: Cement and Concrete Research 35 (2005), pp. 189

[8]   Brown, P.W. and Doerr, A.: Chemical Changes in Concrete Due to the Ingress of Aggressive Species. In: Cement and Concrete Research 30 (2000), pp. 411-418

[9]   Novak, G. A. and Colville, A. A.: Efflorescent Mineral Assemblages Associated with Cracked and Degraded Residential Concrete Foundations in Southern California. In: Cement and Concrete Research 19 (1989), pp. 1-6

[10]   Hartell, J., Boyd, A. J. and Ferraro, C. C.: Sulfate Attack on Concrete: Effect of Partial Immersion. In: Journal of Materials in Civil Engineering, ASCE, May (2011), pp. 572-579

The use of the Stiffness Damage Test "SDT" for the assessment of concrete structures affected by ASR

Leandro Sanchez

CRIB (Research Centre on Concrete Infrastructures)
Laval University, 1065, Avenue de la Medecine, Quebec (G1V 0A6), Canada
Supervisor: Benoit Fournier

Abstract

Alkali-silica reaction (ASR) is one of the main processes reducing the service life of concrete structures worldwide. The objective of this research is to develop an effective and reliable approach to evaluate the damage and the potential for further distress of concrete damaged by ASR, based on selected mechanical (Stiffness Damage Test) and petrographic (Damage Rating Index) test methods performed in the laboratory. To achieve this goal, concrete specimens were cast and exposed to conditions promoting damage due to ASR. At selected expansion levels, they were tested through the procedures mentioned above. Results show that the SDT should be carried out with a percentage of the design strength and parameters such as the hysteresis area, the modulus of elasticity and the plastic deformation provide good correlations with the amount of expansion reached by the concrete.

1   Alkali Silica Reaction (ASR)

Alkali–silica reaction (ASR) is a chemical reaction between the alkali hydroxides in the concrete pore solution and certain reactive siliceous phases present within the aggregates. ASR generates a number of petrographic features of distress within or in the immediate proximity of the reactive aggregate particles (e.g., microcracks, reaction rims, debonding), including an alkali–calcium–silica gel that swells in the presence of water, thus developing pressures on the surrounding cement paste. These pressures are able to exceed the tensile strength of the paste and the aggregates, thus creating microcracks and causing volumetric expansion of concrete [1].

2   Assessment of damage in concretes damaged by ASR

First of all, the word "damage" has been defined in this work as the harmful consequences (measurable ones) of various types of mechanisms (e.g. loadings, shrinkage, creep, ASR, DEF, freezing and thawing, etc.) on the mechanical properties, physical integrity and durability of a concrete element. It is well-established that different deleterious mechanisms affecting the long-term durability of concrete generate different patterns of internal damage whose "signatures" were schematized by BCA [2] and St-John et al. [3]. However, one of the biggest challenges in engineering is to establish the correlation between the above "signatures" and the loss in mechanical properties and durability of the material. Many petrographic methods were developed with this aim but almost all the analyses carried out were described in a qualitative (and narrative) way. Therefore, petrographic analysis is often criticized by engineers as they look for a precise evaluation on the extent of damage of a concrete element [4]. Recent studies dealing with the mechanical responses of damaged materials suggest that the "Stiffness Damage Test (SDT)" could provide a diagnostic evaluation of the damage in concrete due to ASR. Nevertheless, the SDT does not have a standard test procedure so far, and an in-depth evaluation on the input and output parameters of the test should be done for its correct use.

3   The Stiffness Damage Test (SDT)

The concept of the SDT is to quantify the degree of distress in concrete damaged by ASR [5, 6]. The method is based on the cyclic loading (in compression) of concrete samples (cylinders or cores) with diameters greater than 70 mm (length / diameter of 2.5 – 2.75) [4]. Initially, the SDT involved the application of five cycles under a loading stress of 5.5 MPa at a rate of 0.1 MPa/s. Crisp et al. [5, 6] carried out more than 1000 tests with cores extracted from damaged concrete structures and they proposed the following as the diagnostic parameters of the test [5]:

Proc. of the 9th *fib* International PhD Symposium in Civil Engineering, July 22 to 25, 2012,
Karlsruhe Institute of Technology (KIT), Germany, H. S. Müller, M. Haist, F. Acosta (Eds.),
KIT Scientific Publishing, Karlsruhe, Germany, ISBN 978-3-86644-858-2

- Modulus of elasticity (Ec): concrete samples of damaged concretes presented lower modulus of elasticity (average modulus of elasticity value of the last four cycles) than undamaged samples;
- Hysteresis area (H - J/m3): damaged concrete samples showed greater hysteresis areas (average hysteresis area of the last four cycles) than undamaged samples;
- Non linearity index (NLI): it represents the ratio between Ec and the value of the modulus of elasticity when taken on the half of the slope of the stress applied.

Smaoui et al. [1] continued studying the SDT on laboratory concrete samples incorporating a variety of reactive rock types and that had reached differents expansion levels (stored at 38°C at 100% R.H.). After carrying out many tests, the authors found that the best output response for the SDT was the hysteresis area of the first cycle for test specimens loaded at the 10 MPa level; lower stresses did not offer the capacity of diagnosing the degree of damage in the concrete due to ASR. The authors also evaluated other parameters of the test, including plastic deformation during loading/unloading cycles, and they found that the correlation between the expansion and plastic deformation was also fairly satisfactory. They however noted significant variations for either the hysteresis area or the plastic deformation for concretes incorporating different types of reactive aggregates. These differences were possibly associated to the nature of the aggregate (fine or coarse) and differences in the internal pattern of damage, as they can generate their own mode of reaction (i.e. pattern/density/orientation of cracking depending on whether the damage is generated in the fine or the coarse aggregate, or by different rock types, etc.). Based on their findings, Smaoui et al. [1] proposed the following parameters as the best responses of the test:

- Hysteresis area of the first cycle (J/m^3);
- Plastic deformation accumulated during the five cycles of loading/unloading (µstrains).

It is important to mention that the work of Smaoui et al. [1] was based on one single concrete mix-design ($420kg/m^3$), and the proposed value of 10 MPa, as a fixed loading value in the test, worked only for such a type of concrete. It is therefore logical to believe that applying a single load of 10 MPa could result in different responses in the SDT, depending on the characteristics of the concrete analyzed (i.e. $\neq$ mix-designs, $\neq$ types of fine / coarse aggregate, etc.). However, this information has not been studied deeply yet. Without this information, the analysis of the SDT for different mix-designs could result into erroneous estimates of the actual level of damage and the expansion achieved to date. Finally, the SDT was originally developed to assess the effects of ASR on concrete, but the test certainly has the potential to assess damage from other mechanisms, such as freezing and thawing, action of fire, impact loads, DEF, etc. (Crisp et al., [6]). Smaoui et al. [1] also reported a good correlation between the expansion to date of a concrete associable to freezing-thawing cycles and the hysteresis area of the first cycle in the SDT. However, there is currently no data recognizing the signature of a damage mechanism over another when tested by the SDT. In addition, the SDT has the characteristic of being non-destructive, so as the number of samples taken within the structures is often limited (for economic reasons), we can consider using the same cores to perform other tests such as residual expansion, petrographic tests, compressive and tensile strengths, etc. (Crisp et al., [6]).

4 Scope of the work

As indicated in the previous section, the lack of a thorough study on the input/output parameters of the SDT could lead to erroneous interpretation of the test results. Among all of the parameters of the test, the most important ones are as follows:

- Input parameters: Test loading (versus concrete mixture designs);
- Output parameters: Analysis of hysteresis area (1st cycle vs. last four cycles), modulus of elasticity (1st cycle vs. average for cycles 2 and 3) and plastic deformation (over the 5 cycles) of the damaged concretes.

This paper presents the evaluation of these parameters on the responses in the SDT.

5 Materials and methods

5.1 Materials and mixture proportions

The analyses were carried out with two types of concrete (25 MPa and 35 MPa) and two highly-reactive aggregates (New Mexico (NM) gravel and Texas (Tx) sand)[1]. The two concretes were designed to contain the same volume of paste and the same volume of aggregates (i.e. from one mix to another), so one can compare similar systems. A total of 56 samples (cylinders) were cast from each of the mixtures made in the laboratory. After casting, the specimens were cured for 48h in the moist curing room and then stored at 38°C and 100% R.H. until they reach the expansion levels chosen for this research (i.e. 0.05%, 0.12%, 0.20% and 0.30%).

When they reached the above expansion levels, the specimens were wrapped in plastic film and stored at 12°C until testing (because of testing capacity issues). Even though they were wrapped, the specimens were restored for 48h in the moist curing room before stiffness damage testing in order to follow the procedure proposed for concrete cores extracted from real structures (CSA A23.2-14C).

5.2 Methods for assessment and analysis

5.2.1 Stiffness damage test (SDT)

The test samples were subjected to five cycles of loading/unloading at a controlled loading rate (0.15 MPa/s). One set of samples was composed of twelve concrete cylinders with the same mix-design, type of aggregate and expansion level. Each set of samples was divided into 4 sub-sets of three cylinders, in order to evaluate four different loading levels through the test, i.e. 15%, 20%, 30% and 40% of the design (28-day) concrete strength. The results evaluated were the average values of three specimens (sub-sets) tested.

5.2.2 Damage Rating Index (DRI)

The semi-quantitative petrographic analysis of the concrete specimens was carried out using the Damage Rating Index method proposed by Grattan-Bellew & Danay [10] and recently modified by researchers from the Laval University [11]. Petrographic features of deterioration are counted in a grid (minimum of 200, 1 cm by 1 cm squares) drawn on polished concrete sections (16x magnification under the stereomicroscope). The counts of each of the features are then multiplied by selected weighing factors, and those results summed up, to produce the DRI (normalized to 100 cm^2).

The tests were carried out in two ways. First, the DRI was performed on a polished sample that was not subjected to SDT, for verifying the degree of damage at the selected expansion levels. Second, the third sample of each sub-set mentioned before was cut and polished, after completion of the SDT (40% load), for verifying the non-destructive character of the SDT.

6 Results

6.1 Stiffness Damage Test (SDT)

For both aggregates tested in the 25 MPa and the 35MPa concretes, clear differences in the responses with increasing expansion in the test specimens could be seen only at a load corresponding to 40% of the design concrete strength; this can be seen either by the measurements of the hysteresis area or the plastic deformation of the test specimens (first cycle, average value of the last four cycles or over the five cycles). The modulus of elasticity for both mixtures decreases as a function of the expansion level, as expected.

6.2 Damage Rating Index (DRI)

Table 1 shows the results of the petrographic analysis performed on the 25 MPa specimens incorporating the Texas sand (Tx) and New Mexico gravel (NM), before and after the SDT test (40%). One can see that the DRI values are similar from one set to another and the fairly low values of either the standard deviation (SD) or the coefficient of variation (CV) (considering that the DRI is a subjective

---

[1] The 14-day accelerated mortar bar (ASTM C 1260) expansions for the Texas sand and the New Mexico gravel are 0.995% and 1.114%, respectively [9]. The reactive sand (Jobe) and gravel (New Mexico) were used in combination with non-reactive coarse (pure limestone) and fine (granitic) aggregates, respectively.

analysis) suggest that carrying out the SDT up to 40% of the concrete design strength does not introduce additional damage into the sample analyzed, at least at the magnification used in the DRI analysis (16X). This is true for both aggregates investigated, and preliminary results show that this trend is also applicable for the 35 MPa mixtures studied.

6   Discussion

In this study, the SDT was carried out with loading levels ranging from 15 to 40% of the 28-day concrete mix-design strength. The maximum loading (40%) was chosen as one knows that loadings above and beyond this value can introduce new cracks in the sample tested in compression [12]. Table 2 shows that when fixed loading values are used for the SDT, one could easily misinterpret the response of the test, for example based on the hysteresis area values measured. As discussed previously, it seems the best loading to apply in the SDT is 40% of the concrete mix-design strength, as it provides a good diagnosis of the expansion degree of the damaged material and it still maintains a non destructive character in the test, thus allowing other tests like microscopic analysis or compressive strength determination to be carried out on the same specimens, if needed or desired.

The test data obtained in this study confirmed that the hysteresis area, the modulus of elasticity and the plastic deformation are very important responses of the damaged concretes under cyclic loading. But there is a question that still remains: must one use the values measured for first cycle, the average values of the last four cycles or the total value over the test? To answer this question, one needs to analyze each parameter separately. The hysteresis area of a damaged sample corresponds to the energy used to close disseminated cracks during a compressive test. The results obtained in this study indicate that the hysteresis area values for the first cycle are 2 to 3 times greater than the average ones for the last four cycles. Even though both situations could be used for evaluating the damage level in concrete affected by ASR (at the 40% load level), eliminating the response obtained during the first cycle may result in losing important information about the extent of cracking in the test specimens. So, for this parameter, one should consider the first cycle. However, comparing the use of the hysteresis area of the first cycle and the use of the five cycles in terms of correlation against the degree of expansion, it is possible to notice that, for the majority of mixtures, the correlation is better when one uses the five cycles over the SDT (Figures 1A & B). Regarding physically these results, the increase of this correlation could likely be explained by the homogenization of the cracks closure over the test. It seems that maybe one cycle is not sufficient to close all of the cracks and that the use of the four additional ones could help to complete this process. Moreover, the use of the hysteresis area over the five cycles could as well decrease or even mitigate the potential influence of the cage setting (testing equipment "issue") that may occur over the first cycle. Therefore, as all the mixtures with both aggregates were well classified against the expansion level using the hysteresis area over the test, one this parameter is chosen as the first SDT output parameter.

Regarding the plastic deformation in a cyclic testing, it is known that this parameter better distinguishes damaged materials as the numbers of cycles are increased. It is possible to notice that, like to the hysteresis area, the correlation of the plastic deformation against the degree of expansion is better when the five cycles are taken into account (Figures 1E & F). So, it is logical to think that the analyses of the plastic deformation over the five cycles of the test could be more diagnostic and thus, this could be chosen as the second output parameter of the test. Finally, since the five cycles of the SDT will be carried out at 40% of the concrete mix design strength, it is logical that the same parameters commonly used for the ordinary modulus of elasticity determination are maintained. So, the use of the average value of the cycles II and III could be used as the third output parameter of the test (Figures 1C & D). Globally, carrying out the SDT with the above procedure introduces a lot of additional and useful information (hysteresis areas and plastic deformation) to the conventional modulus of elasticity test.

7   Conclusion

Input and output parameters of the Stiffness Damage test (SDT) were discussed for tests carried out using 25 MPa and 35 MPa concrete mixtures incorporating two types of reactive aggregates (Texas sand and New Mexico gravel). The main conclusions are:

- The SDT should be carried out with a percentage of the mix-design strength instead of a fixed loading to analyze damage in different types of concretes;

- Carrying out the SDT with percentages of loading of less than 40% of the concrete mix-design strength, does not make the SDT a diagnostic test against the degree of expansion of concretes damaged by ASR;
- The hysteresis area and the plastic deformation over the five cycles, as well as the average value of the modulus of elasticity obtained in second and third cycles, seem to be the best parameters to use as output responses in the test;
- Even using 40% of the concrete mix-design strength, the test seems to maintain its "non-destructive" character (as one could see from the determination of the microscopic features of deterioration in the samples using the DRI);
- The SDT seems to be a powerful tool but more tests are required with a greater number of samples (different mix-designs and reactive aggregates) to confirm its efficiency;
- Maybe the SDT could be used for other mechanisms as DEF, freezing and thawing, but an in-depth study needs to be carried out.

References

[1]   SMAOUI, N.; BÉRUBÉ, M.A.; FOURNIER, B.; BISSONNETTE, B. & DURAND, B. Evaluation of the expansion attained to date by concrete affected by ASR - Part I: Experimental Study. Canadian Journal of Civil Engineering, Canada, 2004.

[2]   British Cement Association. The diagnosis of alkali silica reaction – report of a working party. England, 1992.

[3]   ST-JOHN, D.A.; POOLE, A.W.; SIMS, I. Concrete Petrography: A handbook of investigative techniques, London, 1998.

[4]   POWERS, L.; SHRIMER, F. H. Quantification of ASR in concrete: An introduction to the Damage-Rating Index method. International Conference on Cement Microscopy (ICMA), 2008.

[5]   CRISP, T. M.; WALDRON, P.; WOOD, J. G. M. Development of a non destructive test to quantity damage in deteriorated concrete. Magazine of Concrete Research, 1993.

[6]   CRISP, T. M.; WOOD, J. G. M.; NORRIS, P. Towards quantification of microstructural damage in AAR deteriorated concrete. International Conference on Recent Developments on the Fracture of Concrete and Rock, 1993.

[9]   FOURNIER, B.; IDEKER, J.H.; FOLLIARD, K.J.; THOMAS, M.D.A.; NKINAMUBANZI, P.C.; CHEVRIER, R. Effect of environmental conditions on expansion in concrete due to alkali–silica reaction (ASR). Materials Characterization Journal, 2010.

[10]   GRATTAN-BELLEW, P. E.; MITCHELL, L. D. Quantitative petrographic analysis for concrete. 8th Symposium on Alkali-Aggregate Reactivity in Concrete, Canada, 2002.

[11]   VILLENNEUVE, V; FOURNIER, B. Determination de l'endommagement du béton par méthode pétrographique quantitative. Annual Progress in concrete Symposium, ACI Québec, 2009.

[12]   MINDESS, S.; YOUNG, J. F.; DARWIN, D. Concrete. Book. Pearson Education Ltda, second edition, 2003.

Table 1.   Microscopic analysis (over the DRI) of the 25 MPa mixtures with the Texas sand and the New Mexico gravel for different degrees of expansion.

Tests		Expansion degrees for all the 25 MPa mixtures						
		0.05% Tx	0.05% NM	0.12% Tx	0.12% Nm	0.20% Tx	0.20% NM	0.30% Tx
DRI values	Standard (without SDT)	226	270	369	396	554	598	724
	SDT 40% + DRI	249	237	360	358	536	599	739
	SD	16.3	23.3	6.4	26.9	12.7	0.7	10.6
	CV (%)	7.2	8.6	1.7	6.8	2.3	0.1	1.5

Table 2.Comparison between loadings over the SDT: fixed vs. percentage of the concrete mix-design.

(%)	fck MPa	Loading (MPa)	Expansion (%)	Hysteresis area (J/m³) - Texas sand	Hysteresis area (J/m³) - New Mexico gravel
40%	25	10	0.12	1142	1883
30%	35	10.5	0.12	691	623
40%	25	10	0.20	1802	2213
30%	35	10.5	0.20	897	913

A - Tx sand

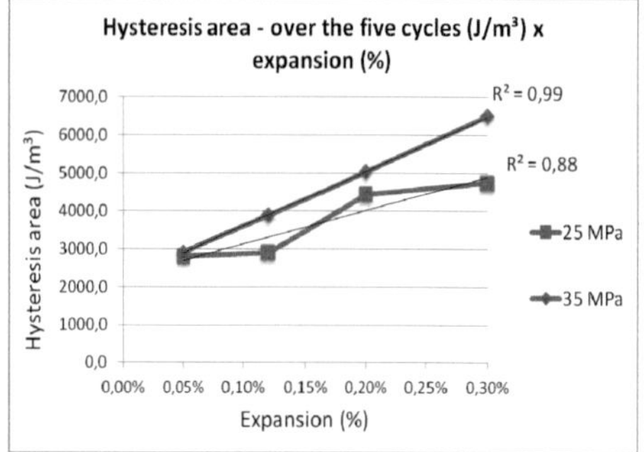

B - NM gravel

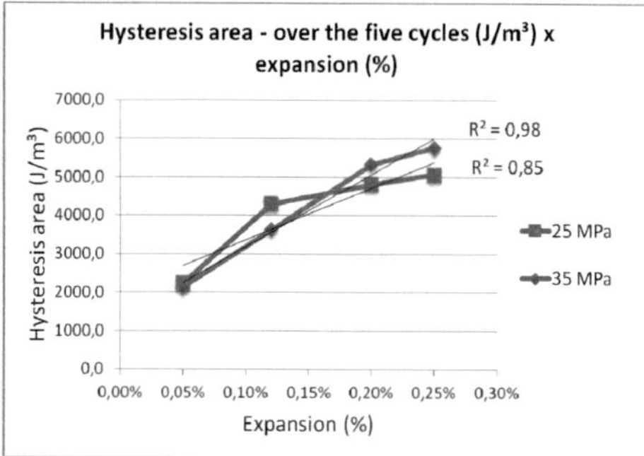

C - Tx sand

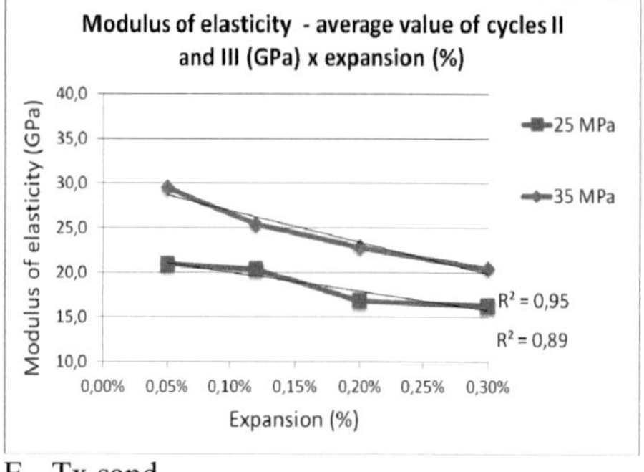

D - NM gravel

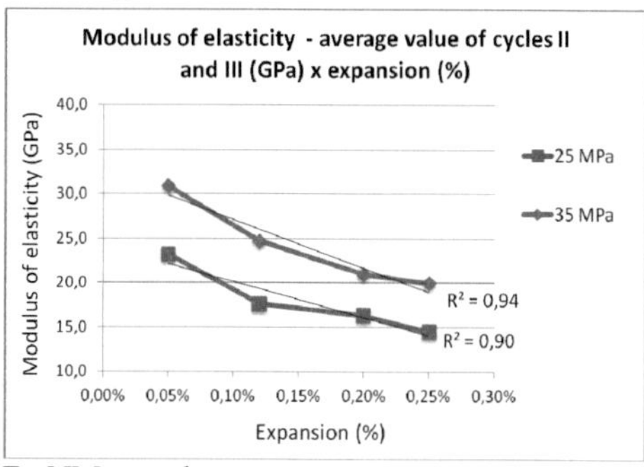

E - Tx sand

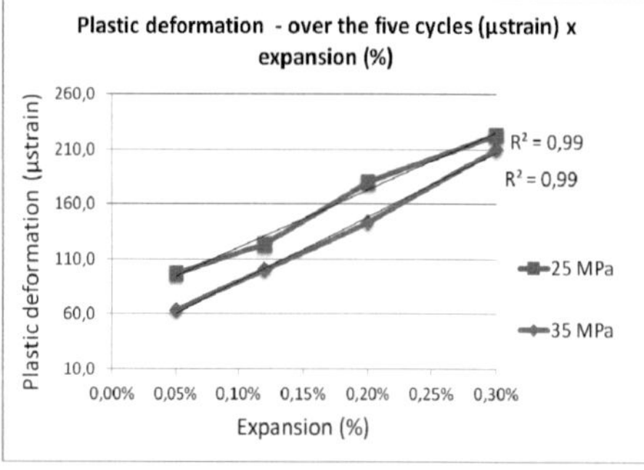

F - NM gravel

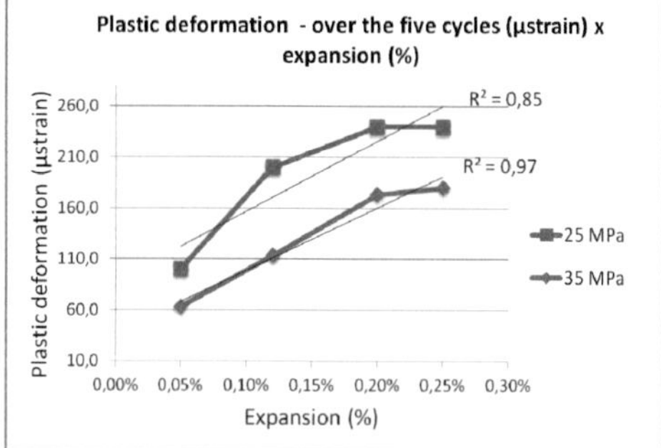

Figure 1.   Analysis of the SDT output parameters when 40% of the mix-design strength is used over the test for the 25 and 35 MPa mixtures with both reactive aggregates.

Deterioration of concrete subjected to hydro-abrasion

Claudia Bellmann

Institute of Construction Materials,
TU Dresden,
Würzburger Str. 46, 01187 Dresden, Germany
Supervisor: Viktor Mechtcherine

Abstract

The concrete surfaces of hydraulic structures can deteriorate severely when continuously or frequently overflowed by water containing solid particles; this process is called hydro-abrasion. In predicting the abrasion resistance and the remaining lifetime of concrete structures subjected to hydro-abrasion, a sound knowledge of the mechanisms leading to the wear of concrete on different levels of observation is needed. Hydro-abrasion tests and morphological analysis of the concrete microstructure were performed on various concrete mixtures in order to characterize the occurring damage. It could be established that wear resistance was affected strongly by both the type and the maximum size of the aggregate. Furthermore, the experimental results show the possibilities and limits of the individual examination methods.

1   Introduction

Surfaces of hydraulic structures made of concrete (e.g. weirs, spillways, plunge pools) are often exposed to flowing water that effects mechanical erosion on the concrete surface. The surface wear referred to as hydraulic abrasion or hydro-abrasion results from the abrasive action of solid particles transported by the water. Essentially, hydraulic abrasion is – along with cavitation and erosion by "pure" moving fluids – just one of the types of wear resulting from flow processes. However, hydro-abrasion by sediments transported in natural waters is the most common form of wear occurring to hydraulic concrete structures. The hydro-abrasive attack comprises a complex exposure which combines sliding, rolling, and saltation effects.

The development of concrete damage due to hydro-abrasion depends on the abrasive regime, which is determined by a variety of influencing factors such as the velocity of the flow, the duration of the exposure as well as the amount and properties (geometry, hardness) of the sediments transported. The main findings of the previous research work in this field can be summarized as follows:

- The higher the flow velocity the higher the erosive-acting kinetic energy of the moving water ([1], [2], [3], [4]);
- the angle of incidence by the solid particles significantly influences the erosion grade ([1], [5]);
- scuffing of the structure increases with increasing grain hardness as well as with increasing sharpness of the grains' edges ([1], [6]).

The meso- and micro-structure of concrete, and herewith, its resistance to abrasion, is influenced by its composition. The main parameters are the water-to-binder ratio, the composition and size of the aggregates, and the quality of the bond between the matrix and aggregates. Although a number of experimental parameter studies have been performed to investigate the hydro-abrasion of concrete, there is very little information on the particular mechanisms of material deterioration due to abrasive wear.

Thus, this study focuses on the methodology for observing and quantifying changes in the surface layer of concrete exposed to hydro-abrasion. The experimental program included both hydro-abrasion tests in various regimes and different methods of the damage characterisations. On the macroscopic level of observation changes in surface roughness, ultra-sound velocity and loss of mass provide a first assessment of the degrees of damage. In order to attain better understanding of the wear mechanisms, morphological analysis of the concrete microstructure is performed, e.g. on the resin-soaked thin-section samples as well as using the environmental scanning electron microscopy (ESEM).

Proc. of the 9th *fib* International PhD Symposium in Civil Engineering, July 22 to 25, 2012,
Karlsruhe Institute of Technology (KIT), Germany, H. S. Müller, M. Haist, F. Acosta (Eds.),
KIT Scientific Publishing, Karlsruhe, Germany, ISBN 978-3-86644-858-2

2   Hydroabrasion tests and concrete composition

For the simulation of planar hydro-abrasive loading and the determination of the abrasion behaviour of concrete test series were accomplished by means of a special testing machine with a rotating drum-like container. The main part of the test facility was an octagonal, horizontally mounted drum controlled via a PC programmable driving unit; cf. Fig. 1 (left). Up to eight slab-type specimens with dimensions of maximum 350mm by 350mm can be mounted in the drum and subjected to dry or wet abrasive loading.

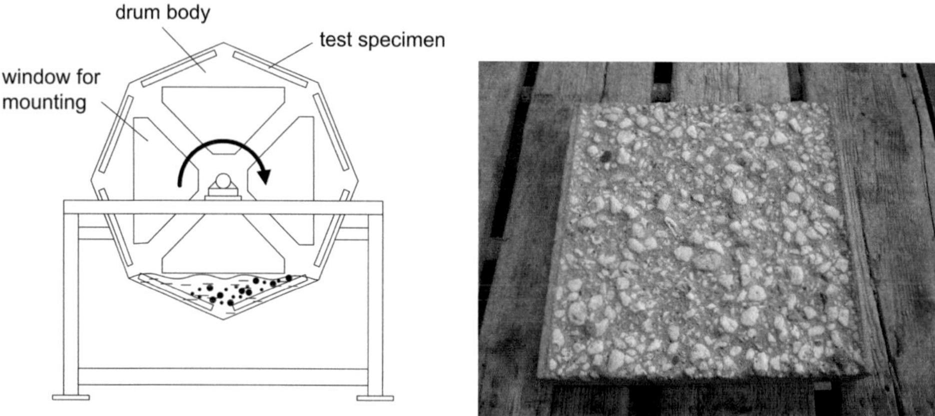

Fig. 1   Left: schematic view of the abrasion test facility [7]; right: testing slab after hydro-abrasive loading [8].

For purposes of this study the dimensions of the slabs were 300x300x50 mm³; cf. Fig. 1 (right) and the abrasive sediment was a mix of steel balls and water. The steel balls-to-water ratio was chosen at 1:1 to coincide with Haroske [1], such that 10 kg of steel balls and 10 kg water were placed into the drum before each experiment. Three characteristic abrasion regimes were examined, which are typical for lower (R1), middle (R2) and upper (R3) river reaches. Therefore the rotational speed of the drum was adjusted to the corresponding, prevalent flow velocity; cf. Table 1. The grain gradation of the sediment was also varied according to the abrasion regime. Representative sediment samples of Elbe, Lower Rhine and Upper Rhine gave an orientation for the selected grading curves (see [8]).

Table 1   Representative conditions of the investigated abrasion regimes [7], [8].

parameter	regime 1 (R1)	regime 2 (R2)	regime 3 (R3)
average grain size: d_m [mm]	4.4	5.0	8.0
rotation speed: u [rot/min]	10.0	13.5	17.0
total loading time of one slab: t_{tot} [h]	15.5	11.5	9.1

Several concrete mixtures were investigated, all of them meeting the basic requirements for concrete subjected to strong abrasion exposure according to the state of the art; cf., e.g., DIN 1045 [9], ZTV-W [10]. A selection of the investigated mixtures is presented in Table 2. A Portland cement CEM I 32.5 R was used as binder and a water-to-cement ratio of 0.45 was kept constant for these mixtures. The fine aggregate was natural quartz sand with a maximum aggregate size of 2 mm. The variation parameter was the type of coarse aggregate (basalt split or quartz gravel) as well as the maximum aggregate size (8 mm or 16 mm); cf. Table 2. Additionally, concrete 2 (cf. Table 2) was pre-damaged by 56 freezing-thawing cycles with plain water prior to the abrasion exposure.

Table 2   Composition of the examined concrete mixtures, data for 1 m³ concrete [7].

		concrete 1	concrete 2	concrete 3	concrete 4	concrete 5
pre-damage by frost			X			
CEM I 32.5 R	[kg]	367.5	367.5	330.8	350.0	315.0
fly ash	[kg]	120.7	120.7	108.7	115.0	103.5
fine sand 0.06/0.2	[kg]	176.9	176.9	139.3	187.0	118.1
sand 0/2	[kg]	675.5	675.5	531.9	714.0	451.0
basalt split 2/5	[kg]	534.2	534.2	420.7		
basalt split 5/8	[kg]	374.7	374.7	295.1		
basalt split 8/11	[kg]			231.9		
basalt split 11/16	[kg]			273.1		
quartz gravel 2/8	[kg]				799.0	504.7
quartz gravel 8/16	[kg]					715.9
water	[kg]	187.1	187.1	168.4	178.2	160.4
super plasticizer	[kg]	2.4	2.4	2.0	3.0	1.1

Further investigated concrete mixtures are presented in [8] and detailed information to the concrete composition can be found in [7].

3   Experimental results

3.1   Weighting of mass loss and laser-optical surface measurement

The hydro-abrasive loading period amounted to altogether 60000 revolutions. After defined test intervals of in total 5000, 15000, 30000 and 60000 drum rotations typical erosion values like mass loss and abrasion depth were determined in order to characterize and quantify the deterioration of the concrete and to analyse the slab's surface condition. The mass loss of the concrete slabs was documented by means of a precision mass weighing and the remaining thickness of the slabs was laser-optically measured. The investigations of the specimens took place always in the water-saturated condition.

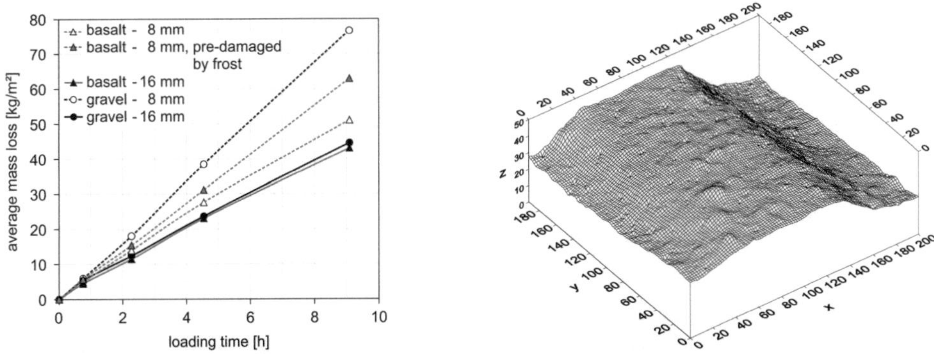

Fig. 2   Left: average mass loss of concrete slabs of all tested concretes in the loading regime R3; right: digitised surface of the concrete slabs (concrete 4) after 9.1 h permanent loading in the abrasion regime R3.

The average absolute mass loss of the slabs stressed in the abrasion regime R3 is presented in Fig. 2 (left) for the concrete mixtures given in Table 2. In comparison to the mixtures, in which smaller coarse aggregates (maximum size of 8 mm) were used, the concrete mixtures with a maximum grain size of 16 mm show a substantially higher resistance to the abrasion, whether the aggregate type is basalt split or quartz gravel. Furthermore, it can be seen that the abrasion resistance was also reduced by the pre-damage of the specimens by the exposure to cyclic freezing and thawing, as had been expected. The linear tendencies presented in Fig. 2 (left) could be observed in all abrasion regimes under investigation (see also [7]).

The values of the laser-based surface measurement indicate similar relations of the material's abrasion resistance. However, this examination method also yields supplementary information, which would remain hidden in the determination of the total loss of mass alone. As the digitised surface of concrete 4 (cf. Fig. 2 (right)) shows, the abrasive wear in the drum was not in every case evenly distributed over the surface of the specimens tested. The rib formation transverse to the direction of movement of the slabs particularly occurred in the abrasion regime R3. This phenomenon has probably its cause in drum geometry and is systems-inherent. The increasing over time of the average abrasion depth can be found in [8] for a variety of concrete mixtures according to the abrasion regimes.

3.2 Ultrasonic measurement

The non-visible, internal deterioration of the concrete was studied by the non-destructive method of ultrasonic measurements during the abrasion process. Fig. 3 clearly shows for exposure in the abrasion regime R1 a reduction of ultrasonic transit times arising from increasing duration of the abrasion process. This behaviour indicates an increase in the internal damage in the concrete slabs. Different to those of mass loss or erosion, the declining values of ultrasonic velocity possess a non-linear character. This indicates that there is no constant increase of damage to the internal structure of the slabs during the drum experiment.

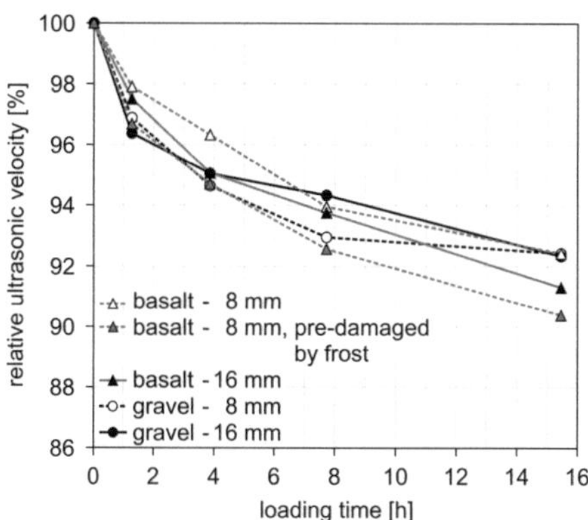

Fig. 3   Relative ultrasonic velocity through concrete slabs of all tested concretes as measured on the side of the specimens exposed to abrasion in the loading regime R1.

Although the values of ultrasonic velocity do not provide quantitative information on the number or size of micro-cracks, the changes in these values resulting from the abrasive loading can be used to characterise the damage to the microstructure. A more detailed presentation of the results of the ultrasonic measurement and an in-depth discussion are presented in [7].

4 Microstructure analysis

The hydro-abrasive loading is a complex procedure, which leads to crack development in concrete. A microscopic analysis of the damaged concrete material by means of an optical microscope and the use of an environmental scanning electron microscope (ESEM) can give information, how this process in detail takes place.

Fig. 4 (left) shows as an example the surface layer of concrete 4 (quartz gravel, maximum grain size of 8 mm) after 9.1 h permanent loading in the abrasion regime R3 on a thin section sample, which had been prepared from the concrete specimen soaked with fluorescent resin. The characteristic damage due to the hydro-abrasive loading can also be noticed by the ESEM-depiction of the stressed slab surface of concrete 3 (basalt split, 16 mm maximum particle size); cf. Fig. 4 (right). Both figures illustrate that the surface of the concrete slabs is uneven and strongly damaged. The aggregate particles at the surface are partially smashed. Further examination should determine whether or not this damage pattern corresponds to damage by natural hydro-abrasion (cf. also Bertram [11], who observed that steel balls caused a higher wear effect then particles of natural stones).

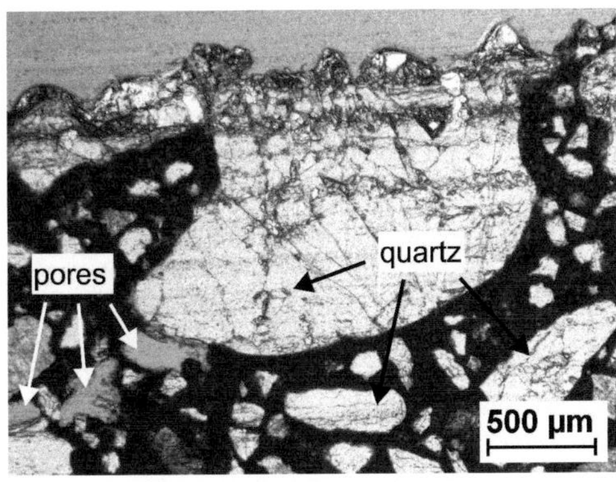

Fig. 4   Microscopic images of the slab surfaces, left: concrete 4 after 15.5 h permanent loading in the abrasion regime R1; right: concrete 3 after 9.1 h permanent loading in the abrasion regime R3 (ESEM).

The spalling of little concrete particles from the slab surfaces is graphically documented in [7]. Further analyses applying environmental scanning electron microscopy (ESEM) are necessary to show micro-cracks and to specify the wear mechanisms in detail.

5   Summary

In this paper the results of an experimental investigation on the effects of wear exposure on the concrete degradation due to hydro-abrasion are presented. Parameters under investigation were the concrete composition, the duration of the wear exposure as well as the degree of pre-damage by freezing and thawing cycles. It was found that wear resistance of concrete was affected strongly by the maximum size of the aggregates. Furthermore, concrete pre-damaged by freezing and thawing cycles shows clearly higher signs of wear than the corresponding concrete mixture without pre-damaging. In all experiments a linear relationship between the duration of the wear exposure and the loss of mass was observed. The laser measurements served as an additional facility for the assessment of the concrete's surface characteristics due to hydro-abrasive loading. The digitisation of the damaged surfaces showed unevenly distributed material losses over the surfaces of the concrete slabs stressed in an intense abrasion regime.

With the determination of the ultrasonic velocity through the concrete a non-destructive measuring procedure was used for the characterization of the internal material damage. Here the relationship between the duration of the exposure and the increase of the relative velocity of the ultrasonic waves was non-linear. At the microscopic level of observation it was discovered by means of an optical microscope and an ESEM that, due to the saltation effect of the steel balls, the aggregate grains were partly destroyed while these were still anchored in the hardened cement. The ongoing microscopic investigations will provide information on the development of micro-cracks due to hydro-abrasion.

The results of the performed experiments show the possibilities and limits of the individual research methods for characterization of concrete damage due to hydraulic abrasion. The determined data will serve among others as a basis for the development of a numeric model, which is based on the Distinct Element Method (DEM), to simulate the hydraulic abrasion of concrete. Further, the experimental data enables also the further development of approximation approaches for the determination of the hydraulic abrasion wear [8].

Acknowledgements

The author acknowledges with gratitude the financial support of this project by the German Research Foundation (DFG). Furthermore, the author gratefully acknowledges the support of the Institute of Hydraulic Engineering and Technical Hydromechanics, TU Dresden.

References

[1]   Haroske, G.: Beitrag zum Hydroabrasionsverschleiß von Betonoberflächen. Dissertation, Universität Rostock, (1998)

[2]   Helbig, U., Horlacher, H.-B., Schmutterer, C., Engler, T.: Möglichkeiten zur Erhöhung der Festigkeit abrasionsbeanspruchter Betonoberflächen bei wasserbaulichen Anlagen. In: Bautechnik 82 (2005) No. 12

[3]   Jacobs, F., Winkler, K., Hunkeler, F., Volkart, P.: Betonabrasion im Wasserbau. Grundlagen – Feldversuche – Empfehlungen. In: Minor, H.-E. (Hrsg.), Mitteilungen der Versuchsanstalt für Wasserbau, Hydrologie und Glaziologie, ETH Zürich (2001) No. 168

[4]   Zanke, U.: Grundlagen der Sedimentbewegung. Berlin/Heidelberg/New York (1982)

[5]   Bania, A.: Bestimmung des Abriebs und der Erosion von Betonen mittels eines Gesteins-stoff-Wassergemisches. Dissertation B, Technische Hochschule Wismar (1989)

[6]   Offermann, P., Horlacher, H.-B. et al.: Erhöhung der Abrasionsfestigkeit von dünnen Mörtelschichten durch textile Strukturen. Schlussbericht zum AiF-Forschungsprojekt Nr. 12872 BR, TU Dresden (2003)

[7]   Bellmann, C., Mechtcherine, V. et al.: Nachbildung der Hydroabrasionsbeanspruchung im Laborversuch Teil 1 – Experimentelle Untersuchungen zu Schädigungsmechanismen im Beton, In: Bautechnik 89 (2012), (in Vorbereitung)

[8]   Helbig, U., Horlacher, H.-B. et al.: Nachbildung der Hydroabrasionsbeanspruchung im Laborversuch Teil 2 – Korrelation mit Verschleißwerten und Prognoseansätze, In: Bautechnik 89 (2012), (in Vorbereitung)

[9]   DIN 1045: Tragwerke aus Beton, Stahlbeton und Spannbeton, Teil 1: Bemessung und Konstruktion, 2008-08 Teil 2: Festlegung, Eigenschaften, Herstellung und Konformität – Anwendungsregeln zu DIN EN 206-1, 2008-08, Teil 3: Bauausführung, 2008-08, Teil 4: Fertigteile, 2001-07

[10]   ZTV-W LB 215: Zusätzliche Technische Vertragsbedingungen Wasserbau, Wasserbauwerke aus Beton und Stahlbeton. Bundesminister für Verkehr, Bau und Stadtentwicklung (2004)

[11]   Bertram, A.: Versuchstechnische Simulation der Hydroabrasion an überströmten Betonoberflächen im Wasserbau. Diplomarbeit, Institut für Wasserbau und Technische Hydromechanik / Institut für Baustoffe, TU Dresden (2007)

Modelling of the saturation behaviour of cement stone during freezing and thawing action

Zorana Djuric

Institute of Concrete Structures and Building Materials,
Karlsruhe Institute of Technology (KIT),
Gotthard-Franz-Str. 3, 76131 Karlsruhe, Germany
Supervisor: Harald S. Müller

Abstract

Modelling of frost damage is a key element in the life cycle management of concrete structures. In case of a pure freeze-thaw cyclic loading, a depth-resolved quantification of the water saturation behaviour of concrete is of key interest. Upon reaching a critical value of saturation, structural damage will occur at only one freeze-thaw cycle (FTC). Because the frost suction cannot be described from single classical transport laws, the challenge in assessing the saturation behaviour of concrete is a continuous, non-destructive and spatially resolved registration of the water uptake during the frost exposure. Based on time- and depth-resolved experimental moisture investigations during the freezing-thawing action on hardened cement paste a new engineering model approach was developed.

1   Introduction

Concrete structures are exposed to various environmental stresses. One of the most relevant stresses in moderate climate zones such as northern Europe is the frost attack. Frost damage occurs in case several unfavourable conditions are met at the same time. The damage characterized by scaling and internal cracking can lead to a reduction of the strength of the concrete so that the intended service life of the building is reduced. The life-time prediction of concrete exposed to frost strongly depends on the accuracy of the model laws. However, due to the complexity of the physical effects observed during freezing and thawing the formulation of time-dependent deterioration laws is extremely difficult. So far, no model with a sufficient accuracy for the service life prediction of the freeze-thaw attack for concrete is available. Hence, there is great need for research.

2   Physical mechanisms of frost damage

The deterioration of concrete caused by freeze-thaw action is directly linked to the formation of ice on the concrete surface and within its porous microstructure. During the phase change from liquid to solid water exhibits a volume dilatation of approx. 9 vol.-%. In case the degree of water saturation in the pores exceeds a defined critical value the pore space is not sufficient for the volume expansion of freezing water [1]. The freezing water causes a burst pressure in the pores which can lead to tensile stresses and so gives rise to an internal cracking [2]. Against this background, the degree of water saturation of the pore system is essential for the formation of damages within the concrete. The critical degree of saturation is reached only by repeated freeze-thaw action with low temperatures [5]. The relevant mechanisms were explained by Setzer with the so-called "micro ice lens model" [4]. Essential for understanding the mechanisms of freeze-thaw damaging is the fact, that the freezing temperature of water is strongly influenced by surface forces. As the surface forces strongly increase with reducing pore size, the freezing behaviour of the pore water thus is a function of the pore size distribution of the concrete [7, 8]. As concrete basically has a very wide range of pore sizes, water exists at normal freezing temperatures in all three aggregate states, i. e. in larger pores in vaporous, liquid and solid state, in small one in vaporous and liquid state [8, 4]. The ice contained in larger micro pores has a much lower vapour pressure than the water contained in the smallest gel pores. This causes a water rearrangement process within the pore system. The degree of saturation in the micro pores increases whereas the much finer gel pores dry out. During the thawing process, the water in larger micro pores is still frozen (micro ice lens) while the gel pores seek to recover their primary saturation state [4]. In case at this time liquid water is available e. g. on the concrete surface, then it will be sucked into the pore system and causes an increase of the degree of saturation [4]. This process is

Proc. of the 9th *fib* International PhD Symposium in Civil Engineering, July 22 to 25, 2012,
Karlsruhe Institute of Technology (KIT), Germany, H. S. Müller, M. Haist, F. Acosta (Eds.),
KIT Scientific Publishing, Karlsruhe, Germany, ISBN 978-3-86644-858-2

described as frost suction [4]. Hence the damage mechanism due to frost attack can be subdivided in a non-damaged, reversible initiation phase and after reaching the critical degree of saturation into a damage phase. The duration of the initiation phase is dominated by the frost suction which cannot be described from classical transport processes such as capillary suction or diffusion (also see [3]). Fig. 1 shows that the water uptake resulting from freeze-thaw action by far outweighs the one from capillary suction. Further results presented by Kruschwitz [3] using a thermodynamic model show, that changes in the water saturation of the concrete are normally limited to the top millimetres of the concrete, see Fig. 1. Experimental evidence for these results is however missing so far.

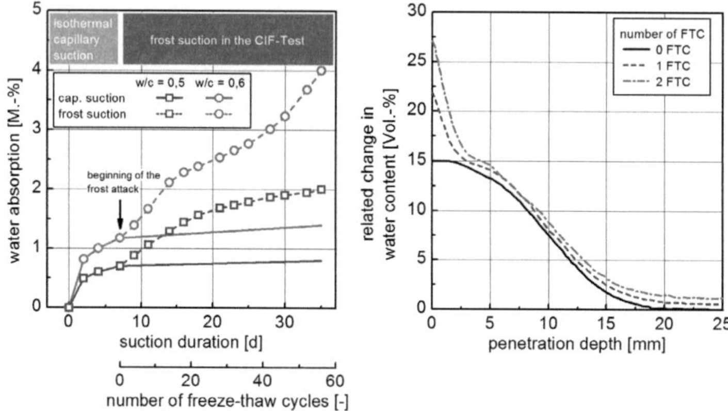

Fig. 1   Left: water absorption of two different concretes with water/cement ratio (w/c) of 0,5 and 0,6 during capillary suction and frost suction (dashed lines: data from [5]). Right: related water content change in dependence of the distance from the hardened cement paste surface before and after the frost stress from a numerical calculation [3].

As damages occur when the critical saturation is reached, modelling the deterioration behaviour of concrete must focus on the description of its saturation behaviour.

3   Deterioration time laws

Deterioration time laws are mathematical models which describe e. g. the scaling of concrete as a function of time. The model can either be chosen on an empiric basis or correctly reproduce the physical processes taking place. The most important ones are discussed below. In case of frost action, this later task turns out to be very difficult. Although with Setzer's micro ice lens model in principle the physical processes of frost damage are understood, only a very restricted prediction of the saturation and damage behaviour of concrete at a given frost action is feasible so far. This is mainly due to the fact of the nearly endless range of possible freezing temperatures, temperature gradients and moisture supply scenarios, which interact via the pore radius freezing point relationship with a very large number of different pore sizes [7, 8].

3.1   Model by Sarja and Vesikari

The deterioration time law of Sarja and Vesikari [9] allows the determination of the annual scaling of the concrete surface based on empirical data, see Equation 1.

$$r(t) = c_{env} \cdot c_{cur} \cdot c_{age} \cdot a^{-0,7} (f_{ck} + 8)^{-1,4} \cdot t \tag{1}$$

where $r(t)$ = concrete scaling [mm]; t = time [year]; c_{env} = coefficient considering the intensity of the frost loading [-]; c_{cur} = coefficient considering the curing quality [-]; c_{age} = coefficient considering the concrete ageing or rather the applied additives [-]; a = air content [%]; and f_{ck} = characteristic compressive cube strength of concrete at 28 days [MPa].

Equation 1 shows that the intensity of frost loading as well as the properties of concrete are included in the model. The climatic conditions are incorporated in the environmental coefficient c_{env}, which is subdivided in four classes [9]. Here a differentiation between a pure frost loading and a combined frost loading with chlorides is possible. Adjacent to the characteristic compressive strength and the air content, the addition of pozzolanic additives can be taken into account by the coefficient c_{age} representing the effective age of the concrete. The curing quality is also considered in the model.

3.2 Model approach by Lowke and Brandes

An additional model for determination of the annual, time-dependant scaling of concrete surfaces is the proposed product approach by Lowke and Brandes [10], see Equation 2.

$$r = k \cdot f_s \cdot f_{T_{min},c} \cdot f_{wc} \cdot f_{bin} \cdot f_{aea} \cdot f_{carb} \tag{2}$$

where r = scaling per one FTC [m]; k = maximum tolerable scaling per FTC [m]; f_S = factor considering the salt concentration [-]; $f_{Tmin,c}$ = factor considering the minimum temperature [-]; f_{wc} = factor considering the w/c-ratio [-]; f_{bin} = factor considering the bonding agent [-]; f_{aea} = factor considering the air-entraining agent [-]; and f_{carb} = factor considering the carbonated edge zone of concrete [-].

The deterioration time law shown in Equation 2 is designated for the description of a combined frost attack with chlorides. The damage (scaling), is calculated as a function of the exposure and the material resistance. Unfortunately the authors do not give any information on the determination of the different factors used in Equation 2. Further it should be kept in mind, that the ongoing saturation process and the concrete deterioration by frost are cumulative processes. The suitability of a product law approach must therefore be questioned until further information regarding the definition of the proposed factors is available.

3.3 Model by Fagerlund

The deterioration time law by Fagerlund is directly based on physical processes which are taking place in the material structure during the freeze-thaw loading [1]. According to Fagerlund, damage occurs when the degree of saturation reaches a defined critical value S_{crit} (see also [6]). Hence Fagerlund focused himself on the modelling of the saturation process. Thereby the saturation process can be separated in two phases. The first phase is characterized by a very fast capillary suction process S_b (see Equation 3), which however is limited in the total saturation it can reach by the gaseous phase in the pores, i. e. the air. For this reason not every pore is completely saturated during capillary suction. The remaining air slowly dissolves within the pore water in the second phase of the saturation process. Consequently the saturation increases also after ending of capillary suction (see Equation 3).

$$S(t) = S_b + a \cdot t^d \leq S_{crit} \tag{3}$$

where $S(t)$ = degree of concrete saturation at time t [-]; S_b = degree of saturation due to fast capillary suction [-]; t = time [days]; a, d = material coefficients; and S_{crit} = critical degree of saturation.

The critical degree of saturation S_{crit} depends on the concrete composition and can be determined in frost-tests by plotting the relative dynamic elastic modulus $E_{dyn,rel}$ as a function of the degree of water saturation. As soon as $E_{dyn,rel}$ reaches a value below 80 % [4], a critical saturation is reached. From Equation 3 it is clear, that with Fagerlund´s model it is not possible to map the real freeze-thaw action. The second additive term in Equation 3 is completely independent of the loading, which however in reality lead to the increased saturation and critical degree of saturation in concrete pores.

3.4 Evaluation of the presented models

So far, no comparative analysis of the presented frost models is existent in the literature. The reason for this is to be seen in the fact, that for some models, the determination of the model parameters is not clearly defined. Further, the definition of frost damage varies significantly, i. e. from the amount of scaling to the degree of saturation, and is thus not directly comparable. The model of Sarja and Vesikari [9] directly describes the damage phase and considers both the intensity of the frost loading as well as the quality and maturity of the used concrete. The large variation in the factor c_{env} however makes it difficult to accurately predict the concrete scaling in the practical use, as information on how to calibrate c_{env} on the actual environmental conditions is missing.. The model approach by Lowke and Brandes [10] represents a refinement of Sarja's and Vesikari's model. In addition, the frost action can be differentiated closer. As already mentioned, unfortunately there is no information regarding the determination of material factors. In contrast to the models mentioned above, Fagerlund's model [1] describes only the initiation phase, but not the damage itself. The duration of the initiation phase ($S < S_{crit}$) is greatly underestimated by the model. This deficiency is attributed to the fact, that the relationship between saturation and intensity of freeze-thaw loading is not covered by the model. Thus the whole mechanism of the micro ice lens pump is not included in this model.

Although with regard to practical applications Fagerlund's model is less suitable, it has to be regarded as a groundbreaking contribution with respect to the depiction of the physical processes causing frost damage. The limit state S_{crit} introduced by Fagerlund closely describes the instance, when damage will start to occur. The cumulative formulation of the model also seems to be suited to describe the unidirectional process of water proceeding into the concrete from the surface.

4   Moisture measurement with NMR

The non-destructive and spatially resolved quantification of the water suction of hardened cement paste caused by repeated freeze-thaw cycling is essential in the context of a physically correct modelling of progressive saturation. A suitable measurement tool for this is nuclear magnetic resonance imaging (NMR). It is primarily known for medical applications. Furthermore, NMR has come into operation in many engineering studies, particularly in building materials. This chapter gives a short introduction to the NMR method and presents preliminary experimental investigations on hardened cement paste. Finally the developed setup for the one-dimensional temperature control within the NMR device is presented.

4.1   Measurement principle of the NMR method

The basic physical concept underlying the NMR method is the interaction of atom nuclei with external magnetic fields. Hereby, hydrogen nuclei can be well detected. The NMR setup in principle is shown in Fig. 2. It consists of two permanent magnets (1), in the centre of which an open-ended cylindrical sample space (2) is located so that the specimen (5) is exposed to a directed magnetic field B_0 through which an equilibrium magnetization is produced. The sample space itself is surrounded by a coil (3), by which the ^{1}H nuclei of the sample can be excited for a short time with a high frequency field B_1, perpendicular to the B_0 field. By the superposition of these two magnetic fields, the nuclear magnetization of ^{1}H nuclei atoms is deflected from its B_0 orientation. After the B_1 field has been turned off, a precession movement of the magnetization in the sample is observed, which induces an electrical response signal in the high frequency coil. The intensity of the response signal is proportional to the existing number of ^{1}H nuclei and thus to the water content of the sample.

Using the NMR measurement technique an integral value for the water content in the sample can be obtained. For a depth-resolved moisture measurement, the experimental setup was equipped with a gradient coil (4) by which the magnetic field can be varied locally in its intensity, thus encoding the measurement signal with a spatial information. The measured intensity profile over the height of the sample is referred to as weighted moisture content.

4.2   Preliminary experimental investigations with the NMR method

To check, whether the low-field NMR methodology is suitable for depth-resolved moisture measurement with high resolution, moisture measurements were performed on hardened cement pastes with different w/c-ratios. For the sample preparation Portland white cement CEM I 42.5 R was used. All hardened cement paste samples were vacuum-treated and then saturated with water. Fig. 2 shows the experimental setup and the results of measurements on saturated hardened cement paste samples with different w/c-ratios.

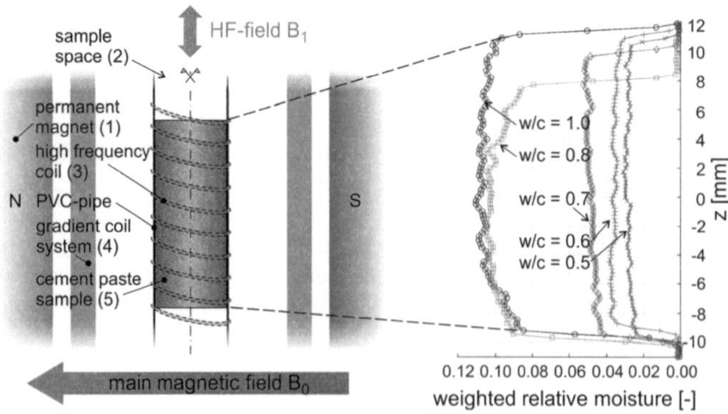

Fig. 2
Left: draft of the setup for NMR measurements on samples. Right: weighted relative moisture distributed over the longitudinal axis of the cement stone samples before drying. (signal is indicated relative to the signal of a homogeneous reference sample)

A spatial resolution of 250 μm between adjacent data points could be achieved in the experiments. The transition from air to the sample reflected in a sudden increase in signal intensity (Fig. 2). Further, Fig. 2 shows that the signal intensity decreases strongly with decreasing w/c-ratio. This is attributed to the decreasing porosity with decreasing w/c-ratio and thus to the decreasing water content in the sample and modified NMR relaxation behaviour.

4.3 In situ one-sided temperature control

For the in situ investigation of moisture profiles in hardened cement paste samples during the frost attack a NMR measurement system with a specially adapted sample head was developed (see Fig. 3).

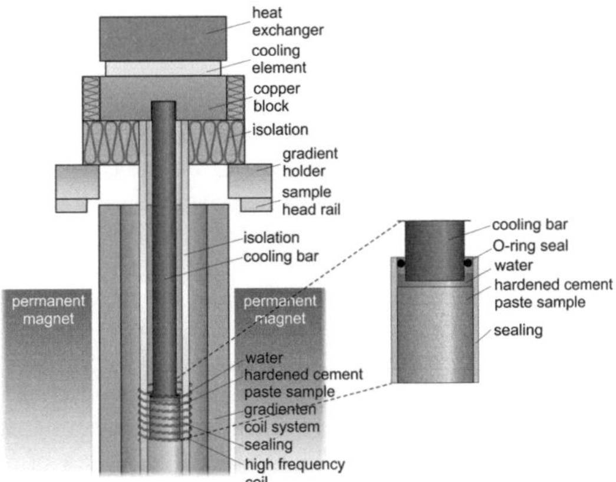

Fig. 3   Setup for the in situ one-sided temperature control.

For the one-sided cooling of the hardened cement paste sample in the NMR system, a suitable thermoelectric cooling element provides the required cooling power, which is transferred via a cooling bar to the sample. The material of the cooling bar has to be a good heat conductor without being an electrical conductor and neither being para- nor ferromagnetic. To ensure a one-dimensional moisture transport in the sample only its circumference is sealed with epoxy resin. In order to allow a unilateral water exposure of the sample from above, the sealant is manufactured longer than the real sample, forming a small water container in which water is inserted and cooled from above by the cooling bar.

4 Modelling of the saturation behaviour of hardened cement paste

The freeze-thaw model presented in the following is based on the Fagerlund's model, which was modified in order to map the principal mechanisms of the micro ice lens model. Therefore the Fagerlund's model was extended by a term that simulates the frost suction. For the engineering-based quantification of a frost event this term should functionally map the incremental increase of water saturation of hardened cement paste during each freezing cycle (see Equation 4). According to Fagerlund, failure of the concrete occurs, in case the water saturation exceeds a critical value S_{crit}. After the determination of the saturation function and the critical degree of saturation it will be possible to estimate the point of time at which the material damage starts to occur. A central feature of the proposed model is the depth-resolved depiction of the attack and the prediction of the resulting change of water saturation. For this, the sample is discretized in layers. At the beginning of a freeze-thaw cycle a defined initial saturation $S_{i,n-1}$ is assigned to each layer i. The current saturation $S_{i,n}$ results from pure capillary suction S_{cap}, increments of previous cycles and from the current freeze-thaw cycle (Equation 4).

$$S_{i,n} = S_{i,n-1} + \Delta S_{i,n} = S_{cap} + \sum_{v=1}^{v=n} \Delta S_{i,v} < S_{crit} \tag{4}$$

where $S_{i,n}$ = degree of saturation after n FTC for the layer i; $S_{i,n-1}$ = degree of saturation after (n-1) FTC for layer i; $\Delta S_{i,n}$ = increase of degree of saturation during the n-th FTC for the layer i; S_{cap} = degree of saturation due to capillary water uptake; $\sum \Delta S_{i,v}$ = summation of all $\Delta S_{i,n}$ (over all FTC) for the layer i; and S_{crit} = critical degree of saturation.

It is assumed that the capillary suction is already completed at the beginning of the freezing action. Thus, S_{cap} is only a function of concrete properties and is assumed to be constant over all layers i and so over the complete attack depth. The increment function $\Delta S_{i,n}$ is dependent on the present environmental conditions. Based on the available literature references and the micro ice lens model it has to be assumed that at pure frost attack the intensity of the suction process is a function of temperature, freezing duration, freezing and thawing rate and number of the freeze-thaw cycles. Moreover, the saturation function or increment function $\Delta S_{i,n}$ has to consider the resistance of the used material, like the w/c-ratio, age of the samples and curing. For the parameterization of the saturation increment function $\Delta S_{i,n}$ each function parameter is to be expressed as a function of the physical parameters given above using the depth-resolved NMR measurement results. The definition of these functions should guarantee the best possible prediction accuracy of the saturation increment function.

6   Summary and Conclusions

An engineering-based durability design allows for a physically based calculation of the concretes behaviour under a given exposure. Hereby the lifetime of the structural element can be estimated. A central role of such a design procedure is formed by deterioration time laws, which model the damage process for the respective exposure. In case of the frost action, design models for the lifetime prediction with acceptable precision do not exist so far.

The presented approach for modelling the deterioration of concrete subjected to a frost attack is based on the physical processes, namely the increasing water uptake during the freezing and thawing process which can lead to a critical saturation and provokes damages. Using non-destructive NMR measurements, the saturation behaviour of the concrete exposed to frost can be quantified in situ with a high spatial resolution. Building up on this data, the influence of both concrete-technological parameters as well as environmental conditions on the saturation behaviour of the concrete could be studied. Using this experimental data, the model presented in this paper will be calibrated. The model allows for a depth resolved prediction of the degree of saturation of the concrete after a certain number of freeze-thaw cycles with varying minimum temperature and temperature gradients.

References

[1]   Fagerlund, G.: A Service Life Model for International Frost Damage in Concrete. Lund Institute of Technology, Division of Building Materials (2004)

[2]   Powers, T. C.: A Working Hypothesis for Further Studies of Frost Resistance of Concrete. In J. of ACI (1945) 41, pp. 245-272

[3]   Kruschwitz, J.: Instationärer Angriff auf nanostrukturierte Werkstoffe – eine Modellierung des Frostangriffs auf Beton. Univ. Duisburg-Essen, Cuvillier, Dissertation (2008)

[4]   Setzer, M. J.: Micro-Ice-Lens Formation in Porous Solid. In J. of Colloid and Interface Science 243 (2001), pp. 193-201

[5]   Müller, H. S., Guse, U.: Zusammenfassender Bericht zum Verbundforschungsvorhaben „Übertragbarkeit von Frostlaborprüfungen auf Praxisverhältnisse". DAfStb (Ed.), Berlin, Beuth, 577 (2010)

[6]   Model Code for Service Life Design. *fib* Bulletin 34, Fédération Internationale du Béton, Lausanne (2006)

[7]   Powers T. C.: Properties of cement paste and concrete. Proc. Fourth Int. Symp. Chemistry of Cement 2 (1960), pp. 577-609

[8]   Setzer, M. J.: Einfluss des Wassergehalts auf die Eigenschaften des erhärteten Betons. In DAfStB, 280 (1977)

[9]   Sarja, A., Vesikari, E.: Durability Design of Concrete Structures. Report of RILEM Techn. Committee 130-CSL (1996)

[10]   Lowke, D., Brandes, C.: Prognose der Schädigungsentwicklung von Betonen bei einem Frost-Tausalz-Angriff. In 8. Münchner Baustoffseminar (2008)

Corrosion of reinforcement steel embedded in concretes manufactured with various cement replacement materials and their gas permeability

Adel Elbusaefi

Cardiff School of Engineering, Cardiff University,
Queen's Building, The Parade, Newport Road,
Cardiff, CF24 1AA, Wales, UK
Supervisors: Robert Lark and Diane Gardner

Abstract

In this paper, the relationship between compressive strength, gas permeability and the corrosion level of steel reinforcement in concretes made with different cement binders is experimentally investigated. These binders included CEM II and blended cements of fly ash (PFA) and ground granulated blast furnace slag (GGBS).The experimental work was conducted by placing 200mm cube test specimens in a saline solution (3.5% NaCl) for a range of exposure times (3, 7, 10 and 14 days) with an applied external current of 10 mA and the gas permeability of concrete was measured using a pressure decay cell apparatus. The results showed that compressive strength increased in the three mixes as function of time. At the end of the exposure time, the corrosion level in CEMII was 21% higher than the blended cements. Also, concrete made with PFA exhibited lower permeability coefficient compared to CEMII and GGBS concrete.

1   Introduction

The surrounding environmental and exposure conditions affect concrete durability. The corrosion of steel reinforcement in concrete is one of the most significant problems that affects the durability and sustainability of reinforced concrete (RC) structures, particularly when the concrete is exposed to a marine environment, de-icing salts and carbonation [1]. The steel bar may become corroded in concrete leading to cracking, spalling of the concrete cover and a reduction in the cross sectional area of the reinforcing steel; all of which, in turn, reduce the service life of the structure[2, 3]. Moreover, concrete permeability is an important factor in the control of the movement of chloride ion through concrete towards the reinforcing bar. It is well known that the volume and size of the interconnected capillary pores in the cement paste affect concrete permeability. Consequently, achieving low permeability of concrete can improve its resistance to the penetration of fluid, chloride ions, alkali ions, carbonation and other causes of chemical attack [4].

The use of supplementary cementitious materials (SCMs), such as ground granulated blastfurnace slag (GGBS), pulverised fuel ash or fly ash (PFA), silica fume (FS) and metakaolin (MK) has become very common in concrete manufacture because of the ability of these materials to significantly improve the durability of concrete by decreasing the permeability (i.e. porosity). An additional advantage is the reduction in the carbon dioxide emissions associated with the manufacture of cement thus, SCMs are considered more environmentally sustainable materials [5].

Many researchers have studied the use of SCMs in reinforced concrete. Yoon-Seok et al [6] found that the partial replacement of cement with PFA led to an improvement in corrosion resistance and decrease in the corrosion rate due to the reduction of permeability and hence an increase in the resistance to movement of chloride ions. Moreover, Tae-Hyun Ha et al [7] investigated the effect of PFA (10-40%) on the corrosion performance of reinforcement in concrete mortar. It was found that up to 30% of PFA replacement enhanced the resistance of the corrosion properties of steel in concrete and improved the permeability characteristics of concrete.

More recently, Hui-sheng et al [8] also observed an improvement in the behaviour of 50% slag concrete against gas permeability for prolonged wet curing and reported that the relationship between compressive strength and gas permeability of high performance concrete (HPC) greatly depends on the water/binder (w/b) ratio and mineral admixture types. According to Torii et al [9] the resistance to chloride penetration of 50% GGBS concrete was equivalent to that of 10% SF concrete. However,

Proc. of the 9th *fib* International PhD Symposium in Civil Engineering, July 22 to 25, 2012,
Karlsruhe Institute of Technology (KIT), Germany, H. S. Müller, M. Haist, F. Acosta (Eds.),
KIT Scientific Publishing, Karlsruhe, Germany, ISBN 978-3-86644-858-2

when Hui-sheng et al [10], studied the influence of the water/cement (w/c) ratio on the gas permeability of HPC, by using different levels of PFA at a 0.30 w/b ratio, it was found that the concrete mixes with PFA exhibited higher gas permeabilities compared to the concrete mixes without PFA.

Currently, there is a growing interest in the use of mineral admixtures in reinforced concrete structures to improve durability and reduced porosity of concrete. The objective of the study is to investigate the influence of the micro-structure of different concrete types, measured via the nitrogen gas relative permeability test, on corrosion level of steel reinforcement.

2 Experimental programme

2.1 Test sample

The dimensions of the cube samples for accelerated corrosion tests were $200 \times 200 \times 200$ mm with a single ribbed bar of 12mm diameter embedded in the centre of the cube. A total of 24 samples of 100 mm diameter by 100 mm long cylinders were cored from concrete slabs of dimensions 350 x 325 x 100 mm for the purpose of gas permeability testing. Each concrete mix comprised eight samples for gas permeability and three 100 mm cubes and cylindrical samples of 100mm diameter and 200 mm height, for compressive strength and splitting strength respectively at 28 days.

2.2 Materials

The cement used in this study was Portland cement siliceous fly ash CEM II/B-V32.5 R containing fly ash up to a maximum of 7 % and manufactured by Lafarge cement UK Ltd, conforming to BS EN 197-1. The GGBS was manufactured according to BS 6699 and supplied by Hanson UK. The PFA used was class F supplied by Ash Resources Ltd. The chemical compositions of these materials are listed in Table 1. The fine aggregate used was a natural marine-dredged sand (0-4 mm), and the maximum size of coarse aggregate used was a 10 mm crushed limestone, both conforming to BS EN 12620.

Table 1   Chemical composition of CEM II, GGBS and PFA

Chemical composition (%)	Sample		
	CEM II	GGBS*	PFA*
Silicon dioxide, SiO_2	29.91	41.78	51.44
Aluminum oxide, Al_2O_3	11.18	10.56	27.03
Iron oxide, Fe_2O_3	4.13	0.27	6.13
Calcium oxide, CaO	45.00	34.65	2.42
Magnesium oxide, MgO	2.43	7.33	1.26
Sodium oxide, Na_2O	0.38	0.50	0.66
Potassium oxide, K_2O	1.45	0.64	2.72

* Chemical composition of GGBS and PFA was obtained from XPD in Cardiff University Laboratories.

2.3 Mix design, casting and curing

The samples were cast from three separate batches of concrete: one control (CEM II) and two mixes containing different percentages of GGBS and PFA. A water-cement ratio of 0.45 was selected to obtain a target 28-day compressive strength of 45 N/mm^2. The mix proportions used for these batches are presented in Table 2. The components were mixed together using 300 kg capacity concrete mixer for 3 minutes. The water was then added in small amounts over a 2 minute period, and the mixing was continued for a further 3 minutes. After the mixing process was complete, a slump test was carried out for each mix according to BS 12350-2:2009 the results of which are given in Table 2. Prior to casting all moulds used for the mix were oiled, fresh concrete was poured and compacted into the wooden moulds in which the reinforcing bar was placed horizontally. The filling process was performed in three layers and the concrete was compacted using a vibrating table after placement of each layer.

After casting, all the samples were left in their moulds overnight at room temperature and then demoulded approximately 24 hours later, at which point the reinforcing steel bar was wrapped in waterproof tape to prevent water from reaching the surface of the exposed sections of the reinforcing bar during the curing period. Finally, all samples were covered with damp hessian which was sprayed with water twice a day for 28 days. Cores were taken from the slab at age of 28 days and stored under wet hessian with the other test samples.

Table 2   Mixture proportions

Reference Mix	Cement content kg/m³	GGBS kg/m³	PFA kg/m³	Water kg/m³	Fine agg. kg/m³	Coarse agg. kg/m³	W/C	Slump mm
CEM II	410			185	710	1070	0.45	125
CEM II+30%PFA	287		123	185	710	1070	0.45	120
CEM II+50%GGBS	205	205		185	710	1070	0.45	75

2.4   Test Procedures

2.4.1   Conditioning regime

At 28 days a 6 mm diameter hole was drilled through the centre of the 100 mm diameter cylindrical cores and the core weighed. These cylinders were dried by placement in an oven at a temperature of 105 °C for 7 days and their weight recorded on a daily basis until no more than 0.02% weight loss was observed between consecutive readings in a 24 hour period as was reported by Gardner [11]. Samples were then kept in the dessicator at 20 °C until the start of the permeability test.

2.4.2   Gas permeability test

The relative gas permeability method was used to investigate the concrete permeability and the full test method is described by Gardner [11]. The coefficient of gas permeability k^{ef} is calculated using equation (1) as follows:

$$k^{ef} = \frac{\ln\left|\frac{(P_{1/2}+P_{atm})(P_{c0}-P_{atm})}{(P_{1/2}-P_{atm})(P_{c0}+P_{atm})}\right|}{2t_{1/2}\pi h P_{atm}} V_0 \mu \ln(r_c / r_h) \tag{1}$$

where k^{ef} = permeability coefficient (m^2); $P_{1/2}$ = half pressure (578250 N/m²); P_{c0}= initial pressure(1053250 N/m²);h = height of sample (m), P_{atm}= atmosphere pressure (103250 N/m²); $t_{1/2}$ = time to reach $P_{1/2}(s)$;r_c = outer radius of cylinder (m); r_h= radius of hole (m); V_0 = the combined volume of the reservoir and the gas surrounding the sample (m^3); μ= dynamic viscosity (1.76 10^{-5} Ns/m²)

2.4.   Accelerated corrosion

In this study, an electrochemical corrosion technique was used to induce artificial accelerated corrosion of the reinforcing bar embedded in the concrete as reported by Fang et al [3] .The direct current supply was applied to the reinforcing steel bar using a series circuit to ensure a constant current (10 mA) through all samples. In order to investigate different corrosion levels, accelerated corrosion was carried out for 3, 7, 10 and 14 days. The corrosion level (C_R) was calculated using equation (2):

$$C_R = \frac{G_0 - G}{g_0 l} \times 100 \tag{2}$$

where G_0 = the initial weight of the reinforcing bar before corrosion; G = the final weight of the steel bar after removal of the corrosion products; g_0 = the weight per unit length of the reinforcing bar (G_0/L), and l = the bond length.

3   Results and discussion

3.1   compressive strength and splitting concrete

The compressive strength results of the concrete mix with CEM II, 30% PFA and 50% GGBS for 7, 28 and 46 days are presented in Table 3. It can be seen that the control concrete (CEM II) exhibited higher value of compressive strength 53.9 N/mm² than the two concrete mixes with pozzolanic materials (PFA, GGBS), which were 50.2 and 37.3 N/mm² at 28-days respectively. It was observed that all concrete mix compressive strengths increased with curing age. This can be explained by the fact that at early age, PFA and GGBS have slower rates of hydration compared to CEM II but with time, the pozzolanic reactions occurring in these blended cements cause an increase in strength.

The splitting tensile strengths of the concrete mixes made with CEM II, 30%PFA and 50%GGBS were also determined at the ages of 28 days, and the test results are shown in Table 3. It can be observed that the higher value of splitting strength was obtained with the CEM II (4.5 N/mm^2) and that, as the compressive strength increases, the split tensile strength increases for all the concretes, irrespective of the type of cement.

Table 3   Strength results of concrete investigated

Cement types	Slump (mm)	Compressive strength (N/mm^2)			Split tensile strength
		7 days	28days	46days	28days
CEM II-control	125	45.0	53.9	61.9	4.5
CEM II + 50% GGBS	85	34.0	50.2	54.6	4.4
CEM II + 30% PFA	75	24.4	37.3	40.2	3.2

It is worth while to note that the obtained results in Table 3 for CEM II + 50 % GGBS are in agreement with findings from other studies by McNally and Sheils [12] who also used CEM II and found the compressive strength of 50%GGBS at 28 and 56 days almost unchanged. Also, Sengul and Tasdemir [13] have concluded that the compressive strength of the concrete samples with 50% slag replacing cement is slightly lower than CEM I concrete for both at 28 and 90 days. Moreover, as for OPC + 30% PFA with a 0.5 w/c ratio, similar trends also have been reported by Saraswathy et al [14] who have found that the compressive strength was lower than control samples.

3.2   Evaluation of corrosion level with time

Figure 1 presents the variation of the corrosion level against with exposure time (3, 7, 10 and 14 days) for three concrete mixes CEM II, 30%PFA and 50% GGBS. From this figure, it can be observed that the level of corrosion of steel bar in CEM II concrete increased throughout the exposure time, with an increasing rate of corrosion at all exposure time considered. This was due to the breakdown of the passive layer surrounding the steel bar as a result of chloride ions penetration. The concrete with 30% of PFA and 50% of GGBS exhibited lower corrosion levels than the CEM II (control) at all exposure times. The lower levels of corrosion in PFA concrete mix compared to the corresponding CEM II samples can be attributed to the lower permeability of the concrete and increased concrete resistivity [15, 16]. It is also worth noting that GGBS exhibited slightly lower levels of corrosion than that PFA for the first 7 days of exposure time but after this the levels of corrosion increased significantly after this to exceed the PFA corrosion levels at 10 days of exposure time. Nevertheless, for more detailed evaluations, different concrete mixes with varying replacement percentages are required.

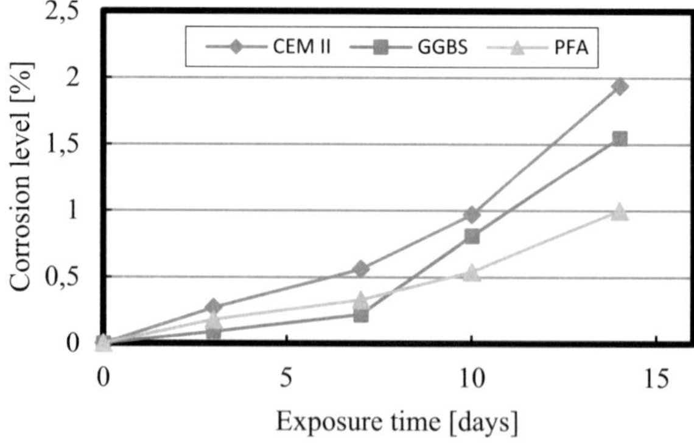

Fig. 1   The corrosion level versus exposure time for CEM II, PFA and GGBS concrete

3.3　Concrete gas permeability and corrosion

Figure 2 shows the relationship between the corrosion modulus, which measures the rate of change of corrosion as a function of exposure time, with the 28 day permeability coefficient for the CEM II, 30%PFA and 50%GGBS concrete mixes cured for 28 days and after 3 and 14 days of corrosion. From figure 2, it can be seen that the mean value of gas permeability coefficient K^{ef} for PFA concrete was 7.58×10^{-17} m^2 and the value of the corrosion modulus (m) was 0.06% for 3 days of exposure time, up to 0.07% of 14 days of exposure time. It can also be observed that the concrete of PFA concrete mix exhibited the lowest permeability coefficient of all concrete mixes. This can be attributed to the formation of the pozzolanic reaction products which produce C-S-H gel leading to enhancement of the pore structure, which ultimately reduces the permeability of the concrete. From literature, 15-30% PFA cement replacement in concrete has been reported to reduce the gas permeability by up to 65% of that of normal cement concrete [17, 10]. In general, the gas permeability coefficient of the CEM II was slightly higher than that of the PFA mix (9.58×10^{-17} m^2), while the values of corrosion modulus were 0.09% for 3day and 0.13% for 14 days. This indicated a reduction in the resistance of CEM II concrete to chloride penetration with increased exposure time, whilst PFA concrete demonstrated enhanced resistance to corrosion, with little deterioration in this resistance over the exposure times considered. On the contrary, the mean value of gas permeability coefficient of the concrete with 50% GGBS was 2.2×10^{-16} m^2 showing the highest gas permeability compared with the other two concrete mixes. Also, the value of the corrosion modulus was 0.03% for 3 days and 0.11% for 14 days of exposure time, which represents about a 3.5 fold increase.

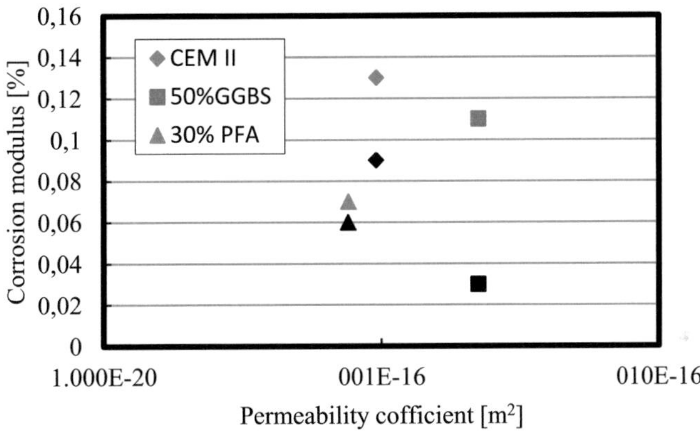

Fig. 2　Relationship between permeability coefficient and modulus of corrosion for CEM II, PFA and GGBS concrete

The inclusion of GGBS in a concrete mix usually leads to reduction in the mobility of chloride ions, as a result of a change in the mineralogy of the cement hydrates, and this may explain why the GGBS concrete possesses the lowest 3-day corrosion modulus of all three mixes [18]. However, once the passive layer protecting the steel reinforcing bar has broken down then the relatively high permeability of the GGBS concrete contributes to a significant increase in the rate of corrosion observed.

4　Conclusions

The following conclusions can be drawn. The compressive strength of the three mixes increased with curing age with that of the CEM II concrete being significantly higher than that of the blended cement (PFA and GGBS) concretes. The splitting tensile strength increased as a function of compressive strength. As far as the corrosion level was concerned, the three mixes exhibited similar trends with exposure time and it was highest in the CEM II concrete. Permeabilities depended on curing time, pozzolanic reaction and cement replacement levels, with the PFA having the lowest permeability compared to the other two mixes. Nevertheless, to fully investigate the corrosion level under a range of exposure times and to identify its relationship with the permeability of the concrete, further experiments using different cement replacement materials and varying replacement percentages are required.

References

[1]   Kivell, A., Palermo, A and Scott, A.: Effects of bond deterioration due to corrosion in rein-forced concrete: Proceeding of the 9th Pacific Conference on Earthquake Engineering (PCEE), Auckland, New Zealand (2011), pp. 081-088

[2]   Cabrera, J. G.: Deterioration of concrete due to Reinforcement steel corrosion: Cement and Concrete Composites 18 (1996), pp. 47-59

[3]   Fang, C., Lundgren, K., Chen, L. and Zhu, C.: Corrosion influence on bond in reinforced Concrete: Cement and Concrete Research 34 (2004), pp. 2159-2167

[4]   Gunevisi, E., Gesoglu, M., Ozturan, T. and Ozbay, E.: Estimation of chloride permeability of concretes by empirical modeling: Considering effects of cement type, curing condition and age: Construction and Building Materials 23 (2009)No.1, pp. 469–481

[5]   Oner, A. and Akyuz, S.: An experimental study on optimum usage of ggbs for the compres-sive strength of concrete: Cement and Concrete Composites 29 (2009), pp. 505-514

[6]   Choi, Y., Kim, J. and Lee, K.: Corrosion behavior of steel bar embedded in fly ash concrete: Corrosion Science 48 (2006) No.7, pp.1733-1745

[7]   Ha, T.H., Muralidharan, S., Ba, J.H., Ha, Y.C., Lee, H.G., Park, K.W. and Kim, D.K.: Accel-erated short-term techniques to evaluate the corrosion performance of steel in fly ash blended concrete: Building and Environment 42(2007), pp.78-85

[8]   Hui-Sheng, S., Bi-wan, X. and Xiao-Chen, Z.: Influence of mineral admixtures on compres-sive strength, gas permeability and carbonation of high performance concrete: Construction and Building Materials 23 (2009), pp.1980-1985

[9]   Torii, K. T., Sasatani, M. and Kawamura.: Effects of Fly Ash, blast Furnace Slag, and Silica Fume on Resistance of Mortar to Calcium Chloride Attack: American Concrete Institute Sp153-49(1995), pp. 931-950

[10]   Hui-Sheng S., Bi-wan, X., Shi, T. and Xiao-Chen, Z.: Determination of gas permeability of high performance concrete containing fly ash: Materials and Structures 44(2008), pp. 1051-1056

[11]   Gardner, D. R.: Experimental and numerical studies of the permeability of concrete: PhD Thesis, Cardiff University, UK (2005).

[12]   McNally, C. and Sheils, E.: Probability-based assessment of the durability characteristics of concretes manufactured using CEM II and GGBS binders: Construction and Building Materi-als 30 (2012),pp. 22–29

[13]   Sengul, O. and Tasdemir, M.A.: Compressive strength and rapid chloride permeability of concretes with ground fly ash and slag: Journal Materials Civil Engineering 21 (2009)No. 9, pp. 434-440

[14]   Saraswathy, V., Muralidharan, S. and Thangavel, S.S.: Influence of active fly ash on corro-sion- resistance and strength of concrete: Cement and Concrete Composites 25(2003),pp. 673-680

[15]   Scott, A. and Alexander, M.G.: The influence of binder type, cracking and cover on corrosion rates of steel in chloride-contaminated concrete: Magazine of Concrete Research 59 (2007) No.7, pp. 495-505

[16]   Arya, C. and Xu, Y.: Effect of cement type on chloride binding and corrosion of steel in concrete: Cement and Concrete Research 25 (1995) No.4, pp. 893-902

[17]   Thomas, M.D.A. and Matthews, J.D.: The permeability of fly ash concrete: Materials and Structures, 25 (1992) No.7, pp. 388-396

[18]   Li, S. and Roy, D.M.: Investigation of relations between porosity, pore structure and chloride diffusion of fly ash blended cement pastes, Cement and Concrete Research 16 (1986) pp.749–759

Session C-3
Behaviour and Performance
of Various Materials and Structures

Direct shear strength test: Analysis of the test influencing parameters

Ildikó Buocz

Department of Construction Materials and Engineering Geology,
Budapest University of Technology and Economics (BUTE),
Műegyetem rakpart 3, 1111 Budapest, Hungary
Supervisors: Nikoletta Rozgonyi-Boissinot and Ákos Török

Abstract

The physical properties of rocks are important input parameters for the modelling and analysis of underground structures. Various test methods are known and are standardised to determine most of the data e.g. uniaxial compression strength or tensile strength. However because of the complexity of the test influencing parameters, such as surface roughness, normal stress, presence of joint infill material, dilatancy…etc., for the analytical prediction of shear strength along discontinuities, no standards are available.

The goal of the research is to investigate the most important shear strength influencing parameters through experiments, and to create a mathematical approach, in order to be able to determine a more exact value for the shear strength of rocks along discontinuities.

The aim of this paper is to give a description of the most important shear strength influencing parameters and to point out their importance through a series of direct shear strength tests, carried out on granite rocks under constant normal load conditions. The values of the surface roughness were determined from 2D profiles that were taken from the 3D images of the surfaces. The results showed that with increasing surface roughness, the shear strength of the joint raises. The same applies to increasing normal stress. On the other hand the presence of the calcite infill material decreased the resistance of the joint.

1   Introduction

The determination of the shear strength of rock joints has been a very popular research topic in the past 50 years. The complexity of the discontinuities and the several factors that influence the test results complicate the analytical description of shear strength. Surface roughness, dilatancy, joint infill material, the magnitude of the applied normal stress, the size of the area that takes part in the shear test (contact area) and the scale effect, are considered to be the most important test influencing parameters of all. A formula that was invented by Barton [1] explains the connection between most of these factors and the peak shear strength, however it has many limitations. Modern technology, with the ability to create 3D images of surfaces opened new doors, which urged the researchers to continue the investigations. Although many series of tests have been delivered on these parameters by Grasselli [2], Indraratna et al. [3], Bahrani and Tannant [4] and others, no such up-to-date and unified formula has been invented that could replace the old one that is still in practise. The aim of this study is to give a short description of the test influencing parameters, to show new test results on the topic taking in consideration the surface roughness and the joint infill material as one of the most important factors, and to give a guideline for the further investigations.

2   Test influencing parameters

To determine the shear strength of rocks along discontinuities analytically, Barton [1] invented a formula (1) that is still in use today.

Proc. of the 9th *fib* International PhD Symposium in Civil Engineering, July 22 to 25, 2012,
Karlsruhe Institute of Technology (KIT), Germany, H. S. Müller, M. Haist, F. Acosta (Eds.),
KIT Scientific Publishing, Karlsruhe, Germany, ISBN 978-3-86644-858-2

$$\tau = \sigma_n \times tg(JRC \times \log_{10}\left(\frac{JCS}{\sigma_n}\right) + \phi_p)$$

(1)

τ shear strength
σ_n normal stress
Φ_p peak internal angle of friction
JRC Joint Roughness Coefficient
JCS Joint Compressive Strength

The results obtained by the formula are only valid for clean joints, neglecting the scale effect and using joint roughness data from 2D profiles. Due to the above mentioned limitations, the formula needs further improvements.

2.1 Surface roughness

The surface roughness of a joint describes the magnitude of the asperities of a discontinuity. It is considered to be the most important shear strength influencing parameter. In 1973 Barton invented a theory on how to specify the surface roughness of rocks numerically [1]. He predefined ten 2D profiles with a length of 10 cm that look the most typical for rock surfaces, and gave a *JRC* value interval (e.g. 8-10) for each. The *JRC* value scale starts at 0 and ends at 20. To record a profile of a rock, the Barton comb method was used [5]. The measured profiles are compared visually to the predefined profiles, and the value of the *JRC* is determined. Although with this method good results are obtained for shear strength, the way of defining the *JRC* is subjective. As technology improved, it offered new opportunities to measure the 2D profiles more precisely. Laser profilometers allowed the profiles to be digitalised [6]. With the evolution of photogrammetry and laser scanning technology, the examination of the entire joint surface became possible. 3D surface models can be produced by several methods e.g. by Advanced Topometric Sensor (ATS) System [2], photogrammetry or laser scanning [7]. Although with 2D profiles the equation of Barton (1) gives reliable results, with these profiles it is impossible to describe the features of the entire surface. To overcome this problem it is necessary either to calculate *JRC* from the 3D surface models of the joints, or develop a new mathematical approach on how to calculate shear strength taking into consideration the whole surface. Although Grasselli [2] proposed an equation for *JRC*, researchers still prefer the usage of the old Barton formula (1) with 2D profiles.

2.2 Dilatancy, normal stress

Dilatancy and normal stress are parameters that depend on each other. The lower the normal load is, the higher the dilatancy becomes. Shear strength measurements can be carried out in two ways. One is under constant normal load conditions (CNL), where the normal load is constant and the dilation is allowed. The other way is under constant normal stiffness (CNS) conditions, when the upper and lower parts of the rock specimen are fixed during the shear test, and the normal load can alter. Both behaviours are present in the nature, CNL conditions occur by landslides, CNS conditions happen when a block would slide from the periphery of a tunnel [8]. Several experiments have been carried out investigating both conditions e.g. by Leichnitz [9] or Vásárhelyi [10].

2.3 Contact area

The magnitude of the contact area during the shear strength test highly influences the test results. The calculation of shear strength according to the suggestion of the International Society for Rock Mechanics [11] is:

$$\tau = \frac{P_s}{A}$$

(2)

τ shear strength
P_s total shear force
A area of shear surface overlap (corrected to account for shear displacement),

from which this high influence of the contact area on the shear strength is visible.

    The determination of the rock surface area that takes part in the shearing process is complex, especially under CNL conditions, when dilation is allowed.

Barton and Choubey [5] already took this phenomenon into consideration when the Barton formula (1) was created and determined that the contact area ratio is related to the JCS/σ_n ratio applied in the shear test. After the 3D modeling of the rock joints became available, Grasselli [2] set up a theory in which he claims that the contact area of the rock joints can be calculated from the triangulated surface area, by knowing the orientation and steepness of these triangles.

2.4   Joint infill material

Most of the experiments and investigations focus on clean joints, but in nature, the presence of joint infill material frequently occurs. Brady and Brown [8] gave a short summary on the effects of infill materials, according to which the presence of infill materials that are weaker than the host rock, decrease the stiffness and shear strength of the rock, whereas the stronger infill materials such as quartz increase it. Also with the increasing thickness of the weaker infill material, the shear strength decreases. In the study of Indraratna et al. [3] the t/a ratio appears, which is the ratio of the thickness of the infill material (t) and the asperity height (a). They determined the critical t/a ratio for clay infills, beyond which no more decrease of shear strength can be measured. Oliveira et al. [12] carried out shear strength tests on clay filled rock joints both under CNL and CNS conditions. They proposed a model, which not only highlights the importance of the t/a ratio with respect to the shear strength, but also shows the interfering and non-interfering zones, and takes into consideration the infill cohesion as well. In 2011 Duriez et al. [13] published an article on the discrete-modeling of infilled rock joints, complemented by experimental data.

2.5   Scale effect

The value of shear strength along discontinuities is an important input parameter for modelling rock stability. However shear strength of rocks is determined under laboratory conditions on samples smaller in size. The application of these test results for rock masses depend on the scale effect. According to the description of Brady and Brown [8] three components change significantly the shear strength of discontinuities by the alteration of scale: the drained residual friction angle, the surface roughness (JRC) and the asperity failure that is controlled by the JCS/σ'_n ratio (σ'_n is the effective normal stress). The bigger the scale is, the smaller the asperity failure component and the surface roughness component gets. A study has been carried out recently on a 45 m high footwall slope in Canada by Bahrani and Tannant [4]. They concluded that the dilation angle for a given shear displacement is not sensitive to the profile length, and the shear displacement needed to reach the peak shear strength can be better estimated with the help of getting high-resolution profile pictures of the slopes.

3   Direct shear strength test

In aware of the test influencing parameters, CNL direct shear strength tests have been carried out on granite rocks according to the suggestion of the ISRM [11]. From the above mentioned test influencing parameters the surface roughness, the normal stress and the infill material were taken into account. However the thickness of the infill material was not taken into consideration. The tests were carried out under laboratory conditions on samples similar in size, therefore the scale effect was irrelevant.

3.1   Sample groups

Two sample groups were formed according to the surface roughness of the granite samples: moderately rough surfaced granites and rough surfaced granites. These samples had clean joints. A third group contained granite specimens with calcite covered surfaces.

Surface roughness was determined from 2D profiles taken from 3D surface images that were created by photogrammetric modelling, with the help of the software ShapeMetriX3D (© 3G Software & Measurement). On Figure 1, profiles of the sections are presented.

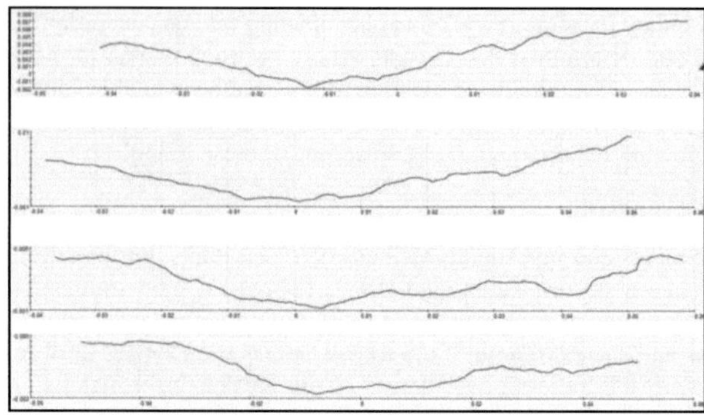

Fig. 1   2D profiles of granite rock surfaces.

The profiles were compared to the predefined Barton profiles, and a *JRC* value was determined for each rock surface. The *JRC* values of the moderately rough surfaced granites fell between 5 and 8, the rough surfaced granites between 8 and 15. For the calcite covered surfaces, the *JRC* value varied between 4 and 11, which means that according to their surface roughness this group contained samples from both of the above mentioned two groups.

3.2   Test results

The direct shear strength tests were carried out in the Laboratory of the Department of Construction Materials and Engineering Geology at Budapest University of Technology and Economics by a Controls shear apparatus, under CNL conditions. Table 1 shows the test results.

Table 1   Direct shear strength test results along discontinuities for granite rocks.

Surface	Specimen	JRC	σ_n [MPa]	τ_p [MPa]
Moderately rough	Gmr1	5	0.44	0.69
	Gmr2	6	0.28	0.63
	Gmr3	6	0.47	0.70
	Gmr4	8	0.44	0.93
Calcite covered	Gcc1	4	0.29	0.34
	Gcc2	4	0.63	0.78
	Gcc3	10	0.24	0.49
	Gcc4	11	0.44	1.21
Rough	Gr1	8	6.82	4.06
	Gr2	10	0.41	0.35
	Gr3	15	2.02	1.24
	Gr4	15	2.01	2.40

Figure 2 shows the direct shear strength test results plotted on a Normal stress – Shear strength diagram. The two clean jointed groups clearly separate from each other, red circles indicate each. Samples with rough surface and high *JRC* values have higher shear strength than the ones with moderately rough surface. Specimen with calcite covered surface behaved more similar to the moderately rough surfaced samples, only one sample with a *JRC* value of 11 fell in the group of the rough surfaced granites. It is also visible that the higher the normal stress value was, the higher the peak shear strength got.

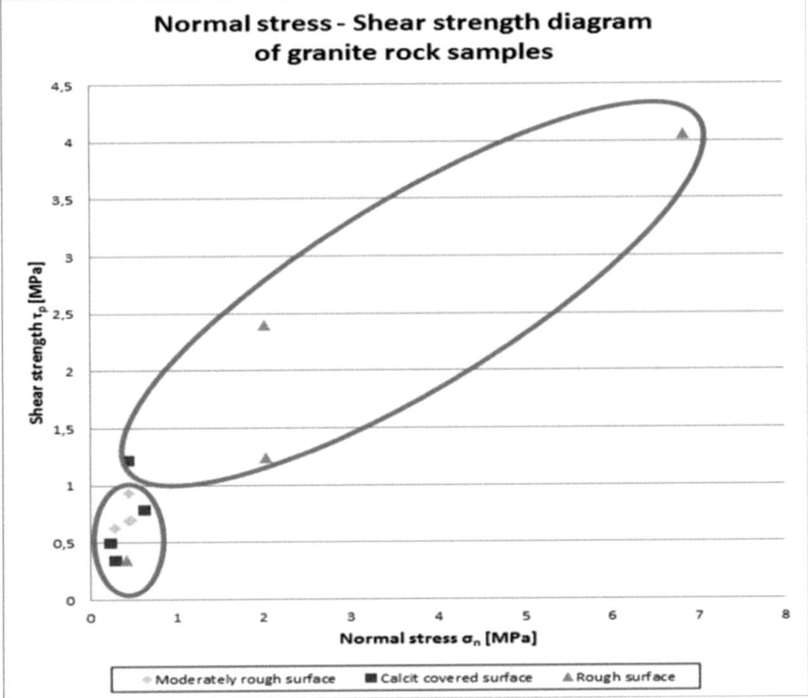

Fig. 2   Normal stress − Shear strength diagram of granite rock samples (yellow rhombi: moderately rough surfaced granite samples, purple squares: granite samples with calcite covered surface, green triangles: rough surfaced granite samples)

4   Conclusions

Direct shear strength tests were carried out on clean and infilled granite joints, under CNL conditions. For the clean joints, sample groups were formed according to the surface roughness (*JRC*) of the discontinuities, determined from 2D profiles that were gained from 3D images.

The test results showed that the samples with calcite covered surface, up to the *JRC* value of 10 had as low or even lower shear strength than the moderately rough surfaced clean jointed granite samples. The calcite as an infill material showed weaker mechanical properties than the granite, therefore what Brady and Brown [8] stated (see 2.4) applied well. However the calcite covered sample having the *JRC* value of 11 fell in the group of the rough surfaces. This could mean that infill materials with high *JRC* values that are not as soft as soil-like infill materials, could influence more the shear strength of a rock joint. Further investigations are recommended on this topic.

The most important shear strength influencing parameters have been introduced above. The topic of the impact of these parameters on shear strength is very popular among researchers, especially today when the technology is even more advanced. Many of the factors have been investigated separately, but a whole extensive solution is needed in order to be able to use the theory in practice. To be able to carry out this project, in aware of the results obtained by the researchers, a big number of experiments should be carried out on the same samples or molds taking in consideration all of the test influencing parameters. All of the test results should be aligned together to be able to set up a global correlation among all the investigated factors, and define such a formula for the shear strength of rock joints that is easily applicable in the engineering practice.

Acknowledgements

Ferenc Deák kindly delivered the photogrammetric measurements. Dr. Nikoletta Rozgonyi-Boissinot, Dr. Ákos Török and Dr. Péter Görög helped and gave their support during the research. The members of the Laboratory of the Department of Construction Materials and Engineering Geology at Budapest University of Technology and Economics provided help in the execution of the experiments.

References

[1]   Barton, N.: Review of a new shear strength criterion for rock joints, In: Engineering Geology 7 (1973) pp. 287-332

[2]   Grasselli, G.: Shear Strength of Rock Joints based on Quantified Surface Description. - EPF-Lausanne, Switzerland (2001).

[3]   Indraratna, B.; Oliveira, D.A.F.; Brown, E.T.; Pacheco de Assis, A.: Effect of soil-infilled joints on the stability of rock wedges formed in a tunnel roof. In: International Journal of Rock Mechanics & Mining Sciences 47 (2010) pp. 739-751

[4]   Baharni, N.; Tannant, D.D.: Field-scale assessment of effective dilation angle and peak shear displacement for a footwall slab failure surface. In: International Journal of Rock Mechanics & Mining Sciences 48 (2011) pp. 565-579

[5]   Barton, N.; Choubey, V.: The shear strength of rock joints in theory and practice. - Rock Mechanics 10 (1977) pp. 1-54.

[6]   Kusumi, H.; Teraoka, K.; Nishida, K.: Study on new formulation of shear strength for irregular rock joints. In: International Journal of Rock Mechanics & Mining Sciences 34:3-4 (1997) No. 168, ISSN 0148-9062

[7]   Buocz, I.; Török, Á.; Rozgonyi-Boissinot, N.; Görög, P.; Deák, F.; Berényi, A.: Összehasonlító módszerek kőzetek felületi érdesség mérésére laboratóriumi körülmények között. (in Hungarian) In: Török Á.; Vásárhelyi B. (edc.): Mérnökgeológia-Kőzetmechanika 2011. Hantken Kiadó, Budapest (2011) pp. 283-289.

[8]   Brady, B. H. G.; Brown, E.T.: Rock Mechanics for underground mining, kluwer, (2004) p. 628.

[9]   Leichnitz, W.: Mechanical properties of rock joints. In: Int. J. Rock mech. Min. Sci. & Geomech. Abstr. (1985) Vol. 22, No. 5, pp. 313-321

[10]   Vásárhelyi, B.: Technical note, Influence of normal load on joint dilation rate. In: Rock mech. Rock Engng 31 (2) (1998) pp. 117-123

[11]   Franklin, J. A.; Kanji, M. A.; Herget, G. and Ladanyi, B.; Drozd, K.and Dvorak, A.; Egger, P.; Kutter, H.and Rummel, F.; Rengers, N.; Nose, M.; Thiel, K.; Peres Rodrigues, F. and Serafim, J. L.; Bieniawski, Z. T. and Stacey, T. R.; Muzas, F.; Gibson, R. E. and Hobbs, N. B.; Coulson, J. H.; Deere, D. U.; Dodds, R. K.; Dutro, H. B.; Kuhn, A. K. and Underwood, L. B.,: Suggested Methods for Determining Shear Strength, International Society for Rock Mechanics Commission on Standardization of Laboratory and Field Tests, (1974) Document No. 1., pp. 131-140.

[12]   Oliveira, D.A.F.; Indraratna, B.; Nemcik, J.: Critical reviw on shear strength models for soil-infilled joints. In: Geomechanics and Geoengineering (2009) Vol. 4, No. 3, pp. 237-244

[13]   Duriez, J.; Darve, F.; Donzé, V.: A discrete modelling-based constitutive relation for infilled rock joints On the formatting of papers for this symposium. In: International Journal of Rock Mechanics & Mining Sciences 48 (2011) pp. 458-468

Numerical modeling of foundation structures with sliding joints

Martina Janulikova and Marie Stará

Department of Building Structures, Faculty of Civil Engineering,
VSB Technical University of Ostrava, Czech Republic,
17.listopadu 2172/15, 708 33 Ostrava, Czech Republic
Supervisors: Radim Cajka and Pavlina Mateckova

Abstract

Rheological sliding joints are preferred for use in reducing friction in foundation structures exposed to effects of horizontal loading (e.g., effects of undermining). Correctness of the rheological sliding joint design depends on knowledge of the mechanical response of materials. The most common material for a sliding joint is an asphalt belt. To better understand the wide spectrum of asphalt-based materials and their mechanical response to long-acting stress, the Faculty of Civil Engineering at VSB - Technical University constructed testing equipment and a temperature-controlled room. Up to this point, a series of measurements has been carried out so the results could be used for numerical modeling. The long-term goal of this research is to contribute to updating the existing design methods for sliding joints.

1   Testing of asphalt belts

Correct design of rheological sliding joints demands the knowledge of the response of used material for long-acting shear load. This was the purpose for which original measuring equipment was constructed at the VSB-Technical University, Faculty of Civil Engineering. In 2007 first measurements were carried out for different types of asphalt belts at laboratory temperature. These measurements followed loosely the measurements of asphalt belts from the 1980s [2]. The basic principle of the test is demonstrated in Figure 1. In 2010, a temperature-controlled room was constructed for investigation into the influence of temperature. The original measurement equipment was placed in this room, Figure 2.

Our measurement methodology is described in detail in [1]. Currently, oxidized asphalt belts have been tested for different temperatures and different load combinations. It is expected that a wide spectrum of modern materials will be measured. Vertical load value corresponds to real stress, horizontal load corresponds to the displacement rate expected in the footing bottom, e.g., from the effect of undermining. Particular results are presented in this paper, Figure 3, and also in the papers [3] and [4].

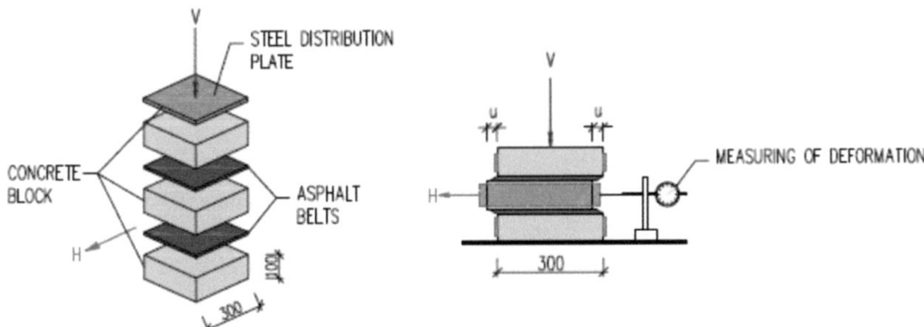

Fig. 1   Basic principle of the test

Proc. of the 9th *fib* International PhD Symposium in Civil Engineering, July 22 to 25, 2012,
Karlsruhe Institute of Technology (KIT), Germany, H. S. Müller, M. Haist, F. Acosta (Eds.),
KIT Scientific Publishing, Karlsruhe, Germany, ISBN 978-3-86644-858-2

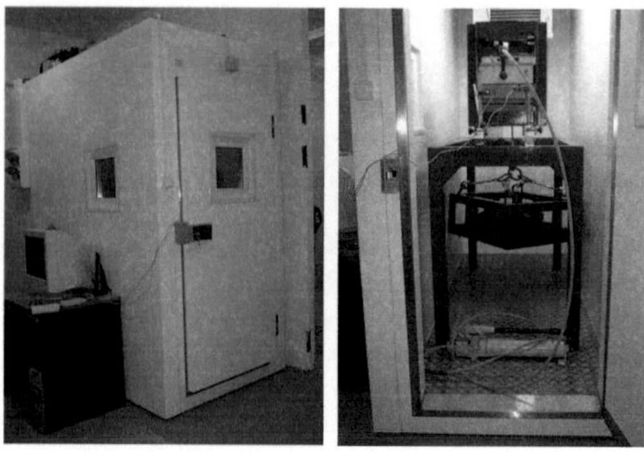

Fig. 2 Measuring equipment in the temperature controlled room

2 Testing results

For the duration of testing, usually 6 days, deformation of the middle concrete block is monitored for arbitrary temperature and load combination (Figure 3). It can generally be stated that at higher temperatures the deformations are greater, and vice versa. From the measured deformation, we can derive the value of the friction parameter, Figure 4, which is possible to utilize in commercial software as well as in design practice.

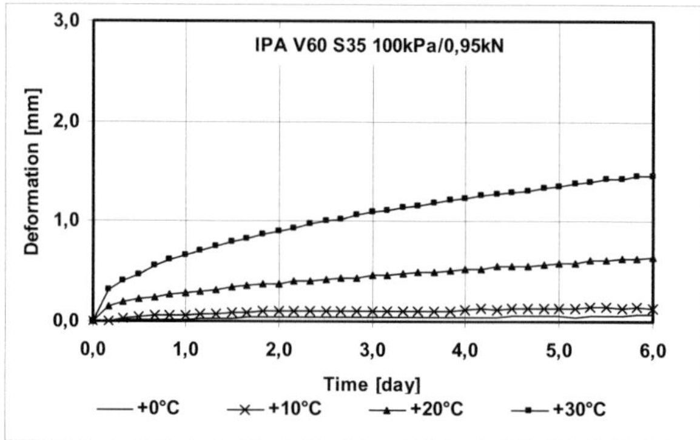

Fig. 3 Measuring results for horizontal force 0,95 kN and vertical pressure 100kPa

3 Friction parameters

There are two methods of the friction parameter calculation – analytical and numerical [5, 6]. Complete derivation of the resulting functions for calculating the friction parameters can be found in [7]

3.1 Analytical calculation of friction parameter

Analytical solutions can only be found for simple one-dimensional problems. In Figure 4 there are demonstrated friction parameters C_{1z}, C_{1x} and C_{1y}, where C_{1z} characterizes vertical resistance and C_{1x} and C_{1y} represent horizontal resistance. From the equilibrium of forces on a differential element (Figure 5), which is loaded with horizontal deformation ε_{max}, a differential equation is derived; whose solution corresponds to the process of axial deformation and the resulting normal force in the foundation.

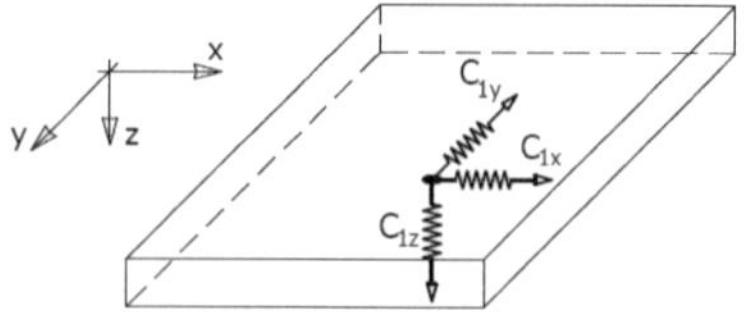

Fig. 4 Scheme of parameters C_z, C_x and C_y

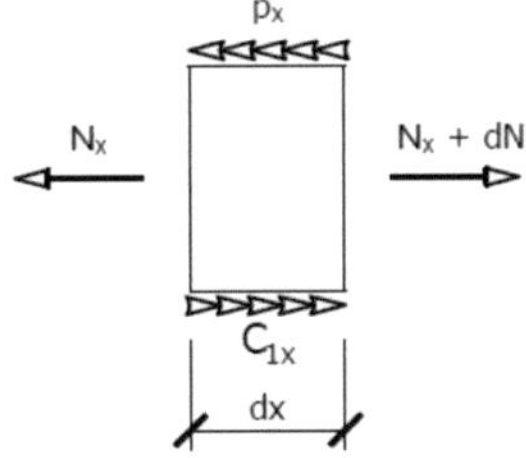

Fig. 5   Equilibrium of diferencial element

If the maximum tensile or compressive force $N_{c,max}$ is known (for example from the simplified solution of internal forces in the foundations according to[8]), the formula for constant friction parameter C_{1x} can be derived:

$$C_{1x} = E_c \cdot A_c \cdot \left[\frac{1}{L} \cdot \arg\cosh\left(\frac{F_x + E_c \cdot A_c \cdot \varepsilon_{max}}{N_{x,max} + E_c \cdot A_c \cdot \varepsilon_{max}} \right) \right]^2$$

3.2   Numerical calculation of friction parameter

If a more detailed solution is necessary to grasp the reality, a numerical calculation could be carried out: in each element a different value of friction parameter C_{1x} is calculated. Friction parameters are determined iteratively, with the aid of FEM analysis.

$$\left[K_g \right] \cdot \left\{ r_g \right\} = \left\{ F_g \right\}$$

where $\left\lfloor K_g \right\rfloor$ is the stiffness matrix of the whole truss structure including the effects of subsoil

$\left\{ r_g \right\}$ is the vector of nodal deformations

$\left\{ F_g \right\}$ is the axial force load vector

The stiffness matrix of the one-dimensional element $[K_e]$, occurring due to friction with the subsoil, has the form

$$[K_e] = E_c \cdot A_c \cdot \begin{bmatrix} \dfrac{1}{L} & -\dfrac{1}{L} \\ -\dfrac{1}{L} & \dfrac{1}{L} \end{bmatrix} + C_{1x} \cdot \begin{bmatrix} \dfrac{L}{3} & \dfrac{L}{6} \\ \dfrac{L}{6} & \dfrac{L}{3} \end{bmatrix}$$

The general nonlinear course of shear force could be expressed with an integral sum and from that, the simplified formula for calculating the friction parameter can be derived. After further derivations and implementation of boundary conditions, one can get this simplified formula for calculating the constant friction parameter C_{1x}.

$$C_{1x} = \frac{\tau_{x,max}}{u_{max}}$$

This constant value does not correspond to the real stress in the foundation bottom. This problem can be partially corrected by using a compact mesh of elements, thus approximating the nonlinear course of the stress.

3.3   Calculation of friction parameter from the tests results

Friction parameters are derived from the test results on the basis of shear stress expected in the foundation joint and the total deformation u. Our goal is to derive shear stress in the sliding joint as a function of time for a particular subsoil deformation. In deriving the friction parameters, we take into account the influence of sliding joint materials, structural deformation and loads, and now also the temperature. The derivation of the friction parameters is treated in more detail in [1].

4　Numerical modeling of sliding joints

In numerical modeling, either of two approaches can be used. The first option is to model an asphalt layer in the "real" thickness and define the characteristics of this layer with viscoelastic material models (Figure 6). The second option is to define a contact layer, (Figure 7) between the foundation structures and a concrete base layer (which is always applied to protect the asphalt belt when applying the sliding joint).

4.1　Modeling by using viscoelastic properties of asphalt

Asphalt is a material with viscoelastic characteristics, which could be replaced with rheological models in a simplified way [9]. To define the viscoelastic behavior of asphalt belts in our case, the ANSYS program was used, in which we prescribed an elementary differential equation of stress in the form

$$\sigma(t) = 2 \cdot \int_0^t G \cdot (t_i - \tau_i) \cdot \frac{d\gamma}{d\tau_i} \cdot d\tau_i + I \cdot \int_0^t K \cdot (t_i - \tau_i) \cdot \frac{d\varepsilon_{vol}}{d\tau_i} \cdot d\tau_i$$

where

G(t)	basic shear relaxation function
K(t)	basic volume relaxation function
t_i	time in solved moment
τ_i	time in moment i-1
γ	shear part of deformation
I	identity matrix
ε_{vol}	volume part of deformation

The basic equation for shear G(t) and volume K(t) function can be expressed in the ANSYS program in one of two ways – using a Maxwell rheological model and its combinations, or using a nonlinear material model using equations of the "Prony series", i.e., equations containing a general exponential expression

$$\sum_{i=1}^{N} \alpha_i \cdot e^{\frac{-t}{\tau_i}}$$

Because the Maxwell model requires us to define a lot of rheological constants, the second possibility was chosen for modeling – to define the material of the sliding joint using equations of the Prony series. In these equations we need to know the rheological constants to define shear modulus (α_i^G) and volume modulus (α_i^K) too, but those can be calculated into the ANSYS program using tools for calculating these constants with results from laboratory tests. The calculation of these constants in selected times was carried out on the basis of our knowing the time evolution of the shear modulus, and calculation of constants was limited only to the shear constants. Volume changes of asphalt belts in shear tests are considered insignificant and are not monitored.

From the results of tests conducted in the temperature-controlled room, shear modulus values at different times and different temperatures are calculated – the methodology used in their calculations is described in [1]. After loading the data file into ANSYS we get the above-mentioned rheological constants in the required time intervals.

Time-dependent material properties of asphalt are implemented as such in the calculation. The whole calculation in ANSYS must then be defined as time-dependent.

The resulting model comprises three materials – concrete, asphalt and subsoil. The asphalt belt should be modeled in its actual thickness, but it is unsuitable since the thickness is small in comparison with other dimensions of the model. In the end it was decided to consider the asphalt layer several times higher and its height was divided into two elements. Another problem is a significant difference in the elastic modulus values of each material. Modulus of elasticity of concrete is about $30 \cdot 10^9$ Pa and modulus of elasticity of asphalt is about $2 \cdot 10^5$ Pa. Figure 6 shows a model of a simple strip foundation with the length of 16 m.

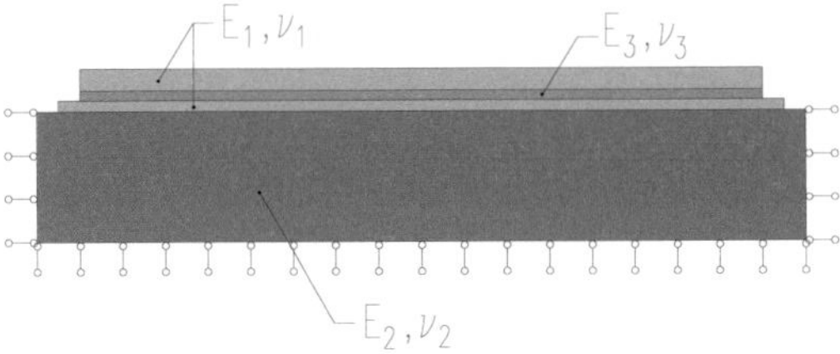

Fig. 6   Model example – with sliding joint, modeling with asphalt layer

4.2   Modeling with use of a contact element

Another possible way to model the friction and shear in a footing bottom is by defining the contact elements. Contact can be defined simply with the friction coefficient depending only on the vertical and horizontal stress according to the Coulomb friction model. The friction coefficient μ for defining the contact elements was in our case calculated depending on the vertical and horizontal stress valid within the laboratory tests. The authors are currently working on a more precisely defined friction parameter with the effect of temperature, for example, which defines the dynamic coefficient of friction. A model with the contact element is in the Figure below.

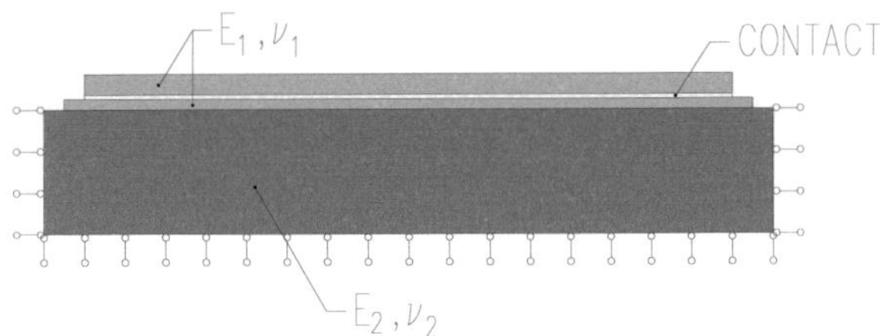

Fig. 7   Model example – with sliding joint, modeling with use of a contact element

4.3   Comparison of results

In Figure 8 there is a comparison of different approaches. In the charts there are shear stress values as a function of a foundation coordinate.

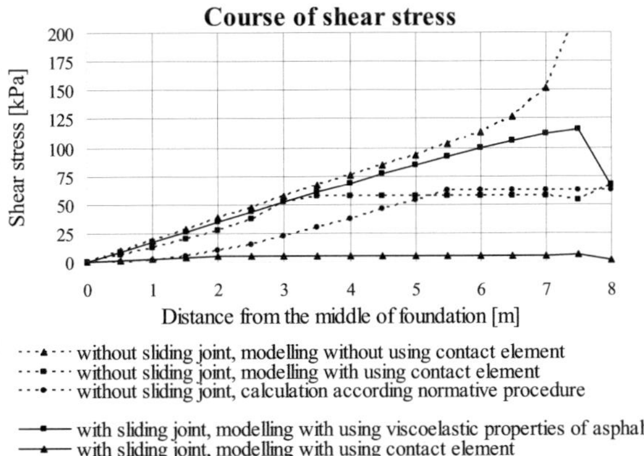

Fig. 8   Comparison of shear stress course

The curve, which shows a course of shear stress without a sliding joint calculated according to current codes [3], is also presented for comparison. The curves in Figure 8 show significant differences between the approaches to modeling.Although the courses of stress cannot be considered as final at all, they show correctly smaller shear resistance when using the sliding joint. This is particularly apparent for the curve that characterizes the course of stress on the model using the contact elements. The question is which model is more suitable for actual calculation. Modeling with an asphalt layer enables us to better express the viscoelastic material properties, including the timing course. As a disadvantage there is the necessity to increase the thickness of asphalt layers to unrealistic proportions. This distorts the resulting distribution of stress in the foundation joint (Figure 8). Due to these inaccuracies and considerable simplification, it seems preferable to model the sliding joint using contact elements. The necessity to find a suitable way to define the contact remains as an open problem.

5   Conclusion

In this paper, the authors introduce experimental testing of shear resistance of bitumen asphalt belts depending on the temperature. Laboratory test results show a significant effect of temperature on the shear deformation of asphalt belts. Further research must be done to implement these test results, including the effect of temperature, using correct mathematical modeling, whether to determine the parameters of contact elements or the material characteristics of the asphalt. On the one hand, it seems suitable to model the sliding joint using the layer with viscoelastic properties; on the other hand, such results can be significantly distorted due to the need to increase the thickness of layer and differences between moduli of elasticity of concrete and asphalt. Models with contact elements do not comprise the viscoelastic properties of asphalt, but in combination with the correct expression of contact parameters this seems to be an appropriate tool for modeling sliding joints. As both approaches have their advantages and disadvantages, both models will be developed in the future.

Acknowledgement

This paper has been achieved with the financial support of the SGS grant, internal no. SP 2012/36.

References

[1]   Maňásek, P.: Foundation structure with sliding point, Ph.D. thesis, VSB TU Ostrava, 2008, in Czech

[2]   Balcárek, V., Bradáč, J.: Utilization of bitumen insulating stripes as sliding joints for buildings on undermined area, Civil Engineering, 1982, in Czech

[3]   Čajka, R., Janulíková, M., Matečková, P., Stará, M.: Laboratory testing of asphalt belts with the influence of temperature, In Transactions of the VŠB - Technical University of Ostrava, 2011, Nr. 2, ISSN 1213-1962

[4]   Čajka, R., Janulíková, M., Matečková, P., Stará, M.: Modeling of foundation structure with slide joint with temperature dependant characteristics. In Thirteenth International Conference on Civil, Structural and Environmental Engineering Computing, Chania, Crete, Greece, 2011, pp. 208. ISBN 978-1-905088-45-4

[5]   Čajka, R.: Soil - structure interaction in case of exceptional mining and flood actions, In COST 12 - Final Conference Proceedings, 20 - 22 January 2005, University of Innsbruck, Austria, ISBN 04 1536 609 7

[6]   Čajka, R., Maňásek, P.: Finite Element Analysis of a structure with a sliding joint affected by deformation loading. In The eleventh International Conference on Civil, Structural and Environmental Engineering Computing. 18-21.9. 2007, St. Julians, Malta, ISBN 978-1-905088-17-1

[7]   Čajka, R.: The influence of friction in the subsoil of the foundation structures, Habilitation, VSB TU Ostrava, 2002, in Czech

[8]   ČSN 730039: Design constructed facilities in mining subsidence areas. Basic requirements, UNMZ Prague, 1990, in Czech

[9]   Čajka, R., Maňásek, P.: Physical and FEM shear load response modeling of viscoelasticity material. In The Eighth International Conference on Computational Structures Technology, (CST 2006), 12-15 September 2006, Las Palmas de Gran Canaria, Spain, ISBN 10-905088-08-6.

Chaboche-based cyclic material model for steel and its numerical application

Viktor Budaházy

Department of Structural Engineering,
Budapest University of Technology and Economics,
Műegyetem rkp. 3., Budapest H-1111, Hungary
Supervisor: László Dunai

Abstract

The subject of the current research is the local ductility of structural members and joints of seismic resistant dissipative steel frames. The main purpose is to develop a model for numerical simulations which can follow the hysteretic behaviour of the dissipative members and by the application of this tool to develop innovative and improve existing structural details. The numerical model should describe the cyclic behaviour of the steel material, the opening and closing phenomena under repeated loading, and the cyclic plate bucking.

In the first part of the paper the research completed on the development of cyclic plasticity material model is presented. A Chaboche model with dynamically updated parameters is applied on the basis of nonlinear kinematic hardening, to follow the cyclic mechanical behaviour of structural steel. In the second part the developed model is applied for the analysis of Buckling Restrained Brace (BRB). BRBs are subjected to a combined loading protocol and the results are verified by experimental tests.

1   Introduction

The global ductility of dissipative steel frame structures is very important in seismic design. The global behaviour highly depends on the local ductility of structural members and joints. The seismic response can be studied by real and virtual experiments. The improvement of computational techniques makes it possible to perform cyclic numerical simulations. In these computational models it is important to describe properly the cyclic behaviour of the material because it has significant effect on the local ductility of the structural members and joints. Despite significant research completed on cyclic material models in the last decades, results are still only partially adapted in finite element programs. The main purpose of the current research is to develop an efficient material model for numerical analysis which takes into account the important characteristics of cyclic behaviour of the steel material and implement it in a finite element program. In the research a model combination (based on nonlinear kinematic and multi-linear isotropic hardening), and a dynamic parameter evaluation method are applied. The model development is done in ANSYS finite element environment [1]. The accuracy and the efficiency of the model are verified by cyclic material test results. On the basis of the developed material model the cyclic simulation of Buckling Restrained Braces (BRB) is completed, subjected to combined loading protocol, in parallel with experimental tests

2   Features of cyclic steel material models

Behaviour under cyclic loading is difficult to describe when plastic deformation appears in the material. The behaviour of an inelastic material can be described by the von Mises yield criterion and the normality yield law.

The primary reason of observed difference between monotonic and cyclic behaviour is the crystallite structure of the metals. Dislocations and other crystal defects are able to modify the behaviour of ideal metals. The velocity of these defects is in the order of $100~\text{ms}^{-1}$, and it causes irregular elastic, and plastic behaviour. This paper focuses on the visible effects, it is not dealing with crystallite translations. The most important physical effects are the elastic after-effect, the Bauschinger effect, the memory of loading history and the effect of strain rate [2]. The monotonic mild steel material behaviour differs from cyclic behaviour, as observed by several experiments (e.g. in [3]). The material model has to take into consideration that as the plastic strain increases, the structure of the crystals becomes more organized, which leads to the decrease of the yielding plateau and saturation of

Proc. of the 9th *fib* International PhD Symposium in Civil Engineering, July 22 to 25, 2012,
Karlsruhe Institute of Technology (KIT), Germany, H. S. Müller, M. Haist, F. Acosta (Eds.),
KIT Scientific Publishing, Karlsruhe, Germany, ISBN 978-3-86644-858-2

the Bauschinger effect. In the case of typical structural steel, the vast of Bauschinger effect is saturating about 1% of the maximum plastic strain.

3 Chaboche model with dynamically updated parameters

Despite numerous Chaboche model based complex material models can be found in the literature, these are not able to describe the decrease of the yielding surface and plateau appropriately [4]. Although increasing complexity is disadvantageous for practical applicability, it is important to take into consideration the aforementioned physical phenomenon since it has significant influence on the global response. The objective is to develop a material model which is able to describe all the governing phenomena, while it can be easily implemented in existing efficient finite element software for numerical applications. The following physical effects are considered during the model development: (i) kinematic hardening, (ii) Bauschinger effect, (iii) decrease of the yielding surface, (iv) disappearance of the yielding plateau, (v) plastic creep, and (vi) strain memory.

3.1 Development strategy

The Chaboche model is selected as the basis of the new cyclic steel material model, due to its several beneficial aspects. Even a simple Chaboche model has fading memory effect, can describe the kinematic hardening and plastic creep. Another advantage is that Chaboche models can be easily superposed with each other and/or combined with other models for isotropic hardening.

The significant differences between static and cyclic behaviour are difficult to describe with the same equations and material constants, therefore a separate static and cyclic parameter set should be created. The developed material model (called PRESCOM – Parameter Refreshed and Strain Controlled Combined Chaboche model) is a combination of five Chaboche models superposed with multi-linear isotropic hardening. The combination of kinematic and isotropic hardening is able to describe the yielding plateau if the isotropic parameters are set as if the material was softening after the yielding point. This approach gives accurate results for monotonic loading, therefore the corresponding parameter set is referred to as static set. A different set of parameters (cyclic set) define a material model that can describe the decrease of the yielding surface, the disappearance of the yielding plateau, and the plastic creep. However, if the plastic strains are small, the PRESCOM model with constant parameters (either static or cyclic) cannot follow the change of yield surface and the transition between the exclusively static and cyclic behaviour. Another parameter setting is defined to describe this phenomenon. This saturated state is followed by the combined model using a function to calculate its parameters on the fly.

3.2 Dynamic model parameter calculation

At the beginning of the loading the static parameter set is used for every steel element in the dissipative zones of the structure. After the first load step, the maximum value of equivalent plastic strain is evaluated, and the material model constants are updated using the following logic: If the value of the plastic strain is zero, the parameter set is kept static. If it is greater than a pre-defined limit (ε_{EQW2}), the cyclic parameter set is switched is loaded. If it is in between 0 and ε_{QEW2}, behaviour can be described by saturated parameters that are described by a function of the equivalent plastic strain mentioned in section 3.1. Equivalent plastic strains are evaluated after every load step. If the new strain value is greater than the previous, the above procedure is repeated, if its lower the parameters stay the same. By this procedure, the hardening memory effect at the critical range of small plastic strains is built into the material model using the maximum plastic strain to describe the change in material behaviour.

The numerical models use the saturated yield stress ($f_{y,num}$), which refers to the decreased yield surface, except for the static parameter set, where the first yielding point corresponds to the virgin state. The decrease of yield surface is the function of maximal strain, and in spite of yielding stress reduction after some plasticity, if the loading reaches the same strain as before, the stress also has be identical, therefore the hardening has to be increased. At the transitional zone the model parameters are changed, in a way to reach the real yielding stress level, if the strain amplitudes reach the maximum.

The material model contains five simple Chaboche models with different parameters (Table 1). The PRESCOM model's initial hardening behaviour depends on first of all the Chaboche models' C and γ parameters (the initial hardening is significant, but the whole hardening saturates at small plastic strain [4]), while the isotropic hardening is relatively small and the other Chaboche models do not influence the hardening behaviour at this strain rate. At the transition range of plastic strain the first three Chaboche models in Table 1 are dominant; the parameters of the other models can be left unchanged. The translation of the yielding surface can be expressed as a function of accumulated plastic strain, as follows:

$$\frac{\Delta\sigma_i}{2} = \frac{C_i}{\gamma_i} \tanh\left(\gamma_i \frac{\Delta\varepsilon_p}{2}\right) \quad \text{where} \quad \gamma_i \frac{\Delta\varepsilon_p}{2} = \beta \text{ and} \tag{1}$$

where C_i, γ_i are material model constants, $\Delta\varepsilon_p$ is the accumulated plastic strain. The expression in parentheses in Eq. 1 (β) controls the saturation of the hardening. If β reaches 2.5, the hardening is 99% saturated. The first Chaboche model reproduces the hardening between the virgin and the saturation tendencies, while the second and third models describe the hardening observed in cyclic experiments. At the end of the transition, the PRESCOM models' behaviour shall be the same as the model with the cyclic parameter set. The parameters of the first model can be calculated as:

$$\gamma_1 = \frac{\beta}{\varepsilon_{EQW}} \quad \text{and} \quad C_1 = \gamma_1 \sigma_\Delta \tag{2}$$

where β=2.5; ε_{EQW} is the maximal von Mises strain, and σ_Δ takes the cyclic degradation of the yield stress into account. The second and third models' parameters are described in Eq. 3:

$$C_i = \left(\frac{\varepsilon_{EQW}}{\varepsilon_{EQW2}}\right)^\alpha C^0_i \quad \text{where } i=1 \text{ or } 2 \text{ and } \alpha=1.5 \tag{3}$$

The stress level at a given strain rate depends on the load history and these results in different stress values for the same strain level. This phenomenon is taken in consideration at transition model, with increasing coefficients at second and third models, and decreasing at the fifth Chaboche model:

$$C_5 = C^0_5 \left(2 - \frac{\varepsilon_{EQW}}{\varepsilon_{EQW2}}\right) \tag{4}$$

If the hardening modulus of the combined models becomes greater than the Young's modulus of steel, it is discarded and the Young's modulus is used instead.

3.3   Model parameter calibration

The model has large number of parameters, therefore *"Trial and error method"* was used for model calibration. The experimental results Youngjiu et. al [5] are used as the basis of the calibration. Two series of experiments with a total of 50 tests on Q235B and Q345B steel specimens were studied by the authors using several different load protocols. The experimental monotonic and hysteresis behaviour, ductility characteristics and cumulative damage degradation are discussed in details. Results of Q345B specimens are used for calibration. The numerical model is a simply supported, cuboids shape, that can describe the material behaviour without the influence of element geometry. The parameters are shown in Table 1.

Table 1   The parameters of PRESCOM model calibrated on the basis of [5]

| $f_{y,num}$=335 MPa | Cyclic material model | | | | | Transitional model | | |
| $f_{y,real}$=423 MPa | Chaboche models | | | Multi-linear | | Chaboche models | | |
Chaboche model no.	C	γ	$\Delta\sigma$	ε_{pl} [-]	σ	C	γ	$\Delta\sigma$
I.	75000	1500	50	0	328	$f(\varepsilon_{EQW}, \sigma_\Delta, \beta)$		σ_Δ
II.	21000	375	56	0.05	339	$f(\varepsilon_{EQW2}, \alpha, C_{02})$		
III.	7000	120	58	0.1	350	$f(\varepsilon_{EQW2}, \alpha, C_{03})$		
IV.	1000	25	40	4.5	451	1000	25	40
V.	300	0				$f(\varepsilon_{EQW})$	0	

3.4   Static, saturated and cyclic behaviour

The above detailed procedure results in a model that can describe the saturation of Bauschinger effect and the decrease of yield surface. Figure 1. shows the model behaviour under static loading, with and without preloading; and the saturation under low strain cyclic loading.

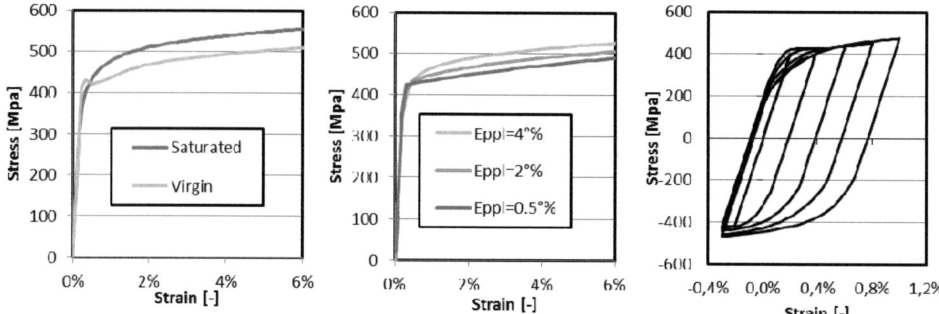

Fig. 1   The pure virgin and cyclic behaviour of PRESCOM model calibrated to Q345 under static loading (left); the behaviour of the model under static loading in the function of maximal plastic strain experienced in preloading (middle); disappearance of the yielding plateau and decrease of the yielding surface during cyclic loading with small amplitudes of plastic strain (right).

Behaviour of the material model under cyclic loading is shown in Fig. 2. The obtained results are sufficiently accurate, especially at the tension side of the hardening. In case of compression the calculated curves shows higher values than the experimental test.

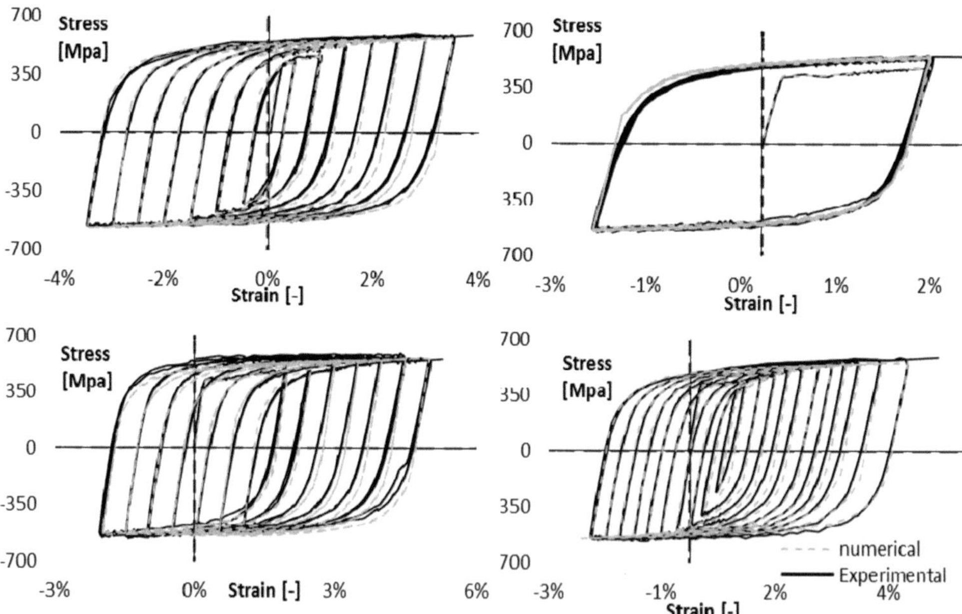

Fig. 2   Cyclic behaviour of the developed model under cyclic loading with proportionally increased amplitudes(left up); constant proportional loading (right up); non-proportionally increased amplitudes at tension side, and decrease at compression side loading, where the difference between the tension-compression is the same (left down); non-proportionally increased amplitude of loading (right down).

4.   Numerical model of Bucking Restrained Braces (BRB)

Using the developed material model a Buckling Restrained Brace (BRB) under cyclic loading is analysed. The BRB has three components: a slender steel core, where the energy can be dissipated; a concrete casing in a steel hollow section, which supports the core laterally under compression; and an air gap separating the former two parts from each other. The purpose of the investigation is to explore the operation of the brace, paying special attention to the additional hardening at large strain levels under compression (*pinching effect*) [6].

4.1   Description of the model

Complexity of the BRB structural member requires the use of several advanced modelling tools, such as cyclic material cyclic plasticity, plastic buckling, contact problem and friction. The four parts of the numerical model are shown in Fig 3.

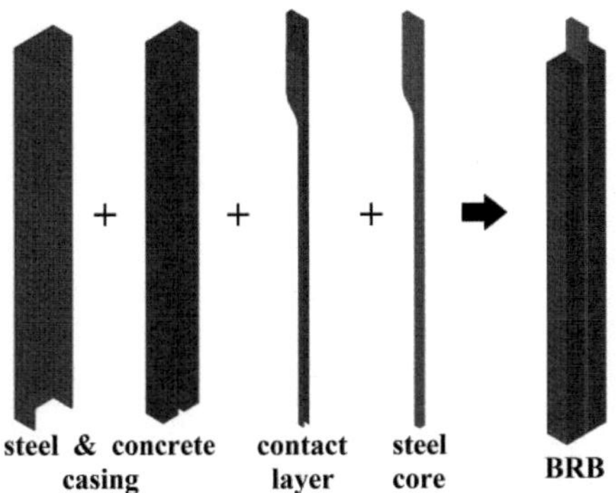

- Steel hollow section: modelled by 20-node-serendipity element and elastic material (E=210 GPa, v=0.3).
- Concrete: modelled by 20-node-serendipity element and elastic material (E=21 GPa, v=0.18).
- Steel core: modelled by 20-node- serendipity element and PRESCOM material.
- Air gap: modelled by 8-node-contact and target element with initial contact offset and friction.

Fig. 3   The numerical model

In order to reduce computation time and improve convergence, symmetry plains are defined and one quarter of the member is modelled, as shown in Fig. 4, [6]. In order to maximize the accuracy of calculation serendipity element type and quadratic mesh have been used. The mesh has 25 mm element size.

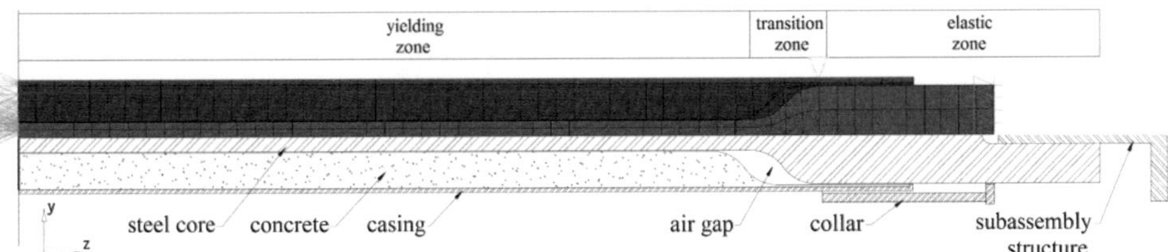

Fig. 4   The mesh and boundary conditions.

Loading is quasi-static with automatic time stepping and displacement control., identical to the laboratory experiments; the applied loading protocol is the combination of ECCS recommendations and EN15129 [6]. In order to improve convergence, the force label convergence criteria were softened, special contact parameters were used, and the normal penalty stiffness factor and the penetration tolerance factor were set to more rigid.

4.2   Numerical analysis

Comparison of the numerical and experimental results is shown in Fig. 5. As it can be seen they are in good agreement and the accuracy of the numerical model is satisfactory. The coefficient of friction $\mu=0.085$ is selected during calibration. The primary reason of the pinching effect is the friction. During the cyclic loading increased plastic compression occurs at the top of the core, causing significant contraction. The contracted core gets trapped and transfers axial forces to the concrete. The pinching effect can be eliminated by simple construction rules that ensure low friction and provide an appropriate air gap size.

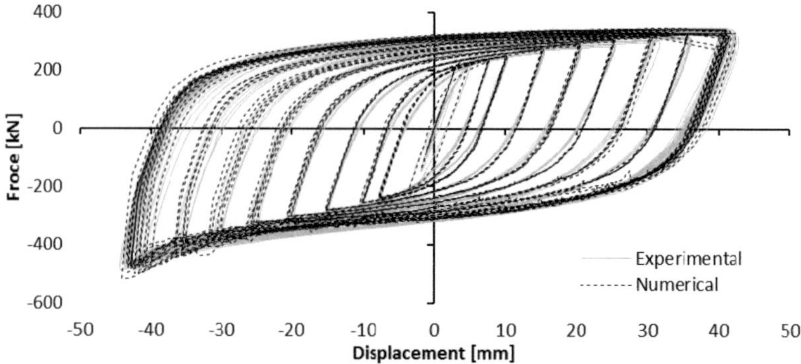

Fig. 5   Result of numerical simulation and experiment.

5.   Concluding remarks

In the paper a parameter refreshed calculation method for Chaboche model is introduced for low carbon steels. Although the model is developed in the ANSYS finite element environment, the procedure can easily be adapted to other software, where the Chaboche model is available anddifferent material models can be combined. The developed model is able to describe the static and cyclic behaviour, the saturation of Bauschinger effect, and the dependence of loading history. The material model calibration is done on the basis of experimental results. The accuracy of the numerical model is satisfactory, the material model can be used in nonlinear time history and cyclic geometrical and material nonlinear imperfect analyses (GMNI). Using the developed material model, a buckling restrained brace element is analysed. It is shown that by the numerical model such a complex structural member can be studied accurately. Based on the result this model can be the basis of innovative structural detail development.

Acknowledgement

The work reported in the paper has been developed in the framework of the project „Talent care and cultivation in the scientific workshops of BME" project. This project is supported by the grant TÁMOP-4.2.2.B-10/1--2010-0009. The authors would also like to express their gratitude to Tsinghua University, Department of Civil Engineering for the steel material laboratory test results.

References

[1]   ANSY Inc, *"Theory reference for ANSYS and ANSYS Workbench 11.0"*, Canonsburg, Pennsylvania, USA (2007).

[2]   Budaházy, V., Dunai, L.: Steel material model development for cyclic analysis. Proc. of 6th European Conference on Steel and Composite Structures, Budapest (2011), Vol. B, pp. 1106-1111.

[3]   Lee, G., Chang, K. C.: The experimental basis of material constitutive laws of structural steel under cyclic and nonproportional loading. Proc of Stability and Ductility of Steel Structures under Cyclic Loading, Osaka (2000), pp. 3-14.

[4]   Chaboche, J.L.: A review of some plasticity and viscoplasticity constitutive theories. International Journal of Plasticity, 24 (2008) pp. 1642-1693.

[5]   Youngjiu, S., Meng, W., Yuanqing, W.: Experimental and constitutive model study of structural steel under cyclic loading. Journal of Constructional Steel Research, 67, (2011) pp. 1185-1197.

[6]   Zsarnóczay, Á., Vigh, L. G.: Experimental Analysis of buckling restrained braces. Proc. of 6th European Conference on Steel and Composite Structures, Budapest (2011), Vol. B, pp. 945-950.

Numerical Simulation Techniques

Behaviour study of single-layer steel grid shells based on numerical analysis

Kitti Gidófalvy

Faculty of Civil Engineering, Department of Structural Engineering,
Budapest University of Technology and Economics,
Műegyetem rkp. 3., 1111 Budapest, Hungary
Supervisor: Levente Katula

Abstract

The goal of the presented research project is to detect the effect of mesh geometry on structural behaviour of free form grid shells by performing systematic numerical analysis. A tool was developed, that is capable of generating an optimal mesh onto a free form surface. The tool performs a shape optimization process on an initial mesh topology by gradually moving its nodal positions along the surface. Optimality criterion is based on maximizing the load bearing capacity. This is determined by nonlinear finite element analysis in which material and geometrical nonlinearities are considered. The paper presents the design methodology and results of optimized structures completed up until writing this paper. Furthermore an imperfection sensitivity analysis is included.

1   Introduction

Free-form single-layer reticulated shells are spatial structures made of steel beams, where the nodes all lie on one double-curved, irregular surface. In the last decade several such structures have been realized, such as the court roof of the British Museum in London. In most cases the network of beams was derived from aesthetic reasons. Previous studies have shown that the behaviour can significantly change only by slightly modifying the mesh. Thus it is investigated, whether performance could be improved by slightly modifying nodal positions of the structure by performing a shape optimization process on given topologies.

In the optimization literature the investigated single-layer structures are mostly domes, and the goal is to minimize weight. In topology or size optimization the geometric description is not general, but rather limited, so not all the geometrically possible structures can be investigated. Therefore the motivation for the presented research was to investigate more complex shapes with various mesh types obtained by a general meshing technique.

A single layer structure is a discrete system of beams. The network made up of beam centrelines is called the mesh; in which five or six centrelines intersect in one node. Adjacent beams form faces, which are all triangles in the current investigation, so that the structure can be covered by flat panels.

2   Mesh generation process

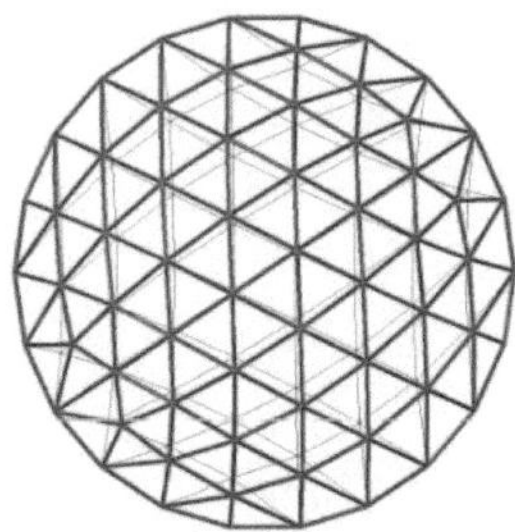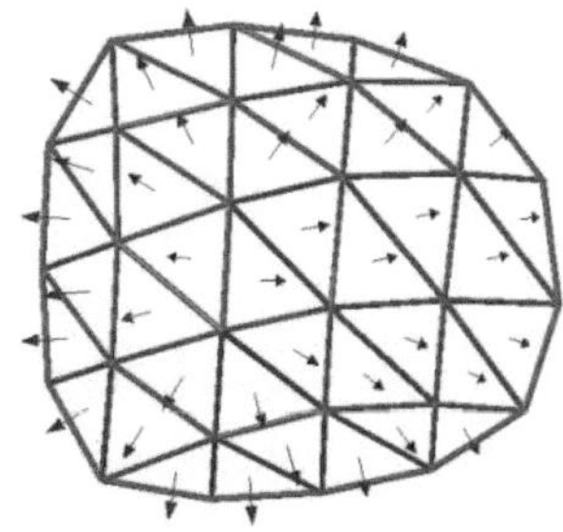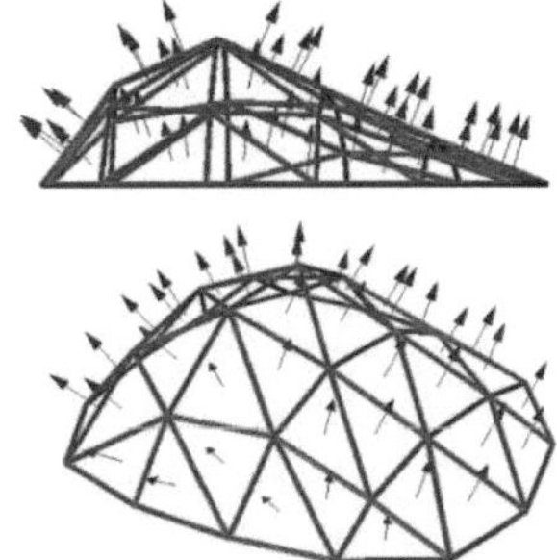

Fig. 1   Automatically generated mesh topologies. Left: mesh generated on a dome (relaxed mesh with thicker line). Middle and right: a mesh generated on a free form surface with face normals.

Proc. of the 9th *fib* International PhD Symposium in Civil Engineering, July 22 to 25, 2012,
Karlsruhe Institute of Technology (KIT), Germany, H. S. Müller, M. Haist, F. Acosta (Eds.),
KIT Scientific Publishing, Karlsruhe, Germany, ISBN 978-3-86644-858-2

As part of the research project, an automated mesh generation process was developed. The tool generates a triangular mesh on any continuous convex free form surface, without singular points and with a convex ground plan. An important property of the mesh is that boundary points are placed equidistantly. The basic concept of the meshing process is to intersect the surface with differently inclined planes, and to determine intersection points of the generated curves. Two basic parameters describe the mesh: one controls the density of the mesh (n), the other controls the angle between the slicing planes (φ). In this paper, mesh is generated on two surface types: on a spherical dome (Fig. 1 left) and on a free form surface (Fig. 1 middle and right).

To improve the generated mesh on a geometrical basic, in certain cases dynamic relaxation was also used [3]. This method is used by architects, because it results in a mesh with smaller variety of beam length and a more harmonic network (Fig. 1 left).

3 Nonlinear analysis

In order to compare the behaviour of each mesh, the load bearing capacity had to be determined. In previous investigations of the authors, load bearing capacity was determined using first order analysis. However investigation on the failure modes of single-layer grid shells shows in many cases the presence of snap-through or global buckling [1, 2]. Therefore it was decided to perform a more exact geometrically nonlinear analysis. However this increases the computational time significantly, therefore, there are only a few accurate numerical investigation in literature [1].

The analysed structures have rigid joints and all boundary nodes are fixed supports. Each straight beam is modelled using 3 beam finite elements. Uniform tubular sections section sizes were chosen, so that element buckling and plasticity is not a typical failure mode, hence the initial mesh shows a global failure mode. The material is perfectly elasto-plastic steel with a yield stress of 23.5 kN/cm^2.

3.1 Determination of Load Bearing Capacity

Nodal loads were applied with one load parameter; this will be detailed later. Load deflection curves were determined by finite element analysis. As single-layer shells may fail due to snap-through, the behaviour is instable. This required a special nonlinear solution method called arc-length method [1]. For each analysed mesh a load bearing capacity was defined. Load-deflection curves of each node - including the end nodes of each finite element - were investigated (Fig. 2). The capacity is defined as the maximum (critical) load, or the load for which deflection is higher than 0.2 meters for any of the nodes.

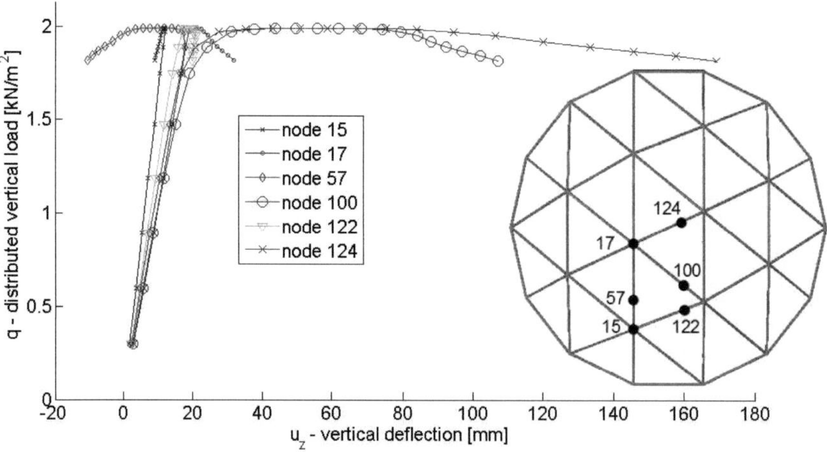

Fig. 2 Load deflection curves of nodes. The load bearing capacity is $q_{cr,0}$=2.08 kN/m^2.

4 Shape Optimization

A shape optimization technique was developed, for which a simplified flowchart on Fig.3 shows how the used tools were linked together. When updating the grid, each inner node is moved along the surface. Each node can be moved as far as the closest adjacent node. During the optimization, the number of variables for a mesh is twice the number of the non-boundary or inner nodes of the mesh.

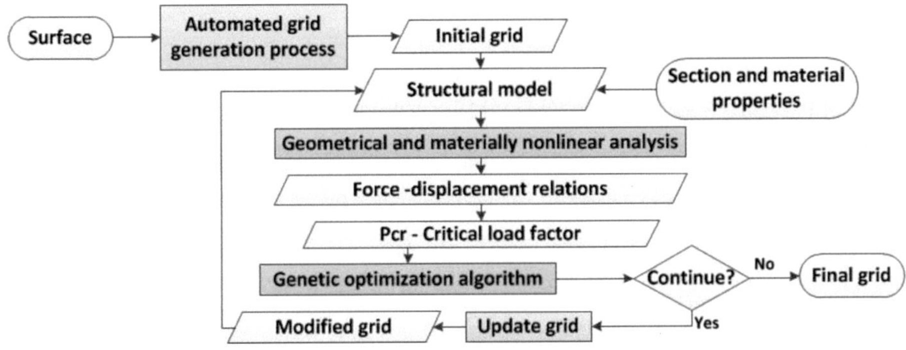

Fig. 3   Flowchart of analysis and optimization procedure.

4.1   Optimization algorithm

The optimization procedure was performed using a genetic algorithm (GA). The algorithm creates a random initial population from a discrete set of possible variables. The variables are represented by a string of numbers, and after computing the fitness value for each individual, strings with the best fitness are selected to construct the next generation. Two-point crossover and mutation are applied in order to obtain a new population. A self-adaptive approach for mutation given by Toğan and Daloğlu [5] is adopted in the design to increase the probability of obtaining the global optimum and to raise the performance of the GA. In this manner, the probability of mutation is chosen according to the fitness value of the solutions. The probability of crossover is fixed during the optimization with the value of 0.85. Elitist strategy is used, which means, that the best two individuals are copied into the new generation directly. In this paper, an optimal or improved mesh refers to the mesh that is the result of the optimization procedure, with the described fitness function, irrelevant of the fact that it is a global or local optimum.

4.2   Load

In order to simplify the optimization procedure the structure is loaded with vertical nodal forces. Loading the mesh with uniform magnitude in all the nodes would bias the results in the optimization procedure. This is demonstrated on the optimization of a dome mesh: Fig. 4 (left) shows the initial and the optimized mesh with a critical uniform nodal load of 225 and 255 kN respectively. The beam on the top of the optimized mesh became longer than the rest. In reality distributed forces act on the surface, hence in the model the nodes of this top beam should be loaded with a higher nodal force.

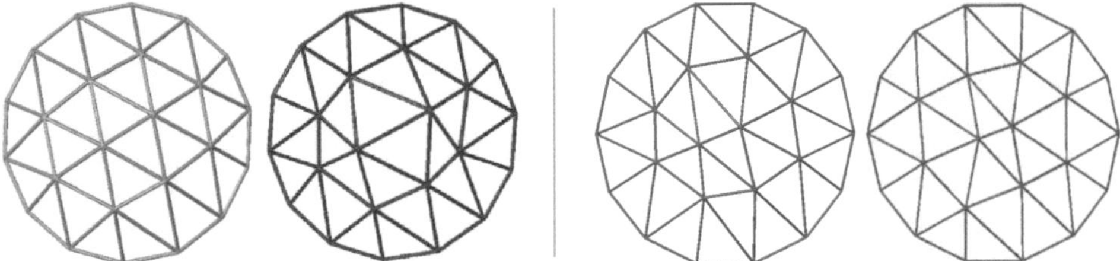

Fig. 4   Left: initial mesh and optimized meshes with uniform vertical nodal loads; Right: Optimized mesh with distributed load with and without scaling.

For this reason a more accurate load model was developed, where nodal force is calculated as if uniform vertical distributed load acted on the surface. It is assumed, that the distributed load on each face is transferred to its three nodes, equally divided between them. Thus the load bearing capacity is the maximum distributed load (q_{max}). If this value was compared for each mesh, then the optimization procedure would result in a mesh, where the faces adjacent to boundary nodes become larger, so that most of the load goes directly to the boundary. To avoid this, the distributed load is scaled by the ratio of the full area (A) of faces, to the area, from where the load is transferred to the inner nodes (A_{inner}). Thus the fitness function is: $q_{max} \cdot A_{inner}/A$. (Fig. 4 right)

4.3   Results

First a mesh on a dome with 10 inner nodes was investigated. The span is 35 meters, and the average beam length is 8 meters, so this is a relatively coarse mesh with only a few inner nodes. Load bearing capacity of the initial mesh was 2.08 kN/m^2 (Fig. 4 left). The fitness of the optimized mesh (Fig. 4 right) is 14% higher than the initial. This does not show a completely symmetric layout, so the algorithm did not find the best solution. The spherical surface - like any surface of revolution - is advantageous, and the initial mesh has a harmonic layout, therefore a higher improvement was not expected.

For a mesh with 42 inner nodes (Fig. 1 left) on the same surface, dynamic relaxation improved the mesh with 4 % increase in load bearing capacity, but the optimization procedure did not find a better solution. Results of the optimization meshes are briefly summarized in Table 1.

The investigated free form mesh has a span of 25 meters. The mesh with 14 inner nodes has an avarage beam length of 6 meters (Fig. 5). In the optimization process the relaxed mesh serves as a better initial mesh, as it saves computational time, while it increased the load bearing capacity by 4%. At the beginning of the optimization process, the nodes close to the very shallow part of the surface were moved away; this way snap through failure mode was prevented. The optimal mesh (Fig. 5 right) is a mesh where one node moved to the peak point of the surface while nodes adjacent to the boundary moved even closer to it. This result reflects the theoretical optimum of a grid with four corner supports, where there is only one inner node, and the grid forms a pyramid [4].

For topologies with more inner nodes, that can be modified, smaller improvement in capacity is possible. The main reason is, that they behave more like shells; hence the typical failure modes are different. Snap through cannot be prevented only by moving away nodes from parts with smaller curvature, but more than one node would snap through. For the free form mesh with 29 inner nodes (Fig. 6), the algorithm found a solution with 19% higer load bearing capacity. An even higher improvement is expected after expanding the set of possible variables. In the current state, the optimization process proved to be rather slow, but converged well.

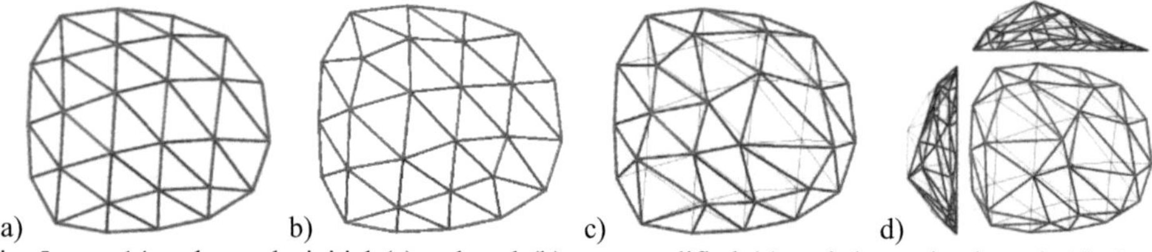

a)   b)   c)   d)

Fig. 5   14 node mesh: initial (a), relaxed (b), one modified (c) and the optimal mesh (d). Load bearing capacity is 2.06; 2.14; 3.20; 4.11 kN/m^2 respectively. Initial mesh with thin line.

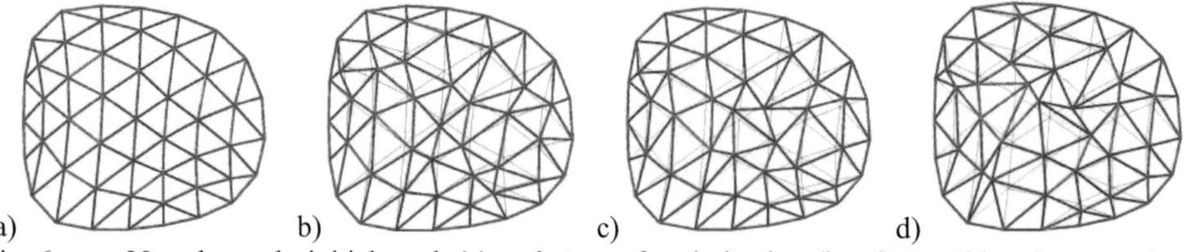

a)   b)   c)   d)

Fig. 6   29 node mesh: initial mesh (a) and steps of optimization (b,c,d). Load bearing capacity is 3.56; 3.96; 4.10; 4.23 kN/m^2 respectively. Initial mesh with thin line.

Table 1   Results of performed optimization

Surface	Number of inner nodes	Load bearing capacity [kN/m^2]		
		Initial	Optimized	Improvement [%]
Dome H/ L=0,2 (Fig. 10)	10	2,08	2,37	14
Dome H/ L=0,2 (Fig. 1 left)	42	7,96	8,25	4
Free form L=25m (Fig. 5)	14	2,06	4,11	100
Free form L=25m (Fig. 6)	29	3,56	4,23	19

5   Imperfection sensitivity analysis

To this point, load bearing capacity was determined based on geometrically and materially nonlinear analysis. However, shell structures are highly sensitive to imperfections. Therefore in future research, imperfection is intended to be included in the optimization process. A parametric analysis was carried out to investigate the effect of imperfection on different mesh types. Imperfection was calculated by scaling the first eigenmode (Fig. 7).

In Fig. 8-11 the ratio of the critical loads (or load bearing capacity) of the perfect and imperfect models of various mesh types are presented. This ratio is also called reduction factor. For an imperfection scaled to a maximum of L/300 (Fig. 11), it varies between 0.4 and 0.9 depending on the eigenmode, surface type, stiffness and mesh coarseness. Fig. 8 shows results for domes with different mesh density parameters. For the densest mesh (n=11) the reduction factor is higher, mainly because the eigenmode is different (Fig. 7). Furthermore the mesh acts more like a shell and is stiffer due to the higher number of nodes. Fig. 9 shows that the reduction factor becomes smaller for a shallow dome, as the eigenmode differs from the other domes. The meshes generated on the investigated free form surface are less imperfection sensitive compared to the dome (Fig. 10-11).

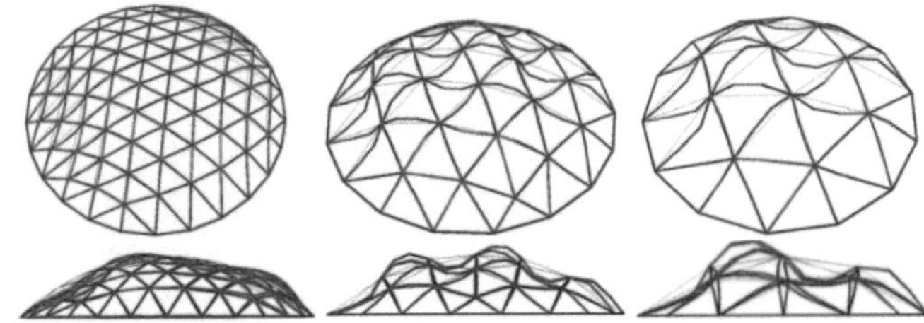

Fig. 7   First eigenmodes of dome meshes. Mesh density parameter varies: n=11, 7 and 5.

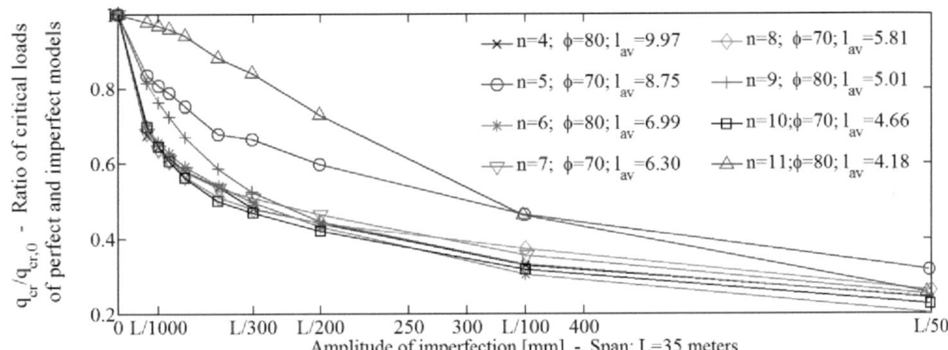

Fig. 8   Effect of imperfection on dome meshes with different mesh density (n). The name of each mesh refers to geometric parameters: n, φ and average beam length respectively.

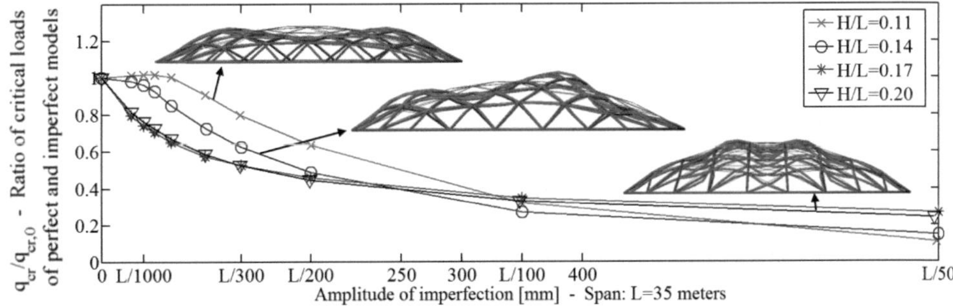

Fig. 9   Effect of imperfection on dome meshes with different height to span ratio.

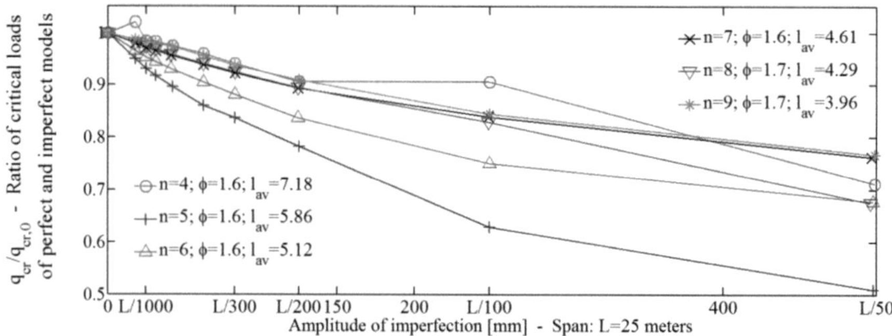

Fig. 10   Effect of imperfection on free form meshes with different mesh density (n). The name of each mesh refers to geometric parameters: n, φ and average beam length respectively.

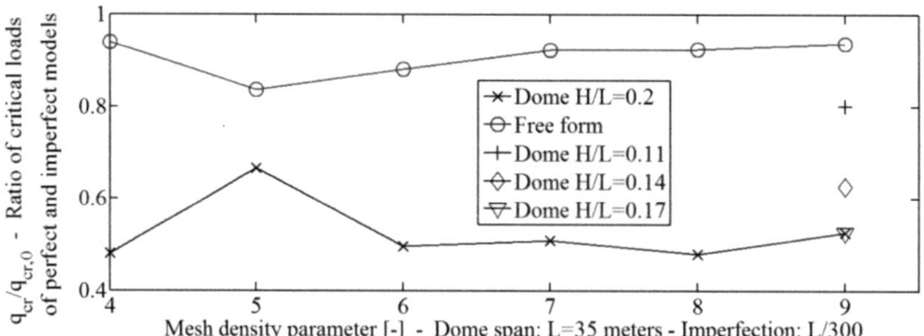

Fig. 11   Effect of imperfection scaled to L/300, on meshes with varying mesh density (n).

6   Conclusions, further investigations

The presented shape optimization process worked well and reached 19% increase in load bearing capacity for a presented free form mesh. It can be concluded that the results are highly dependent on boundary conditions, surface curvature and mesh coarseness: beam length and node number. It was demonstrated that imperfection significantly reduces the load bearing capacity to 40-90%.

In the current paper plasticity occurred only partly in a very few sections and local element buckling was also avoided. In the future, the effect of different section sizes will be investigated. A behaviour study on the effect of nodal stiffness will also be performed.

Acknowledgements

The work presented herein is carried out as part of Ph.D. studies of the first author. The work reported in the paper has been partly developed in the framework of the project „Talent care and cultivation in the scientific workshops of BME" project. This project is supported by the grant TÁMOP-4.2.1/B-09/1/KMR-2010-0002.

References

[1]   Fan, F., Cao, Z., Shen, S.: Elasto-plastic stability of single-layer reticulated shells. In: Thin-Walled Structures (2010) Vol. 48, pp. 827-836

[2]   Kato, S., Fujimoto, M., Ogawa, T.: Buckling Load of Steel Single-Layer Reticulated Domes of Circular Plan. In: Journal of the IASS (2005) Vol. 46, pp. 41-63

[3]   Dimcic, M.: Structural Optimization of Grid Shells based on Genetic Algorithms, Ph.D. dissertation. (2011)

[4]   Kollár, L., Hegedűs, L.: Analysis and Design of Space Frames by the Continuum Method. Akadémiai Kiadó, Budapest, Hungary (1985)

[5]   Toğan V., Daloğlu AT.: Optimization of 3d trusses with adaptive approach in genetic algorithms. In: Engineering Structures (2006) Vol. 28, pp. 1019–1027

3D contact problems with covariant description for large load-steps

Ridvan Izi

Institute of Mechanics,
Karlsruhe Institute of Technology (KIT),
Kaiserstraße 12, 76131 Karlsruhe, Germany
Supervisors: Karl Schweizerhof and Alexander Konyukhov

Abstract

A 3D extension of the so-called large penetration algorithm using a covariant contact description in the local surface coordinate system is presented. As presented in [2] the Node-To-Surface (NTS) approach allows the computation with larger load increments than within a usual consistent linearization. The required specific structure of the tangent matrix for this 3D scheme is found in a straightforward fashion using the covariant contact description. In order to show the efficiency of the scheme for further approaches like e.g. the Surface-To-Analytical-Surface (STAS) examples with contact, bending and sliding of shells modelled with elements of quadratic approximation are presented.

1 Introduction

Contact mechanics as a field within nonlinear computational mechanics requires for all classical methods the consistent linearization in order to reach quadratic convergence. Several approaches to accomplish the linearization of contact expressions are described in [11] and [8]. Within [3], [4] and [7] a special approach allowing the linearization before discretization is developed. This enables the contact formulation in any curvilinear geometry as in [6]. Nevertheless, the consistent linearization converges for all these linearization techniques well only if the initial approximation is close to the searched solution and, therefore, often a large number of load steps is needed to reach the solution.

In order to overcome this difficulty a combination of non-consistent and consistent linearization, the so-called large penetration algorithm, was proposed by Zavarise et al. in [12] for 2D linear geometries. Hereby, the tangent matrix derived according to the consistent linearization is split in a "basic" and a "geometrical" part. Starting only with the "basic" part of the tangent matrix the computation is switching after some iterations to the full tangent matrix. This procedure is nevertheless limited to linear elements in 2D. For 3D problems and arbitrary orders of discretization the required identification of the "basic" and "geometrical" part was already derived in [3]. The matrices therein derived by consistent linearization give a clear understanding which parts have to be incorporated for the large penetration scheme.

A first reliable combination of the covariant description with the large penetration algorithm was presented in [2] for a frictionless Node-To-Surface (NTS) approach. The following contribution is focusing on the large penetration scheme based on the covariant contact description for the Surface-To-Analytical-Surface (STAS) approach. Then 3D-numerical examples concentrating on setups with quadratic approximation of the geometry and non-frictional sliding are presented.

2 Covariant contact formulation

A special 3D local coordinate system as given in Fig. 1 is introduced for the so-called closest point

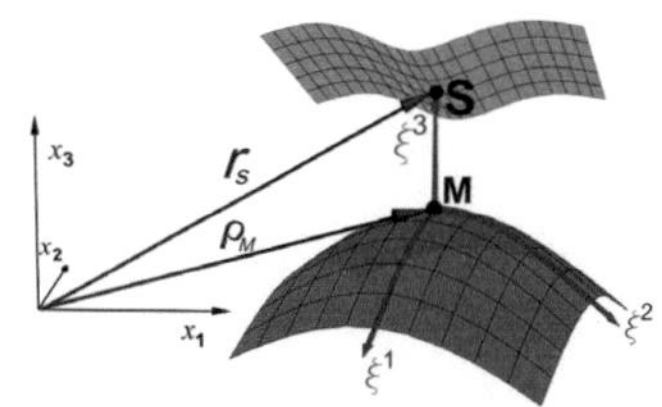

Fig. 1   Definition of a local spatial coordinate system

Proc. of the 9th *fib* International PhD Symposium in Civil Engineering, July 22 to 25, 2012,
Karlsruhe Institute of Technology (KIT), Germany, H. S. Müller, M. Haist, F. Acosta (Eds.),
KIT Scientific Publishing, Karlsruhe, Germany, ISBN 978-3-86644-858-2

projection of the slave point S on the master (M) where the third coordinate ξ^3 has the direction of the normal $\mathbf{n}$. Using $\boldsymbol{\rho}$ for the position vector of the master point M the position of the slave point S yields

$$\mathbf{r}(\xi^1,\xi^2,\xi^3) = \boldsymbol{\rho}(\xi^1,\xi^2)+\xi^3\,\mathbf{n}(\xi^1,\xi^2). \tag{1}$$

Based on this coordinate system contact relies on monitoring the penetration ξ^3 where a non-positive value implies contact. All geometrical and kinematical characteristics of the contact formulation are investigated within this local coordinate system. Regarding contact kinematics with moving and changing contact surfaces all parameters can be considered as time dependent. Therefore, the important operation for obtaining the weak form is to specify the velocity of the slave point. The rate of penetration ξ^3 eminent for the non-frictional formulation is obtained by projecting the slave point velocity $\mathbf{v}_S$ at $\xi^3 = 0$ on $\mathbf{n}$. With the master point velocity $\mathbf{v}_M$ this yields

$$\mathrm{d}/\mathrm{dt}\ \xi^3 = (\mathbf{v}_S - \mathbf{v}_M)\cdot\mathbf{n}\ . \tag{2}$$

2.1 Penalty regularization and consistent linearization

With N denoting the normal component of the traction vector the virtual work for frictionless contact problems can be expressed in the form

$$\delta W_c = \int_s N\delta\xi^3 \mathrm{ds} \tag{3}$$

where the variation $\delta\xi^3$ is derived exploiting the analogy to time derivatives as given in Eq. (2)

$$\delta\xi^3 = (\delta\mathbf{r} - \delta\boldsymbol{\rho})\cdot\mathbf{n}\ . \tag{4}$$

Using a penalty scheme for the normal traction the regularized contact functional is given as

$$\delta W_c = \int_s\varepsilon_N\ \xi^3\ \mathrm{H}(-\ \xi^3)\delta\xi^3 \mathrm{ds} \tag{5}$$

where ε_N is the penalty parameter and $\mathrm{H}(-\ \xi^3)$ is the Heaviside function.

In order to solve this problem within a Newton type solution process a consistent linearization is required with the consequence of exploiting the full material time derivative for the covariant description within the local spatial coordinate system. Each parameter in the contact integral is considered in the spatial local coordinate system of the master surface, therefore, linearization of the slave surface ds is not needed in the process. Thus, ds is assumed to remain constant within linearization. The linearization of the regularized contact functional leads then to

$$D_v(\delta W_c) = \int_s\varepsilon_N\ \mathrm{H}(-\ \xi^3)\ \{(\delta\mathbf{r} - \delta\boldsymbol{\rho})(\mathbf{n}\otimes\mathbf{n})(\mathbf{v}_S - \mathbf{v}_M)- \tag{6}$$
$$-\ \xi^3\ [\delta\boldsymbol{\rho}_{,i}a^{ji}(\mathbf{n}\otimes\boldsymbol{\rho}_{,j})(\mathbf{v}_S - \mathbf{v}_M)+ (\delta\mathbf{r} - \delta\boldsymbol{\rho})(\boldsymbol{\rho}_{,j}\otimes\mathbf{n})a^{ij}\,\mathbf{v}_{M,i}]- \tag{7}$$
$$-\ \xi^3\ (\delta\mathbf{r} - \delta\boldsymbol{\rho})(\boldsymbol{\rho}_{,k}\otimes\boldsymbol{\rho}_{,l})h^{lk}\ (\mathbf{v}_S - \mathbf{v}_M)\}\ \mathrm{ds}\ . \tag{8}$$

While a^{ji} are the contravariant components of the metric tensor, h^{lk} represent the components of the curvature tensor. Obviously, the contact tangent matrix consists of three parts, namely the "main" (Eq. (6)), the "rotational" (Eq. (7)) and the "curvature" (Eq. (8)) part, for details see [3], [4] and [7].

2.2 Finite element discretization

The current contribution is concentrating on the NTS and the STAS contact approaches with quadratic approximation of the master surface. Within finite element formalities using quadratic shape functions $N^{(k)}(\xi^1,\xi^2)$ and the nodal coordinates $\mathbf{x}^{(k)}$ this leads for a master surface point $\boldsymbol{\rho}$ to

$$\boldsymbol{\rho} = \sum N^{(k)}\ (\xi^1,\xi^2)\mathbf{x}^{(k)}\quad\text{with}\quad k = 1,2,...,9. \tag{9}$$

In the case of NTS contact elements the contact surface denoted as the master surface is approximated by surface elements as given in Fig. 2. The slave surface is represented by single nodes directly taken from the structural finite element mesh of the slave body. Thus, a single NTS contact element consists of a pair of 9 nodes for the discretization of the master surface and a further node on the slave surface.

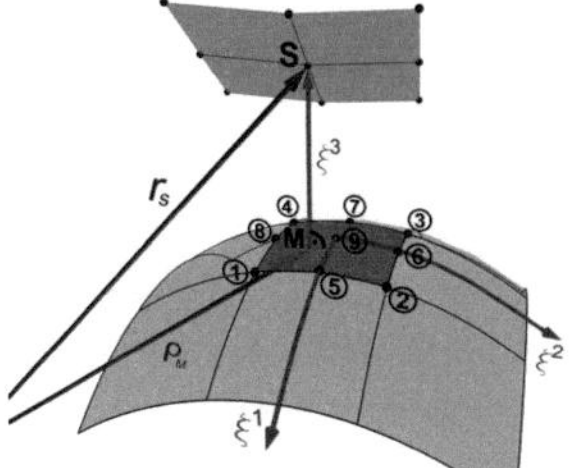

Fig. 2   Quadrilateral NTS element with quadratic approximation

In the case of STAS contact elements the slave surface is represented by an analytically described surface. Thus, a single quadratic STAS contact element consists of only 9 nodes discretizing the master surface. Penetration is checked at integration points as e.g. Gauss or Lobatto points against the analytically described surface and is calculated in a closed form as e.g. for the sphere

$$\xi^3_{ip} = |\rho(\xi^1_{ip},\xi^2_{ip}) - \mathbf{x}^c)| - R \tag{11}$$

with $\mathbf{x}^c$ denoting the center of the sphere and R the radius, while ip goes along with the considered integration points.

Both approaches have been implemented into FEAP-MeKA documented in [9] and [13].

3   Large penetration scheme

Based on the separation of the tangent matrix as given in Eq. (6)-(8) a scheme for large load steps is employed. This scheme was already studied for some examples regarding the NTS approach in [2] and proved to be effective. Whereby, the efficiency is increasing with the complexity of the investigated example as the rotational (Eq. (7)) and curvature parts (Eq. (8)) are activated. This scheme, the so-called 3D large penetration scheme, is directly based on considerations introduced in a recent publication by Zavarise et al., see [12], dealing with 2D linear contact problems.

In the following the general outline of the utilized 3D large penetration scheme for the STAS approach is given, see Fig. 3. The governing assumption, hereby, is to correctly estimate a priori the final resulting maximum contact pressure for a given contact setup.

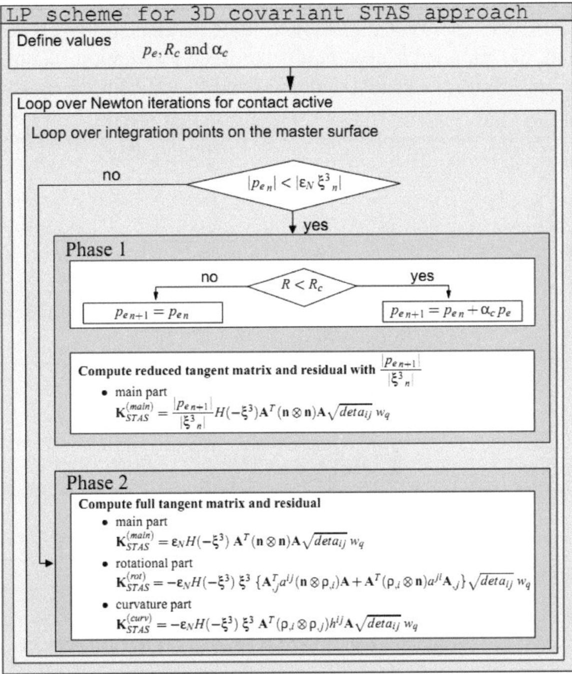

Fig. 3   Outline of applied LP scheme for 3D covariant STAS approach

The a priori estimated contact pressure denoted as p_e is used then as an indicator to switch for each load step between two main phases of the algorithm. Within the first phase the tangent matrix consists only of the main part as Eq. (6) and the residual encounters an upper bound for the maximum contact pressure represented by the a priori estimated contact pressure p_e. The latter modification allows to shift smoothly from phase one to phase two, where the standard consistently linearized terms - the full tangent matrix (Eqs. (6)-(8)) - are used. Furtheron an update scheme for a possibly underestimated a priori contact pressure is used as

$$p_{en+1} = p_{en} + \alpha_c\, p_e \; , \tag{12}$$

where α_c represents a chosen weighting factor. As an update indicator a parameter R is used which considers the penetration within consecutive iteration steps, see [2].

4   Numerical example

The example presented within this contribution represents a setup with very high complexity as sliding is encountered. Nevertheless, the example represents a frictionless contact problem between two bodies and demonstrates the effectiveness of the described large penetration algorithm in combination with the covariant contact approach. This example is taken for a comparison of the effectiveness for both considered approaches (NTS and STAS) regarding the used large penetration scheme.

The efficiency of the used algorithm is measured by the influence on the convergence rate looking at the accumulated number of equilibrium iterations of a Newton algorithm within all load steps. As a reference computation the examples are also computed according to the consistent linearization.

In case of the STAS approach 2×2 Gaussian integration points for each contact element are used and the surface of the rigid bodies is meshed for demonstration purposes.

The setup of the example is as follows. A clamped elastic beam is bended over a double curved body - a rigid sphere. The beam has the dimensions $24.0 \times 4.0 \times 0.25$ and the sphere a radius of 4.0. The beam consists of a St. Venant material with an elasticity modulus of $1.0 \cdot 10^5$ and a Poisson ratio of 0.0. The FE discretization is carried out using 12 "solid-shell" elements with quadratic shape functions. In order to enable sliding on the sphere the edge of the beam is not clamped, but attached to springs with a total spring stiffness of 60. The loading consists of three asymmetric forces $F_1 = 2.5$, $F_2 = 2 \times F_1$ and $F_3 = 4 \times F_1$ applied in vertical direction and $F_4 = 8 \times F_1$, $F_5 = 32 \times F_1$ and $F_6 = 8 \times F_1$ in horizontal direction on the nodes at the free end.

Regarding the contact formulation the bottom surface of the elastic beam is considered as the master surface with quadratic approximation. The contact parameters are chosen as $\varepsilon_N = 500$ for the penalty parameter, and, the parameters specific for the large penetration scheme are given as 1.0 for the cut-off pressure p_e, $\alpha_c = 1$ for the update multiplier and $R_c = 0.1$ for the update indicator. The setup is shown in Fig. 4.

As is indicated in Table 1 the example does not converge for any feasible load step size within the

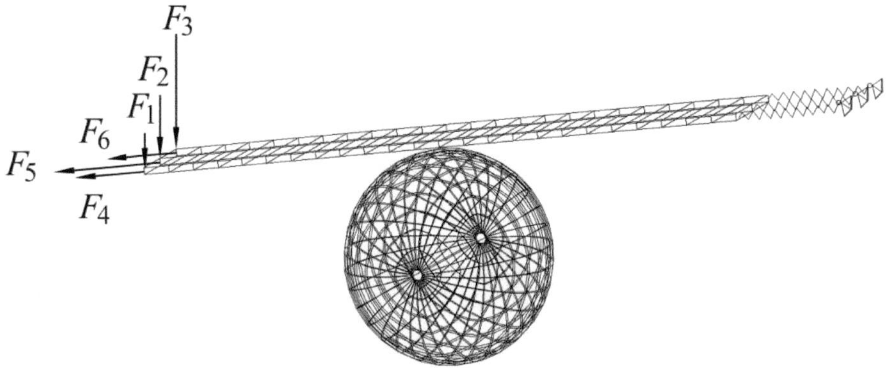

Fig. 4   General setup of sliding and bending of a beam over a rigid sphere

CL scheme. In order to detect the limits of the CL scheme additional computations are carried out for the NTS approach resulting in a maximum load step size of 0.00004 and 75998 iterations for a finally converging computation. However, the LP scheme is capable of dealing with more suitable load step sizes, even applying the full load in one step, thus, the performance is increased tremendously. Another observation within this example is that for some load step sizes no convergence is achieved at all for some load levels, however, convergence is observed again for the next load steps and the computation is then leading to the correct final result. In Table 1 this fact is denoted with $(m \times n)^*$ where m is the number of load steps without convergence and n is the maximum number of iterations for each load step defined within the computation. Finally, a deformation sequence of extreme penetrations allowed during the first iterations of the LP scheme are visualized in Fig. 5 with the contact area developing during the iterations for the load step size of 2.5 (one single load step).

Table 1   Sliding and bending of a beam over a rigid sphere with quadratic approximation of the contact surfaces. Comparison of convergence and accumulated number of iterations necessary for the final load $F_1 = 2.5$ for the NTS and the STAS approach (here: n=100)

scheme	NTS			STAS		
	convergence	load step size	cum. no. it.	convergence	load step size	cum. no. it.
CL	not achieved	2.5	∞	not achieved	2.5	∞
	not achieved	0.5	∞	not achieved	0.5	∞
	not achieved	0.25	∞	not achieved	0.25	∞
	not achieved	0.1	∞	not achieved	0.1	∞
	not achieved	0.02	∞	not achieved	0.02	∞
	not achieved	0.01	∞	not achieved	0.01	∞
LP	achieved	2.5	43	achieved	2.5	38
	achieved	0.5	115	achieved	0.5	$97+(1 \times n)^*$
	achieved	0.25	103	achieved	0.25	$251+(1 \times n)^*$
	achieved	0.1	152	achieved	0.1	$431+(5 \times n)^*$
	achieved	0.02	467	not achieved	0.02	∞
	achieved	0.01	702	achieved	0.01	$2514+(73 \times n)^*$

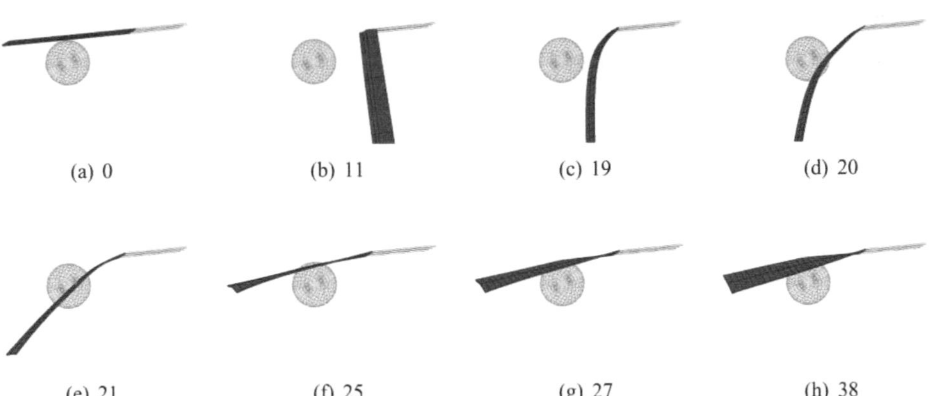

(a) 0   (b) 11   (c) 19   (d) 20

(e) 21   (f) 25   (g) 27   (h) 38

Fig. 5   Sliding and bending of a beam over a rigid sphere with quadratic approximation of the contact surfaces. Development of the contact area at several iteration states for the LP scheme of the STAS approach with load step size 2.5 (one single load step) and 38 iterations necessary for convergence

6   Conclusions

A direct development of a finite element setup for the so-called large penetration scheme was presented via the covariant contact description. An extension of the applicability of the large penetration algorithm to 3D problems with higher order shape functions was found for the NTS and the STAS approach. The restriction of the CL scheme for the STAS and NTS formulation to smaller load step sizes is not present within the LP scheme. Nevertheless, in the STAS approach some difficulties regarding load step sizes with non-converging load steps were found, however, continuing then without a converged solution for the load step the correct final result is achieved in all cases. Finally, it has to be noted that a direct comparison between the NTS and the STAS approach is fairly difficult as the amount of contact points was fairly different using 2×2 Gaussian integration points for the STAS approach.

References

[1]   R. Hauptmann, K. Schweizerhof, A systematic development of 'solid-shell' element formulation for linear and non-linear analysis employing only displacement degrees of freedom, Int. J. Numer. Methods Engrg. 42 (1998) 49-69.

[2]   R. Izi, A. Konyukhov, K. Schweizerhof, Large Penetration Algorithm for 3D Frictionless Contact Problems Based on a Covariant Form, Comput. Methods Appl. Mech. Engrg. 217-220 (2012) 186-196.

[3]   A. Konyukhov, K. Schweizerhof, Contact formulation via a velocity description allowing efficiency improvements in frictionless contact analysis, Comput. Mech. 33 (2004) 165-173.

[4]   A. Konyukhov, K. Schweizerhof, Covariant description for frictional contact problems, Comput. Mech. 35 (2005) 190-213.

[5]   A. Konyukhov, K. Schweizerhof, On the solvability of closest point projection procedures in contact analysis: Analysis and solution strategy for surfaces of arbitrary geometry, Comput. Methods Appl. Mech. Engrg. 197 (2008) 3045-3056.

[6]   A. Konyukhov, K. Schweizerhof, Incorporation of contact for high order FEM in covariant form, Comput. Methods Appl. Mech. Engrg. 198 (2009) 1213-1223.

[7]   A. Konyukhov, Geometrically exact theory for contact interactions, KIT Scientific Publishing, Habilitation thesis (2011) Karlsruhe.

[8]   T.A. Laursen, Computational contact and impact mechanics. Fundamentals of modeling interfacial phenomena in nonlinear finite element analysis, Springer (2002) Berlin.

[9]   K. Schweizerhof and Co-workers. Feap-MeKa, finite element analysis program. Karlsruhe Institute of Technology, based on version 1994 of R. Taylor, 'Feap – A Finite Element Analysis Program', http://www.ce.berkeley.edu/projects/feap. University of California, Berkeley.

[10]   P. Solin, Higher order finite element methods. Chapman & Hall, (2004) CRC.

[11]   P. Wriggers, Computational contact mechanics, John Wiley & Sons (2002).

[12]   G. Zavarise, L. De Lorenzis, R.L. Taylor, A non-consistent start-up procedure for contact problems with large load-steps, Comput. Methods Appl. Mech. Engrg. 205-208 (2012) 91-109.

[13]   O.C. Zienkiewicz, R.L. Taylor, Finite element method: Volume 1 – The basis, 5th edn. Butterworth- Heinemann (2000) New York.

An efficient implementation concept for volumetric and axisymmetric finite shell elements

Christoph Schmied

Institute of Mechanics,
Karlsruhe Institute of Technology (KIT),
Otto-Ammann-Platz 9, 76131 Karlsruhe, Germany
Supervisors: Steffen Mattern and Karl Schweizerhof

Abstract

An implementation concept for highly efficient shell finite elements with linear shape functions is presented. In the context of explicit time integration small time steps in combination with only vector operations on global level lead to a domination of element processing. The presented implementation concept is based on the application of the symbolic programming tool AceGen, an extension to the computer algebra software Mathematica. The presented shell elements including standard (=full) numerical integration in combination with different approaches in order to reduce artificial stiffness effects are implemented into the in-house Finite Element code Feap-MeKa. The specifics of the implementation concept are discussed and the efficiency and functionality of the element formulations are presented on numerical examples including large deformations.

1 Introduction

An explicit time integration method that is suitable for very fast processes and highly nonlinear problems is the central difference scheme [3], often also called VERLET-algorithm [24]. Due to the necessity of very small time steps the element routine is called very often inside the algorithm. For lumped mass matrices the costly solution of coupled linear equations on global level is not necessary. But the time step size is constricted by a critical value that is given by the COURANT-criterion [9]. Consequently the central difference scheme is mostly attractive for highly dynamic problems where small time steps are required anyway. In addition it is preferably used for highly nonlinear problems where convergence with implicit algorithms is hard to achieve.

For analysing thin-walled structures it is adequate to use shell elements that are able to capture thickness effects. For poor aspect ratios of the elements, e.g. for very small thicknesses, the so called geometric locking effect appears that is removed with the method of 'Assumed Natural Strains' (ANS), where the strains are evaluated at specific sampling points and interpolated. The volumetric locking effect that affects simulations near the incompressible limit is reduced with the 'Enhanced Assumed Strain' (EAS) method.

In this contribution the focus lies on an efficient implementation concept that enables the generation of different types of finite elements with high computational performance. The implementation concept is based on the application of the symbolic programming tool AceGen which allows a combination of symbolic operations together with the automatic generation of highly efficient program code. The adaptability of the concept is demonstrated with the solid-shell finite element concept and with an axisymmetric volume shell element. The axisymmetric element formulation can be obtained by assembling axisymmetric constraints on a simple 3D-solid shell element. The implementation of further element procedures like the EAS and ANS method are implemented also applying symbolic programming. The functionality of the element formulations is presented on several numerical examples.

2 Explicit time integration

The implementation of the central difference method is done according to [3] also known as VERLET-algorithm in [24]. The governing equations are shown briefly.

Proc. of the 9th *fib* International PhD Symposium in Civil Engineering, July 22 to 25, 2012,
Karlsruhe Institute of Technology (KIT), Germany, H. S. Müller, M. Haist, F. Acosta (Eds.),
KIT Scientific Publishing, Karlsruhe, Germany, ISBN 978-3-86644-858-2

With a diagonalized mass matrix $\mathbf{M}$, the system damping matrix $\mathbf{C}$ and the time dependent values of the element load vector $\mathbf{f}^n$ at time n and the velocities d at time $n-\frac{1}{2}$ the accelerations can be computed very efficiently as

$$\ddot{\mathbf{d}}^n = \mathbf{M}^{-1}\left(\mathbf{f}^n - \mathbf{C}\dot{\mathbf{d}}^{n-\frac{1}{2}}\right) \tag{1}$$

at time step n. With the time step $\overline{\Delta t}^n = (\Delta t^n + \Delta t^{n-1})/2$ the velocity between two time steps is updated by

$$\dot{\mathbf{d}}^{n+\frac{1}{2}} = \dot{\mathbf{d}}^{n-\frac{1}{2}} + \overline{\Delta t}^n \ddot{\mathbf{d}}^n \tag{2}$$

This leads to the displacements

$$\mathbf{d}^{n+1} = \mathbf{d}^n + \Delta t^n \dot{\mathbf{d}}^{n+\frac{1}{2}} \tag{3}$$

with the current time step size $\Delta t^n = t^{n+1} - t^n$. The time step size is limited by the COURANT-criterion by

$$\Delta t \leq \alpha \cdot \Delta t_{crit} = \alpha \cdot \frac{2}{\omega_{max}} \approx \alpha \cdot (\min_e \frac{l_e}{c_e}) \tag{4}$$

where ω_{max} is the largest eigenfrequency, l_e represents a characteristic element length and c_e the wave propagation velocity. The COURANT-criterion [9] is based on linear problems, so in order to consider non-linearities, the factor $\alpha < 1$ is introduced. For moderately non-linear applications, usually $\alpha \approx 0.9$ is sufficient, see also [10].

3   Element formulation

In this section the governing equations for the solid-shell element and the axisymmetric element formulation are briefly introduced. For more detailed information, it is referred to the literature, e.g. [18,12,17]. The compensation of locking effects by reduced integration is presented in recent developments that are given e.g. in [1,19,7,20] using implicit finite element applications.

First the kinematics for the solid-shell and the axisymmetric formulation are introduced. Then the residual force vector and the mass matrices are presented, where the same implementation concept can be used for each element formulation.

3.1   The axisymmetric solid

A recent formulation of an axisymmetric element is given in [8]. The formulation of an axisymmetric solid element consists according to Fig. 1 (left) of 4 nodal points, framing a plane that is fully rotated around an axis of revolution. This formulation can be obtained by constraining a fully 3D-continuum solid, in the first instance described in cylindrical coordinates, with the condition that the displacement in circumferential direction vanishes. The geometry, material properties, loading and boundary conditions of the problem have to be constant in circumferential direction as a precondition.

Thus the initial geometry can be described as for a plane element only depending on the coordinates that lie in the cross-section of the nodes by

$$\mathbf{X}(\xi,\zeta) = N_i(\xi,\zeta) \cdot \mathbf{X}_i \tag{5}$$

Due to the isoparametric concept the displacements are interpolated with the same shape functions and can be obtained as

$$\mathbf{u}(\xi,\zeta) = N_i(\xi,\zeta) \cdot \mathbf{u}_i \tag{6}$$

where i is the number of the nodal points. For the later described computation of element matrices the kinematics are transformed from cylindrical coordinates to cartesian coordinates.

3.2 The Solid-Shell concept

The concept of Solid-Shell Finite Elements as presented e.g. in [11] provides a shell formulation with full 3D capabilities and displacement degrees of freedom only. Under the assumption of the degenerated shell concept that the normals to the mid-surface remain straight, following the notation given in Fig. 1 (right), the initial geometry is given by

$$\mathbf{X}(\xi,\eta,\zeta) = \frac{1}{2}\big((1+\zeta)\mathbf{X}_u(\xi,\eta) + (1-\zeta)\mathbf{X}_l(\xi,\eta)\big) \tag{7}$$

Linear interpolation of the displacements of the upper and the lower surface leads to

$$\mathbf{u}(\xi,\eta,\zeta) = \frac{1}{2}\big((1+\zeta)\mathbf{u}_u(\xi,\eta) + (1-\zeta)\mathbf{u}_l(\xi,\eta)\big) \tag{8}$$

In this contribution linear isoparametric Solid-Shell elements are used with bilinear interpolation in membrane and linear interpolation in thickness direction. For the discretization of the initial geometry, this leads to

$$\mathbf{X}^{el}(\xi,\eta,\zeta) = \sum_{i=1}^{nip}\left(\frac{1}{2}\mathrm{N}_i(\xi,\eta)\boldsymbol{\Theta}(\zeta)\mathbf{X}_i\right) \tag{9}$$

where nip is the number of in-plane nodes. The upper and lower nodal locations are described by the vector $\mathbf{X}_i = \begin{bmatrix}\mathbf{X}_{iu} & \mathbf{X}_{il}\end{bmatrix}^T$, the interpolation is performed linearly in thickness direction with the interpolation matrix $\boldsymbol{\Theta}(\zeta)$. The in-plane interpolation is achieved in the present case with linear $(nip=4)$ Lagrangian shape functions. According to the isoparametric concept, the displacements are interpolated with the same shape functions. The application of higher order Lagrangian or Serendipity type shapes is also possible, quadratic Lagrange functions e.g. lead to a formulation with 18 element nodes $(nip=8)$.

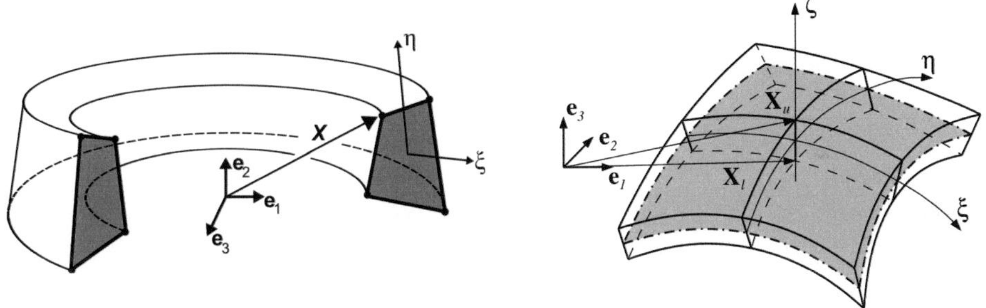

Fig. 1   Geometry of a axisymmetric solid and a solid-shell.

3.3 Residual force vector

The system force vector in Eq. (1) is composed of the globally defined external nodal forces $\mathbf{f}^{ext}$ and the internal nodal forces $\mathbf{f}^{int}$ which are computed on element level. The internal forces can be written as the derivative of the internal energy Π^{int} with respect to the vector of nodal degrees of freedom $\mathbf{d}$. The formulations presented here are implemented for different hyper-elastic material models as e.g. a version of the *Neo-Hooke* material law, defined by the strain energy function

$$\mathrm{W}(\mathbf{C}) = \frac{\mu}{2}(\mathrm{I}_\mathrm{C}-3) - \mu\ln \mathrm{J} + \frac{\lambda}{2}(\ln \mathrm{J})^2 \quad \text{with} \quad \mathrm{J}^2 = \mathrm{III}_\mathrm{C} \tag{10}$$

with the first and third invariant I_C and III_C of the right *Cauchy-Green deformation tensor* and λ and μ as the Lamé parameters. The internal energy and hence the internal nodal force vector is then obtained directly by

$$\Pi^{int} = \int_V \mathrm{W}(\mathbf{C})\mathrm{dV} \quad , \quad \mathbf{f}^{int} = \Pi^{int}_{,d} = \frac{\partial}{\partial \mathbf{d}}\int_V \mathrm{W}(\mathrm{C})\mathrm{dV} \tag{11}$$

3.4 Mass matrices

As mentioned in Section 2, an efficient usage of the central difference method implies diagonalized mass matrices. The entries of the consistent element mass matrix $\mathbf{M}^{el}$

$$\mathbf{M}^{el} = \int_{V} \rho \, \mathbf{N} \cdot \mathbf{N}^{T} \, dV \tag{12}$$

with the shape functions assembled in the matrix $\mathbf{N}$ are therefore e.g. summed up row by row, in order to achieve the diagonalized form

$$M_{ij}^{el,d} = \begin{cases} \sum_{k} M_{ik}^{el} & i = j \\ 0 & i \neq j \end{cases} \tag{13}$$

For element formulations using Lagrangian shape functions, this method leads to identical or very similar mass matrices as other methods as e.g. described in [12] such as '*nodal integration*' or '*scaled diagonals*' which are also implemented in our code. When using Serendipity type shape functions, the '*row-sum*'-technique in Eq. (11) is not applicable, as it leads to negative entries in the diagonalized mass matrix; other schemes are also not perfect. For this reason Serendipity elements are not used here, though they work very well for implicit methods.

3.5 Locking phenomena

A very important issue concerning the implementation of low order shell finite elements is the activation of artificial stresses for different loading situations, the so-called '*locking*' phenomena. Locking can be reduced or even completely removed by several corrections within the element formulation. Proposals for locking-free Solid-Shell elements, using reduced integration rules together with stabilization techniques against artificial kinematics can be found in [1,19,7,20]. In the current contribution, fully integrated element formulations are presented, where different locking phenomena are treated with the well-known methods of '*Assumed Natural Strains* (ANS)' [2, 6] and '*Enhanced Assumed Strains* (EAS)' [23, 22], which have already been applied to Solid-Shell elements for non-linear implicit analyses [12, 11]. For the axisymmetric element formulation *Selective Reduced Integration*' is taken to alleviate transverse shear locking.

4 Implementation concept

4.1 AceGen

In order to achieve an efficient and comfortable implementation of the algorithms on element level, the improved – so called 'automatic' – code generation and optimization tool ACEGEN, a plug-in for the computer algebra software MATHEMATICA is used. The program is developed by the group of KORELC, see [14, 15, 16]. The plug-in uses the symbolic capabilities of MATHEMATICA in order to create automatically optimized program code in FORTRAN. It is possible to enter formulas as written, without bothering about programming issues. A particular advantage is using the differentiation capabilities. Hence matrix operations, summations and differentiation - also with respect to vectors and matrices – can be performed and implemented in a straight forward fashion and programming errors can be reduced significantly. Also the computational speed is increased by using automatic code generation.

For the numerical integration of Eq. (11) at the current time step, the operation

$$\mathbf{f}^{int} = \frac{\partial \Pi^{int}}{\partial \mathbf{d}} = \Pi_{,d}^{int} = \sum_{i=1}^{m} \sum_{j=1}^{m} \sum_{k=1}^{n} \frac{\partial W(\xi_{i}, \eta_{j}, \zeta_{k})}{\partial \mathbf{d}} \det \mathbf{J}(\xi_{i}, \eta_{j}, \zeta_{k}) w_{i} w_{j} w_{k} \tag{14}$$

has to be evaluated for each element using an integration rule with $m \times m$ quadrature points in-plane and n points in thickness direction. For code generation and implementation, the following steps-itemized in Fig. 2 (left) – have to be carried out:

<u>Initialization:</u> The subroutine as well as the input (geometry, current displacements, coordinates and weights of the quadrature points and material parameters) and output variables (internal force vector) are defined. The used ACEGEN commands are `SMSInitialize` and `SMSModule`.

<u>Element matrices:</u> After the import of the element data, the necessary matrices and vectors – Jacobian, convective base vectors, Green-Lagrange strain tensor, etc. – can be evaluated using

MATHEMATICA's symbolic capabilities. Differentiation with respect to variables or tensors is also possible (SMSD), which is used e.g. to evaluate $\Pi^{int}_{,d}$.

Export internal force vector: The internal force vector at the current integration point has to be exported, to be available outside the symbolic subroutine. The command SMSExport is used with the optima "AddIn"=True in order to automatically sum the results for all quadrature points to the global memory field.

Code generation: In the last step, e.g. FORTRAN-code is generated and automatically optimized, using the command SMSWrite. The language (FORTRAN, C, etc.) and the level of optimization depend on options, defined in SMSInitialize. Fig. 3 (right) shows a portion of the automatically generated FORTRAN-code.

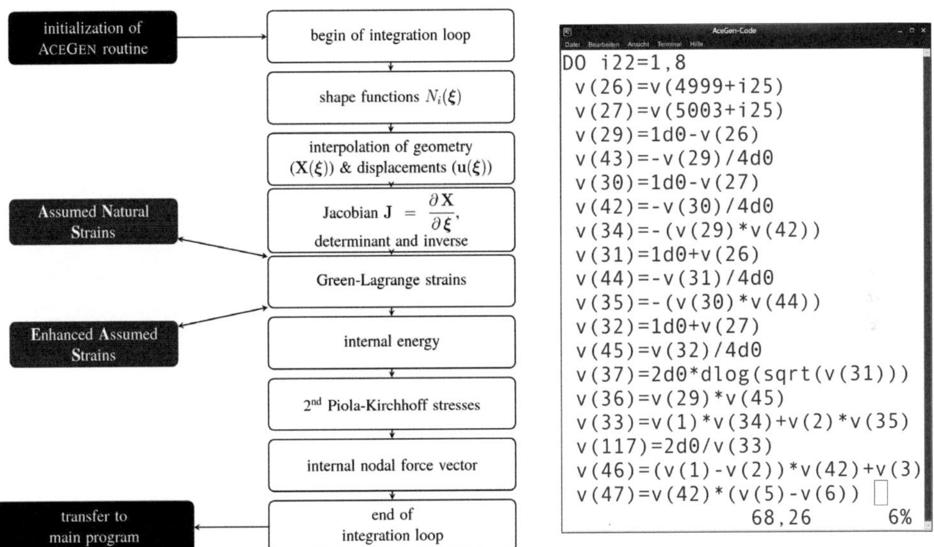

Fig. 2   Implementation concept and exemplary section of automatically generated optimized FORTRAN code

5   Numerical examples

5.1   Bending of a long cylindrical shell by a load uniformly distributed along a circular section

This example is chosen to show how the element formulations behave in the case of volumetric locking by varying the Poisson ratio. The linear elastic solution of the displacement of the mid-section of a long cylindrical shell is given by TIMOSHENKO and KRIEGER and also used in [8].

The uniform load is applied slowly thus quasi-statically according to the load function in Fig. 3 (right), the radius $R = 10.0$ and the thickness $t = 0.04$ describe the geometry and the material is described by $E = 2100 \cdot 10^3$. In this example at first the results of both element formulations are identical until the 3D solid-shell discretization shows – not unexpected as the buckling stress is reached – buckling effects caused by circumferential compression whereas the axi-symmetric element by formulation does not contain this buckling mode. The corresponding buckling mode evolution is displayed in Fig. 4 from left to right.

For different Poisson ratios the displacement of the axi-symmetric element at the mid-section of the cylinder is compared with the analytical solution in Tab. 1 by indicating the ratio of the numerical solution and the analytical result. Due to the use of only one single EAS-Parameter the locking effects could be almost eliminated.

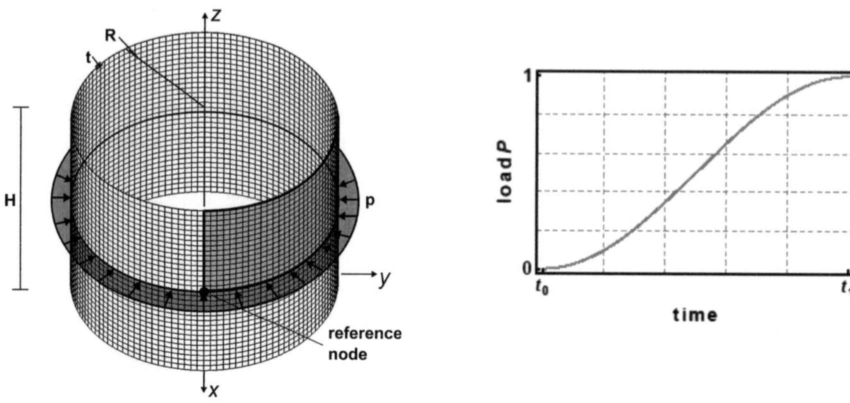

Fig. 3   Geometry of a cylindrical shell and the curve of the applied load

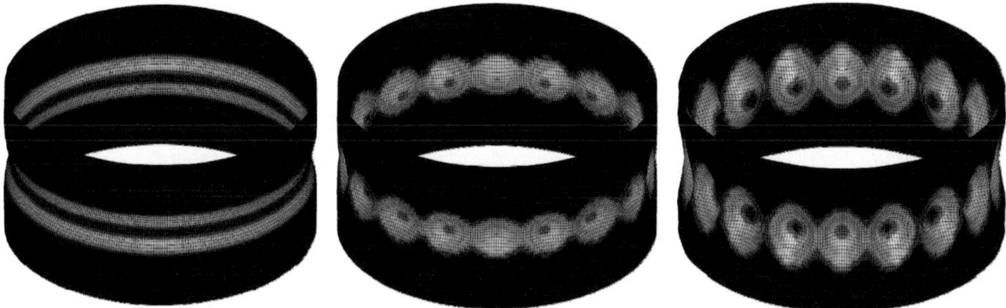

Fig. 4   Buckling mode processing for 3D-solid-shell discretization

Table 1   Displacement ratios at the mid section for different Poisson ratios

$W_{FEM} / W_{analytic}$	Axisymmetric element	
	w/o EAS	w EAS
$\upsilon = 0.3$	0.883	0.948
$\upsilon = 0.4999$	0.181	0.835

5.2   Spherical shell with hole under axisymmetric tip load

This example presents a spherical shell with an *18°* hole on top. Symmetric boundary conditions are constraining all sides of the eighth of the sphere. The geometry is described by radius of $R = 10.0$ and a thickness of $t = 0.04$ and the material is defined by $E = 210 \cdot 10^3$. The prescribed wavelike displacement is given in Fig. 5 (right). The response is captured in radial direction at a reference node that lies on the mid section of the sphere. One eighth of the sphere is discretized with 128×128 elements for the solid-shell formulation, and one fourth of the symmetric sphere section is discretized with 128 axisymmetric elements.

The evolution of the wave is shown in Fig. 6 (left) at the time $t = 3.0 \cdot 10^{-6}$ with tenfold scaled resultant displacement. The response, see Fig. 6 (right), begins with an opposite motion, and then passes into an oscillation. Comparing the solid-shell and axisymmetric element formulation a slightly growing shift of the phase and the amplitude is found. The reasons for this are the minor differences in the element formulations and the error that is automatically embedded by discretizing the curved sphere with bilinear elements. About the time $t = 2.0 \cdot 10^{-5}$ the nearly uniform oscillation is in general disturbed by cumulative reflections of compressional and transversal waves.

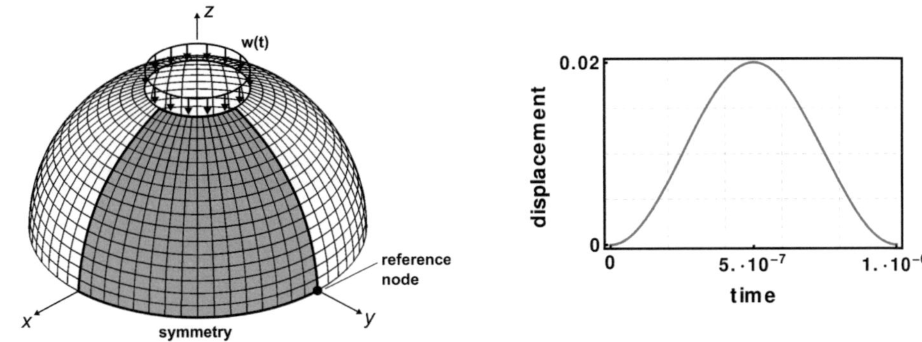

Fig. 5   Geometry of the spherical shell and the wavelike prescribed displacement curve

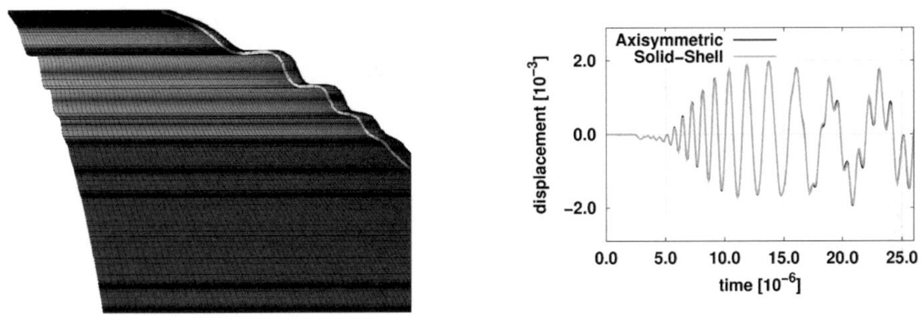

Fig. 6   Propagation of the wave and response at the mid-section of the sphere

6   Conclusions

An implementation concept using the automatic code generation tool AceGen based on the computer algebra program Mathematica is shown for modern shell elements with mixed interpolation and highly nonlinear behaviour. Further the advantages regarding programming and implementing different element formulations are discussed. The correct application of AceGen allows the direct implementation of symbolic algorithms and the fast and error free generation of program code. Also modifications in one element formulation or different element formulations can be quickly implemented without programming errors. A highly efficient program is achieved proven by the numerical examples with explicit time integration.

The application of AceGen for further shell element formulations with quadratic geometry and displacement interpolations has shown similar advantages, as they are especially important to capture the behaviour of arbitrary curved shells in combination with explicit time integration. A very promising application is the implementation of complex material models – currently also underway – as here the advantage of automatic differentiation is particularly interesting.

7   Acknowledgments

The presented work was part of a project (DFG-SCHW307/20-1), currently funded by the German Research Foundation (DFG). The support is gratefully acknowledged.

References

[1]   R. J. Alves de Sousa, R. P. R. Cardoso, R.A. Fontes Valente, J. W. Yoon, R. M. Natal Jorge and J. J. "A new one-point quadrature enhanced assumed strain (EAS) solid shell element with multiple integration points along thickness: Part I – geometrically linear applications.", Int. J. Num. Meth. Eng., 62:952-977, 2005.

[2]   K.-J. Bathe and E. N. Dvorkin. "A formulation of general shell elements – the use of mixed interpolation of tensorial components.", Int. J. Num. Meth. Eng., 22:697-722, 1986.

[3]   T. Belytschko, W. K. Liu and B. Moran. Nonlinear Finite elements for continua and structures. Wiley, 2004.

[4]   P. Betsch and E. Stein. "An assumed strain approach avoiding artificial thickness straining for an non-linear 4-node shell element.", Communications In Numerical Methods In Engineering., 11:899-909, 1995.

[5]   M. Bischoff and E. Ramm. "Shear deformable shell elements for large strains and rotations.", Int. J. Num. Meth. Eng., 40:4427-4449, 1997.

[6]   M. L. Bucalem and K.-J. Bathe. "Higher-order MITC general shell elements.", Int. J. Num. Meth. Eng., 36(21):3729-3754, 1993.

[7]   R. P. R. Cardoso, J. W. Yoon, M. Mahardika, S. Choudhry, R. J. Alves de Sousa and R. A. Fontes Valente. "Enhanced assumed strain (eas) and assumed natural strain (ans) methods for one-point quadrature solid-shell elements.", Int. J. Num. Meth. Eng., 75:156-187, 2008.

[8]   R. P. R. Cardoso, J. W. Yoon, R. Dick. "A new axi-symmetric element for thin walled structures.", Computational Mechanics., 45:281-296, 2009.

[9]   R. Courant, K. O. Friedrichs and H. Lewy. "Über die partiellen Differentialgleichungen der mathematischen Physik.", Mathematische Annalen, 100:32-74, 1928.

[10]   John O. Hallquist. LS-DYNA – Theory manual. LSTC., "http://www.lstc.com /manuals.htm/", 2006.

[11]   R. Hauptmann, S. Doll, M. Harnau and K. Schweizerhof. " 'solid-shell' elements with linear and quadratic shape functions at large deformations with nearly incompressible materials.", Computers & Structures, 79(18),1671-1685, 2001.

[12]   R. Hauptmann and K. Schweizerhof. " 'A systematic development of 'solid-shell' element formulations for linear and non-linear analyses employing only displacement degrees of freedom.", Int. J. Num. Meth. Eng., 42(1):49-69, 1998.

[13]   J. Korelc. "Automatic generation of finite-element code by simultaneous optimization of expressions.", Theoretical Computer Science, 187(1-2):231-248, 1997.

[14]   J. Korelc. "Multi-language and multi-environment generation of nonlinear finite-element codes.", Engineering with Computers, 18(4):312-327, 2002.

[15]   J. Korelc. "http://www.fgg.uni-lj.si/Symech/", 2010.

[16]   J. Korelc and P. Wriggers. "Computer algebra and automatic differentiation in derivation of finite element code.", ZAMM, 79:811-812, 1999.

[17]   C. Miehe. "A theoretical and computational model for isotropic elastoplastic stress analysis in shells at large strains.", Comp. Meth. Appl. Mech. Eng., 155(3-4),193-234, 1998.

[18]   H. Parisch. "A continuum-based shell theory for non-linear applications.", Int. J. Num. Meth. Eng., 38:1855-1883, 1995.

[19]   S. Reese. "A large deformation solid-shell concept based on reduced integration with hourglass stabilization.", Int. J. Num. Meth. Eng., 69(8):1671-1716, 2007.

[20]   M. Schwarze and S. Reese. "A reduced integration solid-shell finite element based on the eas and the ans concept – geometrically linear problems.", Int. J. Num. Meth. Eng., 80(10):1322-1355, 2009.

[21]   K. Schweizerhof and Co-workers. "FEAP-MEKA, finite element analysis program. Karlsruher Institut für Technologie, based on Version 1994 of R. Taylor, 'FEAP – A Finite Element Analysis Program' ", University of California, Berkeley.

[22]   J. C. Simo and F. Armero. "Geometrically non-linear enhanced strain mixed methods and the method of incompatible modes.", Int. J. Num. Meth. Eng., 33:1413-1449, 1992.

[23]   J. C. Simo and M. S. Rifai. "A class of mixed assumed strain methods and the method of incompatible modes.", Computers & Structures, 29:1595-1638, 1990.

[24]   L. Verlet. "Experiments on classical fluids I. Thermomechanical properties of Lennard-Jones molecules.", Physical Review, 159:98-103, 1967.

Flow-induced fibre orientation and distribution in steel fibre reinforced self-compacting concrete

Joren Andries

*Building Materials and Building Technology Section,
KU Leuven,
Kasteelpark Arenberg 40, 3001 Heverlee, Belgium
Supervisor: Lucie Vandewalle, KU Leuve, Ann Van Gysel, Lessius University College*

Abstract

Fibre efficiency in structural elements can be improved by controlling and influencing the final direction of the fibres in order to attain the same post-cracking behaviour with a smaller fibre content. This optimisation will be investigated by experiments. Besides a computational model will be developed to simulate the flow of fresh concrete in a given formwork and to predict the fibre alignment during the flow of SFRSCC.

The results presented in this paper examine the influence of fibre length and flow distance on the fibre orientation and are part of an ongoing experimental programme. Fibre orientation and distribution are determined by manually counting the pulled out fibres in both cracking planes. The post-cracking behaviour is determined from the load-CMOD curve obtained by testing notched prisms, cut from longer beams, under three-point loading (EN 14651).

1 Motivation and aim of the research

The use of Steel Fibre Reinforced Self-Compacting Concrete (SFRSCC) can ease the production process because concrete elements can be made with a reduction or even complete substitution of conventional reinforcement and without vibrating the concrete. Since fibre distribution and orientation are influenced by the fresh concrete properties, casting procedure and mould geometry [1], the assumption of a homogeneous fibre distribution and an isotropic fibre orientation can be questioned.

Fibre efficiency in structural elements can be improved by controlling and influencing the final direction of the fibres so that they have a preferential orientation along the tensile stress direction. In this doctoral study, the flow-induced fibre orientation and distribution in fresh SFRSCC is investigated.

In the first part of the research project the influence of the rheological parameters of the concrete, the flow distance and the fibre geometry on fibre orientation and distribution are empirically investigated. Based on these experimental results, a model will be developed to predict any segregation and to calculate how the fibres orient themselves along the concrete flow. The preliminary results presented in this paper are part of an ongoing experimental programme at the Department of Civil Engineering of the KU Leuven.

Numerical simulations of concrete flow enable to understand the filling behaviour of fresh concrete in a given formwork and provide insight into how fibres align during the flow of SFRSCC. The ability to predict the fibre orientation and a tool to optimise the casting procedure reduce the number of trial-and-error experiments required for the design of new SFRSCC structural elements. To the best of the author's knowledge, this PhD is the first study that will predict fibre orientation and distribution in SFRSCC by use of discrete particle flow simulation in three dimensions.

2 Experimental programme

2.1 Materials used

The SFRSCC composition is designed to examine the flow-induced fibre orientation and distribution. The mixture proportion used in this study is presented in Table 1. The fibre content is kept constant at 30 kg/m³ ($\approx$ 0.38 vol%) of hooked-end steel fibres.

Proc. of the 9th *fib* International PhD Symposium in Civil Engineering, July 22 to 25, 2012,
Karlsruhe Institute of Technology (KIT), Germany, H. S. Müller, M. Haist, F. Acosta (Eds.),
KIT Scientific Publishing, Karlsruhe, Germany, ISBN 978-3-86644-858-2

Table 1   Mixture proportions

Component	Type	kg/m³
River gravel	4/14	695
River sand	0/4	847
Mineral addition	Limestone powder	335
Cement	CEM I 42.5 R HES	335
Water	Tap water	150.8
Superplasticiser	PCE	8.375
Steel fibres	Hooked-end	30.0

The constituent materials are initially mixed without fibres and only 80% of the superplasticiser. The fibres are then gradually added to avoid fibre balling and finally the remaining 20% of superplasticiser is added. Collated hooked-end steel fibres with circular cross-section are used. In order to avoid fibre rupture, steel fibres with a high tensile strength are applied. Their physical properties are presented in Table 2.

Table 2   Properties of the hooked-end steel fibres

Fibre type	Aspect ratio [-]	Length [mm]	Diameter [mm]	Tensile strength [MPa]
RC80/30BP	80	30	0.38	3070
RC80/60BP	80	60	0.71	2600

2.2   Concrete properties

2.2.1   Fresh concrete properties

Slump-flow tests according to EN 12350-8 and V-funnel tests according to EN 12350-9 are performed to measure the flowability and filling ability of the SFRSCC. Additionally the air content is determined by means of the pressure gauge method according to EN 12350-7. For each SFRSCC, the arithmetic mean values of three tests are given in Table 3. The fresh concrete properties are not significantly influenced by the fibre length. The slump-flow test did not indicate any segregation.

Table 3   Fresh concrete properties

Fibre type	Slump-Flow [mm]	t_{500} [s]	t_V [s]	Air content [%]	Fresh concrete density [kg/m³]
RC80/30BP	787	11.3	47.8	1.5	2400
RC80/60BP	772	11.3	50.1	1.5	2400

The product of fibre volume and aspect ratio is called the fibre factor and can be used to describe the effect of the type and content of steel fibres. Fresh mix workability is reduced with increasing fibre amount and aspect ratio of the fibres [2, 3]. Both SFRSCCs have the same flowability because of the constant fibre factor. The length of the fibres may have an influence on the passing ability [4].

2.2.2   Hardened concrete properties

All specimens are cured during four weeks at 20 ± 2 °C and 60 ± 2 %RH. The compressive strength is measured on 150 mm cubes according to EN 12390-3. The average cube compressive strength is equal to 73 N/mm² for SFRSCC80/30 and 75 N/mm² for SFRSCC80/60. SFRSCC80/60 (with a fibre length of 60 mm) has a slightly higher compressive strength due to the fact that this concrete was about 4 days older at the date of testing. The density of both hardened SFRSCCs is 2390 kg/m³.

2.3 Test results and discussion

2.3.1 Flexural tensile strength

For both fibre types, six beams having a length of 2.05 m and a cross-section of 150 mm by 150 mm are cast. These beams are cut in three different prisms and the specimens are numbered as illustrated in Fig. 1.

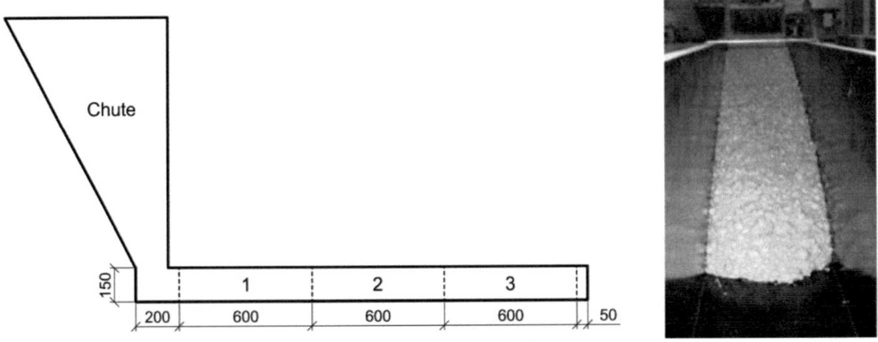

Fig. 1   Casting setup (all dimensions in mm) width = 150 mm

The post-cracking behaviour is determined from the load-crack mouth opening displacement curve obtained by three-point bending tests on notched prisms according to EN 14651. The test is performed under crack mouth opening displacement (CMOD) control.

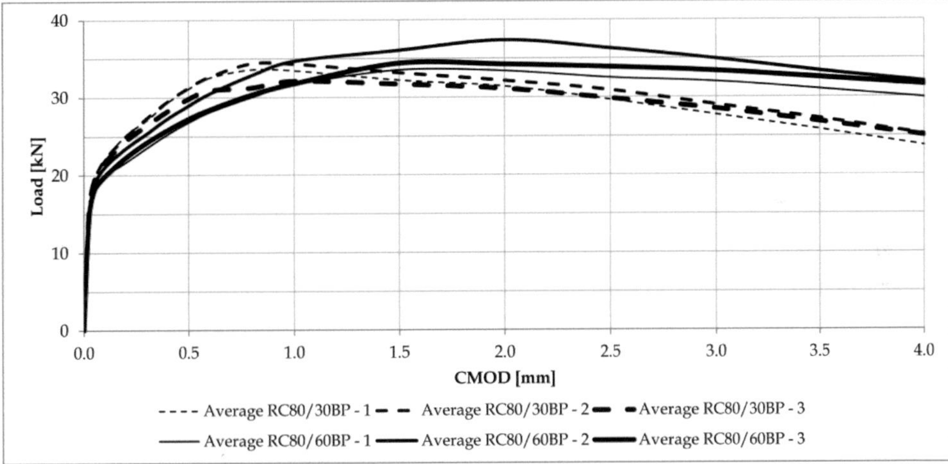

Fig. 2   Average load-CMOD curves

The average load-CMOD curves (Fig. 2) show that the post-cracking behaviour of the prisms is not significantly influenced by the flow distance in the beam. This is probably related to the fact that the fibres already have a preferential orientation due to the flow in the chute. Afterwards, the alignment of the fibres doesn't change much due to the flow in the beam itself.

Complete pull-out of the fibres is the principal failure mode. Fibre fracture would significantly reduce the toughness [5].

The results of three-point bending tests are compared with the number of fibres crossing the active crack and their orientation profile. The results of one beam with RC80/60BP-fibres (Fig. 3) will be used as an example. In this beam, the prism with the greatest flow-distance has the best post-cracking behaviour.

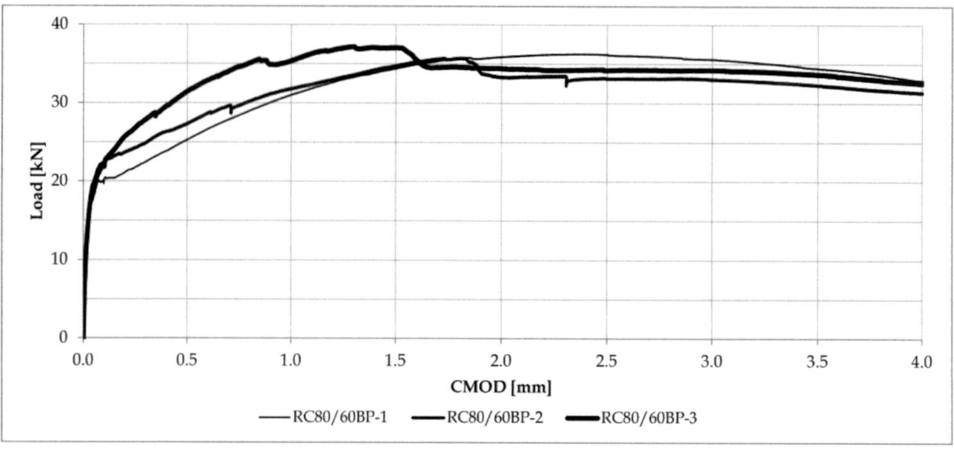

Fig. 3   Load-CMOD curve for one beam with RC80/60BP-fibres

2.3.2   Fibre orientation and fibre distribution

After the bending tests on the notched prisms, the number of fibres present at both cracking planes is manually counted for quantitative measurement of the fibre orientation. The cross section is divided in 36 square areas (Fig. 4) in order to register an orientation profile with different zones and not only an average value. This orientation profile will be used later to quantify the wall effect. From the distribution of the fibres in the cross-section it can be concluded that there is no significant segregation.

The mean value of the sawn fibres in the notch added to the sum of the pulled out fibres in both cracking planes gives the total number of fibres in the cross-section. The average experimental orientation factor is calculated by means of the formula proposed by Krenchel [6]:

$$\alpha = N \cdot \frac{A_f}{V_f \cdot A_c}$$

with: α : average orientation factor (-)
           N : total number of fibres in the cross-section (-)
           A_f : cross section of the fibre (mm²)
           V_f : fibre volume fraction (-)
           A_c : cross section of the concrete beam (mm²)

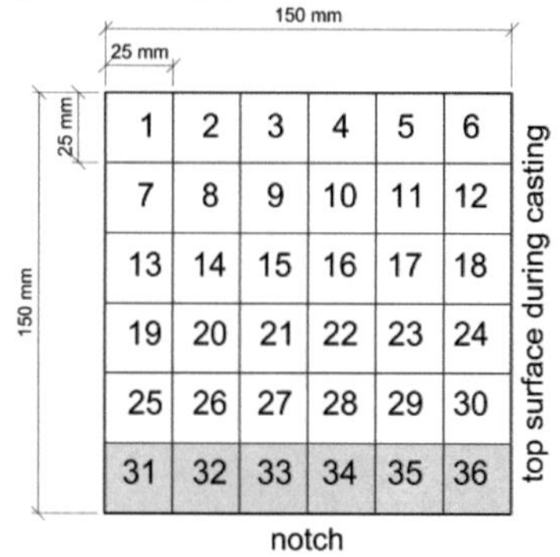

Fig. 4 Template for manual counting

A perfect alignment of all fibres along one direction would result in an average orientation factor along that direction equal to 1. The orientation factor of randomly oriented fibres in bulk is 0.5. In order to take into account the influence of geometrical boundaries on the orientation factor, Dupont and Vandewalle [7] described an averaging calculation procedure for rectangular sections. With a fibre length of 60 mm and a notch depth of 25 mm, this results in a theoretical orientation factor of 0.58.

Table 4   Orientation factor for one beam with RC80/60BP-fibres

Prism	Average orientation factor	
	Experimental	Theoretical
RC80/60BP-1	0.81	
RC80/60BP-2	0.83	0.58
RC80/60BP-3	0.90	

Table 4 compares the experimental with the theoretical orientation factor for the beam from which the load-CMOD curves are given in Fig. 3. Due to the casting procedure, a preferential fibre orientation occurs in the investigated SFRSCC-beam. The actual influence of geometrical boundaries on the orientation factor is greater than the theoretical factor. The prism at the largest distance of the pouring point has the largest average orientation factor and the best post-cracking behaviour.

3   Future research

At this time the orientation coefficients in well defined zones of each sawn surface are being determined by means of image analysis. Besides making use of the formula proposed by Krenchel, the ratio of the diameter of the steel fibre and the actual fibre length in the plane of consideration provides a direct measurement method for the orientation factor. In addition to the average orientation per cross-section, also the orientation factor in each zone will be calculated to quantify the wall effect. The load-CMOD curves will be compared with the results of the manual fibre counts and image analysis. A second SFRSCC with a lower viscosity and comparable flowability is currently in design to compare with the results of the SFRSCC of Table 1. The influence of the fibre aspect ratio will also be investigated.

The computational model that will be developed is based on the Discrete Element Method in order to take into account fibre geometry and predict the fibre orientation profile. DEM is a particle based simulation technique and it allows for a detailed modelling of the moving free surface. Because computation time strongly depends on the number of particles in the simulation, only the largest particles (steel fibres and coarse aggregates) will be modelled separately and the mortar fraction is modelled as a single-phase suspending non-Newtonian fluid. The parameters of the DEM model will be estimated based on small-scale experimental tests such as rheometry and workability tests. Simulations and experiments will then gradually be scaled up to end with a case study.

Based on the experimental findings, the accuracy of the model will be evaluated, and improvements will be made. The research focuses primarily on the fresh state of the concrete, but the fibre orientation and distribution can then be correlated with the mechanical behaviour of a structural element by use of adapted constitutive models designed by other researchers.

4   Conclusions

A viscous steel fibre reinforced self-compacting concrete with high flowability is designed. Two different fibre lengths are applied with all other parameters, apart from the flow distance, kept constant. From the ongoing research program the following conclusions can be drawn:

- Flowability and filling ability are not significantly influenced by the fibre length.
- In beam elements, steel fibres align along the flow direction of the fresh SFRSCC.

Because the fibres already align due to the flow in the chute and the concrete bucket, an additional formwork geometry will be worked out to quantify the influence of the flow distance.

References

[1]   Laranjeira, F. (2010): Design-oriented constitutive model for steel fiber reinforced concrete. PhD-thesis, Universitat Politècnica de Catalunya.

[2]   Groth, P. & Nemegeer, D. (1999): The use of steel fibres in self-compacting concrete. In: First International RILEM Symposium on Self-Compacting Concrete, Stockholm, Sweden. RILEM Publications SARL, pp. 497 - 507.

[3]   Grünewald, S. (2004): Performance-based design of self-compacting fibre reinforced concrete. PhD-thesis, Technische Universiteit Delft.

[4]   Dhonde, H. B., Mo, Y. L., Hsu, T. T. C. & Vogel, J. (2007): Fresh and Hardened Properties of Self-Consolidating Fiber-Reinforced Concrete. ACI Materials Journal 104, pp. 491-500.

[5]   Van Gysel, A. (2000): Studie van het uittrekgedrag van staalvezels ingebed in een cementgebonden matrix met toepassing op staalvezelbeton onderworpen aan buiging. PhD-thesis, Ghent University.

[6]   Krenchel, H. (1975): Fibre spacing and specific fibre surface, In: Fibre Reinforced Cement and Concrete. The Construction Press, UK, pp. 69-79.

[7]   Dupont, D. & Vandewalle, L. (2005): Distribution of steel fibres in rectangular sections. Cement and Concrete Composites 27, pp. 391-398.

Suitable restrained shrinkage test for fibre reinforced concrete: a critical discussion

Adriano Reggia

*DICATA - Department of Civil, Architectural, Environmental and Land Planning Engineering,
University of Brescia,
Via Branze, 43 - 25123 Brescia, Italy
Supervisor: G. A. Plizzari*

Abstract

Concrete performance traditionally refers to compressive strength and workability. Recently, high performance concrete evidenced the possibility of enhancing other material properties. Among these, resistance to shrinkage cracking is gaining more attention among practitioners due to its strict relation to durability requirements. Shrinkage cracks occur in restrained structures: for this reason, material characterisation should be made on the basis of a restrained shrinkage test. The ring test is an easy-to-use tool since one can measure the time-to-cracking of a concrete mix. Focus of this work is to critically discuss via non-linear numerical analyses the actual standard test procedure and then to propose enhancements of the test set-up in order to make the test duration more suitable for practical uses. Furthermore, the effect of fiber reinforcement on shrinkage cracking is presented.

1   Introduction

Recent advances in construction methods and new materials and admixtures have renewed the interest on early age cracking in cement based material due to restrained shrinkage; the latter has been recognized as the main cause of damage for thin structures such as highway pavements, industrial floors, bridge decks but also for partial depth repairs, jacketing and overlays on existing structures.

During the last few years, concrete technologists have been studying the mechanism governing early age cracking in concrete members, especially after the diffusion in the market of High Strength Concrete (HSC), where the autogenous shrinkage plays a major role. The latter occurs initially due to hydration reaction in the cement matrix. At a later stage, when the material is exposed to a low relative humidity environment, drying shrinkage takes place. In Normal Strength Concrete (NSC), where autogenous shrinkage plays a minor role, it can be assumed that the whole shrinkage strain is given by drying shrinkage.

When shrinkage is restrained by a structural element or an internal restrain (rebars), tensile stresses occur and may give rise to crack formation. However, tensile stresses are relaxed by creep phenomena that may delay or prevent cracking [1]. Therefore, shrinkage crack formation and development depends on several factors such as air environmental humidity, creep, restrain conditions, tensile strength and stiffness as well as fracture toughness of concrete.

Currently, there is no general consensus on a standard method to investigate the shrinkage cracking behaviour of concrete. Many studies have been carried out, mainly using three types of test: the linear test [2], the plate test [3] and the ring test according to standard AASHTO PP 34-99 [4].

The linear restrained column type test gives a simple uniaxial stress development but suffers from the disadvantage of not providing a constant degree of restraint that makes this test complex. The plate test provides a biaxial restraint to evaluate both biaxial and plastic shrinkage. The ring test provides a nearly constant degree of reaction through an axi-symmetric specimen geometry. In this type of test, a concrete specimen is cast around an inner steel ring which provides a constant degree of restraint to shrinkage deformation; the steel ring is also used to evaluate the induced tensile stresses in concrete through the measure of the steel compressive strains with strain gauges.

However, this test method can be ineffective: for instance, for a concrete with a high tensile strength, the induced tensile stresses might not be able to generate cracks in the composite. Beside the material properties, the low crack-sensitivity is associated to the specimen geometry.

It is now commonly accepted that fiber reinforcement reduces cracking phenomena but it is not clear how to best quantify this contribution; in other words, the minimum material performance necessary to reduce the crack width below a design value is still an open question.

Proc. of the 9th *fib* International PhD Symposium in Civil Engineering, July 22 to 25, 2012,
Karlsruhe Institute of Technology (KIT), Germany, H. S. Müller, M. Haist, F. Acosta (Eds.),
KIT Scientific Publishing, Karlsruhe, Germany, ISBN 978-3-86644-858-2

This paper provides a critical discussion on the ring test method for assessing the performance of Fiber Reinforced Concrete (FRC) starting from AASHTO requirements and evaluating the influence of different set-up geometries and materials through numerical analyses based on Non Linear Fracture Mechanics (NLFM) [5]. The main phenomena governing the restrained shrinkage behaviour of early-age concrete, such as elastic and post-cracking material properties (with special emphasis on FRC toughness), shrinkage strains and creep are considered. Stress evolution and crack formation is determined through a model based on a discrete crack approach with a time-step procedure.

2 Testing Procedure

As already mentioned, the AASHTO designation PP 34-99 is a testing procedure for the determination of the cracking tendency of ring-shaped concrete specimens. The time-to-cracking of the concrete ring is considered as the age when compressive strains in the steel ring suddenly decrease. This procedure represent a standard method and is not intended to determine the time-to-cracking of any particular structure cast with the same material.

The standard steel ring have a wall thickness of 12.7 mm (1/2 in), an outside diameter of 305 mm (12 in) and an height of 152 mm (6 in). The ring surface in contact with the concrete is coated with a form-release agent to minimize bond between the concrete and the steel. The form is made of a non-absorbent material and have an outside diameter of 457 mm (18 in).

Two or more strain gauges are applied on the inside surface of the steel ring to measure the strain. After a wet curing of 24 hours, specimens are stored in a controlled environment room with a constant air temperature of 21°C (73.4°F) and relative humidity of 50%. During the test, the strains in the steel ring are recorded every 30 minutes; a sudden decrease of the compressive strains in the steel ring higher than 30 µε usually indicates a crack formation.

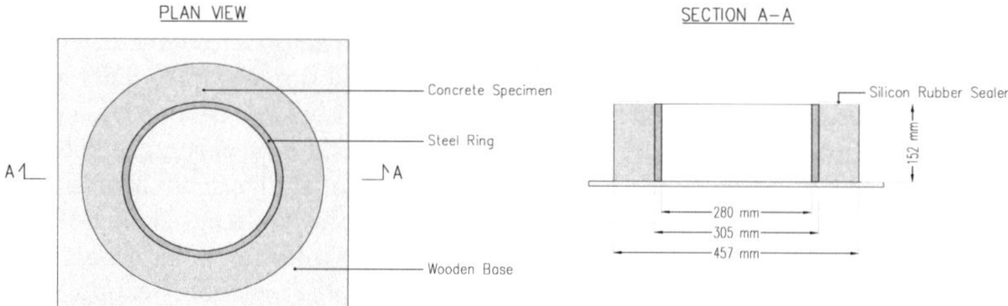

Fig. 1   AASHTO PP34-99 test apparatus.

An essential parameter for this type of test is the degree of restraint that represents a measure of the effectiveness of the restraint provided by the steel ring. The degree of restraint R should be defined as the stiffness of the steel ring over the stiffness of both the concrete and steel ring, that is:

$$R = \frac{A_{st}E_{st}}{A_{st}E_{st} + A_cE_c} \tag{1}$$

where A_{st} and A_c are the cross-sectional areas of the steel and concrete rings, respectively, and E_{st} and E_c are the moduli of elasticity of the steel and concrete, respectively.

3 Numerical Model

A parametric study of the ring test was carried out through several numerical analyses performed with TNO DIANA 9.4 and a discrete crack approach [6]. The main phenomena governing early-age cracking in cement-based materials, such as elastic and post cracking properties, shrinkage strains and tensile creep are considered by a set of features provided by the program.

The inner steel ring and the outer concrete ring are modelled by four-point plane stress elements. Between the two rings, the steel-to-concrete interface is modelled using linear interface elements with a brittle Mode-I behaviour and a Mode-II behaviour, the latter given by a simple bond-slip relationship suitable for smooth rebars [7] and a Coulomb friction criterion. The discrete crack was simulated by zero-thickness interface elements with a bilinear tension softening behaviour (Mode-I). Due to the axi-symmetric geometry the discrete crack is arbitrary placed (Figure 2). Material non-linearity is localized in the discrete crack while the concrete specimen and steel ring are defined as linear elastic.

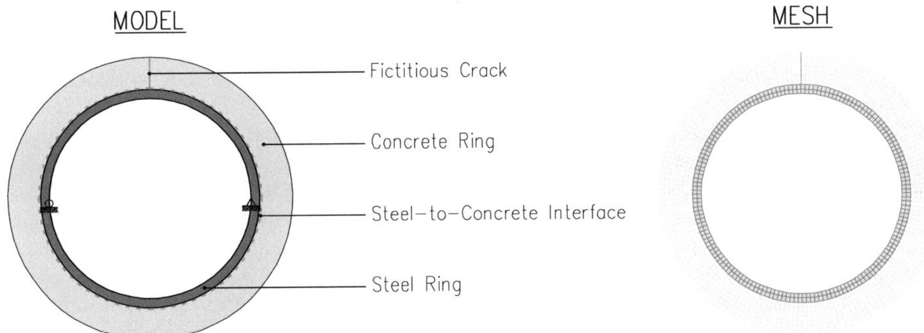

Fig. 2   Ring-test model and mesh.

The numerical analyses were performed with a time step procedure considering the evolution of the tensile stresses induced by shrinkage to be compared with the rising tensile strength of concrete. When the maximum tensile stress exceeds concrete tensile strength, a crack starts to open in the discrete crack. The adopted time step was 1 day; analyses were performed up to 60 days. The time evolution of drying shrinkage εsh(t) (Figure 3a) and compressive strength fc(t) (Figure 3b) were evaluated according to ACI 209R-92 [8]. The time evolution of tensile strength ft(t), elastic modulus Ec(t) and fracture energy Gf(t) were evaluated according to the following relationships:

$$f_t(t) = f_{t28}\sqrt{\frac{f_c(t)}{f_{c28}}} \qquad E_c(t) = E_{c28}\sqrt{\frac{f_c(t)}{f_{c28}}} \tag{2}$$

where ft28, Ec28 and Gf28 are, respectively, tensile strength, elastic modulus and fracture energy after 28 days of curing.

Concerning the viscoelasticity related to tensile creep at early age, it was chosen to develop creep functions in a Taylor series as DIANA does with a Power Law model for the stress calculation [9]. The compliance function for the Power Law model is given by the following relationship:

$$J(t,\tau) = \frac{1}{E(\tau)}\left(1 + \alpha\tau^{-d}(t-\tau)^p\right) \tag{3}$$

For these analyses the power of the creep function p = 0.5, the development point td = 15.0, the coefficient α = 0.16 and the power of the time dependent part of the creep function d = 0.1 have been adopted. These values were chosen in order to best fit the compliance function J in Eq. 3 with that included in ACI [8].

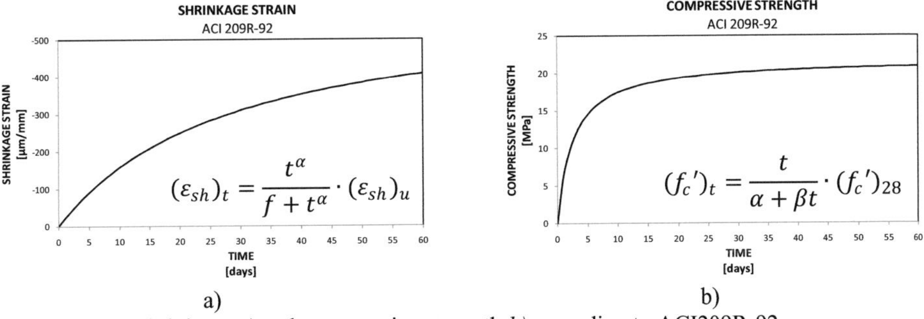

$$(\varepsilon_{sh})_t = \frac{t^\alpha}{f + t^\alpha} \cdot (\varepsilon_{sh})_u$$

$$(f_c')_t = \frac{t}{\alpha + \beta t} \cdot (f_c')_{28}$$

a)   b)

Fig. 3   Free shrinkage *a)* and compressive strength *b)* according to ACI209R-92.

4   Parametric Study

4.1   Effect of Geometry

The parametric study on the specimen geometry is performed by considering as a reference the test setup given by AASHTO PP 34-99 and changing two significant parameters: the steel and the concrete thickness. These parameters are related to the cross-sectional areas of the steel and concrete rings and then to the degree of restraint provided by the test set-up. A variation on the degree of restraint produces a direct effect on the time required to form a crack: an increase of steel thickness or a decrease of concrete thickness reduces the time-to-cracking.

A first set of analyses has been performed with a constant steel ring thickness equal to 12.7 mm and a variable concrete thickness equal to 15, 30, 45, 60 76 and 90 mm, respectively. A second set of analyses has been carried out with a constant concrete ring thickness, equal to 76 mm, and a variable steel thickness equal to 6.3 mm (1/4 in), 12.7 (1/2 in), 25.4 (1 in), 50.8 (2 in) and 76.2 (3 in), respectively. The specimen dimensions have been chosen to favour crack formation in a reasonable time for practical use in a laboratory. For each specimen geometry, three analyses were carried out by considering three different concrete grades: C12, C20 and C30. The mean values of elastic and post-cracking properties were determined according to fib Model Code 2010 [10], as reported in Table I. Table II reports the degree of restraint R for the different specimen geometries considered for concrete C20. In the same table, the steel thickness (t_{st}), the concrete thickness (t_c) and elastic modulus (Ec) are reported. The elastic modulus of steel (Est) was assumed as 210 GPa.

Table 1   Mean values of elastic and post-cracking properties for NSC.
Table 2   Degree of restraint R of ring-test setup for concrete C20.

Table 1		C12	C20	C30
f_{ck}	[MPa]	12.0	20.0	30.0
f_{cm}	[MPa]	20.0	28.0	38.0
f_{ctm}	[MPa]	1.6	2.2	2.9
G_F	[N/m]	125	133	141
E_c	[GPa]	22.9	26.2	29.7
ν	[-]	0.2	0.2	0.2

Table 2		FEM 1	FEM 2	FEM 3	FEM 4	FEM 5	FEM 6	FEM 7	FEM 8	FEM 9	FEM 10
t_{st}	[mm]	12,5	12,5	12,5	12,5	12,5	12,5	6,3	25,4	50,8	76,2
A_{st}	[mm²]	1900	1900	1900	1900	1900	1900	958	3861	7722	11582
t_c	[mm]	15	30	45	60	76	90	76	76	76	76
A_c	[mm²]	2280	4560	6840	9120	11552	13680	11552	11552	11552	11552
E_c	[GPa]	26,2	26,2	26,2	26,2	26,2	26,2	26,2	26,2	26,2	26,2
R	[-]	87%	77%	69%	63%	57%	53%	40%	73%	84%	89%

4.2   Effect of Material Toughness

The parametric study on material toughness focused on the evaluation of cracking behaviour of plain (NSC) and Fibre Reinforced Concrete (FRC) having a post-cracking softening behaviour. A second aim of this study is the evaluation of the effect of the enhanced concrete toughness on crack development. The tensile behaviour of both plain and FRC in tension was described with a linear relationship up to the cracking strength followed by a post-peak bilinear relationship. For plain concrete the latter was assumed according to fib MC2010 [10] (Figure 4) while, for FRC, the two branches of the bilinear law were varied as shown in Figure 5.

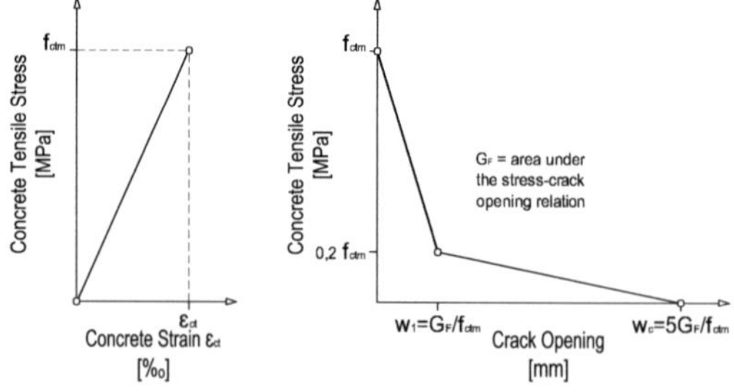

Fig. 4   Constitutive laws for concrete under uniaxial tension.

In particular, a first set of analyses for FRC was performed on Model 3, which was considered the most efficient specimen geometry, with a steel thickness of 12.7 mm and a concrete thickness of 45 mm, by varying the first branch of the softening law (Figure 5a). A second set of analyses was carried out by considering a variation on the second branch of the post-cracking softening law (Figure 5b).

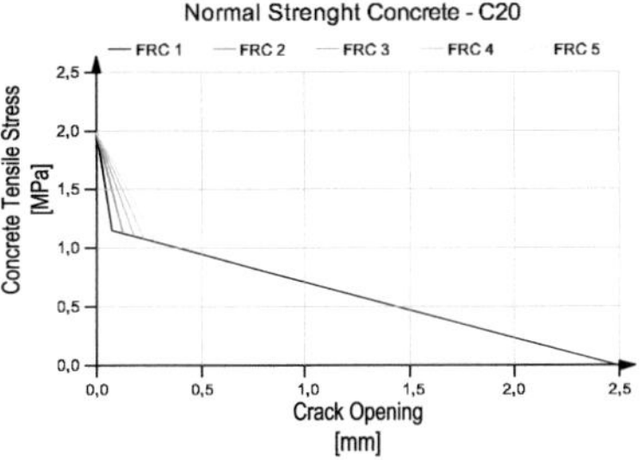

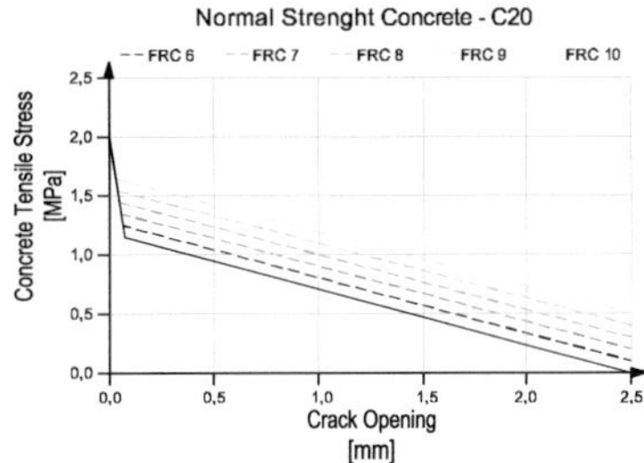

Fig. 5   Bilinear tension softening laws for FRC.

5   Numerical Results

An overview of the results of the parametric study on the effect of different geometries with regard to the time-to-cracking is given in Figure 6. The latter corresponds to a sudden strain drop in the steel ring. From a numerical point of view the time-to-cracking was identified as the age at which the crack interface is completely open and the tensile stress in all interface elements has reached the tensile strength; this assumption is reasonable by considering that a macro-crack has to form for releasing the steel ring, whereas a micro-cracking would probably not have any significant impact on the strain-gauge measurements.

As previously suggested, the time-to-cracking strongly depends on concrete thickness for two principal reasons: a decrease of concrete thickness implies, on one hand, an increase of the degree of restraint R and, on the other hand, an increase of the intensity of drying shrinkage due to a higher moisture exchange with the environment. Figure 6a shows the increase of the time-to-cracking with the concrete thickness for all the concrete grades considered. The results suggest to employ a concrete with a thickness lower than 60 mm (Models 1, 2 and 3) to maintain the time-to-cracking in the first two weeks after initiation of drying. Numerical results also show that the time-to-cracking depends on the steel thickness (Figure 6b): in fact, an initial strong decay is followed by a smoother decrease for a steel thickness larger than 25.4 mm. Therefore, a steel ring thicker than 25.4 mm (Models 9 and 10) does not provide an appreciable decrease of the time-to-cracking and, for this reason, it is not recommended for this test.

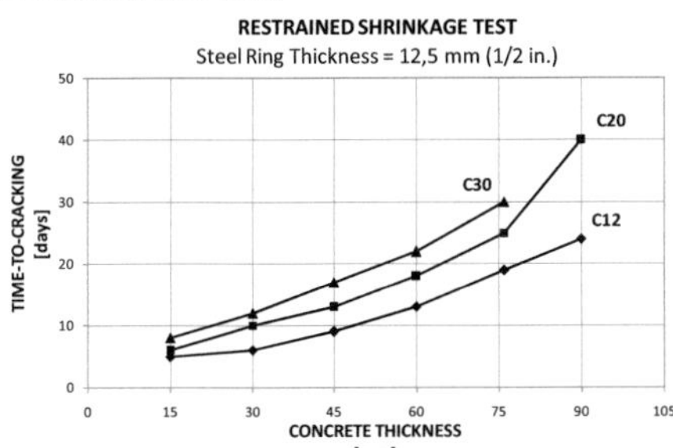

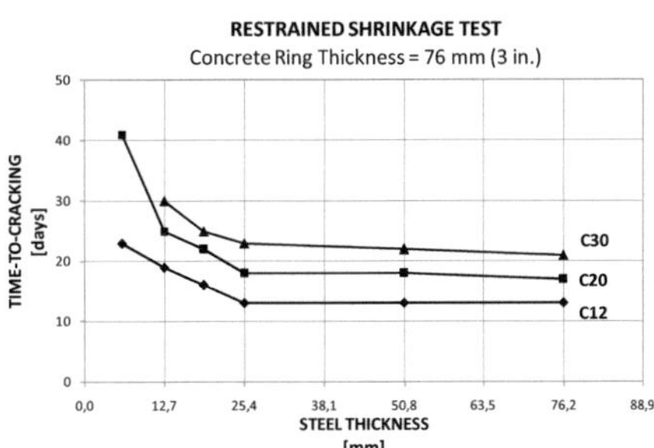

Fig. 6   Effects of concrete and steel thickness on the time-to-cracking.

The influence of fiber reinforcement on shrinkage cracking is shown in Figure 7; both sets of tension softening models adopted were able to prevent a sudden strain drop in the steel ring and allowed to control cracking by transfer tensile stresses after the appearance of the crack. The first set of analyses shows how a different slope of the first branch influences the model response from the early crack formation (Figure 7a). The second set of analyses illustrates how a modification on the second branch changes the numerical response only for considerable crack widths (Figure 7b). The numerical results are consistent with experimental results available into the literature, stating the effectiveness of fiber reinforcement in controlling shrinkage cracking. However, FRC with a higher residual strength for micro-cracking is more suitable in controlling shrinkage cracking.

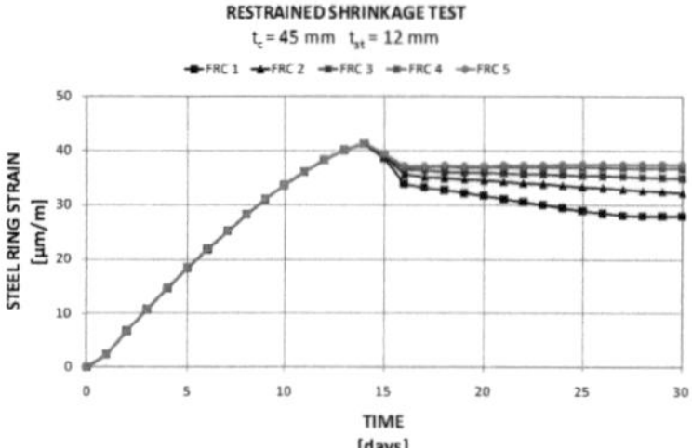

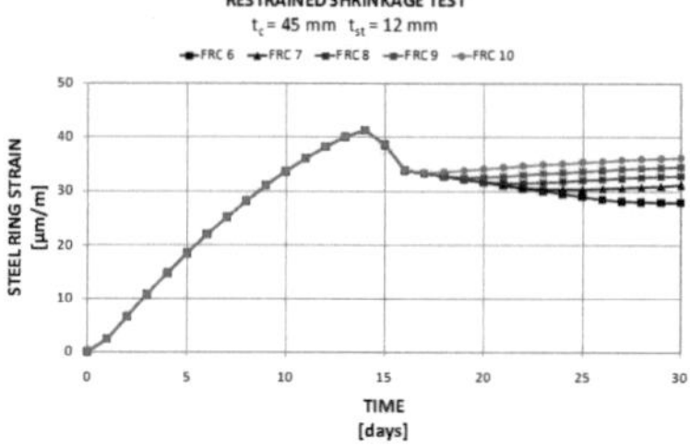

Fig. 7   Influence of different tension softening laws on ring-test response.

6   Concluding Remarks

Shrinkage cracking requires special attention from structural designers for many aspects: with regard to durability, early-age cracking advances the deterioration process of concrete and reduces the service life of a structure; from an aesthetical and psychological point of view, it is not acceptable because makes structure appearance worse and people feel unsafe. Thus, the evaluation of cracking sensitivity of a material is strongly recommended for an effective and durable design of a structure. Based on the numerical study presented herein, the following conclusions can be drawn:

- Time-to-cracking strongly depends on the specimen geometry and on the degree of restraint: an increase of steel ring thickness or a decrease of concrete thickness can accelerate the occurrence of cracking. For the normal concrete considered, starting from the standard AASTHO test, results suggest to employ concrete thickness lower than 60 mm to maintain the time-to-cracking within the first two weeks after drying initiation. It is also recommended the use of a steel ring having a thickness in the range of 12.7÷25.4 mm (0.5÷1 in.) to provide enough stiffness to ensure crack formation.
- Fiber reinforcement has a considerable effect on behaviour of FRC under restrained shrinkage, by preventing sudden drops in the strain of the steel ring, controlling cracking and transmitting tensile stresses after the appearance of the crack. The crack control is mainly influenced by the residual strength for smaller crack opening (first branch of the tension softening law).

References

[1]   See, H.T. Attiogbe, E.K. and Miltenberger, M.A.: Shrinkage Cracking Characteristics of Concrete Using Ring Specimens. In: ACI Materials Journal, V. 100, No. 3, May-June 2003, pp. 239-245.

[2]   Klover, K.: Testing System for Determining The Mechanical Behavior Of Early Age Concrete Under Restrained and Free Uniaxial Shrinkage. In: Materials and Structures, V. 27, No. 170, July 1994, pp. 324-330.

[3]   Kraai, P.P.: Proposed test to determine the cracking potential due to drying shrinkage of concrete. In: Concrete Constr., V. 30, No 9, 1985, pp. 775-778.

[4]   AASHTO: Standard Practice for Estimating the Crack Tendency of Concrete. In: AASHTO Provisional Standards, PP 34-99, Apr. 2000.

[5]   Hillerborg, A. et al.: Analysis of crack formation and crack growth in concrete by means of fracture mechanics and finite elements. In: Cement and Concrete Research, Vol. 6, 1976, pp. 773-782.

[6]   TNO DIANA User's Manual – Release 9.4. TNO Building and Construction Research, Delft, The Netherlands - 2009.

[7]   CEB, 1990, Comité Euro-International du Béton, CEB-FIP Model Code 1990, Bulletin d'Information, No. 195, 480 pp.

[8]   ACI Committee 209, 1992, Prediction of Creep, Shrinkage, and Temperature Effects in Concrete Structures (ACI 209R-92), American Concrete Institute, Farmington Hills, Mich., 47 pp.

[9]   De Borst, R. and van den Boogaard, A. H.: Finite-element modelling of deformation and cracking in early-age concrete. In: J. Eng. Mech. Div., ASCE, Vol. 120, No 12, 1994, pp. 2519–2534.

[10]   Walraven et al., fib Bulletin 55: Model Code 2010 – First complete draft, Volume 1 (chapters 1-6), 2010, 318 pp., ISBN 978-2-88394-95-6.

Effect of fibre properties and embedment conditions on fibre pullout behaviour from concrete matrix

Fanbing Song

Institute for construction materials,
Ruhr University Bochum,
Universitätsstraße 150, 44801 Bochum, Germany
Supervisor: Rolf Breitenbücher

Abstract

Precast tunnel segments are subjected to various loads during production, construction and service stages. Concrete damages, in the form of cracking and spalling, are quite likely to occur due to impact loads and local stress concentrations during construction. By means of adding steel fibres to concrete, the impact resistance and ductility of the material can be improved.

To investigate the fibre-matrix bond characteristics, pullout tests on single steel fibres were performed. Parameters such as fibre shape, strength and inclination angle were studied. From the experimental results, it was observed that the pullout resistance, pullout work and utilization of tensile capacity increased with the increase of the deformation, however, decreased with the increment of the embedment angle; with the increase of embedment angle, the fibre tends to break more easily.

1   Introduction

Concrete, in particular in the case of high-strength concrete, is a brittle material exhibiting low tensile strength and fracture toughness. This brittle material tends to crack under low level of tensile strain. Such kind of tensile characteristics can be improved by the addition of steel fibres into the concrete matrix [1]. The primary contribution of fibres, in terms of structural aspects, is to enhance the post-cracking tensile resistance and ductility, since fibres bridging the cracks can transfer the stresses across these cracks and retard the crack opening or propagation. Furthermore, the fatigue and impact resistance of the material can also be improved.

Thus, fibre reinforced concrete has a distinct advantage in comparison with conventional concrete. This can particularly be used for the tunnel precast segment linings. For the production of tunnel lining segments, concrete with a minimum strength class of C35/45, C45/55 or even higher is applied [2]. The actual value of the concrete compressive strength achieved could reach that for a high-strength concrete. Concrete damages, in the form of cracking and spalling, can occur due to impact loads during demoulding, handling and transportation, especially due to point load from the rams of the Tunnel Boring Machine during the assembly process [3].

As is well known, the bridging efficiency of a given fibre under tensile action depends largely on the fibre-matrix bond characteristics [4]. Therefore, it is important to study the bond characteristics in the fibre-matrix interface for understanding the mechanical behavior of steel fibre reinforced concrete in a real structural scale. Pullout tests on a single fibre are commonly used to investigate the fibre-matrix behavior, where the force for pulling out a fibre embedded in a matrix under uniaxial tension is measured [5].

2   Pullout mechanisms

In general, the overall pullout process of a steel fibre in cementitious composites consists of two stages: debonding and frictional pullout.

In the initial phase, the entire adhesive bonding remains between the fibre and surrounding matrix, if the shear stress, derived from the pullout force, is smaller than the adhesive strength along the fibre-matrix interface. When the shear stress exceeds the adhesive strength, the debonding process, regarded as an interfacial micro cracking process, takes place. The pullout force increases continuously during this process. After the adhesive bonding is fully destroyed, the fibre is being pulled out of the matrix. Subsequently, the frictional resistance controls the pullout behaviour. In this stage the pullout force decreases with the increase of fibre displacement out of the matrix.

Proc. of the 9th *fib* International PhD Symposium in Civil Engineering, July 22 to 25, 2012,
Karlsruhe Institute of Technology (KIT), Germany, H. S. Müller, M. Haist, F. Acosta (Eds.),
KIT Scientific Publishing, Karlsruhe, Germany, ISBN 978-3-86644-858-2

If the fibre is geometrically deformed, the frictional pullout is accompanied by a mechanical bond mechanism and plastic deformation. Different from straight fibres, after fully debonding, the pullout force increases due to the mechanical anchorage. Beginning with the debonding process, the fibre slowly develops its full anchorage and approaches ultimate conditions up to the maximum pullout load [6]. Depending on the properties of fibre and matrix, the deformed fibre can either be pulled out from the matrix or break in the pullout process.

3 Experiments

The experimental program conducted in this research work was to study the bond mechanisms of fibres in high-strength concrete. Parameters controlling the pull-out behaviour, such as fibre shape (straight, crimped, end-hooked and twin-cone), tensile strength (normal and high) and fibre orientation (0°, 15°, 30°, 45° and 60°) with respect to load direction were investigated.

3.1 Materials and specimen preparation

Types and properties of the fibres chosen for the pullout tests are listed in Table 1. In order to perform the pullout tests of straight fibres, the hook end of steel fibre RC-80/60-BN was chopped off with pliers. This allowed a comparison of the same fibre with and without deformation.

A concrete mixture with a compressive strength of 84 MPa was applied as pullout matrix. The proportions and properties are given in Table 2. The compressive strength of the concrete was determined at 28 days using 150 mm-cube in accordance with DIN EN 12390-3: 2009.

A laboratory mortar mixer was used throughout the experiments to prepare the concrete mixture. The following mixing procedure was applied: cement, fly ash and aggregate (0-2 mm Sand, 2-8 mm crushed stone) were first dry-mixed for approximately 30 s. Water mixed with superplastiziser was then added gradually and mixed for another 2 minutes.

Table 1   Types and properties of the steel fibres

Fibre type	Geometric deformation	Length [mm]	Diameter [mm]	Tensile strength [MPa]
Dramix RC-80/60-BP	End-hooked	60	0.71	2600
Dramix RC-80/60-BN		60	0.75	1250
Stratec FWW 60/0.8	Crimped	60	0.80	1250
Arcelormittal Twincone 1/54	Twin-cone	54	1.00	1100

Table 2   Mix proportions and basic properties of the concrete

Constituents					Properties	
Cement [kg/m³]	Fly ash [kg/m³]	Aggregate [kg/m³]	Superplasticizer [kg/m³]	Water [kg/m³]	Density [kg/m³]	Compressive strength [MPa]
400	100	1644	1.5	180	2421	84

A special steel mould shown in Fig. 1 was assembled to produce 6 cylindrical specimens with the dimension 60 mm × 60 mm for each series. During casting, the fibres were fixed with desired embedment length and inclination angle between two bars clamped with two screws. After casting, the specimens were placed in a plastic box and stored for 24 hours at room temperature. After demoulding, the specimens were then placed in a lime saturated water curing tank for 6 days, afterwards they were removed from the curing tank and cured in a climatic chamber with a temperature of 20 ± 2° and a relative humidity of 65 ± 5% for another 21 days. The pullout tests were performed at an age of 28 days.

3.1 Test Setup and test procedure

The pullout tests were performed using an universal testing machine with a 5 kN load cell (Zwick 1435). As shown in Fig. 2, the specimen was glued, using a two part epoxy resin adhesive, to a thick cylindrical steel plate attached to the testing rig with one bolt. The free end of the steel fibre was fixed to the grip which allowed a secure clamping. The pullout displacement of the steel fibre was meas-measured with three LVDTs (HBM WI, linear stroke +/- 5 mm); these three LVDTs were mounted into an aluminium frame which was positioned onto the specimen using three screws. The tip of the LVDT touched the bottom surface of a thin aluminium plate fixed to the fibre with two fine screws. This allowed an exclusion of measuring fibre elongation or displacement at the grip which may occur during the pullout process. Load was applied at a cross-arm travel rate of 0.5 mm/min.

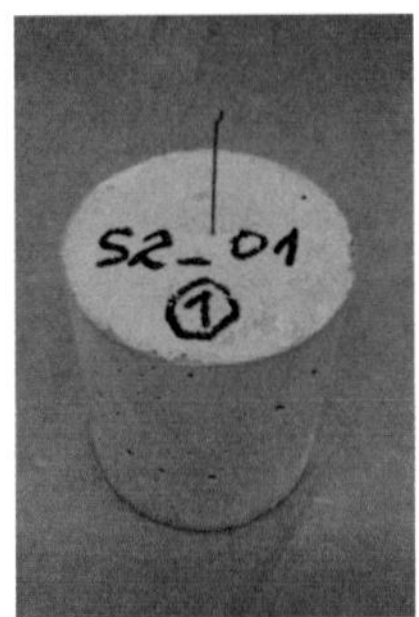

Fig. 1   Mould for casting the specimens (left) and Pullout specimen (right)

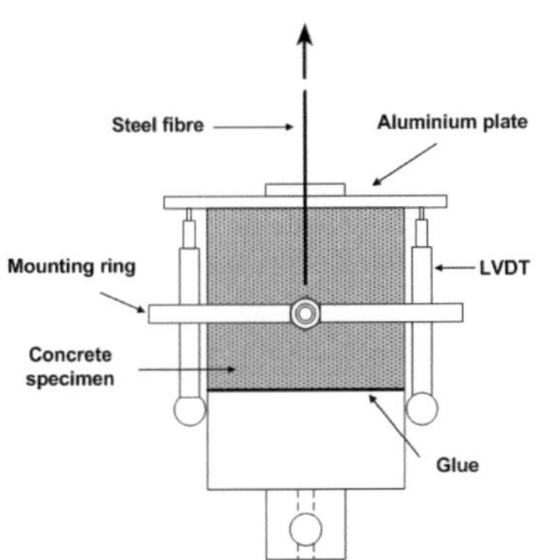

Fig. 2   Schematic of the pullout test (left) and photograph of pullout test in progress (right)

4   Pullout Test results

4.1   Effect of fibre shape

The average load-displacement curves of 6 individual pullout tests up to a pullout displacement of 5 mm for various fibre shapes are depicted in Fig. 3. In this series of tests, the single fibre was embedded 20 mm (L_f = 20 mm) in the matrix aligned with respect to the load direction ($\theta = 0°$).

As shown in Fig. 3, the pullout responses of the fibres differ from each other significantly. For the straight fibres, after fully debonding, the pullout load drops suddenly, which corresponds to an abrupt increase of damage along the fibre-matrix interface. Due to the mechanical bond, the pullout loads of deformed fibres continue to increase until the peak load is reached. Thus, deformed fibres provide significantly higher resistance against pullout than straight fibres. As a result of a better fibre anchoring in the matrix, both crimped and twin-cone fibres broke in all cases. In the case of end-hooked fibre (Dramix RC-80/60-BN), after the hook end is deformed at the maximum pullout load, the mechanical anchorage tends to be progressively mobilized, which causes a decrease of the pullout force and a further increase of the pullout displacement. At a certain fibre displacement, the pullout process is mainly controlled by frictional resistance. Since the hook is usually only partially straightened, it generates an additional frictional resistance during the pullout process. Therefore, end-hooked fibres have higher pullout force in the frictional pullout stage than straight fibres.

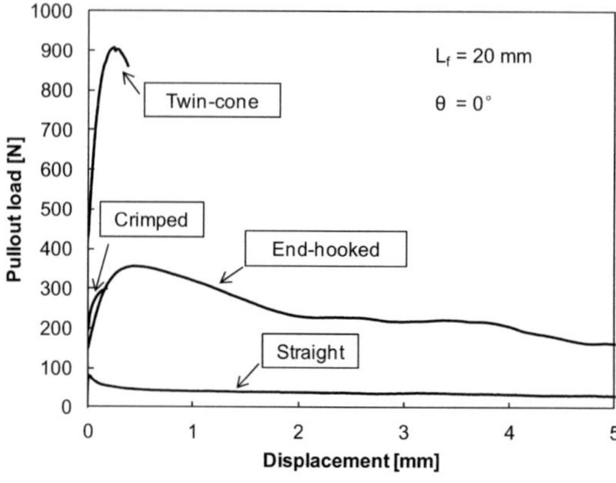

Fig. 3   Influence of various fibre shapes on the pullout behaviour (load-displacement curve)

The load-displacement curves were further analyzed by comparing the peak pullout load (P_{peak}), displacement at peak load (Δ_{peak}), work at peak load (W_{peak}), maximum tensile stress (σ_{max}) and tensile capacity (σ_{max} / f_y). It can be seen in Table 3 that the twin-cone fibres had the highest P_{peak} and W_{peak}, whereas the straight fibres exhibited the lowest corresponding values. Furthermore, the tensile strength of the twin-cone fibres was exceeded, i.e. the tensile capacity is utilized over 100%. Hereby fibre fracture occurred at a very small displacement in all tests. Therefore, its overall absorbed energy is considerably less than that of the hooked-end fibre which underwent a complete pullout without rupture.

Table 3   Average pullout results of various steel fibre shapes

Fibre type	P_{peak} [N]	Δ_{peak} [mm]	W_{peak} [N.m]	σ_{max} [MPa]	f_y [MPa]	σ_{max}/f_y [%]
Straight	83.0	0.03	0.0021	188.0	1250	15.0
Crimped	298.2	0.14	0.0358	593.6	1250	47.5
Hook-ended	356.1	0.45	0.1334	806.4	1250	64.5
Twin-cone	907.0	0.25	0.1893	1155.4	1100	105.0

To obtain a ductile behavior of concrete after cracking, fibre rupture should be avoided at small crack width (i.e. fibre slip), and the fibre should rather be pulled out of the concrete matrix with a high utilization of the tensile capacity. For the application in tunnel linings where the allowable crack openings are relatively large, the end-hooked steel fibre has a distinct advantage over the other three fibre types [6].

4.2   Effect of fibre strength and embedment angle

The average fibre pullout responses, in terms of fibre strength and inclination angle, are shown in Fig. 4 for normal-strength (RC-80/60-BN) and high-strength fibres (RC-80/60-BP) respectively. It can be seen that, for both end-hooked fibres, the configurations of pullout load-displacement curves were similar.

Fibres made of high-strength steel provided expectedly larger pullout load and work in the total pullout process, nevertheless the tensile capacity of fibre RC-80/60-BP (σ_{max}/f_y = 61.6%) was utilized up to almost the same level of fibre RC-80/60-BN(σ_{max}/f_y = 64.5%). In the case of normal-strength steel fibres under complete pullout, the maximum pullout load did not change significantly with the variation of the inclination angles. Only a slight increase was observed at a 30° inclination angle which is followed by a decrease at a 45° inclination angle. On the contrary, a drastic increase of the peak pullout load was ascertained for high-strength fibres at embedment angles of 30° and 45°.

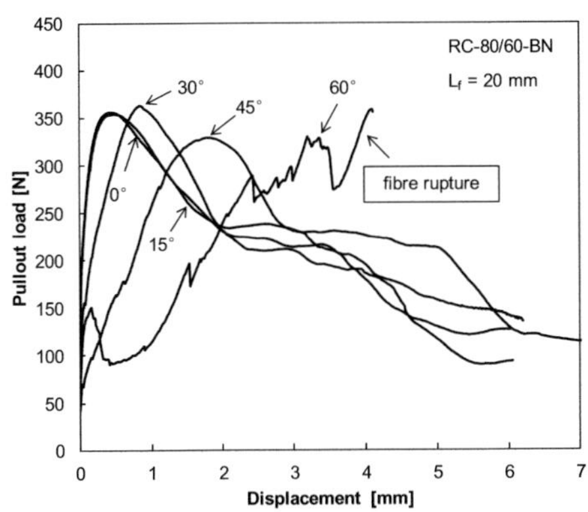

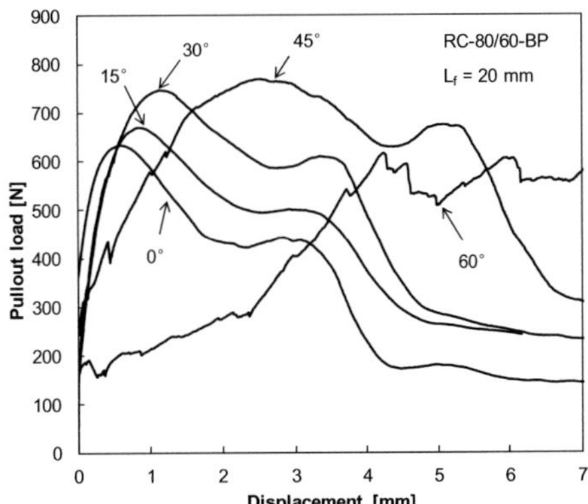

Fig. 4   Influence of fibre inclination angles on pullout behaviour: normal-strength fibre RC-80/60-BN (left) and high-strength fibre RC-80/60-BP (right)

In general, the slope of the pre-peak branch decreases with the increase of fibre orientation angle. That means the fibre displacement at maximum load increases as the embedment angle increases, particularly for an inclination angle greater than 30°. With the increase of the inclination angle, an increase of matrix crushing and spalling at the fibre exit point was also observed during the resting procedure. This phenomena was more pronounced in the case of fibres made of high-strength steel.

Beside a complete fiber pullout, it was also observed that fibre fracture occurred for both inclined fibre types in the first stage of the pullout process. With the increase of inclination angle, the fibre tends to break more easily. Due to its high tensile strength, fibre RC-80/60-BP started to break from an inclination at 45°. Moreover, its number of broken fibres was less than that of fibre RC-80/60-BN.

Table 4   Number of broken fibres at various embedment angles

Fibre type	Inclination angle				
	0°	15°	30°	45°	60°
RC-60/60-BN	0	0	3	4	6
RC-80/60-BP	0	0	0	2	3

In most concrete structures the maximum allowable crack width (i.e. fibre slip) is designed (by code) smaller than 0.4 mm [7]. Since the peak pullout load and the corresponding displacement (in this research work: Δ_{peak} from 0.45 to 4.95 mm) do not say much about the performance of fibres in real structural elements, the tensile stress of inclined fibres should be evaluated at appropriate fibre displacements. Therefore, the ratio between the tensile stresses of aligned and inclined fibres was determined at fibre displacements of 0.1, 0.25, 0.5 mm and 1 mm, shown in Fig. 5. Except for the fibre displacement at 1 mm, aligned fibres (or fibres with very small inclination angle) provided the highest tensile capacity, whereas fibres inclined at 60° possessed the lowest efficiency for all fibre displacements. As a result of embedment inclination, the pullout work also decreased with the increase of inclination angle, shown in Table 5.

Table 5   Pullout work by a fibre displacement of 1 mm at various embedment angles

Fibre type	W_{peak} [N.m]				
	0°	15°	30°	45°	60°
RC-60/60-BN	0,3217	0,3221	0,2896	0,1565	0,1076
RC-80/60-BP	0,5745	0,5445	0,5685	0,4968	0,1949

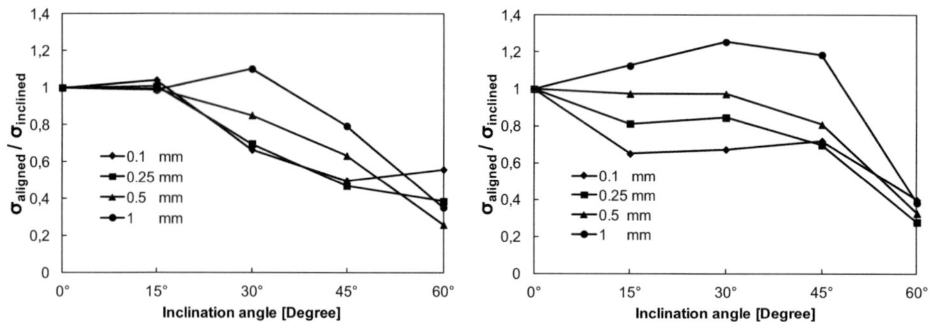

Fig. 5   Influence of inclination angles on the utilization of tensile capacity for 0.1, 0.25, 0.5 and 1 mm fibre displacements : RC-80/60-BN (left) and RC-80/60-BP (right)

5   Conclusions

From the results of the experiments, the following conclusions can be drawn:

Deformed fibres provided a significantly higher pullout resistance and an utilization of tensile capacity than straight fibres. Due to good anchorage and the high matrix strength, crimped and twin-cone fibres broke in all tests at very small fibre displacements, whereas aligned end-hooked steel fibres underwent complete pullout and exhibited a better pullout response. Therefore, it could be more suitable for the application in structures with large allowable crack width. End-hooked fibres made of high-strength steel provided expectably larger pullout load and pullout work, however, its utilization of tensile capacity remained almost the same level of fibres made of normal-strength steel.

For aligned fibres with end-hook, the peak load was obtained at smaller displacements than that of inclined fibres ; furthermore, the utilization of tensile capacity and the absorbed energy were also higher up to a certain fibre displacement. At large inclination angles, fibre fracture was more likely to occur for both normal and high- strength fibres. Therefore, fibres aligned or inclined with small embedment angles exhibited a better pullout response under tensile load.

Consequently, to guarantee a high efficiency of fibres in concrete matrix, i.e. a reliable tensile strength after cracking and a certain ductility in structural elements, fibres (with appropriate geometric deformation) should be distributed aligned or with small inclination angles with respect to the tensile loading direction. In terms of the application in tunnel linings, a homogeneous fibre distribution and an orientation in three dimensions are preferred, since the stresses occurred in the segments during the curing, transport, and assembly process are hardly predictable.

References

[1]   Naaman, A.E.: Fibre Reinforcement for Concretes. In: Concrete International: Design and Construction (1985) No.7, pp. 21-25

[2]   DAUB: Concrete Linings for Tunnels built by underground construction. In: Tunnel (2001) March, pp. 27-43

[3]   Moccichino, M., Romualdi, P., Perruzza, P., Meda, A., Rinaldi, Z.: Experimental Tests on Tunnel Precast Segmental Lining with Fibre Reinforced Concrete. In: Publication: University of Rome "Tor Vergata", Rome, Italy

[4]   Mandel, J., Wei, S. & Said, S.: Studies of the properties of the fibre-matrix interface in steel fibre reinforced mortar. In: ACI Materials J.(1987) No. 84, pp. 101-109

[5]   Robins, P., Austin, S. & Jones, P.: Pull-out behaviour of hooked steel fibres. In: Material and Structures (2002) Vol. 35, pp. 434-442

[6]   Banthia, N., Trottier, J.-F.: Concrete Reinforced with Deformed Steel Fibres, Part I: Bond-Displacement Mechanisms. In: ACI Materials Journal (September-October 1994), pp. 435-445

[7]   Naaman, A.E.: Evaluation of steel fibres for application in structural concrete. In: 6th RILEM Symposium on Fibre-Reinforced Concretes (FRC) (September 2004), pp. 389-399

Load bearing behaviour of textile reinforced concrete with short fibres

Marcus Hinzen

*Institute of Building Materials Research,
RWTH Aachen University,
Schinkelstr. 3, 52062 Aachen, Germany
Supervisor: Wolfgang Brameshuber*

Abstract

Short fibres have a big influence on the load-bearing behaviour of textile reinforced concrete. The first part of the paper deals with the ability of short fibres to increase the first crack strength of textile reinforced concrete under consideration of the pre-storage conditions. The second part describes the transition from the uncracked to the cracked stage and defines a criterion for a suitable fibre content. The main focus is the influence of short fibre volume and stiffness on the post-cracking behaviour. Especially the crack formation is described. The addition of short fibres improves the load-bearing capacity of textile reinforced concrete across the entire area of the stress-strain curve.

1   Introduction

Textile reinforced concrete is a combination of fine grained concrete and technical glass or carbon textiles as reinforcement. A major field of application are façade elements and lightweight construction. The load-bearing behaviour of textile reinforced concrete can be considerably improved by applying short fibres. Investigations show an increase in the first crack load at dry samples, a strain hardening behaviour with a finer distributed crack pattern as well as an increase in the load-bearing capacity mostly across the entire area of the stress-strain curve. The exact mechanisms, which allow for a targeted adjustment of the load-bearing behaviour, are not yet fully understood. Only few researchers [1, 2, 3] have so far been engaged in this field and basically, they made similar observations. In [3] the potential load-bearing behaviour of textile reinforced concrete with short fibres was schematically defined and divided into two parts. The first part describes the contribution of the short fibres to a higher first crack load of the concrete (stage I). The second part describes the phase of crack formation until the textile fails. At first, the influence of short fibres on the crack resistance of concrete is described in this paper. In this context, also different storage conditions are examined. The transition from the uncracked to the cracked stage reveals the need for additional short fibre reinforcement and is described briefly. During the crack formation phase, the short fibres in the widening crack interact with the textile. It is to be expected that this interaction influences the local stiffness of the composite material in state II as well as the load level and the crack formation. All investigations were conducted at RWTH Aachen University within the framework of the Collaborative Research Centre 532.

2   Test specimens, Raw materials and Production

The tensile tests were performed on dumbbell tensile specimens according to [5]. As a rule, all tensile specimens were manufactured with the laminating technique in which fibre concrete as well as textiles are inserted in layers. As basic fine-grained concrete mixes, the mix Fil-10-09, which was especially developed for the application of short fibres, as well as a mix FC [3], which contains more binder, were applied. The textiles used were 1200 tex textiles made of AR-glass with tricot binding. As short fibres, mainly water dispersible micro-fibres made of different materials and with lengths of between 3 and 15 mm were used which are described in detail in the respective chapters. The pre-storage of the specimens influences the load-bearing behaviour. During the investigations, three variations were examined. Wet specimens were stored under water for 28 days until immediately before the test. Permanently watered specimens which were however dry at the test were stored under water for 21 days and in a standard climate at 20°C and relative humidity of 65 % for 7 days. In addition, after demoulding, sealed specimens without moisture exchange were examined. Tensile

Proc. of the 9th *fib* International PhD Symposium in Civil Engineering, July 22 to 25, 2012,
Karlsruhe Institute of Technology (KIT), Germany, H. S. Müller, M. Haist, F. Acosta (Eds.),
KIT Scientific Publishing, Karlsruhe, Germany, ISBN 978-3-86644-858-2   "

strengths and stress-strain curves of tensile specimens were measured with a universal test machine with a displacement ratio of 1 mm/m.

3   First crack strength

The investigations on the increase of the first crack strength of textile reinforced concrete are of special importance because textile reinforced concrete components often have the quality of fair-faced concrete and therefore shall remain without cracks during service life. An increase in the crack resistance of the matrix hence offers additional flexibility for design. The investigations showed a strong dependence on the pre-storage of the samples.

It turned out that fine grained concrete which were permanently kept wet already feature a very high average tensile strength of about 7 MPa. Tensile tests with different types of short fibres (glass, carbon, PVA, PE) revealed that this tensile strength can hardly be further increased by adding short fibres. Crack initiating effects of the short fibres and the air entrainment at higher fibre contents seem to prevent a further increase in the tensile strength of water-stored samples.

The short fibres come into effect, however, when the tensile strength of the fine grained concrete is reduced due to pre-damages induced by drying, which is to be expected under practical conditions. Due to the findings gathered at water stored specimens, an older test series with samples stored sealed was examined again. The plain concrete of this test series had a significantly lower average tensile strength of only 3 MPa. As fibres, glass, carbon and steel fibres were used. All fibre types led to a continuous increase in strength with increasing fibre content. With some fibre types tensile strengths up to 6.5 MPa were reached. But the strength level of the wet concrete was again not exceeded. Thus, the mode of action of the short fibres mainly consists in counteracting the strength losses induced by drying. The result shows that an assessment of the short fibres with regard to the first crack strength must absolutely be made taking into account exact pre-storage conditions. Preliminary results showed that drying of only a few minutes already causes a significant reduction in tensile strength of the plain concrete.

4   Transition state I → state II

A major aim when using short fibres in textile reinforced concrete is to achieve a continuous strain-hardening behaviour after the first matrix crack. Investigations already showed that this can be achieved with several fibre dosages [4]. For practical applications it would be helpful to be able to estimate the necessary fibre volume.

Due to the disturbed bond between textile and matrix in the vicinity of the crack faces and the asynchronous activation of the filaments, the textile needs a certain crack opening displacement to carry the load. This circumstance leads to a sudden load decrease in deformation-controlled tensile tests and to a decrease in stiffness as well.

Fig. 1, left, shows this behaviour for a 1200 tex textile with tricot binding and a 2400 tex textile with chain stitch binding. As expected, the 2400 tex textile shows a bigger load decrease due to lower bond properties.

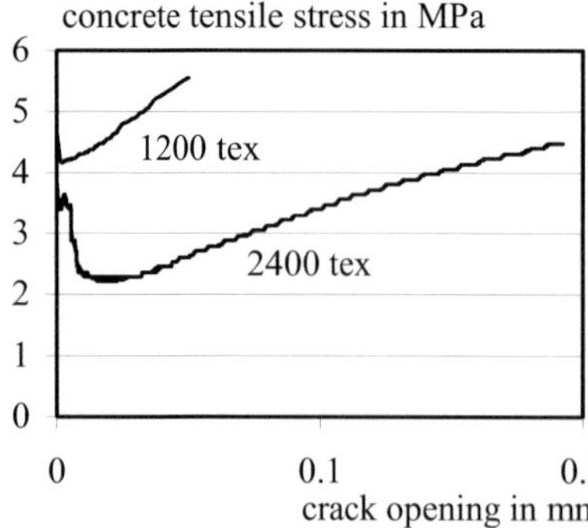

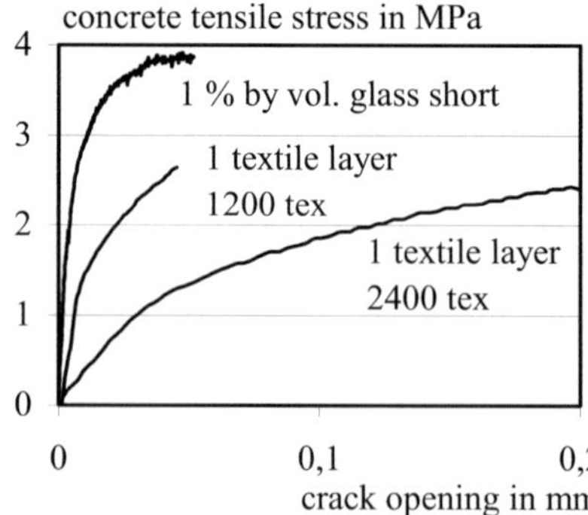

Fig. 1   Stress-crack opening displacement of notched TRC specimens for 2 types of textiles (left), double-sided pull-out behaviour of short fibres and textiles

Fig. 1, right, contains double-sided pull-out curves of either one textile layer or 1 % by vol. of glass short fibres. The comparison shows that 1 % by vol. of short fibres has better crack-bridging abilities in the early stage of the crack-opening compared to textile reinforcement.

The amount of short fibres which is necessary to prevent the loss in strength can be determined by scaling and superimposing the pull-out curve with the stress-crack opening displacement in Fig. 1, left.

5 Post-Crack behaviour

A series of tests on textile reinforced concrete samples with two layers of textile and short fibres were conducted. In these tests, the fibre content and stiffness were varied. Textile reinforced concretes without short fibres with two and three layers of textile served as reference. To illustrate the development of the crack formation, the tests were documented with an optical measuring system (Aramis).

Short fibres with a low stiffness (PP, 15 mm), with a medium stiffness (PVA, 8 mm and glass, 6 mm) as well as with a high stiffness (steel, 12 mm) were applied. Fibre contents of 0.5 and 1.5 % by volume were adjusted per fibre type. The samples were stored in water for 21 days and in laboratory climate for 7 days before the test.

For each fibre type, Fig. 2 illustrates exemplary tensile stress-strain curves of textile reinforced concrete samples with both short fibre contents. For reasons of comparison, each diagram contains the reference samples with only two layers of glass textile without short fibres. Averaged curves were deliberately not displayed so that the curves are not smoothed in the area of the crack formation.

The crack patterns are documented in two ways. During the tests, images of the samples were taken with the optical deformation measuring every one to two seconds. The data was used to evaluate the temporal development of the cracks. Since the procedure is limited in its resolution, the crack pattern was additionally visualised by wetting the samples after the test. Each of the last crack patterns of the optical measuring taken before the sample failed are displayed in Fig. 3. Table contains the mean crack spacing determined on the samples after the test.

5.1 Load level and stiffness in state II

At the evaluation of the results, very different scenarios depending on the short fibre type were observed which will be explained more closely in the following. At first, Fig. 2(a) illustrates the comparison of both reference concretes with two and three layers of textile. The first crack strengths are almost equal. As was to be expected, there is a higher local stiffness in state II due to the higher reinforcement ratio at three layers of textile because, basically, the load is related to the constant cross-sectional area of the concrete.

The following diagrams show how the tensile stress-strain curve changes when short fibres are added instead of a third layer of textile. The volume of the textile equals a short fibre volume of 1.5 %. Fig. 2(b) shows textile reinforced concretes with 0.5 and 1.5 % by volume of PVA short fibres. Both fibre dosages lead to an increase in the first crack strength and thus also to an additionally released energy at the crack formation. This does however not lead to a more abrupt load decrease or to a more reduced stiffness during the crack formation as compared to the reference. Instead, there is a parallel displacement of the stress-strain curve because of the short fibres acting in the crack. It is noticeable that the increase in the short fibre content and thus the increase in the reinforcement ratio does not lead to a higher gradient of the stress-strain curve as displayed in Fig. 2(a). An explanation could be the larger crack formation at higher fibre contents. A distinct multiple crack formation leads to altogether larger deformations and hence to a lower local stiffness. This means that the expected increase in local stiffness due to the higher reinforcement ratio would be counterbalanced by the additional crack formation. This assumption seems possible because, at higher fibre contents, also the load transfer lengths to form further cracks decrease accordingly.

Fig. 2(c) shows a similar situation with integral glass fibres. Although there was no increase in the first crack load at a glass fibre content of 0.5 % by volume, at least a small parallel displacement of the stress-strain curve occurred later since imperfections of samples which lead to an early first crack are no longer relevant after the first cracks have formed.

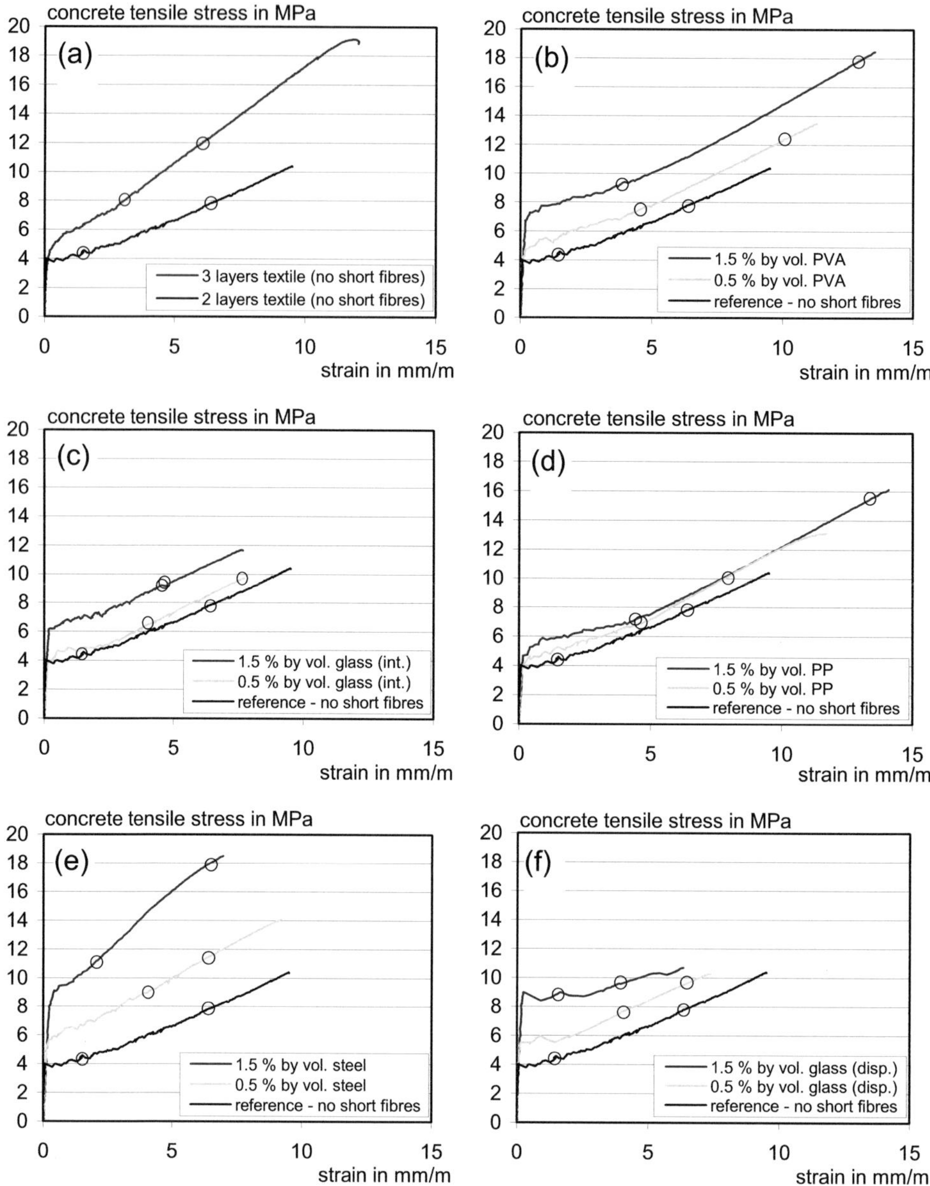

Fig. 2   Influence of different short fibre types and contents on the stress-strain behaviour

Concerning stiffness and load-level, the PP-fibres illustrated in Fig. 2(d) behave in an almost neutral way because of the very low modulus of elasticity. Even at the high fibre content the crack pattern is only slightly improved as compared to the reference without fibres. Likewise, there is hardly any parallel displacement of the curve as explained before because the short fibres can hardly transfer any load due to the low modulus of elasticity. However, the higher ultimate strain with fibre concrete is noticeable. Maybe the slightly higher number of cracks increases the total strain or, more likely, stress concentrations of the textiles at the crack edges are reduced by the short fibres.

The samples containing steel fibres in Fig. 2(e) feature a special behaviour because steel fibres have a considerably higher modulus of elasticity than all other fibre materials. Therefore, there is not only a parallel displacement of the curve but also an increase in the local stiffness. Since the reinforcement ratio and the number of cracks are comparable to the PVA-fibre concretes (cf. Fig. 3), this influence can directly be ascribed to the high stiffness of the steel. Thus, such a strong influence on the local stiffness is not to be expected at the short fibres normally used in textile reinforced concrete.

Another limit case is shown in Fig. 2(f). The first crack strength of concrete containing dispersible glass fibres amounts to up to 9 MPa. Since the crack-bridging reinforcement cannot transfer the high fracture energy, high strains occur in the area of the crack formation. Due to the high load transfer length the crack pattern cannot be sufficiently influenced by the short fibres. In this case, short fibres and textile reinforcement are not correctly adjusted to each other. Therefore, short fibres featuring very good crack-bridging properties should be added when high concrete tensile strengths are achieved.

All in all, it was determined that the advantages of short fibre reinforcement for practical application mainly arise when the first crack strength is increased. The local stiffness of the short fibre reinforced samples immediately after the first crack is slightly lower as compared to a reference sample with three layers of textile. A higher local stiffness in the area of crack formation can only be achieved with higher short fibre volume.

5.2   Crack formation

At conventional textile reinforced concrete, the area of the crack formation can clearly be discerned on the course of the stress-strain curve. At high reinforcement ratios or when short fibres are used, this is not always possible. By approximation, the crack formation is assumed to be completed when the tensile stress-strain curve shows a linear shape in state II. In order to determine exactly until when the crack formation takes place, the crack patterns covering the entire strain area were evaluated.

The examination of the photographed image sequences showed that, in some cases when short fibres were used, the crack formation continued until the samples failed. The marks in the diagrams of Fig. 2 indicate the areas where crack formation occurred. Up to the first mark, continuous cracks formed in quick succession. Since these cracks occur suddenly, there are often major losses in stiffness in this area. In the area between the marks, further (secondary) cracks developed which formed however very slowly and were not continuous. This kind of crack formation cannot always be discerned by means of the tensile stress-strain curves as can be seen from specimens with 1.5 % by volume of PVA-fibres and steel fibres. Many of these secondary cracks developed behind the first mark. Behind the second mark, no further cracks could be visualised with the optical measuring.

A comparison of the fibre types shows that the crack pattern of the reference sample with two layers of textile can be improved with almost all fibre concretes. The dispersible glass fibres are an exception. They are well suited to increase the first crack strength but they are less suited for crack-bridging. It is conceivable that the individual glass filaments are damaged at the fracture and therefore are no longer completely available in the opening crack. All in all, the smallest crack spacings were obtained when using steel and PVA-fibres. As expected, the crack formation does not directly depend on the stiffness of the short fibres. As compared to a third layer of textile, comparable or even significantly smaller crack spacings could be obtained with steel fibres and a higher fibre volume of integral glass, PVA and PP fibres.

Table 1   Crack spacings determined on the samples after the test

Reinforcement		Crack spacing in mm	
Short fibre type	Textile	0.5 % by vol.	1.5 % by vol.
Steel, 12.7 mm		5.3	3.5
Glass, 6 mm (dispersible)		18.0	10.0
Glass, 6 mm (integral)	2 layers	12.0	6.4
PVA, 8 mm		7.2	4.1
PP, 15 mm		7.5	6.9
-	2 layers	9.0	
-	3 layers	6.0	

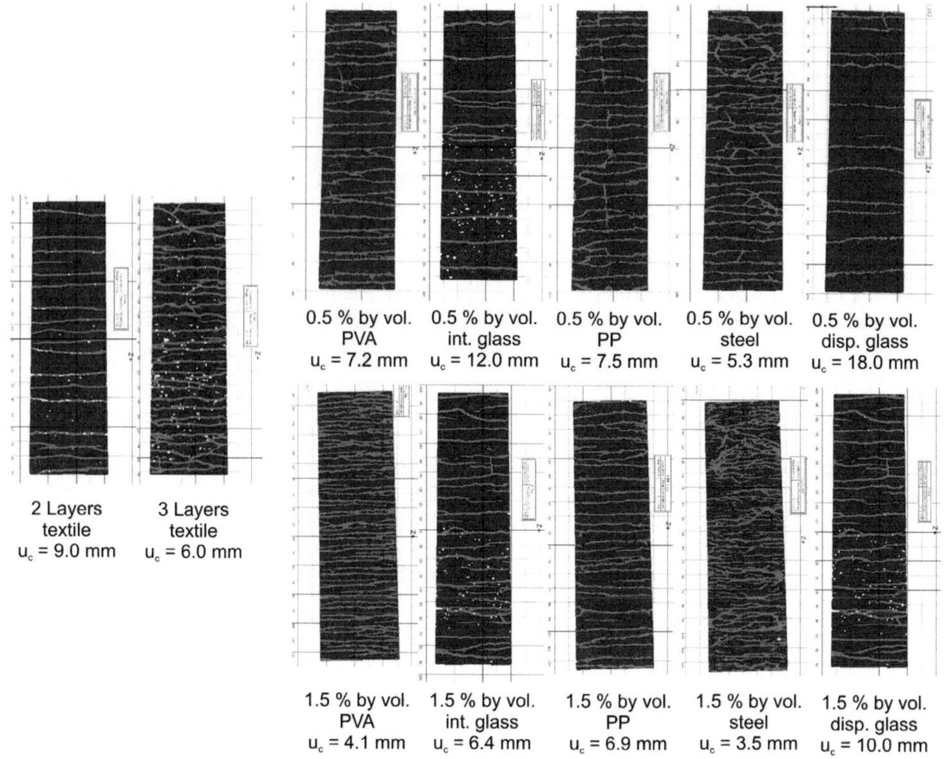

Fig. 3   Visualization of the crack patterns with optical deformation measuring

6   Conclusions

The paper gives a short summary on the load-bearing behaviour of textile reinforced concrete with short fibres. Three stages of the stress-strain curve were defined and the influence of short fibres described. The mode of action of the short fibres mainly consists in counteracting the strength losses induced by drying. The tensile strength of wet-stored specimens could not be further increased. In stress-crack opening displacement tests the load-bearing behaviour of short fibres and textiles were compared. The tests revealed that short fibres take loads at much smaller crack widths which reduces the loss of stiffness during multiple cracking. Finally, the load-bearing behaviour of specimens with different types of short fibres was compared. Polymer fibres increase the strain capacity while dispersible glass fibres mainly increase the first crack strength. It was shown that, when using short fibres, cracking takes place continuously until the textile fails. The crack spacing is dependent on crack strength and type of short fibre but could be reduced with most fibre types.

References

[1]   Barhum, R.; Mechtcherine, V.: Influence of Short Fibres on Fracture Behaviour in Textile Reinforced Concrete. Symposium on Fracture and Damage of Advanced Fibre-reinforced Cement-based Materials. 18th European Conference on Fracture, Dresden, August 30 - September 03, 2010, S. 77-89

[2]   Korb, S.: Untersuchungen zum Zugtragverhalten hochduktiler Faserbetone mit zusätzlicher Textilbewehrung. Kaiserslautern, Technische Universität, Fachbereich Architektur/Raum-und Umweltplanung/Bauingenieurwesen, Dissertation, 2010

[3]   Hinzen, M. ; Brameshuber, W.: Improvement of Serviceability and Strength of Textile Reinforced Concrete by Using Short Fibres. Dresden: Technische Universität Dresden, 2009. - In: Textilbeton Theorie und Praxis. Tagungsband zum 4. Kolloquium zu textilbewehrten Tragwerken (CTRS4) und zur 1. Anwendertagung, Sonderforschungsbereich 528 und 532, Dresden, 3.6.-5.6.2009, (Curbach, M.; Jesse, F. (Eds.)), S. 261-272

[4]   Hinzen, M.; Brameshuber, W.: Influence of Short Fibres on Strength, Ductility and Crack Development of Textile Reinforced Concrete, RILEM Proceedings PRO 53. - In: High Performance Fibre Reinforced Cement Composites (HPFRCC5), Proceedings of the Fifth International RILEM Workshop, Mainz, July 10-13, 2007, (Reinhardt, H.W.; Naaman, A.E. (Eds.)), S. 105-112, ISBN 978-2-35158-046-2

[5]   Brockmann, J. ; Raupach, M.: Durability Investigations on Textile Reinforced Concrete. Durability of Materials and Components, 9th International Conference (CSIRO 2002), Brisbane, Australia, 17-20 March 2002. Paper No. 111

Mechanisms of the interaction between textile reinforcement and short fibres in high-performance cement-based composites subjected to tensile loading

Rabea Barhum

Institute of Construction Materials,
Technische Universität Dresden,
Würzburger Strasse 46, 01187 Dresden, Germany
Supervisor: Viktor Mechtcherine

Abstract

The mechanical performance of textile-reinforced concrete (TRC) in tension can be considerably improved by adding short fibre to its matrix. The goal of the study is to understand the mechanisms of the interactions between continuous and short fibres in cement-based matrix and to develop a multi-scale model for the mechanical behaviour of cement-based composites with such hybrid reinforcement.

Uniaxial tension tests on TRC specimens as well as multifilament-yarn pullout and single-fibre pullout tests were performed to study the fracture behaviour of the composite material and its components and gain better understanding of the crack bridging mechanisms. Visual inspection and microscopic investigation provided deeper insight into the specific phenomena relevant to the characteristic material properties.

This article summarizes experimental findings for some chosen parameters including the type of short fibres and the water-to-binder ratio. Furthermore, the failure mechanisms are discussed providing a basis for a multi-scale model, which will be published elsewhere.

1 Introduction

Textile-reinforced concrete (TRC) is a composite material consisting of a finely grained cement-based matrix and high-performance, continuous multifilament yarns made of alkali-resistant (AR) glass, carbon, or polymer. The major advantages of TRC are its high load carrying capacity and pseudo-ductile behaviour, which is characterised by large deformations due to its tolerance of multiple cracking [1]. Such large deformations prior to material failure are crucial in respect of both structural safety and energy dissipation in the case of impact loading [2]. However, that high strength levels can be only reached at high deformations means that for the service state, where only small deformations are acceptable, the design load-bearing capacity of TRC must be much lower than its tensile strength. Moreover, relatively wide cracks observed at high deformations are undesirable. In recent years researchers have performed several test series to investigate the influence of short fibres on various properties of TRC [3]. However, the mechanisms in the joint action of short fibre and textile reinforcement are still not fully understood. In order to gain more and better insight into the specific material behaviour of the finely grained concrete with such hybrid reinforcement, a new investigative program has been initiated at the TU Dresden. In this ongoing PhD project the influence of adding different types of short fibres on the fracture behaviour of textile-reinforced concrete is studied.

 Uniaxial tension tests (macro-level) on thin, narrow plates made of TRC constitute the core of the experimental program. Special attention is directed at the course of the stress-strain relationship, crack pattern development, and fibre failure behaviour. Furthermore, multifilament-yarn (meso-level) and single-fibre pullout (micro-level) tests were performed to provide detailed insights into the various failure mechanisms observed in the experiments, the bond behaviour between short fibres and the finely grained concrete as well as between the yarn surface and the matrix. Moreover, visual inspections of the specimens' surfaces and microscopic investigation of the fracture surfaces were performed and evaluated.

 Based on the experimental results at the micro-scale physically based, rheological models consisting of simple rheological elements were developed. By means of statistical procedures the

Proc. of the 9ᵗʰ *fib* International PhD Symposium in Civil Engineering, July 22 to 25, 2012,
Karlsruhe Institute of Technology (KIT), Germany, H. S. Müller, M. Haist, F. Acosta (Eds.),
KIT Scientific Publishing, Karlsruhe, Germany, ISBN 978-3-86644-858-2

combination of these models leads to description of the material behaviour at the meso-scale. According to the results of fracture mechanics and phenomenological investigations a material law for the macro-scale is going to be discussed and developed.

The results presented in this paper are limited to the part of the experimental investigation dealing with the influence of the addition of short, dispersed fibres made of alkali-resistant (AR) glass and carbon on the fracture behaviour of TRC.

2 Material

Matrices with slag furnace cement (CEM III) and the addition of pozzolans show favourable properties regarding the durability of glass fibre [4]. Because of the small diameter of both the continuous filaments and the short fibres, the maximum aggregate diameter had to be small as well (<1 mm). One such fine-grained, cement-based concrete was chosen for this investigation. Two designated mixtures M030 and M045, having water-to-binder ratios of 0.30 and 0.45, respectively, were used. Finally, a super-plasticizer was added to achieve sufficient flowability.

One type of polymer-coated, biaxial fabric made of alkali-resistant glass was used as textile reinforcement for TRC specimens as well as for the multifilament-yarn pullout tests. The weft and warp threads had a fineness of 2*640 tex (mass in g of 1km yarn; tex = g/km).

Two types of short fibre were chosen and combined with textile layers for this investigation: dispersed AR glass and dispersed carbon fibres.

More details concerning the properties of the materials used in this PhD-project can be found in [5].

3 Preparations of specimens and test setup

Uniaxial tensile tests were performed on rectangular plates (500 mm x 100 mm x 12 mm) reinforced by 2 layers of textile and produced using the lamination technique explained in [5]. Additionally, specimens of the same dimensions made of plain mortar (matrix) as well of mortar reinforced with short fibre only were tested. The force was introduced to the specimens via non-rotatable steel plates glued to the TRC plates, cf. Fig 1a.

For multifilament yarn pullout tests, rectangular specimens (200 mm x 70 mm x 12 mm) were cut from larger plates produced in the same manner as those for uniaxial tensile tests. Only one layer of textile was used as reinforcement. The anchorage length was determined by specific arrangement of a "should" crack-position using a saw cut on both sides, cf. Fig 1b. In the notched cross-section only one multifilament yarn will be active and connect the two parts of the plates to each other. More details may be found in [6].

Single-fibre pullout tests were performed on specimens prepared according to [7]. This test was performed only on short glass fibres. The specimen was fixed to the clamps, and the fibre was glued to the upper mounting plate of a testing machine, cf. Fig. 1c.

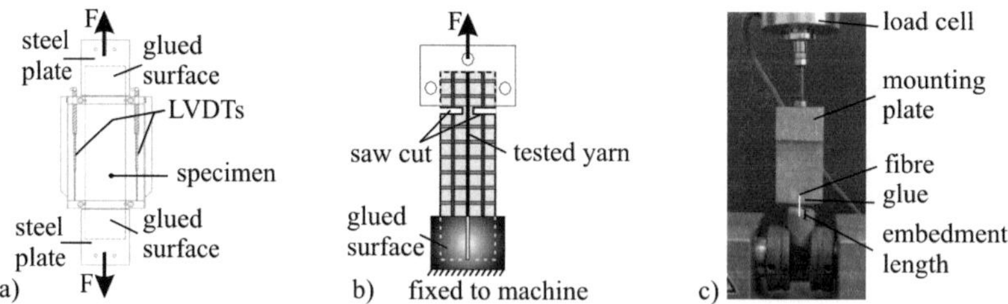

Fig. 1   Schematic view of tests' setups: a) uniaxial tension, b) yarn pullout, c) single-fibre pullout

4 Experimental results

Fig. 2a shows representative stress-strain curves obtained from the tests on TRC specimens with and without the addition of 1.0% by volume of short dispersed glass (SGF) and carbon fibres (SCF). A pronounced increase in first-crack stress could be observed in all experiments with the addition of both types of short fibres in comparison to results obtained with reference TRC plates. The first-crack stress value increased approximately by a factor of 2.5.

The addition of short fibres led to the expansion of the region IIa, where multiple cracks form. This expansion was particularly pronounced in the case of the TRC with short carbon fibres: The width of this region (i.e., stains) more than doubled, cf. Fig. 2a. The observation of the specimens' surfaces showed that this widening resulted from the formation of larger numbers of cracks. To emphasize, observably for the same strain level the TRC specimens with short fibre always exhibited a higher number of cracks in comparison to the TRC specimens without short fibre as shown in Fig. 2b.

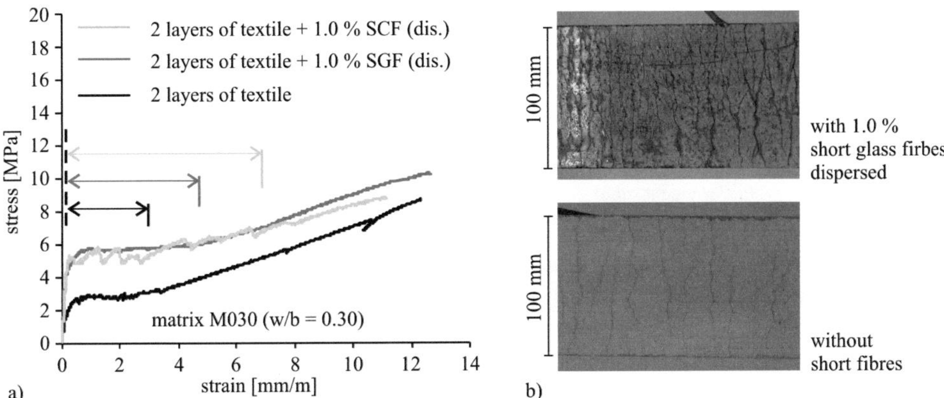

Fig. 2   a) Effect of addition of SGF and SCF on the characteristic stress-strain behaviour of TRC plates subjected to tensile loading, and b) effect of dispersed short AR glass fibres on the cracks pattern on specimen's surface

Moreover, since the stress-strain curves for TRC with short fibre are always above the corresponding curves for TRC without short fibre, it can be concluded that the area under the stress-strain curve, i.e., the energy dissipated when TRC with short fibres is subjected to tensile loading, increases noticeably in comparison to TRC without dispersed fibres. The energy absorption due to the addition of both glass and carbon short fibres increased by approximately 40% for the TRC specimens made of the matrix M030.

Multifilament-yarn pullout tests were performed to investigate the effect of the short fibre on the pullout behaviour. Fig. 3a shows force-displacement curves obtained for the specimens made with matrix M045 and 1.0% of short glass fibre. Obviously, the addition of short fibre led to a higher ultimate pullout force, which indicates a better bond between matrix and yarn due to the addition of short fibres. Furthermore, a higher pullout force was needed when matrix M030 was used in comparison to that achieved for matrix M045.

The single-fibre pullout tests, regarded as micro-level tests, sowed that brittle failure of fibres dominated behaviour when matrix M030 was used. Typically, a vertical drop in the force-displacement curves due to fibre breakage was observed after the relatively high ultimate force was reached, see Fig. 3b. This relatively high ultimate force at fibre failure pointed to the good matrix-fibre bond when matrix M030 was used. The ultimate pullout force of the fibres embedded into matrix M045 was much lower. Most of the fibres provided force-displacement relationships with a pronounced softening branch, which indicated fibre pullout, cf. Fig. 3b.

More details on the obtained results may be found in [8, 9].

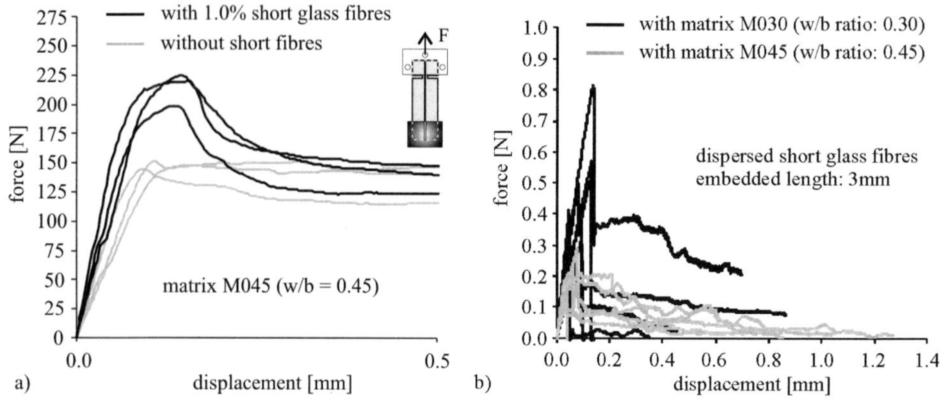

Fig. 3 a) Results of multifilament-yarn pullout tests with the matrix M045 with and without the addition of 1.0% dispersed, short glass fibres, and b) Force-displacement curves obtained from the single glass fibre pullout tests with matrix M030 and matrix M045

5 Discussion

The improvement in the level of first-crack stress of TRC due to the addition of short fibres can be mainly ascribed to the bridging of micro-cracks by short fibres.

The higher number of cracks observed on TRC specimens' surfaces with short fibres can be traced back to the bridging of macro-cracks with short fibres which causes additional stress transfer over the macro-cracks and, thus, less relaxation of the matrix in the cracks' vicinity, as schematically shown in Fig. 4. A new crack can form at a smaller distance from an existing crack, thus more cracks can develop. Microscopic investigation of fracture surfaces by means of ESEM showed that short fibres are frequently linked to multifilament yarns, cf. Fig. 5a. This might improve the bond between textile and matrix, thereby leading to smaller crack widths and higher cracking density.

Microscopic investigation showed, moreover, that short fibres can improve the bond between multifilament yarn and surrounding matrix by means of new, additional cross-links, which lead to a better yarn-matrix bonding. By their random dispersion in the matrix and their positioning on the yarn surface, short fibres provide extra connecting points to the surrounding matrix. Fig. 5b illustrates experimental evidence that such cross-links in connection with short fibre indeed exist.

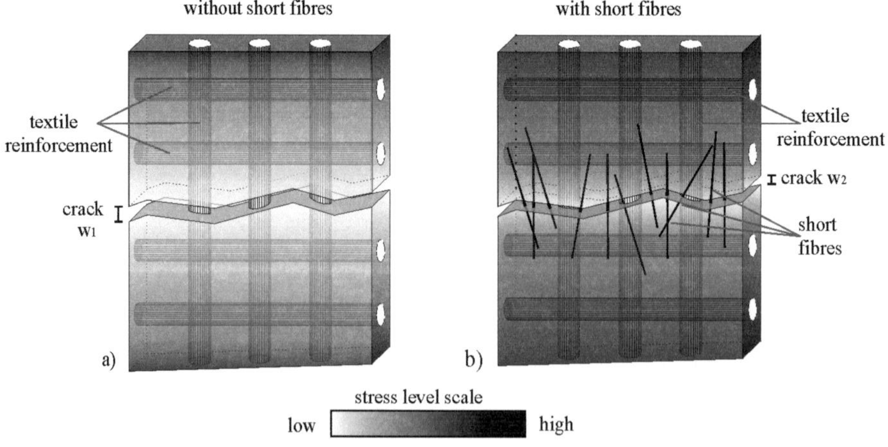

Fig. 4 Stress gradient distribution in the vicinity of a macro-crack in TRC a) without short fibres and b) with the addition of short fibres

Results presented in Fig. 3b showed that the water-to-binder ratio influences the bond between the matrix and single short fibres. In order to explain this, the specimens of single-fibre pullout tests were carefully split in the middle and investigated using ESEM. In most of the specimens made with the matrix M030, a part of the fibre was found, i.e., the part remained in the specimen after fibre breakage, cf. Fig 6a. In contrast, the empty fibre canal in the cases when matrix M045 was used indicated a complete fibre pullout, cf. Fig. 6b.

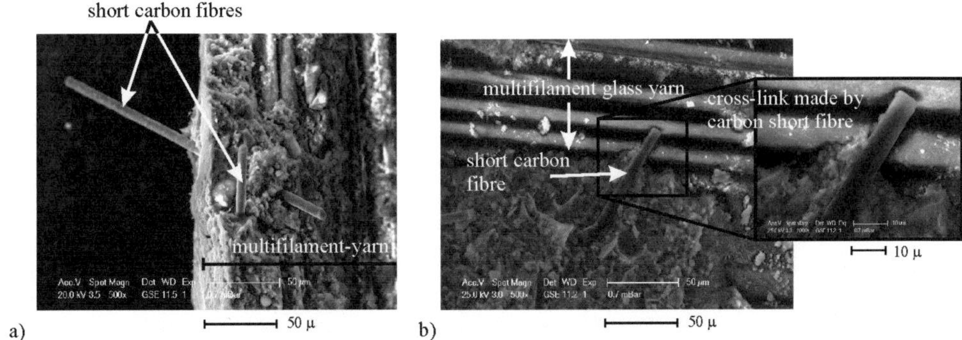

Fig. 5   ESEM images of a) short carbon fibres linked to multifilament yarns of a textile layer, and b) "cross-link" between multifilament-yarn and matrix caused by a short carbon fibre

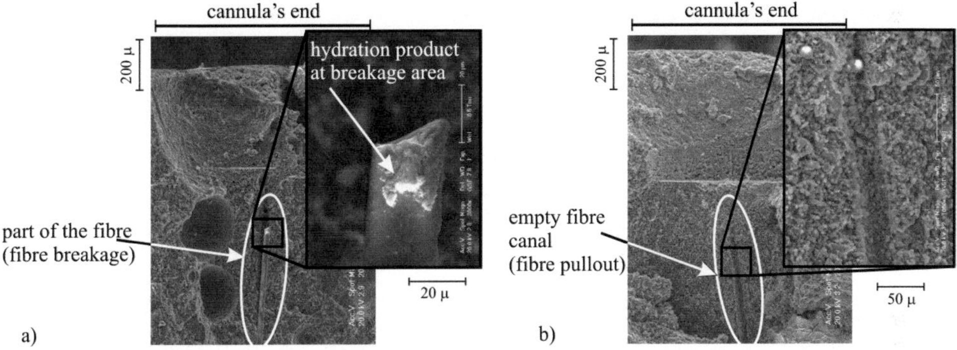

Fig. 6   ESEM images on single-fibre pullout specimen with matrix a) M030 and b) M045

More details of these mechanisms and deeper discussions are presented in [5, 8, 9].

6   Summary and conclusions

In this research project the effects of adding different types of short fibres on the strength, deformation, and failure behaviour of textile-reinforced concrete subjected to tensile loading were investigated. The stress-strain curves obtained from uniaxial tension tests demonstrated clearly the positive influence of short fibre on the mechanical performance of TRC. Furthermore, depending on the water-to-binder ratio of the matrix various degrees of matrix-fibre bond were observed. To clarify the mechanisms leading to the enhancement of the mechanical performance of TRC, multifilament-yarn and single-fibre pullout tests were performed. The morphology of the matrix-fibre interface was studied using an electron microscope and provided some explanation of the phenomena observed. From the findings the following conclusions could be drawn:

- The first-crack stress value of TRC specimens increased by two to three times due to the addition of 1.0% by volume short fibres, more distinctly in the case of the short carbon fibres.
- Expansion of the strain region, where multiple cracks form, was observed for the stress-strain curves due to the addition of short fibres. The visual inspection of the specimens' surfaces showed that a higher number of cracks and finer cracks for given strain levels were formed when short fibres were added to TRC.

- The work-to-fracture of the composite was improved significantly by the addition of short fibre. Here the effect when matrix M030 (water-to-binder ratio of 0.30) was used was more pronounced.
- Short fibres improved the bond between multifilament-yarn and the surrounding matrix. By their random positioning on the yarn's surface; short fibres built new "special" adhesive cross-links which provided extra connecting points to the surrounding matrix.
- The water-to-binder ratio of the matrix influenced bond quality between fibre and matrix and, thus, the effect of short fibre on TRC-behaviour.

Acknowledgement

This project was initiated in the Collaborative Research Centre SFB 528 "Textile Reinforcement for Structural Strengthening and Retrofitting" financed by the German Research Foundation (DFG). The author would like to acknowledge with gratitude the foundation's financial support.

References

[1]   Curbach, M.; Jesse, F. (Eds.): Textile reinforced structures: Proc. of the 4th colloquium on textile reinforced structures (CTRS4), Dresden. Technische Universität Dresden, (2009)

[2]   Silva, F.A.; Butler, M.; Mechtcherine, V.; Zhu, D.; Mobasher, B.: Strain rate effect on the tensile behaviour of Textile-reinforced concrete under static and dynamic loading. Materials Science and Engineering A, 528, (2011), pp. 1727–1734

[3]   Hinzen, M.; Brameshuber, W.: Influence of short fibres on strength, ductility and crack development of textile reinforced concrete. Reinhardt, H.W.; Naaman, A.E. (Eds.): 5th International Workshop on High Performance Fiber Reinforced Cement Composites (HPFRCC 5), Mainz. (2007), pp. 105-112

[4]   Butler, M.; Mechtcherine, V.; Hempel, S.: Experimental investigations on the durability of fibre-matrix interfaces in textile reinforced concrete. Cement and Concrete Composites 31, (2009), pp. 221–231

[5]   Barhum, R.; Mechtcherine, V.: Effect of short fibres on fracture behaviour of textile reinforced concrete. 7th International Conference on Fracture Mechanics of Concrete and Concrete Structures & Post-Conference Workshop (FraMCoS-7). Jeju. (2010), pp. 1498-1503

[6]   Lorenz, E.; Ortlepp, R.: Bond behaviour of textile reinforcements - development of a pull-out test and modelling of the respective bond versus slip relation. Parra-Montesinos, G.J.; Reinhardt, H.W.; Naaman, A.E. (Eds.): 6th International work shop on High Performance Fiber Reinforced Cement Composites (HPFRCC 6), Ann Arbor. (2011), pp. 463-470

[7]   Jun, P.; Mechtcherine, V.: Behaviour of strain-hardening cement-based composites (SHCC) under monotonic and cyclic tensile loading. Cement and Concrete Composites 32, (2010), pp. 801-809

[8]   Barhum, R.; Mechtcherine, V.: Influence of short fibres on fracture behaviour in textile reinforced concrete. The 18th European Conference on Fracture (ECF 18). Dresden. (2010), pp. 77-89

[9]   Barhum, R.; Mechtcherine, V.: Effect of short fibres on the behaviour of textile reinforced concrete under tensile loading. Parra-Montesinos, G.J.; Reinhardt, H.W.; Naaman, A.E. (Eds.): 6th International workshop on High Performance Fiber Reinforced Cement Composites (HPFRCC 6), Ann Arbor. (2011), pp. 487-494

Influence of fibre reinforcement on the fatigue behaviour of concrete

Kerstin Elsmeier

Institute of Building Materials Science,
Leibniz Universität Hannover,
Appelstraße 9A, 30167 Hannover, Germany
Supervisor: Ludger Lohaus

Abstract

One important issue concerning the long-term resistance and durability of concrete structures is the fatigue damage caused by cyclic loading. Several experimental investigations to examine the influence of fibre reinforcement on the fatigue behaviour were carried out at the Institute of Building Materials Science, Leibniz University Hannover. An increase in the ultimate numbers of cycles to failure due to the fibre reinforcement could not be determined. However, the strain development of the fibre reinforced concrete shows an advantageous behaviour concerning the prior indications of failure. In this paper, the experimental and theoretical investigations on the fatigue behaviour of plain and fibre reinforced concretes are described, and the results of the experimental tests and the effects of the fibre reinforcement on the fatigue behaviour observed are presented.

1   Introduction

The fatigue behaviour of plain and fibre reinforced normal strength, high strength and ultra-high strength concretes was investigated in the course of several research projects at the Institute of Building Materials Science, Leibniz University Hannover [1], [2]. Determining the influence of steel fibre reinforcement on the fatigue behaviour was a partial aspect of these investigations [3]. In order to identify the influence of the steel fibres, plain and fibre reinforced concretes were investigated under uniaxial static loading and fatigue compressive loading with different loading amplitudes on cylindrical specimens. Moreover, experimental investigations were conducted on specimens with a varied geometry. One side of the cylindrical specimens was bevelled, resulting in a truncated cone. This varied geometry causes passive transverse tension, thereby increasing the influence of the fibre reinforcement. In this paper, the test results of fatigue investigations on plain and fibre reinforced concrete will be presented and analysed in order to quantify the effect of fibre reinforcement on the fatigue behaviour.

2   Test Programme

2.1   Tested Concrete Composition and Specimen Geometry

The ultra-high strength concrete mixture examined has a maximum grain size of 0.5 mm and a 28-day compressive strength of $f_{c,cube,100} = 160$ MPa, following storage in water. This mixture also contains 2.5 Vol.-% of 9 mm long, smooth, high-strength steel fibres with a l/d ratio of 60. It belongs to the standard mixtures of the priority programme "Sustainable Building with Ultra-High Performance Concrete (UHPC)" of the German Research Foundation (DFG).

The experimental investigations were carried out primarily on cylindrical specimens with dimensions of d/h = 60 mm/180 mm. Furthermore, a variation in geometry has been developed in cooperation with the Institute for Concrete Constructions, Leibniz Universität Hannover [4]. The specimen geometries used are presented in Figure 1.

The formwork of the specimens was removed after 48 hours. All the specimens were then subjected to a two-day heat-treatment of 120° C. The mean value of the static compressive strength of the cylindrical specimens after heat treatment was $f_{c,cyl} = 185$ MPa. Afterwards, the specimens were stored at a standard climate (20°C/65% r.H.). The top surfaces of the specimens were plane-parallel ground and polished before testing.

Proc. of the 9th *fib* International PhD Symposium in Civil Engineering, July 22 to 25, 2012,
Karlsruhe Institute of Technology (KIT), Germany, H. S. Müller, M. Haist, F. Acosta (Eds.),
KIT Scientific Publishing, Karlsruhe, Germany, ISBN 978-3-86644-858-2

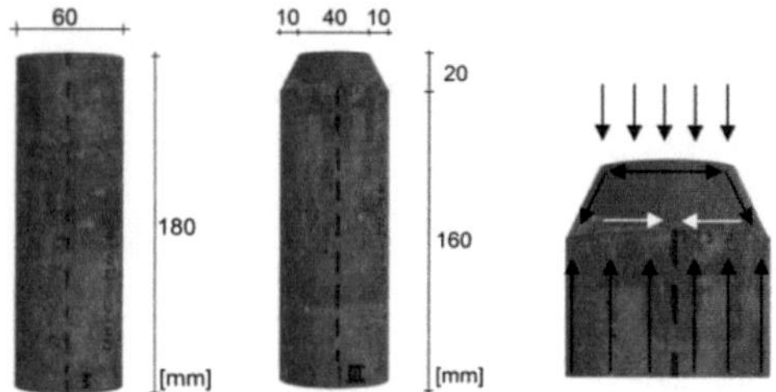

Fig. 1   Specimen geometries and schematically shown force distribution of the truncated cone

Fibre reinforcement is mainly effective under tensile stress. However, direct tensile tests are usually hard to realise due to technical difficulties. The new specimen geometry was developed in order to create a passive transverse tension within the specimen, which should increase the effect of the fibre reinforcement. The force will be applied at the top surface with the reduced diameter, and spreads to the larger cross-sectional area. Due to this force distribution, a transverse tension occurs in the transition area between the cylinder and the truncated cone, as can be seen in Figure 1. In this way, investigations with a combined loading of axial pressure and passive transverse tension can be carried out. According to [5] the occurring passive transverse tension of the presented truncated cones can be determined using the following equation:

$$\sigma_{Z,quer} \cong \frac{-\sigma_{D,\text{längs}}}{12} = -0,0833 \cdot \sigma_{D,\text{längs}} \qquad \text{Eq. 1}$$

Based on this variation in the geometry, the influence of fibre reinforcement can be investigated more effectively.

2.2   Experimental Set-up and Test Programme

The experimental investigations were carried out with a servo-hydraulic testing machine with a 1 MN-actuator. The experimental set-up is shown in figure 2. The axial deformation of the test specimens, the surface temperature, the force applied by the machine, and the corresponding displacement were recorded during the investigations. Three laser distance sensors were positioned at 120° around the specimen in order to measure the axial deformation of the specimen. The surface temperature was measured using a temperature sensor.

Fig. 2   Experimental Set-up

The static compressive strength, which is adopted as the reference strength for the cyclic investigations, was determined based on the mean value of three specimens. The static uniaxial compressive tests were carried out using the displacement-controlled method at a rate of v = 0.2 mm/min prior to the cyclic investigations.

The cyclic investigations were carried out under pure compression loading. At first, the mean load was applied using displacement-controlled method. The sinusoidal force-controlled fatigue load was then applied with a transient oscillation of 100 load cycles. The frequency used in all cyclic investigations had a magnitude of f = 10 Hz.

The cyclic investigations on plain and fibre reinforced concretes using normal cylindrical specimen geometry were carried out by Wefer [1]. The maximum and minimum compressive stress levels for these investigations were calculated based on the mean value of the static compressive strength. The minimum compressive stress level for all cyclic investigations was $S_{c,min} = 0.05$. The fibre reinforced cylindrical specimens were investigated using a maximum compressive stress level $S_{c,max}$ with a value ranging between 70 % and 90 % of the static compressive strength. These tests made use of eight to eleven specimens per stress level. The cylindrical specimens without steel fibres were examined with a maximum stress level which varied between $S_{c,max} = 0.70$ and $S_{c,max} = 0.80$, using six to seven specimens per stress level. The maximum stress level applied to investigate the truncated cones amounted to between 70 % and 90 % of the static compressive strength. Six specimens were used for each stress level for the truncated cones without fibre reinforcement, except the tests with a maximum stress level of $S_{c,max} = 0.70$. For this stress level, only three specimens were investigated. These investigations were stopped after reaching a number of cycles of $N = 2*10^6$. The investigations on the truncated cones with fibre reinforcement made use of four specimens per stress level. The maximum stress level varied between $S_{c,max} = 0.70$ and $S_{c,max} = 0.85$. An overview of the cyclic investigations performed is given in Table 1.

Table 1   Overview of the cyclic investigations

	$S_{c,max}=0.70$	$S_{c,max}=0.75$	$S_{c,max}=0.80$	$S_{c,max}=0.85$	$S_{c,max}=0.90$
Cylinder without fibre reinforcement	7	-	6	-	-
Cylinder with fibre reinforcement	8	-	11	-	9
Truncated cone without fibre reinforcement	$3^{1)}$	6	6	6	6
Truncated cone with fibre reinforcement	4	4	-	4	-

[1)] run-out

3   Experimental Results

3.1   Static Investigations

The mean values of the static compressive strength of the different specimen geometries with and without fibre reinforcement are summarized in Table 2. Furthermore, the ratio between the results with and without fibre reinforcement, and the ratio between the static compressive strength of the different specimen geometries are given. Comparison of the static compressive strength clearly demonstrates that the fibre reinforcement leads to an increase in the static strength for both specimen geometries. This static strength increment for the truncated cones, as a result of the fibre reinforcement, is visibly higher than that of the cylindrical specimens.

Table 2   Comparison of the static strength's values of the specimens with and without fibre reinforcement

		with fibre reinforcement	without fibre reinforcement	ratio [-]
cylinder	[MPa]	192	176	1.1
truncated cone	[MPa]	136	86	1.6
ratio	[-]	1.4	2.0	-

The stress-strain curves of four selected truncated cones with and without steel fibres are presented in figure 3. The compressive strength is calculated based on the larger base area ($d = 60$ mm). As already shown in Table 2, the static strength's values of the truncated cones with fibre reinforcement are about 58 % higher than the static compressive strength of the specimens without steel fibres. This tendency was already observed by Cachim et al. [6] for cylindrical specimens with the dimensions d/h = 150 mm/300 mm. However, the static compressive strength was only about 17 % higher than the corresponding plain concrete compressive strength. Furthermore, it can be observed that the specimens without fibre reinforcement have a brittle material behaviour which results in an abrupt failure. In contrast to this behaviour, the fibre reinforced specimens show certain signs in advance before failure.

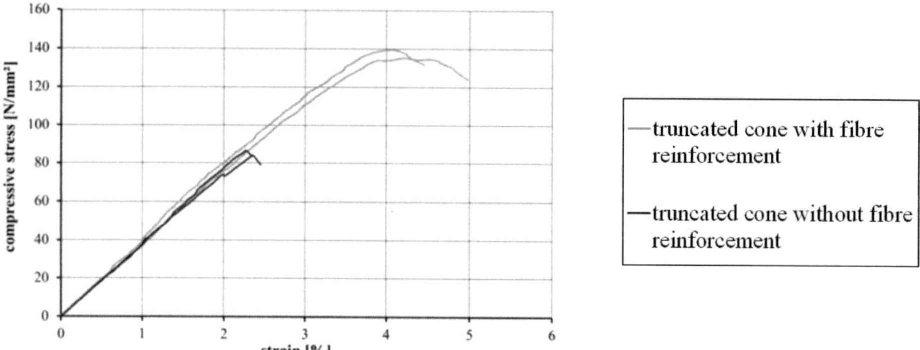

Fig. 3   Comparison of the stress-strain relation of truncated cones with and without fibre reinforcement

3.2   Cyclic Investigations

3.2.1   Ultimate number of cycles to failure

The results of the cyclic investigations on the plain and fibre reinforced concrete are evaluated in relation to the ultimate numbers of cycles to failure N. The mean values of the ultimate numbers of cycles to failure N of the cylindrical specimens with and without fibre reinforcement are presented in figure 4. Furthermore, the S/N-curve for pure compression of the CEB-FIP Model Code 90 [7] and the S/N-curve for pure compression of the fib-Model Code 2010 [8] are shown. As can be seen, the ultimate numbers of cycles to failure of the concrete with and without fibre reinforcement are considerably higher than the specifications in [7] and [8]. Furthermore, the inclination of the S/N-curves of both test series is nearly the same. Apparently, the ultra-high strength concrete without steel fibres reaches slightly higher ultimate numbers of cycles to failure. A comparison of test results on plain and fibre reinforced concrete under fatigue compressive loading by Lee and Barr [9] shows a similar trend concerning the ultimate numbers of cycles to failure. A slight reduction in the fatigue life of the fibre reinforced concrete can be observed compared to the plain concrete under compression loading.

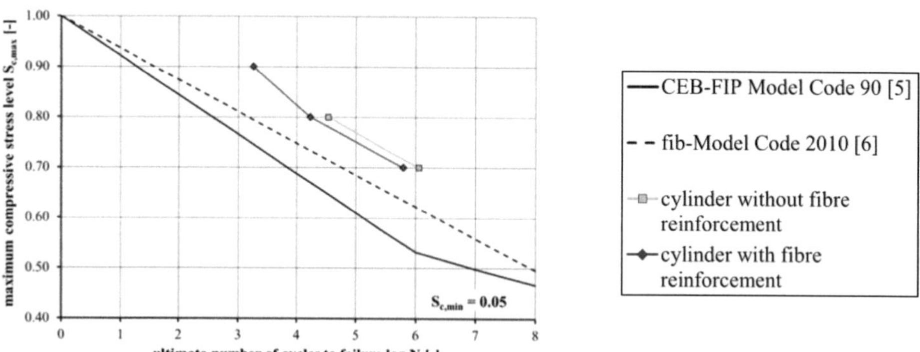

Fig. 4   Ultimate numbers of cycles to failure N of cylindrical specimens

The mean ultimate numbers of cycles to failure of the truncated cones with and without fibre reinforcement are shown in figure 5 and the S/N-curves of CEB-FIP Model Code 90 [7] and fib-Model Code 2010 [8] are presented again. The difference in the ultimate numbers of cycles to failure, observed on the cylindrical specimens, also appears for the truncated cones. The mean values of ultimate numbers of cycles to failure for the fibre reinforced specimens are lower than the results of the plain concrete. Similar to the results of the cylindrical specimens, the inclination of S/N-curves of the truncated cones with and without steel fibre reinforcement is approximately the same.

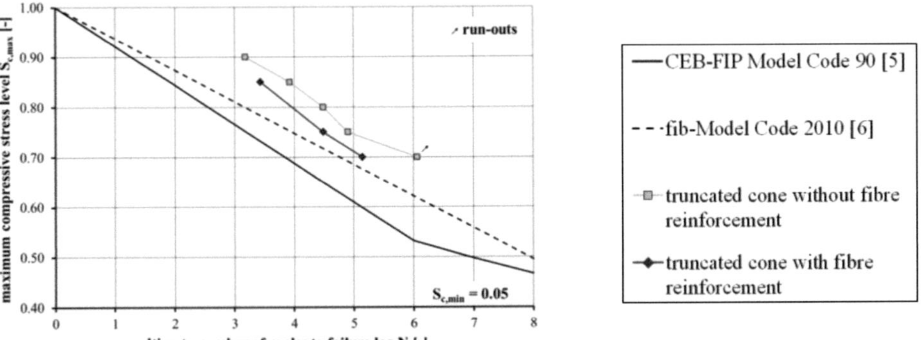

Fig. 5   Ultimate number of cycles to failure N of truncated cones

The mean numbers of cycles to failure of all fibre reinforced specimens are lower than the corresponding results of the plain concrete. With regard to these results, the fibre reinforcement seems to have a negative effect. As the maximum and minimum stress levels are calculated based on the static compressive strength, the influence of the fibre reinforcement is already captured in the static investigations. Based on these results, an assumption can be made that the crack initiating mechanism of the steel fibre reinforcement is dominant. Steel fibres, which are positioned parallel to the direction of loading, could accelerate the damage and, therefore, cause slightly lower ultimate numbers of cycles to failure. This tendency is assisted by unpublished investigations.

3.2.2 Strain Development

In addition to the investigations concerning the influence of fibre reinforcement on the ultimate numbers of cycles to failure, the test results were evaluated with regard to the influence of fibre reinforcement on the strain development due to cyclic loading. Figure 6 shows the strain development of two selected truncated cones with fibre reinforcement (left) and two specimens without steel fibres (right). These tests were carried out with a maximum stress level of 80 % and a minimum stress level of 5 % of the static compressive strength. The progression of the strain is presented for the minimum and the maximum stress level.

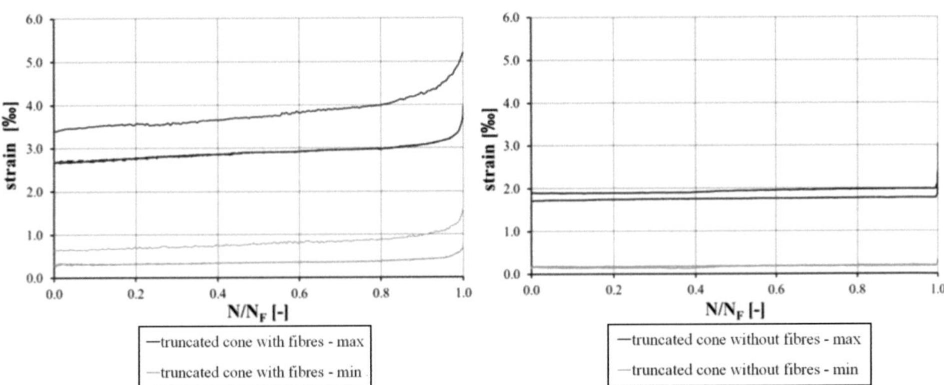

Fig. 6   Comparison of the strain development of specimens with fibres (left) and without fibres (right)

Comparison of the strain development shows clear differences. The second phase of the strain development showed a steeper inclination for the specimens with fibre reinforcement while it remains nearly straight for the plain concrete. This indicates the existence of a continuous crack propagation in the fibre reinforced specimens. In contrast to this, the specimens without fibre reinforcement show merely a limited increase in strain, and therefore, a slight crack propagation. The third phase of the strain development is normally characterized by unstable crack propagation which results into a large increase in strain. As a consequence of this large increase in strain, fractured surfaces occur that lead to failure of the specimens. The third phase of the strain development of the fibre reinforced specimens is more distinctive than the strain development of the plain concrete. An increase in strain can also be seen for the fibre reinforced concrete. Consequently, the failure does not occur abruptly. It rather occurs with a certain prior indications. The strain development of the specimens without fibre reinforcement does not indicate any distinctive pattern with respect to failure. The third phase of the strain development does not really exist. Therefore, the failure of the plain concrete occurs abruptly. In comparison with the results of the ultimate number of cycles to failure N, the fibre reinforcement clearly has a positive influence on the strain development.

4   Conclusions

In this paper, the results of experimental investigations concerning the influence of fibre reinforcement on the static and fatigue behaviour of ultra-high performance concrete are presented. In order to increase the effectiveness of the fibre reinforcement, newly developed specimen geometry was used. The results of the static investigations showed that the static compressive strength can be increased through the usage of steel fibres. The fatigue investigations showed a slight reduction in the ultimate numbers of cycles to failure due to the fibre reinforcement. On the contrary, the plain concrete endures a higher number of load cycles up to failure. As the minimum and maximum stress levels are calculated based on the static compressive strength, the influence of fibre reinforcement is already considered in the static investigations. Therefore, an additional improvement in the fatigue strength under compressive cyclic loading could not be achieved by the inclusion of fibre reinforcement. The fibre reinforcement has shown a positive influence with regard to the strain development. Due to the inclusion of the steel fibres, the failure of the specimens occurs with distinctive signs prior to failure. On the contrary, the failure in plain concrete occurs abruptly.

References

[1]   Wefer, M.: Materialverhalten und Bemessungswerte von ultrahochfestem Beton unter einaxialer Ermüdungsbeanspruchung. Dissertation, Leibniz Universität Hannover, Institut für Baustoffe, 2010

[2]   Lohaus, L.; Wefer, M.; Oneschkow, N.: Ermüdungsbemessungsmodell für normal-, hoch- und ultra-hochfeste Betone. In: Beton- und Stahlbetonbau, Vol. 106, No. 12, pp. 836 - 846, Ernst & Sohn, 2011

[3]   Lohaus, L.; Elsmeier, K.: Fatigue Behaviour of plain and fibre reinforced Ultra-High Performance Concrete, HiPerMat - 3rd International Symposium on Ultra-High Performance Concrete and Nanotechnology for High Performance Construction Materials, Kassel, 2012

[4]   Ertel, C.; Grünberg, J.: Triaxial Fatigue Behaviour of Ultra High Performance Concrete (UHPC). 3rd fib International Congress, Washington D.C., May 29 – June 2, 2010

[5]   Grünberg, J.; Oneschkow, N.: Gründung von Offshore-Windenergieanlagen aus filigranen Betonkonstruktionen unter besonderer Beachtung des Ermüdungsverhaltens von hochfestem Beton. Abschlussbericht zum BMU-Verbundforschungsprojekt, Leibniz Universität Hannover, 2011

[6]   Cachim, P.; Figueiras, J.; Pereira, P.: Fatigue behaviour of fibre-reinforced concrete in compression. In: Cement & Concrete Composites, Vol. 24, pp. 211 - 217, 2002

[7]   CEB – Comité Euro-international du Béton: CEB-FIP Model Code 90. Bulletin d'Information, No. 213/214, Thomas Telford Ltd., London, 1993

[8]   fib – International Federation for Structural Concrete: Model Code 2010, Final draft, September 2011

[9]   Lee, M.; Barr, B.: An overview of the fatigue behaviour of plain and fibre reinforced concrete. Cement & Concrete Composites, Vol. 26, pp. 299 – 305, 2004

Rubberized hybrid fibre-reinforced concrete for special applications

Ana Baricevic

University of Zagreb Faculty of Civil Engineering,
Department for Materials,
Kaciceva 26, Zagreb, Croatia
Supervisor: Dubravka Bjegovic

Abstract

Currently, Croatia is undertaking significant efforts to ensure funds for construction of railway connection between its continental and coastal regions. Construction of high speed railway is planned using concrete blastless track system. Technology and material needed to satisfy requirements for named track system are currently not available in Croatia. Overall goal of the presented research is to prepare concrete which could satisfy prescribed requirements on one side and on the other, provide ecological and economical savings.

Presented rubberized hybrid fibre reinforced concrete (RHFRC) is tailored based on previous experience and available quantities of by-products from mechanical recycling of waste tyres. For purpose of material optimisation, 20 concrete mixtures are prepared with different recycled/industrial (steel and textile) fibre ratio and with addition of fine rubber particles (0.5–2 mm). More than 1 000 concrete specimens are prepared and tested on 16 different standardized tests. In this paper, investigation of compressive strength, modulus of elasticity, toughness, capillary absorption, resistance to wear and freeze-thaw cycles will be represented on selected concrete mixtures.

Obtained results indicate that by partially replacing industrial with recycled fibres together with addition of fine rubber particles (5% on total volume of aggregate) good quality RHFRC can be prepared, assuring significant economic savings due to steel deficiency and its rising prices.

1 Motivation of the research project

Sustainability represents one of the major concerns of contemporary industry; concrete industry with use of million tons of natural resources per year settles between major environmental pollutants. Innovative materials and technologies are more than welcome to help preserving sufficient amount of natural resources as well as to decrease current gas emission. Waste materials are used and incorporated in concrete in order to replace aggregate or cement and in such way increase concretes sustainability. At the same time million tons of waste tyres are inadequately managed and left in the environment. To prevent this global devastation of the environment EU was pursuant to oblige all EU members and its candidates to start applying EU Directive 1999/31/EC which starting from 2006 clearly prohibits any kind of waste tyre disposal in the environment [1]. Recent investigations implied significant decrease of abandoned tyres and currently EU average in their recycling reaches 94%. In Croatia, according to the regulations 70% of all available tyres should be mechanically or in another process recycled, while only 30% can be used for energy recovery. Croatian recycling company has mechanical recycling process and its by-products are currently the only available in Croatia. Previously conducted research on University of Zagreb Faculty of Civil Engineering indicated that all by product obtained during mechanical recycling can be further used in concrete industry [2–15]. Special application of these concrete encompasses structures such as noise barriers, industrial floors, concrete pavements, railway track system and etc.

Although, current economic situation has decreased the amount of construction work, infrastructural projects has to go on. One of these engineering projects with special significance for Croatia and its habitants is modern, high seed railway connecting central Croatia and it capital with western coastal regions. According to the already prepared project documentation, track system will be constructed using concrete due to its capability to absorb large amounts of dynamic loadings and obtained energies. A technology for such systems is currently unavailable in Croatian factories, following that major part of this infrastructural investment should be imported from abroad. Investigations

Proc. of the 9th *fib* International PhD Symposium in Civil Engineering, July 22 to 25, 2012,
Karlsruhe Institute of Technology (KIT), Germany, H. S. Müller, M. Haist, F. Acosta (Eds.),
KIT Scientific Publishing, Karlsruhe, Germany, ISBN 978-3-86644-858-2

and experience in materials design, especially concrete implies that usually applied materials for this purpose (ordinary concrete) shows certain disadvantages when used for named application. Major problem of such concrete structures is disability of ordinary concrete to absorb all energy produced during cycling loadings present in structure life span. Inclusion of short textile and/or steel fibres improves concrete ductility and absorption capacity, but due to the extremely high price of industrial obtained fibres caused by complexity of their production, fibre reinforced concrete is still struggling on the market. Accordingly, alternative materials are investigated whose properties could provide adequate attributes to the concrete at the same time assuring price reduction of fibre reinforced concrete (FRC).

2   Investigation of innovative low cost material

During different recycling processes of waste tyres similar by products are obtained and they mostly consists from rubber particles, steel and textile fibres. Research performed on University of Zagreb Faculty of Civil Engineering indicated that each of obtained by-products can provide special improvements of concrete characteristics, depending on its application.

Rubber particles can be used as replacement of part of the aggregate, while the amount of replaced aggregate depends on the application. For example, small amounts (cca 5% of aggregate volume) are used for increase of concrete ductility and absorption capacity when high strength classes are required; when lightweight concrete is requested larger amounts are incorporated (up to 40-50% of aggregate volume). Example of the lightweight concrete is concrete were expanded clay is replaced with rubber and used for production of RUCONBAR – Rubberized Concrete Noise Barriers [3][12]. Steel fibres produced during mechanical recycling have different lengths and diameters since they origin from different types of tyre reinforcement, and most important they have irregular shapes. Although, they require development of special incorporation technologies due to their tendency to interlock, taking into account their attributes together with price they represent interesting alternative for industrially produced steel fibres [16–18]. Rubber particles are hard to totally dislodge from textile. This small amount of rubber incorporated can be positive for concrete characteristics since it not going to provide significant changes in concrete microstructure but will contribute to its ductility. Current research indicates that textile fibres can be used as a substitute for polypropylene fibres [10].

According to the requirements set in relevant standards for railway track system, applied concrete should have strength class at least C45/55, adequate capillary absorption, wear and freezing resistance, at the same time resisting to the significant dynamic loading present with each train pass. Insight in literature pointed good characteristics of fibre reinforced composite when such dynamic loadings are present. Although, fibre reinforced concrete with only recycled steel fibres has similar structural behaviour as ordinary fibre reinforced concrete [17–19] it is suggested to be used in higher amounts or blended with industrially produced fibres [13–15][16]. Compared to the ordinary concrete, fibres act as crack arrestors, giving substantial increase in toughness and better fatigue performance even when debonding and pulling out [20]. Besides mechanical characteristics durability is determined factor for reaching expected service life of the railway track system since appearance of cracks on the surface reduces service life of structure especially when one is exposed to the environmental and traffic loads [14][15]. Capability of fibres to restrain cracks, presents detrimental factor for longer service life and reduced maintenance. Performed accelerated corrosion investigation [21] indicated good resistance of concrete with only recycled steel fibres since no compressive or flexural behaviour of investigated material was reduced. Disadvantage of recycled steel fibres is certainly their intensive external corrosion causing negative judgments of such and similar materials. It should be pointed that this negative appearance can be reduced by applying different coatings; important is that this external corrosion has no progress in concretes internal microstructure.

Decreased energy absorption capacity of recycled fibres compared to the industrial ones can be further improved with addition of small amount of rubber particles in concrete mixture. Their low modulus of elasticity and high flexural strength additionally enhances concrete ductility and improves toughness and impact capability of recycled fibre reinforced concrete. Investigation of concrete behaviour with addition of steel beads confirms previous explanations [22]. Namely, steel beads consist of small amounts of rubber (cca 15%) and larger amount of recycled steel fibres (cca 75%), obtained during unfinished recycling process were steel fibres were not totally extracted from rubber. According to the results of performed investigation presence of steel beads in mixture increased its toughness and ductility, were cracking is controlled by present steel fibres altering post-cracking behaviour while rubber absorbs produced energy and enhances additionally concrete ductility [22]. Their significant disadvantage is interlocking during mixing process, leaving localized areas of high steel and rubber content. In order to avoid this negative consequence, others [19][20] performed investigation of positive synergy between rubber particles and industrially produced steel fibres. According to their investigation rubber particles act as crack arresters, giving rise to concrete exhibiting high strain capacity before macrocracking localisation while the fibres conferring the crack control mechanisms.

Having in mind the costs of industrially produced steel fibres and already demonstrated capability of recycled steel fibres as their viable alternative, the main objective of this research is to prepare innovative low cost rubberized hybrid fibre reinforced concrete. Insights in literature and previous experience with by-products from mechanical recycling of waste tyres indicated great need for extensive research on possible interaction between industrial and recycled steel fibres with or without incorporation of rubber particles depending of concrete intended application.

3 Design of the experiment

Significant effort was already undertaken to demonstrate the viability and importance of presented research. Results obtained during preliminary phase encompassed 20 different concrete mixtures and 16 tests indicating positive interaction between industrial and recycled steel fibres. Following hypotheses will be verified during the investigation:

- Saturated calcium hydroxide solution provides improvement of rubber adhesion to cement matrix, assuring production of hydrophilic groups on the rubber surface required for adequate degree of hydration on rubber/cement paste interface.
- Presence of mineral admixtures enables consumption of accumulated calcium hydroxide from rubber pre-treatment contributing to the hardening process by providing secondary CSH gel through pozzolanic reaction.
- Interaction of industrial and recycled steel fibres enables production of hybrid fibre reinforced concrete with adequate toughness and impact strength for special applications without influencing on other mechanical and durability properties.
- Addition of rubber particles assures improvement of ductility and post-cracking behaviour of hybrid fibre reinforced composite with blended industrial and recycled steel fibres, especially when air entraining admixtures are used. Other mechanical and durability properties are unchanged.
- Ageing of rubber and steel fibres during material exposure is minimised since they are surrounded with cement matrix and has no influence on the composite chemical and physical behaviour.

Stated hypothesis will be verified using following research procedures:

- Positive influence of rubber pre-treatment with saturated calcium hydroxide solution using: scanning electron microscope and EDS analysis, infrared spectroscopy and contact angle. Testing will be conducted on rubber and rubberized mortar samples, encompassing rubber without pre-treatment, rubber pre-treated with saturated sodium hydroxide and calcium hydroxide solutions. Investigation of mechanical and durability properties will also be conducted to determine pre-treatment influence on concrete properties, encompassing short and long term influence of used pre-treatment.
- Positive interaction of industrial and recycled steel fibres compared to the fibre reinforced concrete with only recycled fibres will be demonstrated using set of mechanical and durability (toughness, impact strength, wear resistance) testing at the age of 28 days.
- Additional enhancement of hybrid composite with rubber particles will be demonstrated using set of mechanical and durability (modulus of elasticity, toughness, impact strength and freezing resistance) testing at the age of 28 days.
- Influence of ageing of recycled fibres, rubber and composite will be determined through long term exposure of the specimens in salt chamber.

4 Preliminary results

Preliminary research was conducted along with project *Concrete track systems – ECOTRACK*. Specimens were due to limited conditions in laboratory, prepared in precast concrete plant (TBP Pojatno, Viadukt dd) which in one way restrained possibility to control all parameters. In order to minimise influences caused by robustness of the technologies used in precast concrete plant, all components were manually added in mixer.

More than 1000 concrete specimens were cast and prepared, in this paper only part of the research is presented (Table 1). Constituents for preparation of concrete incorporate: CEM II/BM SV 42.5 N, combination of crushed and alluvial aggregate, silica fume, superplasticizer (polycarboxylic ether hyperplasticiser) and air entraining admixture. Industrial fibres, 35 mm long with diameter of 0.55 mm and bent ends. Recycled steel fibres obtained during mechanical recycling of waste tires (irregular shape and dimension) and rubber particles (diameter 0.5 – 2 mm). Mixing procedure was adopted due to addition of rubber particles which were initially pre-treaded in saturated calcium hydroxide solution [4][10][12].

Table 1   Mixture composition

Mixture	Cement (kg)	Water (l)	Aggregate (kg)	Superplast. (kg)	Air entr. (kg)	S. fume (kg)	Ind. steel fibres (kg)	By-products (kg)	
								Fibres	Rubber
0I0RAG	420	170	1666	2,31	0,25	21	0	0	18,9
100I0R	420	170	1742	2,31	0	21	30	0	0
100I0RG	420	170	1656	2,31	0,00	21	30	0	18,9
100I0RA	420	170	1742	2,31	0,25	21	30	0	0
100I0RAG	420	170	1656	2,31	0,25	21	30	0	18,9
50I50R	420	170	1742	2,31	0	21	15	15	0
50I50RG	420	170	1656	2,31	0,00	21	15	15	18,9
50I50RA	420	170	1742	2,31	0,25	21	15	15	0
50I50RAG	420	170	1656	2,31	0,25	21	15	15	18,9

Abbreviation: I=industrial steel fibres; R=recycled steel fibres; G=rubber; A=air entraining admixture; 100I0RAG: 100% industrial fibres + 0% recycled fibres + air entraining admixture + rubber granulates

Comparison of different pre-treatments of rubber was performed; without pre-treatment, immersion in saturated sodium hydroxide solution [25][26] and saturated calcium hydroxide solution. No difference in concrete attributes was obtained as result of different pre-treatments, indicating that saturated sodium hydroxide solution could be replaced by calcium hydroxide solution providing similar concrete characteristics. Detailed analysis of interface microstructure and rubber surface (SEM and EDS analyses, infrared spectroscopy and determination of contact angle) will be performed to confirm positive influence of saturated calcium hydroxide on improved hydrophilic properties of rubber surface.

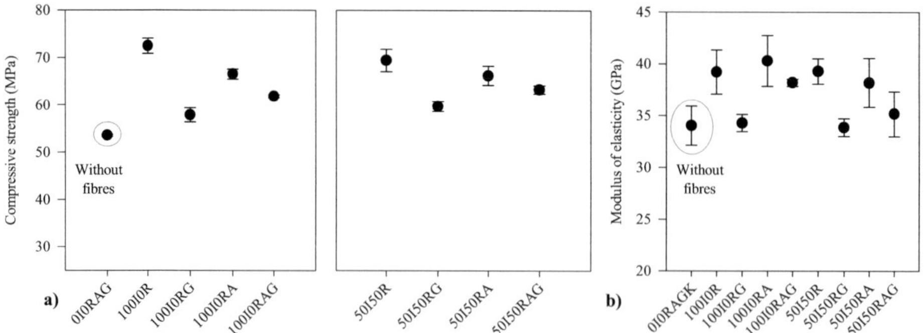

Fig. 1   Influence of different admixtures and fibre ratios of composite: a) compressive strength, b) modulus of elasticity

Positive interaction between industrial and recycled steel fibres was determined during analysis of mechanical properties of the composite. Results of compressive strength (Fig.1 a) indicate slight increase in compressive strength with presence of both fibre types what is in accordance to the literature data [21]. Although, capability of rubber and air entraining admixture to entrap air causes slight decrease of compressive strength, still high strength concretes are prepared (Fig.1 a).

Very low modulus of elasticity of rubber particles, from 25 to 25 000 times lower than modulus of elasticity of the aggregate [27] assures rubber capability to withstand large deformations, acting like springs inside the composite and delaying crack widening as well as catastrophic failures [28]. Although, modulus of elasticity of steel fibres is very high due to their low volume contribution (< 0,4% V) their presence in mixture has no influence on modulus of elasticity of the composite [29]. Even by combining industrial and recycled steel fibres no modifications in modulus were obtained since total volume of fibres in mixture was not changed (Fig.1 b).

Inclusion of fibres assures progress of concrete properties, such as improved energy absorption capacity, ductility and fatigue performance. Their capability to transfer stress, from damaged to un-damaged cross-sections, reduces brittleness of concrete. By reducing crack openings, improving ductility, energy absorption and post-cracking strength steel fibres represent optimum solution for materials used in structures exposed to cyclic dynamic loadings. Neocleous et al. 2011 [17] in their research investigated the use of recycled steel fibres from waste tires and according to the results recycled fibres present viable alternative to the industrially produced steel fibres, if used in higher amounts or blended with industrially produced fibres [16].

Performed investigations of flexural behaviour using third point loading test on specimens with notch, positive synergy between industrial and recycled fibres is demonstrated (Fig.2). Load – displacement curves imply that replacement of 50% of industrial with recycled steel fibres provides decrease of the surface area below the curve, when other admixtures are not present. This investigation is extended to determine positive synergy between hybrid fibre reinforced concrete (HFRC) and rubber particles in order to evaluate rubber contribution to improvement of composite ductility when recycled fibres are present since positive synergy of industrially produced steel fibres and rubber particles was already demonstrated [19][20]. According to the following figure (Fig.2), incorporation of equal content of industrial vs. recycled fibres together with addition of 5% of rubber particles to the total aggregate volume (50I50RG) assures similar composite behaviour as when ordinary FRC is used (100I0R). Thus, depending on requested concrete attributes (compressive strength, resistance to freezing and thawing, value of dynamic loadings etc) hybrid fibre reinforced concrete with or without addition of rubber can be prepared presenting economically and ecologically viable alternative to the ordinary FRC.

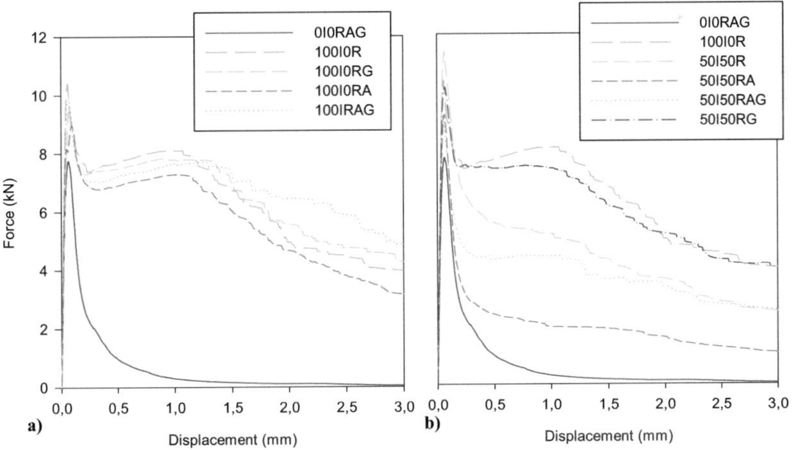

Fig. 2   Load-displacement curves for tested mixtures

5   Conclusions

Inadequate knowledge of by-products from mechanical recycling of waste tyres is currently found to be big obstacle for further use of investigated material in concrete industry. Conducted investigation implies that by additional optimization of concrete constituents, high strength fibre reinforced concrete with adequate properties could be prepared assuring significant cost reductions. Positive synergy between blended industrial and recycled steel fibres together with rubber particles was obtained, confirming that incorporated rubber serves as crack arrestor without affecting steel fibre capability to transfer stress across the crack. This is of special interest when freezing resistance is required and air entraining admixture is present; rubber particles due to their high energy absorption capacity assure better post-cracking behaviour of the composite. Although, testing of durability properties implied that HFRC with or without rubber can comply with all requirements from relevant standards here due to the limited space results are not presented. Following, rubberized hybrid fibre reinforced composite can be used in aggressive environments where its utilization for tailored applications could lead to significant environmental and economic benefits.

References

[1]   Council of the European Union, "Council Directive 1999/31/EC of 26 April 1999 on the landfill of waste," 1999.

[2]   I. Lakušić, S.; Bjegović, D.; Baričević, A.; Haladin, "Betoni visokih uporabnih svojstava kod kolosijeka za velike brzine," in Međunarodni simpozijum o istraživanjima i primeni savremenih dostignuća u građevinarstvu u oblasti materijala i konstrukcija, 2011, pp. 81-88.

[3]   S. Lakušić, D. Bjegović, A. Baričević, I. Haladin, and M. Serdar, "Apsorpcijska svojstva laganog betona s recikliranom gumom - RUCONBAR," in Međunarodni simpozijum o istraživanjima i primeni savremenih dostignuća u građevinarstvu u oblasti materijala i konstrukcija, 2011.

[4]   D. Bjegović, A. Baričević, and M. Serdar, "Durability properties of concrete with recycled waste tyres," in 12th International Conference on Durability of Building Materials and Components, 2011, pp. 1659-1667.

[5]   D. Bjegović, M. Serdar, and K. Opačak, "Primjena reciklirane gume u graditeljstvu," in INDIS 2009 Planiranje, projektovanje, građenje i obnova graditeljstva, 2009.

[6]   S. Lakusic, D. Bjegović, M. Serdar, and A. Baričević, "ECOBAR- betonske barijere za zaštitu od buke – inovativno rješenje," in Ekološki problemi prometnog razvoja, 2011, pp. 123-131.

[7]   S. Lakušić, D. Bjegović, A. Baričević, and I. Haladin, "Ekološke betonske barijere za zaštitu od buke - ECOBAR," in Projektiranje prometne infrastrukture, S. Lakušić, Ed. Zagreb: Građevinski fakultet Sveučilišta u Zagrebu, Zavod za prometnice, 2011, pp. 7-30.

[8]   A. Lakušić, Stjepan; Haladin, Ivo; Baričević, "Primjena reciklirane gume u proizvodnji apsorbirajućih betonskih barijere za zaštitu od buke," in BH KONGRES O ŽELJEZNICAMA, 2011.

[9]   D. Bjegović, S. Lakušić, and M. Serdar, "Primjena reciklirane gume na prometnicama," in Prometnice-nove tehnologije i materijali, Građevinski fakultet Sveučilišta u Zagrebu, Zavod za Prometnice, 2010, pp. 7-46.

[10]   D. Bjegovic, S. Lakusic, M. Serdar, and A. Baricevic, "Properties of concrete with components from waste tyre recycling," in Concrete Structures for Challenging Times, 2010, pp. 134-140.

[11]   Univesity of Zagreb Faculty of Civil Engineering, "Conrete railway tracks - ECOTRACK," 2011.

[12]   Univesity of Zagreb Faculty of Civil Engineering, "Rubberized concrete noise barriers - RUCONBAR," 2011.

[13]   D. Bjegovic, A. Baricevic, and S. Lakusic, "RUBBERIZED HYBRID FIBRE REINFORCED CONCRETE," in 2nd International Conference on Microstructure Related Durability of Cementitious Composites, 2012.

[14]   D. Bjegovic, A. Baricevic, and S. Lakusic, "Innovative low cost fiber-reinforced concrete. Part I: Mechanical and durability properties," in 3rd International Conference on Concrete Repair, Rehabilitation and Retrofitting, 2012.

[15]   J. Krolo, D. Bjegovic, D. Damjanovic, I. Duvnjak, and A. Baricevic, "Innovative low cost fibre-reinforced concrete. Part II: Fracture toughness and impact strength," in 3rd International Conference on Concrete Repair, Rehabilitation and Retrofitting, 2012.

[16]   C. Achilleos, D. Hadjimitsis, K. Neocleous, K. Pilakoutas, P. O. Neophytou, and S. Kallis, "Proportioning of Steel Fibre Reinforced Concrete Mixes for Pavement Construction and Their Impact on Environment and Cost," Sustainability, vol. 3, no. 7, pp. 965-983, Jul. 2011.

[17]   K. Neocleous, H. Angelakopoulos, K. Pilakoutas, and M. Guadagnini, "Fibre- reinforced roller-compacted concrete transport pavements," Proceeding of the ICE - Transport, vol. 164, no. 2, pp. 97-109, 2011.

[18]   M. A. Aiello, F. Leuzzi, G. Centonze, and A. Maffezzoli, "Use of steel fibres recovered from waste tyres as reinforcement in concrete: pull-out behaviour, compressive and flexural strength.," Waste management (New York, N.Y.), vol. 29, no. 6, pp. 1960-70, Jun. 2009.

[19]   K. Pilakoutas, K. Neocleous, and H. Tlemat, "Reuse of tyre steel fibres as concrete reinforcement," Proceedings of the ICE: Engineering Sustainability, vol. 157, no. 3, pp. 131-138, Feb. 2004.

[20]   H. Tlemat, K. Pilakoutas, K. Neocleous, and K. (2006) Tlemat, H., Pilakoutas, K., Neocleous, "Modelling of SFRC using inverse finite element analysis," Materials and Structures, vol. 39, no. 2005, pp. 365-377, 2006.

[21]   A. Graeff, K. Pilakoutas, C. Lynsdake, and K. Neocleous, "Corrosion Durability of Recycled Steel Fibre Reinforced Concrete," Intersections, vol. 6, no. 7, pp. 77-89, 2009.

[22]   C. G. Papakonstantinou and M. J. Tobolski, "Use of waste tire steel beads in Portland cement concrete," Cement and Concrete Research, vol. 36, no. 9, pp. 1686-1691, Sep. 2006.

[23]   A. Turatsinze, S. Bonnet, and J. Granju, "Mechanical characterisation of cement-based mortar incorporating rubber aggregates from recycled worn tyres," Building and Environment, vol. 40, no. 2, pp. 221-226, Feb. 2005.

[24]   A. Turatsinze, J. Granju, and S. Bonnet, "Positive synergy between steel-fibres and rubber aggregates: Effect on the resistance of cement-based mortars to shrinkage cracking," Cement and Concrete Research, vol. 36, no. 9, pp. 1692-1697, Sep. 2006.

[25]   L.-H. Chou, J.-R. Chang, and M.-T. Lee, "Use of waste rubber as concrete additive," Waste Management Resources, vol. 25, pp. 68-76, 2007.

[26]   N. Segre, P. J. M. Monteiro, and G. Sposito, "Surface characterization of recycled tire rubber to be used in cement paste matrix.," Journal of colloid and interface science, vol. 248, no. 2, pp. 521-3, Apr. 2002.

[27]   M. Khorami, E. Ganjian, and A. Vafaii, "Mechanical Properties of Concrete with Waste Tire Rubbers as Coarse Aggregates," in Proceeding of Special sections on International conference on Sustainable Construction Materials and Technologies, 2007, pp. 85-90.

[28]   W. A. Aules, "Utilization of Crumb Rubber as Partial Replacement in Sand for Cement Mortar," European Journal of Scientific Research, vol. 51, no. 2, pp. 203-210, 2011.

[29]   L. Domagała, "Modification of properties of structural lightweight concrete with steel fibres," Journal of Civil Engineering and Management, vol. 17, no. 1, pp. 36-41, 2011.

Monitoring and Repair of Building Structures

Evaluation of the bridge Champangshiel by using static assessment methods

Frank Scherbaum, Jean Mahowald

Faculté des Sciences, de la Technologie et de la Communication,
Université du Luxembourg,
6, rue Richard Coudenhove-Kalergi, L-1359 Luxembourg, Luxembourg
Supervisor: Danièle Waldmann

Abstract

The evaluation of existing buildings regarding their bearing capacity and serviceability is very important today. Especially for bridges, which become increasingly older, the assessment of the remaining useful life and the detection of possible damages become more and more relevant. The investigation of the current state of a structure is normally based on visual inspections and continuous monitoring, which are costly and time consuming.

Therefore, the University of Luxembourg carries out a project to investigate an efficient application of different assessment methods taking into account praxis relevant test conditions. One part of this project is the practicability use of non-destructive testing methods to investigate bridge structures. These results are compared to the evaluation of a structure by in situ loading tests.

Within this paper the results of in situ loading tests of a two span box girder bridge in Luxembourg, which was destined to demolition after the performed tests due to urban transformation, are analysed in order to describe the approach. Different damage scenarios, realized by cutting of a defined number of tendons, were elaborated and, for each damage scenario, static load tests performed. The results of these loading tests will be presented by analysing the load-deformation behaviour.

1 Introduction

Today, more and more buildings come to the end of their calculated utilisation period. But normally the local authority has not enough money to replace these old constructions precautionary with new ones. So bridge inspections and the assessment of the current state of the structure become more and more important today. In the past the current status of a bridge was determinate by periodic visual inspections, by a continuous monitoring and if necessary by an object-related damage analysis together with a static calculation. However this kind of investigation and especially the detailed inspection are time consuming and cost intensive. For that reason, the University of Luxembourg investigates an efficient non-destructive way of bridge inspection by using static, dynamic and non-destructive testing methods.

As a part of this project, the University of Luxembourg investigates efficient use of in-situ load tests to evaluate the condition of a construction. Normally in-situ tests are used to determine the maximum load bearing capacity of a construction. In this project, in-situ load tests are used to detect a damage of a structure. To reach this goal the University got the possibility to damage a structure and investigate its structural behaviour with increasing damage. The here analysed bridge, which is called Champangshiel, was destined to demolition after the performed tests due to urban transformation [1].

1.1 Bridge description

The investigated bridge is a prestressed concrete box girder bridge, which was built from 1965 to 1966 (Fig. 1). It is a straight longitudinal bridge with a total length of 103 m, which is separated in two different spans. The large span is 65 m and the short one 37 m long between the axis of the bearings. For the static calculation a total length of 102 m is used (Fig. 2). Primarily the superstructure was made of prestessed concrete including 32 parabolic, 24 upper straight lined and 20 lower straight lined tendons. Additional to these tendons, 56 external prestressed steel cables were added inside the box girder in the large field in 1987 (Fig. 3).

Proc. of the 9th *fib* International PhD Symposium in Civil Engineering, July 22 to 25, 2012,
Karlsruhe Institute of Technology (KIT), Germany, H. S. Müller, M. Haist, F. Acosta (Eds.),
KIT Scientific Publishing, Karlsruhe, Germany, ISBN 978-3-86644-858-2

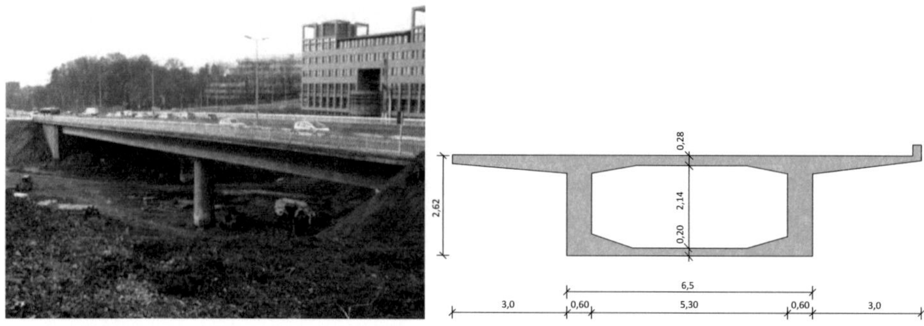

Fig. 1   Side view of the bridge (left) and cross section (right)

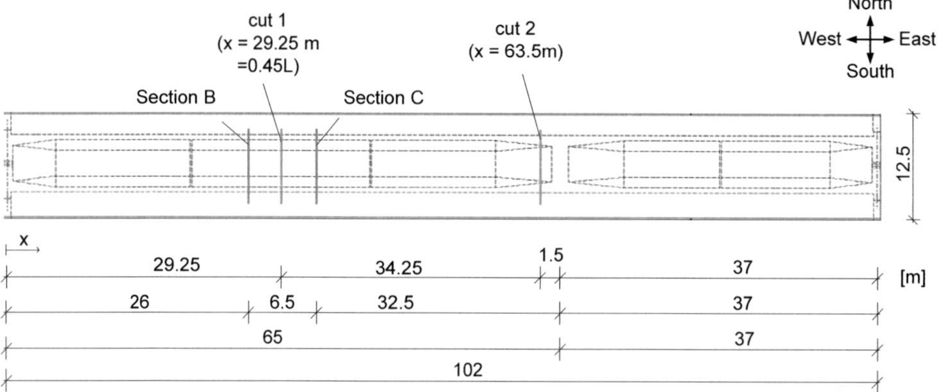

Fig. 2   Longitudinal section of the bridge

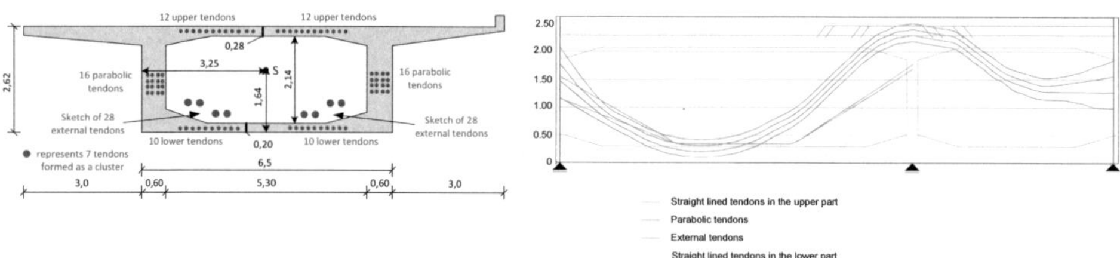

Fig. 3   Cross section of the box girder with the description of the tendons (left) and longitudinal schematic view of the prestressed tendons (right)

2   Executed tests

To evaluate the possibility of damage detection by using an in-situ load test, the superstructure is gradually damaged. The artificial damage is produced by cutting a defined part of the prestressed tendons. This should simulate a local loss of pretension and a local sectional weakening. During the whole test period, four different damage scenarios are realized to observe the reaction of the bridge to an increasing damage (Table 1). These damages are created at different positions of the superstructure to investigate, if this influences the detectability of damage. For each damage scenario an in-situ load test is accomplished. For a constant load in each damage scenario, the superstructure is loaded by 38 steel beams, which means 245 t. The centre of this experimental load is positioned in x = 0.45•L (Figure 4), which corresponds to the section where the first damage is created. The experimental load is removed before the next damage is created.

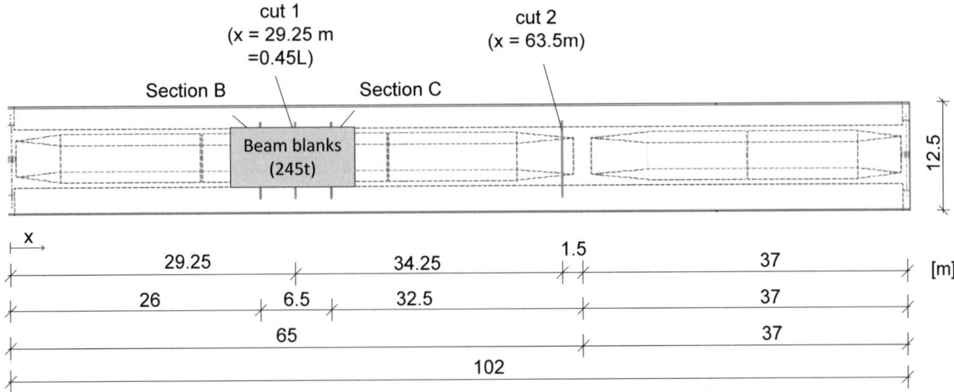

Fig. 4   Position of the steel beams

Before the first damage could be created, the initial state of the superstructure was measured. Therefore, a first in-situ load test is done in the undamaged state (damage scenario #0).

The first damage (damages scenario #1) is created by cutting 20 straight lined tendons in the lower part of the superstructure. The position of this damage is in the large field at x = 0.45•L (cut 1 in Figure 2). As second damage scenario, eight straight lined tendons in the upper part of the superstructure are cut (damage scenario #2). The tendons are cut 1.5 m away from the pylon axis (cut 2 in Figure 2). The third damage state (damage scenario #3) is created by cutting the external tendons. The tendons are cut by a flame cutter from the upper side of the superstructure through a hole in the top plate of the box girder (at the position of cut 2). Finally, a fourth damage scenario (damage scenario #4) is created by cutting the last 16 straight lined tendons in the upper part of the bridge and 8 parabolic tendons. This damage is also produced from the upper side in the section of cut 2. In the following evaluation damage scenario #4 is not considered because the displacement transducers are removed after the third damage scenario (damage scenario 3). All damage scenarios are summarized in table 1.

Table 1   Description of the damage scenarios

Damage scenario	Damage
#0	Undamaged state
#1	Cutting all 20 straight lined tendons in the lower part of the bridge (bottom plate of the box girder) at 0.45•L
#2	Additional cutting of 8 straight lined tendons in the upper part of the bridge (top plate of the box girder) over the pylon
#3	Additional cutting of the external tendons
#4	Additional cutting of 16 straight lined tendons in the upper part of the bridge (top plate of the box girder) and also 8 parabolic tendons

3   In-situ load test

3.1   Load cases

For each damage scenario, the loading and unloading of the superstructure define already two different load cases. Furthermore, respectively a load case is related to the time-dependent behaviour of the bridge after each cutting, loading and unloading. Table 2 describes these five different load cases for each damage state.

Table 2   Description of the load cases

Load case	Damage scenario	Description	Load case	Damage scenario	Description
#0	#0	Initial measurement	#2	#2	Measurement after the second cut
	#0-L	Measurement after the bridge was loaded		#2-Cr	Measurement 19 h after the second cut
	#0-L-Cr	Measurement 19 h after the bridge was loaded		#2-L	Measurement after the bridge was loaded
	#0-U	Measurement after the bridge was unloaded		#2-L-Cr	Measurement 18 h after the bridge was loaded
	#0-U-Cr	Measurement 30 minutes after the bridge was unloaded		#2-U	Measurement after the bridge was unloaded
#1	#1	Measurement after the first cut		#2-U-Cr	Measurement 27 h after the bridge was unloaded
	#1-Cr	Measurement 44 h after the first cut	#3	#3	Measurement after the third cut
	#1-L	Measurement after the bridge was loaded		#3-Cr	Measurement 19 h after the third cut
	#1-L-Cr	Measurement 19 h after the bridge was loaded		#3-L	Measurement after the bridge was loaded
	#1-U	Measurement after the bridge was unloaded		#3-L-Cr	Measurement 20 h after the bridge was loaded
	#1-U-Cr	Measurement 94 h after4 the bridge was unloaded		#3-U	Measurement after the bridge was unloaded
				#3-U-Cr	Measurement 20 h after the bridge was unloaded
			#4	#4	Measurement after the fourth cut

L = loaded state
U = unloaded state
Cr = consideration of the time-dependent behaviour

3.2   Additional cracks in the superstructure

Additional to and as a function of the artificial damage, cracks in the superstructure can be observed during the tests. The first cracks appear in load case #1, during the cutting of the tendons, and are limited to the bottom plate (Figure 5). The first bending crack can be observed after the structure is loaded in this damage state (load case #1-L). In this load case a crack can be seen in both sidewalls of the box girder (Figure 5).

In damage scenario #2 no crack can be visually detected. The next cracks appear in damage scenario #3, during cutting the external tendons. More cracks in the bottom plate and especially in the sidewalls can be seen (Figure 5). The first cracks in the top plate appear in damage scenario #4.

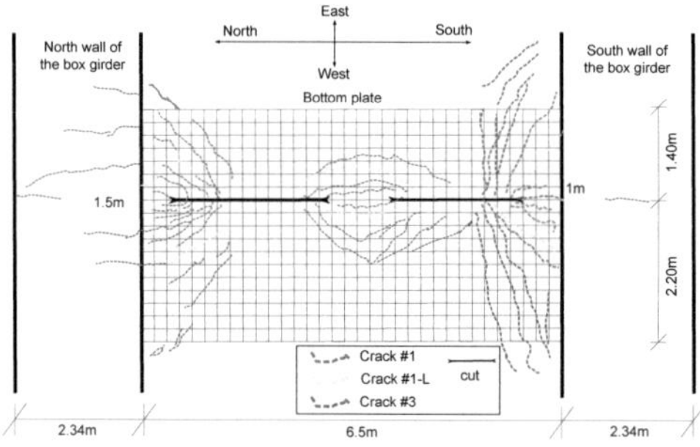

Fig 5   sketch of the cracks after damage scenario #1, #1-L and #3 at 0.45L [1]

3.3   Vertical deflection

To evaluate the structural behaviour, the vertical deflections of the superstructure are measured by two displacement transducers (S_{axisB} and S_{axisC}) in two sections (section B and section C) during the whole test period. Section B is the position of the maximum bending moment at x = 0.40•L. Section C is the middle of the large field (x = 0.50•L). Additionally, the vertical deflection is measured at six sections (section A to F) by a digital levelling (Figure 6). This kind of measurement is executed for several times (one measurement for each load case).

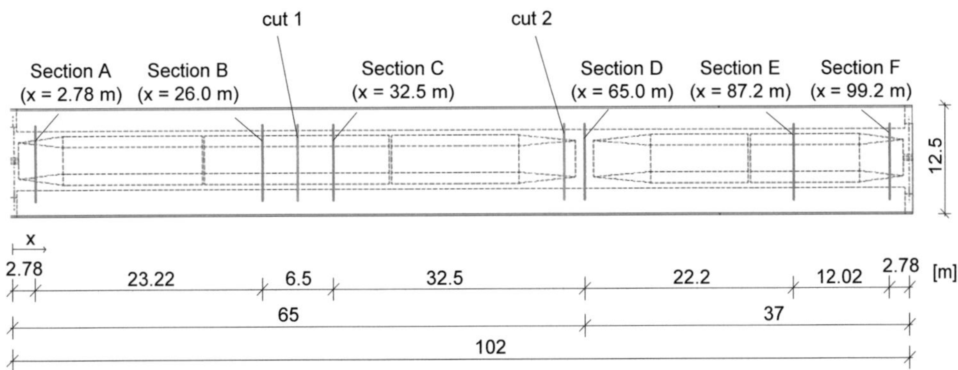

Fig. 6   Position of the measurement sections

Figure 7 illustrates the vertical deflection measured by the displacement transducers for the whole test period including the temperature variations. The two upper lines present the deflection in section B (S_{axisB}) and section C (S_{axisC}). The four lower lines present the variation of the air temperature and the variation of the temperature of the construction measured in section C. The construction temperature is measured in the top plate, the bottom plate and both sidewalls of the box girder.

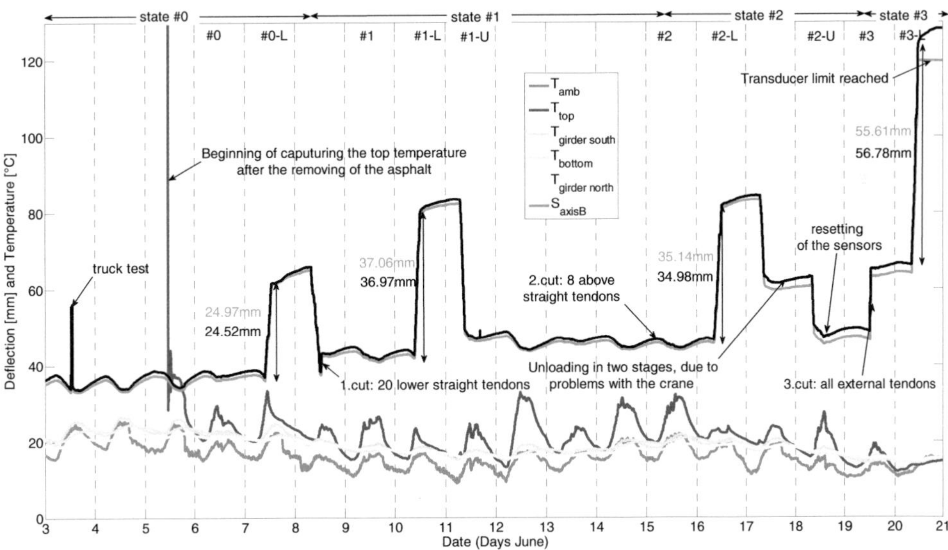

Fig. 7   Vertical deflection measured by a displacement transducer in section B (S_{axisB}) and C (S_{axisC}) and temperature changes during the test period

Figure 7 presents that every loading in each damage scenario lead to an increase of the vertical deflection. In the undamaged state, the loading leads to an increase of 24.97 mm in section B. In damage scenario #1 the vertical deflection increases of 37.06 mm as result of the loading. So, after the first damage and after the first observed cracks, the vertical deflection resulting from the loading is

12.09 mm higher than the deflection in the undamaged state. The vertical deflection in damage scenario #2 (when no crack could be detected) increases of 35.14 mm due to loading, which is less than in damage scenario #1. In damage scenario #3, the loading leads to an increase of the vertical deflection of 55.61 mm. In this scenario more cracks in the bottom plate and in the sidewalls can be observed.

A view to the vertical deflection measured by digital levelling in section A to F shows the same than the displacement transducer. Figure 8 illustrates the vertical deflection curve of the superstructure measured by digital levelling for the loaded structure.

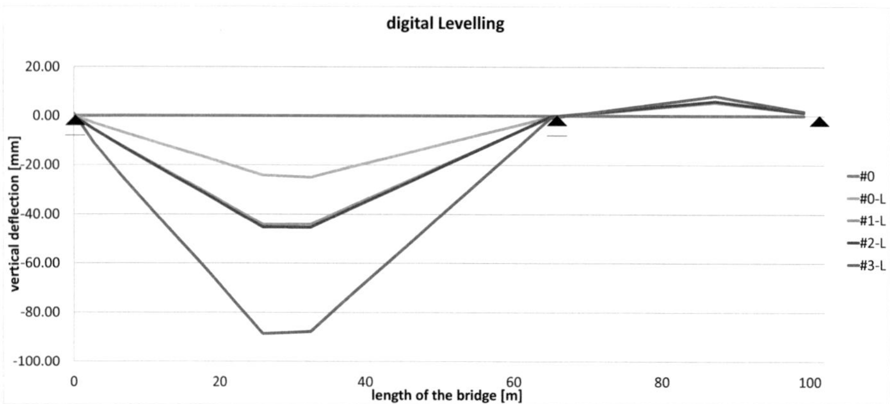

Fig. 8   schematic vertical deflection curve for the loaded superstructure measured by digital levelling

It can be seen that there is nearly no difference between the deflection curve of load case #1-L and #2-L.

The analysis of the vertical deflection in the unloaded state shows a clearly increase due to damage scenario #1 and #3 (Figure 7). In contrast to this, the artificial damage scenario #2 does not lead to an increase of the vertical deflection. Figure 7 also presents that the vertical deflection, which results from temperature changing, is larger than the vertical deflection of such a small damage like damage scenario #2.

4   Conclusions

It has been shown, that high damages like damage scenario #1 or #3, which leads to cracks in the superstructure, could be detected by measuring the vertical deflection. The loading of the superstructure amplify this effect. However, a small damage like damage scenario #2 could not be detected. Here the influence of the temperature changing could be higher, so that it is important to know its influence prior to all test result evaluations. This means, that the knowledge about the behaviour of the structure for different temperatures is necessary.

References

[1]   Scherbaum, F. & Mahowald, J.: Report Bridge Champangshiehl 2. University of Luxembourg (2011)

[2]   Belastungsversuche an Betonbauteilen, DAfStb-Richtlinie im DIN Deutsches Institut für Normung e.V. Berlin

[3]   Bungard, V.; Waltering, M.; Waldmann, D.; Maas, S. & Zürbes, A.: Comparison of static behaviour and nonlinear vibration characteristics of gradually damaged prestressed concrete slabs and reinforced concrete beams. Proceeding of the EVACES 09 conference. Worclaw. Poland. (2009)

A quick in-situ non-destructive test method for external prestressed tendons and stay cables

Steffen Siegel

*Institute of Concrete Structures and Building Materials,
Karlsruhe Institute of Technology (KIT),
Gotthard-Franz-Straße 3, 76131 Karlsruhe, Germany
Supervisor: Lothar Stempniewski*

Abstract

In Germany the use of external tendons is mandatory since 1998 for prestressed concrete bridges with box-type construction in the area of responsibility of the Federation as standard building method. In order to assess the current state of the built-in external tendons and stay cables, various non-destructive testing methods are available with which damage or a decrease in the prestressing force can be detected. All currently existing techniques are so time-consuming that a routine use in situ is not economically feasible. The extent of a routine measurement cannot primarily have the aim to determine the absolute change of the tension force referring to the initial state, especially since the initial state is not known in most cases. Rather, the measurement should provide clear indicators that the condition of the tendon is located inside or outside of the confidence interval. Therefore, a rapid in-situ test has been developed which proceeds only by the following conditions: The initial tension force is not known; the free tendon length is not determined exactly and creep and shrinkage is not of particular interest. Assuming these conditions, it is sufficient to measure only the frequency of the tendons and not make further in-depth considerations. For the pure determination of the frequency, these points are not necessary and above all the time required for these measurements is a few minutes. If you draw the periodic tests in a database, you get a picture of the changes that have occurred in the meantime. In order to classify the results, a classification of the changes has to be done. To this intention, all the frequencies of all tendon sections are normalized to the respective very first value of each section. This allows that all the tendons of the same superstructure can be compared.

1 Research Project

The goal of the research project [10] is to specify criteria and recommendations regarding the selection of testing methods for external tendons and stay cables with parallel strands. Thereby assistance is given to the bridge inspectors, which refers to possible visible damages and suggests – after finding these damages at external tendons and stay cables – in each case to appropriate precautions and necessary work step for further measures. By means of a clear instruction for procedure, an assessment of damages can take place via the bridge inspector at as small expenditure as possible. In dependence of different criteria, like e.g. the existing tendon system, the existing visible damages, the measuring accuracy aspired and not least the expenditure, one is ready to be invested, one or several inspection procedures and measuring procedures as selection are depicted.

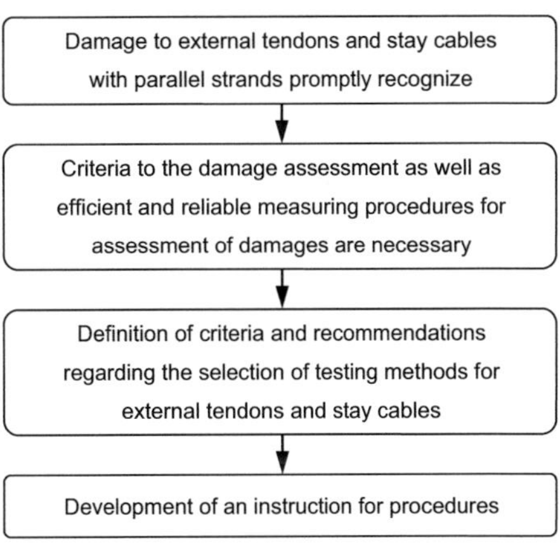

Fig. 1   Research objective.

Proc. of the 9th *fib* International PhD Symposium in Civil Engineering, July 22 to 25, 2012,
Karlsruhe Institute of Technology (KIT), Germany, H. S. Müller, M. Haist, F. Acosta (Eds.),
KIT Scientific Publishing, Karlsruhe, Germany, ISBN 978-3-86644-858-2

2 External tendons used in Germany

The external prestressing acts around tendons for prestressing without bond, which lie outside of the concrete, but within the concrete cross section. The external prestressing was used in 1936 for the first time in Germany after a draft Dischingers, but in the next 50 years the internal prestressing with bond was preferred [1].

In the mid of the eighties the external prestressing was again discovered and continued to develop after discovering not injected tendons at internal tendons. The external prestressing allows a simple visual inspection and can be retightened or replaced if necessary. Table 1 gives an overview of the most important characteristics of the external prestressing systems used in Germany.

Table 1   External tendons used in Germany.

Company	SUSPA	BBV	BBV	BBR VT	VBF	VBT
Typ	Draht EX	Typ E	Typ EMR	CMM D CMM KD	CMMD	BE
Prestressing steel	Wires	Strands	Strands	Strands	Strands	
Cross section						
Corrosion protection	Duct (HDPE), grease / mortar	Duct (HDPE), grease	Strand, duct strand (HDPE), outer duct (HDPE)			
Maximum prestressing force [kN]	2970	6278	3531	3532	2974	

3 Bundles of parallel strands

Contrary to full-locked cables (see fig. 2) bundles of parallel strands are stay cables, which are composed of parallel running mono strands and an additional outside PE spiral coil.

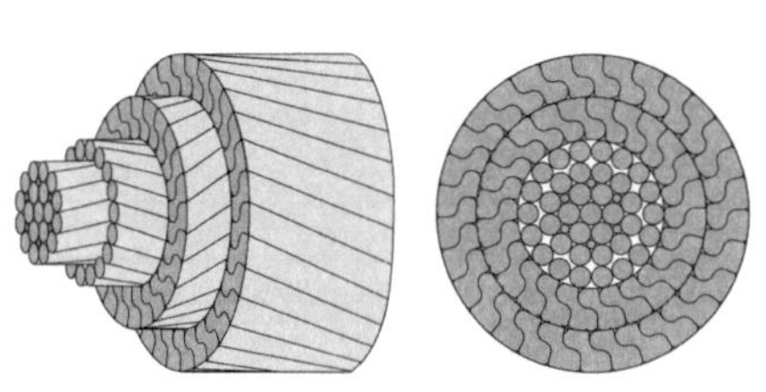

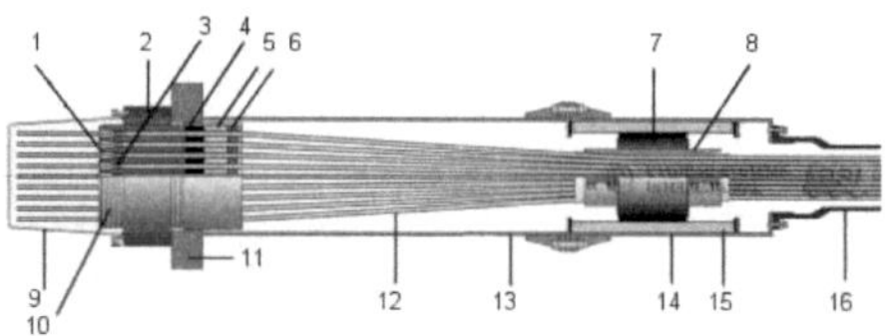

1. Wedges, 2. Ring Nut, 3. Compression Tubes, 4. Sealing Plates, 5. Spacer, 6. Compression Plate, 7. Elastomeric Bearing, 8. Clamp, 9. Cap, 10. Anchor Block, 11. Bearing Plate, 12. Strands, 13. Recess Pipe, 14. Exit Pipe, 15. Filler Material, 16. HDPE Sheathing

Fig. 2   Full locked cable (left), DYNA Grip® anchorage (right)

4 Damage to the prestressing steel

The following kinds of damages can occur to prestressing steels without bond [12]:
- Brittle failure due to local corrosion attack,
- Break due to hydrogen-induced stress corrosion,
- Brittle failure due to a hydrogen brittleness.

The following influences favour the occurrence of defects:

- High notch sensitivity of the prestressing steel,
- Penetration from corrosion-promoting substances,
- Execution error (like e.g. injecting error),
- Foreign effects.

5 Testing methods

In the context of a building examination the individual construction units of the tendons and stay cables including the anchorage and the returning element are to be examined visually for their actual condition. With the occurrence of damages the bridge inspectors have different aids (e.g. endoscope, layer thickness measuring instrument, crack ruler) on-hand, in order to examine the extent of the damage more exactly. If the available aids are not sufficient, then a subject-related damage analysis [6] has to be accomplished. In this context complex technical testing methods become more important constantly. For this, numerous non destructive or destruction-poor testing methods are available [2], [9], [11], which are to be selected for the respective case apart from the aspect of the applicability also regarding the assumed damage form and the developing expenditure.

5.1 Determination of the prestressing force

With the knowledge of the stress in the tendons the condition of the tendons can be judged, before clear damage to the superstructure or the tendon are visible from the outside. In the context of this research project the tested methods are introduced in the following.

5.1.1 Using the tensioning jack

One possibility for the examination of the stress is to apply a tensioning jack. Conditions for this are the free accessibility of the anchor and the presence of a continuous plastic corrosion protection. Setting the tensioning jack at a wire system such as e.g. SUSPA-Draht EX is possible at any time and causes no damage to the tendon system. On the other hand setting of the tensioning jack at strand systems is possible only with a sufficient tendon length behind the anchorage. Apart from a large temporal and financial expenditure the danger of an inadvertent damage of the anchorage or the cable exists due to repeated wedge bite.

Fig. 3   Tensioning jack (left) and jack for several strands (right).

Due to the large expenditure of time to bring the tensioning jack to their place there is the possibility of accomplishing a lift-off-test with several strands. Also a sufficiently large strand projection is necessary for this. The jack is pushed onto every several strand and the necessary pressure is applied with a hand pump, in order to pull the strand out easily.

5.1.2 Vibration measurement

During the stress determination by means of dynamic measurements the oscillations of a tendon are seized and the natural frequencies are determined. With the further knowledge of the diameter, the cable length, the cable mass and the modulus of elasticity of the prestressing steel the tension in the steel can be determined. Considering the conditions of the bearings and sag of the tendon the difference compared with the stress measured by tensioning jack is less than 3% [7]. If the sag of the cables is too large, this procedure is not applicable. This is the case with the main cables from suspension bridges. Usually the tendon is oscillated with a hammer. With an electrical vibration absorber (in general acceleration sensor) and an interpretation device (frequency analyzer) the signals are seized for ascertaining the frequency.

Generally the traction power P in [N] of an ideally tautly strained wire with articulated storage results over the connection between the measured eigenfrequency $f_{k,string}$ with the order k, the oscillation length of the tendon L_S and the mass m per meter of length:

$$P = \left(\frac{2 \cdot f_{k,string} \cdot L_S}{k} \right)^2 \cdot m \tag{1}$$

5.1.3 Magnetoelastic sensors

Over the so-called field coil an alternating current is fed, which induces a tension in the coil. From these values the magnetic characteristics of the tendon can be determined. At measuring instruments, which are based on this method, the tension values or force values of the strands can be read off directly.

5.2 Damage detection

Usual testing methods for the collection of the damage can be divided into optical, acoustic, electromagnetic, magnetic and chemical procedures regarding their operational principles. The tested procedures are represented in the following.

5.2.1 Magnetic detection of prestressing steel breaks

The magnetic scatter field procedure uses the ferromagnetic characteristics of the prestressing steel for the non destructive localization of prestressing steel breaks. The tendon which can be examined is first remanent magnetized. Following the transversal component (orthogonal to the tendon arranged) of the magnetic flow density along the tendon is measured. Tension wire breaks produce a characteristic signal process.

5.2.2 Ultrasound

Sound with frequencies between 20 kHz and 1 GHz is called ultrasonic [11]. With the ultrasonic echo procedure or also ultrasonic pulse echo method ultrasonic waves will transfer from the transmitter into the construction unit which has to be examined. The acoustic waves are reflected, strewn and bent from boundary surfaces of materials of different acoustic impedance (wave velocity multiplied by density). At the transition to air the reflection is the largest (total reflexion). From the intensity and running time of the reflected waves the distance of the reflection place. By scanning the extents of an inhomogeneity can be determined.

5.2.3 High frequency resonance measurement

With the electrical reflectometry high frequency pulsed voltages at opened prestressing steel ends are linked and the signal over an evaluation of the time functions, reflected at damages, is seized [8]. By systematic changes of the frequency resonance features can be determined, from which the entire tendon length is determined. A prestressing steel break shows itself from an increased distance of neighbouring resonances, so that the computational tendon length shortens. In extensive investigations it could be stated among other things that this method cannot be used if prestressing steels are in contact like e.g. SUSPA-Draht EX [5]. In case of mono strands this method is however applicable.

5.2.4 Visual inspection

External tendons inside a bridge (box-type construction) are visually controllable with the naked eye. Stay cables can be examined by using an automatic trolley at which 4 high speed cameras are fixed. Rope diameters between 32 and 350 mm and rope lengths up to 300 m can be examined with a speed of up to 0.4 m/s.

6 Development of an in situ quick test

The existing test methods can only be used in certain sections of a tendon. Therefore several methods must be used to be able to exclude or detect any damage. Only the determination of the tensioning force using vibration measurement covers the entire length of a tendon or cable, if one measures in each field between the deflection points.

To achieve more precise calculation results, it is not sufficient to determine the tensioning force with the formula of the vibrating string (equation 1). It is necessary to use higher natural frequencies and the stiffness of the cable must be considered also (equation 2).

$$f_{k,cable} = f_{k,string}\left(1 + 2 \cdot \sqrt{\frac{E \cdot I}{P \cdot L_S^2}} + \left(4 + \frac{k^2 \cdot \pi^2}{2}\right)\frac{E \cdot I}{P \cdot L_S^2}\right)^2 \tag{2}$$

It is clearly recognizable that the expenditure is relatively high in order to calculate the stiffness of an external tendon and stay cable, which is composed of one to two HDPE ducts, prestressing steel (strands or wires) and a corrosion protection (grease, mortar, etc.). Also, the vibration length has to be determined from deflected tendons complicated, because of the deviation point within the deflection is not known. In addition, the mass of the tendon is required per running meter. Because duct diameter, thickness, prestressing steel weight and corrosion protection varies from bridge to bridge, the determination of the actually existing weight is very difficult or impossible, especially there is not always known according to which admission the tendons are used and for example the dimensions of the duct may vary from admission to admission. To classify the calculated tension force accurately, the tension force applied initially must be known, and after the tensioning a zero measurement has to be performed and the losses due to creep, shrinkage and relaxation must be considered in further measurements.

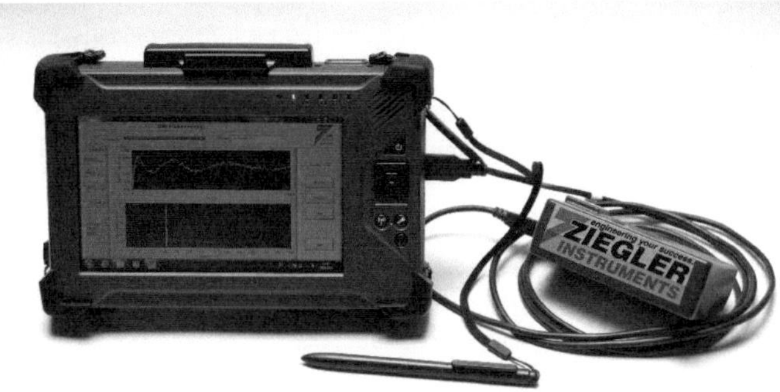

Fig. 4   Measuring instrument.

An invention of the Institute for Concrete and Building Materials (IMB) enables a quick test for monitoring of external tendons. The test is based on the fact that the eigenfrequencies of all the tendons in a bridge can be adapted to each other mathematically. At repeated measurements the tendons are filtered out automatically if an irregular change of state has taken place in the meantime. The analysis of the signals is designed so that the result after the measurement is available on site immediately. With current conventional methods several days or weeks are needed for measurement and evaluation. The new invention reduces the testing time and costs by a factor of five. In addition, the bridge inspector can perform himself the measurements and evaluations and no specialist in measurement and evaluation is required, which is especially in terms of costs a great advantage. The system is universally applicable and also suitable for the inspection of cable-stayed bridges. Currently in a pilot project the external tendons of selected prestressed concrete bridges are examined. Currently, the IMB also develops together with an industrial partner a mobile measuring instrument and a database system for better processing of the measured test data.

7   Conclusions

In laboratory and in-situ tests different non-destructive or destruction-poor testing methods were checked with regard to their practicability. The tested techniques are characterized by different pros and cons. Like that the practical use is given to the existing stress with a regulation by means of dynamic vibration measurement due to the fast measurements with small deviations by far more than those very time and transport-complex determination of the stress by using the tensioning jack.

Ultrasound measurements are a relatively fast method particularly in the range of the anchorages due to finding defects with separate running strands, in order to examine the tendon in the anchor range with suspicion of damage.

At measurements with the magnetic scattering field procedure there is the possibility of examining tendons within executions on cross section reductions. The free length is driven off completely with the magnet and a possibly existing defect is thus detected centimetre-exactly.

The new developed measuring instrument is contrary to the test procedure on the market by the bridge inspector itself applicable and delivers locally a statement about the condition of the stay cables or tendons. With the automated analysis and data storage a measuring instrument is given to the bridge inspector. So tendons and other guyed structures can be checked quickly and inexpensively.

8 References

[1] Baumann, T.: Vorspannung von Brücken – Anwendung interner und externer Spannglieder. Beton- und Stahlbetonbau, Heft 11: 646-656, Berlin: Ernst & Sohn, 2000.

[2] Diem, P.: Zerstörungsfreie Prüfmethoden für das Bauwesen. Bauverlag GmbH, Wiesbaden und Berlin, 1982.

[3] Holst, A. and Wichmann H.-J.: HF-Untersuchungen mit dem Verfahren der Elektromagnetischen Resonanzmessung an externen Spannbändern der Talbrücke Rümmecke, 2007.

[4] Holst, A. and Wichmann H.-J.: Einbau und Kalibrierung magnetoelastischer Kraftsensoren an Spannbändern der Talbrücke Rümmecke, 2007.

[5] Mietz, J. and Fischer, J.: Verifizierung zerstörungsfreier Prüfverfahren zur Detektion von Spannstahlschäden an Spannbetonbauteilen mit nachträglichem Verbund. Abschlussbericht zum Forschungsvorhaben P 32-5-7.207-1015/02, Fraunhofer IRB Verlag, 2005.

[6] OSA. Leitfaden Objektbezogene Schadensanalyse. Bundesanstalt für Straßenwesen (BASt), 2004.

[7] Riad, K.: Überwachung der Vorspannkraft Externer Spannglieder mit Hilfe der Modalanalyse. Dissertation, Fachgebiet Massivbau, Universität Kassel, 2006.

[8] Scheel, H.: Spannstahlbruchortung an Spannbetonbauteilen mit nachträglichem Verbund unter Ausnutzung des Remanenzmagnetismus. Dissertation, Fachbereich 9, Technische Universität Berlin, 1997.

[9] Schickert, G., Krause, M. and Wiggenhauser, H.: ZfP Bau-Kompendium. (www.bam.de), 2004.

[10] Stempniewski, L., Siegel, S., Möller, J., Kiefer, D.: Verfahren zur Prüfung des Zustandes von externen Spanngliedern und Schrägseilen. Bundesanstalt für Straßenwesen (BASt), 2009.

[11] Streicher, D., Wiggenhauser, H., Holst, R. and Haardt, P.: Zerstörungsfreie Prüfung im Bauwesen. Beton- und Stahlbetonbau, Heft 3: 216-224, Berlin: Ernst & Sohn, 2005.

[12] Zilch, K., Zehetmaier, G., and Gläser, C.: Ermüdungsnachweis bei Massivbrücken. Beton-Kalender 2004, Berlin: Ernst & Sohn, 2004.

Reliability analysis of concrete highway bridges for hazard scenarios

Vazul Boros

Institute for Lightweight Structures and Conceptual Design (ILEK),
University of Stuttgart,
Pfaffenwaldring 7, 70569 Stuttgart, Germany
Supervisor: Balthasar Novák

Abstract

The German research project "Protection of Critical Bridges and Tunnels in a Road Network" focused on the identification of critical road infrastructure and the development of protective measures. Therefore a suitable indicator of criticality for concrete highway bridges had to be found, thus enabling the ranking of structures and the evaluation of protection measures. First a complex probabilistic model considering the uncertainties in resistances, actions and mechanical models had been established. Based on this model Monte-Carlo simulations for different limit states and hazard scenarios were carried out, to estimate the reliability index of highway bridges. Thereafter the traffic load bearing capacity of bridges has been investigated as a simplified indicator of criticality and was calibrated by comparison with the results gained in the reliability analysis.

1   Introduction

The road network plays an essential part in a countries economy and social life, as its continuous and undisrupted operation is vital for society. The probably most vulnerable components of the road infrastructure are bridges and tunnels. These structures not only act as bottle-necks in the highway system, but also have to face a number of different hazards. Bridges can be affected by vehicle impact against vital structural components such as cables or pillars. Moreover the effect of climate change on structures has to be taken into account, as it results in higher wind velocities, increased temperatures and rainfall, thus causing storms, floods and other natural disasters. Finally, another ever greater threat to important infrastructure components is represented by terrorist attacks. A research project was carried out to investigate the effects of these hazard scenarios on bridges and tunnels and to develop effective protection measures and strategies. The objective was to be able to rank different types of structures according to their criticality and thus enable traffic administrations to identify their most vulnerable infrastructure components. Three main aspects have been considered regarding the criticality of a structure: effects on road users, influence on the road network and structural aspects. All of these aspects have been investigated in detail in the research project, yet this paper addresses the assessment of the criticality of the structure, in particular of bridges.

2   Reliability analysis

To describe the criticality of a bridge for a certain hazard scenario, a suitable indicator of criticality had to be identified. This indicator had to enable the assessment and ranking of different hazard scenarios and different types of structures. Current standards [1, 2] provide us with such an indicator, namely the reliability of a structure, which can either be expressed by the reliability index (β) or the probability of failure (P_f). Whereas a clear definition of these indicators is given by the standard, there are hardly any instructions regarding the calculation of their values. Therefore a calculation method had to be developed and the required information on the probabilistic models had to be collected and implemented in the calculation. The probability of failure due to a hazard scenario can be divided into the exposure of the structure (P_e - probability of event occurrence) and vulnerability of the structure (P_v - probability of failure if event occurs), where: $P_f = P_e \cdot P_v$. Due to the lack of information and statistical data on the occurrence of disastrous events, it is extremely difficult to determine the exposure of the structure, that is the probability of a hazard scenarios occurrence. Therefore only the vulnerability of the structure was assessed within this research project and it was agreed to set P_e for each scenario equal to 1. Assuming that the different hazard scenarios have similar occurrence probabili-

Proc. of the 9th *fib* International PhD Symposium in Civil Engineering, July 22 to 25, 2012,
Karlsruhe Institute of Technology (KIT), Germany, H. S. Müller, M. Haist, F. Acosta (Eds.),
KIT Scientific Publishing, Karlsruhe, Germany, ISBN 978-3-86644-858-2

ties, this fits the objectives of the project, as the aim was to enable a ranking for different types of hazard scenarios and structures.

2.1 Basic variables

The probabilistic model for the reliability analysis is based on the traditional separation of actions and resistances common to structural engineering. Basic variables have been carefully considered and selected for both sides. Basic variables for resistances mainly include material properties. The distribution functions for the basic variables have been assumed according to various references, for the prestressing steel the permit of the cables [3] provided information regarding the requirements towards material quality. A correlation between the compressive strength and the modulus of elasticity of concrete has been considered. Model uncertainty factors, accounting for random effects neglected in models and simplifications in the mathematical relations [4] have been considered, too. Modelling actions and loads required more advanced considerations. It had to be distinguished between permanent actions, such as self weight and variable actions, as for example traffic loads or temperature differences. The factor for self weight has been applied to each construction section separately. Ground settlements and wind actions, considering the effect of climate change, were applied on pillars. As already indicated, the effects of the hazard scenarios investigated could not be considered as basic variables due to the lack of statistical data, they were introduced as non probabilistic actions instead. An observation period of one year had been chosen, the combination of the different time dependent variable actions within this period has been implemented using the Ferry Borges-Castanheta model [5]. Table 1 presents an overview of the selected basic variables and the main parameters of the corresponding distribution functions with the exception of traffic loads.

In order to reproduce the traffic loads on the bridge, complex traffic simulations were carried out. The simulations were developed on the basis of statistical data acquired in traffic measurements on German highways [10]. This simulation method has also been used to support the development of new traffic load models for the German National Annex to the EN 1991-2 [11]. Some results of the traffic measurements are presented in Fig. 1. Several different traffic scenarios have been investigated. One corresponds to the current traffic on German highways according to the measurements carried out. Other scenarios account for expected changes in the traffic constitution, assuming an increase in vehicle numbers and loading. The traffic simulations required a large number of basic variables including vehicle types, total vehicle weight and distance between vehicles. Traffic congestion was included into the simulation model with certain probability, too.

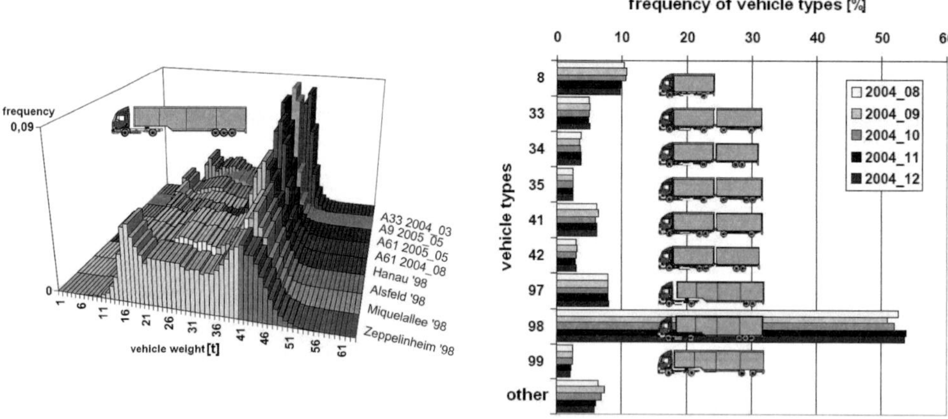

Fig. 1   Results of traffic measurements on German highways: frequency of vehicle types (left) and vehicle weight distribution for truck type 98 (right) according to [10]

Table 1   Basic variables for resistances and actions

Variable	Unit	Distribution	Mean	Standard deviation	Reference
Compressive strength of concrete	[MPa]	Lognormal	43	5	[6]
Modulus of elasticity of concrete	[MPa]	Lognormal	33282	4992	[6]
Yield stress of reinforcement	[MPa]	Normal	560	30	[4]
Modulus of elasticity of reinforcement	[GPa]	Constant	205	-	[4]
Ultimate strength of prestressing steel	[MPa]	Normal	1876	64.5	[3]
Modulus of elasticity of prestressing steel	[GPa]	Constant	195	-	[3]
Eccentricity of load on pillar	[mm]	Normal	0	21	[4]
Factor of model uncertainty for bending	-	Lognormal	1.2	0.15	[4]
Factor of model uncertainty for shear	-	Lognormal	1.0	0.1	[4]
Factor of model uncertainty for normal forces	-	Lognormal	1.0	0.05	[4]
Self weight factor	-	Normal	1.0	0.1	[6]
Temperature difference (positive)	[K]	Weibull	3.62	2.2	[7]
Temperature difference (negative)	[K]	Weibull	-2.59	1.59	[7]
Ground settlements	[cm]	Beta	1.0	0.3	[8]
Wind velocity	[m/s]	Weibull	10.54	3.94	[9]
Factor of model uncertainty for bending	-	Lognormal	1.0	0.1	[4]
Factor of model uncertainty for shear	-	Lognormal	1.0	0.1	[4]

2.2   Monte-Carlo simulation

The Monte-Carlo method has been implemented to determine the reliability indicators in question. In order to improve the accuracy of the Monte-Carlo simulation a special method considering the model uncertainties, has been developed in this research. The technique is similar to the importance sampling method, which utilises prior information about which domain of the possible values of basic variables contributes most to the probability of failure, and achieves a variance reduction by centring the simulation on this area [12]. In an analogous way the applied method considers only those values of the model uncertainties which result in the failure of the structure.

For each of the basic variables with the exception of model uncertainties n different possible realizations according to the corresponding population function are generated. Thereafter the resistance forces R_i and action forces E_i can be calculated for each of these realisations. The limit state function for one realization can be stated as

$$R_i \cdot \theta_R - E_i \cdot \theta_E \geq 0$$

with θ_R and θ_E being the model uncertainty factors for resistances and actions respectively. The model uncertainty factors follow the logarithmic normal distribution, therefore they can be substituted by the exponents of normally distributed variables U_R and U_E with mean value μ_{U_R} and μ_{U_E} along with standard deviation σ_{U_R} and σ_{U_E}.

$$R_i \cdot e^{U_R} - E_i \cdot e^{U_E} \geq 0$$

$$e^{U_R - U_E} \geq \frac{E_i}{R_i}$$

As the exponential function is strictly increasing, the inequality has to be also valid for the logarithm of both sides.

$$U_R - U_E \geq \ln\left(\frac{E_i}{R_i}\right)$$

The variables U_R and U_E being normally distributed, their difference has to be normally distributed also, with mean $\mu_{U_R} - \mu_{U_E}$ and standard deviation $\sqrt{\sigma_{U_R}^2 + \sigma_{U_E}^2}$. Therefore the probability of the limit state function not being fulfilled can be calculated as

$$\hat{P}_{f_i} = \Phi\left(\frac{\ln\left(\frac{E_i}{R_i}\right) - \left(\mu_{U_R} - \mu_{U_E}\right)}{\sqrt{\sigma_{U_R}^2 + \sigma_{U_E}^2}}\right)$$

with $\Phi(x)$ being the distribution function of the standard normal distribution.

By this equation each realization in the simulation provides an estimate of the failure probability for the investigated limit state. The overall failure probability considering all n simulations can be expressed as

$$\hat{P}_f = \frac{1}{n}\sum_{i=1}^{n}\hat{P}_{f_i}$$

The variance of the failure probability can then be calculated in analogy to the importance sampling technique by

$$Var\left(\hat{P}_f\right) = \frac{1}{n-1}\left(\frac{1}{n}\sum_{i=1}^{n}\hat{P}_{f_i}^2 - \hat{P}_f\right)$$

The introduced method proved effective in improving the accuracy of the estimation for the failure probability, as the variances for the dominant limit states of shear and bending could be reduced to less than 0.07.

3   Simplified analysis

The reliability analysis introduced in section 2 is an innovative approach to assess the reliability of a structure for hazard scenarios, yet it requires a large number of simulations and complex calculations, thus it is highly time consuming. In civil engineering practise it is often of importance to get a quick estimate of a structures reliability, without demanding simulations. Therefore a simplified analysis had to be developed to enable a criticality assessment for bridges. As an alternative to the reliability index a simplified criticality indicator γ_L has been introduced, named traffic load factor. The traffic load factor is defined as the factor the traffic load has to be multiplied with, so that for a certain limit state the failure criterion is exactly met. For the persistent and transient design situations of the Eurocodes, the traffic load factor can be determined by solving the following equation:

$$R_{Ed} = E\left\{\sum_{j\geq 1}\gamma_{G,j} \cdot G_{k,j} + \gamma_P \cdot P_k + \gamma_L \cdot \gamma_{Q,1} \cdot Q_{k,1} + \sum_{i>1}\gamma_{Q,i} \cdot \psi_{0,i} \cdot Q_{k,i}\right\}$$

For any structure complying with the specifications of a given standard, the load factor has to be greater than 1.0 for each limit state and action given by the standard. For the extreme hazard scenarios investigated in the project it is of course possible that the calculated load factor is less than 1.0 or cannot be assessed. The load factor is calculated with linear analysis. It has to be remarked, that additional reserves of the structural reliability could be considered taking into account non-linear behaviour and redistribution of internal forces. Two different models for traffic loads have been investigated. The model for current traffic corresponds to the traffic loads defined in the current German standard on actions for bridges [13]. The scenario future traffic is based on the suggestion for a new model according to [10], taking into account prognosticated increase in vehicle numbers and loads.

4 Results

The case studies were carried out on two different structures: a single span prestressed concrete slab and a prestressed concrete continuous beam with five spans and a double-webbed T-beam cross-section. These can be regarded as fairly representative, as beams and slabs account for more than two thirds of bridges in the German highway network. The structures were calculated and designed according to the valid German standards at the time of construction. Both bridges carry three lanes, accommodating traffic of approximately 120.000 vehicles per day, with around 10 % of heavy traffic. As a reference scenario first the reliabilities have been calculated for the undamaged structure. According to the European Standard EN 1990 [1] the target value of the reliability index, for a reference period of one year, should be 4.7 for the ultimate limit states. This criterion is met by both structures for both traffic models. For the concrete slab a partial failure at mid-span and a partial failure at the supports have been examined. For the continuous beam the investigated scenarios included partial failure at supports, partial failure at mid-span and the collapse of one out of a pair of pillars. The results of the Monte-Carlo simulations and the simplified analysis for these structures after the incident are presented in Table 2.

Table 2   Calculated indicators of criticality for different scenarios

Scenario	Reliability index		Traffic load factor	
	Current traffic	Future traffic	Current traffic	Future traffic
Reference scenario (single span slab)	4.85	4.73	1.50	1.14
Partial failure at mid-span (single span slab)	3.04	2.92	0.00	0.00
Partial failure at supports (single span slab)	4.85	4.73	1.05	0.80
Reference scenario (continuous beam)	5.70	5.38	1.39	1.14
Partial failure at mid-span (continuous beam)	4.34	3.97	1.04	0.82
Collapse of a pillar (continuous beam)	5.13	4.84	0.00	0.00
Partial failure at supports (continuous beam)	2.60	2.23	1.26	0.80

The results indicate a clear correlation between the reliability index and the traffic load factor, the sample correlation coefficient being $r = 0.86$. Considering this correlation a formula for the estimation of the reliability index based on the traffic load factor could be developed. If a structure for a certain scenario meets the reliability requirements of the European Standard, the reliability index has to exceed 4.7 and the traffic load factor must not be less than 1.0, therefore for a traffic load factor of 1.0 the formula should provide a reliability index of 4.7.

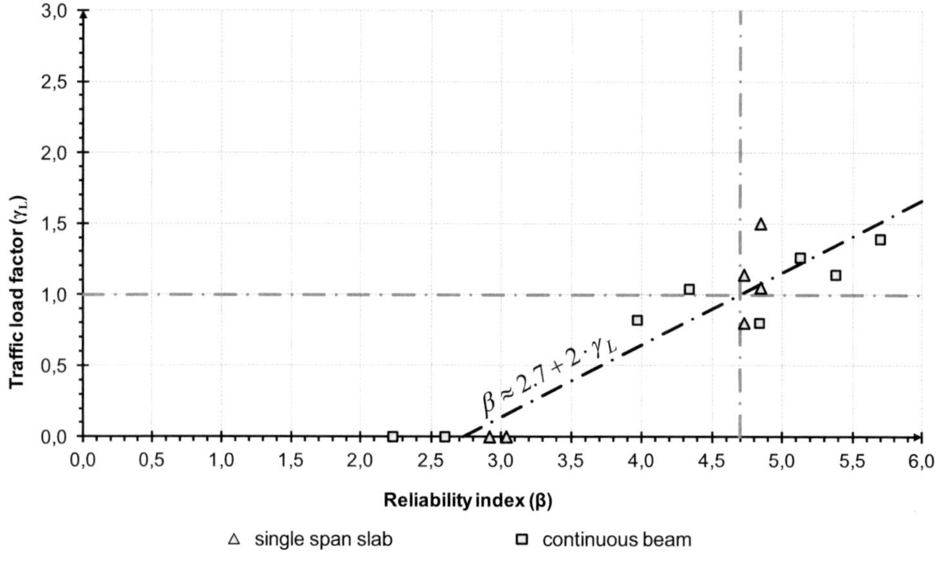

Fig. 2   Comparison of results for reliability index and traffic load factor

A linear equation fulfilling this criterion was fitted to the data presented in Table 2 by applying the method of least squares. As a result the following estimation formula for the reliability index could be developed:

$$\beta \approx 2.7 + 2 \cdot \gamma_L$$

This equation provides for the investigated scenarios an estimation of the reliability index with a maximum error of approximately 15 %. Considering the significantly easier calculation method, this error is acceptable for a first estimate. In Fig. 2 the equation of the estimation is plotted, the corresponding indicators of criticality for each scenario according to Table 2 are represented by individual points as a comparison.

5   Conclusions

The reliability index was identified as a suitable indicator for ranking different types of bridges and hazard scenarios according to their criticality, thus enabling the identification of the most vulnerable structures. A method was developed to estimate the reliability index with Monte-Carlo simulations. The variance of the estimation could be reduced effectively by introducing a technique considering the basic variables for the model uncertainties analytically. A simplified indicator of criticality, named traffic load factor, was introduced and defined. Based on results for case studies a correlation between the two indicators of criticality could be identified and a formula for the estimation of the reliability index was developed.

6   Acknowledgments

This research was conducted within a project titled "Protection of critical bridges and tunnels in a road network (Schutz kritischer Brücken und Tunnel im Zuge von Straßen - SKRIBT)". The project in the context of the programme "Research for Civil Security" was part of the "High-Tech Strategy" of the German Federal Government and was funded by the German Federal Ministry of Education and Research (BMBF). The author would like to thank his supervisor Prof. Balthasar Novák for his guidance and encouragement throughout this research.

References

[1]   EN 1990:2010: Eurocode: Basis of Structural Design (2010)

[2]   DIN 1055-100:2001, Einwirkungen auf Tragwerke, Teil 100: Grundlagen der Tragwerksplanung, Sicherheitskonzept und Bemessungsregeln (2001)

[3]   Allgemeine bauaufsichtliche Zulassung Z-12.3-29: Spannstahllitzen St 1570/1770 aus sieben kaltgezogenen, glatten Einzeldrähten; Deutsches Institut für Bautechnik (2004)

[4]   Joint Committee on Structural Safety, Probabilistic Model Code (2001)

[5]   Ghosn, M. et al.: National Cooperative Highway Research Program Report 489, Design of Highway Bridges for Extreme Events, Transportation Research Board (2003)

[6]   Six, M.: Sicherheitskonzept für nichtlineare Traglastverfahren im Betonbau, Deutscher Ausschuss für Stahlbeton Heft 534 (2003)

[7]   Frenzel, B. et al.: Bestimmung von Kombinationsbeiwerten und – regeln für Einwirkungen auf Brücken, Forschung Straßenbau und Straßenverkehrstechnik Heft 715 (1996)

[8]   Spaethe, G.: Die Sicherheit Tragender Baukonstruktionen (1987)

[9]   FE 89.0232/2009/AP Schlussbericht: Auswirkungen des Klimawandels auf bestehende Spannbetonbrückenbauwerke (2011)

[10]   Kaschner, R. et al.: Auswirkungen des Schwerlastverkehrs auf die Brücken der Bundesfernstraßen, Berichte der Bundesanstalt für Straßenwesen Heft B 68 (2009)

[11]   FE 15.451/2007/FRB Schlussbericht: Anpassung des DIN-Fachberichtes „Einwirkungen auf Brücken" an endgültige Eurocodes und nationale Anhänge einschließlich Vergleichsrechnung (2009)

[12]   Faber, M. H.: Risk and Safety in Engineering, ETH Zürich (2009)

[13]   DIN-Fachbericht 101:2003: Einwirkungen auf Brücken (2003)

Crack Classification and Location Using Acoustic Emission Analysis in Prestressed Concrete Structures

Hisham A. Elfergani

Cardiff School of Engineering, Cardiff University,
Queen's Building, The Parade ,
Cardiff, CF24 3AA, Wales, UK
Supervisor: Rhys Pullin and Karen M. Holford

Abstract

Deterioration, by corrosion of steel wire in reinforced and prestressed concrete is very serious and must be given special consideration; failure may result in the loss of life and a high financial cost. One application is prestressed concrete cylinder pipes which are widely used for conveying water and wastewater. The biggest problem in these pipes is corrosion of the prestressed wires which can lead to catastrophic failure. In the Great Man-mad River in Libya detection of the corrosion in its initial stages is very important to avoid water interruption to homes and industry. This paper reports on the use of the Acoustic Emission (AE) technique to detect and locate the early stages of corrosion and classify different crack types. Preliminary results presented indicate that AE is capable of detecting corrosion in representative structures. In addition, it can identify and classify the crack location and type.

1 Introduction

Corrosion is a significant problem in numerous structures. The cost due to corrosion is estimated at billions of dollars every year. The Department of Transport in the UK evaluated that the cost of repairing concrete structures damaged by corrosion problems as £755 million a year [1]. The risk of corrosion in this type of structure must be given special consideration because failure may result in the worst case a, loss of life, but at a minimum a financial loss. Most studies indicate that the main reason of the failure of bridges and concrete pipes is corrosion during the short period after they were constructed. However, the life of a concrete structure becomes even shorter due to steel corrosion, which may occur by aggressive ion attack from products of chloride or carbonation [1].

The concrete pipes which transport water are one such structure that has suffered from corrosion. For example, the pipes of Great Man-Made River project of Libya have suffered catastrophically from this affect. Five pipe failures due to corrosion have occurred since their installation. However, the substantial challenges which face the engineers, apart from future corrosion protection, is to find the best way to detect the corrosion and prevent the pipes from deteriorating [2]. This project aims to use the Acoustic Emission (AE) technique to detect the early stages of corrosion prior to deterioration and eventual failure of the concrete structures.

AE is defined as the elastic energy released from materials which are undergoing deformation. Also, it can be defined as "the transient elastic waves which are generated by the rapid release of energy from localised sources within a material" [3].The rapid release of elastic energy, the AE event, propagates through the structure to arrive at the structure surface where a piezoelectric transducer is mounted. These transducers detect the displacement of the surface at different locations and convert it into a usable electric signal. By analysis of the resultant waveform in terms of feature data such as amplitude, energy and time of arrival, the severity and location of the AE source can be assessed.

According to many of researches [4, 5, 6, 7] the relationship between RA values (rise time/ Amplitude) and average frequencies (counts/duration) can be used for classification of crack types. They reported that when an AE signal has low average frequency and high RA value it is classified as shear type crack. However when it has a high average frequency and low RA value classified as tensile type crack as shown Fig.1and 2.

Proc. of the 9th *fib* International PhD Symposium in Civil Engineering, July 22 to 25, 2012,
Karlsruhe Institute of Technology (KIT), Germany, H. S. Müller, M. Haist, F. Acosta (Eds.),
KIT Scientific Publishing, Karlsruhe, Germany, ISBN 978-3-86644-858-2

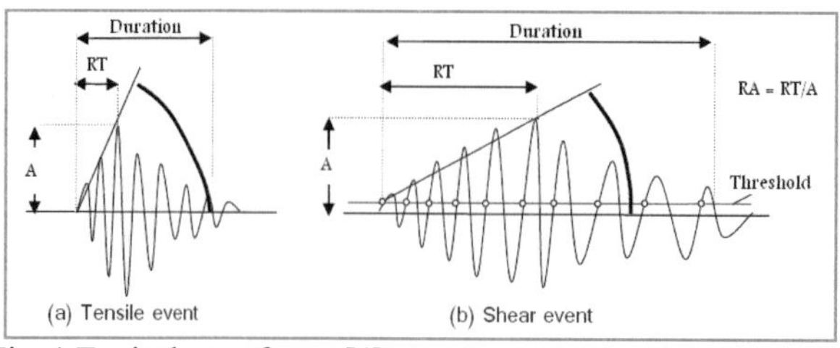

Fig. 1 Typical waveforms [4]

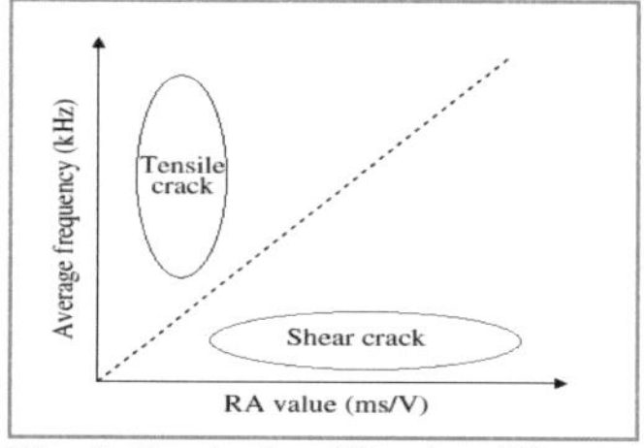

Fig. 2 Crack classification [4]

The Great Man-Made River Project (GMRP) is the one of the major civil engineering projects of the 20th century located in Libya. The project is concerned with water transportation from the aquifers deep in the Sahara desert to the coastal region where over 90% of the population lives. The water is conveyed throughout almost 4000 km of prestressed concrete cylinder pipe (PCCP) networks as shown in Fig. 3. [2, 8, 9]

Pre-stressed concrete cylinder pipes are designed to take the best advantage of the compressive strength and corrosion-inhibiting property of Portland cement concrete and mortar and the tensile strength of prestressing wire. The majority of pre-stressed concrete cylinder pipes are 4.0 m in inner diameter; with a length of 7.5 m, and over 70 tonnes in weight. The concrete pipe consists of a 225 mm thick concrete core within an embedded thin steel cylinder and externally wrapping prestressed wires. The cured concrete core is prestressed by applying over-wrapping with high tensile steel wire at a close pitch under uniform tension. The prestressed wires are coved by a 19mm thick layer of cement mortar to protect the wires against corrosion and mechanical harm. A typical cross-section of the PCCP is shown in Fig. 4.

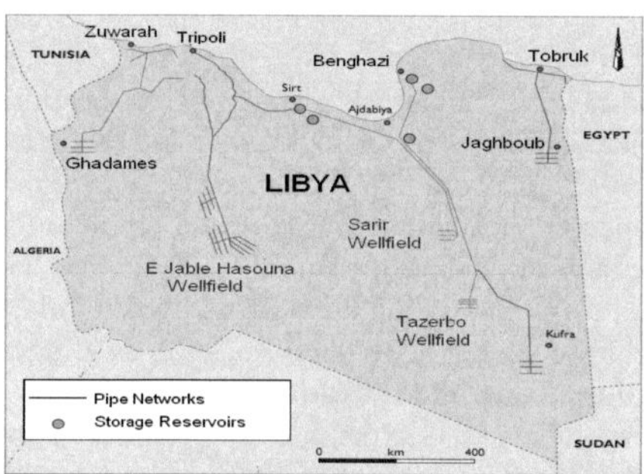

Fig. 3 Pipe Networks of GMRP [8]

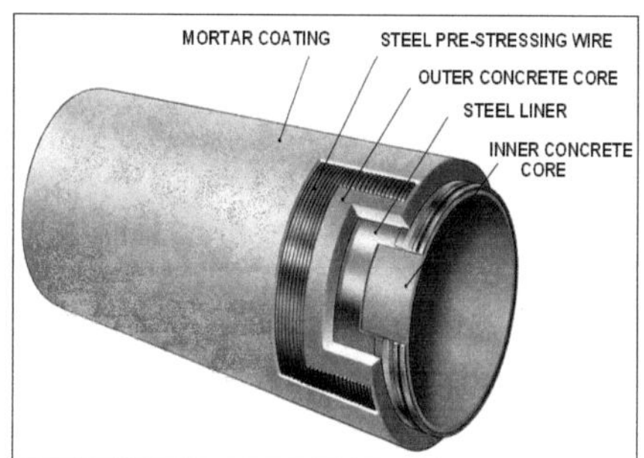

Fig. 4 Typical cross-section

Five catastrophic failures in four metre diameter pipes occurred between 1999 and 2001 after ten years of operation. The main reason for the damage is corrosion of prestressed wires in the pipes due to attack by the chloride ions from the surrounding soil. Detection of the corrosion in the initial stages has been very important to avoid other failures which will interrupt water flows. Even though most of the non-destructive methods which are used in the project are able to detect wire breaks, they cannot detect the presence of corrosion. Hence, in areas where no excavation has been completed, areas of serious damage can go undetected. In this respect, AE has significant advantages compared with other NDT methods because the AE technique is able to reliably detect the very early stages of the corrosion process, before significant damage to the concrete has occurred. Furthermore, it can indicate the level of damage occurring to the concrete. [10, 11, 12]

2 Experimental Procedure

2.1 Tension holding frame

Since it was intended in this work to simulate as close as possible the real physical conditions surrounding the high strength steel wires in concrete pipes, it was important to place and maintain all relevant wire samples under tension equal to 60% of their ultimate tensile strength in PCCP. To achieve this objective a tension frame was designed and fabricated.

The frame consists of two blocks (190mm x 45mm x 45mm) and two threaded steel bars (studding) having a diameter of 20 mm and a length of 500 mm. Two holes (20 mm diameter) and two (6mm) are drilled in each block. Fig. 5 shows a schematic drawing for the tension holding frame. The two blocks are assembled via two threaded bars tightened by means of eight nuts.

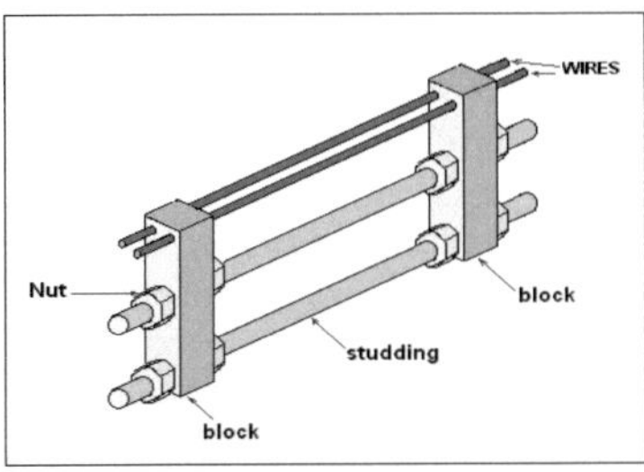

Fig. 5 Tension holding frame

2.2 Wire preparation

The two working high strength steel wires samples were supplied from GMRP manufacturing plant in Libya. The metallurgical composition and mechanical properties as certified by the wire manufactures is summarised as follows: Carbon steel (carbon 0.8-0.84%, 0.85-1.00%Mn, 0.030 %Max S, 0.035% Max P, 0.20-0.35% Si). The tensile strength of the wires is approximately 1738 MPa.

The two working wire samples were passed through 6 mm diameter holes in the steel blocks and then through the two modified bolts and nuts, designed to control the tension load of each wire. Finally a steel cylinder was then threaded over the wire. The cylinder was then compressed in a load machine. In this way the modified nut and bolt could be expanded between the clamped cylinder and the steel block and as a result tension could be introduced into the wire. Each wire was subjected to a tensile force of 20 kN by adjusting the bolts and nuts and the resulting strain was monitored via strain gauges mounted on the wires.

2.3 Concrete and mortar preparation

The concrete specimen (200×200×50mm), representative of the inner pipe was prepared according to the technical specification for PCCP manufacturing used in GMRP, which is in accordance with AWWA C301-92 (Standard for Pre-stressed Concrete Pressure Pipe, Steel Cylinder Type, for Water and Other Liquids)[2]. Three days after casting the concrete specimen, the wires combined with their holding frame were placed on the upper surface of this specimen. Finally the mortar 200×200mm and 20 mm thickness was coated on the upper surface of the concrete. The mortar consists of one part cement to not more than three parts fine aggregate by weight. The final construction is shown in Fig.6.

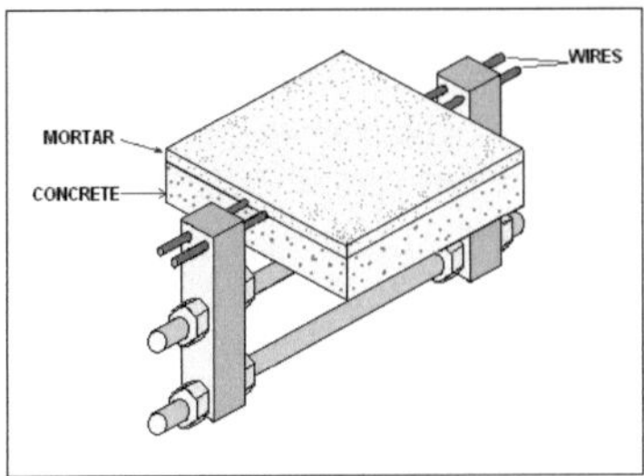

Fig. 6 Concrete and mortar specimen

2.4 Accelerated corrosion technique

To study the effects of corrosion within a realistic time-scale, it is sometimes necessary to accelerate the initiation period and occasionally control the rate of corrosion during the propagation stage. To simulate the corrosion of prestressing steel wires, the corrosion cell was induced by an impressed current ($100\mu A/cm^2$). This is reported as corresponding to the maximum corrosion rate for concrete in laboratory conditions and has been used by several researchers in the laboratory as discussed by Li and Zhang [13]. In this experimental work, the wire corrosion was induced by impressed current $(100\mu A/cm^2)$. The prestressed wires were contacted in an electrical circuit with positive pole of power supplier and the negative pole connected with a stainless steel plate (30*150 mm) resting on the upper mortar. A 4% NaCl solution was poured on the surface of the mortar. Silicon sealant was used to pool the solution on the upper surface.

2.5 Acoustic emission set-up

AE instrumentation typically consists of transducers, filters, amplifiers and analysis software. Four AE sensors (R3I – resonant frequency 30 kHz) were mounted to surface of mortar.The four AE sensors were mounted using silicon sealant and were fixed on the upper surface of mortar with a U shaped plate. The plate was screwed to hold the sensors down and to ensure a good coupling. Then the sensitivity of the sensors was checked by using the Hsu-Neilson source [14].

3 Results and Discussion

Figure 7a is a schematic diagram of the specimen after testing. The figure shows the sensors mounted on the mortar surface, wires position, the stainless steel plate and the crack shape. Fig. 7 b is a photograph of the top mortar surface the after of the test finished, again showing the crack shape

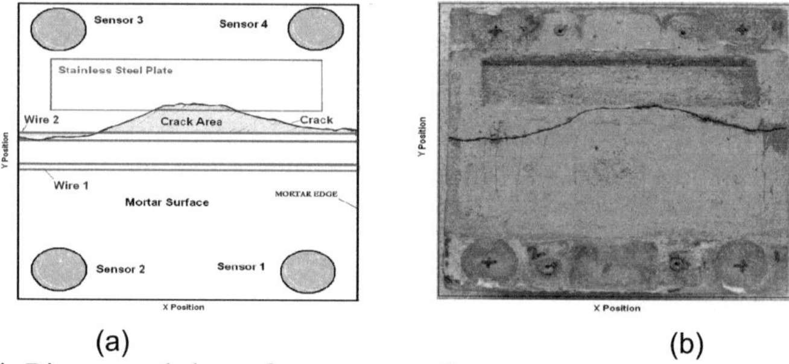

(a)   (b)

Fig. 7 Schematic Diagram and photo of top mortar surface

Fig. 8 shows the location of hits with minimum amplitude 45dB for the whole period of the test. It can be noted that the highest hits concentration and highest energy coincides with maximum wire corrosion and the crack which was visibly observed post test. Four zones have been chosen as examples to differentiate between cracks and non cracked areas. Areas were chosen based on visual observation (Fig. 7b) zone 1 and 2 represent crack areas whereas zone 3 and 4 no cracks were present.

Fig. 9 (a, b, c, and d) Show the AF vs. RA value for different regions of concentration of events on surface of mortar. Fig. 9 (a and b) show the AF vs. RA for zones associated with the crack regions. It can be seen that the most of data points have various AF and low RA value (less than 10 ms/v). Therefore, based on Fig. 7, this indicates that the type of the crack is tensile cracks. However, the Fig. 9 (C and d) show the relationship between RA value and AF, where areas on the location with low concentration of hits. It can be noted that the RA value has a wide distribution (0-50 ms/v).

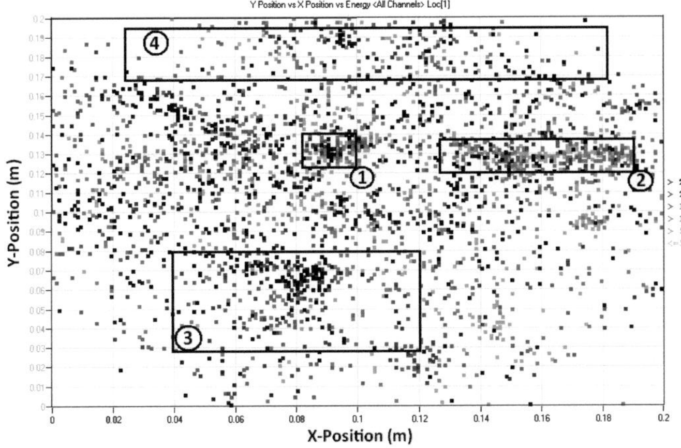

Fig. 8 Source locations for whole test with amplitudes greater than 45dB

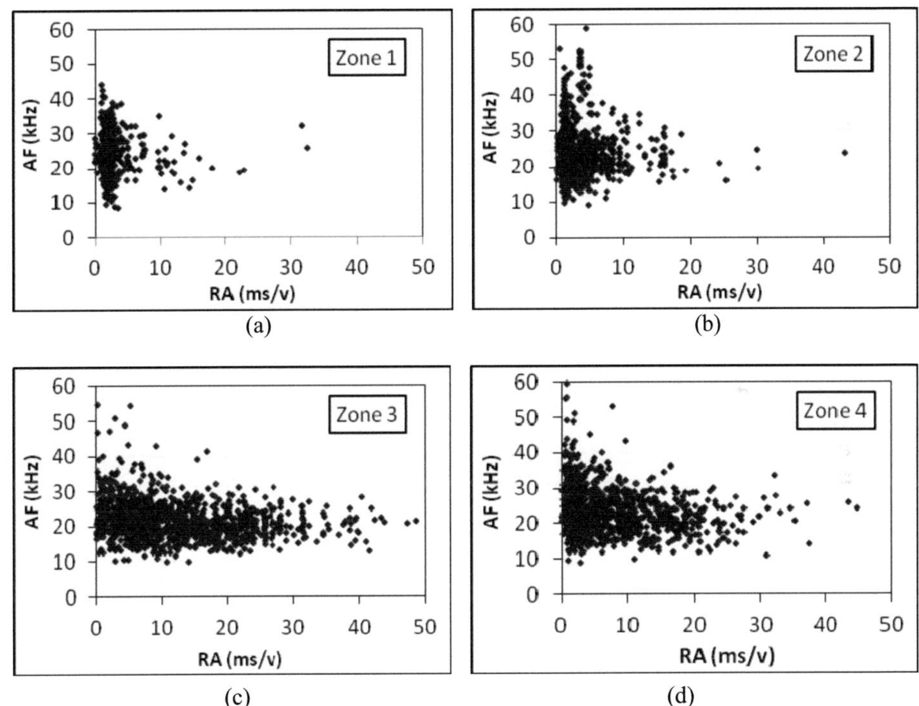

Fig. 9   Relation between the RA value and average frequency of (a) zone1, (b) zone2, (c) zone3 and (d) zone 4

Fig. 10 shows Relation between the RA value and average frequency of zone1 compared with zone 3. The black data points represent the crack region, while the grey represents the no crack region. It can be seen clearly the difference between two zones.

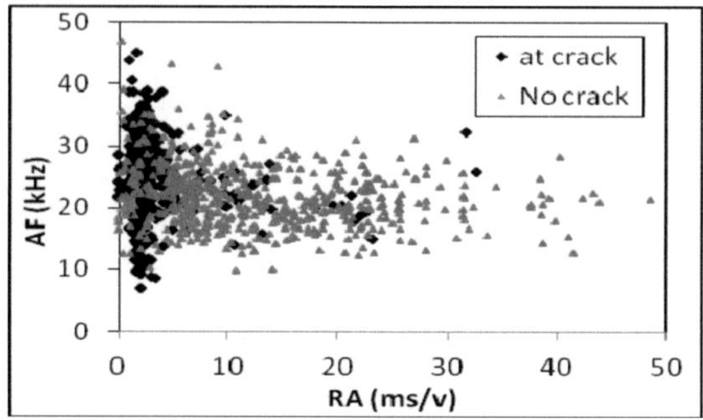

Fig. 10   Relation between the RA value and average frequency of Zone1 and zone 3

It has been demonstrated that by using relationship between RA and AF value, the crack area can be identified (Fig. 10). Hence, it could be possible to provide a corrosion alarm and location to pipe engineers prior to any wire breaks. Furthermore, by knowing the crack types it could be possible to identify the damaged area before the mortar completely fails. Additional analysis has shown that the measured values are not affected by distance making them an ideal approach. Further work is being conducted to investigate the exact mechanisms within this structure that causes differing RA and RF values. Results suggest that using AE techniques as structural health monitoring of the concrete pipes and other concrete structures such as bridges could be achieved. However, applying acoustic emission techniques on site (real pipes line), some limitations need to be considered which include source attenuation, number and location of sensors, noise from the surrounding environment and noise from water flow. Further research studies are needed to overcome these limitations if AE is to be used to monitor pipes.

4   Conclusion

The AE method has been applied to detect the corrosion in small scale pipe samples and a novel analysis approach to this application has been used to evaluate differing crack types. Using the relationship between RA value and AF value, the crack area and crack type can be identified and distinguish.

Acknowledgements

The authors would like to thank to all technical staff of Cardiff School of Engineering who contribute in this work. Also, I thank Dr. Mark Eaton who provided technical assistance

References

[1]   Ann, K.Y. et al.: The importance of chloride content at the concrete surface in assessing the time to corrosion of steel in concrete structures. In: Construction and Building Materials (2009) No. 23, pp239–245.

[2]   Singh, S. K.: Corrosion Studies on Prestressing Steel Wire. In: Ph.D. Thesis (2000) Imperial College, University of London.

[3]   Miller, R. K. et al.: Non-destructive Testing Handbook, Acoustic Emission Testing (2005) Volume 6, USA.

[4]   JCMS-III B5706: Monitoring method for active cracks in concrete by AE. In: Construction Material Standards (2003) Tokyo: Japan.

[5]   Ohtsu, M. and Tomoda, Y.: Corrosion process in reinforced concrete identified by acoustic emission. In: Materials Transactions (2007b) Vol. 48, No. 6, pp1184 to 1189.

[6]   Ohno, K. and Ohtsu, M.: Crack classification in concrete based on acoustic emission. In: Construction and Building Materials (2010) No. 4, pp. 2339-2346.

[7]   Aggelis, D.: Classification of cracking mode in concrete by acoustic emission parameters. In: Mechanics Research Communications (2011) No. 38, pp.153-157.

[8]   Kuwairi, A.: Water mining: the Great Man-made River, Libya. In: Proceedings of ICE Civil Engineering 159 May (2006) No.14382 pp.39-43.

[9]   Essamin, O. and Holley, M.: Great Man Made River Authority (GMRA): The Role of Acoustic Monitoring in the Management of the Worlds Largest Prestressed Concrete Cylinder Pipe Project. In: Proceedings of the ASCE Annual International Conference on Pipeline Engineering and Construction (2004) August 1-4, San Diego, California.

[10]   Ing, M. et al.: Cover zone properties influencing acoustic emission due to corrosion. In: Cement and Concrete Research (2005) No. 35 pp 284– 295.

[11]   Hellier, C. J.: Handbook of Non-destructive Evaluation (2001) McGraw-Hill, USA.

[12]   Carpinteri, A. et al.: Structural damage diagnosis and life-time assessment by acoustic emission monitoring. In: Engineering Fracture Mechanics (2007) No.74, pp273–289.

[13]   Li, X. and Zhang, Y.: Analytical Study of piezoelectric Paint for Acoustic Emission-Based Fracture Monitoring. In: Fatigue and Fracture of Engineering Materials and Structures (2008) No.31, pp 684-694.

[14]   Hsu, N. N. and Breckenridge, F. R.: Characterization and Calibration of Acoustic Emission Sensor. In: Materials Evaluation (1981) No. 39 (1) pp. 60-68.

Time shift method, a new method for early detecting alkali-silica reaction (ASR) damages in concrete structures

Farid Moradi-Marani

Département de génie civil,
Université de Sherbrooke
2500, boul. l'Université, Shebrooke (QC), Canada, J1K 2R1
Supervisors: Patrice Rivard, Serge Apedovi Kodjo and Charles-Philippe Lamarche

Abstract

While a wide variety of methods are available to detect Alkali-Silica Reaction (ASR) damages, recent studies have indicated that nonlinear acoustic techniques appear to be promising for the detection and characterization of early cracks and/or micro-cracks produced by ASR in structural concrete elements. A new method based on nonlinear acoustics, which has been validated in the laboratory, has been developed for the characterization of damages associated with ASR in concrete structures. This method consists in quantifying the influence of an external mechanical disturbance on the propagation of an ultrasonic wave in a solid medium. This paper presents a laboratory test set-up involving three large reactive and nonreactive concrete slabs used to study the nonlinear behaviour of concrete using the recently proposed method. Experimental test results show that this method is suitable for recognizing the ASR damages at early stages.

1　Introduction

Alkali Silica Reaction (ASR), as one of the most destructive reactions in concrete structures, is a complex reaction between silica phases of aggregates and the alkali hydroxides dissolved in the concrete pore solution. This reaction leads to the formation of a swelling and expansive gel that generates a cracking network through aggregate and cement paste. The consequent cracking may reduce concrete stiffness as well as other mechanical properties of concrete.A wide range of test methods, destructive and non-destructive test methods (DT and NDT), are used to assess and monitor ASR in concrete structures. Among NDT methods that have been used to detect cracks in concrete, linear acoustic methods like linear resonant frequency presented by ASTM C 215-02 [1] and ultrasonic pulse velocity (UPV) presented by ASTM C 597-97[2] using stress wave propagation have been utilized several times. However, these methods are less sensitive in detecting early age damages in the concrete [3] whereas it is clear that earlier detecting damages, more effective maintenance and more service life. Recently, nonlinear acoustic techniques have showed good correlation for early age detection of damages due to ASR [4-7]. Nonlinear acoustic techniques take into account the influence of the medium disturbance on propagation of waves. Some strong waves are sent through the medium and reflected waves from the perturbation show any changes in the microstructure of the medium. [4]. Nonlinear approaches, e.g. nonlinear attenuation, harmonic generation, resonant frequency shift, slow relaxation dynamics, have been used for characterizing constituent materials like rocks, but they are new approaches in detection of the ASR-damage in cement based materials [4-7]. Although nonlinear techniques are very effective in early age monitoring, these techniques cannot be used in field work easily and effectively. A complete evaluation of damages in a material takes several minutes. The techniques can be influenced by external sources, e.g. waves generated by vehicles passing in a concrete bridge [5].

　　More than the methods, a new approach in nonlinear acoustic techniques named "Time Shift Method" has been recently developed by Kodjo [4]. This technique has been validated in the laboratory and some unpublished works show the potential of the technique for field applications. The novel developed technique is easily applicable in the field. For monitoring the level of damage in concrete elements, this method uses a source of high frequency-low amplitude waves to probe the material while a disturbance waves with low frequency-high amplitude is apply on the material; the later usually produced by an impact [6]. This source of disturbance opens/closes microcracks, and therefo-

Proc. of the 9th *fib* International PhD Symposium in Civil Engineering, July 22 to 25, 2012,
Karlsruhe Institute of Technology (KIT), Germany, H. S. Müller, M. Haist, F. Acosta (Eds.),
KIT Scientific Publishing, Karlsruhe, Germany, ISBN 978-3-86644-858-2

re, a delay happens in the propagation of the high frequency-low amplitude waves. This delay (or time shift) depends on the quantity of microdefects in the concrete element. In the real concrete structures like pavements or bridges, the impact for producing low frequency and high amplitude waves can be replaced by passing vehicles if the produced waves have enough energy for perturbation of in the material. It may be one of the benefits of the method for monitoring of concrete structures. In this project, the passing vehicle is modeled by a MTS hydraulic press in the laboratory in order to determine the efficiency of the wave source for in-service monitoring of the structure. Initial results confirm the competence of the idea for surviving ASR in three concrete slabs.

2　Time Shift Method

The Time Shift Method consists in quantifying the influence of an external mechanical impact on the propagation of an ultrasonic wave through a medium. The idea is to disturb the material, and then, to follow the behavior of the material to this disturbance. The behavior of the material against this distribution is related to the density of microdefects. With this technique, two groups of signals are used to evaluate damages. First, a blast of the high frequency and low amplitude waves are propagated through concrete. These waves are not strong enough to have an effect on the microdefects in concrete and without external disturbance, they show no changes in the materials, its role is to probe the medium. Then, an impact on surface of the concrete element propagates a low frequency and high amplitude wave through the concrete. This generated wave is strong enough to disturb the medium and induces opening/closing microcracks in the concrete.

This opening/closing phenomenon changes the elastic properties of the material and induces a time delay in the arrival time of the probe waves. The amount of delay in the arrival time is inferred to be representative of density of defects in the material. Normally, a longer delay in arrival times means that there are more defects in the material. It should be mentioned that the energy of the impact can increase the time shift in a same material.

Sometimes there is no direct access for sending the direct waves in real structures and therefore, the transmitter and receiver transducers are placed in the same surface. This method is named indirect transmission and previously is studied by Yaman et al [8] for measuring UPV in the concrete. Indirect transmission determination was more elaborate than the direct transmission due to the uncertainty in wave path length; some propagated signals pass near to surface and some of them scan all of the height of the specimens by reflecting from free surface and by multi-reflection between aggregates before arriving to the receiver. The multi- reflection inter-grain is very important, because for the waves who have wave domain less than aggregate sizes, the waves may be attenuated. The main advantage of the indirect transmission may be that the more parts of the specimens can be monitored when the generated waves reflected several times. Yaman et al [8] mentioned that the differences in calculated UPV decrease and the results converge to a constant value when the distance between transducers increases. In this research, both transducers were placed in the same surface for modeling field experience as well as the distance between the transducers was 60 cm.

3　Experiment to model the vehicle passing in laboratory

In this project, ASR in concrete blocks is monitored using nonlinear acoustic technique of "Time Shift Method" for damage assessment, along with some conventional tools for expansion measurements: Demec measurement pins and an internal strain from vibrant string. By this way, the nonlinear technique and conventional techniques are compared for their sensitivity and efficiency in detecting early age damages associated with ASR. Three reinforced concrete slabs with dimension of $1.40 \times 0.75 \times 0.30$ m³ were fabricated. The first one was made with nonreactive aggregates and the two others were made with reactive coarse aggregates (Spratt limestone). To accelerate the reaction rate, NaOH pellets were added to the mixing water to give a total alkali content of 1.25% Na_2Oeq the nominal cementitious materials content for all concrete mixtures was 400 ± 5 kg/m3 with the water/cement ratio of about 0.50. No chemical admixture was used in this mix design. Moreover, to accelerate the reaction, the slabs were stored at 50 ± 2 °C. Figure 1 shows a view of one of the slabs which is equipped with various instruments for expansion measurements. Moreover, the concrete slabs were reinforced by standard steel bars in order to model real concrete elements of structures. This allows knowing the influence of reinforcement on reflected signals.

Fig. 1   Slab which is equipped by expansion measurement instruments. Left: concrete slabs fabricated in laboratory; middle: internal instrumentation for expansion measuring; right: concrete slabs equiped by pins for expansion measuring by Demec.

To model the real conditions in the laboratory, a MTS hydraulic actuator was used to apply idealized load patterns attempting to simulate vehicle passing. As mentioned above, the time delay depends on both the density of microcracks and the impact energy. In this study, various impacts loading range between 10 and 40 KN with the step of 5 KN was used to know the effect of loading on the time shift. The load range may cover the impact produced by most of the general vehicles when pass on concrete structures like pavements. Also, the time rate of the loading in this research was around between 75 ms and 200 ms for the loading. This is similar to real conditions where different types of the vehicles pass a certain point. Figure 2 shows one of the concrete slabs instrumented to carry out the experiment the test.

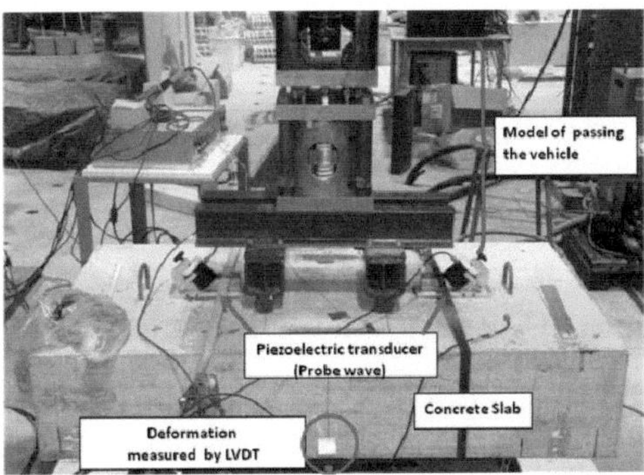

Fig. 2   Test set-up for simulating the real condition (vehicle passing).

A blast of high frequency and low amplitude signals, which is 250 KHz, produced by a signal generator. These waves excite a Panametrics V1012 piezoelectric transducer. The receiving transducer is connected through a receiving channel to the generator where the signals are amplified before being sent to the data acquisition system.

Both transducers are in line with the distance of 60 cm (two time of height of the concrete slab). The distance between the transducers is very important; higher distance yields higher attenuation of high frequency and low amplitude waves. The apparatus between the transducers produces the impact in order to generate low frequency and high amplitude waves. This frequency of the low frequency wave is less than 2000 Hz and it depends on the loading and its rate.

4   Results and discussion

As mentioned previously, the slabs have been monitored by both conventional tests method, like expansion measurement by Demec and vibrating wires gauge, and the new acoustic nonlinear technique of Time Shift. Significant expansions were measured for the two slabs fabricated with reactive aggregate and no clear expansion for nonreactive slab. Figure 3 and 4 shows expansion curves for every slab measured by Demec and vibrating wire, respectively. For nonreactive slab, only internal expansion was measured by vibrating wire. It should be mentioned that Both RT and RQ are reactive concrete and NR is nonreactive one.

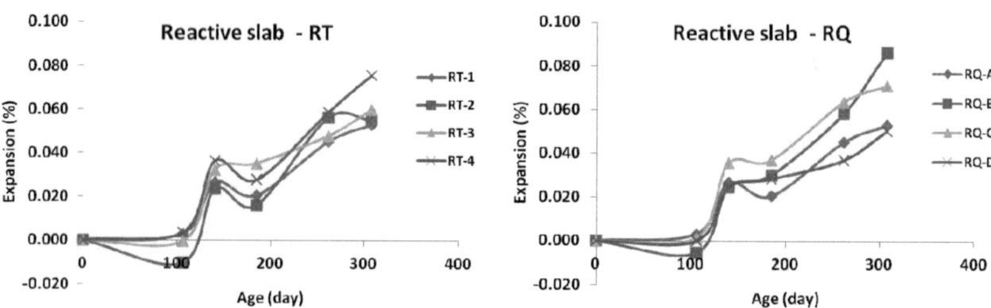

Fig. 3   Expansion measured by Demec in reactive slabs. The position of the measurement in every slab is different.

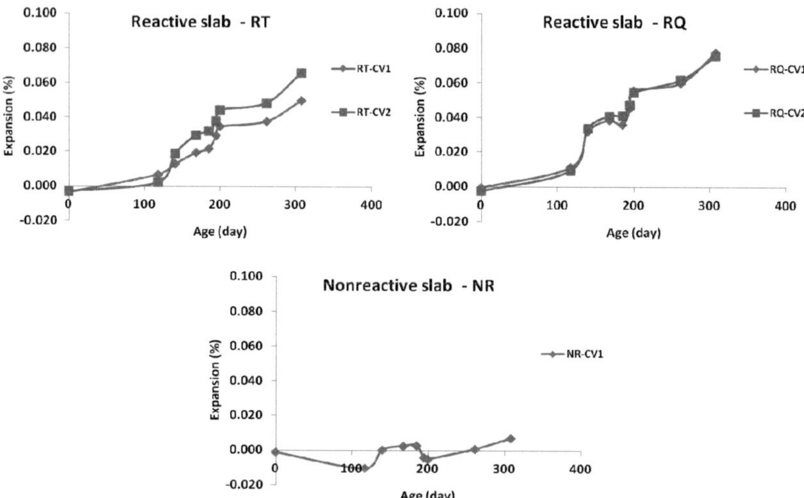

Fig. 4   Expansion measured by vibrating wire in both reactive slabs as well as nonreactive one.

As mentioned previously, the slabs stored for 2 cycles of 15 days in the warm room (50 °C). Measurements after storing in the warm room shows significant growths in expansion curves, which happened after around 120 days. The maximum level of expansion measured by Demec tools and vibrating wires is approximately same.

Development of microcracking by ASR in the slabs was monitored by Time Shift Method. To do this so, a procedure was considered to analyses all received signals. The probe signal record after propagate in the concrete medium is analyzed over a period of 1ms. A rectangular time window with 0.16 ms (160 µs) range from 0.2 ms to 1.0 ms scans the signal from the time corresponding to the first arrival time of the signal. The 0.2 ms is the time where the first signal was recorded from the concrete slabs. The time shift (Δt) was measured by cross-correlation function for each step of the scan, separately. The summation of these time shifts helps to consider the contribution of each part of the signal. Waves which arrive earlier have scanned lower parts of the concrete between two transducers than the waves are received later. The waves who come later are mostly one type of Coda waves. The coda waves are acoustic waves that travel through a medium like concrete, scattered multiple times by medium, being late-arriving waves [9].The Time Shift Method compares by cross correlation technique, the waves before impact (reference wave) and after impact. Indeed, any changes comparing the reference wave shows that there are micro defects in the medium.

Figure 5 presents results of the Time Shift Method in detecting ASR in three concrete slabs of RQ, RT and NR in three time step. Two main points are clear in all subfigures: 1) the amount of the time shift augments when the loading increases; 2) According to the expansion graphs (Figures 5 & 6), the results recorded by both new techniques and expansion instruments are in good agreement. Test 1 is before warming the specimens, test 2 is after first cycle of warming (15 days) and test 3 is after second cycle of warming (15 days).

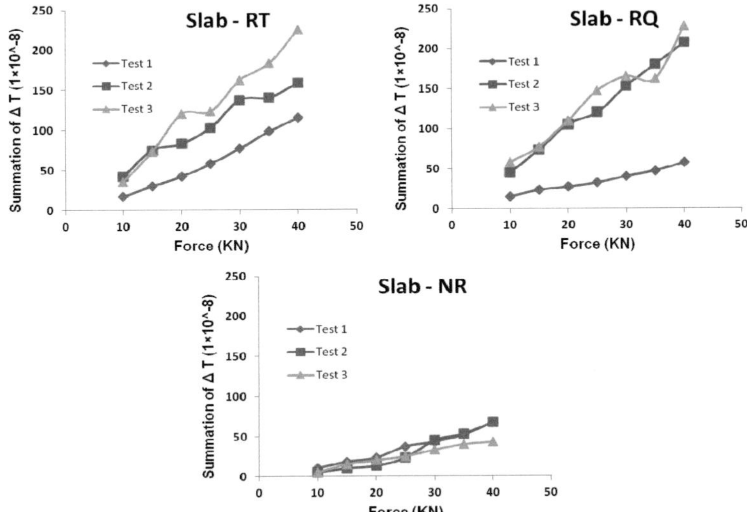

Fig. 5   Summation of time delays measured by Time Shift Method ($\sum \Delta t$) for three tests.

Usually, acoustic linear methods like pulse velocity are not able to detect cracking associated with ASR at early ages. The recent research by Sargolzahi et al. [3] showed no clear changes in ultrasonic pulse velocity of concrete specimens up to 0.09% expansion and after this amount, variations were weak. However, the results recorded by Time Shift method shows reverse conclusion in this study. It means that nonlinear acoustic techniques are very applicable for ASR evaluation of concrete elements and it has a good potential in field measurements for concrete structures. In the case of slabs RT, the amount of time delay increases up to 200% of the primary test for some of the loadings and it is more than 400% for RQ slab. These results are in good correlation with expansion measured by Demec and vibrant strings (Figure 6). On the other hand, the amount of the time shifts was low for nonreactive slab and decrease for third series of the tests, because by passing time, more hydration of cement may increase the density of microstructure of the concrete and fill microdefects into the concrete.

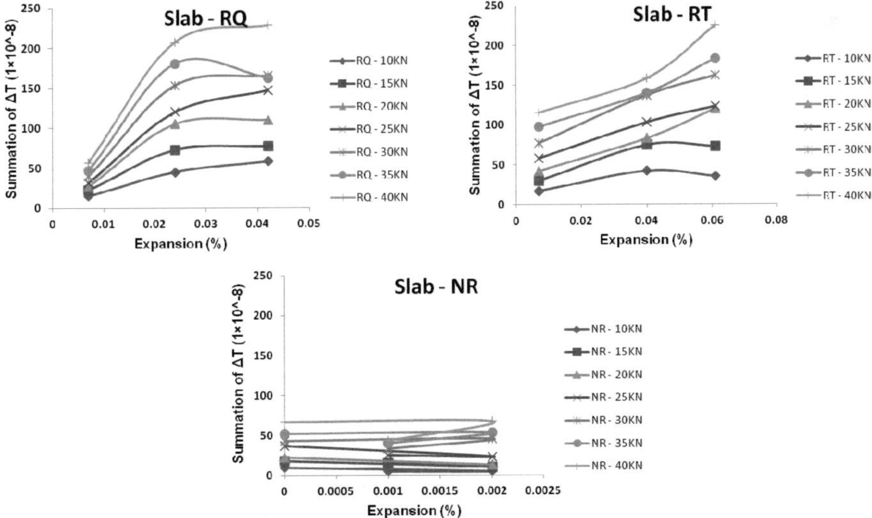

Fig. 6   Comparison of time shifts and expansion for every slab

5   Conclusions

The current research presents application of a new nonlinear acoustic method entitled Time Shift Method in detecting ASR in the field. First, this technique is considered to evaluate ASR in reinforced concrete slabs in the laboratory, and it would be presented field experiences in the near future works. As mentioned previously, two types of the waves are used in the novel method a blast of high frequency and low amplitude waves; an impact by which low frequency and high amplitude wave generated. In real cases like pavements, the later can be generated by vehicles passing. It means that every part of a pavements may be evaluated only by sending high frequency and low amplitude waves when any changes in these waves recorded in a certain time when an impact imposed by passage of a vehicle. This easily shows the efficiency of the method for field surviving. To do this, a research has being run on three concrete slabs; two of them fabricated by alkali reactive aggregates and the rest fabricated by nonreactive aggregates. Expansions measured by Demec and vibrant string gages show the propagation of ASR in reactive slabs and no significant changes in nonreactive slab. Similar results were recorded by the new technique of Time Shift Method. At first when the conventional instruments show no significant expansion, the time shift recorded by this method was relatively low. As long as the expansion instruments showed some observable variations, the amount of the time shifts grew significantly. The recorded results on reinforced concrete slabs confirm the efficiency of the method for damage detections in the field. The novel technique may be used to quantify the concrete damages due to ASR to some extent. In the most of the case, increasing time shift and increasing the time range for relaxation (the time in which microcracks return to initial states before perturbation) means that there is more quantity of damages in the concrete and reverse. It should be considered that the point is not true for all cases and some parameters like amount of humidity of concrete can change both the time shift and the relaxation time.

References

[1]   ASTM C 215-02 (2002): Standard test method for fundamental transverse, longitudinal, and torsional frequencies of concrete specimens. American Society for Testing & Materials, West Construction, Annual Book of ASTM Standards (04.02): Concrete and Aggregates.

[2]   ASTM C 597-97 (1998): Test method for pulse velocity through concrete. American Society for Testing & Materials, West Construction, Annual Book of ASTM Standards (04.02): Concrete and Aggregates.

[3]   Sargolzahi, M., Kodjo, S. A., Rivard, P. and Rhazi, J.: Effectiveness of nondestructive testing for the evaluation of alkali–silica reaction in concrete. In: Construction and Building Materials 24 (2010) No. 8, pp. 1398-1403.

[4]   Kodjo, A. : Contribution à la caractérisation des bétons endommagés par des méthodes de l'acoustique non linéair, application à la réaction alcalis-silice, Ph.D. Thesis in French Université de Sherbrooke/Université Cergy-Pontoise (2008), 127p.

[5]   Payan, C., Garnier, V. and Moysan, J..: Effect of water saturation and porosity on the nonlinear elastic response of concrete. In: Cement and Concrete Research 40 (2010) No. 3, pp. 473-476.

[6]   Kodjo, A. S., Rivard, P., Cohen-Tenoudji, F. and Gallias, J. L.: Evaluation of damages due to alkali-silica reaction with nonlinear acoustics techniques. In: Proceedings of Meetings on Acoustics (POMA), Acoustical Society of America Digital Library 7 (2009) No. 1: 045003-045003-10.

[7]   Moradi-Marani, F. and Rivard, P.: Nondestructive assessment of alkali-silica reaction in concrete: A review. In: International Symposium on Nondestructive Testing of Materials and Structures, May 15-18, Istanbul, Turkey, (2011), 6p.

[8]   Yaman, I. O., Inei, Gokhan, Yesiller, N. and Aktan, H. M.: Ultrasonic pulse velocity in concrete using direct and indirect transmission. In: ACI Material Journal 98 (2001) No. 6, pp. 450-457.

[9]   Snieder, R.: Coda wave interferometry and the equilibration of energy in elastic media. In: Physic Review E 66 (2002), pp. 046615 (1-15).

Freeze-thaw attack on concrete structures

Frank Spörel

Federal Waterways Engineering and Research Institute,
Department of Structural Engineering, Division Construction Materials (B3),
Kussmaulstrasse 17, 76187 Karlsruhe, Germany
Supervisor: Wolfgang Brameshuber

Abstract

The aim of the research was to make a contribution to gathering the relevant parameters of the temperature exposure and the degree of water saturation of the concrete pore structure under different environmental conditions and to evaluate them concerning intensity of freeze-thaw attack on the concrete. Information on the water saturation and temperature exposure of concrete under practical conditions was gained by monitoring of concrete structures. The degree of water saturation was determined by means of resistivity measurements. The results show different degrees of water saturation of concrete members in exposure classes XF1/XF2 respectively XF3/XF4 and a different potential of water to freeze inside the pore structure.

1   Motivation

The descriptive specification in the form of limit values concerning the minimum cement content and the water-to-cement ratio according to the European concrete standard EN 206-1 [1] is based on a classification of the environmental action on concrete. The exposure classes describe the major actions relevant to cause damage on concrete. An assignment of concrete members to exposure classes is based on an evaluation of the regional environmental and operating action. In compliance with the resulting limit values, a proper workmanship and curing a sufficient resistance of the concrete to the action can be assumed for a service life according to EN 206-1 of 50 years.

Concerning freeze-thaw resistance of concrete the descriptive specifications distinguish between "moderate" respectively "high" degree of saturation of the concrete without (exposure classes XF1 respectively XF3) and with chloride exposure (exposure classes XF2 respectively XF4). However there is little data available on the relevant actions concerning freeze-thaw-attack namely the temperature exposure in combination with the degree of water saturation under realistic conditions. The paper focuses on obtaining new findings on these action under different conditions and to evaluate them concerning intensity of freeze-thaw-attack.

2   Laboratory investigations and Monitoring of concrete structures

2.1   Measuring system

To investigate the freeze-thaw-attack on concrete structures two bridges, two tunnels, a lock chamber, a lock economising basin, a quay wall and a waste water treatment plant were equipped with sensors. In each concrete structure several measuring points were installed to observe parts of the construction with a different moisture and temperature exposure. A data logger with a remote control assured an immediate analysis of the data from the building members. Data have been collected for several years.

The non-destructive determination of the degree of saturation of concrete only is possible by indirect measurement methods. A depth-dependent, continuous measurement of the resistivity which can be transferred to the degree of saturation by a calibration in the laboratory can be conducted by the Multiring-Electrode (MRE) [2]. The MRE is a sensor consisting of several rings of stainless steel, each with a thickness of 2.5 mm with an insulating plastic ring in between two steel rings. Cable connections inside the sensor enable AC resistance measurements of the concrete between two adjacent rings of steel using a digital measuring unit at a measuring frequency of 10.8 Hz. The sensors installed consist of nine rings of stainless steel to measure the resistivity in eight steps at a depth between 7 to 42 mm from the concrete surface. Figure 1 shows the set-up of a MRE schematically. By means of a distance piece the measuring depth can be expanded up to 87 mm using two MRE.

Proc. of the 9th *fib* International PhD Symposium in Civil Engineering, July 22 to 25, 2012,
Karlsruhe Institute of Technology (KIT), Germany, H. S. Müller, M. Haist, F. Acosta (Eds.),
KIT Scientific Publishing, Karlsruhe, Germany, ISBN 978-3-86644-858-2

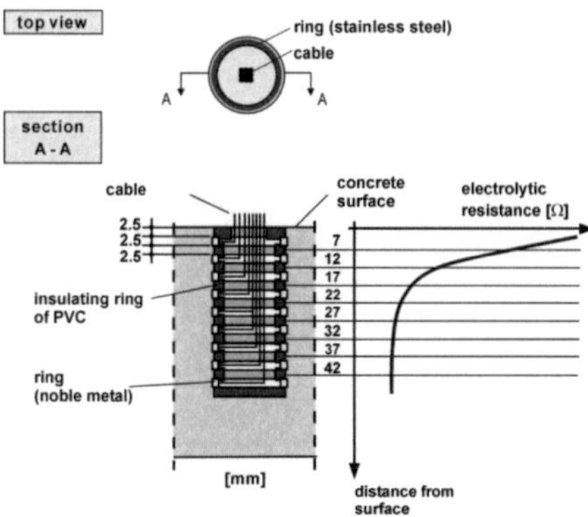

Fig. 1   Schematic set-up of a Multiring Electrode (MRE) [2]

To monitor the temperature exposure a multitemperature probe (MTP) is installed near the MRE. The MTP is equipped with eight PT 1000 probes enabling a temperature measurement in eight different distances from the concrete surface. Furthermore, a compensation of the temperature effects on the resistivity measurement is possible.

2.2   Pore structure

An analysis of the concrete pore structure was performed by means of mercury intrusion and adsorption isotherms. For a comparative evaluation of the results of the mercury intrusion different parameters like porosity, median radius, threshold radius and the fractions of the entire pore volume in different ranges of the pore radii were determined. Adsorption isotherms gave information on the hygroscopic saturation of the different concretes under different environmental humidity situations and allowed to evaluate the results of the measurements of the degree of saturation at the structures. Furthermore information on the pore structure in the range of gel pores is given.

2.3   Calibration

For a conversion of the resistivity into the degree of water saturation of concrete a calibration provided a relation between resistivity and degree of saturation. This relation was determined by a simultaneous measurement of the degree of saturation and the resistivity by means of the Two-Electrode-Method (TEM) at a concrete temperature of 20°C. To minimise the influence of hydration on the resistivity, calibration was started on at least two year old concrete. Different degrees of saturation were adjusted in concrete disks of a diameter of 80 mm and a height of 20 mm.

Water content is the most dominant influencing parameter on the resistivity of concrete. However, besides the concrete specific parameters as e. g. the w/c-ratio, the degree of hydration, the cement type, the content of additions or the carbonation, the resistivity is also influenced by environmental conditions as e. g. temperature variations or the chloride content caused by marine exposure or the application of de-icing agents [2, 3, 4]. Investigations aiming at an indirect determination of the degree of saturation of concrete by means of resistivity measurements have to consider all the influencing parameters to avoid a misinterpretation of the data. Additional calibration tests considered the influence of chlorides and a carbonation of the concrete.

2.4   Consideration of the Influence of temperature on the resistivity

Besides the degree of water saturation concrete resistivity is influenced amongst other parameters by temperature. The calibration functions are only valid for the temperature at which the tests were conducted thus the influence of temperature on the resistivity has to be considered. Whereas concrete specific parameters were considered by a calibration of each investigated concrete, the temperature was taken into account by a compensation routine considering the simultaneously measured resistivity as well as temperature values from the structure. To compensate the temperature influence on the resistivity of concrete the Arrhenius-Equation (equation 1) was used. This equation enables to calculate the resistivity at a temperature of 20°C as in the calibration test from the measured values of temperature and resistivity at the structures. The constant b depends on concrete technological aspects and on the degree of water saturation and can vary within a range of about 2000 to 5000 K [4, 5].

$$\rho_{el} = \rho_{el,0} \cdot e^{\,b \cdot \left(\frac{1}{T} - \frac{1}{T_0} \right)}$$ (equation 1)

ρ_{el} electrolytic resistivity at temperature T in Ωm

$\rho_{el,0}$ electrolytic resistivity at temperature T_0 in Ωm

T, T_0 absolute temperature in K

b constant in K

Because of the above mentioned possible variation range of the decisive constant b data from the structures were evaluated and laboratory test were carried out to account for concrete mix design, degree of water saturation and conditions on site.

Besides TEM-measurements and an evaluation of the data from the structures following [5] MRE-Tests were carried out on partly saturated and sealed concrete cubes with a dimension of 200 mm edge length with installed MREs and MTPs by continuous resistivity measurements at a temperature cycle in a range of +20°C to -15°C. By calculating the degree of saturation by the aid of the calibration functions the correlation of the constant b to the degree of saturation was determined. This way the influence of concrete mix design, concrete age and degree of saturation could be considered. The measurements were carried out at a high concrete age of several years as the data recording for the long term measurement started as a high age as well. Influence of hydration could mostly be neglected.

The application of the compensation of the influence of temperature on the resistivity by now mainly was investigated for temperatures > 0°C, and there are hints at a distinct influence of temperature in a temperature range between 10 and 50 °C [6]. Investigations in temperature ranges below 0 °C are described in [7]. Concerning freeze-thaw attack temperatures below 0 °C are of special interest thus investigations in that direction were performed. Two different states of concrete saturation were investigated whilst the test surface was sealed during the test with cling foil to minimize evaporation or water uptake during the temperature cycle. The first test was performed after storage in climate 20 °C / 65 % relative humidity. After a period of ten days of capillary suction the test was repeated with a higher degree of saturation. In both tests there was a gradient of saturation enabling a measurement over a wide range of concrete saturation.

3   Results

3.1   General

The results are presented using the example of the lock Hohenwarthe near Magdeburg. Fig. 2 shows a sectional view of the lock with four measuring points installed inside the northern side wall of the northern lock chamber. The concrete surface is orientated in southern direction.

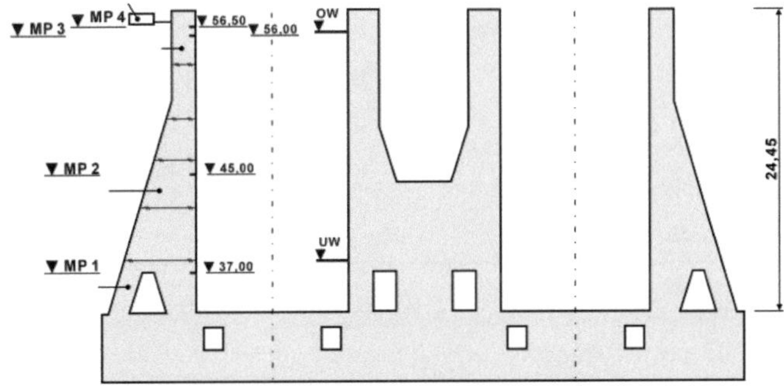

Fig. 2   Measuring points in the side wall of the lock Hohenwarthe

Measuring point MP1 is permanently submerged whereas MP2 and MP3 are located between headwater and tail water and thus in frequent contact to water (XF3). MP4 is unsheltered and exposed to rain and freezing (XF1). The concrete with blast furnace slag cement and fly ash was produced with an equivalent water to cement ratio of 0.47 (k=0.4) without an air entraining agent.

3.2   Pore structure

The results of the mercury intrusion are summarised in table 1 for the concrete of the lock Hohenwarthe. The parameters were determined on cores taken from specimen stored in the laboratory for about three years and on cores taken from the structure (MP1, MP2) at an age of about five years.

Table 1   Results of the mercury intrusion of the Hohenwarthe concrete

Concrete	porosity	Median radius	Threshold radius	Fraction in the range of pore radii			
				r < 5 nm	5 ≤ r < 30 nm	30 ≤ r < 1000 nm	r ≥ 1000 nm
[-]	[Vol.-%]	[nm]	[nm]	[%]			
laboratory	10,4	7	20	35	43	16	6
MP1	11,3	7	17	41	44	9	16
MP2	12,5	7	18	40	40	14	6

The concrete shows a high fraction of pores with a pore radius r < 30 nm which indicates a high fraction of gel pores resulting from the binder combination and high concrete age. The high fraction of gel pores is confirmed by the adsorption isotherm with a high hygroscopic degree of saturation in an environment of 65 to 95 % relative humidity (Table 2).

Table 2   Results of the adsorption isotherm of the Hohenwarthe concrete

hygroscopic saturation in a rel. humidity of … %			
65	76	85	95
%			
51	54	67	79

3.3   Calibration

The calibration function of the concrete of the lock Hohenwarthe is shown in figure 3 for the concrete stored in laboratory conditions and of cores taken from the structure. Differences were observed in the range of degrees of saturation below about 50 %. As the concrete of the cores is most representative the calibration function of the core concrete was used to convert the data from the structure into the degree of saturation. The range of resistivity measured at the structure was covered by the calibration function. At the lock Hohenwarthe the consideration of the influence of chlorides was not necessary.

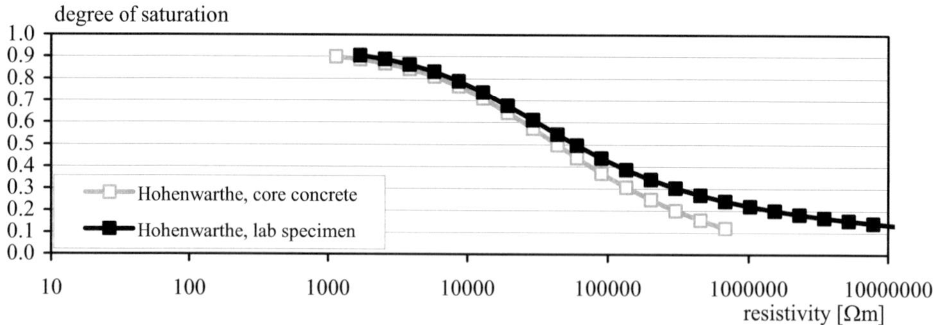

Fig. 3   calibration functions of the concrete of the lock Hohenwarthe

3.4   Influence of temperature on the resistivity

For the application of the calibration function the data from the structure has to be transferred to a temperature of 20 °C by equation 1. A determination of the constant b at temperatures below 0 °C was possible by MRE measurements as described in section 2.4. The constant b decreased with rising degree of saturation. Depending on the different concretes investigated and different degrees of saturation the constant b varied between about 7000 K at low degrees of saturation and about 2500 K at water saturated conditions at concrete ages between about two and five years. The results enabled a sufficient consideration of the influence of temperature on resistivity. The test showed in addition a special effect which occurred at high degrees of saturation at temperatures below about -2 °C. Fig. 4 shows the resistance after compensation of temperature influence according equation 2 during the temperature cycle described in section 2.4.

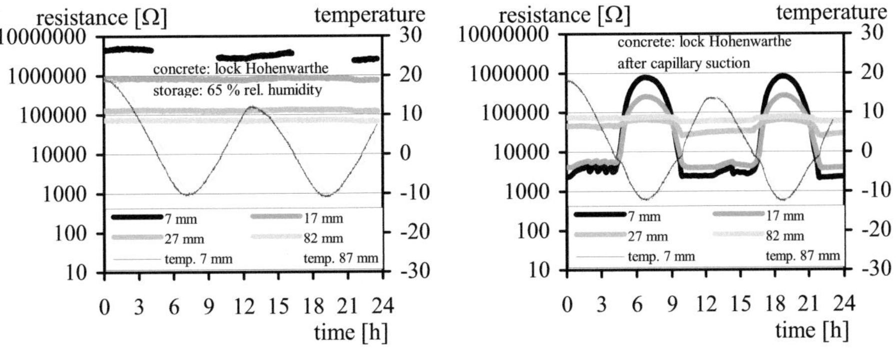

Fig. 4   Effectiveness of the compensation of temperature influence on concrete resistance including temperatures below 0 °C depending on the degree of saturation

Fig. 4 (left) shows an almost linear developing of the resistivity of the concrete during the test after storage in laboratory conditions (20°C/65 % relative humidity). No temperature effects can be detected proving the efficient compensation routine. Fig. 4 (right) shows the test results after a period of capillary suction for about 10 days. At a measuring depth of 82 mm an almost linear developing is observed as well. Closer to the concrete surface an abrupt increase of resistivity is observed when the temperature decreases below about -2°C. The compensation routine according to equation 2 is not able to eliminate these temperature effects at low temperatures for highly saturated concrete. This increase of the resistance is attended by a deviation of the linear temperature decrease in a distance to the surface of 7 mm which is a sign of a phase transformation of water to ice. This means that by resistivity measurements a phase transformation of water to ice can be detected. The fact that the dry concrete (Fig. 4 (left)) does not feature these effects shows that freezing of water in concrete is dependent of the degree of saturation. In rather dry conditions of the concrete during the test mainly fine pores contain water. Freezing, as detectable by resistivity measurements, does not occur under temperatures relevant for practical conditions. By means of the calibration function the minimum degree of saturation at which freezing occurs was determined. The evaluation showed that freezing of water in the pore structure occurred at a degree of saturation in the range of the hygroscopic saturation of the different concretes at a relative humidity of about 95 %.

4   Evaluation of the freeze-thaw-exposure

An evaluation of the data from the structure has shown that at measuring point MP3 of the lock Hohenwarthe high degrees of saturation in the range of capillary saturation of the concrete were calculated during almost the entire observation period (Fig. 5 (left)). A drying of the concrete only took place during an inspection interval of the lock in September 2005. During usual operating conditions of the lock no essential changes of the degree of saturation were observed. At the measuring point MP4 a lower, mainly constant degree of saturation at a distance of more than about 40 mm from the surface was observed corresponding to the hygroscopic saturation of that concrete in a relative humidity of about 95 %. Closer to the surface during winter periods higher medium degrees of saturation and during summer periods lower medium degrees of saturation were determined. During single events also at this XF1 measuring point close to the surface for short periods high degrees of saturation in the order of saturation under atmospheric pressure are possible. Freezing of the water in the pore structure detected by resistivity measurements also occurred under practical conditions marked by arrows in Fig. 5 (right) up to a distance to the surface of 82 mm up to 45 times during one winter.

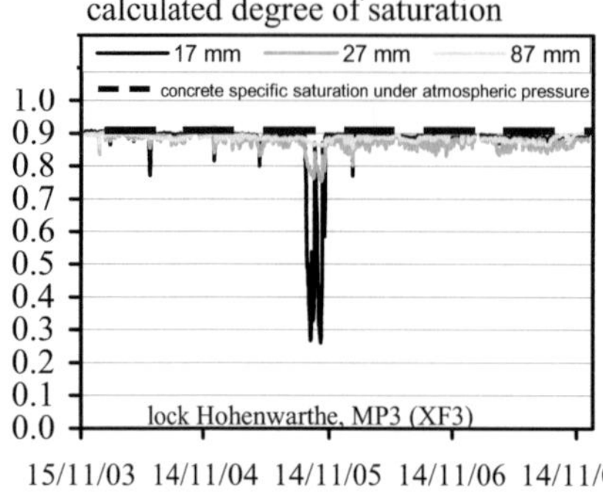 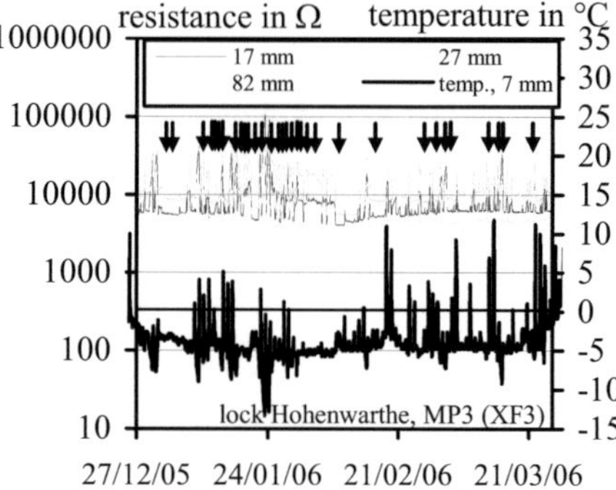

Fig. 5   Calculated degree of saturation and freezing of water in the pore structure as shown by resistivity measurements at MP3 of the lock Hohenwarthe

These effects were observed at other structures at measuring points in exposure class XF3 as well. At measuring points in exposure class XF1 or XF2 similar effects were observed very rarely and less deep than under XF3-conditions.

5   Conclusions

Especially at concrete members under moderate saturation (XF1, XF2) a solely evaluation of freeze-thaw attack by temperature measurements might overestimate the intensity of the freeze-thaw attack on concrete. Damage of concrete by freeze-thaw attack mainly can occurs by freezing of the water in the pore structure in combination with an additional saturation by the micro-ice-lens-pump according to [8]. These effects were observed in the laboratory and under certain conditions resistivity measurements enabled to observed these effects at different distances to the concrete surface at the structure as well under XF3-exposure. The combined measurement of the degree of saturation and the temperature has shown main differences between the freeze-thaw attack on concrete under moderate and high saturation to be considered in future when classifying concrete members to exposure classes.

6   References

[1]   DIN EN-206-1:2001-07; Concrete – Part 1: Specification, performance, production and conformity; German version EN 206-1:2000

[2]   Raupach, M.: Zur chloridinduzierten Makroelementkorrosion von Stahl in Beton. Berlin: Beuth Verlag. – In: Schriftenreihe des Deutschen Ausschusses für Stahlbeton (1992), Nr. 433 (in German)

[3]   Elkey, W., Sellevold, E.J.: Electrical resistivity of concrete, Norwegian Road Research Laboratory, Publication No. 80, 1995

[4]   Bürchler, D.: Der elektrische Widerstand von zementösen Werkstoffen. Modell, Einflussgrößen und Bedeutung für die Dauerhaftigkeit von Stahlbeton. Zürich: Technische Hochschule, Diss., 1996 (in German)

[5]   Schiegg, Y.: Online-Monitoring zur Erfassung der Korrosion der Bewehrung von Stahlbetonbauten. Zürich, Eidgenössische Technische Hochschule, Diss., 2002 (in German)

[6]   McCarter, W. J.: Effects of Temperature on Conduction and Polarisation in Portland Cement Mortar. In: Journal of the American Ceramic Society 78 (1995), Nr. 2, S. 411-415

[7]   Sato, T.; Beaudoin, J. J.: Coupled AC Impedance and thermomechanical Analysis of Freezing Phenomena in Cement Paste. In: Materials and Structures (RILEM) 44 (2011), S. 405-414

[8]   Setzer, M.J.: Mikroeislinsenbildung und Frostschaden. Stuttgart, Ibidem (1999). In: Eligehausen R. (Hrsg.): Werkstoffe im Bauwesen – Theorie und Praxis (Construction Materials – Theory and Application), S. 397-413 (in German)

Behaviour of post-installed expansion anchor under high loading rates

Philipp Mahrenholtz

*Institute of Materials in the Civil Engineering,
University of Stuttgart,
Pfaffenwaldring 4, 70569 Stuttgart, Germany
Supervisor: Rolf Eligehausen*

Abstract

The load transfer mechanism of expansion anchors relies on friction which is sensitive to the rate of loading. Therefore, expansion anchors are potentially at risk to show reduced load capacities when subjected to seismic relevant loading rates. Various experimental tests were carried out to investigate the effects of high loading rates on the frictional behaviour and load capacity in detail. The focus of this paper is put on investigations conducted on the internal friction between the anchor cone and expansion elements as well as on the external friction between expansion elements and the concrete borehole wall. Based on the findings it may be reasonably assumed that in general high loading rates do not have a negative influence on the load capacity of expansion anchors.

1   Introduction

Due to easy handling and straightforward installation, post-installed anchors are very popular and widely used throughout the construction industry. Many products are approved for use in cracked and uncracked concrete and some also for seismic applications. Detailed description of available anchor types, and the associated load transfer mechanisms and failure modes can be found in [3].

As a structure responds to an earthquake, anchors located in the structure are subjected to high loading rates. As known from material sciences, short term loading may influence the material properties in a positive way. Therefore, an anchor generally develops a dynamic load capacity that is higher than the static load capacity ([5], [6]).

However, axially loaded expansion anchors need to be considered in more detail since they also may fail in a pullout mode in which the anchor is pulled out of the borehole without loading the material (concrete or steel) to failure. If the external friction between the expansion elements and the concrete borehole wall is higher than the internal friction between expansion elements and anchor cone (Fig. 1 left), the cone is pulled into the expansion elements and the anchor exhibits a follow-up

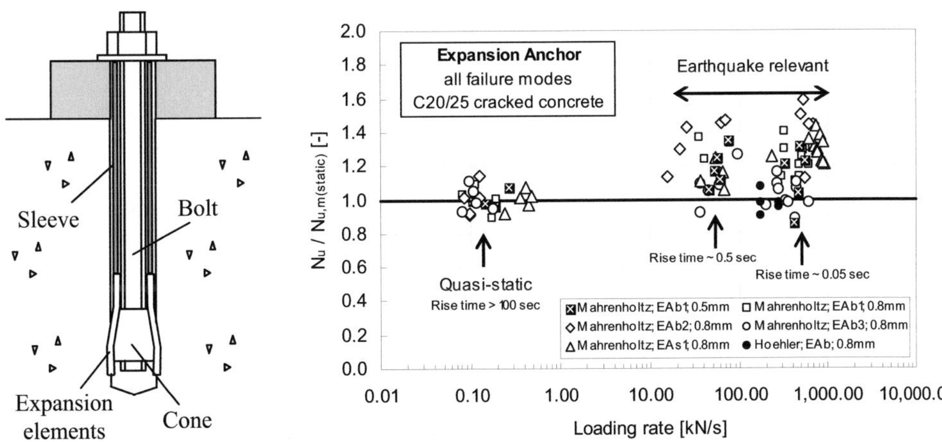

Fig. 1   Left: Schematic of expansion anchor; Right: Results of high loading rate tests on installed expansion anchor pullout tests (after [5] and [10]).

Proc. of the 9th *fib* International PhD Symposium in Civil Engineering, July 22 to 25, 2012,
Karlsruhe Institute of Technology (KIT), Germany, H. S. Müller, M. Haist, F. Acosta (Eds.),
KIT Scientific Publishing, Karlsruhe, Germany, ISBN 978-3-86644-858-2

expansion which is essential for the proper functioning. If the follow-up expansion is insufficient, the anchor will be pulled out and the ultimate load will be reduced substantially.

In order to investigate the effects of high loading rates on the load capacity of expansion anchors, pullout tests on installed anchors with various loading rates were carried out which results are summarized in [6]. The diagram in Fig. 1 (right) depicts the ultimate load normalized with reference to the mean ultimate load under quasi-static loading rate for tests conducted by [5] and [10]. It can be seen that the ultimate load N_u increases for increasing loading rates by trend irrespective of the failure mode.

Although the pullout tests showed that increasing loading rates generally cause an increase in load capacity, increased ultimate anchor loads may also hypothetically indicate the increase of (internal) frictional resistance and, in consequence, the deterioration of the ratio of internal and external friction. For this reason, additional tests were carried out under defined conditions which allowed separate investigations of the effect of loading rate on the internal and external friction. The conducted tests are briefly presented in the following. A detailed report is given in [11].

2 Experimental tests

It is unknown how the external and internal friction are influenced by high loading rates. The literature does not provide load sensitive coefficients of friction for this very specific case. The goal of the investigations was to better understand the load transfer mechanisms of expansion anchors and its effects on the anchor performance, and, ultimately, to decide whether high loading rate tests are required to qualify anchors for seismic applications. Fig. 2 shows the forces acting on an anchor installed in concrete when being pulled out and the corresponding mechanical model.

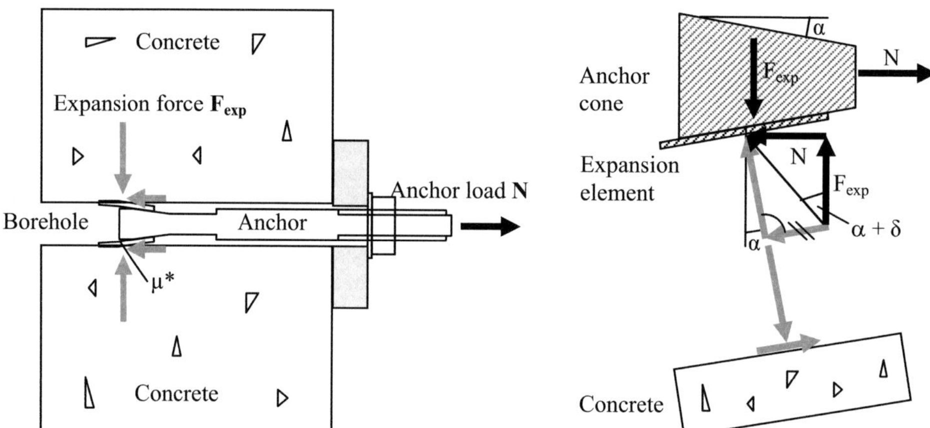

Fig. 2 Left: Forces acting on an anchor installed in concrete; Right: Mechanical model.

2.1 External friction

2.1.1 Experimental procedure and test setup

In [12] the friction behaviour of mechanical anchors was investigated to determine the ratio of N and F_{exp}. Therefore the anchor was installed in between two concrete prisms and pulled out by a manually controlled actuator. The ratio of N and F_{exp} is the tangent of the angle of friction δ^* equal to the coefficient of friction μ^*, where μ^* and δ^* are theoretical values related to a friction plane perpendicular to the axial force N (refer to Fig. 2 left):

$$N / F_{exp} = \tan \delta^* = \mu^*$$

Eq. 2.1

In the following, the equilibrium of forces is considered for the inclined external friction plane of an installed anchor (Fig. 2 right) to determine the coefficient of friction μ acting between the borehole wall and the expansion elements. The expansion force equals to the splitting force integrated over the circumference of the expansion element. With respect to the inclined friction plane (α = angle of the anchor cone), the relation of anchor load and expansion force is after [12]:

$$N / F_{exp} = \pi \cdot \tan (\alpha + \delta_e) \qquad \text{Eq. 2.2}$$

After trigonometric conversion of Eq. 2.2, an equation for the coefficient of friction μ based on μ^* and α can be given:

$$\mu = \frac{\mu^* - \pi \cdot \tan \alpha}{\pi + \mu^* \cdot \tan \alpha} \qquad \text{Eq. 2.3}$$

Eq. 2.3 allows together with Eq. 2.1 the determination of the coefficient of friction μ based on α, N and F_{exp}.

The test setup used in [12] required some substantial changes in order to facilitate high loading rate tests. For this purpose, the test specimen, consisting of the anchor and two concrete prisms, was positioned opposite to a 50 kN servo-controlled actuator. Two braces fixated the concrete prisms. The lower brace was bolted to the strong floor. The upper brace was hold down by means of a traverse and four threaded rods. A load cell was placed in between the upper brace and the traverse. The actuator, fitted with a load cell, was mounted horizontally on a steel abutment which in turn was bolted to a strong floor (Fig. 3).

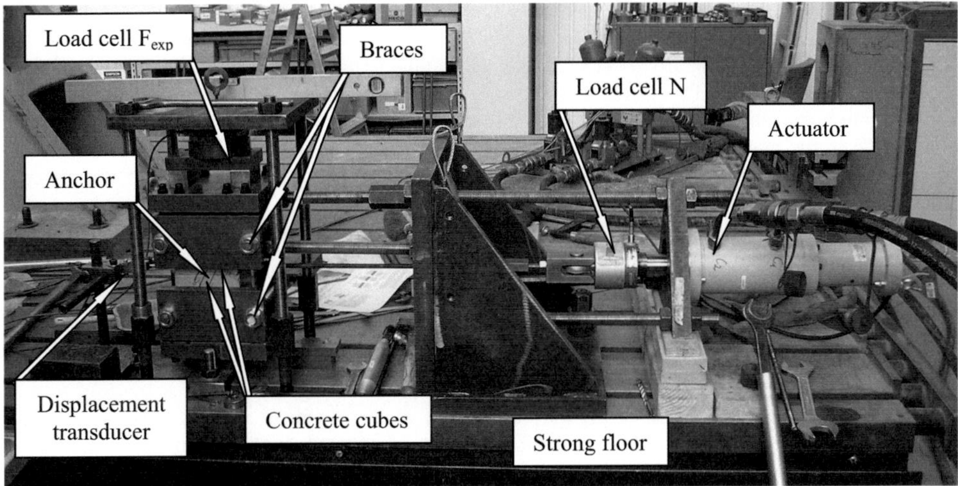

Fig. 3   Test setup for the investigation of anchor load and expansion force.

The expansion elements were expanded by preliminary installation and tack welded to the anchor cone to ensure that the anchor displacement is generated between expansion elements and concrete but not between cone and expansion elements. Next, the anchor and the two concrete prisms were reassembled such that the expansion elements rested exactly in the indentation marks they have been created in the concrete before. The borehole has been drilled through the entire concrete prisms to enable the displacement measurement by a transducer at the unloaded end. By tightening the nuts of the four threaded rods, a preload on the anchor sandwiched between the concrete prisms was created. Adjusting the nuts allowed keeping a uniform and constant gap between the two concrete prisms.

2.1.2 Experimental results and discussion

When the anchor was loaded to failure, the anchor load, the expansion force and the anchor displacement were measured. The external friction was determined for a sleeve-type expansion anchor EAs1 ($\alpha = 13°$) and a bolt-type expansion anchor EAb1 ($\alpha = 11°$) under quasi-static and high loading rates. The anchors were tested in grade C20/25 concrete prisms. Anchor EAb1 was additionally tested in grade C50/60 concrete to investigate the influence of the concrete strength. Another test parameter was the preload acting as a preset expansion force on the anchor. Table 1 comprises the test parameters and the key results regarding the external friction. The friction coefficient was taken as the maximum within the first 3 mm anchor displacement which is in compliance with the procedure applied in [12].

Table 1   Test parameters and key results for experimental tests on external friction

Anchor Type, Angle of Cone	Concrete Strength Class	Preload, kN	Loading Rate	Num. of Tests	Range of Coefficient of External Friction
EAs1; $\alpha=13°$	C20/25	5	quasi-static	3	$0.17 \div 0.37$
			high	3	$0.28 \div 0.45$
EAb1; $\alpha=11°$	C20/25	5	quasi-static	3	$0.30 \div 0.40$
			high	3	$0.22 \div 0.46$
EAb1; $\alpha=11°$	C20/25	15	quasi-static	5	$0.16 \div 0.32$
			high	5	$0.12 \div 0.38$
EAb1; $\alpha=11°$	C50/60	15	quasi-static	5	$0.23 \div 0.33$
			high	5	$0.16 \div 0.37$

As tests on friction are generally associated with substantial scatter, the range of the determined coefficient of friction is large. This effect was already observed by [12] and is more pronounced for high loading rate tests than for quasi-static loading rate tests. Though in theory the coefficient of friction is insensitive to the normal, i.e. expansion force, it is lower for increased preset expansion forces by trend which also complies with the test results presented in [12]. The influence of the concrete strength on the coefficient of friction is not significant.

In conclusion, any potential trend in the influence of loading rate on the external friction coefficient is overcast by the large scatter. Therefore it is reasonable to assume that there is no significant change in the external friction for high loading rates.

2.2   Internal friction

2.2.1   Experimental procedure and test setup

In [7] the theoretical calculation of the pullout resistance of mechanical anchors is shown. The approach for the calculation falls back on indentation tests conducted by [8] to establish a relation of indentation and resulting compressive strength. Increased loading and indentation rates generally results in increased stiffness and strength.

When axially loaded, the anchor cone is pulled through the expansion elements which then indent the concrete due to the cone angle (Fig. 2 right). The resulting compressive force acting on the concrete at that moment equals to the expansion force F_{exp}. The internal angle of friction δ_i related to the friction plane formed by expansion elements and anchor cone and is after [12] given by:

$$\delta_i = \arctan (N / F_{exp}) - \alpha \qquad \text{Eq. 2.4}$$

Comparing N and F_{exp} measured in two test series with quasi-static loading rate (index 0) and high loading rate (index 1) allows evaluating the influence of the loading rate on δ:

$$\frac{\delta_1}{\delta_0} = \frac{\arctan(N_1 / F_{exp1}) - \alpha}{\arctan(N_0 / F_{exp0}) - \alpha} \qquad \text{Eq. 2.5}$$

N / F_{exp} is the ratio of transmission which was investigated in detail for various mechanical anchors in [2]. For the purpose within the scope of this paper it is accurate enough to assume the transmission value to be 1.25 for the EAs1 and EAb1 anchor types. Eq. 2.5 can be further transformed to determine the ratio of δ_1 and δ_0 on the basis of the ratio of anchor load (N_1 / N_0) and ratio of expansion force (F_{exp1} / F_{exp0}):

$$\leftrightarrow \quad \frac{\delta_1}{\delta_0} = \frac{\arctan[((N_1 / N_0)/1.25/(F_{exp1} / F_{exp0}))] - \alpha}{\arctan(1/1.25) - \alpha} \qquad \text{Eq. 2.6}$$

The array of curves in Fig. 4 (left) is the graphical representation of Eq. 2.6 and allows determining the change in the internal angle of friction (δ_1 / δ_0) based on expansion forces evaluated by indentation tests.

For the indentation test, dices of dimensions representing the anchor expansion elements of the EAs1 and EAb1 anchor were pressed by a 250 kN servo-controlled actuator in a concrete prism while the compressive force was measured by the load cell (Fig. 4 right).

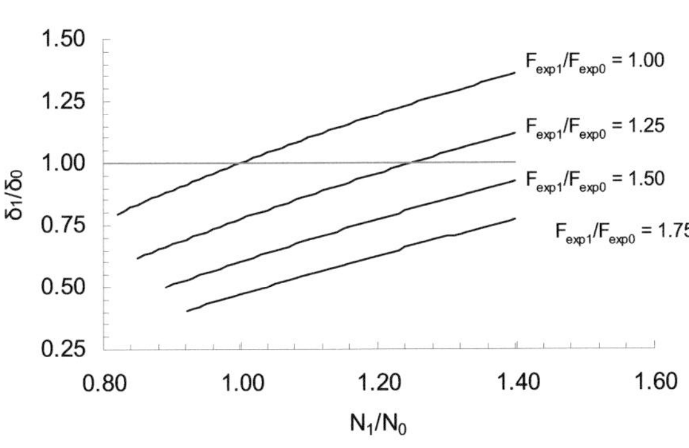

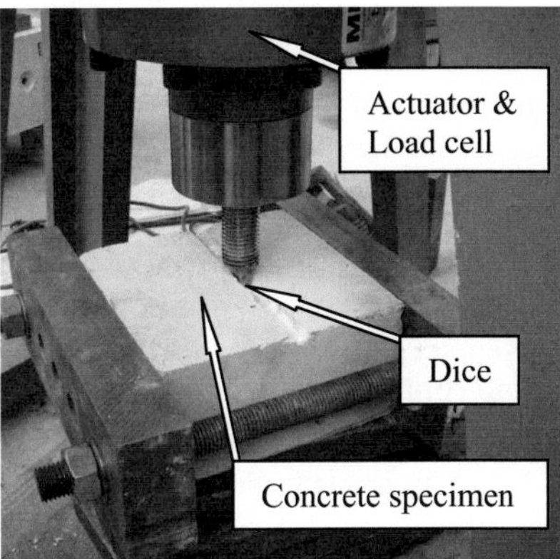

Fig. 4   Left: δ_1 / δ_0 versus N_1 / N_0 for various F_{exp1} / F_{exp0} ; Right: Setup for indentation tests.

2.2.2 Experimental results and discussion

The compressive force measured over indentation increased by approximately 50% for high loading rates if compared to quasi-static loading rates. Therefore, also F_{exp} may be assumed to increase by 50% for high loading rates. With reference to Fig. 1 (right), the increase in anchor load due to high loading rates may be tentatively assumed as 25% on average. With $F_{exp1} / F_{exp0} = 1.50$ and $N_1 / N_0 = 1.25$, the diagram in Fig. 4 (left) reads 0.82 for δ_1 / δ_0 which equals a reduction in internal friction of 18%. The result reveals that the increase in expansion force rather than the hypothetical increase in friction is the reason for increased anchor loads under high loading rates. Table 2 comprises the test parameters and key results regarding the internal friction.

Table 2   Test parameters and key results for experimental tests on internal friction

Anchor Type, Outer Diameter of Dice	Concrete Strength Class	Loading rate	Num. of Tests	Normalized Increase in Compressive Force	Normalized Coefficient of Internal Friction
EAs1;	C20/25	quasi-static	5	-	-
18 mm		high	6	1.53	~0.80
EAb1;	C20/25	quasi-static	5	-	-
12 mm		high	6	1.47	~0.84

3   Summary and conclusions

Considerable number of experimental tests on installed anchors has shown that the mean capacity of anchors tested under high loading rate is generally greater than observed under quasi-static loading. This is true irrespectively of failure mode and type of anchor. However, it was unknown whether this behaviour is accompanied by a potentially adverse change in the ratio of internal and external friction. For this reason, further investigations on expansion anchors were carried out to deepen the understanding of the external and internal frictional behaviour.

For the external friction, the anchor and expansion forces of pulled out anchors were measured by means of a special test setup. Based on the experimental data, the external friction coefficient was determined. The results showed a large scatter overcasting any high loading rate effect. It was not possible to establish a relation between friction and loading rate that was distinguishable from the general scatter in the test data. Therefore, the external friction has to be assumed as relatively constant for variable loading rates.

For the internal friction, indentation tests on concrete specimens were carried out to determine the influence of the loading rate on the concrete resistance when an anchor cone is pulled through the expansion elements. It turned out that the internal friction decreases with increasing loading rate which ensures a proper functioning of the expansion anchor for all loading rates. The reason for increased anchor load capacities despite of reduced frictional resistance is the beneficial effect of increased loading rates on the indentation resistance of the concrete. Increasing expansion forces permit that the anchor load capacity under high loading rates is not reduced in comparison to the capacity under quasi-static loading.

It is concluded that high loading rates are not a critical factor for expansion anchors. In consequence, high loading rate tests are deemed to be not necessary for seismic qualification of anchors. In this respect it is recommended not to incorporate high loading rate tests in anchor qualification guidelines as ETAG 001 [4] or ACI 355.2 [1].

References

[1]   ACI 355.2 (2007): Qualification of post-installed mechanical anchors in concrete, American Concrete Institute (ACI), Farmington Hills, Michigan, 2007.

[2]   Asmus, J. (1999): Bemessung von zugbeanspruchten Befestigungen bei der Versagensart Spalten des Betons (Design of tension loaded fastenings for splitting of the concrete). Dissertation, Institut für Werkstoffe im Bauwesen, Universität Stuttgart, 1999 (in German).

[3]   Eligehausen, R.; Mallée, R.; Silva, J. (2006): Anchorage in concrete construction. Ernst&Sohn, Berlin, 2006.

[4]   ETAG 001 (2006): Guideline for European technical approval of metal anchors for use in concrete, Parts 1 – 6, European Organization of Technical Approvals (EOTA), Brussels.

[5]   Hoehler, M. (2006): Behavior and testing of fastenings to concrete for use in seismic applications. Dissertation, IWB, Universität Stuttgart, 2006.

[6]   Hoehler, M.; Mahrenholtz, P.; Eligehausen, R. (2010): Behavior of anchors in concrete at seismic-relevant loading rates. ACI Structural Journal, Vol. 108, No. 2, 238-247.

[7]   Lehmann, R. (1992): Tragverhalten von Metallspreizdübeln im ungerissenen und gerissenen Beton bei der Versagensart Herausziehen (Load bearing behaviour of mechanical expansion anchors in uncracked and cracked concrete in case of pullout failure), Dissertation, Institut für Werkstoffe im Bauwesen, Universität Stuttgart, 1999 (in German).

[8]   Lieberum, K.-H. (1989): Das Tragverhalten von Beton bei extremer Teilflächenbelastung (Resistance of concrete to external partial area loading), Dissertation, TH Darmstadt 1987.

[10]   Mahrenholtz, P. (2007): High loading rate tests on torque-controlled expansion anchors – Test report. HS II/03-07/03, Institut für Werkstoffe im Bauwesen, Universität Stuttgart, not published.

[11]   Mahrenholtz, P. (2011): Anchor behavior under high loading rates – Summary report and discussion. EII/07-A/01, Institut für Werkstoffe im Bauwesen, University of Stuttgart, 2011, not published.

[12]   Mayer, B. (1990): Funktionsersatzprüfung für die Beurteilung der Eignung von kraftkontrolliert spreizenden Dübeln (Functions substitution test for the assessment of the suitability of torque-controlled expansion anchors), Dissertation, Universität Stuttgart, Institut für Werkstoffe im Bauwesen, 1990 (in German).

Structural behaviours of mortared screw anchors

Andreas Heck

Institute of Concrete Structures - Chair of Building Materials
Technische Universität Darmstadt,
Petersenstraße 12, 64287 Darmstadt, Germany
Supervisor: Harald Garrecht

Abstract

The structural behaviour of metal and composite anchors was subject of numerous scientific projects. Within the scope of those studies a variety of results and design models were developed with which these fixing systems can be described in its various fields of application. For screw anchors installed without mortar there are a lot of experiences and therefore an existing design model was modified to describe the structural behaviour according to the load transfer behaviour. Up to now with mortared screw-anchors only a few experiences have been carried out and there are no proposals to design such a fastening system. The results illustrated within this contribution demonstrate, that in comparison with existing design models an economic design of mortared screw-anchors isn`t possible. Especially with a larger anchorage depth as described in [4] ($h_{ef} \leq 8\ d_0$ - h_{ef} = anchorage depth, d_0 = borehole diameter) significantly higher loads are recorded which couldn´t be predicted according to the currently design model.

1   Introduction

Screw anchors are a common post installing fixing system to anchor tensile and shear loads into concrete constructions. Screw anchors are screwed into a post drilled borehole. They anchor themselves automatically by generating an internal thread by means of a cutting unit. Usual anchor sizes are used to have a diameter of 6 mm to 22 mm and an anchorage depth of 65 mm to 145 mm. Screw anchors can generally be divided into two categories. Screw anchors can be used without special techniques, or they can be mortared. The present elaboration outlines findings and results studying mortared screw-anchor systems.

2   The load transfer behaviour of mortared screw anchors

Mortared screw-anchors are a development of conventional screw anchors. It consists of a screw anchor with or without a connecting thread, a hexagonal head and a mortar cartridge. Before the two-component mortar is injected, the borehole must be cleaned. Afterwards, the anchor is installed with a torque wrench. By screwing in the anchor, the mortar is evenly distributed inside the borehole and around the screw anchor. During the tightening of the anchor the mortar increases along the annular gap. In this way the fit between the thread flanks and the borehole wall is significantly improved (Fig. 1).

Fig. 1   Cross-sectional view of a mortared screw anchor after installation

Proc. of the 9th *fib* International PhD Symposium in Civil Engineering, July 22 to 25, 2012,
Karlsruhe Institute of Technology (KIT), Germany, H. S. Müller, M. Haist, F. Acosta (Eds.),
KIT Scientific Publishing, Karlsruhe, Germany, ISBN 978-3-86644-858-2

As a result of the optimised positive connection there is an increasing by the factor 1.5 of the effective undercut for anchorage compared with standard screw anchors. Thus, the entire thread flank height is available for load application. In case of tensile loads screw-anchors fail with small anchorage depths of 40 mm to 80 mm by a concrete breakout. At greater anchorage depths over 80 mm screw anchors fail by a combination of pull-out and concrete breakout. In this type of failure, the concrete-brackets in the lower area of the anchorage, between the thread pitches, are sheared off (see Fig. 2-section b). After that a reduced concrete breakout is formed out (see Fig. 2- section a).

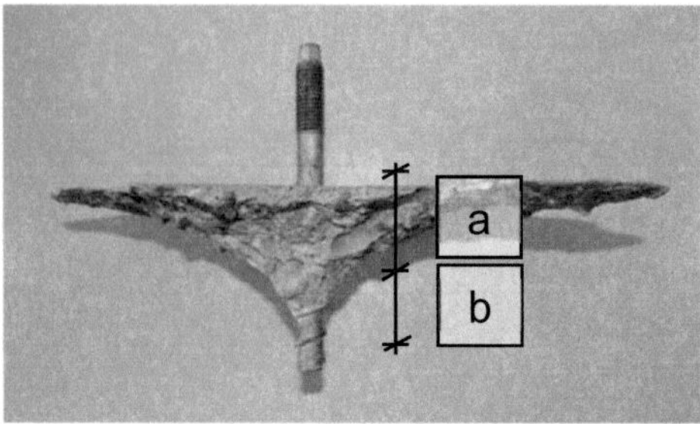

Fig. 2   Typical failure picture of an mortared screw-anchor at a embedment depth of 140 mm

Due to the small size of the concrete breakout it cannot be assumed that the sum of the tensile stresses, based on the lateral surface of the forming eruption cone, is sufficiently for transferring loads as measured in tests., The tensile load is rather introduced by the lower thread flanks (see Fig. 2 -b) into the anchorage base.

3   Studies on the load transfer behaviour of mortared screw anchors

The preliminary experimental program to investigate the structural behaviour of mortared screw-anchors in normal strength concrete C20/25 included experiments with variants of the following parameters:

- screw-in depth of 65 mm to 145 mm
- screw diameter of 8 mm to 16 mm
- reducing the length of thread at a constant depth
- Test specimens of cracked and uncracked concrete (crack width of 0.3 mm)
- mortar A and B

For the provisionally test program a setup was chosen with a wide support, as schematically shown in Fig. 3. With such a structure it can be ensured that the forces, which are introduced by the screw anchors, lead to an anchor typical failure. This device consists essentially of a support device including a hydraulic unit for force transmission. In addition to this an electronic load cell and two inductive displacement transducers are used to record the load-displacement behaviour. To investigate the influence of the mortar, the same device has been modified in the region of the support, to prevent a concrete-breakout. In the diagram in Fig. 4 such an experimental setup with a narrow support is shown. This feature allows to reflect the anchoring-intensity very well. The used measurement technique corresponds to the system which is described in Fig. 3. For the investigations in cracked concrete precast specimens were used. The hairline cracks were generated by mechanical wedge spreaders. To achieve the desired crack width a hydraulic lashing was used. With its help it was possible to control the spreading force which is applied by the wedges. The crack width was checked by two mechanical fine gauge at a distance h_{nom} from the anchor axis. The performance of the tensile test was carried out in the opened crack. For the tests in uncracked concrete the same specimens were used, but without wedges and lashing material.

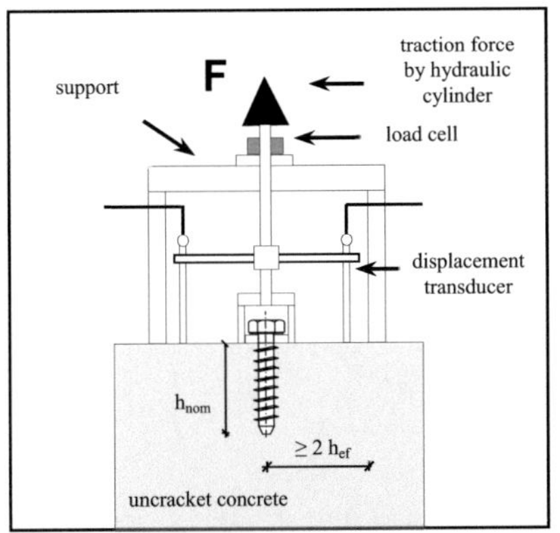
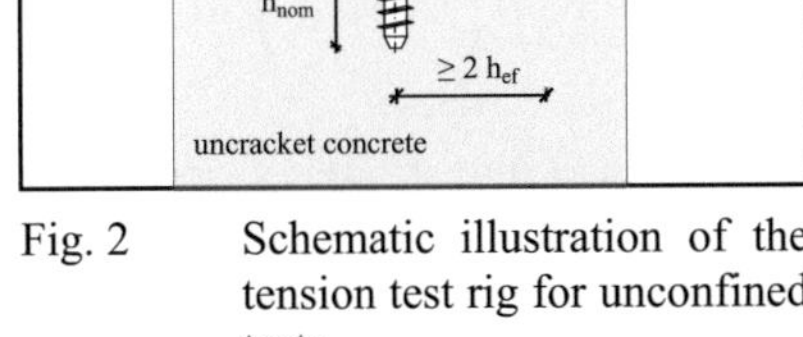

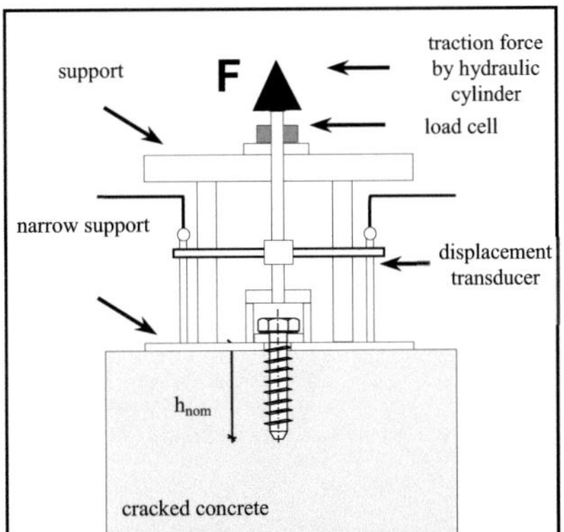

Fig. 2   Schematic illustration of the tension test rig for unconfined tests

Fig. 1   Schematic representation of the tension test rig for confined tests

3.1   The effect of mortar on the load transfer behaviour

Under the original assumption that the mortar used in terms of its composite behaviour has a significant impact on the type of load application, the mortared screw-anchors were tested with two mortars in the experiments, see shown in Table 1. Mortar composition A is a styrene free vinyl ester mortar system, regulated by a technical approval. Mortar-system B is a freely chosen two-component epoxy resin mortar. For the experimental procedure a tension test rig for confined tests (Fig. 4) was used. These tests are primarily used to examine the interactions between the mortar, the anchoring base and the screw-anchor. A total of 40 experiments were carried out. The main values of the test are summarized in Table 1.

Table 1   Influence of mortar composition on the amount of load

Fall	d_{cut}	mortar	execution	$N_{um,Test}$ [kN]
1	16 mm	A	undercut	149
2	16 mm	B	undercut	139
3	20 mm	A	composite	110
4	20 mm	B	composite	62

The investigations have shown, that in case of the undercut of the concrete, the attributes of the mortar have a small influence on the height of the breaking loads. In comparison of case 1 to case 2, the measured failure loads are at a similar level, whereas in a direct comparison of case 3 and case 4, a significant difference can be determined. This leads to the assumption that with the filling of the annular gap with mortar, the contact surface of the concrete consoles is reinforced in the range of undercut. In this way a larger effective area for the introduction of compressive stresses on the flanks can be reached.

3.2   Influence of thread length on the fracture load

To investigate the effects of thread length on the fracture load two series of experiments were performed by using screw-anchors with a diameter of 8 mm, 10 mm, 14 mm and 16 mm. In a series of experiments with a constant screw-in depth, a gradually reduction of the thread length was performed until a significant drop that could be found in the load. The depth of the employed dowel was 85 mm at diameter 8 mm and 125 mm at diameter of 16 mm. The screw thread was reduced from the outlet of thread in steps of 20 mm. The uncut thread lengths were both 80 mm and 100 mm. The evaluation of the investigations has shown, that a thread length of 40 mm is sufficient to transfer the load as a fully threaded screw (Fig. 5).

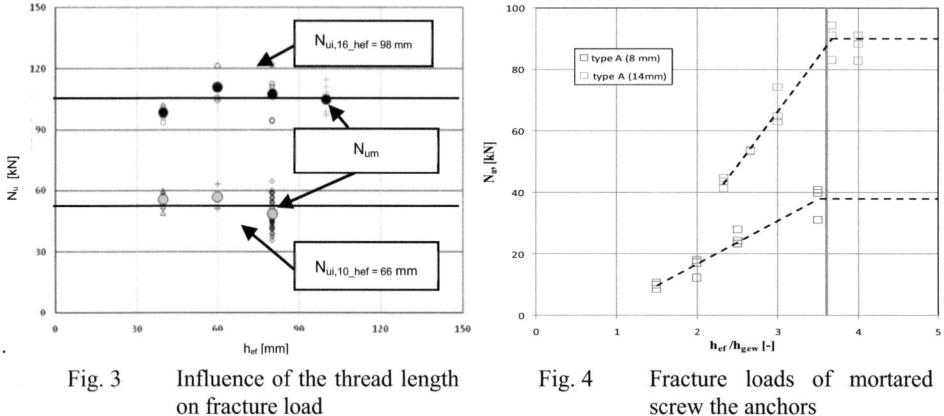

Fig. 3   Influence of the thread length on fracture load

Fig. 4   Fracture loads of mortared screw the anchors

Based on these results a second series of experiments was started by increasing the screw-in depth in a number of steps from 30 mm to 120 mm under retention of the same screw thread length and until no more load increase under tensile was possible. This corresponds to the results of [1] as an enlargement of the screw-in depth related factor of up to 3.6 instead of 1.4 (Fig. 6). From this ratio on there is no significant load increase recognized and it results in pulling out the anchor. In this context, an analogy to the load transfer behaviour of headed studs under the critical maximum bearing pressure can be derived [2] (see Fig. 7). With a decreasing load introduction surface the bearing pressure of the headed studs increases under tensile loading according to [2]. This leads to a larger axial displacement with a simultaneous reduction of the embedment depth and the resulting breaking load. The tensile-load transfer of screw-anchors takes place by the thread flanks. In contrast to headed studs, the maximum breakout cone-depth and the coupled maximum failure load is achieved at a medium load transfer area, in contrast to a fully threaded screw (Fig. 7). For large values of an anchorage depth in relation to threaded-length (h_{ef}/h_{gew}) beyond the size of 3.6 as shown in Fig. 6, it comes with an increasing surface pressure to a significantly increasing displacement. Essentially an increasing of displacement is caused by the local compression of the hardened cement paste matrix in the vicinity of the thread flanks. In this case the surface-pressure of the concrete-brackets is about 10 to 14 x f_c (f_c = 30 N / mm ^{2}). This corresponds approximately to that value, which the critical maximum bearing pressure according to [2] is based on, for ensuring the theoretical failure load (N_u = 15.5 x $h_{ef}^{1,5}$ $f_c^{1/2}$). In this context it can be stated that mortared screw-anchors, having about similar load introduction surfaces as headed studs, are able to form the same concrete breakout loads as headed studs.

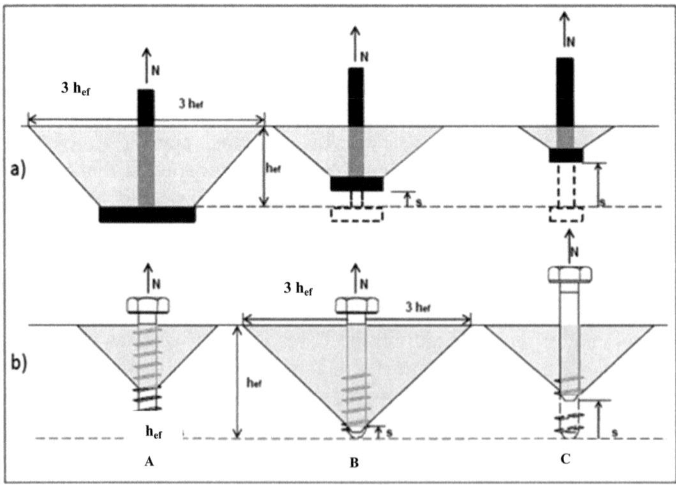

Fig. 7   Concrete breakout cones of headed studs and screw anchors with various load introduction surfaces (A = large, B = medium, C = small)

3.3 Measured and calculated loads in uncracked and cracked concrete

For analysing the comparability of the experimental results with the known approaches of the regulations a total of 373 tests were performed in uncracked and cracked concrete. The experimental setup used within this study corresponds to the schematic depiction shown in Fig. 3. The calculation of the mean concrete breakout load was carried out according to the known formulas of ETAG 001 (see equation 1 and equation 2).

Applied to uncracked concrete:

equation 1[3]: $$F_u^0 = 13,5 * \sqrt{f_{ck,cube}} * h_{ef}^{1,5}$$

Applied to cracked concrete:

equation 2[3]: $$F_{u,c}^0 = 9,5 * \sqrt{f_{ck,cube}} * h_{ef}^{1,5}$$

The effective embedment depth can be calculated using equation 3.

equation 3[2]: $$h_{ef} = 0,85 * (h_{nom} - 0,5 * h_t - h_s)$$

here are:

h_{nom} = embedment depth into the anchorage base material

h_t = slope hight

h_s = height of the crest of thread

The results of pullout tests in uncracked, low strength concrete compared to the calculated values according to equation 1 are shown in Fig. 8. The illustrations of the results are corresponding to the actual embedment depth of the single tests. The evaluations indicate a tendency that the measured concrete breakout loads, performed in tests within the range of small embedment depths (from approx. 50 mm to 70 mm), are tendencially higher than the calculated failure loads in accordance to equation 1. With a further increasing embedment depth (70 mm to 120 mm), the measured values of the investigated screw-anchors are considerably higher in comparison to the relationship line calculated with equation 1. Whereas in cracked concrete the results provide a comparable trend (see Fig. 9). The calculated relationship line, according to equation 2, underestimates the experimental results significantly. It is remarkable that a direct comparison of the measured values in uncracked and cracked concrete is approximately at the same load level. Therefore, the load transfer behaviour of mortared screw-anchors cannot be reproduced by the valid possibilities for the calculation of concrete breakout loads, presented within this contribution.

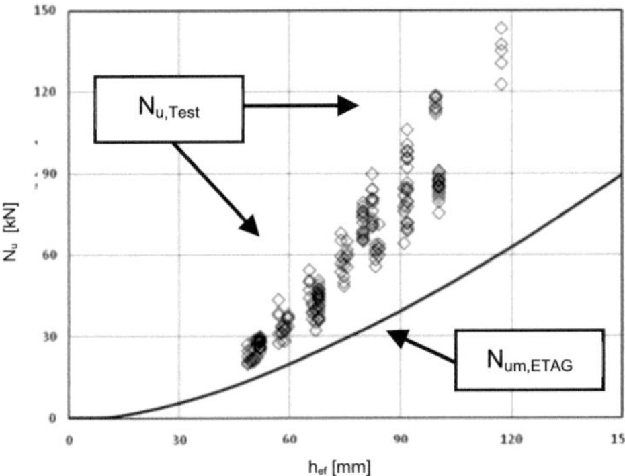

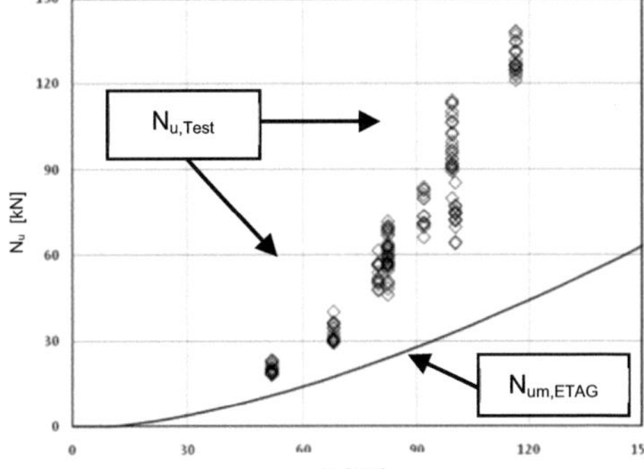

Fig. 8   Comparison of experimental results ($N_{u,Test}$) with calculated values of ETAG 001 Part 3 ($N_{um,ETAG}$) (uncracked concrete)

Fig. 9   Comparison of experimental results (Nu,Test) with calculated values of ETAG 001 Part 3 ($N_{um,ETAG}$) (cracked concrete)

4   Summary and conclusions

The scope of the investigations presented within this contribution, comprised the finding of fundamental arguments to develop a calculation method for the load transfer behaviour of mortared screw-anchors. As the results illustrate it can only partially be concluded from a structural behaviour which is known from guidelines and European Technical Approvals (ETA) to mortared screw-anchors in normal strength concrete. Therefore it is intended to optimise the calculation method for mortared screw-anchors in accordance to the model shown in Fig. 7-b. The optimisation focuses on the extension of the application range of screw-anchors in excess of the current limit of applicability of $h_{ef} \leq 8$ d_0 and the consideration of the load increase, identified in the investigations, towards applicable guidelines.

References

[1]   Küenzlen, J.H.R (2005): Tragverhalten von Schraubdübeln unter statischer Zugbelastung, Institut für Werkstoffe im Bauwesen der Universität Stuttgart, 2005.

[2]   Eligehausen R., Mallée R., Silva J. F. (2006): Anchorage in Concrete Construction, Stuttgart und Waldachtal, 2006.

[3]   European Organisation for Technical Approvals (EOTA): Guidline for European Technical Approval of Metal Anchors for Use in Concrete. Annex C: design methods of anchorages. Edition 1997, 3rd amendment 2010.

[4]   European Organisation for Technical Approvals (EOTA): Guidline for European Technical Approval of Metal Anchors for Use in Concrete. Part 3 – undercut anchors. Edition 1997, amendment 2010.

End anchorage and overlapping of textile reinforcements in textile reinforced concrete

Enrico Lorenz

Institute of Concrete Structures,
Technische Universität Dresden,
George-Bähr-Straße 1, 01062 Dresden, Germany
Supervisor: Manfred Curbach

Abstract

In strengthening existing concrete structures with textile reinforced concrete (TRC), a safe anchoring and transfer of the acting forces is crucial for the functioning of the composite material. This paper deals with the experimental and analytical determination of the end anchorage and overlap lengths of textile reinforcements in TRC members. Based on the experimental research on the bond behaviour between the yarn surface and the fine-grained concrete matrix, an analytical model was developed which can be used to calculate the required end anchorage and overlap lengths in the yarn pull-out limit state. Furthermore, extensive analytical observations were made to predict the probability of failure due to delamination.

1   Introduction

The composite material textile reinforced concrete (TRC) is a new, effective and innovative strengthening method of load bearing concrete structures. It combines the favourable material properties of concrete with those of technical textiles.

With regard to the application of TRC for the strengthening and rehabilitation of existing structures, understanding the load bearing behaviour of the strengthening layers at the construction details is important.

A working bond forms the basis for the static interaction between the individual strengthening components. A load application without damage to this bond is necessary. Hence, extensive experimental and analytical tests have been carried out to examine the decisive failure mechanisms as well as to describe the load bearing behaviour of TRC strengthening in end anchorage and overlap areas.

2   Tests regarding the bond of textile reinforcements in TRC

Knowledge about the bond forces between yarn and matrix is the prerequisite for assessing the bond behaviour of textile reinforcements within TRC. However, due to major influences of textile processing on the bond properties, the results gained from tests on unprocessed single yarns cannot be applied. Consequently, a suitable test setup for asymmetrical pull-out tests on textile reinforcements was developed based on KRÜGER [1] (LORENZ/ORTLEPP [2], [3]). The small-scale slab-shaped specimen used can be manufactured by means of laminating or spraying techniques in a laboratory environment or in the process of quality monitoring of strengthening measures based on TRC. This way, processing induced influences on the bond properties of TRC can be directly observed.

The determination of the multilinear bond stress versus slip relationship (BSR) is based on the analytical modelling of the sample's load carrying behaviour (LORENZ/ORTLEPP [3]). With the help of a simple parameter variation at the supporting points, the BSR can be directly determined using the experimentally determined force-crack-opening relation. With the help of the introduced calculation model, a good approximation to the test results could be reached. In Figure 1, the measured force-crack-opening relation and analytically approximated force-crack opening relationship is exemplarily shown for a biaxial textile fabric consisting of carbon filament yarns in the tested direction.

Subsequently, the evaluation algorithm of the textile pull-out tests developed along with the calculation method enables a reliable investigation of the bond behaviour between the textile reinforcement and the matrix of coated textile fabrics and single yarns.

Proc. of the 9th *fib* International PhD Symposium in Civil Engineering, July 22 to 25, 2012,
Karlsruhe Institute of Technology (KIT), Germany, H. S. Müller, M. Haist, F. Acosta (Eds.),
KIT Scientific Publishing, Karlsruhe, Germany, ISBN 978-3-86644-858-2

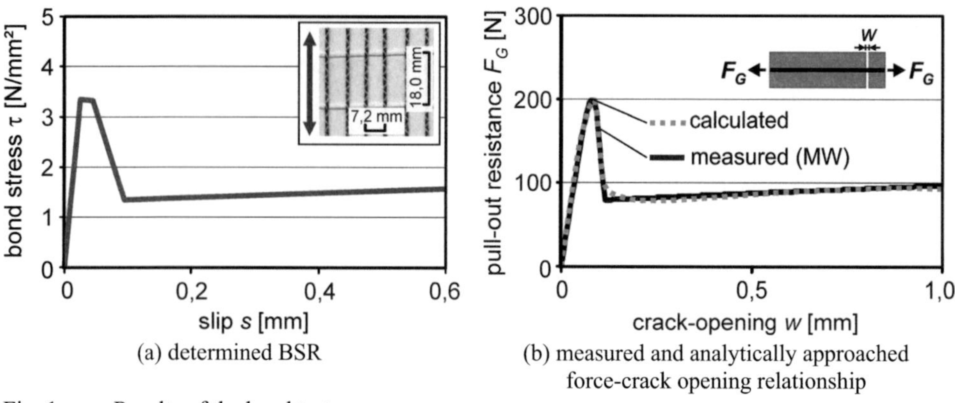

Fig. 1 Results of the bond test

(a) determined BSR

(b) measured and analytically approached force-crack opening relationship

Consequently, the analytically determined BSR forms the basis for the calculation of end anchorage and lap lengths of textile reinforcements in structural members and the strengthening layers of TRC.

3 End anchorage of textile reinforcements in TRC

According to ORTLEPP [4] different failure mechanisms can occur within the anchorage area of TRC. In the course of this work, extensive research concerning the following three failure modes was carried out: delamination failure, collapse of old concrete and destruction of the bond joint between the fine-grained and the old concrete. Owing to their very high tensile strength, especially the load introduction lengths of textile fabrics consisting of carbon fibers are to be determined from the pull-out of the filament yarns of the fine grained concrete. For that reason, further experimental and analytical investigation was necessary to clarify the failure criteria caused by yarn pull-out.

The investigations were carried out by considering different percentages of reinforcement and differently configured coated textile fabrics. The experimental tests were done based on end anchorage tests by ORTLEPP [4]. The test setup is described in LORENZ/ORTLEPP [5]. A comprehensive presentation of the test results can be found in LORENZ/ORTLEPP [2], [5]. Figure 2 illustrates exemplarily the experimentally determined failure loads of single- and two-layered TRC samples with varying end anchorage lengths. The textile reinforcement used was the biaxial textile fabric already presented in Figure 1.

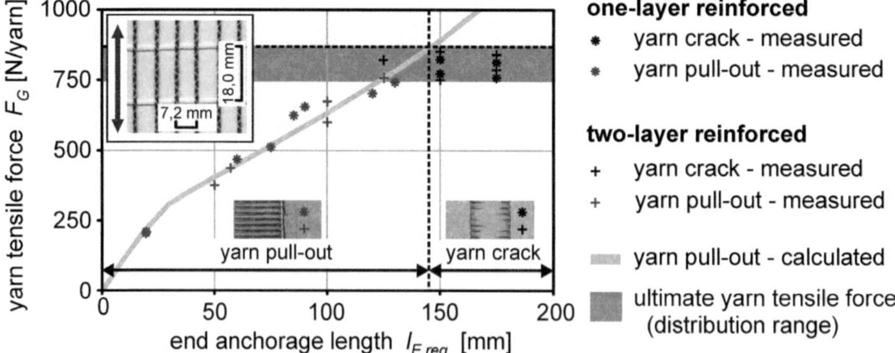

Fig. 2 Comparison of the analytical and experimental tests to determine the in the yarn pull-out limit state end anchorage lengths in the yarn pull-out limit state (LORENZ/ORTLEPP [5])

Two different failure mechanisms were examined. While for anchorage lengths under 125 mm and maximally anchorable yarn tensile forces smaller than the ultimate yarn tensile force yarn pull-out could be observed, anchorage lengths over 145 mm led to a yarn crack on all examined specimens. The maximally anchorable forces vary depending on the scattering of the ultimate yarn tensile force. The end anchorage lengths in the yarn pull-out limit state are independent of the number of textile layers and directly influenced by the magnitude of the bond forces. As has been shown in LO-RENZ/ORTLEPP [6] and [7] an increase in the value of the bond forces between yarn and matrix, for instance by increasing the coating level of the textile reinforcement, reduces the required end anchorage lengths.

Apart from the extensive tests described above, analytical considerations regarding the calculation of end anchorage lengths of textile reinforcements in strengthening layers and structural components made of TRC were made. The tests were conducted applying the analytical calculation method presented in LORENZ/ORTLEPP [5]. In the first step, the bonding behavior of each textile-matrix-system has been experimentally analyzed through the textile pull-out tests presented in LORENZ/ORTLEPP [2] und [3]. In step two follows the analytical modeling of the corresponding incrementally linear BSR, as described in LORENZ/ORTLEPP [3]. In the third step, calculation of the end anchorage lengths required to anchor the relevant yarn- and textile tensile force is done. This can be achieved with the help of a separate model presented in LORENZ/ORTLEPP [5] und [8], which is based on the relations between pull-out force and pull-out length, described in RICHTER [9]. A comparison between the experimentally and analytically determined end anchorage lengths for the examined textile fabric in Figure 2 reveals a good agreement between the failure loads and the corresponding anchorage lengths. The results of the experimental tests confirm the theoretical assumptions. The described relationships were confirmed by multiple additional tests on varyingly configured textile fabrics from carbon and AR-glass. Accordingly, with the help of the relationships presented in LORENZ/ORTLEPP [2], [5], the required end anchorage lengths for textile reinforcements in the yarn pull-out limit state can be determined in dependence of the yarn tensile force. On the whole, with the help of the observed relations and with consideration of the calculation approach regarding delamination and old concrete failure described in ORTLEPP [4], the comprehensive calculation of end anchorage lengths of textile reinforced strengthening layers of fine grained concrete is now possible. Here, the minimum end anchorage length is determined by considering the highest single value resulting from the aforementioned failure modes.

Using the biaxial textile fabric described in Figure 2 as an example, Figure 3 shows the corresponding calculated end anchorage lengths required, under consideration of the bond failure mechanisms. As can be seen, failure by yarn pull-out is relevant for TRC strengthening with one or two layers of textile reinforcements in case of the illustrated carbon textile. However, increased tensile force of the textile reinforced strengthening layer generally necessitates an increase in the required end anchorage lengths to avoid failure by delamination or old concrete failure.

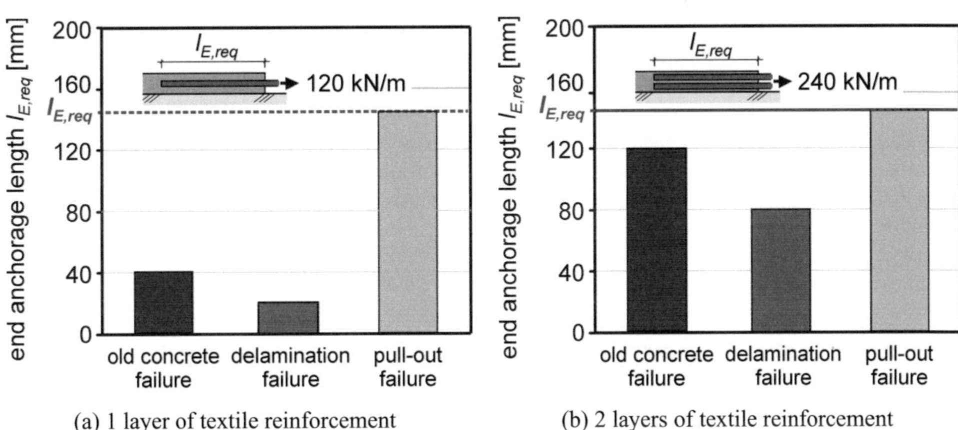

(a) 1 layer of textile reinforcement   (b) 2 layers of textile reinforcement

Fig. 3   Consideration of anchorage failure mechanisms within the overall context (Lorenz/Ortlepp [5])

The displayed relationships present an essential basis for a comprehensive calculation of end anchorage lengths for textile reinforced strengthening layers in fine grained concrete within the overall context.

4 Overlapping of textile reinforcements in TRC

Using strengthening measures with TRC, amongst others described in SCHLADITZ ET AL. [10], lap joints within the textile reinforcement layers usually cannot be avoided.

For the proper functioning the TRC strengthenings, the safe force transmission between the individual composite materials has to be ensured. Consequently, based on uniaxial tension tests by JESSE [11], a uniaxial tension test with a lap joint at the sample centre, as described in LORENZ/ORTLEPP [8], was developed in the framework of tests regarding the load bearing behaviour of TRC samples in the overlapping area. By arranging the reinforcement layers in the sample symmetrically, the lateral tensile stress, which results from the eccentricity of the textile fabrics in the reinforcement layer and contributes to delamination in the lap area, can be taken into account. Experimental and analytical tests on textile-matrix-systems with textile fabrics from carbon and AR-glass in different configurations and varied by the decisive influencing parameters were carried out. Thus, according to LORENZ ET AL. [12], with regard to bond failure different failure modes can occur. As described in LORENZ/ORTLEPP [8], two mechanisms of bond failure are decisive in the lap joint area. On one hand, the bond failure can occur by yarn pull-out from the fine grained concrete matrix, on the other hand, failure due to delamination/longitudinal matrix splitting can happen on the textile layer at the lap joint.

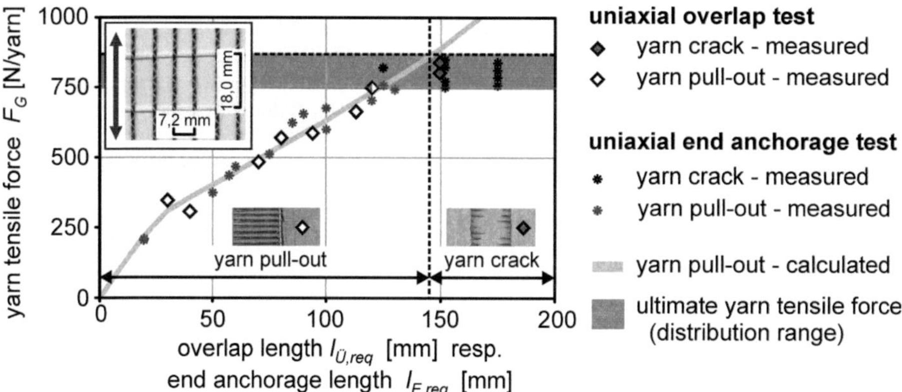

Fig. 4   Comparison of the analytical and experimental tests to determine lap lengths in the in the yarn pull-out limit state (LORENZ/ORTLEPP [8])

Figure 4 illustrates from LORENZ/ORTLEPP [8] the overlap lengths which have been experimentally determined in the yarn pull-out limit state for the biaxial textile fabric, already presented in Figure 1 and 2. The results of the tests for the determination of the end anchorage lengths, described in LORENZ/ORTLEPP [5], are compared. Here, the lap tests and end anchorage tests were carried out in one series based on an identical textile-matrix-system. The results of the tests for the determination of lap lengths in the yarn pull-out limit state are in agreement with the results of the tests for determining end anchorage lengths. The observed failure modes correspond with those of the failure in the end anchorage area. Consequently, with the aim of avoiding yarn pull-out while still making full use of the tensile strength in the exemplarily presented fabric, the required lap and end anchorage lengths can be set to approximately 145 mm. Hence it can be concluded that the distribution of the bond stresses in the lap joint is only marginally influenced by the crack formation compared to the bond stress distribution in the end anchorage area. Thus, it was possible to prove that the lap lengths of the textile reinforcement in the TRC member are 1.0 times the value of the end anchorage length. This relation was confirmed in further tests with differently configured textile fabrics from carbon and AR-glass.

Based on the knowledge that the bond stress distribution in the yarn pull-out limit state is almost identical in the lap- and end anchorage areas of TRC members, the analytical calculation of the lap lengths can be carried out in the same way as has been described for the end anchorage lengths in LORENZ/ORTLEPP [8]. Consequently, the comparison of experimentally determined failure loads with the corresponding analytical calculation values presented in Figure 4 show good agreement.

Accordingly, the required end anchorage- and lap lengths in the yarn pull-out limit state for any multilinear form of the BSR can be determined using the corresponding yarn- or textile tensile force by means of the described relations. Independent numerical tests carried out in ASSAM/RICHTER [13] confirm the presented results.

Failure in the yarn pull-out limit state is usually represented by a significant growth of the crack openings at the beginning and at the end of the lap joint. In contrast to this, sudden failure of the overlap joint can occur in the delamination/longitudinal matrix splitting limit state. The transverse tensile forces trigger failure, which develops through the load introduction and transmission within the lap joint, primarily resulting from the eccentricities of the acting forces, which in turn is a consequence of the bonding forces between yarn and matrix, the distance between the reinforcement layers and the deflection forces on the concrete, as described in LORENZ ET AL. [12] which result from the undulation of the filament yarns in the matrix. If the transverse tensile strength of the concrete is exceeded in the textile reinforcement layer, initially small longitudinal matrix splits occur. These move either abruptly or even under small load increase into the lap area and subsequently lead to failure. Hence, delamination failure of the lap joints should be principally avoided by the purposeful adjustment of the used textile-matrix-system as has been described by LORENZ ET AL. [12].

To guarantee that the results can be transferred to strengthened members, four-point bending tests were carried out on one ply, TRC strengthened RC slabs with the lap joint of a textile reinforcement made of high-performance carbon yarns placed in the middle of the slab. The results of the tests have been presented in Figure 5. The loads which cause textile reinforcement failure obtained in the component tests were calculated based on the experimentally determined ultimate bending resistance by means of the procedure presented in SCHLADITZ, LORENZ & CURBACH [14]. Furthermore, the results of the corresponding uniaxial lap tests have been shown. A good agreement between the results of the uniaxial lap tests and those values determined in the slab tests can be observed.

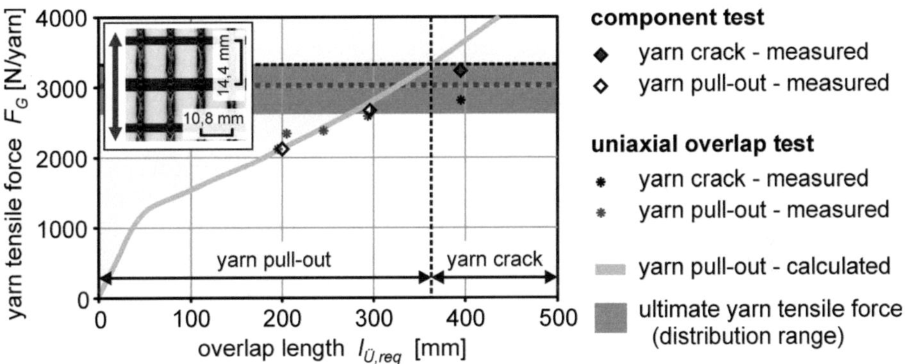

Fig. 5   Results of the component tests to determine the overlap lengths

By using a textile-matrix-system which has been configured based on the tests regarding delamination behavior, described in LORENZ ET AL. [12], the safe transmission of tensile forces in the lap area without longitudinal matrix splitting, delamination and bond damage was proven. Apart from increasing the flexural capacity of structural components reinforced with one layer of TRC in comparison to unstrengthened RC slabs by approximately 150 %, the methods for the determination of lap lengths presented in LORENZ/ORTLEPP [8] were verified.

5   Summary

The carried out experimental and analytical research regarding the load bearing behaviour of end anchorage areas and overlap joints allows the calculation and safe execution of the design details in strengthenings and structural members from textile reinforced concrete. The resulting knowledge for the configuration and improvement of the textile-matrix-system contribute to a more efficient and safe use of the composite material.

References

[1]   Krüger, M.: Vorgespannter textilbewehrter Beton, Universität Stuttgart, Fakultät Bau- und Umweltingenieurwissenschaften, Institut für Werkstoffe im Bauwesen, Dissertation 2004.

[2]   Lorenz, E.; Ortlepp, R.: Berechnungsalgorithmus zur Bestimmung der Verankerungslänge der textilen Bewehrung in der Feinbetonmatrix. In: Curbach, M.; Jesse, F. (Hrsg.): Textile Reinforced Structures: Proceedings of the 4th Colloquium on Textile Reinforced Structures (CTRS4) und zur 1. Anwendertagung, Dresden, 03.-05.06.2009. SFB 528, Technische Universität Dresden, D-01062 Dresden: Eigenverlag - ISBN 978-3-86780-122-5, 2009, pp. 491-502.

[3]   Lorenz, E.; Ortlepp, R.: bond behavior of textile reinforcements - development of a bull-out test and modelling of the respective bond versus slip relation. In: Parra-Montesinos, G.J.; Reinhardt, H. W.; Naaman, A. E. (Hrsg.): Sixth International Workshop on High Performance Fiber Reinforced Cement Composites (HPFRCC6), Ann Arbor, 19.-22.6.2011. Springer, 2011, pp. 463-470.

[4]   Ortlepp, R.: Untersuchungen zur Verbundverankerung textilbewehrter Feinbetonverstärkungsschichten für Betonbauteile. Dresden, Technische Universität, Fakultät Bauingenieurwesen, Institut für Massivbau, Dissertation, 2007.

[5]   Lorenz, E.; Ortlepp, R.: Anchoring Failure Mechanisms of Textile Reinforced Concrete Strengthening of RC Structures. In: Proceedings of the ACI Fall 2010 Convention Pittsburgh (Textile Reinforced Concrete - Modern Developments), 2010.

[6]   Lorenz, E.; Ortlepp, R.: Basic research on the anchorage of textile reinforcement in cementitious matrix. In: 9th International Symposium on Fiber-Reinforced Polymer Reinforcement for Concrete Structures (FRPRCS-9), Sydney, 13.-15.07.2009. Book of Abstracts - ISBN 978-0-9806755-0-4, 2009, pp. 136.

[7]   Ortlepp, R.; Lorenz, E.: Anchoring of textile reinforcements in a fine-grained concrete matrix. In: Kuczma, M.; Wilmanski, K.; Szajna, W. (Hrsg.): Proceedings of the 18th International Conference on Computer Methods in Mechanics CMM2009, Zielona Góra, 18.-21.05.2009. The University of Zielona Góra Press, 2009, pp. 347-348.

[8]   Lorenz, E.; Ortlepp, R.: Untersuchungen zur Bestimmung der Übergreifungslängen textiler Bewehrungen aus Carbon in Textilbeton (TRC). In: Curbach, M.; Ortlepp, R. (Hrsg.): Textilbeton in Theorie und Praxis: Tagungsband zum 6. Kolloquium zu Textilbewehrten Tragwerken (CTRS6) in Berlin am 19. und 20.9.2011. Eigenverlag, 2011, pp. 85-102.

[9]   Richter, M.: Entwicklung mechanischer Modelle zur analytischen Beschreibung der Materialeigenschaften von textilbewehrtem Feinbeton. Technische Universität Dresden, Fakultät Bauingenieurwesen, Institut für Mechanik und Flächentragwerke, Dissertation, 2004.

[10]   Schladitz, F.; Lorenz, E.; Jesse, F.; Curbach, M.: Verstärkung einer denkmalgeschützten Tonnenschale mit Textilbeton. In: Beton- und Stahlbetonbau 104 (2009) 7, pp. 432-437.

[11]   Jesse, F.: Tragverhalten von Filamentgarnen in zementgebundener Matrix. Technische Universität Dresden, Fakultät Bauingenieurwesen, Institut für Massivbau, Dissertation, 2004

[12]   Lorenz, E.; Ortlepp, R.; Hausding, J.; Cherif, C.: Effizienzsteigerung von Textilbeton durch Einsatz textiler Bewehrungen nach dem erweiterten Nähwirkverfahren. Beton- und Stahlbetonbau 106 (2011) 1, pp. 21-30.

[13]   Assam, A.; Richter, M.: Investigation of Stress Transfer Behavior in Textile Reinforced Concrete with Application to Reinforcement Overlapping and Development Lengths. In: Curbach, M.; Ortlepp, R. (Hrsg.): Textilbeton in Theorie und Praxis: Tagungsband zum 6. Kolloquium zu Textilbewehrten Tragwerken (CTRS6) in Berlin am 19. und 20.9.2011. Eigenverlag, 2011, pp. 103-116.

[14]   Schladitz, F.; Lorenz, E.; Curbach, M.: Biegetragfähigkeit von textilbetonverstärkten Stahlbetonplatten. In: Beton- und Stahlbetonbau 106 (2011) 6, pp. 377-384.

Dual confinement of circular concrete columns consisting of CFRP sheets and ties or spirals

Stefan Käseberg

Institute of Concrete Construction,
Leipzig University of Applied Sciences (HTWK Leipzig),
Karl-Liebknecht-Str. 132, Leipzig (04277), Germany
Supervisor: Klaus Holschemacher

Abstract

It is well known, that confinement introduced by CFRP (Carbon Fiber Reinforced Polymer) sheets increases the ultimate compressive strength and ductility of concrete. In past various experimental research programs had been carried out by other researchers to express the increase in strength and strain by the use of CFRP jackets. But in the majority of cases the additional effects of reinforcing elements like ties or spirals are not analyzed very well.

In this paper the investigations on wrapped short concrete columns with and without transverse reinforcement will be shown. In an extensive research program the volumetric ratio of the CFRP jacket as well as the ratio of transverse reinforcement were varied. Thereby, columns with different geometrical shape, different CFRP thickness and with different transverse reinforcement elements (steel ties and spirals) were produced and tested in compression tests.

As a main result it can be shown clearly, that a dual confinement consisting of CFRP and transverse reinforcement increases strongly the ultimate strength and axial strain. The comparison with existing guidelines and models also points out the necessity of new structural models to describe the influence of dual confinement in an exact way.

1 Introduction – CFRP confinement

1.1 Strengthening of members subjected to axial force

Normal transverse reinforcement, like steel ties, specified in design codes for beams and columns has three main functions [1]:

- prevent buckling of longitudinal bars,
- avoid shear failure,
- confine the concrete core.

But this paper addresses only the confinement functions. Confinement is generally applied to members in compression, with the goal to increase their strength and ductility. Thereby, the effective confining pressure f_l, caused by conventional spiral or tie reinforcing steel, can be calculated with equation (1) [3].

$$f_l = \frac{1}{2}\rho_{st} \cdot f_y \quad \text{with} \quad \rho_{st} = \frac{4 \cdot A_{st}}{s \cdot d_s} \tag{1}$$

where ρ_{st} = transverse steel volumetric ratio, f_y = yield stress, A_{st} = cross-section of transverse steel, s = spacing of hoops (spiral), and d_s = diametre of transverse reinforcement.

Besides conventional transverse tie reinforcing steel also advanced FRP (Fiber Reinforced Polymers) materials have only recently recognized as favourable confinement devices. FRP consists of strengthening fibers (for example carbon fibers) in a resign system. The FRP or CFRP confinement appears by orienting the fibers transverse to the longitudinal axis of the concrete member. Through FRP strengthening by confinement, concrete's lateral expansion is efficiently restricted in cases of imposed axial compressive deformation, while the elastic FRP resisting response generates an ever increasing lateral compressive stress state on concrete, leading to structural upgrade of the member core to provide sufficient deformability. For concrete cylinders the confining pressure σ_l can be found from equation (2) [3].

Proc. of the 9th *fib* International PhD Symposium in Civil Engineering, July 22 to 25, 2012,
Karlsruhe Institute of Technology (KIT), Germany, H. S. Müller, M. Haist, F. Acosta (Eds.),
KIT Scientific Publishing, Karlsruhe, Germany, ISBN 978-3-86644-858-2

$$\sigma_1 = \frac{1}{2}\rho_j \cdot \sigma_j = \frac{1}{2}\rho_j \cdot E_j \cdot \varepsilon_j \quad \text{with} \quad \rho_j = \frac{4 \cdot t_j}{D} \tag{2}$$

where ρ_j = volumetric ratio of FRP jacket, σ_j = stress in FRP jacket, E_j = modulus of composite material, ε_j = circumferential strain in FRP jacket (max $f_1 \rightarrow \varepsilon_j = \varepsilon_{ju}$ = ultimate strain of FRP jacket), t_j = FRP thickness, and D = diametre concrete cylinder.

1.2 Stress-strain models of FRP-confined concrete

In past various experimental research programs had been carried out by other researchers to express the increase in strength and strain by the use of FRP or CFRP jacket. Some of them were adopted by design recommendations [4], like the stress-strain model by Lam and Teng in ACI 440.2R-08 [2] or the model by Spoelstra and Monti in technical report by the fib [3]. The stress-strain model by Lam and Teng is illustrated in Fig. 1.

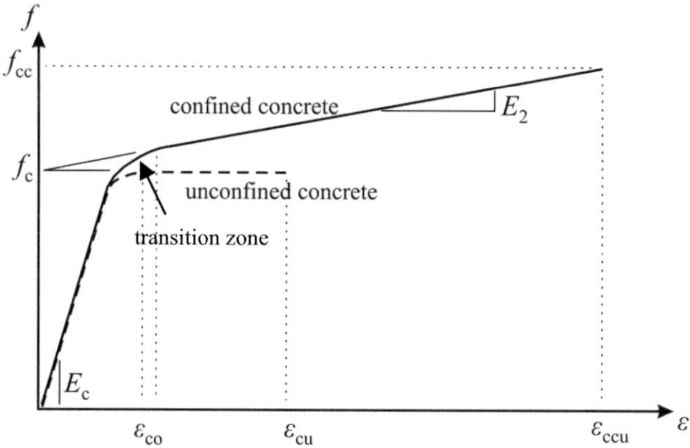

Fig. 1   Stress-strain model for FRP-confined concrete by Lam and Teng [2]

The maximum confined concrete compressive strength f_{cc} can be calculated with the following equations by Lam and Teng (3) or by Spoelstra and Monti (4). Thereby, f_c is the unconfined concrete strength.

$$f_{cc} = f_c + 0.95 \cdot 3.3 \cdot f_1 \tag{3}$$

$$f_{cc} = f_c \left(0.2 + 3\sqrt{\frac{f_1}{f_c}} \right) \tag{4}$$

The ultimate strain ε_{ccu} of the confined concrete member is calculated using equation (5) by Lam and Teng or (6) by Spoelstra and Monti.

$$\varepsilon_{ccu} = \varepsilon_{co} \cdot \left(1.50 + 12 \cdot \frac{f_1}{f_c} \cdot \left(\frac{\varepsilon_{ju}}{\varepsilon_{co}} \right)^{0.45} \right) \tag{5}$$

$$\varepsilon_{ccu} = \varepsilon_{co} \cdot \left(2 + 1.25 \cdot \frac{E_c}{f_c} \cdot \varepsilon_{ju} \sqrt{\frac{f_1}{f_c}} \right) \tag{6}$$

where ε_{co} = unconfined concrete strain at peak stress, and E_c = initial tangent modulus of elasticity of concrete.

Now, like shown in Fig. 1, the second modulus E_2 of the confined concrete can be determined. With the presented models and equations only the contribution of the FRP or CFRP jacket is determined. It is only possible to calculate confined concrete but not reinforced concrete. The contribution of the internal transverse steel reinforcement and other effects like the instability of longitudinal steel reinforcement to the confinement can not be taken into account.

2   Experimental study

2.1   Main goals of research

In the majority of cases the additional effects of reinforcing elements like ties or spirals are not analyzed very well. It results of the limited experimental evidence on the area of FRP confinement of real-size RC columns. Additionally, these limits have not allowed the appropriate implementation of key effects in the current models [4]. Hence the goals of the research work presented in this paper are:

- production of circular columns with different cross sections (from 15 up to 30 cm),
- thereby production of RC elements with different longitudinal und transverse reinforcement,
- wrap with CFRP sheets of different thickness to change the volumetric ratio of FRP jacket,
- increase the knowledge about dual confinement in RC columns.

2.2   Experimental program

In an extensive research program the volumetric ratio of the CFRP jacket (ρ_j) as well as the ratio of transverse reinforcement (ρ_{st}), mainly accountable for effective confining pressure, were varied. Columns with different geometrical shape, different CFRP thickness and with different transverse reinforcement elements (steel ties and spirals) were produced and tested in compression tests. The whole research study is shown in table 1.

During the displacement controlled compression tests two different measurement systems were used. In all the specimens, besides the strain gauges on the FRP jacket (at midheight), two LVDTs were fixed to two opposite sides of each specimen in order to measure the axial shortening. Fig. 2 presents the experimental setup and the stress-strain curves in longitudinal and transverse direction, derived from the compression tests. Thereby, the stress strain curves of series D15 are illustrated. The stress-strain behavior (longitudinal and transverse) of the CFRP confined specimens was bilinear in general, consisting of the three phase behavior reported in Fig. 1. The second modulus E_2 could be observed in longitudinal (E_2) as well as in transverse ($E_{2,q}$) direction (cf. Fig. 2). The failure of CFRP confined plain or steel reinforced specimens was 'explosive' due to the sudden and noisy fracture of CFRP sheets. Furthermore, Fig. 2 also explains the interrelationship between the second modulus and the volumetric ratio of the CFRP jacket (ρ_j). More layers of CFRP produce higher volumetric ratios and higher second modulus. This connection was used for the discussion of the compression tests in chapter 2.3.

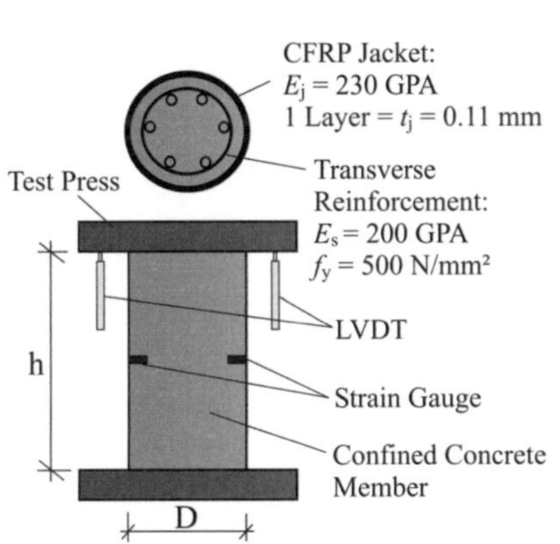

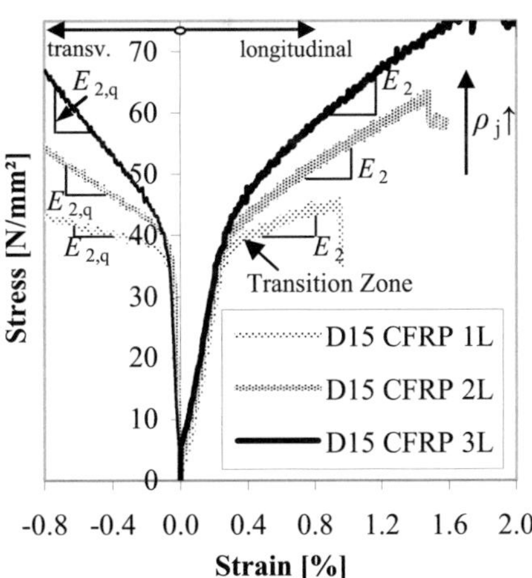

Fig. 2   Experimental setup and derived stress strain curves of CFRP confined concrete (S. D15)

Table 1   Research study on confinement with CFRP wrap and transverse steel reinforcement

1 Series = 3 Specimens	Concrete	D	h	CFRP Wrap (230 GPA)		Transverse Reinforcement		
		[mm]	[mm]	Layer	ρ_i [%]	Ø/s [mm]	ρ_{st} [%]	Type
D15 CFRP 1L	C30/37	150	300	1	0.29	-	-	-
D15 CFRP 2L	C30/37	150	300	2	0.59	-	-	-
D15 CFRP 3L	C30/37	150	300	3	0.88	-	-	-
D15 6/10 CFRP 2L	C35/45	150	300	2	0.59	6 / 100	0.93	steel tie
D15 6/5 CFRP 2L	C35/45	150	300	2	0.59	6 / 50	1.87	steel tie
D20 CFRP 1L	C20/25	200	400	1	0.22	-	-	-
D20 CFRP 2L	C20/25	200	400	2	0.44	-	-	-
D20 CFRP 3L	C20/25	200	400	3	0.73	-	-	-
D20 4/17,5 CFRP 2L	C20/25	200	400	2	0.44	4 / 175	0.17	steel tie
D20 6/17,5 CFRP 2L	C20/25	200	400	2	0.44	6 / 175	0.38	steel tie
D20 6/10 CFRP 2L	C20/25	200	400	2	0.44	6 / 100	0.66	steel tie
D20 6/5 CFRP 2L	C20/25	200	400	2	0.44	6 / 50	1.32	steel tie
D25 CFRP 1L	C20/25	250	500	1	0.18	-	-	-
D25 CFRP 2L	C30/37	250	500	2	0.35	-	-	-
D25 CFRP 3L	C30/37	250	500	3	0.53	-	-	-
D25 CFRP 4L	C25/30	250	500	4	0.70	-	-	-
D25 8/4 CFRP 1L	C25/30	250	500	1	0.18	8 / 40	2.27	steel spiral
D25 8/4 CFRP 2L	C30/37	250	500	2	0.35	8 / 40	2.27	steel spiral
D25 8/4 CFRP 3L	C30/37	250	500	3	0.53	8 / 40	2.27	steel spiral
D25 10/4 CFRP 2L	C25/30	250	500	2	0.35	10 / 40	3.60	steel spiral
D25 6/10 CFRP 2L	C25/30	250	500	2	0.35	6 / 100	0.53	steel tie
D25 6/10 CFRP 2L 1m	C25/30	250	1000	2	0.35	6 / 100	0.53	steel tie
D25 8/4 CFRP 2L 1m	C25/30	250	1000	2	0.35	8 / 40	2.27	steel spiral
D30 CFRP 2L	C25/30	300	600	2	0.29	-	-	-
D30 CFRP 3L	C25/30	300	600	3	0.44	-	-	-
D30 10/4 CFRP 2L	C25/30	300	600	2	0.29	10 / 40	2.93	steel spiral
D30 10/5.5 CFRP 2L	C25/30	300	600	2	0.29	10 / 55	2.13	steel spiral

2.3   Results and discussion

2.3.1   CFRP confined concrete

In this chapter the results on confined plain concrete are explained. Columns with different geometrical shape and different CFRP thickness were produced, to vary the CFRP thickness (t_j) and the diametre of the column (D). Both are responsible for the volumetric ratio of the CFRP jacket.

In the diagrams of Fig. 3 the second modulus E_2 (longitudinal and transverse) is shown as a function of the volumetric ratio of the CFRP jacket. It can be seen clearly, that there is a big influence of the volumetric ratio on the second modulus. In both directions (longitudinal and transverse) it is possible to find a regression curve for mathematical interpretation. Thereby, the volumetric ratio is able to take account of potential size effects. The regression curves of small (D15) and medium specimens (D20 and D25) as well as big specimens (D30) are almost the same. These results permit to conclude that there is no size effect on material behavior of confined specimens.

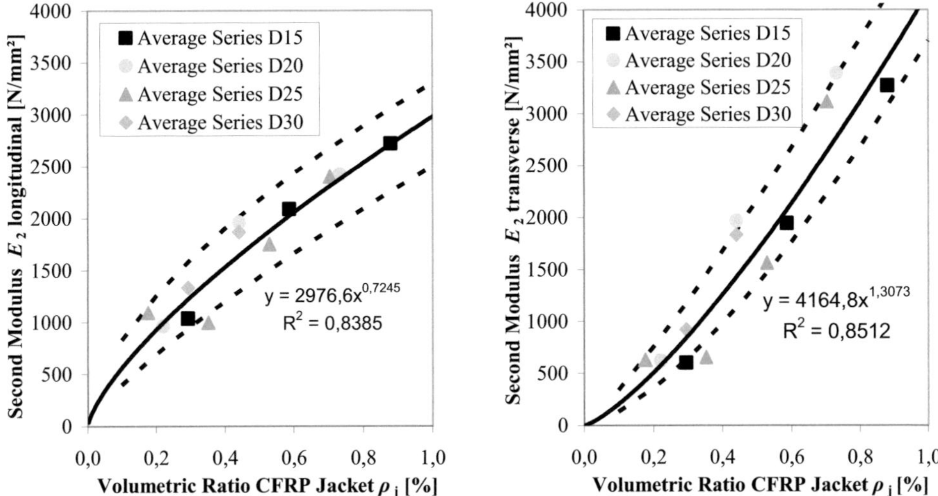

Fig. 3   Second modulus E_2 as a function of the volumetric ratio of CFRP jacket ρ_j.

2.3.2   CFRP-confined reinforced concrete

In this chapter the results on confined reinforced concrete are explained. Thereby, the effect of a dual confinement (consisting of transverse steel reinforcement and CFRP confinement) was the point of interest.

Dual confinement strongly increases the load bearing capacity in general. In transverse direction this is demonstrated in clearly higher second modulus (cf. Fig. 4 a). In doing so the shares of CFRP and steel confinement in E_2 can be summarized. In the diagramm of Fig. 4 b the second modulus E_2 (transverse) is now shown as a function of the transverse steel volumetric ratio. It is possible to find a regression curve for mathematical interpretation. In axial direction in reinforced specimens, an analogous is obvious, where continuous decrease of specimens' axial rigidity occurs. However, this transition zone is more prolonged and smooth than plain FRP confined specimens showed. The following second modulus is similar to E_2 observed with confined plain concrete.

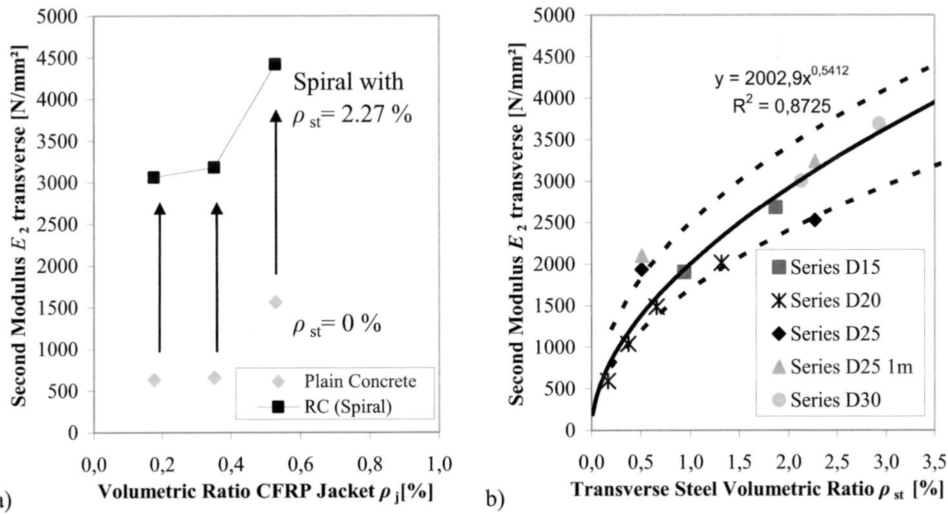

Fig. 4   Influence of transverse steel reinforcement:
a) Second modulus E_2 in transverse direction as a function of the volumetric ratio of CFRP jacket ρ_j for plain concrete and RC with steel spiral
b) Second modulus E_2 as a function of the transverse steel volumetric ratio ρ_{st}

3   Comparison of experimental results and guidelines predictive equations

The material models of Lam and Teng or Spoelstra and Monti only describe the material behavior of FRP confined plain concrete. Hence this chapter only deals with the comparison of the own experimental results on confined plain concrete and results found with the equations (3-6).

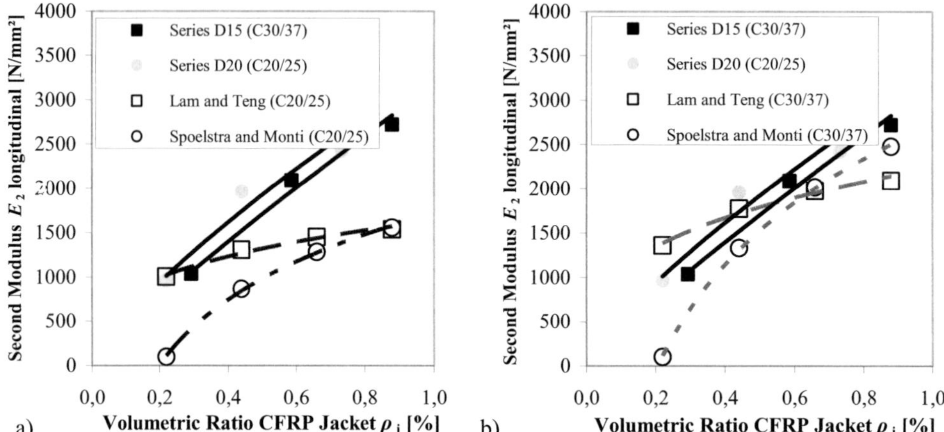

Fig. 5   Second modulus E_2 as a function of the volumetric ratio of CFRP jacket ρ_j:
a) Experimental results versus design equations (calculated for concrete C20/25)
b) Experimental results in opposite to design equations (calculated for concrete C30/37)

Like presented in chapter 1.2 the material models are affected by the concrete properties (f_c, E_c, ε_{cu}). Thus the results found with the equations (3-6) were calculated with the concrete properties of the test series D15 (concrete class C30/37) and D20 (concrete class C20/25). The results for E_2 differ strongly for the different concrete classes (cf. Fig 5 a. concrete C20/25 and Fig 5 b. concrete C30/37). The own experimental results for the series D15 and D20 do not show this influence. In opposite to the theoretical models the regression curves for E_2 as a function of the volumetric ratio of the CFRP jacket do not change because of a different concrete strength or a different concrete modulus of elasticity. Only the volumetric ratios (and confining pressures σ_l) of the CFRP jacket caused changes of the second modulus (cf. chapter 2.3.1).

4   Conclusions

The FRP confinement can significantly increase the strength and ductility of concrete and reinforced concrete. The present study confirms the stress-strain model by Lam and Teng for confined plain concrete. But an influence of concrete properties on the second modulus E_2 could not be observed. Dual confinement effect of steel and CFRP confinement in specimens' transverse E_2 resulted as the total of transverse steel reinforcement and CFRP jacket. Thereby, it is possible to find regression curves, in order to explain mathematically the influence of the volumetric ratio of CFRP and steel confinement on the stress-strain behavior of wrapped concrete.

References

[1]   Paultre, P.; Légeron, F.: Confinement Reinforcement Design for Reinforced Concrete Columns. In: Journal of Structural Engineering 134 (2008) No. 5, pp. 738-749.

[2]   American Concrete Institute (ACI): Guide for the Design and Construction of Externally Bonded FRP Systems for Strengthening Concrete Structures. ACI 440.2R-08 (2008).

[3]   International Federation for Structural Concrete (fib): Design and use of externally bonded fibre reinforced polymer reinforcement for reinforced concrete structures. (2001).

[4]   Rocca, S.; Galati, N.; Nanni, A.: Review of Design Guidelines for FRP Confinement of Reinforced Concrete Columns of Noncircular Cross Sections. In: Journal of Composites for Construction 12 (2008) No. 1, pp. 80-92.

Primary and secondary failure modes in NSM strengthenings

Zsombor K. Szabó

Budapest University of Technology and Economics,
Műegyetem rkp. 3, Budapest 1521, Hungary
Supervisor: György L. Balázs

Abstract

Among all possible strengthening techniques some include the application of fibre reinforced poly-mers (FRP). In near surface mounting (NSM) technique the FRP materials are bonded in grooves cut into the concrete cover. The technique of NSM has many advantages in comparison to externally bonding of reinforcement (EBR). This technique involves a large number of variables. The bond behaviour is of key importance for effective application of the reinforcements.

The main objective of our research was to study the bond behaviour of FRP in NSM applications. An advanced pull-out test setup was developed for that reason. Experiments with an additional double tension-tension test setup were also carried out. This second test setup was used in the framework of an international round robin testing programme on bond. The influence on the bond behaviour of the reinforcements material (fibre materials, adhesives and substrate) and geometrical properties (adhe-sive thickness, reinforcement cross-section, reinforcement surface pattern, edge distance) were in detail researched.

Herein an innovative classification of failure modes of the near surface mounted (NSM) strengthening will be presented based on the results of our experimental studies and the analysis of the available literature data. Based on this primary and secondary failure modes were differentiated. One of the primary failure modes always occur and initiate member failure as it is or as a starting point of secon-dary failure modes. Secondary failure modes are splitting failures of concrete along inclined planes as a result of slip induced by one of the primary failure modes. The presented classification is of key importance for the understanding of the NSM reinforcements bond behaviour.

1 Introductions

In order to preserve our existing structures strengthening is often needed. Among all possible strengthening techniques some include the application of fibre reinforced polymers (FRP). Consider-ably high tensile strength of the FRP material is applied to resist tensile loads. FRP materials can assure proper confinement or flexural resistance of reinforced concrete elements. FRP materials can be used in form of pultruded elements or textiles. Pultruded reinforcements are used in two applica-tion techniques. One is called externally bonding (EBR). This means bonding of strips on the tension side of elements. The other is called near surface mounting (NSM). In NSM technique the FRP mate-rial is bonded into grooves cut into the concrete cover. In order to ensure proper adhesion of the FRP materials to the structural elements these techniques use high strength adhesives. The NSM technique has many advantages in comparison to externally bonding of reinforcement In comparison to other strengthening applications the NSM strengthening is a very complex strengthening method [1][2]. In order to observe the characteristic bond behaviour high number of pull-out test need to be performed owing to the high number of variables. Published experimental data can be used with special regard on the test layout.

2 Experimental background

Our research work started with an extensive literature review of bond test setups and bond influencing parameters of NSM FRP reinforcements. An advanced pull-out test setup was then developed [3]. The advanced L-shaped pull-out specimen (Fig. 1) is formed from a cubic specimen of 250 mm sides with cut-outs. The testing plane is selected to be parallel to one of the diagonal planes in order to have the longitudinal axis of the reinforcement as close as possible to the diagonal plane. The thickness of the

Proc. of the 9th *fib* International PhD Symposium in Civil Engineering, July 22 to 25, 2012,
Karlsruhe Institute of Technology (KIT), Germany, H. S. Müller, M. Haist, F. Acosta (Eds.),
KIT Scientific Publishing, Karlsruhe, Germany, ISBN 978-3-86644-858-2

specimen is the highest in the second diagonal plane (perpendicular to the first diagonal plane). This offers a high stiffness of the specimen in the plane which is weakened by the groove and is usually highly stressed by the bond stresses perpendicular to the reinforcement. The flappers increase the stability and help handling of the specimen. The special L-shaped form of the specimen enables proper view of the supposed failure surface and provides the possibility to measure the displacement on both loaded and unloaded ends.

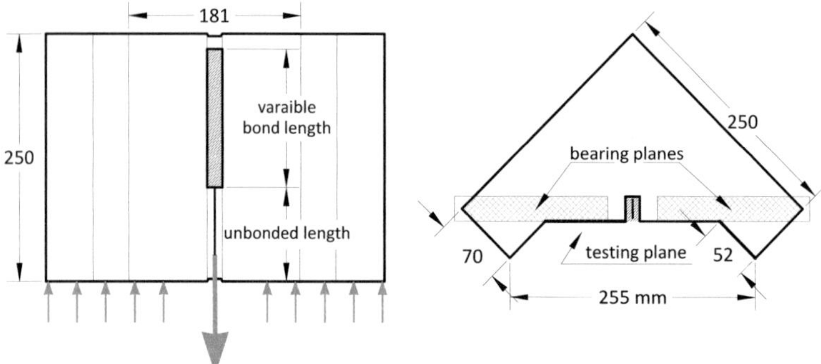

Fig. 1   The newly developed (L-shaped) specimen for bond tests of NSM reinforcements

The tests results obtained by this pull-out test setup were verified using a double tensile-tensile test setup. The test setup used for validation was a double tension-tension test setup [4] used in the Round Robin Test program initiated by the European Network for Composite Reinforcement (EN-CORE) in the European Sixth Framework Program. A double tension-tension test was develeoped to study the bond properties of externally bonded FRP strips. The prismatic specimen was weakened by a steel plate in the middle. FRP strips were mounted on two opposite sides. The gripping of specimens was possible by embedded steel bars at both ends. The specimens were loaded in tension. This double tension-tension bond test setup was adopted by the European Network for Composite Reinforcement (En-core) in Round Robin Test (RRT) programme to investigate bond properties of EBR and NSM reinforcement. The RRT tests program was carried out in several laboratories although herein only results from the Budapest laboratory are presented. The influence of various bond influencing parameters was studied in 65 pull-out tests (L-shaped specimen) and 21 tensile-tensile tests. The influence of bond length, groove width, substrate strength, and adhesive type was studied mostly for strip shaped CFRP reinforcements. In addition to the different rectangular reinforcements round cross-section reinforcements were used with various surface preparations to study the influence of the reinforcement shape and surface preparation on bond behaviour. In addition to pull-out load and slip loaded and unloaded end, in some tests strain along the reinforcement and transverse deformations of the specimens was measured.

3   Failure modes

3.1   Primary failure modes

Based on the results of an extensive experimental study and analyses of available literature data, I suggest the following classification for the failure modes of NSM (Near Surface Mounted) strengthenings. Primary failure modes lead to the global failure or to a secondary failure mode. Primary failure modes are bond failure modes without cracking of the concrete cover. Based on the different failure surfaces three primary failure modes are distinguished.

Two interfaces can be defined first is the FRP-adhesive interface. Bond on this interface is manly influenced by the surface pattern of the bonded reinforcement and the adhesion properties of the adhesives. This interface is the smaller in comparison to the other one and has the main role in the force transfer it will be called generally bond surface. The bond surface to the cross-section ratio is important in case of strip shaped reinforcements or round cross-section reinforcement with small diameter it will be maximized.

The second interface is the so called adhesive-concrete interface. The size of this interface is defined by the groove size .Interfacial failure modes can be separated into two groups: one of pure interfacial failure and one cohesive failure usually close to the interfaces. If failure is taking place in a plane parallel to these interfaces it will be called cohesive failure. Pure interfacial failure is the adhesion failure and it is caused by inadequate adhesive selection or improper surface preparation

A1 FRP failure is the first primary failure mode. In our categorisation failure planes are presented from inside out. The first failure mode is the FRP failure and has two modes. The first one is the **A1a) interlaminar failure** it is caused by the incapability of the FRP materials to transfer shear forces. In externally bonded FRP applications the interlaminar failure is often observed due to the bond stress component perpendicular to fibres which triggers the so called peeling-off failure. This component is balanced in case of the NSM application by the confinement of the concrete.

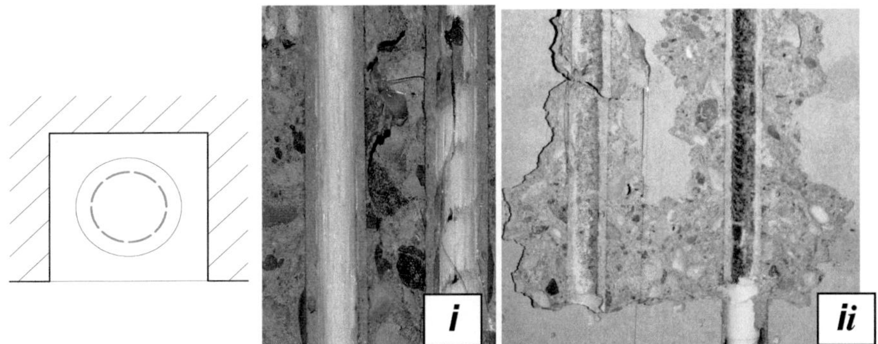

Fig. 2   Failure through the FRP surface pattern (A1b): i) interlaminar shear failure of a GFRP bar surface ribs were sheared-off and ii) shear-off failure of sand coating (BFRP)

The shear **A1b) failure through the FRP surface pattern** (Fig. 2) is considered to be the second FRP failure mode. GFRP bars were tested with surface ribs realized by indentations of the outer layer of the FRP with a helical trajectory. Herby some of the longitudinal fibres were cut, therefore, the effective cross-section of the bar was reduced. Failure of these bars in NSM application (when applied with epoxy adhesive) was a combination of several failure modes including the interlaminar shear failure of the ribs (Fig. 2i). Shear failure of the surface pattern was observed in case of sand coated (BFRP) bars Fig. 2ii).These reinforcements were designed to be applied as internal reinforcements. In case of NSM bars the shear strength of the FRPs contact surface should exceed the shear capacity of adhesives.

A2 FRP-adhesive interface failure is the moust fergvent primary failure mode . Three different failure modes of the adhesive can be defined. **A2a) Adhesion failure** as a pure interfacial failure is one of them. It is rare and it is observed in combination with other failure modes. It was observed in case of plain reinforcements bonded with cement based adhesives where forces should be transferred by mechanical adhesion which was poor due to the plain reinforcement surface. The pure adhesion failure should be avoided by a preparation of the FRP surface .This include removal of any surface impurities (grease and dust) using solvents supplied by the FRP materials supplier. The FRP materials with plain surface can be carefully prepared with a sand-paper to expose external fibres.

A2b) Shear failure of the adhesive with failure of the adhesive between consecutive ribs is shown in Fig. 3i) failure is typical failure mode for embedded steel reinforcement. The failure of consecutive ribs leads to progressive failure. It will be influenced by shear properties of the adhesive and ribs. Such a failure is shown in Fig. 3i) for a GFRP with ribs bar bonded with a low strength cement based adhesive.

As the shear strength and the bond properties of the adhesive increases this failure changes to shear-off failure of ribs (Fig. 2i). The cohesive shear failure in the adhesive was observed with intensive cracking of the adhesive layer with a fishbone pattern (Fig. 3iii). It was critical as the strength of the cement based adhesive was increased for the GFRP bar with ribbs and for CFRP reinforcement with helical wrap and sand coating. The appearance of multiple cracks is a sign of progressive failure with energy dissipation the cracks propagated also to the concrete cover.

The other possible cohesive shear failure of the adhesive with failure plane close to the FRP adhesive interface was observed for plain FRP strips. This was the characteristic failure mode in almost every case when the L-shaped specimen was used to test reinforcement with smooth surface bonded with epoxy base adhesives. Cohesive shear failure for different bond length is shows in Fig. 3ii. The revealed failure surfaces were characterized by a thin layer of adhesive remained on the FRP surface after pull-out. The type of failure was independent form the bond length (35 to 140 mm) and from the groove width. Therefore, we consider that the pull-out load was limited mainly by the adhesive shear strength. The measured residual stress is considered to be the result of friction between the sheared adhesive surfaces. In case of the tension-tension test setup a combination of this failure mode was observed with cracking of the concrete due to the low stiffness of the specimen. The failure is gradually propagating from the loaded end towards the unloaded end as the maximal shear capacity of the adhesive is reached. The failure plane was observed to start from the concrete adhesive interface propagating towards the FRP-adhesive interface. This failure is an indication of a good surface preparation and proper bond at the FRP-adhesive interface.

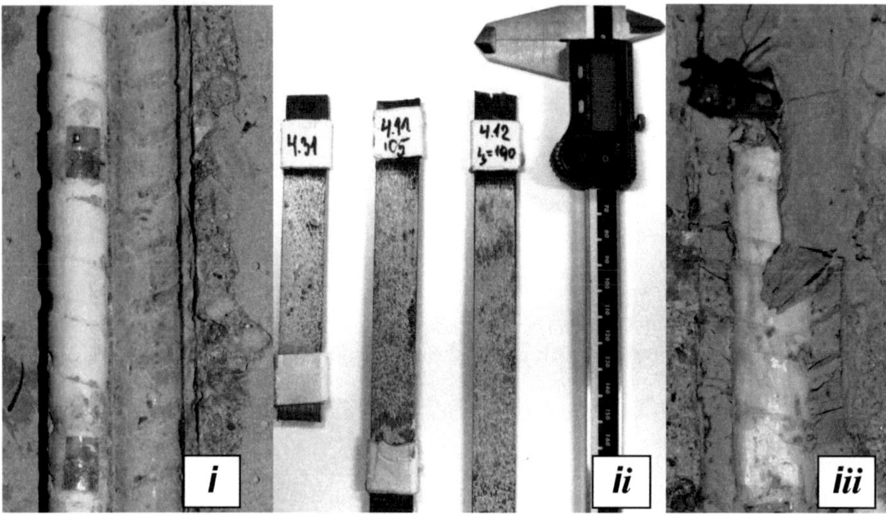

Fig. 3   **FRP-adhesive interface failure:** i) shear failure of adhesive between the reinforcements ribs, ii) cohesive shear failure of the epoxy adhesive near to the FRP surface for strips, iii) fishbone cracking of the cement based adhesive with shear failure of the adhesive

A3. Adhesive-concrete interface failure is the thid primary failure mode it is considered to be rare. This interface is the largest. In case of cut grooves the exposed aggregate surfaces usually guarantee a good bond at concrete adhesive interface especially for epoxy adhesives.

A3 a) Adhesion failure was observed in case of preformed grooves (de Lorenzis, Teng, 2007) for cement and epoxy based adhesives. The concrete surface preparation is very important. Adequate bond requires dry and dust free surfaces. Preformed grooves in situ applications are rather rare because the weaker cement surface should be removed to expose aggregates, or to roughen the adhesive-concrete interface. The surface preparation can be done by water jetting (if epoxy based adheives are used bonding surfaces should be dry) or by bursting.

A3 b) Cohesive adhesive-concrete interface failure was not observed in our experiments. This failure is typical for EBR reinforcement where a thin layer of concrete remains on the adhesive and it is called peeling off failure. It is rare in case of NSM reinforcement, due to the confinement of concrete and teh mechanicak interlocking of the agregates. In EBR applications this failure mode is considered brittle.

3.2   Secondary failure modes

Secondary failure modes are splitting failures along inclined planes (Fig. 4) as a result of slip induced by one of the primary failure modes. The following secondary failure modes are defined based on the starting point of the failure: **B1 Surface splitting (adhesive splitting), B2 Central splitting (mixed adhesive-concrete splitting), B3 Deep splitting (concrete splitting).**

Concrete tensile cracking is caused by the bond stress component perpendicular to the FRP surface. Micro cracks appear in the concrete with fish bone pattern close to the failure load. At ultimate load these cracks join across inclined longitudinal planes. In case of strips cracks are not developing along inclined planes they are running parallel to the concrete surface. Therefore, concrete failure mode was rarely seen in case of rectangular reinforcements with large aspect ratio. In their case the fish bone cracks develop but the bond stresses have a comparatively small components perpendicular to the concrete surface.

The starting level and angle of failure planes is hard to be defined. In case of cement base adhesives (lower strength adhesive) the starting point near the surface of the concrete was observed. Shear failure around the reinforcement was the primary failure, the cover was pushed-off as effect of the bar pull-out. The opening of the inclined planes is comparatively large the cracks are propagating through the concrete surrounding the groove causing reduced damage of the groove corner. Similar failure was observed for plain surfaced CFRP bars (Fig. 4i).

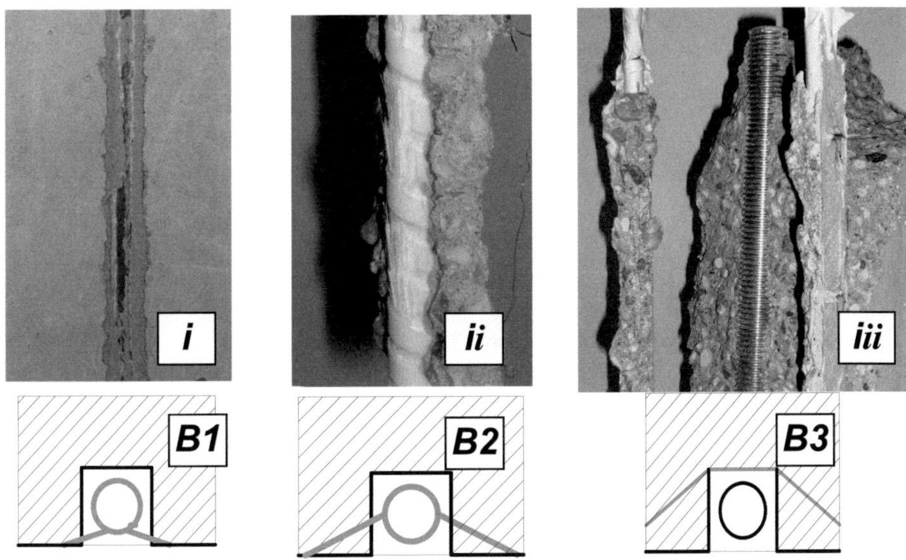

Fig. 4   Secondary failure modes with near surface (B1), central (B2), and deep (B3) starting point

In case of NSM reinforcement the combination of different failure modes is often observed. Primary bond failure starts from the reinforcement level and propagates towards and often through the concrete cover and cause secondary failure modes with concrete cover failure.

As the adhesive strength increased the starting point of inclined planes moved towards the bottom of the groove. In majority of the experiments the failure plane was observed to start from the central level of the reinforcement where the adhesive thickness was the smallest for round bars. This confirms our theory that the shear capacity of the adhesive has an important role in the definition of the inclined planes. Such a failure mode is shown in Fig. 4ii. Another important influencing factor is the surface pattern of the FRP bar which has effect on the radial component of bond stresses. If epoxy adhesives are used their tensile strength is higher than the concrete tensile strength. As the adhesive thickness increases the failure plane shifted with the starting point at bottom of the groove, here this starting level of the failure will be called deep. This failure was not observed in our experiments but a similar failure is shown in Fig. 4iii.

Edge splitting failure is a secondary failure mode with central or deep starting level. It is likely if the reinforcement is close to the member edge. Recommendations minimal edge distance should be defined as a function of the FRP reinforcement splitting tendency and the confinement capacity of the concrete edge. The failure is brittle and is marked by the longitudinal splitting of the concrete edge.

Concrete rip-off failure is likely in case of NSM with high reinforcement ratio. It is a secondary failure mode usually with deep staring level. In comparison to embedded reinforcement radial bond stress component are balanced by a much thinner cover and shear links are not available, unless additional FRP U-wraps are applied [5]. Concrete rip-off is a brittle failure different from the other failure modes due to the sudden and complete loss of composite action. The micro cracks developed around multiple longitudinal reinforcements propagate through the adhesive and join inside the concrete cover before reaching the concrete surface. Cracks join usually in the weakest cross-section for example at the level of internal reinforcement where tension is increased by the bond stress developed at the internal steel bars. Characteristic in case of strip shaped reinforcements where surface perpendicular component of the bond stress is running parallel to the concrete surface. Concrete rip-off and edge splitting failures are likely to develop with strip shaped reinforcements as they have high lateral forces acting on the concrete.

Tensile failure the FRP reinforcement is a material failure. It is considered to be an exceptional failure mode although usually high percentage of the tensile capacity of FRP reinforcement is mobilized for strips with high aspect ratio. This failure is possible in prestressed application were end anchorages and special confinement in the anchorage zone (U-wrap) is applied [6]. This failure is brittle with sudden loss of composite action. It has been rarely observed by non-prestressed strengthening. Structures strengthened with prestressed FRP [7] more frequently fail by fibre rupture because the strain capacity of FRP is used by prestressing.

4   Conclusions

Based on the results of an extensive experimental study and analyses of available literature data, we suggest the following classification for the failure modes of NSM (Near Surface Mounted) strengthenings.

Primary failure modes. Primary failure modes lead to the global failure or to a secondary failure mode. Primary failure modes are bond failure modes without cracking of the concrete cover. Based on the different failure surfaces three primary failure modes are distinguished:

    A1 FRP failure
    A2 FRP to adhesive interface failure
    A3 Adhesive to concrete interface failure.

Secondary failure modes. Secondary failure modes are splitting failures along inclined planes as a result of slip induced by one of the primary failure modes. The following secondary failure modes are defined based on the starting point of the failure:

    B1 Surface splitting (adhesive splitting)
    B2 Central splitting (mixed adhesive-concrete splitting)
    B3 Deep splitting (concrete splitting).

Herein presented classification of failure modes cover the entire range of the failures observed for NSM mounted reinforcements. They can be easily understood therefore are recommended to be used in future scientific publications.

References

[1]   de Lorenzis, L., Rizzo, A. and la Tegola, A. (2002), Modified pull-out test for bond of near-surface mounted FRP rods in concrete. Composites: Part B: Eng., Vol. 33, pp. 589-603.

[2]   Cruz, J. M. S. and Barros, J. A. O. (2002), Bond behavior of carbon laminates strips into concrete by pullout-bending tests. Proceedings of the 3rd International Symposium on Bond in Concrete – From Research to Standards (eds. Balázs et al.), Budapest, pp. 614-622.

[3]   Szabó, K. Zs., Balázs, L. Gy. (2008), Advanced pull-out test for near surface mounted CFRP strips, Proceedings of Challenges for Civil Construction (Eds. Torres et al.), Porto, pp. 92-94, full paper on CD-ROM (8 p.).

[4]   Szabó, K. Zs., Balázs, L. Gy., (2010) Importance of boundary conditions on bond of NSM reinforcement, 3rd fib Congress and PCI Annual Convention, May 29 to June 2, Washington DC, USA, full paper on CD-ROM (12 p.).

[5]   De Lorenzis, L., Teng, J. G. (2007), Near-surface mounted reinforcement: An emerging technique for strengthening structures, Composites: Part B, Vol. 38, pp. 119-143

[6]   Carolin A. (2003), Carbon fibre reinforced polymers for strengthening of structural elements. Technical University of Lulea, PhD thesis, Lulea.

[7]   Nordin, H. and Täljsten, B. (2006), Concrete beams strengthened with prestressed near surface mounted CFRP. Journal of Composite for Construction, Vol. 10, No. 1, pp. 60-68.

Strengthening efficiency of RC beams strengthened with prestressed CFRP laminates

Krzysztof Lasek

Department of Concrete Structures,
Lodz University of Technology,
Al. Politechniki 6, 90-924 Lodz, Poland
Supervisor: Renata Kotynia

Abstract

The paper concerns experimental tests on reinforced concrete (RC) beams strengthened with pre-stressed carbon fiber reinforced polymer (CFRP) laminates. The beams were exhausted to 14%, 25% or 76% of load capacity of non-strengthened beam before strengthening. Strengthening ratio varied in the range from 1,6 to 2,2 of the non-strengthened beam. Research showed high efficiency of strengthening with prestressed strips. Flexural strengthening with prestressed CFRP strips is very effective both for the ultimate and serviceability limit state, especially in case of highly preloaded elements. The active strengthening system causes significant reduction of strains in concrete, mid-span deflections and the increase in stiffness of the tested elements.

1 Introduction

Many research conducted on reinforced concrete members strengthened in flexure with fiber reinforced polymers (FRPs) proved quite low efficiency of strengthening, due to a sudden, premature CFRP strips debonding from the concrete surface induced by flexural cracks [1]. Moreover, it was indicated that the efficiency of the non-active externally bonded strengthening depends on the type of CFRP reinforcement, the distance of the strip's end from the support, the percentage of longitudinal and lateral steel reinforcement as well as on the distribution of bending moments and shear forces in a strengthened element [1, 2]. Although strengthening with non-prestressed composites increased the load capacity of a RC member, it has no significant influence on the serviceability conditions (i.e. cracking moment and deflections). To improve the efficiency of a strengthening on the serviceability limit state and to increase the level of the CFRP tensile strength utilization, prestressing of the CFRP strip has been proposed. This technique allows to reduces deflection of a strengthened element, width of flexural cracks, stresses in longitudinal steel reinforcement, concrete strains and increases the stiffness and the load capacity of a structure.

 Advantages of active strengthening have been confirmed in numerous experimental tests [2, 3, 4, 5, 6, 7, 8], which stated the minimal level of prestressing as 25% of the CFRP tensile strength [3, 4]. In case of CFRP prestressing to the level above 70% of the tensile strength, failure due to a CFRP rupture was observed. On the other hand, prestressing under 60% of the CFRP tensile strength resulted in failure due to the CFRP debonding from a concrete surface [4]. The most effective level of CFRP prestressing has been defined as 50-60% of the CFRP tensile strength, which allows to achieve almost simultaneous CFRP debonding and its rupture.

 The main problem connected with strengthening of RC structures with prestressed CFRP strips is the high shear stress at the end of the strip. The tensile stress in CFRP cannot be transferred to a concrete due to its much lower tensile strength. To overcome this problem, mechanically anchored steel plates should be applied. Pioneer non-mechanically anchored system was developed by Prof. Urs Meier at the Empa laboratory in Zurich [7].

 Strengthening of existing RC members needs consideration of a preloading level of a structure before strengthening. Analysis of experimental test results published so far shows that the effect of the preloading level on the strengthening efficiency has been very rarely taken into consideration. To verify this influence, the author proposes a test program containing the elements strengthened under three different preloading levels, caused by a dead load and/or additional external load equal to 14%, 25% and 75% of the non-strengthened member's load capacity. The next considered parameter of the presented tests is the verification of a bond effect on the strengthening efficiency. Hence, in one case

Proc. of the 9th *fib* International PhD Symposium in Civil Engineering, July 22 to 25, 2012,
Karlsruhe Institute of Technology (KIT), Germany, H. S. Müller, M. Haist, F. Acosta (Eds.),
KIT Scientific Publishing, Karlsruhe, Germany, ISBN 978-3-86644-858-2

the prestressed CFRP laminates were applied as externally bonded reinforcement along their length and entirely unbonded like a bowstring anchored at its ends in the second case.

2   Program description

The research program, carried out in the laboratory of the Department of Concrete Structures at the Lodz University of Technology, contained two series of beams consisted of seven beams in total. Series I contained four beams (P1, P2, P3 and P4) and series II contained two beams (P5 and P6). The main difference between both series was the longitudinal tensile reinforcement (four $\varnothing 12$ mm bars in series I and four $\varnothing 16$ mm bars in series II). All beams were reinforced with four bars $\varnothing 8$ mm at the beam's top. Transverse reinforcement consisted of $\varnothing 8$ mm stirrups at 150mm spacing. The beams were strengthened with a single CFRP strip of 100 x 1.20mm cross-section (Fig. 1). RC members were tested in six point loading as a simply supported beam over a span of 6000mm, with a cross-section of 500 x 220mm.

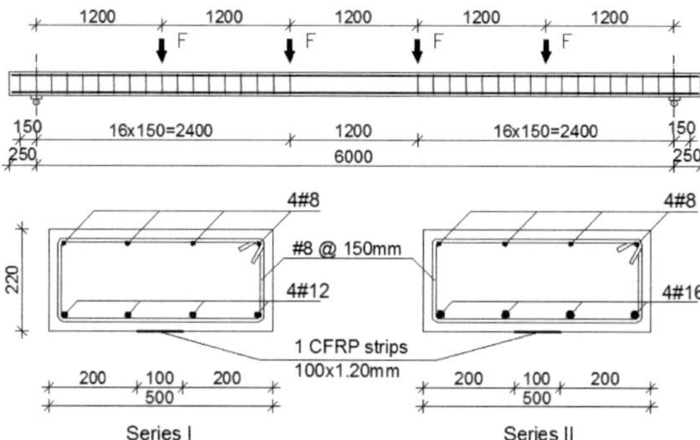

Fig. 1   Static scheme, steel reinforcement and strengthening mode of the beam.

The main variable was the level of the beam's preloading before strengthening. The beams were strengthened under three different preloading levels provided constantly until epoxy resin reached its full strength. To investigate the beam's preloading effect, beams P1, P3 and P5 were strengthened under their dead load (corresponding to the load provided by external forces of $2F_p$=6.1kN), equal to 25% (P1 and P3) or 14% (P5) of the load capacity of the non-strengthened beam. Beams P2 and P4 were strengthened under external load of 2F=13.7kN, the beam P6 under external load of 2F=27.7kN, which with a dead load of 2F=6.1kN equals to $2F_p$=19.8kN (P2 and P4) and $2F_p$=33.8kN (P6), that corresponds to 76% of the non-strengthened beam's load capacity.

The beams were concreted using concrete class C30/37 and strengthened in the set-up. The CFRP strips were bonded to the concrete surface with the epoxy adhesive. Reaching required stress level in the strip allowed to block the anchorage plates, that were removed after next 12 hours (Fig. 2).

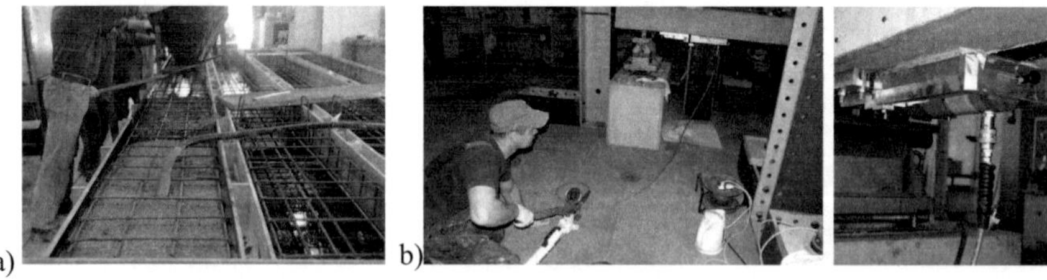

Fig. 2   a) Concreting process   b) Prestressing CFRP laminates.

The CFRP strip was not bonded along the span in beams P3 and P4, behaving like a bowstring by transferring the prestressing force to the beam in the anchorage points.

The following parameters were registered during the test: concrete strains in the tension and compression zone, steel reinforcement strains in the beam's midspan, CFRP strains and vertical displacements of the tested beam. Simultaneously, the crack pattern and the crack widths were registered. Additionally, the vertical reactions were measured with fourload cells, two under each support.

Prestressing levels of the strip for each member are shown in Table 1. In case of the beam P1, after preloading to the level of 2F=25.9kN, six cycles of unloading and loading were performed. Afterwards the beam was loaded until failure.

Fig. 3　View of bottom side after beam failure in the beams: a) P1, b) P4, c) P2.

3　Analysis of tests results

Failure of each beam strengthened with bonded strip (without P3 and P4) was caused by the CFRP strip debonding from the concrete surface, induced by the intermediate crack debonding (ICD) developing to one of the supports. Depending on the anchorage system, failure resulted in the CFRP sliding from the anchorage plate or total CFRP debonding along the full length (Fig. 3a). Failure of the beam P4 strengthed with anunbonded strip occured due to sliding of the strip from the anchorage system (Fig. 3b). After CFRP debonding, the "fishbone-shaped" crack layout in a bottom concrete cover was observed (Fig. 3c).

Table 1　Summary of the test results

Beam		$2F_p$ (kN)	$2F_p$ (-)	$2F_u$ (kN)	$2F_u$ (-)	ε_{fp} (‰)	σ_{fp} (MPa)	$\varepsilon_{f,test}$ (‰)	η_ε (-)	Failure mode
Series I	P1	6.1	0.25	52.6	2.19	5.20	900 0.32 f_{fu}	9.3	0.87	CFRP debonding; strip's end sliding from anchorage system
	P2	19.8	0.76	48.3	1.86	4.75	822 0.29 f_{fu}	6.85	0.69	
	P3	6.1	0.25	46.8	1.95	4.60	796 0.28 f_{fu}	6.9	0.68	Concrete crushing
	P4	19.8	0.76	45.1	1.73	4.40	762 0.27 f_{fu}	5.0	0.56	CFRP end sliding from anchorage system
Series II	P5	6.1	0.14	74.3	1.69	4.80	831 0.29 f_{fu}	8.0	0.76	CFRP debonding; strip's end sliding from anchorage system
	P6	33.8	0.76	71.7	1.63	4.85	840 0.29 f_{fu}	7.15	0.71	

F_p　- preloading load
F_{u0}　- ultimate load of reference beam (cal.),
　　　$2F_{u0} = 24kN$ (for P1 and P3), $2F_{u0} = 26kN$ (for P2 and P4), $2F_{u0} = 44kN$ (for P5 and P6)
F_u　- ultimate load of strengthened beam (test)
ε_{fp}　- strain of prestressed CFRP strip
σ_{fp}　- stress of prestressed CFRP strip
f_{fu}　- ultimate tensile strength of CFRP (test), $f_{fu} = 2857MPa$
$\varepsilon_{f,test}$　- CFRP strain increase (registered in the test for full range of loading)
ε_{fu}　- ultimate tensile strain of CFRP strip (test), $\varepsilon_{fu} = 16.5$ ‰
η_ε　- CFRP strain efficiency, $\eta_\varepsilon = (\varepsilon_{fp} + \varepsilon_{f,test}) / \varepsilon_{fu}$

Independently from the beams' preloading level during application of the strip, CFRP debonding occurred when the strain increase in the CFRP strip reached value in a range from 6.85 to 9.3‰ for the beams'series I which failed finally due to CFRP sliding from the anchorage system and 5‰ for the beam P4 that failed due to concrete crushing. The increase in the CFRP strain in the beams of series II was equal to 7.15 and 8‰ (see Table 1). The maximal strain registered during beam's failure does not depend on the ordinary steel ratio and the level of preloading (compare $\varepsilon_{f,test}$ for corresponding beams

P1 and P5 - strengthened under preloading of 0.25 F_{u0} and beams P2 and P6 - strengthened under 0.76 F_{u0}, see Table 1). Moreover, comparing the CFRP strain (beams P1, P5, P2 and P6) during its debonding from the concrete surface (Table 1), the influence of the prestressing strain for the maximal test CFRP strain is confirmed. For the beams strengthened with higher prestressing force, higher debonding strains were reached, independently of the steel reinforcement ratio.

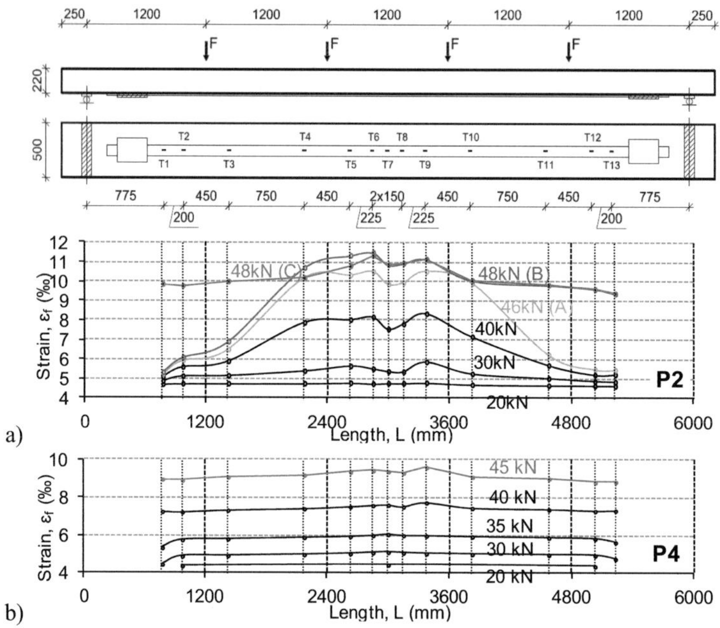

Fig. 4   Strain of CFRP strips at the following load of beam P2 and P4

CFRP debonding process is clearly shown on the load-CFRP strain curve (Fig. 4a). The CFRP strain coresponds to the total strain value calculated as a sum of prestressing and the test strain registered at the beam's failure ($\varepsilon_{fp} + \varepsilon_{f,test}$). Debonding starts under the external load of 2F=48kN ("A") and propagates toward one of the supports ("B"). Shortly after this debonding, the opposite CFRP strip end debonding occurred under the same loading of 2F=48kN ("C"). At the moment, where the CFRP strip debonded at its full length and it was held only at the anchorage platesIt behaved as the external tension bowstring until sliding of the CFRP end from the anchorage system (Fig. 4a), similar to the strip in the beam P4 unbonded for the whole test (Fig. 4b).

Strengthening ratio,defined as the ratio of the ultimate load of the strengthened beam to the calculated ultimate load of the non-strengthened beam ($2F_u / 2F_{u0}$), for the beams' series I (strengthened with bonded CFRP along the span) varied from 1.86 for the beam P2 (strengthened under preloading of 0.76 F_{u0} with the CFRP prestressed to strain ε_{fp}=4.75‰) to 2.19 for the beam P1 (strengthened under dead load equal to 0.25 F_{u0} with the CFRP prestressed to strain of ε_{fp}=5.20‰, see Table 1 and Fig 5a). For the beams' series I with not bonded CFRP strip along the span, the strengering ratio varied from 1.73 for the beam P4 to 1.95 for the beam P3 (Table 1, Fig 5b). While for the beams series II the strengering ratio varied from 1.63 for the beam P6 to 1.69 for the beam P5 (Table 1, Fig 5c). Based on above comparison the strengthening ratio depends on the longitudinal steel reinforcement ratio. For the beams' series I (with 12mm diameter bars) the strengthening ratio was higher than for the beams' series II (with 16mm diameter bars).

The influence of the CFRP prestressing on the stiffness and the ultimate load of the tested members is shown as a function of the average tensile concrete strain and the load. All tested members of series I (Fig. 5a, b) and series II (Fig. 5c) had similar stiffness after yielding of the longitudinal steel reinforcement. Although the beams higher preloaded before strengthening (P2, P4 and P6) indicate higher concrete strain for the same load than the beams less preloaded (P1, P3, P5), using the prestressed strips undeniably increased strengthening efficiency (even for the beams preloaded to 0.76 F_{u0}).

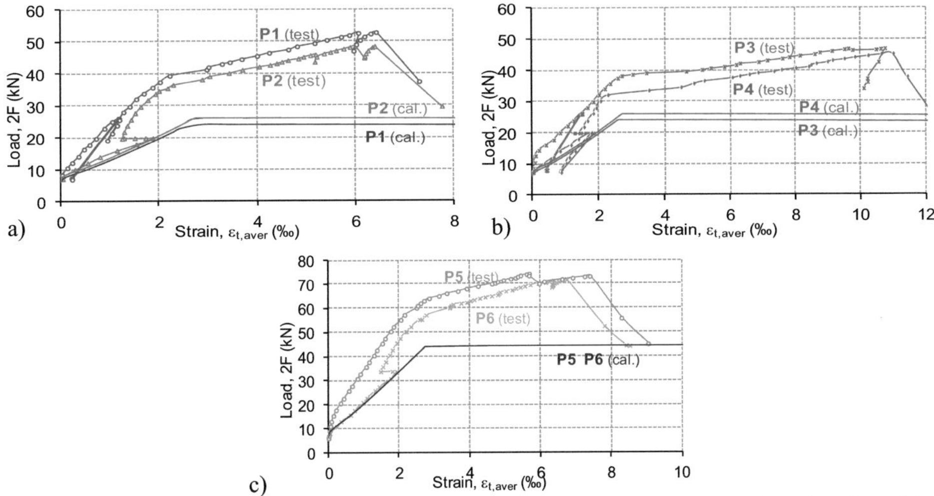

Fig. 5   Average concrete tensile strain at the level of tensile reinforcement in a function of load for beams: a) P1, P2, b) P3, P4 and c) P5, P6.

The capacity of non-strengthened beam was calculated based on a non-linear model of reinforced concrete members [9]. The main assumptions of the model are as follows: non-linear strength-strain relationship for concrete in compression and tension, experimental stress-strain relationships for reinforcing steel, tension stiffening principle and the plane section principle. External load value is calculated based on the equilibrium condition of generalized forces in the cross-section. The load for which a limit strain in one of the material's strength is reached is accepted as the load bearing capacity of the reinforced concrete cross-section. The model has been successfully applied for the analytical verification of the load-strain and load-deflection relationship of the large number of RC members [1].

Vertical displacements of beam P2 before rebar yielding were only 15% higher compared to displacement of the beam P1 strengthened under a dead load (Fig. 6a). The beam P2 showed a minor loss of the stiffness and the load capacity compared to the beam P1 strengthened under only dead load. The same observation confirm the beams P3 and P4 and the beams P5 and P6.

The aim of the CFRP strips prestressing was to improve the serviceability state (deflections and crack width). Comparison of the deflection-load (v-2F) curves (Fig. 6) shows that strengthening beam P2 with prestressed CFRP resulted in a significant reduction of the vertical displacement caused by the preliminary loading. The curves (v-2F) confirm the high influence of the CFRP prestressing on the stiffness of the beam, especially in case of the beams of series I (P1 and P2), strengthened with the CFRP strips prestressed to strains of 5.2‰ and 4.75‰, respectively and the beams of series II (P5 and P6), strengthened with prestressed strips of 4.8‰ and 4.85‰ respectively.

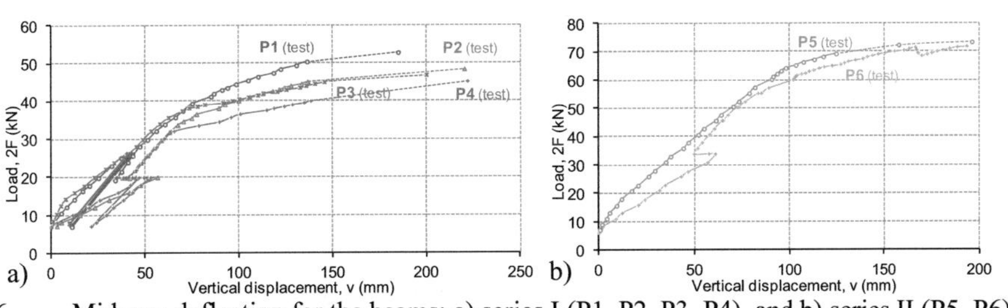

Fig. 6   Midspan deflection for the beams: a) series I (P1, P2, P3, P4) and b) series II (P5, P6).

The test showed that even high beam preloading equal to $0.76F_{u0}$ caused no significant loss of stiffness before and after steel yielding (Fig. 6). High beam preloading caused only minor loss of the load capacity. The load-deflection curve inclinations for all tested members are very similar, both before and after steel yielding. It confirms the high efficiency of strengthening RC members with prestressed CFRP strips. Despite very high exhaustion of the beams (far over the serviceability limit state and even ultimate limit state), application of the prestressed CFRP strip resulted in significant reduction of deflections, strains and the stiffness increase similar to non-preliminary-loaded beam.

The beam P1 showed the highest strengthening ratio ($2F_u / 2F_{u0} = 2.19$) and the maximal CFRP strain reached $0.82\varepsilon_{fu}$ due to the highest prestressing strain of 5.2‰.

3   Conclusions

The analysis of experimental test results of the RC beams strengthened in flexure with prestressed CFRP strips can be summarized with the following conclusions:

- High efficiency of the described strengthening technique is confirmed by strengthening ratio in a range of 1.73 to 2.19 of the non-strengthened beam's ultimate load for the beams of series I and from 1.63 to 1.69 for the beams of series II.
- Strengthening with prestressed CFRP strips is the only alternative for highly exhausted members requiring retrofitting, where the passive strengthening would not prove satisfying results.
- Bond loss between composite and concrete, propagating to one of the supports was the most common failure mode.
- Strengthening of RC members with prestressed CFRP strips resulted in significant improvement of both ultimate limit state and serviceability limit state, especially for highly preloaded elements.
- The band between FRP ship and concrete has on significant effect on beam's strain after steel yielding. The strain of the beams strengthened with bonded strips indicated lower values than the beams strengthened with unbonded strips.
- The longitudinal tensile steel reinforcement ratio has a crucial influence on the strengthening ratio, but no effect on the maximal CFRP strain registered along its debonding.

Acknowledgements

The research was founded by EU Project "Innovative measures and efective methods to improve safety stability for the construction work and transport infrastructure in the suistainable developement strategy", to which the author is grateful.

References

[1]   Kotynia, R., Kamińska, M.E.: Ductility and failure mode of RC beams strengthened for flexure with CFRP. In: Report No. 13, Technical University of Lodz, Poland (2003)

[2]   Deuring, M.: Verstärken von Stahlbeton mit gespannten Faserverbundwerkstoffen. In: Bericht (1993) No. 224, EMPA, Dűbendorf, Switzerland

[3]   Trintafillou, T.C., Deskovic, N., Deuring, M.: Strengthening of concrete structures with prestressed fiber reinforced sheets. In: ACI Structures Journal (1992) No. 89(3), pp. 235–244

[4]   Meier U.: Strengthening of structures using carbon fibre/epoxy composites. In: Construction Building Materials (1995) No. 9(6), pp. 341-351

[5]   Teng, J.G., Chen, J.F., Smith, S.T., Lam, L.: FRP strengthened RC structures. In:Wiley(2002)

[6]   Wight, R.G., Green, M.F., Erki, M.-A.: Prestressed FRP sheets for post-strengthening reinforced concrete slabs. In: Journal of Composites for Construction (2001) No. 5(4), pp.214-220

[7]   Stocklin, I., Meier, U.: Strengthening of concrete structures with prestressed and gradually anchored CFRP strips. In: Proceedings of 6th International Symposium on FRP Reinforcement for Concrete Structures (2003), pp. 1321-1330

[8]   Kotynia, R., Walendziak, R., Stoecklin, I., Meier, U.: RC Slabs Strengthened with Prestressed and Gradually Anchored CFRP Strips under Monotonic and Cycling Loading. In Journal of composites for constructions (2011) No 4-5

[9]   Czkwianianc, A., Kamińska, M.E.: Method of nonlinear analysis of one-dimensional reinforced concrete members. In: KILiW PAN IPPT, Warsaw, (1993), (in Polish)

Numerical modelling of RC wall panels strengthened with FRP composites

Carla Todut

Civil Engineering Department
"Politehnica" University of Timisoara,
2A Traian Lalescu str., 300223 Timisoara, Romania
Supervisor: Valeriu Stoian

Abstract

This work belongs to the field of earthquake engineering and it refers to the seismic behaviour of the precast reinforced concrete walls panels in terms of strengthening and repair. Numerical models have been developed in order to predict the behaviour of reinforced concrete wall panels with cut-out openings, strengthened by fiber reinforced composites. Three cut-out opening types were considered, in order to assess the lateral load capacity loss. The Precast Reinforced Concrete Large Panel (PRCLP) buildings are characterised by the dense distribution of load bearing walls. During the life of a building may happen new cut-out requirements or enlarged openings in the structural walls due to the change in functional destination of the building or other reasons, but these changes cause stress redistribution in the proximity of the cut-out intervention and sometimes it can affect the structure's global behaviour. The type of analysis that will be conducted on these models is a material nonlinear analysis. The main objectives of these numerical modelling is to provide data on the failure criteria of the analysed models, provide comparison between the analysed models and experimental specimens tested in the Civil Engineering Department by Istvan Demeter, and present the effect of fiber type. Parametric studies conducted with respect to the analysed models indicate shear failure in all cases.

1 Introduction

A significant percentage of the population of Romania live in precast reinforced concrete large panel buildings built in the period of 1960-1990. In this paper, a numerical model was developed in order to predict the behaviour of shear walls and compare the results with the tested specimens.

2 Experimental program description

The seismic performance of the precast large wall panels was evaluated considering the outrigger effect of the adjacent structural members, assessing the weakening of doorway cut-outs and investigating the performance of the CFRP-EBR strengthening method.

The experimental walls, consisting of a bare solid wall (S), a wall with a narrow door cut-out (S/E1) and a post-damage strengthened wall with narrow door cut-out (S/E1-TR) were already tested by I. Demeter in the Laboratory of Reinforced Concrete Structures, Politehnica University of Timisoara. The walls reinforcement consists of PC52 type (S355 grade) for Φ 8÷16 mm longitudinal bars, and STPB type (S490) for Φ 3÷4 mm welded wire mesh. The ultimate cubic compressive strength of concrete for the tested and modelled wall panels is 17.5 MPa [1].

3 Numerical analysis

Simulating the non-linear behaviour of Reinforced Concrete (RC) walls subjected to severe earthquake ground motion is an important problem for the engineering community. These models must describe essential geometrical and material characteristics as well as the basic mechanisms that control the behaviour of reinforced concrete structures [2]. The nonlinear analysis was performed in a 2D model using the ATENA software [3]. A number of three RC shear walls were modelled and analysed in order to compare the behaviour, critical parts and failure mode with the tested ones.

The tested specimen had a lateral loading, reversed cyclic- displacement controlled, with increasing displacement amplitude load/displacement history. The control displacement was the horizontal drift calculated as the difference between the horizontal displacements measured at the top and bottom edge of the tested specimen. In comparison with the tested wall, in the modelled wall was not

Proc. of the 9th *fib* International PhD Symposium in Civil Engineering, July 22 to 25, 2012,
Karlsruhe Institute of Technology (KIT), Germany, H. S. Müller, M. Haist, F. Acosta (Eds.),
KIT Scientific Publishing, Karlsruhe, Germany, ISBN 978-3-86644-858-2

applied a reversed cyclic lateral loading. The lateral load were displacement controlled increments of 0.1 mm. The axial loads for the tested specimen were composed of two parts, namely a constant and a variable part. In addition to the constant level, alternating axial loads were imposed (taking into account the uplifting of the loaded end of the upper beam) in order to restrain the rocking rotation of the laterally loaded walls. For the modelled wall the axial loads consist of a constant part (95 kN + 95 kN) and an additional part composed of an axial load imposed on the left side by adding 15 kN each step performed by the analysis in order to restrain the rocking rotation of the laterally loaded walls. The value of 15 kN was adopted considering the total load applied in the case of the tested specimen and the number of steps. The boundary conditions consist of restrained rotation in case of the tested wall while for the modelled wall the bottom side of the lower beam is fixed in both horizontal and vertical directions.

4 Results

4.1 Bare solid wall results

The primary results for the tested bare solid wall (1-S-T) are shown in Fig.1. In the present paper, the actual drift values of the tested specimen were reduced to 2/3 of the measured values by neglecting the rotational component. The first inclined cracks developed at $\delta = 2.87$ mm drift level, extending from the top to the bottom edges. Due to the testing facility, the solid wall was not loaded beyond 1210 kN and 8.6 mm drift level [2]. According to the analysis performed using the ATENA software, the first inclined cracks developed at 0.7 mm drift level and it had a width of 0.25 mm (Fig. 3a). Inclined cracks recorded at 2.87 mm drift level had a 0.48 mm width (Fig.3b). Crack state at 7.43 mm drift level had a width of 1.71 mm and is shown in Fig. (3c). Cracking was most intensive near the joint between the lower beam and the panel at the right side. In comparison with the experimental results where the first inclined cracks developed at $\delta =2.87$ mm drift level, the numerical modelling may record thin crack width not easily seen experimentally. Concrete ultimate compressive strain was not recorded in the case of the tested solid wall and the ultimate compressive strength of concrete was not reached either. In the case of the modelled solid wall, the ultimate compressive strain of concrete was reached at 7.4 mm drift level (Fig. 4a) and the ultimate compressive strength of concrete was attained at 7.4 mm drift level on the left side of the specimen between the unconfined wing element and panel (Fig. 4b). Figure 4 c) presents the compressive strength of concrete at 7.43 mm, representing the point on the load displacement curve where it decreases suddenly. Reinforcement yielding takes place at 4.97 mm drift level at the upper right corner of the specimen. Failure of the specimen takes place at 7.87 mm drift level by loosing 20% of its bearing capacity.

Table 1   Bare solid wall comparison results

BARE SOLID WALL RESULTS			
	First inclined cracks	Maximum lateral load imposed	Drift level corresponding to max lateral load imposed
Tested	$\delta = 2.87$ mm drift level	1210 kN	8.6 mm
Modelled	$\delta = 0.70$ mm drift level	1380 kN	7.43 mm

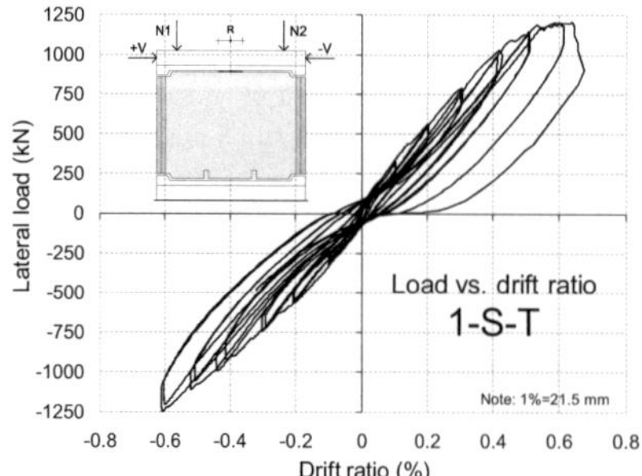

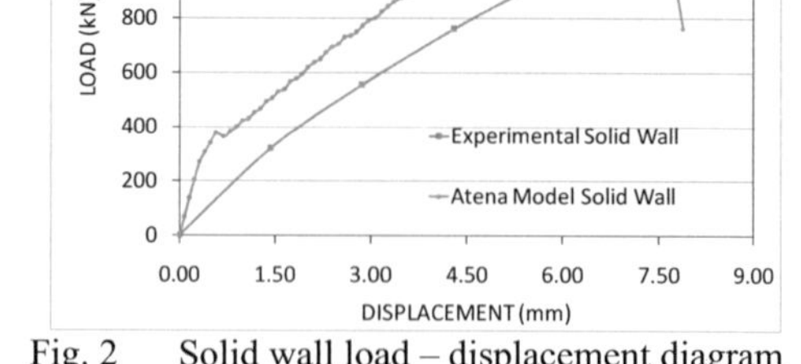

Fig. 1   Primary results for the solid wall [1]   Fig. 2   Solid wall load – displacement diagram

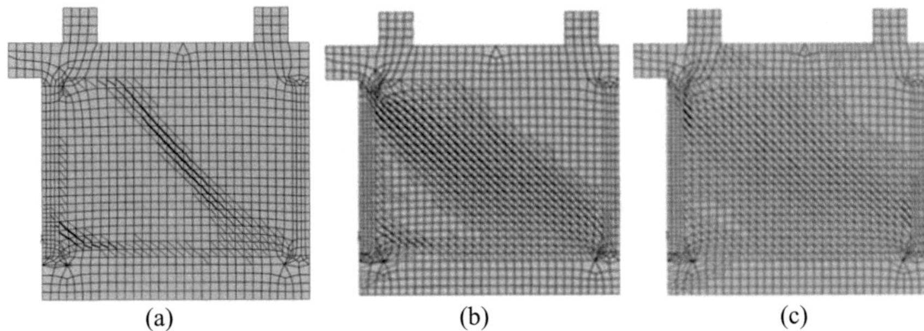

(a)   (b)   (c)

Fig. 3   Crack state at (a) 0.70 mm, (b)2.87 mm and (c) 7.43 mm drift level

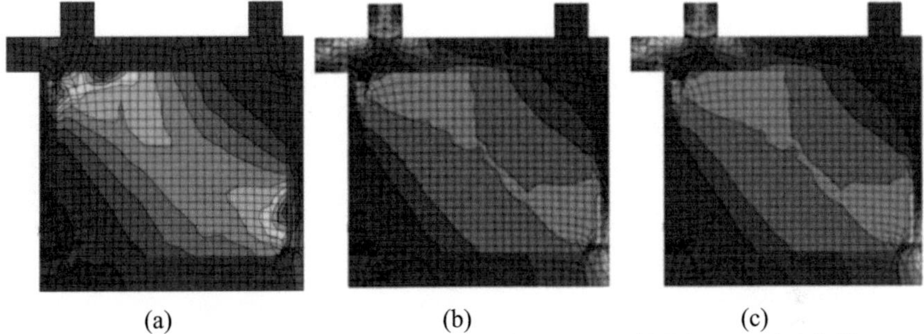

(a)   (b)   (c)

Fig. 4   Compressive strain at 7.4 mm (a), compressive stress at b) 7.4 mm and c) 7.43 mm

4.2   Wall with narrow door cut-out results

The loading procedure for the modelled wall was applied in a similar manner with the tested wall, except the axial load imposed on the left side, by adding 10 kN at each step performed by the analysis in order to restrain the rocking rotation of the laterally loaded walls. The value of 10 kN was adopted considering the total loading applied in the case of the tested specimen and the number of loading steps. The results of the tested S/E1 are depicted in Fig. 5. The cracking initiated at the base and top corners of the opening, while concrete spalling initiation was observed at the inside toe (near the opening) of pier 2. The failure occurred at δ=10.98 mm in the positive loading direction by severe crushing and spalling of the concrete at the spandrel to pier 2 connection and along the top of pier 2.

According to the numerical analysis the first diagonal cracks developed at 1.23 mm drift level and it had a width of 0.39 mm (Fig. 7a). Inclined cracks recorded at 4.67 mm drift level, where the load-displacement curve starts to decrease, and had a 0.92 mm width (Fig. 7b). Inclined cracks recorded at 6.29 mm drift level had a width of 0.65 mm (Fig. 7c) and in comparison with the tested element here is a severe state of cracking. In the case of the modelled S/E1, the ultimate compressive strain of concrete was reached at 5.02 mm drift level (Fig.8a) and the ultimate compressive strength of concrete was attained at 5.54 mm drift level between the upper left side of the spandrel and the loading beam (Figure 8b). Reinforcement yielding takes place at 6.55 mm drift level at the bottom right side of the cut-out. Figure 8c represents the stress state of the specimen at failure δ=9.68 mm.

Table 2   Wall with narrow cut-out door comparison results

WALL WITH NARROW DOOR CUT-OUT RESULTS			
Element	First diagonal crack	Maximum lateral load imposed	Drift level corresponding to max lateral load imposed
Tested	δ = 6.28 ÷ 7.85 mm drift level	581.8 kN	9.42 mm
Modelled	δ = 1.23 mm drift level	541.8 kN	4.7 mm

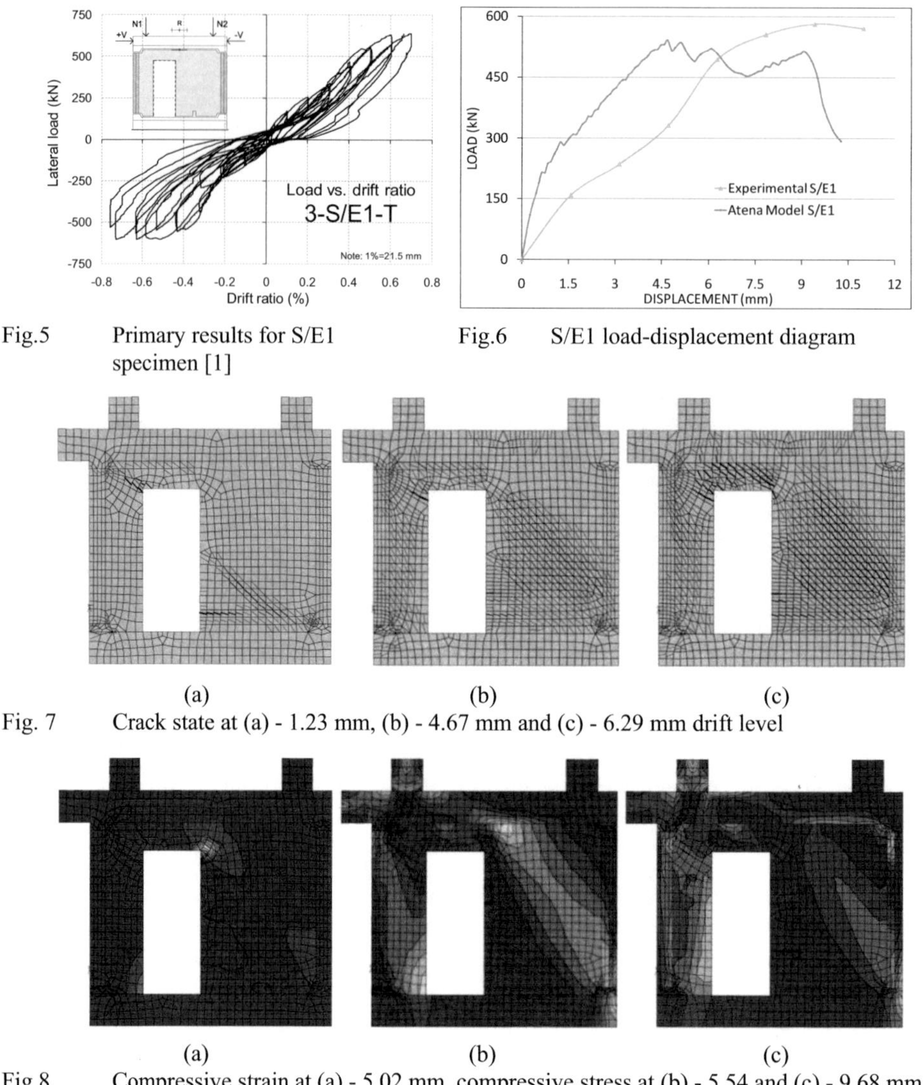

Fig.5   Primary results for S/E1 specimen [1]

Fig.6   S/E1 load-displacement diagram

(a)   (b)   (c)

Fig. 7   Crack state at (a) - 1.23 mm, (b) - 4.67 mm and (c) - 6.29 mm drift level

(a)   (b)   (c)

Fig.8   Compressive strain at (a) - 5.02 mm, compressive stress at (b) - 5.54 and (c) - 9.68 mm

4.3   Post-damage strengthened wall with narrow door cut-out results

In the case of post-damage strengthened solid wall with narrow door cut-out (S/E1-TR), the strengthening was performed by means of Carbon Fiber Reinforced Polymer (CFRP) using the Externally Bonded Reinforcement (EBR) technique. The overall objectives of the strengthening were to counterbalance the weakening incurred in a solid wall as a result of a door cut-out. The strengthening strategy was divided into three directions: (1) to offer flexural capacity along the vertical and horizontal edges of the cut-out opening, (2) to increase the shear capacity of the wall piers, and (3) to provide confinement effect at the cut-out opening corners (Fig.11). The use of near-surface mounted (NSM) fiber-reinforced polymer (FRP) reinforcement is currently emerging as a new technology for increasing flexural and shear strength of reinforced concrete elements. In the modelled specimens were applied the technique of NSM- FRP reinforcement, instead of EBR-FRP and the characteristics for the modelled FRP system are given below (Table 3):

Table 3   Properties of FRP system [4]

FRP system	Resin or adhesive	Fibre type and content	Design thickness (mm)	Tensile strength (Mpa)	Elastic modulus (Gpa)	Elongation at failure (%)
SikaWrap 230 C	Sikadur 330	carbon 230g/m^2	0.381	715	61	1.09

The primary results of the tested post-damage S/E1-TR are presented in Figure 9. The cracking of the specimen was initially (from δ =3.14 mm) characterised by existing crack opening in pier 2. Some flexural FRPs around the top corners of the opening fractured at δ =6.28 mm and 12.56 mm drift level, whereas at the inside toe of pier 2 compression bulging of the vertical FRP strips was observed between the horizontal confinement strips. The specimen failed between 14.13 ÷ 15.7 mm by concrete crushing and FRP confinement fracture at the compression toes of the wide pier [2]. In the case of the modelled specimen, the first diagonal cracks developed at 1.53 mm drift level and it had a width of 0.48 mm (Fig. 12a). Inclined cracks recorded at 3.13 mm drift level had a 0.696 mm width (Fig. 12b) and in comparison with the tested element here we deal with a severe state of cracking and not with a crack opening in pier 2. Inclined cracks recorded at 6.48 mm drift level had a width of 0.94 mm (Fig. 12c), where the load-displacement diagram starts decreasing. Debonding process can not be modelled in ATENA 2D mode. In the case of the NSMR – FRP yielding of reinforcement was reached at 6.48 mm drift level at the left upper corner of the opening. Concrete crushing appears in the modelled specimen at 6.38 mm, where the ultimate concrete strain reaches 3.68 ‰ (Fig.13a) and the ultimate concrete strength is attained (Fig.13b). Fig.13c represents the state of stress in specimen at failure.

Table 4   Post-damage strengthened wall with narrow door cut-out comparison results

POST-DAMAGE STRENGTHENED WALL WITH NARROW DOOR CUT-OUT RESULTS				
Element	First diagonal crack	FRP failure initiation	Maximum lateral load imposed	Drift level corresponding to max lateral load imposed
Tested	δ = 3.14 mm drift level	6.28 mm	625.3 kN	14.13 mm
Modelled	δ = 1.53 mm drift level	6.48 mm	642 kN	6.48 mm

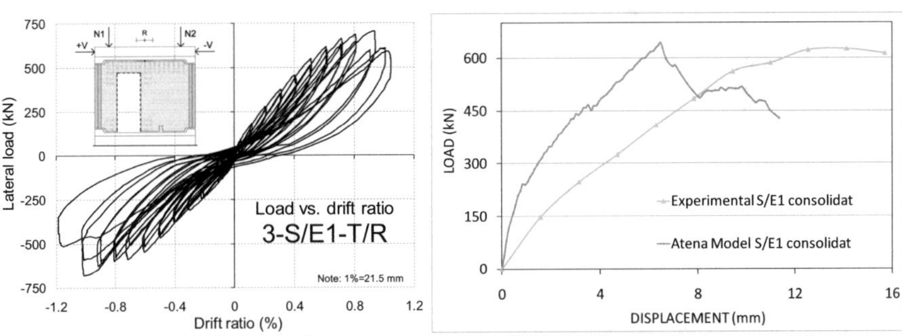

Fig. 9   Primary results for S/E1-TR [1]   Fig. 10   S/E1-TR load– displacement diagram

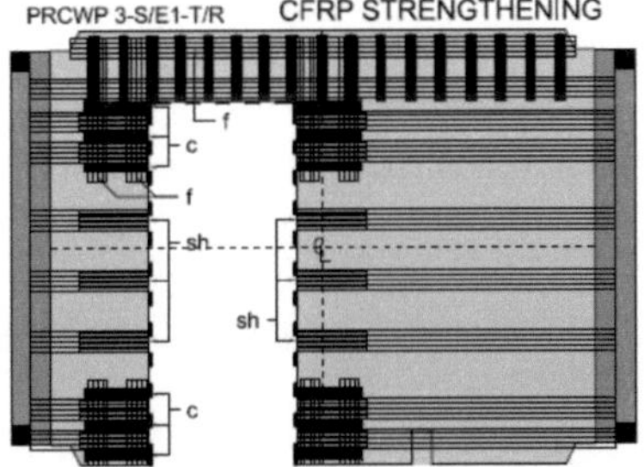

Fig.11   CFRP strengthening strategy for S/E1-T/R specimen [1]

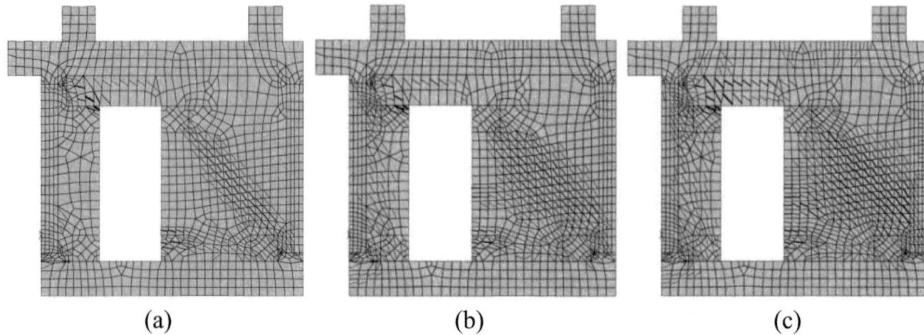

(a)   (b)   (c)

Fig. 12   Crack state at (a) - 1.53 mm, (b) - 3.13 mm and (c) - 6.48 mm drift level

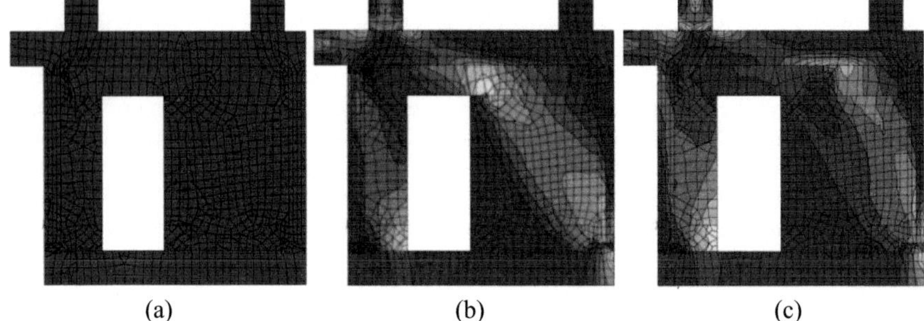

(a)   (b)   (c)

Fig.13   Compressive strain at (a) - 6.38 mm, compressive stress at (b) - 6.38 and (c) - 7.8 mm

5   Conclusions

Post-processing of the numerical results revealed that extensive cracking, reinforcement yielding and concrete crushing was taking place in the specimen. Under the left-side loading numerous diagonal cracks formed, similar to that reported in the experiment. The analysis predicted well the tendency of the experimental curves, sharing the same critical regions of the specimens. However, the failure appears to be brittle after the peak. In terms of the maximum load supported by the element, the modelled wall with a narrow door cut-out recorded 541.8 kN while the tested specimen had 581.8 kN. The modelled post-damage strengthened wall with narrow door cut-out recorded 642 kN and the tested specimen had 625.3 kN. The maximum drift level obtained by the modelled wall with a narrow door cut-out was 9.68 mm while for the tested specimen was 10.98 mm. For the modelled post-damage strengthened wall with narrow door cut-out the maximum drift level was 7.8 mm and for the tested specimen was 15.7 mm. Failure of the modelled wall with narrow door cut-out occurred due to concrete crushing in pier 2 and for the modelled post-damage strengthened wall with narrow door cut-out due to FRP reinforcement yielding at the left upper corner of the opening and concrete crushing in pier 2. Concluding upon the performance of the strengthening system this was efficient from the point of view of the load bearing capacity. The failure mode was similar with the tested elements. Also from the recorded data available one can remark that the loss of bearing capacity caused by a door cut-out with respect to the solid wall is more than significant.

References

[1]   Demeter, I.: Seismic retrofit of precast RC walls by externally bonded CFRP composites. PhD Thesis, Politehnica University of Timisoara, (2011).

[2]   Kheyroddin, A., Naderpour, H., and Hoseini Vaez, S.R.: Hysteretic evaluation of seismic behaviour of RC shear walls strengthened with FRP sheets. In: The 14th World Conference on Earthquake Engineering, October 12-17(2008), Beijing, China .

[3]   ATENA – Advanced tool for engineering nonlinear analysis, Technical specifications 2010, Cervenka Consulting Ltd., Prague, Czech Republic.

[4]   Demers, M., Neale, K.W., and Sheikh, S.: FRP rehabilitation of reinforced concrete structures. Design Manual No.4, Version 2, ISIS Canada Corporation, Winnipeg, Manitoba, Canada, (2008).

Performance evaluation of sodium resistant mortars as sacrificial layer in fast breeder reactors

K. Mohammed Haneefa

Department of Civil Engineering
Indian Institute of Technology Madras, (IITM), Chennai, India
Supervisor: Manu Santhanam

Abstract

In fast breeder nuclear reactors, sodium is used as a coolant. Accidental spillage of sodium at 550°C and above can cause damage to the structural concrete. For this reason, a sacrificial layer of limestone aggregate concrete is provided over the conventional siliceous aggregate concrete. In spite of extensive understanding of the chemistry of interactions, limited information is available on the microstructural description of deterioration, including creation of mineralogical polymorphs and growth of microcracks. To study and improve the performance of the sacrificial layer in the presence of hot liquid sodium, a research project is taken up at IIT Madras. This paper presents overview of the state of the art, research needs and motivations supplemented by results and analysis of first phase of work performed at IIT Madras, which includes thermal performance evaluation of limestone mortars. This study will be followed by an evaluation of sodium attacked mortars.

1 Introduction

In fast breeder reactors, concrete is used as structural material for constructing foundations, containment, radiation and support equipment cells. The alkali metal sodium is used as a coolant in fast breeder nuclear reactors. Accidental spillage of sodium at high temperature from pipes in inert equipment cells or inside reactor cavity can lead to various safety issues in fast breeder reactors due to sodium fire [1]. A sacrificial layer of concrete is employed to protect the structural concrete from interacting with hot liquid sodium at around 550°C and above to reduce the risk factors related to structural integrity, functional efficiency and radiation safety of nuclear reactors [2]. The choice of materials for sacrificial layer may depend on economy and ease with which the sacrificial layer can be constructed, demolished, repaired, rehabilitated or disposed. From the various studies carried out for several decades, it is apparent that concrete with limestone aggregate has gained applicability in this field due to its high reaction temperature threshold [3]. The accidental spillage of liquid sodium at high temperature and its interaction with concrete are realistic phenomena in sodium cooled fast breeder reactors. Hence, the development and construction of a sacrificial layer of concrete to protect the structural concrete from thermo-chemical damages is of importance from the perspective of the safety and efficient functioning of a reactor. Numerous studies have been carried out on such concretes called 'sodium resistant concrete', but limited information is available in public domain possibly due to the closed nature of the nuclear industry. In India, the Indira Gandhi Centre for Atomic Research (IGCAR), Kalpakkam, has conducted various studies on sodium resistant concrete using limestone aggregates [1-4]. All studies conducted thus far on sodium-concrete interaction are limited to the macro-scale damage due to chemical, mechanical and thermal phenomena. Amount of hydrogen release from the concrete-sodium interactions based on different ages of concrete and sodium penetration into the concrete are well documented. However, the fundamental mechanisms of concrete degradation during liquid sodium-concrete interaction have not been still addressed at a microstructural level. It is believed that research at this scale will help in development of better sodium resistant concrete while optimizing the compositions and ingredients of the concrete.

2 Properties of concrete

Concrete is a composite material. Strength and other properties of concrete are governed by mix design, which controls type and amount of aggregates, cement, water, and chemical or mineral admixtures. The mix gains strength upon hydration of cement with time.

Proc. of the 9th *fib* International PhD Symposium in Civil Engineering, July 22 to 25, 2012,
Karlsruhe Institute of Technology (KIT), Germany, H. S. Müller, M. Haist, F. Acosta (Eds.),
KIT Scientific Publishing, Karlsruhe, Germany, ISBN 978-3-86644-858-2

The major product of the cement hydration is calcium silicate hydrate (CSH), which is the main strength imparting phase in hydrated cement (50 to 60% of total volume of solids). Calcium hydroxide (20 to 25%), calcium sulfoaluminates (15 to 20%) namely ettringite and monosulphate, and calcium aluminate hydrates (C-A-H) are the other products of cement hydration. Microstructural characterization and chemistry of hydrated cement have been well documented over the years [5]. In cementitious materials, stress concentrations under load occur due to microstructural defects and discontinuities [6]. Mechanical properties exhibit influence of individual phases according to their concentration, distribution and spatial configurations. The heterogeneous nature of concrete is typified by the presence of an interfacial transition zone (ITZ) [5-8].

3 Performance of concrete at elevated temperature

Wang [9] explains the performance of concrete at elevated temperature as "Unlike wood, concrete does not burn and unlike steel, it does not lose a substantial degree of its rigidity at moderately high temperatures". Changes in the basic properties of concrete such as strength, modulus of elasticity and volume stability due to elevated temperature can challenge the structural integrity and even cause failures. Performance of concrete at elevated temperature is governed by types of aggregate and cement, temperature and duration of exposure, moisture content of concrete and size of the structural elements.

3.1 Alterations in paste phase

Temperatures from ambient up to $1000\,°C$ or more bring about numerous changes in the hydrated cement paste structure, and this aspect has been studied by a number of researchers. Lee et al. [10] summarize the processes of decomposition depending on the temperature regime (Table 1) assuming that aluminate phases are non-reactive substances at elevated temperatures.

Table 1   Process of decomposition depending on the temperature regime

Temperature	Decomposition
20–120 °C	Evaporation of free water, dehydration of C-S-H and ettringite
120–400 °C	Dehydration of C-S-H
400–530 °C	Dehydration of C-S-H, dehydration of calcium hydroxide
530–640 °C	Dehydration of C-S-H, decomposition of poorly crystallized $CaCO_3$
640–800 °C	Dehydration of C-S-H, decomposition of $CaCO_3$

Cement paste shrinks at elevated temperatures and eventually cracks when the tensile stress induced crosses the tensile capacity of the paste [9]. Figure 1 pictorially depicts crack formation in cement paste at different ranges of elevated temperature [11]. Cracks are formed beyond 400°C and grow with respect to increment in temperature. Extremely severe cracks are observed beyond 600°C.

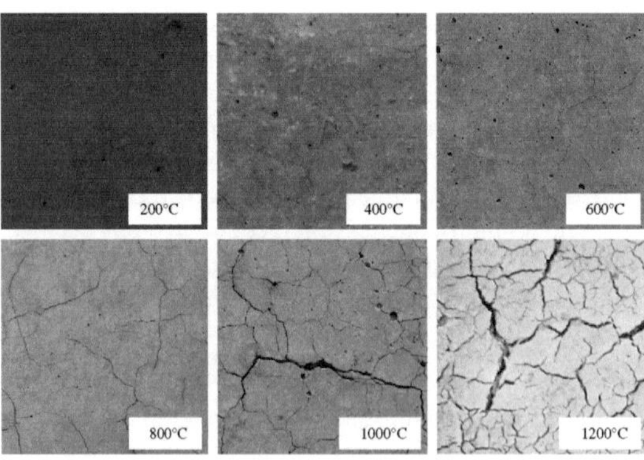

Fig. 1   Crack patterns of cement pastes exposed to elevated temperatures [11]

3.2 Effect of interfacial transition zone

Elevated temperatures up to 800°C for 1 hour reduce strength and durability of concrete by weakening the ITZ, concurrent coarsening of hydrated cement paste pore structure and increase in pore diameter. Microhardness test conducted by Hossain [12] revealed that ITZ width increased when the concrete was exposed to 200°C for 1 hour. Significant drop in microhardness at 400°C, increased total porosity and average pore diameter at 600°C were the other observations from the same study.

3.3 Role of mineral admixtures

Poon et al. [13] performed a comprehensive performance evaluation of concrete incorporating pozzolanic materials like silica fume, fly ash and blast furnace slag. Concrete with fly ash and blast furnace slag exhibited increase in strength for a temperature range of 20-200°C. Up to 400°C high performance concrete retained its strength, while normal concrete with ordinary Portland cement showed an average loss in strength of 20%. Beyond 400°C concrete started losing strength rapidly and rate of loss was more for high strength concrete. Fly ash and blast furnace slag concrete performed well compared to ordinary Portland cement concrete at 600°C. Study reveals that spalling or splitting is the associated risk with silica fume concrete at elevated temperatures. Fly ash and blast furnace slag concretes showed network of fine cracks without any spalling or splitting. Figure 2 gives some typical crack patterns for concrete using OPC and pozzolanic materials at 800°C.

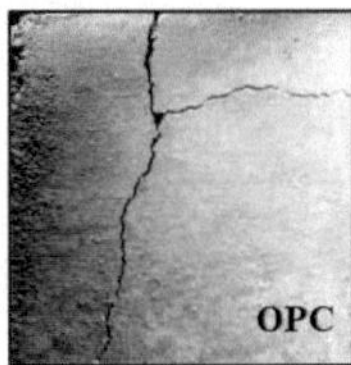

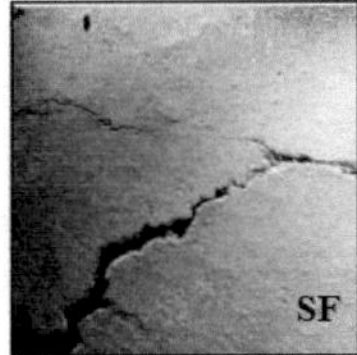

Fig. 2   Typical crack patterns observed at 800°C for OPC, silica fume (SF), fly ash (FA) and slag (GGBS) concretes [13]

3.4 Influence of aggregate type

Concrete constitutes around 70% aggregates by volume, which provide dimensional stability and stiffness. Performance of aggregate at elevated temperature is relatively better compared to cement paste phase [14]. For the most part, carbonate aggregates like limestone or siliceous aggregate like granite are used for making concrete. Granite is a siliceous rock with quartz and feldspar as essential minerals. Granite may also contain dark coloured minerals like muscovite, biotite, hornblende, augite and magnetite [15]. Limestone is a sedimentary rock consisting of more than 50 % of carbonates of calcium or magnesium [16]. Calcite is the predominant mineralogical component of limestone. Zing et al. [17] studied the influence of the nature of aggregates on the behavior of concrete subjected to elevated temperature. Thermal conductivity of concrete is affected by thermal properties of aggregates. According to different mineralogical composition and microstructure, aggregates exhibit different thermal conductivities. Decrease in compressive strength of concrete with siliceous aggregate is more compared to carbonate aggregate, because of higher thermal expansion and consequent increase in volume due to phase transition. Specific heat capacity of calcareous aggregate is 10 times more compared to siliceous aggregates, for rise in temperatures up to 600°C. In carbonate aggregates, $CaCO_3$ decomposes into CaO and CO_2 at 800–900°C and further expands at elevated temperatures associated with volume changes causing destruction [9]. Re-hydration of carbonates during cooling may cause spalling of concrete made from calcareous aggregates.

4 Sodium - concrete interactions

In fast breeder nuclear reactors, sodium is used as a coolant. These coolant conduits are in near vicinity of the sacrificial concrete layers on the reactor structure. The sodium can spill accidentally and get in contact with the sacrificial concrete. The thermo-chemical interactions of liquid sodium at 550°C and above with concrete can result in degradation of concrete in several ways. These interactions include fifteen exothermic and four endothermic reactions [1]. The chemical reactions occurring as a result of this spillage have been studied for several years. In short, the steam from the evaporated water reacts with liquid sodium and produces sodium hydroxide, sodium monoxide, and gaseous hydrogen.

Due to the thermo-chemical effects, the sodium hydroxide can erode the concrete surface. Eventually, sodium carbonate, calcium oxide, magnesium oxide, sodium metasilicate, sodium aluminates, and carbon are produced – resulting in the deterioration of concrete. Cracking in concrete during sodium fire occurs because of large thermal gradients caused due to rapid increase in temperature, and production of steam [3, 18]. Interaction of sodium with silicate phase of cement, or with silica in the aggregate, can possibly lead to alkali silica reaction (ASR), especially when temperatures increase significantly (Conversion of quartz to cristobalite occurs at ~ 1000°C). Alkali silica reaction refers to the expansion causing reaction between the alkalis from cement and reactive siliceous aggregate [19]. In this case, the alkalis would be directly available from the liquid sodium. However, there are very little details in literature with respect to this possibility. Although limestone is not prone to ASR, it can have several impurities, including various forms of reactive silica such as chert and opal. Furthermore, the presence of clay can lead to a specific deleterious texture of calcite (or dolomite) crystals that could make the aggregate prone to alkali-carbonate reaction [19]. For concrete with high w/c, the presence of a large amount of free water can cause a considerable increase in the amount of hydrogen that is released, apart from the increased risk of spalling caused by steam pressure. Additionally, since early age concrete has more free water, younger concrete would tend to release more hydrogen in a sodium fire [20]. Special cement such as slag cement and high alumina cement could be employed for their beneficial properties. Fritzke and Schultheiss [21] tested the option of using high alumina cement along with Al_2O_3 fine aggregate in the protective layer over quartzitic concrete. They concluded that using mortar with increased porosity could reduce the damage to some extent. Various researchers have studied liquid sodium and concrete interactions either in air or in inert argon atmosphere to simulate various accident scenarios in sodium cooled fast breeder reactors. Sodium exposure experiments are equipped with oxygen monitor, hydrogen monitor, hygrometer, pressure transducers and thermocouples. Bae et al. [20] reported that possibility of hydrogen explosion reaction cannot be neglected, because, lower flammable limit of hydrogen is 4.0 mol%, whereas the experimental maximum concentration reached up to 31 mol% in several cases. Hot sodium and concrete interactions can lead to various endothermic and exothermic reactions, resulting in dehydration and erosion of concrete along with production of hydrogen gas. Degradation of concrete strength and accumulation of hydrogen gas associated with hydrogen burning could challenge the integrity of the structure due to over pressurization. It could possibly lead to release of radioactive materials which are extremely harmful. Thus, it is necessary to understand sodium-concrete reactions to predict thermal energy, hydrogen gas release and degradation of concrete. Figure 3 lists parameters which govern hot liquid sodium and concrete interactions, along with phases formed during the interactions and reaction products [1-4, 18, 20-22].

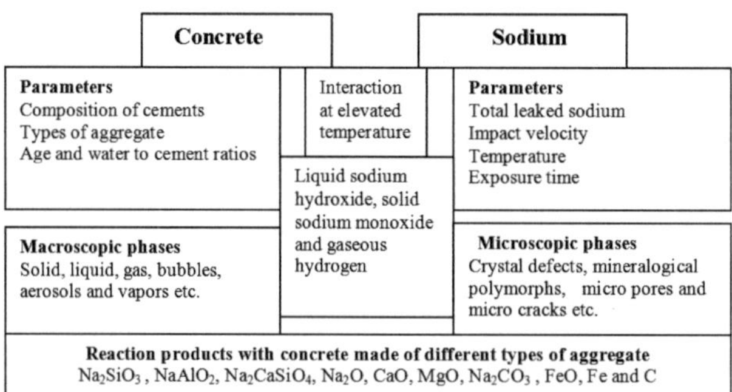

Fig. 3   Parameters which govern hot liquid sodium and concrete interactions, along with phases formed during the interactions and reaction products [1-4, 18, 20-22].

5 Methodologies

The research work can be divided into three phases, namely, comprehensive characterization of materials used, thermal performance evaluation and sodium interaction studies

1st Phase: Comprehensive Characterization of materials used- Physical, chemical and mineralogical characterization of ordinary Portland cement (OPC), Portland pozzolana cement (PPC), Portland slag cement (PSC) and high alumina cement (HAC), limestone aggregate, river sand and granite.

2nd Phase: Thermal performance evaluation- Mineralogical characterization of limestone, granite and river sand after exposure to 550°C for 10,20 and 30 minutes durations and thermal performance of limestone mortars (1:2.75) after exposure to 550°C for 10, 20 and 30 minutes duration. Compressive strength, flexural strength, mass loss due to thermal exposure and abrasion are considered for performance evaluation and development of performance indices. All studies on limestone mortars are repeated on river sand mortar also. Different types cement are used with the same water to cement (w/c) ratio of 0.55 and effect of water cement ratio on thermal performance is studied for a range of 0.40 -0.60 using ordinary Portland cement.

3rd Phase: Sodium interaction studies- Sodium interaction studies of limestone mortars for different durations of 10, 20 and 30 minutes.

6 Results and discussions

After performing characterization of materials, limestone mortars were prepared and comprehensive performance evaluation was carried out based on the performance in the compressive strength, flexural strength, mass loss and abrasion tests. The eight mixes of limestone mortars have been ranked in Table 2 separately for the performance with respect to mass loss, compressive strength, flexural strength and abrasion. The ranking has been done on a scale from 1 to 10, 1 being the best and 10 being the worst. This allows a direct comparison of the performances of all eight mortars – M5 with a w/c ratio of 0.60 with OPC is the worst, while the mix M8 with OPC for a w/c ratio 0.40 is the best, which indicates that the effect of water to cement ratio is significant. Among the mortars with different types of cement for same water to cement ratio, mortar with Portland pozzolana cement performed well.

Table 2 Performance indices based on compressive strength, flexural strength, mass loss and abrasion

Mix	w/c ratio/ Type of cement	PIBCS[1]	RBCS[2]	PIBFS[3]	RBFS[4]	PIBML[5]	RBML[6]	PIBA[7]	RBA[8]
M1	0.55/OPC	1.92	7	1.86	7	2.07	6	1.146	5
M2	0.55/PPC	1.53	4	1.36	3	1.62	4	1.025	3
M3	0.55/PSC	1.8	5	1.76	6	2.03	5	1.223	7
M4	0.55/HAC	1.57	6	1.56	4	2.29	7	1.524	8
M5	0.60/OPC	2.07	8	2.07	8	2.76	8	1.208	6
M6	0.50/OPC	1.37	3	1.6	5	1.42	3	1.044	4
M7	0.45/OPC	1.08	2	1.07	2	1.26	2	1.02	2
M8	0.40/OPC	1	1	1	1	1	1	1	1

PIBCS[1] : Performance index based on compressive strength after 30 minutes exposure = (Strength Loss)i/(Strength Loss)min
RBCS[2] : Ranking based on compressive strength after 30 minutes exposure
PIBFS[3] : Performance index based on flexural strength after 30 minutes exposure= (Strength Loss)i/(Strength Loss)min
RBFS[4] : Ranking based on flexural strength after 30 minutes exposure
PIBML[5] : Performance index based on mass loss after 30 minutes exposure = (Mass Loss)i/(Mass Loss)min
RBML[6] : Ranking based on mass loss after 30 minutes exposure
PIBA[7] : Performance index based on abrasion after 30 minutes exposure = (Abrasion)i/(Abrasion)min
RBA[8] : Ranking based on abrasion after 30 minutes exposure

7 Possible research outcomes

The proposed research is aimed at understanding fundamental degradation mechanisms of concrete at elevated temperatures and in the presence of sodium. Better understanding of these mechanisms will contribute to development concrete with higher resistance to hot sodium. Development of performance indices will help in identifying beneficial use of best combinations of materials and water to cement ratios to achieve good resistance against hot liquid sodium. The data obtained can be effectively used to perform safety analysis in fast breeder reactors for establishing safety regulations in the case of a hypothetical sodium spillage accident.

References

[1]   Das, S.K., Sharma, A.K., Parida, F.C., and Kashinathan, N. (2009). "Experimental study on thermo-chemical phenomena during interaction of limestone concrete with liquid sodium under inert atmosphere", *Construction and building materials, 23*, pp. 3375-3381

[2]   Chawla, T.C., and Pederesen, D.R. (1985. "A review of modeling concepts for sodium-concrete reactions and a model for liquid sodium transport to the un reacted concrete surface", *Nuclear engineering and design, 88*, pp. 85-91

[3]   Pramila, M., Sivasubramanian, K., Amarendra, G., and Sundar, C.S. (2008). "Thermo chemical degradation of limestone aggregate concrete on exposure to sodium fire", *Journal of nuclear materials, 375*, pp. 263-269

[4]   Parida, F.C., Das, S.K., Sharma, A.K., Rao, P.M., Ramesh, S.S., Somayajalu, P.A., Malarvizhi, B., and Kashinathan, N. (2006). "Sodium exposure tests on limestone concrete used as sacrificial protection layer in FBR", *Proceeding of ICON 14, 14th International conference on Nuclear Engineering, July 17-20*, Miami, Florida, USA

[5]   Mehta, P.K., and Moteiro, P.J.M.(2006). "Concrete- microstructure, properties and materials", Third Edition, *Tata McGraw Hill, New Delhi*

[6]   Prokopski, G and Halbiniak, J. (2000). "Interfacial transition zone in cementitious materials", *Cement and concrete research, 30*, pp. 579-583

[7]   Diamond, S and Haung, J. (2001), "The ITZ in the concrete- A different view based on image analysis and SEM observations", *Cement and concrete composites, 23*, pp. 179-188

[8]   Liao, K.Y., Chang, P.K., Peng, Y.N., and Yang, C.C. (2004). "A study on characteristics of interfacial transition zone in concrete", *Cement and concrete research, 34*, pp. 977-989

[9]   Wang, H.Y. (2008). "The effects of elevated temperature on cement paste containing GGBFS", *Cement & concrete composites, 30*, 992–999

[10]   Lee, J., Xi., Y., William, K and Jung, Y.(2009). "A multi scale model for modulus of elasticity of concrete at high temperatures", *Cement and concrete research 39*, pp. 754–762

[11]   Arioz, O., (2007). "Effect of elevated temperature on properties of concrete", *Fire safety journal, 42*, pp.516-522

[12]   Hossain., K.M.A. (2006)."High strength blended cement concrete incorporating volcanic ash: Performance at high temperatures", *Cement & concrete composites, 28*, pp. 535–545

[13]   Poon, C.S., Azhar, S., Anson, M., and Wong, Y.L. (2001). "Comparison of the strength and durability performance of normal- and high-strength pozzolanic concretes at elevated temperatures", *Cement and concrete research , 31*, pp.1291–1300

[14]   Vodak, F., Trtık, K., Kapickova, O., Hoskova, S and Demo, P. (2004). "The effect of temperature on strength – porosity relationship for concrete", *Construction and building materials, 18*, pp.529–534

[15]   Garg, S. (1999). "Physical and engineering geology", 3rd Edition, KHANNA PUBLISHER, Delhi

[16]   Gribble, C.D. (1988). "Rutley's elements of mineralogy" , *27th edition, UNWIN HYMAN,* London

[17]   Xing, Z., Beaucou, A., L., Hebert, R., Noumowe, A., and Ledesert, B. (2011). "Influence of the nature of aggregates on the behavior of concrete subjected to elevated temperature", *Cement and concrete research , 41*, pp.392–402

[18]   Chasanov, M.G and Staahl, G.E., Sr, (1977). High Temperature Sodium Concrete interactions, *Journal of Nuclear materials, N0.66, pp.*217 -220

[19]   Mantuani, L.D.(1983), Hand book of Concrete Aggregates: A petrographic and technological evaluation, Noyes Publication , 345

[20]   Bae, J.H., Shin, M.S., Min, B.H and Kim, S.M. (1998). "Experimental study on sodium-concrete reactions", *Journal of the Korean nuclear society* (1998), pp. 568-580

[21]   Fritzke, H.W., and Schultheiss, G.F. (1983). "An experimental study on sodium concrete interaction and mitigating protective layers, *7th international conference on structural mechanics in reactor technology,*pp. 135-142

[22]   Muhlestien, L.D and Postma , A.K. (1984). "Application of sodium-concrete reaction data on breeder reactor safety analysis", *Nuclear safety*, Vol 25, No.2, , pp.212-220

Previous *fib* International PhD Symposia

1st *fib* International PhD Symposium in Civil Engineering

May 1996, Technical University of Budapest, Budapest, Hungary
order info: Balázs, György (Ed.)
               Sokszorositotta a Muegyetemi Kiado
               Risograph technologiaval
               Fedelos vezeto: Veress Janos
               Munkaszam: 96-Mk10
               E-Mail: fib@eik.bme.hu
               ISBN 963 420 504 6

2nd *fib* International PhD Symposium in Civil Engineering

August 1998, Technical University of Budapest, Budapest, Hungary
order info: Balázs, György (Ed.)
               ISBN 963 420 560 7
               Publishing Company of Technical University of Budapest
               Head: Janos Veress
               Offset printing No: 98-0134
               E-Mail: fib@eik.bme.hu

3rd *fib* International PhD Symposium in Civil Engineering

October 2000, University of Agricultural Sciences Vienna, Vienna, Austria
order info: IKI - Institute for Structural Engineering,
               BOKU - University of Agricultural Sciences,
               Peter Jordanstr. 82, A-1190 Vienna, Austria
               Phone: +43 1 47654 5250
               Fax: +43 1 47654 5292
               E-mail: lilo@iki.boku.ac.at
               ISSN 1028-5334

4th *fib* International PhD Symposium in Civil Engineering

September 2002, Technische Universität München, Universität der Bundeswehr München, München, Germany
order info: Springer-VDI-Verlag GmbH & Co. KG
               Düsseldorf, Germany
               ISBN 3-935065-09-4

5th *fib* International PhD Symposium in Civil Engineering

2004, Delft University of Technology, Delft, The Netherlands
order info: W. Sutjiadi
               Faculty of Civil Engineering and Geosciences
               Delft Univ. of Technology
               Stevinweg 1, 2628 CN Delft, The Netherlands
               Tel: +31 15 278 4665
               Fax: +31 15 278 7313
               E-mail: w.sutjiadi@ct.tudelft.nl
               ISBN Vol 1: 90 5809 677 7
               ISBN Vol 2: 90 5809 678 5

6th *fib* International PhD Symposium in Civil Engineering

August 2006, ETH Zurich, Zurich, Switzerland
order info: *fib* secretariat
        Case Postale 88,
        CH-1015 Lausanne, Switzerland
        Phone: + 41 21 693 2747
        Fax: + 41 21 693 6245
        E-mail: *fib*@epfl.ch

7th *fib* International PhD Symposium in Civil Engineering

September 2008, University of Stuttgart, Stuttgart, Germany
order info: PhD Symposium Secretariat
        IWB Universität Stuttgart
        Pfaffenwaldring 4
        70569 Stuttgart, Germany
        Phone: +49 711 685 63320
        Fax: +49 711 685 67681
        E-mail: stumpp@iwb.uni-stuttgart.de

8th *fib* International PhD Symposium in Civil Engineering

June 2010, Technical University of Denmark, Kgs. Lyngby, Denmark
order info: Technical University of Denmark
        Department of Civil Engineering, Brovej, Building 118,
        DK-2800 Kgs. Lyngby, Denmark
        Phone: + 45 45 25 17 00
        Fax: + 45 45 88 32 82
        E-Mail:byg@byg.dtu.dk

9th *fib* International PhD Symposium in Civil Engineering

July 2012, Karlsruhe Institute of Technology (KIT), Karlsruhe, Germany
order info: KIT Scientific Publishing
        Straße am Forum 2
        D-76131 Karlsruhe, Germany
        Phone: + 49 721 608 43104
        Fax: + 49 721 608 44886
        E-mail: info@ksp.kit.edu
        available as download at: http://www.ksp.kit.edu